T0122060

WILD CARNIVORES *of* NEW MEXICO

Photograph: © Hari Viswanathan

WILD CARNIVORES
of NEW MEXICO

Edited by

Jean-Luc E. Cartron

and **Jennifer K. Frey**

UNIVERSITY OF NEW MEXICO PRESS · ALBUQUERQUE

© 2023 by University of New Mexico Press
All rights reserved. Published 2023
Printed in China

ISBN 978-0-8263-5151-7 (cloth)
ISBN 978-0-8263-5153-1 (electronic)

Library of Congress Control Number: 2023947216

Founded in 1889, the University of New Mexico sits
on the traditional homelands of the Pueblo of Sandia.
The original peoples of New Mexico—Pueblo, Navajo,
and Apache—since time immemorial have deep
connections to the land and have made significant
contributions to the broader community statewide.
We honor the land itself and those who remain stew-
ards of this land throughout the generations and also
acknowledge our committed relationship to Indigenous peoples. We gratefully recognize our history.

Cover photographs by Doug Burkett, Sally King, Ed
MacKerrow, Mark Watson, and Hari Viswanathan

Cover design by Felicia Cedillos
Text design by Lisa C. Tremaine

Text composed in Alegreya, designed by Juan Pablo
del Peral for Huerta Tipográfica.

CONTRIBUTING PHOTOGRAPHERS

Kalon Baughan	**Jared A. Grummer**	**Sole Marittimi**
Dennis Buchner	**Roger Hogan**	**John McClure**
Douglas W. Burkett	**Sally King**	**Robert Shantz**
Mike Dunn	**Evan Kipp**	**Geraint Smith**
Pat Gaines	**Don MacCarter**	**Hari Viswanathan**
Kari Greer	**Ed MacKerrow**	**Mark L. Watson**

Carol A. Anderson	Harrison Frazier	Stewart Liley	Diane Rocchia
Corey Anderson	Gordon French	Ulli Limpitlaw	Gary W. Roemer
Romain Baghi	Jennifer K. Frey	Jenny Lisignoli	Diego Romero
Marc Baldwin	John Gallagher	Travis Livieri	Britt Runyon
Jonathan Batkin	Della Garelle	Mike Lockhart	Leif Saul
Louis C. Bender	Keith Geluso	Ken Logan	Jake Schoellkopf
Scott C. Bender	Joel Gilb	Brian Jay Long	Tim Shortell
David L. Bergman	Warner Glenn	Dustin Long	Christopher Sichko
Kevin Bixby	Rick Gonzales	Robert C. Lonsinger	Greg Silsby
Craig E. Blakemore	Matthew J. Gould	Barbara Magnuson	Jamie Simo
Robin Bonner	Rick Grothe	Robert K. Mark	Meg Sommers
Robyn Bortner	Art H. Harris	Mark H. Meier	Ken Steiner
Daniel Boyes	Robert L. Harrison	Rüdiger Merz	Gak Stonn
Kenneth F. Brinster	Christine C. Hass	Ella Meyer	James N. Stuart
Darren L. Bruning	Clint Henson	David Mikesic	Steve Sunday
Jessica K. Buskirk	Patty Hoban	Denise Miller	Leslie Sutton
Miranda Butler-Valverde	William Horton	Jennifer Miyashiro	Brad Taylor
Jean-Luc Cartron	Bob Inman	Victoria Monagle	Christopher Taylor
Jo Castillo	Bennette Jenkins-Tanner	Samuel Mulder	Myles Traphagen
Cassidi Cobos	Don Jones	Robert Muller	Jack Triepke
Steve Collins	Jeff Kaake	Alissa Diane Mundt	Stella F. Uiterwaal
Rodney H. Cosper	Brendon Kahn	Sarah Nelson	Pat (Ranger) Ward
Cecily Costello	Daniel Kalal	Christopher Newsom	Jannelle Weakly
Michael D. Cox	Jan F. Kamler	Lori Nixon	Matthew M. Webb
Ana D. Davidson	Rebekah Karsch	Eric A. Odell	Jim Webber
Robert Deans	Tom Kennedy	Ryan P. O'Donnell	Mara E. Weisenberger
Jerry W. Dragoo	Brianne Kenny	Albert Ortega	Ruth Wheeler
Colin R. Dunn	Surasit Khamsamran	Mark Packila	Scott Wilber
David Eads	Larry Kimball	Marty Peale	Dan J. Williams
Katherine Eagleson	Tamar Krantz	Martin Perea	Greg D. Wright
Jon Dunnum	Larry Lamsa	Travis W. Perry	Jill Terese Wussow
Trey Flynt	Andrew J. Lawrence	Clifford G. Powell	J. Judson Wynne
Bernard R. Foy	Alma Leaper	J. T. Radcliff	James E. Zabriskie
Kimberly Fraser	David C. Lightfoot	Jason Roback	

Maps by Ryan Trollinger
and Kenneth C. Calhoun

SPONSORS

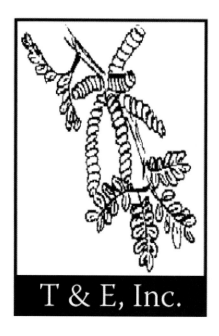

T & E, Inc.

Daniel B. Stephens & Associates, Inc.

SWCA®
ENVIRONMENTAL CONSULTANTS

Sound Science. Creative Solutions.

Friends of
Bosque del Apache

DePAULI ENGINEERING & SURVEYING LLC
CIVIL ENGINEERS & LAND SURVEYORS
307 S. 4TH STREET GALLUP, NM 87301
Phone: 505 863 5440
www.depauliengineering.com

**New Mexico
Wildlife Center**

Connecting People and Wildlife for an Abundant Tomorrow

CONTENTS

Carnivores have fascinated humans throughout our existence for a variety of reasons. For the earliest humans in New Mexico, some carnivores represented dangerous predators. Later visitors and inhabitants trapped several species as valuable furbearers. Eventually, as the descendants of European immigrants began raising domestic livestock, predator control became the norm. Our knowledge of carnivores grew during these eras mainly through direct observation. When Vernon Bailey summarized knowledge of the mammals of New Mexico in 1931, his carnivore accounts were rich in details gleaned from those accomplished early hunters and trappers.

Bailey and his colleagues from the US Bureau of Biological Survey initiated scientific surveys of the mammals of the state and laid the groundwork for subsequent workers. In 1955, James S. Findley began a program of modern investigations of mammals at the University of New Mexico, and by the time I arrived in 1965 to study with him, he had a cadre of graduate students working on many aspects of mammalian biology. Findley and his students concentrated more on small mammals than on carnivores because at that time we knew less about most rodents, bats, and shrews than we did about carnivores. Findley and his students Art Harris (a contributing author to this volume), Clyde Jones, and I produced an updated *Mammals of New Mexico* in 1975, but the carnivore accounts were mostly limited to taxonomic and distributional information, with limited natural history information.

Studies of the carnivores in New Mexico blossomed during the latter half of the last century and the first two decades of the current one. This happened concurrently with an explosion of new techniques and methods for studying them. Chief among those is the development of a revolutionary array of molecular techniques that allows modern researchers to compare the DNA of individual animals in detail. Our understanding of the evolutionary relationships of virtually every species of carnivore has grown apace. The use of radio and GPS telemetry and modern camera traps, both still and video, yield a wealth of information on animal movements and behavior.

New Mexico, the "Land of Enchantment," is also the land of diversity in so many ways. This volume explores the diversity of carnivores in New Mexico and helps to explain it by outlining the diversity in the fossil record, in modern-day landscapes and their history, and in how we as humans have treated carnivores throughout our own history in the state. Early fur trappers viewed carnivores as a natural resource to be exploited, and settlers demanded predator control to protect increasing livestock production. Today, wildlife managers search for a happy coexistence of humans and carnivores, and modern conservation biologists work hard to better understand human-carnivore interactions with a view toward maintaining all extant species in smoothly functioning ecosystems.

To set the stage for summarizing what we know about each of the 28 species of native, wild carnivores currently found or known to have occurred recently in New Mexico, the early chapters explore each of the topics mentioned above in some detail. The fossil record of carnivores in New Mexico is rich and Chapter 1's summary of the ice-age carnivore fauna nicely documents what we know of each of the many species. The Pleistocene megafauna included many iconic carnivores from sabertooths to dire wolves, most of which became extinct around 11,000 years ago. Although many smaller carnivore species survived into the Holocene, the only remnants of the megafauna are the jaguar, cougar, gray wolf, black bear, and brown bear.

Climate, landscape setting, geology, and soil all contribute to the diversity of ecosystems in New Mexico. That vegetative diversity in turn allows for the great diversity in the carnivore fauna. Chapter 2 examines the ecosystems of the state through the lens of climate and fire ecology. This rich summary of the life zones will grow in value as we face the onslaught of climate change in coming decades. The next three chapters take us through the history of human-carnivore interactions in the state. Chapter 3 summarizes the controversy surrounding predator control and documents much of the history of predator management activities in New Mexico. Modern-day management of wild carnivores in New Mexico is described in Chapter 4. Chapter 5 brings much of this human-carnivore interaction activity into focus by using the threat of climate change to summarize what we know about the present and future of the order Carnivora in New Mexico. Clearly, climate change is the single most serious, overarching threat to animal and plant populations everywhere.

The meat of the book, so to speak, is found in the individual chapters dedicated to each species. From white-nosed coatis to wolverines, the current carnivore fauna of New Mexico is as varied as the landscapes and ecosystems they inhabit. We owe it to future generations to do all we can to ensure that this fauna survives intact for our grandchildren to enjoy. To do that, we need to understand the biology of these unique animals in as much detail as possible. The individual accounts presented here summarize what we know and what we do not know about each species.

To provide a hint of the wealth of information included in each species account, perhaps a few sentences chosen from different accounts will serve:

The Canada lynx (*Lynx canadensis*; hereafter lynx) is a medium-sized felid, standing ~50 cm (~1 ft 7 in) tall at the shoulders.

Bobcats are creatures of shadows and stealth.

Taxonomists classify the cougar along with other meat-eaters in the order Carnivora and with all other cats in the family Felidae.

Based on the fossil record, the coyote made its appearance during the Pleistocene's late Irvingtonian (0.4–0.250 mya).

Presently, gray foxes are found in northern South America, in both Venezuela and Columbia, north throughout Central America—though curiously absent from some Caribbean watersheds along the Atlantic coast—and Mexico, including the Baja Peninsula, and across the United States from Florida to Oregon, but absent in some parts of the Great Plains and the northern Rocky Mountains.

Although there are no data documenting continental-scale shifts in the distribution of the kit fox, there is evidence of a regional decline due to agricultural and industrial habitat conversion and urban development in the past century.

There have been a number of swift fox reintroductions in an effort to restore populations to unoccupied yet suitable areas within the species' historical range.

The genetic data, while acknowledging that hybridization did historically occur in localized areas of the eastern United States, also have demonstrated that red foxes in North America are a distinct species that has not been displaced or subjected to extensive hybridization with European red foxes. Put simply, North American red foxes are native and represent a distinct species, *V. fulva*.

In New Mexico, black bears are highly selective of closed-canopy forest and woodland communities that provide visual and thermal cover, and a variety of plant and animal foods

While the skulls of weasels and martens are delicate, smooth, and armed with sharp pointed teeth for catching and slicing small rodents, the skull of the wolverine is massive, with prominent ridges for the insertion of powerful muscles that work a jaw armed with large, robust teeth that can crush large bones of ungulates.

River otters do not hibernate and are active year-round.

New Mexico ermines are not only the smallest known mustelid but also the smallest known carnivore!

In 2012, New Mexico became the latest state to attempt the repatriation of the black-footed ferret to its former range, though it remains to be seen whether the species' return to the "Land of Enchantment" succeeds.

In New Mexico, many specimens of long-tailed weasel have come from places far from any water, such as prairie dog towns on the Llano Estacado or White Sands.

Minks are generally solitary except during the mating season and except for females with young.

While playing, young, excited [western spotted skunks] will let out a loud, ear-piercing, high-pitched screech.

Besides humans, known and suspected predators of coatis in the American Southwest include the jaguar (*Panthera onca*), cougar (*Puma concolor*), bobcat (*Lynx rufus*), ocelot (*Leopardus pardalis*), black bear (*Ursus americanus*), domestic dog (*Canis familiaris*), spotted owl (*Strix occidentalis*), red-tailed hawk (*Buteo jamaiscensis*), and golden eagle (*Aquila chrysaetos*).

During the 2014–2015 fiscal year, the New Mexico Department of Game and Fish received 535 nuisance wildlife complaints statewide, with the main species involved consisting of the black bear (*Ursus americanus*), followed by the raccoon.

The final chapter is a summary of information on feral dogs and cats in New Mexico, a problem of increasing proportions in modern times. In summary, you are holding a volume that provides a wealth of information on the current carnivore fauna of New Mexico. Modern syntheses like this provide invaluable sources of information to help us manage natural resources in difficult times.

Don E. Wilson
Washington, DC
November 2021

More than 10 years have passed since the publication of the *Raptors of New Mexico* (UNM Press, 2010). The book—all 700 pages of it—was well received, and almost immediately the idea of a companion volume presented itself. Just like raptors, members of the mammalian order Carnivora (hereafter referred to as carnivores) tend to exhibit complex social behaviors and play critically important roles in natural ecosystems. Both raptors and carnivores are charismatic animals, and some of them even serve as flagship species for garnering support around conservation issues. Like raptors, which are often described as the embodiment of freedom, strength, and bravery, carnivores are important symbols—of power, stealth, or cunningness—including in Native American culture. With its wide latitudinal spread, its low deserts and high mountains, and the vastness of its public lands and natural ecosystems, New Mexico harbors large populations of raptors and carnivores alike and a diverse fauna of both groups. Our state also lends itself to much ecological research. As was the case with several birds of prey—the northern goshawk, Harris' hawk, and Mexican spotted owl, long-term studies of cougars and black bears in New Mexico have led to significant advances in our understanding of these species' population structure and dynamics, social organization, competition, and/or conservation needs in the Southwest and beyond. With all their uncertainty and logistic complexity, species reintroductions have been attempted in New Mexico for the aplomado falcon among raptors, and the Mexican gray wolf, northern river otter, and black-footed ferret among carnivores. The large body of ecological research conducted in New Mexico, together with all the important management issues and fascinating conservation and recovery stories, seemed highly deserving of a companion volume to the *Raptors of New Mexico*, dedicated this time to the state's carnivores.

Carnivores are typically elusive and rarely seen by humans, and a face-to-face encounter with one of them in the wild can provide one of those rare, heart-stopping moments in life. It happened to one of us in the spring of 2008 around 1:00 a.m. in Sugarite Canyon State Park near Raton in Colfax County. Having completed owl surveys along the mountain slopes above Lake Maloya, I (Cartron) reached my car parked by the lake and started driving slowly down the road through the canyon. Near the visitor center, I eased to a complete stop as right in front of me, not one but four cougars (one adult female and her three full-grown cubs) started crossing the road, seemingly unhurried and unbothered. Too stunned to even attempt grabbing my camera, I simply took in the moment and watched the four cougars make their way to the other side of the road, realizing in that instant that I perhaps should not be conducting owl surveys at night by myself. Later, news of my late-night encounter reached other biologists. Apparently, one of the park rangers had seen the same cougar and her cubs in hot pursuit of a mule deer during the previous winter. Although I was trained as an ornithologist, some of my most memorable experiences in the field have consisted of encounters with carnivores, not just cougars in Sugarite Canyon and elsewhere but also bears, foxes, skunks, and ringtails. Those encounters also inspired the idea of this book project, in which I was immediately joined by my colleague Jennifer Frey.

I (Frey) grew up with an intense fascination for animals, especially mammals. As a child, I developed my own natural history museum based on the bones, feathers, shells, and other animal remains I found in backyards, campgrounds, and other out-of-the-way places. I labeled the specimens with seemingly pertinent information, but I wanted to know more about what they were. After a lucky find at a garage sale, I became the new owner of a book that seemed tailor-made for me—Burt and Grossenheider's *Field Guide to the Mammals*. I poured over the pages finally able to understand what species was what, how you identified it, and where it occurred. The first Latin name I memorized was *Procyon lotor*, the raccoon. I loved how learning the scientific name, seeing its distribution portrayed on a map, and finding the raccoon arranged under the heading Procyonidae, along with its close relatives the ringtail and coati, helped me understand the raccoon, in a deeper way than my casual encounters with campground-raiding raccoons ever could. Shortly after acquiring my mammal field guide, I checked out a new book from the library that described mammals that were at risk of extinction. Extinction? Appalled by the idea these animals could disappear, I carefully cross-referenced

them into my field guide highlighting their names in red crayon and marker—grizzly bear, black-footed ferret, wolverine, gray wolf, kit fox, mountain lion, and lynx. Even though a child, I resolved that I would do something to help prevent extinctions of mammals. I did not yet know the word *conservation*. And later, mirroring my fascination for animals and especially mammals, I focused my educational training in the fields of zoology, wildlife management, and mammalogy.

It has taken us more than 10 years to write and edit the *Wild Carnivores of New Mexico*. All that time, the loss of biodiversity has accelerated and climate-driven impacts have become more pronounced everywhere. These two global environmental crises have heightened the stakes for some of the already complex and controversial issues surrounding the management of carnivores. It is our hope that the information in this volume can help wildlife management agencies and conservation organizations to navigate the perhaps difficult times ahead and promote our continued coexistence with carnivores.

ACKNOWLEDGMENTS

In the course of editing this volume, it has been our privilege to work with many university, government-agency, and independent researchers, whose published studies have considerably aided the advancement of knowledge concerning carnivores and their habitats in New Mexico. We are also deeply grateful to the many photographers who contributed their wonderful images to the book, with special thanks extended in particular to Hari Wiswanathan, Mark Watson, Sally King, Mike Dunn, Larry Lamsa, Jim Stuart, Jon Dunnum, Alissa Diane Mundt, and Kari Greer. We thank all of the institutions that kindly granted us the use of their photograph collections through authorized representatives, including the Mexican wolf reintroduction project's Interagency Field Team (Colby Gardner), White Sands Missile Range Garrison Environmental Division and ECO Inc. (Cristina Rodden, Greg Silsby), Sevilleta National Wildlife Refuge (Jon Erz), New Mexico Department of Game and Fish (Dan Williams and Nick Forman), Sandia Mountain Natural History Center (Paul Mauermann and Fiana Shapiro), New Mexico Wildlife Center (Alissa Mundt), Gila Wildlife Rescue (Dennis Miller), Santa Fe New Mexican (Allison Dellinger), Northern Jaguar Project (Megan "Turtle" Southern), San Andres National Wildlife Refuge (Mara Weisenberger), and California Academy of Sciences (Seth Cotterell). An important contribution was made by Ryan Trollinger, who expertly prepared the species' occurrence maps, and by all those who reviewed the chapters of the book, including Jerry Apker, Erin Barding, Michael A. Bogan, Wesley Brashear, Steven W. Buskirk, James W. Cain III, Ivonne Cassaigne, Robert Dowler, David A. Eads, Adam W. Ferguson, James S. Findley, Stanley D. Gehrt, Matthew E. Gompper, Matthew J. Gould, Kevin Hansen, Larisa Harding, Bart J. Harmsen, Jan F. Kamler, Robert C. Lonsinger, Gary S. Morgan, Este Muldavin, Richard E. Richards, Meghan D. Riley, Harley G. Shaw, Bob Sivinski, Mark Statham, Todd D. Steury, and Mark Vieira. We are also indebted to Sandra West (editorial assistance), Ivana Mali and Darren A. Pollock (ENMU specimen spreadsheets), and Don E. Wilson, who did us the honor of writing the foreword of the book. We acknowledge the great generosity of our sponsors, without whom the production of the *Wild Carnivores of New Mexico* would not have been possible, particularly the New Mexico Department of Game and Fish's Ecological and Environmental Planning Division and Tom and Eleanor Wootten's foundation, T&E Inc. Thank you, finally, to Zubin Meer for his meticulous and thorough copyediting, Felicia Cedillos and Lisa C. Tremaine for the beautiful cover and text design, James Ayers for expertly overseeing all stages of the book's design and production, and Clark Whitehorn, Michael Millman, and the University of New Mexico Press for their trust in the "Wild Carnivores of New Mexico" book project.

INTRODUCTION

Jean-Luc E. Cartron and Jennifer K. Frey

The order Carnivora (the carnivorans, commonly referred to as carnivores) is a phylogenetic (evolutionary) classification of placental mammals, not to be confused with the ecologically defined carnivores, organisms that primarily or exclusively consume the flesh of animals. Distributed across nearly all the continents and oceans, the Carnivora is monophyletic (i.e., derived from a single common ancestor) yet ecologically diverse, occupying a wide range of terrestrial and aquatic environments from the poles to the tropics. Many carnivoran species are predators that eat other vertebrates, but others are omnivores, while the aardwolf (*Proteles cristata*) specializes on a diet of termites and the walrus (*Odobenus rosmarus*) prefers mollusks. As for the kinkajou (*Potos flavus*), it thrives on a diet of fruit, while the red panda (*Ailurus fulgens*) and even more so the giant panda (*Ailuropoda melanoleuca*) are bamboo eaters. Thus, not only does carnivory exist in other animals outside the order Carnivora, such as the killer whale (*Orcinus orca*; order Cetacea), snakes, crocodiles, or birds of prey; it is also not a defining characteristic of modern carnivorans.

The early evolutionary history of the Carnivora has not been fully resolved (Solé et al. 2016; Hassanin et al. 2021; Chapter 1). Several Northern Hemisphere clades (i.e., evolutionarily groups) of placental meat-eating mammals are now believed to have evolved independently during the Paleocene (~66 to 56 million years ago [mya]) (Solé et al. 2016). Filling the ecological land niche vacated by the dinosaurs, they made the transition from an insectivorous diet to being primarily meat-eaters. That transition was made possible by a number of morphological adaptations ranging from strong jaws and dagger-like canines (used for killing prey) to the development of the carnassials, paired upper and lower cheek teeth modified to act as scissor blades (Colbert and Morales 1991). When the jaw closes, the sharp edges at the rear of the upper carnassial and front of the following lower carnassial slide past each other, tearing flesh (Van Valkenburgh 2007). In the Carnivora, there is only one pair of carnassials consisting of the last upper premolar and the first lower molar on each side of the jaw. In contrast, the carnassials of other early carnivorous mammals (e.g., oxyaenids and hyaenodontids, two unrelated clades originally classified together as Creodonta) did not involve premolars. Instead, they were all modified molars, sometimes occurring as more than one pair (Ungar 2010). Non-carnivoran carnivores were the dominant predators during the Eocene, but they were all extinct by the end of the Miocene just over 5 mya (Goswami 2010).

In the warm greenhouse climate of the early-middle Eocene, forests dominated across all the continents, and carnivorans likely consisted of small arboreal species (Hassanin et al. 2021). As inferred from a recent molecular study with multiple fossil time calibration points, the two carnivoran suborders Caniformia (dog-like carnivorans) and Feliformia (cat-like carnivorans) diverged around 50 mya, during or just after the Early Eocene Climatic Optimum (Hassanin et al. 2021). Within the Caniformia, a further split occurred approximately 42–48 mya between the dog lineage and the clade Arctoidea, the latter giving rise to the weasel, skunk, raccoon, bear, seal, sea lion, and walrus families. Within the Feliformia, the cat, hyena, and mongoose lineages began to diverge perhaps as late as the Oligocene (Hassanin et al. 2021). The history of the Carnivora (and Creodonta) is marked by recurring morphological convergence and the repeated iteration of similar forms and ecomorphs across different lineages

(Martin 1989; Werdelin 1996; Van Valkenburgh 1999). For example, sabertooth short-faced predators evolved independently not just in the Felidae and the Nimravidae (a now extinct family within the Feliformia), but also in the Creodonta, not to mention among dasyurid marsupials (Van Valkenburgh 2007). Skull homoplasies (i.e., similar features in unrelated organisms) in particular may be indicative of developmental constraints that are most pronounced in hypercarnivores—animals with a diet that is > 70% meat—and related to biomechanics, such as bite force and degree of mouth opening (Figueirido et al. 2011; Tamagnini et al. 2021).

By the latest count, the world's carnivorans consist of 302 extant species in 16 families (Mammal Diversity Database, version 1.2). Excluding the domestic cat (*Felis catus*) and the domestic dog (*Canis familiaris*), New Mexico is, or was until recently, home to 28 of these species, but several of them are now extirpated including the brown bear (*Ursus arctos*) sometime during the first half of the 20th century (Chapter 17), the wolverine (*Gulo gulo*) likely earlier, by the end of the 19th century (Chapter 18), and perhaps more recently also the American mink (*Neogale vison*; Chapter 24) (Table i.1). The largest—and most diverse—family of carnivorans, both globally and in New Mexico, is the Mustelidae, which includes weasels, ferrets, minks, martens, wolverine, badgers, and otters (Chapters 18–25). No longer included in the Mustelidae are the skunks, elevated to the status of family, the Mephitidae (Dragoo and Honeycutt 1997) and represented in New Mexico by at least five species in three genera (*Conepatus, Mephitis, Spilogale*; Chapters 26–29). With few exceptions, all of New Mexico's species first appeared in the Northern Hemisphere, whether in North America or Eurasia. However, the Carnivora entered South America as part of the Great American Biotic Interchange (GABI) that began approximately 2.7 mya, or even earlier, before the full emergence of the Isthmus of Panama

(e.g., the Procyonidae; Tarquini et al. 2020). Having originated in the Old World, either the jaguar (*Panthera onca*) itself or an ancestral Eurasian jaguar-like pantherine (*P. gombaszoegensis*) reached North America from Siberia through the Beringia Land Bridge, then expanded its geographic range south into Central and South America (Kurtén and Anderson 1980; Hemmer et al. 2001; Mol et al. 2001; Arroyo-Cabrales 2002; O'Regan 2002; Marciszak 2014; Stimpson et al. 2015). Although the jaguar was extirpated from North America during the Quaternary Extinction Event that ended about 11,650 years ago, the remnant population that subsisted in South America subsequently expanded its distribution, this time to the north, and recolonized Mexico, reaching the southern United States (Kurtén and Anderson 1980). The geographic origin of the cougar (*Puma concolor*) is unclear, but until recently, the prevailing theory was that the species had originated in North America, then spread southward into South America like *Panthera* (Culver 1999; Culver et al. 2000). The cougar had then disappeared for some time from North America, before the arrival of a small number of individuals migrating from South America about 10,000 to 12,000 years ago (see Chapter 9). However, based on the recent discovery of a fossil from Argentina's early-middle Pleistocene (1.2–0.8 mya), the cougar might have evolved instead in South America then expanded its distribution northward into North America (Chimento and Dondas 2017).

The history of competition and conflict between humans and carnivorans on a scale large enough to cause population declines and species extirpations might date back to the late Pleistocene glacial period (~110,000 to 11,650 years ago). On several continents, but particularly in North America, the end of the Pleistocene was marked by the disappearance of many large mammals, an event variously attributed to rapid climate and vegetation change, over-hunting by humans, or a combination of both (Nyhus 2016; Surovell et al. 2016). With

Table I.1. New Mexico's extant and recently extirpated native carnivoran species.

Family	Common Name	Scientific Name	Population Status
Felidae	Canada lynx	*Lynx canadensis*	Reintroduced in Colorado, with individuals ranging into New Mexico
	Bobcat	*Lynx rufus*	Extant
	Jaguar	*Panthera onca*	Last verified occurrence in 2006
	Cougar	*Puma concolor*	Extant
Canidae	Coyote	*Canis latrans*	Extant
	Gray wolf	*Canis lupus*	Reintroduced
	Gray fox	*Urocyon cinereoargenteus*	Extant
	Kit fox	*Vulpes macrotis*	Extant
	Swift fox	*Vulpes velox*	Extant
	North American red fox	*Vulpes fulva*	Extant
Ursidae	Black bear	*Ursus americanus*	Extant
	Brown bear	*Ursus arctos*	Extirpated
Mustelidae	Wolverine	*Gulo gulo*	Extirpated
	Northern river otter	*Lontra canadensis*	Reintroduced
	Pacific marten	*Martes caurina*	Extant
	American ermine	*Mustela richardsonii*	Extant
	Black-footed ferret	*Mustela nigripes*	Reintroduced
	Long-tailed weasel	*Neogale frenata*	Extant
	American mink	*Neogale vison*	Possibly extirpated
	American badger	*Taxidea taxus*	Extant
Mephitidae	White-backed hog-nosed skunk	*Conepatus leuconotus*	Extant
	Hooded skunk	*Mephitis macroura*	Extant
	Striped skunk	*Mephitis mephitis*	Extant
	Western spotted skunk	*Spilogale gracilis*	Extant
	Rio Grande spotted skunk	*Spilogale leucoparia*	Extant
Procyonidae	Ringtail	*Bassariscus astutus*	Extant
	White-nosed coati	*Nasua narica*	Extant
	Raccoon	*Procyon lotor*	Extant

increasing evidence that *Homo sapiens* contributed directly to some of these megafaunal extinctions in North America (e.g., Broughton and Weitzel 2018), the disappearance of the dire wolf (*Canis dirus*), the saber-toothed cat (*Smilodon fatalis*), and other large predators could have been in part the result of competition with newly arrived humans (Ripple and Van Valkenburgh 2010). Over-hunting of large herbivores such as camels (*Camelops*), mammoths (*Mammuthus*), or mastodons (*Mammut americanum*) would have reduced or eliminated the primary prey base of those carnivores.

New Mexico has been home to many indigenous human cultures, including the ancient Ancestral Pueblo and Mogollon, and the present-day Pueblo, Navajo, Apache, Kiowa, Comanche, and Ute peoples (Dutton 1991). Wildlife, including carnivorans, were important to these people as sources of food, clothing, tools, and trade (Weber 1971). However, customs varied among groups. For instance, while bears were hunted by the Jemez, they were not killed by the Isletans (Beals 1935). Carnivorans were also objects of cultural and religious reverence. For instance, the Puebloans ceremonially associated the cougar with hunting as it was thought to make a hunter more powerful. Thus, a hunter might carry a cougar fetish or a quiver made of the skin of a cougar (Beals 1935). Similarly, images of carnivores are often depicted in petroglyphs and on pottery (see Chapter 31).

New Mexico was first settled by the Europeans in 1598, well before the establishment of Jamestown in 1607. During the 17th century, primary trade was coarse fur (pelts of ungulates such as deer [*Odocoileus hemonius* and *O. virginianus*] and bison [*Bison bison*]; Weber 1971). However, in 1821 the region that is now New Mexico gained independence from Spain as part of newly established Mexico. This geopolitical transition had enormous implications for natural resources of the region as it resulted in the opening up of New Mexico to foreigners, who under Spanish rule had not been allowed to enter.

Thus, the Santa Fe trail was opened as a commercial route in 1821 to allow trade between (New) Mexico and the United States, which at that time had its western limits in Missouri. The chief economic draw of the new frontier was its abundant wildlife, particularly "fine furs" including beaver and several species of carnivores such as foxes and the Canada lynx (*Lynx canadensis*), bobcat (*L. rufus*), wolverine, Pacific marten (*Martes caurina*), American mink, and northern river otter (*Lontra canadensis*). Taos became the hub of the fur trade for the entire Southern Rocky Mountain and Southwest region. Trapping was intensive, much of it conducted by companies employing large numbers of well-provisioned trappers. Beavers were especially numerous and valuable, and the so-called Mountain Men trapped enormous quantities of these animals, moving from one location to the next as trapping grounds were exhausted. Even as early as 1830, before changing fashions caused an end to the "Mountain Man" era in the next decade, some rivers previously teaming with beavers were now depleted. During that same period, it seems likely that many of the most valuable carnivorans were greatly reduced in numbers in the state.

During the latter half of the 19th century, two important events shaped the next phase of the relationship between humans and carnivorans. The Homestead Act of 1862 brought new settlers to nearly every region of the state, while railroads allowed for the development of a robust open range livestock industry (Love 1916). These conditions resulted in conflict between humans and the larger carnivorans—the so-called predators— that could kill livestock. To protect the interests of settlers and ranchers, there was an effort to remove those predators, with a special focus on wolves (*Canis lupus*), bears (*Ursus* spp.), and cougars, though many control methods used at the time were indiscriminate (e.g., applying toxicants to carcasses to poison scavengers). By that time, populations of game species had also been dramatically reduced due to the overharvesting of prior decades as well as the sustenance needs of homesteaders. In 1906, the US Forest Service launched a predator control program. It ended in 1915 when congress instead appropriated funding and directed the Bureau of Biological Survey to oversee predator control activities (Hawthorne 2004). J. Stokley Ligon, whose name is mentioned in several chapters of this volume, was appointed New Mexico–Arizona District inspector within the Biological Survey's Predatory Animal and Rodent Control (PARC) program in 1915 (Chapter 17). Reflecting the prevailing views of the time that the only good predator was a dead one, he wrote the following lines in his 1917 annual report, as quoted in Bailey (1931:307).

> The successful hunting of wolf dens is a science that few have mastered; by successful, I mean involving the destruction of both adults and pups. As a general thing, it is an easy matter to get the pups provided they are found and taken without delay—before the parent animals have a chance to move them after being disturbed. The getting of the adult animals is the problem, and failure to do this has been the occasion, in the past, of much criticism of the bounty system, not always because the adults could not be taken, but too often because there was no effort made—for selfish reasons—to get them.

Through all the early predator control campaigns, many biologists nonetheless gained insight and experience with carnivorans, including not just J. Stokley Ligon but also and more prominently Aldo Leopold and Vernon Bailey. Aldo Leopold, who participated in predator eradication efforts in New Mexico in the early 20th century, was trained as a forester and held several positions within the US Forest Service (including that of Forest Supervisor on the Carson National Forest). He came to distance himself from the idea that the wilderness is a human hunting ground and predators need to be shot down to the last one. Instead, he developed a holistic philosophy emphasizing the integrity of the land, where large carnivores play an important role (Lorbiecki 1998). That New Mexico did not lose more of its carnivoran fauna in the 20th century is based largely on Leopold's influence in shaping land ethics and wildlife management. In his classic of environmental philosophy, *A Sand County Almanac*, Leopold (1949:129) reminisces about a key moment that helped him develop a new understanding of nature and mammalian predators, after he had fired on a female wolf and her pups.

> We reached the old wolf in time to watch a fierce green fire dying in her eyes. I realized then, and have known ever since, that there was something new to me in those eyes— something known only to her and to the mountain. I was young then, and full of trigger-itch; I thought that because fewer wolves meant more deer, that no wolves would mean hunters' paradise. But after seeing the green fire die, I sensed that neither the wolf nor the mountain agreed with such a view.

Among all animals, the world's carnivorans have now experienced the most pronounced anthropogenic range contractions, the result of habitat loss and degradation, a declining prey base, persecution, and over-harvesting for traditional medicine or sport hunting (e.g., Di Minin et al. 2016). In New Mexico, most populations of Carnivores appear more secure—for now—than they were during much of the 20th century, with the gray wolf (Chapter 11), northern river otter (Chapter 19), and black-footed ferret (*Mustela nigripes*; Chapter 22) even reintroduced in the state. As described in this volume, however, significant challenges lie ahead, and climate change in particular may require a rethinking of conservation planning, policies, and practice.

LITERATURE CITED

Arroyo-Cabrales, J. 2002. Registro fósil del jaguar. In *El jaguar del nuevo milenio*, ed. R. A. Medellín, 343–54. Fondo de Cultura Económica, Universidad Nacional Autónoma de México, Mexico City, Mexico.

Bailey, V. 1931 (=1932). *Mammals of New Mexico*. North American Fauna 53. Washington, DC: US Department of Agriculture, Bureau of Biological Survey.

Beals, R. L. 1935. *Preliminary report on the ethnography of the Southwest*. Berkeley, CA: US Department of the Interior National Park Service, Field Division of Education.

Broughton, J. M., and E. M. Weitzel. 2018. Population reconstructions for humans and megafauna suggest mixed causes for North American Pleistocene extinctions. *Nature Communications* 9, article 5441. https://doi.org/10.1038/s41467-018-07897-1.

Chimento, N. R., and A. Dondas. 2018. First record of *Puma concolor* (Mammalia, Felidae) in the early-middle Pleistocene of South America. *Journal of Mammalian Evolution* 25:381–89.

Colbert, E. H., and M. Morales. 1991. *Evolution of the vertebrates*. 4th ed. Wiley-Liss: New York.

Culver, M. 2010. Lessons and insights from evolution, taxonomy, and conservation genetics. In, *Cougar: ecology and conservation*, ed. M. Hornocker and S. Negri, 27–40. Chicago: University of Chicago Press.

Culver, M., W. E. Johnson, J. Pecon-Slattery, and S. J. O'Brien. 2000. Genomic ancestry of the American puma (*Puma concolor*). *Journal of Heredity* 91:186–197.

Di Minin, E., R. Slotow, L. T. B. Hunter, F. Montesino Pouzols, T. Toivonen, P. H. Verburg, N. Leader-Williams, L. Petracca, and A. Moilanen. 2016. Global priorities for national carnivore conservation under land use change. *Scientific Reports* 6:23814. doi: 10.1038/srep23814

Dragoo, J. W., and R. L. Honeycutt. 1997. Systematics of mustelid-like carnivores. *Journal of Mammalogy* 78:426–43.

Dutton, B. P. 1991. *American Indians of the Southwest*. Albuquerque: University of New Mexico Press.

Figueirido, B., N. MacLeod, J. Krieger, M. De Renzi, J. A. Perez-Claros, and P. Palmqvist. 2011. Constraint and adaptation in the evolution of carnivoran skull shape. *Paleobiology* 37:490–518.

Goswami, A. 2010. Introduction to Carnivora. In *Carnivoran evolution: new views on phylogeny, form and function*, ed. A. Goswami and A. Friscia, 1–24. Cambridge: Cambridge University Press.

Hassanin, A, G. Veron, A. Ropiquet, B. Jansen van Vuuren, A. Lécu, S. M. Goodman, J. Haider, and T. T. Nguyen. 2021. Evolutionary history of Carnivora (Mammalia, Laurasiatheria) inferred from mitochondrial genomes. *PLOS One* 16(2): e0240770. https://doi.org/10.1371/journal.pone.0240770

Hawthorne, D. W. 2004. The history of federal and cooperative animal damage control. *Sheep and Goat Research Journal* 19:13–15.

Hemmer, H., R.-D. Kahlke, and A. K. Vekua. 2001. The Jaguar—*Panthera onca gombaszoegensis* (Kretzoi, 1938) (Carnivora: Felidae) in the late lower pleistocene of Akhalkalaki (south Georgia; Transcaucasia) and its evolutionary and ecological significance. *Geobios* 34:475–86.

Kurtén, B., and E. Anderson. 1980. *Pleistocene mammals of North America*. New York: Columbia University Press.

Leopold, A. 1949. *A Sand County almanac*. New York: Oxford University Press.

Lorbiecki, M. 1998. The land makes the man: New Mexico's influence on the conservationist Aldo Leopold. *New Mexico Historical Review* 73:235–52.

Love, C. M. 1916. History of the cattle industry in the Southwest. *Southwestern Historical Quarterly* 19:370–99.

Marciszak, A. 2014. Presence of *Panthera gombaszoegensis* (Kretzoi, 1938) in the late Middle Pleistocene of Biśnik cave, Poland, with an overview of Eurasian jaguar size variability. *Quaternary International* 326:105–13.

Martin L. D. 1989. Fossil history of the terrestrial Carnivora. In *Carnivore behavior, ecology, and evolution*, ed. J. L. Gittleman, 536–68. Ithaca, NY: Cornell University Press.

Mol, D., W. van Logchem, and J. de Vos. 2001. New record of the European jaguar, *Panthera onca gombaszoegensis* (Kretzoi, 1938), from the Plio-Pleistocene of Langenboom (The Netherlands). *Cainozoic Research* 8:35–40.

Nyhus, P. J. 2016. Human-wildlife conflict and coexistence. *Annual Review of Environment and Resources* 41:143–71.

O'Regan, H. J. 2002. A phylogenetic and palaeoecological review of the Pleistocene felid *Panthera gombaszoegensis*. PhD dissertation, Liverpool John Moores University, Liverpool, UK.

Ripple, W. J., and B. Van Valkenburgh. 2010. Linking top-down forces to the Pleistocene megafaunal extinctions. *BioScience* 60:516–26.

Solé, F., T. Smith, E. De Bast, V. Codrea, and E. Gheerbrant. 2016. New Carnivoraforms from the latest Paleocene of Europe and their bearing on the origin and radiation of Carnivoraformes (Carnivoramorpha, Mammalia). *Journal of Vertebrate Paleontology* 36:e1082480. doi: 10.1080/02724634.2016.1082480.

Stimpson, C. M., P. S. Breeze, L. Clark-Balzan, H. S. Groucutt, R. Jennings, A. Parton, E. Scerri, T. S. White, and M. D. Petraglia. 2015. Stratified Pleistocene vertebrates with a new record of a jaguar-sized pantherine (*Panthera* cf. *gombaszogensis*) from northern Saudi Arabia. *Quaternary International* 382:168–80.

Surovell, T. A., S. R. Pelton, R. Anderson-Sprecher, and A. D. Myers. 2016. Test of Martin's overkill hypothesis using radiocarbon dates on extinct megafauna. *Proceedings of the National Academy of Sciences* 113:886–91.

Tamagnini, D., C. Meloro, P. Raia, and L. Maiorano. 2021. Testing the occurrence of convergence in the craniomandibular shape evolution of living carnivorans. *Evolution* 75:1738–52. doi:10.1111/evo.14229

Tarquini, J., L. H. Soibelzon, R. Salas-Gismondi, and C. De Muizon. 2020. *Cyonasua* (Carnivora, Procyonidae) from Late Miocene of Peru shed light on the early dispersal of Carnivorans in South America. *Journal of Vertebrate Paleontology* 40(5): e1834406. doi: 10.1080/02724634.2020.1834406

Ungar, P. S. 2010. *Mammal teeth: origin, evolution, and diversity*. Baltimore: Johns Hopkins University Press.

Van Valkenburgh, B. 1999. Major patterns in the history of carnivorous mammals. *Annual Review of Earth and Planetary Sciences* 27:463–93.

———. 2007. Déjà vu: the evolution of feeding morphologies in the Carnivora. *Integrative and Comparative Biology* 47:147–163. doi:10.1093/icb/icm016

Weber, D. J. 1971. *The Taos trappers: the fur trade in the far Southwest, 1540–1846*. Norman: University of Oklahoma Press.

Werdelin, L. 1996. Carnivoran ecomorphology: a phylogenetic perspective. In *Carnivore Behavior, Ecology and Evolution*, ed. L. Gittleman, 582–624. Ithaca, NY: Cornell University Press.

AADT	average annual daily traffic
ADC	Animal Damage Control
AMNH	American Museum of Natural History
APHIS	Animal and Plant Health Inspection Service
ASM	Arizona State Museum
ASNHC	Angelo State Natural History Collections
AVMA	American Veterinary Medical Association
BBS	Bureau of Biological Survey
BLM	Bureau of Land Management
BMZ	Bear Management Zone
BP	before present
CAVE	Carlsbad Caverns National Park
CDC	Centers for Disease Control and Prevention
CDOW	Colorado Division of Wildlife
CDV	canine distemper virus
CITES	Convention on International Trade in Endangered Species of Wild Fauna and Flora
CLAWS	Colorado Lynx and Wolverine Strategy
CMIP5	Coupled Model Intercomparison Project, Phase 5
COAs	Conservation Opportunity Areas
CPV	canine parvovirus
CPV-2	canine parvovirus type 2, a.k.a. *Carnivore protoparvovirus 1*
CPW	Colorado Parks and Wildlife
DDE	dichlorodiphenyldichloroethylene
DDT	dichlorodiphenyltrichloroethane
DMNS	Denver Museum of Nature and Science
DOR	dead on the road
enFeLV	endogenous feline leukemia virus
ENMUNHM	Eastern New Mexico University, Natural History Museum
EPA	Environmental Protection Agency
ESA	Endangered Species Act
ESP	Endangered Species Program
FCoV	feline coronavirus
FCV	feline calicivirus
FeLV	feline leukemia virus

FFV	feline foamy virus
FHSM	Fort Hays State University, Sternberg Museum of Natural History
FIV	feline immunodeficiency virus
FMNH	The Field Museum of Natural History
FPV	feline panleukopenia virus (feline distemper)
FY	fiscal year
GABI	Great American Biotic Interchange
GAM	General Additive Model
GBIF	Global Biodiversity Information Facility
GIS	Geographic Information System
GISD	Global Invasive Species Database
GPA	Game Protective Association
HIV	human immunodeficiency virus
HSU	Humboldt State University Vertebrate Museum
IG	intraguild
IGP	intraguild predation
IHS	Indian Health Service
IPCC	Intergovernmental Panel on Climate Change
IUCN	International Union for the Conservation of Nature and Natural Resources
K	nutritional-carrying capacity
KU	University of Kansas, Natural History Museum and Biodiversity Research Center
kya	thousand years ago
LPC	livestock protection collar
MACA	Multivariate Adaptive Constructed Analogs
MCPs	minimum convex polygons
MOU	Memorandum of Understanding
MRGCOG	Middle Rio Grande Council of Governments
MSB	University of New Mexico, Museum of Southwestern Biology
mtDNA	mitochondrial DNA
MVZ	University of California, Berkeley, Museum of Vertebrate Zoology
MWEPA	Mexican Wolf Experimental Population Area
mya	million years ago
NASS	National Agriculture Statistics Service
NCSM	North Carolina Museum of Natural Sciences
NEPA	National Environmental Policy Act
NMDA	New Mexico Department of Agriculture

NMDGF	New Mexico Department of Game and Fish
NMDOH	New Mexico Department of Health
NMDOT	New Mexico Department of Transportation
NMMNH	New Mexico Museum of Natural History and Science
NMSA	New Mexico Statutes Annotated
NMSU	The Vertebrate Museum, New Mexico State University
NMSUVWM	New Mexico State University Vertebrate Wildlife Museum (a.k.a. NMSU Wildlife Museum)
NRCS	Natural Resources Conservation Service
NSRL	Natural Science Research Laboratory
NWR	National Wildlife Refuge
NYSM	New York State Museum
OMNH	University of Oklahoma, Sam Noble Oklahoma Museum of Natural History
PARC	Predatory Animal and Rodent Control
PCBs	polychlorinated biphenyls
PCEs	primary constituent elements
PDM	predator damage management
PIT	passive integrated transponder
RCP	Representative Concentration Pathway
RMSF	Rocky Mountain spotted fever
SEMARNAT	Ministry of Environment and Natural Resources of Mexico
SFCT	Swift Fox Conservation Team
SGCN	Species of Greatest Conservation Need
SNPs	single nucleotide polymorphisms
SPV	sylvatic plague vaccine
SSP	Species Survival Plan
SWE	Snow Water Equivalent
TCWC	Biodiversity Research and Teaching Collection (formerly Texas Cooperative Wildlife)
TESF	Turner Endangered Species Fund
TNR	trap-neuter-return
TTU-NSRL	Texas Tech University–Natural Science Research Laboratory
UA	Collection of Mammals, University of Arizona
UCM	University of Colorado Museum of Natural History
UNM	University of New Mexico
UMMZ	University of Michigan Museum of Zoology
USDA	United States Department of Agriculture

USFS	United States Forest Service
USFWS	United States Fish and Wildlife Service
USGCRP	United States Global Change Research Program
USGS	United States Geological Survey
USNM	United States National Museum of Natural History
UTEP	University of Texas, El Paso, Biodiversity Collections, Mammals Division
UWYMV	University of Wyoming Museum of Vertebrates
VCNP	Valles Caldera National Preserve
VHF	very high frequency
WGFD	Wyoming Game and Fish Department
WNMU	Western New Mexico University
WNV	West Nile virus
WRI	Wildlife Rescue Inc.
WS	Wildlife Services
ybp	years before present
YPM	Yale University, Peabody Museum of Natural History

SKULL MORPHOLOGY

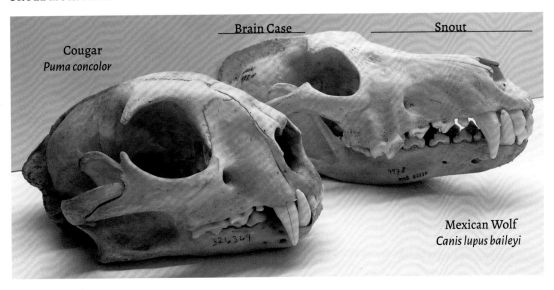

Brain Case   Snout

Cougar
Puma concolor

Mexican Wolf
Canis lupus baileyi

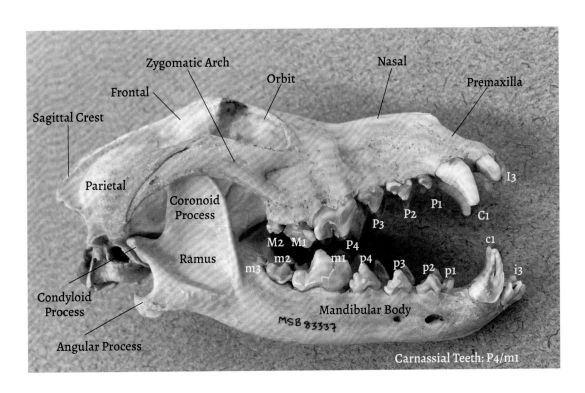

Zygomatic Arch

Orbit

Nasal

Premaxilla

Frontal

Sagittal Crest

Parietal

Coronoid Process

Ramus

Condyloid Process

Angular Process

I3

P1

P2

C1

P3

c1

M2 M1

P4

M2 m1

m2

m1 p4

m3

p3

p2 p1

i3

Mandibular Body

MSB 83337

Carnassial Teeth: P4/m1

PAW MORPHOLOGY

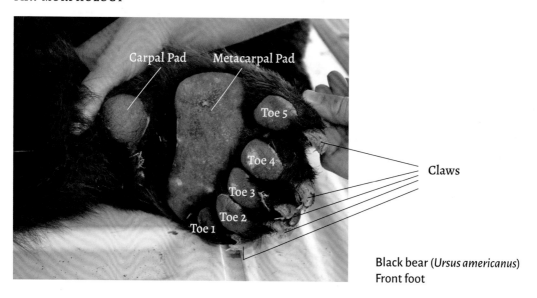

Carpal Pad Metacarpal Pad

Toe 5

Toe 4

Toe 3

Toe 2

Toe 1

Claws

Black bear (*Ursus americanus*)
Front foot

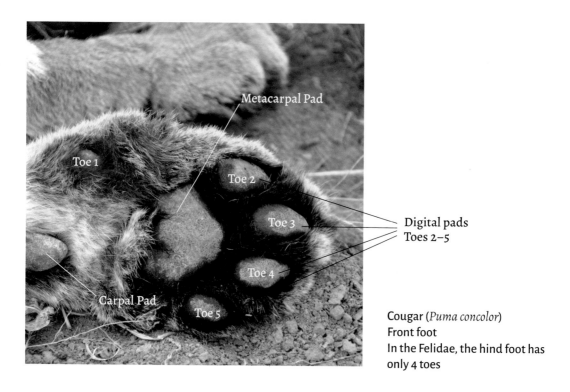

Metacarpal Pad

Toe 1

Toe 2

Toe 3

Digital pads
Toes 2–5

Toe 4

Carpal Pad

Toe 5

Cougar (*Puma concolor*)
Front foot
In the Felidae, the hind foot has
only 4 toes

MORPHOMETRICS

Body Length

Tail Length (fur beyond
the last tail vertebra is
excluded)

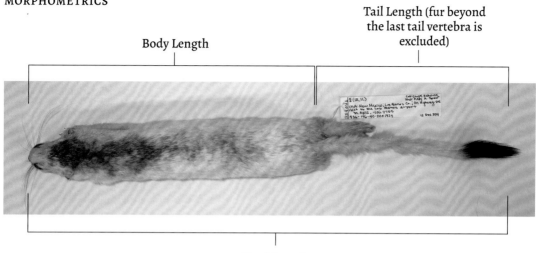

Total Length

Photo credits. Skull and specimen photos: ©Jon Dunnum; black bear and cougar paws: © Greg D. Wright.

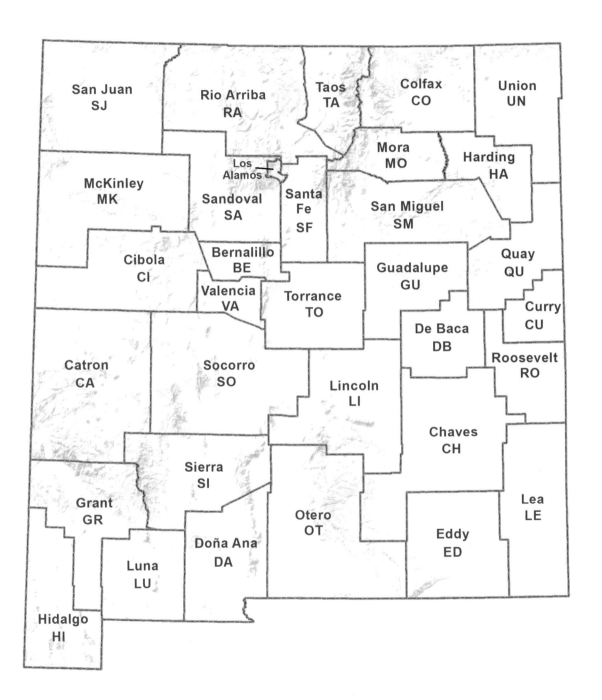

Map i.1. New Mexico's counties.

GENERAL NEW MEXICO MAPS

With a total area of 314,917 km² (121,590 mi²), New Mexico is the fifth largest US state. Five main rivers flow through the state, the Rio Grande, Pecos, Canadian, San Juan, and Gila. Elevation ranges from 866 m (2,842 ft) at Red Bluff Reservoir in Eddy County to 4,013 m (13,167 ft) on Wheeler Peak, located in Taos County in the Sangre de Cristo Mountains. New Mexico is divided into 33 counties (designated by acronyms on the species maps) and harbors seven national wildlife refuges, two national parks, and five national forests. More than 920,000 people live in the Greater Albuquerque metropolitan area. Based on the 2020 Census Redistricting dataset, Las Cruces now holds a population of 111,385. No other town in New Mexico harbors a population of more than 100,000.

SPECIES DISTRIBUTION MAPS

Whereas in the *Raptors of New Mexico* range boundary maps were drawn for most of the 37 birds of prey regularly occurring in the state, in this volume we rely instead mainly on point location maps. Each map type has potential shortcomings. Range boundary maps can overestimate the true distribution of a species (commission errors); point location maps tend to underestimate it (omission errors). Despite some spatial and temporal biases, specimen records are the preferred distributional data used to generate point location maps. Species specimens present the advantage of having been taxonomically validated by experts. They are preserved in the natural history collections of museum institutions, where they remain available for any re-examination in the event of taxonomic changes. Most specimen records are also now widely available through online searchable databases.

In some cases, specimen records can be supplemented with other verified distributional information. For example, researchers are now increasingly using remote cameras equipped with motion triggers to assess the distribution and habitat associations of species in certain areas. Remote camera photos from those studies can augment the occurrence data for a larger-scale distribution map. Other verified distributional data consists of trapping data obtained by researchers.

Not all species are well represented in museum collections. For example, no Canada lynx specimens were ever collected in New Mexico, despite the likelihood that the species was present historically in the state. In this volume, the lynx distribution map is based on telemetry data, which requires animals to be fitted with GPS collars.

Range boundary maps (or polygon range maps) are generated based on known occurrence data, which are then typically interpolated to draw range boundaries. The resulting polygons assume a homogeneous species distribution, with the potential for commission errors. A range boundary map is presented in the swift fox (*Vulpes velox*) chapter, based on extensive surveys within the historical distribution of the species in New Mexico. Specimen records exist from outside the range boundaries, but they are now dated as the distribution of the swift fox has contracted in the state.

Range boundary maps can also be predictive in nature, using land-cover characteristics and other variables to model habitat. The results of the habitat model must then be validated with occurrence data, such as locations of hunter-killed animals.

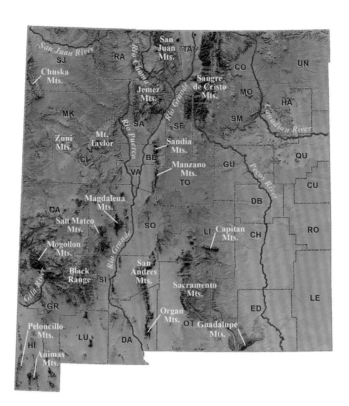

Map i.2. Main physical features of New Mexico's geography.

Map i.3. New Mexico's national parks, national forests, and national wildlife refuges.

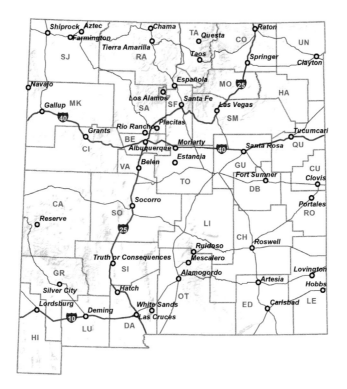

Map i.4. New Mexico's main roads and towns.

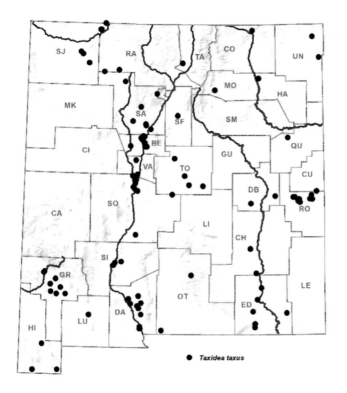

Map i.5. Example of point location map based on geo-referenced specimen records: the badger (*Taxidea taxus*).

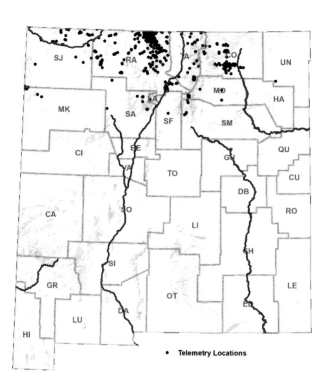

Map i.6. Point location map based on geo-referenced specimen records, physical capture locations, and remote camera photos: the gray fox (*Urocyon cinereoargenteus*).

Map i.7. Point location map based on telemetry data: the lynx (*Lynx canadensis*). Note that the symbols on the map represent positional fixes rather than individual animals.

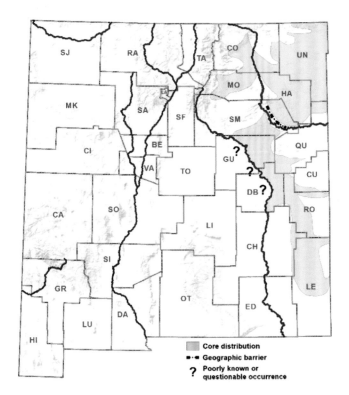

Map i.8. Example of range boundary map: the swift fox (*Vulpes velox*).

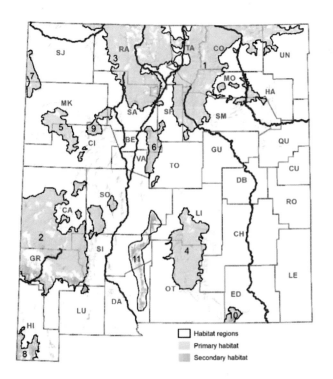

Map i.9. Example of predictive range boundary map: the black bear (*Ursus americanus*).

CARNIVORES IN THE ICE AGE OF NEW MEXICO

Arthur H. Harris

The fossil record of carnivores in New Mexico starts in the Paleocene, near the time when the Carnivora can first be recognized, and continues through the recently ended Pleistocene—the ice age of popular usage (Fig. 1.1). The modern-day carnivores of New Mexico form an impoverished fauna descended directly from the more diverse fauna of the Pleistocene.

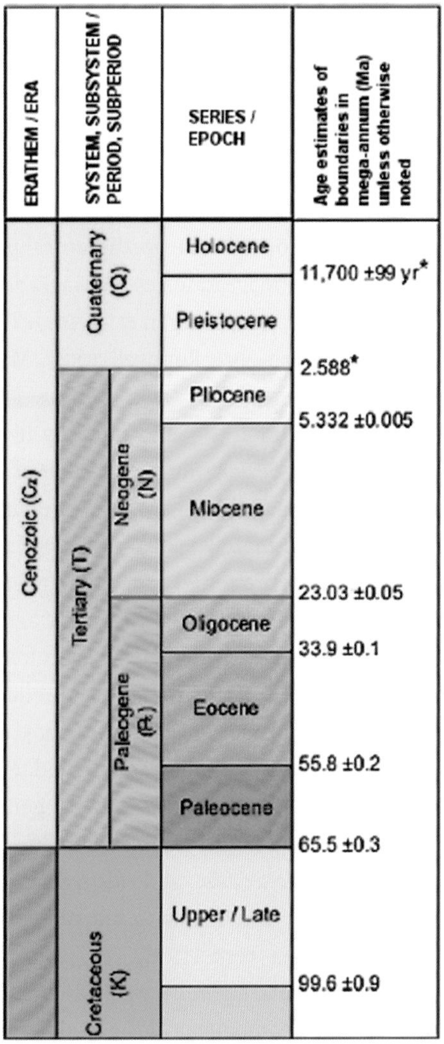

EARLY DAYS

Although the focus of this review is on relatively recent geologic time, a brief overview of the earlier New Mexican fossil record of carnivores is in order.

Members of the order Carnivora are characterized by, among other features, a pair of teeth, the carnassials, that allow their recognition in the fossil record. The carnassials primitively are specialized for slicing and consist of the fourth (last) premolar of the upper jaw and the first molar of the lower jaw.

The earliest recognizable carnivores, members of the extinct family Viverravidae, appeared in the early Paleocene. In New Mexico, the family appears in the Paleocene of the San Juan Basin. Genera include *Protictis* and *Intyrictis*. Goswami (2010:5) described them as "small- to medium-sized terrestrial animals that incorporated insects as a large part of their diet." In the past they often have been considered as ancestral to living carnivores, but it now appears that the

ERATHEM / ERA	SYSTEM, SUBSYSTEM / PERIOD, SUBPERIOD		SERIES / EPOCH	Age estimates of boundaries in mega-annum (Ma) unless otherwise noted
Cenozoic (Cz)	Quaternary (Q)		Holocene	
				11,700 ±99 yr*
			Pleistocene	
				2.588*
	Tertiary (T)	Neogene (N)	Pliocene	
				5.332 ±0.005
			Miocene	
				23.03 ±0.05
		Paleogene (Pe)	Oligocene	
				33.9 ±0.1
			Eocene	
				55.8 ±0.2
			Paleocene	
				65.5 ±0.3
	Cretaceous (K)		Upper / Late	
				99.6 ±0.9

Figure 1.1. Late Cretaceous and post-Cretaceous geologic time, courtesy of the US Geological Survey.

(*opposite page*) New Mexico Museum of Natural History and Science mural by Kent Pendleton. Photograph: © Jean-Luc Cartron.

Viverravidae was a side lineage without descendants (Polly et al. 2006).

Although named as a family, the "Miacidae" is a "waste basket" group of primitive carnivores whose relationships to one another are unclear, though they are somewhat advanced over the Viverravidae. The most recent common ancestor of all the living Carnivora apparently is somewhere within this group. The miacid *Uintacyon* has been identified from the Eocene of the Galisteo Basin (Stearns 1943), and five genera have been identified from Eocene deposits in the San Juan Basin (Lucas et al. 1981).

Fossiliferous deposits of the Oligocene are poorly represented in New Mexico, but the Miocene has large deposits in the north-central portions of the state (MacFadden 1977; MacFadden et al. 1979; Tedford 1981; Morgan et al. 1997; Tedford and Barghoorn 1997; Rothwell 2001). Modern families and even some modern genera are recorded, though none is represented by living species. The genus *Bassariscus* is a member of the raccoon family (Procyonidae) and is the genus to which the modern ringtail belongs. The dog family (Canidae) was represented by the living genus *Vulpes*, represented today by such canids as the North American red fox (*Vulpes fulva*), swift fox (*V. velox*), and kit fox (*V. macrotis*). The cats (Felidae) showed up in the form of *Pseudaelurus*, a possible ancestor to later cats, and the bears (Ursidae) as *Hemicyon ursinus* and *Cephalogale*. Several genera represent the weasel family (Mustelidae), including *Promartes*, *Pliotaxidea*, and *Plesiogulo*, the names suggesting similarities to martens, badgers, and wolverines, respectively.

Not all Miocene carnivores have close modern relatives. A major component of the canid family is now an extinct subfamily, the Borophaginae, the so-called bone-crushing dogs (Wang et al. 1999). Also appearing then was the Amphicyonidae, the extinct bear-dogs, creatures displaying some characteristics of both bears and dogs (Hunt 1998).

Morgan and Lucas (2003) summarized the New Mexican mammalian record for the time span between about 4.9 and 1.8 million years ago (mya), an interval known as the Blancan North American Land Mammal Age. The borophagine dogs continued into the early Pleistocene before becoming extinct, and *Canis* appeared in the form of the extinct, coyote-sized *C. lepophagus* that survived into the early Pleistocene. Both cat subfamilies (Machairodontinae and Felinae) were present, but unidentified to lower taxonomic level. The Procyonidae was still represented by *Bassariscus*, and the American badgers (*Taxidea*) represented the Mustelidae. Bears were present, but unidentifiable other than to family.

THE PLEISTOCENE

The Pleistocene is the geologic epoch immediately preceding our present Holocene epoch. It was long considered to have started about 1.8 mya, but the International Union of Geological Sciences has now officially recognized (though not without controversy) the beginning of the Pleistocene as approximately 2.6 mya. Thus the Pleistocene has been extended significantly back into what was the late Pliocene. The new date reflects the ramping up of glacial activity in the Northern Hemisphere, with approximately a score of cold glacial and warm interglacial cycles of various magnitudes up to our present interglacial. Although fossil carnivores are known from throughout the Pleistocene deposits of New Mexico, the best record is from the latest of the major glacial ages, the Wisconsin (also known as the Wisconsinan). The most severe glacial episode of the Wisconsin climaxed about 20 thousand years ago (kya) and, in contrast to the slow buildup of glacial conditions, rapidly reached arguably early interglacial conditions by around 15 kya. The end of the Rancholabrean and the Pleistocene is marked by a major extinction event, when most large North American mammals (those over 45 kg [~100 lbs]) became extinct (the Pleistocene megafaunal extinction). A number of carnivores were included in the extinction. The Pleistocene epoch is replaced by the Holocene at about 11,700 calendar years ago (10,000 radiocarbon years ago).

Asia and North America currently are separated from each other only by the narrow Bering Strait. During glacials, sea levels fell due to the large amount of water tied up in terrestrial ice caps, and Siberia and Alaska were connected by a broad land bridge. Eastern Siberia, the land bridge, and non-glaciated central Alaska and vicinity formed a geographic entity called Beringia. Beringia was mostly or entirely cut off from the more southerly regions of North America by ice sheets during glacial ages. With each advance of the ice sheets, vegetation was forced southward, though there is little direct evidence of the nature of the vegetation until the late Pleistocene. During the relatively warm interglacial conditions, the climate and vegetation became distributed somewhat similarly to today, allowing faunal interchange between Beringia and regions of North America farther south. Carnivoran movements during the Pleistocene were predominantly southward, made by animals originating in Asia.

Numerical dates often are not available for fossils. Chronological information may be given by referring to climatic events (Marine Isotope Stages; Glacial/Interglacial Ages) or biological events (North American Land Mammal Ages). The ages of the latter are characterized by particular suites of mammals. Of concern here are the last three. The Blancan lies between about 4.9 and 1.8 mya, with only the late Blancan (ca. 2.6 to 1.8 mya) lying within the Pleistocene; the Irvingtonian encompasses the time span from about 1.8 mya to about 250 kya; the Rancholabrean finishes up the Pleistocene, ending with the mass extinction of large mammals (the Pleistocene megafaunal extinction).

Although the last glacial age, the Wisconsin, was characterized throughout by large terrestrial ice sheets, the mid-Wisconsin was somewhat milder than the early Wisconsin and the severe late Wisconsin, and other fluctuations occurred within these major divisions of the Wisconsin. It is only for the mid- and late Wisconsin that the fossil record is outstanding, and thanks to the sparsity of cave faunas in the northern sector, the fossil record appears to be nearly continuous only for the southern portions of New Mexico.

Based on a wide variety of disciplines, including paleontology, geology, palynology, and preserved vegetation in woodrat (*Neotoma*) middens, it is clear that New Mexico was ruled by far different climatic conditions than at present during the last glacial age. Cool summers and increased effective moisture allowed northern and high-elevation plants to invade areas currently barred to them by warm temperatures and aridity. Northern coniferous forest replaced much of the more temperate vegetation, thus allowing movement of forest species between now-isolated mountain ranges; the desert and desert grasslands were replaced with woodlands and grasslands more typical of areas to the north. Big sagebrush (*Artemisia tridentata*), so typical now of the Great Basin, apparently came far south, probably even into extreme northern Mexico (Harris 1990a). Along with shifting plants came the herbivores and omnivores dependent on them, and following along, their predators—the carnivores.

From the point of view of carnivores, there was a wealth of prey—but also an abundance of competitors. Representatives of most of today's groups of medium and small prey thrived. By the Wisconsin, both living and extinct (marked by the dagger symbol †) larger prey included various species of horses (*Equus*, extinct in the New World at the end of the Pleistocene), Shasta ground sloth (†*Nothrotheriops shastensis*), two kinds of camels (†*Camelops hesternus* and †*Hemiauchenia macrocephala*), at least three kinds of pronghorn (†*Capromeryx furcifer*, †*Stockoceros conklingi*, and *Antilocapra americana*), mammoth (†*Mammuthus columbi*), extinct species of bison (†*Bison*),

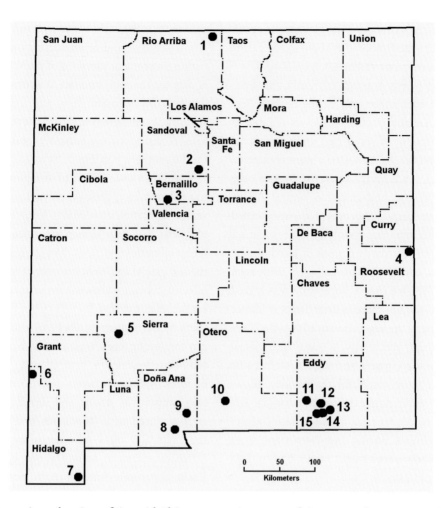

Map 1.1. Approximate locations of sites with Pleistocene carnivoran records in New Mexico. 1 San Antonio Cave (SAM Cave), 2 Sandia Cave, 3 Isleta Caves, 4 Blackwater Draw, 5 Palomas Creek Cave, 6 Virden, 7 U-Bar Cave, 8 Mesilla Basin (La Union, Adobe Ranch), 9 Conkling Cavern, 10 Pendejo Cave, 11 Burnet Cave, 12 Dry Cave, 13 Dark Canyon Cave, 14 Big Manhole Cave, and 15 Muskox Cave.

bighorn sheep (*Ovis canadensis*), Harrington's mountain goat (†*Oreamnos harringtoni*), and deer (†*Navahoceros, Odocoileus*), along with a few other large prey types. But also present were the dire wolf (†*Canis dirus*), the American lion (†*Panthera atrox*), the huge short-faced bear (†*Arctodus simus*) and, at least marginally, the sabertooth (†*Smilodon fatalis*)—these in addition to nearly all of the predators currently inhabiting New Mexico.

Extinctions, both of prey and of predator, demanded wholesale reorganization of biological communities toward the end of the Pleistocene. As many of the large prey animals became extinct, the larger carnivores lost much of their subsistence base and in turn disappeared, leaving only the historic remnants of the once-larger carnivoran fauna. With the loss of the herbivorous megafauna, there also must have been major vegetational adjustments to that loss; the details, and even the major features, of such are largely unknown, but presumably changes in vegetative cover, density, and taxonomic makeup must have affected the hunting strategies of the surviving predators.

PLEISTOCENE SITES

Although there are many Pleistocene sites in New Mexico, many record single species or a very limited number of kinds of mammals, mostly large forms such as mammoth, horse, and camel. Most carnivoran records come from cave sites, and most from a relatively small number of such. A complete list with descriptions of Pleistocene sites in the state can be found in Harris (2012b; and see Map 1.1). A few of the more important ones for records of carnivores follow.

Big Manhole Cave, Eddy County. This cave is just outside of the northern boundary of Carlsbad Caverns National Park. It currently consists of a domed chamber with the sole opening in the ceiling about 15 m above the chamber floor. A large Holocene fauna was excavated from the floor of the chamber (Lear and Harris 2007), and an important Pleistocene fauna was recovered by spelunkers from an exploratory shaft and deeper chambers revealed by the shaft (Harris 2012a).

Burnet Cave, Eddy County. One of the many caves associated with the Guadalupe Mountains and vicinity. Excavated in the early 1930s, it produced a large, important fauna (Schultz and Howard 1935).

Dark Canyon Cave, Eddy County. A large rock shelter near the mouth of Dark Canyon close to the edge of the Pecos River floodplain near Carlsbad. Various collections over a number of years have resulted in a very large fauna (Howard 1971; Tebedge 1988; Harris 2012a). It apparently dates to the late mid-Wisconsin (a radiocarbon date is 21,120 ± 420 [Tebedge 1988], but on bone apatite, often considered untrustworthy).

Dry Cave, Eddy County. Some 24 different sites, ranging in age from mid- or early Wisconsin to Holocene, lie within this extensive cave system. Major collections have been made over several decades (Harris 2012a), including by two field schools from the University of Texas at El Paso.

Isleta Cave No. 1, Isleta Cave No. 2, Bernalillo County. Caves in lava formed from gases or draining of liquid lava from below solidified lava. Isleta Cave No. 2 is a pitfall, self-baiting trap, luring in carnivores by the presence of trapped small mammals such as rabbits and prairie dogs; once in, most mammals could not access the ceiling entrance. The major excavation was in 1946 by an archaeological field school from the University of New Mexico. Faunal remains include Holocene material as well as late Wisconsin fossils (Harris and Findley 1964; Harris 2012a). Apparently, much mixing has occurred.

Muskox Cave, Eddy County. The cave, within the boundaries of Carlsbad Cavern National Park, reportedly includes material from the Sangamon interglacial (the interglacial previous to our current one) to the Holocene. Radiocarbon dates available are at about 25 and 18 kya (Logan 1981). Unfortunately, dates associated with specific fossils are unavailable.

Pendejo Cave, Otero County. This small cave is located on Fort Bliss land on the western breaks of Otero Mesa, on the east side of the Tularosa Basin. Deposits range in age from modern to >55,000 radiocarbon years; a large number of radiocarbon dates are available (Harris 2003; MacNeish and Libby 2003).

U-Bar Cave, Hidalgo County. Located in the "bootheel" of extreme southwestern New Mexico near the Mexican border, U-Bar Cave has produced large mid- and late Wisconsin faunas and, together with Dry Cave sites and Pendejo Cave, has allowed reconstruction of mid- and late Wisconsin faunas and vegetation across southern New Mexico (Harris 1987, 1989).

FAUNA

In the following accounts, the relatively sparse earlier Pleistocene fossil record receives mention, but concentration is on the far better record of the late Wisconsin. Published Pleistocene occurrences of carnivores in New Mexico are mapped by Harris (2012a).

Felidae—Cat Family

Pleistocene cats fall into two subfamilies: the Machairodontinae with flattened, elongated upper canines and reduced lower canines; and the Felinae with conical upper and lower canines more or less subequal in size (Kurtén and Anderson 1980). Both are represented in New Mexico, though the Felinae is by far the more common.

MACHAIRODONTINAE
†*Smilodon gracilis*. Gracile Sabertooth.

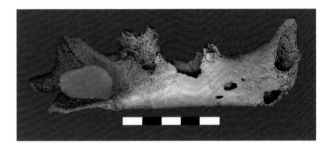

Photo 1.1. Right lower jaw of *Smilodon gracilis* (UTEP 97-2) from the La Union fauna. Scale bar = 5 cm. Photograph: © Art H. Harris.

This sabertooth was roughly the size of modern jaguars at 55–100 kg (121–220 lbs) by the estimate of Christiansen and Harris (2005). It is represented in New Mexico by a single lower jaw (Photo 1.1) from early Pleistocene deposits of the Mesilla Basin south of Las Cruces (Morgan and Lucas 2003).

†*Smilodon fatalis*. Sabertooth.
Smilodon fatalis is the animal most commonly thought of when *sabertooth* is mentioned (unfortunately, often as *sabertooth tiger*, though tigers are in the other subfamily and share few characteristics other than those possessed by all cats). According to the study by Christiansen and Harris (2005), body mass ranged from 160 kg (352 lbs) to 280 kg (616 lbs), comparable to that of the largest living cat, the Siberian tiger, *Panthera tigris altaica*. New Mexican records include lower jaws from the Blackwater Draw Fauna in Roosevelt County (Lundelius 1972) and the 25 Mile Stream site in Eddy County (Morgan and Lucas 2001).

There has been much speculation about how sabertooths utilized their formidable upper incisors to kill their prey. A recent study by McHenry

Photo 1.2. Reconstruction of *Smilodon fatalis* by Charles R. Knight for the American Museum of Natural History, 1905. Public domain.

et al. (2007) found that the jaws were relatively weak, suggesting that prey was brought to ground and immobilized before the powerful neck muscles were utilized to drive the canines into the neck region.

In general, predator size tends to vary roughly with the preferred prey size. Thus, likely prey were in the general size range of Pleistocene camels and horses, though perhaps with emphasis on subadults and young.

FELINAE

Lynx rufus. Bobcat.

The bobcat is nearly ubiquitous today, and Pleistocene fossils are widespread in cave deposits. Most bobcat fossils are from the late Pleistocene, but a possibly aberrant specimen is known from the latest Blancan La Union fauna, and another individual is from the Irvingtonian Adobe Ranch deposits (Photo 1.3). Presumably, prey selection was similar to that of present-day bobcats.

†Panthera atrox. American Lion.

This cat is a huge problem, literally and figuratively. It represented one of the largest members of the subfamily Felinae *sensu lato* (Christiansen and Harris 2009) but has been shunted around taxonomically. There are possible problems in separating New Mexican specimens of this species from fossil jaguars (see the jaguar account in Chapter 8).

Kurtén and Anderson (1980), in common with the times, considered this to be a large sized member of *Panthera leo*, commonly known to the general public as the African lion. A morphological study, however, has concluded that it has jaguar affinities and that *P. leo*, though occurring in Beringia, never made it to the lower United States (Christiansen and Harris 2009). However, a DNA study, also published in 2009, found DNA from *P. atrox*, the extinct Old World cave lion (*P. spelaea*), and modern *P. leo* formed

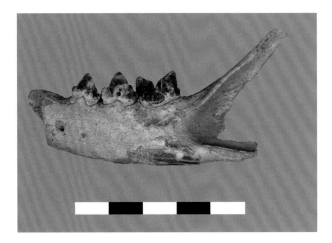

Photo 1.3. Left lower jaw of *Lynx rufus* from the Irvingtonian Adobe Ranch fauna. UTEP 97-3; scale bar = 5 cm. Photograph: © Art H. Harris.

a grouping of three clusters separate from the jaguar, negating any link with the jaguar line (Barnett et al. 2009). There was no evidence of gene flow between the three lion groups, which would suggest that the groupings denote three different species. Nevertheless, further study was deemed necessary to clarify the specific status. Thus, conflicting evidence renders both the specific status and the relationships within the Felinae of this beast problematical. On the basis of the DNA evidence, Barnett et al. (2009) estimated the split between ancestral Beringian *P. spelaea* and the southern populations that would become *P. atrox* occurred about 337 kya.

Pleistocene jaguars were larger than present-day northern jaguars, but are said to be distinguishable from *P. atrox* by size, the jaguar (*Panthera onca*) being smaller (Kurtén and Anderson 1980). Some New Mexican specimens, however, are intermediate in size, judging by published measurements. These are interpreted here as pertaining to *P. atrox*.

Four sites in New Mexico have produced remains of this lion, from the Isleta Caves in the north to three sites in the southeastern part of the state (Harris 2012b).

Panthera onca. Jaguar.

There is one site with sure identifications of Pleistocene jaguar in New Mexico. Dry Cave is about halfway between the Guadalupe Mountains proper and the Pecos River (Harris 1970). The older deposits of Dry Cave are mid-Wisconsin or older and have produced a fragmentary skull clearly representing the jaguar.

Rancholabrean records of *P. onca* reveal a geographic range in North America far greater than today's. Sites in Oregon, Nebraska, and Tennessee suggest the northern limits, though the northern portions of the range were retreating to the south by Wisconsin times (Kurtén and Anderson 1980).

Puma concolor. Cougar.

The Pleistocene cougar, or mountain lion, is the same species as present through much of North and South America today. It is thought that an Old World ancestor moved into North America about 6 mya, with its descendants splitting into the cougar and the American cheetah-like cat about 3.2 mya (Barnett et al. 2005). Genetic data from North American populations show very little variation compared with the variation among South American populations. This may indicate that the North American populations died out at about the same time as so many other large mammals near the end of the Pleistocene; if this is the case, then the present North American cougars are descended from a relatively few individuals repopulating the area from the south (Culver et al. 2000; and see Chapter 9).

†Miracinonyx trumani. American Cheetah-like
    Cat.

This cat, roughly the size of a cougar (*Puma concolor*), was originally described as a new species of *Felis* (Orr 1969), the genus to which the cougar was then assigned; then later as a subgenus of the African cheetah (Adams 1979); and yet later as a cheetah-like cat most closely related to the cougar (Barnett et al. 2005). In part at least, the similarities to the cheetah appear to be due to convergent evolution for a highly cursorial ecological niche (Martin et al. 1977; Barnett et al. 2005), implying open country for fast running. Chorn et al. (1988:137) noted that "Pronghorns are the most likely candidates for being the preferred prey of the American cheetah, *Miracinonyx trumani*."

The single record from New Mexico is from Muskox Cave on the eastern side of the Guadalupe Mountains (Logan 1981); unfortunately, nothing about the nature of the specimen(s) has been published.

Canidae—Dog Family

The Canidae includes wolves and foxes. Unlike felids, which sometimes are characterized as hypercarnivores because they are highly adapted for a meat diet with dentition unsuitable for processing plant material, canids have a somewhat intermediate dentition suitable both for meat and plant material (Photo 1.4). Whereas the felids have lost most of the teeth behind the meat-slicing carnassials, our canids retain all but the posteriormost upper molar.

†Canis armbrusteri. Armbruster's Wolf.

Armbruster's wolf was a large canid of Irvingtonian age. It has been recovered from the early Irvingtonian Adobe Ranch deposits along the Rio Grande Valley (Vanderhill 1986). The animal is similar to the gray wolf (*Canis lupus*), but with enough differences in skull and dentition that most paleontologists recognize it as a separate species. Kurtén and Anderson (1980) speculate that its extinction may have been prodded along by the entry of the gray wolf into the New World. Nowak (2002) places this species as leading to the late Pleistocene dire wolf (*Canis dirus*).

†Canis dirus. Dire Wolf.

The dire wolf appears in a number of sites in

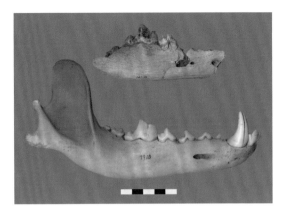

Photo 1.5. Right lower jaw of *Canis armbrusteri* from the Rio Grande Valley (*top*) contrasted with a modern *Canis lupus* from the northern United States (*bottom*). Scale bar = 5 cm. Photograph: © Art H. Harris.

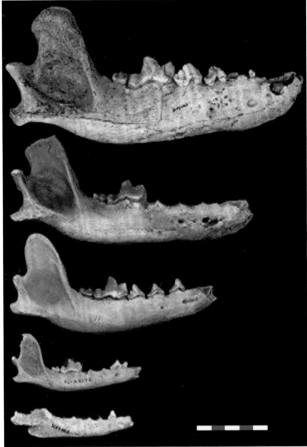

Photo 1.4. Comparison of fossil lower jaws (dentaries) of five species of canids. Top to bottom: *Canis dirus* (left dentary, reversed), *Canis lupus*, *Canis latrans*, *Vulpes velox*, and *Urocyon cinereoargenteus*. After Harris (2012b). Photograph: © Art H. Harris.

New Mexico, including SAM Cave, Sandia Cave, Blackwater Draw, U-Bar Cave, and several of the sites in the Guadalupe Mountains region, widespread enough to suggest that virtually all of the state was inhabited. This was a large canid, roughly the size of a gray wolf, but with a heavier build and shorter lower leg elements. It possessed the most powerful dentition of any *Canis*, but had a somewhat smaller braincase than found in the gray wolf (Kurtén and Anderson 1980).

The oldest probable fossil remains of the dire wolf are from about 252 kya, though the next older remains appear to date from or after the last interglacial (Dundas 1999). The species apparently disappeared at about the same time as the end-Pleistocene megafaunal extinction.

Conkling (1932:15), in discussing canids identified from Conkling Cavern, noted that "at least one individual is represented here in the large dire wolf, whose remains were found in the 'Wolf Den' with gnawed horse, camel, and human bones."

Canis latrans. Coyote.
The coyote is widespread in New Mexican cave deposits, though seldom preserved in open

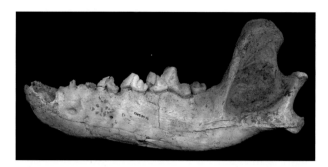

Photo 1.6. Left mandible of a dire wolf (*Canis dirus*) from U-Bar Cave, dated between 13 and 15 kya. Scale bar = 5 cm. Photograph: © Art H. Harris.

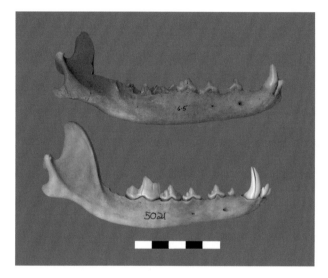

Photo 1.7. Fossil right mandible of coyote (*Canis latrans*) from Harris' Pocket, Dry Cave, approximately 15,000 years old (top), compared to a modern specimen from Hudspeth County, Texas. Scale bar = 5 cm. Photograph: © Art H. Harris.

sites. In at least one case, Isleta Cave No. 2, the appetites of coyotes and swift foxes resulted in disaster for these carnivores. This small cave is a lava bubble, more or less bell-shaped, with the entrance in the ceiling. The distance between the floor and the entrance is just sufficient as to allow entry without injury, but too high to allow exit for medium-sized carnivores, thus forming what is known as a self-baiting trap. Careless prairie dogs and rabbits falling into the cavity

and trapped within must surely have appeared to carnivores as free lunch. Except, of course, they in turn found themselves trapped. Over 130 individual bones of coyotes and more than 450 fox elements have been identified from the site.

Kurtén and Anderson (1980) note decreasing size from Irvingtonian-age coyotes to the present. One possible cause might be competition between larger forms of the coyote and the gray wolf, with the latter winning out.

†*Canis lepophagus.* Johnson's Coyote.
Johnson's coyote is recorded from the Virden fauna (latest Blancan) in southwestern New Mexico (Morgan and Lucas 2003; Morgan et al. 2008), and Tedford et al. (2009) identified a lower jaw of Blancan age from the Mesilla Basin in Doña Ana County south of Las Cruces. Kurtén and Anderson (1980) considered that this species led to the modern coyote and possibly to the wolves. It was coyote-sized, but with some dental differences, and the limb proportions indicate that it was less cursorial than the coyote.

Canis lupus. Gray Wolf.
Canis lupus possibly originated in the Old World from *Canis mosbachensis* (Nowak 2002), invading the New World via Beringia in late Rancholabrean times (Nowak 2002) or, according to Kurtén and Anderson (1980), surely identified in the New World in the very late Irvingtonian. In New Mexico, it occurs from early or early-middle Pleistocene deposits to those of Holocene age and, of course, occurred in the region historically. Although as yet unrecorded from west of the Rio Grande Valley, this likely is a combination of sampling error and scarcity of excavated Pleistocene cave deposits in much of that area. Records are available from as far north as Sandia Cave in the Sandia Mountains, as far east as Blackwater Draw near the Texas line, and to a number of sites in southeastern and south-central New Mexico.

The relatively small size of southwestern wolves

compared with northern wolves is manifest in the specimen shown in Photo 1.8; the specimen is very late Wisconsin or possibly Holocene. The large carnassial tooth indicates wolf rather than large domestic dog.

Canis rufus. Red Wolf.

The red wolf is reported from Irvingtonian deposits of SAM Cave in extreme north-central New Mexico (Rogers et al. 2000). Identification was on the basis of a single element, a lower second molar. There has been considerable uncertainty regarding the legitimacy of this canid. Some have considered it a hybrid form between the gray wolf and the coyote, but others have considered it a good species in its own right (Nowak 1979, 2002). Nowak (1979) originally suggested that the red wolf is relatively unmodified from a line of North American small wolves appearing in the early Pleistocene. He later (2002) indicated, however, that the red wolf possibly descended from an invasion of an Old World small wolf (*Canis mosbachensis*) that became isolated in North America by glaciation and evolved into *Canis rufus*. Wilson et al. (2000) presented genetic data indicating that *C. rufus* is separate from the gray wolf, instead being closely related to the eastern wolf (*Canis lycaon*).

There remains the question as to how diagnostic is the single tooth from SAM Cave, especially in light of the environmental differences from the modern red wolf habitat "of southern forests and marshes" (Nowak 1979:90). The distribution of the red wolf during the Recent period consists of southeastern North America and does not include New Mexico.

Urocyon cinereoargenteus. Gray Fox.

As a creature of rough terrain and wide ecological amplitude, it is not surprising to have a number of records from cave deposits in New Mexico. Skulls and lower jaws are easily identified, but post-cranial elements and various skull fragments can easily be confused with those of

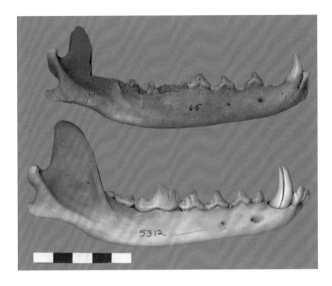

Photo 1.8. *Canis lupus*. Lower jaw above is from Isleta Cave No. 2; lower jaw below is of a large subspecies from Montana. Scale bar = 5 cm. Photograph: © Art H. Harris.

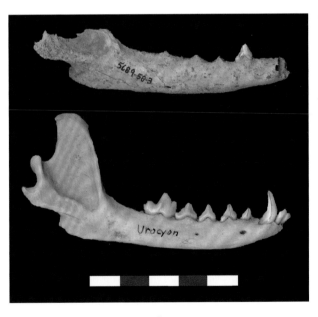

Photo 1.9. Mid-Wisconsin gray fox (*Urocyon cinereoargenteus*) jaw from U-Bar Cave (top). The distinctive "step" in the lower border (also seen in the jaw of the modern gray fox specimen [bottom]) easily separates this from other foxes. Scale bar = 5 cm. Photograph: © Art H. Harris.

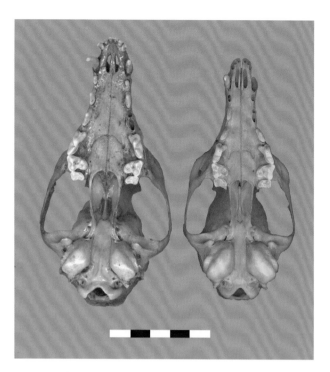

Photo 1.10. Comparison of the skulls of *Vulpes velox* (*left*) and *V. macrotis*. Note especially the size differences and the enlarged auditory bullae (bulbous structures toward the back of the skull) of *V. macrotis*. Scale bar = 5 cm. Photograph: © Art H. Harris.

the swift fox. New Mexican records are mid- and late Wisconsin, but it is known elsewhere from Blancan and Irvingtonian sites (Kurtén and Anderson 1980).

The adaptability of this fox is notable—for example, it is not uncommonly seen on the campus of the University of Texas at El Paso and has raised young in a landscaping boulder complex on campus (and see Chapter 12).

Vulpes macrotis/Vulpes velox. Kit Fox/Swift Fox.
The small foxes of the genus *Vulpes* (*V. macrotis* and *V. velox*) pose problems. They have been considered to be subspecies of a single species (*V. velox*) or as separate species. Rohwer and Kilgore (1973) produced evidence of limited interbreeding where the two meet in the Pecos

Valley of southeastern New Mexico. Dragoo et al. (1990) concluded that the genetic evidence, based on electrophoretic data, indicates that a single species (*V. velox*, since that was the earlier name) should be recognized, with morphological data indicating two subspecies, *V. v. velox* and *V. v. macrotis*. On the other hand, Mercure et al. (1993) found mtDNA evidence of two distinct groups of genotypes, mainly separated by the Rocky Mountains, consistent with the named species *V. velox* and *V. macrotis*. Although they recognized the existence of a hybrid zone, they concluded that the degree of interbreeding was so limited as to allow continued evolutionary differentiation and thus two species were recognized. Also playing a role in treating *V. velox* and *V. macrotis* as species was the recognition that the arctic fox (*V. lagopus*) is closely related to these arid-land foxes and is separated from them by about the same degree as the two foxes under consideration are from each other (see Chapters 13 and 14).

Fossils recognizable as belonging to the swift/kit fox group go back to the late Irvingtonian, and Kurtén and Anderson (1980) rated the species as indistinguishable in fossil material. However, working with data assembled by Jasso (2001), it appears that there is one fossil element from our area surely assignable to *V. macrotis* (Harris 2012b). This element may date to the late mid-Wisconsin and is from Dry Cave, near the area of present hybridization; *V. velox* is recorded from the same deposit (Harris 2012b). There are several other fossils identified as *V. macrotis*, but all come from sites that include Holocene material.

Although the Pleistocene fossils from New Mexico mostly appear to be *V. velox*, there is some possibility that we are seeing a single species responding to environmental change in the Pleistocene since larger size and smaller bullae are characters expected in colder, moister climates. Alternatively, these climatic conditions may have resulted in *V. velox* expanding westward. There is little indication as to where *V. macrotis* might have been during this time.

Vulpes velox is common in sites in southeastern New Mexico and in central New Mexico (Isleta Caves). Farther to the west and south in New Mexico, however, the genus is represented only by a single element (probably *V. velox*) from Palomas Creek Cave in west-central New Mexico.

Vulpes fulva. North American Red Fox.
North American red foxes are now believed to have originated from two colonization events involving ancestral red fox populations from Eurasia, the first one during (or before) the Illinoian Glacial period some 130,000–300,000 years ago, the second one during the Wisconsin glaciation (Aubry et al. 2009; Chapter 15). As expected, Jasso (2001) found the North American red fox to be larger than *V. velox* and *V. macrotis* in virtually all cranial and post-cranial features. Of the two subspecies traditionally recognized in New Mexico (but see Chapter 15), the one found in the eastern parts of the state was thought for a time to be descended from foxes introduced from the Old World (Frey 2004). That population now is believed to be native (Statham et al. 2012; Frey 2013; see Chapter 15). The other subspecies is primarily montane, though it also occurs in the San Juan Basin (Frey 2004).

Fossil remains are scanty. SAM Cave, near the Colorado line in north-central New Mexico, records it from the medial Irvingtonian (Rogers et al. 2000). A number of specimens are from Isleta Cave No. 2; though both late Pleistocene and Holocene remains occur at the site, preservation suggests most of the red fox remains are Wisconsin in age. Other records are from Wisconsin-age deposits in the Guadalupe Mountains region. Tebedge (1988) records a number of specimens from Dark Canyon Cave. Otherwise, specimens are rare in the region (Burnet Cave: Schultz and Howard 1935) or uncertain (Dry Cave, Blackwater Draw: Slaughter 1975; Harris 2012b).

Ursidae—Bear Family

†*Arctodus pristinus.* Lesser Short-faced Bear.
This smaller relative of the giant short-faced bear is tentatively known from fragmentary material from the late Blancan of the Mesilla Valley (Morgan and Lucas 2005). It is considered more primitive than its larger relative (Kurtén and Anderson 1980).

†*Arctodus simus.* Giant Short-faced Bear.
This extremely large bear was more closely related to the extant spectacled bear (*Tremarctos ornatus*) of South America than to the living

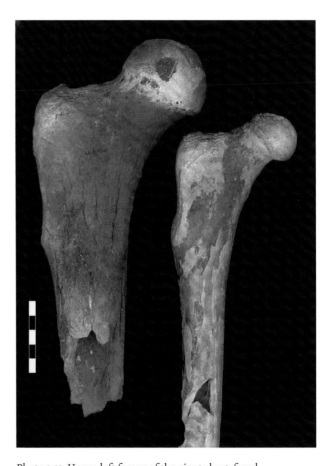

Photo 1.11. Upper left femur of the giant short-faced bear (*Arctodus simus*) from U-Bar Cave (left) compared to the same element of the black bear from Isleta Cave No. 2. Scale bar = 5 cm. Photograph: © Art H. Harris.

North American and Old World bears. Kurtén (1967) gave a weight estimate for a large male of 590–630 kg (1,298–1,386 lbs). Kurtén (1967:50) envisioned it as "a predominantly carnivorous form, by far the most powerful predator in the Pleistocene fauna of North America." However, a more recent study based on skull morphology indicates that it was omnivorous, having skull characteristics intermediate between highly carnivorous bears such as the polar bear (*Ursus maritimus*) and those primarily herbivorous such as the panda (*Ailuropoda melanoleuca*; Figueirido et al. 2009). Thus, it seems likely that the food habits were similar to those of the omnivorous brown bear (*Ursus arctos*, also known as the grizzly). The brown bear entered North America south of Alaska near the end of the Pleistocene, leading to speculation that competition between the brown bear and giant short-faced bear may have been instrumental in the extinction of the latter (Kurtén and Anderson 1974).

There are seven sites for New Mexico, scattered through the central and southern parts of the state, including the Isleta Caves, Conkling Cavern, Dry Cave, and U-Bar Cave. Absence of giant short-faced bear records from north of Albuquerque is almost certainly a reflection of site distribution.

Ursus americanus. Black Bear.
Black bears occur in cave deposits across the southern tier of New Mexico as well as at Blackwater Draw between Clovis and Portales and at the Isleta Caves. The majority of recovered elements come from Isleta Cave No. 2, suggesting that it may have served as a hibernaculum. These bears are large enough to likely have been able to exit the cave with little problem.

Mustelidae—Weasel Family
Lontra canadensis. River Otter.
In agreement with its historical rarity in New Mexico, only a single fossil is recorded. This is

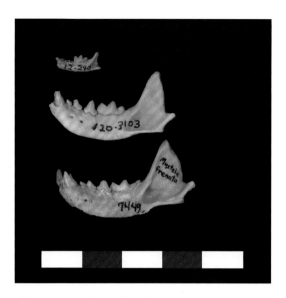

Photo 1.12. Fossil *Mustela richardsonii*. Left mandible from Dry Cave (top) compared to a fossil long-tailed weasel (*Neogale frenata*) from Big Manhole Cave (middle) and a modern female long-tailed weasel (*N. f. neomexicana*) from El Paso, Texas (bottom). Scale bar = 2 cm. Photograph: © Art H. Harris.

from late Wisconsin or Holocene deposits of SAM Cave in the Rio Grande drainage near the Colorado border (Rogers et al. 2000).

Mustela richardsonii. American Ermine.
The American ermine is recorded from the medial Irvingtonian of SAM Cave near the Colorado border, within its general current range (Rogers et al. 2000), which today is limited in New Mexico to the northern mountains (Chapter 21). *Mustela richardsonii* is recorded from the late Wisconsin of Dry Cave (Harris 1993) and tentatively, based only on an innominate (one side of the pelvic girdle), from the mid-Wisconsin or full-glacial of Pendejo Cave (Harris 2003).

Mustela nigripes. Black-footed Ferret.
Although nowhere near as common as the long-tailed weasel (*Neogale frenata*), *M. nigripes* is represented by several individuals from the Isleta Caves, one skull from Big Manhole Cave, and a probable individual from Pendejo Cave. Prairie

Photo 1.13. Left mandible of *Mephitis mephitis* from U-Bar Cave. Scale bar = 5 cm. Photograph: © Art H. Harris.

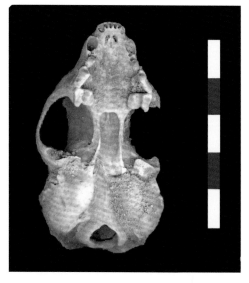

Photo 1.14. Spotted skunk (*Spilogale* sp.) skull from Big Manhole Cave. Scale bar = 5 cm. Photograph: © Art H. Harris.

dogs are the mainstay of modern black-footed ferret diet, and Wisconsin-age prairie dog remains are widespread.

Neogale frenata. Long-tailed Weasel.
Long-tailed weasels occur at present in river bottomlands, montane forests, and grasslands. Conditions undoubtedly were amenable in the Wisconsin, also, judging from the number of fossil specimens and their distribution. They are known from 13 sites, ranging in age from mid-Wisconsin to terminal Pleistocene. One especially rich cave deposit (Big Manhole Cave) produced 42 specimens, including 14 skulls and skull parts.

Neogale vison. American Mink.
The sole record from the Pleistocene of New Mexico rests on a fragment of a lower jaw from Isleta Cave No. 2. American minks are associated with riparian habitats; the Isleta Caves are just west of the Rio Grande Valley that undoubtedly

supported a healthy riparian flora throughout the Wisconsin.

Mephitidae—Skunk Family
Conepatus leuconotus. White-backed Hog-nosed Skunk.
Only two Pleistocene records of the genus occur in our area, both in the Guadalupe Mountains region. This may be somewhat of an artifact, since post-cranial elements are very similar to those of the much more common striped skunk and likely to be so identified.

Mephitis mephitis. Striped Skunk.
Just like today, the striped skunk was common in New Mexico in the late Pleistocene. Occurrences are from near the Colorado border (SAM Cave) to near the Mexican border in extreme southwestern New Mexico (U-Bar Cave) to east-central New Mexico near the Texas border. Nearly every cave site with more than a few fossil remains is represented by the species.

Spilogale Species. Western/Eastern Spotted
   Skunk Clades.

The phylogeny and taxonomy of spotted skunks remain in flux (Chapter 29). Western spotted skunks (*Spilogale gracilis* and *S. leucoparia* in New Mexico) differ from eastern spotted skunk forms (until recently referred to as *S. putorius*; see Chapter 29) primarily by features of reproductive physiology and morphological aspects of the reproductive system, though western forms are also somewhat smaller (Mead 1968). Without an extensive database of measurements (which might or might not separate the species), the fossils from New Mexico are treated as indistinguishable between the two clades. The likelihood, based on the modern distribution, however, is that the western form is represented. Fossil distribution is similar to that of the striped skunk.

Procyonidae—Raccoon Family

Bassariscus astutus. Ringtail.

Ringtails are well known to frequent caves and rock shelters, so as expected, their remains are common in such sites, though fossil remains are limited to the extreme southern parts of the state.

Skinner (1942) described *B. sonoitensis* from the Pleistocene of southern Arizona. However, recovery of a skull (Photo 1.15) from U-Bar Cave led to a study resulting in synonymizing it with *B. astutus* (Harris 1990b). The limited number of specimens of *B. sonoitensis*, however, indicate that specimens average somewhat larger and more robust than modern specimens, possibly representing a temporal subspecies.

END TIMES

The end of the Pleistocene marked the end of the megacarnivores, with the exception only of the jaguar, cougar, gray wolf, black bear, and brown bear. Fiedel (2009) indicated that the extinction

Photo 1.15. Skull and lower jaw of *Bassariscus astutus* from U-Bar Cave. Scale bar = 5 cm. Photograph: © Art H. Harris.

of at least a major portion (and perhaps all) of the Pleistocene megafauna occurred in less than a 400-year window around 11,000 radiocarbon years ago. Since about that time, New Mexican survivors—both predator and prey—have had to adjust to increasing temperature and aridity and the concomitant reorganization of plant communities. Numerous prey items retreated north and upslope along with the formerly continuous areas of forest and savannah.

Despite the vast climatic and biotic changes at the end of the Pleistocene, New Mexico's remaining carnivores seem to have taken all in stride. A few additions in the form of raccoon (*Procyon lotor*) and coati (*Nasua narica*) eventually wandered into the state and some geographic ranges shifted (e.g., the American ermine became restricted to the northern mountains, and the kit fox replaced the swift fox in western areas). Otherwise, the post-Pleistocene record shows little change from glacial time, even surviving the booming populations of the superpredator, humans.

LITERATURE CITED

Adams, D. B. 1979. The cheetah: Native American. *Science* 205:1155–58.

Aubry, K. B., M. J. Statham, B. N. Sacks, J. D. Perrine, and S. M. Wisely. 2009. Phylogeography of the North American red fox: vicariance in Pleistocene forest refugia. *Molecular Ecology* 18:2668–86.

Barnett, R., I. Barnes, M. J. Phillips, L. D. Martin, C. R. Harington, J. A. Leonard, and A. Cooper. 2005. Evolution of the extinct sabretooths and the American cheetah-like cat. *Current Biology* 15(15):R589–R590.

Barnett, R., B. Shapiro, I. Barnes, S. Y. W. Ho, J. Burger, N. Yamagudhi, T. F. G. Higham, H. T. Wheeler, W. Rosendahl, A. V. Sher, M. Sotnikovsa, T. Kuznetsova, G. F. Brayshnikov, L. D. Martin, C. R. Harington, J. A. Burns, and A. Cooper. 2009. Phylogeography of lions (*Panthera leo* ssp.) reveals three distinct taxa and a late Pleistocene reduction in genetic diversity. *Molecular Ecology* 18:1668–77.

Chorn, J., B. A. Frase, and C. D. Frailey. 1988. Late Pleistocene Pronghorn, *Antilocapra Americana*, from Natural Trap Cave, Wyoming. *Transactions of the Nebraska Academy of Sciences* 16:127–39.

Christiansen, P., and J. M. Harris. 2005. Body size of *Smilodon* (Mammalia: Felidae). *Journal of Morphology* 266:369–84.

———. 2009. Craniomandibular morphology and phylogenetic affinities of *Panthera atrox*: implications for the evolution and paleobiology of the lion lineage. *Journal of Vertebrate Paleontology* 29:934–45.

Conkling, R. P. 1932. Conkling Cavern: the discoveries in the bone cave at Bishop's Cap, New Mexico. *West Texas Historical and Scientific Society Bulletin* 44:6–19.

Culver, M., W. E. Johnson, J. Pecon-Slattery, and S. J. O'Brien. 2000. Genomic ancestry of the American puma (*Puma concolor*). *Journal of Heredity* 91:186–97.

Dragoo, J. W., J. R. Choate, T. L. Yates, and T. P. O'Farrell. 1990. Evolutionary and taxonomic relationships among North American arid-land foxes. *Journal of Mammalogy* 71:318–32.

Dundas, R. G. 1999. Quaternary records of the dire wolf, *Canis dirus*, in North and South America. *Boreas* 28:375–85.

Fiedel, S. 2009. Sudden deaths: the chronology of terminal Pleistocene megafaunal extinction. In *American megafaunal extinctions at the end of the Pleistocene*, ed. G. Haynes, 21–37. Springer Science + Business Media.

Figueirido, B., P. Palmqvist, and J. A. Pérez-Claros. 2009. Ecomorphological correlates of craniodental variation in bears and paleobiological implications for extinct taxa: an approach based on geometric morphometrics. *Journal of Zoology* 277:70–80.

Frey, J. K. 2004. Taxonomy and distribution of the mammals of New Mexico: an annotated checklist. *Occasional Papers, Museum of Texas Tech University* 240:1–32.

———. 2013. Re-evaluation of the evidence for the importation of red fox from Europe to colonial America: Origins of the southeastern red fox (*Vulpes vulpes fulva*). *Biological Conservation* 158:74–79.

Goswami, A. 2010. Introduction to Carnivora. In *Carnivoran evolution: new views on phylogeny, form and function*, ed. A. Goswami and A. Friscia, 1–24. Cambridge: Cambridge University Press.

Harris, A. H. 1970. The Dry Cave mammalian fauna and late pluvial conditions in southeastern New Mexico. *Texas Journal of Science* 22:3–27.

———. 1987. Reconstruction of mid-Wisconsin environments in southern New Mexico. *National Geographic Research* 3:142–51.

———. 1989. The New Mexican late Wisconsin—east versus west. *National Geographic Research* 5:205–217.

———. 1990a. Fossil evidence bearing on southwestern mammalian biogeography. *Journal of Mammalogy* 71:219–29.

———. 1990b. Taxonomic status of the Pleistocene ringtail *Bassariscus sonoitensis* (Carnivora). *Southwestern Naturalist* 35:343–46.

———. 1993. A late-Pleistocene occurrence of ermine (*Mustela erminea*) in southeastern New Mexico. *Southwestern Naturalist* 38:279–80.

———. 2003. The Pleistocene vertebrate fauna from Pendejo Cave. In *Pendejo Cave*, ed. R. S. MacNeish and J. G. Libby, 36–65. Albuquerque: University of New Mexico Press.

———. 2012a. *Pleistocene vertebrates of New Mexico and Trans-Pecos Texas*. Vol. 3, *Fossil sites*. UTEP Biodiversity Collections, University of Texas at El Paso. (E-book)

———. 2012b. *Pleistocene vertebrates of New Mexico and Trans-Pecos Texas*. Vol. 2, *Mammals*. UTEP Biodiversity Collections, University of Texas at El Paso. (E-book)

Harris, A. H., and J. S. Findley. 1964. Pleistocene-recent fauna of the Isleta Caves, Bernalillo County, New Mexico. *American Journal of Science* 262:114–20.

Howard, H. 1971. Quaternary avian remains from Dark Canyon Cave, New Mexico. *Condor* 73:237–40.

Hunt, R. M., Jr. 1998. Amphicyonidae. In *Evolution of tertiary mammals of North America*. Vol. 1, *Terrestrial carnivores, ungulates, and ungulatelike mammals*, ed. C. M. Janis, K. M. Scott, and L. L. Jacobs, 196–227. Cambridge: Cambridge University Press.

Jasso, H. A. 2001. Systematics of Pleistocene *Vulpes* in New Mexico. MS thesis, University of Texas at El Paso.

Kurtén, B. 1967. Pleistocene bears of North America. 2. Genus *Arctodus*, short-faced bears. *Acta Zoologica Fennica* 117:1–60.

Kurtén, B., and E. Anderson. 1974. Association of *Ursus arctos* and *Arctodus simus* (Mammalia: Ursidae) in the late Pleistocene of Wyoming. *Breviora* 426:1–6.

———. 1980. *Pleistocene mammals of North America*. New York: Columbia University Press.

Lear, L. L., and A. H. Harris. 2007. The Holocene fauna of Big Manhole Cave, Eddy County, New Mexico. *Southwestern Naturalist* 52:110–15.

Logan, L. E. 1981. The mammalian fossils of Muskox Cave, Eddy County, New Mexico. *Proceedings of the Eighth International Congress Speleology* 1:159–60.

Lucas, S. G., R. M. Schoch, E. Manning, and C. Tsentas. 1981. The Eocene biostratigraphy of New Mexico. *Geological Society of America Bulletin* 92:951–67.

Lundelius, E. L., Jr. 1972. Vertebrate remains from the Gray Sand. In *Blackwater Draw Locality No. 1: a stratified early man site in eastern New Mexico*, ed. J. J. Hester, 148–63. Fort Burgwin Research Center, Southern Methodist University, Publication no. 8.

MacFadden, B. J. 1977. Magnetic polarity stratigraphy of the Chamita Formation stratotype (Mio-Pliocene) of north-central New Mexico. *American Journal of Science* 277:769–800.

MacFadden, B. J., N. M. Johnson, and N. D. Opdyke. 1979. Magnetic polarity stratigraphy of the Mio-Pliocene mammal-bearing Big Sandy Formation of western Arizona. *Earth and Planetary Science Letters* 44:349–64.

MacNeish, R. S., and J. G. Libby, eds. 2003. *Pendejo Cave*. Albuquerque: University of New Mexico Press.

Martin, L. D., B. M. Gilbert, and D. B. Adams. 1977. A cheetah-like cat in the North American Pleistocene. *Science* 195:981–82.

McHenry, C. R., S. Wroe, P. D. Clausen, K. Moreno, and E. Cunningham. 2007. Supermodeled sabercat, predatory behavior in *Smilodon fatalis* revealed by high-resolution 3D computer simulation. *Proceedings of the National Academy of Sciences* 104:16010–15.

Mead, R. A. 1968. Reproduction in western forms of the spotted skunk (genus *Spilogale*). *Journal of Mammalogy* 49:373–90.

Mercure, A., K. Ralls, K. P. Koepfli, and R. K. Wayne. 1993. Genetic subdivisions among small canids: mitochondrial DNA differentiation of swift, kit, and arctic foxes. *Evolution* 47:1313–28.

Morgan, G. S., and S. G. Lucas. 2001. The sabertooth cat *Smilodon fatalis* (Mammalia: Felidae) from a Pleistocene (Rancholabrean) site in the Pecos River valley of southeastern New Mexico/southwestern Texas. *New Mexico Geology* 23:130–33.

———. 2003. Mammalian biochronology of Blancan and Irvingtonian (Pliocene and Early Pleistocene) faunas from New Mexico. *Bulletin of the American Museum of Natural History* 279:269–320.

———. 2005. Pleistocene vertebrate faunas in New Mexico from alluvial, fluvial, and lacustrine deposits. *New Mexico Museum of Natural History and Science Bulletin* 28:185–248.

Morgan, G. S., P. L. Sealey, and S. G. Lucas. 2008. Late Pliocene (late Blancan) vertebrate faunas from Pearson Mesa, Duncan Basin, southwestern New Mexico and southeastern Arizona. *New Mexico Museum of Natural History and Science Bulletin* 44:141–88.

Morgan, G. S., P. L. Sealey, S. G. Lucas, and A. B. Heckert. 1997. Pliocene (latest Hemphillian and Blancan) vertebrate fossils from the Mangas Basin, southwestern New Mexico. In *New Mexico's Fossil Record 1*, ed. S. G. Lucas, J. W. Estep, T. E. Williamson, and G. S. Morgan, 97–128. *New Mexico Museum of Natural History and Science Bulletin* 11.

Nowak, R. M. 1979. North American quaternary *Canis. Monograph of the Museum of Natural History, University of Kansas*, 6:1–154.

———. 2002. The original status of wolves in eastern North America. *Southeastern Naturalist* 1:95–130.

Orr, P. C. 1969. *Felis trumani*, a new radiocarbon dated cat skull from Crypt Cave, Nevada. *Santa Barbara Museum of Natural History, Department of Geology Bulletin* 2:1–8.

Polly, P. D., G. D. Wesley-Hunt, R. E. Heinrich, G. Davis, and P. Houde. 2006. Earliest known carnivoran auditory bulla and support for a recent origin of crown-group Carnivora (Eutheria, Mammalia). *Palaeontology* 49, part 5:1019–27.

Rogers, K. L., C. A. Repenning, F. G. Luiszer, and R. D. Benson. 2000. Geologic history, stratigraphy, and paleontology of SAM Cave, north-central New Mexico. *New Mexico Geology* 22:89–117.

Rohwer, S. A., and D. L. Kilgore Jr. 1973. Interbreeding in the arid-land foxes, *Vulpes velox* and *V. macrotis. Systematic Zoology* 22:157–65.

Rothwell, T. 2001. A partial skeleton of *Pseudaelurus* (Carnivora:Felidae) from the Nambé Member of the Tesuque Formation, Española Basin, New Mexico. *Novitates* 3342:1–31.

Schultz, C. B., and E. B. Howard. 1935. The fauna of Burnet Cave, Guadalupe Mountains, New Mexico. *Proceedings of the Academy of Natural Sciences of Philadelphia* 87:273–98.

Skinner, M. F. 1942. The fauna of Papago Springs Cave, Arizona, and a study of *Stockoceros*; with three new antilocaprines from Nebraska and Arizona. *Bulletin of the American Museum of Natural History* 80:143–220.

Slaughter, B. H. 1975. Ecological interpretation of the Brown Sand Wedge local fauna. In *Late Pleistocene environments of the southern high plains*, ed. F. Wendorf and J. J. Hester, 179–92. Rancho de Taos, NM: Fort Burgwin Research Center Publication no. 9.

Statham, M. J., B. N. Sacks, K. B. Aubry, J. D. Perrine, and S. M. Wisely. 2012. The origins of recently established red fox populations in the United States: translocations or natural range expansions? *Journal of Mammalogy* 93:52–65.

Stearns, C. E. 1943. The Galisteo Formation of north-central New Mexico. *Journal of Geology* 51:301–19.

Tebedge, S. 1988. Paleontology and paleoecology of the Pleistocene mammalian fauna of Dark Canyon Cave, Eddy County, New Mexico. PhD dissertation, University of Texas at Austin.

Tedford, R. H. 1981. Mammalian biochronology of the late Cenozoic basins of New Mexico. *Geological Society of America Bulletin, Part I* 92:1008–22.

Tedford, R. H., and S. Barghoorn. 1997. Miocene mammals of the Española and Albuquerque basins, north-central New Mexico. In *New Mexico's Fossil Record 1*, ed. S. G. Lucas, J. W. Estep, T. E. Williamson, and G. S. Morgan, 77–95. *New Mexico Museum of Natural History and Science Bulletin* 11.

Tedford, R. H., X. Wang, and B. E. Taylor. 2009. Phylogenetic systematics of the North American fossil Caninae (Carnivora: Canidae). *Bulletin of the American Museum of Natural History* 325:1–218.

Vanderhill, J. B. 1986. Lithostratigraphy, vertebrate paleontology, and magnetostratigraphy of Plio-Pleistocene sediments in the Mesilla Basin, New Mexico. PhD dissertation, University of Texas at Austin.

Wang, X., R. H. Tedford, and B. E. Taylor. 1999. Phylogenetic systematics of the Borophaginae (Carnivora: Canidae). *Bulletin of the American Museum of Natural History* 243:1–391.

Wilson, P. J., S. Grewal, I. D. Lawford, J. N. M. Heal, A. G. Granacki, D. Pennock, J. B. Theberge, M. T. Theberge, D. R. Voigt, W. Waddell, R. E. Chambers, P. C. Paquet, G. Goulet, D. Cluff, and B. N. White. 2000. DNA profiles of the eastern Canadian wolf and the red wolf provide evidence for a common evolutionary history independent of the gray wolf. *Canadian Journal of Zoology* 78: 2156–66.

VEGETATION, FIRE ECOLOGY, AND CLIMATE OF NEW MEXICO'S ECOSYSTEMS

F. Jack Triepke and Timothy K. Lowrey

The impressive diversity of land features and climate zones found in New Mexico translates into a vast array of ecosystem types, which in turn supports a rich carnivore fauna (Frey and Yates 1996). The coyote (*Canis latrans*), cougar (*Puma concolor*), and kit fox (*Vulpes macrotis*), to name just a few species, represent quintessential carnivores that find ample habitat in the state. Less known is the fact that with their deep snow cover and subalpine plant communities, New Mexico's northern high mountains are home to the southernmost populations of several boreal carnivores, including the Pacific marten (*Martes caurina*) and the American ermine (*Mustela richardsonii*) (see Chapters 20 and 21). To the south, enclaves of Madrean vegetation correspond to the northernmost distribution of the white-nosed coati (*Nasua narica*), a tropical and subtropical species (Chapter 31).

New Mexico's climate is generally cold-temperate in mountainous areas, plains, and grasslands in the north, in the upper Gila and San Francisco basins in the southwest, and in the Sacramento Mountains of south-central New Mexico (Map 2.1). The climate transitions from cold- to warm-temperate over a large mild zone found in the southern half of the state including the lower Rio Grande basin. With both winter precipitation and summer monsoon rains, bi-modal precipitation exists in much of the state, adding to the complexity of environmental conditions and plant habitat. New Mexico's

Photo 2.1. Badlands and hoodoos of the US Bureau of Land Management Bisti Wilderness Area (De-Na-Zin) in northwestern New Mexico. Photograph: © Larry Lamsa.

geology is a mixture of tertiary volcanics, middle-age sedimentary rocks (e.g., table lands), and ancient igneous basement rock that underlies the sedimentary mountain ranges. Areas of volcanic history are represented by vast expanses of un-eroded lava flows (malpais), and volcanic masses, cones, and calderas (Dick-Peddie 1993). Igneous mountain-building accounts for many peaks over 3,000 m (10,000 ft), which, together with several river systems and the erosion of extensive sedimentary strata, has given New Mexico's landscape its badlands and large areas of steep topography (Photo 2.1). Regional vegetation patterns have responded accordingly to the range of geological and climatic conditions,

(*opposite page*) Photograph: © Geraint Smith.

Table 2.1. List of major ecosystems found in New Mexico and depicted in this chapter, cross-referenced to Merriam's life zones (Merriam 1890) and Biotic Communities (Brown and Lowe 1974).

Merriam's Life Zones	Biotic Communities	Major Ecosystem Types
Arctic-Alpine	Rocky Mountain and Great Basin Tundra (111.6), Rocky Mountain Alpine and Subalpine Scrub (131.4)	Alpine and Tundra
Hudsonian	Rocky Mountain and Great Basin Subalpine Conifer Forest (121.3)	Spruce-Fir Forest
Canadian	Rocky Mountain Montane Conifer Forest (122.6)	Mixed Conifer with Aspen
		Mixed Conifer-Frequent Fire
Transition		Ponderosa Pine Forest
Hudsonian, Canadian, Transition	Rocky Mountain Alpine and Subalpine Grassland (141.2), Rocky Mountain Montane Grassland (142.4)	Montane / Subalpine Grassland
Canadian, Transition, Upper Sonoran	Great Basin Montane Scrub (132.1)	Gambel Oak Shrubland
		Mtn Mahogany Mixed Shrubland
Transition, Upper Sonoran	Great Basin Desert Scrub (152.1)	Sagebrush Shrubland
Upper Sonoran	Great Basin Conifer Woodland (122.7)	Pinyon-Juniper Woodlands
	Madrean Evergreen Forest and Woodland (123.3)	Madrean Woodlands
	Great Basin Shrub-Grassland (142.2)	Colorado Plateau/Great Basin Grassland
	Chihuahuan (Semi-desert) Grassland (143.1)	Semi-desert Grassland
	Plains Grassland (142.1)	Great Plains
Lower Sonoran	Chihuahuan Desert Scrub (153.2)	Chihuahuan Desert Scrub
All	Various	Riparian (various types)

together with the continual influences of fire and other natural and human processes to form a distinct variety and geography of ecosystem types and plant species (Sivinski et al. 1996).

This chapter provides an overview of New Mexico's ecosystems with an emphasis on geographic distribution, floristic composition, and fire regime relevant to the conservation of carnivores and their habitats, along with some discussion of the effects of contemporary and future climate conditions (and see Chapter 5).

ECOSYSTEM TYPE CONCEPT

Throughout his *Mammals of New Mexico*, Vernon Bailey (1931) used C. Hart Merriam's (1890) life zones when describing distribution and habitat, tying the American marten to the Canadian zone forests of the Sangre de Cristo and San Juan mountains, the white-backed hog-nosed skunk (*Conepatus leuconotus*) to the Lower Sonoran Zone, or the gray fox (*Urocyon cinereoargenteus*) primarily to the Upper Sonoran Zone. Thus, the use of classification and mapping systems to describe the spatial distribution of the vegetation, ecosystems, or land cover is nothing new. After Merriam (1890), various additional classification and mapping schemes have been applied (e.g., Brown and Lowe 1974; Dick-Peddie 1993; Muldavin et al. 2000; Robbie 2004), which not only help to characterize wildlife habitat but also serve as important decision-making tools for environmental planning and management.

For our purpose here, we adopt the US Department of Agriculture (USDA) Forest Service approach of ecosystem types implemented regionally since 2006 (TNC 2006; Moreland et al. 2017; Triepke et al. 2017). The underpinnings of these ecosystem types are found in the Biotic Communities of Brown and Lowe (1974) and in the analysis framework used by the Forest Service and other organizations in the Southwest. Table 2.1 provides a crosswalk between the ecosystem types of this chapter and legacy classification schemes. In time, the National Vegetation Classification System (Jennings et al. 2009) will be adopted across the United States as more refinements are made to the classification and as successional relationships among classes become established.

While driven mainly by climate, these ecosystem types are also shaped by soils and represent areas with similar plant succession, disturbance regime, and dominant plant species. They are mapped and described using many technical references (e.g., Brown and Lowe 1974; Dick-Peddie 1993; Muldavin et al. 2000; Comer et al. 2003; USDA Forest Service 2006), sometimes as groupings of finer vegetation classes with similar site properties and ecology. The ecosystem types in this chapter represent units of land that are similar in site potential *and* historical fire regime, delineated and characterized for purposes of habitat analysis and management. The descriptions to follow discuss these key elements along with changes to the contemporary landscape of New Mexico brought about in the last century or so by land use and climate change. For many of the ecosystem types where natural processes have been substantially altered, current habitat conditions stand in stark contrast to those of historical landscapes.

MAJOR CLIMATE AND LIFE ZONE STRATIFICATION

Knowledge of the state's climate patterns is essential to understanding the geographic distribution of New Mexico's major ecosystem types. The climate is characterized by subregional climatic zones that have been delimited based on temperature and precipitation, as shown in Map 2.1. As mentioned, the state is broadly divided into cold and mild climates between the north and south, respectively, based on a mean annual soil temperature threshold of 11° C (~52° F) at 50 cm (~20 in) depth. These temperature zones are further defined by the time of the year that receives the most precipitation (Carlton and Brown 1983)—either winter precipitation zones or summer/monsoonal zones. A semi-arid climatic zone exists where the Great Plains extend into the northeastern corner of the state, depicted by low annual precipitation of 300–500 mm (12–20 in), hot and dry summers, cold winters with some snowfall, and considerable day-night temperature swings (up to 20° C) (McKnight and Hess 2000). Other climate zones, not included in Map 2.1., occur over minor areas and represent mixed temperature conditions. For example, in northeastern New Mexico some areas meet the soil temperature threshold for a mild climate yet have winters that are very cold relative to the summer. Due to the cold-limiting effects of harsh winters, these plant communities are dominated by the shortgrass prairie vegetation of the Great Plains rather than the vegetation of semi-desert grasslands of mild climate areas to the south that have similar levels of precipitation. The mountainous areas of the state likewise lend themselves to mixed climate conditions of cold winters with hot summers and a thermic

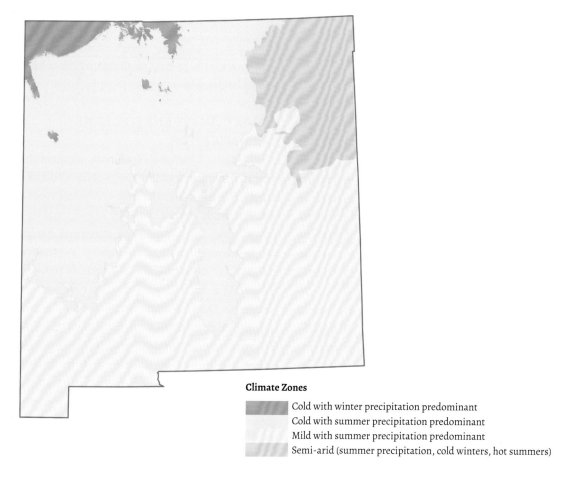

Climate Zones

Cold with winter precipitation predominant
Cold with summer precipitation predominant
Mild with summer precipitation predominant
Semi-arid (summer precipitation, cold winters, hot summers)

Map 2.1. New Mexico's major climate zones based on the concepts of Carlton and Brown (1983) and climate surface data of Rehfeldt (2006).

soil temperature regime where mean annual soil temperatures vary from 15 to 22° C.

For the mountain ranges of New Mexico, it is also important to consider life zone patterns in conjunction with the major climate zones (Map 2.1), given the indirect effect of altitude on climate and vegetation. In the mountainous areas of the north-central, south-central, and southwestern parts of the state, topography and elevation lend themselves to life zone stratification of vegetation associated with foothill, montane, subalpine, and alpine settings (Lowrey 2010). Life zones were conceptualized and described beginning in the southwestern United States by C. Hart Merriam (1890), who recognized belts of vegetation that were distinct in appearance and dominant plant species. To a greater or lesser degree, animal and plant diversity similarly changes with increasing altitude. Merriam's basic scheme of six different life zones (Table 2.1) remains in use, and related concepts have been refined to account for the compensatory effects of other environmental variables such as slope and aspect. For instance, in the Northern Hemisphere the life zones will be lower on northern exposures than on southern exposures, all else being equal, reflecting the greater sun energy on south aspects. Note also that some life zones can contain more than one ecosystem type, given their ecological and environmental variability.

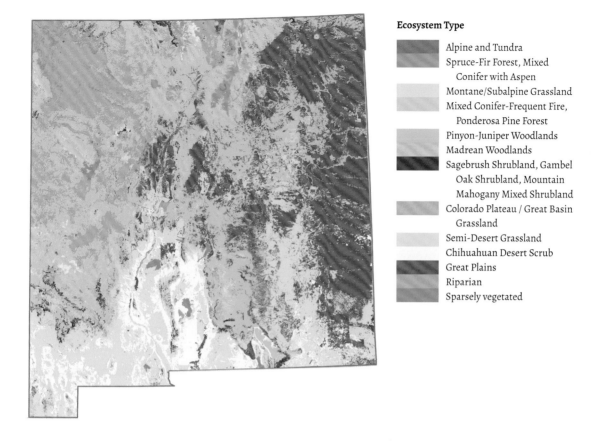

Ecosystem Type

- Alpine and Tundra
- Spruce-Fir Forest, Mixed Conifer with Aspen
- Montane/Subalpine Grassland
- Mixed Conifer-Frequent Fire, Ponderosa Pine Forest
- Pinyon-Juniper Woodlands
- Madrean Woodlands
- Sagebrush Shrubland, Gambel Oak Shrubland, Mountain Mahogany Mixed Shrubland
- Colorado Plateau / Great Basin Grassland
- Semi-Desert Grassland
- Chihuahuan Desert Scrub
- Great Plains
- Riparian
- Sparsely vegetated

Map 2.2. New Mexico's major ecosystem types based on a map of Forest Service Ecological Response Units.

The Alpine life zone makes up the uppermost life zone in New Mexico. At some of the highest altitudes of the state, such as Sierra Blanca in the Sacramento Mountains or in the Sangre de Cristo Mountains in the north, alpine communities occur in nearly treeless and climatically extreme settings above subalpine forests. The subalpine forests comprise the "spruce-fir" zone, immediately above montane forests of mixed conifer and ponderosa pine (shown together in Map 2.2). Mixed conifer is composed especially of Douglas-fir (*Pseudotsuga menziesii*), white fir (*Abies concolor*), southwestern white pine (*Pinus strobiformis*), and ponderosa pine (*Pinus ponderosa*). At lower altitudes and warmer exposures, the dryer end of mixed conifer grades into the ponderosa pine zone, which is generally situated just above a belt of low-stature woodlands. The woodland zone is distinguished by coniferous pinyon (*Pinus* spp.) and juniper (*Juniperus* spp.) trees and, in southern New Mexico, evergreen oak (*Quercus* spp.) tree types that occur alone or in combination with pinyon-juniper. With decreasing elevation, woodlands grade into grassland zones. In many areas of the mild climate zone to the south, desert plant communities are downslope of grassland zones. New Mexico's mountain ranges typically reflect this zonation of vegetation types, and are often surrounded by expanses of grassland or desert systems, making up "sky island" formations.

New Mexico's regional climate zones in conjunction with the life zone stratification in mountainous areas help explain the geographic distribution of major ecosystem types regionally and locally (Map 2.2). Fire ecology is another primary influence on the distribution of ecosystems and is discussed in the individual characterizations to follow. Attributes of climate and life zone can likewise help explain ecosystem types in terms of their vegetation composition and physiognomy (appearance, structure).

PAST AND CONTEMPORARY CONDITIONS

New Mexico's land cover has changed dramatically since the last ice age. Approximately 18,000 years ago in the late Pleistocene, lowland New Mexico's climate was cooler, more humid, and overall comparable to that found today in southern Oregon (see Dick-Peddie 1993). The dominant vegetation type throughout the state consisted of montane coniferous forest with Douglas-fir, Southwestern white pine, and white fir. The upper tree line could be found as low as 2,500 m (8,250 ft) in the northwestern part of the state, or about 900 m (2,970 ft) lower than its current elevation. Approximately 28,000 years ago, alpine tundra occurred not just in the Sangre de Cristo Mountains and on Sierra Blanca but in all the major mountain ranges of the state. Subalpine coniferous forests were also more widespread, whereas now they are restricted to the few mountainous areas of the state reaching above 3,000 m (~9,850 ft). Grassland and desert scrub did not occur in the state 18,000 years ago (Dick-Peddie 1993).

The abundance and location of ecosystem types found today in New Mexico reflect a warmer and drier climate than that 18,000 years ago, the result of natural change over time. However, today's ecosystem types also share many of the uncharacteristic and undesirable conditions associated with relatively recent land use patterns of fire suppression, livestock grazing, water diversions, aberrant timber practices, and other contemporary system perturbations. As discussed in this chapter, notable changes in fire-adapted forest and woodland ecosystems typically include increased tree densities and increased patch size (aggregation) as a result of fire suppression, and the simplification of the vertical canopy structure that comes with the ingrowth of many small trees and the high-grading of larger, more merchantable trees (Brown et al. 2001; Sánchez Meador et al. 2011). In grassland systems, fire suppression has favored the encroachment of trees and shrubs, in turn limiting forage potential and altering plant composition (Jameson 1967; Kramer et al. 2015). Woody encroachment in combination with livestock grazing and the reduction of fine fuels (Yanoff et al. 2008) further limits the capacity for wildfire to spread, exacerbating the detrimental effects of fire suppression policy. In some grasslands, intense livestock grazing of the previous decades has reduced the amount of perennial grass cover, simplified plant communities, and favored invasive herb species that thrive under chronic press disturbance (Arnold 1950; Clary 1975; Milchunas and Lauenroth 1989; Ambos et al. 2000; Milchunas 2006). Modern game management, with the promotion of exceptionally large wild ungulate populations, has similarly impacted ecosystem structure, composition, and process. Impacts of both native and non-native grazers are common in riparian and wetland systems of the Southwest (Krueper 1995), where introduced grasses such as Kentucky bluegrass (Poa pratensis) better withstand trampling as they lack the thick fibrous root matting of native sedges and grasses (Milchunas 2006). Unlike fire-adapted systems in the uplands, where woody encroachment is a ubiquitous issue, in riparian zones shrub and tree cover are often reduced from past levels

Table 2.2. New Mexico's 16 major ecosystem types with associated climate and life zones.

Ecosystem Type	Precipitation	Temperature
Alpine and Tundra	Summer[1]	Cold
Spruce-Fir Forest	Summer or winter	Cold
Mixed Conifer with Aspen	Summer or winter	Cold
Mixed Conifer–Frequent Fire	Summer or winter	Cold
Ponderosa Pine Forest	Summer or winter	Cold
Montane/Subalpine Grassland	Summer or winter	Cold
Pinyon-Juniper Woodlands	Summer or winter	Cold or mild
Madrean Woodlands	Summer[1]	Mild
Gambel Oak Shrubland	Summer or winter	Cold
Mtn Mahogany Mixed Shrubland	Summer or winter	Cold or mild
Sagebrush Shrubland	Winter	Cold
Colorado Plateau/Great Basin Grassland	Summer or winter	Cold
Semi-Desert Grassland	Summer[1]	Mild
Chihuahuan Desert Scrub	Summer[1]	Mild
Great Plains	Summer	Cold[2]
Riparian (various types)	Summer or winter	Cold or mild

1. Also occurs in winter precipitation zones in Arizona

2. Occurs in semi-arid climate, with very hot summers and cold winters

due to livestock (Kauffman and Krueger 1984). Together with changes in understory composition, the reduction of woody vegetation can lower stream bank stability leading to bank sloughing and the eventual widening or downcutting of the stream channel (Krueper 1995; Neary and Medina 1996). Where stream channels are downcut, the associated effects include de-watering (drying) of nearby riparian communities and changes in plant composition from wetland and aquatic species to more mesic upland species. Degradation of the tree and shrub components is also associated with increased stream temperatures and decreased cover for wildlife habitat.

MAJOR ECOSYSTEM TYPES

Among New Mexico's 16 major ecosystem types described in this chapter (Table 2.2), some occur in more than one climate regime or life zone. Further, the Great Plains are recognized here as one of the state's major ecosystem types, but they do not lend themselves as seamlessly to the concept of life zone and are better subdivided into mainly Shortgrass Prairie, Sandsage, and Shinnery Oak. These three types can co-occur in areas of similar climate and, as a result, must instead be differentiated at the site level based on setting and soil (edaphic) properties. Within New Mexico, the climate of the Great Plains is one of hot summers, cold winters, and relatively brief spring and fall seasons, with the majority of the precipitation occurring from April to September. Finally, riparian ecosystems occur throughout all climate and life zones, driven principally by local climate conditions, hydrology, and soil properties. Many different riparian systems occur in New Mexico, with some described later in the chapter.

Though landscape contrast among ecosystem types is occasionally stark, the units listed in Table 2.2 typically occur along continua of climate variables and soil conditions. It is nevertheless useful to impose classification concepts and map unit boundaries to help highlight points along these gradients that denote the physiological limits of major ecosystem types (Daubenmire 1968; Kormondy 1969). The following narratives provide a cursory overview of the vegetation and ecology of New Mexico's major ecosystems. The application of scientific and common names is based on the USDA Plants Database (USDA NRCS 2016) or Allred and Ivey (2012).

Alpine and Tundra

The Alpine and Tundra ecosystem type is limited in extent to only the highest elevations, above approximately 3,800 m (~12,500 ft), all of it in north- and south-central New Mexico (Brown 1982) including on Wheeler Peak and other high peaks in the Sangre de Cristo Mountains, and on Sierra Blanca. It can be found not just on peaks but also along gradual to steep slopes in valleys and basins and on flat ridges.

Alpine areas are low in productivity and biomass, but have a rich and unique diversity of low-growing shrubs, forbs, graminoids, mosses, and lichens. Extreme cold, exposure to high winds and desiccation, unstable surfaces, and a short growing season limit vegetation to all but the hardiest plant species with specific adaptations. Alpine shrubs are few but include alpine willow (*Salix petrophila*). Prostrate and mat-forming vegetation with thick taproots or rootstocks typify the forb component, while rhizomatous sod-forming sedges are the dominant graminoids. Forbs include Ross's avens (*Geum rossii*), phlox (*Phlox pulvinata*), and alpine clover

(*Trifolium dasyphyllum*), while graminoids include tufted hairgrass (*Deschampsia caespitosa*), Bellardi bog sedge (*Kobresia myosuroides*), and several sedge species of the genus *Carex*, along with fescue grasses (*Festuca* spp.). Avens, phlox, and Bellardi bog sedge also occur in less stable settings and open fell-fields along with twinflower sandwort (*Minuartia obtusiloba*), moss campion (*Silene acaulis*), creeping sibbaldia (*Sibbaldia procumbens*), nailwort (*Paronychia pulvinata*), and black and white sedge (*Carex albonigra*). Fires are rare (Moir 1993), as they were historically, most often creeping among patches of vegetation in a mixed severity pattern. Wind, desiccation, grazing and trampling, and instability are far more significant as disturbances in fragile alpine settings (Dick-Peddie 1993). As a stressor, climate change and temperature increases pose a particular problem for alpine vegetation, where no additional area exists for upward plant migration.

Spruce-Fir Forest

The Spruce-Fir Forest ecosystem type occurs only at the highest elevations in mountain ranges of north- and south-central New Mexico, and on the Mogollon Plateau in the southwest. The Spruce-Fir Forest ranges in elevation from about 2,700 to 3,500 m (~8,850 to 11,500 ft), depending on climate zone and aspect, and occurs on both steep and gentle mountain topography. This type is dominated mostly by Engelmann spruce (*Picea engelmanii*) and corkbark fir (*Abies arizonica*), but at lower elevation can be co-dominated by tree species more prevalent in the mixed conifer zone including Douglas-fir, white fir, southwestern white pine, and limber pine (*Pinus flexilis*). Aspen (*Populus tremuloides*) are concentrated in the lower spruce-fir, sometimes forming their own forest cover type in a mosaic with conifer-dominated stands. Common understory species include currants (*Ribes* spp.), maples (*Acer* spp.), honeysuckle (*Lonicera* spp.), huckleberry (*Vaccinium* spp.), red baneberry (*Actaea rubra*), alpine clover, fleabane (*Erigeron* spp.), twinflower (*Linnaea borealis*), and sedges. The characteristic fire regime is one of stand replacement fires at long intervals of 300 or more years (Grissino-Mayer and Swetnam 1995), though mixed-severity fires also play a role (Vankat 2013). Tree insect outbreaks and blowdown are other significant disturbances that are natural to this ecosystem. Snag and downed wood, both products of disturbance, are important habitat features of the Spruce-Fir Forest for carnivores including the Pacific marten. While younger post-fire tree stands are often dense, they tend to thin with age and become structurally diverse, both horizontally and vertically, with some communities developing into large stands of old growth. Old-growth components include old trees, snags, downed wood (coarse woody debris), and multistory conditions, with the location of these features shifting on the landscape over time as a result of disturbance and succession. Today's disturbance regimes and associated patch patterns are similar to historical conditions in many parts of the region in terms of patch size and patch size diversity. Like alpine settings, spruce-fir ecosystems are susceptible to warmer temperatures, given the limited opportunities for upward expansion in New Mexico.

Mixed Conifer with Aspen

The Mixed Conifer with Aspen type represents the moist-mesic constituent of the mixed conifer zone (Photos 2.2 and 2.3), situated between Ponderosa Pine Forest below and Spruce-Fir Forest above. Mixed Conifer–Frequent Fire, discussed in the next section, represents the opposing warm-dry theme of the mixed conifer zone. At opposite extremes, the two types—Mixed Conifer with Aspen and Mixed Conifer–Frequent Fire—differ substantially in structure and fire regime, but much of the mixed conifer zone exists in gradation without strong affinities

to the two extremes. Mixed Conifer with Aspen occurs mostly at elevations between 1,950 and 3,050 m (~6,400 and 10,000 ft) and has a geographical distribution similar to spruce-fir in New Mexico, though it extends farther south into the Guadalupe Mountains bordering Texas to the south, and to the Animas Mountains in the far southwestern Bootheel region of the state. Tree species dominance is driven by environmental conditions and the sequence of successional stages following fire and insect events. Seral plant communities are dominated by aspen, southwestern white pine, and occasionally limber pine. It is noteworthy that ponderosa pine (*Pinus ponderosa*) occurs only as a co-dominant element within some communities, contrary to Mixed Conifer–Frequent Fire where the species is a major element. Late succession stands are represented by Douglas-fir, white fir, and blue spruce (*Picea pungens*), and less frequently by bigtooth maple (*Acer grandidentatum*). Important subordinate woody species include New Mexico locust (*Robinia neomexicana*) and Rocky Mountain maple (*Acer glabrum*), with an understory made up of a wide variety of shrubs, forbs, and grasses whose presence and abundance depend on aspect, soil properties, and other site factors. Some classic mixed conifer shrub taxa include Gambel oak (*Quercus gambelii*), oceanspray (*Holodiscus discolor*), thimbleberry (*Rubus parviflorus*), five-petal cliffbush (*Jamesia americana*), mountain ninebark (*Physocarpus monogynus*), and kinnikinnick (*Arctostaphylos uva-ursi*). The herbaceous stratum may be dense or sparse and dominated by either forbs or graminoids including Fendler's meadow-rue (*Thalictrum fendleri*), Nevada pea (*Lathyrus lanszwertii*), Canadian white violet (*Viola canadensis*), elkweed (*Frasera speciosa*), paintbrush (*Castilleja* spp.), yarrow (*Achillea millefolium*), and several species of grasses and sedges.

Stand composition and structure are shaped mostly by the ecosystem's fire regime but insect

Photo 2.2. Mixed Conifer with Aspen in the Valles Caldera preserve of northcentral New Mexico, Sandoval County, showing the characteristic components of mixed conifers and patches of quaking aspen. Photograph: © Jack Triepke.

Photo 2.3. Mixed Conifer with Aspen in the Jemez Mountains of north-central New Mexico. Photograph: © Tom Kennedy.

and pathogen agents affecting trees also play an important role in Mixed Conifer with Aspen (USDA Forest Service 2013). Fires typically occur either as large infrequent events, particularly stand replacement fires, or as smaller disturbances alongside (or in combination with) insects, disease outbreaks, or strong winds. Disturbances, in turn, lead to the development

of downed wood and snag habitat, which are typically plentiful in this ecosystem type. While younger post-fire tree stands can be dense, tree thinning will occur naturally to create communities that are vertically and horizontally diverse, with some stands developing into old growth. Historically, the fire regime was one of mixed-severity and stand replacement fires (Romme et al. 2009; O'Connor et al. 2014), with fire severity now increasing on some contemporary landscapes of the region. Stand replacement fire is important in triggering the regeneration of large continuous patches of aspen, and there is some concern that aspen cover has declined with the onset of fire suppression in combination with other factors such as browsing by deer and elk (*Cervus canadensis*) (Jones et al. 2005; Smith et al. 2016).

Mixed Conifer–Frequent Fire

As already mentioned, this ecosystem represents the opposing theme to the moist-mesic mixed conifer type, Mixed Conifer with Aspen. The Mixed Conifer–Frequent Fire type may be found at elevations from approximately 1,800 to over 3,000 m (~5,900 to 9,850 ft), existing in settings that are predisposed to frequent fire. Historically, such settings would have been dominated by ponderosa pine, given its specific adaptations to frequent fire, and to a lesser extent by Douglas-fir, southwestern white pine, and limber pine. White fir was a minor component in contrast to contemporary plant communities. Aspen is present in many stands but as a subordinate feature, achieving dominance only in the moist-mesic mixed conifer where aspen cover types are a signature trait of the ecosystem. The understory vegetation is comprised of many of the same constituents as the moist-mesic type, though the cover of grasses averages higher, in turn favoring the frequent-fire regime that is inherent to this type. In southwestern New Mexico, at the northern extent of the Madrean influence, Mixed

Conifer–Frequent Fire communities may have an evergreen oak component, notably silverleaf oak (*Quercus hypoleucoides*).

Stand composition and structure are shaped mostly by the ecosystem's fire regime but tree disease from insects and parasitic plants, especially bark beetles and dwarf mistletoe, also play an important role in forming key habitat characteristics of snags, downed wood, dead limbs, and broken treetops. Historically, these old-growth features would have occurred individually or in small clumps, in contrast to the stand-level dynamics of Spruce-Fir Forest and Mixed Conifer with Aspen. Fires were frequent and of low severity (Ahlstrand 1980; Baisan and Swetnam 1990), favoring open communities with trees of all sizes and ages. Here, succession would have occurred in small clumps or individual trees rather than as stands as with the moist-mesic forest systems.

Due to fire suppression and other causes, fires today occur much less frequently and are much more severe (Photos 2.4 and 2.5), associated with stand conditions that are denser, more even-aged, and prone to insect outbreaks and uncharacteristic fire behavior. For carnivores, many of the plant communities found today in the Mixed Conifer–Frequent Fire ecosystem type have taken on habitat conditions of upper, more-mesic, mixed conifer stands. Where stand replacement fires occur, there may be long-term type conversions to herbaceous and shrub-dominated plant communities with the lack of seed source for tree regeneration in unnaturally large fire openings (Savage and Mast 2005).

Ponderosa Pine Forest

The Ponderosa Pine Forest ecosystem type is widespread in forested areas of New Mexico and represents the classic fire-adapted system of the western United States. It occurs at elevations ranging from about 1,800 to 2,300 m (~5,900 to 7,500 ft), and as its name indicates is dominated by ponderosa pine, with other trees such

Photo 2.4. Uncharacteristic fire in the Mixed Conifer–Frequent Fire type; from the 2012 Little Bear Fire on the Lincoln National Forest in south-central New Mexico. Photograph: © Kari Greer/US Forest Service.

Photo 2.5. Resulting stand replacement in the Mixed Conifer–Frequent Fire type after the 2011 Las Conchas Fire in north-central New Mexico. Photograph: © Larry Lamsa.

Photo 2.6. Ponderosa Pine Forest in the upper Gila River watershed in southwestern New Mexico, Grant County, showing characteristic multi-age stand structure, intact as a result of wilderness fires that have been allowed to burn repeatedly. Photograph: © Jack Triepke.

Photo 2.7. Ponderosa pine stand in Bandelier National Monument, north-central New Mexico. Photograph: © Larry Lamsa.

as pinyon, juniper, and Gambel oak (tree form) in lesser abundance (Brown 1994). Shrub density varies according to local environment and land use. The abundance of shrub-form Gambel oak or, conversely, bunchgrasses in the understory helps to define two important subclasses of the Ponderosa Pine Forest (Ponderosa Pine/Gambel Oak, Ponderosa Pine/Bunchgrass). Shrubs also include New Mexico locust, and common grass species are Arizona fescue (*Festuca arizonica*), mountain muhly (*Muhlenbergia montana*), pine dropseed (*Blepharoneuron tricholepis*), muttongrass (*Poa fendleriana*), and blue grama (*Bouteloua gracilis*).

In Ponderosa Pine Forest, stand composition and structure are shaped especially by fire regime but also by tree disease and insect infestations (USDA Forest Service 2013). As with all forest systems, these processes are important in creating habitat features such as snags, downed wood, dead limbs, and broken treetops. Historically, wildfires were frequent and of low severity (Swetnam and Dieterich 1985; Muldavin et al. 2003), favoring open communities with trees of all sizes and ages (Photos 2.6 and 2.7). Seasonal climate patterns, the plant physiology of Ponderosa Pine Forest, a thick fire-resistant bark, and the mild topography on which much of the Ponderosa Pine Forest occurs are some of the key variables that mutually promote a system of frequent fire and uneven-aged structure. In the last century, however, fire suppression and land use have led to less frequent fires that are considerably more severe, in turn favoring denser and more evenly aged conditions (Moore et al. 2004). As with the Mixed Conifer–Frequent Fire type, the combined effects of altered stand structure and climate change can further lead to fires of greater severity, and to the long-term conversion of previously forested communities to shrub- and grass-dominated systems (Savage and Mast 2005).

Ponderosa Pine–Evergreen Oak is a minor system, related to the Ponderosa Pine Forest, known from mild climate zones and the mountain ranges of southwestern and south-central New Mexico, and with characteristics of the Madrean province extending north from Mexico. Like the Ponderosa Pine Forest, this system occurs below mixed conifer and above the pinyon-juniper zone, but is co-dominated by evergreen oak trees such as silverleaf oak, netleaf oak (*Quercus rugosa*), gray oak (*Q. grisea*), and Arizona white oak (*Q. arizonica*) (Dick-Peddie 1993). This ecosystem was also one of frequent low-severity fires (Baisan and Swetnam 1990; Kaib 2001), but with the added variability of mixed-severity fires at long intervals in some settings. In recent decades Ponderosa Pine–Evergreen Oak has been impacted by the combined effects of fire suppression and land use, resulting in denser stands and a contemporary disturbance regime of less frequent but higher severity fires.

Montane/Subalpine Grassland

The Montane/Subalpine Grassland ecosystem type of the mountains of New Mexico (Photo 2.8) spans elevations from about 2,400 to 3,350 m (~7,900 to 11,000 ft), representing a variety of plant associations and flora (Moir 1967). The ecology of these grasslands is tied closely to snowmelt and seasonal wetness. In valley bottoms

Photo 2.8. Montane/Subalpine Grassland in the Valles Caldera National Preserve in Sandoval County, New Mexico. Photograph: © Jack Triepke.

and basins the Montane/Subalpine Grassland type is often interspersed with herbaceous wetlands, sometimes forming belts of grasslands surrounding riparian and wetland communities of lower settings. Characteristic graminoids include Thurber's fescue (*Festuca thurberi*), Arizona fescue, Parry's oatgrass (*Danthonia parryi*), pine dropseed, and various sedges (Robbie 2004). In communities that have been grazed by livestock, Kentucky bluegrass can be abundant, occasionally forming large patches of sod vegetation. Forb diversity is often high in these grasslands, and can include shooting star (*Dodecatheon* spp.), lupine (*Lupinus* spp.), Rocky Mountain iris (*Iris missouriensis*), larkspur (*Delphinium* spp.), Parry's bellflower (*Campanula parryi*), Porter's licorice root (*Ligusticum porteri*), and California false hellebore (*Veratrum californicum*) among others.

Historically this ecosystem type was subject to frequent surface fires (Dick-Peddie 1993), which limited shrub- and tree-cover and ensured regular nutrient cycling. With the onset of fire suppression, the vigor of the herb layer has declined in most of the ecosystem type's plant communities, and trees have encroached along forest edges, represented by ponderosa pine, Douglas-fir, blue spruce, and other conifers (Allen 1989; White 2002).

Pinyon-Juniper Woodlands

Pinyon-Juniper Woodlands make up the most common forested ecosystem type, not just in New Mexico but overall in the Southwest. They occur mostly at elevations between 1,300 and 2,300 m (~4,250 to 7,550 ft), covering vast areas of plateaus, foothills, and surrounding plains throughout the entire state except southeastern New Mexico. Despite the ecosystem type's common appearance (Photo 2.9), Pinyon-Juniper Woodlands represent over a dozen tree species not counting the oaks that sometimes co-dominate. Depending on the type of pinyon-juniper, the historical fire regime ranged from frequent, low-severity fires to infrequent, stand replacement events, with a commensurate diversity of structural conditions, from open communities with trees of all sizes to more closed and even-aged conditions. Pinyon-Juniper Woodlands can be divided into six general subtypes based on climate, fire regime, and structural attributes (Moir and Carlton 1986; Romme et al. 2009; Moreland et al. 2017) (Table 2.3). The Pinyon-Juniper subtypes can be further divided to express differences in vegetation based on temperature and precipitation regimes, with a comparable diversity of shrub, forb, and grass species (Dick-Peddie 1993).

As Table 2.3 suggests, some types of Pinyon-Juniper Woodlands have been more affected by fire suppression than others. On contemporary landscapes, the frequent-fire ecosystem types exhibit the most obvious impacts of stand densification and increased fire severity. Of the Southwest ecosystem types, Pinyon-Juniper Woodlands have understandably received much of the scrutiny associated with climate change, with the widespread dieback of trees from warmer summers

Photo 2.9. Pinyon-Juniper Woodlands on the Kiowa National Grassland in northeastern New Mexico, Harding County (Juniper Grass subclass). Photograph: © Jack Triepke.

Table 2.3. List of major subclasses of the Pinyon-Juniper Woodlands (or PJ Woodlands) and their associated climate and historical fire regime.

Subclass	Temperature	Historical Fire Regime[1]
PJ Woodland (persistent)	Cold or mild	V, III
PJ Sagebrush	Cold	V, III
PJ Deciduous Shrub	Cold	III, IV
PJ Evergreen Shrub	Mild	III, IV
PJ Grass	Cold or mild	I
Juniper Grass	Cold or mild	I

1. I (frequent, non-lethal), II (frequent, stand replacement), III (moderately frequent, mixed-severity), IV (moderately frequent, stand replacement), V (infrequent, stand replacement) (Barrett et al. 2010).

and higher moisture deficits (Allen 2009; Williams et al. 2013).

Madrean Woodlands

Madrean Woodlands occur in areas of mild climate primarily in southern New Mexico on foothills extending out onto piedmonts (bajadas), and also on plateaus and in canyons. This ecosystem type is at the northern extension of the Madrean floristic province of Mexico and it provides habitat for Madrean carnivores like the coati. Plant communities akin to Madrean Woodlands in physiognomy and dynamics can be found as far north as the Sandia Mountains near Albuquerque, and as far east as the Guadalupe Mountains on the border with Texas (Dick-Peddie 1993). Like Pinyon-Juniper Woodlands, Madrean Woodlands occur in the zone sandwiched between Ponderosa Pine Forest above and grassland systems below, roughly between 1,200 and 2,100 m (~3,950 to 6,900 ft). Intergradation with neighboring ecosystem types is common so that boundaries among related units are not always obvious. Madrean Woodlands can be conceptualized as two more precise units, either Madrean Encinal Woodland or Madrean Pinyon-Oak (Brown et al. 1998), but for our purposes here are described as one system.

Madrean Woodlands are dominated by evergreen oaks including Arizona white oak, Emory oak (*Quercus emoryi*), gray oak, and Mexican blue oak (*Q. oblongifolia*), along with alligator juniper (*Juniperus deppeana*), pinyon species, and Chihuahua pine (*Pinus leiophylla* var. *chihuahuana*). Hybridization among oak species is common, making species identification difficult. Pines have low representation in the subtype of Madrean Encinal Woodland, but are dominant or co-dominant in the Madrean Pinyon-Oak subtype, where the large pines of the montane zone above, such as ponderosa or Arizona pine (*Pinus arizonica*), are uncommon. In the Guadalupe Mountains, Texas madrone (*Arbutus xalapensis*) can co-dominate Madrean Woodlands. Understory constituents include various deciduous and evergreen shrubs, including shrub-form oaks of some of the tree species mentioned above. A strong grass component is common and includes several species of grama (*Bouteloua* spp.), threeawns (*Aristida* spp.), Arizona cottontop (*Digitaria* spp.), muhly grasses (*Muhlenbergia* spp.), plains lovegrass (*Eragrostis intermedia*), vine mesquite (*Panicum obtusum*), and Texas bluestem (*Schizachyrium cirratum*).

The historical fire regime is generally thought of as frequent and low severity (Baisan and Swetnam 1990; Kaib et al. 1996), though a component of mixed-severity fire was likely, especially on steeper slopes that favored more intense fire behavior. Madrean Woodlands may intergrade with and resemble, at least in early succession, surrounding shrubland ecosystems. As with other fire-adapted types, modern fire suppression and land use practices have changed the dynamics and the resulting stand structures of this ecosystem type. Today's Madrean Woodlands have been substantially altered with more severe fires and trend toward denser and more homogenous tree structure, along with increased shrub cover and decreased grass cover. Climate change may also be playing a role in elevating tree dieback particularly in species of pine (Allen 2007).

Gambel Oak Shrubland

The Gambel Oak Shrubland ecosystem type is dominated by shrub-form Gambel oak, and to a lesser extent by other deciduous shrubs, often occurring in continuous patches of relatively homogenous structure. In New Mexico this type occurs from about 2,000 to 2,900 m (~6,600 to 9,500 ft), generally on all aspects but mostly on southern exposures especially at the highest elevations. The Gambel Oak Shrubland spans the montane forest and upper woodland zones, often expressed as a fire disclimax system on steep topography subjected to repeat stand replacement fire (Vankat 2013). Its occurrence can also be edaphically promoted by soil properties, often in combination with steep topography and high severity fire.

Little is known directly about historical stand dynamics and fire patterns in Gambel Oak Shrubland. However, plant physiology, topography, soil properties, and fire behavior all suggest that historically, fires were moderately frequent and of high severity. Such fires are followed by rapid re-sprouting of Gambel oak and other shrubs from live root crowns, to form dense thickets or clumpy patterns that often resemble the preexisting plant community. In this manner, the Gambel Oak Shrubland is relatively stable in space and time in contrast to some woodland and forest systems that include Gambel oak cover types only as a temporary seral condition (Dick-Peddie 1993). Occasionally, stands of Gambel oak escape fire for significant amounts of time, self-thin, and acquire more fire resistance and understory plant diversity. Unlike other ecosystem types, Gambel Oak Shrubland may have changed little from historical times, with the exception of local conifer encroachment.

Mountain Mahogany Mixed Shrubland

The Mountain Mahogany Mixed Shrubland ecosystem type is distributed across all mountainous regions of the state, but has particular affinity to the mountain ranges adjacent to the Great Plains of eastern New Mexico. It occurs in foothills, on lower mountain slopes, and in canyons (Photo 2.10), in association with rocky substrates, well-drained soils, and exposed topography. As with Gambel Oak Shrubland, recurring stand replacement fires promote shrub growth through re-sprouting while limiting tree encroachment. The constant of this ecosystem type is alderleaf mountain mahogany (*Cercocarpus montanus*), while co-dominants can include skunkbush sumac (*Rhus trilobata*), serviceberry (*Amelanchier utahensis*), cliffrose (*Purshia stansburiana*), and, in some areas, Gambel oak and desert ceanothus (*Ceanothus pauciflorus*) (Dick-Peddie 1993). Small inclusions of grassland or tree cover may be present, but the characteristic physiognomy of the system is one of large continuous shrub patches.

The Mountain Mahogany Mixed Shrubland spans the upper woodland and lower montane zones, often intergrading with Ponderosa Pine Forest and Pinyon-Juniper Woodlands. Historical fires were of high severity and moderate frequency and, as in other New Mexico shrublands, favored relatively stable floral composition and vegetation structure over time. Inferences of fire behavior, plant response, and environmental setting

Photo 2.10. Mountain Mahogany Mixed Shrubland in Mills Canyon, Harding County, northeastern New Mexico. Photograph: © Jack Triepke.

corroborate the assumed historical fire regime in lieu of more direct evidence. The presence of trees in Photo 2.10 largely reflects the effects of 20th-century fire suppression in an ecosystem type that otherwise appears to be unchanged.

Sagebrush Shrubland

The Sagebrush Shrubland ecosystem type is distributed across northwestern and north-central New Mexico, in areas of winter precipitation and cold climate, often on well-drained soils of plateaus and basin-bottoms (Photo 2.11). Dick-Peddie (1993) explains that big sagebrush (*Artemisia tridentata*), the signature dominant plant of this ecosystem type, also occurs in Great Basin grasslands, but only as a component subordinate to grasses. Grass cover is less substantial in Sagebrush Shrubland, though blue grama and needle-and-thread (*Hesperostipa comata*) are common in the understory. In this ecosystem type, other shrubs include silver sagebrush (*Artemisia cana*), black sagebrush (*A. nova*), rubber rabbitbrush (*Ericameria nauseosa*), and Bigelow sage (*A. bigelovii*). In New Mexico, Sagebrush Shrubland occurs at elevations between about 1,450 and 1,800 m (~4,750 and 5,900 ft), often adjacent to

Colorado Plateau/Great Basin Grassland and Pinyon-Juniper systems. Modern fire exclusion may play a role in the encroachment of conifers into Sagebrush Shrubland ecosystems, though other local factors can impose greater limitations to tree growth than fire. Information on the historical fire regime of Sagebrush Shrubland is sparse, but nonetheless suggests that fires were infrequent, with stand replacement carrying through shrub crowns only in extreme conditions of wind and low fuel moisture. It can be hypothesized that the cover of shrubs has increased in the last century, less as a result of fire suppression than grazing practices that favor shrub and tree growth.

Colorado Plateau/Great Basin Grassland

The Colorado Plateau/Great Basin ecosystem type is the cold-climate counterpart to the Semi-desert Grassland of mild climate zones described below, and it assumes the same zone position below woodlands. As the name implies, the Colorado Plateau/Great Basin Grassland is made up of the two grassland subclasses that have been effectively differentiated based on floristics and recent vegetation classification (USNVC 2016)

Photo 2.11. Sagebrush shrubland southwest of Taos, Taos County with the Picuris Mountains in the background on the left. Photograph: © Jean-Luc Cartron.

along with ecological mapping (Robbie 2004). But here the two are combined based on similar ecosystem dynamics and habitat features. This type is concentrated in the northwestern part of the state where precipitation falls mostly in winter and early spring, but its grasslands can be found south to the upper Gila and San Francisco river basins of southwestern New Mexico, and east to the eastern front of the Sangre de Cristo Mountains.

Historically the vegetation of this ecosystem type consisted mostly of grasses including galleta (*Pleuraphis jamesii*), Indian ricegrass (*Achnatherum hymenoides*), western wheatgrass (*Pascopyrum smithii*), needle-and-thread, sideoats grama (*Bouteloua curtipendula*), and blue grama, with intermittent patches of shrubs. On

contemporary landscapes shrubs have increased substantially in cover due especially to grazing and fire suppression (Yanoff et al. 2008), in a system that likely witnessed frequent fires before European settlement (Wright and Bailey 1982). The shrub stratum is dominated especially by members of the sunflower and goosefoot families including big sagebrush, shadscale (*Atriplex confertifolia*), winterfat (*Krascheninnikovia lanata*), and greasewood (*Sarcobatus vermiculatus*) (Lowrey 2010). The warmer temperatures forecast for the Southwest (Gutzler and Robbins 2010) imply conditions more favorable to the vegetation of mild ecosystems such as the Semi-Desert Grassland, but perhaps the most likely change in the immediate future is an increased abundance of scrub species.

Semi-Desert Grassland

The Semi-Desert Grassland ecosystem type (Photo 2.12) is the mild regime counterpart to the Colorado Plateau/Great Basin Grassland of the cold regions of northern New Mexico, holding a similar zone position, typically below the woodlands and above Chihuahuan Desert Scrub concentrated in the southern third of the state. Semi-Desert Grassland is generally considered a frequent-fire type (Bahre 1985; McPherson 1995; Kaib et al. 1996). It is represented by several subclasses, which are differentiated by floristics, topographic settings, and soils (Muldavin et al. 2004), yet are treated together here based on similarity in habitat, structure, and ecosystem processes. Semi-Desert Grassland is distributed at elevations ranging from about 900 to 1,350 m (~2,950 to 4,450 ft) across the mild southern third of the state, where precipitation is concentrated in the summer monsoon rains. Characteristic grass species include black grama (*Bouteloua eriopoda*), blue grama, tobosagrass (*Pleuraphis mutica*), big sacaton (*Sporobolus wrightii*), vine mesquite, bush muhly (*Muhlenbergia porteri*), and burrograss (*Scleropogon brevifolius*) (Robbie 2004; Lowrey 2010). Shrubs include mesquite (*Prosopis velutina*), creosote bush (*Larrea tridentata*), tarbush (or American tarwort; *Flourensia cernua*), turpentine bush (*Ericameria laricifolia*), desert ceanothus, and soaptree yucca (*Yucca elata*).

The exact boundaries between the Semi-Desert Grassland and Chihuahuan Desert Scrub ecosystem types can be ambiguous owing to several factors. First, the two ecosystem types share many of the same shrub species, including those listed above, to the point that Semi-Desert Grassland and Chihuahuan Desert Scrub are sometimes intermingled (Robbie 2004) (Photo 2.13). Also, grazing practices and fire suppression have promoted an increase in shrub cover (Fletcher and Robbie 2004) at the expense of the grass component, giving many Semi-Desert Grassland communities the appearance of desert scrub (Dick-Peddie 1993). Vast expanses of former grassland are now mesquite coppice dunes. Nevertheless, soil properties along with the lack of tarbush and the presence/absence of other floristic indicators can be used to distinguish scrub communities from what may have been grassland. At the same time, there are Semi-Desert Grassland sites that are naturally high in shrub cover, sometimes referred to as "hot steppe" or Semi-Desert Shrub systems. Finally, adding to the

Photo 2.12. Semi-Desert Grassland near Deming in Luna County, New Mexico. Photograph: © Jack Triepke.

Photo 2.13. Scrub-dominated Semi-Desert Grassland intermingled with Chihuahuan Desert Scrub at Oliver Lee State Park south of Alamogordo, New Mexico. Photograph: © Tom Kennedy.

difficulty in perfectly delineating the two types, there is some evidence to suggest that climate change and drought conditions in the Southwest are becoming less conducive to grasses and could promote shrub dominance (Báez et al. 2013).

Chihuahuan Desert Scrub

The Chihuahuan Desert Scrub ecosystem type occurs in the mild and summer precipitation zone of southern New Mexico (Map 2.1), extending north from the greater Chihuahuan Desert of Mexico (Brown 1994). This type includes large expanses of open-canopied scrub lands at elevations below approximately 1200 m (~3,950 ft), in somewhat warmer-dryer settings than Semi-Desert Grassland. Chihuahuan Desert Scrub is distributed on the edges of basin floors, on alluvial fans, and up the foothills of mesas and desert mountain ranges. While several subtypes of Chihuahuan Desert Scrub have been described (e.g., Muldavin et al. 2004), creosote bush and tarbush are diagnostic elements across much of the spectrum (Photo 2.14). Other shrubs and subshrubs include whitethorn acacia (*Vachellia constricta*), viscid acacia (*V. neovernicosa*), ocotillo (*Fouquieria splendens*), lechuguilla (*Agave lechuguilla*), Wright's beebrush (*Aloysia wrightii*), and many species of cactus including cactus apple (*Opuntia engelmanii*) (Photo 2.15). Chihuahuan Desert Scrub has the highest diversity of cacti of any desert province in the Southwest (Lowrey 2010). Some areas are barren with less than 1% vegetation cover, as with plant communities in and around the White Sands in the south-central part of the state. While grasses and forbs are common, their collective cover is low, a key factor in the rarity of fires owing to the lack of fine fuels needed to facilitate fire spread. Herbaceous species include black grama, tobosagrass, and burrograss (LANDFIRE 2010).

Great Plains Systems

The Great Plains in New Mexico are represented by multiple ecosystem subtypes, including Shortgrass Prairie, Sandsage, and Shinnery Oak, with the former being the most common by far. These subtypes occupy the same climate zones with differences in niche driven by setting and edaphic factors. The Great Plains exist principally in northeastern and east-central New Mexico, in areas continuing south from the Kiowa-Rita Blanca National Grasslands. Shortgrass Prairie extends west and south, with shortgrass elements existing in the central Rio Grande Valley and as far west as the San Rafael Valley in southeastern Arizona. Shortgrass Prairie typically occurs on broad plains (Photos 2.16 and 2.17) and flat to gently rolling uplands and mesa tops, and is represented by the signature taxa of blue grama and buffalograss (*Buchloe dactyloides*), as well as sideoats grama, New Mexico feathergrass (*Hesperostipa neomexicana*),

needle-and-thread, purple three-awn (*Aristida purpurea*), sand dropseed (*Sporobolus cryptandrus*), and other grasses (Robbie 2004). Along with the other Great Plains subtypes, Shortgrass Prairie is particularly adapted to large ungulate herbivory and is more resistant to grazing pressure than the other ecosystem types of New Mexico. As with all Great Plains systems, historical fire in Shortgrass Prairie is assumed to have been frequent (Wright and Bailey 1982). And like other grassland systems, Shortgrass Prairie suffers from the same symptoms caused by fire suppression and land use, as evidenced by the increase in woody vegetation including juniper and oak (Fletcher and Robbie 2004).

Sandsage shares much of the same distribution as Shortgrass Prairie, though it spans farther west into northwestern New Mexico and beyond. This type occurs mainly on sand dunes, areas where sediment has blown in and deposited as in the case of the plant communities established following the Dust Bowl. Dune formation and sandsage (*Artemisia filifolia*) development continue to this day as a result of both natural and human processes. The vegetation is of low stature, with patches of the low-growing sandsage and other

Photo 2.14. Creosote bush flat within the Chihuahuan Desert Scrub ecosystem type near Hatch in Doña Ana County, New Mexico. Photograph: © Jack Triepke.

Photo 2.15. Chihuahuan succulent desert scrub foothills (Chihuahuan Desert Scrub ecosystem type) in southwestern New Mexico, Hidalgo County. Photograph: © Tom Kennedy.

Photo 2.16. Shortgrass Prairie in the Kiowa National Grassland of northeastern New Mexico, Harding County. Photograph: © Jack Triepke.

Photo 2.17. Shortgrass Prairie in Roosevelt County in southeastern New Mexico. Photograph: © James N. Stuart.

shrubs, all of which are nevertheless important as hiding cover for wildlife. Although the chief constituent is sandsage (sand sagebrush), other characteristic plant species include mid and tall-grass species such as sideoats grama, little bluestem (*Schizachyrium scoparium*), sand bluestem (*Andropogon hallii*), mesa dropseed (*Sporobolus flexuosus*), and needle-and-thread. The historical fire regime is largely unknown, but assumed similar to Shortgrass Prairie (frequent fire), perhaps with a greater propensity for mixed-severity fires due to the lower continuity of fine fuels.

In New Mexico, Shinnery Oak represents the other main shrub-dominated subtype of the Great Plains. Of the Great Plains subtypes discussed, its range is the most limited, occurring in the far east-central part of the state. Shinnery Oak often exists in complexes with Sandsage, Shortgrass Prairie, and other Great Plains subtypes that form mosaics of structurally and biologically diverse land cover. It occurs primarily on sandy soils with shinnery oak (*Quercus havardii*) as the characteristic dominant, accompanied by other shrubs such as mesquite, catclaw acacia (*Senegalia greggii*), sandsage, and species of yucca. The grass component includes little bluestem, Indiangrass (*Sorghastrum nutans*), and dropseeds

(*Sporobolus* spp.). This system is thought to be one of frequent fires, with shinnery oak re-sprouting vigorously following each fire (Peterson and Boyd 1998). With fire suppression, the assumption is that the frequency of fire is much reduced and shrub cover is uncharacteristically high on some natural extents of this ecosystem type, at least where herbicides are not applied.

Riparian

Riparian areas are specialized plant communities associated with water that exhibit high productivity and biological diversity. In terms of habitat for wildlife, they rank among the most important vegetation communities across the landscape (Price et al. 2005), even though they occupy less than 1% of New Mexico's land cover and are barely perceptible in Map 2.2. The majority of all southwestern vertebrate species use riparian systems for at least half of their life cycles, with the majority characterized as riparian dependent species (Chaney et al. 1993; Krueper 1995). The nearby aquatic habitats and the biota they support are likewise dependent on the functioning of riparian plant communities. Riparian areas exist interstitially among all previous ecosystem types discussed in this chapter, usually forming linear corridors through upland settings

(Photo 2.18), though riparian ecosystems can also exist adjacent to lakes, ponds, springs, and even human impoundments (Photo 2.19). The plant communities that develop in riparian ecosystems include wetland obligates that require sub-irrigation for their reproduction and sustenance, as well as upland vegetation of larger more prolific growth forms. By their productivity and diversity, riparian ecosystems naturally concentrate trophic systems, an ecosystem's means of bringing energy and nutrients through food chains to much of New Mexico's biota. Of course, different types of riparian communities support different plant and animal diversity.

New Mexico riparian types can be divided into several general categories (Table 2.4). These subcategories will not be discussed further except to say that, like the state's upland ecosystems, riparian ecosystems are similarly diverse owing to the mix of climate conditions, geomorphological features, edaphic qualities, and past disturbance histories.

Common riparian trees include cottonwood (*Populus* spp.), willow (*Salix* spp.), Arizona alder (*Alnus oblongifolia*), walnut (*Juglans* spp.), and boxelder (*Acer negundo*) (Dick-Peddie 1993; Cartron et al. 2008; Lowrey 2010). Arizona sycamore (*Platanus wrightii*) occurs in southwestern New Mexico. Shrubs include mountain alder (*Alnus tenuifolia*), red osier dogwood (*Cornus sericea*), seepwillow (*Baccharis glutinosa*), shrub form willows (*Salix* spp.), and desert willow (*Chilopsis linearis*). The herb layer is dominated by sedges and grasses and, to a lesser extent, by species of rush (Juncaceae family).

Flooding is the chief natural disturbance factor in riparian areas and constitutes an important ecosystem process for riparian obligates that depend on periodic floods for their spread and reproduction. Fire was likely low frequency in many of New Mexico's riparian areas given their moisture content, landscape position, and the lack of fire adapted species (Stuever 1997). For other areas, the frequency and role of fire

Photo 2.18. (*left*) Riparian area within the Guadalupe Mountains of southcentral New Mexico, Eddy County. Photograph: © Jack Triepke.

Photo 2.19. (*above*) Rattlesnake Springs at Carlsbad Caverns National Park in southeastern New Mexico. Photograph: © Jean-Luc Cartron.

Table 2.4. List of general riparian ecosystem types found in New Mexico with approximate elevation ranges (Triepke et al. 2014).

General Riparian Type	Approximate Elevation Range
Herbaceous/Wetland	900–3,700 m (~2,950–12,150 ft)
Desert Willow	900–2,100 m (~2,950–6,900 ft)
Cottonwood	1,000–3,000 m (~3,300–9,850 ft)
Cottonwood-Evergreen Tree	1,900–3,300 m (~6,250–10,850 ft)
Montane-Conifer Willow	1,100–3,600 m (~3,600–11,800 ft)
Walnut-Evergreen Tree	1,400–3,000 m (~4,600–9,850 ft)

remain uncertain, but overall they could have varied considerably among subtypes, reflecting the influence of fire regimes in the surrounding upland systems (Stromberg et al. 2009). Modern riparian fire patterns vary considerably but are generally influenced by domestic and native herbivory and other land use practices, as well as by the presence of invasive vegetation and by stream-flow regulation. The lack of flooding in many regulated river systems, sometimes in combination with increased fire severity, favors invasive vegetation over native flora. Regardless of disturbance history, invasive plants such as saltcedar (*Tamarix* spp.) and Russian olive (*Elaeagnus angustifolia*) represent a major modification to riparian habitats in New Mexico (Cartron et al. 2008). Finally, climate change may affect riparian ecosystems significantly through changes in the amount and timing of precipitation and stream flow (Gutzler 2013), decreases in groundwater levels, and indirectly through the added frequency and severity of fires (Price et al. 2005). Perhaps of all New Mexico ecosystem types, riparian and wetland communities have witnessed the most degradation and loss (Mitsch and Gosselink 2007), with a commensurate response in the rarity and federal listing of obligate plants and animals. As with upland systems, understanding the interactions between land use and climate change is key to the sustainable management of ecosystems.

CONCLUSION

Climate, landscape setting, geology and soil properties, and other variables contribute to the distribution and abundance of New Mexico's varied ecosystems. Within each of the major ecosystem types (Table 2.2), disturbance history, land use, invasive plants, and climate change all express themselves in the quality of habitat conditions at multiple scales. Each of the major types provides habitat for carnivores and a range of flora and fauna in the form of space, energy, nutrition, and other resources necessary for continued viability. The sustainability of these resources and the habitats dependent on them will require careful and expedient measures to restore and maintain ecosystem resilience while simultaneously considering measures to address climate change.

Observations of Southwest vegetation based on remeasurements of long-term sampling transects already indicate the effects of the changing climate (Bell et al. 2014, Brusca et al. 2013; Guida et al. 2014) and testify to the rapid rate of change and upward shifts. Climate projections for the region suggest that summer temperature averages will continue to increase and even exceed the historical range of variation as early as mid-century (Williams et al. 2013). It is reasonable to assume that individual ecosystem types will change in their abundance and geographic distribution (see Chapter 5), and that novel types of unfamiliar species combinations may emerge in the longer term (Notaro et al.

2012; Rehfeldt et al. 2012). With shifting ecosystem types and changes to the relative abundance of forested, shrubland, and grassland systems, a net loss of forest cover is expected (Parks et al. 2019), along with increases in the amount of grassland and desert types. For a time, there may be an overall decrease in vegetation cover if ecosystems struggle to realign and keep pace with the rate of climate change. The ecological sustainability of New Mexico's ecosystems, and the carnivore species they support, rely on the effective planning, management, and monitoring of ecosystems and populations.

LITERATURE CITED

Ahlstrand, G. M. 1980. Fire history of a mixed conifer forest in Guadalupe Mountains National Park. In *Proceedings of the Fire History Workshop, 20–24 October 1980, Tucson AZ*, ed. M. A. Stokes and J. H. Dieterich, 4–7. USDA Forest Service Gen. Tech. Rep. RM-81. Ogden, UT: Rocky Mountain Forest and Range Experiment Station.

Allen, C. D. 1989. Changes in the landscape of the Jemez Mountains, New Mexico. PhD dissertation, University of California, Berkeley.

———. 2007. Interactions across spatial scales among forest dieback, fire, and erosion in northern New Mexico landscapes. *Ecosystems* 10:797–808.

———. 2009. Climate-induced forest dieback: an escalating global phenomenon? *Unasylva* 60:43–49.

Allred, K. W., and R. D. Ivey. 2012. *Flora Neomexicana III: an illustrated identification manual.* Raleigh, NC: Lulu.

Ambos, N., G. Robertson, and J. Douglas. 2000. Dutchwoman Butte: a relict grassland in central Arizona. *Rangelands* 22:3–8.

Arnold, J. F. 1950. Changes in ponderosa pine-bunchgrass ranges in northern Arizona resulting from pine regeneration and grazing. *Journal of Forestry* 48: 118–26.

Báez, S., S. L. Collins, W. T. Pockman, J. E. Johnson, and E. E. Small. 2013. Effects of experimental rainfall manipulations on Chihuahuan Desert grassland and shrubland plant communities. *Oecologia* 172:1117–27.

Bahre, C. J. 1985. Wildfire in southeastern Arizona between 1859 and 1890. *Desert Plants* 7:190–94.

Bailey, V. 1931 (=1932). *Mammals of New Mexico.* North American Fauna 53. Washington, DC: US Department of Agriculture, Bureau of Biological Survey.

Baisan, C. H., and T. W. Swetnam. 1990. Fire history on a desert mountain range: Rincon Mountain Wilderness, Arizona, USA. *Canadian Journal of Forest Research* 20:1559–69.

Barrett, S. W., D. Havlina, J. L. Jones, W. J. Hann, C. K. Frame, D. Hamilton, K. Schon, T. E. DeMeo, L. C. Hutter, and J. P. Menakis. 2010. *Interagency fire regime condition class (FRCC) guidebook*, Version 3.0. USDA Forest Service, US Department of the Interior, and The Nature Conservancy. Technical guide available online at www.frcc.gov.

Bell, D. M., J. B. Bradford, and W. K. Lauenroth. 2014. Early indicators of change: divergent climate envelopes between tree life stages imply range shifts in the western United States. *Global Ecology and Biogeography* 23:168–80.

Brown, D. E. 1982. Biotic communities of the American Southwest—United States and Mexico. *Desert Plants* 4:3–341.

———, ed. 1994. *Biotic communities: southwestern United States and northwestern Mexico.* 2nd ed. Salt Lake City: University of Utah Press.

Brown, D. E., and C. H. Lowe. 1974. A digitized computer-compatible classification for natural and potential vegetation in the Southwest with particular reference to Arizona. *Journal of the Arizona-Nevada Academy of Science* 9:3–11.

Brown, D. E., F. Reichenbacher, and S. E. Franson. 1998. *A classification of North American biotic communities.* Salt Lake City: University of Utah Press.

Brown, P. M., M. W. Kaye, L. S. Huckaby, and C. H. Baisan. 2001. Fire history along environmental gradients in the Sacramento Mountains, New Mexico: influences of local patterns and regional processes. *Ecoscience* 8:115–26.

Brusca, R. C., J. F. Wiens, W. M. Meyer, J. Eble, K. Franklin, J. T. Overpeck, and W. Moore. 2013. Dramatic response to climate change in the Southwest: Robert Whittaker's 1963 Arizona Mountain plant transect revisited. *Ecology and Evolution* 3:3307–19.

Carlton, O. J., and H. G. Brown. 1983. Primary climate gradients of Region 3. In *Proceedings of the workshop on southwestern habitat types, 6–8 April 1983, Albuquerque, NM*, ed. W. H. Moir and L. Hendzel, 85–90. Albuquerque: USDA Forest Service, Southwestern Region.

Cartron, J.-L., D. C. Lightfoot, J. E. Mygatt, S. L. Brantley, and T. K. Lowrey. 2008. *A field guide to the plants and animals of the middle Rio Grande bosque*. Albuquerque: University of New Mexico Press.

Chaney, E., W. Elmore, and W. S. Platts. 1993. *Managing change: livestock grazing on western riparian areas*. July 1993. Northwest Resource Information Center Inc. Resource report available online at the United States Environmental Protection Agency website, https://nepis.epa.gov/.

Clary, W. P. 1975. *Range management and its ecological basis in the ponderosa pine type of Arizona: the status of our knowledge*. USDA Forest Service Res. Pap. RM-158. Fort Collins, CO: Rocky Mountain Forest and Range Experiment Station.

Comer, P., D. Faber-Langendoen, R. Evans, S. Gawler, C. Josse, G. Kittel, S. Menard, M. Pyne, M. Reid, K. Schulz, K. Snow, and J. Teague. 2003. *Ecological systems of the United States: a working classification of US terrestrial systems*. NatureServe technical guide available online at http://www.natureserve.org.

Daubenmire, R. F. 1968. *Plant communities*. New York: Harper & Row.

Dick-Peddie, W. A, ed. 1993. *New Mexico vegetation: past, present, and future*. Albuquerque: University of New Mexico Press.

Fletcher, R., and W. A. Robbie. 2004. Historic and current conditions of southwestern grasslands. In *Assessment of grassland ecosystem conditions in the southwestern United States*, ed. D. M. Finch, 120–129. USDA Forest Service Gen. Tech. Rep.

RMRS-GTR-135-vol. 1. Fort Collins, CO: Rocky Mountain Research Station.

Frey, J. K., and T. L. Yates. 1996. Mammalian diversity in New Mexico. *New Mexico Journal of Science* 36:4–37.

Grissino-Mayer, H. D., and T. W. Swetnam. 1995. Fire history in the Pinaleño Mountains of southern Arizona: effects of human-related disturbances. In *Biodiversity and management of the Madrean Archipelago: the Sky Islands of southwestern United States and northwestern Mexico*, ed. L. H. DeBano, P. H. Ffolliott, A. Ortega-Rubio, et al., 399–407. USDA Forest Service Gen. Tech. Rep. RM-GTR-264. Fort Collins, CO: Rocky Mountain Research Station.

Guida, R. J., S. R. Abella, W. J. Smith, H. Stephen, and C. L. Roberts. 2014. Climatic change and desert vegetation distribution: assessing thirty years of change in southern Nevada's Mojave Desert. *Professional Geographer* 66:311–22.

Gutzler, D. S. 2013. Regional climatic considerations for borderlands sustainability. *Ecosphere* 4:1–12.

Gutzler, D. S., and T. O. Robbins. 2010. Climate variability and projected change in the western United States: regional downscaling and drought statistics. *Climate Dynamics* 37:835–49.

Jameson, D. A. 1967. The relationship of tree overstory and herbaceous understory vegetation. *Journal of Range Management* 20:247–49.

Jennings, M. D., D. Faber-Langendoen, O. L. Loucks, R. K. Peet, and D. Roberts. 2009. Standards for associations and alliances of the US National Vegetation Classification. *Ecological Monographs* 79:173–99.

Jones, B. E., T. H. Rickman, A. Vazquez, Y. Sado, and K. Tate. 2005. Removal of encroaching conifers to regenerate degraded aspen stands in the Sierra Nevada. *Restoration Ecology* 13:373–79.

Kaib, J. M. 2001. *Fire history reconstructions in the Mogollon Province ponderosa pine forests of the Tonto National Forest Central Arizona*. US Fish and Wildlife Service technical report available online at www.fireleadership.gov. Region 2, Albuquerque, NM.

Kaib, M., C. H. Baisan, H. D. Grissino-Mayer, and T. W. Swetnam. 1996. Fire history of the gallery pine-oak forests and adjacent grasslands of the Chiracahua Mountains of Arizona. In *Effects of fire* on *Madrean Province Ecosystems—a symposium proceedings, March 11–15, 1996, Tucson AZ*, ed. P. F. Ffolliott, L. F. DeBano, M. B. Baker, G. J. Gottfried, G. Solis-Garza, C. B. Edminster, D. G. Neary, L. S. Allen, and R. H. Hamre, 253–64. USDA Forest Service Gen. Tech. Rep. RM-GTR-289. Fort Collins, CO: Rocky Mountain Research Station.

Kauffman, J. B., and W. C. Krueger. 1984. Livestock impacts on riparian ecosystems and streamside management implications: a review. *Journal of Range Management* 37:430–38.

Kormondy, E. J. 1969. *Concepts of ecology*. Englewood Cliffs, NJ: Prentice Hall.

Kramer, D. W., G. E. Sorensen, C. A. Taylor, R. D. Cox, P. S. Gipson, and J. W. Cain. 2015. Ungulate exclusion, conifer thinning, and mule deer forage in northeastern New Mexico. *Journal of Arid Environments* 113:29–34.

Krueper, D. J. 1995. Effects of livestock management on Southwestern riparian ecosystems. In *Desired future conditions for southwestern riparian ecosystems: bringing interests and concerns together*, ed. D. W. Shaw and D. M. Finch, 281–301. USDA Forest Service Gen. Tech. Rep. RM-GTR-272. Fort Collins, CO: Rocky Mountain Research Station.

LANDFIRE. 2010. LANDFIRE 1.1.0 vegetation dynamics models and biophysical setting descriptions. Model files and reports available online at www.landfire.gov. USDA Forest Service and US Department of the Interior.

Lowrey, T. K. 2010. An introduction to New Mexico's floristic zones and vegetation communities. In *Raptors of New Mexico*, ed. J.-L. E. Cartron, 17–27. Albuquerque: University of New Mexico Press.

McKnight, T. L., and D. Hess. 2000. *Physical geography: a landscape appreciation*. Upper Saddle River, NJ: Prentice Hall.

McPherson, G. R. 1995. The role of fire in the desert grasslands. In *The desert grassland*, ed. M. P. McClaran and T. R. Van Devender, 130–51. Tucson: University of Arizona Press.

Merriam, C. H. 1890. Results of a biological survey of the San Francisco Mountains region and desert of the Little Colorado in Arizona. *North American Fauna* 3:1–136.

Milchunas, D. G. 2006. Responses of plant communities to grazing in the southwestern United States. USDA Forest Service Gen. Tech. Rep.

RMRS-GTR-169. Fort Collins, CO: Rocky Mountain Research Station.

Milchunas, D. G., and W. K. Lauenroth. 1989. Three-dimensional distribution of plant biomass in relation to grazing and topography in the shortgrass steppe. *Oikos* 55:82–86.

Mitsch, W. J., and J. G. Gosselink. 2007. *Wetlands*. 4th ed. Hoboken, NJ: John Wiley.

Moir, W. H. 1967. The subalpine tall grass, *Festuca thurberi*, community of Sierra Blanca, New Mexico. *Southwestern Naturalist* 12:321–28.

———. 1993. Alpine tundra and coniferous forest. In *New Mexico vegetation: past, present, and future*, ed. W. A. Dick-Peddie, Chapter 5. Albuquerque: University of New Mexico Press.

Moir, W. H., and J. O. Carlton. 1986. Classification of pinyon-juniper (P-J) sites on National Forests in the Southwest. In *Proceedings—Pinyon-Juniper conference, January 13–16, 1986*, ed. R. L. Everett, 216–26. USDA Forest Service Gen. Tech. Rep. INT-215. Ogden, UT: Intermountain Research Station.

Moore, M. M., D. W. Huffman, P. Z. Fulé, W. W. Covington, and J. E. Crouse. 2004. Comparison of historical and contemporary forest structure and composition on permanent plots in southwestern ponderosa pine forests. *Forest Science* 50:162–76.

Moreland, J. C., W. A. Robbie, F. J. Triepke, E. H. Muldavin, and J. R. Malusa. 2017. Ecological response units: ecosystem mapping system for the southwest US (abstract). In *Proceedings of the sixth natural history of the Gila symposium, 25–27 February 2016, Silver City, New Mexico, USA*, ed. K. Whiteman and W. Norris, 29. https://wnmu.edu/events/6th-natural-history-of-the-gila-symposium/.

Muldavin, E. P., C. Baisan, T. Swetnam, L. DeLay, and K. Morino. 2003. *Woodland fire history studies in the Oscura and northern San Andres Mountains, White Sands Missile Range, New Mexico*. Natural Heritage New Mexico publication no. 03-GTR-256, White Sands Missile Range document no. 92F018. University of New Mexico, Albuquerque.

Muldavin, E. P., P. Durkin, M. Bradley, M. Stuever, and P. Mehlhop. 2000. *Handbook of wetland vegetation communities of New Mexico*. Vol. 1, *Classification and community descriptions*. New Mexico Natural

Heritage Program technical guide available online at www.allaboutwatersheds.org/library/general-library-holdings. University of New Mexico, Albuquerque.

Muldavin, E. H., G. Harper, P. Neville, and S. Wood. 2004. A vegetation classification of the Sierra del Carmen, U.S.A. and México. In *Proceedings of the sixth symposium on the natural resources of the Chihuahuan Desert region, October 14–17, 2004*, ed. C. A. Hoyt and J. Kargesd, 117–50. Fort Davis, TX: Chihuahuan Desert Research Institute.

Nature Conservancy, The (TNC). 2006. *Southwest forest assessment project: historical range of variation and state and transition modeling of historical and current landscape conditions for potential natural vegetation types of the southwestern U.S.* The Nature Conservancy technical report available online at http://azconservation.org/projects/southwest_forest_assessment, November 2013. TNC Arizona Chapter, Tucson.

Neary, D. G., and A. Medina. 1996. Geomorphic response of a montane riparian habitat to interactions of ungulates, vegetation, and hydrology. In *Desired future conditions for southwestern riparian ecosystems: bringing interests and concerns together, September 18–22, 1995*, ed. D. W. Shaw and D. M. Finch, 143–47. USDA Forest Service Gen. Tech. Rep. RM-GTR-272. Fort Collins, CO: Rocky Mountain Forest and Range Experiment Station.

Notaro, M., A. Mauss, and J. W. Williams. 2012. Projected vegetation changes for the American Southwest: combined dynamic modeling and bioclimatic-envelope approach. *Ecological Applications* 22:1365–88.

O'Connor, C. D., D. A. Falk, A. M. Lynch, and T. W. Swetnam. 2014. Fire severity, size, and climate associations diverge from historical precedent along an ecological gradient in the Pinaleño Mountains, Arizona, USA. *Forest Ecology and Management* 329:264–78.

Parks, S. A., S. Z. Dobrowski, J. D. Shaw, and C. Miller. 2019. Living on the edge: trailing edge forests at risk of fire-facilitated conversion to non-forest. *Ecosphere* 10(3):e02651. https://doi.org/10.1002/ecs2.2651

Peterson, R. S., and C. S. Boyd. 1998. *Ecology and management of sand shinnery communities: a literature review*. USDA Forest Service Gen. Tech. Rep.

RMRS-GTR-16. Fort Collins, CO: Rocky Mountain Research Station.

Price, J., C. H. Galbraith, M. Dixon, J. Stromberg, T. Root, D. MacMykowski, T. Maddock, and K. Baird. 2005. *Potential impacts of climate change on ecological resources and biodiversity in the San Pedro Riparian National Conservation Area, Arizona*. American Bird Conservancy technical report to the US Environmental Protection Agency available online at https://cfpub.epa.gov/si/si_public_record_report.cfm?dirEntryId=180083&Lab=NCEA.

Rehfeldt, G. L. 2006. *A spline model of climate for the western United States*. USDA Forest Service Gen. Tech. Rep. RMRS-GTR-165. Fort Collins, CO: Rocky Mountain Research Station.

Rehfeldt, G. E., N. L. Crookston, C. Saenz-Romero, and E. M. Campbell. 2012. North American vegetation model for land-use planning in a changing climate: a solution to large classification problems. *Ecological Applications* 22:119–41.

Robbie, W. A. 2004. Grassland assessment categories and extent. In *Assessment of grassland ecosystem conditions in the southwestern United States*, ed. D. M. Finch, 11–17. USDA Forest Service Gen. Tech. Rep. RMRS-GTR-135-vol. 1. Fort Collins, CO: Rocky Mountain Research Station.

Romme, W. H., C. D. Allen, J. D. Bailey, W. L. Baker, B. T. Bestelmeyer, P. M. Brown, and P. J. Weisberg. 2009. Historical and modern disturbance regimes, stand structures, and landscape dynamics in piñon-juniper vegetation of the western United States. *Rangeland Ecology and Management* 62:203–22.

Sánchez Meador, A. J., P. F. Parysow, and M. M. Moore. 2011. A new method for delineating tree patches and assessing spatial reference conditions of ponderosa pine forests in northern Arizona. *Restoration Ecology* 19:490–99.

Savage, M., and J. N. Mast. 2005. How resilient are southwestern ponderosa pine forests after crown fires? *Canadian Journal of Forest Research* 35:967–77.

Sivinski, R., T. K. Lowrey, and P. Knight. 1996. New Mexico vascular plant diversity. *New Mexico Journal of Science* 36:60–78.

Smith, D. S., S. M. Fettig, and M. A. Bowker. 2016. Elevated Rocky Mountain elk numbers prevent positive effects of fire on quaking aspen (*Populus tremuloides*) recruitment. *Forest Ecology and Management* 362:46–54.

Stromberg, J. C., T. J. Rychener, and M. D. Dixon. 2009. Return of fire to a free-flowing desert river: Effects on vegetation. *Restoration Ecology* 17:327–38.

Stuever, M. C. 1997. Fire induced mortality of Rio Grande cottonwood. MS thesis, University of New Mexico, Albuquerque.

Swetnam, T. W., and J. H. Dieterich. 1985. Fire history of ponderosa pine forests in the Gila Wilderness, New Mexico. In *Proceedings, symposium and workshop on wilderness fire, 15–18 November 1983, Missoula MT*, ed. J. E. Lotan, B. M. Kilgore, W. C. Fischer, and R. W. Mutch, 390–97. USDA Forest Service Gen. Tech. Rep. INT-182. Ogden, UT: Intermountain Forest and Range Experiment Station.

Triepke, F. J., W. A. Robbie, E. W. Taylor, and M. M. Wahlberg. 2017. A framework for the analysis, planning, and management of ecosystems in the Southwest (abstract). In *Proceedings of the sixth natural history of the Gila symposium, 25–27 February 2016, Silver City, New Mexico, USA*, ed. K. Whiteman and W. Norris, 33. https://wnmu.edu/events/6th-natural-history-of-the-gila-symposium/

Triepke, F. J., M. M. Wahlberg, D. C. Cress, and R. L. Benton. 2014. *RMAP—Regional riparian mapping project*. USDA Forest Service project report available online at www.fs.usda.gov/main/r3/landmanagement/gis. Southwestern Region, Albuquerque, NM.

United States Department of Agriculture (USDA) Forest Service. 2006. *Terrestrial ecosystem survey of the Cibola National Forest and National Grasslands*. Technical report on file. Southwestern Region, Regional Office, Albuquerque, NM.

———. 2013. Forest insect and disease conditions in the Southwestern Region, 2013. Technical report PR-R3–16–12 available online at www.fs.usda.gov/detail/r3/maps-pubs. Southwestern Region Forest Health, Regional Office, Albuquerque, NM.

United States National Vegetation Classification (USNVC). 2016. The US NVC hierarchy explorer. Detailed descriptions of vegetation types in the US with ecological context and geographic ranges, available online at http://usnvc.org/explore-classification. North Carolina State University, Raleigh.

USDA Natural Resources Conservation Service (NRCS). 2016. The PLANTS database, available online at http://plants.usda.gov. National Plant Data Team, Greensboro, NC.

Vankat, J. L. 2013. *Vegetation dynamics on the mountains and plateaus of the American Southwest*. New York: Springer.

White, M. R. 2002. Characterization of and changes in the subalpine and montane grasslands, Apache-Sitgreaves National Forests. PhD dissertation, Northern Arizona University, Flagstaff.

Williams, A. P., C. D. Allen, A. K. Macalady, D. Griffin, C. A. Woodhouse, D. M. Meko, T. W. Swetnam, S. A. Rauscher, R. Seager, H. D. Grissino-Mayer, J. S. Dean, E. R. Cook, C. Gangodagamage, M. Cai, and N. G. McDowell. 2013. Temperature as a potent driver of regional forest drought stress and tree mortality. *Nature Climate Change* 3:292–97.

Wright, H. A., and A. W. Bailey. 1982. *Fire ecology: United States and southern Canada*. New York: John Wiley.

Yanoff, S., P. McCarthy, J. Bate, L. M. Wood-Miller, A. Bradley, and D. Gori. 2008. *New Mexico rangeland ecological assessment: findings and application*. The Nature Conservancy technical report available online at www.nmconservation.org. New Mexico Chapter, Santa Fe.

PREDATOR DAMAGE MANAGEMENT IN NEW MEXICO

J. Alan May and Stewart Breck

Predator damage management (PDM) is a concept that elicits strong emotional responses from people on all sides of a very polarizing issue. Some of the public views PDM as a misguided effort. Killing carnivores, it is argued, is morally wrong, even for the sake of livestock and crop protection or big-game production. Perhaps in an ideal world free of conflicts between wildlife and ranchers, farmers, and even pet owners, there would be little or no need for PDM. The reality, however, is that predation and depredation can have devastating impacts on the livelihoods of ranchers and farmers. While animal welfare activists and others demand that only non-lethal methods be used, ranchers and farmers just want the most effective forms of PDM, and they alone carry the extra financial burden of protecting their livestock, their orchards, or their beehives. Even urban and suburban residents who want their pets protected from coyotes (*Canis latrans*) might in the end resort to lethal methods as few other tools are available to them. These and other arguments form the crux of the debate on whether PDM is justified and necessary, and how it should be implemented. In practice, wildlife managers who seek a balanced approach must weigh impacts to predator populations and ecosystems as well as the economic cost, humaneness, and efficacy of each method used. Indeed, the job of PDM has become a very challenging endeavor. Our goal in this chapter is

to provide a brief history of PDM in New Mexico; to quantify the damage caused by coyotes and other carnivores on the state's livestock industry and big-game production in particular; and to provide details of contemporary PDM as implemented by the US Department of Agriculture's Wildlife Services.

Federal government involvement in predator damage management in New Mexico intensified in 1914 when the early naturalist J. Stokley Ligon became an "inspector" for the Bureau of Biological Survey (BBS) and was granted authority to hire hunters and trappers. In 1940, the BBS was combined with the Bureau of Fisheries to form the US Fish and Wildlife Service (USFWS), and predator control was conducted by Animal Damage Control (ADC), within the USFWS under the authority of the Animal Damage Control Act of 1931. In 1985 ADC was transferred from the US Department of Interior to the US Department of Agriculture (USDA), and today federal wildlife damage management is conducted by the USDA/Animal Plant Health Inspection Service (APHIS)/Wildlife Services (WS). Early agency documents for New Mexico indicate that the goal of BBS work was to prevent loss of game and domestic livestock. And in a letter dated

(*opposite page*) Photograph: © Mark L. Watson.

Photo 3.1 (*top left*). Henry W. Henshaw, Bureau of Biological Survey (BBS) Chief 1910–1916. The BBS was the precursor of the US Fish and Wildlife Service, and in 1914 it opened its Predatory Animal and Rodent Control (PARC) branch. Photograph: © Don Hawthorne Collection (NWRC 0033) USDA/APHIS/WS National Wildlife Research Center Archives.

Photo 3.2 (*top right*). Coyote control by poisoning, Armendaris Ranch, Fra Cristobal Mountains in north-central Sierra County, 1918. Photograph: © Denver Public Library, Conservation Collection 92; Stokley Ligon photographer.

Photo 3.3 (*bottom left*). Early trapper and cougar. Photograph: © New Mexico Department of Game and Fish.

Photo 3.4 (*bottom right*). Benjamin "Ben" Vernon Lilly (1856–1936) was a famous mountain man and big game hunter of the American West. In 1916, he was hired by J. Stokley Ligon, who at the time was an inspector for the PARC branch of the Bureau of Biological Survey, to help eradicate wolves, bears, and other predators. He is credited with killing the last grizzly bear in the Gila Wilderness. Photograph: © New Mexico Department of Game and Fish.

20 September 1951, from J. Stokley Ligon to then State Game Warden Elliott Barker, Mr. Ligon stated, "As you no doubt know, I have always strongly advocated complete elimination of coyotes in wooded and mountainous regions." In that same letter, he also advocated eradication of the "lobo wolf" because it was "incompatible with intensified civilization." Eradication campaigns were carried out throughout New Mexico, and they included broad-scale application of poison in baits and carcasses, as well as intensive trapping and hunting of the state's predators. These efforts played a role in the eventual extirpation of gray wolves (*Canis lupus*) and grizzly bears (*Ursus arctos*) in New Mexico and reduced populations of other predators (see Chapters 11 and 17).

As social values evolved and knowledge regarding biology, ecology, wildlife behavior, and other aspects of wildlife damage management accumulated, the mission of the agency that eventually became USDA/Wildlife Services changed. In 1972, President Richard M. Nixon banned the use of toxicants by federal employees and on federal land (Conover 2001). This ban was later overturned by President Ronald Reagan to allow targeted use of some pesticides for predators. More importantly, the efforts moved away from large-scale suppression of predator populations to greater use of non-lethal techniques and more targeted and limited use of lethal methods, and the way the agency mission was implemented also changed dramatically. Today, Wildlife Services is a national, cooperatively funded program with operational programs in each state. Funding is often provided by multiple entities, including state agencies, county governments, local associations, and major stakeholder groups, and provides for increased service delivery to protect agricultural resources. For example, in New Mexico, approximately 50% of the annual budget is composed of money provided by agricultural producers and local, state, and federal agencies. The current mission of Wildlife Services is "to provide Federal leadership and expertise to resolve wildlife conflicts, thereby allowing people and wildlife to coexist."

The Wildlife Services Program is a service-delivery program within the USDA. As such, Wildlife Services is not a regulatory agency, meaning that it has no legal authority to manage wildlife. Therefore, decisions made to implement specific management actions are reached at the request, or under the legal guidelines, of agencies with management authority such as the New Mexico Department of Game and Fish (NMDGF). Wildlife Services works with land and wildlife management agencies through Memoranda of Understanding (MOU) and Joint Powers Agreements. At the state level, Wildlife Services has agreements with the New Mexico Department of Agriculture (NMDA) and NMDGF that specify roles and functions. National-level MOUs between Wildlife Services and the Bureau of Land Management and the US Forest Service establish the common understanding of the agencies on how Wildlife Services will conduct Predator Damage Management in response to requests from permittees on their lands, as appropriate. Under the agreement with the NMDA, Wildlife Services responds to damage from coyotes involving crops and livestock. Wildlife Services refers complaints from the public involving other predators, including furbearers and game species, to NMDGF because the individual must first obtain a depredation complaint number or permit for a problem to be resolved. Native American tribes are responsible for wildlife management on their own lands and can request assistance directly from Wildlife Services.

Wildlife damage management is defined as the alleviation of damage or other problems caused by wildlife (Leopold 1933; Wildlife Society 1990; Berryman 1991). Wildlife Services uses an "Integrated Wildlife Damage Management" approach as defined in an Environmental Impact Statement (USDA 1997). Depending on the damage scenario, management options include non-lethal strategies such as the modification of the habitat or offending animal's behavior, and control of the offending animal or local population of the offending species with lethal or non-lethal methods (USDA 1997). For any particular management scenario, a variety of management options may be used to minimize loss of resources due to predators.

Wildlife Services works to prevent and reduce wildlife predation on livestock through education, technical assistance to producers (i.e., making recommendations on how they can help reduce wildlife conflicts without involvement by Wildlife Services), and direct predation damage management. In the modern West, non-lethal measures available to reduce livestock predation include fencing, guarding animals, shed lambing, moving livestock to pastures where livestock are less vulnerable, using devices to frighten predators away from livestock, and using herders or range riders to watch over livestock. Lethal techniques include use of traps, snares, M-44 sodium cyanide ejectors, the livestock protection collar, aerial hunting, and calling and shooting. The best approach to reduce livestock predation often involves a combination of these non-lethal and lethal methods. While minimizing damage links the past and present, the mission of Wildlife Services as carried out today focuses on reducing damage with minimal impact on wildlife populations.

NEED FOR PREDATOR DAMAGE MANAGEMENT

The total value of agriculture production in New Mexico was nearly $3.5 billion in 2021, and total cash receipts from livestock products reached just over $2.4 billion (National Agriculture Statistics Service [NASS] 2022). Cattle and sheep production contributed substantially to local economies as range livestock production was the second largest sector of the state's agriculture industry behind milk production. As of 1 January 2022, livestock inventories included 1,300,000 cattle and calves, 90,000 sheep and lambs, and 8,000 goats (NASS 2022). Sheep inventories in New Mexico have declined sharply over the last 35+ years from a high of 660,000 in 1980, partially due to a declining market.

Nationwide, predators accounted for more than $66 million worth of livestock losses in 2015 (USDA 2015a and b); this is with predation management programs in place throughout most of the nation. The impact of predators on livestock without predator management would undoubtedly be higher (Nass 1977, 1980; Howard and Shaw 1978; Howard and Booth 1981; Bodenchuk et al.

2002), but the degree to which it would increase likely varies depending upon a host of unstudied variables. Research indicates that coyote predation on sheep tripled in areas without predator management in Utah and Montana (O'Gara et al. 1983; Wagner and Conover 1999), but whether these rates can be extrapolated to other states or to different predators and livestock species is unknown.

According to APHIS National Animal Health Monitoring System, 70,000 adult cattle and calves valued at over $57 million were lost to all causes in 2015 in New Mexico (USDA 2015a). Non-predator losses due to causes including digestive, respiratory, metabolic, and calving problems, poisoning, mastitis and other diseases, weather, and theft totaled 60,500 head in that same year. Predator losses involving adult cattle and calves in New Mexico during 2015 totaled 9,500 head (USDA 2015a). USDA (2015b) reported that predators killed 4,813 sheep and lambs valued at $833,000 in the state during 2015. By comparison, 8,187 sheep and lambs in New Mexico valued at $1,420,000 were lost to all other causes during this same year.

Cattle and calves are vulnerable to predation, especially at calving (NASS 1992, 1996, 2001, 2006, 2011). Sheep, goats, and poultry are highly susceptible year-round to predation (Henne 1975; Nass 1977, 1980; Tigner and Larson 1977; O'Gara et al. 1983; NASS 1991, 1995, 2000, 2005, 2010). In New Mexico, USDA/Wildlife Services data indicate reported and verified damage caused by carnivores averaged over $462,500 between 1 October 2011 and 30 September 2021 (Table 3.1). Through 2017, coyotes were responsible for about 88% of the losses stemming from predation on calves, cougars (*Puma concolor*) for about 2.8% of those same losses (USDA Wildlife Services, unpubl. data). These losses occurred despite PDM efforts by producers, who must bear the additional costs for these activities (see Jahnke et al. 1987).

A variety of predators prey on livestock, but the

Table 3.1. Value of reported and verified damage caused by carnivores in New Mexico.

Species	FY 12	FY 13	FY 14	FY 15	FY 16	FY 17
Coyote	$395,299	$343,552	$322,813	$334,615	$222,004	$231,725
Cougar	$27,166	$18,159	$13,197	$8,714	$6,045	$7,269
Gray Wolf	$9,606	$11,522	$33,197	$59,822	$56,005	$14,317
Black bear	$5,885	$11,391	$12,006	$1,225	$4,187	$5,700
Bobcat	$13,957	$13,762	$2,484	$1,421	$7,659	$1,994
Feral/Free Ranging Dog	$6,582	$24,569	$2,254	$36,949	$1,665	$5,902
Gray fox	0	$1,754	$350	$80	$511	0
Raccoon	$333	$335	$240	$310	$287	$2,600
Striped Skunk	$7,755	$473	$110,794[*]	$21,810	$1,438	$3,043
Total	**$466,583**	**$425,487**	**$497,335**	**$464,946**	**$299,801**	**$272,550**

[*]includes two incidents totaling $104,544 total damage to sod grass.

Data from USDA Wildlife Services Management Information System. Reported damage includes losses reported to Wildlife Services personnel that were not verified. Verified damage is that damage which is verified by USDA Wildlife Services staff. During FYs 12–17, there was no recorded damage value associated with other carnivores including feral/free ranging cats (*Felis catus*), North American red foxes (*Vulpes fulva*), swift foxes (*V. velox*), and kit foxes (*V. macrotis*), to name a few.

one that inflicts the highest percentage of damage is almost always the coyote. For example, in New Mexico coyotes are the predominant predator on livestock both in terms of number killed and damage monetary value (Tables 3.1 and 3.2). Studies in Montana and Utah (O'Gara et al. 1983; Palmer et al. 2010) show coyote predation has routinely accounted for over 90% of the predation loss. Other predators of cattle, calves, sheep, and lambs in New Mexico are wolves, bobcats (*Lynx rufus*), black bears (*Ursus americanus*), cougars, and feral or free-ranging dogs (Table 3.2; and see Chapter 33). Palmer et al. (2010) found that predation by carnivores other than coyotes had increased since the 1970s with more threats originating from black bears and cougars. Breck et al. (2011) reported that in Arizona and New Mexico cougars were the dominant threat to calves but that this threat varied depending upon habitat.

Surplus killing is a term referring to an event when many livestock are killed in a short time period (i.e., a day or two), but not all of the animals are eaten or only selected tissues or parts are consumed. In a study in New Mexico, DeLorenzo and Howard (1977) found that more than 43% of lambs killed by coyotes were not fed upon. Bears, cougars, and wolves have all similarly been implicated in surplus killing with most incidents occurring with sheep and lambs (Mysterud 1977; Shaw 1987). Bears or cougars may also frighten an entire flock of sheep as they attack, resulting in a mass stampede. This can result in many animals suffocating as they pile up on top of one another in a confined area, such as against a thicket of dense vegetation or in corrals. In a single documented incident, over one hundred sheep were killed by a cougar in Texas (Wade and Bowns 1982). Although these events are fairly rare, they can have important economic impacts to a few producers.

Confirmation of the cause of livestock death is a vital step toward establishing any need for PDM and the response necessary to resolve the problem. Wildlife Services specialists use physical

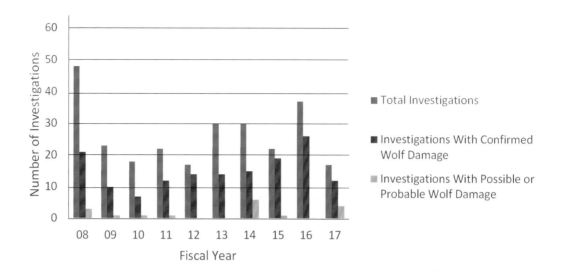

Figure 3.1. Mexican wolf depredation investigations by US Department of Agriculture, Wildlife Services in New Mexico: From the 2008 fiscal year to the 2017 fiscal year.

Table 3.2. Damage occurrences by species as recorded by New Mexico Wildlife Services from the 2015 fiscal year to the 2017 fiscal year.

	Agriculture			Property			Human Health & Safety			Natural Resources		
	FY 15	FY 16	FY 17	FY 15	FY 16	FY 17	FY 15	FY 16	FY 17	FY 15	FY 16	FY 17
Badger	2	3	6	3	5	1	0	0	0	0	0	0
Black Bear	4	2	5	0	0	1	0	0	1	0	0	0
Bobcat	12	19	10	3	1	1	1	1	1	0	0	0
Feral Cat	0	0	0	0	0	0	0	0	1	0	0	0
Coyote	1,087	878	444	61	77	91	28	57	25	11	14	8
Feral Dog	9	4	2	0	1	1	0	1	2	0	0	0
Gray Fox	2	0	2	6	0	7	5	0	9	0	0	0
Cougar	13	11	4	1	1	5	2	1	4	0	0	0
Raccoon	1	6	0	6	4	10	1	2	2	0	1	0
Striped Skunk	5	4	2	47	23	38	172	173	151	0	0	0
Totals	1,135	927	475	127	112	155	209	235	196	11	15	8

Agriculture includes crops and livestock, irrigation systems.
Property includes pets, buildings, vehicles, etc.
Human Health and Safety includes disease or injury in humans including injuries on airport runways.
Natural Resources include wildlife species.

evidence to confirm the predators responsible for losses when possible. This physical evidence may include bite marks, hemorrhaging under the skin, and scat and tracks in the area. The condition of the hooves of a newborn calf or lamb may also indicate whether the animal stood and walked or was simply stillborn.

Between the federal 2015 fiscal year (1 October 2014–30 September 2015) and the 2017 fiscal year (1 October 2016–30 September 2017), New Mexico Wildlife Services received requests for assistance averaging over 900 annually for coyotes, 200 for striped skunks (*Mephitis mephitis*), 16 for bobcats, 14 for cougars, and four for black bears. The majority of requests in New Mexico are associated with livestock depredation (Table 3.2). Between FY 2008 and FY 2017, the average annual number of Mexican wolf depredation investigations conducted by Wildlife Services in New Mexico was 26, the average number with confirmed wolf damage was 15, and the average number of investigations that resulted in a determination of possible or probable depredation was two (Figure 3.1). Wildlife Services confirmed livestock depredation caused by wolves in New Mexico varied from seven to 26 head of livestock annually between the 2008 and 2017 fiscal years. However, a study in Idaho suggested that many more losses likely go undetected, thus unreported (Oakleaf et al. 2003).

Connolly (1992) determined that only a fraction of the total predation attributable to coyotes is reported to or confirmed by Wildlife Services. Breck et al. (2011) found similar results and reported that detection rates of livestock killed by carnivores can vary dramatically across sites in New Mexico. These results indicate that in many areas throughout the state the number of livestock killed by predators is likely underestimated if this figure is based only on reported and confirmed losses. Connolly (1992) determined that based on scientific studies and livestock loss surveys generated by the National Agriculture Statistics Service, Wildlife Services only confirms

about 19% of the total adult sheep and 23% of the lambs actually killed by predators. Larger livestock typically have higher detection rates than suggested by the sheep and lamb data, but cattle carcasses in rugged terrain may also be difficult to locate.

METHODS FOR CARNIVORE MANAGEMENT AND CONTROL

A wide range of methods are available for resource owners and Wildlife Services personnel. Predator damage management methods fall into different categories including cultural practices (i.e., shed lambing and guard animals), habitat and behavior modification (i.e., exclusion, chemical repellents, and hazing with pyrotechnics), protection with dogs, translocation, and lethal control including shooting, calling and shooting, aerial hunting, foothold traps, cage traps, snares, M-44 sodium cyanide spring-loaded ejector, the livestock protection collar, chemical immobilization and euthanasia, denning, gas cartridges, decoy and tracking dogs, and hand-capture. Wildlife Services specialists use lethal methods as a last resort and often educate producers about predator behavior and effective long-term non-lethal solutions to minimize conflict. Most non-lethal methods associated with livestock production involve changes in husbandry practices or ranch management, and thus it is the responsibility of the producer to implement them. Livestock producers often have already implemented multiple methods before they even contact Wildlife Services for assistance. Over 83% of New Mexico livestock producers in a 1994 survey had implemented at least one non-lethal method to protect their livestock, and over 94% of the methods tried were still in use (May 1996). Properly trained and well-managed livestock guardian dogs can be a very effective method to control predation (Texas A&M Agrilife Extension 2015). Another potentially promising method for coyote predation management is reproductive inhibition (Knowlton et al.

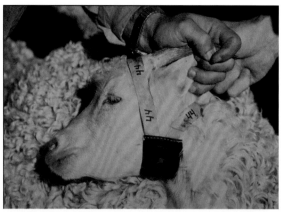

Photo 3.5 (*top left*). Warning sign for spring-activated M-44 sodium cyanide ejectors, which are staked to the ground. When the target animal, often a coyote or a fox, pulls on the baited capsule holder, the M-44 device is activated, releasing the plunger and delivering a dose of cyanide power into the animal's mouth. Photograph: © USDA/APHIS/WS National Wildlife Research Center Archives.

Photo 3.6 (*top right*). Sheep wearing a livestock protection collar (LPC). The LPC is a tool used for selectively killing coyotes that prey on sheep and goats. The collars contain two vessels holding low concentration doses of sodium monofluoroacetate (Compound 1080), which is fatal to predators after ingestion. Coyotes attack sheep and goats with a crushing bite to the neck, puncturing one of the reservoirs in the process. Photograph: © USDA/APHIS/WS National Wildlife Research Center Archives.

Photo 3.7 (*bottom left*). Aerial gunning in helicopter over the mountains in New Mexico, 2017. Photograph: © Justin Hendricks/New Mexico Wildlife Services, USDA/APHIS/WS National Wildlife Research Center Archives.

Photo 3.8 (*bottom right*). Livestock protection dog guarding lambs. Photograph: © USDA/APHIS/WS National Wildlife Research Center Archives.

1999; Bromley and Gese 2001; Mitchell et al. 2004). Studies have shown that predation on livestock increases during spring and early summer and this increase has been attributed to the need for the mating pair to feed their young. Bromley and

Gese (2001) demonstrated that sterilizing coyotes stopped production of young and reduced the amount of depredation. Furthermore, they showed that sterilized coyotes maintained their territories, thus potentially creating a longer-term solution for minimizing depredation on sheep. In 2017 NMDGF sterilized coyotes on Fort Stanton, New Mexico to attempt to reduce coyote predation of neonate pronghorns (*Antilocapra americana*). This field trial was judged to be unsuccessful because hunters shot many of the sterilized coyotes (N. Tatman, NMDGF, pers. comm.).

Wildlife Services implements PDM based upon requests from resource owners experiencing damage and from state agencies such as NMDGF when human health or safety or natural resources threats exist. PDM is implemented throughout New Mexico to help protect penned livestock as well as free-range livestock, backyard chicken flocks, pets, and damage to crops such as watermelons. Requests for cattle protection often coincide with calving periods. Calves and smaller livestock are more vulnerable to predation than adult cattle or larger livestock. When lethal control is implemented, effort is focused either on targeting individual animals or on impacting the small portion of the predator population that is causing damage (e.g., on and immediately adjacent to lambing or calving grounds). In New Mexico, coyotes are subjected to lethal control more than any other carnivore species, primarily because they are the most abundant and pose the biggest threat to livestock producers. The coyote is the only species subjected to routine aerial shooting, a technique that is very selective and can effectively suppress coyote numbers on a local scale where depredation is prevalent. However, the total number of coyotes taken where aerial gunning is routinely practiced is relatively small, thus the impact of aerial gunning combined with all other lethal removal is likely not having any long-term impacts on coyote populations. Connolly and Longhurst (1975) examined the effect of control on coyote populations using a simulation model and determined that a minimum annual removal of 75% of the breeding population was needed to consistently lower coyote

density. Sterling et al. (1983) found that control programs inflicting less than 50% annual mortality could not be expected to reduce coyote populations. Only a 10% success rate in offspring survivorship to breeding age is necessary to maintain most coyote populations (Knowlton et al. 1999). The combined (or cumulative) impact of take by Wildlife Services and sportsman harvest on the coyote in New Mexico (Table 3.3) is arguably minimal given how common that species is in the state (Chapter 10). Annual Wildlife Services take is much lower—and in some cases negligible—for all other carnivore species. Except for the striped skunk, its impact is lower by an order of magnitude or two compared with sportsman harvest (Table 3.3).

MEASURING THE BENEFITS OF PREDATOR DAMAGE MANAGEMENT

The best measure of success of a Predator Damage Management program is the number of livestock saved from predation. However, estimating that number with precision is difficult because finding control areas (i.e., where livestock producers are willing to allow operations to proceed without predator control for the sake of research) that can be compared to treatment areas (with predator control) is challenging. In a rare study based on that experimental design (Ogara et al. 1983), coyote predation rates on sheep were so high in the control area that the producer requested that the study be modified. The best available evidence indicates that areas without PDM can incur predation losses as high as 8.4% and 29.3%, for adult sheep and lambs respectively, of the total number of head (Henne 1975; Munoz 1977; O'Gara et al. 1983; Wagner and Conover 1999). Other studies have indicated that sheep and lamb losses are significantly lower where Predator Damage Management is applied (Nass 1977; Tigner and Larson 1977; Howard and Shaw 1978; Howard and Booth 1981; Wagner and Conover 1999). Bodenchuk et al. (2002) estimated more conservatively

Table 3.3. Cumulative impacts on carnivore species in New Mexico from the 2014 fiscal year to the 2017 fiscal year.

	Mean Annual Wildlife Services	Mean Annual Sportsman	Cumulative
	Take[1]	Harvest[2]	Total
Coyote	3,291	5,600	8,891
Gray fox	19	2,353	2,372
Kit fox	7	170	177
Swift fox	8	61	69
Striped skunk	546	400	946
Hog-nosed skunk	1	30	31
Black bear	3	535	538
Raccoon	3	312	315
Cougar	5	278	283
Bobcat	17	1,858	1,875
Badger	12	215	227

1. Based on USDA New Mexico Wildlife Services Management Information System data.
2. From New Mexico Department of Game and Fish harvest reports. Estimated sportsman coyote and skunk harvest from prior year data.

that livestock losses in the absence of predator damage management would be 3%, 5.6%, and 17.5% for calves, sheep, and lambs, respectively. Using the estimates from Bodenchuk et al. (2002) and applying them to data from the 2015 to 2017 fiscal years, the total estimated annual average value of livestock saved in New Mexico during that time was $2,117,350 (Table 3.4). With PDM in place, annual calf losses averaged 270 head, adult sheep losses averaged 42 head, and lamb losses averaged 372 head. This analysis implies that without PDM, the average annual estimated

Table 3.4. Average annual resources protected by New Mexico Wildlife Services and the average annual estimated value of livestock saved from predation during the 2015–2017 fiscal years.

Type	No. Protected	Total Value	Value per Head	No. Reported and Confirmed Predation Losses	% Protected Lost to Predation	Est. % Lost without PDM	Est. no. Lost without PDM	Est. No. Saved	Est. Value of Livestock Saved
Cattle (calves)	128,289	$58,696,462	$457	270	0.21%	3%	3,849	3,579	$1,635,603
Adult sheep	28,804	$2,367,011	$82	42	0.15%	5.6%	1,613	1,571	$128,822
Lambs	23,352	$2,229,827	$95	372	1.59%	17.5%	4,087	3,715	$352,925
Total	180,445	$63,293,300	—	684	—	—	9,549	8,865	$2,117,350

Number of livestock protected, values, and loss data from USDA Wildlife Services Management Information System. Percentages for livestock lost without predation management were used from Bodenchuk et al. (2002).

livestock losses would have been approximately 3,849 calves, 1,613 adult sheep, and 4,087 lambs.

In evaluating the cost effectiveness of Predator Damage Management, USDA (1997) concluded that benefits, in terms of avoided sheep and lamb losses plus price benefits to consumers, are 2.4 times the cost of providing Wildlife Services PDM services for sheep protection in the 16 western states. That analysis did not address the value of calf protection, which is a substantial component of services provided by Wildlife Services in New Mexico.

CROPS, PROPERTY, HUMAN HEALTH AND SAFETY, AND NATURAL RESOURCES

In New Mexico, carnivores impact a number of resources other than livestock. Typically, the values of those losses are less than that of the losses to livestock, yet they can be significant. Other resources that carnivores sometimes damage include crops, property, human health and safety, and natural resources. Field crops, such as melons (watermelons and cantaloupes), milo, sweet and field corn, and wheat, have been damaged by predators, such as coyotes, feral/free-roaming dogs, American badgers (*Taxidea taxus*), and raccoons (*Procyon lotor*). Fruit and nut crops have been damaged by raccoons in New Mexico. And occasionally, improved or planted pasture is damaged by badger and coyote burrowing which leaves the ground uneven. This can hamper the use of planting and mowing equipment or even result in damage to the equipment.

Occasionally, carnivores cause property damage. Coyotes, cougars, or raccoons may kill pets, black bears cause damage to cabins or homes, coyotes and bears damage drip irrigation systems by biting holes in the pipe, raccoons and skunks burrow under homes to den, and feeding activities of badgers, skunks, or raccoons sometimes damage landscaping, gardens, or golf courses. Generally, this type of damage by carnivores is a fairly insignificant component of Wildlife Services' day-to-day operations, but addressing these conflicts is still important when they occur.

Wildlife Services, NMDGF, New Mexico Cooperative Extension Service, and several other agencies regularly answer calls for assistance regarding urban wildlife. In addition, there are a few private individuals and companies offering predator control services. While it is difficult to estimate how often these services are used, private predator control providers must have a permit from the New Mexico Department of Game and Fish to handle any protected species and are required to submit monthly reports to the NMDGF.

People in urban areas are often surprised to see coyotes in their neighborhoods. Coyotes are extremely adaptable and do well in city environments surviving on a variety of natural and anthropogenic food sources readily available (see Chapter 10). Most coyotes in urban environments do not pose a problem. However, urban environments create ideal conditions (i.e., plenty of food and little if any human persecution) for coyotes to develop significant changes in their behavior toward people and their pets. Urban coyotes are certainly more tolerant of humans but also can become bolder and more aggressive toward people (Breck et al. 2019). This can lead to the development of problem individuals that can cause significant conflict primarily in the form of attacks on pets (Poessel et al. 2013) and the need for effective management solutions. Preventative use of non-lethal methods (e.g., scaring or hazing coyotes, and reducing accidental and purposeful feeding of coyotes) will likely reduce the potential for development of problem individuals but can be challenging to implement in urban environments. A long-term continuous educational strategy is likely the most effective method for preventing conflict from developing. When problem individuals do occur, reports of pet attacks and aggressive encounters with people tend to increase within certain geographical areas. At this point non-lethal methods become ineffective and lethal control measures should be implemented. Targeted removal using traps or professional "sharp-shooters" who are trained to operate in urban environments are likely the most effective strategies for stopping conflict and minimizing impacts to the coyote population.

Carnivores can also pose a threat to human health and safety in certain instances, including human attacks from cougars or bears that result in injuries or death; disease threats from rabies and plague outbreaks where predators act as reservoirs for the disease; odor and noise nuisances from skunks and raccoons in attics and under houses; and airstrike hazards from coyotes crossing runways at airports or airbases. In response to confirmed rabies cases near Wink, Texas in 2009, Wildlife Services staff distributed oral rabies vaccine baits in a >360 km^2 (140 mi^2) area including southern Lea County in southeastern New Mexico to vaccinate coyotes and foxes during 2010 and 2011. No new cases of coyote or fox rabies have occurred in the area since (despite animal rabies cases in nearby Roosevelt County as recently as 2020). Although it is rare, both black bears and cougars have killed or injured people in New Mexico. In 2001, a bear broke into a home and killed a woman in Cleveland, Mora County. In 2008, a cougar killed a man near Pinos Altos, Grant County. Increasing human populations and spread of residential areas into cougar habitat have resulted in an increased number of human encounters with cougars (Shaw et al. 2007). These include sightings, actual approaches by cougars, attacks on pets, and attacks on humans (Beier 1991; Fitzhugh et al. 2003; see Chapter 9). At the request of the NMDGF, Wildlife Services has responded to complaints involving cougars that were perceived as threats to public safety. Baker and Timm (1998), after several human-coyote

interactions in an area, concluded that the use of leghold traps to capture and euthanize a few coyotes would be the best method to resolve the problem and have the longest lasting effects. Breck et al. (2017) found that removal of problem coyotes in urban environments decreased the recurrence of severe conflict events and that removal of problem individuals (1–2% of population) had no significant effect on the size of urban coyote populations. After a child was killed by a coyote in Glendale, California, city and county officials trapped 55 coyotes in an 80-day period from within one-half mile of the home, an unusually high number for such a small area (Howell 1982). Wildlife Services assists many residents, especially in urban areas such as Albuquerque, concerned about coyote attacks on their pets and their apparent loss of fear for humans.

Under certain conditions predators can impact prey populations causing wildlife managers to deem predation management a viable means to boost prey populations. This is a controversial topic, but research can shed light on when and how predators impact prey populations. A good synthesis of this topic focused on predator impacts to deer is by Ballard et al. (2001). They reviewed published predator-deer relationship studies since the mid-1970s and found that predators (coyote, cougar, and wolf) could cause significant mortality, but Predator Damage Management may or may not result in higher deer populations and increased harvest levels for hunters. They found that PDM benefitted deer mostly when herds were well below forage carrying capacity, predation was identified as a limiting factor, PDM efforts sufficiently reduced the predator population, PDM efforts were timed correctly (prior to fawning and denning), and PDM was focused on a relatively small area (<~670 km^2 [~260 mi^2]). PDM was not effective when the above conditions were not

Photo 3.9. Two mule deer (*Odocoileus hemionus*) at Bandelier National Monument, 4 April 2014. Published research suggests that predator damage management (PDM) may benefit deer herds well below forage carrying capacity when the limiting factor is predation. Photograph: © Larry Lamsa.

Photo 3.10. Mule deer (*Odocoileus hemionus*) freshly killed by a cougar (*Puma concolor*) in the spring of 1993 in Doña Ana County. Photograph: © Jenny Lisignoli.

Photo 3.11. Coyote (*Canis latrans*) chasing pronghorn (*Antilocapra americana*) at the Sevilleta National Wildlife Refuge. Photograph: © Sevilleta National Wildlife Refuge, US Fish and Wildlife Service.

met, but, additionally, Ballard et al. (2001) suggested that the experimental design of research being conducted on PDM effectiveness needed to be improved because it was unclear in several studies if PDM had a significant effect protecting deer herds. The most convincing evidence of deer population increases as a result of PDM were from studies conducted in small enclosures (<~40 km² [15 mi²]) because predator populations were much easier to regulate in smaller

areas. In New Mexico, management of predators (primarily coyotes and cougars) occurs occasionally to enhance populations of mule deer (*Odocoileus hemionus*), pronghorn, and bighorn sheep (*Ovis canadensis*). Decisions to carry out such actions are made based on published literature, local knowledge of predator and game species populations, and input from local, state, and federal wildlife management agencies.

Some would argue that wildlife managers

Photo 3.12. Neonatal bighorn sheep (*Ovis canadensis*) lamb captured in the spring of 2012 in the Peloncillo Mountains of southwestern New Mexico. It has just been fitted with an ear tag and expandable VHF radio collar for monitoring during a survival study. Photograph: © Rebekah Karsch.

Photo 3.13. Coyote (*Canis latrans*) kill documented in early 2013 in the Peloncillo Mountains during a 2011–2013 study of bighorn sheep (*Ovis canadensis*) lamb survival and specific causes of mortality. Coyotes were the second most common source of predator-caused mortality during the study. Photograph: © Rebekah Karsch.

Photo 3.14. Cougar (*Puma concolor*) kill discovered in the summer of 2012 in the Peloncillo Mountains. A total of 12 lambs were captured and collared during the 2011–2012 season, 14 in the 2012–2013 season. Cougars appeared to be the main predators of desert bighorn lambs with five documented kills. Note the bite marks on the neck on the lamb. Photograph: © Rebekah Karsch.

should just let nature take its course; that the "balance of nature" has been upended by predator damage management and hunting. But we have altered the environment for our benefit to produce food and fiber for the world. While predator damage management remains controversial in the eyes of many, it helps maintain lower production costs for producers and less expensive food supplies for consumers, helps to protect vulnerable threatened or endangered wildlife from predation by other wildlife (including non-native invasive species), and provides a partial defense against disease outbreaks such as rabies. USDA Wildlife Services has made great strides in addressing humaneness and non-target wildlife issues since the days of J. Stokley Ligon and the Bureau of

Biological Survey. EPA restrictions on the use of "predacides" are designed to help protect both people and the environment. Wildlife managers use administrative tools such as the National Environmental Policy Act (NEPA) to consider public concerns and analyze program impacts on wildlife. USDA Wildlife Services managers also consult with the US Fish and Wildlife Service to minimize impacts on threatened and endangered species. And further development of cultural tools such as livestock guarding animals, exclusion techniques, and electronic devices to scare predators away from vulnerable livestock, help wildlife managers balance the need to protect livestock, property, human health and safety, and natural resources, with the desire to maintain minimal impacts on the fantastic diversity of wildlife that we all enjoy.

ACKNOWLEDGMENTS

The authors wish to thank Taliva Ortega for preparing two of the tables using Wildlife Services' annual Program Data Reports.

LITERATURE CITED

Baker, R. O., and R. M. Timm. 1998. Management of conflicts between urban coyotes and humans in southern California. *Proceedings of the Vertebrate Pest Conference* 18:299–312.

Ballard, W. B., D. Lutz, T. W. Keegan, L. H. Carpenter, and J. C. deVos Jr. 2001. Deer-predator relationships: a review of recent North American studies with emphasis on mule and black-tailed deer. *Wildlife Society Bulletin* 29:99–115.

Beier, P. 1991. Cougar attacks on humans in the United States and Canada. *Wildlife Society Bulletin* 19:403–12.

Berryman, J. H. 1991. Animal damage management: responsibilities of various agencies and the need for coordination and support. *Proceedings of the Eastern Wildlife Damage Control Conference* 5:12–14.

Bodenchuk, M. J., J. R. Mason, and W. C. Pitt. 2002. Economics of predation management in relationship to agriculture, wildlife, and human health and safety. In *Human Conflicts with Wildlife: Economic Considerations, Proceedings of the Third NWRC Symposium, Fort Collins, Colorado, August 1–3, 2000*, ed. L. Clark, J. Hone, J. A. Shivik, R. A. Watkins, K. C. Vercauteren, and J. K. Yoder, 80–80.

Breck, S. W., B. M. Kluever, M. Panasci, J. Oakleaf, D. L. Bergman, W. Ballard, and L. Howery. 2011. Factors affecting predation on calves and producer detection rates in the Mexican wolf recovery area. *Biological Conservation* 144:930–36.

Breck, S. W., S. A. Poessel, and M. A. Bonnell. 2017. Evaluating lethal and non-lethal management option for urban coyotes. *Human Wildlife Interactions* 11:133–45.

Breck, S. W., S. A. Poessel, P. Mahoney, and J. A. Young. 2019. The intrepid urban coyote: a comparison of bold and exploratory behavior in coyotes from urban and rural environments. Scientific Reports 9, 2104. https://doi.org/10.1038/s41598-019-38543-5.

Bromley, C., and E. M. Gese. 2001. Surgical sterilization as a method of reducing coyote predation on domestic sheep. *Journal of Wildlife Management* 65:510–519.

Connolly, G. E. 1992. Coyote damage to livestock and other resources. In *Ecology and management of the eastern coyote*, ed. A. H. Boer, 161–69. Fredericton: University of New Brunswick.

Connolly, G. E., and W. M. Longhurst. 1975. The effects of control on coyote populations. *Bulletin of the Division of Agricultural Sciences, University of California, Davis* 1872:1–37.

Conover, M. R. 2001. *Resolving human-wildlife conflicts: the science of wildlife damage management*. Boca Raton, FL: CRC Press.

DeLorenzo, D. G., and V. W. Howard Jr. 1977. Evaluation of sheep losses on a range lambing operation in southeastern New Mexico. Agricultural Experiment Station, New Mexico State University, Las Cruces.

Fitzhugh, E. L., S. Schmid-Helmes, M. W. Kenyed, and K. Etling. 2003. Lessening the impact of cougar attack on a human. In *Proceedings of the Seventh Mountain Lion Workshop*, ed. S. A. Becker, D. D. Bjornlie, F. G. Lindzey, and D. S. Moody, 89–103. Lander: Wyoming Game and Fish Department.

Henne, D. R. 1975. Domestic sheep mortality on a western Montana ranch. In *Proceedings of the 1975 predator symposium*, ed. R. L. Phillips and C. Jonkel, 133–49. Missoula: Montana Forest and Conservation Experiment Station, University of Montana.

Howard, V. W., Jr., and T. W. Booth. 1981. *Domestic sheep mortality in southeastern New Mexico*. Bulletin 683, Agricultural Experiment Station, New Mexico State University, Las Cruces.

Howard, V. W., Jr., and R. E. Shaw. 1978. Preliminary assessment of predator damage to the sheep industry in southeastern New Mexico. Agricultural Experiment Station, New Mexico State University, Las Cruces.

Howell, R. G. 1982. The urban coyote problem in Los Angeles County. *Proceedings of the Vertebrate Pest Conference* 10:21–23.

Jahnke, L. J., C. Phillips, S. H. Anderson, and L. L. McDonald. 1987. A methodology for identifying sources of indirect costs of predation control: a study of Wyoming sheep producers. *Vertebrate Pest Control and Management Materials* 5, ASTM International Special Technical Publication 974:159–69.

Knowlton, F. F., E. M. Gese, and M. M. Jaeger. 1999. Coyote depredation control: an interface between biology and management. *Journal of Range Management* 52:398–412.

Leopold, A. S. 1933. *Game management*. New York: Charles Scribner.

May, J. A. 1996. Results of a non-lethal survey and report to the New Mexico legislature. *Proceedings of the Seventeenth Vertebrate Pest Conference, Rohnert Park, California, March 5–7, 1996*, ed. R. M. Timm and A. C. Crabb, 225–29.

Mitchell, B. R., M. M. Jaeger, and R. H. Barrett. 2004. Coyote depredation management: current methods and research needs. *Wildlife Society Bulletin* 32:1209–18.

Munoz, J. R. 1977. Cause of sheep mortality at the Cook Ranch, Florence, Montana. 1975–1976. MS thesis, University of Montana, Missoula.

Mysterud, I. 1977. Bear management and sheep husbandry in Norway, with discussion of predatory behavior significant foe evaluation of livestock losses. *International Conference on Bear Research and Management* 4:233–41.

Nass, R. D. 1977. Mortality associated with range sheep operations in Idaho. *Journal of Range Management* 30:253–58.

———. 1980. Efficacy of predator damage control programs. *Proceedings of the Vertebrate Pest Conference* 9:205–8.

National Agricultural Statistics Service (NASS). 1991. *Sheep and goat predator loss*. USDA, NASS, Washington, DC.

———. 1992. *Cattle and calves death loss*. USDA, NASS, Washington, DC.

———. 1995. *Sheep and goat predator loss*. USDA, NASS, Washington, DC.

———. 1996. *Cattle predator loss*. USDA, NASS, Washington, DC.

———. 2000. *Sheep and goat predator loss*. USDA, NASS, Washington, DC.

———. 2001. *Cattle predator loss*. USDA, NASS, Washington, DC.

———. 2005. *Sheep and goat predator loss*. USDA, NASS, Washington, DC.

———. 2006. *Cattle predator loss*. USDA, NASS, Washington, DC.

———. 2010. *Sheep and goat predator loss*. USDA, NASS, Washington, DC.

———. 2011. *Cattle predator loss*. USDA, NASS, Washington, DC.

———, and New Mexico Department of Agriculture. 2022. *2021 New Mexico Agricultural Statistics Annual Bulletin*. USDA, NASS, Las Cruces, NM.

Oakleaf, J. K., C. Mack, and D.L. Murray. 2003. Effects of wolves on livestock calf survival and movements in central Idaho. *Journal of Wildlife Management* 67:299–306.

O'Gara, B. W., K. C. Brawley, J. R. Munoz, and D. R. Henne. 1983. Predation on domestic sheep on a western Montana ranch. *Wildlife Society Bulletin* 11:253–64.

Palmer, B. C., M. R. Conover, and S. N. Frey. 2010. Replication of a 1970s study on domestic sheep losses to predators on Utah's summer rangelands. *Rangeland Ecology and Management* 63:689–95.

Poessel S. A., S. W. Breck, T. L. Teel, S. Shwiff, K. R. Crooks, and L. Angeloni. 2013. Patterns of human coyote conflicts in the Denver Metropolitan Area. *Journal of Wildlife Management* 77:297–305.

Shaw, H. G. 1987. *A mountain lion field guide.* Federal Aid in Wildlife Restoration Project W-87-R, 3rd, Special Report Number 9. Arizona Game and Fish Dept., Phoenix.

Shaw, H. G., P. Beier, M. Culver, and M. Grigione. 2007. *Puma field guide.* Concord, MA: Cougar Network.

Sterling, B., W. Conley, and M. R. Conley. 1983. Simulations of demographic compensation in coyote populations. *Journal of Wildlife Management* 47:1177–81.

Texas A&M Agrilife Extension. 2015. *Livestock guardian dogs.* EWF-028. 9/15.

Tigner, J. R., and G. E. Larson. 1977. Sheep losses on selected ranches in southern Wyoming. *Journal of Range Management* 30:244–52.

United States Department of Agriculture (USDA). 1997. *Animal Damage Control Program Final Environmental Impact Statement.* (Revision) USDA-APHIS-WS, Operational Support Staff, 6505 Belcrest Rd, Room 820 Federal Bldg, Hyattsville, MD 20782.

———. 2015a. *Cattle and Calves Death Loss in the United States Due to Predator and Nonpredator Causes, 2015.* USDA, APHIS, VS, CEAH, Fort Collins, CO. #745.1217.

———. 2015b. *Sheep and Lamb Predator and Nonpredator Death Loss in the United States, 2015.* USDA, APHIS, VS, CEAH, Fort Collins, CO. #721.0915.

Wade, D. A., and J. E. Bowns. 1982. Procedures for evaluating predation on livestock and wildlife. Texas Agricultural Experiment Station, Texas A&M University/US Dept. Interior-USFWS Pub. B-1429.

Wagner, K. K., and M. R. Conover. 1999. Effect of preventive coyote hunting on sheep losses to coyote predation. *Journal of Wildlife Management* 63:600–612.

Wildlife Society, The. 1990. *Conservation Policies of the Wildlife Society.* Washington, DC: Wildlife Society.

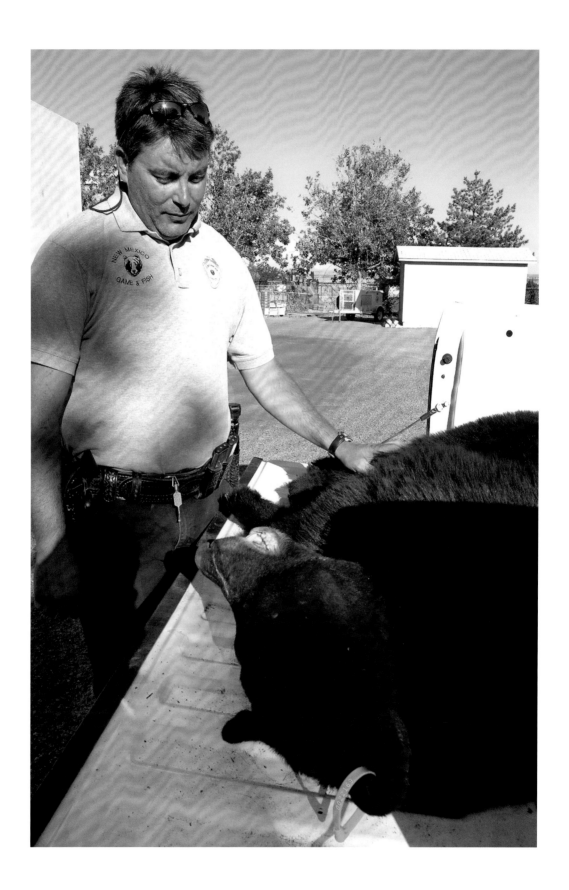

THE NEW MEXICO DEPARTMENT OF GAME AND FISH AND CARNIVORE MANAGEMENT

Elise Goldstein

Wildlife management in New Mexico has a long and active history, stemming from the need to regulate the number of animals killed by humans and thus sustain wildlife populations into the future. Formed in 1912, the New Mexico Department of Game and Fish (hereafter "Department") is the agency responsible for managing wildlife, including implementing and enforcing fishing and hunting regulations set by the State Game Commission, a citizen board created in 1921. With the 1995 amendments to New Mexico's 1974 Wildlife Conservation Act, the Department not only maintains a state-specific list of endangered and threatened species but also now manages recovery and conservation efforts for these species. The Department has authored multiple recovery plans for species listed under the Wildlife Conservation Act, and most recently, it released its 2016 State Wildlife Action Plan for New Mexico, which identifies needs and opportunities for wildlife conservation. For several decades, the Department has found itself at the center of public controversies regarding predator management, trapping on public lands (now outlawed, with some exceptions for management and research), the use of traditional leghold traps, or more generally, what is viewed by part of the public as the Department privileging consumptive over non-consumptive values and users. These controversies should not eclipse the important successes in the history of the Department, from restoring New Mexico's

elk (*Cervus canadensis*) population—now crucial to Mexican wolf (*Canis lupus baileyi*) recovery efforts (Chapter 11)—to the more recent reintroduction of the northern river otter (*Lontra canadensis*; see Chapter 19). Before describing some of the current hunting and trapping regulations and conservation initiatives regarding New Mexico's carnivores, I provide an overview of the history of wildlife management and the New Mexico Department of Game and Fish. In the words of John Crenshaw (2002:7), former Department employee, "Along the trail the Department has assumed new responsibilities as the public's desire to retain its wildlife heritage embraced species once believed less than desirable."

TERRITORIAL WILDLIFE MANAGEMENT

Early colonial settlers in New Mexico found a vast richness of wildlife upon arriving in the territory (Barker 1970). As the railroad brought more people to the area, increasing strain was put on natural resources. Wanton slaughter of bison (*Bison bison*), over-hunting native white-tailed and mule deer (*Odocoileus virginianus, O. hemionus*), elk, pronghorn (*Antilocapra americana*), and bighorn sheep (*Ovis canadensis*) populations, competition with livestock for forage, and disease transmission from domestic sheep and cattle dramatically reduced game herds in the mid- to late 1800s (Bailey 1931; Lee 1962; Barker 1970; Crenshaw 2002). Carnivores were perceived as threats to livestock and wildlife,

(*opposite page*) Photograph: © New Mexico Department of Game and Fish.

Photo 4.1. Coyote hunters on New Year's Day, 1909. Photograph: © New Mexico Department of Game and Fish.

and there was a focus on reducing their populations (e.g., Barker 1970; see also Chapters 10 and 17). New Mexico's first law pertaining to carnivores was passed in 1867 by the Territorial Legislative Assembly, offering a bounty for coyotes (*Canis latrans*), lynx (*Lynx lynx*), bobcats (*L. rufus*), cougars (*Puma concolor*), and grizzly and black bears (*Ursus arctos*, *U. americanus*) as a way to protect livestock production (Barker 1970). This paradigm of predators as "varmints" to be exterminated would last for many years. It then took until 1880 before the first game protection law was passed by the Territorial Legislative Assembly, closing the hunting season for big game between May and August as a means to limit the number of animals being killed (Crenshaw 2002). Although the existence of such a law marked the dawning of wildlife management, it had very few provisions and even fewer people enforcing it. Not only did many people hunt to provide their families with food or supplemental income, but market hunters put a lot of additional pressure on the game resource. The territorial game laws of 1895, 1897, and 1899 further restricted the hunting season, created

wanton-waste laws, made it illegal to dump pollutants into waterways, and banned market hunting (Barker 1970; Crenshaw 2002).

As bison, bighorn sheep, pronghorn, elk, and deer populations declined or were extirpated, an unsuccessful attempt was made to establish a game and fish department in the territory in 1901. Page Otero, an early conservationist and brother of Territorial Governor Miguel A. Otero II (1897–1906), led this effort because he recognized the need to protect big game from high hunting pressure (Barker 1970). New laws closed the season for most ungulates, but there remained no way to enforce these laws, and public sentiment did not generally support them. In 1903, Page Otero was appointed the first Territorial Game Warden in New Mexico (Barker 1970). He was authorized to appoint deputies to enforce the laws with their compensation being one half of the fines they collected. Early work by territorial game and fish wardens focused on advertising the laws and, in 1909, establishing a funding base for wildlife conservation from license fees that would be deposited in the Game Protection Fund for the exclusive purpose of wildlife conservation (Barker 1970; Crenshaw 2002).

HISTORY OF THE NEW MEXICO DEPARTMENT OF GAME AND FISH

In 1912, New Mexico became the 47th state in the Union and the Department of Game and Fish was officially created in the same year (Barker 1970; Jones and Schmitt 1997). Governor William McDonald appointed Trinidad C. de Baca as the first State Game Warden. Volunteer field deputies made up the majority of the workforce, with approximately 340 deputies around the state (see Barker 1970). Use of newspapers to inform people of the new laws, as well as attorneys and judges taking wildlife law more seriously, gave enforcement some traction. In the 1920s the Department established not only bag limits but also legal weapons types, and continued to strengthen laws, make

Photo 4.2. Trinidad C. de Baca, first State Game Warden, 1912–1914. In 1914, de Baca commented about the thankless work of his deputies: "At least three-quarters of the wardens receive no pay whatever, and yet their work is decidedly of the most difficult and strenuous character. There is no glory in the service, either, and all that a successful warden generally receives is criticism and abuse. If he is slow and timid, he is ridiculed. If he enforces the law rigidly he is abused" (Crenshaw 2003d:8). Photograph: © New Mexico Game and Fish.

Photo 4.3. Grace Melaven, State Game Warden in 1923. The wife of a Santa Rosa banker, she was not a hunter, nor a member of the Game Protective Association (GPA), and she had no background in wildlife conservation. But she had campaigned for James F. Hinkle, who defeated his republican opponent in the 1922 gubernatorial race. Her appointment was met at first with hostile reactions by the GPA, and as a result of lobbying, the New Mexico Legislature transferred the authority to hire state game wardens from the governor to the State Game Commission in 1925. Regulation authority was conferred to the State Game Commission six years later, in 1931. Photograph: © New Mexico Game and Fish.

Photo 4.4. Elliott Barker and his staff at the New Mexico Department of Game and Fish. Elliott Barker led the Department as State Game Warden from 1931 to 1953. In 1955, the State Legislature changed the title of State Game Warden to that of Director of the Department. Photograph: © New Mexico Game and Fish.

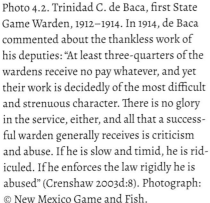

hunting seasons shorter, and increase license fees (Barker 1970). The large military presence in the state led to a high poaching rate among military personnel. Military presence was largely composed of people new to the state who were likely unfamiliar with game laws, often living in remote areas with access to game. This added to the difficulty in enforcing early wildlife protection laws.

The 1910s gave rise to several local sportsmen's organizations around the state (Crenshaw 2003). From their union, a statewide umbrella organization was formed in March 1916, the Game Protective Association (GPA, still in existence today but renamed New Mexico Wildlife Federation), thanks largely to the efforts of conservationists Aldo Leopold and Miles W. Burford (Lorbiecki 1998). Their main focus was to support better law enforcement, predator control, and the establishment of game refuges. The GPA lobbied for legislation that was enacted to restrict Game Protection Fund expenditures, the fund established by the legislature that contained revenue from hunting and fishing license sales and penalty assessments, to just fish and wildlife management (Barker 1970). The GPA also successfully worked toward establishing a three-member State Game Commission (which was later expanded to five, and eventually seven members), in the end transferring authority to appoint the State Game Warden from the Governor to the State Game Commission, and granting the State Game Commission authority to determine hunting seasons, bag limits, and manner and method of take, define game animals, and set these in regulation (Barker 1970; Crenshaw 2003).

Under a philosophy that certain areas should be protected from hunting to provide source populations that would slowly disperse and repopulate areas open to hunting, state wildlife refuges were established on National Forests (Bailey 1928; Barker 1970). This was a basic tenet of wildlife management in New Mexico for many years. By 1926 there were 57 wildlife refuges, increasing to 169 refuges by 1939 covering 11,314 km² (2,795,704

acres) (Barker 1970; Pederson 1987). However, by the early 1950s deer overpopulation and associated negative impacts prompted abolishing many of the wildlife refuges so that there were only about 1,214 km² (300,000 acres) protected in state refuges by the end of the decade (Barker 1970; Pederson 1987). The Department continued to increase the number of employees on staff, enabling increased law enforcement activities, higher emphasis on wildlife education, and big-game transplants to re-establish and augment populations (Barker 1970; Pederson 1987).

By the late 1960s the Department took an increasing role in broader ecosystem protection, recognizing the need to protect habitat and non-game animals, curb pollution, and educate the public on the important role hunting plays in funding wildlife conservation and providing funding for management actions; it also began funding habitat and wildlife restoration projects (Pederson 1987). This was followed by an increase in endangered species work, prompted by the legislature allocating general tax funds for this purpose; it was the first time general funds were allocated to the Department. In general, the 1970s witnessed an increase in national awareness—reflected in Congress—about the need to protect non-game animals from extinction. The federal Endangered Species Act was passed in 1973, with Section 6 providing the mechanism for funding programs for the management of threatened and endangered species by state wildlife agencies. New Mexico's Wildlife Conservation Act was passed in 1974 to provide state protection for endangered species, additional biologists were hired for research and recovery work, and the first list of endangered species was adopted by the State Game Commission (Jones and Schmitt 1997). Among the species first listed under the 1974 Wildlife Conservation Act were the Mexican gray wolf and the black-footed ferret (*Mustela nigripes*), though the latter species, believed to be extinct throughout its historical distribution, was delisted in 1988 (Jones and Schmitt 1997; see Chapter 22). Later, in 1995, important amendments to the Wildlife Conservation Act were passed, requiring the development of recovery plans for state endangered and threatened species (replacing endangered groups 1 [G1] and 2 [G2], respectively) and the listing of species based only on biological information (Jones and Schmitt 1997).

The Department bases its current wildlife management policy on concepts presented in the North American Model of Wildlife Conservation as first described by Valerius Geist (1995). This is not a rigid set of rules; rather, it is a paradigm describing the major tenets of wildlife conservation unique to the United States and Canada compared with other conservation programs around the world. These tenets are summarized by the following seven principles: wildlife resources are a public trust; markets for game are eliminated; allocation of wildlife is by law; wildlife can be killed only for a legitimate purpose; wildlife is considered an international resource; science is the proper tool to discharge wildlife policy; and democracy of hunting is standard (Geist 1995). Exceptions have been made for the active market for furbearer pelts. This market is highly regulated and consistent with conservation principles because state regulations keep harvest levels consistent with maintaining sustainable populations, help manage human-furbearer conflicts, and support habitat conservation and restoration efforts (Organ et al. 2012). Harvest regulations in New Mexico continue to be refined to maximize hunting opportunity while maintaining sustainable populations and upholding the tenets of the North American Model of Wildlife Conservation.

In 2021, the New Mexico Legislature passed the Wildlife Conservation and Public Safety Act (NMSA 1978 § 17-11-1). Also known as Roxy's Law, this new piece of legislation went into effect on April 1, 2022. It prohibits using traps, snares, or poison to kill, capture or injure animals on New Mexico's public lands. The provisions of the law do not apply to certain situations including scientific research and wildlife management.

Photo 4.5. Unknown bear hunter. Photograph: © New Mexico Department of Game and Fish.

Photo 4.6. Left to right, Elliott Barker, New Mexico Governor Thomas J. Mabry (1947–1951), and Roy Snyder hunting cougars (*Puma concolor*). Roy Snyder (1893–1974) joined the New Mexico Department of Game and Fish in 1944 and became the "premiere lion hunter" in the state. Photograph: © New Mexico Game and Fish.

CARNIVORE MANAGEMENT

With the onset of World War I, the Council of Defense encouraged conserving food resources. As such, big-game and livestock conservation in the form of predator control was promoted as patriotic and a matter of national security (Sayre 2017). The US Biological Survey had a very active

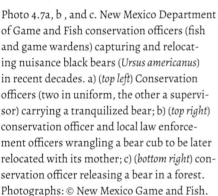

Photo 4.7a, b , and c. New Mexico Department of Game and Fish conservation officers (fish and game wardens) capturing and relocating nuisance black bears (*Ursus americanus*) in recent decades. a) (*top left*) Conservation officers (two in uniform, the other a supervisor) carrying a tranquilized bear; b) (*top right*) conservation officer and local law enforcement officers wrangling a bear cub to be later relocated with its mother; c) (*bottom right*) conservation officer releasing a bear in a forest. Photographs: © New Mexico Game and Fish.

predator control program, and though by 1923 the bounty system was no longer in place, there was a lot of pressure to eradicate predator species such as the gray wolf (*Canis lupus*) and grizzly bear. The grizzly bear was extirpated from New Mexico sometime during the 1930s (Chapter 17), the Mexican gray wolf sometime thereafter (Chapter 11). Predator control was a principal focus of the Department, and it was generally held that high predation rates were limiting game populations (Barker 1970). In contrast, black bears were protected as a game animal in 1927 and laws were enacted requiring a license to kill nuisance bears (Costello et al. 2001).

Furbearer trapping became regulated in 1939 when the legislature defined several species as non-predator fur-bearing animals (later called "protected furbearers"), including the American badger (*Taxidea taxus*), white-nosed coati (*Nasua narica*), black-footed ferret, American mink (*Neogale vison*), Pacific marten (*Martes caurina*), raccoon (*Procyon lotor*), and ringtail (*Bassariscus astutus*), in addition to all species of foxes (*Vulpes* spp., *Urocyon cinereoargenteus*), weasels (*Mustela richardsonii*, *Neogale frenata*), and the muskrat (*Ondatra zibethicus*) (see NMDGF 2020). A license was required to trap these species, and a season was set from 1 October to 31 March. The legislature added the nonnative nutria (*Myocastor coypus*) and the northern river otter to the list of protected furbearers in 1955. Exact season dates have been modified over the years, and seasons for some species have been closed periodically in response to declining numbers. In particular, the season was closed for the muskrat, mink, and marten for the first time in 1947. Open seasons were held periodically for these animals, in some instances because the population could support it, and in others as a means to determine if the species was still present. Seasons on American mink, northern river otter, and Pacific marten were permanently closed in 1975, when these species were placed on the New Mexico endangered species list. The mink and the river otter were delisted in 1985 due to their presumed extinction (Jones and Schmitt 1997), and the seasons remain closed today.

The 1930s and 1940s saw a large increase in predator control work conducted by the Department. The program started with three part-time predator control employees, and by 1947 the number had increased to 20 employees (Barker 1970). The intensive effort resulted in a dramatic increase in the number of predators removed annually by the Department. Compound 1080, a highly effective poison introduced in 1945 also contributed to increased predator removal (Barker 1970). Concurrently, the statewide deer population began increasing in the 1940s and reached a high in the mid-1950s, and trappers and houndsmen reported sharp declines in predator numbers. In the mid-1950s, the Department began shifting its focus away from predator control and by the end of the decade there were only four trappers and houndsmen on payroll (Barker 1970). This did not necessarily reflect a changing attitude regarding predator control; rather, it reflected a shift in the responsibility for predator control away from the Department and toward the federal government and private landowners (see Chapter 3).

A paradigm shift, from regarding predators as "varmints" to recognizing them as having a vital role in the ecosystem and in agricultural life, began to emerge in the next decade. By the mid-1960s the Department had adopted a policy of trapping predators in specific instances with a clear wildlife objective, and no longer engaged in indiscriminate predator killing. Department trappers stopped using poisons in 1972 when it was outlawed by the federal government, and

an Animal Control Section was created within the Department to address nuisance wildlife in a more systematic manner. Cougar numbers appeared to be declining and an emphasis was placed on protecting cougar populations while allowing for a mechanism for ranchers to protect their livestock. The Department was concerned that if federal protections were granted for cougars under the Endangered Species Act it would limit its ability to manage New Mexico's population of this species, and it promoted implementing protections to preclude the need for federal listing. In 1971 the New Mexico legislature added the cougar to the list of protected game animals, which resulted in the State Game Commission regulating season, bag limits, and manner and method of take. New pelt tagging requirements allowed the Department to begin tracking the number of cougars harvested, as well as other harvest statistics (e.g., harvest location, age, sex, method of take), and make changes to the harvest rules to modify harvest numbers as needed.

In 1978, the Department implemented a mandatory pelt-tagging requirement for harvested bobcats in response to the International Union for Conservation of Nature's (ICUN) prohibition on export of any bobcat pelt from the United States without a permit from the Department of Interior. This permit, or pelt tag, would confirm that the pelt belonged to a bobcat and not to one of several spotted cats of similar appearance that are considered endangered in other countries (e.g., the Iberian lynx [Lynx pardinus]). Pelt tags would not be issued to the state until it submitted substantial data demonstrating that export would not adversely affect the state bobcat population. The Department pelt tagging requirement also allowed for biological data collection to monitor total harvest, harvest distribution throughout the state, and bobcat population demographics. However, a 1979 lawsuit seeking to prevent bobcat exportation resulted in export prohibition in seven states, including New Mexico. In 1980, the New Mexico legislature added bobcats to the list of protected furbearers and the State Game Commission set trapping regulations, including limits on trap size, minimum distances that traps can be set from roads, campgrounds, and developed trails, restrictions on visual baits, and mandatory 48-hour trap checks. With this system in place, bobcat pelt export was approved for the following season, and the bobcat pelt tagging requirement remains in place today (Evans 1981). An average of approximately 2,200 bobcats are harvested every year in New Mexico (Figure 4.1). Among the carnivores, only the gray fox (*Urocyon cinereoargenteus*) is harvested in comparable numbers. All other protected furbearers are harvested in much smaller numbers in the state. The number of furbearer licenses has increased somewhat since the 1990s (Figure 4.2).

New policies overhauling the depredation system were implemented in the mid-1980s. Previous policies granting landowners unlimited authority to kill nuisance or depredating bears and cougars were set aside in favor of a more regulated system. Landowners who actively improve habitat for wildlife can apply for a permit allowing limited cougar take on their property, and the director grants or denies requests on a case-by-case basis. In the event of an immediate threat posed by a predator to human life or livestock, a person may kill the animal and is required to report it to the Department within 24 hours. For cougar sport harvest, New Mexico has put in place limits on both total number and number of females that can be harvested within each cougar management zone (see Chapter 9).

FUNDING

From its inception, the Department has been a self-funded agency with revenues generated from the sale of hunting and fishing licenses and penalty assessments for violations (Barker 1970). No general funds are typically received. Funding does not automatically go to the Department;

rather, it is allocated by the state legislature which passes a bill each session authorizing revenue to the Department. The Wildlife Restoration Act passed by Congress in 1937 created an 11% excise tax on sport arms, ammunition, and archery equipment. This act, also known as the Pittman-Robinson Act, in recognition of its sponsors, provides states with funding for the purchase and construction of public recreation areas, big-game transplants, scientific research, and other wildlife conservation activities. For every dollar the Department spends on qualifying activities, it receives a 75-cent reimbursement from these federal funds. In 2021 this reimbursement totaled approximately $6.5 million, enabling the Department to implement numerous projects to benefit wildlife.

Increasing the scope of work performed by the Department led to a need for additional funding. In the 1980s the Share with Wildlife Program was created to fund non-game projects through a voluntary contribution system. In 2021, the contributions totaled approximately $130,000, and were matched with approximately $115,000 in federal grant funds. The Habitat Stamp Program was initiated in 1986 under the federal Sikes Act, requiring all hunters and fishermen on Bureau of Land Management (BLM) lands or national forests to purchase a $5 stamp to be used for habitat improvement projects on those federal public lands. This fee was raised to $10 in 2022. In 2021, this program provided approximately $1 million for artificial wildlife water sources, prescribed burns, and thinning. Matching federal funds can be applied to individual projects when they qualify. Created in 2006, the Habitat Management and Access Validation Stamp provides funds to lease access to private lands for public use, public access to landlocked public land, as well as the improvement, maintenance, development, and operation of State Game Commission property for fish and wildlife habitat management. This program provided approximately $1 million in

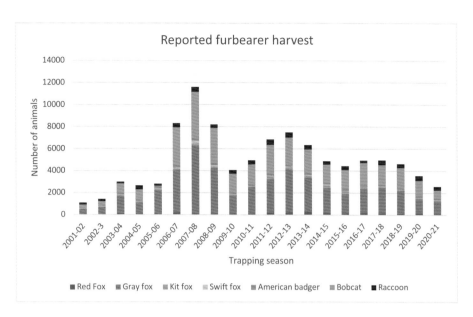

Figure 4.1. Reported harvest for seven carnivores classified as protected furbearers in New Mexico, from the 2001–2002 through the 2020–2021 trapping season. The gray fox (*Urocyon cinereoargenteus*) and the bobcat (*Lynx rufus*) represented most of the reported harvest during the 20-year period (note that no bobcat harvest data is available for 2005–2006). Protected furbearers can be taken only during a designated open season, which currently extends from 1 November through 15 March for most species. There is no open season for the black-footed ferret (*Mustela nigripes*), Pacific marten (*Martes caurina*), American mink (*Neogale vison*), northern river otter (*Lontra canadensis*), and white-nosed coati (*Nasua narica*), as their populations in the state are considered too small to support take from hunting or trapping. Some of the regulations related to hunting and trapping furbearers are part of the Furbearer Rule, which is updated every four years.

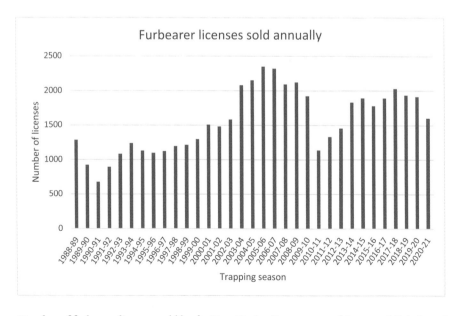

Figure 4.2. Number of furbearer licenses sold by the New Mexico Department of Game and Fish from the 1988–1989 through the 2020–2021 trapping season (a trapping license allows both trapping and hunting for furbearers).

Photo 4.8. Cougar (*Puma concolor*) on Mesa Prieta, Santa Fe National Forest north of Coyote. Licensed hunters are currently limited to killing a maximum of two cougars per year. Taking of spotted kittens or females with spotted kittens is illegal. Photograph: © New Mexico Department of Game and Fish.

Photo 4.9. Black bear (*Ursus americanus*) near Rio del Oso on the Santa Fe National Forest west of Ohkay Owingeh. Except for research or management, black bears cannot be trapped or baited in New Mexico. Photograph: © New Mexico Department of Game and Fish.

Photo 4.10. Bobcat (*Lynx rufus*) photographed by a trail camera on Brokeoff Mountain on the Carson National Forest northwest of Tres Piedras. The Convention on International Trade in Endangered Species of Wild Fauna and Flora (CITES) is an international agreement that went into effect on 1 July 1975. Ratified by the United States and 182 other nations, it offers protection to those species whose populations could be threatened by international trade. Pelt tags are required for bobcats harvested in New Mexico before they can be sold or transported out of state, not because the species is declining or under threat, but to confirm that the pelt belongs to a bobcat and not to one of several spotted cats of similar appearance and considered endangered in other countries (e.g., Iberian lynx [*L. pardinus*]). Photograph: © New Mexico Department of Game and Fish.

Photo 4.11. American badger (*Taxidea taxus*) photographed by a trail camera near the Rio Nutrias northwest of Cebolla in Rio Arriba County. The American badger is classified as a protected furbearer in New Mexico, with an open season currently extending from 1 November through 15 March. Harvest of badgers requires the purchase of a trapping license (which covers both trapping and hunting for furbearers). Photograph: © New Mexico Department of Game and Fish.

Photo 4.12. Coyote (*Canis latrans*) photographed by a trail camera on Brokeoff Mountain, on the Carson National Forest northwest of Tres Piedras. Coyotes are unprotected furbearers and a non-game species, and as such they may be taken throughout the year with no bag limits. No license is needed for residents to hunt or trap non-game species. Non-residents must have a non-resident trapper license to hunt protected furbearers or trap any animal (protected furbearer and non-game). A non-resident hunting license allows hunting (and the possession) of non-game species but not setting traps or snares (a non-resident trapper license is required for the latter). Note that trapping of coyotes and other wildlife is now illegal on public lands in New Mexico, as a result of the Wildlife Conservation and Public Safety Act (also known as Roxy's Law) taking effect on 31 March 2022. Photograph: © New Mexico Department of Game and Fish.

2021, and federal funds are used as match for qualifying projects. Taken together, these funds support the approximately $45 million annual budget spent by the Department to maintain sustainable wildlife populations and support the habitats in which they live.

RULE DEVELOPMENT

In 1953, the New Mexico Statutes (NMSA 1953) formalized that the state is responsible for, "providing an adequate and flexible system for the protection of the game and fish of NM and for their use and development for public recreation and food supply, and to provide for their propagation, planting, protection, regulation and conservation to the extent necessary to provide and maintain an adequate supply of game and fish within the State of New Mexico." One of the primary ways the Department accomplishes this goal is to set harvest regulations for hunted and trapped species through the rule and regulation development process. Final rules can be found on the Department's website or summarized in the Rules and Information Booklets that are published annually.

The foundation of all harvest rules is the biology and status of the species, and over the years Department harvest levels have become more data- or evidence-based. The Department now collects a variety of data through the mandatory harvest reporting system, including mandatory pelt tag reports for bears, cougars, and bobcats, research, field observations, and interactions with hunters and trappers. Population estimates are often held up by the public as the gold standard upon which to base harvest levels. However, population estimates can be difficult and expensive to obtain for rare or elusive species and other metrics may be just as effective in setting harvest levels. For example, data on location, age, and sex of harvested animals can provide insights into population dynamics. If the ratio of various age and sex components of the population in the harvest changes over time, it may indicate an increase or decrease in the population and the Department may choose to modify harvest regulations accordingly. Although this does not provide insight into why the population is changing, the Department can increase or decrease harvest opportunity to help maintain a sustainable population. Other metrics such as catch-per-unit effort, success rate, type of weapon used, etc., can help wildlife managers understand how the population might respond to specific changes in the hunting rule. Harvest statistics have proved to be a valuable management tool for many years in the absence of precise population estimates.

All game and furbearer species are classified by the state legislature as "protected species," meaning that it is illegal to kill them without a license. Some protected species such as the northern river otter and the American mink do not have open seasons because their populations are not large enough to support it (Table 4.1.). The State Game Commission has the authority to set rules (sometimes referred to as regulations) for species regarding, for example, season dates, bag limits, and weapon types legally allowed. Rules for most species such as those designated as furbearers, in addition to the black bear and cougar, automatically open for amendment every four years. Rule development is initiated when the Department makes a presentation to the State Game Commission during a public meeting announcing that a rule is open. This initial presentation includes an overview of population and harvest trends, other harvest statistics, and current research being conducted in the state. A general timeline for the process is presented along with amendments the Department is considering and the locations of public meetings. The public has a variety of opportunities to comment on the rule including sending comments via email or standard mail, attending a public meeting, and making comments during State Game Commission meetings. The Department posts information

Table 4.1. Protection status and hunting/trapping rules for New Mexico carnivore species during the 2022–2023 hunting season.

Note that the Wildlife Conservation and Public Safety Act (also known as Roxy's Law) went into effect in April 2022. With some exceptions, it prohibits using traps, snares, or poison to kill, capture or injure animals on public lands in New Mexico.

Common Name	Species	Classification	Open Season	Bag Limits	Comments
Coyote	*Canis latrans*	Unprotected	No closed season	None	
Mexican gray wolf	*Canis lupus baileyi*	Non-essential experimental population (federal); State endangered	No open season	-	Any incidental capture of a Mexican wolf must be reported as soon as possible to the USFWS Interagency Field Team to arrange for radio-collaring and release of the wolf.
Gray fox	*Urocyon cinereoargenteus*	Protected furbearer	1 November–15 March	None	
Kit fox	*Vulpes macrotis*	Protected furbearer	1 November –15 March	None	
Swift fox	*Vulpes velox*	Protected furbearer	1 November–15 March	None	
North American red fox[1]	*Vulpes fulva*	Protected furbearer	1 November–15 March	None	
Canada lynx	*Lynx canadensis*	Federal endangered species	No open season	-	
Bobcat	*Lynx rufus*	Protected furbearer	1 November–15 March	None	Every bobcat harvested in New Mexico and sold or transported out of state must have a pelt tag.
Jaguar	*Panthera onca*	Federal endangered species	No open season	—	
Cougar	*Puma concolor*	Protected game animal	1 April 1–31 March	Two	Trapping and baiting are not allowed. Harvested cougars must be both carcass tagged and pelt tagged. Proof of sex is required. No spotted kitten(s) or females with spotted kittens may be taken.
Northern river otter	*Lontra canadensis*	Protected furbearer	No open season	—	
Pacific marten	*Martes caurina*	State threatened and Protected furbearer	No open season	—	
American ermine	*Mustela richardsonii*	Protected furbearer	1 November–15 March	None	
Long-tailed weasel	*Neogale frenata*	Protected furbearer	1 November–15 15 March	None	
Black-footed ferret	*Mustela nigripes*	Federal endangered and Protected furbearer	No open season	—	

Common Name _continued_	Species	Classification	Open Season	Bag Limits	Comments
American mink	_Neogale vison_	Protected furbearer	No open season	—	
American badger	_Taxidea taxus_	Protected furbearer	1 November–15 March	None	
White-backed hog-nosed skunk	_Conepatus leuconotus_	Unprotected	No closed season	None	
Hooded skunk	_Mephitis macroura_	Unprotected	No closed season	None	
Striped skunk	_Mephitis mephitis_	Unprotected	No closed season	None	
Western spotted skunk[2]	_Spilogale gracilis_	Unprotected	No closed season	None	
Ringtail	_Bassariscus astutus_	Protected furbearer	1 November–15 March	None	
White-nosed coati	_Nasua narica_	Protected furbearer	No open season	—	
Raccoon	_Procyon lotor_	Protected furbearer	1 September–15 May, 16 May–31 August (restricted harvest methods[3] only)	None	
Black bear	_Ursus americanus_	Protected game animal	Varies by bear management zone and sporting arm, but generally from August into or through November	One	Trapping and baiting are not allowed. Harvested bears must be both carcass tagged and pelt tagged. Proof of sex is required. No cubs, one-year old, or females with cubs may be taken

1. See Chapter 15. New Mexico's red fox was previously treated as _Vulpes vulpes_.
2. See Chapter 29 for a proposed taxonomic split.
3. Restricted harvest methods include cage traps, foot-encapsulating traps, and hunting.

on its website to keep the public informed of the process, opportunities to be involved, and proposed changes to the rules. It also issues press releases that may appear in local newspapers or broadcasts.

At subsequent State Game Commission meetings, the Department provides further detail on public participation, recommendations that are being considered, and the science underlying the recommendations. The Department presents final draft recommendations and posts them on the Department's website for public consideration. At the final meeting, the State Game Commission votes whether to approve the recommendations as presented. The new rule goes into effect during the subsequent season. Department biologists continue to monitor populations to understand how new rules impact populations, and present the data at subsequent State Game Commission meetings when the rule is opened again for amendment.

PROTECTION STATUS AND CONSERVATION INITIATIVES

New Mexico's carnivores benefit from varying levels of protection. Those listed under the

Photo 4.13. Mexican wolf (*Canis lupus baileyi*) photographed in 2014 by a trail camera in New Mexico's Gila National Forest. There are currently three full-time New Mexico Department of Game and Fish biologists assisting with wolf releases, cross-fostering pups, depredation, hazing problem animals, and interacting with the public. Photograph: © Interagency Field Team.

federal Endangered Species Act of 1973 and/or the New Mexico Wildlife Conservation Act of 1974 (as amended in 1995) are protected from "take," with some exceptions (Table 4.2). Under the Endangered Species Act, "take" means to harass, harm, pursue, hunt, shoot, wound, kill, trap, capture, or collect, or to attempt to engage in any such conduct (16 US Code § 1532). Under the Wildlife Conservation Act, it is defined as harassing, hunting, capturing, or killing any wildlife or attempting to do so (NMSA 1978 § 17-2-38). Under both pieces of legislation, illegal take can result in civil or criminal penalties. However, the Endangered Species Act provides for habitat protection whereas the Wildlife Conservation Act does not, other than the acquisition of land or water rights (Table 4.2).

The Department has been actively involved in much of the federal Mexican Wolf reintroduction program under the Endangered Species Act, led by the US Fish and Wildlife Service (see Chapter 11). As a sister agency, the Department has helped support reintroduction efforts, cross-fostering pups, surveys, captures, adverse conditioning, and interacting with landowners. The Department currently has three full-time wolf biologists collaborating with the US Fish and Wildlife Service in southwestern New Mexico.

Other than coyotes and skunks (family Mephitidae), carnivores are officially categorized as protected game animals or protected furbearers by the state of New Mexico. Harvest rules determine the degree of regulation based on the length of the open season and bag limits (Table 4.1). Finalized in 2016, with minor revisions in 2019, the New Mexico Statewide Action Plan identifies Species of Greatest Conservation Need (SGCNs), including those known to be represented by declining populations (Table 4.3). Although the Statewide Action Plan is a non-regulatory planning document, species categorized as SGCNs qualify for funding under the State Wildlife Grant program administered by the US Fish and Wildlife Service, which amounts to approximately $200,000 in federal funds annually.

Table 4.2. Comparison of provisions contained in the federal Endangered Species Act of 1973, as amended, and New Mexico's Wildlife Conservation Act of 1974, as amended in 1995.

	Endangered Species Act	Wildlife Conservation Act[1]
Status designations	Endangered and Threatened	Endangered and Threatened
Listing process	Petition or self-initiated process. Final Rule published in the *Federal Register*, preceded by a Proposed Rule and a public comment period.	Listing recommendation of the Department's director to the State Game Commission based on investigation with the input of a peer-review panel. Public hearings required in affected areas of the state. The separate Biennial Review process is to up- or downlist between threatened and endangered only.[2] It does not address new listings or delisting.
Powers to regulate	US Fish and Wildlife Service and National Marine Fisheries Service	State Game Commission
Protection against take	"Take" is prohibited with few exceptions (e.g., incidental take permit, national security issue)	"Take" is prohibited with exceptions (e.g., special use permits, Native American religious purposes)
Habitat protection provisions	Designation of Critical Habitat; Section 7 Consultation process (federal nexus required); and acquisition of habitat and creation of refuges (Section 5)	Under Section 17-2-44, the Department's director may acquire land for the conservation and protection of listed species.
Species recovery tools	Recovery plans (non-regulatory)	Recovery plans (non-regulatory)
Penalties	Citizen suits; civil and criminal penalties	Civil and criminal penalties

1. New Mexico also has the Endangered Plant Species Act, passed in 1985 (NMSA 1978 § 75-6-1).

2. *Downlisting* means that the species' designation changes from endangered to threatened; *uplisting* occurs when a threatened species becomes listed as endangered.

RESEARCH EFFORTS

The first major biological undertaking in New Mexico was conducted in 1927 in a statewide game survey by J. Stokley Ligon (Ligon 1927). This was the beginning of incorporating science into game management, and included having game wardens and predator control personnel record data on animals harvested. It was followed a few years later by the publication of Vernon Bailey's *Mammals of New Mexico*, a detailed species account sponsored by the US Biological Survey (Bailey 1931).

The first cougar study was conducted in the mid-1930s to obtain basic life history information in northern and western New Mexico. A more intensive effort was initiated in 1971 in the southwestern quadrant of the state. Cougars were captured, marked, and tracked for three

Table 4.3. Carnivores listed as Species of Greatest Conservation Need (SGCN) in the 2016 Statewide Wildlife Action Plan. Note that species legally harvestable with statutory protection as game animals are not eligible for SGCN designation, unless they are simultaneously designated as Threatened or Endangered by the State (NMDGF 2019).

Common Name	Scientific Name	Federal and State Status	Criteria for Inclusion as SGCN	Monitoring
Jaguar	*Panthera onca*	E	De, V, K	CT, M(o)
Gray wolf	*Canis lupus*	NEP, SE	De, V, K	CT, M(a)
Northern river otter	*Lontra canadensis*	—	V	M(p)
Pacific marten	*Martes caurina*	ST	V	M(p)
Black-footed ferret	*Mustela nigripes*	E	De, V	CT, M(a)
American mink	*Neogale vison*	-	V	M(o)

Abbreviations for federal status are as follows: E = Endangered; NEP = Non-essential experimental population. State status (2020 Biennial Review): SE = Endangered; ST = Threatened. Criteria for Inclusion as SGCNs: De = Declining; K = Keystone; V = Vulnerable. Monitoring: CT = Conservation/recovery team; M = Monitoring (a) = at least once per year; (p) = periodically (but less than annually); (o) = opportunistically.

years to estimate population size and to evaluate diet, home range, predator-prey relationships, and other life history characteristics. Department biologists recognized a lot of uncertainty in their population estimate of approximately 500 cougars in southwestern New Mexico, and cautioned against extrapolating results to the rest of the state. At the same time, they suggested that the previous statewide estimate of 350, based on intuition and not data, was probably quite low. A subsequent study in the early 1980s was funded by the National Park Service to investigate cougar population dynamics in Carlsbad Caverns and Guadalupe Mountains national parks.

Recognizing the need for better population and life history data on which to base harvest regulations, the Department initiated a 10-year cougar study in 1987 in conjunction with the Hornocker Wildlife Institute at the University of Idaho. This study quantified many life history attributes such as survival, mortality, recruitment, age and sex structure, home range size, reproduction, predator-prey interaction, and population density in the San Andres Mountains in south-central New Mexico (Logan et al. 1996; Chapter 9). Data were gathered by radio-collaring and tracking individual cougars, and by experimental removal of cougars on half of the study area. Data collected during this study have been used as the foundation for cougar harvest regulations for many years. More recently, the Department has been conducting research to more accurately estimate the size of individual cougar populations around the state. This involves radio-collaring cougars with GPS collars and deploying remote cameras to "recapture" animals when they come within range of the motion sensor. The technique, known as spatial mark-recapture, provides a population density estimate. When combined with movement and habitat use data from the radio collars and metrics from hunter harvest reporting, an integrative population model can be developed to understand population dynamics more fully.

In 1988, a black bear study was initiated at the Philmont Scout Ranch to evaluate efficacy of management actions to reduce human/bear conflicts (see Chapter 16). Permanent cables from which campers could suspend their food were installed in campgrounds, as were bear-proof garbage containers. These improvements resulted in reducing the number of bear complaints received by the Department in the area by just over 50%, and the number of nuisance bears removed from the area by over 60%. In addition, location data from 11 radio-collared bears was used to identify home range size and extent of overlap with the campgrounds (Jones 1991). In 1992, the Department collaborated with the Hornocker Wildlife Institute and New Mexico State University to initiate a comprehensive eight-year bear study. Research occurred in two distinct habitat types, in the xeric Mogollon Mountains of west-central New Mexico and the mesic Sangre de Cristo Mountains in the northern part of the state (see Chapter 16). Impacts of hunting were evaluated by closing one area to hunting during the majority of the study period. The study estimated reproduction and survival rates, validated a technique for aging bears based on tooth rings, described home range size, habitat use, and minimum population densities among sex-age categories, and developed a GIS model to predict bear habitat across New Mexico. In addition, a model relating sex-age composition of harvested bears and population dynamics was developed. Data were collected by radio-collaring and tracking individual bears and investigating den sites (Costello et al. 2001; Chapter 16).

Modern developments in statistical methods provided an opportunity for the Department to implement a study to collect more accurate, average (as opposed to minimum) population estimates in some areas where data had previously been collected, and in other areas in which there was a particular management need. In collaboration with New Mexico State University, a spatial mark-recapture study was initiated in 2012 in the Sangre de Cristo Mountains in northern New Mexico, the Sacramento Mountains in southeastern New Mexico, and the Sandia Mountains in central New Mexico. Substantial support for the Sandia Mountains research

effort was provided by the Sandia Mountains Bear Collaborative. Hair-snare stations were set up to obtain genetic data from individual bears who left hair resulting in "recapture" of the individual. The number of "recaptures" and their spatial distribution over the landscape were used to estimate population sizes in each study area (Gould et al. 2016). Non-invasive genetic monitoring of black bear population dynamics continues using hair-snare stations in different parts of the state (Bear Management Zone [BMZ] 1 in 2019, BMZ 10 in 2020–2021, and BMZs 5 and 7 in 2022). Radio collars were also deployed in BMZs 5 and 7 in 2022. Data collected on movement and habitat use will be incorporated, along with harvest statistics, in an integrated population model to understand bear ecology more fully. In recent years, the Department deployed trail cameras as part of more than 10 monitoring studies, primarily species specific. For example, the Department conducted a survey for Pacific martens in the Sangre de Cristo Mountains from the fall of 2019 to the spring of 2020. During that time, a total of 64 sites were monitored with the use of remote, motion-activated cameras. During the fall of 2019, additional coverage was provided by secondary cameras that operated for two to three weeks at a time before being rotated from one site to the next. The camera monitoring effort resulted in several marten detections (see Chapter 20). Other camera monitoring has focused again on the cougar and black bear, but also on the bobcat and swift fox (*Vulpes velox*). These research efforts are aimed at collecting additional information on distribution and population dynamics, as well as developing monitoring methods. The Department views the use of remote cameras as a viable long-term monitoring method for nearly all of New Mexico's carnivores, combining camera "captures" with data from GPS-collared individuals or non-invasive genetic sampling (scat or hair), where feasible.

Following the northern river otter's initial reintroduction in New Mexico's upper Rio Grande watershed, population augmentations are ongoing with the release of otters from Louisiana (see Chapter 19). The Department funded a non-invasive capture-recapture genetic study based on the analysis of scats (Cox and Murphy 2019). That study yielded estimates of current otter population size and annual population growth rate in the upper Rio Grande. Although it provided evidence of continued reproduction in the reintroduced population, it also documented a bottleneck effect and concluded that population size lies below the minimum necessary to prevent inbreeding depression (Cox and Murphy 2019). Cox and Murphy's (2019) study provides the justification for the current population augmentations, and it is the goal of the Department to rely on this and similar research to inform future carnivore management decisions.

LITERATURE CITED

Bailey, F. M. 1928. *Birds of New Mexico*. Santa Fe: New Mexico Department of Game and Fish.

Bailey, V. 1931 (=1932). *Mammals of New Mexico*. North American Fauna 53. Washington, DC: US Department of Agriculture, Bureau of Biological Survey.

Barker, E. 1970. A history of the New Mexico Department of Game and Fish. Unpublished manuscript, New Mexico Department of Game and Fish.

Costello, C. M., D. E. Jones, K. A. Green-Hammond, R. M. Inman, K. H. Inman, B. C. Thompson, R. A. Deitner, and H. B. Quigley. 2001. *A study of black bear ecology in New Mexico with models for population dynamics and habitat suitability*. Final Report, Federal Aid in Wildlife Restoration Project W-131-R. Santa Fe: New Mexico Department of Game and Fish.

Cox, J. J., and S. M. Murphy. 2019. Final report to the New Mexico Department of Game and Fish Share with Wildlife Program. Project # 171012. Unpublished report to New Mexico Department of Game and Fish, Santa Fe, NM.

Crenshaw, J. 2002. Making tracks: a century of wildlife management, Part 1. *New Mexico Wildlife* 47(4):7–10.

———. 2003. Making tracks: a century of wildlife management, Part 3. *New Mexico Wildlife* 48(2):7–10.

———. 2004. Making tracks: a century of wildlife management, Part 9. *New Mexico Wildlife* 49(4):7–10.

Evans, W. 1981. *Bobcat study*. Final report. New Mexico Department of Game and Fish, Federal Aid in Wildlife Restoration Project W-124-R-5, Job 1.

Geist, V. 1995. North American policies of wildlife conservation. In *Wildlife conservation policy*, ed. V. Geist and I. McTaggart-Cowan, 77–129. Calgary: Detselig Enterprises.

Gould, M. J., J. W. Cain, G. W. Roemer, and W. R. Gould. 2016. *Estimating black bear density in New Mexico using noninvasive genetic sampling coupled with spatially explicit capture-recapture methods*. Final Report. New Mexico Department of Game and Fish, Federal Aid in Wildlife Restoration Project W-93-R-56, Job 2.

Jones, C., and C. G. Schmitt. 1997. Mammal species of concern in New Mexico. In *Life among the muses: papers in honor of James S. Findley*, ed. T. L. Yates, W. I. Gannon, and D. E. Wilson, 179–205. Albuquerque: University of New Mexico Museum of Southwestern Biology.

Jones, D. 1991. Philmont bear project. Unpublished report. New Mexico Department of Game and Fish, Santa Fe, NM.

Lee, L. 1962. 50 years of hunting. *New Mexico Wildlife* 7(1):6–8.

Ligon, J. S. 1927. *Wild life of New Mexico: its conservation and management*. Santa Fe, NM: State Game Commission, Department of Game and Fish.

Logan, K. A., L. L. Sweanor, T. K. Ruth, and M. G. Hornocker. 1996. *Cougars of the San Andres Mountains, New Mexico*. Final Report, Federal Aid in Wildlife Restoration Project W-128-R. New Mexico Dept. Game and Fish, Santa Fe.

Lorbiecki, M. 1998. The land makes the man: New Mexico's influence on the conservationist Aldo Leopold. *New Mexico Historical Review* 73:235–52.

New Mexico Department of Game and Fish (NMDGF). 2019. State Wildlife Action Plan for New Mexico. New Mexico Department of Game and Fish, Santa Fe, New Mexico, USA.

———. 2020. *New Mexico furbearer law and species identification course*. New Mexico Trapper Education Program, New Mexico Department of Game and Fish.

Organ, J. F., V. Geist, S. P. Mahoney, S. Williams, P. R. Krausman, G. R. Batcheller, T. A. Decker, R. Carmichael, P. Nanjappa, R. Regan, R. A. Medellin, R. Cantu, R. E. McCabe, S. Craven, G. M. Vecellio, and D. J. Decker. 2012. *The North American model of wildlife conservation*. The Wildlife Society Technical Review 12–04. Bethesda, MD: Wildlife Society.

Pederson, J. 1987. A look at 75 years of game and fish history. *New Mexico Wildlife* 32(5):25–34.

Sayre, N. F. 2017. *The politics of scale: a history of rangeland science*. Chicago: University of Chicago Press.

FUTURE CLIMATE-DRIVEN IMPACTS AND THE CONSERVATION OF CARNIVORES IN NEW MEXICO

Jean-Luc E. Cartron, F. Jack Triepke, David S. Gutzler,
Kassidy M. Steckbeck, and Kenneth C. Calhoun

Throughout northern North America, wildlife conservation has had many successes due to the enactment of important government legislation, as well as the creation of protected areas, inter-jurisdictional, national, and international cooperation, and a set of principles guiding the sustainable management of natural resources (Geist 1995; Geist et al. 2001; Organ et al. 2012). In all US states and Canadian provinces, wildlife is considered a public trust for the benefit of all, markets for game species (though not furbearers) have been eliminated, and science, despite important shortcoming, informs management decisions (Bean 1983; Batcheller et al. 2010; Organ et al. 2012). In New Mexico in particular, gone are the days of intensive, unregulated trapping and hunting; bounties placed on predators of game species and livestock; severe overstocking on rangelands; and clearcutting in the mountains for fuelwood consumption and for the railroad and mining industries (see Bailey 1931; Brown 1983, 1985; Scurlock 1998; Chapters 3 and 4).

With the arrival and spread of European settlers, overexploitation, persecution, and habitat loss cost New Mexico some of its native carnivores and other fauna. In addition, although New Mexico has seen more limited human population growth compared to neighboring states (especially during this last decade), many of its ecosystems today have lost their resilience to

Photo 5.1. Wolf Program veterinarian Dr. Susan Dicks assists tribal biologist Theo Guy with a blood draw on a sedated Mexican gray wolf (*Canis lupus baileyi*) in January 2016. The reintroduction of the Mexican wolf in New Mexico and Arizona has relied on inter-jurisdictional cooperation involving the Arizona Game and Fish Department, the New Mexico Department of Game and Fish, the US Department of Agriculture–Animal and Plant Health Inspection Service–Wildlife Services, the US Forest Service, the US Fish and Wildlife Service, and the White Mountain Apache Tribe. Photograph: © Interagency Field Team.

natural and man-made disturbances (e.g., Buffington and Herbel 1965; Hunter et al. 1988; Allen 1989; Dahl 1990; Dick-Peddie 1993; Scurlock 1998; Gibbens et al. 2005; Petrakis et al. 2017). On the one hand, the current human footprint and "legacy effects" from past land use practices in

(*opposite page*) Photograph: © Kari Greer/US Forest Service.

Photo 5.2. Reintroduced recently in New Mexico, the black-footed ferret (*Mustela nigripes*) is listed as Endangered, both under the Endangered Species Act and on the IUCN Red List of Threatened Species. New Mexico has restored once-extirpated species using both wild-to-wild translocations and releases of captive-bred animals. Photo taken on 20 September 2019 in Mora County, New Mexico. Photograph: © James N. Stuart.

New Mexico pale in comparison with the ongoing, industrial-scale deforestation in the tropics, where a collapse of biodiversity is imminent (Barlow et al. 2018). On the other hand, New Mexico is no different from the rest of the world in that it faces new and deepening climate-driven threats likely to affect all wildlife, and in particular the distribution, population status, and ecology of carnivores.

CONSERVATION THREATS AND THE GLOBAL STATUS OF CARNIVORES

Referring to the ongoing, global loss of biological diversity, Diamond (1984, 1989) famously introduced the "Evil Quartet" (also dubbed the "Four Horsemen of the Ecological Apocalypse") as 1) habitat loss and fragmentation; 2) overexploitation and persecution; 3) introduced species (predators, competitors, and vectors or hosts of exotic diseases); and 4) the indirect effects of these threats on ecological interactions (e.g., co-extinctions). To these drivers of biodiversity loss, two others have

been added more recently: 5) climate change (e.g., Graham and Grimm 1990; Gates 1993; Hughes et al. 1997; Kappelle et al. 1999; Hughes 2000; McCarty 2001; McLaughlin et al. 2002), and 6) extinction synergies (Brook et al. 2008; Laurence and Useche 2009; Liess et al. 2016). Extinction synergies can be defined as amplifying feedback loops among pervasive stressors, as, for example, when habitat loss is the initial driver of a population decline but may also facilitate other threats such as hunting access and invasions by non-native species (Brook et al. 2008). Climate change in particular is increasingly reshaping ecosystems by interacting with the more direct impacts of human activities and an altered fire regime (Brook 2008; O'Connor et al. 2011; Le Page et al. 2017; Roos and Guiterman 2021; Hicke et al. 2022).

Of the world's 250 or so extant species of the Carnivora (Wilson and Mittermeier 2009; Hunter 2019), more than 40% are listed as Threatened or Near Threatened on the International Union for the Conservation of Nature (IUCN) Red List, with decreasing population trends and also, in many cases, severely contracted geographic ranges (IUCN 2021). Declining populations are reported in particular for 24 (77%) of the 31 largest carnivores, which are especially at risk as a result of wide-ranging movements (large home ranges and long-distance dispersal), low reproductive rates, low population densities, and a reliance on larger prey to meet their high energy requirements (Ripple et al. 2014). These life history traits both make large carnivores more prone to conflicts with humans and increase their vulnerability to habitat loss, persecution, overexploitation, and declining prey populations (Carbone et al. 1999; Cardillo et al. 2004, 2005). The ability to identify and set aside the large tracts of land needed for the conservation of carnivores and other biodiversity is complicated by climate change–related projections, which despite some uncertainty (e.g., variation in detail among different models and scenarios), include range shifts, contractions, and fragmentations especially in a context of increasing land-cover changes (Parmesan 2006; Mason et al. 2015; Zanin et al. 2021). Climatic niche modeling in the Felidae (cat family) in particular projects range centroid displacements of up to 1,067 km (663 mi) and range contractions of up to 460 km² (178 mi²) by 2080–2100 (Zanin et al. 2021).

GLOBAL CLIMATE CHANGE

The global models used for our analysis of impacts on carnivores were generated as part of the CMIP5 modeling effort that contributed to the fifth assessment (AR5) of global climate change carried out by the Intergovernmental Panel on Climate Change (IPCC 2013). These global simulations have been used for hundreds of research papers and operational climate projections since then. The general picture of anticipated climate change over the next several decades includes several robust, well-understood features that tend to recur in projections of 21st-century climate change forced by increasing greenhouse gases, as confirmed in newer CMIP6 model projections in the recently released Sixth IPCC Assessment (AR6, IPCC 2021).

- Increasing temperature. Increasing greenhouse gas concentration (carbon dioxide CO_2, methane CH_4, and several other gases) increases the atmospheric absorption of heat emitted from Earth's surface. The atmosphere re-radiates this energy in all directions, including downward, leading to increasing surface temperature. This effect has been understood since the 19th century, and it is simulated in all climate models. CO_2 and CH_4 are long-lived and well-mixed in Earth's atmosphere, so the increases in their concentrations due to human-caused emissions tend to be globally distributed. Less well-understood feedback effects modify the direct surface-warming effect of increasing greenhouse gases, thereby introducing uncertainty in the magnitude and geographical variability of the climate response.

- Amplified temperature increase over continental interiors. Oceans have a very large heat capacity, because energy absorbed near the surface is mixed through a relatively deep layer of water, and the molecular properties of water itself (its specific heat) are such that more energy input is required to raise the water temperature compared to a land surface. So while higher greenhouse gas concentrations induce global warming, the changes in surface temperature are generally largest on land, far from the moderating effects of the ocean. Temperature in New Mexico, far from any ocean, is already observed to be increasing at roughly double the rate of the global average warming (e.g., Tebaldi et al. 2012; Hoerling et al. 2013; USGCRP 2017; IPCC 2013, 2021).

- Diminished snow and sea ice. As temperatures rise globally, mid-latitude storms from fall through spring drop a greater fraction of their precipitation as rain instead of snow, the elevation of the snow line rises, and the snowpack and ice that have formed on the surface melt earlier in spring. Diminished snowpack in subtropical and middle latitude regions, such as the western United States, has already been observed and is one of the most confident projections derived from climate models as greenhouse gases increase and temperature rises (USGCRP 2017; IPCC 2013, 2021).

- Poleward expansion of tropical and subtropical climate zones. A robust feature of a globally warming climate is the poleward expansion of the tropics and subtropics, and associated contraction of cold polar regions. In addition to being warm regions, subtropical latitude zones (generally between about 20° and 35° latitude) include many of the world's desert regions. As the subtropics of the Northern Hemisphere expand, the zone of winter storm development along the boundary separating tropical and polar air masses tends to shift northward (Wang et al. 2017). Across North America, this means that many simulations of future climate exhibit a northward shift in winter precipitation, so that northern Mexico gets less, while southern Canada gets more (IPCC 2013, 2021). New Mexico is located near the transition zone of this northward shift, but a majority of CMIP5 climate model simulations exhibit a decrease in winter and spring season precipitation across New Mexico (Dunbar et al. 2022).

- Sensitivity to the assumed future emissions scenario. The rate of future climate change will depend on international energy policy choices that would modify greenhouse gas emissions. For the first two decades of the 21st century, global atmospheric CO_2 concentration has been increasing at a rate averaging about 0.57% per year to its present (2020) annually averaged value of 412 ppm. The recent observed rate of emissions increase corresponds to something close to a mid-range scenario for future projections. In a higher emissions future, greenhouse gas emissions continue to increase significantly. A lower emissions scenario can be envisioned based on the presumption of future declines in emissions, due to effective international mitigation policy and more aggressive development of non-fossil fuel energy. The projections shown here (described in more detail in Appendix 5.1) are developed using global models driven by a high-emissions scenario (RCP 8.5) that assumes minimal emissions mitigation, so these simulations represent pessimistic projections of future climate change. The simulations can be interpreted as a worst-case scenario that project the highest risks to humans and wildlife associated with rapid, high-magnitude 21st-century climate change. Global model output has been downscaled statistically to finer horizontal resolution to provide more realistic estimates of temperature and precipitation change appropriate for New Mexico's sharp topographic gradients (see Appendix 5.1).

CLIMATE CHANGE IN NEW MEXICO

Temperature has been increasing rapidly, considerably exceeding global or continental average rates of change as noted above. Over the past half-century annual-average temperature has risen across New Mexico by approximately 3° F, or between 1° and 2° C (Dunbar et al. 2022). Warmer future temperature, a direct physical consequence of an increase in the Greenhouse Effect (amplified by less well-constrained feedback effects), is the most robust projection of climate change in the 21st century (Maps 5.1a and b; and see Appendix 5.1).

Snowpack has declined in the high-elevation headwaters of major rivers flowing through New Mexico. As discussed above, diminishing snowpack is an expected consequence of rising temperature in many parts of the world. New Mexico and southern Colorado, at the southern edge of continental snowpack in North America, are particularly vulnerable to temperature-related snowpack declines as climate gets warmer over the next half-century. Pronounced decreases in snowpack in the headwaters of the Rio Grande and throughout the western United States have already been thoroughly documented (Fig. 5.1; Mote et al. 2018; Chavarria and Gutzler 2018).

Precipitation is extremely variable from one year to the next across southwestern North America. Most simulations exhibit some rate of long-term winter precipitation decrease toward the southern part of the state, with less consistency among simulations closer to Colorado (IPCC 2013, 2021; USGCRP 2017; Map 5.2). Declines in spring season precipitation are shown in most simulations (IPCC 2013, 2021; USGCRP 2017; Dunbar et al. 2022). Significant long-term trends in historical total precipitation (rain and snow) are difficult to detect amid high

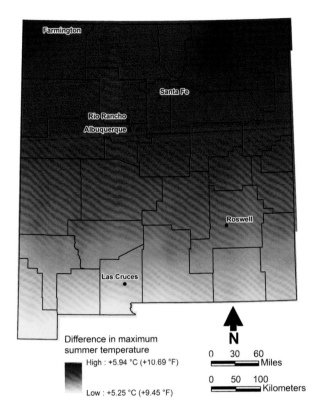
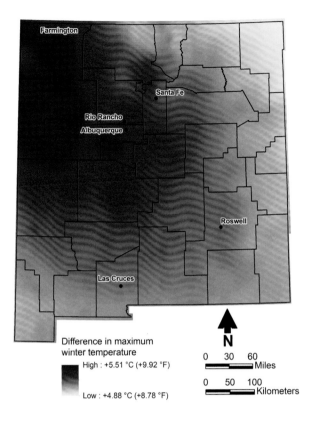

Maps 5.1a and b (*left to right*). Temperature change projections for New Mexico by approximately 2085 (period 2070–2099), averaged over multiple climate model simulations and compared to 1971–2000. (a) Mean maximum summer (June–August) temperatures are expected to rise by 5.2° C (9.4° F) in the southwestern quadrant of the state and by up to 5.9° C (10.6° F) in the northwest and northeast. (b) Mean maximum winter temperature is projected to increase by up to 5.5° C (9.9° F) in the northwestern quadrant of New Mexico. Note that although warmer temperature across New Mexico is a general feature of 21st-century climate change, the quantitative rate of change and the geographical and seasonal distributions of temperature change exhibit considerable variation from one simulation to another.

inter-annual variability in observed time series, though the past two decades have been unusually dry (Williams et al. 2020). Flows in New Mexico rivers and streams have been low during this period, a consequence of generally low (but quite variable) precipitation and exceptionally warm temperatures. Under all emissions scenarios, streamflow reductions are projected in the Rio Grande watershed (Fig. 5.2) when averaged over many independent future model simulations, though the range of projected change among different model simulations is quite large.

The combination of increasing temperature

and increasing variability of precipitation means that *extended drought episodes*, which in the past have been caused by persistent precipitation deficits, are projected to become more severe in future decades (Williams et al. 2022). Even during periods of near-average precipitation, warmer temperatures from late spring through early autumn promote increased evaporation from moist surface or open water, resulting in diminished water resources and making the landscape more arid (Dunbar et al. 2022). The trend toward aridity (increased evaporation and surface dryness) is exacerbated in mountainous

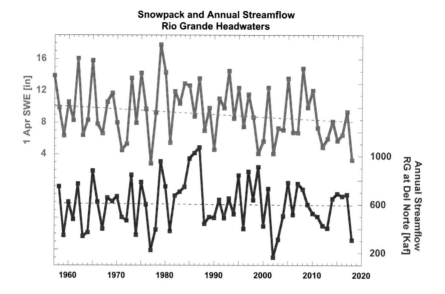

**Snowpack and Annual Streamflow
Rio Grande Headwaters**

Figure 5.1. (top, green curve) Annual values of observed snowpack on 1 April in the Rio Grande headwaters basin, southern Colorado. Values are expressed as Snow Water Equivalent or SWE (inches of liquid water). The first of April is generally close to the annual peak of snowpack in the basin. (bottom, blue curve) Annual observed volume of water flowing past the stream gage on the Rio Grande at Del Norte, CO, at the outlet of the headwaters basin (thousands of acre-feet). Linear trend lines, calculated over the period 1958–2018, are shown as light dashed lines. The downward trend in snowpack is statistically significant; the trend in streamflow is not significantly different from zero. Adapted from Chavarria and Gutzler (2018).

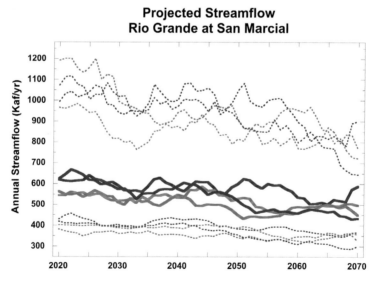

**Projected Streamflow
Rio Grande at San Marcial**

Figure 5.2. Time series of simulated future annual streamflow (thousands of acre-feet/year) on New Mexico's middle Rio Grande (between Cochiti Dam and Elephant Butte Reservoir) at San Marcial, derived using climate models coupled to a surface hydrologic model (Reclamation 2013). Annual simulated flow values have been adjusted to account for the water removed for beneficial use upstream from San Marcial. Each of the four colored curves represents an average of 16 simulations, forced by one of four greenhouse gas emissions scenarios (blue = lowest, red = highest emissions). Each thick line is the average of 16 simulations. The thin dashed lines show the low and high extremes of annual streamflow among the 16 simulations, indicating the natural variability among the simulations. Each plotted line has been smoothed to emphasize decadal and longer variability. Data from Townsend and Gutzler (2020).

regions by the reduction in both the length of the snow season and the snow-covered area, because snow cover keeps the underlying surface moist.

Another consequence of warmer temperatures and a trend toward aridity is the likelihood of *more frequent and intense wildfires*, especially during the spring dry season. Forests in New Mexico are projected to undergo intense drought stress in the 21st century, especially if a high-emissions future (with associated rapid warming) comes to pass. Nearly complete elimination of many stands of pine and fir forests in New Mexico are a feature of high-emissions projections (Williams et al. 2012; see below).

Some important features of the climate system exhibit only modest or uncertain systematic changes in CMIP5 global model simulations. Changes in total summer precipitation across southwestern North America are uncertain in CMIP5 (USGCRP 2017) and CMIP6 (IPCC 2021) simulations; a somewhat more consistent projection is that the onset of summer precipitation at the end of the spring dry season will be delayed in a warmer climate (Seth et al. 2011; Cook and Seager 2013).

ECOSYSTEM VULNERABILITY AND CLIMATE-DRIVEN IMPACTS

Warming trends and increased aridity are likely to change the amount and distribution of New Mexico's ecosystem types along with the manner in which such types structurally respond to disturbance and the services they provide (Gutzler 2013). Chapter 2 provides an overview of the state's major ecosystem types, their characteristics and geography, and their fire regimes. Here we look at groupings of these types and predictions of range contractions and expansion under 21st-century climate trends (Table 5.1). While it is possible to provide mapped predictions of individual ecosystem types, greater certainty can be provided for more general vegetation types.

Map 5.3 shows the current and projected distribution of general vegetation types in New Mexico.

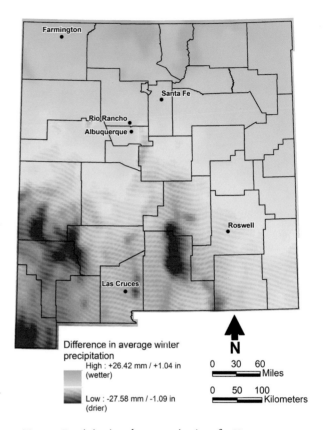

Map 5.2 Precipitation change projections for New Mexico by approximately 2085 (period 2070–2099), averaged over multiple climate model simulations and compared to 1971–2000. Most simulations show some rate of winter precipitation decrease toward the southern part of the state.

The current distribution represents a generalized 2020 map rendering of ecosystem types (Wahlberg et al. 2020), while projected conditions were based on a 2090 forecast of a 30-year climate model (2061–2090) and the climate vulnerability estimated for each plant community (Triepke et al. 2019). Differences between the current and future distributions represent areas of significant departure from the historic climate envelope for a given ecosystem type at a given location. The contrast in distribution suggests remarkable change in familiar ecosystem patterns including significant loss in the extent of forests and woodlands and a substantial increase in desert shrub (Table 5.1). This

Photo 5.3. Aerial view of the Thompson Ridge fire in the Jemez Mountains in June 2013. Photograph: © Bernard R. Foy

Photo 5.4. The Whitewater-Baldy Complex fire burned 120,534 ha (297,845 acres) in 2012, mostly within the Gila Wilderness. It replaced the 2011 Las Conchas fire as the largest wildfire in New Mexico state history. In 2022, the Whitewater Baldy Complex fire was eclipsed by two even larger fires: the Black fire (131,577 ha [325,133 acres]) and the Calf Canyon Hermit's Peak fire (138,188 ha [341,471 acres]). Photograph: © Kari Greer/US Forest Service.

Photo 5.5. Burn scar photographed in 2019 near Emory Pass in the Black Range, six years after the Silver Fire, which burned 56,129 hectares (138,698 acres) of forest land in 2013. Some trees survived the fire, but note the lack of trees sprouting in the burn scar. High burn severity promotes conversion to non-forested type such as grassland, oak scrub, and weedy, herbaceous-dominated vegetation. Photograph: © Jean-Luc Cartron.

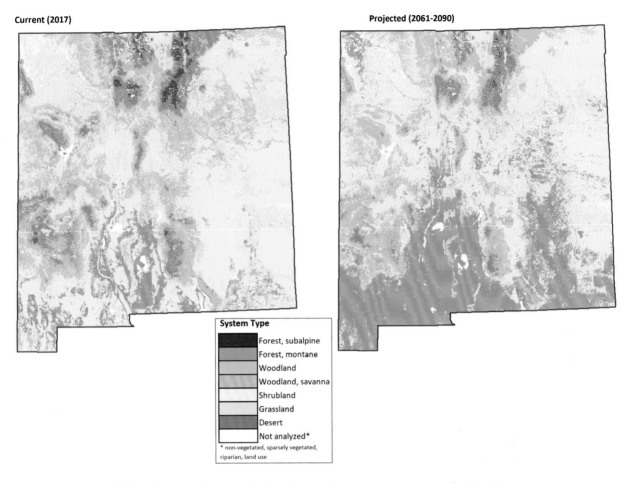

Current (2017)

Projected (2061-2090)

System Type

- Forest, subalpine
- Forest, montane
- Woodland
- Woodland, savanna
- Shrubland
- Grassland
- Desert
- Not analyzed*

* non-vegetated, sparsely vegetated, riparian, land use

Map 5.3 Current (left) and projected 2090 (right) distribution of general vegetation types for New Mexico.

forecast does not factor disturbance agents, which could accelerate anticipated changes.

Map 5.4 shows the geographic pattern of uncertainty in New Mexico that was estimated for the projections in Table 5.1 and Map 5.3. This uncertainty assessment represents the level of agreement in climate vulnerability predictions among multiple global climate model forecasts (Triepke et al. 2019). Uncertainty tends to be highest in mountainous terrain, and the least uncertainty occurs in the basins and plains of the state.

Shifts in vegetation types are a concern due to the potential for abrupt change with severe disturbance such as high-severity wildfire and the associated impacts to wildlife habitat and other services (Keyser et al. 2020). Of particular concern are the individual, combined, and synergistic effects of wildfire, drought, and warmer conditions on forests and woodlands (Breshears et al. 2009; Allen et al. 2010). Concerns are heightened in New Mexico and the Southwest due to increased fire activity and the compounding effect of fire exclusion and unnaturally dense forests and abundant fuels (Westerling et al. 2006). Estimates in Table 5.1 suggest a late-century loss in forest and woodland cover of over 30%, consistent with other reporting indicating losses in excess of 34% (Parks et al. 2019).

Table 5.1. List of general system types found in New Mexico and depicted in Figure 5.5, cross-referenced to the ecosystem types of Chapter 2.

General Vegetation Type	Ecosystem Type (Ch. 2)	Current		Projected 2090	
		Area (ha)	%	Area (ha)	%
Shrubland/mixed (alpine)	Alpine and Tundra	4,012	0.01%	0.0%	0.0%
Forest, subalpine	Spruce-Fir Forest	321,456	1.0%	41,546	0.1%
Forest, montane	Mixed Conifer Ponderosa Pine Forest	2,891,483	9.2%	1,431,648	4.5%
Woodland	Pinyon-Juniper Woodlands Madrean Woodlands	3,813,615	12.1%	2,171,456	6.9%
Woodland, savanna	Juniper Grass (savanna)	2,998,672	9.5%	3,329,152	10.6%
Shrubland	Gambel Oak Shrubland Mtn Mahogany Mixed Sagebrush Shrubland	2,959,730	9.4%	2,021,937	6.4%
Grassland	Montane/Subalpine Grassland Colo Plateau/Great Basin Grassland Semi-desert Grasslands Great Plains (Shortgrass Prairie)	15,679,136	49.7%	12,942,966	41.1%
Shrubland, desert	Chihuahuan Desert Scrub	2,455,798	7.8%	9,185,198	29.1%
Other (not analyzed)	Riparian, other	393,673	1.2%	393,673	1.2%
		31,517,574		31,517,74	

PROJECTED CLIMATE-DRIVEN IMPACTS ON CARNIVORES IN NEW MEXICO

Worldwide, climate-driven impacts on carnivores are expected to include 1) poleward and upslope distribution shifts; 2) contraction and disruption of habitat; 3) lower availability of water and other key resources; 4) reduced prey populations; 5) increased interspecific competition; and 6) higher human-wildlife conflict (e.g., Elmhagen et al. 2017; Trouwborst and Blackmore 2020). Many wildlife species are already shifting their distributions toward the poles and to higher elevations in response to climate change (Parmesan and Yohe 2003; Tingley et al. 2009; Scheffers et al. 2016), though range boundaries are moving at rates that are extremely variable (Lenoir et al. 2010; Angert et al. 2011; Crimmins et al. 2011; MacLean and Beissinger 2017; Freeman et al. 2018). Because New Mexico's life zones are structured primarily on the basis of elevation rather than latitude (Chapter 2), upslope displacements of mammal species distributions in the state might arguably prove a more measurable biological response to climate change compared to northward range shifts.

The outlook for many species may hinge on

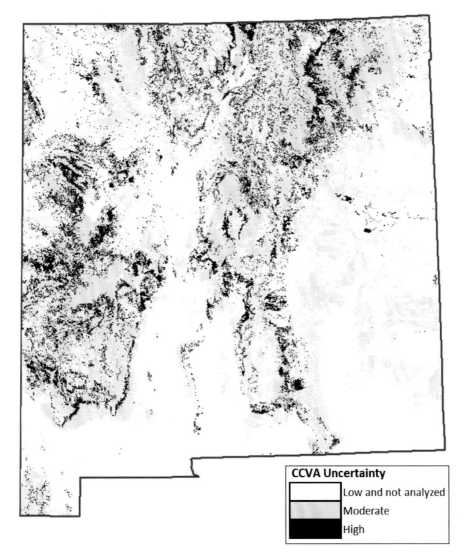

CCVA Uncertainty

Low and not analyzed

Moderate

High

Map 5.4 Uncertainty in the 2090 projection of ecosystem type distribution in New Mexico.

whether 1) they are habitat generalists vs. specialists; 2) their dispersal abilities will allow them to fully track their respective climatic niches; and 3) they can sustain the level of predation pressure and interspecific competition in their new geographic ranges. Hof et al. (2012) found that under a full dispersal-ability scenario in the subarctic regions of Europe, range contractions were projected only in a few mammals, primarily the cold-climate specialists such as the wolverine

(*Gulo gulo*). For the others, ranges were predicted to expand, though predation pressure and interspecific competition were likely to increase.

With many species moving poleward or upslope, it is not clear whether New Mexico's fauna would actually gain any new carnivore species arriving from the south, even in the absence of a border wall preventing cross-border wildlife migration (discussed below). López-González and García-Mendoza's (2012) inventory of mammals

Table 5.2. Examples of projected changes in the extent of preferred/primary habitat for several carnivores in New Mexico. Species associated with riparian areas were not analyzed but are likely to be seriously impacted by climate change.

Species	Ecosystem Types Representing Preferred or Primary Habitat	Approx. 2017 Extent (ha)	Approx. Projected 2090 Extent (ha)	Projected 2090 % Change
Canada lynx (reintroduced population)	Subalpine Forest	305,000	40,000	-87%
Jaguar	Madrean Woodlands	270,000	125,000	-53%
Wolf (historical population)	Madrean Woodlands and Montane Forest	3,065,000	1,545,000	-49%
Gray fox	Madrean Woodlands, Pinyon-Juniper Woodland, Ponderosa Pine Forest, and Ponderosa Pine– Evergreen Oak	3,240,000	1,025,000	-68%
Kit fox	Deserts and Valley and Basin Shrublands and Grasslands	15,390,000	18,535,000	20%
Swift fox	Shortgrass Prairie	5,965,000	780,000	-87%
Black bear	Subalpine and Montane Forests (primary habitat)	2,945,000	820,000	-72%
Wolverine (extirpated)	Subalpine Forest and Alpine and Tundra	270,000	40,000	-85%
Pacific marten	Subalpine and Mixed Conifer Forests	660,000	465,000	-30%
White-nosed coati	Madrean Woodlands and Mogollon Chaparral	270,000	125,000	-53%

in Chihuahua, just across the border with Mexico, does not list any carnivores not already present in New Mexico. The ocelot (*Leopardus pardalis*) has been recorded both in Arizona (wandering males only) and Texas (Haines et al. 2006; Avila-Villegas and Lamberton-Moreno 2013; Culver et al. 2016). It has never been documented in New Mexico (and neither has the jaguarundi [*Puma yagouaroundi*]); all Arizona ocelots are dispersers from a breeding population located ~50 km (30 mi) south of the Mexico border in northern Sonora (Rorabaugh et al. 2020). While most ocelot records in Arizona have originated from oak and pine-oak woodland (López-González et al. 2003), ecosystem types also found in New Mexico, the likelihood of the species establishing itself in the Land of Enchantment seems low. Not only do the tropical vegetation communities preferred by the ocelot not occur in

New Mexico (more generally in the Southwest), but pine-oak and pine woodlands are likely to see their extent diminish as a result of the changing climate (see Table 5.1).

Based on the above projections of higher temperatures, increased aridity, reduced streamflow, northward expansion of deserts, and contraction of woodlands and forests, we can make a number of predictions about New Mexico's carnivores and their respective habitats. Some projected trends are based directly on future climatic conditions, others on river flows and on the forecast of 2090 ecosystem type extent described above. As with the rest of the world, land-cover changes (e.g., increased urbanization) and land management are likely to also play an important role. Feedback loops also exist, as with the case of urbanization: urban areas act as heat sinks, which can amplify warming-related environmental effects at a local scale (see Santamouris 2015; Zhao et al. 2016; Aflaki et al. 2017; Mohajerani et al. 2017).

Habitat generalists such as the striped skunk (*Mephitis mephitis*) were not analyzed, but they may rank among the projected "winners" in the state, particularly as they might be able to range farther up in the mountains. Another species that might benefit from climate change in New Mexico is the kit fox (*Vulpes macrotis*), uniquely adapted to deserts, projected to increase in extent (Table 5.2). Conversely, the swift fox (*Vulpes velox*), found only on the eastern plains of New Mexico, would see its shortgrass and mixed-grass prairie habitat reduced drastically in the state. Table 5.2 lists some of the carnivores historically or currently present in New Mexico and projected changes in the extent of their preferred or primary habitat using climate-driven ecosystem type projections. For those species, ecosystem types are believed to represent prime indicators for assessing habitat quality or suitability.

Snow-adapted, Northern Species

New Mexico lies at the extreme southern end of the distribution of four snow-adapted carnivores. Two of them, the Canada lynx (*Lynx canadensis*; Chapter 6) and the wolverine (Chapter 18), are primarily associated with boreal forests (taigas) across Alaska and Canada. The American ermine (*Mustela richardsonii*; Chapter 21) occurs in the taiga biome, while the Pacific marten (*Martes caurina*; Chapter 20) occupies Pacific coastal areas and Sierra Nevada and Rocky Mountain coniferous forests. With its very tall peaks and prolonged winter snowpack, the north-south trending Rocky Mountains allow all four of these species (in addition to the Rocky Mountains clade of North American red fox [*Vulpes fulva*], also snow-adapted) to extend their distribution far to the south while still remaining within their climatic niche and finding vegetation both structurally similar to the boreal forest and related to it at the generic level (*Picea* and *Abies* spp.).

Photo 5.6. Wolverine (*Gulo gulo*) photographed at McNeil River State Wildlife Sanctuary in Alaska. The wolverine is a snow-adapted species that mostly occurs at high latitudes. Historically, it reached the southern edge of its distribution in the southern Rocky Mountains. Photograph: Drew Hamilton; © Alaska Department of Fish and Game.

A reduced snowpack, the upward shift of lower montane forest types, an altered fire regime, and increased competition with generalist predators moving upslope all represent climate-related threats to the Canada lynx, wolverine, American ermine, and Pacific marten. For the wolverine in particular, the extent of snow cover persisting through the end of the spring denning period may represent the single most important habitat requirement (see Chapter 18), and trends in the observed length of the snow cover season are a source of concern in the United States (Figure 5.3).

According to our projections of ecosystem type change, alpine tundra and subalpine forest—the main vegetation communities associated with wolverines in the southern Rocky Mountains—will be reduced by a combined 85% by 2090 in New Mexico's Sangre de Cristo and San Juan mountains. Lynx reintroduced in Colorado have ranged into the Sangre de Cristo, San Juan, Jemez, and Chuska mountains of northern New Mexico (see Chapter 6), and within those mountains, subalpine forest (where the lynx mainly occurs) would

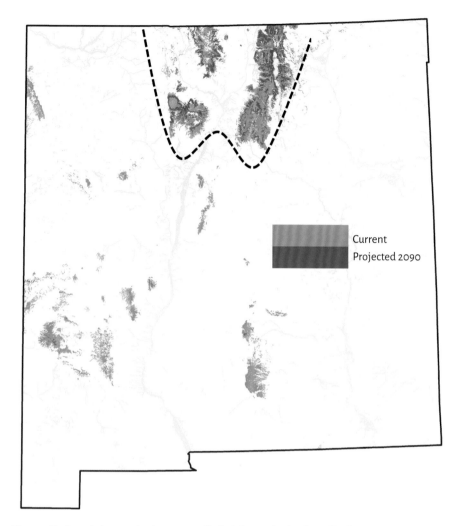

| | Current |
| | Projected 2090 |

Map 5.5 Projected changes in the extent of subalpine and mixed conifer forests in New Mexico. The dashed line represents the southern limit of the verified or probable distribution of both the Pacific marten (*Martes caurina*) and the American ermine (*Mustela richardsonii*), two species restricted to those two forest types.

similarly decrease by 85%. Within the Pacific marten's known current range in New Mexico (Sangre de Cristo and San Juan Mountains; the species is also suspected to occur in the Jemez Mountains), the extent of spruce-fir and mixed conifer forests (the main vegetation communities where the species occurs) would be reduced by a combined 30% (Map 5.5). Although the ermine's narrower ecological niche does not allow for projections based on changes in ecosystem type extent, its habitat

(with winter snow cover and proximity to perennial water as key constituents), would likely also contract.

Forest and Woodland-adapted Species

In New Mexico, black bear (*Ursus americanus*) populations are likely to experience declines as their primary forest habitat shrinks by a projected 70% by the late 21st century (Map 5.6). However, population declines are likely to vary by region (see

Snow Cover Season in the United States, 1972–2013

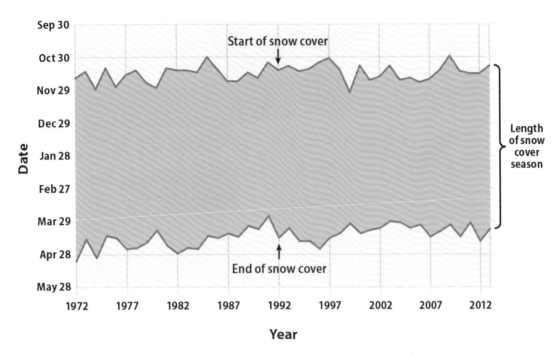

Figure 5.3. Length of the snow cover season in the United States, 1972–2013. Note the trend toward earlier dates for the end of the snow cover season. A reduced snow cover season is likely to impact snow-adapted species such as the wolverine (*Gulo gulo*), Canada lynx (*Lynx canadensis*), Pacific marten (*Martes caurina*), and American ermine (*Mustela rechardsonii*), in addition to the Rocky Mountain clade of the North American red fox (*Vulpes fulva*). Data source: National Oceanic and Atmospheric Administration 2015, ftp://eclipse.ncdc.noaa.gov/pub/cdr/snowcover.

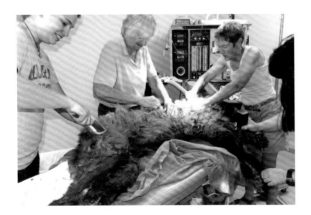 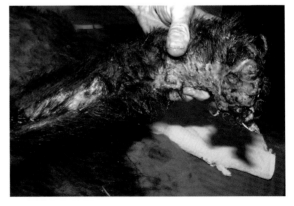

Photos 5.7a and b (*left to right*). Adult female black bear (*Ursus americanus*) treated by veterinarian Dr. Kathleen Ramsay at the New Mexico Wildlife Center in Española in 2011. The bear sustained second and third-degree burns during the Las Conchas fire, which burned a total of 624.6 km² (154,349 acres) in northern New Mexico. Photographs: © Katherine Eagleson.

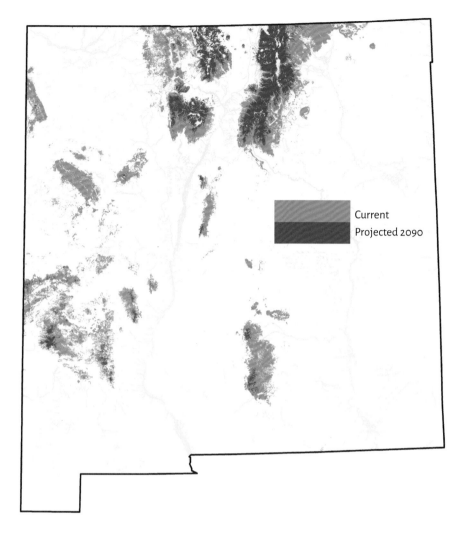

Current

Projected 2090

Map 5.6 Projected changes in the extent of montane forests within the habitat regions of the black bear (*Ursus americanus*) in New Mexico.

Chapter 16) and in relation to the occurrence of high-severity fires acting as a catalyst in the conversion of forest into non-forest type vegetation. A study in the southern Mazatzal Mountains of central Arizona (Cunningham and Ballard 2004) offers a preview of potential negative impacts on black bear demographics from direct fire mortality and from post-fire food and vegetation cover shortages. Cunningham and Ballard's (2004) research was conducted immediately after a large

wildfire (the Lone Fire) destroyed >90% of the vegetation in an area totaling 237 km² (58,564 acres) in April and May 1996. The local black bear population had been studied in a prior study (LeCount 1982), allowing for comparisons between pre- and post-fire (in addition to within vs. outside the burn area) demographics. A sharp decrease in the number of adult females and female and male subadults was documented post-fire (Cunningham and Ballard 2004). The drop in the number

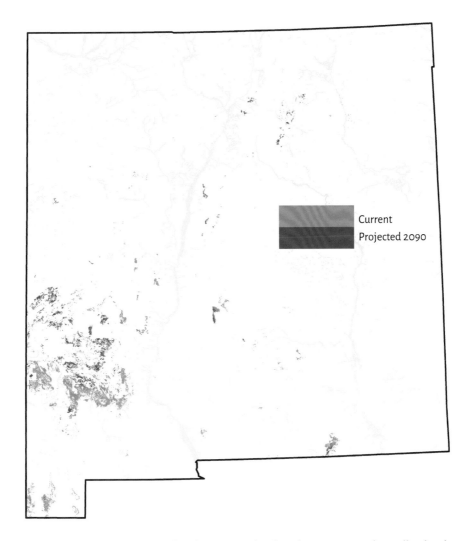

Current
Projected 2090

Map 5.7 Projected changes in the extent of Madrean encinal and Madrean pinyon-oak woodland and Mogollon chaparral—ecosystem types representing a large portion of the white-nosed coati's (*Nasua narica*) habitat—in New Mexico.

of adult females—known to exhibit strong site fidelity (see Chapter 16)—indicated either direct mortality from the fire (see Photos 5.7a and b) or emigration prompted by a shortage of food or other resources. The lack of subadults was linked to low post-fire recruitment to the yearling age class (cub production was unchanged, but cub survival appeared to have plummeted post-fire).

Other carnivores showing some degree of preference for woodlands and forests include the gray fox (*Urocyon cinereoargenteus*; Chapter 12) and the coati (*Nasua narica*; Chapter 31). The gray fox occurs throughout most of New Mexico, where the extent of pinyon-juniper and Madrean encinal, as well as Madrean pinyon-oak woodlands, ponderosa pine, and ponderosa pine-evergreen oak forest—representing preferred habitats—is projected to decrease by approximately 65%. The coati may occur primarily in Madrean encinal and Madrean

Photo 5.8. Two northern river otters (*Lontra canadensis*) at the Taos Junction Bridge in the Orilla Verde (in the Rio Grande del Norte National Monument), New Mexico on 20 October 2015. Photograph: © Geraint Smith.

pinyon-oak woodland and Mogollon chaparral within the southwestern quadrant of the state. Those ecosystem types are projected to decrease collectively by 50% (Map 5.7).

Semi-aquatic Species and Wetland Obligates

The northern river otter (*Lontra canadensis*) likely became extirpated in New Mexico sometime in the mid-to-late 20th century (Chapter 19). However, reintroductions in New Mexico and surrounding states have resulted in the river otter making a comeback in two watersheds, the upper Rio Grande and the San Juan River. In the upper Rio Grande Basin, where the reintroduced river otter population seems firmly re-established (Chapter 19), climate model simulations project declining streamflow (Figure 5.4), which could in turn negatively impact the river otter, a semi-aquatic carnivore.

Based on projections derived from CMIP5 climate models, Bjarke and Gutzler (2023) estimated that the average (median) upper Rio Grande spring season flow would decline by 14% over the 21st century, as defined by a straight-line trend fit to the thick blue line in Figure 5.4. There is a very wide envelope of uncertainty among model simulations surrounding the overall average decline quoted above. The 14% median decline is affected by a relatively small number of simulations in which headwaters precipitation increases strongly, compensating for the decline in flow associated with increasing temperature that is ubiquitous in these simulations. These higher-flow simulations, represented by the upper thin blue line, exhibit no 21st-century trend in headwaters streamflow (insignificant 5% decline). Models that simulate unchanging or declining average winter precipitation tend to exhibit much more sharply declining streamflow, as shown by the lower thin blue line (representing

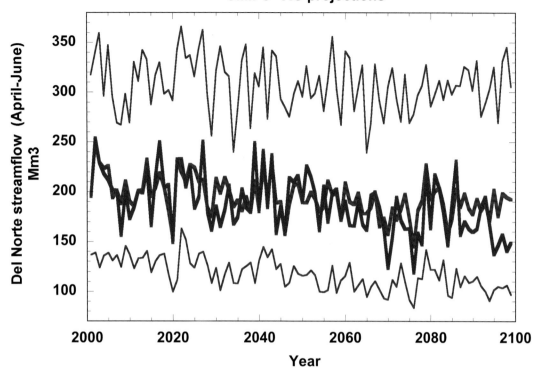

**Rio Grande Headwaters Snowmelt Runoff
CMIP5+VIC projections**

Figure 5.4. Time series of model-simulated streamflow (mm³) out of the Rio Grande headwaters during the snow-melt runoff season (April–June) for years 2001–2099, adapted from Bjarke (2019). Simulations were produced by the US Department of Interior, Bureau of Reclamation (Reclamation 2013) using an ensemble of 97 individual CMIP5 global climate model simulations driven by a wide variety of greenhouse gas forcing scenarios ranging from low-emissions (RCP2.6) to high-emissions (RCP8.5) futures. The thick blue line denotes the median snow-melt runoff season flow among all 97 simulations. Thin blue lines represent the 25% and 75% limits of the distribu-tion of flow values among these simulations, an indication of the spread of the 97 simulations from lower flows to higher flows. The thick red line denotes the median flow among the subset of 45 simulations that are based on just the higher-emissions futures (RCP6.0 and RCP8.5).

lower-flow simulations) that declines by 25% over the 21st century.

If we consider just the simulations associated with a high-emissions future, the median decline in streamflow is 23% (thick red line), similar to the lower-flow simulations derived from the entire ensemble of future simulations. So projected Rio Grande streamflow is sensitive to whether a low-emissions or high-emissions scenario is pre-scribed, declining more in a future with higher greenhouse gas emissions and greater global climate change. Spring season snowmelt run-off comprises the largest component of annual streamflow in the Rio Grande headwaters, and it shows the clearest climate change signal (because rising temperature leads to reduced snowpack). Simulation of the other months is less certain and does not affect trend estimates much, but the projections become somewhat noisier.

Furthermore, projected peak snowmelt runoff

in a warmer climate occurs earlier in the spring season, as more spring precipitation falls as rain rather than snow and the sparser snowpack melts earlier (Moeser et al. 2021). The trend toward earlier snowmelt runoff is already observed (IPCC 2021) and is projected to cause the spring dry season in New Mexico to become considerably drier in future decades, especially in high-elevation catchments where snowpack declines.

Overall, climate change may lead to a 10% to 16% expansion of the northern river otter's geographic range as more areas become climatically suitable at high latitudes in North America (Cianfrani et al. 2018). At the southern end of the species' distribution, however, some areas (e.g., Gulf Coast region) may be lost as they are climatically no longer suitable (Cianfrani et al. 2018). A future reduction in spring streamflow of 23%–25% in the upper Rio Grande, as suggested above, would likely impact river otters at a time when they are giving birth to their young and raising them. Everything else being equal, lower fish abundance would be expected in the spring in the upper Rio Grande simply from lower habitat volume, with concomitant elevated temperatures (due to reduced thermal capacity; e.g., Webb et al. 2003; Van Vliet et al. 2011) and lower dissolved oxygen (the solubility of oxygen in water decreases as temperature rises), likely representing additional impacts. The potential for conflict between river otters and fishermen seeking to protect fish stocks would also likely increase (see Chapter 19).

Worldwide, wetlands have already been severely impacted by water diversions and the regulation of river flows (Kingsford 2009). New Mexico in particular has already lost over one-third of its historical wetlands from a combination of factors ranging from agricultural conversion and water diversions to overgrazing and urbanization (Yuhas 1996), with trapping of beavers another important factor along smaller rivers and headwater streams (Chapter 24). The American mink (*Neogale vison*), last recorded in 1987 in New Mexico, is a wetland obligate (Chapter 24). Although the return of minks to New Mexico cannot be ruled out, increasing temperatures and decreasing snowmelt runoff will further place wetlands (and riparian areas) under stress, with additional impacts possible from changes in the timing, magnitude, and frequency of precipitation (see Bates et al. 2008; Milliman et al. 2008; Klausmeyer and Shaw 2009; Palmer et al. 2009; Viers and Rheinheimer 2011). Although not wetland obligates, both the American ermine (Chapter 21) and the raccoon (*Procyon lotor*; Chapter 32) represent two additional species that could be affected by reduced waterflow and a decline in the extent of wetlands and riparian vegetation. The ermine may well represent a poster child for climate change impacts in New Mexico given that it requires both snow and riparian zones.

Pathogens, Vectors, and Transmission Dynamics

Wild carnivore populations including those in New Mexico are likely to face increasing threats from climate-induced changes in host-pathogen-vector interactions and the always-important risk of cross-species spillovers from domestic dogs (*Canis familiaris*) and domestic cats (*Felis catus*) (Costanzi et al. 2021; and see Chapter 33). Pathogens (e.g., bacteria, viruses, and protozoa) and vectors (e.g., mosquitoes, ticks, fleas) are small organisms unable to regulate their internal temperature, and their survival and development thus tend to be highly dependent on local climate (Patz et al. 2003; Lata et al. 2018). With some infectious diseases already shifting their distribution in response to warmer temperatures and other drivers, new epidemic infections (e.g., dispersal of pathogen into new areas, emergence of new strains) and patterns of transmission (e.g., host shifts) are increasingly likely as climate change accelerates (Patz et al. 2003; Rosenthal 2009; Lata et al. 2018; Rohr and Cohen 2020). In the case of the West Nile virus (WNV), known to infect many bird species but also mammals, warmer temperatures are associated with higher viral replication rates and growth rates of mosquito vector populations (Paz 2015; Hicke et al. 2022). Warmer temperatures also correlate with shorter intervals between blood meals and incubation time in mosquitoes. Precipitation plays another important (yet more complex) role, as it influences mosquito densities likely through the availability of standing water (Paz 2015; Hicke et al. 2022). Recent climatic changes (warmer temperatures, more frequent and more intense heatwaves) have been implicated in the establishment of WNV in new parts of the world including North America (Paz 2015). Mammals are considered dead-end hosts of WNV (Ahlers and Goodman 2018), but exposure to the virus has nonetheless been documented in a number of mammals including the black bear, raccoon, and striped skunk in North America, with infections also reported from other parts of the world in the genera *Vulpes* and *Martes*, both represented in North America (including in New Mexico) (Root and Bosco-Lauth 2019). WNV seroprevalence (the proportion of individuals showing exposure to the disease based on the presence of antibodies) was high in a small number of raccoons tested during a study in Louisiana (Dietrich et al. 2005), and severe disease is possible in carnivores (Root and Bosco-Lauth 2019). A polar bear (*Ursus maritimus*) held captive at the Toronto Zoo had to be euthanized after developing sudden onset paraparesis (partial paralysis of leg muscles) and testing positive for WNV infection (Dutton et al. 2009). Much remains to be discovered on the impact of WNV on carnivores, but the virus (already established in New Mexico) represents a source of concern in the context of climate change with possible continued dispersal into new areas and amplification (Paz 2015; Hicke et al. 2022).

Ticks represent the second most important vectors of infectious diseases after mosquitoes (Brites-Neto et al. 2015), and they can transmit a wide variety of pathogens consisting of viruses, bacteria, protozoa, and roundworms (Weaver et al. 2022). While temperature determines the development rates of ticks, survival in the free-living stages in addition to hatching and moulting are conditioned by humidity. Tick-borne bacterial infections affecting domestic and/or wild carnivores include those caused by *Borrelia burgdorferi* (Lyme disease), *B. turicatae* (tick-borne relapsing fever), *Anaplasma phagocytophilum* (granulocytic anaplasmosis), all detected in New Mexico and all subject to changes in their prevalence (see Chapter 33).

Sylvatic plague is a flea-borne infectious disease that is caused by the bacterium *Yersinia pestis*. It has been detected in more than 200 mammal species with carnivores being exposed through different infection routes, mainly from bites by fleas carrying the bacterium and through the ingestion of infected prey (Brinkerhoff et al. 2009). It can be highly pathogenic, especially in the black-footed ferret (*Mustela nigripes*; Chapter 22), but also in other carnivores such as the Canada lynx (Wild et al. 2006; see Chapter 6). Rocke et al. (2020) studied the potential impact of climate change on sylvatic plague transmission dynamics by collecting fleas on prairie dogs (*Cynomys* spp.), also highly susceptible to the disease. Ambient temperature was found to correlate positively with flea development rates, numbers of fleas on their hosts, and prairie dog burrow temperatures (Rocke et al. 2020). According to Rocke et al. (2020), climate change may lead to significant increases in flea populations by shortening flea development time, but the impact on transmission dynamics remains unclear. According to Nakazawa et al. (2007), the distribution of plague outbreaks may simply shift northward with areas to the south exceeding a threshold temperature and becoming unfavorable to flea development. Here again, the impact on carnivores from sylvatic plague within a context of climate change is unclear for New Mexico, but changes in prevalence are possible.

Climate change is also likely to affect non-vectored infectious diseases such as canine

Photo 5.9. Gray fox (*Urocyon cinereoargenteus*) in a parking lot in Las Cruces, Doña Ana County on 20 February 2020. Photograph: © Colin R. Dunn.

parvovirus type 2 (CPV-2; also known as *Carnivore protoparvovirus 1*), found in New Mexico and not just extremely contagious but also highly pathogenic in many species of wild carnivores (Jiang 2018; and see Chapter 33). Most infections are acquired through contact with feces, vomit, or saliva from infected animals, or contaminated water or food. CPV-2 was first recognized around 1976 in Europe in domestic dogs (as the result of two or three genetic mutations in feline panleukopenia virus [FPV] infecting domestic cats) and within a few years had spread around the world, with new genetic and antigenic variants also detected (Jiang 2018). The most important predictors of CPV-2 establishment include annual mean temperature, isothermality, November precipitation, maximum temperature during the warmest month of the year, and precipitation during the warmest quarter (Jiang 2018). Although the southwestern US may not lie in an area of high risk for increased transmission of CPV-2, canine parvovirus infections illustrate the potential for a distribution shift and/or amplification of non-vectored infectious diseases interacting with cross-species spillovers from domestic animals. Much remains to be learned about the connection between climate change and infectious diseases, and research is needed to characterize and project trends in

transmission, host-pathogen interactions, and the risk of cross-species spillovers under diverse climate change scenarios (Rupasinghe et al. 2022).

Urban Carnivore Populations

Although most of New Mexico's counties reported a loss of human population during the last decade, the main urban (and exurban) areas of the state are still projected to grow. In the middle Rio Grande region, for example, the human population is expected to continue growing and reach 1.6 million people by 2050 (MRGCOG 2000). Worldwide, the total urban human population has increased more than fourfold over the last six decades, and it is expected to reach nearly 5 billion people by 2030 (UNFPA 2007; Gehrt 2010; Seto et al. 2012). Urbanization is a leading cause of habitat loss and fragmentation, yet city parks, golf courses, cemeteries, riparian corridors, and even private yards serve as spatial refugia exploited by carnivores, particularly mesocarnivores, and other wildlife (McKinney 2002; Belaire et al. 2014; Gallo et al. 2017). Mesocarnivores known to occur in urban areas in New Mexico and surrounding states include coyotes (*Canis latrans*), North American red foxes (*Vulpes fulva*), gray foxes, bobcats (*Lynx rufus*), striped skunks, raccoons, and ringtails (*Bassariscus astutus*) (Grubbs and Krausman 2009; Brashear 2013; Lombardi 2014; Poessel et al. 2013, 2016; Magle et al. 2019). Conflicts between city residents and coyotes in particular have been well documented (e.g., Poessel et al. 2013), as have those involving black bears, who are not permanent residents of urban areas but forage in them, attracted by unsecured garbage and other anthropogenic foods (Lewis et al. 2015; Chapter 16). Synergies between climate change and the growth of cityscapes as new ecosystems are not well understood, but worldwide climate change is likely to exacerbate the problem of carnivores increasingly relying on anthropogenic foods (Newsome et al. 2017), with cascading effects on population densities, home ranges, movements, life history traits, and interspecific interactions (Newsome et al. 2015). Gray foxes,

Photos 5.10a and b (*left to right*). Wildlife corridor and road-crossing structure south of Cuba in Sandoval County. (a) New Mexico Department of Transportation employees inspect one of two large bridges over the Rio Puerco on US Highway 550 south of Cuba, a known wildlife-vehicle collision hotspot. Construction of eight-foot-tall woven wire fencing along the highway right-of-way has been completed, and the fence is forcing wildlife to move under the highway at the two bridge locations. Photograph: © Mark L. Watson. (b) Cougar (*Puma concolor*) crossing under the bridge after it likely followed the fence along the right-of-way; it is photographed by a motion-triggered, remote camera installed for monitoring the success of this wildlife-vehicle collision mitigation project. Photograph: © Arizona Game and Fish Department and New Mexico Department of Transportation Research Bureau.

Photo 5.11. Adult female white-nosed coati (*Nasua narica*) and her five kits along New Mexico Highway 152 in Hillsboro, Sierra County. As shown in neighboring Arizona (Gagnon et al. 2020), the coati could benefit from the construction of wildlife road-crossing structures in New Mexico (Cramer et al. 2022). Advanced driver assistance systems should help reduce wildlife mortality on roads in the coming years, but unlike road-crossing structures they do not address the issue of road permeability. High-volume traffic can form a complete barrier to animal movement once average annual daily traffic (AADT) reaches around 10,000 vehicles (see Charry and Jones 2009; Cramer et al. 2022). In the southern part of the state, I-10 at Steins in Hidalgo County (one of the high-priority wildlife corridors identified in the Wildlife Corridors Action Plan [Cramer et al. 2022]) has an average annual daily traffic (AADT) of 12,440 vehicles per day (projected to grow to 18,433 vehicles per day by 2038). Along I-25 north of Albuquerque, AADT ranges from 40,000 to 58,000 vehicles per day, forming a virtual wall of vehicles and severely limiting habitat connectivity in an important wildlife corridor between the Sandia and Jemez mountains (see Cramer et al. 2022). Photograph: © Jake Schoellkopf/NMDOT.

Photos 5.12a and b. More than 730 km (455 mi) of border wall were built between the US and Mexico (about 160 km [100 mi] in New Mexico) from 2017 to 2021. The border wall negatively impacts habitat connectivity on a regional scale for many animal species including the cougar (*Puma concolor*), bobcat (*Lynx rufus*), coyote (*Canis latrans*), jaguar (*Panthera onca*) and gray wolf (*Canis lupus*) (Traphagen 2021). (a) View from the Whitewater Mountains in Hidalgo County looking east toward the Antelope Wells port of entry along the international border. Visible in the background is a ~9.5 km (6 mi) long section of border wall constructed in 2020 and sealing off the Playas Valley on the US side (left) from Chihuahua (right). (b) view to the west in Doña Ana County, showing the ~9 m (30 ft) tall bollard wall. A border marker (Monument 16) is visible through a small gap between the steel bollards. Photographs: © Myles Traphagen.

first reported in New Mexico's residential rural areas by Harrison (1997), appear to have increasingly habituated to humans and human development in the state. They now occur (and breed) in residential neighborhoods in cities such as Albuquerque, Santa Fe, and Las Cruces (Photo 5.9), where the availability of water likely allows them to thrive on a diet of mammals with less reliance on fleshy fruit (see Harrison 1997). The gray fox is only one of several species whose numbers could grow or continue to grow in New Mexico's urban areas as a result of increasingly arid conditions.

CLIMATE CHANGE MITIGATION AND ADAPTATION

In addition to addressing the root causes of climate change (i.e., curbing greenhouse gas emissions), a number of mitigation and adaptation measures have gained attention including those aimed at boosting the resistance (the capacity to remain unaffected) or resilience (capacity to absorb and recover) of species affected by climate-driven impacts (e.g., Hansen et al. 2003).

Wildlife Movements and Habitat Linkages

Habitat linkages represent an increasingly important tool to promote species resilience in a context of global climate change (Nuñez et al. 2013; McGuire et al. 2016; Costanza et al. 2020). With many animals expected to move in search of new habitats, perhaps using altered dispersal or migration routes (Heller and Zavaleta 2009; Finch et al. 2012; Schloss et al. 2012), landscape permeability will likely become crucial. Roads in particular create barriers that impede wildlife movements, isolate populations, and result in direct mortality from collisions with vehicles (e.g., Forman and Alexander 1998; Trombulak and Frisell 2000; Forman 2003; Van der Ree et al. 2011). In the western United States, numerous road-crossing structures have already been built to promote habitat linkages, reduce wildlife mortality from collisions with vehicles, and increase motorist safety (e.g., Gagnon et al. 2011; Dodd et al. 2012; Cramer 2012, 2014; Cramer and McGinty 2018; Kintsch et al. 2021; Cramer and Hamlin 2021). Pre- and post-mitigation monitoring is key to determining the effectiveness of new road-crossing structures and the associated fencing that funnels wildlife toward those structures, both in terms of reduced wildlife mortality and higher road permeability (e.g., Gagnon et al. 2011; Cramer and Hamlin 2019).

In 2019, the New Mexico legislature passed the Wildlife Corridors Act mandating the development of an action plan to 1) identify both the state's top wildlife-vehicle collision hotspots and the wildlife corridors most affected by roads; and 2) propose specific road-crossing structures for wildlife in those priority conflict areas. The newly released Wildlife Corridors Action Plan (Cramer et al. 2022) focuses mostly on large mammals including two carnivores, the cougar (*Puma concolor*) and the black bear, but also takes into account other species known to be ecologically impacted by roads in New Mexico or elsewhere (e.g., American badger [*Taxidea taxus*] and swift fox). Within the top wildlife-vehicle collision hotspots and the wildlife corridors most affected by roads (identified through modeling and other tools), road-crossing structures (e.g., overpasses, bridge underpasses, culverts) are being proposed at specific locations (i.e., mileposts) for future implementation (Cramer et al. 2022). Already in New Mexico, ten main road mitigation projects have been implemented (Photos 5.10 a and b). Another project, along I-25 between Raton and Raton Pass, is currently in its construction phase (Cramer et al. 2022). It should be beneficial to black bears in particular as they are often involved in wildlife-vehicle collisions along that segment of the freeway. Funding now needs to be secured for the implementation of the Action

Plan, which is intended as a "living document," with updates and modifications as needed every few years (Cramer et al. 2022). As the effects of climate change degrade current core and connective habitats, identifying future dispersal corridors and areas of wildlife-vehicle conflict will likely become key.

More than 730 km (455 mi) of US-Mexico border wall were built from 2017 to early 2021 (Traphagen 2021; Photos 5.12a and b). The border wall fragments habitat, closes important cross-border dispersal corridors for the jaguar (*Panthera onca*) and many other species, and precludes gene flow between some now disjunct wildlife populations. It also undermines the bi-national collaboration of government agencies, non-government organizations, and private landowners to create and expand networks of protected areas (see Peters et al. 2018). The Illegal Immigration Reform and Immigrant Responsibility Act of 1996 was amended in 2005 to authorize the secretary of the Department of Homeland Security to waive all legal environmental requirements in order to expedite the construction of barriers and roads related to border security. Executive Order no. 13767 issued by the Trump administration in January 2017 for the construction of a border wall was then followed by waivers of environmental laws and regulations. As a result, no environmental impact analysis was conducted under the National Environmental Policy Act (NEPA) and none of the Endangered Species Act provisions applied. Recommendations by a coalition of more than 2,500 scientists from 43 countries (Peters et al. 2018) were ignored as none of the bollard border wall sections was designed for maximum wildlife permeability even in ecologically sensitive areas or known important wildlife corridors. With Traphagen (2021) reporting a gap between the bollards measuring only ~7–10 cm (3–4 in), the border wall thus represents a largely continuous, complete, and perhaps permanent barrier to cross-border dispersal of carnivores. How to mitigate this impact is far beyond the scope of this chapter but likely involves as a starter recognizing the link between climate change and increased human migration (see Chemnick and Matthews 2019) and possibly removing harmful wall sections. Significant time and effort will likely need to be invested in inventorying and studying the ecological impacts of the border wall on local and regional scales with a consideration of climate change increasingly displacing species. Alternative designs to the current wall design should be explored.

Adjustment of Harvest Levels

If not properly adjusted, harvest could potentially compound climate-driven impacts and accelerate the decline of regional populations for some of the species mentioned above. A demographic modeling study of polar bears (Regehr et al. 2017) shows that sustainable harvest remains possible even in the face of habitat loss. A prerequisite is that population estimates must be updated, and that harvest should be reduced with lower certainty in the demographic data. Survival rates from one age group to the next in both males and females are tied to environmental carrying capacity in density dependent relationships, and a management framework is then developed that reflects goals and risk tolerance (Regehr et al. 2017). A similar model could be developed for black bear populations in New Mexico, also at risk currently and in the next several decades from habitat loss. In Regehr et al.'s (2017) study, a proxy metric for the carrying capacity is derived using sea-ice extent as documented by satellite data. A similar approach could be used for black bears in New Mexico using instead forest extent. It would require a significant investment in collecting data such as age-specific survival, breeding probabilities, and age structure, all quite difficult to obtain but necessary for high model output precision and thus confidence in the management decisions that would follow.

Serosurveillance of Infectious Diseases

Targeted surveillance and seroprevalence studies are important tools allowing researchers to investigate whether a pathogen is a source of concern for the conservation of carnivores (and other wildlife; see Root and Bosco-Lauth 2019). As emerging infectious diseases or new transmission patterns may first appear in developing countries where financial resources are particularly limited, a continued investment should be made into global electronic surveillance networks, and readily accessible databases and predictive maps, as well as the dissemination of syndromic surveillance (i.e., based on symptoms) and rapid diagnostic methods (see Institute of Medicine 2007).

Increasing the Resistance and/or the Resilience of Montane Forests

Historically, southwestern ponderosa pine and dry mixed conifer forests were subject to natural fire disturbance occurring at intervals of a few years to several decades (Safford and Stevens 2017). These regular and lower-severity surface fires burned through the understory, maintaining a more open forest structure and reducing competition for the larger, older trees, most of which were left unaffected by the heat and the flames (Taylor and Skinner 2003; Knapp et al. 2013; Collins et al. 2015; Stephens et al. 2015). As a result of more than a century of fire suppression, however, frequent-fire forests have become denser and trees now suffer from competitive stress, which coupled with drought, makes them more vulnerable to forest health agents such as bark beetle infections (Kolb et al. 2016). Drought conditions and the larger amounts of fuels in the understory have also resulted in the occurrence of large-sized, high-severity fires accompanied not only by massive tree mortality (Mallek et al. 2013, Stevens et al. 2017) but also the permanent conversion of forest stands to non-forest type (e.g., Walker et al. 2018).

With the realization that fire suppression

posed an important threat to forests, forest managers shifted to a paradigm of ecological sustainability, with the use of prescribed fire and mechanical thinning of small to medium diameter trees proving to be effective tools to increase resilience in frequent-fire forests (Collins et al. 2014). However, the operating budgets of land management agencies limit the ability for extensive treatments in forested areas even in a context of increasingly urgent needs. Innovative solutions are needed, possibly including complete fuel breaks around the highest risk or highest value settings (Millar et al. 2007). Decision-making tools are also important as they show for example that for optimum results, tree replanting efforts could focus on moist and cool forests along with north- or east-facing aspects and slopes under 45% gradient (Stevens-Rumann and Morgan 2019). Also, planting could focus on areas where the distance to the nearest living tree (natural seed source) exceeds 200 m.

Without change agents such as severe wildfire, rapid type conversions (e.g., forest to grassland) are unlikely in the next immediate decades due to plant-climate disequilibrium and the survivability of older plant cohorts delaying vegetation type shifts. Also, changes at the trailing warm-dry end of a given ecosystem type are expected to occur at a faster pace than the advancing cool-moist end (Brusca et al. 2013). But the potential for disturbance to trigger rapid conversion underscores the need for biologists and land managers to consider where type conversions are most likely, and to consider the interactions of disturbance and climate in the design, prioritization, and implementation of land management aimed at conservation. Identifying areas where change is likely and then narrowing to those stand conditions and landscape positions at greatest risk to high severity fire can be helpful in determining elevated risk to the delivery of ecosystem services. Spatial analysis can further aid in focusing on zones important for the habitat resilience of carnivores and other wildlife. Climate adaption that is aimed at *resilience* can help in forestalling the impacts of a changing climate or, when coupled with a *transformation* approach, can assist in the realignment of an ecosystem to gradually changing conditions (Peterson St-Laurent et al. 2021).

Climate projections presented in this chapter are all based on a high-emissions (RCP 8.5, also known as "business as usual") scenario that assumes nearly unmitigated, continued fossil fuel use. This high-emissions scenario, which yields rapid climate change in model projections (especially during the later decades of this century), is now considered somewhat unlikely in the long-term. However, it remains highly relevant until fossil fuel use is actually curtailed, and remains realistic under existing energy policies through the middle of the century (Schwalm et al. 2020). As emphasized by IPCC (2021), sustained reductions in greenhouse gas emissions would limit future climate change within several years of their implementation. To a large extent, carnivores and their habitat will be affected by the changing climate in New Mexico as elsewhere, but the ecological impacts of future climate change can be lessened by bold and innovative management and conservation actions.

LITERATURE CITED

Abatzoglou, J. T. 2011. Development of gridded surface meteorological data for ecological applications and modelling. *International Journal of Climatology* 31:1135–42. https//doi: 10.1002/joc.3413.

Aflaki, A., M. Mirnezhad, A. Ghaffarianhoseini, A. Ghaffarianhoseini, H. Omrany, Z. Wang, and H. Akbari. 2017. Urban heat island mitigation strategies: a state-of-the-art review on Kuala Lumpur Singapore and Hong Kong. *Cities* 62:131–45.

Ahlers, L. R. H., and A. G. Goodman. 2018. The immune responses of the animal hosts of West Nile virus: a comparison of insects, birds, and mammals. *Frontiers in Cellular and Infection Microbiology* 8, article 96. doi: 10.3389/fcimb.2018.00096.

Allen, C. D. 1989. Changes in the landscape of the Jemez Mountains, New Mexico. PhD dissertation, University of California, Berkeley.

Allen, C. D., A. K. Macalady, H. Chenchouni, D. Bachelet, N. McDowell, M. Vennetier, T. Kitzberger, A. Rigling, D. D. Breshears, E. T. Hogg, and P. Gonzalez. 2010. A global overview of drought and heat-induced tree mortality reveals emerging climate change risks for forests. *Forest Ecology and Management* 259:660–84.

Angert, A. L., L. G. Crozier, L. J. Rissler, S. E. Gilman, J. J. Tewksbury, and A. J. Chunco. 2011. Do species' traits predict recent shifts at expanding range edges? *Ecology Letters* 14:677–89.

Avila-Villegas, S., and J. A. Lamberton-Moreno. 2013. Wildlife survey and monitoring in the Sky Island region with an emphasis on neotropical felids. In *Merging science and management in a rapidly changing world: biodiversity and management of the Madrean Archipelago*, ed. G. J. Gottfried, P. F. Ffolliott, B. S. Gebow, L. G. Eskew, and L. C. Collins, 441–47. RMRS-P-67. Fort Collins, CO: US Department of Agriculture, Forest Service, Rocky Mountain Research Station.

Bailey, V. 1931 (=1932). *Mammals of New Mexico*. North American Fauna 53. Washington, DC: US Department of Agriculture, Bureau of Biological Survey.

Barlow, J., F. Franca, T. A. Gardner, C. C. Hicks, G. D. Lennox, E. Berenguer, L. Castello, E. P. Economo, J. Ferreira, B. Guénard, C. Gontijo Leal, V. Isaac, A. C. Lees, C. L. Parr, S. K. Wilson, P. J. Young, and N. A. J. Graham. 2018. The future of hyperdiverse tropical ecosystems. *Nature* 559:517–26.

Batcheller, G. R., M. C. Bambery, L. Bies, T. Decker, S. Dyke, D. Guynn, M. McEnroe, M. O'Brien, J. F. Organ, S. J. Riley, and G. Roehm. 2010. *The public trust doctrine: implications for wildlife management and conservation in the United States and Canada*. Technical Review 10-1. Bethesda, MD: Wildlife Society.

Bates, B., Z. W. Kundzewicz, S. Wu, and J. Palutikof. 2008. *Climate change and water*. Technical paper of the Intergovernmental Panel on Climate Change. Geneva: Intergovernmental Panel on Climate Change.

Bean, M. J. 1983. *The evolution of national wildlife law*. New York: Praeger.

Belaire, J. A., C. J. Whelan, and E. S. Minor. 2014. Having our yards and sharing them too: the collective effects of yards on native bird species in an urban landscape. *Ecological Applications* 24:2132–43.

Bjarke, N. R., and D. S. Gutzler. 2023. Use of observed hydroclimatic trends to constrain projections of snowmelt season runoff in the Rio Grande headwaters. *Journal of the American Water Resources Association* 59. https://doi.org/10.1111/1752-1688.13118.

Brashear, W. A. 2013. An assessment of the genetic structure of a striped skunk (*Mephitis mephitis*) population across an urban landscape. MS thesis, Angelo State University, TX.

Breshears, D. D., O. B. Myers, C. W. Meyer, F. J. Barnes, C. B. Zou, C. D. Allen, N. G. McDowell, and W. T. Pockman. 2009. Tree die-off in response to global change-type drought: mortality insights from a decade of plant water potential measurements. *Frontiers in Ecology and the Environment* 7:185–89.

Brinkerhoff, R. J., S. K. Collinge, Y. Bai, and C. Ray. 2009. Are carnivores universally good sentinels of plague? *Vector-Borne and Zoonotic Diseases* 9:491–97.

Brites-Neto, J., K. M. R. Duarte, and T. F. Martins. 2015. Tick-borne infections in human and animal population worldwide. *Veterinary World* 8:301–15. doi: 10.14202/vetworld.2015.301–315.

Brook, B. W. 2008. Synergies between climate change, extinctions and invasive vertebrates. *Wildlife Research* 35:249–52.

Brook, B. W., N. S. Sodhi, and C. J. A. Bradshaw. 2008.

Synergies among extinction drivers under global change. *Trends in Ecology and Evolution* 23:453–60.

Brown, D. E. 1983. *The wolf in the Southwest: the making of an endangered species*. Norman: University of Oklahoma Press.

———. 1985. *The grizzly in the Southwest: documentary of an extinction*. Norman: University of Oklahoma Press.

Brusca, R. C., J. F. Wiens, W. M. Meyer, J. Eble, K. Franklin, J. T. Overpeck, and W. Moore. 2013. Dramatic response to climate change in the Southwest: Robert Whittaker's 1963 Arizona Mountain plant transect revisited. *Ecology and Evolution* 3:3307–19.

Buffington, L. C., and C. H. Herbel. 1965. Vegetational changes on a semidesert grassland range from 1858 to 1963. *Ecological Monographs* 35:139–64.

Carbone, C., G. M. Mace, S. C. Roberts, and D. W. Macdonald. 1999. Energetic constraints on the diet of terrestrial carnivores. *Nature* 402:286–88.

Cardillo, M., G. M. Mace, K. E. Jones, J. Bielby, O. R. P. Bininda-Emonds, W. Sechrest, C. D. L. Orme, and A. Purvis. 2005. Multiple causes of high extinction risk in large mammal species. *Science* 309:1239–41.

Cardillo, M., A. Purvis, W. Sechrest, J. L Gittleman, J. Bielby, and G. M. Mace. 2004. Human population density and extinction risk in the world's carnivores. *PLOS Biology* 2:e197. doi: 10.1371/journal.pbio.0020197; pmid: 15252445.

Charry, B., and J. Jones. 2009. Traffic volume as a primary road characteristic impacting wildlife: a tool for land use and transportation planning. In *Proceedings from the 2009 International Conference on Ecology and Transportation*, 159–72. University of North Carolina, Raleigh. Center for Transportation Research. https://escholarship.org/uc/item/4fx6c79t#main.

Chavarria, S. B., and D. S. Gutzler. 2018. Observed changes in climate and streamflow in the upper Rio Grande Basin. *Journal of the American Water Resources Association* 54:644–59. doi.org/10.1111/1752-1688.12640.

Chemnick, J., and M. K. Matthews. 2019. Trump's border wall highlights the climate–migration connection. *E&E News*, 22 January 2019.

Cianfrani, C., O. Broennimann, A. Loy, and A. Guisan. 2018. More than range exposure: global otter vulnerability to climate change. *Biological Conservation* 221:103–113.

Collins, B. M., A. J. Das, J. J. Battles, D. L. Fry, K. D. Krasnow, and S. L. Stephens. 2014. Beyond reducing fire hazard: fuel treatment impacts on overstory tree survival. *Ecological Applications* 24:1879–1886.

Collins, B. M., J. M. Lydersen, R. G. Everett, D. L. Fry, and S. L. Stephens. 2015. Novel characterization of landscape-level variability in historical vegetation structure. *Ecological Applications* 25:1167–74.

Cook, B. I., and R. Seager. 2013. The response of the North American Monsoon to increased greenhouse gas forcing. *Journal of Geophysical Research: Atmospheres* 118:1690–99. doi:10.1002/jgrd.50111.

Costanza, J. K., J. Watling, R. Sutherland, C. Belyea, B. Dilkina, H. Cayton, D. Bucklin, S. S. Romañach, and N. M. Haddad. 2020. Preserving connectivity under climate and land-use change: no one-size-fits-all approach for focal species in similar habitats. *Biological Conservation* 248:108678. https://doi.org/10.1016/j.biocon.2020.108678.

Costanzi, L., A. Brambilla, A. Di Blasio, A. Dondo, M. Goria, L. Masoero, M. Silvia Gennero, and B. Bassano. 2021. Beware of dogs! Domestic animals as a threat for wildlife conservation in Alpine protected areas. *European Journal of Wildlife Research* 67:70. doi: 10.1007/s10344-021-01510-5.

Cramer, P. 2012. *Determining wildlife use of wildlife crossing structures under different scenarios*. Final report to the Utah Department of Transportation, Salt Lake City.

———. 2014. Wildlife crossings in Utah: Determining what works and helping to create the best and most cost-effective structure designs. Report to the Utah Division of Wildlife Resources, Salt Lake City, UT.

Cramer, P., J.-L. E. Cartron, K. C. Calhoun, J. W. Gagnon, M. B. Haverland, M. L. Watson, S. A. Cushman, H. Y. Wan, J. A. Kutz, J. N. Romero, T. J. Brennan, J. A. Walther, C. D. Loberger, H. P. Nelson, T. D. Botkin, and J. G. Hirsch. 2022. *Wildlife corridors action plan*. Daniel B. Stephens and Associates, New Mexico Department of Transportation, and New Mexico Department of Game and Fish.

Cramer, P., and R. Hamlin. 2019. *US 89 Kanab-Paunsaugunt Wildlife Crossing and existing structures research project*. Final report to Utah Department of Transportation, Salt Lake City. Available at https://rosap.ntl.bts.gov/view/dot/63544.

Cramer, P., and R. Hamlin. 2021. *US 160 Dry Creek wildlife research*. Final report to the Colorado Department of Transportation, Denver, CO.

Cramer, P., and C. McGinty. 2018. *Prioritization of wildlife-vehicle conflict in Nevada*. Final report to the Nevada Department of Transportation.

Crimmins, S. M., S. Z. Dobrowski, J. A. Greenberg, J. T. Abatzoglou, and A. R. Mynsberge. 2011. Changes in climatic water balance drive downhill shifts in plant species' optimum elevations. *Science* 331:324–27.

Culver, M., S. Malusa, J. L. Childs, K. Emerson, T. Fagan, P. M. Harveson, L. E. Haynes, J. G. Sanderson, J. H. Sheehy, T. Skinner, N. Smith, K. Thompson, and R. W. Thompson. 2016. Jaguar surveying and monitoring in the United States. US Geological Survey Open-File Report 2016–1095. http://dx.doi.org/10.3133/ofr20161095.

Cunningham, S. C., and W. B. Ballard. 2004. Effects of wildfire on black bear demographics in central Arizona. *Wildlife Society Bulletin* 32:928–37.

Dahl, T. E., 1990. *Wetlands-Losses in the United States, 1780's to 1980's*. US Fish and Wildlife Service Report to Congress, Washington, DC.

Diamond, J. 1989. Overview of recent extinctions. In *Conservation for the twenty-first century*, ed. D. Western and M. C. Pearl, 37–41. New York: Wildlife Conservation International.

Diamond, J. M. 1984. "Normal" extinction of isolated populations. In *Extinctions*, ed. M. H. Nitecki, 191–246. Chicago: University of Chicago Press.

Dick-Peddie, W. A, ed. 1993. *New Mexico vegetation: past, present and future*. Albuquerque: University of New Mexico Press.

Dietrich, G., J. A. Montenieri, N. A. Panella, S. Langevin, S. E. Lasater, K. Klenk, J. C. Kile, and N. Komar. 2005. Serologic evidence of West Nile virus infection in free-ranging mammals, Slidell, Louisiana, 2002. *Vector-borne and Zoonotic Diseases* 5:288–92.

Dodd, N. L., J. W. Gagnon, S. Boe, K. Ogren, and R. E. Schweinsburg. 2012. Wildlife-vehicle collision mitigation for safer wildlife movement across highways: State Route 260. Arizona Department of Transportation, Phoenix, AZ. FHWA-AZ-12–603.

Dunbar, N. W., D. S. Gutzler, K. S. Pearthree, F. M. Phillips, C. D. Allen, D. DuBois, M. D. Harvey, J. P. King, L. D. McFadden, B. M. Thomson, and A. C. Tillery, 2022. *Climate Change in New Mexico Over the Next 50 Years: Impacts on Water Resources*. New Mexico Bureau of Geology and Mineral Resources, Bulletin 164, 218 pp. https://geoinfo.nmt.edu/publications/monographs/bulletins/164/home.cfm.

Dutton, C. J., M. Quinnell, R. Lindsay, J. DeLay, and I. K. Barker. 2009. Paraparesis in a polar bear (*Ursus maritimus*) associated with West Nile virus infection. *Journal of Zoo and Wildlife Medicine* 40:568–71. doi: 10.1638/2008-0121.1.

Elmhagen, B., D. Berteaux, R. M. Burgess, D. Ehrich, D. Gallant, H. Henttonen, R. A. Ims, S. T. Killengreen, J. Niemimaa, K. Norén, T. Ollila, A. Rodnikova, A. A. Sokolov, N. A. Sokolova, A. A. Stickney, and A. Angerbjörn. 2017. Homage to Hersteinsson and Macdonald: climate warming and resource subsidies cause red fox range expansion and Arctic fox decline. *Polar Research* 36:sup1. doi: 10.1080/17518369.2017.1319109.

Finch, D. M., D. M. Smith, O. LeDee, J.-L. E. Cartron, and M. A. Rumble. 2012. Climate change, animal species, and habitats: adaptation and issues. In *Climate change in grasslands, shrublands, and deserts of the interior American West: a review and needs assessment*, ed. D. M. Finch, 60–79. General Technical Report RMRS-GTR-285. Fort Collins, CO: US Department of Agriculture, Forest Service, Rocky Mountain Research Station.

Forman, R. T. T. 2003. *Road ecology: science and solutions*. Washington, DC: Island Press.

Forman, R. T. T., and L. E. Alexander. 1998. Roads and their major ecological effects. *Annual Review of Ecology, Evolution, and Systematics* 29:207–31.

Freeman, B. G., J. A. Lee-Yaw, J. M. Sunday, and A. L. Hargreaves. 2018. Expanding, shifting and shrinking: The impact of global warming on species' elevational distributions. *Global Ecology and Biogeography* 27: 1268–76. doi: 10.1111/geb.12774.

Gagnon, J. W., C. A. Beach, S. C. Sprague, C. D. Loberger, and C. Rubke. 2020. *Evaluation of measures to reduce wildlife-vehicle collisions and promote connectivity in a Sonoran Desert environment—State*

Route 77 Santa Catalina—Tortolita Mountain Corridor. Prepared for the Pima County Regional Transportation Authority.

Gagnon, J. W., N. L. Dodd, K. S. Ogren, and R. E. Schweinsburg. 2011. Factors associated with use of wildlife underpasses and importance of long-term monitoring. *Journal of Wildlife Management* 75:1477–87.

Gallo, T., M. Fidino, E. Lehrer., and S. B. Magle. 2017. Mammal diversity and metacommunity dynamics in urban green spaces: implications for urban wildlife conservation. *Ecological Applications* 27(8):2330–41. doi: 10.1002/eap.1611.

Gates, D. M. 1993. *Climate change and its biological consequences*. Sunderland, MA: Sinauer Associates.

Gehrt, S. D. 2010. The urban ecosystem. In *Urban Carnivores: ecology, conflict, and conservation*, ed. S. D. Gehrt, S. P. D. Riley, and B. L. Cypher, 3–11. Baltimore: Johns Hopkins University Press.

Geist, V. 1995. North American policies of wildlife conservation. In *Wildlife conservation policy*, ed. V. Geist and I. McTaggart-Cowan, 77–129. Calgary: Detselig Enterprises.

Geist, V., S. P. Mahoney, and J. F. Organ. 2001. Why hunting has defined the North American model of wildlife conservation. *Transactions of the North American Wildlife and Natural Resources Conference* 66:175–85.

Gibbens, R. P., R. P. McNeely, K. M. Havstad, R. F. Beck, and B. Nolen. 2005. Vegetation changes in the Jornada Basin from 1858 to 1998. *Journal of Arid Environments* 61:651–668.

Graham, R. W., and E. C. Grimm. 1990. Effects of global climate change on the patterns of terrestrial biological communities. *Trends in Ecology and Evolution* 5:289–92.

Grubbs, S. E., and P. R. Krausman. 2009. Use of urban landscape by coyotes. *Southwestern Naturalist* 54:1–12.

Gutzler, D. S. 2013. Regional climatic considerations for borderlands sustainability. *Ecosphere* 4:1–12.

Haines, A. M., J. E. Janecka, M. E. Tewes, L. I. Grassman Jr., and P. Morton. 2006. The importance of private lands for ocelot *Leopardus pardalis* conservation in the United States. *Oryx* 40:90–94. doi:10.1017/S0030605306000044.

Hansen, L. J., J. L. Biringer, and J. R. Hoffman. 2003. *Buying time: a user's manual for building resistance and resilience to climate change in natural systems*. Berlin: World Wildlife Fund Climate Change Program.

Harrison, R. L. 1997. A comparison of gray fox ecology between residential and undeveloped rural landscapes. *Journal of Wildlife Management* 61:112–22.

Heller, N. E., and E. S. Zavaleta. 2009. Biodiversity management in the face of climate change: a review of 22 years of recommendations. *Biological Conservation* 142:14–32.

Hicke, J.A., S. Lucatello, L. D. Mortsch, J. Dawson, M. Domínguez Aguilar, C. A. F. Enquist, E. A. Gilmore, D. S. Gutzler, S. Harper, K. Holsman, E. B. Jewett, T. A. Kohler, and K. A. Miller. 2022. North America. In *Climate Change 2022: Impacts, Adaptation and Vulnerability. Contribution of Working Group II to the Sixth Assessment Report of the Intergovernmental Panel on Climate Change* (H.-O. Pörtner, D.C. Roberts, M. Tignor, E. S. Poloczanska, K. Mintenbeck, A. Alegría, M. Craig, S. Langsdorf, S. Löschke, V. Möller, A. Okem, B. Rama [eds.]). Cambridge University Press, Cambridge, UK, and New York, NY, USA, 1929–2042, doi:10.1017/9781009325844.016.

Hoerling, M. P., M. Dettinger, K. Wolter, J. Lukas, J. Eischeid, R. Nemani, B. Liebmann, and K. E. Kunkel. 2013. Present weather and climate: Evolving conditions. In *Assessment of climate change in the Southwest United States: a report prepared for the National Climate Assessment*, ed. G. Garfin, A. Jardine, R. Merideth, M. Black, and S. LeRoy, 74–97. A report by the Southwest Climate Alliance. Washington, DC: Island Press.

Hof, A. R., R. Jansson, and C. Nilsson. 2012. Future climate change will favour non-specialist mammals in the (Sub)Arctics. *PLOS ONE* 7(12): e52574. doi:10.1371/journal.pone.0052574.

Hughes, J. B., G. C. Daily, and P. R. Ehrlich. 1997. Population diversity: its extent and extinction. *Science* 278:689–92.

Hughes, L. 2000. Biological consequences of global warming: is the signal already apparent? *Trends in Ecology and Evolution* 15:56–61.

Hunter, L. 2019. *Carnivores of the World*. 2nd ed. Princeton, NJ: Princeton University Press.

Hunter, W. C., R. D. Ohmart, and B. W. Anderson. 1988. Use of exotic saltcedar (*Tamarix chinensis*) by birds in arid riparian systems. *Condor* 90:113–23.

Institute of Medicine. 2007. *Global Infectious Disease*

Surveillance and Detection: Assessing the Challenges— Finding Solutions. Institute of Medicine (US) Forum on Microbial Threats, Workshop Summary. Washington, DC: US National Academies Press.

Intergovernmental Panel on Climate Change (IPCC). 2013. *Climate Change 2013: The Physical Science Basis.* Contribution of Working Group I to the Fifth Assessment Report of the Intergovernmental Panel on Climate Change [eds. T. F. Stocker, D. Qin, G.-K. Plattner, M. Tignor, S. K. Allen, J. Boschung, A. Nauels, Y. Xia, V. Bex, and P. M. Midgley]. Cambridge: Cambridge University Press.

———. 2021. *Climate Change 2021: The Physical Science Basis.* Contribution of Working Group I to the Sixth Assessment Report of the Intergovernmental Panel on Climate Change [ed. V. Masson-Delmotte, P. Zhai, A. Pirani, S. L. Connors, C. Péan, S. Berger, N. Caud, Y. Chen, L. Goldfarb, M. I. Gomis, M. Huang, K. Leitzell, E. Lonnoy, J. B. R. Matthews, T. K. Maycock, T. Waterfield, O. Yelekçi, R. Yu and B. Zhou]. Cambridge: Cambridge University Press.

International Union for the Conservation of Nature (IUCN). 2021. The IUCN red list of threatened species. Version 2021-1. https://www.iucnredlist.org.

Jiang, F. 2018. Bioclimatic and altitudinal variables influence the potential distribution of canine parvovirus type 2 worldwide. *Ecology and Evolution* 8:4534–43.

Kappelle, M., M. M. I. Vuuren, and P. Baas. 1999. Effects of climate change on biodiversity: a review and identification of key research issues. *Biodiversity and Conservation* 8:1383–97.

Keyser, A. R., D. J. Krofcheck, C. C. Remy, C. D. Allen, and M. D. Hurteau. 2020. Simulated increases in fire activity reinforce shrub conversion in a southwestern US forest. *Ecosystems* 23:1702–13.

Kingsford, R. T. 2009. Conservation management of rivers and wetlands under climate change—a synthesis. *Marine and Freshwater Research* 62:217–22.

Kintsch, J., P. Cramer, P. Singer, and M. Cowardin. 2021. *State Highway 9 wildlife crossings monitoring final report, Study number 115.01.* Report to the Colorado Department of Transportation, Denver, CO.

Klausmeyer, K. R., and M. R. Shaw. 2009. Climate change, habitat loss, protected areas and the climate adaptation potential of species in Mediterranean ecosystems worldwide. *PLOS ONE* 4(7):e6392. doi:10.1371/JOURNAL.PONE.0006392.

Knapp, E. E., C. N. Skinner, M. P. North, and B. L. Estes. 2013. Long-term overstory and understory change following logging and fire exclusion in a Sierra Nevada mixed-conifer forest. *Forest Ecology and Management* 310:903–14.

Kolb, T. E., C. J. Fettig, M. P. Ayres, B. J. Bentz, J. A. Hicke, R. Mathiasen, J. E. Stewart, and A. S. Weed. 2016. Observed and anticipated impacts of drought on forests insects and diseases in the United States. *Forest Ecology and Management* 380:321–34.

Lata, K., G. Das, R. Verma, and R. P. S. Baghel. 2018. Impact of climate variability on occurrence and distribution of vector and vector-borne parasitic diseases. *Journal of Entomology and Zoology Studies* 6:1388–93.

Laurence, W. F., and D. C. Useche. 2009. Environmental Synergisms and Extinctions of Tropical Species. *Conservation Biology* 23:1427–37.

LeCount, A. L. 1982. Population characteristics of Arizona black bears. *Journal of Wildlife Management* 46:861–68.

Lenoir, J., J.-C. Gégout, A. Guisan, P. Vittoz, T. Wohlgemuth, N. E. Zimmermann, S. Dullinger, H. Pauli, W. Willner, and J.-C. Svenning. 2010. Going against the flow: potential mechanisms for unexpected downslope range shifts in a warming climate. *Ecography* 33:295–303.

Le Page, Y., D. Morton, C. Hartin, B. Bond-Lamberty, J. M. Cardoso Pereira, G. Hurtt, and G. Asrar. 2017. Synergy between land use and climate change increases future fire risk in Amazon forests. *Earth System Dynamics* 8:1237–46.

Lewis, D. L., S. Baruch-Mordo, K. R. Wilson, S. W. Breck, J. S. Mao, and J. Broderick. 2015. Foraging ecology of black bears in urban environments: guidance for human-bear conflict mitigation. *Ecosphere* 6(8), article141. http://dx.doi.org/10.1890/ES15-00137.1.

Liess, M., K. Folt, S. Knillmann, R. B. Schafer, and H. D. Liess. 2016. Predicting the synergy of multiple stress effects. *Scientific Reports* 6:32965. doi: 10.1038/srep32965.

Lombardi, J. V. 2014. Ecology of mesopredators within a small urban area in east Texas. MS thesis, Stephen F. Austin State University, Nacogdoches, TX.

López-González, C., and D. F. García-Mendoza. 2012. A checklist of the mammals (Mammalia) of Chihuahua, Mexico. *Check List* 8:1122–33.

López-González, C. A., D. E. Brown, and J. P. Gallo-Reynoso. 2003. The ocelot *Leopardus pardalis* in north-western Mexico: ecology, distribution and conservation status. *Oryx* 37: 358–64. doi:10.1017/S0030605303000620.

MacLean, S. A., and S. R. Beissinger. 2017. Species' traits as predictors of range shifts under contemporary climate change: a review and meta-analysis. *Global Change Biology* 23:4094–105.

Magle, S. B., M. Fidino, E. W. Lehrer, T. Gallo, M. P. Mulligan, M. J. Ríos, A. A. Ahlers, J. Angstmann, A. Belaire, B. Dugelby, A. Gramza, L. Hartley, B. MacDougall, T. Ryan, C. Salsbury, H. Sander, C. Schell, K. Simon, S. St Onge, and D. Drake. 2019. Advancing urban wildlife research through a multi-city collaboration. *Frontiers in Ecology and the Environment* 17:232–39. doi:10.1002/fee.2030.

Mallek, C, H. Safford, J. Viers, and J. Miller. 2013. Modern departures in fire severity and area vary by forest type, Sierra Nevada and southern Cascades, California, USA. Ecosphere 4(12), article 153. doi: 10.1890/ES13-00217.1.

Mason, S. C., G. Palmer, R. Fox, S. Gillings, J. K. Hill, C. D. Thomas, and T. H. Oliver. 2015. Geographical range margins of many taxonomic groups continue to shift polewards. *Biological Journal of the Linnean Society* 115:586–97.

McCarty, J. P. 2001. Ecological consequences of recent climate change. *Conservation Biology* 15:320–31.

McGuire, J. L., J. J. Lawler, B. H. McRae, T. A. Nuñez, and D. M. Theobald. 2016. Achieving climate connectivity in a fragmented landscape. *Proceedings of the National Academy of Sciences* 113:7195–200. doi: 10.1073/pnas.1602817113.

McKinney, M. L. 2002. Urbanization, biodiversity, and conservation. *BioScience* 52:883–90.

McLaughlin, J. F., J. J. Hellmann, C. L. Boggs, and P. R. Ehrlich. 2002. Climate change hastens population extinctions. *Proceedings of the National Academy of Sciences* 99: 6070–74.

Meinshausen, M., S. J. Smith, K. Calvin, J. S. Daniel, M. L. T. Kainuma, J-F. Lamarque, K. Matsumoto, S. A. Montzka, S. C. B. Raper, K. Riahi, A. Thomson, G. J. M. Velders, and D. P. P. van Vuuren. 2011. The RCP greenhouse gas concentrations and their extensions from 1765 to 2300. *Climatic Change* 109:213–41.

Middle Rio Grande Council of Governments (MRGCOG). 2000. *Focus 2050 Regional Plan*. Report SPR-278. Albuquerque: Middle Rio Grande Council of Governments.

Millar, C. I., N. L. Stephenson, and S. L. Stephens. 2007. Climate change and forests of the future: managing in the face of uncertainty. *Ecological Applications* 17:2145–51.

Milliman, J. D., K. L. Farnsworth, P. D. Jones, K. H. Xu, and L. C. Smith. 2008. Climatic and anthropogenic factors affecting river discharge to the global ocean, 1951–2000. *Global and Planetary Change* 62:187–94. doi:10.1016/J.GLOPLACHA.2008.03.001.

Moeser, C. D., S. B. Chavarria, and A. M. Wootten. 2021. Streamflow response to potential changes in climate in the upper Rio Grande Basin. US Geological Survey Scientific Investigations Report 2021-5138.

Mohajerani, A., J. Bakaric, and T. Jeffrey-Bailey. 2017. The urban heat island effect, its causes, and mitigation, with reference to the thermal properties of asphalt concrete. *Journal of Environmental Management* 197:522–38.

Mote, P. W., S. Li, D. P. Lettenmaier, M. Xiao, and R. Engel. 2018. Dramatic declines in snowpack in the western US. *npj Climate and Atmospheric Science* 1, article 2. https://doi.org/10.1038/s41612-018-0012-1.

Nakazawa, Y., R. Williams, A. T. Peterson, P. Mead, E. Staples, and K. L. Gage. 2007. Climate change effects on plague and tularemia in the United States. *Vector-Borne and Zoonotic Diseases* 7:529–40.

Newsome, T. M., J. A. Dellinger, C. R. Pavey, W. J. Ripple, C. R. Shores, A. J. Wirsing, and C. R. Dickman. 2015. The ecological effects of providing resource subsidies to predators. *Global Ecology and Biogeography* 24:1–11.

Newsome, T. M., P. J. S. Fleming, C. R. Dickman, T. S. Doherty, W. J. Ripple, E. G. Ritchie, and A.

J. Wirsing. 2017. Making a new dog? *BioScience* 4:374–81.

Nuñez, T. A., J. J. Lawler, B. H. McRae, D. J. Pierce, M. B. Krosby, D. M. Kavanagh, P. H. Singleton, and J. J. Tewksbury. 2013. Connectivity planning to address climate change. *Conservation Biology* 27:407–16. doi: 10.1111/cobi.12014.

O'Connor, C. D., G. M. Garfin, D. A. Falk, and T. W. Swetnam. 2011. Human pyrogeography: a new synergy of fire, climate and people is reshaping ecosystems across the globe. *Geography Compass* 5/6:329–50.

Organ, J. F., V. Geist, S. P. Mahoney, S. Williams, P. R. Krausman, G. R. Batcheller, T. A. Decker, R. Carmichael, P. Nanjappa, R. Regan, R. A. Medellin, R. Cantu, R. E. McCabe, S. Craven, G. M. Vecellio, and D. J. Decker. 2012. *The North American model of wildlife conservation*. The Wildlife Society Technical Review 12–04. Bethesda, MD: Wildlife Society.

Palmer, M. A., D. P. Lettenmaier, N. L. Poff, S. L. Postel, B. Richter, and R. Warner. 2009. Climate change and river ecosystems: protection and adaptation options. *Environmental Management* 44:1053–68. doi: 10.1007/S00267-009-9329-1.

Parks, S. A., S. Z. Dobrowski, J. D. Shaw, and C. Miller. 2019. Living on the edge: trailing edge forests at risk of fire-facilitated conversion to non-forest. *Ecosphere* 10(3):e02651. doi: 10.1002/ecs2.2651.

Parmesan, C. 2006. Ecological and evolutionary responses to recent climate change. *Annual Review of Ecology, Evolution, and Systematics* 37:637–69.

Parmesan, C., and G. A. Yohe. 2003. A globally coherent fingerprint of climate change impacts across natural systems. *Nature* 421:37–42.

Patz, J. A., A. K. Githeko, J. P. McCarty, S. Hussein, U. Confalonieri, and N. de Wet. 2003. Climate change and infectious diseases. In *Climate Change and Human Health: Risks and Responses*, ed. A. J. McMichael, et al., 103–32. Geneva: World Health Organization.

Paz, S. 2015. Climate change impacts on West Nile virus transmission in a global context. *Philosophical Transactions of the Royal Society B: Biological Sciences* 370(1665): 20130561. doi: 10.1098/rstb.2013.0561.

Peters, R., W. J. Ripple, C. Wolf, M. Moskwik, G. Carreón-Arroyo, G. Ceballos, A. Córdova, R. Dirzao, P. R. Ehrlich, A. D. Flesch, R. List, T. E. Lovejoy, R. F. Noss, J. Pacheco, J. K. Sarukhán, M. E. Soulé, E. O. Wilson, and J. R. B. Miller. 2018. Nature divided, scientists united: US-Mexico border wall threatens biodiversity and binational conservation. *BioScience* 68:740–43.

Peterson St-Laurent, G., L. E. Oakes, M. Cross, and S. Hagerman. 2021. R–R–T (resistance–resilience–transformation) typology reveals differential conservation approaches across ecosystems and time. *Communications Biology* 4:1–9.

Petrakis, R. E., W. J. van Leeuwen, M. L. Villarreal, P. Tashjian, R. Dello Russo, and C. A. Scott. 2017. Historical analysis of riparian vegetation change in response to shifting management objectives on the Middle Rio Grande. *Land* 6(29). doi:10.3390/land6020029

Poessel, S. A., S. W. Breck, and E. M. Gese. 2016. Spatial ecology of coyotes in the Denver metropolitan area: influence of the urban matrix. *Journal of Mammalogy* 97:1414–1427.

Poessel, S. A., S. W. Breck, T. L. Teel, S. Shwiff, K. R.Crooks, and L. Angeloni. 2013. Patterns of human-coyote conflicts in the Denver Metropolitan Area. *Journal of Wildlife Management* 77:297–305.

Reclamation. 2013. *Downscaled CMIP3 and CMIP5 Climate and Hydrology Projections: Release of Downscaled CMIP5 Climate Projections, Comparison with preceding Information, and Summary of User Needs*. Prepared by the US Department of the Interior, Bureau of Reclamation, Technical Services Center, Denver, CO.

Regehr, E. V., R. R. Wilson, K. D. Rode, M. C. Runge, and H. L. Stern. 2017. Harvesting wildlife affected by climate change: a modeling and management approach for polar bears. *Journal of Applied Ecology* 54:1534–43.

Ripple, W, J. J. A. Estes, R. L. Beschta, C. C. Wilmers, E. G. Ritchie, M. Hebblewhite, J. Berger, B. Elmhagen, M. Letnic, M. P. Nelson, O. J. Schmitz, D. W. Smith, A. D. Wallach, and A. J. Wirsing. 2014. Status and ecological effects of the world's largest carnivores. *Science* 343, article 1241484. doi: 10.1126/science.1241484.

Rocke, T. E., R. Russell, M. Samuel, R. Abbott, and J. Poje. 2020. *Effects of climate change on plague* exposure pathways and resulting disease dynamics. Final report SERDP Number 16 RC01-012.

Rohr, J. R., and J. M. Cohen. 2020. Understanding how temperature shifts could impact infectious disease. *PLOS Biology* 18(11):e3000938. https://doi.org/10.1371/journal.pbio.3000938.

Roos, C. I., and C. H. Guiterman. 2021. Dating the origins of persistent oak shrubfields in northern New Mexico using soil charcoal and dendrochronology. *Holocene* 31:1212–20.

Root, J. J., and A. M. Bosco-Lauth. 2019. West Nile virus associations in wild mammals: an update. *Viruses* 11(5), article 459. doi: 10.3390/v11050459.

Rorabaugh, J. C., J. Schipper, S. Avila-Villegas, J. A. Lamberton-Moreno, and T. Flood. 2020. Ecology of an ocelot population at the northern edge of the species' distribution in northern Sonora, Mexico. *PeerJ* 8:e8414. https://doi.org/10.7717/peerj.8414.

Rosenthal, J. 2009. Climate change and the geographic distribution of infectious diseases. *Ecohealth* 6:489–95. doi: 10.1007/s10393-010-0314-1.

Rupasinghe, R., B. B. Chomel, and B. Martínez-López. 2022. Climate change and zoonoses: a review of the current status, knowledge gaps, and future trends. *Acta Tropica* 226:106225. doi: 10.1016/j.actatropica.2021.106225.

Safford, H. D., and J. T. Stevens. 2017. *Natural range of variation (NRV) for yellow pine and mixed conifer forests in the Sierra Nevada, southern Cascades, and Modoc and Inyo National Forests, California, USA*. US Department of Agriculture, Forest Service, Pacific Southwest Research Station. General Technical Report no. PSW-GTR-256.

Santamouris, M. 2015. Analyzing the heat island magnitude and characteristics in one hundred Asian and Australian cities and regions. *Science of the Total Environment* 512:582–98.

Scheffers, B. R., L. De Meester, T. C. L. Bridge, A. A. Hoffmann, J. M. Pandolfi, R. T. Corlett, S. H. M. Butchart, P. Pearce-Kelly, K. M. Kovacs, D. Dudgeon, M. Pacifici, C. Rondinini, W. B. Foden, T. G. Martin, C. Mora, D. Bickford, and J. E. M. Watson. 2016. The broad footprint of climate change from genes to biomes to people. *Science* 354. doi: 10.1126/science.aaf7671.

Schloss, C. A., T. A. Nuñez, and J. J. Lawler. 2012. Dispersal will limit ability of mammals to track climate change in the Western Hemisphere.

Proceedings of the National Academy of Sciences 109:8606–11. doi: 10.1073/pnas.1116791109.

Schwalm, C. R., S. Glendona, and P. B. Duffya. 2020. RCP8.5 tracks cumulative CO_2 emissions. *PNAS* 117:19656–57. https://doi.org/10.1073/pnas.2007117117.

Scurlock, D. 1998. *From the rio to the sierra: an environmental history of the Middle Rio Grande Basin*. General Technical Report RMRS-GTR-5. Fort Collins, CO: US Department of Agriculture, Forest Service, Rocky Mountain Research Station.

Seth, A., S. A. Rauscher, M. Rojas, A. Giannini, and S. J. Camargo. 2011. Enhanced spring convective barrier for monsoons in a warmer world? *Climatic Change* 104:403–14. doi:10.1007/s10584-010-9973-8.

Seto, K. C., B. Güneralp, and L. R. Hutyra. 2012. Global forecasts of urban expansion to 2030 and direct impacts on biodiversity and carbon pools. *Proceedings of the National Academy of Sciences* 109:16083–88.

Stephens, S. L., J. M. Lydersen, B. M. Collins, D. L. Fry, and M. D. Meyer. 2015. Historical and current landscape-scale ponderosa pine and mixed conifer forest structure in the Southern Sierra Nevada. *Ecosphere* 6, article 79:1–63. doi: 10.1890/ES14-00379.1.

Stevens, J. T., B. M. Collins, J. D. Miller, M. P. North, and S. L. Stephens. 2017. Changing spatial patterns of stand-replacing fire in California conifer forests. *Forest Ecology and Management* 406:28–36.

Stevens-Rumann, C. S., and P. Morgan. 2019. Tree regeneration following wildfires in the western US: a review. *Fire Ecology* 15:15. 10.1186/s42408-019-0032-1.

Taylor, A. H., and C. N. Skinner. 2003. Spatial patterns and controls on historical fire regimes and forest structure in the Klamath Mountains. *Ecological Applications* 13:704–19.

Taylor, K. E., R. J. Stouffer, and G. A. Meehl. 2012. An overview of CMIP5 and the experiment design. *Bulletin of the American Meteorological Society* 93:485–98.

Tebaldi, C., D. Adams-Smith, and N. Heller. 2012. *The heat is on: U.S. temperature trends*. Princeton, NJ: Climate Central.

Tingley, M. W., W. B. Monahan, S. R. Beissinger, and C. Moritz. 2009. Birds track their Grinnellian niche through a century of climate change. *Proceedings of the National Academy of Sciences of the United States of America* 106:19637–43.

Townsend, N. T., and D. S. Gutzler. 2020. Adaptation of climate model projections of streamflow to account for anthropogenic flow impairment. *Journal of the American Water Resources Association* 56:586–98.

Traphagen, M. 2021. The Border Wall in Arizona and New Mexico—July 2021. *Wildlands Network*. Available at https://storymaps.arcgis.com/stories/8532c503c2084293bb8847407245228d. Accessed 15 January 2022.

Triepke, F. J., E. H. Muldavin, and M. M. Wahlberg. 2019. Using climate projections to assess ecosystem vulnerability at scales relevant to managers. *Ecosphere* 10(9):e02854. doi: 10.1002/ecs2.2854

Trombulak, S. C., and C. A. Frisell. 2000. Review of ecological effects of roads on terrestrial and aquatic communities. *Conservation Biology* 14:18–30.

Trouwborst, A., and A. Blackmore. 2020. Hot dogs, hungry bears, and wolves running out of mountain—International wildlife law and the effects of climate change on large carnivores. *Journal of International Wildlife Law and Policy* 23:212–38.

United Nations Population Fund (UNFPA). 2007. 2007 *State of World Population (UNFPA)*.

US Global Change Research Program (USGCRP). 2017. *Climate Science Special Report: Fourth National Climate Assessment*, Vol.1 [ed. D. J. Wuebbles, D. W. Fahey, K. A. Hibbard, D. J. Dokken, B. C. Stewart, and T. K. Maycock]. Washington, DC: US Global Change Research Program. doi: 10.7930/J0J964J6.

Van der Ree, R., J. A. G. Jaeger, E. van der Grift, and A. P. Clevenger. 2011. Effects of roads and traffic on wildlife populations and landscape function. *Ecology and Society* 16:48–57.

Van Vliet, M. T. H., F. Ludwig, J. J. G. Zwolsman, G. P. Weedon, and P. Kabat. 2011. Global river temperatures and sensitivity to atmospheric warming and changes in river flow. *Water Resources Research* 47:W02544. doi:10.1029/2010WR009198.

Viers, J. H., and D. E. Rheinheimer. 2011. Freshwater conservation options for a changing climate in California's Sierra Nevada. *Marine and Freshwater Research* 62:266–78.

Wahlberg, M., W. Robbie, S. Strenger, J. Triepke, D. Vandendriesche, E. Muldavin, J. Malusa, P. Shahani, J. Moreland, R. Crawford, and C. Bogart. 2020. *Ecological response units of the southwestern United States*. USDA Forest Service technical report, Southwestern Region, Regional Office, Albuquerque, NM.

Walker, R. B., J. D. Coop, S. A. Parks, and L. Trader. 2018. Fire regimes approaching historic norms reduce wildfire-facilitated conversion from forest to non-forest. *Ecosphere* 9(4): e02182. https://doi.org/10.1002/ecs2.2182.

Wang, J., H.-M. Kim, and E. K. M. Chang. 2017. Changes in Northern Hemisphere winter storm tracks under the background of arctic amplification. *Journal of Climate* 30:3705–24.

Weaver, G. V., N. Anderson, K. Garrett, A. T. Thompson, and M. J. Yabsley. 2022. Ticks and tick-borne pathogens in domestic animals, wild pigs, and off-host environmental sampling in Guam, USA. *Frontiers in Veterinary Science* 8, article 803424. doi: 10.3389/fvets.2021.803424.

Webb, B., P. Clack, and D. Walling. 2003. Water–air temperature relationships in a Devon river system and the role of flow. *Hydrological Process* 17:3069–84.

Westerling, A. L., H. G. Hidalgo, D. R. Cayan, and T. W. Swetnam. 2006. Warming and earlier spring increase western US forest wildfire activity. *Science* 313: 940–43.

Wild, M. A., T. M. Shenk, and T. R. Spraker. 2006. Plague as a mortality factor in Canada lynx (*Lynx canadensis*) reintroduced to Colorado. *Journal of Wildlife Diseases* 42:646–50.

Williams, A. P., C. D. Allen, A. K. Macalady, D. Griffin, C. A. Woodhouse, D. M. Meko, T. W. Swetnam, S. A. Rauscher, R. Seager, H. D. Grissino-Mayer, J. S. Dean, E. R. Cook, C. Gangodagamage, M. Cai, and N. G. McDowell. 2013. Temperature as a potent driver of regional forest drought stress and tree mortality. *Nature Climate Change* 3:292–97.

Williams, A. P., B. I. Cook, and J. E. Smerdon. 2022. Rapid intensification of the emerging southwestern North American megadrought in 2020–2021. *Nature Climate Change* 12: 232–34.

Williams, A. P., E. R. Cook, J. E. Smerdon, B. I. Cook,

J. T. Abatzoglou, K. Bolles, S. H. Baek, A. M. Badger, and B. Livneh. 2020. Large contribution from anthropogenic warming to an emerging North American megadrought. Science 368: 314–18. https://doi.org/10.1126/science.aaz9600.

Wilson, D. E., and R. A. Mittermeier, eds. 2009. *Handbook of the mammals of the world*. Vol. 1, *Carnivores*. Barcelona: Lynx Edicions.

Yuhas, R. H. 1996. *Loss of wetlands in the southwestern United States*. US Geological Survey Water-Supply Paper 2425, National Water Summary on Wetland Resources.

Zanin, M., F. Palomares, and A. L. Mangabeira Albernaz. 2021. Effects of climate change on the distribution of felids: mapping biogeographic patterns and establishing conservation priorities. *Biodiversity and Conservation* 30:1375–94.

Zhao, S., D. Zhou, and S. Liu. 2016. Data concurrency is required for estimating urban heat island intensity. *Environmental Pollution* 208:118–24.

CANADA LYNX (*LYNX CANADENSIS*)

Jacob S. Ivan

The Canada lynx (*Lynx canadensis*; hereafter lynx) is a medium-sized felid, standing ~50 cm (~1 ft 7 in) tall at the shoulders. Its pelage is thick and soft, usually gray to gray brown in color (though it can take on an auburn tinge during summer months), with subtle mottling. It has a short muzzle, a prominent ruff of dark-tipped fur that extends outward from the cheeks and below the chin, and conspicuous dark ear tufts that are usually as long as the ears themselves. Lynx are well adapted to life in areas that receive deep, soft snow. As such, they have a compact body but elongated legs (especially the rear) and unusually large feet, giving them an overall appearance that is simultaneously solid yet lanky. Lynx have a bobbed tail with a black tip.

Lynx can be easily confused with bobcats (*Lynx rufus*) as they are about the same size and share some distinctive features including ear tufts, a pronounced facial ruff, and a bobbed tail (see Chapter 7). However, bobcats are generally reddish tan rather than mottled gray, and they usually have some level of black spotting on their dorsum and flanks. Bobcats that lack distinctive spotting on their dorsum almost always have dark spots and/or dark bars on the inside of their legs, whereas lynx generally lack these marks. Also, the legs and feet of a bobcat are proportional to the body (i.e., bobcats are built like large house cats), and though their tail is bobbed, it is relatively longer than a lynx tail and the black tip covers only the top of the tail; the underside

Photo 6.1. Large male Canada lynx trotting through the snow after having been trapped, collared, and released near Leadville, Colorado. Note the gray-brown coloration, large snowshoe-like feet, long ear tufts, and ruff of fur that extends from ear to ear just behind the jaw. Photograph: © Steve Sunday/US Forest Service.

is white (Chapter 7). The ear tufts of a bobcat are much less pronounced (often less than half the size of lynx tufts), and bobcats often have a very distinctive white patch behind their ears that is more muted, or absent, in lynx.

On average, native-born adult lynx in the Southern Rockies—all descended from Canada and Alaska lynx (see below)—weigh about 9 kg (~20 lbs). The species is sexually dimorphic as males are generally larger (9.7 kg, range = 7.8–11.8 kg) than females (8.3 kg, range = 7.4–9.5 kg;

(*opposite page*) Photograph: Gerald and Buff Corsi © California Academy of Sciences.

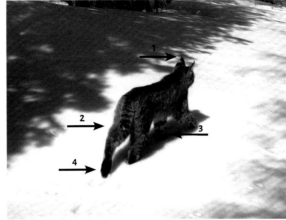

Photos 6.2a and b (*left to right*). Distinctive field marks delineate (a) Canada lynx from (b) bobcats. 1) Ear tufts are relatively longer (i.e., as long as the ear itself) in lynx compared to bobcats, and bobcats have a distinctive white patch flanked in black behind their ear; this patch is muted or absent in Canada lynx. 2) The tail of the bobcat is relatively longer than that of a lynx, and the underside is white whereas lynx have a shorter tail that is completely tipped in black. (3) Bobcats nearly always have dark bars on the inside of their legs, whereas lynx rarely exhibit this characteristic (a faint bar may be discernible on juveniles). (4) The legs and hind feet of a bobcat are proportional to the body, like a large housecat. Canada lynx have long hind legs and oversized feet. Photographs: © Colorado Parks and Wildlife.

Colorado Parks and Wildlife [CPW], unpubl. data). Like all felids, lynx have four toes, retractable claws, and digitigrade foot posture, meaning that they walk exclusively on their toes with their "heel" off the ground. Dense fur lines the space between their toes. Thus, tracks registered in the snow (and other substrates) generally lack distinctive outlines of digital and interdigital pads typical of other cats. Lynx prints are generally as wide (9–12 cm [3.5–4.7 in]) as they are long (Forrest 1988). The intergroup distance between prints (i.e., the distance between each single print) is usually 20–71 cm (7.9–28 in) for walking gaits; the straddle (i.e., the width of the trail or distance between the outermost edges of the prints) is roughly 15–23 cm (5.9–9.1 in) (Forrest 1988).

Lynx leave a unique signature in the compacted snow beneath their prints. Given typical winter conditions (i.e., cold weather and deep, soft snow), snow under the track of any animal will be compacted by the weight of the individual. This compacted snow then hardens in the cold. When tracks are discovered, one can excavate a given print by digging around and underneath it. Removing all loose snow from the print reveals the compressed, hard-packed snow underneath. Because lynx have disproportionately large feet (i.e., natural snowshoes), they tend to leave a unique compressed snow signature that is wide but relatively shallow (up to 7.6 cm [3 in]), like a pan. Other carnivores that overlap the lynx's range (e.g., bobcats, coyotes [*canis latrans*]) have much greater foot-loading than lynx and thus leave an "inverted cone" or columnar compression signature that is relatively narrow and extends deeper into the snow column.

Taxonomists currently consider Canada lynx to be part of the genus *Lynx* as per Wozencraft (1993; historically it had been considered as belonging to the genus *Felis*). As such, congeners only include bobcats in North America, Iberian Lynx (*Lynx pardinus*), which are restricted to a few remnant populations in southern Spain, and

 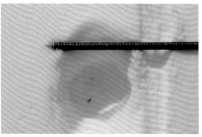 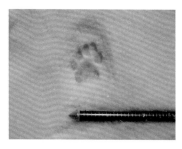

Photos 6.3a, b, and c (*left to right*). Canada lynx print with characteristic lack of distinct edges for toe and inter-digital pads (even in shallow snow) owing to the dense fur between the toes (a). Lynx tracks (b) can be more than twice the width of bobcat tracks (c), which often have more distinct pad patterns. Photographs: © Colorado Parks and Wildlife.

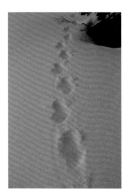  

Photos 6.4a, b, and c (*left to right*). Typical track patterns for Canada lynx walking through (a) deep, (b) shallow, and (c) warm snow conditions. Photographs: © Colorado Parks and Wildlife.

Photos 6.5a and b (*left to right*). (a) Shallow, disc-shaped compression signature formed from compacted snow under tracks left by Canada lynx. b) Deep, columnar-shaped compression signature left by a coyote. Photographs: © Colorado Parks and Wildlife.

the Eurasian lynx (*L. lynx*), which is distributed across boreal Europe and Asia. Canada lynx are descended from an ancestral felid lineage that likely came to North America across the Bering Strait approximately 8 Ma ago (mya) (Johnson et al. 2006). Bobcats separated from this lineage about 3 Ma; Canada lynx, along with the precursor to Eurasian and Iberian lynx, separated 1.6 Ma (Johnson et al. 2006). Canada lynx and bobcats have been known to hybridize in the wild (Schwartz et al. 2004), though such events are uncommon.

DISTRIBUTION

Historically, Canada lynx ranged throughout the boreal forest biome in North America from Newfoundland to Alaska, north to the tundra, and south as far as climatic and topographic conditions combine to create boreal-like conditions (McKelvey et al. 2000). Thus, in the contiguous United States, lynx were known to occur in peninsular extensions of boreal-like habitat protruding south out of Canada into the Cascades, Rocky Mountains, Great Lakes, and New England. In the Southern Rocky Mountains, verified historical records of lynx occurred as far south as southwestern Colorado (McKelvey et al. 2000).

There are no known verified records of lynx in New Mexico (Frey 2006). However, verified records in Colorado occur only a few kilometers north of the border in the San Juan and Sangre de Cristo mountain ranges. These ranges and the subalpine forest habitat (i.e., boreal-like habitat) contained within them extend continuously into northern New Mexico. Also, other subalpine forest species that are sympatric with the lynx in other parts of its range (e.g., snowshoe hares, *Lepus americanus*) occur in northern New Mexico. Furthermore, lynx are naturally rare, widely dispersed, and occur in high-elevation areas less likely to be visited by people. Thus, it is plausible, or even probable, that Canada lynx occurred in northern New Mexico historically but were never

officially documented there prior to their extirpation (Frey 2006).

Colorado Lynx Reintroduction

For most of the 20th century, the Canada lynx was considered an "unprotected predator" in Colorado. In 1970, however, harvest of lynx became prohibited, and in 1973 the species was designated as "state endangered." The last known lynx in Colorado (and the Southern Rocky Mountains in general) was trapped illegally in 1974 near Vail Mountain in the central portion of the state (McKelvey et al. 2000). Subsequently, the Colorado Division of Wildlife (CDOW, now Colorado Parks and Wildlife) completed 12 snow-tracking surveys aimed at documenting the remaining presence of lynx in the state, but none of these surveys produced conclusive evidence that the species was still extant (Seidel et al. 1998). Given that Colorado was isolated from the nearest known lynx population (in western Montana) by hundreds of kilometers of unsuitable habitat, CDOW concluded that reintroduction was the most practical means of re-establishing the species (Seidel et al. 1998). A reintroduction program was given a high probability of succeeding because many of the factors that likely led to the local demise of lynx had been mitigated. That is, since the extirpation of lynx in Colorado in the 1970s, the species' legal status had changed to fully protected, the indiscriminate predator poisoning of the early 20th century had ceased, more stringent hunting and trapping regulations had been adopted, wildlife and habitat conservation was being given more attention in political and management decisions, and in at least some places the landscape had recovered from past uses inconsistent with lynx conservation (Shenk 2003).

Initially, CDOW felt that reintroducing ~100 individuals to the state would suffice to produce a viable population that would survive well, reproduce, and eventually populate the species' historical range. Thus, 96 individuals were released into southwestern Colorado starting in 1999 and 2000.

Their subsequent monitoring revealed that initial (8-month) survival was lower than anticipated and that several individuals starved or were killed while crossing highways. High mortality especially affected both lynx that were hard-released (i.e., released into the wild as soon as was practical after transport from source populations in Canada and Alaska) and pregnant females. In response to these findings, release protocols were modified. Individuals were held longer in captivity and released in spring rather than winter; source animals were trapped before the breeding season to prevent the release of pregnant females.

The changes in release protocols led to adequate survival among reintroduced lynx. However, there was no evidence that reproduction occurred during the initial years of the project, and without reproduction to offset mortality, the goal of producing a viable population could not be attained. Thus, during the winter of 2002–2003, CDOW obtained permission to release more animals over the next several years in an effort to boost densities of lynx in what was now known as the Core Research Area (i.e., the high-elevation zones within the San Juan Mountain Range). The goal of the additional releases was to reach at least the lower end of the range of lynx densities documented in well-established northern populations. An additional 122 individuals were released from 2003 through 2006. Survival of these individuals was high, and in summer of 2003, six dens (sites where females give birth and raise litters) were discovered (Shenk 2003). Another 31 dens were discovered from 2004 to 2006. Also in 2006, CDOW documented the first kittens born to lynx that had themselves been born in Colorado.

Unfortunately, the number of dens declined in 2006, and in 2007 and 2008 CDOW did not document any evidence of kitten production despite monitoring 28–34 females each season (Shenk 2006). Coincident with this apparent lack of reproduction in Colorado's lynx population, densities of hares (the primary prey of lynx) experienced a clear and dramatic decline (Ivan 2011; Ivan et al. 2014), at least in central Colorado. However, by winter of 2009, hare populations in central Colorado began to recover (Ivan et al. 2014), and 5 dens were discovered during the summer of 2009. The number of kittens produced per female that year was lower than previously recorded, but another milestone, third-generation kittens, was documented. In 2010, another five dens were discovered and the rate of kitten production per female approached that documented during 2003–2005. Thus, lynx in Colorado appeared to have successfully weathered a downturn in density of their primary prey (snowshoe hares), mirroring a cycle of events that have played out in northern populations for centuries (Krebs et al. 2001a)

By June 2010, the Canada lynx population in Colorado had reached all the benchmarks for success outlined by CDOW at the outset of the program: 1) release protocols had been developed that led to high initial post-release survival; 2) long-term survival of lynx in Colorado was high; 3) reintroduced lynx showed fidelity to areas of good habitat; 4) reintroduced lynx bred successfully; 5) breeding led to successful reproduction and survival of kittens; 6) lynx born in Colorado reached breeding age and reproduced successfully; and 7) on average, over a decade, recruitment outpaced mortality. In September 2010, CDOW officially concluded that the reintroduction program had been successful. If future generations of lynx in Colorado were to replicate the levels and patterns of survival and productivity measured during the 10 years of the reintroduction and monitoring program, the trajectory of the Canada lynx population would remain positive.

Distribution in New Mexico

Many lynx released into Colorado traveled extensively either immediately after their release, as part of exploratory movements during summer (after which they often returned to their winter home range), or during dispersal (in which case they did not return to their former home range). Reintroduced lynx often ranged into New Mexico from the beginning of the reintroduction program in 1999 through 2011, when efforts to locate telemetered animals ceased. In total, 61 individuals (40 females, 21 males) were located in New Mexico at least once, and several individuals either spent significant time in that state, or spent time there periodically over the course of several years. Even when the data are truncated by excluding information from the first six months post-release (assuming that during this period, lynx movements may not have been representative of their normal activities as they explored their new environment), at least 35 individuals used New Mexico from time to time.

Most lynx activity was concentrated in the northern portion of the state where the San Juan Range extends from Colorado into Rio Arriba County (Map 6.1). Lynx also occurred, to a lesser extent, in the Sangre de Cristo Range in Taos and Colfax counties. The occurrence of monitored lynx in New Mexico coincided with subalpine habitat generally associated with lynx activity in other parts of the Rocky Mountains (Squires et al. 2010; Ivan et al. 2011). These movements are also consistent with the range of snowshoe hares, the main prey for lynx, in New Mexico (Frey and Malaney 2006; Malaney and Frey 2006). A few individual lynx made brief incursions into San Juan County; single individuals also explored the Jemez and Chuska mountains, and made more southerly forays down the Sangre de Cristo Mountains into Mora, Sante Fe, and San Miguel counties (Map 6.1). The lengthiest venture into New Mexico was made by a female released into Colorado in 2000. She lived in south-central Colfax County nearly continuously from July 2006 through July 2007, and traveled the farthest south into New Mexico of any monitored lynx: she was located 3 times in western San Miguel County in June 2007.

Most lynx activity in New Mexico occurred during summer and early fall when lynx are most likely to make large, exploratory movements (Buderman et al. 2018), especially if they are not reproductively active. However, between 2001 and 2011, seven individuals were known to winter in New Mexico. Four wintered in the San Juan Mountains, two in the Sangre de Cristos (one of these far to the east in lower-elevation forest), and one wintered just east of the Sangre de Cristo Mountains in open country west of Springer. Shenk (2009:15) noted that "New Mexico and Wyoming have been used continuously by lynx since the first year lynx were released in Colorado" and that "although no reproduction has been documented in New Mexico or Utah to date . . . the continuous presence of lynx within these states for over six years does suggest the potential for year-round residency of lynx and reproduction in those states."

In summary, Canada lynx likely occurred historically in northern New Mexico, at least intermittently. Following reintroduction of the species into Colorado, lynx are known to have occurred in New Mexico annually. Both males and females were found in the state at all times of year. Thus, there is potential for breeding activity, but whether any reproduction has truly occurred is unknown.

HABITAT ASSOCIATIONS

Regardless of the region in which lynx occur, they are strongly associated with forested habitats containing dense understory (Aubry et al. 2000; Mowat et al. 2000) and where winter snow is deep and soft. Their primary prey, snowshoe hares, are tightly tied to such habitats as well (see below). More specifically, in northern boreal forests, lynx select forest stands regenerating from fire or timber harvest, and prefer stands that are >20 years post-disturbance (Mowat et al. 2000). Lynx in these areas may also be found in mature stands, but use is in proportion to availability and such stands are thought to be of secondary

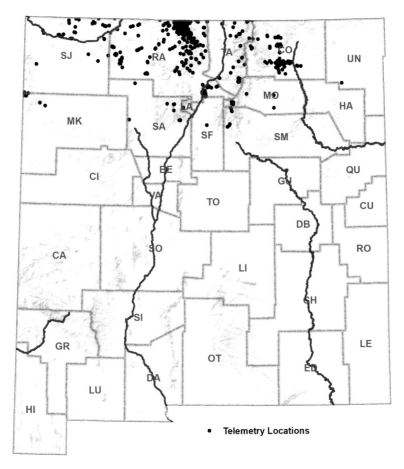

Map 6.1. Canada lynx (*Lynx canadensis*) occurrence data points for New Mexico during the period 1999–2011, from telemetered animals belonging to the population reintroduced in Colorado. A total of 61 individuals (40 females, 21 males) were located in New Mexico at least once. Several lynx either spent significant time in New Mexico or spent time there periodically over the course of several years.

Photos 6.6a and b (*left to right*). Within their home range, lynx select areas with dense cover (conifer saplings, conifer branches, down wood) near the ground as these are places snowshoe hares select for both cover and foraging. Photographs: © Colorado Parks and Wildlife

importance (Mowat et al. 2000). Similarly, in (broadly defined) boreal forests of the Pacific Northwest, lynx preferred 20-year-old regenerating stands of densely stocked lodgepole pine (*Pinus contorta*) (Koehler 1990; Aubry et al. 2000) and only used mature Engelmann Spruce (*Picea engelmannii*)-subalpine fir (*Abies concolor*) stands in rough proportion to their availability. Likewise, in boreal regions of eastern North America, lynx were associated with early seral forests, using mature forests less than, or equal to, expectation (Aubry et al. 2000; Hoving et al. 2004).

Compared to the rest of the species' range, habitat associations differ somewhat in the US Rocky Mountains. During winter in Montana, lynx selected mature stands composed primarily of Engelmann spruce and subalpine fir. Such stands were old enough that their structure was complex and multistoried, including a well-developed understory layer that was dense enough to support snowshoe hares (Squires et al. 2010). During summer, lynx broadened their preference to also include early seral stands with dense understory, similar to those described above for lynx in Washington and in Canada (Squires et al. 2010). The highest quality home ranges, as indexed by propensity of female lynx to produce kittens, contained large blocks of well-connected mature forest interspersed with regenerating stands (Kosterman et al. 2018).

Habitat selection by Canada lynx in Colorado is similar to that described for Montana. Relative use of the landscape by lynx during winter months was positively associated with high-elevation areas comprised mostly of Engelmann spruce-subalpine fir and aspen (*Populus tremuloides*). Lynx also selected areas with relatively steep slopes (i.e., mountainous terrain) and high topographic wetness (i.e., areas where aspect and topography tend to result in high soil moisture). Relative use of the landscape during winter was negatively associated with distance to large (>50 ha [>124 acres]) forest patches and

shrubland. Habitat use was positively associated with willow (*Salix* spp.), Douglas-fir (*Pseudotsuga menziesii*), and distance to road segments with heavy traffic, but these associations were weak (Ivan et al. 2011). In Colorado forests recently impacted by spruce beetle, lynx selected home ranges (and areas within home ranges) that were characterized by large diameter trees, live spruce-fir canopies and subcanopies, and an abundance of dense understory cover (Squires et al. 2020).

During summer, lynx habitat use in Colorado included a wider diversity of forest stands dominated by lodgepole pine. Also, use of landscapes dominated by aspen increased compared to winter, and high-volume road segments were avoided more strongly. However, the Douglas-fir forest–types were negatively associated with lynx use during summer (Ivan et al. 2011). Habitat associations have not been studied in New Mexico but would likely be similar to those documented in Colorado.

LIFE HISTORY

The life history of lynx has not been documented in New Mexico. However, individuals ranging into the state likely exhibit similar life history habits as in Colorado.

Diet and Foraging

Wherever Canada lynx have been studied, the majority of their diet consists of snowshoe hares (Aubry et al. 2000; Mowat et al. 2000). In Canada and Alaska, where lynx and hares undergo famous 10-year population cycles (Fig. 6.1), snowshoe hares make up 52–100% (average = 84%) of prey items in the diet during increases in the hare population and 20–91% during declines (average = 59%; Aubry et al. 2000). In more southern montane habitats, percent occurrence of snowshoe hares in the diet is intermediate between the highs and lows reported for northern populations (Aubry et al. 2000; Squires and Ruggiero 2007; Ivan and Shenk 2016). There is

Photos 6.7a and b (*left to right*). Canada lynx are heavily reliant on snowshoe hares (*Lepus americanus*) (a) throughout their range, including in the Southern Rockies where over a 10-year period, snowshoe hares made up an average of 70% of the winter diet of lynx. b) Red squirrels (*Tamiasciurus* spp.) constitute most of the remaining diet, and are an important alternative food source during years when snowshoe hare abundance declines. Photographs: © Colorado Parks and Wildlife.

significant annual variability in the proportion of hares in the diet in southern habitats, just as there is in more northerly populations. For example, from 1999 to 2009 in Colorado, the percent occurrence of hares in the diet of lynx ranged from 26 to 90% depending on the year (average = 70%, Ivan and Shenk 2016).

Red squirrels (*Tamiasciurus hudsonicus* and *T. femonti*) are usually the alternative prey of choice

(Aubry et al. 2000; Mowat et al. 2000), and they constituted an average of 27% of the prey items consumed by lynx in Colorado across 11 winters (Ivan and Shenk 2016). Generally, when the prevalence of hares in the diet declines (usually as hare abundance declines), that of red squirrels increases, and vice versa (Mowat et al. 2000). In Colorado, hare numbers in sprucefir habitat declined during the latter part of

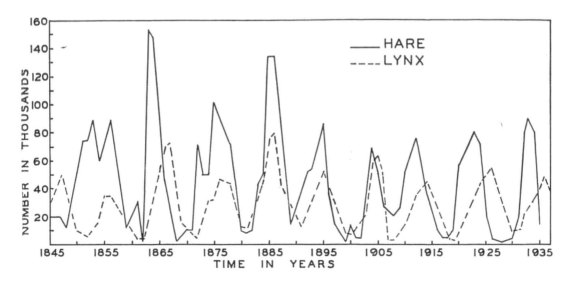

Figure 6.1. Ten-year snowshoe hare (*Lepus americanus*) and lynx population cycles in Canada. Reproduced with permission from E. P. Odum's (1953:134) *Fundamentals of Ecology* (Philadelphia: W. B. Saunders).

the reintroduction research (Ivan et al. 2014), at which time percent occurrence of red squirrels increased to 39–72% in the diet. Aside from snowshoe hares and red squirrels, lynx in Colorado fed on an assortment of additional species: mountain cottontails (*Sylvilagus nuttallii*), Pacific marten (*Martes caurina*), mice (*Peromyscus* spp.), white-tailed jackrabbits (*Lepus townsendii*), mule deer (*Odocoileus hemionus*), weasels (Mustelidae), white-tailed ptarmigan (*Lagopus leucura*), gray jay (*Perisoreus canadensis*), dusky grouse (*Dendragapus obscurus*), and woodpeckers (*Picoides* spp.). However, most of these species were rarely observed in the lynx's diet in Colorado, no more than once or twice through 11 winters of monitoring and across 4,600 km (2,860 miles) of tracking. Collectively they comprised <3% of recorded prey items (Ivan and Shenk 2016). Among all those species, the mountain cottontail represented the only slight exception, with 10 cottontails taken over that period (Ivan and Shenk 2016).

Nearly all food habit research has occurred during winter months. The few summer diet studies that have been conducted indicate that snowshoe hares are still the prey of choice when available, but alternative prey sources may be more important compared to winter (Aubry et al. 2000). Perhaps hares are more difficult to catch during summer, alternative prey are more abundant, or some combination of these and other factors contributes to the shift (Aubry et al. 2000; Mowat et al. 2000).

The density of snowshoe hares tends to be higher in areas with high horizontal cover (i.e., dense understory; Hodges 2000a, b; Ivan et al. 2014). Despite this, however, evidence suggests that lynx may be reluctant to hunt in areas with very dense understory because the high stem density and associated diminished visibility decrease their capture efficiency. Instead, lynx often choose to hunt in areas with intermediate cover and intermediate densities of hares. Thus, they seem to balance the opportunity to capture

hares with accessibility and increased probability of success (Mowat et al. 2000; Fuller et al. 2007; Fuller and Harrison 2010; Ivan and Shenk 2016). In Montana, however, lynx appear to simply hunt where hares are most abundant (Squires and Ruggiero 2007).

Lynx hunt by actively stalking prey or by ambushing them from a stationary position (O'Donoghue et al. 1998). Which strategy is employed likely depends on the abundance and behavior of prey, the physical condition of the predator, and the vegetative cover in which the lynx is hunting (Murray et al. 1995; Mowat et al. 2000), and as a result there is considerable variation in hunting mode across the range. O'Donoghue et al. (1998) found that lynx in the Yukon Territory turned to ambushing during years when hare abundance declined, postulating that hunting from ambush beds became more energy-efficient, and/or that lynx were more successful capturing alternative prey such as red squirrels using this approach. However, in southern portions of the lynx's range where hares occur at relatively low densities, evidence does not support this hypothesis. In Montana, lynx chased hares more often than they ambushed them (55% chased, 45% ambushed; Squires and Ruggiero 2007), and in Colorado 93% of chases were initiated while moving (i.e., stalking); only 7% were initiated from beds (i.e., ambushes; Ivan and Shenk 2016).

Murray et al. (1995) noted that lynx stalked prey when hunting in more open habitat, and opted for ambushing in denser cover. Perhaps the propensity for lynx to stalk hares in the southern portion of their range is due to the inherent patchiness of southern boreal-like forests. Lynx may have to travel farther—and consequently spend more time—to find a patch of good foraging habitat and therefore spend less time lying in wait at any suitable location. Or perhaps habitat patches in the Southern Rockies sustain too few hares to warrant an ambush strategy. Indeed, lynx in Colorado took hares at a rate of only 0.08 kills/km traveled (i.e., one hare every 12.5 km [7.8

mi]; Ivan and Shenk 2016), which is lower than rates reported for lynx in central Alberta (0.15–0.55 kills/km; Brand et al. 1976), Nova Scotia (0.13 kills/km; Parker et al. 1983), and Montana (0.12 kills/km; Squires and Ruggiero 2007).

O'Donoghue et al. (1998) did not detect any difference in hunting success between the two modes of capture, but Murray et al. (1995) found that lynx had greater success ambushing prey (46% success rate) compared to stalking prey (27% success rate). Lynx in Colorado were more successful at stalking hares (31%) and red squirrels (53%) compared to when they ambushed either prey species from a bed (16% and 31%, respectively; Ivan and Shenk 2016).

Reproduction and Social Behavior

Canada lynx generally breed from March through April. They are polygamous, meaning that each female is thought to breed with only one male, but males can breed with multiple (usually one to three) females (Armstrong et al. 2011). Gestation is approximately 70 days and the kittens are born between mid-May and early June (the earliest date a den was discovered in Colorado was 21 May). Kittens are born altricial; their eyes are shut and ears are closed until about two weeks of age (Mowat et al. 2000; Armstrong et al. 2011). They reach adult size by late winter of their first year and typically become independent around the time their mother breeds again, just before their first birthday (Armstrong et al. 2011). Yearlings generally disperse away from their natal territory, but female yearlings may carve out a home range adjacent to their mother as well. Male yearlings tend to disperse farther from their natal territory (Mowat et al. 2000).

In northern populations, females can become pregnant as yearlings when hares are abundant; otherwise both males and females first breed as two-year-olds (Mowat et al. 2000). During the increase and peak phases in the snowshoe hare cycle, pregnancy rates (number of females

Photos 6.8a, b, and c (*top to bottom*). Canada lynx kittens are born between mid-May and early June. They are born altricial, meaning they are underdeveloped at birth and require extensive parental care. However, they develop quickly. (a) The eyes and ears of newborn lynx kittens are closed, but they can more than double in size after a month (b), and they are fully grown by the winter following their birth (c: collared female with her yearling kitten). Photographs a–b: © Colorado Parks and Wildlife; Photograph c: © Diane Rocchia.

that become pregnant) and birth rates (number of females that give birth) are high, potentially approaching 100% (Mowat et al. 2000). The average litter size is four to five kittens per female. When hare populations crash, pregnancy rates of lynx become highly variable but overall decline. Birth rates decrease dramatically and may be near zero in some areas. Likewise, kitten survival also declines such that few, if any, young are recruited into the adult population when hares are scarce (Mowat et al. 2000).

In Colorado, successful reproduction was first observed in 2003, four years after the first lynx were released into the state and the same year as the original release population was augmented with more individuals. Between 2003 and 2010, Colorado Parks and Wildlife discovered 48 litters (CPW, unpubl. data). The average litter consisted of 2.75 kittens, which is consistent with litter sizes recorded during cyclical hare lows in northern populations. When the snowshoe hare population apparently declined in Colorado following the winter of 2007 (hare density declined precipitously in one portion of the study area where it was measured intensively and the decline was presumed to be widespread based on anecdotal information), no dens were found and no evidence of kitten recruitment was noted the following winter. However, denning resumed and kittens were produced again following a rebound in snowshoe hare abundance a few years later (Shenk 2009; Ivan et al. 2014).

Of the 48 dens discovered from 2003 to 2010, 47 (98%) were located in the southwestern portion of the state and one was located in Wyoming just over the Colorado border. All dens occurred in mature Engelmann spruce-subalpine fir forests on steep (mean slope = 30°), north-facing slopes at high (mean = 3,354 m [11,004 ft]) elevations. Most were situated in areas with extensive, thick downfall of trees (Shenk 2006, 2009). Some den sites occurred within hollow logs or stumps or in crevices created by rock outcroppings, but most often lynx placed their kittens directly on the ground in well-protected areas under downed trees (lynx do not tunnel or excavate or otherwise create a "den" as do other species). All dens occurred within the home range used by the female during the previous winter, but on average den sites were located on higher and steeper terrain compared to areas used routinely for traveling, bedding, and hunting (Shenk 2006). Few other studies have located and described den characteristics for lynx, but results were similar in all cases—lynx den sites are generally characterized as having extensive, complex structure (i.e., downed trees or other features that provide protection; Koehler 1990; Slough 1999; Mowat et al. 2000; Squires et al. 2008).

The home range size of the Canada lynx varies with season, sex, age, and geographic locality. In general, males have larger home ranges than females, lynx ranges become larger when snowshoe hares are scarce, and home ranges in the southern portion of the species' range are comparable to those in the northern part of the range during snowshoe hare declines (Aubry et al. 2000; Mowat et al. 2000). In Colorado, the average home range size of reproductively active individuals was 103 km² (~40 sq mi) for males, 75 km² (29 sq mi) for females, and often encompassed one to three major drainages (Shenk 2006). These average home range sizes are well within the range of values reported elsewhere (Aubry et al. 2000). Lynx exhibit a form of intrasexual territoriality in which male home ranges typically do not overlap, at least not the core home ranges (males may overlap other males at the fringes of their home range), but male and female home ranges can overlap completely. Females seem more likely to tolerate other females within their home range. It is thought that related females are especially tolerant of each other as daughters have been known to establish home ranges within or immediately adjacent to their mother. Related females may even hunt together on occasion or den near each other (Mowat et al. 2000).

Photos 6.9a, b, and c (*clockwise from top left*). Lynx do not excavate dens. Instead, the young are usually born and raised directly on the ground in well-protected areas beneath extensive downfall such as those depicted in photos a and b. Photograph 6.9c shows a kitten in a den under a thick cover of downed trees. Photographs: © Colorado Parks and Wildlife.

Lynx are known to make large exploratory movements outside of their usual home range (Squires and Laurion 2000; Squires and Oakleaf 2005; Shenk 2006; Buderman et al. 2018). Increased movements are more likely during summer months, and for individuals that are not reproductively active (Buderman et al. 2018). On average, Colorado lynx engaged in just over two exploratory movements each, for a mean distance of 107 km (66 mi) per movement bout, though some distances exceeded 1000 km (~620 mi). Individuals often resettled in the area from which they started. Lynx from Colorado made exploratory movements that took them to Kansas, South Dakota, Nebraska, Iowa, Nevada, Utah, Wyoming, Montana, and New Mexico. One individual that was translocated to Colorado from British Columbia even made its way back to Canada after living in the San Juan Mountains for four years.

Photo 6.10. A collared female keeps a close eye on her den and kittens while researchers work quickly to take measurements and mark the kittens with passive integrated transponder (PIT) tags. Photograph: © Colorado Parks and Wildlife.

Photos 6.11a, b, and c (*clockwise from top left*). The home range of a Canada lynx in southern Colorado often includes several large drainages with large patches of mature spruce-fir forest, such as those pictured here. Photographs: © Colorado Parks and Wildlife.

These movements may not be considered "normal" for lynx if they were somehow induced or precipitated by the reintroduction process. Nevertheless, they demonstrate the incredible ability lynx have to cover large distances through a variety of vegetation types (including poor lynx habitat) over relatively short periods of time. The purpose of these movements is unknown, but perhaps they provide opportunity to assess habitat quality elsewhere in the event an individual is forced to disperse due to competition or local collapse of a prey population (Aubry et al. 2000).

Dispersal (permanent movement away from a home range with the intent to establish a new home range elsewhere) most often occurs with yearling lynx when the mother-kitten unit is broken up and kittens leave their natal range (i.e., mother's home range) to establish their own. However, older individuals may also disperse when prey sources collapse or when they are pushed out of their home range via competition (Mowat et al. 2000). Dispersal events can result in an individual moving to an area adjacent to its previous home range (i.e., as can be the case with dispersal from the natal range by female kittens), but movements of 500 km (~310 mi) or more have been recorded for adults during declines in hare abundance (Mowat et al. 2000).

Interspecific Interactions

The interaction between Canada lynx and snowshoe hares is well documented and famous for the regular 10-year population cycle exhibited by the two species in Canada (Hodges 2000a; Krebs et al. 2001a, b; Fig. 6.1). Large-scale field experiments in the Kluane Region of the Yukon Territory indicated that the cycle is largely driven by predation on snowshoe hares, though food availability for the hares, production of secondary compounds by plants that are browsed heavily by hares, and disease may all play varying roles as well (Krebs et al. 2001a).

In simplified terms, the cycle (Fig. 6.1) proceeds as follows: As hare numbers increase, lynx and other predator populations respond by increasing as well (i.e., the numerical response, lagged by a few years). The functional response of predators also follows suit—lynx and other predators focus more on hares as opposed to alternate prey. The reproductive output of hares then begins to decline during the increase phase of the population cycle (Krebs et al. 2001a), perhaps due to chronic, sublethal stress from increased predation pressure (Sheriff et al. 2009) and/or from stress due to malnutrition as food resources become more scarce. Following the loss of reproductive output, the survival of hares declines rapidly as increased numbers of lynx and other predators prey heavily on them (malnutrition may play a role in increasing susceptibility to predation). This causes a crash in the hare population, which precipitates a subsequent crash in the lynx population a few years later. Once predation pressure is relieved, hares survive at a higher rate and begin to reproduce more effectively, and the cycle repeats itself (Krebs et al. 2001a). Thus, the hare-lynx cycle is mediated strongly by predation, but food availability likely plays a role as well.

Whether snowshoe hare (and lynx) populations truly cycle in the southern portions of these two species' ranges is debatable. Hodges (2000b) collected historical harvest, pellet transect, and track transect data for hares from various jurisdictions in the southern portion of the snowshoe hare's range and concluded that southern populations are cyclical, but the amplitude of the cycle is diminished compared to northern populations. However, Dolbeer and Clark (1975) and Wolff (1980) both contend that hare populations in the western United States are relatively stable. At a minimum, hare and lynx populations in the southern portion of their range do not exhibit the typical, high-amplitude cycle observed farther north. This may be because habitat in the southern part of either species' range is patchy and discontinuous compared to that of northern populations. When hare populations begin to increase, excess hares are forced to disperse into and across poor habitat where survival is poor (Dolbeer and Clark 1975; Wolff 1980; Hodges 2000b). Also, a larger suite of generalist predators in southern systems may prevent hare populations from ever becoming so exceedingly large that they then crash. At the same time, southern systems provide a wider variety of prey available to lynx and other predators in their southern range. Switching to these alternative prey rather than focusing almost exclusively on hares may disrupt or diminish cyclicity as well (Wolff 1980; Hodges 2000b).

Canada lynx are well-adapted to life in subalpine forests that receive deep, soft snow whereas most other carnivores in the Southern Rockies are not. Thus, in one sense, lynx are relatively isolated from competition during winter, at least from similar sized competitors such as bobcats, coyotes, and cougars (*Puma concolor*), though less so the Rocky Mountain lineage of the red fox (*Vulpes fulva macroura*) also adapted for life on snow. However, lynx may compete with this suite of species for food and/or space at the lower elevational limits of their range, where winter conditions are not as severe, or during summer months (Buskirk et al. 2000a, 2000b). The fragmented, high-relief nature of the landscape in the Southern Rockies likely exacerbates this competition because open, dry forests and shrub-steppe habitats are often closely juxtaposed to good lynx habitat (Buskirk et al. 2000b). Road networks and timber management may further fragment subalpine forests and facilitate occupancy by bobcats, cougars, and coyotes. Additionally, snow compaction by snowmobiles and other winter sports activities may encourage movement of these species up into the subalpine zone during winter, when lynx would normally have a competitive advantage in the soft snow. However, the only research conducted on this topic concluded that the overall influence of snowmobile trails on coyote movements appeared minimal, at least in Montana (Kolbe et al. 2007).

Because cougars are known to prey on lynx and they are much larger and more aggressive (Aubry et al. 2000; Buskirk et al. 2000b), their presence directly interferes with the ability of lynx to persist in an area. Similarly, bobcats can be slightly larger than lynx and are generally the more dominant competitor (Buskirk et al. 2000b), and they tend to exclude lynx in areas where the two species interact closely (Parker et al. 1983; Peers et al. 2013). Coyotes and North American red foxes (*Vulpes fulva*; Chapter 15) may represent strong competition for lynx in the Southern Rockies as they can exploit a wide variety of habitats and prey items (including subalpine forests and snowshoe hares, respectively), and both species have a high reproductive potential (Byrne 1998; Buskirk et al. 2000b).

STATUS AND MANAGEMENT

The Canada lynx is currently listed as a federally Threatened species in the contiguous United States (US Fish and Wildlife Service [USFWS] 2000). The "Distinct Population Segment" that is specifically listed is composed of lynx occurring in six geographic units: 1) northern Maine, 2) northeastern Minnesota, 3) northwestern Montana/northeastern Idaho, 4) north-central Washington, 5) the Greater Yellowstone Area, and 6) western Colorado (USFWSe 2017). However, the protections of the Endangered Species Act extend to lynx wherever they occur, including New Mexico (USFWS 2013). The main factor leading to the decision to list the Canada lynx as a Threatened species was the inadequacy of regulatory mechanisms and the lack of guidance for the conservation of lynx on public lands administered by the US Forest Service and the Bureau of Land Management (USFWS 2000). In response to listing, National Forest plans in Wyoming and Colorado were

Photo 6.12. A Canada lynx caught on camera in the San Juan Mountains in southern Colorado as part of ongoing, non-invasive monitoring to determine population status in the area. Photograph: © Colorado Parks and Wildlife.

amended to include specific recommendations for managing lynx habitat where appropriate (Ruediger et al. 2000; USFWS 2008). In general, these changes involve specific recommendations for creating and/or maintaining dense understory cover in subalpine forests. Public lands in New Mexico are not required to follow these recommendations. Because the lynx is a listed species, no hunting or fur harvest is allowed. In southwestern Colorado, special regulations are in place to minimize the impact that recreational trapping for other species has on lynx (https://cpw.state.co.us/Documents/RulesRegs/Regulations/Ch03.pdf).

Fifteen lynx from Colorado are known to have died in New Mexico. Of these, four were shot, two were hit by a vehicle, one starved, and eight died of unknown causes (CPW, unpubl. data). During the Colorado reintroduction project, gunshots and vehicle collisions represented the most common known sources of mortality (Devineau et al. 2010). However, both seem to have waned as a mortality source in the decade since the project ended. Underreporting could be a factor now that individuals are no longer radio-collared. However, recent monitoring of lynx that have home ranges near state highways in Colorado suggests that they cross these highways often (every other day on average), most often at night (i.e., 1 am) when traffic volume is lowest (Baigas et al. 2017).

Many of the monitored individuals (62%) did not show any avoidance behavior toward roads within their home range (Baigas et al. 2017). Thus, current evidence suggests that lynx may be relatively adept at managing two-lane highways and highway crossings.

Canada lynx are susceptible to common felid pathogens and other North American wildlife diseases. Of the 52 lynx mortalities documented between 1999 and 2003 in Colorado, six involved individuals that tested positive for sylvatic plague (*Yersinia pestis*), a bacterial disease carried by fleas (Wild et al. 2006). Plague is known to be maintained within small mammal populations in North America, and it is likely that lynx contracted this disease via contact with infected prey (Wild et al. 2006). Eventually plague was listed as the cause of death for nine (6%) of the 148 mortalities recorded during the Colorado reintroduction (CPW, unpubl. data). An outbreak of canine distemper virus (CDV) was documented in six lynx collected on Cape Breton Island in Nova Scotia in the late 1990s (Daoust et al. 2009). Infected individuals eventually died due to encephalitis induced by the virus. Biek et al. (2002) sampled lynx from six study areas, ranging from Montana to central Alaska, for antibodies to feline panleukopenia virus (FPV), feline coronavirus, CDV, feline calicivirus, feline herpesvirus, sylvatic plague, tularemia (*Francisella tularensis*), and feline immunodeficiency virus (FIV). Except for FIV, they found evidence of exposure to each pathogen in at least one study area, but overall prevalence was low for all pathogens (<8% of the 215 individuals tested for any given disease). Thus, while lynx are apparently susceptible to a variety of infectious diseases, Biek et al. (2002) concluded that the species rarely encounters the underlying pathogens in the wild.

The outlook for the lynx in the Southern Rockies (including northern New Mexico) remains good, an assessment somewhat tempered by some of the projections linked to climate change (see Chapter 5). Ongoing non-invasive monitoring efforts in the San Juan Mountains of Colorado indicate that the population of lynx there is stable (Odell et al. 2020). However, because Canada lynx are highly adapted to boreal-like regions that receive deep, soft snow, they are naturally a species of concern relative to climate change. The spruce-fir forests that lynx prefer in the Southern Rockies are expected to persist in the next 50–100 years when matched against expected changes in future conditions. Without considering the influence of wildfires (see below), the range of this forest type may slowly shift upward in elevation, and overall extent of these types of forests may decline, but current assessments suggest that spruce-fir forests in the Southern Rockies are only moderately vulnerable to climate change (Decker and Fink 2014; Rice et al. 2018). Extensive spruce beetle outbreaks (natural disturbances that have been amplified by climate change) have impacted much of the lynx's range in the Southern Rockies, with significant tree mortality throughout (Bentz et al. 2009). However, recent research suggests that lynx still persist in these areas and select habitat in much the same way they did before the beetle outbreak (Squires et al. 2020). This may be because neither snowshoe hare occupancy (Ivan et al. 2018) nor density (CPW, unpubl. data) seems to have been impacted significantly by beetle-induced changes to overstory vegetation. Red squirrel occupancy and density have both declined, however, likely due to significant loss of mature, overstory trees and much of the associated seed crop (cones) that red squirrels rely on (Ivan et al. 2018; Latif et al. 2020). This raises the potential threat that lynx may be limited in secondary food resources the next time snowshoe hares density declines. Perhaps the greatest risk to lynx persistence in the Southern Rockies is wildfire (see Chapter 5). Wildfire occurrence, and severity, is likely to increase with predicted future climate conditions (Rocca et al. 2014), and recent fire seasons have precipitated some of the largest, most severe wildfires on record in the Southern Rockies. Such large, high intensity fires can quickly eliminate large swaths of habitat for Canada lynx; regeneration of these forests to the mature, multistoried conditions that support lynx would likely take centuries.

LITERATURE CITED

Armstrong, D. M., J. P. Fitzgerald, and C. A. Meaney. 2011. *Mammals of Colorado*. 2nd edition. Boulder: University Press of Colorado.

Aubry, K. B., G. M. Koehler, and J. R. Squires. 2000. Ecology of Canada lynx in southern boreal forests. In *Ecology and conservation of lynx in the United States*, ed. L. F. Ruggiero, K. B. Aubry, S. W. Buskirk, G. M. Koehler, C. J. Krebs, K. S. McKelvey, and J. R. Squires, 373–96. Fort Collins, CO: Department of Agriculture, Forest Service, Rocky Mountain Research Station.

Baigas, P. E., J. R. Squires, L. E. Olson, J. S. Ivan, and E. K. Roberts. 2017. Using environmental features to model highway crossing behavior of Canada lynx in the Southern Rocky Mountains. *Landscape and Urban Planning* 157:200–13.

Bentz, B. J., J. A. Logan, J. A. Macmahon, C. D. Allen, M. P. Ayres, E. Berg, A. Carroll, M. Hansen, J. Hicke, L. Joyce, W. Macfarlane, S. Munson, J. Negron, T. Paine, J. Powell, K. Raffa, J. Regniere, M. Reid, B. Romme, S. J. Seybold, D. Six, D. Tomback, J. Vandygriff, T. Veblen, M. White, J. Witcosky, and D. Wood. 2009. *Bark beetle outbreaks in western North America: causes and consequences*. Bark Beetle Symposium, Snowbird, Utah. Salt Lake City: University of Utah Press.

Biek, R., R. L. Zarnke, C. Gillin, M. Wild, J. R. Squires, and M. Poss. 2002. Serologic survey for viral and bacterial infections in western populations of Canada lynx (*Lynx canadensis*). *Journal of Wildlife Diseases* 38:840–45.

Brand, C. J., L. B. Keith, and C. A. Fischer. 1976. Lynx responses to changing snowshoe hare densities in central Alberta. *Journal of Wildlife Management* 40:416–28.

Buderman, F. E., M. B. Hooten, J. S. Ivan, and T. M. Shenk. 2018. Large-scale movement behavior in a reintroduced predator population. *Ecography* 41:126–39.

Buskirk, S. W., L. F. Ruggiero, K. B. Aubry, D. E. Pearson, J. R. Squires, and K. S. McKelvey. 2000a. Comparative ecology of lynx in North America. In *Ecology and conservation of lynx in the United States*, ed. L. F. Ruggiero, K. B. Aubry, S. W. Buskirk, G. M. Koehler, C. J. Krebs, K. S. McKelvey, and J. R. Squires, 397–417. Fort Collins, CO: Department of Agriculture, Forest Service, Rocky Mountain Research Station.

Buskirk, S. W., L. F. Ruggiero, and C. J. Krebs. 2000b. Habitat fragmentation and interspecific competition: implications for lynx conservation. In *Ecology and conservation of lynx in the United States*, ed. L. F. Ruggiero, K. B. Aubry, S. W. Buskirk, G. M. Koehler, C. J. Krebs, K. S. McKelvey, and J. R. Squires, 83–100. Fort Collins, CO: Department of Agriculture, Forest Service, Rocky Mountain Research Station.

Byrne, G. 1998. *Core area release site selection and considerations for a Canada lynx reintroduction in Colorado*. Glenwood Springs: Colorado Division of Wildlife.

Daoust, P. Y., S. R. McBurney, D. L. Godson, M. W. G. Van De Bildt, and A. D. M. E. Osterhaus.

2009. Canine distemper virus-associated encephalitis in free-living lynx (*lynx canadensis*) and bobcats (*lynx rufus*) of Eastern Canada. *Journal of Wildlife Diseases* 45:611–24.

Decker, K., and M. Fink. 2014. *Colorado Wildlife Action Plan enhancement: climate change vulnerability assessment*. Fort Collins: Colorado Natural Heritage Program.

Devineau, O., T. M. Shenk, G. C. White, P. F. Doherty Jr., P. M. Lukacs, and R. H. Kahn. 2010. Evaluating the Canada lynx reintroduction programme in Colorado: patterns in mortality. *Journal of Applied Ecology* 47:524–31.

Dolbeer, R. A., and W. R. Clark. 1975. Population ecology of snowshoe hares in the central Rocky Mountains. *Journal of Wildlife Management* 39:535–49.

Forrest, L. R. 1988. *Field guide to tracking animals in snow*. Mechanicsbury, PA: Stackpole Books.

Frey, J. K. 2006. Inferring species distributions in the absence of occurrence records: an example considering wolverine (*Gulo gulo*) and Canada lynx (*Lynx canadensis*) in New Mexico. *Biological Conservation* 130:16–24.

Frey, J. K., and J. L. Malaney. 2006. Snowshoe hare (*Lepus americanus*) and mountain cottontail (*Sylvilagus nuttallii*) biogeography at their southern range limit. *Journal of Mammalogy* 87:1175–82.

Fuller, A. K., and D. J. Harrison. 2010. Movement paths reveal scale-dependent habitat decisions by Canada lynx. *Journal of Mammalogy* 91:1269–79.

Fuller, A. K., D. J. Harrison, and J. H. Vashon. 2007. Winter habitat selection by Canada lynx in Maine: prey abundance or accessibility? *Journal of Wildlife Management* 71:1980–86.

Hodges, K. E. 2000a. The ecology of snowshoe hares in northern boreal forests. In *Ecology and conservation of lynx in the United States*, ed. L. F. Ruggiero, K. B. Aubry, S. W. Buskirk, G. M. Koehler, C. J. Krebs, K. S. McKelvey, and J. R. Squires, 117–61. Fort Collins, CO: Department of Agriculture, Forest Service, Rocky Mountain Research Station.

———. 2000b. Ecology of snowshoe hares in southern boreal and montane forests. In *Ecology and conservation of lynx in the United States*, ed. L. F. Ruggiero, K. B. Aubry, S. W. Buskirk, G. M. Koehler, C. J. Krebs, K. S. McKelvey, and J. R. Squires, 163–206. Fort Collins, CO: Department of Agriculture, Forest Service, Rocky Mountain Research Station.

Hoving, C. L., D. J. Harrison, W. B. Krohn, W. J. Jakubas, and M. A. McCollough. 2004. Canada lynx *Lynx canadensis* habitat and forest succession in northern Maine, USA. *Wildlife Biology* 10:285–94.

Ivan, J. S. 2011. Density, demography, and seasonal movement of snowshoe hares in central Colorado. PhD dissertation, Colorado State University, Fort Collins.

Ivan, J. S., M. Rice, P. M. Lukacs, T. M. Shenk, D. M. Theobald, and E. Odell. 2011. *Predicted lynx habitat in Colorado*. Colorado Division of Parks and Wildlife.

Ivan, J. S., A. E. Seglund, R. L. Truex, and E. S. Newkirk. 2018. Mammalian responses to changed forest conditions resulting from bark beetle outbreaks in the southern Rocky Mountains. *Ecosphere* 9:e02369. https://doi.org/10.1002/ecs2.2369

Ivan, J. S., and T. M. Shenk. 2016. Winter diet and hunting success of Canada lynx in Colorado. *Journal of Wildlife Management* 80:1049–58.

Ivan, J. S., G. C. White, and T. M. Shenk. 2014. Density and demography of snowshoe hares in central Colorado. *Journal of Wildlife Management* 78:580–94.

Johnson, W. E., E. Eizirik, J. Pecon-Slattery, W. J. Murphy, A. Antunes, E. Teeling, and S. J. O'Brien. 2006. The late Miocene radiation of modern Felidae: a genetic assessment. *Science* 311:73–77.

Koehler, G. M. 1990. Population and habitat characteristics of lynx and snowshoe hares in north central Washington. *Canadian Journal of Zoology* 68:845–51.

Kolbe, J. A., J. R. Squires, D. H. Pletscher, and L. F. Ruggiero. 2007. The effect of snowmobile trails on coyote movements within lynx home ranges. *Journal of Wildlife Management* 71:1409–18.

Kosterman, M. K., J. R. Squires, J. D. Holbrook, D. H. Pletscher, and M. Hebblewhite. 2018. Forest structure provides the income for reproductive success in a southern population of Canada lynx. *Ecological Applications* 28:1032–43.

Krebs, C. J., R. Boonstra, S. Boutin, and A. R. E. Sinclair. 2001a. What drives the 10-year cycle of snowshoe hares? *Bioscience* 51:25–35.

Krebs, C. J., S. Boutin, and R. Boonstra. 2001b.

Ecosystem Dynamics of the Boreal Forest. New York: Oxford University Press.

Latif, Q. S., J. S. Ivan, A. E. Seglund, D. L. Pavlacky, and R. L. Truex. 2020. Avian relationships with bark beetle outbreaks and underlying mechanisms in lodgepole pine and spruce-fir forests of Colorado. *Forest Ecology and Management* 464:118043.

Malaney, J. L., and J. K. Frey. 2006. Summer habitat use by snowshoe hare and mountain cottontail at their southern zone of sympatry. *Journal of Wildlife Management* 70:877–83.

McKelvey, K. S., K. B. Aubry, and Y. K. Ortega. 2000. History and distribution of lynx in the contiguous United States. In *Ecology and conservation of lynx in the United States*, ed. L. F. Ruggiero, K. B. Aubry, S. W. Buskirk, G. M. Koehler, C. J. Krebs, K. S. McKelvey, and J. R. Squires, 207–64. Fort Collins, CO: Department of Agriculture, Forest Service, Rocky Mountain Research Station.

Mowat, G., K. G. Poole, and M. O'Donoghue. 2000. Ecology of lynx in northern Canada and Alaska. In *Ecology and conservation of lynx in the United States*, ed. L. F. Ruggiero, K. B. Aubry, S. W. Buskirk, G. M. Koehler, C. J. Krebs, K. S. McKelvey, and J. R. Squires, 265–306. Fort Collins, CO: Department of Agriculture, Forest Service, Rocky Mountain Research Station.

Murray, D. L., S. Boutin, M. O'Donoghue, and V. O. Nams. 1995. Hunting behaviour of a sympatric felid and canid in relation to vegetative cover. *Animal Behaviour* 50:1203–10.

Odell, E., J. S. Ivan, and S. Wait. 2020. Canada lynx monitoring in Colorado. In *Wildlife Research Report—Mammals*, 2–6. Fort Collins: Colorado Parks and Wildlife. https://spl.cde.state.co.us/artemis/nrserials/nr616internet/nr616201920internet.pdf.

O'Donoghue, M., S. Boutin, C. J. Krebs, D. L. Murray, and E. J. Hofer. 1998. Behavioural responses of coyotes and lynx to the snowshoe hare cycle. *Oikos* 82:169–83.

Odum, E. P. 1953. *Fundamentals of Ecology.* Philadelphia: W. B. Saunders.

Parker, G. R., J. W. Maxwell, L. D. Morton, G. E. J. Smith, and S. Morton. 1983. The ecology of the lynx (*Lynx canadensis*) on Cape Breton Island. *Canadian Journal of Zoology* 61:770–786.

Peers, M. J. L., D. H. Thornton, and D. L. Murray. 2013. Evidence for large-scale effects of competition: Niche displacement in Canada lynx and bobcat. *Proceedings of the Royal Society B: Biological Sciences* 280:20132495.

Rice, J. R., C. Regan, D. Winters, R. L. Truex, and L. A. Joyce. 2018. Subalpine spruce-fir ecosystems: vulnerability to nonclimate and climate stressors in the U.S. Forest Service Rocky Mountain region. In *Climate change vulnerability assessment of aquatic and terrestrial ecosystems in the US Forest Service Rocky Mountain region*, ed. J. R. Rice, L. A. Joyce, C. Regan, D. Winters, and R. Truex, 81–111. Gen. Tech. Rep. RMRS-GTR-376. Fort Collins, CO: US Department of Agriculture, Forest Service, Rocky Mountain Research Station.

Rocca, M. E., P. M. Brown, L. H. MacDonald, and C. M. Carrico. 2014. Climate change impacts on fire regimes and key ecosystem services in Rocky Mountain forests. *Forest Ecology and Management* 327:290–305.

Ruediger, B., J. Claar, S. Gniadek, B. Holt, L. Lyle, S. Mighton, B. Naney, G. Patton, T. Rinaldi, J. Trick, A. Vendehey, F. Wahl, N. Warren, D. Wenger, and A. Williamson. 2000. *Canada lynx conservation assessment and strategy.* 2nd ed. Missoula, MO: US Department of Agriculture, Forest Service, US Department of Interior, Fish and Wildlife Service, Bureau of Land Management, National Park Service.

Schwartz, M. K., K. L. Pilgrim, K. S. McKelvey, E. L. Lindquist, J. J. Claar, S. Loch, and L. F. Ruggiero. 2004. Hybridization between Canada lynx and bobcats: genetic results and management implications. *Conservation Genetics* 5:349–355.

Seidel, J., B. Andree, S. Berlinger, K. Buell, G. Byrne, B. Gill, D. Kenvin, and D. Reed. 1998. *The conservation and reestablishment of lynx and wolverine in the Southern Rocky Mountains.* Colorado Division of Wildlife.

Shenk, T. M. 2003. *Post-release monitoring of lynx (Lynx canadensis) reintroduced to Colorado.* Colorado Division of Wildlife.

———. 2006. *Post-release monitoring of lynx (Lynx canadensis) reintroduced to Colorado.* Colorado Division of Wildlife.

———. 2009. *Post-release monitoring of lynx (Lynx canadensis) reintroduced to Colorado.* Colorado Division of Wildlife.

Sheriff, M. J., C. J. Krebs, and R. Boonstra. 2009. The sensitive hare: sublethal effects of predator stress on reproduction in snowshoe hares. *Journal of Animal Ecology* 78:1249–58.

Slough, B. G. 1999. Characteristics of Canada lynx, *Lynx canadensis*, maternal dens and denning habitat. *Canadian Field-Naturalist* 113:605–8.

Squires, J. R., N. J. DeCesare, J. A. Kolbe, and L. F. Ruggiero. 2008. Hierarchical den selection of Canada lynx in western Montana. *Journal of Wildlife Management* 72:1497–506.

———. 2010. Seasonal resource selection of Canada lynx in managed forests of the northern Rocky Mountains. *Journal of Wildlife Management* 74:1648–60.

Squires, J. R., J. D. Holbrook, L. E. Olson, J. S. Ivan, R. W. Ghormley, and R. L. Lawrence. 2020. A specialized forest carnivore navigates landscape-level disturbance: Canada lynx in spruce-beetle impacted forests. *Forest Ecology and Management* 475:10.1016/j.foreco.2020.118400

Squires, J. R., and T. Laurion. 2000. Lynx home range and movements in Montana and Wyoming: preliminary results. In *Ecology and conservation of lynx in the United States*, ed. L. F. Ruggiero, K. B. Aubry, S. W. Buskirk, G. M. Koehler, C. J. Krebs, K. S. McKelvey, and J. R. Squires, 337–49. Fort Collins, CO: Department of Agriculture, Forest Service, Rocky Mountain Research Station.

Squires, J. R., and R. Oakleaf. 2005. Movements of a male Canada lynx crossing the Greater Yellowstone Area, including highways. *Northwest Science* 79:196–201.

Squires, J. R., and L. F. Ruggiero. 2007. Winter prey selection of Canada lynx in northwestern Montana. *Journal of Wildlife Management* 71:310–315.

United States Fish and Wildlife Service (USFWS). 2000. Endangered and threatened wildlife and plants: determination of threatened status for the contiguous U.S. distinct population segment of the Canada lynx and related rule, final rule. *Federal Register* 65:16052–86 (24 March 2000).

———. 2008. Final Environmental Impact Statement: Southern Rockies Lynx Management Direction.

———. 2013. Endangered and threatened wildlife and plants; revised designation of critical habitat for the contiguous U.S. distinct population segment of the Canada lynx and revised distinct population segment boundary, proposed rule. *Federal Register* 78:59430–74 (26 September 2013).

———. 2017. *Species status assessment for the Canada lynx (Lynx canadensis) contiguous United States distinct population segment.* Version 1.0, October 2017. Lakewood, Colorado.

Wild, M. A., T. M. Shenk, and T. R. Spraker. 2006. Plague as a mortality factor in Canada lynx (*Lynx canadensis*) reintroduced to Colorado. *Journal of Wildlife Diseases* 43:646–50.

Wolff, J. O. 1980. The role of habitat patchiness in the population dynamics of snowshoe hares. *Ecological Monographs* 50:111–30.

Wozencraft, W. C. 1993. Order Carnivora. In *Mammal species of the world: a taxonomic and geographic reference*, Vol. 1, ed. D. E. Wilson and D. M. Reeder, 286–346. 2nd ed. Washington, DC: Smithsonian Institution Press.

BOBCAT (*LYNX RUFUS*)

Robert L. Harrison and Jean-Luc E. Cartron

Bobcats (*Lynx rufus*) are medium-sized members of the cat family, Felidae. Also known as wildcats, they often symbolize the ferocity of wild animals, willing and able to unleash pure savagery in any desperate fight for survival. In a trap, a bobcat will lie still, warily watching its captor's every move while waiting for the right moment to explode with a violence startling in its suddenness and too fast for the human eye to follow. But despite the fierce reputation of bobcats, unprovoked attacks on humans by healthy bobcats are extremely rare (see under "Status and Management").

Bobcats are creatures of shadows and stealth. They burst from cover, using their explosive power to surprise and overwhelm prey. Both the ruff of facial hair and the spotted or streaked fur coats of bobcats provide excellent camouflage, especially in shrubby habitats where the vision of prey is obscured and sunlight and shadows are intermixed. Extending downward from the ears and often compared to wide sideburns, the bobcat's ruff in particular helps conceal the outline of the head. Blending in with shadows and breaking up the outline of the bobcat's body, dark brown or black spots and bars typically occur throughout the coat. Coat colors have been described as reddish brown, yellowish brown, gray brown, gray, or buff (moderate orange yellow) (e.g., Peterson and Downing 1952; Larivière and Walton 1997; Anderson and Lovallo 2003). Spot patterns on the inner portions of legs are

nearly always present and have been used to identify individual bobcats (Heilbrun et al. 2003). The bellies of bobcats are white, again with black spots (Anderson and Lovallo 2003), and it is this fur that is of prime interest to trappers (Hansen 2007). Although true color phases do not occur, there is considerable variation between individuals seemingly reflecting use of different habitats (Larivière and Walton 1997; Anderson and Lovallo 2003). Semiannual molting may also result in shorter, more reddish hair in summer and longer, more grayish hair in winter in some areas. Variation in the width of the white ventral fur and in the distinctness of ventral black spots influence the value of pelts, those from the western United States bringing the highest prices (Fur Harvesters Auction Inc. 2020). Melanistic individuals have been documented, mostly in Florida (e.g., Ulmer 1941; Regan and Maehr 1990; Hutchinson and Hutchinson 2000). In general, bobcats in the Pacific Northwest are considered to be more colorful with more distinct spots (McCord and Cardoza 1982), whereas those in the Southwest tend to be more pallid (Koehler 1987). Bobcat coloration in New Mexico has not been adequately documented– Bailey's (1931) description of the species is based on type specimens collected elsewhere—but varies across the state. Bobcat coloration in the Chihuahuan Desert near Truth or Consequences in Sierra County, New Mexico varied from light gray to buff (R. Harrison, unpubl. data).

(*opposite page*) Photograph: © Ed MacKerrow/In Light of Nature Photography.

Photo 7.1. Adult bobcat in Santa Fe County on 6 October 2013. In dense vegetation the bobcat is camouflaged by the combination of dark bars, spots, and rosettes breaking up the outline of the body. The ruffs of facial hair beneath the ears make the face appear wider. Tufts of black hair extend upward from the pointy ends of the ears. Photograph: © Bernard Foy.

Photo 7.2. Adult female bobcat and her kitten check out a scent station under a log pile at the Bosque del Apache National Wildlife Refuge on 31 August 2008. The tail of the bobcat is short with dark bands and a dark tip on the dorsal side; the underside of the tail is white. Camera trap project by Matthew Farley, Jennifer Miyashiro, and James N. Stuart.

Bobcats are named for their short, "bobbed" tail, a distinctive characteristic they share with the closely related Canada lynx (*Lynx canadensis*; Chapter 6). The two species may be mistaken for one another, as they are very similar in general appearance, including long legs relative to body length and tufts of hair extending upward from ears that are black dorsally with white spots (more conspicuous in bobcats) (Anderson and Lovallo 2003). The bobcat and the Canada lynx overlap in northern New Mexico where a population of lynx reintroduced into southwestern Colorado has spread into New Mexico (Shenk 2009; see Chapter 6). However, lynx may be distinguished from bobcats by their feet, which appear quite large and hairy for support in deep snow. Lynx tracks measure more than 8.9 cm (3.5 in) wide, compared to less than 6.4 cm (2.5 in) for bobcats. The ear tufts of lynx are at least 3.8 cm

(1.5 in) long, compared to less than 2.5 cm (1.0 in) for bobcats. The tail of the lynx is brownish or buff to the tip, which is black both dorsally and ventrally. In contrast, the tail of the bobcat has dark bands, including a dark tip limited to the dorsal side; the underside of the tail is white. Bobcats, both adults and kittens, may also be mistaken for lost or wandering domestic cats (*Felis catus*) especially in areas of exurban development and along the wildland-urban interface. However, despite significant geographic variation in body size (see below), adult bobcats are typically about twice as big as domestic cats. Bobcat kittens are usually born with spots or leopard-like rosettes in their coats, whereas only rare breeds of domestic cats are spotted. In addition, the black ear tufts of bobcat kittens, noticeable by the age of 3–4 weeks, are nearly diagnostic, as is the stubby tail. Occasionally, bobcats are even confused with cougars (*Puma concolor*), even though adult cougars average nearly twice as long in body length and up to five times or more in body weight (see Chapter 9). Contrary to adults, however, cougar cubs show conspicuous dark spots, which begin to fade at the age of about three months but do not disappear completely until after the first year of life. Compared to bobcats, cougar cubs nonetheless have much longer tails.

Considerable geographic variation has been documented in the size of adult bobcats, together with pronounced sexual dimorphism (Anderson 1987; Larivière and Walton 1997; Wigginton and Dobson 1999; Anderson and Lovallo 2003). Wigginton and Dobson (1999) examined geographic variation in bobcat body size (in individuals ≥ 2 years of age) across western North America including New Mexico. About 40% of the

Photo 7.3. Four New Mexico bobcat specimens housed at the University of New Mexico's Museum of Southwestern Biology. Note the short tail and the dorsal pelage coloration ranging from gray to gray brown or brown. The bobcat kitten specimen (left) shows spots and leopard-like rosettes (jagged black circles). Photograph: © Jon Dunnum.

Photo 7.4 (*top*). Adult female bobcat with gray-brown fur in Santa Fe County on 27 April 2010. Although there is considerable individual variation in pelage coloration (largely related to habitat), no true color phases are recognized in bobcats. Photograph: © Dan J. Williams.

Photo 7.5 (*bottom*). Bobcat with reddish brown pelage in Helm's Valley on White Sands Missile Range, 16 September 2014. Photograph: © WSMR Garrison Environmental Division and ECO Inc., Ecological Consultants.

Photo 7.6. Bobcat kitten found and rescued on White Sands Missile Range, and later released back into the wild after rehabilitation by Gila Wildlife Rescue in Silver City. Although bobcat kittens may be mistaken for domestic cats, they can easily be identified by their stubby tail and black ear tufts. Spots or leopard-like rosettes may or may not be present on their coats. Photograph: Denise Miller.

variation in mean body size of males and nearly half the variation in mean body size of females could be attributed to latitude and elevation combined, and overall bobcats of both sexes were larger in those environments that were colder and drier and/or experienced greater seasonal temperature variation (Wigginton and Dobson 1999). Rangewide, male bobcats have also been found to average 10% longer and 25–80% heavier than females (Anderson 1987). Interestingly, however, sexual dimorphism seems most prominent in mountainous regions whereas it is less noticeable in areas of flat topography (Sikes and Kennedy 1993). The largest weight ever verified was that of a male bobcat from Minnesota, or 17.6 kg (38.7 lb; Berg 1979). The only reported weights of adult New Mexico bobcats are from the Armendaris Ranch in the Chihuahuan Desert of Sierra County, where four males and six females averaged 12.6 kg (27.7 lb) and 9.0 kg (19.8 lb), respectively (Harrison 2010). By comparison, Zezulak and Schwab (1980) documented lower bobcat weights (an average of 8.0 kg [17.6 lb] for seven males; 6.2 kg [13.6 lbs] for a single female) from another desert environment, at Joshua Tree National Monument in the Mojave Desert of California. More surprisingly, average weights reported from the Chihuahuan Desert by Harrison (2010) are higher than those calculated by Banfield (1987) for bobcats in Canada (9.6 kg [21 lbs] for males and 6.8 kg [15 lbs] for females). However, bobcats do not achieve full adult body size until sometime during their second year of life—as late as 93 weeks of age in females (Crowe 1975a)—complicating any comparison of body size among geographic locations or studies. In areas where they are not trapped, such as on the Armendaris Ranch, bobcats would be expected to live longer on average (see Anderson and Lovallo 2003; Kamler and Gipson 2004) and therefore reach larger weights, despite living in a warm or hot climate that lacks significant temperature seasonality.

Bobcats are known to disperse over long distances, at times over 150–200 km (~100–125 mi; Knick and Bailey 1986; Johnson et al. 2010), likely creating considerable gene flow between populations and habitats and thus potentially obscuring boundaries between subspecies. Bailey (1931) recognized two bobcat subspecies in New Mexico, *uinta* in the major mountain ranges, and *baileyi* everywhere else. Based largely on Young (1958), Hall (1981) listed 12 subspecies of bobcat rangewide and remains the most frequently cited authority. He designated all bobcats in New Mexico as belonging to the subspecies *baileyi*. However, based upon a study of cranial variation among 956 specimens collected in Texas and surrounding states, Schmidly and Read (1986) instead assigned New Mexico bobcats east of the continental divide to the subspecies *texensis*, a conclusion accepted in Frey's (2004) annotated checklist of mammalian taxa in New Mexico. According to Schmidly and Read (1986) *L. rufus texensis* can be differentiated from the subspecies *baileyi* by a larger body size, a more rounded and higher skull, and a more colorful (reddish brown or grayish) pelage.

The intraspecific taxonomy of the bobcat nonetheless remains in flux as recent molecular studies (Croteau 2009; Reding 2011) suggest the existence of two distinct phylogeographic groups, eastern and western, that formed as a result of glacial vicariance during the Pleistocene. When uninhabitable continental ice sheets covered much of North America, the distribution of bobcats was reduced to two disjunct refugia, one in the southeast, the other one along the west coast. The two groups experienced a post-glacial range expansion but remained largely separated by the Great Plains acting as a geographic and ecological barrier (Reding 2011; see under "Habitat Associations"). Kitchener et al. (2017:38–40) recognizes them as the only two subspecies north of Mexico, with the western subspecies, *Lynx rufus fasciatus*, subsuming both *baileyi* and *texensis*.

Bobcats use a variety of behaviors to communicate with conspecifics, including urine spraying, claw marking, body rubbing, deposition of feces, and vocalizations (Allen et al. 2015). The vocal repertoire of bobcats is similar to that of domestic cats, although much louder. Bobcats purr and mew, especially when with young. When threatened, they may hiss and snarl. The deep rumbling growl of a trapped bobcat is especially impressive. Yowling or caterwauling occurs during fights or times of mating.

DISTRIBUTION

Bobcats occur in central British Columbia and elsewhere in southern Canada in a narrow strip along the border with the United States, throughout the contiguous United States (except in Delaware), and in Mexico as far south as Oaxaca. Historically, bobcat numbers in midwestern states from Iowa and southern Minnesota to western Pennsylvania and New York were severely reduced as a result of habitat changes and predator control, both related to intensive agriculture (Hansen 2007). However, midwestern populations have bounced back in recent decades (Woolf and Hubert 1998; Roberts and Crimmins 2010). Roberts and Crimmins (2010) also reported an approximate doubling of the bobcat's population in Canada since the early 1980s, while in Quebec in particular the species has expanded its range northward by ~105 km (~70 mi) during roughly the same time period (Lavoie et al. 2009).

Specimen records exist from nearly all New Mexico counties but suggest greater bobcat numbers in the northern third of the state and in the western half, particularly in association with rugged topography; by comparison, few specimen records originate from New Mexico's eastern plains (Map 7.1). According to the New Mexico Department of Game and Fish (NMDGF) unpublished fur harvest records, bobcats have been trapped in all counties of New Mexico (Table 7.1; Map 7.2). County harvest numbers are useful for management purposes, but the relationship between numbers trapped and bobcat population size is not known. Trapping harvest by county is influenced by not only bobcat population size but also the number of trappers, the level of trapping effort by individual trappers, and the size of the county. Nevertheless, reported trapping harvest might give a general idea of regional population densities. Bobcat harvest appears to be fairly evenly distributed throughout New Mexico, with the exception of lesser numbers in the eastern regions, notably in prairie and agricultural environments (Table 7.1; Map 7.2).

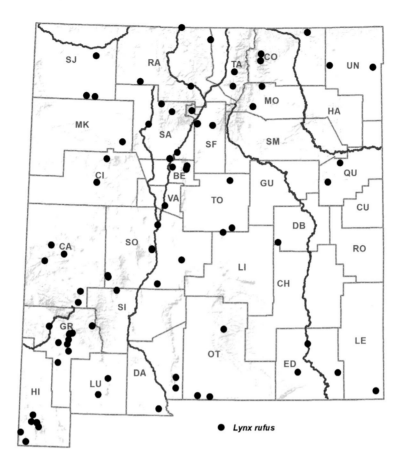

Map 7.1. Distribution of the bobcat (*Lynx rufus*) in New Mexico based on museum specimen records.

Bailey (1931) described New Mexico's bobcats as occurring primarily at elevations ranging from ~ 1,370 m (4,500 ft) to ~ 2,590 m (8,500 ft). Bailey (1931) mentioned records of bobcats occurring higher up in the mountains, including one individual he accidentally flushed from brushy cover in the Gallinas Mountains in Rio Arriba County at an elevation of 3,170 m (10,400 ft). Occurrence in river valleys at the lowest elevations (<1,370 m [<4,500 ft]) of the state (e.g., lower Rio Grande Valley and Pecos River Valley south of Santa Rosa) was deemed less common, though today bobcats are regularly observed in the Las Cruces area in Doña Ana County (J. Frey, pers. comm.).

HABITAT ASSOCIATIONS

Rangewide, bobcats occur in a wide variety of forested and more open vegetation types ranging from mature second-growth deciduous forest, mixed upland hardwoods, montane coniferous woodlands and forests, white-cedar swamps, and longleaf pine savanna to subtropical scrub, canyon lands, deserts, and grasslands (e.g., Kamler and Gipson 2000; Harrison 2010; Espinosa-Flores and López-González 2017; Little et al. 2018; Young et al. 2019; Jones et al. 2020). Although typically uncommon in areas densely populated by humans, bobcats may nonetheless also occupy natural spaces within urban landscapes (Poessel et al. 2014; Young et al. 2019). Given certain conditions, agricultural

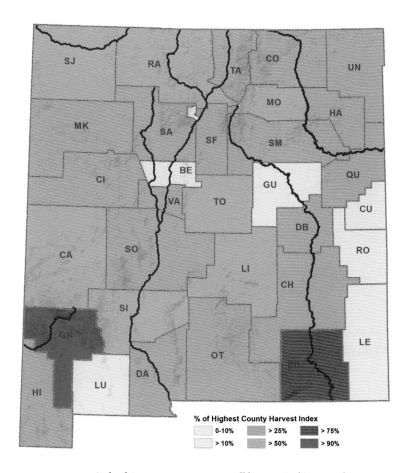

Map 7.2. Harvest index by county, 2010–2020. All harvest indices are shown as percentages of the highest harvest index (in Grant County). See Table 7.1 for details on the calculation of harvest indices.

row crop fields can even be suitable for bobcats, as documented by Conner et al. (1992) in east-central Mississippi, where numerous brushy fence rows and drainages appeared to support a high abundance of prey. Adaptable though they may be, bobcats nonetheless tend to avoid landscape-scale, intensive row crop agriculture such as seen today in the Midwest's "Corn Belt" (Tucker et al. 2008), one main reason that eastern and western bobcat phylogeographic groups might be kept separated on either side of the northern Great Plains (Reding 2011). In Iowa, for example, row crop fields intersected some bobcat home ranges, but on average the largest row crop patch did not exceed 3% of the home range (Tucker et al. 2008).

Sufficient prey and brushy or other cover for stalking or ambushing them are the primary bobcat habitat requirements (Kleiman and Eisenberg 1973; McCord and Cardoza 1982; Anderson and Lovallo 2003). However, other factors also influence habitat selection at the home range scale (second-order selection) or within the home range (third-order selection) during at least part of the year, including the depth of winter snow cover and the availability of dens (e.g., dense brush, small caves, rocky ledges, or even abandoned buildings; for breeding females and their kittens) and concealment or protective cover (Bailey 1974; Hamilton 1982; Koehler and Hornocker 1989; Kolowski and Woolf 2002). Depth of winter

Table 7.1. Bobcats trapped by county from 2010 to 2020, based upon unpublished New Mexico Department of Game and Fish fur harvest records*. The reported number of bobcats trapped is based upon reports from trappers, which averaged 92.8% of CITES (the Convention on International Trade in Endangered Species of Wild Fauna and Flora) total tagged. Harvest index is calculated as the reported number of bobcats trapped times 1000/county area.

County	Region of New Mexico	Reported Number of Bobcats Trapped	County Area (km²)	Harvest Index	Mean Regional Harvest Index
McKinley	northwestern	627	14113	44.43	47.84
San Juan	northwestern	732	14281	51.26	
Catron	west-central	1061	17943	59.13	48.87
Cibola	west-central	454	11759	38.61	
Grant	southwestern	1031	10272	100.37	49.32
Luna	southwestern	158	7679	20.58	
Hidalgo	southwestern	241	8925	27.00	
Santa Fe	north-central	131	4944	26.50	38.63
Rio Arriba	north-central	1012	15172	66.70	
Taos	north-central	231	5706	40.48	
Sandoval	north-central	401	9609	41.73	
Los Alamos	north-central	5	282	17.73	
Bernalillo	central	26	3020	8.61	38.63
Valencia	central	106	2766	38.32	
Torrance	central	442	8664	51.02	
Socorro	central	683	17216	39.67	
Lincoln	central	695	12512	55.55	
Doña Ana	south-central	476	9860	48.28	47.28
Otero	south-central	512	17164	29.83	
Sierra	south-central	690	10826	63.74	
Mora	northeastern	252	5001	50.39	44.17
Colfax	northeastern	427	9731	43.88	
San Miguel	northeastern	578	12217	47.31	
Union	northeastern	480	9920	48.39	
Harding	northeastern	170	5506	30.88	
Curry	east-central	2	3642	0.55	20.23
Quay	east-central	362	7394	48.96	
Roosevelt	east-central	12	6343	1.89	
Guadalupe	east-central	144	7850	18.34	
DeBaca	east-central	189	6022	31.38	
Eddy	southeastern	963	10831	88.91	43.51
Chaves	southeastern	599	15724	38.09	
Lea	southeastern	40	11378	3.52	

* No data were available for the 2011–2012 and 2014–2015 trapping seasons.

Photo 7.7. Bobcat in dense, brushy vegetation at the Bosque del Apache National Wildlife Refuge on 17 November 2009. In New Mexico as elsewhere, the species is typically found in areas with brushy or other cover for stalking or ambushing prey. Photograph: © Samuel Mulder.

Photo 7.8. Bobcat visiting a wildlife drinker in June 2016 in the north-eastern corner of the Sevilleta National Wildlife Refuge. The dominant vegetation is juniper savanna. Photograph: © Sevilleta National Wildlife Refuge, US Fish and Wildlife Service.

Photo 7.9. Bobcat in ponderosa pine forest in the Manzano Mountains on 4 September 2009. Photograph: © David C. Lightfoot.

Photo 7.10. Bobcat below Aguirre Springs in the Organ Mountains on 15 October 2006. Contrary to earlier reports, bobcats occur in desert scrub in New Mexico. Photograph: © Jeff Kaake.

Photo 7.11. Bobcats visiting a wildlife drinker at the San Andres National Wildlife Refuge on 16 January 2009. Photograph: © San Andres National Wildlife Refuge.

Photo 7.12. Adult female and her three kittens in a backyard in the Northeast Heights of Albuquerque. Bobcats occur in suburban areas and on the edge of larger urban centers in New Mexico. Photograph: © Kenneth F. Brinster.

Photo 7.13. Bobcat photographed on 7 April 2009 about 6 m (20 ft) up in the crook of a mulberry tree in the center of the very busy main cantonment area at White Sands Missile Range. Bobcats are frequently observed on and around the main post and often become relatively tolerant of human activity. Photograph: © Douglas W. Burkett.

snow is an important limiting factor at more northern latitudes or in high mountainous areas. Snow deeper than 15 cm (6 in) is difficult for bobcats to traverse, forcing them to seek trails, roads, logs, and other pathways more often than they would otherwise (McCord 1974; see Chapter 6 for a comparison with the lynx). Rocky outcrops may be especially important in areas that lack brushy or tree cover (Zezulak and Schwab 1980; Hamilton 1982; Koehler and Hornocker 1989; Anderson 1990). Theodore Bailey (1974:435) studied a bobcat population in southeastern Idaho in what was deemed "ideal habitat," with rugged topography and large numbers of caves, lava flows, and volcanic craters. Bailey (1974) found and investigated 50 fecal marking locations (or latrines), 45 (90%) of which were located on broken, rocky terrain, often in proximity to natal or temporary dens of adult females with kittens, and/or to rocky ridges and rims of craters used as travel routes.

In New Mexico, Vernon Bailey (1931) emphasized two factors that in his view determined habitat quality in New Mexico: rugged, rocky topography and an abundance of prey. According to Bailey (1931), it was in the canyonlands and on other broken terrain of the state that bobcats were most abundant, with the presence of rocky outcrops and canyons enhancing habitat in some of the more open pinyon-juniper woodlands and even allowing bobcats to venture out far onto the eastern plains. In Bailey's (1931) view, canyons and cliffs sustained higher bobcat densities by both providing dens for bobcats and supporting large small-prey populations (Bailey 1931). Similarly, Hooper (1941) linked what appeared to be high numbers of bobcats in the volcanic and mountainous landscape in and around today's El Malpais National Monument (in west-central New Mexico) to an abundance of potential den sites in the form of caves and ledges. Thompson et al. (1992) attempted to map the extent of bobcat habitat in New Mexico and predict the occurrence of the species within ecological zones

as defined by Dick-Peddie (1993). They identified coniferous and mixed woodland, juniper savanna, and mixed coniferous forest as the primary occupied habitats and subalpine coniferous forest and Great Basin Desert scrub as secondary habitats for bobcats. Environments deemed not suitable by Thompson et al. (1992) consisted of sand dunes, alpine tundra, urban areas, farmland, subalpine montane grassland, plains-mesa grassland, desert grassland, Chihuahuan desertscrub, closed basin scrub, and plains-mesa sand scrub. However, the methodology used by Thompson et al. (1992) is questionable and the results must be viewed with caution. Thompson et al. (1992) relied extensively on a literature review of habitat associations in other states, and no systematic field surveys were conducted to validate and calibrate the results. Some areas deemed not suitable appear to be used by bobcats, though perhaps in smaller numbers. For example, Harrison (2010) studied a bobcat population in Sierra County that occurred in Chihuahuan desertscrub. In another study at White Sands National Park, bobcats were photographed in fourwing saltbush (*Atriplex canescens*), mesquite (*Prosopis* spp.), and pickleweed (or quinine bush, *Allenrolfea occidentalis*) shrublands (Robinson et al. 2014; G. Roemer, in litt). Thompson et al. (1992) also excluded sand dunes from bobcat habitat, but remote cameras at White Sands National Park photographed bobcats in barren and vegetated gypsum duneland, vegetated gypsum outcrop, and gypsum interdune swale grassland (G. Roemer, in litt.).

On the Armendaris Ranch, bobcats used Chihuahuan desertscrub, which occurred on hilly areas bisected by arroyos, with creosotebush (*Larrea tridentata*), honey mesquite (*Prosopis glandulosa*), and desert sumac (*Rhus microphylla*) as some of the dominant shrubs (Harrison 2010). Unusually large home ranges documented by Harrison (2010) for male bobcats may have been the result of large areas of desert grassland also being included in those home ranges. Desert grassland occurred on flat areas and was dominated by black grama (*Bouteloua eriopoda*), alkalai sacaton (*Sporobolus airoides*), galleta (*Hilaria jamesii*), with some stands of honey mesquite and desert sumac. Within Harrison's (2010) study area, bobcats usually occurred in areas of moderate-to-dense shrub cover, which became more fragmented in desert grassland (Harrison 2010).

The value of grasslands as habitat for bobcats has been assessed with seemingly conflicting results. In Kansas, Kamler and Gipson (2000) documented bobcat occupancy of a tallgrass and mixed-grass prairie ecosystem. Intermixed with prairie grasslands were riparian woodlands, which were also included in the home ranges of resident bobcats. During winter, bobcats were mainly associated with woodlands, but during summer months they preferentially used grasslands (Kamler and Gipson 2000). In arid central Arizona, however, Lawhead (1984) found that bobcats preferred riparian areas but avoided extensive grassland areas. Although a 0.5 km² (149-acre) area of continuous desert grassland was partially or totally contained within the home ranges of three bobcats, only once was a bobcat located within this area. Lawhead (1984) attributed the avoidance of extensive desert grasslands to lack of stalking cover, insufficient prey densities, or both. Thus it may be that tallgrass and mixed-grass prairies provide enough stalking cover, but only during the growing season (in winter, grasses are compacted by snow); desert grasslands never do.

Boyle and Fendley (1987) recognized that water availability could influence bobcat habitat associations in parts of the western United States. Boyle and Fendley (1987) mention in particular Lawhead's (1984) research documenting the preference of bobcats for riparian areas in the Sonoran Desert, which could be due in part to the availability of standing water in an otherwise arid landscape. In an unpublished report, Shafer (2016) analyzed about 10 years of photographic evidence from 13 remote cameras installed at water sources on a ranch in San Benito County, California. Shafer (2016) reported that of a total of 2,694 visits when water was in the field of view, bobcats drank 685 (25.4%) times. Drinking occurred mainly during summer months, and the observed monthly rate of drinking increased as a direct function of ambient temperature, while the observed mean duration of visits was longer during spring and summer (Shafer 2016). In the Chihuahuan Desert of western Texas, in an area with no permanent natural water sources, remote cameras examined bobcat—as well as coyote (*Canis latrans*) and gray fox (*Urocyon cinereoargenteus*)—visitation of stock tanks and earthen impoundments through time (Attwood et al. 2011). The rate of bobcat visitation was best predicted by just two variables, time elapsed since last rainfall and distance to the next nearest water source. These studies collectively suggest that bobcats need access to water in hot, arid, or semi-arid environments and that, at a minimum, droughts and/or hot temperatures influence bobcat activity patterns. Bobcats have been photographed visiting water sources in the Chihuahuan Desert in south-central New Mexico, including on the San Andres National Wildlife Refuge (Photo 7.11). In the San Andres Mountains, Bender et al. (2017) found bobcat occupancy of the landscape to be correlated with small prey availability as well as proximity to water.

In New Mexico as elsewhere, habitat flexibility has enabled bobcats to survive in the presence of increasing human encroachment. Bobcats can live on the fringes of some of the state's urban areas, and within suburban development if sufficient cover and prey remain. For example, they are often seen within the communities of Placitas, Tijeras, Cedar Crest, and Sandia Heights adjacent to Albuquerque. Over 70% of sightings reported to a mail survey were within 25 m (~ 80 ft) of houses (Harrison 1998). Before development, Sandia Heights consisted primarily of open juniper (*Juniperus* sp.) grassland but subsequent landscaping, especially with chamisa (also known as rabbitbrush, *Ericameria* sp.), substantially increased stalking cover. The addition of small pets and bird feeders, which attract not only birds but also small mammals, likely augmented the prey base. Now bobcats are frequently seen within Sandia Heights. However, sightings are most common adjacent to large undeveloped areas, suggesting that while bobcats may use residential neighborhoods, their core home range areas are likely not within developed areas, as found by Nielsen and Woolf (2001) in Illinois.

LIFE HISTORY

Diet and Foraging

In contrast to its congener the lynx (Chapter 6), the bobcat preys upon many different species range-wide, as would be expected for a species occupying a wide variety of habitats. Overall, lagomorphs are the most frequently consumed prey (Dearborn 1932; Bailey 1979; Parker and Smith 1983), but rodents, ungulates, and game and nongame birds can all be important in some areas, with shrews, raccoons (*Procyon lotor*), skunks, reptiles, fish, amphibians, and insects also occasionally taken, among others (e.g., McCord 1974; Bailey 1979; Fritts and Sealander 1978; Knick et al. 1984; López-Vidal et al. 2014). Most mammalian prey weigh between 700 g and 5.5 kg (~1.5–12 lbs; Rosenzweig 1966), though bobcats take prey as small as pocket mice (*Chaetodipus* spp., ~10 g [~0.4 oz]) and kangaroo rats (*Dipodomys* spp., ~60 g [2.1 oz]), as well as much larger species, such as deer (*Odocoileus* sp.) and pronghorn (*Antilocapra americana*; see Anderson and Lovato 2003 and Hansen 2007; weights from López-Vidal et al. 2014). Deer (mostly fawns and does) are especially vulnerable in deep snow, and they represent an important portion of the bobcat's diet in northern areas where, importantly, the species is larger (Anderson and Lovallo 2003). In Massachusetts, for example, McCord (1974) reported white-tailed deer (*Odocoileus virginianus*) in 79% of bobcat scats during winter. Most

Photos 7.14a and b (*top to bottom*). Bobcat in a tree with a mallard (*Anas platyrhynchos*) at the Bosque del Apache, 5 March 2011. Photographs: © Steve Collins.

predation attempts by bobcats involved white-tailed deer as the intended prey, often while they were bedded down in the snow (McCord 1974). Male bobcats are more likely to kill larger prey than females, likely the result of bobcat sexual-size dimorphism (Knick et al. 1984; Litvaitis et al. 1984; Anderson 1987). The availability of prey influences which species are taken. When the population of one species of favored prey declines, bobcats will switch to other prey (e.g., Bailey 1981; Maehr and Brady 1986; Knick 1990).

Photo 7.15. Bobcat with desert cottontail (*Sylvilagus audubonii*) in Doña Ana County on 6 May 2017. Although bobcats prey on a wide variety of vertebrates, lagomorphs are particularly important in their diet. Note also the surrounding desertscrub. Photograph: © WSMR Garrison Environmental Division and ECO Inc., Ecological Consultants.

Photo 7.16. Bobcat with rock squirrel (*Otospermophilus variegatus*) on 7 September 2020 in Sandoval County. Most mammalian prey weigh between 700 g and 5.5 kg (~1.5–12 lbs). Photograph: © Mark L. Watson.

The only study of bobcat diets in New Mexico remains that of Harrison (2010) in the Chihuahuan Desert near Truth or Consequences in Sierra County. Desert cottontail (*Sylvilagus audubonii*) remains were found in 60% of 143 bobcat scats. Black-tailed jackrabbits (*Lepus californicus*), white-throated woodrats (*Neotoma albigula*), and unidentified birds were identified less frequently, in slightly more than 10% of scats each. Other species of small mammals, in addition to reptiles and invertebrates, were also consumed. Altogether, lagomorphs represented an estimated 58% of prey items found, rodents an additional 29% (Harrison 2010). López-Vidal et al. (2014) also studied the diet of bobcats in the Chihuahuan Desert, but farther south, in the Mapimí Biosphere Reserve of the state of Durango in northern Mexico. Here again, the diet of bobcats was dominated by lagomorphs followed by rodents. Rather than desert cottontails, however, black-tailed jackrabbits were found to be the most important prey species based on both biomass and frequency of detection in bobcat scats. Lagomorphs in general and, within lagomorphs, desert cottontails in particular, appeared to be consumed more often than expected based on their availability (compared to other prey). López-Vidal et al. (2014) concluded that bobcats in their study area acted as specialist foragers rather than generalists.

On the National Rifle Association Whittington Center near Raton in Colfax County, New Mexico, bobcats were responsible for two (5%) of 33 mule deer fawn predation events (Taylor 2013). Also on the National Rifle Association Whittington Center, predation accounted for 22 documented cases of mule deer doe mortality (Sorensen 2015). None of the predation was attributed to bobcats, however, and instead cougars (*Puma concolor*, Chapter 9) were determined to be the main predators of mule deer does (Sorensen 2015). Other information on the food habits of bobcats in New Mexico tends to be dated and anecdotal. Bailey (1931) reported predation on prairie dogs (*Cynonys* spp.), as well as a propensity for bobcats to depredate domestic livestock, particularly chickens and lambs. Elsewhere, bobcats have indeed been shown to kill sheep, goats, and chickens, but their impact is generally minor and localized (Young 1958; Anderson and Lovallo 2003; see also Chapter 3).

Bobcats usually forage alone, although kittens learn to hunt with their mothers. Hunting occurs in a variety of ways, usually involving a slow, stealthy approach and a short chase (Hansen 2007). An approach may take over an hour, with the chase covering over 100 m (328 ft, Biggins and Biggins 2006). Bobcats take advantage of all types of covers, including trees, shrubs, rocks, logs, topography, and even suburban landscaping and fences. Bobcats may also wait by a trail or the entrance of a rodent burrow. When hunting for mice or voles in long grass, careful stalking ends with a high, curved jump and a pounce. Bobcats may use ledges and other advantageous spots as "lookouts" for prey.

Activity Patterns

Bobcats are traditionally considered nocturnal, though in reality diel activity patterns vary considerably between regions (Hansen 2007). By operating 308 camera stations for a total of 5,844 trap-nights, Fedriani et al. (2000) found that bobcats were active at all times of day and night in the Santa Monica Mountains in California, with 40% of camera survey captures during daytime. Other studies (e.g., Hall and Newsom 1976; Buie et al. 1979) have reported that most bobcat activity occurs around crepuscular hours at sunrise and sunset. According to Zezulak and Schwab (1980), bobcats in the Mojave Desert exhibited a crepuscular activity pattern in the winter but became more nocturnal during the spring months. In the Chihuahuan Desert in south-central New Mexico, Harrison (2010) found daytime activity (18.8%) to be less than in other states: the percentage of activity was highest from sunset

Photo 7.17. Bobcat photographed at night at White Sands National Park on 31 August 2009. Bobcats may be active at all times of day and night, though studies in New Mexico documented most activities during crepuscular hours and at night. Photograph: © Gary W. Roemer.

Photo 7.18. Two bobcats after sunset at the Bosque del Apache National Wildlife Refuge on 2 January 2009. Camera trap project by Matthew Farley, Jennifer Miyashiro, and James N. Stuart.

through two hours before sunrise; bobcats were least active three to six hours before sunset. From 2007 to 2011 in the San Andres Mountains, Bender et al. (2017) operated camera traps for at least 9–13 weeks annually from early to mid-February until late April or May (through July in 2011). As recorded photographically, only 10%

of all bobcat activity occurred during the day. Most of the observed activity was split between nighttime (48%) and crepuscular hours (42%). To a large extent, bobcat activity patterns may be linked to the times of peak activity of their primary prey (Anderson and Lovallo 2003), though ambient temperature may also play a role and some movements are not related to hunting (e.g., change of den). In the Mojave Desert, Zezulak and Schwab (1980) found that bobcat activity began and ended when cottontail and jackrabbit observations peaked in both winter and spring, as determined by roadside surveys. In the spring, at the time that bobcats became more nocturnal, Zezulak and Schwab (1980) observed an increase of spring prey activity during the middle portion of the night that was largely due to the abundance of rodents. Human activity may also play a role in determining bobcat activity patterns. In California, for example, Wang et al. (2015) found that bobcats reduced their diurnal activity in areas with a higher human presence.

Reproduction and Social Behavior

No studies have been published on the reproductive biology of bobcats in New Mexico. Based on research in other states, kittens may be born at any time of year though seemingly only from March to October at more northern latitudes, and in most cases from April through June (Gashwiler et al. 1961; Crowe 1975a) following a gestation period of about 65 days (e.g., Ewer 1973; Mehrer 1975; Stys and Leopold 1993). Litters usually consist of two to four kittens (up to six; Gashwiler et al. 1961; Anderson 1987). An average litter size of 2.5 kittens was documented in Arkansas (Fritts and Sealander 1978), 2.8 in Idaho (Bailey 1972) and Wyoming (Crowe 1975a), and 3.5 in Utah (Gashwiler et al. 1961). Litters of three and four kittens have been documented in New Mexico (e.g., Photo 7.19). Bailey (1931) in particular mentioned a litter of four kittens taken on 9 April 1908 along the headwaters of "Sapello Creek," likely

Photo 7.19 (*top left*). Three kittens photographed in June 2006 after being found and rescued in Columbus, Luna County. Litters of three and four have been documented in New Mexico. Photograph: © Denise Miller.

Photo 7.20 (*top right*). Bobcat kitten (ca. four weeks old) found in the Albuquerque Heights and later rehabilitated. Photograph: © Mark L. Watson.

Photo 7.21 (*left*). Bobcat adult female and her two kittens on 10 December 2014 on the Armendaris Ranch in southern New Mexico. Kittens start accompanying their mothers at the age of three months and usually remain with her until the next breeding season. Photograph: © Robert L. Harrison.

Photo 7.22. An adult female (probably) with her two young-of-year juveniles, lounging at a scent station at the Bosque del Apache National Wildlife Refuge on 10 December 2008. Camera trap project by Matt Farley, Jennifer Miyashiro, and James N. Stuart.

Sapillo Creek in Grant County—the collector, H. Hotchkiss, worked in the southwestern part of the state. Kittens do not open their eyes until 9–18 days after birth (Pollack 1950). They begin to accompany their mothers away from the den at three months (Bailey 1979) and usually stay with her until the next breeding season (Anderson and Lovallo 2003).

Female bobcats were formerly thought to be induced ovulators (see Asdell 1946), requiring the act of copulation in order to trigger ovulation. Instead, however, they appear to be seasonally polyestrus, experiencing up to three estrous cycles per breeding season (or year) if not impregnated or if the litter is resorbed or aborted (Pollack 1950; Gashwiler et al. 1961; Crowe 1975a; Mehrer 1975). If sufficient prey is available, females may become sexually mature within their first year (e.g., Rolley 1985), although the rate of pregnancies in yearlings is less than in adults (Fritts 1973; Crowe 1975a, Knick et al. 1985; Anderson 1987; Stys and Leopold 1993). Males do not become sexually active until their second year (Anderson and Lovallo 2003).

After leaving their mothers, bobcats usually become transients while searching for an unoccupied area of habitat in which to establish their home range (Johnson et al. 2010). In populations that suffer heavy mortality from trapping, vacancies are common and young bobcats may become residents within a few months (Crowe 1975b; Griffith and Fendley 1986). In other areas, where bobcat densities are locally high and/or habitat fragmentation has occurred, bobcats may not settle for two years or longer or may even remain transients (Johnson et al. 2010). Dispersal distances vary greatly, from occupying an area adjacent to the natal home range to movements of over 200 km (120 mi; Johnson et al. 2010). Females usually disperse later and over shorter distances than males, and may even occupy a portion of their mother's home range; longer dispersal distances and periods of transience are more typical of males (e.g., Robinson and Grand 1958; Knick and Bailey 1986; Johnson et al. 2010). Johnson et al. (2010) estimated that dispersing bobcats spent 39–80% of their nomadic life occupying temporary home ranges in Indiana, where the landscape was fragmented by row crop agriculture; time spent in a temporary home range, continuously or intermittently, varied from about a month to nearly half a year. In Idaho, transients did not spend more than a few days in the home ranges of resident adults and simply passed through (Bailey 1974).

Once established, bobcats usually remain in the same home ranges throughout their lives (Anderson and Lovallo 2003). They may move if an adjacent individual with a better home range dies (Bailey 1974; Anderson 1988; Lovallo and Anderson 1995), or if prey populations decline severely within their home range (e.g., Knick 1990). In the latter case, bobcats become transients again.

In keeping with their hunting style, bobcats are primarily solitary (see Anderson and Lovallo 2003). Males and females travel together only during times of mating. Where resources are adequate and fairly evenly distributed, the home ranges of adult females show little to no overlap (e.g., Bailey 1974; Hall and Newsom 1976; Buie et al. 1979; Lawhead 1984; Anderson 1987). Male territories often overlap those of several females and may overlap those of other males to varying degrees (Anderson and Lovallo 2003). Where overlap occurs, bobcats maintain their separation by using common areas at different times through the use of scent marking and visual cues. Females mark their territories using feces, urine, and anal gland secretions (Provost et al. 1973; Bailey 1974). Males are thought to mark their territories less than females. Fights are uncommon, the result of what has been called a land tenure system, established based on prior residence and maintained by scent marking (Bailey 1974; Anderson 1988; Lovallo and Anderson 1995; Benson et al. 2004).

Bobcat home ranges vary considerably in size

among geographic areas (see Anderson and Lovallo 2003). Home range size tends to be much larger toward the northern end of the species' distribution. It is believed to be influenced mainly by prey densities but also reflects other habitat quality parameters (e.g., snow and amount of escape cover), as well as energy requirements and local bobcat densities. Male home ranges are two to five times larger than those of females (Hall and Newsom 1976; Major 1983; Witmer and DeCalesta 1986). The annual size of home ranges for bobcats in the Chihuahuan Desert in Sierra County, New Mexico was estimated at 128.9 km² (49.8 mi²) for males and 28.5 km² (11.0 mi²) for females (Harrison 2010), or among the largest reported and comparable to those in northern states (see Anderson and Lovallo 2003).

Interspecific Interactions

Rangewide, bobcats incur mortality mainly from humans (see under "Status and Management") but also from cougars, coyotes, and dogs (Ackerman et al. 1984; Lembeck 1986; Knick 1990; Koehler and Hornocker 1991; Fedriani et al. 2000; Hass 2009; see also Chapter 33). Among competing wild carnivores, it can be difficult to distinguish between true predation and interference competition, the latter involving two species aggressively excluding one another from habitats or scarce food resources. For example, cougars kill, but do not always consume, bobcats. In Idaho, killed but uneaten bobcats were found near food caches and indicated interference competition where cougars were defending or usurping those feeding sites (Koehler and Hornocker 1991). In central New Mexico, Prude (2020) documented five cases of cougar intraguild predation (true predation or interference competition) on bobcats.

The impact of coyotes as both dominant competitors and predators of bobcats has been well documented in the western United States (Fedriani et al. 2000, Hansen 2007). Coyotes are larger, have a higher reproductive output, and overlap extensively with bobcats in their diet, with a heavy reliance on lagomorphs (Chapter 10). Bobcat numbers tend to respond directly and positively to the amount of coyote control (Nunley 1978). Robinson (1961) in particular reported changes in population numbers of several carnivores in apparent response to coyote control in seven areas spread out between Colorado, Wyoming,

Photo 7.23. A bobcat encounters a snake on 13 July 2012 at a wildlife drinker on the Sevilleta National Wildlife Refuge. Photograph: © Sevilleta National Wildlife Refuge, US Fish and Wildlife Service.

and New Mexico. In the control areas in Wyoming, the total number of coyotes trapped—as a measure of their abundance—in 1960 fell to 8% of 1940–1941 levels as a result of continuous use of thallium stations and later Compound 1080 and shooting from airplanes (see Chapter 3 for details on control methods). An opposite trend was observed in bobcats, with 25 times more individuals trapped in 1960 compared to 1940–1941 in the same control areas in Wyoming. In New Mexico and Colorado, where coyote control measures were curtailed and/or less effective, coyote numbers did not decline steadily as they did in Wyoming, and no concurrent increase in bobcat numbers was observed (Robinson 1961).

Encounters with sympatric competitors do not always result in aggression by the dominant competitor nor avoidance by the subordinate competitor. In western Texas, Attwood et al. (2011) recorded seven encounters between coyotes and bobcats at water sites. None of those encounters resulted in agonistic behavior, and from observed patterns of activity, bobcats did not appear to be avoiding coyotes when visiting the water sites (Attwood et al. 2011). In New Mexico's San Andres Mountains, Bender et al. (2017) did not detect any avoidance of cougars by bobcats and instead documented a high similarity in the spatial and temporal occupancy of the landscape by the two species. Bobcats did not appear to associate with cougars in order to benefit from suppression of coyote activity where cougars were present: bobcat occupancy was also correlated with coyote occupancy (Bender et al. 2017). In California, bobcat detections were higher by 39% at sites with both low housing development and puma occupancy, whereas coyotes were less likely to be detected at those same sites (Wang et al. 2015). However, higher bobcat detections were also found in association with greater coyote activity, both at sites with low housing development and in more developed areas, the latter being characterized with reduced cougar detections (Wang et al. 2015). Thus, despite the

risk of intraguild predation, bobcats do not appear as a rule to avoid either cougars or coyotes.

Bobcats likely also compete with foxes in New Mexico as elsewhere. In western Texas, gray foxes behaved as subordinate competitors to not just coyotes but also bobcats, based on patterns of water source visitations of all three species (Attwood et al. 2011). No agonistic interactions between bobcats and gray foxes were captured on camera in western Texas (Attwood et al. 2011), but in northern California Riley (2001) reported clear evidence of interference competition between the two species. Riley (2001) found that bobcats killed but did not consume gray foxes, despite some spatial segregation between the two species, and one bobcat was seen chasing a gray fox. Also in California, Disney and Spiegel (1992) reported that at least 44% of mortalities of kit foxes (*Vulpes macrotis*) in a developed area were caused by bobcats.

Bobcats and lynx have largely segregated latitudinal and altitudinal distributions (see Chapter 6), but in areas where they co-occur, the two species can hybridize (Schwartz et al. 2004). Bobcats tend to be larger where the two species are sympatric, and on Cape Breton Island, Nova Scotia, they are known to have displaced lynx from the island's lower elevations (Parker et al. 1983). However, because lynx alone are adapted for deep snow cover, they are much less likely to be outcompeted by bobcats at higher elevations (Anderson and Lovallo 2003). Snowshoe hares (*Lepus americanus*) reach the southern end of their broad geographic range in New Mexico, in a narrow elevational band corresponding to subalpine coniferous forest (Frey and Malaney 2006). Whether bobcats and lynx compete for snowshoe hares in the high mountains of northern New Mexico is unknown.

Diseases and Parasites
Like other carnivores, bobcats are susceptible to many pathogens, but rangewide the effects of

infectious diseases on the species are unclear. Among all infectious diseases affecting bobcats, rabies has received the greatest amount of attention, but this is due simply to human public health implications. Rabies results in acute, progressive encephalomyelitis with associated behavioral changes such that infected animals lose their fear of people (Aiello and Moses 2016). In the case of bobcats, infected individuals can wander around residential areas during daytime and have been known to attack people. A total of 488 cases of rabies were reported in bobcats from 1960 to 2000 in the United States., including just five in New Mexico (Krebs et al. 2003). The New Mexico Department of Health did not report any cases of rabies in bobcats in 2019 or 2020, but three rabid bobcats were found in 2022, all in the southwestern part of the state. A gray fox rabies virus strain is predominant among rabid bobcats in Texas and Arizona.

Infections with feline panleukopenia virus (FPV, also known as feline distemper) result in severe, often fatal hemorrhagic gastroenteritis in bobcats. During an outbreak of FPV in a Florida population, 11 (61%) of 18 radio-collared bobcats died over a three-month period (Wassmer et al. 1988). In the San Francisco Bay Area, proximity to urban areas seemingly increased the risk of feline calicivirus (FCV) exposure from domestic cats carrying the virus (Riley et al. 2004; see Chapter 33). FCV infection typically manifests with mild symptoms in domestic cats except in such cases where mutant strains spontaneously arise, causing much more serious disease (FCV-associated virulent systemic disease or FCV-VSD) with multiple organ damage or even death. Bobcats are susceptible to additional viruses including feline immunodeficiency virus (FIV) (e.g., Franklin et al. 2007; Lagana et al. 2013).

Bacterial diseases such as sylvatic plague and tularemia can also affect bobcats (see Anderson and Lovallo 2003). Additionally, the bobcat serves as a host to numerous endoparasites such as *Toxoplasma gondii*, which rarely causes clinical toxoplasmosis, but for which the prevalence of antibodies has been reported as ranging from 34% to 73% by Dubey and Beattie (1988). Ectoparasites include the notoedric mite *Notoedres cati*, which burrows into the skin and can cause mange (Pence et al. 1982). In a previously unreported combination of human and non-human factors, Riley et al. (2007) reported that exposure to rodent poison in developed areas was strongly correlated with advanced notoedric mange. Annual survival declined to 28% two years after the beginning of the mange epizootic. Cytauxzoonosis is an emerging disease affecting both domestic cats and wild felids and transmitted by the bite of an infected tick (e.g., *Dermacentor variabilis*) (Wang et al. 2017). The bobcat serves as the most common, natural host of *Cytauxzoon felis*, and typically experiences a non–life threatening illness when infected, though fatal cases have been reported (see Wang et al. 2017).

STATUS AND MANAGEMENT

The bobcat is one of the relatively few felid species that are not considered endangered, threatened, or vulnerable (Nowell and Jackson 1996). The International Union for the Conservation of Nature and Natural Resources (IUCN) lists the status of the species as of "Least Concern" due to its abundance and wide distribution and despite the persistence of local challenges to bobcat populations including habitat loss, poisoning, and fur trapping (Kelly et al. 2016). This should be viewed as somewhat surprising, as during most of the period of contact between the species and European Americans, bobcats have been the object of both exploitation and deliberate persecution. Settlers saw them as threats to both livestock and wild game and established bounties for dead bobcats. Bounties continued until the 1970s (Hansen 2007). The Animal Damage Control program (now USDA APHIS Wildlife Services, see Chapter 3) was created in 1931 by the federal

Photo 7.24. Dead bobcat draped over a fence near Hachita in Grant County. The species continues to be the object of deliberate persecution in New Mexico and elsewhere in North America. Photograph: © Mark L. Watson.

government to control rodent pests and to eradicate livestock predators, including bobcats. During the time that bobcats were classified as predators, no limit was imposed on the number that could be killed. Between 1937 and 1970, the peak number of bobcats killed annually throughout the United States by Animal Damage Control was 26,000 in 1960 (Hansen 2007).

The bobcat is listed in Appendix II of the Convention on International Trade in Endangered Species of Wild Fauna and Flora (CITES; Nowell and Jackson 1996; CITES 2021). Although bobcats have been the most heavily harvested and traded cat species (Nowell and Jackson 1996), they were only listed because of the similarity between their pelts and those of endangered felids such as the Iberian lynx (*L. pardinus*). The concern is that illegally obtained endangered cats of other species may be labeled as bobcats and enter the fur trade. Listing requires that exporting countries show that export is not detrimental to the survival of the species. All bobcat pelts, whether for sale or other purposes, must be tagged by government officials, usually state game management officers. A benefit of CITES listing is that tagging produces relatively accurate counts of the numbers of bobcats killed in each US state, in contrast to voluntary reports by trappers (see below).

In most areas where bobcats occur, the primary source of mortality is trapping (Anderson and Lovallo 2003). Trappers kill bobcats in order to sell their fur pelts to the fashion industry, which then uses the pelts primarily as trim for women's coats. The largest markets are in Russia, China, and eastern Europe, and most pelts obtained in the United States are exported. The level of trapping activity is strongly influenced by the prices trappers obtain for pelts. When prices are high, the number of active trappers increases dramatically and so does harvest (Rolley 1987; Nowell and Jackson 1996; Anderson and Lovallo 2003). Since prices are determined by the fashion industry and are not related to bobcat population size, the risk always exists that demand for bobcat pelts will exceed sustainable levels.

Fur trading began in the 1600s and continues today. From 1920 to 1980, approximately 920,000 bobcats were killed nationally (Obbard et al. 1987). The peak year between 1970 and 2003 was 1987–1988, when 85,000 bobcats were taken by trappers (Hansen 2007). At that time the average price of a bobcat pelt was about $110. Today prices for prime western pelts (including New Mexico) may be as high as $1,400 each, but average around $430 (Fur Harvesters Auction Inc. 2020).

State regulations for bobcat trapping and/or hunting vary significantly between US states,

Photos 7.25a, b, c, d, and e. Research on the Armendaris Ranch in southern New Mexico focused on bobcat habitat associations, diet, home range, dispersal, survival, patterns of activity, body length and weight, and exposure to disease. a) (*top left*) Diet studies rely upon collecting scats in the field or obtaining stomachs from carcasses of trapped bobcats. Scats can be difficult to find, especially where bobcat densities are low or where bobcats do not follow obvious trails or roads. Harrison (2006) found detector dogs to be very helpful in locating bobcat scats. b) (*top right*) Locating radio-collared bobcats on the vast Armendaris Ranch requires extra-large antennas. c) (*middle right*) Hair snares are normally placed on trees; but in habitats with no trees, logs may be used. d) (*bottom left*) Scent-station surveys were the method of choice for bobcat surveys prior to the advent of automatic cameras. e) (*bottom right*) Cage traps are cumbersome to use but cause much less injury to trapped animals than leghold traps. Photographs: © Robert L. Harrison.

Photos 7.26a and b (*left to right*). Bobcats on a power pole in Socorro County on 14 November 2014. Electrocutions on utility poles are a documented source of bobcat mortality. Photographs: © WSMR Garrison Environmental Division and ECO Inc., Ecological Consultants.

with some, such as Texas, allowing unlimited take (Texas Parks and Wildlife Department 2023), whereas others like California have banned all trapping and hunting of bobcats (Woolf and Hubert 1998; Hansen 2007; Loftus-Farren 2019). Differences among US and Mexican states and Canadian provinces also revolve around the length of the trapping season, devices permitted, and bag limits. As reported by Hansen (2007), harvesting is allowed in seven of the eight Canadian provinces where bobcats occur and hunting is allowed in five Mexican states.

The fact that bobcats have survived such intense harvest pressure may be attributed to several factors: They occur over a very large geographic area and are flexible in their diet and habitat choices. Bobcats may live 12 years or longer in the wild (Hansen 2007), or long enough to reproduce multiple times. They are small enough to live close to humans without causing excessive alarm. Although they do kill sheep (Bailey 1931; USDA 2015), they are not a significant threat to cattle and thus not a major retaliatory focus of the most influential livestock industry. Demand for pelts fluctuates, potentially allowing overexploited populations to recover. Finally,

though rabid bobcats will attack people, attacks by bobcats are rare. The only attacks on people by healthy bobcats of which we are aware occurred at Death Valley National Park, and were likely the result of habituation to humans who had been feeding the bobcats (Repanshek 2007).

Radiotelemetry studies cited by Anderson and Lovallo (2003) found legal and illegal trapping and poaching to account for at least 50%, and as high as 100%, of bobcat mortality. In areas that are heavily trapped, annual survival of adults may be as low as 8% (Fuller et al. 1985). By contrast, annual adult survival in unexploited populations is usually 80–97% (Anderson and Lovallo 2003). For example, on the Armendaris Ranch near Truth or Consequences, where trapping is not allowed, Harrison (2010) found adult male and female survival to be 87.5% and 91.7%, respectively. Although kittens are not trapped as often as adults, loss of females directly affects their survival (Knick 1990). By comparing two bobcat populations in southeastern Idaho, one harvested (trapped) and the other unharvested, Knick (1990) calculated that if annual survival rates exceed 80%, populations remain relatively stable. But if average mortality due to all factors exceeds 20%, then large declines in population density may occur. Populations under heavy exploitation have higher proportions of younger animals and the ratio of males to females becomes biased toward females as males tend to move more extensively than females and are more readily trapped (Anderson and Lovallo 2003).

In New Mexico, bounties for bobcats were paid from 1897 until 1924 and bobcats were classified as predators from 1897 until 1980, when they became protected furbearers (Evans 1981). Regulations were then enacted limiting the times of year that trapping could occur and the number of animals that could be taken. Currently in New Mexico, bobcats may be killed with traps, snares, dogs, firearms, and bows and arrows from 1 November to 15 March each year (NMDGF 2020). There is no limit upon the number which may be taken. However, in early 2021 the state of New Mexico banned the use of leghold traps and snares on public land beginning with the 2022–2023 trapping season (State of New Mexico 2021; leghold traps are steel traps that are buried beneath the soil and close upon an animal's leg when stepped upon). Because leghold traps and snares are the most commonly used devices to capture bobcats, a significant mortality factor will be removed. However, the ban (known as the Wildlife Conservation and Public Safety Act, or Roxy's Law) passed the legislature by a very narrow margin. It could be overturned in the future if political trends shift, allowing trapping to resume as in the past.

Roberts and Crimmins (2010) presented information on bobcat population trends obtained from 46 contiguous US states (Delaware has no bobcat population while Colorado did not report any information). Populations were reported as increasing in 32 states (including New Mexico), stable or increasing in one other, and stable in eight more. Population trends were unknown in the last five of the 46 reporting states. A total of 27 states were able to provide an actual population estimate (Roberts and Crimmins 2010).

Despite New Mexico being included among them, reliable data have been lacking to accurately estimate bobcat population size in the state. Use of data from outside New Mexico led Thompson et al. (1992) to calculate a statewide population of between 6,084 and 328,503 bobcats and a sustainable harvest level of 1,156 to 62,416 bobcats, which Thompson et al. (1992) themselves described as absurd. In 2006, the NMDGF estimated the number of bobcats in New Mexico as 36,269 to 54,403, again based upon population density estimates from outside New Mexico (NMDGF unpubl. Furbearer Population Assessment and Harvest Management Matrix). It then estimated sustainable annual harvest limits as no more than 10% of the estimated population, or 3,627 to 5,440 bobcats per year. In New Mexico, CITES tagging records indicate that between 1980 and 2020, the average number of bobcats killed each year was 1,855 (range: 65–4,240). An average of 1,603 trapping licenses were sold annually between 1980 and 2020. On its face, it would appear that bobcat harvest at recent levels is sustainable in New Mexico, but any such conclusion hinges on the reliability of the 2006 population estimate. The applicability of out-of-state bobcat density estimates to New Mexico is highly questionable, and, in our opinion, likely led to a substantial overestimate.

The most recent published analysis of the effects of trapping upon New Mexico bobcat populations was that of Haussamen (1989), who examined harvest data collected from 1969 through 1987, a period of unusually high prices and trapper effort (trapping and shooting are actually not differentiated, with predator calling an important recreational activity in the state; J. K. Frey, pers. comm.). The largest harvest was in the 1976–1977 season, when an estimated 5,077 bobcats were killed. Haussamen (1989) found that the average age of bobcats trapped declined and that the proportion of females harvested increased, indicative of an overall population decline. No corrective action was taken by the NMDGF. One of us (R. L. Harrison, unpubl. data) found no evidence of any obvious decline in the numbers of bobcats trapped per trapping license sold between 1980 and 2020. Despite variations in fur prices and the number of licenses sold, the number of bobcats trapped per license sold was remarkably constant, at an average of one bobcat per license (range: 0.1–2.1). Although these statistics imply that the rate of bobcat harvest in New Mexico has been sustainable, in fact the actual bobcat population size is not known and not monitored. Therefore, no conclusion regarding changes in population size due to trapping or any other factor in New Mexico can be drawn at this time.

In addition to trapping (and predator control; see Chapter 3), bobcats also suffer direct mortality from the human footprint on the landscape, particularly collisions with road traffic (e.g., Johnson et al. 2010), attacks by domestic dogs (Lembeck 1986; see Chapter 33) and, to a lesser extent, electrocutions on power poles (Bailey 1974). None of these potential sources of mortality has been studied in New Mexico, but their impact on populations is likely negligible in the state.

State wildlife management agencies and researchers have used a variety of methods to monitor or survey bobcat populations, including capture-mark-release, surveys for tracks on natural or prepared surfaces with or without a scented lure (a.k.a. scent stations or scent posts), scat surveys, the use of motion-triggered automatic remote cameras, hair snares, trapper questionnaires, prey surveys, examination of annual harvests (for trapper success and effort and sex and age distribution of harvested bobcats), counts of license sales, and CITES tagging (Knowlton and Tzilkowski 1979; Anderson and Lovallo 2003; Harrison 2006; Hansen 2007; Larrucea et al. 2007; Long et al. 2008; Ruell et al. 2009; Mahaard et al. 2016). In recent years, automatic cameras have become the method of choice. Cameras are cost-efficient and easy to deploy, and they provide documentable evidence for multiple species. Cameras may be placed along trails or other likely travel routes, or where bobcat tracks or scats have been found. An attractant such as a scented lure or a feather or shiny object like a pie tin suspended over the site may also be used.

At present there is no formal management plan for the bobcat or any other furbearer in New Mexico. The NMDGF does require trappers to report the numbers of bobcats caught each season. Before 2010, when reporting was voluntary, only 39.7% of trappers reported. After reporting was made mandatory in 2010, that percentage increased to 82.3% (NMDGF, unpub. fur harvest records). However, CITES tagging is required to sell bobcat pelts, and more accurate numbers are thus obtained (see above). The NMDGF also records the annual number of trapping license sales. By comparing trapper harvest with license sales, a rough assessment of bobcat population trends is possible, but any trends such as a population decline may take several years to detect (Rolley 1987).

To obtain a more direct evaluation of population trends for the bobcat and other species, before 2010 the NMDGF conducted scent-station surveys in selected portions of New Mexico. However, the effort was not extensive enough to be successful. Currently, the NMDGF uses motion-triggered remote cameras in selected portions of the state to monitor populations (N. Forman, NMDGF, pers. comm.). This effort is directed primarily at swift foxes (*Vulpes velox*), but bobcat photographs are obtained as well.

Bobcats are adaptable and have survived many challenges. The outlook for the bobcat in New Mexico seems promising. New Mexico is a large state with a relatively small human population. It also has numerous areas which are protected or inaccessible to trappers. Prey will have to remain wary of explosions of teeth and claws from the shadows in New Mexico for a long time to come.

LITERATURE CITED

Ackerman, B. B., F. G. Lindzey, and T. P. Hemker. 1984. Cougar food habits in southern Utah. *Journal of Wildlife Management* 48:147–55.

Aiello, S. E., and M. A. Moses, eds. 2016. *Merck Veterinary Manual*. 11th ed. Hoboken, NJ: Wiley.

Allen, M. L., C. F. Wallace, and C. C. Wilmers. 2015. Patterns in bobcat (*Lynx rufus*) scent marking and communication behaviors. *Journal of Ethology* 33:9–14. doi 10.1007/s10164-014-0418-0.

Anderson, E. M. 1987. *A critical review and annotated bibliography of literature on bobcat*. Special Report No. 62. Colorado Division of Wildlife.

———. 1988. Effects of male removal on spatial distribution of bobcats. *Journal of Mammalogy* 69:637–41.

———. 1990. Bobcat diurnal loafing sites in southeastern Colorado. *Journal of Wildlife Management* 54:600–602.

Anderson, E. M., and M. J. Lovallo. 2003. Bobcat and Lynx. In *Wild mammals of North America: biology, management, and economics*, 2nd ed., ed. G. A. Feldhamer, B. C. Thompson, and J. A. Chapman, 758–786. Baltimore: Johns Hopkins University Press.

Asdell, S. A. 1946. Patterns of mammalian reproduction. Ithaca, NY: Comstock Press.

Attwood, T. C., T. L. Fry, and B. R. Leland. 2011. Partitioning of anthropogenic watering sites by desert carnivores. *Journal of Wildlife Management* 75:1609–15.

Bailey, T. N. 1972. Ecology of bobcats with special reference to social organization. PhD dissertation, University of Idaho, Moscow.

———. 1974. Social organization in a bobcat population. *Journal of Wildlife Management* 38:435–446.

———. 1979. Den ecology, population parameters and diet of eastern Idaho bobcats. In *Proceedings of the 1979 bobcat research conference*, ed. P. C. Escherich and L. Blum, 62–69. Science and Technology Series 6. Washington, DC: National Wildlife Federation.

———. 1981. Factors of bobcat social organization and some management implications. In *Proceedings of the worldwide furbearer conference*, ed. J. A. Chapman and D. Pursley, 984–1000. Frostburg, MD.

Bailey, V. 1931 (=1932). *Mammals of New Mexico*. North American Fauna 53. Washington, DC: US Department of Agriculture, Bureau of Biological Survey.

Banfield, A. W. F. 1987. *The mammals of Canada*. Toronto: University of Toronto Press.

Bender, L. C., O. C. Rosas-Rosas, and M. E. Weisenberger. 2017. Seasonal occupancy of sympatric larger carnivores in the southern San Andres Mountains, south-central New Mexico, USA. *Mammal Research* 62:323–29.

Benson, J. F., M. J. Chamberlain, and B. D. Leopold. 2004. Land tenure and occupation of vacant home ranges by bobcats (*Lynx rufus*). *Journal of Mammalogy* 85:983–88.

Berg, W. E. 1979. Ecology of bobcats in northern Minnesota. In *Proceedings of the 1979 bobcat research conference*, ed. P. C. Escherich and L. Blum, 55–61. Science and Technology Series 6. Washington, DC: National Wildlife Federation.

Biggins, D. E., and D. M. Biggins. 2006. Bobcat attack on a cottontail rabbit. *Southwestern Naturalist* 51:119–22.

Boyle, K. A., and T. T. Fendley. 1987. Habitat suitability index models: bobcat. U.S. Fish and Wildlife Service Biological Report 82. Washington, DC.

Buie, D. E., T. T. Fendley, and H. McNab. 1979. Fall and winter home ranges of adult bobcats on the Savannah River Plant, South Carolina. In *Proceedings of the 1979 bobcat research conference*, ed. P. C. Escherich and L. Blum, 42–46. Science and Technology Series 6. Washington, DC: National Wildlife Federation.

Conner, L. M., B. D. Leopold, and K. Sullivan. 1992. Bobcat home range, density, and habitat use in east-central Mississippi. *Proceedings of the annual conference of the Southeastern Association of Fish and Wildlife Agencies* 46:147–58.

Convention on International Trade in Endangered Species of Wild Fauna and Flora (CITES). 2021. Checklist of CITES Species Appendix II. https://checklist.CITES.org. Accessed 26 May 2021.

Croteau E. K. 2009. Population genetics and phylogeography of bocats (*Lynx rufus*) using microsatellites and mitochondrial DNA. PhD dissertation, Southern Illinois University, Carbondale.

Crowe, D. M. 1975a. Aspects of ageing, growth, and reproduction of bobcats from Wyoming. *Journal of Mammalogy* 56:177–98.

———. 1975b. A model for exploited bobcat populations in Wyoming. *Journal of Wildlife Management* 39:408–15.

Dearborn, N. 1932. Food of some predatory furbearing animals of Michigan. *Conservation Bulletin No. 1*, 1–52. University of Michigan School of Forestry.

Dick-Peddie, W. A, ed. 1993. *New Mexico vegetation: past, present, and future*. Albuquerque: University of New Mexico Press.

Disney, M., and L. K. Spiegel. 1992. Sources and rates of San Joaquin kit fox mortality in western Kern County, California. *Transactions of the Western Section of the Wildlife Society* 28:73–82.

Dubey, J. P., and C. P. Beattie. 1988. *Toxoplasmosis of animals and man*. Boca Raton, FL: CRC Press.

Espinosa-Flores, M. E., and C. A. López-González. 2017. Landscape attributes determine bobcat (*Lynx rufus escuinapae*) presence in Central Mexico. *Mammalia* 81:101–105.

Evans, W. 1981. Bobcat status report. Final report. Pittman-Robertson Project W-124-R-5, Job 1. Unpublished report. Santa Fe: New Mexico Department of Game and Fish.

Ewer, R. E. 1973. *The Carnivores*. Ithaca, NY: Cornell University Press.

Fedriani, J. M., T. K. Fuller, R. M. Sauvajot, and E. C. York. 2000. Competition and intraguild predation among three sympatric carnivores. *Oecologia* 125:258–70.

Franklin S. P., J. L. Troyer, J. A. Terwee, L. M. Lyren, W. M. Boyce, S. P. D. Riley, M. E. Roekle, K. R. Crooks, and S. VandeWoude. 2007. Frequent transmission of immunodeficiency virus among bobcats and pumas. *Journal of Virology* 81:10961–69.

Frey, J. K. 2004. Taxonomy and distribution of the mammals of New Mexico: an annotated checklist. *Occasional Papers, Museum of Texas Tech University* 240:1–32.

Frey, J. K., and J. L. Malaney. 2006. Snowshoe hare (*Lepus americanus*) and mountain cottontail (*Sylvilagus nuttallii*) biogeography at their southern range limit. *Journal of Mammalogy* 87:1175–82.

Fritts, S. H. 1973. Age, food habits, and reproduction in the bobcat (*Lynx rufus*) in Arkansas. MS thesis, University of Arkansas, Fayetteville.

Fritts, S. H., and J. A. Sealander. 1978. Diets of bobcats in Arkansas with special reference to age and sex differences. *Journal of Wildlife Management* 42:533–39.

Fuller, T. K., W. E. Berg, and D. W. Kuehn. 1985. Survival rates and mortality factors of adult bobcats in north-central Minnesota. *Journal of Wildlife Management* 49:292–96.

Fur Harvesters Auction Inc. 2020. Auction Results. www.furharvesters.com/auctionresults.html. Accessed 27 January 2021.

Gashwiler, J. S., W. L. Robinette, and O. W. Morris. 1961. Breeding habits of bobcats in Utah. *Journal of Mammalogy* 42:76–84.

Griffith, M. A., and T. T. Fendley. 1986. Pre and post dispersal movement behavior of subadult bobcats on the Savannah River Plant. In *Cats of the world: biology, conservation and management*, ed. S. D. Miller and D. Everet, 277–89. Washington, DC: National Wildlife Federation.

Hall, E. R. 1981. *The mammals of North America*. 2nd ed. Vol. 2. New York: John Wiley.

Hall, H. T., and J. D. Newsom. 1976. Summer home ranges and movement of bobcats in bottomland hardwoods of southern Louisiana. *Proceedings of the Annual Conference of the Southeastern Association of Fish and Wildlife Agencies* 30:427–36.

Hamilton, D. A. 1982. Ecology of the bobcat in Missouri. MS thesis, University of Missouri, Columbia.

Hansen, K. 2007. *Bobcat: master of survival*. New York: Oxford University Press.

Harrison, R. L. 1998. Bobcats in residential areas: distribution and homeowner attitudes. *Southwestern Naturalist* 43:469–75.

———. 2006. A comparison of survey methods for detecting bobcats. *Wildlife Society Bulletin* 34:548–52.

———. 2010. Ecological relationships of bobcats (*Lynx rufus*) in the Chihuahuan Desert of New Mexico. *Southwestern Naturalist* 55:374–381.

Hass, C. 2009. Competition and coexistence in sympatric bobcats and pumas. *Journal of Zoology* (London) 278:174–80.

Haussamen, W. 1989. *Bobcat final report*. Rev. Project No. W-124-R-11. Santa Fe: New Mexico Department of Game and Fish.

Heilbrun, R. D., N. J. Silvy, M. E. Tewes, and M. J. Peterson. 2003. Using automatically triggered cameras to individually identify bobcats. *Wildlife Society Bulletin* 31:748–55.

Hooper, E. T. 1941. *Mammals of the lava fields and adjoining areas in Valencia County, New Mexico.* Museum of Zoology, University of Michigan, Miscellaneous Publications 51:1–52.

Hutchinson, J. T., and T. Hutchinson. 2000. Observation of a melanistic bobcat in the Ocala National Forest. *Florida Field Naturalist* 28:25–26.

Johnson, S. A., H. D. Walker, and C. M. Hudson. 2010. Dispersal characteristics of juvenile bobcats in south-central Indiana. *Journal of Wildlife Management* 74:379–85.

Jones, L. R., P. A. Zollner, R. K. Swihart, E. Godollei, C. M. Hudson, and S. A. Johnson. 2020. Survival and mortality sources in a recovering population of bobcats (*Lynx rufus*) in south-central Indiana. *American Midland Naturalist* 184:222–32.

Kamler, J. F., and P. S. Gipson. 2000. Home range, habitat selection, and survival of bobcats, *Lynx rufus*, in a prairie ecosystem in Kansas. *Canadian Field-Naturalist* 114:388–94.

———. 2004. Survival and cause-specific mortality among furbearers in a protected area. *American Midland Naturalist* 151:27–35.

Kelly, M., D. Morin, and C. A. Lopez-Gonzalez. 2016. Bobcat. *Lynx rufus.* The IUCN Red List of Threatened Species 2016:e.T12521A50655874. http://dx.doi.org/10.2305/IUCN.UK.2016-1.RLTS.T12521A50655874.en

Kitchener, A. C., C. Breitenmoser-Würsten, E. Eizirik, A. Gentry, L. Werdelin, A. Wilting, N. Yamaguchi, A. V. Abramov, P. Christiansen, C. Driscoll, J. W. Duckworth, W. Johnson, S.-J. Luo, E. Meijaard, P. O'Donoghue, J. Sanderson, K. Seymour, M. Bruford, C. Groves, M. Hoffmann, K. Nowell, Z. Timmons, and S. Tobe. 2017. A revised taxonomy of the Felidae: the final report of the Cat Classification Task Force of the IUCN Cat Specialist Group. *Cat News Special Issue* 11:1–80.

Kleiman, D. G., and J. F. Eisenberg. 1973. Comparisons of canid and felid social systems from an evolutionary perspective. *Animal Behaviour* 21:637–59.

Knick, S. T. 1990. Ecology of bobcats relative to exploitation and a prey decline in southeastern Idaho. *Wildlife Monographs* 108:1–42.

Knick, S. T., and T. N. Bailey. 1986. Long-distance movements by two bobcats from southeastern Idaho. *American Midland Naturalist* 116:222–23.

Knick, S. T., J. D. Brittell, and S. J. Sweeney. 1985. Population characteristics of bobcats in Washington State. *Journal of Wildlife Management* 49:721–28.

Knick, S. T., S. J. Sweeney, J. R. Alldredge, and J. D. Brittell. 1984. Autumn and winter food habits of bobcats in Washington State. *Great Basin Naturalist* 44:70–74.

Knowlton, F. F., and W. M. Tzilkowski. 1979. Trends in bobcat visitation to scent-station survey lines in western United States, 1972–1978. In *Bobcat Research Conference Proceedings*, 8–12. National Wildlife Federation Scientific and Technical Series 6.

Koehler, G. 1987. *The bobcat.* Audubon Wildlife Report 1987. New York: Academic Press.

Koehler, G. M., and M. G. Hornocker. 1989. Influences of seasons on bobcats in Idaho. *Journal of Wildlife Management* 53:197–202.

———. 1991. Seasonal resource use among mountain lions, bobcats, and coyotes. *Journal of Mammalogy* 72:391–96.

Kolowski, J. M., and A. Woolf. 2002. Microhabitat use by bobcats in southern Illinois. *Journal of Wildlife Management* 66:822–32.

Krebs, J. W., S. M. Williams, J. S. Smith, C. E. Rupprecht, and J. E. Childs. 2003. Rabies among infrequently reported mammalian carnivores in the United States, 1960–2000. *Journal of Wildlife Diseases* 39:253–61.

Lagana, D. M., J. S. Lee, J. S. Lewis, S. N. Bevins, S. Carver, L. L. Sweanor, R. McBride, C. McBride, K. R. Crooks, and S. VandeWoude. 2013. Characterization of regionally associated feline immunodeficiency virus (FIV) in bobcats (*Lynx rufus*). *Journal of Wildlife Diseases* 49:718–22.

Larivière, S., and L. R. Walton. 1997. *Lynx rufus.* *Mammalian Species* 563:1–8.

Larrucea, E. S., G. Serra, M. M. Jaeger, and R. H. Barrett. 2007. Censusing bobcats using remote cameras. *Western North American Naturalist* 67:538–48.

Lavoie, M., P. Collin, F. Lemieux, H. Jolicoeur, P. Canac-Marquis, and S. Larivière. 2009. Understanding fluctuations in bobcat harvest at the northern limit of their range. *Journal of Wildlife Management* 73:870–75. doi:10.2193/2008-275.

Lawhead, D. N. 1984. Bobcat *Lynx rufus* home range,

density and habitat preference in south-central Arizona. *Southwest Naturalist* 29:105–113.

Lembeck, M. 1986. Long term behavior and population dynamics of an unharvested bobcat population in San Diego County. In *Cats of the world: biology, conservation and management*, ed. S. D. Miller and D. Everet, 305–10. Washington, DC: National Wildlife Federation.

Little, A. R., L. M. Conner, M. J. Chamberlain, N. P. Nibbelink, and R. J. Warren. 2018. Adult bobcat (*Lynx rufus*) habitat selection in a longleaf pine savanna. *Ecological Processes* 7, article 20. https://doi.org/10.1186/s13717-018-0129-5.

Litvaitis, J. A., C. L. Stevens, and W. W. Mautz. 1984. Age, sex, and weight of bobcats in relation to winter diet. *Journal of Wildlife Management* 48:632–35.

Loftus-Farren, Z. 2019. California just became the first state to ban fur trapping. *Earth Island Journal*. 5 September 2019. www.earthisland.org/journal/index.php/articles/entry/California-first-state-to-ban-fur-trapping.

Long, R. A., P. MacKay, W. J. Zielinski, and J. C. Ray. 2008. *Noninvasive survey methods for carnivores.* Washington, DC: Island Press.

López-Vidal, J. C., C. Elizalde-Arellano, L. Hernández, J. W. Laundré, A. González-Romero, and F. A. Cervantes. 2014. Foraging of the bobcat (*Lynx rufus*) in the Chihuahuan Desert: Generalist or specialist? *Southwestern Naturalist* 59:157–66.

Lovallo, M. J., and E. M. Anderson. 1995. Range shift by a female bobcat (*Lynx rufus*) after removal of neighboring female. *American Midland Naturalist* 134:409–12.

Maehr, D. S., and J. R. Brady. 1986. Food habits of bobcats in Florida. *Journal of Mammalogy* 67:133–38.

Mahaard, T. J., J. A. Litvaitis, R. Tate, G. C. Reed, and D. J. A. Broman. 2016. An evaluation of hunter surveys to monitor relative abundance of bobcats. *Wildlife Society Bulletin* 40:224–32.

Major, J. T. 1983. Ecology and interspecific relationships of coyotes, bobcats, and red foxes in western Maine. PhD dissertation, University of Maine, Orono.

McCord, C. M. 1974. Selection of winter habitat by bobcats (*Lynx rufus*) on the Quabbin Reservation, Massachusetts. *Journal of Mammalogy* 55:428–37.

McCord, C. M., and J. E. Cardoza. 1982. Bobcat and

lynx (*Felis rufus* and *F. lynx*). In *Wild mammals of North America: biology, management, and economics*, ed. J. A. Chapman and G. A. Feldhamer, 728–66. Baltimore: Johns Hopkins University Press.

Mehrer, C. F. 1975. Some aspects of reproduction in captive mountain lions (*Felis concolor*), bobcats (*Lynx rufus*), and lynx (*Lynx Canadensis*). PhD dissertation, University of North Dakota, Grand Forks.

New Mexico Department of Game and Fish (NMDGF). 2020. 2020–2021 New Mexico Furbearer Rules and Info. Santa Fe: New Mexico Department of Game and Fish.

Nielsen, C. K., and A. Woolf. 2001. Bobcat habitat use relative to human dwellings in southern Illinois. In *Proceedings of a symposium on current bobcat research and implications for management*, ed. A. Woolf, C. K. Nielsen, and R. D. Bluett, 40–44. The Wildlife Society 2000 Conference, Nashville, TN.

Nowell, K., and P. Jackson. 1996. *Wild cats status survey and conservation action plan.* Gland, Switzerland: International Union for the Conservation of Nature and Natural Resources.

Nunley, G. L. 1978. Present and historical bobcat population trends in New Mexico and the West. In *Proceedings of the Vertebrate Pest Conference.* 8:77–84.

Obbard, M. E., J. G. Jones, R. Newman, A. Booth, A. J. Satterthwaite, and G. Linscombe. 1987. Furbearer harvests in North America. In *Wild furbearer management and conservation in North America*, ed. M. Novak, J. A. Baker, M. E. Obbard, and B. Malloch, 1007–33. Ontario, Canada: Ministry of Natural Resources.

Parker, G. R., J. W. Maxwell, L. D. Morton, and G. E. J. Smith. 1983. The ecology of the lynx (*Lynx canadensis*) on Cape Breton Island. *Canadian Journal of Zoology* 61:770–86.

Parker, G. R., and G. E. J. Smith. 1983. Sex- and age-specific reproductive and physical parameters of the bobcat (*Lynx rufus*) on Cape Breton Island, Nova Scotia. *Canadian Journal of Zoology* 61:1771–82.

Pence, D. B., F. D. Matthews, and L. A. Windberg. 1982. Notoedric mange in the bobcat, *Felis rufus*, from south Texas. *Journal of Wildlife Diseases* 18:47–50.

Peterson, R. L., and S. C. Downing. 1952. Notes on the bobcat (*Lynx rufus*) of eastern North America with the description of a new race. *Contributions of the Royal Ontario Museum* 33:1–23.

Poessel, S. A., C. L. Burdett, E. E. Boydston, L. M. Lyren, R. S. Alonso, R. N. Fisher, and K. R. Crooks. 2014. Roads influence movement and home ranges of a fragmentation-sensitive carnivore, the bobcat, in an urban landscape. *Biological Conservation* 180:224–32.

Pollack, E. M. 1950. Breeding habits of the bobcat in northeastern United States. *Journal of Mammalogy* 31:327–330.

Provost, E. E., C. A. Nelson, and D. A. Marshall. 1973. Population dynamics and behavior in the bobcat. In *The world's cats*. Vol. 1, ed. R. L. Eaton, 42–67. Winston, OR: World Wildlife Safari.

Prude, C. H. 2020. Influence of habitat heterogeneity and water sources on kill site locations and puma prey composition. MS thesis, New Mexico State University, Las Cruces.

Reding D. M. 2011. Patterns and processes of spatial genetic structure in a mobile and continuously distributed species, the bobcat (*Lynx rufus*). PhD dissertation, Iowa State University, Ames.

Regan, T. W., and D. S. Maehr. 1990. Melanistic bobcats in Florida. *Florida Field Naturalist* 18:84.

Repanshek, K. 2007. Bobcats attack two at Death Valley National Park. National Parks Traveler. 31 December 2007. www.nationalparkstraveler.com/2007/12/bobcats-attack-two-death-valley-national-park.

Riley, S. P. D. 2001. Spatial and resource overlap of bobcats and gray foxes in urban and rural zones of a national park. In *Proceedings of a symposium on the ecology and management of bobcats*, ed. A. Woolf and C. K. Nielsen, 32–39. Champaign: Illinois Department of Natural Resources.

Riley, S. P. D., C. Bromley, R. H. Poppenga, F. A. Uzal, L. Whited, and R. M. Sauvajot. 2007. Anticoagulant exposure and notoedric mange in bobcats and mountain lions in urban southern California. *Journal of Wildlife Management* 71:1874–84.

Riley, S. P. D., J. Foley, and B. Chomel. 2004. Exposure to feline and canine pathogens in bobcats and gray foxes in urban and rural zones of a national park in California. *Journal of Wildlife Diseases* 40:11–22.

Roberts, N. M., and S. M. Crimmins. 2010. Bobcat population status and management in North America: evidence of large-scale population increase. *Journal of Fish and Wildlife Management*. 1:169–74. doi:10.3996/122009-JFWM-026.

Robinson, Q. H., D. Bustos, and G. W. Roemer. 2014. The application of occupancy modeling to evaluate intraguild predation in a model carnivore system. *Ecology* 95:3112–23.

Robinson, W. B. 1961. Population changes of carnivores in some coyote-control areas. *Journal of Mammalogy* 42:510–15.

Robinson, W. B., and E. F. Grand. 1958. Comparative movements of bobcats and coyotes as disclosed by tagging. *Journal of Wildlife Management* 22:117–22.

Rolley, R.E. 1985. Dynamics of a harvested bobcat population in Oklahoma. *Journal of Wildlife Management* 49:283–92.

———. 1987. Bobcat. In *Wild furbearer management and conservation in North America*, ed. M. Novak, J. A. Baker, M. E. Obbard, and B. Malloch, 671–81. Ontario, Canada: Ministry of Natural Resources.

Rosenzweig, M. L. 1966. Community structure in sympatric Carnivora. *Journal of Mammalogy* 47:602–12.

Ruell, E. W., S. P. D. Riley, M. R. Douglas, J. P. Pollinger, and K. R. Crooks. 2009. Estimating bobcat population sizes and densities in a fragmented urban landscape using noninvasive capture-recapture sampling. *Journal of Mammalogy* 90:129–35.

Schmidly, D. J., and J. A. Read. 1986. Cranial variation in the bobcat (*Felis rufus*) from Texas and surrounding states. *Occasional Papers, Museum of Texas Tech University* 101:1–39.

Schwartz, M. K., K. L. Pilgrim, K. S. McKelvey, E. L. Lindquist, J. J. Claar, S. Loch, and L. F. Ruggiero. 2004. Hybridization between Canada lynx and bobcats: genetic results and management implications. *Conservation Genetics* 5:349–55.

Shafer, E. P. 2016. Monitoring bobcat (*Lynx rufus*) activity at watering sites via camera traps. Unpublished senior's thesis, University of California at Berkeley.

Shenk, T. 2009. *Post release monitoring of lynx (Lynx canadensis) reintroduced to Colorado*. Wildlife Research Report. Colorado Division of Wildlife.

Sikes, R. S., and M. L. Kennedy. 1993. Geographic variation in sexual dimorphism of the bobcat (*Felis rufus*) in the United States. *Southwestern Naturalist* 38:336–44.

Sorensen, G. E. 2015. Ecology of adult female Rocky Mountain mule deer (*Odocoileus hemionus hemionus*) following habitat enhancements in north-central New Mexico. PhD dissertation, Texas Tech University, Lubbock.

State of New Mexico. 2021. Wildlife Conservation & Public Safety Act (Senate Bill 0032). New Mexico Legislature. www.nmlegis.gov/Legislation/Legislation?Chamber=S&LegType=B&LegNo=32&Year=21. Accessed 18 May 2021.

Stys, E. D., and B. D. Leopold. 1993. Reproductive biology and kitten growth of captive bobcats in Mississippi. *Proceedings of the Southeastern Association of Fish and Wildlife Agencies* 47:80–89.

Taylor, C. A. 2013. Behaviour and cause-specific mortality of mule deer (*Odocoileus hemonius*) fawns on the National Rifle Association Whittington Center of north-central New Mexico. MS thesis, Texas Tech University, Lubbock.

Texas Parks and Wildlife Department. 2023. Nongame, Exotic, Endangered, Threatened and Protected Species. Valid 1 September 2022 through 31 August 2023. https://tpwd.texas.gov/regulations/outdoor-annual/hunting/nongame-and-other-species. Accessed 1 February 2023.

Thompson, B. C., D. F. Miller, T. A. Doumitt, and T. R. Jacobson. 1992. *Ecologically-based management evaluation for sustainable harvest and use of New Mexico furbearer resources*. Santa Fe: New Mexico Department of Game and Fish.

Tucker, S. A., W. R. Clark, and T. E. Gosselink. 2008. Space use and habitat selection by bobcats in the fragmented landscape of south-central Iowa. *Journal of Wildlife Management* 72:1114–24.

Ulmer, F. A., Jr. 1941. Melanism in the Felidae, with special reference to the genus *Lynx*. *Journal of Mammalogy* 22:285–88.

United States Department of Agriculture (USDA). 2015. Sheep and lamb predator and nonpredator death Loss in the United States, 2015. USDA–APHIS–VS–CEAH–NAHMS Fort Collins, CO #721.0915.

Wang, J.-L., T.-T. Li, G.-H. Liu, X.-Q. Zhu, and C. Yao. 2017. Two tales of *Cytauxzoon felis* infections in domestic cats. *Clinical Microbiology Reviews* 30:861–85.

Wang, Y., M. L. Allen, and C. C. Wilmers. 2015. Mesopredator spatial and temporal responses to large predators and human development in the Santa Cruz Mountains of California. *Biological Conservation* 190:23–33.

Wassmer, D. A., D. D. Guenther, and J. N. Layne. 1988. Ecology of the bobcat in south-central Florida. *Bulletin of the Florida State Museum, Biological Sciences* 3:159–228.

Wiggington, J. D., and F. S. Dobson. 1999. Environmental influences on geographic variation in body size of western bobcats. *Canadian Journal of Zoology* 77:802–13.

Witmer, G. W., and D. S. deCalesta. 1986. Resource use by unexploited sympatric bobcats and coyotes in Oregon. *Canadian Journal of Zoology* 64:2333–38.

Woolf, A., and G. F. Hubert Jr. 1998. Status and management of bobcats in the United States over three decades: 1970s–1990s. *Wildlife Society Bulletin* 26:287–94.

Young, J. K., J. Golla, J. P. Draper, D. Broman, T. Blankenship, and R. Heilbrun. 2019. Space use and movement of urban bobcats. *Animals* 9:275. doi:10.3390/ani9050275

Young, S. P. 1958. The bobcat of North America: its history, life habits, economic status and control, with list of currently recognized subspecies. Washington, DC: Wildlife Management Institute.

Zezulak, D. S., and R. G. Schwab. 1980. *Bobcat biology in a Mojave Desert community*. California Department of Fish and Game Report. Federal Aid Wildlife Restoration Project W-54-R-12, job IV-4.

JAGUAR (*PANTHERA ONCA*)

James N. Stuart and Charles L. Hayes IV

Probably the most striking of New Mexico's "charismatic megafauna" species, the jaguar (*Panthera onca*) is among our state's largest and most elusive carnivores. While very few people will ever have the unique good fortune of seeing this rare cat in the wild, the jaguar has nonetheless captured the imagination of many throughout the New World, including in the southwestern United States, where the species is as controversial as it is enigmatic.

The jaguar is the only New World representative of the "roaring cats" or "big cats" (genus *Panthera*), which also include the African lion (*P. leo*), tiger (*P. tigris*), and leopard (*P. pardus*). In much of the older scientific literature, it was assigned to the genus *Felis* along with most other cat species (Seymour 1989). It is the largest New World cat and third largest in the world, exceeded only by the African lion and the tiger (Valdez 2000). The jaguar and the more common cougar (*Puma concolor*), the only two large, long-tailed cats in the New World, might be confused due to overlap in geographic range and similar body size, especially at the northern end of the jaguar's distribution (Emmons 1999). Compared to a cougar, however, the jaguar has a more powerful, more compact appearance; the head is massive and supported by a thickly muscled neck, and the legs and tail are relatively short in comparison to the torso (see also Chapter 9). Throughout its distribution, the jaguar varies substantially in size (Valdez 2000). Individuals from various parts of

the species' range are from 111 to 185 cm (43.7 to 72.8 in) in body and head length combined, with a tail length of 44–75 cm (17–30 in); body mass is 31–158 kg (68–348 lbs) (Leopold 1959; Emmons 1999). Adult jaguars exhibit sexual size dimorphism with males typically 10–25% larger than adult females. In northwestern Mexico, jaguars are relatively small compared to those in the Neotropics, averaging 54.5 kg (120 lbs) for males and 36 kg (80 lbs) for females (Brown and López González 2001). Body masses of three jaguars killed in Arizona were 62.7 kg (138 lbs) for an adult male and 47.7 kg (105 lbs) and 50 kg (110 lbs) for two adult females (Brown and López González 2000). As suggested by Brown and López González (2001), body size in any given population may be less a function of latitude and habitat than the types and sizes of prey that are available.

The distinctive pelage of jaguars is boldly marked by many large black rosettes (irregular circles of smaller blotches), each of which surrounds a lighter colored center with one or several black spots in the middle (Photos 8.1–8.2). The arrangement of the rosette pattern is unique to each individual jaguar and has proven to be a useful identification tool for studies of this species using photographs (e.g., McCain and Childs 2008). This pattern is superimposed on a background color of orange or yellow, and like other carnivores the most richly colored individuals occur in the tropics, whereas paler animals are found in the arid regions near the US-Mexico

(*opposite page*) Photograph: © Northern Jaguar Project.

Photo 8.1. Jaguar photographed in Sonora, Mexico in May 2009. The distinctive markings on the jaguar's body are called rosettes, which correspond to irregular circles made of smaller blotches surrounding a lighter colored center with smaller, dark spots. On the head, neck, legs, and tail are large, dark spots. The arrangement of the rosettes and spots forms a unique pattern that allows researchers to identify individual jaguars using photographs taken by motion-detection wildlife cameras. Photograph: © Northern Jaguar Project.

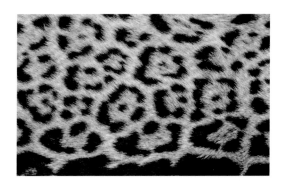

Photo 8.2. Close-up of a jaguar pelt from Mexico donated to the New Mexico Department of Game and Fish. Note the blotchy rosettes that often enclose one or several smaller, dark spots. Photograph: © Martin Perea.

border (Brown and López González 2001; Caro 2013). The dark spots on the tail, limbs, and head are generally solid (not in the form of rosettes) and a series of elongate dark spots along the midline of the back may merge into a solid line (Nowak 1991). The chin, throat, and underside of the torso are white or whitish-gray (especially in jaguars in the northern extent of their range). Although the overall pelage pattern sets the jaguar apart from any other mammal in the New World, with the possible exception of the much smaller ocelot (*Leopardus pardalis*), it is worth noting that immature cougars also have darkish spots, though much more diffuse and not consisting of rosettes. Similarly, the bobcat (*Lynx rufus*) also can exhibit a pattern of irregular dark spots. It is likely that some unverified reports from the Southwest of cats that were identified by observers as jaguars were actually either cougars or smaller native felids whose size was estimated incorrectly.

Some jaguars in populations from southern Mexico and Central and South America have melanistic pelage which appears uniformly blackish or dark smoky gray, though the darker rosettes are still evident in the pelt pattern upon close inspection. These individuals occur infrequently alongside jaguars with typical pelage, primarily in dense tropical forests where the dark coloration is possibly adaptive (Brown and López González 2001). Although no melanistic jaguar has ever been verified in North America, and melanism is unknown in the cougar, "black panthers" are still occasionally reported from many parts of the United States, including the Southwest (Childs and Childs 2008), and have figured in American cryptozoology and folklore (e.g., Willis 2007; see also Johnson et al. 2011). It is likely that all "black panthers" reported from the United States, including New Mexico, are normally colored cougars or other large animals seen under poor light conditions. In one unusual case from New Mexico, a road-killed specimen that was initially reported to be of a large black cat was later determined to be the carcass of a cougar covered in road tar (NMDGF 2016a).

Geographic variation in the jaguar has long been a topic of debate among mammalogists. Nelson and Goldman (1933) recognized 16 geographical races of the jaguar from throughout the species' range, whereas both Pocock (1939) and Emmons (1999) described only eight. Bailey (1931) assigned jaguars from New Mexico to the subspecies "Felis onca hernandesii" (West Mexican jaguar), which was originally described from Mazatlán, Sinaloa, Mexico (Nelson and Goldman 1933). After Bailey's publication, Goldman (1932) considered jaguars from northeastern Sonora, Arizona, and New Mexico to be larger and distinctive in skull characteristics compared with the small West Mexican jaguar and recognized the northern form, "Felis onca arizonensis" (Arizona jaguar), based on a type specimen from Cibecue, Navajo County, Arizona. However, more recent studies of geographic variation based on morphology or genetics have found remarkably little evidence to support the recognition of any subspecies in Panthera onca, despite its widespread distribution, the variety of habitats it occupies, and genetic isolation of some populations (Larson 1997; Eizirik et al. 2001; Johnson et al. 2006; Culver and Hein 2016). For this reason, many modern-day jaguar researchers recognize P. onca as a monotypic species or consider the geographic races described in the past as too poorly defined to be useful (Kitchener et al. 2017). Although the names "northern jaguar" and "borderland jaguar" have been applied to the species in northwestern Mexico and the southwestern United States (e.g., Brown and López González 2001; Boydston and López González 2005), they are only terms of convenience and do not imply any taxonomic distinctiveness.

DISTRIBUTION

The jaguar is a widely distributed but primarily neotropical cat that has been documented from 21 countries in the Americas (Wozencraft 2005). The species ranges from Brazil and Argentina northward through northern South America, Central America, and much of Mexico along the Atlantic and Pacific slopes. In the latter country, it may have always been rare or absent on the central Mexican Plateau and all or most of Baja California (Swank and Teer 1989; Seymour 1989; Sanderson et al. 2002b). In Mexico and Central and South America, populations of the jaguar persist in many areas but have been greatly reduced and isolated by ongoing habitat loss and fragmentation and from persecution by humans (Swank and Teer 1989; Sanderson et al. 2002b; Rabinowitz and Zeller 2010).

Historically, the species reached its northernmost range limits in the United States in two

Photo 8.3. The jaguar breeding population closest to the southwestern United States occurs along the west slope of the Sierra Madre Occidental in east-central Sonora, Mexico, near the confluence of the Yaqui and Aros rivers, about 200 km (125 mi) due south of the US-Mexico border where Arizona and New Mexico meet. The Río Aros is the last free-flowing river in northern Mexico. Jaguars documented in Arizona and New Mexico during recent decades likely originate from the east-central Sonora population. The jaguar in the photograph is one of several known animals belonging to the east-central Sonora population and monitored through the use of remote cameras in the 223 km² (86 mi²) Northern Jaguar Reserve. Photograph: © Northern Jaguar Project.

regions: 1) southern and central Texas, and possibly Louisiana; and 2) Arizona, New Mexico, and possibly southern California (Nelson and Goldman 1933; Brown and López González 2001). Although the jaguar was probably always uncommon in the United States, confirmed records since the early 1900s have nonetheless both declined in number and become more sporadic. The only verified occurrences in the United States since the 1960s have been in Arizona and New Mexico (Brown and López González 2000, 2001).

Today, the known breeding population of jaguars that is closest to the southwestern United States is on the west slope of the Sierra Madre

Occidental in east-central Sonora, Mexico, in the general vicinity of the towns of Huasabas, Nácori Chico, and Sahuaripa and near the confluence of the Yaqui and Aros rivers (Brown and López González 2001; Rosas-Rosas et al. 2008; Rosas-Rosas and Bender 2012). This population, which is about 225 km (140 mi) due south of where Arizona and New Mexico meet at the US-Mexico border, is the most likely source of jaguars recently verified in the southwestern United States (Brown and López González 2001; Johnson et al. 2011).

Reliable records of the jaguar in Arizona are more numerous than those in New Mexico. Brown and López González (2000) compiled records of jaguars that were killed or photographed in the Southwest during 1900–1999 and listed 51 from Arizona but only 6 from New Mexico. The last jaguar known to have been killed in Arizona was in 1986, whereas in New Mexico the last one taken was in 1909. Perhaps this disparity in numbers of records reflects more suitable habitat in Arizona or that state's proximity to dispersal corridors and occupied habitat in adjacent Sonora (Barber 1902; Menke and Hayes 2003). McCain and Childs (2008) and Culver (2016) provided evidence of at least one male jaguar residing in southern Arizona from the 1990s up near present time. No such evidence exists for New Mexico, perhaps in part due to the lack of comparable survey efforts.

In New Mexico, both the historical and current distributions of the jaguar remain poorly understood, and opinions differ significantly among researchers on what constitutes the species' geographic range in the state (Map 8.1). This lack of consensus stems from disagreements concerning the reliability of various jaguar records from years past (NMDGF 2016a; see under "Status and Management," below). Despite the fact that a few jaguars were killed in New Mexico from the late 1800s through the early 1900s and their pelts retained, it is remarkable that no preserved specimen from the state is known to exist today

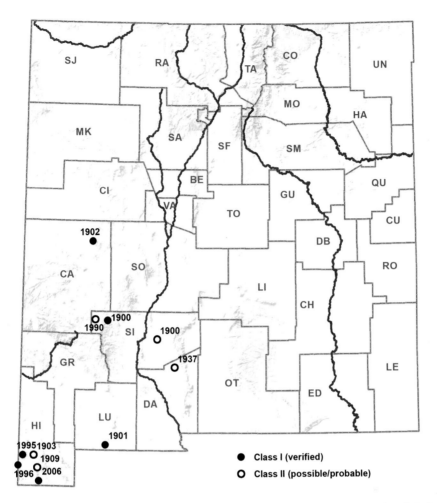

Map 8.1. Confirmed records and other reports of the jaguar (*Panthera onca*) in New Mexico. Closed circles indicate detections that are accepted as Class I (verified jaguar records based on physical evidence such as a photograph or specimen). Open circles indicate reports considered Class II (observations, mainly from the scientific literature, that probably or possibly represent valid detections but for which physical evidence is lacking). The year of observation is indicated next to the symbol. Other reports that are considered Class III (observations that are unreliable, lack sufficient detail to be fully evaluated, or have been rejected) are not mapped; some of these have been published in the scientific literature and are discussed under "Distribution." Locations of symbols on the map are based on available data and in some cases are approximate. The Class I–III ranking system for evaluating observations is discussed under "Status and Management."

(Findley et al. 1975; NMDGF 2016a). If residency of a species is based on evidence of reproduction, then New Mexico was perhaps always marginal territory as no female or juvenile jaguar has ever been confirmed from the state (Brown and López González 2000, 2001; Boydston and López González 2005), contrary to statements by

Grigione et al. (2007). In contrast, several female jaguars, including a few with kittens, were reported from Arizona from the late 1800s up to 1963 (Brown and López González 2000; Valdez 2000; Johnson et al. 2011), which suggests that a small breeding population existed in that state well into the 20th century.

Evidence of jaguars in New Mexico during pre-Columbian times is equivocal. The species seems to have figured little in the imagery and legends of Native American peoples of the Southwest. Daggett and Henning (1974) suggested that jaguar artistic motifs are found at various Native American archeological sites throughout the contiguous United States, while Brown and López González (2001) discussed some iconographs from the Southwest that are suggestive of jaguar and may have been influenced by artwork from Mexico. In New Mexico, one such site is the Shrine of the Stone Lions at Bandelier National Monument in the Jemez Mountains, which includes a sculpted image of a cat that could be interpreted as a jaguar (Brown and López González 2001). Nonetheless, the paucity of unambiguous artifacts in the Southwest would seem to indicate that the species was not familiar to the indigenous peoples in the region (Rabinowitz 1999).

The earliest possible observation of a jaguar in New Mexico by European explorers was by the Spaniard Francisco Vásquez de Coronado, who reported both "*tigres*" and "*ounces*" during his trek north to Zuni Pueblo in western New Mexico in 1540 (Whipple et al. 1856, in Bailey 1931). However, it is impossible to tell from this brief mention in Coronado's account if any of the cats his party encountered, or were told about, were indeed jaguars.

Most of the historical reports from New Mexico that are considered reliable or at least probable were summarized by Bailey (1931) and are primarily from the southwestern quarter of the state during the late 19th and early 20th centuries. A member of the Mexican Boundary Survey party reportedly saw a jaguar in Guadalupe Canyon in extreme southwestern Hidalgo County around 1855, but details are lacking (Baird 1859, in Bailey 1931). Barber (1902:192) provided a more detailed early account of a "*Felis hernandesii*" from the Black Range:

Mr. Nat Straw, hunter and trapper, informed me that he trapped a jaguar near Grafton on Taylor Creek, Socorro County [now in Sierra County], New Mexico in May, 1900. He gave its length as 8 feet and 3 inches [~ 2.45 m]. I saw the skin made up into a rug. I have heard of several others being seen or killed.

In 1903, rancher W. P. Burchfield shot a jaguar that was feeding on a bull it had killed in Clanton Canyon west of the Gray Ranch in Hidalgo County; the skin apparently was taken but was not available to Bailey (1931) for examination at the time he interviewed the rancher. Famous hunter Ben Lilly reportedly killed a jaguar in semi-desert grassland at Dog Springs near the Animas Mountains in 1909, but no specimen is known to exist (Carmony 1998). This record was nonetheless accepted as valid by Brown and López González (2000).

Bailey (1931:284) provided a fairly detailed account by Ned Hollister of a jaguar that had been killed in the Datil Mountains, Catron County in 1902. Hollister's photograph of the preserved specimen (in Bailey 1931:Plate 16A; Photo 8.4) provides a rare verified record of a jaguar from the state.

In 1905 Hollister saw and photographed a skin that had been mounted as a rug and was in the possession of O. Reddeman, at Magdalena [New Mexico]. The original skull was mounted in the skin and showed the animal to be an adult with well-worn teeth. Reddeman had purchased the skin from a Mr. Manning, whose wife poisoned the animal in the Datil Mountains in August, 1902. A little later, when in the Datil Mountains, Hollister visited Manning and obtained an account of the killing of the animal. Mrs. Manning had been in the habit of putting out poison to kill the predatory animals about their ranch, in the mountains 12 miles [19.3 km] northwest

Photo 8.4. A rare historical, verified jaguar record from New Mexico: a pelt mounted as a rug and photographed by the early mammalogist Ned Hollister in 1905. The pelt was obtained from ranchers in the Datil Mountains in western New Mexico. The ranchers poisoned the jaguar 19 km (12 mi) northwest of Datil in present-day Catron County in August 1902. Photograph reproduced from Bailey (1931).

of Datil, and among the victims of the poisoned baits was this jaguar, which had been killing stock on the ranch for some time. It had killed 17 calves near the house during a short period before it was secured. The ranch was located at about 9,000 feet [2,743 m] altitude in the pine and spruce timber of this exceedingly rough range of mountains. At the time Hollister was there another jaguar was supposed to be at large in the general neighborhood.

Other historical reports suggest that the jaguar might once have ranged across much of southern New Mexico, including perhaps in the San Andres and Caballo mountains of Doña Ana and Sierra counties. Halloran (1946) provided a second-hand account of a jaguar killed by Bob Burch, foreman of the Goldberg Ranch, in the Caballo Mountains in the late 1800s. In 1903, Bailey (1931) examined and measured a rug made from a jaguar pelt that had been presented to New Mexico Governor Miguel Otero; the animal reportedly had been shot the previous year somewhere in Otero County, though other evidence from newspaper articles suggests that it was actually taken in 1901

in Luna County at the US-Mexico border (Fig. 8.1). New Mexico State Game Warden, Page B. Otero, brother of the governor, also reported to Bailey in the early 1900s that the jaguar was known from the San Andres and Sacramento mountains, but without supporting evidence.

Bailey (1931) also reported that around 1904 or 1905, a jaguar was killed by a hunter named Morris on the west slope of the Caballo Mountains. A more detailed newspaper account of this incident places it in December 1900 and indicates that the cat's pelt was preserved (Fig. 8.2), though the current location of the specimen is unknown. Halloran (1946:160) described an observation by a Biological Survey hunter named Bannermann in 1937 of a jaguar in the San Andres Mountains. The cat was "jumped" by his hunting dogs but "would not 'tree' as does a lion [cougar]" and managed to escape.

Jaguars also have been reported in the past from the northern half of New Mexico and even as far north as Colorado (Seton 1920). However, all the available records are problematic due to the lack of physical evidence and often questionable veracity. Despite their unreliability, these old reports have been cited by some as evidence that

"John Cravens, a prominent cattleman south of Deming, recently killed a jaguar on the line of Mexico. Mr. Cravens preserved the skin, and the citizens of Deming will buy the same, send it to Denver and have it mounted in first class style, after which it will be presented to a lady in the northern part of this territory."

(*Deming Headlight*, May 25, 1901, p. 5)

Beautiful Jaguar Skin.

W. H. Greer, manager of the Victoria Cattle company in Luna county, is in Santa Fe. He will remain over Monday on account of business before the territorial land board. This afternoon, on behalf of citizens of Luna county, he presented to Mrs. Ortega at the executive mansion, the most beautiful jaguar skin ever seen in this section. The jaguar was caught some time ago in Luna county after a hard fight and after it had killed six dogs. The skin measures seven and a half feet from the tip of the tail to nose, and it is most beautifully marked. – New Mexican.

(*Albuquerque Daily News*, May 7, 1902, p. 5)

Figure 8.1. Transcriptions of newspaper articles on a jaguar reportedly taken in Luna County, New Mexico during the early 1900s.

the jaguar might once have enjoyed an extensive range throughout most of New Mexico (e.g., Seton 1929; Findley et al. 1975; Robinson et al. 2006). In 1902 or 1903, State Game Warden Page B. Otero reported to Bailey (1931) that he had been told of jaguar sightings at Cow Creek near Fulton, San Miguel County, and Ute Creek in the Canadian River drainage, but no details were provided. Hill (1942) briefly mentioned a report of a jaguar killed near Springer, Colfax County, some time prior to 1938. The pelt of the Springer cat, which apparently was not examined by Hill, was supposedly preserved but apparently has never been verified, and the record was considered unlikely by Brown and López González (2001). A remarkable Spanish narrative from 1825 describes an incident in which a jaguar reportedly entered a convent in Santa Fe, New Mexico and killed four men before being shot (Baird 1859). Seton (1929:28) discussed this report in a chapter titled "The Murderous Jaguar of Santa Fe," and though he asserted that "there can be no question of its accuracy," he did allow, based on the opinions of others, that the location of the incident might have been at Peña Blanca, Sandoval County, south of Santa Fe. On the other hand, Bailey (1931:284) considered the Santa Fe incident described by Baird as "very improbable." In retrospect, the most significant problem with the story is that it seems to have originated not in New Mexico but in Santa Fé, Argentina (O'Mara 1997; Brown and López González 2001)! As noted by Brown and López González (2001:130), "Separating myth from reality has never been easy for would-be jaguar researchers."

We know of no reliable reports of jaguars observed in New Mexico from the 1940s through the 1980s, though in Arizona the species was still being encountered occasionally during this period (Brown and López González 2000, 2001; NMDGF 2016a). However, during the last decade of the 20th century, the jaguar made a dramatic reappearance in New Mexico. In 1990, a biologist and

AN IMMENSE JAGUAR.

Killed in Mountains Near Engle, New Mexico.

TRACKS OF OTHERS SEEN.

A special correspondent, writing to the El Paso Herald from Engle, N. M., under date of Dec. 10, says: Yesterday, Sunday, Dec. 9, Louis O. Morris, being camped in the hills doing some assessment work, was walking over the hills, near the camp, when he discovered an immense jaguar or American tiger coming straight toward him. He opened fire on him with a 30-30 rifle, and after firing eleven shots, the monster lay dead, with three mortal wounds in his head and body, not more than fifteen feet from where young Morris was standing.

…

The measurements are as follows: Length from tip to tip, seven feet, nine inches; height, thirty-four inches; around head twenty-five inches; length of tusk, two inches; around fore foot, nine inches; length of foot, eight inches; length of hind foot, ten inches.

The hide was taken to Engle to be shipped immediately to a taxidermist in Kansas City to be dressed, and while there will be exhibited at the Manual Training high school, and will then be returned to New Mexico.

While such ferocious animals are roaming around these mountains, it will be well for prospectors, and visitors traveling through these parts, to be well armed.

The immense tracks that have been seen so often in these mountains, and attributed to mountain lions, have finally revealed their identity, as the killing of this animal will show.

The worst feature of this animal is that he did not wait to be attacked, but when he first saw young Morris, started after him with leaps from ten to fifteen feet at a time, and the slayer said he lost no time in working the lever of his gun.

Probably hereafter, no one will prowl around these hills without his gun and plenty of cartridges, and his eyes well opened.

Later, Dec. 11 – To-day Dan O'Shea while passing from Engle to Las Palomas, just west of the foot of the mountains, encountered a second tiger, about two and one half miles from where the one was killed last Sunday.

He fired two shots at the animal, but his horse and burro seeing it, took fright and ran away, jerking O'Shea over a steep bluff, badly bruising and skinning his left arm. The burro, going over the bluff, smashed his pack, so in the fray, O'Shea lost sight of the animal.

At first, the idea prevailed that the animal had escaped from a show that passed through the country some months ago, but it [is] now known that it was not alone.

During a light snow about two weeks ago, the tracks of three were seen in the snow within less than a mile of where the one was killed.

Undoubtedly cattlemen have suffered loss from these animals and have been charging it to the account of wolves and mountain lions.

(*Albuquerque Daily Citizen*, December 17, 1900, p. 3)

Figure 8.2. Transcription of an article published in the Albuquerque Daily Citizen about a jaguar reportedly shot on the west slope of the Caballo Mountains during the early 1900s.

his wife observed what they described in detail as a jaguar crossing a dirt road east of Beaverhead in extreme northwestern Sierra County, New Mexico, though, as no photograph or other physical evidence was obtained, this sighting remains unverified (NMDGF 2016a). In April 1995, an Arizona researcher found and photographed a large felid track in Clanton Canyon in Hidalgo County, New Mexico. The photograph was later examined by personnel of the Arizona Game and Fish Department who determined that the track was of a jaguar. Then, in March 1996, Arizona rancher Warner Glenn encountered an adult male jaguar in the Peloncillo Mountains on the Arizona-New Mexico state line (south of where the track was found the previous year) during a cougar hunt and obtained a series of amazing photographs of this animal (Glenn 1996; Arritt 1997; Photo 8.5). Although some authors (e.g., Brown and López González 2000, 2001) place this record in Cochise County, Arizona rather than in Hidalgo County, New Mexico, the exact location in the Peloncillos apparently has never been pinpointed (even by Glenn himself) and thus can be claimed by both states. A few months after Glenn's encounter, a different male jaguar was photographed by Jack Childs and his hunting party in the Baboquivari Mountains of southern Arizona (Brown and López González 2001; Childs and Childs 2008). This animal would be later nicknamed "Macho B" and eventually became the subject of a long-term monitoring study (McCain and Childs 2008). Remarkably, Glenn encountered a second male jaguar in southern Hidalgo County near the US-Mexico border a decade later in February 2006 (Photo 8.6). This jaguar, like the two in 1996, was bayed by hunting hounds during a cougar hunt and was well photographed before being allowed to escape. Glenn's (1996) illustrated book of his first jaguar encounter provides a remarkable first-hand account of what he assumed then would be a once in a lifetime event. His subsequent encounter with another jaguar in 2006 was described in a few popular articles (Runyan

2006; Kahn 2007; Mahler 2010). These highly publicized jaguar sightings had a significant impact on both wildlife conservationists and the public that is still being felt today (see under "Status and Management").

A review of jaguar records from New Mexico would be incomplete without some mention of the case of Curtis Prock, a hunting guide from Idaho and Belize. During the early 1970s, Prock illegally imported a number of jaguars to New Mexico from Central America or Mexico for use in so-called canned hunts—guided hunts in which the prey is released in the vicinity of the hunter, who in some cases does not know that his quarry was recently caged, virtually guaranteeing that the animal can be bagged. Based on investigations by the New Mexico Department of Game and Fish and US Fish and Wildlife Service, Prock arranged at least nine of these hunts, at up to $3,500 apiece, in the vicinity of Apache Creek, Catron County, New Mexico during 1972–1973 (Jones 1974; Johnson et al. 2011). Could a few jaguar records in New Mexico and Arizona have resulted from illegally imported animals? Although it has been suggested that some Arizona records from the late 1950s to the early 1970s might be traced to canned hunts (Hoffmeister 1986; Girmendonk 1994; Brown 1991, in Valdez 2000), we know of no jaguar reports from New Mexico during the time period when these hunts were being conducted (NMDGF 2016a). Therefore, it appears unlikely that any of our New Mexico records considered as reliable were the result of jaguars being transplanted from elsewhere.

Unlike in Arizona, where at least one jaguar has been documented almost continuously from 1996 to the present (Childs and Childs 2008; McCain and Childs 2008; Johnson et al. 2011; Culver 2016; Photos 8.7 and 8.8), no verified reports in New Mexico have surfaced since Warner Glenn's sightings in (or very near) Hidalgo County of two different animals in 1996 and 2006. If a jaguar has been resident in New Mexico in recent years, it has eluded detection.

Photo 8.5. Adult male jaguar photographed by Arizona rancher Warner Glenn during a cougar (*Puma concolor*) hunt in the Peloncillo Mountains on the Arizona–New Mexico state line in March 1996. Only a year earlier (in April 1995), a large felid track had been discovered and photographed just to the north in Clanton Canyon in Hidalgo County, New Mexico. The track was determined to be that of a jaguar by the Arizona Game and Fish Department. No reliable records exist from New Mexico from the 1940s through the 1980s, but the 1990s witnessed the return of a few jaguars to the southwestern United States. Photograph: © Warner Glenn.

Photo 8.6. Male jaguar during another cougar (*Puma concolor*) hunt with dogs—involving again Warner Glenn—in February 2006 in southern Hidalgo County near the US-Mexico border. After it was well photographed, the jaguar was allowed to escape. Photograph: © Warner Glenn.

Photo 8.7. Male jaguar photographed by motion-detection wildlife cameras in the Santa Rita Mountains southeast of Tucson, Arizona on 30 July 2015 as part of a Citizen Science jaguar monitoring project conducted by the University of Arizona, in coordination with the US Fish and Wildlife Service. More reliable records of jaguar occurrence exist from Arizona compared to New Mexico, where detection efforts are far more limited. Photograph courtesy of the University of Arizona/US Fish and Wildlife Service.

Photo 8.8. Male jaguar photographed by motion-detection cameras in southeastern Arizona's Dos Cabezas Mountains on 16 November 2016. This same male, only the seventh jaguar to be documented in the United States since 1996, was later photographed on 16 April 2017 in the Chiricahua Mountains area. The Dos Cabezas Mountains represent a northwestern extension of the Chiricahua Mountains beyond Apache Pass. Jaguars use rugged terrain, canyons, and creek beds as travel corridors in the northernmost part of their distribution. Photograph courtesy of the Bureau of Land Management.

HABITAT ASSOCIATIONS

As is often characteristic of large carnivores, the jaguar occupies a variety of habitats throughout its broad geographic range. In Central and South America, it inhabits tropical forests with closed canopies including evergreen and deciduous communities and dense riparian woodlands, avoiding areas of more open vegetation (Brown and López González 2001). Throughout most of their range in Mexico, jaguars exhibit flexible associations with ecological communities but seem to be "mainly associated with tropical rain forests, high prey-species richness and regularly flooded vegetation, with a clear avoidance [of] arid vegetation, higher elevations and grassland" (Rodríguez-Soto et al. 2011:353). However, in northwestern Mexico, jaguars occur in drier areas primarily vegetated by Sinaloan thornscrub and Madrean woodlands of oak (*Quercus* spp.), juniper (*Juniperus* spp.), and pine (*Pinus* spp.). Most populations throughout the species' range are at elevations below 1,200 m (~ 3,950 ft), though at least some individuals occur regularly at much higher elevations such as above 1,800 m (~ 5,900 ft) in pine-oak forests of central Mexico (Monroy-Vilchis et al. 2009). In the southwestern United States, the locations of the 62 jaguars known to have been killed since 1900 range in elevation from 500 m (~ 1,650 ft) in lowland desert areas to over 3,000 m (~ 9,850 ft) in temperate coniferous forests, with most over 1,500 m (~ 4,900 ft) (Brown and López González 2000). The variety of habitats in which jaguars have been found north of the US-Mexico border suggests that occurrence in New Mexico may not be limited by the availability of suitable vegetation communities. All biomes present in New Mexico, except for expanses of plains grasslands or tundra, could potentially support jaguars if other required habitat features were present, with the Sierra Madrean–type woodlands having the strongest association with jaguar activity (Brown and López González 2001). Whether dense wet forest in the Neotropics or drier woodlands associated with the rugged topography of northern Mexico and the United States, the availability of cover and hard-to-access retreats provided by dense vegetation, often in association with rough terrain and a diverse prey base, seem to be key components of jaguar habitat.

Various researchers or research institutions have attempted to characterize jaguar habitat in Arizona and New Mexico, including Rabinowitz (1999), the Sierra Institute Field Studies Program in Arizona (2000), Brown and López González (2001), Menke and Hayes (2003), Hatten et al. (2005), Boydston and López González (2005), Robinson et al. (2006), Sanderson and Fisher (2013), Sloan (2019), and Sanderson et al. (2021). This task is complicated by different interpretations of 1) what constitutes a valid record of jaguar occurrence; 2) whether that record is indicative of a vagrant, dispersing, or resident animal; and 3) the importance of that locality and its associated habitat to persistence of the species. There is also a temporal aspect in such evaluations because historical habitat may no longer exist at a location.

Recent studies of "northern jaguars" in northwestern Mexico and Arizona provide some insights into what constitutes habitat in the Southwest, including New Mexico (see Menke and Hayes 2003). At the northern limits of their range, jaguars are found on rugged montane terrain and in foothills, typically in Madrean pine-oak or oak-juniper associations. This geographic area in the Southwest has been characterized as the "Sky Islands" or Madrean Archipelago as it includes disjunct patches of upland forest habitat that are floristically and faunally more typical of the Sierra Madre Occidental of western Mexico than the surrounding deserts and desert-grasslands that predominate in the region (Brown 1994; Brown and López González 2001). Access to water, especially in the Southwest, is a likely limiting factor for the jaguar, and springs and

Photo 8.9. Overview of the Río Aros and surrounding mountainous jaguar habitat in the Northern Jaguar Reserve of east-central Sonora, Mexico. Only about 200 km (125 mi) south of the US-Mexico border, the area is characterized by rugged topography and a heterogeneous mosaic of mostly xerophilous and subtropical thornscrub, relics of tropical deciduous forest, and riparian vegetation. Oak woodlands and natural grasslands are present at upper elevations. This vegetation is interspersed within large areas of non-native buffel grass (*Pennisetum ciliare*). Photograph: © Brendon Kahn/Northern Jaguar Project.

Photo 8.10. Two jaguars in typical habitat in the Northern Jaguar Reserve of east-central Sonora, Mexico: a dry stream bed lined with xeroriparian vegetation with mesquites and native palm trees. Jaguars also use xerophilous and semi-tropical thornscrub, oak woodlands, and tropical deciduous forest in the reserve. Photograph: © Northern Jaguar Project.

livestock ponds may be the only available surface water in much of this region. The presence of large areas of undeveloped land is also important. Although jaguar observations from more densely forested habitats in the tropics indicate the species exhibits some tolerance for human development (such as by using logging roads as travel corridors), there is general agreement among researchers that high-traffic roads associated with urbanized and other developed areas in the Southwest are avoided (Menke and Hayes 2003; Hatten et al. 2005; Rabinowitz and Zeller 2010; Sanderson and Fisher 2013; Sloan 2019).

Compared to areas used by the jaguar in Central and South America, the habitat at the northern extremes of the species' distribution has often been described as marginal due to its aridness and much sparser tree cover. For a top carnivore, prey availability—which itself is determined by the quality of the habitat—is a primary consideration in determining if an area can support a population (see under "Diet and Foraging"). This has led some (e.g., Rabinowitz 1999) to assert that the southwestern United States has never been ideal jaguar habitat, and hence the scarcity of this animal here. More generally, a large body of ecological research points to increasing abiotic and biotic stress away from a species' range core and/or toward its geographic boundaries, resulting in reduced abundance and higher susceptibility to extirpation from natural and anthropogenic factors (e.g., Brown 1984; Sagarin and Gaines 2002; Yancovitch Shalom et al. 2020).

Despite differing opinions about the suitability of the Southwest for this species, the US Fish and Wildlife Service (USFWS 2014) designated Critical Habitat for the jaguar in portions of southern Arizona and southern Hidalgo County, New Mexico (see under "Status and Management," below). Criteria for defining Critical Habitat were based on several earlier habitat studies including those of Menke and Hayes (2003) for New Mexico, Hatten et al. (2005) for Arizona, and

Sanderson and Fisher (2013) for both states and northwestern Mexico. This designation, made under the authority of the federal Endangered Species Act, identified several primary constituent elements (PCEs) or necessary components of the environment deemed crucial for allowing the species to persist in the Southwest, all of which must be present to constitute Critical Habitat. The PCEs defined by the US Fish and Wildlife Service (2014:12587) are

expansive open spaces in the southwestern United States of at least 100 km² (38.6 mi²) in size which: (1) Provide connectivity to Mexico; (2) Contain adequate levels of native prey species . . . ; (3) Include surface water sources available within 20 km (12.4 mi) of each other; (4) Contain from greater than 1 to 50 percent canopy cover within Madrean evergreen woodland . . . or semidesert grassland vegetation communities . . . ; (5) Are characterized by intermediately, moderately, or highly rugged terrain; (6) Are below 2,000 m (6,562 feet) in elevation; and (7) Are characterized by minimal to no human population density, no major roads, or no stable nighttime lighting over any 1-km² (0.4 mi²) area.

These criteria for Critical Habitat have been criticized by some as either too restrictive or too broad, depending on one's perspective or experience with the species. This difference of opinion further emphasizes the difficulty in developing conservation measures at the edge of a species' distribution.

The future effect of climate change on ecosystems in the Southwest is a topic of concern among conservation biologists, and the possible impact of our changing environment on the jaguar is still largely speculative (see Chapter 5). Nonetheless it remains a consideration in any long-term conservation planning for this species (USFWS 2014, 2018). Predicted shifts in precipitation patterns, rising temperatures, increased xerification, and the likely increase in major wildfires (USGCRP 2017) potentially could have an adverse effect on both the jaguar and the prey animals it relies upon. Given that the habitat for the jaguar in New Mexico is already considered marginal by many researchers, climate change potentially could make the state even less suitable for this species (Chapter 5).

LIFE HISTORY

The jaguar is arguably one of the least studied of the large cats (Valdez 2000). Most of what we do know about the species in the Southwest was reviewed by Brown and López González (2001). However, no comprehensive study of its life history has been conducted in the United States, where our knowledge of its habits is fragmentary at best (Emmons 1999). Although jaguar life history has been studied in Central and South America, it is unclear how relevant much of that information is to the species at the northern extreme of its range where its environment is considerably different. Studies of jaguar populations in northwestern Mexico likely provide the best data that can be applied to the American Southwest.

Diet and Foraging

Jaguars are entirely carnivorous and top predators throughout their range (López González and Miller 2002). They hunt either using a sit-and-wait strategy, in which the prey animal is ambushed, or by stalking and then making a short dash when attacking. Like many cat species, jaguars rely on stealth and cover, combined with their cryptic coloration, to capture prey. Most prey, especially larger ungulates (hooved mammals), are killed by a single piercing bite to the skull and less often by seizing the throat. Whereas smaller prey are usually consumed in one sitting, large prey typically are dragged to a sheltered and secluded spot and fed upon over several days. Unlike cougars, jaguars do not typically consume the long bones of large prey nor do they scrape debris over a prey carcass to cover it. Additionally, jaguars may take some parts of the animal preferentially, such as the tongue and meat from the neck, shoulders, and ribs, while avoiding internal organs (Seymour 1989; Childs 1998; Johnson et al. 2011). Although carrion may be consumed, it apparently is not an important food item (Seymour 1989).

Jaguars are known to prey upon a wide diversity of animal species. Leopold (1959:467) described their taste in prey as "almost catholic." Their preferred prey has been described as terrestrial mammals greater than 1 kg (2.2 lbs) in body mass (Seymour 1989). When adapted to the available or abundant prey animals within a given area, this description provides a reasonable approximation of observed diets across a broad range of habitats. In Central and South America, medium-sized to large animals recorded in the diet include armadillos (family Dasypodidae), capybara (*Hydrochoerus hydrochaeris*), tapir (*Tapirus* spp.), peccaries (family Tayassuidae), deer (family Cervidae), birds, turtles, and caimans (family Alligatoridae; Rabinowitz and Nottingham 1986; Crawshaw and Quigley 1991). López González and Miller (2002) provided evidence that populations of jaguar closest to the equator take more medium-sized prey than expected, whereas at higher latitudes (e.g., Jalisco, Mexico) jaguars rely on large prey as a greater component of the diet. However, Cassaigne et al. (2016) suggested that jaguars in Sonora might rely more on prey items that weigh less than 15 kg (33 lbs). Simoni (2012) proposed that the robust build of the jaguar is more suitable for subduing potentially dangerous prey, such as peccaries, than is the more gracile body plan of the cougar, which is suitable for pursuing agile, lighter animals. She suggested that some degree of prey selection is evident where the two cat species co-occur, with jaguars taking more peccaries and cougars

Photo 8.11. Jaguar with collared peccary (*Dicotyles tajacu*) in east-central Sonora, Mexico. Collared peccaries (also called javelinas) likely rank among the most important prey of jaguars in southwestern North America, together with mule deer (*Odocoileus hemionus*) and white-tailed deer (*O. virginianus*). Photograph: © Northern Jaguar Project.

hunting more deer, but with substantial overlap in prey consumed.

Little detailed information is available on prey species in Arizona or New Mexico. In the southwestern United States and northwestern Mexico, the most important native prey likely includes the collared peccary or javelina (*Dicotyles tajacu*; Photo 8.11), mule deer (*Odocoileus hemionus*), and white-tailed deer (*O. virginianus*), while medium-sized mammals such as the white-nosed coati (*Nasua narica*), the opossum (*Didelphis* spp.), skunks (*Mephitis* and *Spilogale* spp.), the northern raccoon (*Procyon lotor*), and hares (*Lepus* spp.) are also likely taken but are possibly of lesser importance (Hatten et al. 2005; Rosas-Rosas et al. 2008). No data are available from New Mexico, but Brown and López González (2000:51) summarized some of the information obtained from

the stomachs of jaguars killed in Arizona during the 1900s; contents included meat of domestic horses (*Equus caballus*) and cattle (*Bos taurus*), elk (*Cervus canadensis*) carrion, white-tailed deer, and white-nosed coati, and one stomach was "full of frogs." The jaguar encountered by Glenn (1996) in the Peloncillo Mountains smelled strongly of skunk, suggesting the cat might have preyed upon, or otherwise had a close encounter with, one of these animals (Brown and López González 2001). While none of these alternative food sources may serve as staples for borderland jaguars in the same manner as deer and peccary, it is possible that any locally abundant prey species could contribute to the diet.

As documented in Arizona, jaguars prey upon introduced domestic livestock when available, particularly if native prey animals have been depleted and livestock (especially young animals) are pastured in the vicinity of usual foraging areas. In many parts of Central and South America, jaguars are known to take cattle, horses, burros (*Equus asinus*), and pigs (*Sus scrofa*) and this naturally results in conflict with ranchers who often kill any large cat that poses an actual or potential threat to their animals (Leopold 1959; Rosas-Rosas et al. 2008). Leopold (1959) suggested that some individual jaguars in Mexico may become regular livestock killers while others do not. In northeastern Sonora, cow calves constituted 58% of the prey biomass for all jaguars in one study area, primarily due to predation by only three jaguars. However, overall calf survival in the study area remained high and most losses could be attributed to causes other than jaguars or cougars, though ranchers generally believed that any calf mortality was due to cat predation (Rosas-Rosas et al. 2008). The concentration of cattle in riparian areas during the dry season was identified as a major contributing factor to jaguar predation in Sonora as these areas were also frequented by native prey such as deer and peccaries (Rosas-Rosas et al. 2010). A proposed solution to minimize this conflict with ranching interests was to locate watering sources for cattle in upland areas, away from areas frequented by native prey species and where stalking cover for predators was limited (Rosas-Rosas et al. 2008, 2010).

In the Southwest, predation on introduced livestock is not as well-documented as it is in Mexico, presumably due to the rarity of jaguars. Although reports of jaguars taken as "stock killers" occurred in several areas of Arizona through the 1960s (Brown and López González 2001), detailed confirmation of livestock predation events is generally lacking. More recently (2007), one cow calf was confirmed killed by a jaguar in Arizona (McCain and Childs 2008). In New Mexico, Bailey (1931) discussed reports of a jaguar feeding on a bull in Hidalgo County and another allegedly taking livestock in the Datil Mountains, both in the early 1900s.

Another non-native ungulate species in the United States, the feral domestic pig, has been long established in southern Hidalgo County where it occurs in the same habitat as the collared peccary (Findley et al. 1975). It seems likely that feral pigs, similar in biology and habits to peccaries, would be suitable prey for a borderland jaguar.

Jaguar predation on humans is rare and, unlike in Old World species of *Panthera*, there is scant evidence of any jaguar becoming a habitual "man-eater," even in the Neotropics where the species is more common (Leopold 1959; Brown and López González 2001). Nonetheless, as with any large predator, it can be difficult to separate fact from folklore. The account of the Santa Fe jaguar, discussed above, in which a jaguar allegedly dispatched four men before being subdued (O'Mara 1997; Brown and López González 2001:40–41) seems more fanciful than credible. Well-known naturalist John James Audubon may have also promoted the view of jaguars as man-eaters, referring to the species with the following description: "Alike beautiful and ferocious, the Jaguar is of all American animals unquestionably the most to be dreaded, on account of its combined strength, activity, and courage, which not only give it a vast physical power over other wild creatures, but enable it to frequently destroy man" (Audubon and Bachman 1854:3). Despite the fantastical nature of these descriptions, some reports of jaguar preying on humans or hunters being attacked by a cornered or wounded jaguar are likely true (Brown and López González 2001).

Although jaguars are most often characterized as nocturnal animals (e.g., McCain and Childs 2008; Monroy-Vilchis et al. 2009), some studies from the Neotropics show that these cats can also be most active during crepuscular periods (dawn and dusk) or even during daylight (Rabinowitz and Nottingham 1986; Crawshaw and Quigley 1991; Cavalcanti and Gese 2009). Such variability in activity, which may change seasonally, is perhaps related to the presence or lack of dense cover in foraging areas and the activity patterns of the particular prey animals being pursued.

Reproduction and Social Behavior

Most information on reproduction comes from captive jaguars or populations in the Neotropics and very little is known from northwestern Mexico or the United States. Studies in Brazil suggest that both males and females are promiscuous, mating with more than one individual of the opposite sex (Cavalcanti and Gese 2009), though extension of this finding to very low-density populations such as those of the Borderlands may be questionable. Jaguars breed at various times of the year throughout their range, but in Mexico the birth of young may be concentrated during mid- to late summer, possibly corresponding with the rainy season and shifts in prey availability (Leopold 1959; Nowak 1991). In Brazil, female jaguars may come into estrus at various times, which suggests there is no defined breeding season (Cavalcanti and Gese 2009). Estrus lasts for 6 to 17 days, during which time the female may wander from her usual home range and be briefly accompanied by one or more males. Data mainly from captive animals indicate that gestation lasts from 93 to 105 days and the typical litter is two kittens, rarely three or four (Nowak 1991; Emmons 1999). The young weigh 700–900 grams (1.5–2.0 lbs) at birth and open their eyes after 3–13 days. Young jaguars are weaned after about six months but may stay with their mother for 1.5–2 years before dispersing, suggesting that a female may not have a litter every year (Nowak 1991; Emmons 1999).

Adult size and sexual maturity are reached as early as 2 years old. In Belize, jaguars two to three years old were considered to be young adults, those four to ten years old were mature adults, and those over 11 years of age were old adults (Rabinowitz and Nottingham 1986). Although captive jaguars have been reported to live to 22 years of age, the usual lifespan in the wild is approximately 11 years (Seymour 1989; Nowak 1991). The male jaguar studied in Arizona, called Macho B (see McCain and Childs 2008), was estimated to be 15–16 years old when it was captured in failing health and euthanized in early 2009 (Johnson et al. 2011).

The sex ratio of reproducing jaguar populations is close to the expected 1:1 in Sonora (Boydston and López González 2005) but may be skewed toward males elsewhere (e.g., up to 1.5 males for every female in Brazil; Cavalcanti and Gese 2009). Males greatly outnumber females in particular at the northern extreme of the range where most jaguars seem to be dispersing individuals from farther south. The great majority of jaguars that were detected in the southwestern United States and could be reliably sexed have been males (Brown and López González 2001). Although eastern Sonora and perhaps parts of Arizona can support both sexes, at least historically, the more marginal environment in New

Mexico and Chihuahua may currently be suitable only for males, which disperse farther than females (Boydston and López González 2005).

Information on home ranges in jaguars is mainly from populations in the tropics and probably has little relevance to northern jaguars which, again, occupy very different environments. Home ranges vary from 10 to over 400 km² (~ 4 to over 150 mi²), with smaller ranges reported for females, from rain forest (compared to open habitats), and from studies that did not use Global Positioning System collars (Emmons 1999; de la Torre et al. 2017). In the Pantanal of Brazil, home ranges were typically smaller in the wet season when widespread flooding occurred, and larger in the dry season, presumably due to changes in prey density (Crawshaw and Quigley 1991), though Cavalcanti and Gese (2009) found little variation in core foraging areas for either sex across seasons in this same region.

Jaguars are generally considered to be solitary animals, though as noted by Cavalcanti and Gese (2009), this does not exclude social behavior. Some tolerance for members of their own species, including between sexes outside of breeding periods, is evident from studies in Brazil. Territoriality appears to be flexible. Males may have overlapping territories, which in turn overlap those of multiple females (Emmons 1999; Cavalcanti and Gese 2009). This pattern differs from those seen in other solitary cat species, though the degree to which neighbors may be tolerated is possibly influenced by the abundance of prey (Rabinowitz and Nottingham 1986; Crawshaw and Quigley 1991). A well-studied jaguar population in Sonora was described as "fluid" as population size varied year-to-year and turnover of individuals within the population was high, perhaps due to resource variability or the disruptive effect of illegal killing (Rosas-Rosas and Bender 2012). The authors of that same study noted that jaguar density in Sonora (1.1/100 km²) was lower than reported densities elsewhere in Mexico such as Jalisco (2/100 km²), the Yucatan Peninsula (3–7/100 km²), and Campeche (3.7/100 km²), or in Guatemala (2/100 km²) or Belize (9/100 km²), perhaps due to marginal habitat conditions and poaching.

Individual jaguars may "passively" maintain separation from each other through communication such as scent marking. Territorial scent-marking behavior includes urine-spraying and cheek-rubbing against objects. Jaguars also claw-rake trees and logs, but rarely mark the ground in this way as do cougars (Emmons 1999; McCain and Childs 2008; see Chapter 9). Individuals also communicate vocally through deep, resonant grunts that can carry over long distances. Despite these methods of communication and avoidance, aggressive encounters do occur occasionally between males, sometimes resulting in death (Cavalcanti and Gese 2009).

Interspecific Interactions

Humans are the main predators of jaguars, and, other than members of their own species, very few other animals pose a threat to an adult. Although jaguars and cougars are broadly sympatric from South America to the southwestern United States, there seems to be little evidence that the two species regularly interact or actively exclude each other despite similar habitat and life habits. Some partitioning of prey type and size and temporal avoidance may allow these two cat species to coexist (Rabinowitz and Nottingham 1986; de la Torre et al. 2017), though it is possible that aggressive encounters occasionally do occur. Childs and Childs (2008) discuss a very large male cougar, taken in the same general area of Arizona where they observed a jaguar in 1996, that had wounds consistent with having been in a fight with another large cat. The authors speculated that this cougar may have had an encounter with the local jaguar. Another large carnivore, the grizzly bear (*Ursus arctos horribilis*), may have once posed a threat to one of our small northern

jaguars, perhaps in competition for prey carcasses, but this species no longer exists anywhere within the range of the jaguar (see Chapter 17). Some large and potentially dangerous prey animals, such as peccaries, can reportedly pose a risk of injury or occasionally even death to a jaguar (Simoni 2012).

STATUS AND MANAGEMENT

The question of whether the jaguar is, or ever was, a resident species during historical times in New Mexico remains a subject of debate, and this uncertainty colors any discussion of its conservation status and management needs within the state. Brown and López González (2000) articulated the two competing viewpoints concerning jaguars in the southwestern US: 1) individuals detected north of the US-Mexico border have always been transients (mainly dispersing or displaced males) from northwestern Mexico where breeding populations persist today; versus 2) the species was, until the 1900s, a resident of Arizona and New Mexico, and individuals entering this region today from Mexico can be considered "re-colonizers" that may be capable of re-establishing a population north of the border. An additional hypothesis that has been proposed is that, since the late 1800s, the jaguar's presence in our area can be attributed to the establishment of an abundant, anthropogenic source of prey (domestic cattle) that facilitated the survival of jaguars as "subsidized predators" (animals that thrive with help from human-provided food sources) in Arizona and New Mexico. However, Brown (1983) discounted this last idea as jaguars had been detected in Arizona prior to widespread stocking of cattle.

Although the jaguar was perhaps always uncommon in the southwestern United States, it is clear from the available evidence that the species declined in numbers regionally during the 20th century, likely due to persecution by humans (Brown and López González 2000). Today, the species occurs in Arizona and New

Photo 8.12. Colonel Theodore Roosevelt kneeling besides the jaguar he just shot in 1913 during a hunt on the Taquari River in Brazil. Although sport hunting and poaching still occur in some areas, loss of habitat and conflicts with livestock ranchers now seemingly represent the main threats to the jaguar. Photo by Kermit Roosevelt, courtesy of Charles Scribner's Sons and the American Museum of Natural History.

Mexico as either an infrequent visitor from Mexico or a very rare resident. Nonetheless, the species has been prominent in the mind of the public in recent years. The jaguar has been considered by wildlife conservationists as both a "flagship species," a charismatic animal that can raise awareness of biodiversity in the general public, and an "umbrella species," a wide-ranging animal whose protection indirectly benefits many other species (Simberloff 1997). As such, the species has served for many as an icon for wildlife conservation in the Southwest, despite its rather tenuous position as a member of our state's mammal fauna.

The jaguar was listed as Endangered in the United States in 1972 and retained this status following the passage of the federal Endangered Species Act (ESA) in 1973, thus making it one of the first animals to receive formal protection under federal law. Around the same time, the Convention on International Trade in Endangered Species of Wild Fauna and Flora (CITES)

included the jaguar in its Appendix 1 (a list of those species threatened with extinction) due to the extensive commercial trade in pelts throughout the species' range (Swank and Teer 1989). In its original listing under the ESA, the jaguar's range in the United States was inadvertently omitted and so protection under that federal act did not include any jaguars north of the US-Mexico border. This omission remained uncorrected for the next two decades—justifiably, in the views of some, given that there was almost no evidence of jaguars in the United States during this time period. It was not until 1997, following the discovery of two jaguars in Arizona and New Mexico the previous year, that the Endangered status was extended to wild individuals found within the historical range in the United States (USFWS 1997). In Mexico, the jaguar was first protected in 1986 and hunting was banned in 1987; it is currently listed as Endangered under that country's General Wildlife Law (SEMARNAT 2010; Johnson et al. 2011).

Although the global demand for pelts has diminished since the 1970s, illegal hunting continues within the species' range outside of the United States and persecution of jaguars as threats to livestock persists in many parts of rural Mexico and elsewhere in Latin America (Swank and Teer 1989; Rosas-Rosas et al. 2008; Rosas-Rosas and Valdez 2010). Sanderson et al. (2002b) reviewed the various complexities of conserving the jaguar across its broad range in the Americas based on often limited information.

Protection at the state level in New Mexico has been relatively minimal. In 1975, the jaguar was listed as Endangered under the newly authorized New Mexico Wildlife Conservation Act but was removed from this same category in 1985 due to a lack of evidence that the species still occurred in the state (Anonymous 1989; Jones and Schmitt 1997). Through the 1990s and until 2012, the jaguar was categorized as a "Restricted" species in New Mexico, along with six other (non–New

Mexican) wild cats with high-value pelts, in accord with its status as a CITES species. In 1999, a New Mexico state law was passed that established new penalties for the illegal killing of a jaguar that would go into effect should the species ever be removed from protection under the ESA. According to that law, the penalties that could be assessed would exceed those for the killing of a state-protected wildlife species in New Mexico, though they would not be as high as the federal penalties set forth under the ESA. Another state law passed in 2006 authorized the creation of new regulations by the New Mexico Game Commission by which the state could assess significantly higher civil damages for the illegal killing of any wildlife species designated as a trophy animal. To date, no action has been taken to include the jaguar within such regulations (Johnson et al. 2011). The jaguar is also considered a Species of Greatest Conservation Need under New Mexico's Wildlife Action Plan (NMDGF 2016b), though this designation does not confer any additional regulatory protection.

Many conservation and research activities related to jaguars in Arizona and New Mexico were coordinated through the Arizona-New Mexico Jaguar Conservation Team, which was formed in 1997 following the detection of two jaguars in those states the previous year. The "Jag Team," as it has been called, was originally organized as a means to develop a state-led collaborative and conservation agreement that could serve as an alternative to including jaguars north of Mexico under ESA protection. Although federal protection of the jaguar was extended to the United States in 1997, meetings of the Jag Team continued to serve as a forum for discussing information on jaguars and their habitat, presenting results of conservation and research efforts, and for coordination between state and federal agencies, municipalities, non-profit organizations, and the general public (Van Pelt and Johnson 1998; Johnson et al. 2011).

In 2018, a federal recovery plan was prepared by the US Fish and Wildlife Service and the Jaguar Recovery Team for the "northern jaguar" of northwestern Mexico and the southwestern United States (USFWS 2018). The Recovery Team, comprised of members involved in the implementation of the plan, includes many of the same representatives from the former Jag Team, in addition to jaguar biologists from the United States and Mexico. The primary focus for jaguar recovery is on the known breeding sites in northwestern Mexico, especially in Sonora, which are the apparent population source of all jaguars that have entered the United States in recent decades. Thus, recovery efforts for this species in the United States are somewhat atypical in that they focus heavily on protection of breeding populations located outside of the country to ensure future persistence north of the border. Such protections require close coordination among US and Mexico wildlife managers, as well as the support of rural communities on both sides of the border, particularly within jaguar habitat in Mexico where conflicts between livestock ranchers and jaguars pose major challenges to the species' conservation (Valdez 2000; Rosas-Rosas and Valdez 2010). Despite many challenges, progress has been made in Sonora to establish landowner-based conservation efforts for the jaguar (Rosas-Rosas and Valdez 2010).

In 2014, the US Fish and Wildlife Service designated Critical Habitat for the species in parts of the southwestern United States based on verified occurrences of jaguars since the 1990s (USFWS 2014). The Critical Habitat areas include all or parts of the Atascosa, Baboquivari, Huachuca, Patagonia, Santa Rita, and Whetstone mountains and other nearby ranges in southern Arizona; the southern Peloncillo Mountains along the Arizona-New Mexico state line; and the northern San Luis mountains in southern Hidalgo County, New Mexico (USFWS 2014). Critical Habitat has been a topic of much controversy among landowners, land management agencies, and the general public in Arizona and New Mexico and also has been opposed in the past by some jaguar researchers (e.g., Rabinowitz 1999) who view it as counterproductive to effective management of the species. Critical Habitat in the United States does not extend protection to habitat and travel corridors south of the border, where jaguars in the United States presumably originate, but requires management actions taking place on federal lands or requiring federal funding or permitting to be evaluated for potential impacts to jaguar habitat. The effect of Critical Habitat on private landowners is limited as the designation only applies to federal actions. Still, this controversy led to a lawsuit and federal court of appeals decision (*New Mexico Farm and Livestock Bureau v. US Department of Interior* 2020) that reversed the US Fish and Wildlife Service's designation of Critical Habitat in New Mexico, based on a lack of documented evidence that the state's Critical Habitat units (Peloncillo Mountains and San Luis Mountains) were occupied by jaguars at the time the species was listed as Endangered in 1972 (USFWS 2021).

Although jaguars are far less likely to be killed today in the United States as livestock predators or for trophies, the changing human environment in the Southwest poses many new challenges to the regional conservation of the species. Increasing infrastructure along the US-Mexico border, combined with human activity, both legal and unauthorized, is a significant obstacle for jaguars and conservation planners alike. The >1,100 km (>680 mi) of pedestrian and vehicle barrier fencing that was mandated under the US Secure Fence Act of 2006 (Haddal et al. 2009) serves to restrict human traffic in the borderlands region but also can prevent transboundary movement of large terrestrial animals like jaguars (Flesch et al. 2010; Photos 8.13a and b). Multilane highways and urban development can similarly obstruct natural corridors for wildlife movement and reduce

the quality of habitat (Theobald et al. 1997; Sanderson et al. 2002a; Beckmann and Hilty 2010). Protection and, in some cases, re-establishment of travel corridor habitat and refugia for this species in the region is further complicated by a mosaic of land ownership, requiring large-scale planning efforts among agencies and landowners in both countries (King and Wilcox 2008; McCain and Childs 2008; Grigione et al. 2009; Rabinowitz and Zeller 2010; Rodríguez-Soto et al. 2011; USFWS 2018). A major element of retaining the species in the Borderlands is the maintenance of genetic variability within the population and elsewhere in northwestern Mexico and the possible establishment of females where only males have been detected in recent decades (Miller 2013; USFWS 2014, 2018). To what extent jaguars are currently being impacted by changes in the borderland environment—including recently expanded sections of border wall—is open to speculation, although it is reasonable to conclude that any large animal requiring vast areas of wild land to persist must be affected.

As already noted, management of the jaguar is also impeded by a lack of information. A major obstacle in understanding the current status of jaguars is their rarity, elusiveness, and the difficulty of detecting animals within the rugged and remote areas they occupy (Silver et al. 2004). Because jaguars are rarely seen, many surveys have relied on sign that these cats leave behind such as tracks, scats, hair samples, and prey remains (Haynes et al. 2005). Although jaguar tracks can be confused with those of cougars where the two co-occur, careful measurements of key features in clear tracks on smooth dirt substrates can be used with high reliability to differentiate between the two species (Childs 1998; Rosas-Rosas and Bender 2012). In addition, the dimensions and unique features in some tracks also can be used to distinguish individual cats (Rosas-Rosas and Bender 2012). Identification of felid scats to species has been conducted

Photos 8.13a and b (*top to bottom*). a) View to the east along a portion of the border wall under construction in southeastern Arizona's Guadalupe Canyon, with the United States on the left and Mexico on the right, 30 November 2021. Guadalupe Canyon is located in the southern Peloncillo Mountains and is a known jaguar corridor allowing northward movements from Sonora, Mexico. b) Another section of the border wall, photographed on 3 September 2021 in the Carrizalillo Hills in Luna County, New Mexico from the US side looking southeast. Photographs: © Myles Traphagen/Wildlands Network.

using bile acid patterns (Fernández et al. 1997) and more recently through DNA extraction and analysis, the latter of which can also allow identification of individual animals based on unique genetic signatures (Culver and Hein 2016).

The increased availability of camera traps (motion- and heat-triggered wildlife cameras) has contributed greatly to studies of elusive animals, including jaguars (Childs and Childs 2008; Crooks et al. 2008). The technique has been effective in detecting and monitoring jaguars in Arizona and northwestern Mexico, especially when conducted in combination with track and scat surveys that can detect individual cats not captured on camera (e.g., McCain and Childs 2008; Rosas-Rosas and Bender 2012). Because each jaguar has a unique spotting pattern, good photographs often can be used to identify individual cats. From the late 1990s until recently, the Borderlands Jaguar Detection Project deployed up to 50 camera traps in southern Arizona and obtained photographic evidence of two, possibly three, jaguars in the survey area in addition to many other wildlife species (McCain and Childs 2008; Childs and Childs 2008; Johnson et al. 2011). However, this survey and monitoring method can be expensive, laborious, and logistically difficult, and field personnel and camera equipment are also at some risk in the more remote parts of the US-Mexico borderlands region where both jaguars and illegal activities occur (Childs and Childs 2008; Johnson et al. 2011). Despite such obstacles, a broad-scale jaguar detection project was implemented during 2012–2015 in the Southwest by the University of Arizona's Wild Cat Research and Conservation Center that used an array of camera traps placed in suitable habitat. The effort also employed survey teams accompanied by dogs that have been trained to detect jaguar scat. During its first year, the Jaguar Survey and Monitoring Project obtained photographs of a jaguar that had been previously detected in the Santa Rita Mountains of Arizona. The project area mainly focused on mountains in southeastern Arizona but also included the Peloncillo Mountains of Hidalgo County, New Mexico. Unfortunately, no additional jaguars were detected other than the Santa Rita Mountains' individual (Culver 2016).

In addition to the non-invasive methods mentioned above, another "high-tech" method to study jaguars is radio-telemetry, which has proved to be a useful tool in wildlife research and management for many species, including large cats (e.g., Rabinowitz and Nottingham 1986; Cavalcanti and Gese 2009). The technique can be difficult, however, as it requires the capture and handling of the animal to equip it with a radio-transmitter collar. Life history information that can be provided by a radio-collared animal includes daily and seasonal movements, habitat use, and locations of denning and feeding sites. The only jaguar to be radio-collared in the United States was "Macho B," a resident male that had been documented via camera traps or direct observation in southern Arizona since at least 1996 before it was captured by Arizona Game and Fish Department personnel in a foothold snare southwest of Tucson in February 2009. The capture, radio-collaring, and release of Macho B was initially viewed as a major accomplishment in jaguar conservation because of the expectation that the data could significantly improve our regional understanding of the species. However, after showing signs of ill health 12 days following his capture, Macho B was recaptured and euthanized (Johnson et al. 2011). The ramifications of the Macho B incident have made it unlikely, at least for the time being, that any further efforts to live-capture and radio-telemeter a jaguar in the United States will be attempted.

An additional source of information on jaguars is sighting reports from the public. Although alleged sightings in the southwestern United States are not uncommon, most turn out to be cases of misidentification of other animal species

Photo 8.14. Jaguar and her melanistic cub at the Big Cat Sanctuary in the UK. Global captive breeding programs aim to maintain genetic diversity within captive populations for possible future supplementation of existing populations in the wild. Note that melanistic jaguars have never been documented in the northernmost part of the species' range. Photograph: © Alma Leaper, Lead Photographer at The Big Cat Sanctuary UK.

or are too lacking in detail to be considered useful. A surprising number of such reports, especially from Arizona, are of "black panthers," none of which has yielded any credible information on the jaguar (Brown and López González 2001; Johnson et al. 2011). Despite the many false or at least dubious reports, wildlife agencies in both Arizona and New Mexico continue to solicit such information in hopes that a jaguar will be detected. Since the 1990s, both the New Mexico Department of Game and Fish and the Arizona Game and Fish Department have collected putative jaguar reports in their respective states using a consistent standardized ranking system for evaluating report reliability (e.g., Girmendonk 1994; Van Pelt and Johnson 1998; NMDGF 2016a). When possible, the observer is interviewed and the location of the sighting is visited to look for tracks or other evidence. Reports ranked as "Class I" are those for which some physical evidence (e.g., carcass, photograph, scat, or track) was preserved and therefore are considered verified records of jaguars. "Class II" reports lack physical evidence but are considered either "probable" or "possible," depending on the information provided by the observer. The lowest category of reports, "Class III," consists of those that lack sufficient details to be accepted, are confirmed cases of mistaken identity, or are otherwise considered unreliable. Included in Class III are some older records in the scientific literature that have insufficient documentation to be considered credible. The vast majority of reports received from the public in recent years in both Arizona and New Mexico have been assigned to the Class III category.

Despite the infrequency of verified occurrences in the southwestern United States, the jaguar will remain a high-profile species for the foreseeable future, due both to its Endangered status and prominence in the public mind as a symbol of a wilder past in the Southwest. Through the protection of individual cats, public information campaigns about the species, and the careful management of its habitat by resource agencies and landowners alike on both sides of the US-Mexico border, the jaguar may continue to be counted among New Mexico's carnivores—even if only as an occasional visitor—for years to come.

ACKNOWLEDGMENTS

We thank J. K. Frey, J.-L. E. Cartron, I. Cassaigne, and S. M. Murphy for comments on earlier versions of the manuscript; and Megan "Turtle" Southern and the Northern Jaguar Project for granting us permission to use their photos.

LITERATURE CITED

Anonymous. 1989. Jaguar: gone but remembered. *New Mexico Wildlife* 34(3):11.

Arritt, S. 1997. Jaguar! *New Mexico Partners Conserving Endangered Species (New Mexico Department of Game and Fish)* 2(3):8–13, 16.

Audubon, J. J., and J. Bachman. 1854. *The quadrupeds of North America*. Vol. 3. New York: V. G. Audubon.

Bailey, V. 1931 (=1932). *Mammals of New Mexico*. North American Fauna 53. Washington, DC: US Department of Agriculture, Bureau of Biological Survey.

Baird, S. F. 1859. Mammals of the boundary. In *Report on the U.S. Mexican boundary survey, under order of Lieut. Col. W. H. Emory, Major First Cavalry, and United States Commissioner*, Vol. 2, Pt. 2, 1–62. Washington, DC: Department of Interior.

Barber, C. M. 1902. Notes on little-known New Mexican mammals and species apparently not recorded from the territory. *Proceedings of the Biological Society of Washington* 15:191–93.

Beckmann, J. P., and J. A. Hilty. 2010. Connecting wildlife populations in fractured landscapes. In *Safe passages: highways, wildlife, and habitat connectivity*, ed. J. P. Beckmann, A. P. Clevenger, M. P. Huijser, and J. A. Hilty, 3–16. Washington, DC: Island Press.

Boydston, E. E., and C. A. López González. 2005. Sexual differentiation in the distributional potential of northern jaguars (*Panthera onca*). *USDA Forest Service Proceedings*, RMRS-P-36:51–56.

Brown, D. E. 1983. On the status of the jaguar in the Southwest. *Southwestern Naturalist* 28:459–60.

———. 1991. Revival for el tigre? *Defenders* 66:27–35.

———, ed. 1994. *Biotic communities: southwestern United States and northwestern Mexico*. 2nd ed. Salt Lake City: University of Utah Press.

Brown, D. E., and C. A. López González. 2000. Notes on the occurrences of jaguars in Arizona and New Mexico. *Southwestern Naturalist* 45:537–46.

———. 2001. *Borderland jaguars: tigres de la frontera*. Salt Lake City: University of Utah Press.

Brown, J. H. 1984. On the relationship between abundance and distribution of species. *American Naturalist* 124:255–79.

Carmony, N. B., ed. 1998. *Ben Lilly's tales of bears, lions, and hounds*. Silver City, NM: High-Lonesome Books.

Caro, T. 2013. The colours of extant mammals. *Seminars in Cell and Developmental Biology* 24:542–52.

Cassaigne, I., R. A. Medellín, R. W. Thompson, M. Culver, A. Ochoa, K. Vargas, J. L. Childs, J. Sanderson, R. List, and A. Torres-Gómez. 2016. Diet of pumas (*Puma concolor*) in Sonora, Mexico, as determined by GPS kill sites and molecular identified scat, with comments on jaguar (*Panthera onca*) diet. *Southwestern Naturalist* 61:125–32.

Cavalcanti, S. M. C., and E. M. Gese. 2009. Spatial ecology and social interactions of jaguars (*Panthera onca*) in the southern Pantanal, Brazil. *Journal of Mammalogy* 90:935–45.

Childs, J. L. 1998. *Tracking the felids of the borderlands*. El Paso, TX: Printing Corner Press.

Childs, J. L., and A. M. Childs. 2008. *Ambushed on the jaguar trail: hidden cameras on the Mexican border*. Tucson, AZ: Rio Nuevo Publishers.

Crawshaw, P. G., Jr., and H. B. Quigley. 1991. Jaguar spacing, activity and habitat use in a seasonally flooded environment in Brazil. *Journal of Zoology (London)* 223:357–70.

Crooks, K. R., M. Grigione, A. Scoville, and G. Scoville. 2008. Exploratory use of track and camera surveys of mammalian carnivores in the Peloncillo and Chiricahua mountains of southeastern Arizona. *Southwestern Naturalist* 53:510–17.

Culver, M. 2016. *Jaguar surveying and monitoring in the United States*. US Geological Survey Open-File Report 2016-1095.

Culver, M., and A. O. Hein. 2016. *Jaguar taxonomy and genetic diversity for southern Arizona, United States, and Sonora, Mexico*. US Geological Survey Open-File Report 2016-1109.

Daggett, P. M., and D. R. Henning. 1974. The jaguar in North America. *American Antiquity* 39:465–69.

De la Torre, J. A., J. M. Núñez, and R. A. Medellín. 2017. Spatial requirements of jaguars and pumas in Southern Mexico. *Mammalian Biology* 84:52–60.

Eizirik, E., J.-H. Kim, M. Menotti-Raymond, P. G. Crawshaw, S. J. O'Brien, and W. E. Johnson. 2001. Phylogeography, population history and conservation genetics of jaguars (*Panthera onca*, Mammalia, Felidae). *Molecular Ecology* 10:65–79.

Emmons, K. L. H. 1999. Jaguar, *Panthera onca*. In *The Smithsonian Book of North American Mammals*, ed. D. E. Wilson and S. Ruff, 236–237. Washington, DC: Smithsonian Institution Press.

Fernández, G. J., J. C. Corley, and A. F. Capurro. 1997. Identification of cougar and jaguar feces through bile acid chromatography. *Journal of Wildlife Management* 61:506–10.

Findley, J. S., A. H. Harris, D. E. Wilson, and C. Jones. 1975. *Mammals of New Mexico*. Albuquerque: University of New Mexico Press.

Flesch, A. D., C. W. Epps, J. W. Cain III, M. Clark, P. R. Krausman, and J. R. Morgart. 2010. Potential effects of the United States—Mexico border fence on wildlife. *Conservation Biology* 24:171–81.

Girmendonk, A. L. 1994. *Ocelot, jaguar and jaguarundi sightings report: Arizona and Sonora, Mexico*. Arizona Game and Fish Department, Nongame and Endangered Wildlife Program, Technical Report 35.

Glenn, W. 1996. *Eyes of fire: encounter with a borderlands jaguar*. El Paso, TX: Printing Corner Press.

Goldman, E. A. 1932. The jaguars of North America. *Proceedings of the Biological Society of Washington* 45:143–46.

Grigione, M. M., K. Menke, C. Lopez-Gonzalez, R. List, A. Banda, J. Carrera, R. Carrera, A. J. Giordano, J. Morrison, M. Sternberg, R. Thomas, and B. Van Pelt. 2009. Identifying potential conservation areas for felids in the USA and Mexico: integrating reliable knowledge across an international border. *Oryx* 43:78–86.

Grigione, M., A. Scoville, G. Scoville, and K. Crooks. 2007. Neotropical cats in southeast Arizona and surrounding areas: past and present status of jaguars, ocelots and jaguarundis. *Mastozoologia Neotropical* 14:189–99.

Haddal, C. C., Y. Kim, and M. J. Garcia. 2009. *Border security: barriers along the U.S. international border*. Washington, DC: Library of Congress, Congressional Research Service.

Halloran, A. F. 1946. The carnivores of the San Andres Mountains, New Mexico. *Journal of Mammalogy* 27:154–61.

Hatten, J. R., A. Averill-Murray, and W. E. Van Pelt. 2005. A spatial model of potential jaguar habitat in Arizona. *Journal of Wildlife Management* 69:1024–33.

Haynes, L., Z. Hackl, and M. Culver. 2005. Wild cats of the Sky Islands: a summary of monitoring efforts using noninvasive techniques. In *Connecting mountain islands and desert seas: biodiversity and management of the Madrean Archipelago II*, ed. G. J. Gottfried, B. S. Gebow, L. G. Eskew, and C. B. Edminster, 185–88. Fort Collins, CO: US Department of Agriculture, Forest Service, Rocky Mountain Research Station. Proceedings RMRS-P-36.

Hill, J. E. 1942. Notes on mammals of northeastern New Mexico. *Journal of Mammalogy* 23:75–82.

Hoffmeister, D. F. 1986. *Mammals of Arizona*. Tucson and Phoenix: University of Arizona Press and Arizona Game and Fish Department.

Johnson, T. B., W. E. Van Pelt, and J. N. Stuart. 2011. *Jaguar conservation assessment for Arizona, New Mexico and northern Mexico*. Final report (31 January 2011) prepared for the Arizona-New Mexico Jaguar Conservation Team by Arizona Game and Fish Department and New Mexico Department of Game and Fish.

Johnson, W. E., E. Eizirik, J. Pecon-Slattery, W. J. Murphy, A. Antunes, E. Teeling, and S. J. O'Brien. 2006. The late Miocene radiation of modern Felidae: a genetic assessment. *Science* 311(5757):73–77.

Jones, C., and C. G. Schmitt. 1997. Mammal species of concern in New Mexico. In *Life among the muses: papers in honor of James S. Findley*, ed. T. L. Yates, W. I. Gannon, and D. E. Wilson, 179–205. Albuquerque: University of New Mexico, Museum of Southwestern Biology. *Special Publication of the Museum of Southwestern Biology, University of New Mexico* 3:1–290.

Jones, R. F. 1974. The man who loved cat killing. *New Mexico Wildlife* 19(3):9–11.

Kahn, J. 2007. On the prowl. *Smithsonian Magazine* 38(8):84–90, 92.

King, B., and S. Wilcox. 2008. Peace parks and jaguar trails: transboundary conservation in a globalizing world. *GeoJournal* 71:221–31.

Kitchener, A. C., C. Breitenmoser-Würsten, E. Eizirik, A. Gentry, L. Werdelin, A. Wilting, N. Yamaguchi, A. V. Abramov, P. Christiansen, C. Driscoll, J. W. Duckworth, W. Johnson, S.-J. Luo, E. Meijaard, P. O'Donoghue, J. Sanderson, K. Seymour, M. Bruford, C. Groves, M. Hoffmann, K. Nowell, Z. Timmons, and S. Tobe. 2017. A revised taxonomy of the Felidae: the final report of the Cat Classification Task Force of the IUCN Cat Specialist Group. *Cat News Special Issue* 11:1–80.

Larson, S. E. 1997. Taxonomic re-evaluation of the jaguar. *Zoo Biology* 16:107–20.

Leopold, A. S. 1959. *Wildlife of Mexico: the game birds and mammals*. Berkeley: University of California Press.

López González, C. A., and B. J. Miller. 2002. Do jaguars (*Panthera onca*) depend on large prey? *Western North American Naturalist* 62:218–22.

Mahler, R. 2010. Where the wild things are. *New Mexico Magazine* 88(10):31–35.

McCain, E. B., and J. L. Childs. 2008. Evidence of resident jaguars (*Panthera onca*) in the southwestern United States and the implications for conservation. *Journal of Mammalogy* 89:1–10.

Menke, K. A., and C. L. Hayes. 2003. Evaluation of the relative suitability of potential jaguar habitat in New Mexico. Unpublished report to New Mexico Department of Game and Fish, Santa Fe, New Mexico.

Miller, P. S. 2013. *Population viability analysis for the jaguar (Panthera onca) in the northwestern range*. Final report to US Fish and Wildlife Service, Arizona Ecological Services, Tucson, Arizona.

Monroy-Vilchis, O., C. Rodríguez-Soto, M. Zarco-González, and V. Urios. 2009. Cougar and jaguar habitat use and activity patterns in central Mexico. *Animal Biology* 59:145–57.

Nelson, E. W., and E. A. Goldman. 1933. Revision of the jaguars. *Journal of Mammalogy* 14:221–40.

New Mexico Department of Game and Fish (NMDGF). 2016a. Reports of jaguar (*Panthera onca*) in New Mexico, with evaluations of their reliability. Revised 24 August 2016. Unpublished report. New Mexico Department of Game and Fish, Wildlife Management Division, Santa Fe.

———. 2016b. *State wildlife action plan for New Mexico*. Santa Fe: New Mexico Department of Game and Fish.

Nowak, R. M. 1991. *Walker's mammals of the world*. 2 Vols. Baltimore: Johns Hopkins University Press.

O'Mara, R. 1997. Jaguar. *Virginia Quarterly Review* 73:307–13.

Pocock, R. I. 1939. The races of jaguar (*Panthera onca*). *Novitates Zoologicae* 41:406–22.

Rabinowitz, A. R. 1999. The present status of jaguars (*Panthera onca*) in the southwestern United States. *Southwestern Naturalist* 44:96–100.

Rabinowitz, A. R., and B. G. Nottingham. 1986. Ecology and behaviour of the jaguar (*Panthera onca*) in Belize, Central America. *Journal of Zoology (London)* 210:149–59.

Rabinowitz, A., and K. A. Zeller. 2010. A range-wide model of landscape connectivity and conservation for the jaguar, *Panthera onca*. *Biological Conservation* 143:939–45.

Robinson, M. J., C. Bradley, and J. Boyd. 2006. *Habitat for jaguars in New Mexico*. Final report, Center for Biological Diversity, Tucson, Arizona.

Rodríguez-Soto, C., O. Monroy-Vilchis, L. Maiorano, L. Boitani, J. C. Faller, M. A. Briones, R. Núñez, O. Rosas-Rosas, G. Ceballos, and A. Falcucci. 2011. Predicting potential distribution of the jaguar (*Panthera onca*) in Mexico: identification of priority areas for conservation. *Diversity and Distribution* 17:350–61.

Rosas-Rosas, O. C., and L. C. Bender. 2012. Population status of jaguars (*Panthera onca*) and pumas (*Puma concolor*) in northeastern Sonora, Mexico. *Acta Zoológica Mexicana (n. s.)* 28:86–101.

Rosas-Rosas, O. C., L. C. Bender, and R. Valdez. 2008. Jaguar and puma predation on cattle calves in northeastern Sonora, Mexico. *Rangeland Ecology and Management* 61:554–60.

———. 2010. Habitat correlates of jaguar kill-sites of cattle in northeastern Sonora, Mexico. *Human-Wildlife Interactions* 4:103–11.

Rosas-Rosas, O. C., and R. Valdez. 2010. The role of landowners in jaguar conservation in Sonora, Mexico. *Conservation Biology* 24:366–71.

Runyan, C. 2006. Return of a native cat: wild jaguar spotted in New Mexico. *Nature Conservancy* 56(3):12–13.

Sagarin, R. D., and S. D. Gaines. 2002. The "abundant centre" distribution: to what extent is it a biogeographical rule? *Ecology Letters* 5:137–47.

Sanderson, E. W., and K. Fisher. 2013. *Jaguar habitat modeling and database update*. Final report to the US Fish and Wildlife Service in response to Solicitation F12PS00200, submitted March 12, 2013.

Sanderson, E. W., K. Fisher, R. Peters, J. P. Beckman, B. Bird, C. M. Bradley, J. C. Bravo, M. M. Grigione, J. R. Hatten, C. A. Lopez-Gonzalez, K. Menke, J. R. B. Miller, P. S. Miller, C. Mormorunni, M. J. Robinson, R. E. Thomas, and S. Wilcox. 2021. A systematic review of potential habitat suitability for the jaguar *Panthera onca* in central Arizona and New Mexico, USA. *Oryx* 56(1):1–12. doi:10.1017/S0030605320000459.

Sanderson, E. W., M. Jaiteh, M. A. Levy, K. H. Redford, A. V. Wannebo, and G. Woolmer. 2002a. The human footprint and the last of the wild. *BioScience* 52:891–904.

Sanderson, E. W., K. H. Redford, C.-L. B. Chetkiewicz, R. A. Medellin, A. R. Rabinowitz, J. G. Robinson, and A. B. Taber. 2002b. Planning to save a species: the jaguar as model. *Conservation Biology* 16:58–72.

SEMARNAT (Ministry of Environment and Natural Resources of Mexico). 2010. *NORMA Oficial Mexicana NOM-059-SEMARNAT-2010, Protecciónambiental-Especies nativas de México de flora y fauna silvestres-Categorías de riesgo y especificaciones para su inclusión, exclusión o cambio-Lista de especies en riesgo*. Diario Oficial de la Federación (Mexico's federal register), 30 December 2010.

Seton, E. T. 1920. The jaguar in Colorado. *Journal of Mammalogy* 1:241.

———. 1929. *Lives of game animals*. Vol. 1, Pt. 1, *Cats, wolves, and foxes*. Garden City, NJ: Doubleday, Doran & Company.

Seymour, K. L. 1989. *Panthera onca*. *Mammalian Species* 340:1–9.

Sierra Institute Field Studies Program in Arizona. 2000. *Jaguar habitat in southern Arizona and New Mexico*. Report to Arizona-New Mexico Jaguar Conservation Team from the University of California Extension, Santa Cruz.

Silver, S. C., L. E. T. Ostro, L. K. Marsh, L. Maffei, A. J. Noss, M. J. Kelly, R. B. Wallace, H. Gomez, and G. Ayala. 2004. The use of camera traps for estimating jaguar *Panthera onca* abundance and density using capture/recapture analysis. *Oryx* 38:148–54.

Simberloff, D. 1997. Flagships, umbrellas, and keystones: is single-species management passé in the landscape era? *Biological Conservation* 83:247–57.

Simoni, L. S. 2012. Living with large carnivores: insights from diet choice, habitat use, and the ecology of fear. MS thesis, University of Illinois, Chicago.

Sloan, S. 2019. Use of least-cost path analysis to identify potential movement corridors for jaguars across the U.S.-Mexico border. MS thesis, University of Southern California.

Swank, W. G., and J. G. Teer. 1989. Status of the jaguar—1987. *Oryx* 23:14–21.

Theobald, D. M., J. R. Miller, and N. T. Hobbs. 1997. Estimating the cumulative effects of development on wildlife habitat. *Landscape and Urban Planning* 39:25–36.

United States Fish and Wildlife Service (USFWS). 1997. Endangered and threatened wildlife and plants; final rule to extend endangered status for the jaguar in the United States. *Federal Register* 62(140):39147–57 (22 July 1997).

———. 2014. Endangered and threatened wildlife and plants; designation of critical habitat for jaguar; final rule. *Federal Register* 79(43):12572–654 (5 March 2014).

———. 2018. Jaguar recovery plan (*Panthera onca*). US Fish and Wildlife Service, Southwest Region (Region 2), Albuquerque, New Mexico.

———. 2021. Endangered and threatened wildlife and plants; revision of the critical habitat designation for the jaguar in compliance with a court order. *Federal Register* 86(138):38570–2 (22 July 2021).

United States Global Change Research Program (USGCRP). 2017. *Climate science special report: fourth national climate assessment*, Vol.1 [ed. D. J. Wuebbles, D. W. Fahey, K. A. Hibbard, D. J. Dokken, B. C. Stewart, and T. K. Maycock]. Washington, DC: US Global Change Research Program.

Valdez, R. 2000. Jaguar. In *Ecology and management of large mammals in North America*, ed. S. Demarais and P. R. Krausman, 378–88. Upper Saddle River, New Jersey: Prentice Hall.

Van Pelt, W. E., and T. B. Johnson. 1998. *Annual report on the Jaguar Conservation Agreement for Arizona and New Mexico*. Nongame and Endangered Wildlife Program, Technical Report 132. Arizona Game and Fish Department, Phoenix.

Whipple, A. W., T. Ewbank, and W. W. Turner. 1856. Report upon the Indian tribes. In *Reports of explorations and surveys to ascertain the most practicable and economical route for a railroad from the Mississippi River to the Pacific Ocean*, 1–127. Made under the direction of the Secretary of War, in 1853–4. Vol. 3, Pt. 3. Washington, DC: War Department.

Willis, A. 2007. *Black panthers: little known North American treasure*. Petersburg, VA: Dietz Press.

Wozencraft, W. C. 2005. Order Carnivora. In *Mammal species of the world: a taxonomic and geographic reference*, Vol. 1, ed. D. E. Wilson and D. M. Reeder, 532–628. 3rd ed. Baltimore: Johns Hopkins University Press.

Yancovitch Shalom, H., I. Granot, S. A. Blowes, A. Friedlander, C. Mellin, C. E. L. Ferreira, A. C. J. Ernesto, M. Kulbicki, S. R. Floeter, P. Chabanet, V. Parravicini, and J. Belmaker. 2020. A closer examination of the "abundant centre" hypothesis for reef fishes. *Journal of Biogeography* 47:2194–209.

COUGAR (*PUMA CONCOLOR*)

Kenneth A. Logan and Linda L. Sweanor

Of the two big cat species found in the Western Hemisphere, the cougar alone occurs widely in New Mexico, where people also know it as the mountain lion, león, puma, painter, catamount, and panther. Unlike the jaguar (*Panthera onca*; Chapter 8), rare in the state, the cougar joins the list of carnivores—along with the coyote (*Canis latrans*; Chapter 10), Mexican wolf (*Canis lupus baileyi*; Chapter 11), and black bear (*Ursus americanus*; Chapter 16)—that can limit the growth of New Mexico's ungulate populations. Cougars compete with these other carnivores, as well as with humans, for some of the same ungulate and other prey animals. Moreover, humans hunt cougars and can impact them to the point of reducing their numbers and altering the age and sex structure of their populations. Because the species thrives in diverse habitats and on a variety of prey, wild and domestic, it often comes into proximity with humans. Beautiful, dangerous, graceful, powerful, intelligent, cunning, efficient, wasteful, wild, adaptable, skulking, bold, good, and evil-—the cougar has been all these things to people. Public attitudes toward the cougar defined the history of the species in New Mexico during the past two centuries. They still shape its fate in the present and will determine its future.

DESCRIPTION

The cougar is one of the "big cats," the second largest felid in New Mexico (only the jaguar is larger), the fourth largest cat in the world (after the tiger [*P. tigris*], African lion [*P. leo*], and jaguar, respectively), and is comparable in size to the African leopard (*P. pardus*) and snow leopard (*P. uncia*). Cougars are sexually dimorphic, meaning that males are larger than females. On average, adult males weigh about 40% more than adult females (Anderson 1983). In our 10-year study of cougars on the San Andres Mountains in southern New Mexico (Logan and Sweanor 2001:18, 21, Appendix 1), we found that adult male cougars averaged 56.3 kg (123.9 lbs), whereas the average weight of females was 33.1 kg (72.8 lbs). Total lengths (tip of nose to tip of last vertebra in tail) of adult males ranged from 200 to 227 cm (79–89 in), whereas those of females ranged from 172 to 205 cm (68–81 in). Height from the bottom of the forefoot to top of the shoulder ranged from 54 to 67.5 cm (21–27 in) for adult males, and 43 to 61 cm (17–24 in) for adult females. The feature of the cougar that most captures the attention of people is the long, round, stout tail that comprises about one-third of the total length of the animal.

Cougars have a round-shaped head, a short neck, and a muscular lithe torso on powerful, robust legs. Forelegs are somewhat shorter than the aft and have larger paws on supine wrists. Hind legs are proportionately the longest in large felids (Gonyea 1976). The cougar's legs are adapted for traversing rugged terrain, climbing, subduing and handling strong prey, and self-defense.

(*opposite page*) Photo © Ken Logan.

The long muscular tail seems to function as a counterbalance while making precise acrobatic movements such as those used while attacking prey, defending against competitors, traversing ledges, and climbing trees. It might also be a part of postures in up-close communications with other cougars.

The pelage of an adult cougar is principally of one color on the exposed sides and back, with varying hues of tawny, reddish brown, and grayish brown. Whitish hair covers the muzzle, underparts of the neck, chest, abdomen, legs, and tail. There are black accents around the sides of the muzzle, backs of the ears, on the inside of the forelegs, and tip of the tail. Despite seemingly ubiquitous reports of "black cougars," we are not aware of any confirmed, preserved melanistic specimens. Kittens are born with black spots on tawny-colored coats. However, those spots fade to light brown dapples by the time kittens reach the age of nine months. Eye color in young kittens is blue, but it changes to the brown or amber color of the adult within the first six months. Sex can be distinguished by external organs. Males have a spot of black hair that encircles the opening of the penis sheath in adolescents and adults, and which is separated from the anus by the visible scrotum. The female's vulva is directly below the anus and is sometimes ringed with black hair. Sex can be accurately determined this way when an animal is treed with the use of hounds by hunters or researchers. Cougars have thirty teeth with an upper/lower dental formula of 3/3 incisors, 1/1 canines, 3/2 premolars, 1/1 molars. Fewer teeth compared to other mammals are associated with a short muscle-bound jaw that allows for a powerful bite strong enough to break the neck of a mule deer (*Odocoileus hemionus*) or subdue a full-grown elk (*Cervus canadensis*) in a matter of seconds. These characteristics also enable cougars to eat bone.

Like most cats, cougars have retractable claws, so claw marks generally do not show in footprints. Only where cougars use their claws for

Photo 9.1. Adult male cougar at Good Fortune Spring in Sierra County in July 2012. Cougars are large, slender cats with a muscular torso and powerful legs. The long, muscular tail seemingly functions as a counterbalance to a cougar's movements during the pursuit of a prey. Photograph: © WSMR Garrison Environmental Division and ECO Inc., Ecological Consultants.

Photo 9.2. Adult male cougar in New Mexico's San Andres Mountains on 24 June 1988. Cougars have relatively small, round heads. Black accents are found around the sides of the muzzle and on the backs of the ears. Whitish hair covers the muzzle. Photograph: © Ken Logan.

Photo 9.3. Six-week-old cougar cubs showing their characteristic spotted pelage and blue eyes. Photograph: © Ken Logan.

Photo 9.4. Hunters that tree a cougar with hounds can usually determine its sex. Males are characterized by a ring of black hair around the opening of the prepuce, which is anterior to the scrotum. Photograph: © Ken Logan.

extra traction in muddy and icy conditions will claw marks sometimes appear in tracks. Forefeet and hind feet have five and four toes each, respectively. But, because the pollex (analogous to our thumb) on the forefoot is raised, only four toes register in a forefoot track. The plantar pad (the heel pad behind the toes) has three distinct rear lobes, a common feature in cats. The forefeet are larger than the hind feet. Usually the hind foot steps onto the forefoot print when a cougar is walking, and it is the hind footprint that registers clearly on the ground. When a cougar is trotting, however, typically all four footprints register on the ground. Tracks with large almond-shaped toes showing no claw impressions, and with plantar pads measuring about 43 to 56 mm (~1.7–2.2 in) wide and with three lobes at the rear, are sign of a cougar. Except for the jaguar in the Southwest, no other large mammal makes tracks similar to those of the cougar.

Taxonomists classify the cougar with other meat-eaters in the order Carnivora and with all other cats in the family Felidae (Wozencraft 1993). The cougar is also given two Latin names, consistent with the binomial system of biological nomenclature founded by Swedish zoologist

Photos 9.5a and b (*top to bottom*). Two treed female cougars. Females can be identified by the absence of a ring of black hair or scrotum; the vulva is directly below the anus, and both are usually obscured by the base of the tail. Photographs: © Ken Logan.

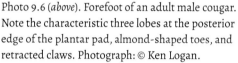

Photo 9.6 (*above*). Forefoot of an adult male cougar. Note the characteristic three lobes at the posterior edge of the plantar pad, almond-shaped toes, and retracted claws. Photograph: © Ken Logan.

Photo 9.7 (*top right*). Tracks of both a forefoot and hind foot of an adult cougar cast in dried mud. The print at the top is of the hind foot; the print at the bottom is of the forefoot. Note that the forefoot is larger than the hind foot. Photograph: © Ken Logan.

Photo 9.8 (*right*). Cougar tracks in snow. Photograph: © Ken Logan.

Carl Linnaeus, who named the cougar *Felis concolor* in 1771, meaning cat of one color. A third name, indicating a subspecies, is attached when there is evidence of sufficient localized variation to warrant more specialized groups. In the early 1900s as many as 30 cougar subspecies were recognized based on shapes and sizes of skulls and colors of skins (Goldman 1946). At the turn of the 21st century, however, the use of much more advanced molecular techniques distinguished only six so-called phylogeographic groupings or subspecies (Culver 1999; Culver et al. 2000). The New Mexico cougar belonged to the subspecies occurring from Canada south to Guatemala and Belize and named *Puma concolor couguar*. The genus *Puma* recognizes the cougar lineage, which evolved from a common ancestor together with the cougar's closest extant relatives, the cheetah (*Acinonyx jubatus*) and jaguarundi (*Puma yaguarondi*) (Culver 2010). An ancestor of the cougar arrived in North America from the Old World about 6 mya, then expanded its geographic range southward into South America via the Panamanian land bridge. However, the modern cougar probably originated in eastern South America then spread to other regions of the Americas. The modern North American cougar is likely descended from a founder event involving a small number of cougars that migrated from South America about 10,000 to 12,000 years ago (Culver 1999; Culver et al. 2000; Chapter 1; and see Young and Goldman 1946).

Our knowledge about the cougar has grown tremendously since the original peoples of the

Americas, and later the first European immigrants, formed their own, original impressions of the animal from mere chance observations (Young 1946). In North America, the early 1900s marked the initial stages of modern ecology, at which time a few men on horses and with hunting dogs went to the field to search for cougar sign, and attempted to formally describe the species' habits, habitat, and diet. One of those men was Frank C. Hibben in Arizona and New Mexico during 1934 and 1935 (Hibben 1937). During the modern environmental movement of the late 1960s and early 1970s in America, pioneering researchers in Idaho, Maurice Hornocker (Hornocker 1969, 1970) and John Seidensticker (Seidensticker et al. 1973), began gathering data on cougar behavior and population ecology by using new technologies including drugs to anesthetize cougars and radio-telemetry to track animal movements. From their innovative work emerged an initial theoretical framework about cougar population limitation and ungulate population responses to predation. Also in the 1970s, Harley Shaw used similar techniques to examine cougar abundance and effects of cougar predation on mule deer and cattle in Arizona (Shaw 1977, 1980). Other cougar studies followed during the 1980s and 1990s. One such study occurred from 1985 to 1995 on the San Andres Mountains in south-central New Mexico, where we studied over 200 individual cougars (Logan and Sweanor 2001). Since then, the technology to study cougars has become even more advanced and now includes 1) a variety of safe-capture and handling methods; 2) Global Positioning System (GPS) collars that locate the animals via satellites on predetermined schedules and with an accuracy of within a few meters, 3) laboratory molecular genetic techniques; and 4) faster and more powerful computers for data storage and analyses. Together with more complex study designs and sophisticated mathematical and statistical models, all of the new available technology has rapidly accelerated the science on cougars

Photo 9.9 (*left*). Linda Sweanor with a four-week-old cub in the San Andres Mountains, where the authors studied over 200 individual cougars from 1985 to 1995. The cub in the photo was marked with an eartag for data on survival. Photograph: © Ken Logan.

Photo 9.10 (*right*). Adult male cougar M7 recaptured in a foot-hold snare as part of a study of cougar population dynamics, behavior, and predation on mule deer and desert bighorn sheep on the San Andres Mountains, New Mexico, 10 March 1991. Photograph: © Ken Logan.

Photo 9.11. Adult female and her two older cubs in a canyon in Los Alamos County on 16 August 2021. Cougars are found throughout most of New Mexico, primarily in mountainous areas and canyonlands. There are no reliable estimates of cougar population size in New Mexico. Photograph: © Hari Viswanathan.

through the early 2000s. Yet, at the same time, there is still much more to learn. This chapter represents an overview of what we know about the cougar in New Mexico, and, where needed, also includes information about the species from elsewhere.

DISTRIBUTION AND ABUNDANCE

Cougars once had the largest geographic distribution of any land-dwelling wild mammal in the Western Hemisphere, occurring from the Atlantic to the Pacific coast, and extremes from southern Alaska to the tip of South America. The species adapted to diverse environments along an elevational gradient ranging from sea level to as high as 3,960 m (13,000 ft) in South America. In North America, cougars occurred from the east to the west coast and from its northern limits south to Mexico (Young 1946; Wilson and Reeder 1993).

European settlers to North America threatened cougar survival. Direct killing, reductions in native prey from over-hunting, and habitat

loss (e.g., through conversion of forest and woodland into farmland or growth of urban centers) all severely reduced cougar numbers and the species' distribution. Cougars were mostly gone from eastern North America by the late 1800s and diminished in the West by the early 1900s (Young 1946; Cahalane 1964). Beginning in the 1960s, however, conditions changed again, which allowed cougars to rebound in numbers and recolonize former parts of their range, except in eastern North America, where only one, small breeding population can be found in southern Florida (Pierce and Bleich 2003; see under "Status and Management"). The most important factors in this recent change were 1) state- and province-mandated regulations that limited the killing of cougars by people, along with the institution of hunting seasons with bag limits and protections of females and kittens, and 2) increases in prey animals, such as white-tailed deer (*Odocoileus virginianus*), mule deer, and elk, with all these species benefitting from professional wildlife management in North America.

Although cougars occur in all parts of New Mexico, their distribution is not uniform throughout the state. Instead, they are recorded primarily in mountainous areas and canyon lands and are comparatively rare on New Mexico's eastern plains (map 9.1). No one knows how many cougars occur in the state. An educated guess by the New Mexico Department of Game and Fish is all that is currently available (see below). Statewide surveys for reliable data on cougar abundance have so far proven impractical for two main reasons. First, not enough trained biologists and financial resources are available to survey cougar populations throughout New Mexico; thus, available surveys are limited to individual mountain ranges. Second, reliable, cost-effective methods were only very recently developed to survey cougar populations in meaningful management-size areas as large as 15,000 km^2 (~5,800 sq mi) in New Mexico (Alldredge et al. 2019; Murphy et al. 2019;

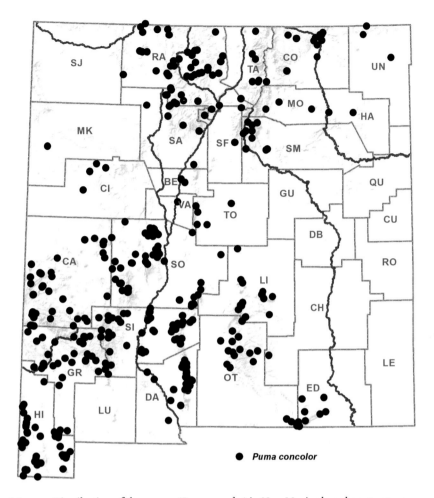

Map 9.1. Distribution of the cougar (*Puma concolor*) in New Mexico based on museum specimen records. Note that the density of records in some areas of the state may reflect cougar management actions rather than density.

Murphy and Augustine 2019; Proffitt et al. 2020). As a substitute for statewide surveys, Perry (2010) used 1) the movements of 10 GPS-collared cougars from 2005 to 2010 in south-central New Mexico; and 2) 1,684 geo-referenced, statewide hunter-killed cougar records to test the predictive power of habitat models developed with environmental variables including topographic ruggedness and vegetation type. Perry (2010) then categorized cougar habitat as excellent, good, moderate, and fair. The New Mexico Department of Game and Fish, which funded the study, surmised density levels for independent cougars

(both adults and subadults) in each of the four habitat categories: excellent = 3 to 4 cougars/100 km² (8–10 cougars/100 mi²); good = 1.2 to 1.7 cougars/100 km² (3.1–4.4 cougars/100 mi²), moderate = 0.6 to 0.9 cougars/100 km² (1.6–2.3 cougars/100 mi²), and fair = 0.4 to 0.5 cougars/100 km² (1–1.3 cougars/100 mi²) (NMDGF 2010, 2020). These densities of cougars were extrapolated to regions of the mapped cougar habitat categories, and summed for a statewide population projection of 3,100 to 4,300 adult cougars (NMDGF, data on file). Today, the New Mexico Department of Game and Fish still uses the same assumptions of population size and abundance scaled to all but one of the cougar management zones (see under "Status and Management"). In north-central New Mexico, field research conducted by Murphy et al. (2019) produced density estimates as low as 0.84 independent cougars per 100 km², much lower than expected given the quality of the habitat, perhaps in part the result of heavy hunting pressure on cougars in that region. Thus, the New Mexico Department of Game and Fish now uses a density of 1.1 cougars/100 km² across that entire zone to reflect the new information (NMDGF 2020). The cougar population in the state is currently assumed to be 3,512 independent cougars by the New Mexico Department of Game and Fish (NMDGF 2020).

The New Mexico Department of Game and Fish's extrapolated cougar numbers in other management zones are only as reliable as the data used to produce them. The highest assumed densities of three to four independent cougars per 100 km² are not supported by field data gathered on cougars from New Mexico. The highest cougar densities estimated from field data in New Mexico were 2.1 and 2.4 independent cougars per 100 km² in the Black Range and in the Rio Grande bosque, respectively (Pitman 2010; Perry et al. 2011). In the Chihuahuan Desert in south-central New Mexico, the highest observed density reached 2.1 independent cougars per 100 km² for a population protected from human exploitation (Logan and Sweanor 2001). As more sophisticated methods of estimating cougar density are applied in more regions of New Mexico, more reliable data should emerge to help guide management decisions.

HABITAT ASSOCIATIONS

At the scale of their entire distribution, cougars inhabit a wide variety of habitats including foothills woodlands, high mountain forests, boreal forests, deserts, riparian woodlands, and tropical forests. Cougars occur in the xerophytic deciduous woodlands of central Argentina, tropical semi-deciduous and secondary forests of South and Central America, and the hammock and swamp forests of southern Florida (Belden et al. 1988; Maehr and Cox 1995; Negrões et al. 2010; De las Guerisoli et al. 2019; Ávila-Nájera et al. 2020). In Mexico, cougars are found in all vegetation types, though they prefer forests with dense cover (Lira and Naranjo 2003; Chávez 2010; Rodríguez-Soto et al. 2013). In central Argentina, cougar abundance was positively related with greater structural landscape complexity, vegetation cover, and larger numbers of prey species, but negatively related with the amount of human disturbance (De las Guerisoli et al. 2019). In Belize, the likelihood of cougar occurrence was higher in a large block of secondary rainforest than in an adjacent, more fragmented landscape with higher human disturbance (Foster et al. 2010).

Currently in New Mexico, cougars may occur in most natural settings where there is 1) sufficient prey; and 2) adequate rugged terrain or vegetation cover for stalking prey, raising young, and avoiding predators that may otherwise kill their young. For the most part this means the foothills, mountains, canyonlands, and riparian forests of the state. One invalid assumption about cougar abundance made by the New Mexico Department of Game and Fish (2010) is that cougar habitat quality is static when it is largely influenced by prey availability. Far from being uniformly distributed and static, cougar prey abundance varies not only across regions but also through time (NMDGF 2021). The population of mule deer, the cougar's most common prey in the state, fluctuated between a high of 300,000 in the 1960s to a low of 80,000 in 2019 (NMDGF, data on file; Western Association of Fish and Wildlife Agencies 2019), with population growth driven primarily by the influence of the amount and timing of precipitation on habitat quantity and quality (Western Association of Fish and Wildlife

(*clockwise from top left*) Photo 9.12. Cougar habitat in the San Andres Mountains in south-central New Mexico, within the Chihuahuan Montane Woodlands ecoregion (Griffith et al. 2006). In this ecoregion, mule deer are the main prey of cougars. Photograph: © Ken Logan. Photo 9.13. Cougar habitat in the Black Range in southwestern New Mexico, within the Madrean Lower Montane Woodlands ecoregion (Griffith et al. 2006). In this ecoregion, mule deer, white-tailed deer, elk, and collared peccary are the main prey of cougars. Photograph: © Ken Logan. Photo 9.14. Cougar at Goat Draw on the Sevilleta National Wildlife Refuge on 18 September 2009. Rugged terrain or vegetation cover are important habitat features, in addition to prey availability. Cougar habitat quality can vary through time with changes in prey abundance. Photograph: © Sevilleta National Wildlife Refuge, US Fish and Wildlife Service. Photo 9.15. Adult female cougar at the Bosque del Apache National Wildlife Refuge in March 2010. High-quality habitat in New Mexico includes riparian areas according to Dickson et al.'s (2013) modeling study. Photograph: © Greg D. Wright.

Agencies 2019; see also under "Effects of Cougar Predation on Prey" below).

Landscape ecologists recognize the importance of both the physical continuity of core habitat patches and the functional connectivity among those habitat patches to animal fitness and inter- and intrapopulation dynamics (e.g., Clevenger et al. 2002; Bennett, 2003; Crooks and Sanjayan 2006; Ewers and Didham, 2006), attributes that certainly apply to the cougar. Functional connectivity is often described in the form of habitat linkages (Wan et al. 2019), which are threatened in many parts of the world by anthropogenic disturbance and by human development including roads. Research in southern California revealed how human development was associated with

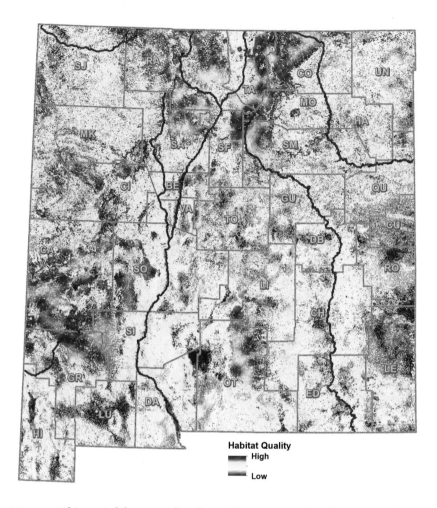

Habitat Quality
High
Low

Map 9.2. Habitat suitability (or quality) derived from scores assigned by experts to land cover, terrain ruggedness, barrier, topographic position, road density, and human population density classes. Adapted with permission from Dickson et al. 2013.

cougars dying mainly from vehicle strikes and depredation control, but also from illegal shootings and human-caused wildfire, which resulted in survival rates comparable to those found in heavily hunted populations (Ernest et al. 2003; Vickers et al. 2015; Gustafson et al. 2019). Dense development and multilane highways created barriers to dispersal that resulted in restrictions to emigration and immigration, and genetic population subdivision (Ernest et al. 2003; Vickers et al. 2015; Gustafson et al. 2019). Dickson et al. (2013) developed maps of cougar habitat and potential movement connections for New Mexico (and Arizona) to assist in conserving not just the integrity of large blocks of habitat but also protecting the smaller patches that can act as linkages or "stepping-stones" for dispersers to maintain movement interactions among population segments (Maps 9.2 and 9.3).

LIFE HISTORY

The Cougar as Predator

Cougars are consummate carnivores in the sense that, once weaned, they eat meat and bone for

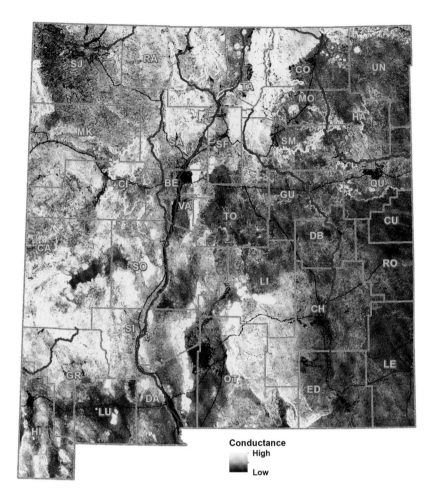

Conductance
High
Low

Map 9.3. Mean cougar habitat conductance (or permeability) derived from scores assigned by experts to land cover, terrain ruggedness, barrier, topographic position, road density, and human population density classes. Adapted with permission from Dickson et al. 2013.

survival. They are both opportunistic and selective predators and occasional scavengers (Logan and Sweanor 2001; Bauer et al. 2005; Krumm et al. 2009; Knopff et al. 2010). Cougars sometimes ingest vegetation accidentally while feeding on prey, or intentionally to obtain micronutrients, scour the digestive tract of parasites, or help expel ingested hair. A cougar stays hydrated by drinking free water and probably by consuming the blood and other moist tissues of its prey.

Although cougars prey on animals ranging in size from hares to adult elk (Iriarte et al. 1990),

they are adapted to killing animals that are greater than 45% of their weight (Carbone et al. 1999). In fact, of the large, solitary felids, the cougar most frequently kills the largest prey relative to its own body size (Packer 1986). Prey are sometimes much larger, as much as 5 times the cougar's mass (Ross and Jalkotzy 1996), and may have antlers, hooves, or canines that can inflict serious injury or cause death (Gashwiler and Robinette 1957; Hornocker 1970; Lindzey et al. 1988; Ross and Jalkotzy 1992; Logan and Sweanor 2001). Hunting tactics and physical adaptations,

Photo 9.16. Mule deer buck that was killed and eaten by a cougar. Note that the cougar plucked swaths of hair from the carcass, and consumed most of the ribs along with almost all of the muscular tissue and internal organs. However, cougars normally do not consume the stomachs and intestines of their prey; the cougar usually removes these organs and covers them with ground debris. Photograph: © Ken Logan.

Photo 9.18. Remains of an elk bull killed and consumed by a cougar. In addition to the soft tissues, all the ribs were consumed. Note also the chewed edges of the pelvis, scapula, and skull, and the plucked hair. Uncompahgre Plateau, Colorado. Photograph: © Ken Logan.

Photo 9.17. Another mule deer buck that was killed and partially consumed by a cougar. After the cougar fed on the carcass, it covered the remains with snow and hair plucked from the deer. This behavior is thought to reduce deterioration of the prey's tissues and detection from scavengers. The cougar returns to further feed on the deer until it is thoroughly consumed. Note the marks on the right side of the deer where the cougar used its forepaws to pull snow onto the carcass. Photograph: © Ken Logan.

including large eyes with binocular and night-time vision, sharp retractable claws, strong short jaws with robust canines, powerful legs, a flexible torso, and the ability to move swiftly over short distances and rough terrain, allow the cougar to dispatch prey efficiently (Logan and Sweanor 2001; Murphy and Ruth 2010).

Cougars typically hunt in habitats that afford stalking cover and security while feeding, including brushy, forested, and rugged terrain (Logan and Irwin 1985; Laing and Lindzey 1991; Jalkotzy et al. 2000; McKinney et al. 2006). Cougars apparently need sufficient visibility to observe prey but also adequate cover for stalking (Laundré and Hernandez 2000). They employ two basic tactics while hunting. In some cases, they use a slow-walking gait to traverse complex terrain while using their senses to detect prey. This approach enables cougars to cover a lot of ground while hunting, but still maintain strength and energy for attacking and killing the prey. In other cases, they hide and wait at locations where they expect to encounter prey, ready to ambush them. These tactics are not mutually exclusive in the sense that the slow-walk and hide-and-wait approaches may be combined when cougars are fatigued or they detect prey (Williams et al. 2014). During hunting activities, more experienced cougars probably employ mental maps of their home ranges and memories of locations where they were successful in killing prey in the past. Upon detection of prey, a cougar may wait in ambush or stalk to within a few meters before attempting a quick rush and attack (Murphy and Ruth 2010). Attack distances of less than 20 m (~65 ft) seem to be the norm. Moreover, cougars precisely match the required pouncing force to the size of the prey (Williams et al. 2014). Long-distance pursuits appear to be rare and may be because of the cougar's lower stamina capacity. The cougar evolved a physique to deploy a tremendous surge of strength and quickness, not endurance. Comparatively, the cougar's maximum speed is about one-fifth that of the cheetah (Williams et al. 2014).

Because the cougar can take advantage of the most abundant and vulnerable prey in a wide range of habitats, it can be considered more of a generalist than specialized predator. However, where prey varies widely in size, cougars of a particular sex, age, size, or experience may specialize on certain prey (Elbroch and Quigley 2019), such as in Canada where adult male cougars focused on large prey (>200 kg or >440 lbs) and adult females on smaller prey (<100 kg or <220 lbs) (Knopff 2010). Within the same general prey size category, cougars may also select prey species over others, as in Washington where cougars preferentially hunted mule deer over white-tailed deer (Cooley et al. 2008). The size and type of prey cougars kill may also be dependent on the cougar's reproductive status, as well as the ease of finding and consuming the prey, the energetic value of the prey, and the risk of injury during pursuit and handling (Pierce et al. 2000; Murphy and Ruth 2010; Knopff 2010).

A question often asked by naturalists, hunters, biologists, and wildlife managers alike is, "How frequently will a cougar kill a deer?" Based on some of the more quantitative data obtained from GPS-collared cougars, an average of about one ungulate per week is a reliable estimate. In Wyoming, for example, cougar kill rates of mule deer and elk per week for adult males, adult females, subadult males, and subadult females were 0.91, 1.01, 0.74, and 0.97, respectively (Anderson and Lindzey 2003). In northern Arizona, independent cougars of both sexes killed mule deer and elk every 6.0–9.2 days (Mattson et al. 2007).

A cougar will often drag or carry its prey to vegetative cover to cache the animal some distance from the kill site, probably to reduce exposure to other animals and to provide security while eating (Beier et al. 1995; Mattson et al. 2007). Large prey that allow for multiple feedings may also be covered with available ground material around the cache, including soil, sticks, leaves, and snow, or hair plucked or sheared from the prey (Beier et al. 1995). Cougars achieve this by raking the material over the carcass with their forepaws. Sometimes they remove and cover the rumen and intestines separately. Caching serves to reduce loss from insects and microbes, and it prevents desiccation during dry, hot conditions, as well as detection by avian and mammalian scavengers (Bischoff-Mattson and Mattson 2009). Some scavengers may be dangerous competitors, such as other cougars, bears, and wolves, as they may usurp the carcass and also harm or kill the cougar involved in killing the prey (White and Boyd 1989; Ruth and Buotte 2007; Murphy and Ruth 2010).

Competition for prey between cougars and

other carnivores can also affect cougar population growth in a density-dependent manner. As the density of cougars and of other carnivores increases (but the availability of prey does not), competition for food also increases and reduces the availability of food per capita. This can lead to declining survival and reproduction rates in the cougar population. Conversely, as the density of cougars and of other carnivores decreases, so does competition, and the rates of these population parameters are expected to increase (Logan and Sweanor 2001; Logan 2019; Ruth et al. 2019; Elbroch et al. 2020a).

Interactions with other carnivores can be especially costly to cougar females and families. Because of their smaller size, females are less able to compete with larger carnivores they encounter, and as a result entire families can be lost. In the San Andres Mountains, we observed the aftermath of encounters with interloping males. In some instances, males killed lone females in competition for prey carcasses. In others, a male cougar killed all family members, or males killed the mothers and the cubs starved to death afterward. In one particular instance, we directly observed a lone coyote usurp a mule deer buck freshly killed by a subadult female cougar by barking incessantly at her. After a while, the cougar moved off the kill and laid down about 70 m (~230 ft) away. The coyote then moved in and started eating the deer, periodically pausing and lifting its head out of the carcass to look in the direction of the cougar.

The Cougar Diet

In New Mexico, as in most of North America, cougars rely on deer as their principal source of food (Anderson 1983; Logan and Sweanor 2001; Murphy and Ruth 2010). The mule deer is the most common and most widely distributed wild ungulate in the state (see NMDGF 1999), and it is indeed the most frequently recorded prey species in the diet of New Mexico's cougars (Table 9.1). Where locally abundant, white-tailed deer and elk are also important prey. To a lesser extent, cougars also rely on collared peccary (*Dicotyles tajacu*) and both Rocky Mountain (*Ovis canadensis canadensis*) and desert bighorn sheep (*O. canadensis mexicana*) where they are available (Logan and Sweanor 2001; Goldstein and Rominger 2006; Perry and Upton 2011; Williams 2011; Prude 2020). Pronghorn (*Antilocapra americana*) are vulnerable when they move into dense cover and on rugged terrain (Shaw 1977; Ockenfels 1994; Logan and Sweanor 2001; Prude 2020). On the San Andres Mountains, some pronghorn entering canyons to gain access to perennial spring water were ambushed by cougars. Porcupines (*Erethizon dorsatum*), skunks (family Mephitidae), raccoons (*Procyon lotor*), rabbits and hares (family Leporidae) may be frequently preyed upon as well, especially where they are abundant (Robinette et al. 1959; Spalding and Lesowski 1971; Ackerman et al. 1984; Maehr et al. 1990; Beier and Barrett 1993; Bender and Rosas-Rosas 2011; Perry and Upton 2011). In New Mexico, Prude (2020) documented 32 different prey species at cougar kill sites along the Rio Grande and adjacent Chihuahuan Desert uplands, including the ungulates previously mentioned, and beaver (*Castor canadensis*), common carp (*Cyprinus carpio*), and waterfowl. Although small prey animals may typically comprise less than 10% of the cougar's diet (e.g., Logan and Sweanor 2000), along the Rio Grande River they represented over half of the diet (Prude 2020). Carnivore species sometimes killed by cougars, but not always eaten, include coyote, gray fox (*Urocyon cinereoargenteus*), bobcat (*Lynx rufus*), and badger (*Taxidea taxus*), perhaps the result of interspecific interference competition, as they typically occur at cougar prey caches (Koehler and Hornocker 1991; Logan and Sweanor 2001; Murphy and Ruth 2010; Prude 2020). Cougars can also be cannibalistic or scavenge other

Table 9.1. Percent frequency of occurrence of prey identified from cougar scats and kills from 1934 to 2020, New Mexico.

Species	Western NM Scats[a]	South-western NM Kills[b]	South-eastern NM Scats[c]	San Andres Mts. Scats[d]	San Andres Mts. Kills[d]	San Andres Mts. Scats[e]	San Andres Mts. Scats[e]	Ladder Ranch, South NM Kills[f]	Bosque, South NM Kills[f]	Chihuahua Desert and Rio Grande, South NM Kills[g]
Mule deer	62.7	77.9	82	85.6	91.2	17	31	60.8	17.1	28
White-tailed deer	26.9	2.6								
Desert bighorn sheep				0.6	1.9	9	3			4
Elk		1.3	<1					30.7	15.9	2
Pronghorn				0.7	1.0	2				3
Collared peccary		2.6		0.1			3	3.3		3
Gemsbok (=oryx)				0.4	0.6	17	3			5
Cattle	1.0	9.1	2						1.2	1
Sheep			6							
Goat			1							<0.5
Lagomorphs	3.6		7	4.2	0.2	28	48		1.2	<1.5
Porcupine	4.2	1.3	15		1.0					<0.5
Beaver		1.3							14.6	
Muskrat										<0.5
Small rodents			3	5.5		13	3			
Cougar		1.3		2.0	2.0					
Bobcat										1
Badger				1.9	0.2				2.4	1
Skunk	0.5			0.7	0.6	11	7	0.6		2
Coyote				0.1	0.6			1.1	12.2	
Fox	1.0	2.6	<1					1.1	1.2	<3.5
Raccoon								0.6	9.8	7
Ringtail				0.1	0.2					<0.5
Other carnivore						12	7			<0.5
Wild turkey								1.1	3.7	1
Sandhill crane									4.9	
Waterfowl										1
Golden eagle					0.2					
Owl									1.2	
American crow										<0.5

(continued on next page)

Species (continued)	Western NM Scats[a]	South-western NM Kills[b]	South-eastern NM Scats[c]	San Andres Mts. Scats[d]	San Andres Mts. Kills[d]	San Andres Mts. Scats[e]	San Andres Mts. Scats[e]	Ladder Ranch, South NM Kills[f]	Bosque, South NM Kills[f]	Chihuahua Desert and Rio Grande, South NM Kills[g]
Unidentified birds				0.7						1
Box turtle				0.1						
Spiny softshell turtle										2
Common carp										7
Channel catfish										<0.5
Vegetation				20						
Total sample	**193**	**77**	**318**	**832**	**525**	**89**	**29**	**281**	**686**	

[a] Hibben 1937. Hibben assumed each scat represented a single cougar meal. Total sample excludes 3 scats that contained grass.
[b] Donaldson 1975; Johnson 1982.
[c] Smith et al. 1986.
[d] Elmer 1997; Logan and Sweanor 2001. Cougar kills included animals that were killed or probably killed and consumed by a cougar. Cougars also killed but did not eat 14 other cougars, 4 gray foxes, 1 coyote, and 1 long-eared owl.
[e] Bender and Rosas-Rosas 2011.
[f] Perry and Upton 2011.
[g] Prude 2020. Some species were combined into prey types (e.g., Lagamorphs include cottontails and jackrabbits).

cougars that have died from causes other than predation (Logan and Sweanor 2001; Murphy and Ruth 2010).

Domestic livestock usually make up far less than 10% of the cougar's diet in New Mexico (Table 9.1). Consequently, there is no evidence in the state that domestic prey "subsidize" cougar diets to the extent of artificially boosting cougar population growth, a notion promoted without rigorous evaluation (Rominger et al. 2004, 2005; Rominger 2018; see below). This phenomenon, however, may occur in parts of southern Arizona where cattle calf remains were found in up to 34% of cougar scats representing 44% of the biomass consumed by cougars (Cunningham et al. 1995, 1999). At the same time, it should be noted that adult females, not mature males, drive cougar population growth. Yet, mature males may be responsible for most of the cougar's predation on cattle where it occurs (Cunningham et al. 1999; McKinney et al. 2010).

Effects of Cougar Predation on Prey

Our research in the San Andres Mountains (Logan and Sweanor 2001) suggested that cougar predation on mule deer and desert bighorn sheep was partly compensatory and partly additive. Compensatory mortality by one cause substitutes for other causes of mortality and tends to have little influence on animal population abundance and growth. However, additive mortality adds to other sources of mortality and can contribute to lower population growth and even population reduction. If predators selectively target prey in poor condition, then the loss of those individuals might be largely compensatory to mortality from starvation, disease, or a higher likelihood of accident. Although in our study cougar predation was the most important proximate cause of mortality in mule deer and desert bighorn sheep, the effect of cougar predation on these ungulates was linked to variation in weather and its associated impact on habitat, in addition to disease in the case of the bighorn sheep (Boyce and Weisenberger 2005). An unusually wet weather pattern from 1985 to 1991 coincided with rapidly increasing mule deer and cougar populations. During those wet years, cougar predation seemed to slow the growth rate of the deer population, and as a result cougar predation was probably mostly additive. Ample deer in the environment likely reduced cougar predation on the rare bighorn sheep. Cougar predation on the bighorn sheep was sporadic and partially compensatory because scabies disease (caused by the *Psoroptes* mite) was widespread in the bighorn sheep population; a higher number of the bighorn sheep died of non-predation causes (16) than from cougars (10); and six of the 10 bighorn sheep killed by cougars had clinical scabies.

Conditions changed in the San Andres Mountains ecosystem beginning in 1992 when drought struck. Mule deer fawn production and survival declined, further slowing and even reversing deer population growth. Cougar predation increased on reproduction-aged deer and hastened the deer population decline. During the drought, cougar predation was probably mostly compensatory, meaning that it substituted for other causes of mortality and did not add to them. Consequently, the deer population would have declined anyway in a habitat with now less capacity to sustain its relatively large size. The decline in deer, the cougars' principal prey, apparently exposed the few remaining bighorn sheep to greater predation by cougars, as the cougars had to search their home ranges more assiduously to encounter prey. The increase in cougar predation on the bighorn sheep was the key factor that caused the bighorn population to continue to decline, leaving only one adult ewe on the San Andres Mountains by 1997 (Logan and Sweanor 2001). All this occurred while the cougar population was still at a relatively high density. This created a time lag before the cougar population would also decline in response to the declining prey base. Such lags in cougar population responses to a reduced prey base has been documented in studies in Utah and Idaho (Laundré et al. 2007), and California (Pierce et al. 2012).

Our study in the San Andres Mountains showed the importance of studying cougar predation impacts within a larger context involving population growth and density in both the cougar and main prey species, as well as other environmental factors. Other predators further complicate the impact of cougars on prey populations. While studying a mule deer population in a semi-mesic pinyon-juniper and ponderosa pine-fir forest in Colfax County in northeastern New Mexico, Sorensen (2015) found that a low-density deer herd increased at a rate of about 7% per year in 2012 and 2013. This occurred during drought conditions and while cougar predation accounted for 59% of radio-collared doe mortalities (n = 16/27). Predation by cougars seemingly represented partly compensatory and partly additive mortality in adult does, whose annual survival was nonetheless greater than 88%. In a study conducted simultaneously on the same deer population, Taylor (2013) reported low fawn survival, again, largely from predation. Predators of fawns, however, were mainly black bears, coyotes, and bobcats, in decreasing order of frequency. No predation of fawns by cougars was documented in this study. The two studies combined (Taylor 2013; Sorensen 2015) suggested that though cougars were the main predators of adult does, the main factor slowing but not preventing the growth of the local deer population was predation on fawns by other carnivores.

As was found in our study in the San Andres Mountains, increased predation on bighorn sheep can result from cougars switching to alternative prey when experiencing declines in their primary prey. In Arizona, cougars caused declines in translocated bighorn sheep populations after switching from mule deer to sheep four years after deer numbers declined (Kamler et al. 2002). In California, the impact that cougar predation had on a small population of bighorn sheep was linked to the degree of winter range overlap with mule deer, the main prey of cougars (Johnson et al. 2013). Rominger et al. (2004, 2005) found that cougar predation was a major limiting factor in desert bighorn sheep in the Ladrón Mountains and elsewhere in New Mexico. Citing in particular Cunningham et al.'s (1999) finding of cattle constituting an important portion of the cougar's diet in southeastern Arizona, Rominger et al. (2004, 2005) hypothesized that cougar populations in isolated mountain ranges of the Chihuahuan Desert occur at artificially enhanced densities because nearby cattle augment the cougars' food supply. So far, cougar diet studies conducted in New Mexico and cited previously (Table 9.1) do not support Rominger et al.'s (2004, 2005) hypothesis. Moreover, rigorous experimental investigations are needed to show the link between variation in diet composition involving cattle consumption and cougar population abundance. However, it now seems well established that cougar predation is a major proximate cause of mortality in New Mexico's desert bighorn sheep, listed as a state endangered species in 1980 (Rominger and Goldstein 2008). In 2001, fewer than 170 desert bighorn sheep remained in all of New Mexico. Three of seven desert bighorn sheep populations went extinct, including in the Alamo Hueco, Animas, and San Andres mountains. To recover the desert bighorn sheep populations, the New Mexico Department of Game and Fish implemented both augmentation of sheep numbers through translocation and cougar culling on four historical desert bighorn sheep ranges: the Peloncillo, Ladrón, Hatchet, and San Andres mountains. Between October 1999 and September 2007, 149 bighorn sheep were translocated from the Red Rock Wildlife Area in southwestern New Mexico, 44 more from the Kofa National Wildlife Refuge in southwestern Arizona. Concurrently, a total of 98 cougars were killed. Average cougar predation rates on radio-collared bighorn sheep declined from 0.17 to 0.04. The statewide desert bighorn sheep population increased to over 400 by 2007 from the combination of population augmentation and increased survival (Rominger and Goldstein 2008). This recovery resulted in the New Mexico Game Commission removing desert bighorn sheep from the state's list of threatened and endangered species in November 2011. Desert bighorn sheep numbers continued to increase, and by 2018, there were an estimated minimum of 1,100–1,300 desert bighorn sheep in New Mexico (NMDGF 2019).

Cougars and Domestic Animals

Livestock constitute a minor percentage of the cougar's diet in New Mexico (see Table 9.1). Diet studies in the western or southwestern part of the state and along the Rio Grande better typified areas where livestock are available as potential cougar prey in the state. Studies in the Guadalupe Mountains (Smith et al. 1986) and San Andres Mountains (Logan and Sweanor 2001) represented areas where livestock were largely excluded. The Guadalupe Mountains study area was comprised mainly of Carlsbad Caverns and Guadalupe Mountains national parks. On the San Andres Mountains, cattle occurred only along the western boundary of the study area on private and Bureau of Land Management and State-Trust lands, and trespass cattle infrequently ranged into the interior of the mountains.

Statewide in New Mexico, predators reportedly killed 1,290 head of cattle weighing over 500 pounds, and 8,210 calves according to records compiled in 2015 (USDA, APHIS 2017a; see also Chapter 3). Meanwhile, 23,710 and 36,790 cattle in those respective categories died of non-predator causes. Of cattle taken by predators, cougars killed approximately 10.7%, or 1,017, with the remainder killed by wolves, black bears, coyotes, and domestic dogs. Predators killed 2,152 sheep and 2,661 lambs in New Mexico in 2014 (USDA,

Photo 9.19. A scrape made by a cougar by pushing its hind feet backward in alternating motions. A cougar will usually also urinate or defecate on the scrape. This visual and olfactory mark is thought to signal to other cougars in the area the presence and reproductive condition of the individual that made the scrape, and may also attract prospective mates. The ruler in the scrape is ~15 cm (6 in) long. Photograph: © Ken Logan.

Photo 9.20. Typically, cougar mothers choose nursery sites with vertical and overhead cover, such as rock outcrops, boulders, and dense shrubs or woody downfall to hide from predators and provide shelter from inclement weather. In this photo from western Colorado, the nursery is located in the rock outcrop below the conifer tree on the upper left. Photograph: © Ken Logan.

Photo 9.21. A mother cougar's nursery consisting of a deep hole in the base of rim rock where she raised her two nurslings during the first month of their lives, 25 November 1990, San Andres Mountains, New Mexico. Photograph: © Ken Logan.

Photo 9.22. A cougar nursery under the base of a granite boulder with three three-week-old nurslings, San Andres Mountains, New Mexico, 13 September 1990. The observed peak in births occurred from July to September during our research from 1985 to 1995. Photograph: © Ken Logan.

Photo 9.23. One of two nurslings in a nursery under a rock ledge on 5 July 1994. Photograph: © Jenny Lisignoli.

Photo 9.24. A mother cougar in her nursery where she was raising nurslings, Mockingbird Mountains, New Mexico, May 1989. Photograph: © Ken Logan.

Photo 9.25. The average litter size is three cubs, with roughly an equal number of males and females born in a population. These cubs are about six weeks old. San Andres Mountains, New Mexico, 14 August 1993. Photograph: © Ken Logan.

Photo 9.26. Severe head wounds suffered by an adult male cougar from fighting with another cougar, Bighorn Mountains, Wyoming, 16 March 1983. Photograph: © Ken Logan.

APHIS 2015; see Chapter 3). Non-predator causes of death befell to 3,848 sheep and 4,339 lambs that same year. Cougars were reportedly responsible for 5.6 percent of all adult sheep and 4.5 percent of all lambs killed by predators, equating to 121 adult sheep and 120 lambs, respectively. The remainder of the predation was attributed to bears, bobcats, lynx (*Lynx canadensis*), coyotes, dogs, foxes, eagles, wolves, feral pigs, and other known or unknown predators including ravens (*Corvus* spp.) and vultures. Similar statistics on predation of goats and kids in New Mexico were not reported to "avoid potential disclosure of respondents" (USDA, APHIS 2017b).

Cougar Reproduction and Social Behavior

During the early years of cougar biology, it was thought that cougars did not cooperate with one another to forage, rear young, achieve matings, or defend against predators, and thus, exhibited a solitary lifestyle (Sandell 1989). The typical picture was that the largest social units were composed of a mother with dependent kittens, with social interactions also including the short-term bond of a breeding pair and the brief confrontations between males as they competed for access to estrus females. Raising young was the sole responsibility of the female. For the most part, all this is probably still accurate.

Our intensive research of cougars in the San Andres Mountains using the older technology of the time—very high frequency radio collars requiring simultaneous ground and aerial locations of cougars—started to reveal that cougars associated more than was previously thought. We logged 643 associations involving pairs of independent cougars and observed 269 of those and tried to discern their meanings (Logan and Sweanor 2001). A majority of the associations were between adult females and adult males. The next most frequent associations occurred between adult females with kittens and adult males. The third most frequent type was between adult males. We surmised that the vast majority of the associations pertained to reproduction—courtship and breeding, strengthening pair-bonds between mates, and competition between males for access to mates. The latter included fights sometimes resulting in the death of one of the combatants. In some cases, cougars also fought over a deer carcass, which in some cases resulted in the death of the smaller cougar. The total number of associations we logged comprised only about 5 percent of the total number of locations we acquired on independent cougars, thus seemingly confirming the general view that the species is solitary.

New technology of GPS-collars, digital wildlife cameras, and molecular genetics have since provided new insights into cougar behavior. From our own research we have learned that cougars may socialize even more frequently and for other reasons than the old technology previously revealed. Sometimes cougars socialize with other cougars of the same or opposite sex at food caches of large prey such as elk. In such cases, sharing a large kill may be less costly than fighting over it and risking injury or death. Some of these interactions involve mothers with their kittens or their grown offspring, and even males of unknown relatedness. As many as eight or more cougars may associate at one carcass. Sometimes genetically related females with overlapping home ranges congregate along with their offspring for extended periods, almost like a pride, except for the absence of males (K. Logan, unpubl. data). We have observed sires associating with mothers and kittens, and sires with kittens but without the mothers present for more than a day. Some researchers focus much attention to these observations, but such associations are nonetheless rare. The proximate and ultimate motivators for particular associations are likely complex, requiring more research into the competitive and cooperative tradeoffs for the involved individuals (Elbroch et al. 2017; Elbroch and Quigley 2017; Logan 2019). Pending the results of such research, we should perhaps characterize cougars as *conditionally social* and not strictly solitary. That is, cougars are social depending upon conditions that may affect their immediate individual survival and long-term reproductive success.

Cougars exhibit a polygamous and promiscuous mating system (Seidensticker et al. 1973; Anderson 1983; Logan and Sweanor 2000, 2001). This means that a male may breed with multiple females that live within his territory. Likewise, a female may breed with more than one male during single or subsequent estrus cycles (Seidensticker et al. 1973; Anderson 1983; Logan and Sweanor 2000, 2001). In non-hunted cougar populations, more adult males usually exist than are needed to breed with the available females that come into estrus, resulting in intense male-on-male competition for access to mates (Logan and Sweanor 2001). In the San Andres Mountains, cougars became sexually mature at about two years of age, though we documented conception in females as young as 18 months. Females in estrus advertised their breeding condition by laying down scent usually at cougar scrape sites, vocalizing their caterwaul-like calls, and seeking out males. Meanwhile, adult males, which were physically prepared to breed throughout the year, continually patrolled their territories looking for breeding opportunities, similarly laying down scent at scrape sites, vocalizing, and competing with other males. We found that breeding pairs stayed together for 1–10 days; in comparison, reported estrus in captive cougars ranged from 3 to 12 days (Anderson 1983; Logan and Sweanor 2001). The females' estrus cycles lasted on average 24.5 days (range = 14–35 days), while the interval between births of litters averaged 17.4 months (range =12.6–22.1 months; Logan and Sweanor 2001).

Because female cougars continue to cycle until becoming pregnant, litters can be born during any month of the year. On the San Andres Mountains, litters were born during every month of the year except February, with a peak in births from July to September. The peak coincided with the monsoon season and birth period for mule deer fawns. Other studies have shown cougar births peak in summer (Nevada, Ashman et al. 1983; Alberta, Ross and Jalkotzy 1992; South Dakota, Jansen and Jenks 2012; Yellowstone National Park, Wyoming, Ruth et al. 2011; Wyoming, Elbroch et al. 2015a; Colorado, Logan and Runge 2021), summer-fall (British Columbia, Spreadbury 1989; Utah, Lindzey et al. 1994; Utah and Idaho, Laundré and Hernández 2007), and spring-summer (Florida, Maehr 1997). The timing of births in North America seems influenced by weather interacting with variations in prey abundance and distribution, all of which likely affect kitten survival. Kittens born from spring to fall are expected to have higher survival because they are not exposed to freezing winter temperatures and there is greater abundance and diversity of prey (Logan and Sweanor 2001; Laundré and Hernández 2007; Jansen and Jenks 2012).

Some of the most quantitative data on cougar litters in the western United States come from our research on the San Andres Mountains. We observed an average of 3.0 kittens (range = two to four kittens) per litter, and 53 nursling litters aged 9 to 49 days old still consisted most frequently of three kittens. The size of litters with older, weaned cubs 52 to 427 days old was lower, averaging 2.2 kittens and likely reflected post-weaning mortality. The sex ratio for 148 nursling kittens in 50 litters was essentially 1:1 (75 males, 73 females; Logan and Sweanor 2001).

Cougar young typically remain dependent on their mother for one to two years. Forty-three young born in the San Andres Mountains averaged 13.7 months old (range = 11.1–16.0 months) at the time that they became independent. We observed littermates sometimes remaining together for a short period after independence, a behavior also documented in Alberta (Ross and Jalkotzy 1992). Although some female cougars born on the San Andres dispersed post-independence, most of them later established a home range adjacent to or overlapping their natal range (the area where they were born), a behavior referred to as philopatry. In contrast, all surviving male progeny dispersed, traveling, on average, 8.1 times farther than females away from their natal ranges. Male dispersal distances ranged from 47 to 215 km (29 to 134 mi), whereas female dispersal distances ranged from 1 to 118 km (0.6 to 73 mi). Dispersal occurred prior to the age of puberty and at an average age of 15.2 months (Logan and Sweanor 2001).

Dispersal likely confers one chief benefit for the

dispersing individual, that of avoiding competition for limiting resources (see below). It also may lead to reduced competition in the natal range, the ability for the species to reoccupy vacant habitat, augmentation of existing population segments from immigration, enhanced gene flow, and the avoidance of deleterious inbreeding effects. In our study, dispersing cougars tended to follow the north-south axis of the San Andres Mountains. However, some also left the mountain range and traversed the wide, adjacent desert basins relatively quickly, using small habitat patches (i.e., rugged terrain with increased vegetation and available prey) as stepping stones where they lingered a variable number of days before continuing (Sweanor et al. 2000). Cougars born on the San Andres Mountains dispersed up to 215 km (134 mi) from their natal ranges, reaching as far as the Guadalupe Mountains to the southeast, the Gallo Mountains to the northwest, the Black Range to the west, and beyond Gallinas Peak to the northeast. The ability for males to move long distances was further illustrated by the return of two adult male cougars to their territories in the San Andres Mountains after we translocated them 465 to 490 km (279 to 294 mi) in northern New Mexico (Ruth et al. 1998). Extreme movements of over 1,000 km (>660 miles) have been documented in the case of two males leaving their native population in South Dakota (Thompson and Jenks 2005; The Wildlife Professional 2011).

Dispersal in subadult cougars may occur for different reasons in males and females. For males, dispersal seems almost obligatory and independent of the density of adult males already in residence (Seidensticker et al. 1973; Logan and Sweanor 2001), and it may primarily be to avoid combat with resident adult males, which are usually larger, physically mature, and more experienced, and that probably view any younger and other interloping males as competitors for mates. Unlike in males, competition for food is probably a motivator for female dispersal,

with hunger likely a proximate factor. Direct and exploitative competition among cougars and between cougars and other carnivores occurs where and when the same prey are being sought (Logan and Sweanor 2001; Laundré and Hernández 2003; Logan and Sweanor 2010; Ruth et al. 2019; Elbroch et al. 2020a).

Adult female cougar life history strategies gear toward survival and successfully rearing offspring. Females do not typically exhibit territorial behavior such as defending an area to the exclusion of other cougars (Logan and Sweanor 2001, 2010). Earlier research on cougar behavior seemingly established that females were territorial, a notion that persisted for over 40 years among some biologists (e.g., Hornocker 1969, 1970; Beausoleil et al. 2013; Elbroch et al. 2015c). We found no evidence of adult female cougars fighting one another during our study in the San Andres Mountains. Nor, to our knowledge, has this been reported in other cougar populations in the science literature. Instead, it may be more accurate to state that adult female cougars exhibit strong fidelity to home ranges, where they can grow familiar with compatible mates, other females using the same area, and food and shelter for their survival and the successful raising of young.

Adult female home ranges are roughly about one-third the size of adult male ranges. Sizes of female home ranges are variable and change over time based on surrounding vegetation and terrain; prey abundance, distribution, and movements; interactions with other cougars; and the female's reproductive stage and associated energy needs (Logan and Sweanor 2010). Females with kittens tend to use larger areas, especially as their kittens grow larger and require more food. Annual home range size for adult females in the San Andres Mountains averaged 70 km^2 (27 mi^2), but with wide variation (range = 13 km^2 to 287 km^2 or 5 mi^2 to 111 mi^2). A study in the Guadalupe Mountains in New Mexico and Texas reported an average home range size for adult females of 39 km² (15 mi²; Smith et al. 1986), within the range of what we observed in the San Andres Mountains.

An adult female will usually have a home range that overlaps the home ranges of several other adult females, including relatives and non-relatives (Logan and Sweanor 2001, 2010), another indication of non-territorial behavior in females. In non-hunted populations in particular, overlapping female home ranges can belong to closely related individuals such as mothers, daughters, sisters, and aunts, making up groups referred to by us as *matrilines* (Logan and Sweanor 2001). Matrilines are expected to be disrupted, reduced, and possibly eliminated in hunted populations, depending upon the extent of hunting pressure. In the San Andres Mountains, the area shared between adult female home ranges averaged approximately 40 to 60%, though there was less overlap between individual core use areas (10–30%). Females generally practiced mutual avoidance with other cougars by utilizing shared areas at different times, probably to reduce direct competition with other females and males, and to avoid aggressive males that could threaten their kittens.

As with other cougar populations, each adult female in the San Andres Mountains shared her home range with one or more adult males, and one or more of these males sired her offspring. Adult females visited cougar scrape sites and sometimes scraped themselves. Sometimes they also left their scent by urinating and defecating, and we observed some of them vocalizing at scrape sites. These behaviors that advertised presence, and perhaps breeding condition, were presumably efforts to assess and attract prospective mates (Logan and Sweanor 2001; Allen et al. 2014).

Adult male cougar life history strategies similarly gear toward survival and reproductive success. To obtain food and maximize mating opportunities, the male maintains fidelity to an area that overlaps a number of female home ranges.

He exhibits territorial behavior through fighting or intimidation and indirectly by patrolling and advertising his presence to other males. The male's territory is often two to three times larger than the home range of a female, thereby enabling him to encounter multiple adult females. On the San Andres Mountains, territories for adult males averaged 193 km² (74 mi²), and ranged from 59 km² to 640 km² (23 mi² to 247 mi²). Territories of male cougars in the Guadalupe Mountains were similar in size to San Andres Mountains males, averaging 228 km² (88 mi², Smith et al. 1986). Unlike adult females, adult males with overlapping territories are likely not genetically closely related, tend to avoid each other, and normally do not share kills of prey. Males patrol their territories and advertise their presence by leaving visual and olfactory markers including scrapes that contain their urine or feces while searching for estrus females and interloping males. Consequently, territories are dynamic and change spatially in response to the distribution and reproductive status of females, relative dominance of other males, and variations in the abundance and spatial distribution of prey. Motivations and behaviors of territorial males, along with those of adult females, likely contribute to the structuring of local social interactions that Elbroch et al. (2017:4–5) termed "social communities" (Logan 2019).

We hypothesize that a male's territorial behavior maximizes his reproductive success by procuring mates and discouraging the activity of competing males, which could otherwise threaten his mates and offspring. If territoriality increases an adult male's reproductive success through *sexual selection*, it will become the norm in the population (Logan and Sweanor 2010). In this case, territorial behavior in the adult male cougar is a result of successful breeding and higher survival of resulting offspring while in competition with other males for breeding success.

Intraspecies killing has been documented in North American cougars (Logan and Sweanor

2001, 2010). During our study, males were responsible for all the intraspecific killing we observed in the San Andres Mountains. Males killed both adult males and females, killed and cannibalized kittens, and killed subadult males and females. Motivations for these killings could not be determined in every case. However, based on both observed behaviors of the cougars involved and field evidence, we surmised that males usually killed other males during fights over access to mates. When males committed infanticide and killed all the kittens in a litter, this act reset the reproductive cycle of the females and the males could then breed with them. Consequently, females would raise the offspring of that male and not another one. Infanticide thus appears to be an indirect strategy used by males to compete for mates and increase their own lifetime reproductive success, and is referred to as *sexually selected infanticide* (Hrdy 1979). Males also sometimes killed females, which were trying to defend their kittens. We suspected that most intraspecific killing of kittens and adult females was by males who were not previous mates of the females and thus were not sires of the kittens. In other instances, males killed females and males while directly competing for cached prey.

Our cougar research on the San Andres Mountains revealed that male cougars do not always attempt to harm females and kittens they encounter (Logan and Sweanor 2001:265–66). Several observed interactions involving adult males and females with kittens nine to 12 months old resulted in copulations between the adult cougars and, later, in pregnancies. Clearly, the females were in estrus at the time of those interactions, and because the older kittens were also radio-collared, we knew they survived. We documented several other, apparently amicable associations between pregnant females and resident males, some known to be the sires of the kittens. On other occasions, mothers with nursing kittens associated with known sires of the kittens

or other males for up to four days, and without any of the kittens harmed. When we were able to follow the tracks of the adult pair, the cougars traveled together away from the nurseries, sometimes laying and rolling on the ground, probably contacting one another, possibly playing. Some tracks indicated the pair might have copulated. We speculated that these associations reinforced individual recognition and pair-bonds among mates, perhaps also revealing qualities of learning, memory, and mate choice. Furthermore, associations involving non-sires may confuse paternity and reduce chances of infanticide by the involved males. Our observations of tolerant social associations among cougars were not the first or the last ones recorded. Similar associations were observed in cougars in Idaho in the early 1970s (Seidensticker et al. 1973), and much later than ours in Colorado (K. Logan, unpublished data) and Wyoming (Elbroch et al. 2014; Elbroch and Quigley 2017).

Cougar Population Dynamics

During our San Andres Mountains cougar study, we gathered data on reproduction, mortality, dispersal, social behavior, and relationships with prey, main elements that define the short- and long-term changes in cougar population growth and structure (Logan and Sweanor 2001). We classified cougars as adults, subadults, and kittens (or cubs). Adults were the breeding-age individuals, usually two years old and older. Subadults were independent of their mothers, but not yet breeding, usually one to two years old. Subadults also comprised the dispersing segment of the population. Kittens were offspring still dependent upon the care of their mothers. The San Andres Mountains cougar population was comprised roughly of one-quarter adult males, one-third adult females, one-tenth subadults (males and females combined), and one-third kittens (Logan and Sweanor 2001:77).

On the San Andres Mountains, we estimated

the growth of the adult portion of the cougar population in a reference area without cougar removal and in a treatment area where we experimentally reduced the population. The observed average annual growth rate for the adult population in the reference area was 0.11 (i.e., 11% growth per year) for an entire seven-year period. However, the average annual growth rate declined from 0.17 (years 1–4) to 0.05 (years 4–7), with the lowest rate occurring during the period of highest population density. In the treatment area we observed average annual growth rates of 0.21 and 0.28 for two successive four-year periods when cougar population density was substantially reduced, first by a previous culling action before our research, and again, after we experimentally removed up to 53% of adults, 58% of independent cougars. It is important to emphasize that the highest rates of population growth occurred after substantial reduction in cougar population density was followed by complete protection of the cougar population from human-caused mortality. Our research was the first to reveal that cougar population growth is probably density dependent; that is, as density increases, the rate of population growth declines (Logan and Sweanor 2001).

Findings from our research enabled us to develop a hypothesis explaining cougar population regulation (Logan and Sweanor 2001, 2010; Logan 2019). As mentioned previously, adult female cougars are not territorial like their male counterparts. We expect growth of the adult female component of a population to be regulated by competition for food. That is, as the density of adult females and other competing carnivores increase in relation to the food supply, in this case available prey, competition for food also increases, and this slows the rate of growth in the female component. We expect increasing competition is also a factor motivating subadult females to disperse from natal areas at higher rates. In the adult male component of the cougar population,

however, territorial behavior is the result of competition for access to mates. Male-male competition for mates is hypothesized to be the factor regulating the rate of growth of the male component of the population (Logan and Sweanor 2001, 2010; Logan 2019). As the density of adult males increases and competition between them also increases, the growth rate of the male component of the population declines. Because the rate of increase in the adult male component is linked to the density of adult females, we expect that the natural ultimate limiting factor to the abundance of cougars in a population is food, the amount of available prey.

However, in cougar population segments that are hunted annually, population growth rates are expected to decline as a function of the amount of hunting pressure, with the attendant mortality becoming additive rather than compensatory beyond a certain threshold (Anderson and Lindzey 2005; Cooley et al. 2009; Robinson and DeSimone 2011; Logan and Runge 2021). Researchers in Colorado learned that when 22% of all independent cougars (i.e., adults and subadults) were harvested on average annually, the result was a 35% decline in cougar abundance after four years (Logan and Runge 2021). In this case, neither reduction in other causes of mortality nor increased rates of reproduction and recruitment compensated for hunting mortality. Other cougar population segments in Utah and Wyoming declined when average hunting mortality rates among independent cougars were 10% and 22%, respectively (Wolfe et al. 2016; Anderson and Lindzey 2005, respectively). Such cougar population segments are no longer limited by natural processes but by hunting mortality.

Source-sink Metapopulation Dynamics

Our research in southern New Mexico revealed that the regional landscape consisting of mountain ranges separated by desert basins naturally fragments the cougar population into smaller population segments. However, each population segment was linked to other segments through the movements of young, dispersing cougars, which crossed broad desert basins and reached a different mountain range. We learned just how important dispersal was to help maintain individual segments and for the resiliency of the cougar population as a whole. Approximately three-quarters of the males and about one-third of the females recruited into the San Andres Mountains population segment were immigrants from elsewhere. On the other hand, we found that about three-quarters of male and about one-half of female surviving kittens born on the San Andres Mountains emigrated and were potential immigrants to areas occupied by other cougar population segments in New Mexico (Sweanor et al. 2000). The movements of immigrants and emigrants were just as important as the direct recruitment of kittens born locally for the overall maintenance of a population segment. Moreover, both the rate of emigration and population trends within any particular cougar population segment were dependent upon local changes in habitat productivity and the extent of human-caused mortality. Some population segments would be expected to act as *sources* because local cougar numbers are stable or increasing and they produce a net supply of dispersers that emigrate to other segments. Conversely, other population segments would represent *sinks* because survival, reproduction, immigration, and recruitment are insufficient for cougar numbers to remain stable or grow.

We concluded that cougar population segments in New Mexico, and probably the entire Southwest, exhibited a *source-sink metapopulation structure* (Sweanor et al. 2000; Logan and Sweanor 2001). Consequently, we recommended that any management actions targeting a particular population segment take into consideration potential impacts to other segments and the overall resiliency of the state population. More recently, a study of cougar population genetics in Nevada revealed five genetically distinct population segments in that state. These population segments probably formed from a combination of factors including the physiographic structure of the region, with mountain ranges separated by desert basins; distances between population segments occupying the mountain ranges; and asymmetrical dispersal movements of cougars between segments apparently caused by different cougar harvest rates (Andreasen et al. 2012). Results from that study further supported the *source-sink* structure that we hypothesized applies to cougars in the Southwest.

Interspecific Interactions

Interactions between cougars and people are the factor most important to the survival and persistence of cougar populations today. Cougar interactions with other carnivores nonetheless exist and consist of competition and predation (Ruth and Murphy 2010). In New Mexico, cougars likely compete for prey with black bears, Mexican wolves, coyotes, foxes (*Urocyon cinereoargenteus*, *Vulpes fulva*), and bobcats if they consume the same foods. In the Rio Grande valley, badgers, bobcats, gray foxes, and coyotes were all documented as prey, some of them killed at cougar caches (Perry and Upton 2011; Prude 2020). In the San Andres Mountains, cougars killed coyotes and gray foxes, but did not always eat them (Logan and Sweanor 2001). Direct interactions resulting in the death of a fox or a coyote, but not its consumption, likely represented cases of interference competition, especially as we documented sign of these species at many cougar prey caches. However, competition for the same food resources was not the only possible motivator. Cougars sometimes may kill other carnivores to protect their offspring. Coyotes are capable of killing young cougars, while foxes and bobcats may be able to take nursling kittens, especially when the mother is away hunting

(Logan and Sweanor 2001). During our study in the San Andres Mountains, a cougar killed a long-eared owl (*Asio otus*) but did not eat it. One golden eagle (*Aquila chrysaetos*) was consumed after apparently being killed trying to scavenge at a deer killed by a cougar (Logan and Sweanor 2001). Cougars may be larger than many other carnivores, but they do not always win encounter contests where competition involves smaller species. As indicated earlier, we directly observed a coyote usurping a mule deer buck killed by a sub-adult female cougar. In most cases, coyotes and bobcats nonetheless likely risk injury or death whenever they attempt to scavenge cougar kills. Other studies also have documented cougars killing bobcats at cache sites (e.g., Koehler and Hornocker 1991; Beier and Barrett 1993). In Arizona, coyotes even comprise a substantial amount of the cougar diet (Mattson et al. 2007).

Human activities can affect the likelihood of encounters between cougars and some of the other carnivores. In Colorado, bobcats avoided wild, natural areas for up to three days where cougars had just visited (Lewis et al. 2015). In areas of residential development, however, no such avoidance of cougars occurred. The authors speculated that this difference might be because bobcats and cougars altered their behavior to avoid human disturbance instead by funneling their movements through smaller patches of habitat, and this increased the potential for more frequent interactions between the felids (Lewis et al. 2015).

Conversely, adult black bears are larger than cougars (weights of female bears in New Mexico range from 55 to 70 kg [120 to 155 lbs]; males are larger, with a minimum reported weight of 90 kg [200 lbs]; Chapter 16), and they may dominate cougars at cache sites. Some bears may rely on cougar-killed prey during the lean times in early spring and prior to denning in the fall. In California, black bears usurping ungulate carcasses from cougars caused the cougars to compensate for the loss of food by increasing their kill rates of ungulates (Elbroch et al. 2015d). In Yellowstone and Glacier national parks, black and grizzly (*Ursus arctos horribilis*) bears visited cougar kills and often displaced the cougars (Murphy et al. 1998).

Relationships between cougars and Mexican wolves have yet to be investigated. Farther north, however, the cougar appears to be a subordinate competitor where gray wolves (*Canis lupus*) were restored. In the Middle Rocky Mountains, for example, cougars tend to select home ranges farther from gray wolf packs (Lendrum et al. 2014). Wolves sometimes usurp cougar kills and kill cougars (Kunkel et al. 1999; Kortello et al. 2007; Elbroch et al. 2015b; Ruth et al. 2019). The incidence of cougar starvation also rises as wolves increase in numbers, probably the result of greater direct and exploitation competition for food sources (Ruth 2004; Kortello et al. 2007; Elbroch et al. 2020a). In the Greater Yellowstone Ecosystem, competition among cougars and with wolves and bears appears to be a major factor in regulating cougar population growth (Ruth et al. 2019; Elbroch et al. 2020a).

STATUS AND MANAGEMENT

Cougars in New Mexico first received legal recognition in 1867 when the Territorial Legislature passed a predatory bounty law providing for the payment of $5.00 for each dead cougar (NMDGF Operational Plan, 1987–1995). The bounty was suspended in 1923 (Nowak 1976), at the time when the federal government's Bureau of Biological Survey became active in predator control in New Mexico (Brown 2013). The general view of the late 19th- and early 20th-century settlers to New Mexico was to eradicate the cougar through predator control activity of bounty hunters, trappers, ranchers, and federal and state agents with an objective to make the ranges safe for livestock and game animals (Ligon 1927; Bailey 1931). In addition, intense recreational hunting of cougars

Photo 9.27. Cougar warning sign in the Zuni Mountains in May 2013. Although still extremely rare, attacks on humans have increased since the 1950s in North America. Photograph: © Jean-Luc Cartron.

was gaining popularity at the beginning of the 20th century and considered to have "some effect on their control" (Ligon 1927:52). A sign of the early 20th-century strong utilitarian ideology, renowned US Biological Survey Agent J. Stokley Ligon (1927:52) wrote, "Depredations by mountain lions are so serious to both domesticated animals and game, particularly deer, that their complete control, which is only a matter of continued systematic hunting, is now essential." The cougar was excluded from Ligon's (1927:66) list of economically important mammals, and he did not consider it to be a big-game animal. Ligon (1927) viewed cougars as an obstacle to wildlife conservation.

The resolute and unregulated killing of cougars in New Mexico apparently took a toll. In a May 1917 report, Ligon speculated there were 400 cougars remaining alive in New Mexico (Bailey 1931:287). By 1927, Ligon thought that although cougars were "still widely distributed throughout the state, they were nowhere common and commit far less damage to game than was formerly the case" (Ligon 1927:52). In 1931, Bailey (1931:287) then wrote that cougars had been "common over practically all of New Mexico, but are rapidly decreasing in numbers." More precisely he stated that their distribution was at the time "probably most common in the Mogollon Mountain region, and in the Animas, San Luis, and Sacramento mountain ranges, with a few scattered through some of the small desert ranges over the southern and western parts of the State." An impetus for the first biological investigation of cougars by Frank C. Hibben (1937:6) in New Mexico "was the growing alarm, either real or imagined, over the scarcity of lions. It was felt that the lion was in danger of extinction in several areas of his former range. The number of lions which the modern hunter encountered was alarmingly small compared to that reported two or three decades ago." Later, in his 1946 description of the cougar's distribution in New Mexico, Young (1946:28) stated, "The puma is widely distributed throughout the State but of late years, due to intensified hunting, is not as common as it was at the beginning of the present century. The animals may now be said to be confined mainly to the rougher mountainous sections west of the Rio Grande." Even later, Berghofer (1967) offered that there could be as few as 350 cougars in New Mexico, thus agreeing with previous assessments of lower numbers for the species in the state. The human-caused decline of cougars was not limited to New Mexico. By the late 19th century, the species had been extirpated from the eastern half of North America, except for a remnant population in Florida. By the 1920s the cougar was restricted to the mostly inaccessible western mountains (Nowak 1976). In the early 1960s, the wildlife management agencies of California, Colorado, Idaho, New Mexico, Oregon, Utah, and Washington issued reports that suggested cougars now numbered in the hundreds in each of those western states (Cahalane 1964). Furthermore, cougars were thought to be

"nearly extirpated in Wyoming" (Long 1965:705). Although none of these states performed rigorous surveys of cougar abundance, they all consistently contributed to the general notion of low cougar numbers in much of the West.

In the 1960s, the status of the cougar in the western United States began shifting from that of "vermin" in need of killing by any means anytime, to that of game animal with regulations on the killing of cougars by people. This change coincided with the beginning of the American environmental revolution (Gill 2010), and more specifically with both public concern that cougars were disappearing from some areas and a greater appreciation for recreational and aesthetic values for all large carnivores (Nowak 1976). Colorado and Nevada were the first states to regulate the killing of cougars in 1965. In New Mexico, the legal status of the cougar changed in 1971 when the species was placed on the state list of protected wildlife; management authority was then given to the New Mexico Department of Game and Fish to establish hunting seasons, bag limits, and resolve depredation on livestock (Evans 1983).

Cougar hunting seasons in New Mexico have varied substantially since the cougar was designated a protected species in 1971. Except for the most recent changes (see below), the trend over time has been to move from conservative management (i.e., greater protection) to more relaxed hunting harvest and to population control of cougars in some jurisdictions (NMDGF hunting proclamations, 1971–1995, 2004–2015). In 1971, three-quarters of New Mexico was closed to cougar hunting. The southwestern quarter was open for four months with a bag limit of one cougar per hunter, and thereafter, females followed by kittens and kittens less than one year old were protected. Dogs could legally be used to hunt cougars. In subsequent years, more areas of New Mexico were progressively opened to cougar hunting, and the length of the season was extended to 11 months. From 1979 to 1983, almost all of New Mexico was opened to an 11-month hunting season with an increased bag limit of two cougars per hunter. After 1979, hides of all cougars killed by hunters had to be inspected and tagged by the New Mexico Department of Game and Fish. Harvest data from 1981 to 1994 indicated that 70 to 150 cougars were killed per year (NMDGF 1995).

Beginning in the early 1980s in New Mexico, the legal protection status of the cougar was frequently contested mainly by members of the agricultural and hunting communities; their challenges sometimes reached the state legislature for consideration. A 1983 challenge resulted in a thorough investigation of information on cougar management in New Mexico. Members of the agriculture industry, concerned about predation on livestock, attempted to return the cougar to its former status as "varmint" by introducing a bill to the New Mexico House of Representatives. The bill was tabled in committee, but the legislature requested more information from the State Game Commission and the New Mexico Department of Game and Fish. The New Mexico Department of Game and Fish responded with the first in-depth report on cougars in the state titled, *The cougar in New Mexico: biology, status, depredation of livestock, and management recommendations*. In the report, Evans (1983) concluded that cougar numbers probably had declined during the previous 11 years (1972–1983) based on three separate statistics: 1) The number of cougars killed per year by hunters had doubled; 2) The number of hunters had quadrupled; and 3) The hunter success ratios had declined from 0.65 to 0.19. His recommendations, supported by influential public sentiment, resulted in more conservation-oriented cougar hunting regulations in most of the state, together with targeted efforts to control cougar abundance in specified areas.

In 1984, the cougar hunting season was three months long throughout almost all of New Mexico. However, there were five hunt units (two in the southwest, three in the southeast) with harvest quotas of 10 to 17 cougars and where the season was extended two additional months. The objective in those units was to obtain a higher cougar kill where depredation on livestock was problematic. From 1985 to 1998, cougar hunting regulations were uniform. Almost all areas of New Mexico were open to cougar hunting for four months, from 1 December to 31 March, with a bag limit of one cougar per hunter.

Cougar management changed again in 1999 when the New Mexico Department of Game and Fish adopted a zone management structure. Cougar hunting was open for six months from 1 October to 31 March, and a quota limited the number of cougars that could be killed by hunters in each zone. There were 18 management zones in total, all of which were hunted (Map 9.4). This structure included the cougar hunting and cougar population control zones, but excluded refuge zones (i.e., closed to hunting) as originally designed for New Mexico (NMDGF 1997; Logan and Sweanor 2001:385–88). Harvest objectives in each zone were set to result in either the increase, decline, or stability of the corresponding cougar population segment, as determined by public sentiment and the State Game Commission (Beausoleil 2000). Moreover, on private land, cougars could be killed year-round.

In 2010, the New Mexico Department of Game and Fish completely rethought its cougar hunting management (NMDGF, data on file). The agency argued that New Mexico was home to 40 to 50% more independent cougars (i.e., ≥18 months old, subadults and adults) than previously assumed based on a new habitat map and a set of untested values representing cougar densities across habitat categories (NMDGF, data on file; see Map 9.4). The resulting population projections were used to justify substantial increases in harvest quotas. In April 2010, before the new cougar kill quotas took effect, the projected sustainable total

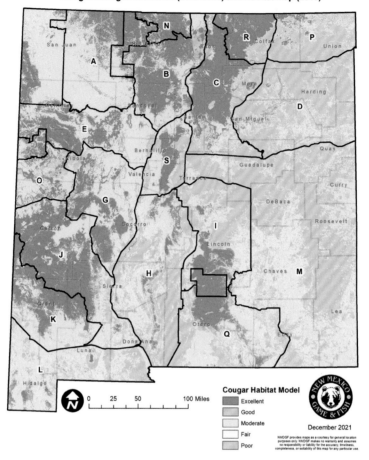

New Mexico Department of Game and Fish

Cougar Management Zones (2019-2023) and Habitat Map (2010)

Cougar Habitat Model
- Excellent
- Good
- Moderate
- Fair
- Poor

December 2021

NMDGF provides maps as a courtesy for general location purposes only. NMDGF makes no warranty and assumes no responsibility or liability for the accuracy, timeliness, completeness, or suitability of this map for any particular use.

Map 9.4. The New Mexico Department of Game and Fish's cougar management zones, overlaid on top of the 2010 habitat model. Note the five categories of cougar habitat.

annual mortality was 487 cougars and the female sublimit was 140. In harvest management, males are targeted to protect females, the reproductive component in polygynous or promiscuous species, because only a subset of males inseminate females. However, six months later, in October 2010, the agency implemented its new quotas and adopted a cougar mortality limit of 745 with a female sublimit of 299. Female sublimits were implemented presumably to more stringently regulate the killing of the female component of the population. The newly assigned cougar

densities, however, might have rendered quotas practically non-effective as a way to limit the cougar kill; that is to say, the quotas might never be reached if they were unrealistically inflated. Consequently, the factors that might limit the annual number of cougars killed were the regulations on the length of the hunting season and bag limits on public land, the actual abundance and vulnerability of the cougars, and the numbers and efforts of hunters. On the other hand, the New Mexico Department of Game and Fish instituted a mandatory Cougar Education and

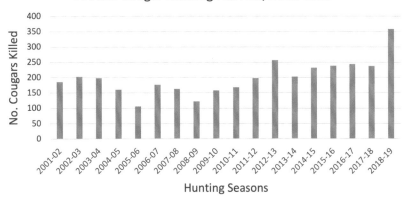

Figure 9.1. Numbers of cougars killed by hunters each year in New Mexico from the 2001–02 hunting season to the 2018–19 hunting season, New Mexico Department of Game and Fish data on file, Santa Fe, New Mexico.

Identification Course for all hunters that purchased a license to hunt cougars from 2010 forward. The purpose of the program was to inform hunters about the natural history of cougars and to teach them how to determine the sex and reproductive stage of female cougars they catch while hunting.

In April 2012 the New Mexico Department of Game and Fish further relaxed its rules with year-long hunting on public lands, and annual bag limits increased to two cougars per hunter. Only cougars killed by hunters would be counted toward harvest limits; cougars killed by vehicles or by game wardens and control agents for management purposes would be excluded (NMDGF, data on file). New Mexico's legislators again in 2015 proposed a bill that would strip the cougar of its protected status and return it to varmint status where its killing would be unregulated. The bill was tabled in committee after strong public opposition. Later that same year, the New Mexico Department of Game and Fish adopted even more relaxed regulations with a bag limit of up to four cougars on all hunt zones where quotas had not been met during the two of three prior years, while also now allowing trapping and snaring of

cougars on private lands and state-trust lands (NMDGF 2015a). No regulation required hunters to prepare edible parts of the carcass for human consumption; the meat of the cougar could be discarded.

The New Mexico Department of Game and Fish issued new assumptions of 3,123 to 4,269 independent cougars, including adults and subadults, in New Mexico in 2015 (NMDGF 2015b). The limit placed on annual hunter harvest for the years 2012 to 2016 was 749 cougars (NMDGF, data on file). Based on the New Mexico Department of Game and Fish numbers on cougar abundance, that harvest limit would represent 17.5% to 23.9% of the statewide population of independent cougars. However, despite the assumed high number of cougars, high hunter harvest quotas, and lax regulations, the number of cougars killed by hunters each year from 2001 to 2018 ranged from 122 to 257, though it increased to 358 during the 2018–2019 harvest year, for an average annual harvest of 201 (Fig. 9.1). There was an apparent increasing trend in the number of cougars killed by hunters from 2009–2010 to 2018–2019. During approximately that same period (i.e., 2009–2018), an additional 135 cougars were killed by vehicle traffic while attempting to cross roads in New Mexico (NMDOT, unpubl. data; Map 9.5).

Hunting harvest quotas changed again in 2020 after the New Mexico Department of Game and Fish conducted a field-based study of cougar density in north-central New Mexico (Murphy et al. 2019). Based on new estimated cougar densities ranging from 0.84–1.65 cougars per 100 km^2 in high-quality habitat, harvest quotas were reformulated for habitats represented in that part of the state. Cougar abundances in the remainder of New Mexico remained undetermined. This resulted in a lower adjusted statewide cougar population projection and a reduction in the total mortality limit to 580 with a female sublimit of 181 (NMDGF, data on file). The bag limit per individual hunter was set at two cougars. Trapping of cougars was again prohibited. The hunting season was year-round, from 1 April to 31 March of the following year. Females accompanied by spotted kittens and kittens were protected from hunting. To improve cougar management in the state, reliable field-based cougar population density estimates are needed from other mountain ranges and habitat types.

An Alternate Cougar Management Model

A cougar management model called *zone management* was developed for New Mexico in the late 1990s and early 2000s (Logan and Sweanor 2001; Logan et al. 2003). This model was intended to address the full spectrum of New Mexico citizens' interests in cougars through broad democratic representation and a science-based management framework. Zone management handled cougar-related issues by recognizing three different types of management zones.

1. *Control zones* are specified regions with the objective to control or reduce the cougar population segment, for the purpose of protecting a local livestock industry, an endangered animal species, or human safety. Cougar removal rates would be set to exceed the natural rate of population increase to cause a population decline. This could be achieved with high sport-hunting pressure and direct culling actions if necessary. Control zones would function as sink cougar population segments.

2. *Hunting zones* are regions managed to provide sustainable sport-hunting opportunities. Harvest rates of cougars in hunting zones would be regulated to not exceed the natural rate of population increase. Hunting zones might function as source or sink population segments depending upon actual harvest rates and dispersal capacity from individual hunting zones.

3. *Refuge zones* are regions where cougars would not be subject to sport-hunting or population control. Individual cougars that cause depredation on private property and threaten human safety could still be killed. Refuge zones would be designated in regions with good to excellent quality cougar habitat, and where cougars pose little or no threat to wildlife of concern (e.g., a threatened or endangered desert bighorn sheep herd), livestock, or human safety. These zones could be created in association with national and state parks, conservation areas, and military reservations where cougar hunting is already non-existent or highly limited, and with a sufficient buffer around them so that the refuge zones are biologically functional. At least two such zones could be created (one in northern New Mexico, the other in the southern half of the state), each covering a minimum size of 3,000 sq km (~1,860 sq mi). Refuge zones would protect the overall capacity of the statewide cougar population to be resilient to all forms of human-caused and natural mortality. They would function as primary source population segments, because

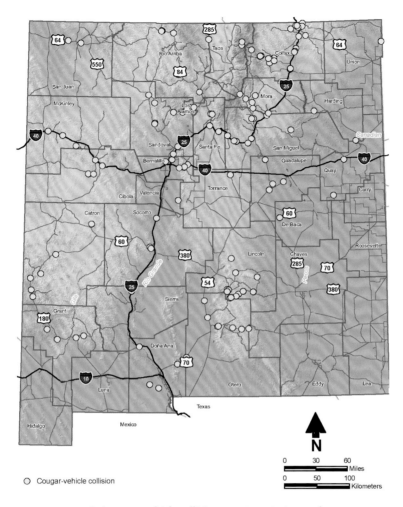

Map 9.5. Recorded cougar-vehicle collisions on New Mexico roads, 2009–2018. A total of 135 cougars were killed from vehicle strikes throughout the state during the 10-year period (NMDOT and NMDGF, unpubl. data).

○ Cougar-vehicle collision

successfully dispersing cougars would prevent extirpation in the control zones and augment the hunting zones. Refuge zones would also provide the right conditions for natural and sexual selection to continue to act on individual cougars. From those population segments exposed to minimal human impacts, cougar genes would move, via dispersing individuals, into the control and hunting zones.

Zones of each type would be managed with an adaptive management approach (Johnson et al. 2015). Each zone would have clearly written objectives and management prescriptions developed with a reliance on sound biological information and on judgments formed from both cougar research in New Mexico and the technical published literature from elsewhere. A sample representative of each zone type would closely track important parameters such as cougar density, population structure, mortality rates, predation rates, and conflict rates to learn how management in the zones performs toward achieving the stated objectives. Adjustments by wildlife managers would be made as necessary. Moreover, the control, hunting, and refuge zones would provide the complete spectrum of human pressure on cougars to further study the effects of hunting on the species, impacts on prey populations, and the effectiveness of mitigating management for human safety. Information from the adaptive management process and research in the zones would also inform the public and stakeholders to generate understanding and support for management options.

Cougar Attacks on People

Although cougars generally avoid humans, encounters with them may increase as human development expands into cougar habitat (Wang et al. 2019). Attacks on humans remain extremely rare in North America, but have become more frequent since the 1950s. Between 1890 and 2008, 29 deaths were recorded, in addition to 171 non-fatal attacks (Beier 1991; Mattson et al. 2011). From 1974 to 2020, the New Mexico Department of Game and Fish has recorded all cougar attacks on people in the state (R. Winslow, NMDGF, pers. comm.). According to those records, two people have been killed by cougars and two people injured to the extent that medical attention was needed. Three other attacks involving physical contact were recorded but either no information on injuries was documented or no injuries occurred. The first fatality was an eight-year-old boy in Arroyo Seco in north-central New Mexico on 20 January 1974. The cougar, a subadult female that appeared to be emaciated, was tracked down and killed. The second fatality was a 55-year-old man who was killed and partially consumed near Piños Altos in Grant County on 16 or 17 June 2008. Two cougars were tracked down and killed in the vicinity; at least one of them was involved in the fatality. Of the two attacks that resulted in injury, the first took place on 17 May 2008 in the Sandia Mountains, and the victim was a five-year-old boy who suffered numerous injuries to the head, back, neck, and shoulders. The second attack resulting in an injury occurred on 26 September 2008 in Taos Ski Valley. The victim (undescribed) was scratched, and stitches were required.

Analyses of recorded cougar attacks on humans reveal characteristics of involved cougars and humans, potential ways of reducing the severity of injuries, and preventing or defending against them. One such study (Mattson et al. 2011) was based on 386 threatening encounters between humans and cougars, including 200 cougar attacks, in North America between 1890 and 2008. It indicated that young (i.e., ≤2.5 years old) or unhealthy cougars were more likely than any others to be involved in close encounters with people (usually within 5 meters [~16 ft.]; Mattson et al. 2011). The likelihood of a younger cougar population age-structure and of animals in poor physical condition theoretically could be exacerbated by high hunting mortality in older adults or nutritional stress caused by local shortages of prey. Furthermore, Mattson et al. (2011) found that female cougars were more likely than males to attack, and adults were more likely to kill the victim than young cougars. Victims of cougar attacks were more likely to be children ≤10 years old rather than single adults. However, when people of any age intervened in defending against an attack, this halved the odds of a child's death; thus, there was an advantage to people being in groups when in proximity to cougars. People who were moving quickly or erratically, such as running or playing, during an encounter were more likely to be attacked. Aggressive behavior toward a cougar by people, including yelling, throwing objects, charging, looming large, and discharge of a firearm, lessened the odds of an attack. The presence of a dog on trails during daylight also reduced the odds of a cougar attacking a person (Mattson et al. 2011).

During our research on the San Andres Mountain, we approached radio-collared cougars at distances of 2–50 meters (~2–55 yards) on 172 occasions to gather data on behavior and reproductive status (Sweanor et al. 2005). In 114 of those instances the cougar chose to avoid us by leaving the area, most likely an expression of its fear of approaching humans. In another 42 instances, the cougar stayed where it was, and did not give any threat response. Threat responses by cougars occurred in only 16 (9%) of all cases, and included vocalizations such as hissing and growling, spitting, and deliberate approaches or charges without making contact. Of those 16 last cases, 14 involved mothers that were probably trying to defend their cubs.

Diseases and Parasites

Epizootic infections occur in cougars in the western United States (Foley 1997), though the population effects of many are not clearly known. Some diseases cause death in individual cougars, of which plague is one. Plague is a disease caused by the bacteria *Yersinia pestis*, and it is zoonotic, meaning that the disease can pass from animals to humans. A first possible route of infection involves the bite of a flea carrying the bacterium. A cougar can also catch the disease from an infected animal, either from exposure to body fluids or by consuming its tissues. Plague caused the death of a wildlife biologist in Arizona who apparently inhaled aerosols originating from a dead cougar's infected tissues (Wong et al. 2009). Plague has been responsible for some cougar mortality in New Mexico (Logan and Sweanor 2001) and in Wyoming (Elbroch et al. 2020b). In Wyoming, antibodies to *Y. pestis* occurred in eight of 17 (47%) of sampled cougars (Elbroch et al. 2020b). Likewise, in Colorado, 32 of 69 (46%) sampled cougars tested positive for the bacteria. Prevalence of the disease ranged from 0–7% among cougars from three California counties (n = 4, 5, 38; Bevins et al. 2009).

Other diseases of cougars include the feline immunodeficiency virus (FIV), the feline analogue to the human immunodeficiency virus (HIV), and occurs in some populations including in the northern Rocky Mountains (Wheeler et al. 2010; Bevins et al. 2012). This virus, which weakens the immune system, is spread through saliva, usually from bites or scratches during mating or while fighting, and from transmission from infected mothers to their kittens (Biek et al. 2003; Wheeler et al. 2010). Feline foamy virus (FFV) can rapidly be disseminated to cougar populations when cougars prey on domestic cats (*Felis catus*), but with no apparent impact to the cougars (Kraberger et al. 2020, Malmberg et al. 2021). The so-called Florida panther (*Puma concolor coryi*), which is characterized by low genetic diversity, is highly susceptible to the feline leukemia virus (FeLV) from exposure to infected domestic cats, which the panthers sometime eat (Chiu et al. 2019). FeLV can cause high mortality rates in domestic cats, and it has been implicated in the deaths of some Florida panthers and infections of several others. Panthers that were necropsied showed signs of anemia, moderate to severe dehydration, lymphadenopathy, septicemia, bronchointerstitial pneumonia, abscesses, and puncture wounds (Cunningham 2005). The bacterium, *Bartonella* spp., which is transmitted by fleas and commonly infects domestic cats—and sometimes their human owners—also has infected cougars in California and Colorado. *Bartonella* infections in felids are often minor; infected felids are usually asymptomatic, but can cause fever, lethargy, uveitis, endocarditis, urinary tract disease, and neurological disease (Chomel et al. 2006, Bevins et al. 2012).

Other than ticks, cougars apparently harbor few external parasites (Currier 1983; Pierce and Bleich 2003). Internal parasites include protozoans, nematodes, cestodes, and trematodes (Anderson 1983; Forrester et al. 1985; Waid 1990). Felids are the only known definitive host of the protozoan *Toxoplasma gondii*, which can be ubiquitous in the local environment. Although it can infect cougars, it does not cause serious disease in otherwise healthy individuals (Bevins et al. 2012). The roundworm *Trichinella* is common in wild carnivores and occurs in cougars in western North America. In Colorado, 17 of 39 (44%) sampled cougars were infected with *Trichinella* larvae (Reichard et al. 2016). Humans can be infected if they ingest raw or undercooked muscle tissue from an infected animal. An outbreak of the disease trichinosis (or trichinellosis) occurred in Idaho in 1995. During that outbreak, 10 people were infected after eating uncooked jerky made from a harvested cougar. Of those, seven persons became ill with symptoms including myalgia, fever, chills, headache, rash, fatigue, facial edema, and arthralgia (Dworkin et al. 1996).

Cougar Control

The New Mexico Department of Game and Fish has managed cougar depredation on livestock since 1971. From 1971 to 1987 the Department employed two full-time predator control officers (see also Chapter 4). The officers were permitted to remove offending cougars after depredation incidents were verified. The officers killed the cougars themselves or they gave ranchers permission to kill the animals. Some exceptions to this process occurred when ranchers killed cougars that were directly threatening or in the process of killing livestock, or when depredating cougars were killed by licensed hunters during the sport-hunting season (Evans 1983). Since 1987, federal predator control agents (currently United States Department of Agriculture, Animal and Plant Health Inspection Service, Wildlife Services; Chapter 3) have worked in collaboration with New Mexico Department of Game and Fish officials to handle depredating animals. Aside from the issue of depredation on livestock, any person may also kill a cougar in defense of private property and human life.

Special cougar control programs have been authorized through the years by the New Mexico Department of Game and Fish. From September 1980 to May 1984, the Department supported a cougar control action involving state and federal predator control agents on the San Andres and Oscura mountains to protect a remnant population of desert bighorn sheep. During this control effort, 42 cougars were killed, including 34 adults (Logan and Sweanor 2001). Cougar control on the San Andres Mountains was resumed again in 2002 to protect reintroduced desert bighorn sheep. Similarly, the New Mexico Department of Game and Fish has authorized cougar control programs since 1999 in the Caballo, Fra Cristobal, Hatchet, Ladrónes, and Peloncillo mountains to protect desert bighorn sheep. By 2010, the cougar control program to protect desert bighorn sheep was tailored to factors that included the sheep population size and rate of cougar predation on sheep in each population. As indicated previously, the cougar control program to protect desert bighorn sheep in addition to sheep augmentation from transplants had contributed to the increase in numbers of this ungulate in New Mexico (NMDGF 2010). To protect domestic sheep on ranches in the Guadalupe Mountains, the New Mexico Department of Game and Fish authorized cougar control from 1985 to 2008, during which over 300 cougars were killed.

According to New Mexico Department of Game and Fish summary records for the period of 2001 to 2015 (NMDGF, data on file), the number of cougars reported killed annually statewide for depredation control ranged from four to 33 and averaged 20 cougars. Females outnumbered males 144 to 128 (five were of undetermined sex). The number of additional cougars killed annually to protect the desert bighorn sheep in particular ranged from 11 to 30 and averaged 16 cougars. Males outnumbered females 140 to 88.

Outlook for the Cougar

In the years that the two of us have studied cougars in the West, we have observed these cats well over 1,000 times during our efforts to capture and track them with radio-telemetry. In that same time, however, we have seen cougars by accident less than a dozen times, even though we were living and working among them. Seeing a cougar by chance makes a person question one's senses. They appear as an apparition—not there, then there, then not there, moving purposefully and silently among the shadows and light of the landscape, of which they are the same colors. On rare occasions, we have heard their mesmerizing caterwauls carried by the wind in the twists and turns of canyons, or emanating from valleys and ridgetops. Evidence of cougars on the landscape is more about the signs they leave in their wake: their diagnostic tracks, scrapes to communicate with other cougars, scraps of prey they killed, consumed, and transmuted into the next form, cougar flesh. Although rarely seen, cougars are an integral part of New Mexico's natural history, ecosystems, and folklore.

The last several decades have seen greater public tolerance of cougars and a more flexible *managed coexistence* with the species. The cougar has certainly adapted and responded well. However, a current of a reversal to primitive human instincts, dogmatic interpretations of old lore and science, and an expanding human population threaten the species again. The extent to which New Mexicans will continue to tolerate or even protect cougars and other big carnivores remains unknown. The fate of the cougar is in our hands.

LITERATURE CITED

Ackerman, B. B., F. G. Lindzey, and T. P. Hemker. 1984. Cougar food habits in southern Utah. *Journal of Wildlife Management* 48:147–55.

Alldredge, M. W., T. Blecha, and J. H. Lewis. 2019. Less invasive monitoring of cougars in Colorado's Front Range. *Wildlife Society Bulletin* 43:222–30.

Allen, M. L., H. U. Wittmer, and C. C. Wilmers. 2014. Puma communication behaviours: understanding functional use and variation among sex and age classes. *Behavior* 151:819–40.

Anderson, A. E. 1983. *A critical review of literature on puma (Felis concolor)*. Colorado Division of Wildlife Special Report No. 54, Denver.

Anderson, C. R., and F. G. Lindzey. 2003. Estimating cougar predation rates from GPS location clusters. *Journal of Wildlife Management* 67:307–16.

———. 2005. Experimental evaluation of population trend and harvest composition in a Wyoming cougar population. *Wildlife Society Bulletin* 33:179–88.

Andreasen, A. M., K. M. Stewart, W. S. Longland, J. P. Bechmann, and M. L. Forister. 2012. Identification of source-sink dynamics in mountain lions of the Great Basin. *Molecular Ecology* 21:5689–701.

Ashman, D., G. C. Christensen, M. L. Hess, G. K. Tsukamoto, and M. S. Wickersham. 1983. *The mountain lion in Nevada*. Nevada Fish and Game Department, Federal Aid in Wildlife Restoration Final Report, Project W-48-15.

Ávila-Nájera, D. M., C. Chávez, S. Pérez-Elizalde, J. Palacios-Pérez, and B. Tigar. 2020. Coexistence of jaguars (*Panthera onca*) and pumas (*Puma concolor*) in a tropical forest in south-eastern Mexico. *Animal Biodiversity and Conservation* 43:55–66.

Bailey, V. 1931 (=1932). *Mammals of New Mexico*. North American Fauna 53. Washington, DC: US Department of Agriculture, Bureau of Biological Survey.

Bauer, J. W., K. A. Logan, L. L. Sweanor, and W. M. Boyce. 2005. Scavenging behavior in puma. *Southwest Naturalist* 50:466–71.

Beausoleil, R. A. 2000. Status of the mountain lion in New Mexico, 1971–2000. In *Proceedings of the sixth mountain lion workshop*, ed. L. A. Harveson, P. M. Harveson, and R. W. Adams, 14–21. Austin, TX.

Beausoleil, R. A., G. M. Koehler, B. T. Maletzke, B. N. Kertson, and R. B. Wielgus. 2013. Research to regulation: cougar social behavior as a guide for management. *Wildlife Society Bulletin* 37:680–88.

Beier P. 1991. Cougar attacks on humans in the United States and Canada. *Wildlife Society Bulletin* 19:403–12.

Beier, P., and R. H. Barrett. 1993. The cougar in the Santa Ana Mountain Range, California. In *Final Report, Orange County Cooperative Mountain Lion Study*, 1–103. Department of Forestry and Resource Management, University of California, Berkeley.

Beier, P., D. Choate, and R. Barrett. 1995. Movement patterns of mountain lions during different behaviors. *Journal of Mammalogy* 76:1056–70.

Belden, R. C., W. B. Frankenberger, R. T. McBride, and S. T. Schwikert. 1988. Panther habitat use in southern Florida. *Journal of Wildlife Management* 52:660–63.

Bender, L. C., and O. C. Rosas-Rosas. 2011. *Predator-prey relationships of pumas in a priority recovery area for the endangered desert bighorn sheep: San Andres National Wildlife Refuge*. Annual report to US Fish and Wildlife Service, San Andres National Wildlife Refuge, Las Cruces, NM.

Bennett, A. F. 2003. *Linkages in the landscape: the role of corridors and connectivity in wildlife conservation*. Gland, Switzerland: IUCN.

Berghofer, L. 1967. *New Mexico wildlife management*. Santa Fe: New Mexico Department of Game and Fish.

Bevins, S. N., S. Carver, E. E. Boydston, L. M. Lyren, M. Alldredge, K. A. Logan, S. P. D. Riley, R. N. Fisher, T. W. Vickers, W. Boyce, M. Salmon, M. R. Lappin, K. R. Crooks, and S. VandeWoude. 2012. Three pathogens in sympatric populations of pumas, bobcats, and domestic cats: implications for infectious disease transmission. *PLOS One* 7:e3140. doi:10.1371/journal.pone.0031403

Bevins, S. N., J. A. Tracey, S. P. Franklin, V. L. Schmit, M. L. MacMillan, K. L. Gage, M. E. Schriefer, K. A. Logan, L. L. Sweanor, M. W. Alldredge, C. Krumm, W. M. Boyce, W. Vickers, S. P. D. Riley, L. M. Lyren, E. E. Boydston, R. N. Fisher, M. E. Roelke, M. Salman, K. R. Crooks, and S. VandeWoude. 2009. Wild felids as hosts for human plague, Western United States. *Emerging Infectious Diseases* 15:2021–24.

Biek R., A. G. Rodrigo, D. Holley, A. Drummond, C. R. Anderson Jr., H. A. Ross, et al. 2003. Epidemiology, genetic diversity, and evolution of endemic feline immunodeficiency virus in a population of wild cougars. *Journal of Virology* 77:9578–89.

Bischoff-Mattson, Z., and D. Mattson. 2009. Effects of simulated mountain lion caching on decomposition of ungulate carcasses. *Western North American Naturalist* 69:343–50.

Boyce, W. M., and M. E. Weisenberger. 2005. The rise and fall of Psoroptic scabies in bighorn sheep in the San Andres Mountains, New Mexico. *Journal of Wildlife Diseases* 41:525–31.

Brown, D. E. 2013. *Bringing back the game: Arizona wildlife management, 1912–1962*. Phoenix: Arizona Wildlife History Series, Arizona Game and Fish Department.

Cahalane, V. H. 1964. *A preliminary study of the distribution and numbers of cougar, grizzly and wolf in North America*. Bronx, New York: New York Zoological Society.

Carbone, C., G. M. Mace, S. C. Roberts, and D. W. MacDonald. 1999. Energetic constraints on the diet of terrestrial carnivores. *Nature* 402:286–88.

Chávez, C. 2010. Ecología y Conservación del jaguar (*Panthera onca*) y Puma (*Puma concolor*) en la región de Calakmul y sus implicaciones para la conservación de la Península de Yucatán. PhD dissertation, Universidad de Granada, Granada, Spain.

Chiu, E. S., S. Kraberger, M. Cunningham, L. Cusack, M. Roelke, and S. VandeWoude. 2019. Multiple introductions of domestic cat Feline Leukemia Virus in endangered Florida panthers. *Emerging Infectious Diseases* 25:92–101.

Chomel, B. B., R. W. Kasten, J. B. Henn, and S. Molia. 2006. *Bartonella* infection in domestic cats and wild felids. *Annals of the New York Academy of Sciences* 1078:410–15.

Clevenger, A. P., J. Wierzchowski, B. Chruszcz, and K. Gunson. 2002. GIS-generated, expert-based models for identifying wildlife habitat linkages and planning mitigation passages. *Conservation Biology* 16:503–14.

Cooley, H. S., H. S. Robinson, R. B. Wielgus, and C. S. Lambert. 2008. Cougar prey selection in a white-tailed deer and mule deer community. *Journal of Wildlife Management* 72:99–106.

Cooley, H. S., R. B. Wielgus, G. M. Koehler, H. S. Robinson, and B. T. Maletzke. 2009. Does hunting regulate cougar populations? A test of the compensatory mortality hypothesis. *Ecology* 90:2913–21.

Crooks, K. R., and M. Sanjayan, eds. 2006. *Connectivity conservation*. Cambridge: Cambridge University Press.

Culver, M. 1999. Molecular genetic variation, population structure, and natural history of free-ranging pumas (*Puma concolor*). PhD dissertation, University of Maryland, College Park.

———. 2010. Lessons and insights from evolution, taxonomy, and conservation genetics. In *Cougar: ecology and conservation*, ed. M. Hornocker and S. Negri, 27–40. Chicago: University of Chicago Press.

Culver, M., W. E. Johnson, J. Pecon-Slattery, and S. J. O'Brien. 2000. Genomic ancestry of the American puma (*Puma concolor*). *Journal of Heredity* 91:186–97.

Cunningham, M. W. 2005. Epizootiology of feline leukemia virus in the Florida panther. MS thesis, University of Florida, Gainesville.

Cunningham, S. C., C. R. Gustavson, and W. B. Ballard. 1999. Diet selection of mountain lions in southeastern Arizona. *Journal of Range Management* 52:202–7.

Cunningham, S. C., L. A. Haynes, C. Gustavson, and D. D. Haywood. 1995. *Evaluation of the interaction between mountain lions and cattle in the Aravaipa-Klondyke area of southeast Arizona*. Arizona Game and Fish Department Technical Report 17, Phoenix.

Currier, M. J. P. 1983. *Felis concolor*. *Mammalian Species* 200:1–7.

De las Guerisoli, M. M., N. Caruso, E. M. Luengos Vidal, and M. Lucherini. 2019. Habitat use and activity patterns of *Puma concolor* in a human-dominated landscape of central Argentina. *Journal of Mammalogy* 100:202–11.

Dickson, B. G., G. W. Roemer, B. H. McRae, and J. M. Rundall. 2013. Models of regional habitat quality and connectivity for pumas (*Puma concolor*) in the southwestern United States. *PLOS One* 8:1–11.

Donaldson, B. 1975. *Mountain lion research*. Job progress report, P-R Project W-93-R-17, Work Plan 15, Job 1. New Mexico Department of Game and Fish.

Dworkin, M. S., H. R. Gamble, D. S. Zarlenga, and P. O. Tennican. 1996. Outbreak of Trichinellosis associated with eating cougar jerky. *Journal of Infectious Diseases* 174:663–66.

Elbroch, L. M., J. M. Ferguson, H. Quigley, D. Craighead, D. J. Thompson, and H. U. Wittmer. 2020. Reintroduced wolves and hunting limit the abundance of a subordinate apex predator in a multiuse landscape. *Proceedings of the Royal Society B: Biological Sciences* 287. http://dx.doi.org/10.1098/rspb.2020.2202

Elbroch, L. M., P. E. Lendrum, P. Alexander, and H. Quigley. 2015a. Cougar den site selection in the Southern Yellowstone Ecosystem. *Mammal Research* 60:89–96.

Elbroch, L. M., P. E. Lendrum, M. L. Allen, and H. U. Wittmer. 2015b. Nowhere to hide: pumas, black bears, and competition refuges. *Behavioral Ecology* 26:247–54.

Elbroch, M. L., P. E. Lendrum, J. Newby, H. Quigley, and D. J. Thompson. 2015c. Recolonizing wolves influence the realized niche of resident cougars. *Zoological Studies* 54, article 41. doi:10.1186/s40555-015-0122-y

Elbroch, M. L., P. E. Lendrum, H. Quigley, and A. Caragiulo. 2015d. Spatial overlap in a solitary carnivore: support for the land tenure, kinship or resource dispersion hypothesis? *Journal of Animal Ecology* 85:487–96.

Elbroch, M. L., M. Levy, M. Lubell, H. Quigley, and A. Caragiulo. 2017. Adaptive social strategies in a solitary carnivore. *Science Advances* 3:e1701218. https://www.science.org/doi/full/10.1126/sciadv.1701218.

Elbroch, M. L., and H. Quigley. 2017. Social

interactions in a solitary carnivore. *Current Zoology* 63:357–62.

———. 2019. Age-specific foraging strategies among pumas, and its implications for aiding ungulate populations through carnivore control. *Conservation Science and Practice* 1. doi:10.1111/csp2.23

Elbroch, L. M., H. Quigley, and A. Caragiulo. 2014. Spatial associations in a solitary predator: using genetic tools and GPS technology to assess cougar social organization in the Southern Yellowstone Ecosystem. *Acta Ethologica* 18:127–36.

Elbroch, L. M., T. W. Vickers, and H. B. Quigley. 2020b. Plague, pumas and potential zoonotic exposure in the Greater Yellowstone Ecosystem. *Environmental Conservation* 47:75–78.

Elmer, M. 1997. Cougar food habit dynamics in the San Andres Mountains, New Mexico. MS, University of Idaho.

Ernest, H. B., W. M. Boyce, V. C. Bleich, B. May, S. J. Stiver, and S. G. Torres. 2003. Genetic structure of mountain lion (*Puma concolor*) populations in California. *Conservation Genetics* 4:353–66.

Evans, W. 1983. *The cougar in New Mexico: biology, status, depredation of livestock, and management recommendations*. Response to House Memorial 42. Santa Fe: New Mexico Game and Fish Department.

Ewers, R. M., and R. K. Didham. 2006. Confounding factors in the detection of species responses to habitat fragmentation. *Biological Reviews* 81:117–42.

Foley, J. E. 1997. The potential for catastrophic infectious disease outbreaks in populations of mountain lions in the western United States. In *Proceedings of the fifth mountain lion workshop*, W. D. Padley, 29–36. San Diego, CA.

Forrester, D. J., J. A. Conti, and R. C. Belden. 1985. Parasites of the Florida panther (*Felis concolor coryi*). *Proceedings of the Helminthological Society of Washington* 52:95–97.

Foster, R. J., B. J. Harmsen, and C. P. Doncaster. 2010. Habitat use by sympatric jaguars and pumas across a gradient of human disturbance in Belize. *Biotropica* 42:724–31.

Gashwiler, J. S., and W. L. Robinette. 1957. Accidental fatalities of the Utah cougar. *Journal of Mammalogy* 38:123–26.

Gill, R. B. 2010. To save a mountain lion: evolving philosophy of nature and cougars. In *Cougar: ecology and conservation*, ed. M. Hornocker and S. Negri, 3–16. Chicago: University of Chicago Press.

Goldman, E. A. 1946. Classification of the races of the puma: part 2. In *The puma: mysterious American cat*, ed. S. P. Young and E. A. Goldman, 177–302. Washington, DC: American Wildlife Institute.

Goldstein, E. J., and E. M. Rominger. 2006. Cause-specific average annual mortality in low-elevation Rocky Mountain bighorn sheep. *Northern Wild Sheep and Goat Council Proceedings* 15:57.

Gonyea, W. J. 1976. Adaptive differences in the body proportions of large felids. *Acta Anatomica* 96:81–96.

Gustafson, K. D., R. B. Gagne, T. W. Vickers, S. P. D. Riley, C. C. Wilmers, V. C. Bleich, B. M. Pierce, M. Kenyon, T. L. Drazenovich, J. A. Sikich, W. M. Boyce, and H. B. Ernest. 2019. Genetic source-sink dynamics among naturally structured and anthropogenically fragmented puma populations. *Conservation Genetics* 20:215–27.

Hibben, F. C. 1937. A preliminary study of the mountain lion (*Felis Oregonensis* sp.). *University of New Mexico Bulletin*, Whole Number 318, Biological Series 5, No. 3.

Hornocker, M. G. 1969. Winter territoriality in mountain lions. *Journal of Wildlife Management* 33:457–64.

———. 1970. An analysis of mountain lion predation upon mule deer and elk in the Idaho Primitive Area. *Wildlife Monographs* 21:3–39.

Hrdy, S. B. 1979. Infanticide among animals: a review, classification, and examination of the implications for the reproductive strategies of females. *Ethology and Sociobiology* 1:13–40.

Iriarte, J. A., W. L. Franklin, W. E. Johnson, and K. H. Redford. 1990. Biogeographic variation of food habits and body size of the American puma. *Oecologia* 85:185–90.

Jalkotzy, M. G., P. I. Ross, and J. Wierzchowski. 2000. Regional scale cougar habitat modeling in southwestern Alberta, Canada (abstract). In *Proceedings of the sixth mountain lion workshop*, ed. L. A. Harveson, P.M. Harveson, and R. W. Adams, 62. Austin, TX.

Jansen, B. D., and J. A. Jenks. 2012. Birth timing for mountain lions (*Puma concolor*); testing the prey availability hypothesis. *PLOS One* 7:9:e44625. https://journals.plos.org/plosone/article?id=10.1371/journal.pone.0044625

Johnson, F. A., G. S. Boomer, B. K. Williams, J. D. Nichols, and D. J. Case. 2015. Multilevel learning in the adaptive management of waterfowl harvests: 20 years and counting. *Wildlife Society Bulletin* 39:9–19.

Johnson, H. E., M. Hebblewhite, T. R. Stephenson, D. W. German, B. M. Pierce, and V. C. Bleich. 2013. Evaluating apparent competition in limiting the recovery of an endangered ungulate. *Oecologia* 171:295–307.

Johnson, J. F. 1982. *Mountain lion research, final report.* P-R Project W-124-R-4, Job 1. New Mexico Game and Fish Department.

Kamler, J. F., R. M. Lee, J. C. deVos Jr., W. B. Ballard, and H. A. Witlaw. 2002. Survival and cougar predation of translocated bighorn sheep in Arizona. *Journal of Wildlife Management* 66:1267–72.

Knopff, K. H. 2010. Cougar predation in a multi-prey system in west-central Alberta. PhD dissertation, Biological Sciences, University of Alberta, Edmonton.

Knopff, K. H., A. A. Knopff, A. Kortello, and M. S. Boyce. 2010. Cougar kill rate and prey composition in a multiprey system. *Journal of Wildlife Management* 74:1435–47.

Koehler, G. M., and M. G. Hornocker. 1991. Seasonal resource use among mountain lions, bobcats, and coyotes. *Journal of Mammalogy* 72:391–96.

Kortello, A. D., T. E. Hurd, and D. L. Murray. 2007. Interactions between cougars (*Puma concolor*) and gray wolves (*Canis lupus*) in Banff National Park, Alberta. *Ecoscience* 14:214–22.

Kraberger, S., N. M. Fountain-Jones, R. B. Gagne, J. Malmberg, N. G. Dannemiller, K. Logan, M. Alldredge, A. Varsani, K. R. Crooks, M. Craft, S. Carver, and S. VandeWoude. 2020. Frequent cross-species transmissions of foamy virus between domestic and wild felids. *Virus Evolution* 6(1): vez058. https://doi.or/10.1093/ve/vez058

Krumm, C. E., M. M. Conner, N. T. Hobbs, D. O. Hunter, and M. W. Miller. 2009. Mountain lions prey selectively on prion-infected mule deer. *Biology Letters* 6(2). doi:101098/rsb1.2009.0742

Kunkel, K. E., T. K. Ruth, D. H. Pletscher, and M. G. Hornocker. 1999. Winter prey selection by wolves and cougars in and near Glacier National Park, Montana. *Journal of Wildlife Management* 63:901–10.

Laing, S. P., and F. G. Lindzey. 1991. Cougar habitat selection in south-central Utah. In *Mountain lion-human interaction symposium and workshop, April 24–26, 1991*, ed. C. E. Braun, 27–37. Denver: Colorado Division of Wildlife.

Laundré, J. W., and L. Hernández. 2000. Habitat composition of successful kill sites for lions in southeastern Idaho and northwestern Utah (abstract). In *Proceedings of the sixth mountain lion workshop*, ed. L. A. Harveson, P.M. Harveson, and R. W. Adams, 24–25. Austin, TX.

———. 2003. Factors affecting dispersal in young male pumas. *Proceedings of the seventh mountain lion workshop*, ed. S. A. Becker, D. D. Bjornlie, F. G. Lindzey, and D. S. Moody, 151–60. Lander, WY.

———. 2007. Do female pumas (*Puma concolor*) exhibit a birth pulse? *Journal of Mammalogy* 88:1300–1304.

Laundré, J. W., L. Hernández, and S. G. Clark. 2007. Numerical and demographic responses of lions to changes in prey abundance: testing current predictions. *Journal of Wildlife Management* 71:345–55.

Lendrum, P. E., L. M. Elbroch, H. Quigley, D. J. Thompson, M. Jimenez, and D. Craighead. 2014. Home range characteristics of a subordinate predator: selection for refugia or hunt opportunity? *Journal of Zoology* 294:59–67.

Lewis, J. S., L. L. Bailey, S. VandeWoude, and K. R. Crooks. 2015. Interspecific interactions between wild felids vary across scales and levels of urbanization. *Ecology and Evolution* 5:5946–61. doi:10.1002/ece3.

Ligon, J. S. 1927. *Wild life of New Mexico: its conservation and management.* Santa Fe, NM: State Game Commission, Department of Game and Fish.

Lindzey, F. G., B. B. Ackerman, D. Barnhurst, and T. P. Hemker. 1988. Survival rates of mountain lions in southern Utah. *Journal of Wildlife Management* 52:664–67.

Lindzey, F. G., W. D. Van Sickle, B. B. Ackerman, D.

Barnhurst, T. P. Hemker, and S. P. Laing. 1994. Cougar population dynamics in southern Utah. *Journal of Wildlife Management* 58:619–24.

Lira, T. I., and E. Naranjo. 2003. Abundancia, preferencia de hábitat e impacto del ecoturismo sobre el puma y dos de sus presas en la Reserva de la Biósfera El Triunfo, Chiapas, México. *Revista Mexicana de Mastozoología* 7:20–39.

Logan, K. A. 2019. Puma population limitation and regulation: what matters in puma management? *Journal of Wildlife Management* 83:1652–66.

Logan, K. A., and L. L. Irwin. 1985. Mountain lion habitats in the Bighorn Mountains, Wyoming. *Wildlife Society Bulletin* 13:257–62.

Logan, K. A., and J. P. Runge. 2021. Effects of hunting on a puma population in Colorado. *Wildlife Monographs* 29:1–35.

Logan, K. A., and L. L. Sweanor. 2000. Puma. In *Ecology and management of large mammals in North America*, ed. S. Demarais and P. R. Krausman, 347–77. Upper Saddle River, New Jersey: Prentice Hall.

———. 2001. *Desert puma: evolutionary ecology and conservation of an enduring carnivore*. Washington, DC: Island Press.

———. 2010. Behavior and social organization of a solitary carnivore. In *Cougar: ecology and conservation*, ed. M. Hornocker and S. Negri, 105–17. Chicago: University of Chicago Press.

Logan, K. A., L. L. Sweanor, and M. G. Hornocker. 2003. Reconciling science and politics in the puma management in the West: New Mexico as a template (abstract). *Proceedings of the seventh mountain lion workshop, May 15 to 17, 2003*, ed. S. A. Becker, D. D. Bornlie, F. G. Lindzey, and D. S. Moody, 146. Lander, WY.

Long, C. A. 1965. *The mammals of Wyoming*. Lawrence: University of Kansas.

Maehr. D. S. 1997. *The Florida panther: life and death of a vanishing carnivore*. Washington, DC: Island Press.

Maehr, D. S., R. C. Belden, E. D. Land, and L. Wilkins. 1990. Food habits of panthers in southwest Florida. *Journal of Wildlife Management* 54:420–423.

Maehr, D. S, and J. A. Cox. 1995. Landscape features and panthers in Florida. *Conservation Biology* 9:1008–19.

Malmberg, J. L., L. A. White, and S. VandeWoude. 2021. Bioaccumulation of pathogen exposure in top predators. *Trends in Ecology and Evolution* 2812:1–10.

Mattson, D. J., J. Hart, M. Miller, and D. Miller. 2007. Predation and other behaviors of mountain lions in the Flagstaff Uplands. In *Mountain lions of the Flagstaff Uplands 2003–2006* (progress report), ed. D. J. Mattson, 31–42. Washington, DC: US Department of the Interior, US Geological Survey.

Mattson, D., K. Logan, and L. Sweanor. 2011. Factors governing risk of cougar attacks on humans. *Human-Wildlife Interactions* 5:135–58.

McKinney, T. T., J. C. deVos, W. B. Ballard, and S. R. Boe. 2006. Mountain lion predation of translocated desert bighorn sheep in Arizona. *Wildlife Society Bulletin* 34:1235–63.

McKinney, T. T., B. F. Wakeling, and J. C. O'Dell. 2010. Mountain lion depredation harvests in Arizona, 1976–2005. In *The Colorado plateau*. Vol. 4, *Shaping conservation through science and management*, ed. C. van Riper III, B. F. Wakeling, and T. D. Sisk, 303–12. Tucson: University of Arizona Press.

Murphy, K., and T. K. Ruth. 2010. Diet and prey selection of a perfect predator. In *Cougar: ecology and conservation*, ed. M. Hornocker and S. Negri, 118–37. Chicago: University of Chicago Press.

Murphy, K. M., G. S. Felzien, M. G. Hornocker, and T. K. Ruth. 1998. Encounter competition between bears and cougars: some ecological implications. *Ursus* 10:55–60.

Murphy, S. M., and B. C. Augustine. 2019. Toward a cohesive framework for large-scale spatially explicit monitoring of puma populations. *Wild Felid Monitor* 2:14–16.

Murphy, S. M., D. T. Wilkins, B. C. Augustine, M. A. Peyton, and G. C. Harper. 2019. Improving estimation of puma (*Puma concolor*) population density: clustered camera-trapping, telemetry data, and generalized spatial mark-resight models. *Scientific Reports* 9, article 4590. https://www.nature.com/articles/s41598-019-40926-7

Negrões, N., P. Sarmento, J. Cruz, C. Eira, E. Revilla, C. Fonseca, R. Sollmann, N. M. Tôrres, M. M.

Furtado, A. T. A. Jácomo, and L. Silveira. 2010. Use of camera-trapping to estimate puma density and influencing factors in central Brazil. *Journal of Wildlife Management* 74:1195–203.

New Mexico Department of Game and Fish (NMDGF). 1995. *Big Game Surveys*. Santa Fe, New Mexico. Project W-93-R-36, Job 9.

———. 1997. *Draft action plan for the management of cougar in New Mexico*. Sante Fe: New Mexico Department of Game and Fish.

———. 1999. *Mule deer of New Mexico*. New Mexico Department of Game and Fish brochure. https://www.wildlife.state.nm.us/download/publications/wildlife/Mule-Deer-of-New-Mexico%20.pdf. Accessed 10 January 2022.

———. 2010. *Management strategy for cougar control to protect desert bighorn sheep*. Santa Fe: New Mexico Department of Game and Fish.

———. 2015a. New Mexico State Game Commission meeting minutes, August 27, 2015. Santa Fe, New Mexico.

———. 2015b. Fiscal year 2015 annual report. Santa Fe, New Mexico.

———. 2019. *2019 spring desert bighorn helicopter surveys*. Santa Fe, New Mexico.

———. 2020. *Cougar population and harvest management matrix 2020–21 through 2023–24*. https://www.wildlife.state.nm.us/download/hunting/species/cougar/Cougar-Management-Strategy-Hunting-Seasons-2020_2024.pdf. Accessed 10 January 2022.

———. 2021. *2021–2022 New Mexico deer hunting prospects*. https://www.wildlife.state.nm.us/download/hunting/species/deer/New-Mexico-Deer-Hunting-Prospects-2021_2022.pdf. Accessed 10 January 2022.

Nowak, R. M. 1976. *The cougar in the United States and Canada*. Washington, DC and New York: US Department of the Interior, Fish and Wildlife Service, and New York Zoological Society.

Ockenfels, R. A. 1994. Factors affecting adult pronghorn mortality rates in central Arizona. *Arizona Game and Fish Department Wildlife Digest* 16:1–11.

Packer, C. 1986. The ecology of sociality in felids. In *Ecological aspects of social evolution*, ed. D. I Rubenstein and R. W. Wrangham, 429–51. Princeton, NJ: Princeton University Press.

Perry, T. 2010. Mountain lion habitat model and population estimate for New Mexico 2010. New Mexico Department of Game and Fish, Santa Fe.

Perry, T., E. Anderson, M. Pittman, B. Talwar, and B. Upton. 2011. Exclusive (nearly) use of riparian habitat by a population of puma in central New Mexico. Poster presented at the 10th Mountain Lion Workshop, 2–5 May 2011, Bozeman, MT.

Perry, T. W., and B. Upton. 2011. *Puma concolor* predation rates and prey selection in riparian and piedmont habitats in New Mexico. Poster presented at the Annual Meeting of the Southeastern Association of Biologists, Huntsville, AL.

Pierce, B. M., and V. C. Bleich. 2003. Mountain lion. In *Wild mammals of North America: biology, management, and economics*, 2nd ed., ed. G. A. Feldhamer, B. C. Thompson, and J. A. Chapman, 744–57. Baltimore: Johns Hopkins University Press.

Pierce, B. M., V. C. Bleich, and R. T. Bowyer. 2000. Selection of mule deer by mountain lions and coyotes: effects of hunting style, body size, and reproductive status. *Journal of Mammalogy* 81:462–72.

Pierce, B. M., V. C. Bleich, K. L. Monteith, and R. T. Bowyer. 2012. Top-down versus bottom-up forcing: evidence from mountain lions and mule deer. *Journal of Mammalogy* 93:977–88.

Pitman, M. E. 2010. Developing a management tool to estimate unmarked puma (*Puma concolor*) populations with a remote camera array. MS thesis, Clemson University.

Proffitt, K. M., R. Garrott, J. A. Gude, M. Hebblewhite, J. Jimenez, J. T. Paterson, and J. Rotella. 2020. Integrated carnivore-ungulate management: a case study in west-central Montana. *Wildlife Monographs* 206:1–208.

Prude, C. H. 2020. Influence of habitat heterogeneity and water sources on kill site locations and puma prey composition. MS thesis, New Mexico State University, Las Cruces.

Reichard, M. V., K. Logan, M. Criffield, J. E. Thomas, J. M. Paritte, D. M. Messerly, M. Interisano, G. Marucci, and E. Pozio. 2016. The occurrence of *Trichinella* species in the cougar *Puma concolor couguar* from the state of Colorado and other regions of North and South America. *Journal of Helminthology* 91:320–25.

Robinette, W. L., J. S. Gashwiler, and O. W. Morris. 1959. Food habits of the cougar in Utah and Nevada. *Journal of Wildlife Management* 23:261–73.

Robinson, H., and R. DeSimone. 2011. *The Garnet Range mountain lion study: characteristics of a hunted population in west-central Montana*. Final Report. Federal Aid in Wildlife Restoration Project W-154-R. Montana Fish, Wildlife and Parks, Helena.

Rodríguez-Soto, C., M. Hernández-Téllez, and O. Monroy-Vilchis. 2013. Distribución y uso del hábitat de *Puma concolor* en la Sierra Nanchititla, México. In *Ecología de Puma concolor en la Sierra Nanchititla, México*, ed. O. Monrroy-Vilcis and L. Soria-Díaz, 47–63. Mexico City, Mexico: Universidad Autónoma del Estado de México.

Rominger, E. M. 2018. The Gordian knot of mountain lion predation and bighorn sheep. *Journal of Wildlife Management* 82:19–31.

Rominger, E. M., and E. J. Goldstein. 2008. Evaluation of an 8-year mountain lion removal management action on endangered desert bighorn sheep recovery. New Mexico Department of Game and Fish, Santa Fe.

Rominger, E. M., H. A. Whitlaw, D. L. Weybright, W. C. Dunn, and W. B. Ballard. 2004. The influence of mountain lion predation on bighorn sheep translocations. *Journal of Wildlife Management* 68:993–99.

Rominger, E. M., F. S. Winslow, E. J. Goldstein, D. W. Weybright, and W. C. Dunn. 2005. Cascading effects of subsidized mountain lion populations in the Chihuahuan Desert (abstract). *Desert Bighorn Council Transactions* 48:65.

Ross, P. I., and M. G. Jalkotzy. 1992. Characteristics of a hunted population of cougars in southwestern Alberta. *Journal of Wildlife Management* 56:417–26.

———. 1996. Cougar predation on moose in southwestern Alberta. *Alces* 32:1–8.

Ruth, T. K. 2004. Ghost of the Rockies: the Yellowstone cougar project. *Yellowstone Science* 12:13–24.

Ruth, T. K., and P. C. Buotte. 2007. *Cougar ecology and cougar-carnivore interactions in Yellowstone National Park*. Final technical report. Bozeman, MT:

Hornocker Wildlife Institute/Wildlife Conservation Society.

Ruth, T. K., P. C. Buotte, and M. G. Hornocker. 2019. *Yellowstone cougars: ecology before and during wolf restoration*. Louisville: University Press of Colorado.

Ruth, T. K., M. A. Haroldson, K. M. Murphy, P. C. Buotte, M. G. Hornocker, and H. B. Quigley. 2011. Cougar survival and source-sink structure on Greater Yellowstone's Northern Range. *Journal of Wildlife Management* 75:1381–1398.

Ruth, T. K., K. A. Logan, L. L. Sweanor, M. G. Hornocker, and L. J. Temple. 1998. Evaluating cougar translocation in New Mexico. *Journal of Wildlife Management* 62:1264–1275.

Ruth, T., and K. Murphy. 2010. Competition with other carnivores for prey. In *Cougar: ecology and conservation*, ed. M. Hornocker and S. Negri, 163–72. Chicago: University of Chicago Press.

Sandell, M. 1989. The mating tactics and spacing patterns of solitary carnivores. In *Carnivore Behavior, Ecology and Evolution*, ed. L. Gittleman, 164–82. Ithaca, NY: Cornell University Press.

Seidensticker, J. C., M. G. Hornocker, W. V. Wiles, and J. P. Messick. 1973. Mountain lion social organization in the Idaho Primitive Area. *Wildlife Monographs* 35:3–60.

Shaw, H. G. 1977. Impact of the mountain lion on mule deer and cattle in northwestern Arizona. In *Proceedings of the 1975 predator symposium, June 16–19, 1975*, ed. R. L. Phillips and C. Jonkel, 17–32. Missoula: University of Montana.

———. 1980. *Ecology of the mountain lion in Arizona*. Arizona Game and Fish Department Final Report. Federal Aid Project W-78-R, WP2, J13.

Smith, T. E., R. R. Duke, M. J. Kutelek, and H. T. Harvey. 1986. *Mountain lions (Felis concolor) in the vicinity of Carlsbad Caverns National Park, New Mexico, and Guadalupe Mountains National Park, Texas*. Final Report. Santa Fe: US Department of Interior, National Park Service.

Sorensen, G. E. 2015. Ecology of adult female Rocky Mountain mule deer (*Odocoileus hemionus hemionus*) following habitat enhancements in north-central New Mexico. PhD dissertation, Texas Tech University, Lubbock.

Spalding, D. J., and J. Lesowski. 1971. Winter food of

the cougar in south-central British Columbia. *Journal of Wildlife Management* 35:378–81.

Spreadbury, B. 1989. Cougar ecology and related management implications and strategies in southeastern British Columbia. Master's thesis (Environmental Design), University of Calgary, Alberta.

Sweanor, L. L., K. A. Logan, and M. G. Hornocker. 2000. Cougar dispersal patterns, metapopulation dynamics, and conservation. *Conservation Biology* 14:798–808.

———. 2005. Puma responses to close approaches by researchers. *Wildlife Society Bulletin* 33:905–13.

Taylor, C. A. 2013. Behavior and cause-specific mortality of mule deer (*Odocoileus hemionus*) fawns on the National Rifle Association Whittington Center of north-central New Mexico. MS Thesis, Texas Tech University, Lubbock.

The Wildlife Professional. 2011. State of wildlife, northeast North America. *Wildlife Professional* 5:18.

Thompson, D. J., and J. A. Jenks. 2005. Long distance dispersal by a subadult male cougar from the Black Hills, South Dakota. *Journal of Wildlife Management* 69:818–20.

United States Department of Agriculture (USDA), Animal Plant Health Inspection Service (APHIS). 2015. *Sheep and lamb predator and nonpredator death loss in the United States, 2015*. Fort Collins, CO: United States Department of Agriculture, Animal and Plant Health Inspection Service.

———. 2017a. *Death loss in U.S. cattle and calves due to predator and nonpredator causes, 2015*. Fort Collins, CO: United States Department of Agriculture, Animal and Plant Health Inspection Service.

———. 2017b. *Goat and kid predator and nonpredator death loss in the United States, 2015*. Fort Collins, CO: United States Department of Agriculture, Animal and Plant Health Inspection Service.

Vickers, T. W., J. N. Sanchez, C. K. Johnson, S. A. Morrison, R. Botta, T. Smith, B. S. Cohen, P. R. Huber, H. B. Ernest, and W. M. Boyce. 2015. Survival and mortality of pumas (*Puma concolor*) in a fragmented, urbanizing landscape. *PLOS One* 10(7):e0131490. doi:10.1371/journal.pone.0131490

Waid, D. D. 1990. Movements, food habits, and helminth parasites of mountain lions in southwestern Texas. PhD dissertation, Texas Tech University, Lubbock.

Wan, H. Y., S. A. Cushman, and J. L. Ganey. 2019. Improving habitat and connectivity model predictions with multi-scale resource selection functions from two geographic areas. *Landscape Ecology* 34:503–19.

Wang, Y. Y., T. G. Weiser, and J. D. Forrester. 2019. Cougar (*Puma concolor*) injury in the United States. *Wilderness and Environmental Medicine* 30:244–55.

Western Association of Fish and Wildlife Agencies. 2019. *Range-wide status of black-tailed and mule deer*. Boise, ID: Mule Deer Working Group, Western Association of Fish and Wildlife Agencies.

Wheeler, D. C., L. A. Waller, and R. Biek. 2010. Spatial analysis of feline immunodeficiency virus infection in cougars. *Spatial and Spatio-Temporal Epidemiology* 1:151–61.

White, P. A., and D. K. Boyd. 1989. A cougar (*Felis concolor*) kitten killed and eaten by gray wolves (*Canis lupus*) in Glacier National Park, Montana. *Canadian Field-Naturalist* 103:408–9.

Williams, D. 2011. Surviving the wild. *New Mexico Wildlife* 56:1, 14–15.

Williams, T. M., L. Wolfe, T. Davis, T. Kendall, B. Richter, Y. Wang, C. Bryce, G. H. Elkaim, and C. C. Wilmers. 2014. Instantaneous energetics of puma kills reveal advantage of felid sneak attacks. *Science* 346:81–85.

Wilson, D. E., and D. M. Reeder, eds. 1993. *Mammal species of the world: a taxonomic and geographic reference*. Washington, DC: Smithsonian Institution Press.

Wolfe, M. L., E. M. Gese, P. Terletzky, D. Stoner, and L. M. Aubry. 2016. Evaluation of harvest indices for monitoring cougar survival and abundance. *Journal of Wildlife Management* 80:27–36.

Wong, D., M. A. Wild, M. A. Walburger, C. L. Higgins, M. Callahan, L. A. Czarnecki, E. W. Lawaczeck, C. E. Levy, J. G. Patterson, R. Sunenshine, P. Adem, C. D. Paddock, S. R. Zaki, J. M. Petersen, M. E. Schriefer, R. J. Eisen, K. L. Gage, K. S. Griffith, I. B. Weber, T. R. Spraker, and P. S. Mead. 2009. Primary pneumonic plague contracted from a mountain lion carcass. *Clinical Infectious Diseases* 49:33–38.

Wozencraft, W. C. 1993. Order Carnivora. In *Mammal species of the world: a taxonomic and geographic reference*, Vol. 1, ed. D. E. Wilson and D. M. Reeder, 286–346. 2nd ed. Washington, DC: Smithsonian Institution Press.

Young, S. P. 1946. History, life habits, economic status, and control: part 1. In *The puma: mysterious American cat*, ed. S. P. Young and E. A. Goldman, 1–173. Washington, DC: American Wildlife Institute.

Young, S. P., and E. A. Goldman, ed. 1946. *The puma: mysterious American cat*. Washington, DC: American Wildlife Institute.

COYOTE (*CANIS LATRANS*)

Jean-Luc E. Cartron and Eric M. Gese

America's native wild "song dog," the coyote (*Canis latrans*) conjures up images of the Old West, filling the night air with its long, mournful wails and staccato yelps. Often cast in the dual role of the "Bad" and the "Ugly," it figures prominently in the tales and legends of Native Americans—including New Mexico's Zuni and Navajo—typically being portrayed as a trickster and troublemaker. The American writer Agnes Morley Cleaveland (1874–1958), who grew up on a ranch in the Plains of San Agustin in western New Mexico, referred to the coyote as a "cowardly beast." And she never forgot the night she rode nervously to White Lake east of Datil, followed "not ten yards behind" by "the largest pack [. . . she] had ever seen [. . . and which] sounded like the wailing of all the tormented in hell" (Morley Cleaveland 1941:87–88). The scourge of many generations of ranchers, the coyote to this day accounts for more livestock losses than any other predator in New Mexico (see Chapter 3). Efforts to control its populations are controversial, but unlike other larger carnivores under human pressure, the coyote has not plummeted in numbers. On the contrary, it continues to thrive in our state as elsewhere, proof of its resilience and adaptability. It ranks among the most important "game" species in New Mexico, where it is hunted and trapped, both for its fur and for recreation. In many areas, coyotes also play an important role in regulating local ecosystems and structuring faunal communities (Henke and Bryant 1999; Thompson and Gese 2007; Berger et al. 2008).

A member of the dog family (Canidae), the coyote resembles other canids in its general appearance, with the slender, deep-chested body, long legs, and digitigrade locomotion—the weight is on the digits or toes, while the heels are raised off the ground—all pointing to adaptations for running. Also characteristic of the Canidae are the long snout and the ears held erect above the head, reflecting the importance of the senses of smell and hearing, and the non-retractable claws, worn down during an animal's normal activity. About the size of a medium-sized domestic dog (*Canis familiaris*), a coyote is larger than a fox but smaller than the wolf (*Canis lupus*). The pelage is variable, but typically a blend of gray, tan, white, black, and fulvous brown, the result of individual hairs being banded with multiple colors (as in most wolves). The tail, long and usually black-tipped (more rarely white-tipped), is usually carried low to the ground, in contrast to that of a domestic dog, wolf, or fox, all of which hold their tails up high or straight out when running. The ears are triangular and pointed, and, especially when compared to the wolf, large, relative to the size of the head. The muzzle is narrow compared to that of the wolf, with a

(*opposite page*) Photograph: © Roger Hogan.

Photo 10.1. Coyote at the Bosque del Apache National Wildlife Refuge on 3 November 2006. Note the gray, tan, white, black, and fulvous brown pelage and the black-tipped tail. Thicker and longer guard hairs and undercoat characterize the winter pelage. Photograph: © Samuel Mulder.

Photo 10.2. Coyote at Bandelier National Monument on 5 February 2008. A coyote's ears are pointed, triangular, and larger than those of wolves relative to the size of the head. Also compared to wolves, the muzzle is narrower, with a smaller nose pad. Note the pupils, round instead of vertically slit as in foxes. Photograph: © Sally King.

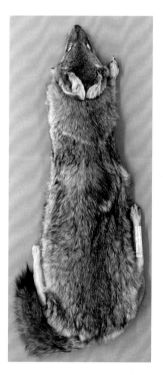

Photo 10.3a, b, and c (*left to right*). Variation in the coloration of winter dorsal pelage in New Mexico's coyotes based on three Museum of Southwestern Biology (MSB) specimens from Sandoval, Hidalgo, and Sierra counties (MSB:Mamm:21263, 89012, and 92554). Coyotes from the higher elevations of northern New Mexico (left) tend to exhibit more black and gray coloration compared to their southern counterparts (center and right), which are more brown and fulvous. Photograph: © Jon Dunnum.

smaller nose pad. Banks et al. (2015) have shown a correlation between the shape of a terrestrial animal's pupils and the ecological niche the species occupies. Foxes have vertically slit pupils, a trait that is functionally adaptive for nocturnal, ambush predators. In contrast, coyotes are typically more diurnal and they use a wider variety of hunting tactics; the eyes of an adult coyote, typically yellow, have black, round pupils.

Based on weight data from New Mexico, male coyotes are on average 9–22% heavier than females. Windberg et al. (1997a) reported average weights of 11.6 kg (25.6 lbs) for males (n = 21) and 9.5 kg (20.9 lbs) for females (n = 13) in Doña Ana and Sierra counties; Young and Jackson (1951) reported averages of 11.1 kg (24.5 lbs) for males (n = 446) and 10.1 kg (22.3 lbs) for females (n = 383). Body length averaged 88.0 cm (34.6 in) for males and 83.1 cm (32.7 in) for females (Windberg et al. 1997a). By comparison, Mexican wolves (*Canis lupus baileyi*) weigh on average twice to four times more than coyotes and their bodies are twice as long (see Chapter 11).

As is typical of canines, coyote paw prints are oval-shaped, with four toes and a triangular heel pad that consists of three lobes, two along the

Photo 10.4. Two coyotes at Panther Seep in Sierra County (south-central New Mexico). Note the extent and brightness of the orange-fulvous to cinnamon rufous coloration, most notable on the fore and hind legs, the top of the head including the ears, and the muzzle. Photograph: © WSMR Garrison Environmental Division and ECO Inc., Ecological Consultants.

trailing edge and one at the front (in contrast to feline tracks). Except on firm soils, the claws register, and they point forward. As opposed to most dogs, an X can be drawn in the negative space between the toes and the heel pad. As in other canid (and felid) species, the fore print is usually larger than the hind print.

Photos 10.5a and b (*left to right*). Coyote tracks observed in mud along the Rio Grande in Bernalillo County (4 and 26 August 2017, respectively). The tracks of coyotes (and other canids) are oval-shaped with four toes and a more or less triangular heel pad. The claws, which may or may not register in soft substrate, are sharp and pointy, instead of blunt in most domestic dogs. Coyote tracks can also be differentiated from most dog tracks when an "X" can be drawn in the negative spaces (between toes and pad, as shown in Photo 10.5b). Photographs: © Michael D. Cox.

The vocal repertoire of a coyote is quite extensive (Lehner 1978). It includes barks, growls, huffs, whines, yelps, and several types of howls. Lone and group howls, which may be preceded by "herald barks," are drawn-out, quavering cries used to communicate location to others in the social group. Also familiar is the group yip-howl, a chorus that is thought to promote group bonding and announces territorial occupancy (Lehner 1982). Often compared with cackling or maniacal laughter, it begins with high-pitched yips and grows in intensity as more members of the group join in.

Based on the fossil record, the coyote made its appearance during the Pleistocene's late Irvingtonian (0.4–0.250 mya; Telford et al. 2009). Coyotes of the Pleistocene were larger and more similar to wolves than those found today, and showed adaptations for eating larger prey (Meachen and Samuels 2012). Within 1,000 years following the megafaunal extinction (the disappearance of many giant ice age mammals in the Americas and northern Eurasia at the end of the last ice age), the morphology of coyotes shifted toward smaller body size (see also Chapter 1).

Of the 19 extant subspecies of the coyote recognized by Wilson and Reeder (2005), four occur in New Mexico (Frey 2004). A medium-sized subspecies, the Plains coyote (*C. l. latrans* Say 1823; *pallidus* Merriam and *nebracensis* Merriam are synonyms) is distinguished by its pale yellowish coloring and whitish winter coat. Described by Clinton Hart Merriam (1855–1942) as the "handsomest of coyotes," *C. l. mearnsi* Merriam 1897 is small and brightly colored, with a cinnamon rufous muzzle and a bright orange fulvous tint of the hind and forelegs all contrasting with the buffy ochraceous upperparts overlaid with black. The Texas Plains coyote (*C. l. texensis* Bailey 1905) resembles the Plains coyote, but averages slightly smaller and exhibits brighter colors, together with a heavier overlay of black-tipped hairs on the front legs and on the back. Nearly as large as *C. l. latrans*, the mountain coyote (*C. l. lestes* Merriam 1897) is distinguished by a large tail and particularly large ears (see Merriam 1897).

DISTRIBUTION

The coyote's remarkable range expansion since pre-settlement times has been well documented (Moore and Parker 1992; Hidalgo-Mihart et al. 2004; Hody and Kays 2018). Recently, however, Hidalgo-Mihart et al. (2004) and Hody and Kays (2018) both established that coyotes were not historically restricted to the Great Plains and southwestern North America as once believed. Historical records show instead that the species also occurred farther south, in southern Mexico and Costa Rica during the Pleistocene-Holocene and in Belize during pre-Columbian time (Hidalgo-Mihart et al. 2004; see also Chapter 1 for geological time and age boundaries); as well as farther west, reaching the Pacific coast of northern North America throughout the Holocene prior to European colonization (Hody and Kays 2018). Historical records thus cast doubt on earlier statements that in northern North America, coyotes first expanded their range northward and westward during the 19th century (see Moore and Parker 1992). The coyote's actual range expansion occurred instead during the 20th century, both northward and eastward, into the eastern deciduous forest and northern boreal forest biomes, to the point that the species is now found in all of the continental United States and in most of Canada. The coyote had invaded mainland Alaska by 1915–1917 (Sherwood 1981). It expanded its distribution around the Great Lakes along a northern and a southern front before establishing itself in southwestern Quebec and in northeastern US states including New York in the 1940s (Young and Jackson 1951; Williams et al. 1985; Weeks et al. 1990; Fener et al. 2005; Kays et al. 2010). In the southeastern United States the coyote gradually expanded its distribution from the 1940s through the 1970s (Gipson 1978). In the early 1970s, coyotes also made their first appearance in the Bas-Saint-Laurent region of southeastern Quebec (Georges 1976). In many areas, the coyote likely benefitted from a combination of factors, primarily deforestation leading to an opening of the landscape, together with the extirpation of wolf (and other top predator) populations (e.g., Richer et al. 2002). The so-called northeastern coyotes—whose ancestry and taxonomy have been much debated—have been shown to have larger body size than found in other populations (Way 2007). Kays et al. (2010) have argued that during the eastward spread of the species north of the Great Lakes (i.e., the north front through Ontario), coyotes came into contact with remnant Great Lakes wolf populations and hybridization occurred. Introgression of wolf genes into northeastern coyotes led to larger body size and the ability to hunt larger prey such as deer, thus accelerating range expansion. In Mexico and Central America (south to northern Panama), coyotes likely expanded their range in areas where the clearing of vast tracts of tropical forest occurred for agricultural development (Hidalgo-Mihart et al. 2004).

Coyote fossil deposits from the Pleistocene have been documented in all parts of New Mexico (Harris 2012; Chapter 1), including at Pendejo Cave in Otero County during the mid-Wisconsin (Harris 2003); Dark Canyon Cave in Eddy County during the mid- to late Wisconsin (Tebedge 1988); Big Manhole Cave, Eddy County (Harris 1993), Blackwater Loc. No. 1 (Lundelius 1972), Pendejo Cave (Harris 2003), and U-Bar Cave, Hidalgo County (13–14 ka, 14–15 ka, 15–18 ka; Harris 1989) during the late Wisconsin; and Burnet Cave, Eddy County (Schultz and Howard 1935); Conkling Cavern, Doña Ana County (Conkling 1932); Isleta Cave No. 1, Bernalillo County (UTEP); Isleta Cave No. 2 (UTEP); Pendejo Cave (Harris 2003); and San Antonio (SAM) Cave in Rio Arriba County (cf. Rogers et al. 2000) during the late Wisconsin/ Holocene (see Figure 1.1 in Chapter 1 for details on geologic time and age boundaries).

In historical times, coyotes were recorded throughout New Mexico at all elevations (but see below), so much so in fact that the word *coyote* appears in many place names in the state: Coyote Peak (2,546 m [8,354 ft]) in Catron County, Coyote Mountain (1,799 m [5,903 ft]) in Sierra County, Coyote Mesa in Lincoln County, Coyote Creek (a tributary of the Mora River) in the Sangre de Cristo Mountains, the village of Coyote in Rio Arriba County, Coyote Canyon in McKinley County, Arroyo Coyote in Sandoval County, and Coyote Springs in the foothills of the Sandia Mountains in Bernalillo County (Julyan 1996). The subspecies *C. l. mearnsi* is found west of the Rio Grande and the Rio Chama in the western part of the state (Frey 2004). Most of the state east of the Rio Grande is occupied by the Texas Plains coyote (*C. l. texensis*), north to about Española in Rio Arriba County, Cowles along the Pecos River in San Miguel County, and northwestern Quay County. The last two subspecies have more restricted distributions in New Mexico. The Plains coyote (*C. l. latrans*) is known only from Union County in the extreme northeastern part of the state. The Mountain coyote (*C. l. lestes*) occurs in and around the San Juan and Sangre de Cristo mountains in north-central New Mexico (Frey 2004; and see Map 10.1).

Among some of the more interesting records from New Mexico is one mentioned by Bailey (1931), a coyote seen in August near Pecos Baldy in the Sangre de Cristo Mountains, at an elevation of 3,500 m (11,500 ft) near the timberline. According to Bailey (1931), the coyote was likely an adult provisioning young, as it was carrying off a freshly killed "brown weasel." In Bailey's mind, the young—if there were any—could have been born in the area, or else they had climbed up the mountain after leaving their den farther downslope. A contemporary of Vernon Bailey, J. Stokley Ligon (1946) argued that coyotes were new invaders of New Mexico's high mountainous areas, including the Truchas Peaks, where

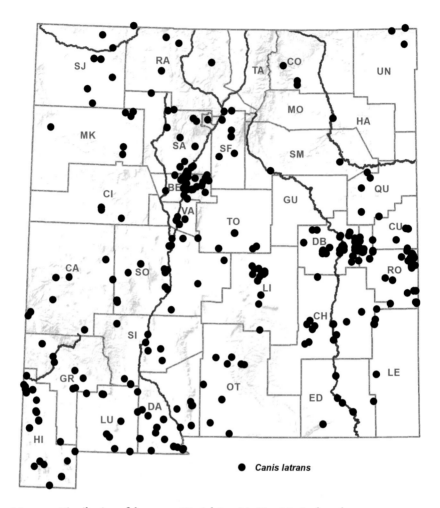

Map 10.1. Distribution of the coyote (*Canis latrans*) in New Mexico based on museum specimen records.

previously coyotes did not reach the timberline and above. Ligon (1946:69) was remarkably precise in his analysis:

The coyote was formerly generally referred to as the "prairie wolf," but now it is just as much at home in the forest. The opening of trails and roads into and through the mountains, and the progressive movement of livestock into the high country, together with the constant availability of game, including turkeys no doubt, combined to aid the coyote greatly in its invasion of new range . . . The coyote

began to invade and remain in the higher forested mountains of New Mexico from 1912 to 1916. Since then its occupation has become general and complete. When the author first started to explore the Black Range and San Mateo Mountains in 1910, there were some wolves but few or no coyotes, and turkeys and deer were abundant.

In 1910, however, Ligon was a newcomer to New Mexico, working mainly as a windmill installer with his uncle on the Jornada del Muerto (Shaw and Weisenberger 2011). His travels into the San

Mateo Mountains consisted of hunting trips with fairly large parties of either family or equally young peers. His later views of the changing distribution of coyotes may or may not be true, but one has to wonder if it might just as easily represent the increased time spent and area covered by Ligon as he matured as a naturalist, rather than changes in coyote range. Warren's (1910) *Mammals of Colorado* reported coyotes from above the timberline at a time when the montane forests in our neighboring state would have been just as pristine as they were in New Mexico.

HABITAT ASSOCIATIONS

Rangewide, coyotes are known from nearly all types of environments including prairies, deserts, high mountains, deciduous forests, swamps, and urban areas (e.g., Bekoff and Gese 2003). In Mexico and Central America, however, coyotes are absent from the more extensive patches of tropical moist forests (Hidalgo-Mihart et al. 2004). In southeastern Quebec, they inhabit deciduous forests, but in lower densities and carrying lower fat reserves compared to coyotes in surrounding rural areas (i.e., woodlots and cultivated fields, together with old fields and regenerating forest), despite a greater abundance of potential prey in the forest landscape (Richer et al. 2002; but see also Kays et al. 2008 for a more nuanced view). For coyotes, which are likely better adapted for hunting in open areas, prey availability in dense forests does not as readily translate into hunting success.

In New Mexico, coyotes are habitat generalists in the sense that they use many biotic community types, but they somewhat avoid forests both at the landscape and home range scales. Their numbers are presumed lowest in the high-elevation forests of the state, as seems to be the case also in neighboring Colorado (Fitzgerald et al. 1994). In the Valles Caldera National Preserve in the Jemez Mountains of northern New Mexico, coyote packs were found to favor some of the more open vegetation communities over the forested areas within their home ranges (Gifford 2013). Average use of herbaceous riparian areas (12% of the time) by coyote packs was greater than expected based on availability (4.4% of home ranges). Wet and dry

Photo 10.6. Coyotes photographed by a motion-activated remote camera (camera trap) at the Rio Grande Nature Center in Albuquerque on 14 December 2008. Coyotes are found in nearly every vegetation type across the state but are better adapted for hunting in open areas. Photograph: © Matt Farley, Jennifer Miyashiro, and James N. Stuart.

Photo 10.7. Coyotes resting in the snow at the Valles Caldera National Preserve on 5 November 2011. In Valles Caldera, coyote packs have been found to use open vegetation communities more than forested areas. Photograph: © Dan J. Williams.

Photo 10.8. Coyote at White Rock, Los Alamos County on 15 September 2009. In some areas, coyotes may use rugged terrain for den sites and access to medium-sized prey. Photograph: © Sally King.

meadows were used most often in proportion to their availability. Forest patches were frequented only 16% of the time, or less than expected based on their availability (22.9% of home ranges) (Gifford 2013).

Overall, the coyote's use of New Mexico's contemporary, arguably more fragmented landscapes can be tied to a few environmental factors, chief among them the availability of prey. In the San Andres Mountains, Bender et al. (2017) found that occupancy of a site by coyotes was influenced by the ruggedness of the terrain and by the presence of rabbits and hares and other medium-size prey, but not by proximity to water. At White Sands National Monument (redesignated as a national park in 2019), coyotes avoided gypsum duneland, pickleweed playas, and interdunal grasslands, where lagomorph abundance as well as rodent densities and biomass were very low. Instead, they selected mesquite, gypsum outcrop, and fourwing saltbush (*Atriplex canescens*) shrublands where indicators of prey availability tended to be much higher (Robinson et al. 2014). Visitation to natural and artificial sources of water was monitored by cameras at the Sevilleta National Wildlife Refuge in central New Mexico (Harris et al. 2015). Predictors of coyote site visitation included reduced weekly precipitation, reduced relative humidity, and higher maximum temperature, and, especially, visitation by mule deer (*Odocoileus hemionus*), though not lagomorphs (Harris et al. 2015). In exurban areas—low density developments outside of urban centers—near Las Cruces, coyote densities were found to be higher than in non-developed, adjacent sites. All sites, non-developed and exurban, were dominated by creosote (*Larrea tridentata*) and mesquite (*Prosopis glandulosa*), but the exurban sites had higher likelihood of occupancy by desert cottontail (*Sylvilagus audubonii*) (DaVanon et al. 2016).

Mesopredators are medium-sized mammalian carnivores that can be preyed on by larger carnivores representing the top (or apex) predators. Recent studies point to mesopredators adapting their use of the landscape to minimize the risk of encountering a top predator. Trade-offs between food availability and perceived risks of predation—discussed under the "landscape of fear" conceptual framework (e.g., Bleicher 2017)—may apply to some coyote populations, particularly where wolves have been reintroduced. In particular, Flagel et al. (2017) found that in a Great Lakes forest, coyotes selectively utilized low-wolf-use areas twice as much as high-wolf-use areas. Mexican wolves were reintroduced in 1998 to the Gila National Forest in western New Mexico. In that part of the state, research is needed to determine what influence—if any—wolves may have on the coyote's use of habitat. Carrera et al. (2008) only provided a comparison between coyote and wolf diet in west-central New Mexico and east-central Arizona, but some of their results suggest another

mechanism through which habitat use may be affected (see under "Diet and Foraging").

LIFE HISTORY

Coyotes are highly adaptable and intelligent animals that display both a wide range of abilities and high intraspecific variability in life histories and social organization. Much research has been conducted on them in New Mexico, particularly on many aspects of their diet, which has important management implications. Of special interest are the results of studies conducted 1) in the Valles Caldera National Preserve, where coyotes are not hunted or trapped; and 2) in west-central New Mexico, where the reintroduction of the wolf—both a competitor and predator—might result in added ecological pressure on coyote populations.

Diet and Foraging

Coyotes have traditionally been described as opportunistic foragers, with a diet that consists largely of small and medium-sized mammals, fruit, and insects, but often also includes wild and domestic ungulates (hoofed animals)—typically calves and fawns but also adults—in addition to carrion (Gese and Grothe 1995). In his time, Bailey (1931) emphasized the high incidence of predation on turkeys and other domestic fowl, goats, and especially sheep, but he also recorded a wide variety of other food items in the diet of New Mexico's coyotes. Jackrabbits and cottontails were described then as staple food in the eastern part of the state, along with small rodents, insects, cactus fruit, juniper berries, and mesquite beans. In the Chuska Mountains in northwestern New Mexico, Bailey (1931) watched a coyote forage along the edge of a lake and catch salamanders (*Ambystoma* sp.). The stomach contents of another coyote apparently shot at the same location—or at least in the Chuska Mountains—also contained a salamander, in addition to a frog and pocket gophers (*Thomomys* sp.).

Photo 10.10. Coyote with mallard (*Anas platyrhynchos*) at the Bosque del Apache National Wildlife Refuge on 10 December 2004. Photograph: © Gordon French.

Photo 10.11. Coyote holding a common carp (*Cyprinus carpio*) at the Bosque del Apache National Wildlife Refuge in 2003. Photograph: © Rüdiger Merz.

Photos 10.9a and b (*top and bottom*). Coyote hunting and swallowing a mouse at the Bosque del Apache National Wildlife Refuge in November 1997. Coyotes are opportunistic predators generally known for preying on small and medium-size prey, though they can also shift their diet to include much larger animals. The coyote in the photo located the mouse burrowed in a cut bank along a ditch. After tracking it through the grass, it appeared to toy with it for a few minutes before finally eating it. Photographs: © Barbara Magnuson and Larry Kimball.

Photo 10.12. Three coyotes feeding on a snow goose (*Chen caerulescens*) at the Bosque del Apache National Wildlife Refuge with sandhill cranes (*Antigone canadensis*) in the background (undated). Photograph: © New Mexico Department of Game and Fish.

Photo 10.13. Elk (*Cervus canadensis*) at Bandelier National Monument on 20 December 2014. Research has shown that two or three coyotes can bring down prey as large as an adult elk, though perhaps only one in poor nutritional condition. In some parts of New Mexico, coyote scats show that elk is consumed seasonally or even year-round, but some or most of it is likely carrion. Photograph: © Larry Lamsa.

Photo 10.14. Coyote on an elk carcass (undated photograph likely from the Valle Calderas National Preserve). As in the Valles Caldera, carcasses are available to coyotes where hunters fatally wound elk but are unable to retrieve them. In areas where wolves (*Canis lupus*) have been reintroduced, elk eaten by coyotes may also represent carcasses left by wolves after predation events. Photograph: © Don MacCarter/New Mexico Department of Game and Fish.

Photo 10.15. Coyote feeding on elk roadkill near Los Alamos, Los Alamos County on 29 January 2011. Photograph: © Sally King.

Photo 10.16. Elk (*Cervus canadensis*) calf that was just captured and fitted with a radio transmitter as part of a mortality study at the Valle Vidal and Vermejo Park Ranch in northern New Mexico (June 2010). Predation on calves may limit elk population recruitment and growth (Tatman et al. 2018). Photograph: © Stewart Liley.

Photo 10.17. Elk calf that survived an attack by coyotes in May 2009 (note torn skin on front leg). Research conducted at Valle Vidal and Vermejo Park Ranch in northern New Mexico found coyotes to be among the top predators of elk calves, together with black bears (*Ursus americanus*) and cougars (*Puma concolor*). Photograph: © Stewart Liley.

Photo 10.18. One of two coyotes circling a female pronghorn (*Antilocapra americana*) in January 2009 at the Jornada Experimental Range in southern New Mexico. The pronghorn is North America's fastest native land mammal, and successful coyote predation on adults appears to be uncommon in New Mexico. Photograph © Greg D. Wright.

Photo 10.19. Two coyotes attack a mule deer (*Odocoileus hemionus*) doe on 8 August 2017 east of Wagon Mound in Mora County. © Louis C. Bender.

Is the coyote really an opportunistic predator, or is it instead an optimal forager, or both? Tests of the coyote's foraging strategy have been conducted throughout the species' range including in New Mexico. During a seven-year study in Socorro County, coyotes were found to prey mainly on lagomorphs, rodents, and arthropods, especially grasshoppers (Hernández et al. 2002). The proportion of lagomorphs, the energetically more profitable prey, in the diet did not fluctuate as a function of availability (jackrabbits and cottontails were not differentiated, a limitation of the study). The proportions of rodents and arthropods consumed did not change in relation to availability, when lagomorph densities remained constant. The ratio of lagomorphs consumed to their availability was negatively correlated with availability. All three of those patterns corresponded to predictions of the optimal foraging model (Hernández et al. 2002), with coyotes appearing to maximize their net rate of energy intake by striking a balance between two opposite strategies: 1) devoting a long time and using more energy in search of highly profitable prey, versus 2) spending minimal time and using less energy hunting more common but less profitable food items. In northern Mexico, coyote activity (as determined with use of GPS collars) peaked in the morning between 7:00 and 10:00 and at dusk around 19:00 (Arias-Del Razo et al. 2011). Coyotes remained active at night and were least active during the afternoon hours. Black-tailed jackrabbits (*Lepus californicus*) and desert cottontails were both crepuscular and nocturnal, with peaks of activity detected (with game cameras installed at feeding stations in coyote activity areas) at dawn and dusk. This partial overlap in temporal peak activity patterns suggested a complex interplay between prey seeking and predator avoidance factors. According to Arias-Del Razo et al. (2011), coyotes may be attempting to match lagomorph activity patterns, but they also rely primarily on their sense of vision to pursue their prey. Their crepuscular vision is excellent, but not their night vision, resulting in a trade-off between maximizing lagomorph prey availability and rate of capture. As for jackrabbits and cottontails, they can afford having peaks of activity during crepuscular hours, at a time when coyotes are active (dawn) or very active (dusk), because those periods are relatively brief and the higher risk of predation can be mitigated by heightened vigilance (see Lima and Bednekoff 1999).

The coyote is also an opportunistic predator, shifting its diet to take advantage of the most available or most vulnerable prey. In some areas of New Mexico, the diet of coyotes is predominantly ungulates rather than smaller prey, at least for parts of the year. In west-central New Mexico (and adjacent east-central Arizona), Carrera et al. (2008) reported that coyote scats contained elk (*Cervus canadensis*), in addition to red squirrel (*Tamiasciurus fremonti*), ground squirrel, skunk, chipmunk (*Neotamias*), shrew (*Sorex*), mouse, collared peccary (*Dicotyles tajacu*), porcupine (*Erethizon dorsatum*), birds, reptiles, and insects,

Photo 10.20. Mule deer (*Odocoileus hemionus*) doe with fawn chasing a coyote at the Bosque del Apache National Wildlife Refuge in November 1997. Photograph: © Barbara Magnuson and Larry Kimball.

Photo 10.21. A coyote and a mule deer (*Odocoileus hemionus*) buck (capable of killing the coyote with its antlers) together at a wildlife drinker on the Sevilleta National Wildlife Refuge on 24 January 2011. The proverbial "water truce" between predators and prey species is considered anomalous. Photograph: © Sevilleta National Wildlife Refuge.

among other prey remains. Year-long, the largest numbers of remains found in scats were those of ungulates, both adults and young, representing 36% of all observed remains. Carrera et al. (2008) did not have diet information preceding the introduction of Mexican wolves in their study area, but they suggested that some of the elk eaten by coyotes could be carcasses left by wolves after predation events. Seasonal and annual variation was observed in the diet of coyotes, with a higher proportion of native ungulates in autumn compared to other seasons (Carrera et al. 2008). In the Valles Caldera National Preserve in northern New Mexico, elk remains were present in coyote scats during all months of the study (May 2005–November 2008; Gifford 2013). Two peaks of elk consumption tended to be observed every year, one in spring or early summer (usually June and representing calf consumption), and another in winter (November–January).

In Valle Vidal and on the adjacent Vermejo Park Ranch in the northeastern part of the state, coyotes are among the main predators of elk calves, together with black bears (*Ursus americanus*) and cougars (*Puma concolor*) (Quintana 2016; Table 10.1). Coyote predation on elk calves has been observed on several occasions. When hunting alone, a coyote may detect a bedded calf (0–10 days old) but is unable to approach it without being chased by the mother (S. Liley, pers. comm.). That chase can last up to half an hour as the cow elk attempts to kick the coyote and kill it. Observed successful predation events all involved coyotes in groups of two to four. As the elk cow chased one coyote, the bedded calf was left unprotected and the other coyote(s) moved in for the kill. After the calf had been killed, it often took many hours before the coyotes were given an opportunity to feed on their prey, as the elk cow continued to chase them off (S. Liley, pers. comm.).

Coyotes also prey on mule deer and pronghorn (*Antilocapra americana*) fawns and on desert bighorn sheep lambs (*Ovis canadensis nelsoni*) in New Mexico. In Colfax County, Lomas and Bender (2007) documented eight cases of neonate mule deer fawn mortality due to coyote predation (more than half of all predation-related mortalities). In the Peloncillo Mountains of southwestern New Mexico, predation by coyotes was also a cause of mortality among desert bighorn sheep lambs 5–47 days old (Cain et al. 2019). In a six-year study in southeastern New Mexico, only one case of pronghorn fawn mortality was attributed to coyote predation (Montoya 1972), though fawn survival rates were higher in a coyote control versus non-control area in two of three years (Larsen 1970). The results of that study stand in sharp contrast to findings from other states, with up to 71% and 90–93% fawn mortality attributed to predation by coyotes in southeastern Colorado (Gese et al. 1988) and Montana (Corneli 1979), respectively. On Anderson Mesa in neighboring Arizona, low fawn survival was due primarily to coyotes (Neff et al. 1985), and subsequent coyote control efforts were credited for increasing fawn survival (Smith et al. 1986).

Predation on adult pronghorn has been documented in New Mexico at both White Sands Missile Range in the southern part of the state and at the Coronado Range and Livestock Research Center just east of Corona, Lincoln County (Bender et al. 2013, 2019). Unlike findings in other states (e.g., Bright and Hervert 2005; Barnowe-Meyer et al. 2009; Keller et al. 2013), predation rates on adult pronghorn by coyotes (and other predators) appear low in New Mexico, with drought-caused malnutrition instead representing the primary cause of mortality. Bender et al. (2019) determined that coyote predation on adult pronghorn appeared to be compensatory rather than additive, with important implications for management (see under "Status and Management").

The presence of a dominance hierarchy in the coyote pack dictates the level of resources acquired by individual members of the pack

Table 10.1. Causes of mortality among elk (*Cervus canadensis*) calves captured and radio-collared from 2009 to 2012 in Valle Vidal and on Vermejo Park Ranch in northern New Mexico. Coyotes accounted for 25% of all calf mortality attributed to predation and at least 22% of all cases of mortality. Adapted from Quintana (2016).

Mortalities	2009	2010	2011	2012	Total
Predation					
Black bear	7	16	10	24	57
Coyote	**3**	**11**	**8**	**9**	**31**
Cougar	3	2	7	8	20
Golden eagle	1	0	0	0	1
Unknown predator	1	4	5	4	14
Other (e.g., legal harvest, drowning, disease)	1	1	1	4	7
Unknown cause	2	0	3	8	13
TOTAL	**18**	**34**	**34**	**57**	**143**

during winter (Gese et al. 1996a), with resource partitioning (i.e., division of resources to avoid competition) occurring within packs, particularly during a time of restricted abundance of food resources during winter. During winter in Yellowstone National Park, as snow depth increased, access to small mammals declined (Gese et al. 1996b). However, as this snow cover limited access to the small mammal prey base for coyotes, it equally made foraging for plant material more difficult for ungulates. As winter progressed and ungulates became nutritionally stressed, animals succumbed to malnutrition and died, or these weakened individuals could be killed by coyotes (Gese and Grothe 1995). Surprisingly, only two to three coyotes were needed to kill an adult elk, but these adult elk were generally in extremely poor nutritional condition, though showing no evidence of injuries (Gese and Grother 1995). Gese and Grothe (1995) reported several instances of coyote predation on elk and found that predation attempts on ungulates almost always involved the alpha pair (the alpha male was the main aggressor) and the remainder of the pack did not participate in the attack

but was often observed to be watching the attack (perhaps the younger animals were too inexperienced). Once a kill had been made or an ungulate succumbed to winter stress, the resident pack began feeding on this resource. However, rather than all members of the pack feeding equally, resource partitioning was observed (Gese et al. 1996a). Apparently, pups were restricted from feeding on the carcass by the older members of the pack. The carcass was monopolized by the alpha pair first, then the higher-ranking beta animals, then the lower-ranking individuals, and lastly the pups (Gese et al. 1996a). Even though these pups were the offspring of the alpha pair and usually related to the older betas in the pack, this restriction of access to the carcass indicated the pups were no longer provided food by their parents or other pack members, and therefore had to fend for themselves. Parent-offspring conflict, where at some point in time the parents must stop all help to and care of their present offspring and invest in their own survival and future offspring, seemed to be in effect within these coyote packs as food resources became restricted during winter. In response to this

Photos 10.22a and b (*left to right*). Coyote den with four to five pups found on 5 May 2014 along the side of a shinnery oak dune blow-out in eastern Chaves County. Based on research outside of New Mexico, litter size in coyotes averages six. Photographs: © Andrew J. Lawrence.

Photo 10.23 (*left*). Coyote pups photographed near Santa Fe by a motion-activated remote camera on 27 May 2008. The pups are possibly a couple months old with a den nearby. Photograph: © Matt Farley, Jennifer Miyashiro, and James N. Stuart.

Photos 10.24a and b (*left to right*). Approximately six-and-a-half-week (a) and eight-week-old (b) coyote pups from New Mexico. Pups reach adult weight at the age of about nine months. Photographs: © James N. Stuart.

resource partitioning, pups adopted a different foraging strategy and spent more time hunting small mammals even when conditions were poor (Gese et al. 1996b, c). In desert environments, the restriction of food resources may occur in a different season than winter due to the lack of snow. However, the social system remains the same in either desert or mountainous ecosystems.

Reproduction

Females are seasonally monoestrus, showing one period of heat per year between January and March, depending on geographic locale (Kennelly 1978). Proestrus lasts two to three months and estrus up to 10 days. Mating ends with a copulatory tie lasting up to 25 minutes, during which the male and female remain locked together. Juveniles (animals < one year old) of both sexes are able to breed by nine months of age, but often do not reproduce their first year. The percentage of females breeding each year varies with local conditions and food supply (Knowlton et al. 1999). Usually, about 60–90% of adult females and 0–70% of female yearlings (one to two years of age) produce litters (Knowlton et al. 1999). Gestation lasts about 63 days. Litter size averages about six pups (range = 1–9 pups) and may be affected by population density and food availability during the previous winter (Knowlton et al. 1999). In the Intermountain West, coyote litter size changed in response to cycles in black-tailed jackrabbits (*Lepus californicus*) (Knowlton and Stoddart 1992). Gese et al. (1996c) found an increase in litter size after cold, snowy winters had increased the number of ungulate carcasses available to ovulating females. Litter sex ratio is generally 1:1 (Knowlton 1972).

Coyotes dens have been documented in a wide variety of settings, including in brush-covered slopes, steep banks, thickets, and hollow logs, and under rock ledges. Coyotes can dig their own dens, but will also use those of other animals.

Some dens have more than one entrance and interconnecting tunnels. In a study in Kansas (Gier 1968), the entrance to the den was typically oriented to the south to maximize solar radiation. The same den can be used from year to year. Denning and pup rearing are the focal point for coyote families for several months until the pups are large and mobile (Bekoff and Wells 1986). The pups are born blind and helpless in the den. Birth weight is 240–275 g (8.5–9.7 oz); length of the body from tip of head to base of tail is about 160 mm (6.3 in) (Gier 1968). Eyes open when the pups are about 14 days old, and the pups emerge from the den when they are about three weeks old. The young are cared for by the parents and other associates, usually siblings from a previous year (Bekoff and Wells 1986). Pups are weaned at about five to seven weeks of age, and reach adult weight by about nine months.

Social Structure and Behavior

Coyotes are considered less social than wolves (but see Gese et al. 1996a, c). The basic social unit is the adult breeding male-female pair, referred to as the "alpha pair," plus "beta" animals (next-ranked individuals, often previous offspring), and pups of the year. Alpha coyotes form pair-bonds that may persist for several years, but not necessarily for life. Coyotes up to 10–12 years of age may maintain pair-bonds and whelp pups; reproductive senescence (i.e., reduced reproductive ability) generally occurs around 9–10 years of age. Associate animals (i.e., betas) may remain in the pack and possibly inherit or displace members of the breeding pair and become alphas themselves. Testing of their social status occurs infrequently and alphas maintain their dominant position in the pack often by intimidation and suppression of lower ranking animals. This social suppression also includes suppressing reproduction among lower ranking individuals in the pack. However, alphas can lose their social position to a younger, stronger beta, and may elect to remain in the pack at a lower social status or leave the pack. Associates participate in territorial maintenance (i.e., active defense of the territory, plus contributing to scent marking) and rearing the pups, but not to the extent of the alpha pair. They receive the benefits of increased access to food and higher survival compared to animals that disperse from the pack.

Additionally, other coyotes exist outside of the resident packs as transient/nomadic individuals. These transients hold a nomadic existence traveling alone over larger areas and do not breed, but will move into territories when vacancies occur. The transient cohort is often young or very old (>8 years old) coyotes of either sex. Sometimes older alphas that are displaced from the pack by a younger, stronger coyote will be forced into a transient existence after losing their social position in the pack.

Coyotes are territorial with a dominance hierarchy within each pack that resides in the territory (Bekoff 1982; Bekoff and Gese 2003; Gese et al. 1996a, c). A dominance hierarchy is defined as the ranking of individuals by their social status. Territoriality mediates the regulation of coyote numbers as packs space themselves across the landscape in relation to available food and habitat (Knowlton et al. 1999). A territory is defined as a defended space that is exclusively occupied and defended by the pack. The territory is actively defended by the pack via direct confrontation of intruding coyotes, and passively defended by scent marking and howling. The dominance hierarchy influences access to food resources within the pack (Gese et al. 1996a, c).

On the one hand, coyotes are social carnivores that often form packs (Camenzind 1978; Bekoff and Wells 1980; Bowen 1981; Andelt 1985; Gese et al.1988). On the other hand, prey size influences social organization, and coyotes may occur in pairs or trios (instead of larger packs) where small or medium mammals represent the bulk of their diet (Bekoff and Gese 2003). The formation of packs, as well as their size (up to 10 individuals; Gese et al. 1996a, c), may be related to hunting elk and deer (typically not prevalent in New Mexico outside mountainous areas), as well as being able to defend ungulate carcasses from other coyotes. Gifford et al. (2017) studied the influence of diet on social organization of coyotes at the Valles Caldera National Preserve in New Mexico. The authors of the study did not find any relationship between coyote social cohesion (i.e., the amount of time the pack members spend together) and the proportion of elk in the diet. They concluded instead that social organization remains relatively constant year-round, despite seasonal differences in diet.

Dispersal of coyotes generally occurs in the fall and winter, and consists of the lowest-ranking individuals in the pack. Typically, some young of the year disperse in their first year of life, while some pups may remain if food resources allow. Dispersal may be into adjacent packs or long distances in search of a vacant territory. Dispersal distances exceeding 100 km are not uncommon (Bekoff and Gese 2003).

It has long been thought that coyote populations respond to indiscriminate lethal control with higher pregnancy rates and with larger litters that have higher survival rates (Knowlton 1972; Connolly and Longhurst 1975; Connolly 1978; Sterling et al. 1983; Knowlton et al. 1999). This mechanism, which is known as compensatory reproduction, is density dependent and mediated by higher food availability in coyote control areas (in controlled areas, more food is available because coyote densities are lower). In South Carolina, however, Kilgo et al. (2017) did not find any significant increase in pregnancy rates, litter size, or fecundity of adult females in response to newly applied trapping pressure. Coyote numbers remained stable, but the age structure of the population changed, with relatively more juveniles and fewer adults represented. Because

Photo 10.25. Coyote and badger in a black-tailed prairie dog (*Cynomys ludovicianus*) town at Maxwell National Wildlife Refuge in the summer of 2010. The coyote and badger were observed running together, the badger following the coyote. When the coyote stopped, it would look behind as if expecting the badger to catch up. The prairie dogs stood at the edge of their burrows and watched. Hunting associations between coyotes and badgers are short-lived and typically initiated by coyotes. Photograph: © Bennette Jenkins-Tanner.

Photo 10.26. Badger chasing after a coyote in New Mexico's Valles Caldera National Preserve during the summer of 2013. The hunting association between the two species is wholly beneficial to the coyote, but its benefit to the badger is more uncertain. In the photograph, the coyote (right) is running and holding a Gunnison's prairie dog (*Cynomys gunnisoni*) in its mouth; the badger (far left) is attempting to keep up with the coyote. The badger nearly caught the prairie dog above ground, but it is the coyote that made the kill. Coyote and badger hunting associations have been observed frequently in New Mexico's Valles Caldera National Preserve. Photograph: © Miranda Butler-Valverde.

juveniles rarely breed, the overall population fecundity decreased, and it was the immigration of juveniles from neighboring areas, rather than compensatory reproduction, that allowed coyote numbers to remain stable (Kilgo et al. 2017).

Interactions with Other Carnivores

According to Native American folklore, coyotes have long been known to form temporary hunting associations with badgers (*Taxidea taxus*) (see Goodwin 1939; Dobie 1950; Ramsey 1977). Observations of that hunting association transcend the scientific literature and in one such instance served as the inspiration for the children's book *Coyote and badger: desert hunters of the Southwest* (Hiscock 2001). Set in Chaco Canyon, San Juan County, the story of Badger and Coyote hunting together is based on a real-life observation by the author while working for the Sierra Club at that location. In his author's note, Bruce Hiscock (2001) relates that he "saw a coyote and badger catch a rabbit . . . the two animals worked as a team, always keeping in touch. They acted like friends in the same way that dogs pal around together." Other observations of coyotes and badgers hunting together have been recorded from other locations, both in New Mexico—including at Maxwell NWR and the Valles Caldera National Preserve—and elsewhere (Photos 10.25–10.26). The two species possess skills that are complementary—for example, the coyote is better adapted to chase prairie dogs above ground while badgers are better suited for hunting them in their burrows. As pointed out by Minta et al. (1992), however, it is more appropriate to refer to a hunting association between badger and coyote than cooperative hunting (see also Chapter 25).

Interactions between carnivore species are common and occur in most major ecological communities worldwide (Johnson et al. 1996; Creel et al. 2001). These interactions generally bring about resource partitioning between the competing species in terms of spatial and temporal avoidance, and may result in direct mortality of the smaller species (Johnson et al. 1996; Creel et al. 2001). As a result of human activities, coyotes find themselves now in the position of a top predator in many ecosystems, while they remain mesopredators in others. Coined by Soulé et al. (1988), the mesopredator-release hypothesis states that mesopredators become more prevalent where populations of larger mammalian carnivores have collapsed. Mesopredator release or its opposite, mesopredator suppression, may cause changes among various species in different trophic levels and reshape ecosystem and community dynamics (Soulé et al. 1988; Crooks and Soulé 1999; Ripple and Beschta 2004). Mesopredator release is a phenomenon in which populations of medium- or small-sized predators rapidly increase in an ecosystem after removal of larger, top carnivores. In New Mexico, coyotes can suppress populations of swift foxes (*Vulpes velox*), kit foxes (*V. macrotis*), North American red foxes (*V. fulva*), and bobcats (*Lynx rufus*). Reintroduction of Mexican wolves into New Mexico may now be suppressing coyote abundance with a corresponding release of small carnivores. The coyote is also not the top predator on the Armendaris Ranch and the Sevilleta National Wildlife Refuge (in south-central New Mexico), with Prude and Cain (2021) documenting a total of 84 cases of cougar predation on coyotes at those two sites combined from 2014 to 2018. Coyotes represented a total of 12% of all cougar kills detected, only second to mule deer (Prude and Cain 2021).

Researchers have documented various impacts of larger canids interacting with smaller co-occurring canids (e.g., Kitchen et al. 1999; Schauster et al. 2002; Kamler et al. 2003) as well as smaller sympatric felid species (e.g., Major and Sherburne 1987; Litvaitis and Harrison 1989; Thornton et al. 2004). In North America, studies of red foxes have found that they avoid coyote territories and persist in boundary areas between coyote territories (Voigt and Earle 1983; Sargeant et al. 1987; Harrison et al. 1989). Theberge and Wedeles (1989) documented how red foxes persisted among coyotes because they utilized more alternative prey during a snowshoe hare (*Lepus americanus*) decline. In a more human-modified landscape, Gosselink et al. (2003) reported how red foxes used farms and urban areas as refugia from coyotes and avoided habitats used by coyotes (see also Chapter 15).

The interaction between coyotes and wolves has been documented in various studies. Results of these interactions were variable, ranging from complete extinction of coyotes by wolves to localized changes in coyote numbers. On Isle Royale, Michigan, Krefting (1969) reported that the coyote population went from an estimated 150 animals on the main island in 1948 to no evidence of coyotes after 1958 following the arrival of wolves to the island. On the Kenai Peninsula, Alaska, Thurber et al. (1992) found coyotes and wolves coexisting with minimal exploitation competition and coyote home ranges overlapping with wolf territories. Similarly, in Riding Mountain National Park, Manitoba, Paquet (1991) documented spatial and temporal overlap of coyotes and wolves. In northwestern Montana, Arjo and Pletscher (1999) reported an increase in coyote pack size following wolf recolonization as coyotes scavenged wolf-killed ungulates. Paquet (1992) similarly found coyotes following wolf tracks and scavenging wolf-killed ungulates.

In Yellowstone National Park, Wyoming, wolves were reintroduced in 1995 and 1996, after an absence of over 60 years. Before wolf reintroduction, coyotes were the top canid in the park and possibly impacted smaller sympatric carnivores. With the return of the wolf, coyote numbers were observed to have declined in the Lamar River Valley (Crabtree and Sheldon 1999). This decline was precipitated by both direct mortality (wolves killing coyotes) and possibly by displacement of pack members (dispersal of coyotes in

areas of high use by wolves). Behaviorally, coyotes responded to wolves by increasing the use of wolf-killed ungulate carcasses, increasing the amount of time spent traveling, and decreasing the amount of time they spent resting (Switalski 2003). Other mesocarnivores, particularly North American red foxes, bobcats, and Canada lynx (*Lynx canadensis*), may benefit from a decline in the coyote population following wolf reintroduction; coyotes may compete for shared prey species, limit their distribution, and inflict direct mortality on these three sympatric species. The potential impacts of wolf reintroduction on the other mesocarnivores in Yellowstone, namely badgers, raccoons (*Procyon lotor*), Pacific martens (*Martes americana*), fishers (*M. pennanti*), and wolverines (*Gulo gulo*), are less well understood. These species may benefit from mesopredator release and increased scavenging opportunities at wolf-killed carcasses. Cascading effects among the trophic levels (i.e., sequential levels in the food chain) are complex and may take many years to be realized.

STATUS AND MANAGEMENT

Coyotes are somewhat informally designated as "unprotected furbearers" in New Mexico, with no closed season or bag limit, and no license required for state residents to trap or hunt them. As a result, there are also no reports available on annual coyote harvest in the state. Altogether, coyotes are killed ("taken") in New Mexico as elsewhere as a result of trapping, hunting, poisoning, and aerial gunning for real or perceived depredation control (see Chapter 3); recreational or commercial trapping for pelts; and opportunistic hunting. The "Coyote Calling Method" of take, whereby a coyote is lured and then shot, may be practiced for depredation control, recreational killing, or personal use or commercial sale of the pelt. Coyote calling is popular in many areas of New Mexico. However, some New Mexico residents have transitioned to a form

Photo 10.27. Coyote at a wildlife drinker near Santa Fe on 19 April 2010. No license is needed for New Mexico residents to trap or hunt coyotes in the state, where the species is informally classified as an unprotected furbearer. Photograph: © Matt Farley, Jennifer Miyashiro, and James N. Stuart.

of coyote calling that is non-consumptive and instead, they just take photographs of lured coyotes (J. Frey, pers. comm.). Coyote-killing contests (described below) are ramped up versions of coyote calling with prizes for the largest coyote or highest number killed. Commercial harvest and use of coyotes for pelts are a recreational pursuit in the state for trappers and hunters. The price of a coyote pelt varies based in part on geography and pelt quality. "Prime" pelts, often used for full-length coats, as the fur hood trim on down coats, and blankets, tend to be those from farther north in the coyote's distribution, from western Montana east to North and South Dakota and north

Photo 10.28. Thirty-nine dead coyotes dumped outside Las Cruces, Doña Ana County in December 2014, following a coyote killing contest held by a local predator hunting club. Photograph: © Kevin Bixby.

Photo 10.29. Coyotes hung over a fence on 30 March 2010 along NM 172 north of Maljamar in Lea County. Coyotes are still regarded as vermin and a threat to livestock in many ranching communities including in New Mexico. Photograph: © Mark L. Watson.

to Alberta. But New Mexico coyote pelts may still reach $50 and are advertised as particularly silky and smooth.

In New Mexico, large-scale hunting, trapping, and poisoning of coyotes began in the late 19th century, at a time when the sheep industry was flourishing (numbers of sheep far exceeded those of cattle until about 1915). In 1907, the Forest Service reported a tally of 510 coyotes killed on just New Mexico's national forest lands, while over a six-month period in 1918, Charles Springer took 438 coyotes on the Bartlett Ranch in Colfax County. The threat to sheep herds from coyote predation is captured by some of James Stokley Ligon's words as quoted by Vernon Bailey (1931:315).

I have always had the idea that the abundance or scarcity of natural food supply governed to a great extent the amount of damage to domestic livestock by coyotes. This idea has been greatly modified by facts learned at the McKenzie ranch in the Pecos Valley. During the past spring the loss in cattle has been very heavy, probably 1,000 out of 6,000 head having died of starvation, thus spreading a feast that one would think sufficient for coyotes. Rabbits are also extremely abundant. Even with this supply of food so easily obtainable, the coyotes rarely eat the dead cattle, but continue to take lambs and even older sheep from the flock that is kept on the McKenzie place . . . It appears that the coyote's appetite for sheep is too strong to be resisted. He cannot forego the temptation of a fat lamb. That he follows sheep from range to range can hardly be doubted.

Beginning in the 1940s, New Mexico's sheep industry declined steadily, as a result of various factors including conversion to the more profitable raising of cattle but also, at least locally, high predation rates (Pearson 1975). A 1974 survey among commercial sheep producers reported that rates of coyote predation on lambs may reach over 20%

in New Mexico (Gee et al. 1977). At that time, much research was being conducted in the state on depredation of domestic livestock and whether all coyotes cause damage, as well as the effectiveness of various control methods (e.g., Robinson 1961; Balser 1964; Littauer 1983; Conolly 1988). Windberg et al. (1997b), for example, studied coyote predation on three small experimental, free-ranging flocks of Angora goats on the Jornada Experimental Range in Doña Ana County. Twelve cases of predation were documented, involving eleven of eighteen territorial and non-territorial transient coyotes with home ranges that extended over the locations of the flocks. Twelve of the goats killed by coyotes were kids (most of them very young), one was a doe; coyotes appeared more prone to attack small groups of goats (Windberg et al. 1997b). Given that 41% of Angora kids were killed in just 12 days of exposure, Windberg et al. (1997b) proposed protecting livestock by controlling local coyote populations rather than simply removing a few "problem animals," when and where lambs or kids are most vulnerable. In contrast, other research outside New Mexico (e.g., Timm and Conolly 1980; Shivik et al. 1996; Conner et al. 1998; Sacks et al. 1999) suggests that many or most coyotes do not attack livestock. Sheep losses are typically caused by older, breeding adults with large litters of pups, whose territories overlap with pastures harboring sheep (especially lambs), in areas lacking sufficient natural prey (Jaeger et al. 2001). Jaeger et al. (2001) recommended removing only the problem animals in areas experiencing significant livestock losses due to predation (see Chapter 3).

Today, management of coyotes for reducing livestock damage or increasing big-game populations is prevalent throughout the state. Protecting livestock from coyotes is a complex endeavor, with each situation requiring an assessment of the legal, social, economic, biological, and technical aspects with no one technique solving the problem in all circumstances (Knowlton et al. 1999). Resolution of conflicts involves an analysis of the efficacy, selectivity, and efficiency of various management scenarios (Knowlton et al. 1999, Young and Green 2015), with an integration of opportunities to empower the local public to protect their private property. Techniques are either corrective (after a depredation event) or preventive (before the event). Selectivity is important when attempting to solve the depredation problem. Techniques that selectively remove the offending individual are preferred over non-selective techniques that coyotes learn to avoid, or which may create warier animals (Knowlton et al. 1999). Identifying the "problem" animal is difficult.

An array of techniques (non-lethal and lethal) is used to prevent or deter depredations on livestock and poultry (Knowlton et al. 1999; Young and Green 2015). Unfortunately, many of these do not carry over to protecting wildlife resources, but see Conner et al. (2016) for benefits of coyote exclusion on fawn survival. Seidler et al. (2014) documented surgical sterilization of coyotes increased pronghorn fawn survival. Some techniques developed for protection of domestic commodities (e.g., fencing) may reduce depredations on natural resources, but are generally limited to small-scale applications. Equally challenging is implementing lethal or non-lethal management techniques for problem coyotes in urban settings where these techniques will be viewed and scrutinized by the general public.

In parts of New Mexico, coyotes are regarded as vermin to the point that they are routinely killed then draped over fences as a supposed deterrent against depredation, or they may be the focus of killing contests. In January 2015, the discovery of 39 coyote and two fox carcasses just outside Las Cruces (Photo 10.28) made headlines statewide and provoked public outcry. Discovered by joggers near the airport, the carcasses had been dumped near the county airport following a coyote-killing contest. In the wake of the discovery, it took less than a month for a bipartisan measure banning coyote-killing contests in New Mexico to be approved by the State Senate. Later that month, however, the measure was tabled by the House of Representatives' Agriculture, Water, and Wildlife Committee. As reported in the *Albuquerque Journal*, opponents of the measure—which did not prevent landowners to kill nuisance animals—had argued that coyote-killing contests represent a key tool for predator management (a claim that is disputed by the Wildlife Society in its March 2019 issue statement regarding killing contests). Animal-rights activists and conservation organizations did not abandon their stance, and in April 2019, New Mexico Governor Michelle Lujan Grisham signed bipartisan legislation that prohibits organizing, sponsoring, or participating in coyote-killing contests in the state. The banning of killing contests is in alignment with the North American Model of Wildlife Conservation, co-developed by the Wildlife Society (see Organ et al. 2012). That model is articulated around seven tenets, including the need to kill wildlife only for a legitimate purpose, the latter defined as "food, fur, self-defense, and the protection of property (including livestock)." The North American Model of Wildlife Conservation has not been universally endorsed (e.g., Peterson and Nelson 2016), but it helps distinguish clearly unethical forms of wildlife use such as killing contests. There are some non-profit organizations that do not subscribe to the North American Model of Wildlife Conservation but advocate "compassionate" conservation and coexistence with native carnivores.

Besides shooting, trapping, and poisoning by humans, other sources of mortality have been documented. For example, serological analyses for antibodies in coyotes show that the species has been exposed to many diseases (e.g., Gier and Ameel 1959; Thomas et al. 1984; Williams et al. 1988; Gese et al. 1991). Generally, the effects of diseases on coyote populations are unknown, but disease can be a substantial mortality factor, especially a parvovirus infection among pups (e.g., Gese et al. 1997). Prevalence of antibodies against canine parvovirus, canine distemper, and canine infectious hepatitis varies geographically (Bekoff and Gese 2003). The prevalence of antibodies against *Yersinia pestis* (plague) ranges from <6% in California (Thomas and Hughes 1992) to levels >50% (Gese et al. 1997); prevalence of antibodies against *Francisella tularensis* (tularemia) ranges from 0% in coyotes in Texas (Trainer and Knowlton 1968) to 88% in Idaho (Gier et al. 1978). Serologic evidence of exposure to brucellosis and leptospirosis varies across locales (Bekoff and Gese 2003). Coyotes in an urban area are equally exposed to pathogens (Grinder and Krausman 2001). Coyotes are also inflicted with a variety of parasites, including fleas, ticks, lice, cestodes, roundworms, nematodes, intestinal worms, hookworms, heartworms, whipworms, pinworms, thorny-headed worms, lungworms, and coccidia fungus (Gier and Ameel 1959; Gier et al. 1978; Bekoff and Gese 2003). The observed prevalence of *Trichinella* (a parasitic roundworm) larvae is generally low in coyotes in Oklahoma and Texas, and the only coyote examined from New Mexico tested negative for *Trichinella* (Pozio et al. 2001; Reichard et al. 2011). Coyotes may carry rabies and suffer from mange, cancer, cardiovascular diseases, and aortic aneurysms (Bekoff and Gese 2003). In February 1994, $474,000 in federal emergency funds were allocated to help fight an outbreak of canine rabies in southern Texas. Infected coyotes were spreading the outbreak northward through Texas toward New Mexico and other states. Control of the outbreak required a massive effort to develop an effective oral canine rabies vaccine and air dropping hundreds of thousands of baits laden with the vaccine (see Chapter 4).

Coyotes adapt to human environs and occupy most habitats, including urban areas. Hybridization with dogs may be a local threat near urban areas, and genetic contamination between dogs, coyotes, and wolves may be occurring in the northeastern United States (Kays et al. 2010). Hybridization between coyotes and red wolves is problematic in the southeastern United States for the red wolf (*Canis rufus*) recovery (Gese et al. 2015; Gese and Terletzky 2015). However, to the extent that any hybridization occurs between coyotes and dogs or wolves in New Mexico, it is unlikely to be the source of significant genetic contamination (e.g., Wayne and Vilà 2003; see also National Academies of Sciences, Engineering, and Medicine 2019 and Chapter 33).

As demonstrated by recent research on the Sevilleta National Wildlife Refuge in central New Mexico (Seamster et al. 2014), coyotes can serve as an indicator of environmental changes. In Seamster et al.'s (2014) study, changes in the structure and composition of local food webs (due to woody shrub encroachment) were tracked through a stable carbon and nitrogen isotope analysis of hair removed from scats of 45 different coyotes. Whether an indicator or keystone species, a game animal, a predator of domestic livestock, or the victim of human persecution, the coyote is clearly many things to many people. As recognized by Knowlton (1972), the key is to balance the need for management of coyote depredations with the aesthetic and ecological values of the species.

Photo 10.30. Coyote with demodectic mange (demodicosis) on 30 April 2021 in Roosevelt County. Demodectic mange is a non-communicable, parasitic skin disease caused by *Demodex canis*, a mite that lives in the hair follicles of canines. It can be exacerbated by times of chronic stress or immune system dysfunction. Photograph: © Mark L. Watson.

LITERATURE CITED

Andelt, W. F. 1985. Behavioral ecology of coyotes in south Texas. *Wildlife Monographs* 94:1–45.

Arias-Del Razo, I., L. Hernández, J. W. Laundré, and O. Myers. 2011. Do predator and prey foraging activity patterns match? A study of coyotes (*Canis latrans*), and lagomorphs (*Lepus californicus* and *Sylvilagus audobonii*). *Journal of Arid Environments* 75:112–18.

Arjo, W. M., and D. H. Pletscher. 1999. Behavioral responses of coyotes to wolf recolonization in northwestern Montana. *Canadian Journal of Zoology* 77:1919–27.

Bailey, V. 1931 (=1932). *Mammals of New Mexico*. North American Fauna 53. Washington, DC: US Department of Agriculture, Bureau of Biological Survey.

Balser, D. S. 1964. Management of predator populations with antifertility agents. *Journal of Wildlife Management* 28:352–58.

Banks, M. S., W. W. Sprague, J. Schmoll, J. A. Q. Parnell, and G. D. Love. 2015. Why do animal eyes have pupils of different shapes? *Science Advances* 1:e1500391. doi: 10.1126/sciadv.1500391

Barnowe-Meyer, K. K., P. J. White, and T. L. Davis. 2009. Predator-specific mortality of pronghorn on Yellowstone's northern range. *Western North American Naturalist* 69:186–94.

Bekoff, M. 1982. Coyote, *Canis latrans*. In *Wild mammals of North America: biology, management, and economics*, ed. J. A. Chapman and G. A. Feldhamer, 447–59. Baltimore: Johns Hopkins University Press.

Bekoff, M., and E. M. Gese. 2003. Coyote (*Canis latrans*). In *Wild mammals of North America: biology, management, and economics*, 2nd ed., ed. G. A. Feldhamer, B. C. Thompson, and J. A. Chapman, 467–81. Baltimore: Johns Hopkins University Press.

Bekoff, M., and M. C. Wells. 1980. The social ecology of coyotes. *Scientific American* 242:130–48.

———. 1986. Social ecology and behavior of coyotes. *Advances in the Study of Behavior* 16:251–338.

Bender, L. C., J. C. Boren, H. Halbritter, and S. Cox. 2013. Factors influencing survival and productivity of pronghorn in a semiarid grassland-woodland in east-central New Mexico. *Human Wildlife Interactions* 7:313–24.

Bender, L. C., O. C. Rosas-Rosas, M. J. Hartsough, C. L. Rodden, and P. C. Morrow. 2019. Effect of predation on adult pronghorn *Antilocapra americana* (Antilocapridae) in New Mexico, Southwestern USA. *Mammalia* 83:248–54.

Bender, L. C., O. Rosas-Rosas, and M. E. Weisenberger. 2017. Seasonal occupancy of sympatric larger carnivores in the southern San Andres Mountains, south-central New Mexico, USA. *Mammal Research* 62:323–29.

Berger, K. M., E. M. Gese, and J. Berger. 2008. Indirect effects and traditional trophic cascades: a test involving wolves, coyotes, and pronghorn. *Ecology* 89:818–28.

Bleicher, S. S. 2017. The landscape of fear conceptual framework: definition and review of current applications and misuses. *PeerJ* 1:1–12. https://peerj.com/articles/3772/#

Bowen, W. D. 1981. Variation in coyote social organization: the influence of prey size. *Canadian Journal of Zoology* 59:639–52.

Bright, J. L., and J. J. Hervert. 2005. Adult and fawn mortality of Sonoran pronghorn. *Wildlife Society Bulletin* 33:43–50.

Cain, J. W., III, R. C. Karsch, E. J. Goldstein, E. M. Rominger, and W. R. Gould. 2019. Survival and cause-specific mortality of desert bighorn sheep lambs. *Journal of Wildlife Management* 83:251–59.

Camenzind, F. J. 1978. Behavioral ecology of coyotes on the National Elk Refuge, Jackson, Wyoming. In *Coyotes: biology, behavior, and management*, ed. M. Bekoff, 267–94. New York: Academic Press.

Carrera, R., W. Ballard, P. Gipson, B. T. Kelly, P. R. Krausman, M. C. Wallace, C. Villalobos, and D. B. Wester. 2008. Comparison of Mexican wolf and coyote diets in Arizona and New Mexico. *Journal of Wildlife Management* 72:376–81.

Conkling, R. P. 1932. Conkling Cavern: the discoveries in the bone cave at Bishops Cap, New Mexico. *West Texas Historical and Scientific Society Bulletin* 44:6–19.

Conner, L. M., M. J. Cherry, B. T. Rutledge, C. H. Killmaster, G. Morris, and L. L. Smith. 2016. Predator exclusion as a management option for increasing white-tailed deer recruitment. *Journal of Wildlife Management* 80:162–70.

Conner, M. M., M. M. Jaeger, T. J. Weller, and D. R. McCullough. 1998. Impact of coyote removal on sheep depredation in Northern California. *Journal of Wildlife Management* 62:690–99.

Connolly, G. E. 1978. Predator control and coyote

populations: a review of simulation models. In *Coyotes: biology, behavior, and management*, ed. M. Bekoff, 327–45. New York: Academic Press.

Connolly, G. E., and W. M. Longhurst. 1975. The effects of control on coyote populations: a simulation model. *Division of Agricultural Science Bulletin 1872, University of California, Davis*:1–37.

Corneli, P. S. 1979. Pronghorn fawn mortality following coyote control on the National Bison Range. MS thesis, University of Montana, Missoula.

Crabtree, R. L., and J. W. Sheldon. 1999. The ecological role of coyotes on Yellowstone's northern range. *Yellowstone Science* 7:15–22.

Creel, S., G. Spong, and N. Creel. 2001. Interspecific competition and the population biology of extinction-prone carnivores. In *Carnivore Conservation.*, ed. J. L. Gittleman, S. M. Funk, D. MacDonald, and R. K. Wayne, 35–60. Cambridge: Cambridge University Press.

Crooks, K. R., and M. E. Soulé. 1999. Mesopredator release and avifaunal extinctions in a fragmented system. *Nature* 400:563–66.

DaVanon, K. A., L. K. Howard, K. E. Mabry, R. L. Schooley, and B. T. Bestelmeyer. 2016. Effects of exurban development on trophic interactions in a desert landscape. *Landscape Ecology* 31:2343–54.

Dobie, J. F. 1950. *The voice of the coyote*. Lincoln, NE: Bison Books.

Fener, H. M., J. R. Ginsberg, E. W. Sanderson, and M. E. Gompper. 2005. Chronology of range expansion of the Coyote, *Canis latrans*, in New York. *Canadian Field-Naturalist* 119:1–5.

Fitzgerald, J. P., C. A. Meaney, and D. M. Armstrong. 1994. *Mammals of Colorado*. Niwot, CO: Denver Museum of Natural History and University Press of Colorado.

Flagel, D. G., G. E. Belovsky, M. J. Cramer, D. E. Beyer Jr., and K. E. Robertson. 2017. Fear and loathing in a Great Lakes forest: cascading effects of competition between wolves and coyotes. *Journal of Mammalogy* 98:77–84.

Frey, J. K. 2004. Taxonomy and distribution of the mammals of New Mexico: an annotated checklist. *Occasional Papers, Museum of Texas Tech University* 240:1–32.

Gee, C. K., R. S. Magleby, W. R. Bailey, R. L. Gum, and L. M. Arthur. 1977. *Sheep and lamb losses to predators and other causes in the western United States*. US Department of Agriculture (USDA) Economic Research Center, Agricultural Economic Report No. 369.

Georges, S. 1976. A range extension of the coyote in Quebec. *Canadian Field-Naturalist* 90:78–79.

Gese, E. M., and S. Grothe. 1995. Analysis of coyote predation on deer and elk during winter in Yellowstone National Park, Wyoming. *American Midland Naturalist* 133:36–43.

Gese, E. M., F. F. Knowlton, J. R. Adams, K. Beck, T. K. Fuller, D. L. Murray, T. D. Steury, M. K. Stoskopf, W. T. Waddell, and L. P. Waits. 2015. Managing hybridization of a recovering endangered species: the red wolf *Canis rufus* as a case study. *Current Zoology* 61:191–205.

Gese, E. M., O. J. Rongstad, and W. R. Mytton. 1988. Relationship between coyote group size and diet in southeastern Colorado. *Journal of Wildlife Management* 52:647–53.

Gese, E. M., R. L. Ruff, and R. L. Crabtree. 1996a. Foraging ecology of coyotes (*Canis latrans*): the influence of extrinsic factors and a dominance hierarchy. *Canadian Journal of Zoology* 74:769–83.

———. 1996b. Intrinsic and extrinsic factors influencing coyote predation of small mammals in Yellowstone National Park. *Canadian Journal of Zoology* 74:784–97.

———. 1996c. Social and nutritional factors influencing dispersal of resident coyotes. *Animal Behaviour* 52:1025–43.

Gese, E. M., R. D. Schultz, M. R. Johnson, E. S. Williams, R. L. Crabtree, and R. L. Ruff. 1997. Serological survey for diseases in free-ranging coyotes (*Canis latrans*) in Yellowstone National Park, Wyoming. *Journal of Wildlife Diseases* 33:47–56.

Gese, E. M., R. D. Schultz, O. J. Rongstad, and D. E. Andersen. 1991. Prevalence of antibodies against canine parvovirus and canine distemper virus in wild coyotes in southeastern Colorado. *Journal of Wildlife Diseases* 27:320–23.

Gese, E. M., and P. A. Terletzky. 2015. Using the "placeholder" concept to reduce genetic introgression of an endangered carnivore. *Biological Conservation* 192:11–19.

Gier, H. T. 1968. *Coyotes in Kansas*. Manhattan, KS: Kansas Agricultural Experiment Station, Kansas State University.

Gier, H. T., and D. J. Ameel. 1959. *Parasites and diseases of Kansas coyotes*. Technical Bulletin 91.

Manhattan, KS: Agricultural Experiment Station, Kansas State University.

Gier, H. T., S. M. Kruckenburg, and R. J. Marler. 1978. Parasites and diseases of coyotes. In *Coyotes: biology, behavior, and management*, ed. M. Bekoff, 37–71. New York: Academic Press.

Gifford, S. J. 2013. Ecology of coyotes on the Valles Caldera National Preserve, New Mexico: implications for elk calf recruitment. MS thesis, Utah State University, Logan.

Gifford, S. J., E. M. Gese, and R. R. Parmenter. 2017. Space use and social ecology of coyotes (*Canis latrans*) in a high-elevation ecosystem: relative stability in a changing environment. *Journal of Ethology* 35:37–49.

Gipson, P. S. 1978. Coyotes and related *Canis* in the southeastern United States with a comment on Mexican and Central American *Canis*. In *Coyotes: biology, behavior, and management*, ed. M. Bekoff, 191–208. New York: Academic Press.

Goodwin, G. 1939. *Myths and tales of the White Mountain Apache*. New York: American Folklore Society.

Gosselink, T. E., T. R. Van Deelen, R. E. Warner, and M. G. Joselyn. 2003. Temporal habitat partitioning and spatial use of coyotes and red foxes in east-central Illinois. *Journal of Wildlife Management* 67:90–103.

Grinder, M., and P. R. Krausman. 2001. Morbidity-mortality factors and survival of an urban coyote population in Arizona. *Journal of Wildlife Diseases* 37:312–17.

Harris, A. H. 1989. The New Mexican late Wisconsin—east versus west. *National Geographic Research* 5:205–17.

———. 1993. Quaternary vertebrates of New Mexico. *Vertebrate Paleontology in New Mexico, New Mexico Museum of Natural History, Bulletin* 2:179–97.

———. 2003. The Pleistocene vertebrate fauna from Pendejo Cave. In *Pendejo Cave*, ed. R. S. MacNeish and J. G. Libby, 36–65. Albuquerque: University of New Mexico Press.

———. 2012. Pleistocene vertebrates of New Mexico and Trans-Pecos Texas. Vol. 3, *Fossil Sites*. UTEP Biodiversity Collections, University of Texas at El Paso. (E-book)

Harris G., J. G. Sanderson, J. Erz, S. E. Lehnen, and M. J. Butler. 2015. Weather and prey predict mammals' visitation to water. *PLOS ONE* 10(11):e0141355. https://journals.plos.org/plosone/article?id=10.1371/journal.pone.0141355

Harrison, D. J., J. A. Bissonette, and J. A. Sherburne. 1989. Spatial relationships between coyotes and red foxes in eastern Maine. *Journal of Wildlife Management* 53:181–85.

Henke, S. E., and F. C. Bryant. 1999. Effects of coyote removal on the faunal community in western Texas. *Journal of Wildlife Management* 63:1066–81.

Hernández, L., R. R. Parmenter, J. W. Dewitt, D. C. Lightfoot, and J. W. Laundré. 2002. Coyote diets in the Chihuahuan Desert, more evidence for optimal foraging. *Journal of Arid Environments* 51:613–24.

Hidalgo-Mihart, M. G., L. Cantú-Salazar, A. González-Romero, and C. A. López-González. 2004. Historical and present distribution of coyote (*Canis latrans*) in Mexico and Central America. *Journal of Biogeography* 31:2025–38.

Hiscock, B. 2001. *Coyote and badger: desert hunters of the Southwest*. Honesdale, PA: Boyds Mills Press.

Hody, J. W., and R. Kays. 2018. Mapping the extension of coyotes (*Canis latrans*) across North and Central America. *ZooKeys* 759:81–97. https://doi.org/10.3897/zookeys.759.15149

Jaeger, M. M., K. M. Blejwas, B. N. Sacks, J. C. C. Neale, M. M. Conner, and D. R. McCullough. 2001. Relative vulnerability of coyotes to removal methods on a Northern California ranch. *California Agriculture* 55:32–36.

Johnson, W. E., T. K. Fuller, and W. L. Franklin. 1996. Sympatry in canids: a review and assessment. In *Carnivore Behavior, Ecology and Evolution*, Vol. 2, ed. L. Gittleman, 189–218. Ithaca, NY: Cornell University Press.

Julyan, R. 1996. *The place names of New Mexico*. Albuquerque: University of New Mexico Press.

Kamler, J. F., W. B. Ballard, R. L. Gilliland, P. R. Lemons, and K. Mote. 2003. Impacts of coyotes on swift foxes in northwestern Texas. *Journal of Wildlife Management* 67:317–23.

Kays, R., A. Curtis, and J. J. Kirchman. 2010. Rapid adaptive evolution of northeastern coyotes via hybridization with wolves. *Biology Letters* 6:89–93.

Kays, R. W., M. E. Gompper, and J. C. Ray. 2008. Landscape ecology of eastern coyotes based on large-scale estimates of abundance. *Ecological Applications* 18:1014–27.

Keller, B. J., J. J. Millspaugh, C. Lehman, G. Brundige, and T. W. Mong. 2013. Adult pronghorn (*Antilocapra americana*) survival and cause-specific mortality in Custer State Park, South Dakota. *American Midland Naturalist* 170:311–22.

Kennelly, J. J. 1978. Coyote reproduction. In *Coyotes: biology, behavior, and management*, ed. M. Bekoff, 73–93. New York: Academic Press.

Kilgo, J. C., C. E. Shaw, M. Vukovich, M. J. Conroy, and C. Ruth. 2017. Reproductive characteristics of a coyote population before and during exploitation. *Journal of Wildlife Management* 81:1386–93.

Kitchen, A. M., E. M. Gese, and E. R. Schauster. 1999. Resource partitioning between coyotes and swift foxes: space, time, and diet. *Canadian Journal of Zoology* 77:1645–56.

Knowlton, F. F. 1972. Preliminary interpretations of coyote population mechanics with some management implications. *Journal of Wildlife Management* 36:369–82.

Knowlton, F. F., E. M. Gese, and M. M. Jaeger. 1999. Coyote depredation control: an interface between biology and management. *Journal of Range Management* 52:398–412.

Knowlton, F. F., and L. C. Stoddart. 1992. Some observations from two coyote-prey studies. In *Ecology and management of the eastern coyote*, ed. A. H. Boer, 101–21. Fredericton: Wildlife Research Unit, University of New Brunswick.

Krefting, L. W. 1969. The rise and fall of the coyote on Isle Royale. *Naturalist* 20:24–31.

Larsen, P. 1970. A six year study of antelope productivity and survival in southern New Mexico. *Proceedings of the Antelope States Workshop* 4:97–103.

Lehner, P. N. 1978. Coyote vocalizations: a lexicon and comparisons with other canids. *Animal Behaviour* 26:712–22.

———. 1982. Differential vocal response of coyotes to "group howl" and "group yip-howl" playbacks. *Journal of Mammalogy* 63:675–79.

Ligon, J. S. 1946. *History and Management of Merriam's Wild Turkey*. Albuquerque: New Mexico Game and Fish Commission.

Lima, S. L., and P. A. Bednekoff. 1999. Temporal variation in danger drives antipredator behavior: the predation risk allocation hypothesis. *American Naturalist* 153:1172–81.

Littauer, G. A. 1983. Rancher use of Compound 1080 toxic collars in New Mexico. *Proceedings of the Great Plains Wildlife Damage Control Workshop* 6:92–104.

Litvaitis, J. A., and D. J. Harrison. 1989. Bobcat-coyote niche relationships during a period of coyote population increase. *Canadian Journal of Zoology* 67:1180–88.

Lomas, L. A., and L. C. Bender. 2007. Survival and cause-specific mortality of neonatal mule deer fawns, north-central New Mexico. *Journal of Wildlife Management* 71:884–94.

Lundelius, E. L., Jr. 1972. Vertebrate remains from the gray sand. In *Blackwater Draw Locality No. 1: a stratified early man site in eastern New Mexico*, ed. J. J. Hester, 148–63. Fort Burgwin Research Center, Southern Methodist University, Publication no. 8.

Major, J. T., and J. A. Sherburne. 1987. Interspecific relationships of coyotes, bobcats, and red foxes in western Maine. *Journal of Wildlife Management* 51:606–16.

Meachen, J. A., and J. X. Samuels. 2012. Evolution in coyotes (*Canis latrans*) in response to the megafaunal extinctions. *Proceedings of the Academy of Sciences of the United States of America* 109:4191–4196.

Merriam, C. H. 1897. Revision of the coyotes or prairie wolves, with descriptions of new forms. *Proceedings of the Biological Society of Washington* 11:19–33.

Minta, S. C., K. A. Minta, and D. F. Lott. 1992. Hunting associations between badgers (*Taxidea taxus*) and coyotes (*Canis latrans*). *Journal of Mammalogy* 73:814–20.

Montoya, W. O. 1972. *Antelope study area population estimates*. Federal Aid Project W9 3R-14:WP4, J5. Santa Fe: New Mexico Department of Game and Fish.

Moore, G. C., and G. R. Parker. 1992. Colonization by the eastern coyote (*Canis latrans*). In *Ecology and management of the eastern coyote*, ed. A. H. Boer, 23–38. Fredericton: Wildlife Research Unit, University of New Brunswick.

Morley Cleaveland, A. 1941. *No Place for a Lady*. Boston: Houghton Mifflin.

National Academies of Sciences, Engineering, and Medicine. 2019. *Evaluating the taxonomic status of the Mexican gray wolf and the red wolf*. Washington,

DC: National Academies Press. https://doi.org/10.17226/25351

Neff, D. J., R. H. Smith, and N. G. Woolsey. 1985. Pronghong antelope mortality study. Final Report, Arizona Game and Fish Department. Federal Aid Project W-78-R.

Organ, J. F., V. Geist, S. P. Mahoney, S. Williams, P. R. Krausman, G. R. Batcheller, T. A. Decker, R. Carmichael, P. Nanjappa, R. Regan, R. A. Medellin, R. Cantu, R. E. McCabe, S. Craven, G. M. Vecellio, and D. J. Decker. 2012. *The North American model of wildlife conservation*. The Wildlife Society Technical Review 12–04. Bethesda, MD: Wildlife Society.

Paquet, P. C. 1991. Winter spatial relationships of wolves and coyotes in Riding Mountain National Park, Manitoba. *Journal of Mammalogy* 72:397–401.

———. 1992. Prey use strategies of sympatric wolves and coyotes in Riding Mountain National Park, Manitoba. *Journal of Mammalogy* 73:337–43.

Pearson, E. W. 1975. Sheep-raising in the 17 western states: populations, distribution, and trends. *Journal of Range Management* 28:27–31.

Peterson, M. N., and M. P. Nelson. 2016. Why the North American model of wildlife conservation is problematic for modern wildlife management. *Human Dimensions of Wildlife* 22:43–54.

Pozio, E., D. B. Pence, G. La Rosa, A. Casulli, and S. E. Henke. 2001. *Trichinella* infection in wildlife of the southwestern United States. *Journal of Parasitology* 87:1208–10.

Prude, C. H., and J. W. Cain. 2021. Habitat diversity influences puma *Puma concolor* diet in the Chihuahuan Desert. *Wildlife Biology* 4:wlb.00875. https://doi.org/10.2981/wlb.00875

Quintana, N. T. 2016. Predator-prey relationships between Rocky Mountain elk and black bears in northern New Mexico. PhD dissertation, Texas Tech University, Lubbock.

Ramsey, J. 1977. *Coyote was going there: Indian literature of the Oregon country*. Seattle: University of Washington Press.

Reichard, M. V., K. E. Tiernana, K. L. Paras, M. Interisano, M. H. Reiskind, R. J. Panciera, and E. Pozio. 2011. Detection of *Trichinella murrelli* in coyotes (*Canis latrans*) from Oklahoma and North Texas. *Veterinary Parasitology* 182:368–71.

Richer M.-C., M. Crête, J.-P. Ouellet, L.-P. Rivest, and J. Huot. 2002. The low performance of forest versus rural coyotes in northeastern North America: inequality between presence and availability of prey. *Ecoscience* 9:44–54.

Ripple, W. J., and R. L. Beschta. 2004. Wolves and the ecology of fear: can predation risk structure ecosystems? *Bioscience* 54:755–66.

Robinson, Q. H., D. Bustos, and G. W. Roemer. 2014. The application of occupancy modeling to evaluate intraguild predation in a model carnivore system. *Ecology* 95:3112–23.

Robinson, W. B. 1961. Population changes of carnivores in some coyote-control areas. *Journal of Mammalogy* 42:510–15.

Rogers, K. L., C. A. Repenning, F. G. Luiszer, and R. D. Benson. 2000. Geologic history, stratigraphy, and paleontology of SAM Cave, north-central New Mexico. *New Mexico Geology* 22:89–117.

Sacks, B. N., K. M. Blejwas, and M. M. Jaeger. 1999. Relative vulnerability of coyotes to removal methods on a Northern California ranch. *Journal of Wildlife Management* 63:939–49.

Sargeant, A. B., S. H. Allen, and J. O. Hastings. 1987. Spatial relations between sympatric coyotes and red foxes in North Dakota. *Journal of Wildlife Management* 51:285–93.

Schauster, E. R., E. M. Gese, and A. M. Kitchen. 2002. Population ecology of swift foxes (*Vulpes velox*) in southeastern Colorado. *Canadian Journal of Zoology* 80:307–19.

Schultz, C. B., and E. B. Howard. 1935. The fauna of Burnet Cave, Guadalupe Mountains, New Mexico. *Proceedings of the Academy of Natural Sciences of Philadelphia* 87:273–98.

Seamster, V. A., L. P. Waits, S. A. Macko, and H. H. Shugart. 2014. Coyote (*Canis latrans*) mammalian prey diet shifts in response to seasonal vegetation change. *Isotopes in Environmental and Health Studies* 50:3, 343–60.

Seidler, R. G., E. M. Gese, and M. Conner. 2014. *Using sterilization to change predation rates of wild coyotes: a test case involving pronghorn fawns*. USDA National Wildlife Research Center—Staff Publications 1541.

Shaw, H., and M. E. Weisenberger. 2011. *Twelve hundred miles by horse and burro: J. Stokley Ligon and New Mexico's first breeding bird survey*. Tucson: University of Arizona Press.

Sherwood, M. 1981. *Big game in Alaska*. New Haven, CT: Yale University Press.

Shivik, J. A., M. M. Jaeger, and R. H. Barrett. 1996. Coyote movements in relation to the spatial distribution of sheep. *Journal of Wildlife Management* 60:422–30.

Smith, R. H., D. J. Neff, and N. G. Woolsey. 1986. Pronghorn response to coyote control: a benefit: cost analysis. *Wildlife Society Bulletin* 14:226–31.

Soulé, M. E., D. T. Bolger, A. C. Alberts, J. Wright, M. Sorice, and S. Hill. 1988. Reconstructed dynamics of rapid extinctions of chaparral-requiring birds in urban habitat islands. *Conservation Biology* 2:75–92.

Sterling, B., W. Conley, and M. R. Conley. 1983. Simulations of demographic compensation in coyote populations. *Journal of Wildlife Management* 47:1177–81.

Switalski, T. A. 2003. Coyote foraging ecology and vigilance in response to gray wolf reintroduction in Yellowstone National Park. *Canadian Journal of Zoology* 81:985–93.

Tatman, N. M., S. G. Liley, J. W. Cain, and J. W. Pitman. 2018. Effects of calf predation and nutrition on elk vital rates. *Journal of Wildlife Management* 82:1417–28. https://www.jstor.org/stable/26610352.

Tebedge, S. 1988. Paleontology and paleoecology of the Pleistocene mammalian fauna of Dark Canyon Cave, Eddy County, New Mexico. PhD dissertation, University of Texas at Austin.

Telford, R. H., X. Wang, and B. E. Taylor. 2009. Phylogenetic systematics of the North American fossil Caninae (Carnivora: Canidae). *Bulletin of the American Museum of Natural History* 325:CP1–218.

Theberge, J. B., and C. H. R. Wedeles. 1989. Prey selection and habitat partitioning in sympatric coyote and red fox populations, southwest Yukon. *Canadian Journal of Zoology* 67:1285–90.

Thomas, C. U., and P. E. Hughes. 1992. Plague surveillance by serological testing of coyotes (*Canis latrans*) in Los Angeles County, California. *Journal of Wildlife Diseases* 28:610–13.

Thomas, N. J., W. J. Foreyt, J. F. Evermann, L. A. Windberg, and F. F. Knowlton. 1984. Seroprevalence of canine parvovirus in wild coyotes from Texas, Utah, and Idaho (1972–1983). *Journal of the American Veterinary Medical Association* 185:1283–87.

Thompson, C. M., and E. M. Gese. 2007. Food webs and intraguild predation: community interactions of a native mesocarnivore. *Ecology* 88:334–46.

Thornton, D. H., M. E. Sunquist, and M. B. Main. 2004. Ecological separation within newly sympatric populations of coyotes and bobcats in south-central Florida. *Journal of Mammalogy* 85:973–82.

Thurber, J. M., R. O. Peterson, J. D. Woolington, and J. A. Vucetich. 1992. Coyote coexistence with wolves on the Kenai Peninsula, Alaska. *Canadian Journal of Zoology* 70:2494–98.

Timm, R. M., and G. E. Connolly. 1980. How coyotes kill sheep. *National Wool Grower* 70:14–15.

Trainer, D. O., and F. F. Knowlton. 1968. Serologic evidence of diseases in Texas coyotes. *Journal of Wildlife Management* 32:981–83.

Voigt, D. R., and B. D. Earle. 1983. Avoidance of coyotes by red fox families. *Journal of Wildlife Management* 47:852–857.

Warren, E. R. 1910. *The mammals of Colorado: an account of the several species found within the boundaries of the state, together with a record of their habits and of their distribution*. New York: G. P. Putnam.

Way, J. G. 2007. A comparison of body mass of *Canis latrans* (coyotes) between eastern and western North America. *Northeastern Naturalist* 14:111–24.

Wayne, R. K., and C. Vilà. 2003. Molecular genetic studies of wolves. In *Wolves: behavior, ecology, and conservation*, ed. L. D. Mech and L. Boitani, 218–37. Chicago: Chicago University Press.

Weeks, J. L., G. M. Tori, and M. C. Shieldcastle. 1990. Coyotes (*Canis latrans*) in Ohio. *Ohio Journal of Science* 90:142–45.

Williams, E. S., E. T. Thorne, M. J. G. Appel, and D. W. Belitsky. 1988. Canine distemper in black-footed ferrets (*Mustela nigripes*) from Wyoming. *Journal of Wildlife Diseases* 24:385–98.

Williams, S. L., S. B. McLaren, and M. A. Burgwin. 1985. Paleo-archaeological and historical records of selected Pennsylvania mammals. *Annals of the Carnegie Museum* 54:77–188.

Wilson, D. E., and D. M. Reeder, eds. 2005. *Mammal species of the world: a taxonomic and geographic reference*. 3rd ed. Baltimore: Johns Hopkins University Press.

Windberg, L. A., S. M. Ebbert, and B. T. Kelly. 1997a. Population characteristics of coyotes (*Canis latrans*) in the northern Chihuahuan Desert of New Mexico. *American Midland Naturalist* 138:197–207.

Windberg, L. A., F. K. Knowlton, S. M. Ebbert, and B. T. Kelly. 1997b. Aspects of coyote predation on Angora goats. *Journal of Range Management* 50:226–30.

Young, J. K., and J. S. Green. 2015. Predator damage control. In *Sheep production handbook*, Vol. 8, 901–44. Fort Collins, CO: American Sheep Industry.

Young, S. P., and H. H. T. Jackson. 1951. *The clever coyote*. Harrisburg, PA: Stackpole.

GRAY WOLF (*CANIS LUPUS*)

John K. Oakleaf, Phillip S. Gipson, Colby M. Gardner, Stewart Breck, and Tracy Melbihess

The gray wolf (*Canis lupus*) is undoubtedly the most controversial and highly publicized carnivore species in New Mexico, eliciting passionate emotions from across a broad spectrum of people holding different viewpoints. For them, wolves symbolize either wilderness and wild places or dangerous, harmful animals to be controlled or eliminated out of fear for human safety and land use economic interests. In truth, wolves are neither the vicious, wanton killers they are sometimes portrayed to be, nor innocuous predators that prey only on sick and weak wild ungulates (hoofed mammals), with no impact on either livestock production or elk (*Cervus canadensis*) and deer (*Odocoileus* spp.) population densities. The wide range of human perceptions on the wolf belies the rather straightforward biology of *Canis lupus* as an apex predator and instead is indicative of the deeply rooted, complex ties, both positive and negative, that bind our two species together. As social carnivores, for example, wolves rely on a wide variety of vocal and physical forms of communication, including the howl, which allows pack members spread out over large distances to coordinate social activities and movements, maintain territories, signal the nearby presence of a predator, and locate and provision wolves too young to hunt (Harrington and Mech 1978; Harrington and Asa 2003; McIntyre et al. 2017). From a human standpoint, the long, low tones of the howl represent the most notable, distinguishing, and perhaps polarizing feature of the

wolf (Harrington and Asa 2003). Some of the authors' most memorable, enriching experiences have indeed come from listening to wolves howling in Arizona, Idaho, Montana, Minnesota, Wyoming, and New Mexico. Depending on an individual's perspective, however, wolves howling can either be a celebrated song highlighting a backpacking experience or a foreboding tale warning of potential danger.

The history of the gray wolf in North America, including New Mexico, illustrates the long journey of the gray wolf, from pre-human colonization times through a prolonged period of control and eradication due to conflict with livestock production and other human activities during European settlement of the American West to more recent protection and recovery efforts. In 1978, the gray wolf was federally listed as Endangered in 47 contiguous states and Threatened in Minnesota. Since then, the clash of public opinions and interests has resulted in a fierce legal and political battle that culminated with the delisting of the gray wolf in January 2021, followed in February 2022 by the reinstatement of Endangered Species Act (ESA) federal protections for the species throughout much, though not all, of its distribution in the contiguous United States (see below). Although less publicized

(*opposite page*) Photograph: © Mark L. Watson.

than the reintroduction and recovery of the gray wolf in Yellowstone and Idaho that began in 1995 in the Northern Rockies, reintroduction efforts in the southwestern United States for the Mexican wolf (*Canis lupus baileyi*) embarked on a parallel but more difficult track starting in 1998. The Mexican wolf recovery effort in the southwestern United States provides a unique genetic and geographic extension of gray wolf restoration efforts in North America, and as such the Mexican wolf is listed as an endangered subspecies separately from other gray wolves (USFWS 2014a).

Gray wolves are the largest extant member of the dog family (*Canidae*), and they resemble most other canids with the same basic body plan for cursorial locomotion, or the ability to run over fairly large distances (see Chapters 10 and 12–15). In addition to an agile body, wolves have elongated distal limbs to increase stride length. The metapodial bones (i.e., the long bones of the feet that connect the digits to the larger lower leg bones), in particular, are longer in wolves and their proximal end is elevated off the ground in the adoption of the digitigrade (i.e., walking- or running-on-toes) posture. Wolves exceed the size threshold (~20kg [45 lbs]), above which a predator must tackle prey as large or larger than itself to meet its energetic requirements (Wang and Tedford 2008). As a result, they also have very strong jaws capable of applying high bite pressure and teeth designed to tear large pieces of flesh and crush bone (Peterson and Ciucci 2003).

Wolves are generally described as either black, gray, or white in pelage; however, most of the wolves historically present in New Mexico and Arizona (all subspecies) were gray with some appearing lighter or darker but only a few specimens of complete black or white coloration (Brown 1983; Gipson et al. 2002). All Mexican gray wolves (*C. l. baileyi*, hereafter Mexican wolves), the subspecies that currently inhabits New Mexico and Arizona, are gray with red, brown, black, and white parts of the body or tips of the hairs (Photo

11.1). Based on historical data, adult wolves in the Southwest (all subspecies) measured 1.5 to 1.8 m (5–6 ft) in total length and 63–81 cm (25–32 in) in height at the shoulder, and they weighed between 23 and 41 kg (50–90 lbs), with males larger and heavier than females (Brown 1983). Sexual size dimorphism has been well documented in the reintroduced population of Mexican wolves in the Southwest. Adult females weighed on average 26 kg (58 lbs, n = 78, SD = 7.3 lbs) whereas adult males were nearly 20% heavier, averaging 31 kg (68 lbs, n = 75, SD = 8.7 lbs) among Mexican wolves captured in Arizona and New Mexico from 1998 to 2020 (USFWS, unpubl. data). Total length (including tail) averaged 1.49 m (4.89 ft, n = 39, SD = 0.13 m) in adult females and 1.6 m (5.25 ft, n = 40, SD = 0.12 m) in adult males, while mean height was 66 cm (26 in, n = 37, SD = 3.5 cm) and 70 cm (28 in, n = 41, SD = 3.8 cm) for adult females and males, respectively (USFWS, unpubl. data). Other characteristics of the wolf (e.g., paw prints, diurnal behavior, smell, hearing) are similar to those described for the coyote (*Canis latrans*; Chapter 10) and not discussed in this chapter (but see Photo 11.3).

The gray wolf ranks among the most extensively studied carnivores. However, even for well-studied species, wildlife science is an inherently imperfect effort, informed at times by inconsistent or conflicting data due to variability in the natural world. Throughout this review, we seek to illuminate areas of agreement and disagreement among researchers and at times develop a coalesced opinion from the available literature.

EVOLUTIONARY HISTORY AND TAXONOMY

While the description of gray wolves presents no great difficulties, the species' evolutionary history and, perhaps more importantly because of the direct conservation implications, its taxonomic status, are at the center of a long-standing

Photo 11.1. The Mexican wolf (*Canis lupus baileyi*) is the smallest, southern-most occurring, rarest, and most genetically distinct subspecies of gray wolf (*C. lupus*) in North America. Mexican wolves typically weigh ~23–36 kg (50–80 lbs), measure up to nearly 170 cm (~5 1/2 ft) from nose to tail, and stand up to 81 cm (32 in) high at the shoulder. They have a distinctive, richly colored coat of buff, gray, rust, and black, often with distinguishing facial patterns. Solid black or white variations do not exist as with other subspecies of North American gray wolves. Photograph: © Mark L. Watson.

Photos 11.2a and b (*top and right*). Skulls of an adult male Mexican gray wolf collected by A. R. Bayne on 16 May 1947 in the Animas Mountains in Hidalgo County (MSB 160138; left) and an adult male coyote from Socorro County (*Canis latrans*; MSB 324248; right). Note the more massive wolf skull, with a heavier maxilla, a broader palate, and more widely spreading zygomatic arches, all attributable to wolves preying on larger-size prey. Compared to the massive rostrum, the wolf's brain case is small. Photographs: © Jon Dunnum.

Photo 11.3. Adult wolf tracks in the mud on the left. Wolf pup (center) and black bear (*Ursus americanus*; right) tracks are also present. Adult Mexican wolf tracks are generally twice the size of adult western coyote (*Canis latrans*) tracks. They usually measure ~9 cm wide by 11.5 cm long (3.5 in × 4.5 in), whereas coyote tracks are ~ 5 cm wide by 6.5 cm long (2 in × 2.5 in). Photograph: © Interagency Field Team.

scientific debate. The Canidae originated about 40 million years ago (mya) in North America, where it underwent an early burst of evolutionary diversification during the early Oligocene and soon split into three subfamilies, the Hesperocyoninae, Borophaginae, and Caninae (Wang 1994; Wang et al. 2004, 2008; Wang and Tedford 2008). The Hesperocyoninae and Borophaginae remained confined (or almost exclusively so) to North America until their extinction during the middle Miocene and early Pleistocene, respectively (Wang and Tedford 2007; Wang et al. 2008). At first represented by small, fox-like generalists, the third subfamily, the Caninae, spread from North America to Eurasia approximately 6 to 7 mya through the Beringian land bridge and reached South America beginning about 3 mya as the Isthmus of Panama formed (Wozencraft 2005; Lucherini and Vidal 2008; Wang and Tedford 2008; Prassack and Walkup 2022). The evolutionary history of the *Caninae* and

of the genus *Canis* in particular is quite convoluted and remains obscured by a conservative anatomy and/or ecomorphological convergence, hybridization, and phenotypic plasticity across extensive distributions (see Krofel et al. 2021; Perri et al. 2021). However, it seems now established that the genus *Canis* (traditionally known as wolves, dogs, and jackals; but see Krofel et al. 2021) made its appearance in North America sometime during the late Miocene to early Pliocene and became widespread in both North America and Eurasia (Nowak 2003; Wang and Tedford 2008; Tedford et al. 2009). In Eurasia, it gave rise to modern-day gray wolves, first recorded ~1 mya (Wang and Tedford 2008). As revealed by a recent genomic study (Perri et al. 2021), the dire wolf (*C. dirus*; Chapter 1) is not a close relative of the gray wolf, and instead its lineage diverged from the other canids ~ 5.7 mya in North America. Characterized by its large head, massive teeth, and relatively short limbs, the dire wolf became extinct about 8,000 years ago, likely due to the loss of megafauna prey during the time period and being outcompeted for nimbler prey by gray wolves (Nowak 2003). By comparison, the gray wolf and coyote lineages may have split only 1.1 mya ago, and both have their origin in Eurasia (Perri et al. 2021). *Canis latrans* evolved in North America and first appears in the fossil record during the late Irvingtonian, between 0.4 and 0.25 mya (Nowak 1978; Tedford et al. 2009). The arrival of gray wolves in North America from Eurasia dates back approximately 500,000 years ago (Nowak 1979).

Foundational biological concepts, such as the definition of species, are not so straightforward in the genus *Canis*. For instance, gray wolves have been documented to breed with domestic dogs (*Canis familiaris*; USFWS, unpubl. data), coyotes (Wilson et al. 2000; Leonard et al. 2005; Hailer and Leonard 2008; Sinding et al. 2018), eastern wolves (*Canis lycaon*; Wilson et al. 2000; Leonard and Wayne 2008; Wheeldon and White 2009; Wheeldon et al. 2010), and red wolves (*Canis rufus*;

see Wayne and Jenks 1991). As reflected by their genomic composition, present-day red wolves and eastern wolves do not fit well into the traditional narrative of unique lineages with distinct evolutionary histories, in part due to relatively recent hybridization occurring between coyotes and red wolves, and between gray wolves, coyotes, and eastern wolf populations, respectively (Sacks et al. 2021). Considerable scientific debate has centered on the exact number of distinct wolf species in North America (two versus three or four), and whether red and eastern wolves evolved separately from coyotes or represent the same species (Wilson et al. 2000; Sacks et al. 2021), or even whether they are simply the result of hybridization between gray wolves and coyotes (VonHoldt et al. 2011, 2016). The scientific debate is not merely esoteric because the ESA does not include any provisions for the protection of hybrids. Although the wolf ancestry debate has not been entirely resolved in the scientific literature, the hypothesis that red and eastern wolves split from coyotes ~50,000 to 100,000 years ago and that interbreeding was only the result of anthropogenic impacts and the corresponding eastward expansion of coyotes in the 1900s appears to receive the most recent support (National Academies of Sciences, Engineering, and Medicine 2019; Sacks et al. 2021). Further, the reintroduction of gray wolves in the Greater Yellowstone Ecosystem, central Idaho, Arizona, New Mexico, and Mexico, all harboring large coyote populations, without any interbreeding recorded, provides ancillary evidence that gray wolves do not readily hybridize with coyotes absent an intermediary species. Gray wolves appear to only interbreed with eastern wolves and coyotes in areas where the three species are present (Wilson et al. 2000). While the scientific debate will likely continue around the origins of the red wolf and eastern wolf, conservation efforts (red wolves are currently listed under the ESA as Endangered) should continue to protect the unique genome associated with remnant populations of these species (National Academies of Sciences, Engineering, and Medicine 2019).

With hybridization having occurred among North American canine species, one can easily imagine the mired imbroglio at the subspecies level, further compounded by the fact that gray wolf subspecies undoubtedly interbred across broad intergradation zones that could extend several hundred kilometers (based on the potential for wolf dispersal over large distances). Early authors (Young and Goldman 1944; Hall 1981) described 24 subspecies of gray wolves in North America (Map 11.1). Subsequent morphological investigation and taxonomic interpretation (Nowak 1979, 1995, 2003) reduced the number of subspecies to five, two of which occurred in New Mexico (see under "Distribution").

Gray wolves may have dispersed from Eurasia in three separate waves during the Pleistocene's glaciation events (Vilà et al. 1999; Nowak 2003; VonHoldt et al. 2011; Fan et al. 2016). According to that scenario, the three waves of invasion (or colonization) are each represented by one of the three most widely recognized subspecies of gray wolves in North America (*C. l. baileyi*, *C. l. nubilus*, and *C. l. occidentalis*). Recently, however, some authors have suggested that gray wolves arrived in a single wave of invasion (Koblmüller et al. 2016; Sinding et al. 2018). Mexican wolves were therefore either the result of the earliest wave of colonization (Nowak 1995; VonHoldt et al. 2011; Fan et al. 2016) or diverged early from other gray wolves during the single wave of colonization (Koblmüller et al. 2016; Sinding et al. 2018). *C. l. baileyi* was first described as a distinct subspecies by Nelson and Goldman (1929) on the basis of having a small body size, a darker, more reddish pelage, and a small, narrow, and arched skull with a slender and depressed rostrum. There is still broad scientific consensus on the uniqueness of Mexican wolves as a subspecies using both morphometric (Bogan and Melhop 1983; Nowak 1995; Nowak 2003; Chambers et al. 2012) and genetic data (Hedrick et

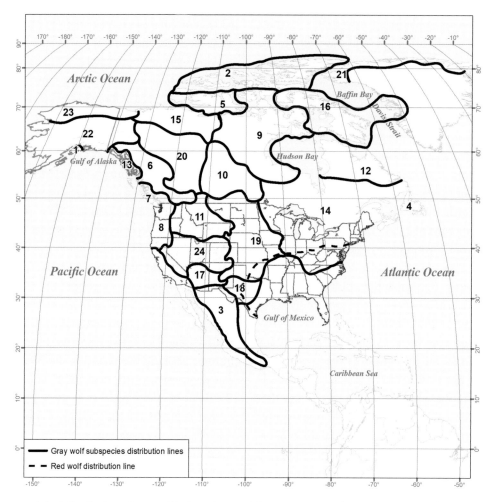

Map 11.1. Original distribution of gray wolf subspecies (*Canis lupus* spp.) in North America (adapted from Hall 1981). The red wolf's (*Canis rufus*) distribution is south of the dashed line. The gray wolf subspecies are represented by numbers, identified as: 1) *C. l. alces*, 2) *C. l. arctos*, 3) *C. l. baileyi*, 4) *C. l. beothucus*, 5) *C. l. bernardi*, 6) *C. l. columbianus*, 7) *C. l. crassodon*, 8) *C. l. fuscus*, 9) *C. l. griseoalbus*, 10) *C. l. hudsonicus*, 11) *C. l. irremotus*, 12) *C. l. labradorius*, 13) *C. l. ligoni*, 14) *C. l. lycaon*, 15) *C. l. mackenzii*, 16) *C. l. manningi*, 17) *C. l. mogollonensis*, 18) *C. l. monstrabilis*, 19) *C. l. nubilus*, 20) *C. l. occidentalis*, 21) *C. l. orion*, 22) *C. l. pambasileus*, 23) *C. l. tundrarum*, and 24) *C. l. youngi*. All these subspecies were later merged into just five: *C. l. arctos* (roughly represented by 2, 5, and 21 on this map); *C. l. baileyi* (represented by 3); *C. l. lycaon* (a small area around Lake Erie and Lake Ontario in 14); *C. l. nubilus* (roughly represented by 4, 7–13, most of 14, 16–19, and 24); and *C. l. occidentalis* (roughly represented by 1, 6, 15, 20, 22, and 23) by Nowak (1995).

al. 1997; Leonard et al. 2005; VonHoldt et al. 2011; Fredrickson et al. 2015) likely reflecting their early divergence from other gray wolves in North America. Today, gray wolves in New Mexico are represented by that one subspecies, the Mexican wolf. It was almost extinct in the wild—only a few animals remained in Mexico and the southwestern United States—at the time that it was listed as Endangered on 28 April 1976 (USFWS 1976a), with none thought to exist in the wild prior to its reintroduction beginning in 1998. The US Fish and Wildlife Service has managed Mexican wolves under flexible management measures (under the designation of a non-essential experimental population)

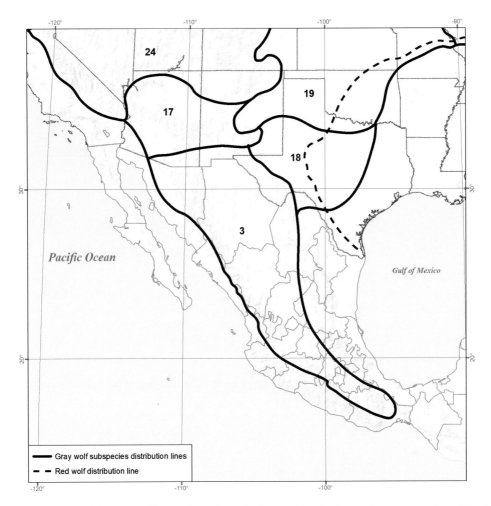

Map 11.2. Subspecies of the gray wolf (*Canis lupus*) historically occurring in the southwestern section of North America (according to Hall 1981; but see Nowak [1979, 1995, 2003]): 3) *C. l. baileyi*, 17) *C. l. mogollonensis*, 18) *C. l. monstrabilis*, 19) *C. l. nubilus*, and 24) *C. l. youngi*. Scientists have disagreed whether *C. l. mogollonensis* and *C. l. monstrabilis* should be synonymous with *C. l. baileyi* (Bogan and Melhop 1983; Chambers et al. 2012) or with *C. l. nubilus* (Nowak 1995; Nowak 2003). This area is likely a transition zone where characteristics of each subspecies are present, with males being morphologically closer to *C. l. baileyi*, and females being closer to *C. l. nubilus* (Bogan and Melhop 1983). The US Fish and Wildlife Service established areas previously identified as *C. l. mogollonensis* and *C. l. monstrabilis* as within the probable historical range of Mexican wolves (*C. l. baileyi* [USFWS 1998; see Map 11.3]).

to facilitate reintroduction into Arizona and New Mexico (USFWS 2015a)

DISTRIBUTION

The gray wolf once ranked among the most widely distributed wild mammals on Earth as it occupied nearly all of Eurasia and North America, except for extreme desert environments where prey populations are too limited to support wolves (Mech and Boitani 2003). In the Old World, the historical distribution of gray wolves stretched from the Artic to the Indian peninsula (~13°N latitude), and it might have included all of Eurasia and northern Africa (Nowak 2003), though animals previously

Photo 11.4. Mexican wolf (*Canis lupus baileyi*) collected near Hatch in Doña Ana County sometime before 1940 and now preserved as a specimen (MSB:Mamm:4078) at the University of New Mexico's Museum of Southwestern Biology. According to Nowak (1979, 1995, 2003), two subspecies of the gray wolf (*Canis lupus*) occurred historically in New Mexico, the Mexican wolf in the southwestern part of the state, the Great Plains wolf (*C. l. nubilus*) elsewhere. Railroad development brought large numbers of livestock to the Southwest during the 1870s and 1880s, resulting in overstocking of rangelands and conflicts between ranchers and predators such as the gray wolf. Bounties and later control programs led by the US government resulted in the eradication of Mexican and Great Plains wolves in New Mexico. Photograph: © Jon Dunnum.

classified as *C. lupus* in northern Africa now appear to be separate species based on genetic data (see review by Hatlauf et al. 2021). Similarly, gray wolf populations stretched across most of North America from polar regions to approximately Mexico City (~15°N latitude) in the south, with the possible exceptions of areas in the northeastern United States and Canada and in the southeastern United States, where as discussed above the eastern wolf (identified as subspecies 14 in Map 11.1, based on Hall [1981]) and red wolf (south of the dashed line in Map 11.1), respectively, could both represent separate species (Sacks et al. 2021).

Of the 24 subspecies described by Young and Goldman (1944) and Hall (1981), five were found in New Mexico (Maps 11.1 and 11.2). Of the five subspecies recognized by Nowak (1979, 1995, 2003), two had distributions overlapping New Mexico (*C. l. baileyi* in the southwestern part of the state

and the Great Plains wolf [*C. l. nubilus*] elsewhere). Today, controversy remains as to the exact historical distribution of these two subspecies (Leonard et al. 2005; Hendricks et al. 2016, 2017; Heffelfinger et al. 2017a, b). The issue is germane to Mexican wolf recovery efforts in Arizona and New Mexico under the ESA because implementing ESA regulations that provide more management flexibility (e.g., the ability of private individuals to take [i.e., kill, injure, or harass] wolves caught in the act of attacking livestock on private land) are based on whether the reintroduction is taking place within the historical range of the species (or subspecies in this case) (see USFWS 1998a and 2015a for more information).

Following the listing of the Mexican wolf as Endangered under the ESA on 28 April 1976 (USFWS 1976a), another one of Hall's (1981) subspecies (*C. l. monstrabilis*; not recognized by Nowak [1979, 1995,

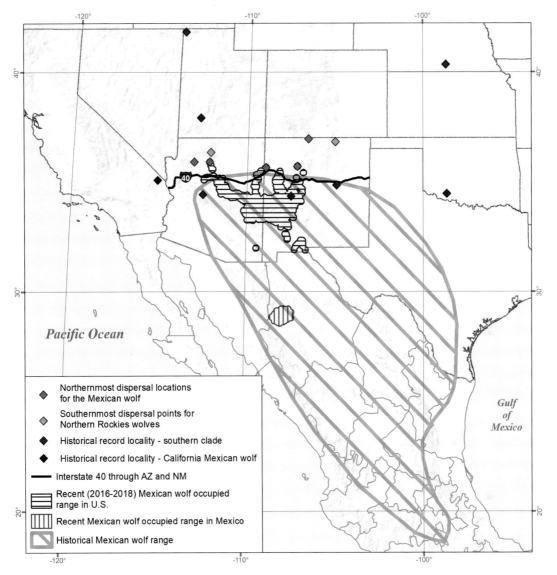

Map 11.3. Mexican wolf historical range (hashed-line) as adopted by the US Fish and Wildlife Service (Parsons 1996; USFWS 1998) in North America. The authors of this chapter delineated the current occupied range by generating a 10-mile buffer around known wolf locations from 2016 to 2018. Blue and black diamonds indicate historically collected specimens that appear to have Mexican wolf ancestry (Leonard et al. 2005; Hendricks et al. 2016), though the interpretation relative to historic range is disputed (Hendricks et al. 2016; Hefflefinger et al. 2017a; Hefflefinger et al. 2017b; Hendricks et al. 2017). It is clear that Mexican wolves (gray diamonds) are dispersing to the north and absent manipulation would likely exchange genetic material with wolves dispersing from northern populations of gray wolves (orange diamonds).

2003]) was listed on 14 June 1976 (USFWS 1976b) in the Southwest. This subspecies was likely listed because two dispersing animals from Mexico were documented in 1970 in Texas (Scudday 1972), where previously *C. l. monstrabilis* (the Texas gray wolf) had been described (Young and Goldman 1944). However, these animals were determined to be Mexican wolves based on morphological characteristics (Scudday 1972; Nowak 1995), and *C. l. monstrabilis*, once also found in southeastern New

Mexico according to Hall (1981) (Maps 11.1 and 11.2), is now considered to have been extinct since 1942 (Wozencraft 2005). Another dispersing Mexican wolf documented in Concho, Arizona was within the northern historical range of yet another one of Hall's (1981) subspecies, *C. l. mogollonensis* (Nowak 1995). The US Fish and Wildlife Service has consistently included the broader area occupied by *C. l. mogollonensis* and *C. l. monstrabilis* (Map 11.2) within the historical range of *C. l. baileyi*. This broader area (Map 11.3) is consistent with multiple interpretations of the subspecies' historical range (USFWS 1982, 1998a, 2014a, 2015a; Bogan and Melhop 1983; Parsons 1996; Chambers et al. 2012), a range at the time of listing in 1976 (specimens used in support of a *C. l. mogollonensis* and *C. l. monstrabilis* listing corresponded to *C. l. baileyi*), and the reintroduction of *C. l. baileyi* into central Arizona and New Mexico). While the location of the northern boundary of *C. l. baileyi*'s historical range has been debated, the scientific consensus is that *C. l. baileyi* was the only subspecies of gray wolf in Mexico, and it occurred south to approximately Mexico City (Nowak 1995).

Numerous authors (Leonard et al. 2005; Hedrick 2013; Carroll et al. 2014; Hendricks et al. 2016, 2017, 2019) have argued for an expanded interpretation of Mexican wolf historical range boundaries based on 1) the presence of closely related members of a southern gray wolf clade to the north of the traditionally recognized Mexican wolf's historical distribution (Leonard et al. 2005; Map 11.3); 2) genetic analyses revealing a unique Mexican wolf mtDNA haplotype in a specimen collected in 1922 in San Bernardino County, California (MVZ:MAMM:33389) (Hendricks et al. 2016); and 3) conservation needs and vacant wolf habitat available to the north, coupled with the idea that highly mobile species would exhibit extensive admixture with other subspecies of gray wolf (Leonard et al. 2005; Carroll et al. 2014; Hendricks et al. 2016, 2019). However, other authors disagree with this expanded interpretation of the Mexican wolf's range because of the inadequate sampling of genetic markers, morphological differences, and ecological relationships (Heffelfinger et al. 2017a, b).

Definitions of historical range boundaries may become less important in the future, as dispersals by Mexican wolves in both Arizona and New Mexico have resulted in a northward expansion of the subspecies' occupied range within each of these two states (Map 11.3). Northern movements by Mexican wolves reaching the south rim of the Grand Canyon and southern dispersal by a northern gray wolf (*C. l. occidentalis*) from Wyoming to the north rim of the Grand Canyon before returning to Utah suggest that the Grand Canyon may form a natural filter for dispersal or interaction between subspecies (Map 11.3), though Mexican wolves were removed before completion of the dispersal due to current regulations stipulating that Mexican wolves must remained confined within the Mexican Wolf Experimental Population Area (MWEPA), which lies south of I-40 (USFWS 2015a). The two subspecies (*C. l. baileyi* and *C. l. occidentalis*) could interact even more frequently in New Mexico than in Arizona (Map 11.3). A presumed southward dispersal by a northern gray wolf and a northward dispersal of a Mexican wolf to northern New Mexico suggest that this part of the state would be an intergradation zone if wolves were allowed to naturally recolonize the area (Map 11.3). Individual northern wolves from core breeding populations in Wyoming that formed as a result of reintroductions into Yellowstone may continue to disperse into northern Arizona and New Mexico. But absent the reintroduction proposed in Colorado, more flexible Mexican wolf regulations, or substantial natural recolonization in Colorado (one breeding pair has been documented in that state) or Utah, wolves are unlikely to form packs in northern Arizona and New Mexico because of low survival rates for dispersing wolves (Smith et al. 2010) and the low likelihood of finding mates far from

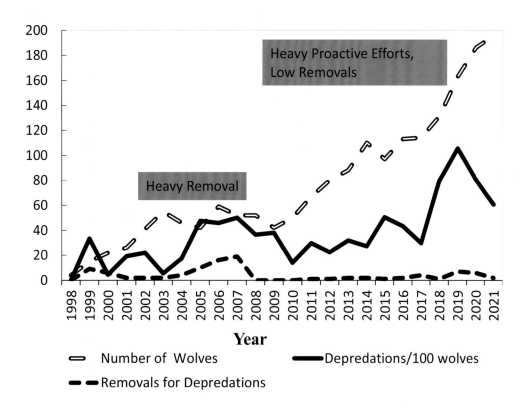

Figure 11.1. Minimum population count (number of wolves), removals, and depredation rates (depredations/ 100 wolves) in the southwestern United States from 1998 to 2021. Changes in management strategies allowed for an increase in the growth of the Mexican wolf population from 2009 to 2021, but recent increases in depredation rates are a cause for concern and reconsideration of management methods.

established populations (Mech and Boitani 2003; Jimenez et al. 2017).

Historically, gray wolves in the Southwest appeared to be most prevalent in the mountainous areas above 1,800 m (6,000 ft; Brown 1983). They nonetheless occurred throughout New Mexico prior to European colonization, except for low desert areas where wolves were likely only transient (Brown 1983). In contrast, wolves in the southwestern United States and Mexico currently exist only in areas where releases of Mexican wolves from captivity have occurred to establish populations. In Mexico, a small population of wolves (~30) has been established to the southwest of Casa Grandes, Chihuahua, in one of

the last reported areas where wolves existed prior to extirpation (McBride 1980). Individual wolves released in Mexico have traveled north to approximately Las Cruces in New Mexico and the Chiricahua Mountains in Arizona. The Mexican wolf population in Arizona and New Mexico is composed entirely of animals born and raised in the wild, except for some pups from captivity placed into wild dens to improve the genetic diversity of the population. This population is increasing in size and distribution (Map 11.3 and Figure 11.1). It consisted of approximately 196 animals at the end of 2021 in the United States (Figure 11.1), with breeding packs occurring between White River, Arizona and Socorro, New Mexico. Individual

Photo 11.5. Mountain stream running in late winter in the Mexican Wolf Experimental Population Area (MWEPA). The MWEPA is a defined geographic area that encompasses Arizona and New Mexico from I-40 south to the international border with Mexico. Most Mexican wolves can be found in areas of evergreen pine-oak woodlands and mixed conifer forests where elk (*Cervus canadensis*) and mule deer (*Odocoileus hemonius*) also occur. Most wolf habitat in the MWEPA is located at higher elevations in the Apache-Sitgreaves National Forest and in portions of the Fort Apache Indian Reservation in Arizona, and in the Gila National Forest and the Magdalena Ranger District of the Cibola National Forest in New Mexico. Photograph: © Interagency Field Team.

Photo 11.6. Yearling Mexican wolf (Luna pack) in February 2011 in the Gila National Forest, New Mexico. The Luna pack utilized the central portion of the Gila National Forest, which is dominated by mixed conifer and ponderosa pine woodlands. Photograph: © Interagency Field Team.

Photo 11.7. Three eight-month-old wolf pups (Seco pack) together in December 2021 on the Ladder Ranch in south-central New Mexico. The Seco pack utilizes a mix of vegetation types ranging from large open areas of pinyon-juniper grassland ecotone at the lower elevations to mixed conifer at the higher elevations. Photograph: © Cassidi Cobos/ Turner Endangered Species Fund.

radio-collared animals from the re-established population have been documented as far north as the south rim of the Grand Canyon (Cataract Canyon) in Arizona, the Chuska Mountains along the New Mexico and Arizona border, and north of Albuquerque in New Mexico (Map 11.3). Individual Mexican wolves have also dispersed as far south as the US-Mexico border to the west of Las Cruces, New Mexico. In the foreseeable future Mexican wolves will likely expand their range throughout most of the quality habitat above 1,800 m (6,000 ft) south of I-40 (Map 11.3).

HABITAT ASSOCIATIONS

Wolf habitat utilization patterns and densities can best be predicted by ungulate density and anthropogenic influences (e.g., road and human densities). Ungulate density is the primary factor regulating wolves in areas without high anthropogenic impacts (Fuller 1989; Fuller et al. 1992, 2003; Cariappa et al. 2011; McRoberts and Mech 2014). Thus, wolf habitat associations both historically and currently mimic those of the primary prey in a particular area. Habitat relationship models typically estimate ungulate habitat either through an index of primary productivity, the greenness index (Carroll et al. 2003, 2006, 2014), or a direct assessment of ungulate densities (Mladenoff et al. 1995; Larsen and Ripple 2006; Oakleaf et al. 2006; Martinez-Meyer et al. 2021), all of which predict wolf productivity (Carroll et al. 2003, 2006), densities (Mladenoff et al. 1995, Larsen and Ripple 2006), colonization probabilities (Oakleaf et al. 2006), or relative habitat suitability across potential recovery areas (Martinez-Meyer et al. 2021).

Negative relationships between anthropogenic influences and wolf use of an area have been observed by numerous authors (Thiel 1985; Jensen et al. 1986; Mech et al. 1988; Fuller et al. 1992; Mladenoff et al. 1995; Carroll et al. 2003; Larsen and Ripple 2006; Oakleaf et al. 2006). These relationships likely change through time as wolves colonize reduced quality habitat (e.g., wolves select habitat closer to anthropogenic influences, but likely suffer higher mortality in these areas) (Mech 2006a, b; Mladenoff et al. 2006), but they have remained useful for predicting wolf landscape recolonization and distribution (Mladenoff et al. 1995, 1999, 2006). We generally consider the most important current habitat attributes needed for wolves to persist and succeed to be high forest cover, high native ungulate density, and low livestock density; in contrast, unsuitable areas are characterized by low forest cover and high human density (sometimes modeled as road density because road and human densities are correlated) and land use (Oakleaf et al. 2006). Suitable areas for wolves are associated with minimal road development, as human access to areas inhabited by wolves can result in increased wolf mortality (due to illegal killing, vehicular mortality, or other causes). Similarly, areas with minimal forest cover also likely lead to increased wolf mortality because wolves are more observable and susceptible to human-caused mortality, while high livestock density typically leads to removal of the wolves by management agencies. Public lands, such as national forests, are considered to offer more suitable conditions for wolf reintroduction and recovery efforts in the United States than other land ownership types because they typically are characterized by minimal human development and appropriate habitat characteristics (Fritts and Carbyn 1995).

In the southwestern United States and in Mexico, habitat selection by wolves is somewhat predictable because prey animals are too rare to support wolves in the low-lying, extreme desert areas whereas they are more abundant in mountains. Thus, wolf habitat models tend to identify very similar areas in Arizona and New Mexico (see Figure 5 in USFWS 2014a). Indeed, this pattern continues to be reinforced through recent investigations in Mexico and the United States (USFWS 2017; Martinez-Meyer et al. 2021). Historically, wolves in the Southwest were known to occupy areas ranging from foothills characterized by evergreen oaks (*Quercus* spp.) or pinyon (*Pinus edulis*) and juniper (*Juniperus* spp.) to higher elevation pine (*Pinus* spp.) and mixed conifer forests (Brown 1983). Factors making these habitats attractive to wolves likely included an abundance of prey, the availability of water, and the presence of hiding cover and suitable den sites. Wolves also likely occurred in the northeastern plains and the Llano Estacado plains of southeastern New Mexico in association with bison (*Bison bison*). However, both of these areas are unlikely to be utilized in the future due to the nearly complete regional disappearance of bison (note: elk and deer are present but in moderate numbers) and the relative openness of the vegetation, which today indicates high risks of high human-caused mortality. Thus, the simple description of mountainous areas over 1,800 meters (6,000 feet) in elevation (Brown 1983) as wolf habitat is an easy way to understand the areas where most wolves in Arizona and New Mexico occur and are likely to persist into the future.

LIFE HISTORY

The gray wolf is an obligate carnivore whose morphology, social organization, and behavior have been shaped by its evolutionary history as a cursorial (running/chasing) predator of large herbivores (Peterson and Ciucci 2003). Wolves are social animals born into a family unit referred to as a pack, which consists of a breeding pair and its offspring from the last year or the last few years (Mech and Boitani 2003). A wolf pack establishes a territory that is defended from other packs, and pack members hunt, breed, den, and raise pups within the territory (Mech and Boitani 2003). Wolf pack territory size varies depending on both prey density and pack size (Fuller et al. 2003). Much of the wolf's ecology and life history are determined by the predominant species of ungulate present locally.

Diet and Foraging

Throughout their distribution, wolves prey mainly on wild ungulates, but they are also known to consume a wide variety of animals ranging in size from small rodents, rabbits (*Sylvilagus* spp.), and jackrabbits (*Lepus* spp.) to large bison and moose (*Alces alces*) (Peterson and Ciucci 2003; Carrera et al. 2008), and they can even switch to a diet consisting primarily of garbage or domestic animals (Peterson and Ciucci 2003). Wolves have a narrower diet than coyotes (see Chapter 11), and they focus a large proportion of their foraging activities just hunting large ungulates. They lack the supinating forearms or retractable claws of felids, and the jaw's functionality is a balancing act with teeth designed to tear, grasp, and hold (Peterson and Ciucci 2003). Thus, wolves are relatively ineffective predators on a per-encounter basis and are successful at killing prey <20% of the time that they come in contact with a prey item the size of an elk or bigger (MacNulty et al. 2020). One can imagine, then, that wolves spend a lot of time searching for vulnerable prey and tend to encounter potential prey items relatively frequently when moving through the landscape (MacNulty et al. 2020). When wolves encounter a potential prey item, they want to force the ungulate(s) to start running, which then makes them more vulnerable to a wolf attack (MacNulty et al. 2020). The wolves bite hind quarters and flanks as they pursue a potential prey item and often grasp the nose or neck during the final stages of a kill (Peterson and Ciucci 2003; MacNulty et al. 2020). Mexican wolves do not deviate from this strategy when hunting elk in the Southwest (USFWS, unpubl. data) but may be less successful due to their smaller size relative to northern wolves (MacNulty et al. 2020).

Ungulates in turn have evolved a wide variety of defensive traits and mechanisms to avoid or lessen the risk of wolf predation, such as antlers and hooves, the lack of scent in neonates, grouping and vigilant behavior, the following of adults at a young age, and use of specialized habitats such as steep and rocky terrain (Mech and Peterson 2003). Because wolves are coursing predators and select prey animals from a herd, a larger proportion of their kills is thought to represent compensatory (animals that would die in a given year from other causes) rather than additive mortality (animals that would otherwise have survived in a given year) compared to ambush predators such as cougars (*Puma concolor*) (Husseman et al. 2003; Wright et al. 2006). Indeed, wolves typically kill a higher proportion of juveniles and older adults than adults in the prime of their life in an ungulate population (Mech et al. 1995; Husseman et al. 2003; Smith et al. 2004; Stahler et al. 2006).

Historically, the diet of wolves in the Southwest likely reflected the distribution of the primary ungulate prey species. Thus, in the isolated mountains in southern Arizona and New Mexico as well as in Mexico, wolves preyed principally on Coues white-tailed deer (*Odocoileus virginianus*

couesi) (Brown 1983). Mule deer (*O. hemionus*) were mainly located at low densities in low-lying, open basins and valleys and thus unlikely served as primary prey of wolves in the southern part of their distribution (Brown 1983). Wolves likely focused on elk and mule deer historically in the Mogollon Rim in central Arizona and New Mexico and areas farther to the north, as well as the in Sacramento and Guadalupe mountains in southern New Mexico until elk were extirpated or greatly reduced in numbers in the early 1900s (Truett 1996). Currently, elk appear to be the favored prey of Mexican wolves in the reintroduction area in Arizona and New Mexico, with deer and cattle being secondary prey (Mexican Wolf Interagency Field Team and Mexican Wolf Adaptive Management Oversight Committee 2005; Reed et al. 2006; Merkle et al. 2009; USFWS 2018). Many other species have been documented in the diet of Mexican wolves in Arizona and New Mexico since the reintroduction of Mexican wolves, including

Photo 11.8. Mexican wolves feed on an elk (*Cervus canadensis*) carcass in August 2011 on the Apache-Sitgreaves National Forest in Arizona. Elk make up approximately 80% of the diet of Mexican wolves in the United States. Photograph: © Interagency Field Team.

porcupine (*Erethizon dorsatum*), squirrels (*Sciurus and Tamiasciurus* spp.), and lagomorphs (*Lepus and Sylvilagus*) (Reed et al. 2006; Merkle et al. 2009). However, these items combine for less than 3% of the biomass consumed (killed or scavenged) by Mexican wolves, whereas elk comprised 75 to 80% of the observed diet, followed by cattle at 8 to 17%, and deer at 0.2 to 3% (Reed et al. 2006; Merkle et al. 2009).

Impact On Prey Populations

One of the most controversial aspects of wolf predation relates to whether, and by how much, wolves reduce ungulate numbers. Severe reductions in ungulate numbers by wolves could impact human hunting opportunities, but control of wolves to maintain ungulate populations is controversial (National Research Council 1997). The real impact of wolves on ungulate population dynamics is related to other environmental factors, such as weather, browse and forage conditions, and the number and type of predator and prey species present in the ecosystem. Wolf predation likely accelerates declines in ungulate populations already negatively affected by stressors such as drought or disease (Gasaway et al. 1983; Fuller 1990; Boertje et al. 1996, 2010). Wolf predation is also likely to slow the rate of population recovery after declines in ungulate numbers (Gasaway et al. 1983; Fuller 1990; Boertje et al. 1996, 2009, 2010). In this case wolf predation can become an additive mortality factor and may limit the ability of a prey population to increase (Klein 1995; Boertje et al. 2009). In extreme situations, localized extinction of prey has been observed as a result of wolf predation (Mech and Karns 1977). In Yellowstone, elk populations decreased by 50% following wolf reintroduction (Creel et al. 2011), but other factors (e.g., weather, human hunting on the outskirts of Yellowstone, and mortality from other predators) also play important roles; thus, the wolf's true role in the

elk's decline is debated (Creel 2005; Vucetich et al. 2005; Eberhardt et al. 2007).

The ratio of wolves to ungulates in an area is another metric that researchers have used to describe predator-prey interactions. In the northern Rockies, elk herds that exist in a multi-predator system (wolves, grizzly bears [*Ursus arctos*] and black bears [*U. americanus*], cougars, and coyotes) and are exposed to human harvest may be impacted by wolves at ratios of 4:6 wolves per 1,000 elk (White and Garrott 2005; Hamlin and Cunningham 2009). Wolf-to-elk ratios of 9:35 wolves per 1,000 elk in a multi-predator system were associated with elk population declines and significant impacts to human hunting opportunities (Dekker et al. 1995; Ballard et al. 2001; White and Garrott 2005; Hamlin and Cunningham 2009; Beschta and Ripple 2010). However, other authors caution that these same wolf-to-elk ratios may produce very different effects in areas characterized by different climates (e.g., less severe winters may reduce vulnerability of elk to wolf predation and allow elk populations to withstand higher densities of wolves without experiencing declines), hunting harvest levels (e.g., lower levels of female harvest may again allow elk population to withstand higher densities of wolves without any decline), and availability of prey refugia (e.g., areas near human populations or between packs where predation is reduced) (Hamlin and Cunningham 2009; Vucetich et al. 2011; Hebblewhite 2013). Winter mortality is comparatively low, and baseline survival higher, for elk populations in the areas wolves occupy in the Southwest (Ballard et al. 2000). For this reason, the relationship between wolf-to-elk ratios and elk population trends, as established in ecosystems with grizzly bears and more severe winters, is less useful for predicting wolves' impact on elk in the Southwest. Wolf-to-elk ratios in Arizona and New Mexico do not appear to be high enough to impact prey populations at a large scale because the Mexican wolf population is still recovering, but research should examine whether localized impacts are occurring in areas with high wolf densities.

Reproduction and Social Behavior

The basic social unit of gray wolves is a pack that is typically composed of one breeding pair and their offspring from one or more years (Mech and Boitani 2003). Wolves are generally considered monogamous, but there are exceptions (see below), with the breeding pair producing a litter of pups, usually four to six, each year (Mech 1970). Gray wolves reach sexual maturity just before two years of age and have one reproductive cycle per year (Mech and Boitani 2003). Most wolves disperse from their natal pack by the age of three (Mech and Boitani 2003), while those that remain following reproductive maturity do not move into a breeding role unless one of their parents dies and an unrelated individual of the opposite sex joins the pack, thus avoiding inbreeding (Packard 2003).

Several instances of multiple litters in the same pack have been documented but generally relate to a new breeding male joining the pack and mating with multiple unrelated females in the pack (Stahler et al. 2002; UFWS, unpubl. data). Mexican wolf pups are generally born between early April and early May and remain inside the den for three to five weeks. In the Southwest, dens are generally dug out holes typically under rock ledges or in piles of large rocks, but some have also been found in hollow downfall trees or in shallow depressions in the ground. Once pups are born the den site becomes the center of the pack's activity as all members provision the litter (Mech and Boitani 2003). Pups are generally moved from den sites to rendezvous or home sites in the summer at approximately two months of age. Rendezvous sites are places where the pack's adults leave the pups prior to hunting. The pups are not strong enough yet to participate in hunting activities, but they do not hide and instead readily explore the area surrounding the rendezvous site until the adults return. The pack generally utilizes one or several rendezvous sites until the pups reach five to six months of age and they can travel larger distances. After the pups are older than six months, members of the pack then only localize for a short period of time at kill sites while they feed on a carcass, until pups are born the following year.

Mortality among young pups occurs frequently. In fact, approximately half of all pups die before reaching one year of age in Arizona and New Mexico (USFWS 2017). Even with high mortality rates, however, pups generally represent the greatest component of a pack in the winter, followed by yearlings. Offspring that are two to three years old are occasionally located with the pack, but relatively few remain as most within the cohort have died or dispersed. Adults that have established territories have a much higher annual survival rate (>80% in the Southwest) and generally persist as the breeding pair

Photo 11.9. Eight members of the Bluestem pack investigate a food cache just after sunset in Arizona's Apache-Sitgreaves National Forest in October 2014. In 2014, the Bluestem pack consisted of 12 wolves, the alpha pair, five yearlings, and five pups. Usually, Mexican wolf packs are smaller, consisting of only four to six wolves. Wolf pack size can be influenced by multiple factors, such as habitat quality, the level of experience of the adult wolves, and human tolerance. Photograph: © Interagency Field Team.

Photo 11.10 (*top left*). Rim pack den on the Apache-Sitgreaves National Forest in Arizona in June 2011. Rim pack AF858 (barely discernable here) remains hidden in the den, while biologists document pup survival following the Wallow Fire that burned over significant portions of the Rim pack territory. Mexican wolves utilize a wide variety of den types, ranging from deep underground dens of more than 4.5 m (15 ft) in length to aboveground dens located in rocky hillsides or cavities under large logs or fallen trees. Photograph: © Interagency Field Team.

Photo 11.11 (*bottom left*). Two-week-old, Iron Creek wolf pups huddle together in their den on the Gila National Forest in New Mexico in May 2016. Mexican wolves usually give birth to four to eight pups per litter. Mexican wolves breed once a year, usually in mid to late February and whelp approximately 60 days later in mid to late April. Photograph: © Interagency Field Team.

Photo 11.12 (*top right*). Twelve-week-old Seco Creek pack wolf pups remain at the rendezvous area in July 2021 on the Ladder Ranch, New Mexico, while adult wolves hunt elk nearby. Wolf pups tend to begin exploring outside the den around 4 weeks of age. Once the pups become mobile at 8–12 weeks of age, the pack will generally move a short distance from the den site to a rendezvous area. The pups remain in that area while the older wolves hunt and forage. When the pups reach four to six months of age, they begin accompanying the pack as it travels throughout its home range. Photograph: © Cassidi Cobos/Turner Endangered Species Fund.

and produce pups during each annual cycle. The average number of members in the pack varies as a function of major prey item, prey density, and terrain (Fuller et al. 2003). In the southwestern United States and Mexico, pack size ranges from two to 14 with an average of four to five wolves (USFWS 2017).

Territoriality and Dispersal

Daily and seasonal movements of wolves within a pack vary in response to care of young and the distribution, abundance, and availability of prey. A territory must contain enough prey to maintain the pack. Gray wolf territory sizes range from under 100 km^2 to 1600 km^2 (39 to 618 mi^2; Fuller et al. 2003). Bednarz (1988) predicted that reintroduced Mexican wolves would likely occupy territories ranging from approximately 200 to 400 km² (78 to 158 mi²) and hypothesized that Mexican wolf territories were comparable or slightly larger in size to those of small packs of northern gray wolves due to habitat patchiness (mountainous terrain that included areas of unsuitable lowland areas) and lower prey densities associated with the regional arid environment. Between 1998 and 2015, territory size for 138 denning packs in the MWEPA population averaged 510 km² +/- 324 km² (197 mi² +/- 125mi²; USFWS 2013, 2014b, 2015b), or much larger than predicted by Bednarz (1988) but consistent with other gray wolf territory sizes where the primary prey is elk (Fuller et al. 2003; Oakleaf et al. 2006).

Gray wolves exhibit a keen ability to disperse, find mates, and locate vacant territories. An individual wolf will disperse from the natal pack's territory in search of vacant habitat or a mate, typically when between nine and 36 months of age. Each pack can be considered a dispersal pump that sends offspring into the landscape to colonize new areas or replace a breeding animal in an established territory (Mech and Boitani 2003). Dispersals may consist of short trips to a neighboring territory, or a long-distance journey of hundreds of kilometers (Ballard et al. 1983; Fritts 1983; Boyd et al. 1995; Packard 2003; Jimenez et al. 2017). For instance, a wolf dispersed from around Cody, Wyoming to the north rim of the Grand Canyon in Arizona (a straight-line distance of ~965 km or 600 mi). Wolves that disperse and locate a mate and an unoccupied patch of habitat establish a new territory (Rothman and Mech 1979; Fritts and Mech 1981) and generally have a higher lifetime reproductive output than individuals joining existing packs to replace breeding animals (Jimenez et al. 2017). However, dispersing wolves tend to incur a higher risk of mortality relative to wolves that are resident in a pack (Smith et al. 2010). Regardless of the risks associated with dispersal, the phenomenal capacity of wolves to disperse over large distances explains how they can rapidly recolonize vacant habitat and maintain populations despite high levels of human-caused mortality (Mech and Boitani 2003; Adams et al. 2008; Jimenez et al. 2017).

Similar to gray wolves elsewhere, Mexican wolves have shown that they can travel hundreds of kilometers during dispersal. For instance, a wolf dispersed from around Nutrioso, Arizona in a southeasterly direction to the New Mexico–Mexico border wall to the west of El Paso, Texas (a straight-line distance of ~321 km or 200 mi) and, heading back to the northwest, was able to reach the Gila National Forest. Movements over such large distances show that Mexican wolves could disperse to any location in Arizona and New Mexico from the centrally occupied range (Map 11.3). However, Mexican wolf dispersal and occupancy outside the boundaries of the Gila and Apache national forests (USFWS 1998a) and Fort Apache Indian Reservation have been limited by regulation, resulting in 12% of dispersal events from 1998 to 2015 ending with the removal or translocation of the wolf due to the I-40 boundary rule. Another 17% of Mexican wolf dispersal events during 1998–2015 resulted in mortality. However, over half (55%) of the dispersal events documented between 1998 and 2015 ended with the wolf successfully locating a mate (USFWS 2017), which was similar to the rate observed in the northern Rockies (Jimenez et al. 2017). Long-distance dispersal continues to be limited by the regulation that prevents occupancy north of I-40 in Arizona and New Mexico (USFWS 2015a), which is why a true representation of the dispersal capabilities of Mexican wolves may depend on movements to the south through areas where a border wall does not exist (see Chapter 5).

Interspecific Interactions

Large carnivores such as wolves can have complex interactions with other species. As predators, wolves may kill and consume individuals of other species (see the section on diet and foraging, above). They may also be involved in aggressive interactions, some of them resulting in mortality, with competing predators such as cougars or coyotes. Physical exclusion of other predators from resources through aggressive behavior is referred to as "interference competition" (Ballard et al. 2003; see Chapters 7 and 9). Wolves can also interact with other carnivore species through competition that results in reduced shared resources, or "exploitative competition." A host of other species are indirectly affected, positively or negatively, by wolves regulating prey numbers and/or behavior (Hebblewhite and Smith 2010; Ripple et al. 2014b), providing carrion (Wilmers et al. 2003), and/or reducing other carnivore species, which in turn may have been limiting the populations of smaller animals (Hebblewhite and Smith 2010; Ripple et al. 2014b).

Wolves regulate wolf-ungulate ecosystems through a "top–down" (predator-driven) process rather than through a "bottom-up" (food-limited) trophic cascade (Vucetich and Peterson 2004; Hebblewhite and Smith 2010). A "top-down" process of regulation suggests that carnivores occupying the highest trophic level will exert a significant influence on their prey through direct or indirect interactions (Ripple et al. 2014b). Trophic cascades can occur when trophic levels are limited by predation (e.g., because they are reduced by wolf predation, elk populations have less of an impact on willows [*Salix* spp.]), which then provides more nesting habitat for songbirds and results in corresponding increases in songbird populations). Thus, wolves can influence many species by limiting ungulate abundance or changing ungulate grazing patterns.

"Top-down" processes generally occur when wolves and a host of other predators (cougars, grizzly and black bears, coyotes, humans) limit ungulates below the nutritional-carrying capacity (K) of the environment (Brodie et al. 2013). K varies throughout the year (e.g., it is higher in the summer than the winter), between years (e.g., particularly hard winters or drought conditions can limit K), and with landscape or long-term climate change (e.g., increasing human development or hotter and drier conditions). Thus, K is generally only useful as a concept (MacNab 1985) except in low variability environments where it can be modeled (McLeod 1997). Predation that occurs on populations that are below K at any particular time of year may be considered additive mortality (e.g., mortality that would not have otherwise occurred in a given year).

Conversely, a "bottom-up" trophic regulation would suggest that ecosystems are driven by resource abundance and weather without significant influence from predation (Strong 1992). In this instance, most predation would be considered compensatory, meaning that total mortality in the prey population remains unchanged, as, for example, if the animal killed by the predator would still have died from starvation or disease. Nature is constantly changing, so "top-down" and "bottom-up" processes are constantly interacting, with one dominating depending on the area (e.g., number and types of predators and prey) and/or specific weather conditions in a given year (e.g., drought or severe winters). For instance, wolf predation combined with severe winters may limit elk populations (Brodie et al. 2013). Further complications with the simple classification of "top-down" versus "bottom-up" processes occur in the form of anthropogenic impacts (e.g., fire control, cattle grazing, hunting, habitat alteration) on areas outside of national parks (Mech 2012) and prey refugia near human settlements (Hebblewhite and Smith 2010).

The effect of the Mexican wolf reintroduction on ecosystems in Arizona and New Mexico is not yet observable. For example, severe declines of aspen (*Populus tremuloides*) have been observed in both states with a wide variety of potential contributing factors such as fire management, climate change, and extensive elk herbivory preventing aspen regeneration (Beschta and Ripple 2010). However, an investigation into the possible relationship between Mexican wolf presence (and a behavioral or numerical impact on ungulate populations) and aspen regeneration did not detect any pattern. Beschta and Ripple (2010) hypothesized that Mexican wolves had not reached an ecologically effective population size at the time of their research. The impact of wolves on aspen through behavioral or numerical impacts on elk remains a tenuously established and highly debated case of trophic cascade (Hebblewhite and Smith 2010; Kauffman et al. 2010; Winnie Jr. 2012, 2014; Beschta et al. 2014) that might be associated with a longer lag time and thus is not yet discernable in the Southwest (Hebblewhite and Smith 2010). By contrast, the link between wolves and willow regeneration and growth appears to be better established (Hebblewhite and Smith 2010; Ripple et al. 2014a). However, the extent of willow is extremely limited in areas where Mexican wolves are present in the Southwest, and provisions are in place to prevent numeric impacts on ungulates by wolves (USFWS 2015a), which would also prevent the wolves' ability to limit elk impacts on willow or aspen through reduction of elk populations. Thus, we do not anticipate that impacts to vegetation will occur in Arizona and New Mexico from Mexican wolves (USFWS 2014a). Further, the entire Southwest, except for small national parks and some wilderness areas, would be characterized as having dominant anthropogenic influences (e.g., fire control, cattle grazing, hunting, habitat alteration) on prey populations and vegetation, thus overriding any ecosystem-level impact of wolves in the region (Mech 2012; Brodie et al. 2013).

Wolves can also create trophic cascades through impacts on other predators. Wolves have caused reductions in coyote populations in Yellowstone, primarily due to wolves killing them at carcasses (Crabtree and Sheldon 1999;

Ballard et al. 2003; Berger and Gese 2007; see Chapter 11). The reduction in coyote numbers can cause an increase in the abundance of other mesopredators such as the North American red fox (*Vulpes fulva*) through the reduction of interference (Mech and Boitani 2003) and exploitative competition that allows more small mammals to be present in a given area after coyote numbers decline (Ripple et al. 2014a). Although impacts to coyotes from Mexican wolves reintroduced to the Southwest were predicted to occur (Bednarz 1988) and the diets of the two carnivores overlap to a large extent (Carrera et al. 2008), research has not yet been conducted to determine Mexican wolf impacts on coyotes. Although coyotes can derive some benefit from scavenging wolf-killed carcasses, the risk of injury or mortality to coyotes is significant if interference competition is high. Mexican wolves and coyotes have been observed simultaneously feeding on wolf-killed carcasses. In a few cases, no aggressive interactions were observed, but in others, wolves have aggressively chased coyotes (USFWS, unpubl. data; Photos 11.13a and b).

Wolves interact with bears primarily at carcasses through interference competition, with grizzly bears dominant in Yellowstone and wolves dominant in Banff (Ballard et al. 2003; Hebblewhite and Smith 2010). Black bears are generally dominated by wolves (Ballard et al. 2003). Much of these interactions depend on the number of wolves present (greater numbers allow wolves to be more successful) and whether the bear has cubs. Female bears with cubs are rarely observed at carcasses (Ballard et al. 2003), but generally retreat in the presence of wolves. In the Southwest, no intensive study of interactions between black bears and wolves has taken place. However, trail cameras suggest that male black bears are dominant over Mexican wolves (USFWS, unpubl. data; Photos 11.14a and b).

Wolves tend to usurp or scavenge cougar kills, whereas cougars rarely scavenge or usurp

Photos 11.13a and b (*top to bottom*). A coyote (*Canis latrans*) squares off with an eight-month-old wolf pup over a recently stocked food cache on the Ladder Ranch, New Mexico in December 2021. Both the coyote and the wolf pup are exhibiting similar defensive or threat postures to scare the other animal away from the food cache, all without risking injury. Additional photos documented the wolf pup successfully defending the food cache from the coyote. Since reintroducing Mexican wolves into the Southwest, biologists have rarely documented wolf-coyote interactions. The few interactions that have been observed have not resulted in mortality or injury to either species. Photographs: © Cassidi Cobos/Turner Endangered Species Fund.

wolf kills (Kortello et al. 2007). Wolves appear to impact cougars through interference and exploitative competition, with cougars not only being killed by wolves but also dying of starvation (Hebblewhite and Smith 2010). Cougars adapt to the presence of wolves by shifting prey resources and habitat utilizations (Hebblewhite and Smith 2010). One lone Mexican wolf was killed by a cougar in the Southwest (Mexican Wolf Interagency Field Team and Mexican Wolf Adaptive Management Oversight Committee 2005). However, little information is available on Mexican wolf–cougar interactions.

Interspecific interactions between wolves and various other species are also largely unstudied in the Southwest. In general, Mexican wolves appear to experience less mortality from intraspecific strife (i.e., wolves killing other wolves) than in other wolf populations (Smith et al. 2010; USFWS 2017), and perhaps this extends to interspecific interactions. Mexican wolves are present in areas without grizzly bears or significant winter die-offs of ungulates, which lessens the likelihood of numerical decline of ungulates caused by wolves and corresponding trophic cascades. The Southwest harbors a substantial wintering bald eagle (*Haliaeetus leucocephalus*) population, and wolves may be providing a significant resource to scavenging eagles based on photos and observations of a large number of eagles at wolf kills (USFWS, unpubl. data). Overall, the growing population of Mexican wolves provides a distinct

Photos 11.14a and b (*top to bottom*). A Black bear (*Ursus americanus*) displaces Mexican wolves from an elk (*Cervus canadensis*) carcass in July 2015 on the Apache-Sitgreaves National Forest. Typical interactions between the two species in the Mexican Wolf Experimental Population Area may involve the black bear usurping a carcass and wolves having to wait for their turn again. Photographs: © Interagency Field Team.

opportunity for researchers to study trophic cascades in a unique environment.

STATUS AND MANAGEMENT

The primary limitation to the wolf's distribution and persistence is another widely distributed mammal, humans. Indeed, the organized persecution of wolves by humans from the Middle Ages until the mid-1900s in Europe and from the 1800s through mid-1900s in North America eliminated gray wolves from broad swaths of their former range (Boitani 2003). As a result, the southern boundary of the gray wolf's distribution in North America was pushed northward approximately 30 degrees of latitude or 3,300 km (2,000 mi) from Mexico City to northern Minnesota. The species was also eliminated from parts of southern Canada, Alaska, and extreme northern

Minnesota, but Canada still harbored large numbers of wolves and thus could provide source populations for natural recolonization or reintroductions. At the nadir of the wolf's recorded history in Europe, small populations persisted in southern Europe (the Iberian Peninsula and Italy), but the species had otherwise been eliminated from the British Isles, France, Denmark, Switzerland, Scandinavia, and the Rhine valley and greatly reduced in numbers in areas farther to the east (Boitani 2003). At the same time, vast numbers of wolves remained in eastern Europe (Poland, Romania, and the Balkans) and Russia and thus, here again, source populations existed for natural recolonization in some areas. Little is known about the history of gray wolves in eastern and southern Asia, but important reductions in both

Photo 11.16 (*above*). A male Mexican wolf drinking water from a cast iron frying pan after it was caught on 22 October 1915, northwest of Chloride, Sierra County, New Mexico. The wolf's right front paw is caught in an animal trap. J. Stokley Ligon estimated that only 103 Mexican wolves remained in New Mexico in 1917, 45 in 1918 (Findley et al. 1975). Small numbers of Mexican wolves persisted in southwestern New Mexico through the 1930s, perhaps into the 1940s, with the last wolf of record dating to 1970, from the Peloncillo Mountains (Brown 1983). Photograph: J. S. Ligon. J. Stokley Ligon papers and photographs, Denver Public Library Special Collections CONS92-2017-54.

Photo 11.15. J. Stokley Ligon on horseback with the carcass of a dead Mexican wolf laid across the pommel of his saddle, returning to his camp at East Water in the former Datil National Forest, New Mexico on 19 October 1917. At the time, J. Stokley Ligon worked as Predatory Animal Inspector for the New Mexico–Arizona District within the US Biological Survey (see also Chapter 17). The gray wolf was distributed throughout most of New Mexico before persecution led to its extirpation from the state. J. Stokley Ligon papers and photographs, Denver Public Library Special Collections CONS92-2017-52.

numbers and distribution have been observed in large areas of China and India. These ongoing reductions are likely the result of persecution and/or limited prey because of competition with humans in that part of the world (Boitani 2003).

Wolves were essentially eliminated from New Mexico by the late 1920s with the exception of dispersing wolves from Mexico in the mountains of Hidalgo County (Brown 1983). The last wolf in New Mexico was probably killed in the Peloncillo Mountains in 1970 (Brown 1983). Exports of techniques for controlling wolves (e.g., poison) from the United States to Mexico likely severely reduced the number of wolves in Mexico from 1950 to 1980 such that few wolves persisted in Mexico in the late 1970s (McBride 1980), and Mexican wolves were likely extinct in the wild by the mid-1990s (Parsons 1998). Numerous books describe in detail the reduction of wolf numbers following the European colonization of North America (Lopez 1978; Brown 1983; Robinson 2005). Fortunately, a handful of Mexican wolves were captured in Mexico prior to extinction in the wild (McBride 1980)

A first turning point occurred in North America and Europe when laws were established to protect gray wolves by the 1960s (Boitani 2003). In both Europe and North America today, human attitudes are also more favorable toward wolves and other predators (Musiani et al. 2009). As a result, wolf reintroductions and recovery programs are being implemented in a variety of locations on both continents (Musiani et al. 2009). In Europe, most gray wolf populations appear to be increasing or stable and wolves are expanding their geographic range to the west and north (Boitani and Ciucci 2009). In North America, reintroductions have occurred in central Idaho, the Greater Yellowstone Ecosystem, Arizona, New Mexico, and Mexico, while natural recovery of wolves has occurred in northwestern Montana and the Great Lake states (Minnesota, Michigan, and Wisconsin). Reintroductions in central Idaho and Yellowstone eventually resulted in wolves expanding their distribution in adjacent areas and establishing robust breeding populations in the states of Washington and Oregon, with a single breeding pair recorded—for now—in California and Colorado each. Wolves in the southwestern United States and Mexico have followed the general historical pattern with persecution at the hands of humans and extinction in the wild, followed by the recent reintroduction of Mexican wolves (Brown 1983; Parsons 1998). The wolves captured in Mexico were subsequently bred with two other captive lineages to launch a captive breeding program with seven founders for Mexican wolves (Hedrick et al. 1997). The success of the captive breeding program, managed through the Species Survival Plan (SSP) with the participation of numerous zoos in Mexico and the United States, allowed the captive population

to serve as the source for reintroductions into the southwestern United States and Mexico.

Mexican Wolf's Reintroduction

The presence of Mexican wolves today in the wild lands of New Mexico is the result of a remarkable conservation effort against what many biologists considered as overwhelming odds. The original team charged with developing a recovery plan for the US Fish and Wildlife Service (USFWS 1982) saw no possibility of complete recovery and delisting of the Mexican wolf. Instead, the prime objective of the original recovery plan was "to conserve and ensure the survival of *Canis lupus baileyi* by maintaining a captive breeding population and re-establishing a viable, self-sustaining population of at least 100 Mexican wolves in the middle to high elevations of a 5,000-square-mile area [~13,000 km²] within the Mexican wolf's historic range" (USFWS 1982:23). At the time, the captive breeding stock consisted of only ten animals including three founders (called the "certified lineage" captured in Mexico in the late 1970s and early 1980s). Considering that few options to capture more wolves existed in Mexico, the prime objective of the 1982 *Mexican wolf recovery plan* was ambitious. Further, genetic fitness and inbreeding depression from the accumulation of deleterious mutations were concerning with only three founders (theoretically unrelated individuals). Eventually, however, genetic results confirmed that two additional, pure Mexican wolf lineages existed, stemming from four additional founders living in captive facilities (Hedrick et al. 1997). All three lineages were combined through breeding in captivity to improve the genetic variability of the subspecies. Even so, the 1982 *Mexican wolf recovery plan* remained focused on how to recover a subspecies with only a few remaining wolves in captivity. Clearly, Mexican wolf recovery could not depend on either the natural dispersal of wolves recolonizing available, unoccupied areas unlike what had happened with the Great Lakes

Photo 11.17. A tethered mule carrying two Aspen pack wolves in metal panniers in June 2005. The specially trained mule is about to transport the two wolves into the Gila Wilderness for release. Photograph: © Interagency Field Team.

Photo 11.18. A wolf release in January 2010 in the Gila National Forest in New Mexico. Mexican wolves were first released in New Mexico in 2000. Photograph: © Interagency Field Team.

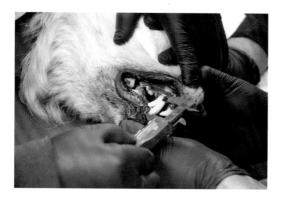

Photo 11.19. Biologists examine the teeth and mouth of a sedated Mexican wolf during wolf handling procedures. They also check for injuries and record the amount of wear and staining observed on the teeth. Measurements of canine teeth may be recorded for data purposes. Photograph: © Interagency Field Team.

Photo 11.20. A wolf from the Maverick pack tries to evade capture from the helicopter-dart crew during the annual winter population survey on the Apache-Sitgreaves National Forest, Arizona in January 2012. Photograph: © Interagency Field Team.

population (Wydeven et al. 2009), or the translocation of wild wolves, as in the reintroduction of wild Canadian wolves into Yellowstone and central Idaho (Fritts et al. 2001). The establishment of a healthy captive population of Mexican wolves was the first successful milestone in the Mexican wolf's recovery, as it kept the subspecies from going extinct and provided hope for reintroduction to the wild. Over several decades, the Association of Zoos and Aquariums Society's SSP for the Mexican wolf resulted in a sufficient number of animals for the US Fish and Wildlife Service to consider reintroduction (Parsons 1998).

After decades of planning and captive breeding

efforts and buoyed by the success of the gray wolf reintroduction programs in Yellowstone and central Idaho, the US Fish and Wildlife Service made plans to reintroduce Mexican wolves back into the mountains of central Arizona and New Mexico. Yet, the reintroduction project faced numerous hurdles compared to the more celebrated reintroductions of gray wolves in Yellowstone and central Idaho. First and foremost, Mexican wolves had to be released from captivity. Captive wolves were conditioned in remote, pre-release facilities and evaluated to identify suitable candidates based on behavior (e.g., high fear of humans) and other criteria (past reproductive success, pups present with the pack, and other pack composition considerations) for release. However, even the most suitable candidates would prove that establishing a wild population from a captive stock is not without challenges. Mexican wolves sometimes exhibited nuisance behavior (e.g., moving within close proximity of houses or towns, exhibiting a lack of fear of humans) for several months following release. Many released animals died or were removed for behavioral reasons, one main reason that only about 30% of Mexican wolves from captivity have successfully bred and produced pups in the wild. In addition, the Mexican wolf reintroduction area was not located in a large national park or designated wilderness with vast expanses of high-quality habitat without anthropogenic influences (i.e., the Gila Wilderness is small compared with the Greater Yellowstone Ecosystem or the Central Idaho Wilderness). Rather, the reintroduction occurred on a multiple-use landscape where wolves would more consistently encounter livestock. Mexican wolves were required to remain within the 18,130 km² (7,000 sq mi) Recovery Area (consisting of the Primary and Secondary Recovery zones; Map 11.4), whereas gray wolves in the northern Rocky Mountains could roam across an area of approximately 1,000,000 km² (385,000 sq mi), encompassing multiple national parks and remote wilderness areas where cattle were

absent. Finally, many biologists suggested that Mexican wolves were primarily predators of the diminutive Coues white-tailed deer, and being of relatively small size, they would be unable to prey upon elk, the primary wild ungulate that existed in the area where the reintroduction was to take place (Brown 1983 and subsequent editions). Thus, on 29 March 1998 when Mexican wolves were first released in the wild, the odds were decidedly stacked against them, and many predicted the failure of the reintroduction effort.

The first groups of Mexican wolves released in 1998 were not successful. By the end of the year only four (from releases in December) remained in the wild (USFWS 1998b). Thus, the possibility of failure continued to hover over reintroduction efforts. Release efforts in 1999 were more successful compared to 1998, but the population at the end of the year still only consisted of 15 animals despite the release of 33 Mexican wolves from captivity in 1998 and 1999 (USFWS 1999). By the end of 1999, biologists determined that releases conducted in areas where deer were the primary prey and year-round cattle grazing occurred were not successful. Thus, from 2000 onward, wolves were released in areas where elk were the primary prey available and cattle grazing was absent or seasonal in the summer months (USFWS 2000a, 2000b). This shift of focus for release areas helped to establish a growing population and a minimum of 55 Mexican wolves in the wild by the end of 2003 (USFWS 2003). At the same time, difficulties continued to persist with human-caused mortality, depredations, and maintaining a population within a relatively small area.

Depredations

Depredation on cattle and sheep has been, and continues to be, one of the most controversial issues surrounding wolves and wolf conservation efforts. Wolves preying on livestock, after the over-harvest or elimination of native prey by settlers, was a primary cause of the federal

Map 11.4. The geography of the Mexican wolf non-essential experimental population rule from 1998 to 2014. Mexican wolves were only allowed to occupy the areas in gray (the Primary and Secondary Recovery zones, later, also the Fort Apache Indian Reservation) during this period. Wolves that established territories wholly outside of these zones were required to be removed. Wolves from captivity, without wild experience, were only allowed to be released in the Primary Recovery Zone (dark gray). However, wolves with wild experience were allowed to be released in the Secondary Recovery Zone.

eradication campaign against wolves (Lopez 1978). Wolves in Arizona and New Mexico were hunted to extinction because of high and often exaggerated depredations on livestock (Brown 1983; Gipson et al. 1998). The last remaining wolves were given names such as "Ol' Three Toes" or "Ol' One Toe," indicative of them having survived from being trapped before and had legendary and inflated depredation numbers attributed to them (Lopez 1978; Brown 1983; Gipson et al. 1998).

Today, wolf depredations of livestock are rare relative to historical accounts (Bangs et al. 2005). The ability to recover wolves in the United States began with the regulation of hunting and the recovery of native ungulate populations starting in the early 1900s (Bangs et al. 2005). The increased presence of wild prey in areas selected

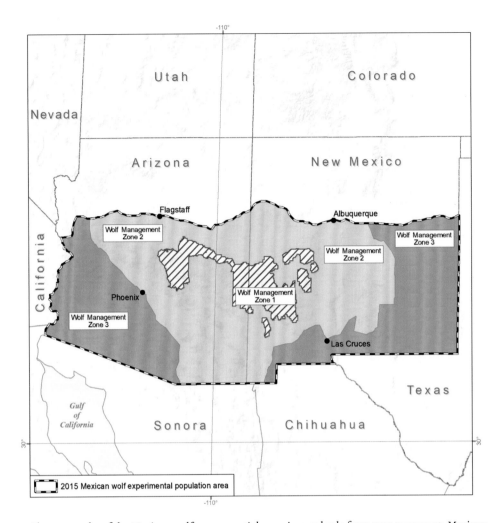

Map 11.5. The geography of the Mexican wolf non-essential experimental rule from 2015 to present. Mexican wolves are allowed to persist anywhere to the south of I-40, but are removed if they are found north of the interstate. Wolves from captivity without wild experience can be released in Wolf Management Zone 1 (hatched area) and in Zone 2 in limited circumstances (as pups). Wolves with wild experience can be translocated into Wolf Management Zone 2. Wolves cannot be translocated or released in Management Zone 3 and in that zone may be more actively managed to reduce human conflict because of the lack of habitat.

for recovery efforts has allowed wolves to develop search images (selective focus on one type of prey even when others are present, based on previous hunting success) and predation on native ungulates with limited impact to cattle (Bangs et al. 1998). Thus, while wolf depredation may result in localized impacts on individual ranchers, there is no statewide impact on the livestock industry (USFWS 2014a).

In areas where wolves prey on livestock, compensation for wolf depredations confirmed by government agencies has long been a standard method to offset a portion of the economic impact to ranchers (Fischer 1989). However, not all livestock killed by wolves are found by ranchers, or the remains are too old to be confirmed as wolf kills by government professionals (Bjorge and Gunson 1985; Oakleaf et al. 2003; Sommers

Photo 11.21. Mexican wolf near cattle in July 2003 on the Gila National Forest in New Mexico. Mexican wolf depredation of livestock led to a federal eradication campaign against wolves in the early 20th century. Today, Mexican wolves and livestock exist on public lands in the Mexican Wolf Experimental Population Area year-round. Compared to historical accounts, depredation of livestock is relatively rare based on the number of livestock and amount of overlap in distribution between wolves and livestock. The success of the wolf's recovery hinges in large part on maintaining healthy native ungulate populations and preventing wolves from depredating on livestock. Photograph: © Interagency Field Team.

et al. 2010; Breck et al. 2011). Additional impacts difficult to measure include stress-related weight loss in herds that experience regular depredation (Sommers et al. 2010; Ramler et al. 2014).

In the Southwest, depredation rates (i.e., depredations per 100 wolves) have been greater than in other areas of the United States (USFWS, unpubl. data). The pattern is understandable because livestock are often left year around on large public land allotments in the Southwest. In northern areas where wolves and free-ranging livestock overlap, most livestock are removed from public lands during late fall, produce calves and are fed on small private pastures through

spring, and ultimately return to public areas in the early summer with calves that are already three to six months old. The small, private, calving pastures are close to home ranches and effective non-lethal tools (e.g., constant human presence, hazing, fladry, light and siren scare devices) can more easily be used for wolf deterrence (Shivik 2006). In contrast, livestock in the Southwest calve across hundreds of square kilometers in allotment pastures with calving peaking in February–April, well before parturition of native ungulates, which makes beef calves the most vulnerable, most available ungulate for wolf predation during that time of year. Altogether, many cattle in the Southwest persist in Mexican wolves' territories year around, with pulses of calving occurring throughout the year in areas too large for effective use of non-lethal techniques. Clustered cattle grazing scenarios and herding can be effective non-lethal measures to reduce wolf depredation, but in the Southwest their use is limited because water and grass are in short supply. Nevertheless, non-lethal tools such as range riders (personnel hired by non-government organizations to prevent depredations by increasing the amount of human presence around livestock), and diversionary food caches during denning and the rendezvous season, appear to be effective in the Southwest based on the authors' experience.

Removal of wolves (either through lethal means or capture and translocation) has been the primary method for reducing impacts of wolves on livestock (Bangs et al. 2005). Removals were heavily utilized in Arizona and New Mexico during the early phases of the Mexican wolf recovery, but this limited population growth (Figure 11.1). From 2003 to 2007 in particular, the wolf reintroduction team conducted fewer releases and adopted a more aggressive management response to depredations (e.g., more removals of wolves). With findings that the reintroduced population was not growing, the focus shifted toward removing fewer wolves, while applying new management techniques to reduce conflict with livestock. The project focused on diversionary food caches near den and rendezvous sites to prevent depredations during April–September. In addition, Defenders of Wildlife and the Mexican Wolf Conservation Fund increased their funding for range riders and other non-lethal techniques (e.g., hazing of wolves). Combined, these efforts appeared to keep depredation rates at the same level as the time period when the project was conducting aggressive wolf removals, and the population increased from 42 wolves at the end of 2009 to 196 wolves at the end of 2021 (Figure 11.1).

However, depredation rates rose during 2018 and 2019 (Figure 11.1), illustrating the need for adaptive management in the face of changing conditions. Mexican wolves are expanding to new areas with a high potential for wolf depredation on livestock in Arizona and New Mexico (Goljani Amerkhiz et al. 2018). Thus, impacts on livestock will continue to be at the center of significant controversy in the Southwest and will remain an important focus of management as Mexican wolf recovery efforts continue.

Cross-Fostering and the New Recovery Goals

While adaptive management efforts continued in the field to address wolf conflicts, ongoing concern over the genetic health of the population was also present. Unfortunately, due to the cessation of releases from captivity and the disproportionate success of only a few wolves, the wild population at the end of 2014 did not represent the full genetic diversity available in the captive population. However, releases from captivity were inordinately controversial, and again the reintroduction adapted to potentially accomplish genetic goals with less impact to local ranchers. Cross-fostering (placing young pups [<14 days old] produced in captivity or wild dens into non-natal wild dens with similarly aged pups) has been utilized since 2014, with some promising initial results. The pups are raised in the wild by experienced packs and have not displayed the nuisance behavior or vulnerability to human-caused mortality that plagued releases of captive adults. In addition, five (23%) of the 22 cross-fostered pups released in 2014 and 2016–2018 have bred and produced pups in the wild. Three other cross-fostered pups from these releases are alive but have yet to produce pups. An additional 34 cross-fostered pups released in 2019 and 2020 are too young to have produced pups at this time with seven of them being monitored. Regardless of the relative success of cross-fostering, the project has resumed the release of animals

Photos 11.22a and b (*top left to right*). Wolf pack (the "Seco Creek pack") in crates, prior to its release on the Ladder Ranch in New Mexico in 2021. The adult wolf in the crate (a) is the alpha male AM 1693. The six pups together in their separate crate (b) are ~six weeks old. The Seco Creek pack is composed of AM1693 and the alpha female AF1728 and six pups born in 2021. In March 2021, AM1693 and AF1728 were temporarily removed from an area of high livestock conflict in the Gila National Forest and held at the Sevilleta Wolf Management Facility. In June 2021, the pack, now consisting of the alpha pair and six pups born in May, were moved into a remote release pen on the Ladder Ranch in south-central New Mexico. The pack was then released back to the wild in July and has since been closely monitored by Turner Endangered Species Fund biologists. Photographs: © Cassidi Cobos/Turner Endangered Species Fund.

Photo 11.23 (*left*). Six-week-old wolf pup receiving a health check by the project veterinarian at the Sevilleta Wolf Management Facility, New Mexico in June 2021. The pup was later released to the wild with the Seco Creek pack. Photograph: © Interagency Field Team.

(pups) from captivity because cross-fostering is more socially acceptable (e.g., local human communities are more accepting of small pups being placed in an existing pack versus establishing new packs in areas where wolves do not exist). The incorporation of cross-fostered pups into the wild population of Mexican wolves will increase its genetic diversity.

With wolves successfully established on the ground, it was clear that the US Fish and Wildlife Service needed to reassess its goals. After all, the Mexican wolf population in Arizona and New Mexico had fully reached the original Recovery Team's prime objective. A new non-essential experimental rule in 2015 allowed Mexican

wolves to occupy all areas south of I-40 in New Mexico and Arizona (Maps 11.4 and 11.5), and the 2017 recovery plan established delisting goals for a population in the southwestern United States (i.e., Arizona and New Mexico south of I-40) and Mexico. In addition, our colleagues in Mexico have started to establish their own population of Mexican wolves. Mexico will face similar and, in some cases, more severe challenges during their reintroduction project. Doubts will be voiced, but the story of the Mexican wolf is one of overcoming tremendous odds and the ability of biologists and wolves to adapt to challenges. Indeed, the dedicated and adaptable staff of the Arizona Game and Fish Department, New Mexico Department

Photo 11.24. Wolf den (Elk Horn pack) on the Apache-Sitgreaves National Forest, Arizona in April 2016. Two captive born Mexican wolf pups from the Brookfield Zoo were placed in the Elk Horn den alongside five wild born pups. Four of the pups were documented surviving to the end of the year. In 2018, one of the cross-foster pups (M1471) became the alpha male of the Prime Canyon pack. Biologists have documented AM1471 and his mate producing a minimum of 19 genetically valuable pups in addition to serving as a cross foster pack in 2019 and 2020. Photograph: © Interagency Field Team.

Photo 11.25. Cross fostering of a Mexican wolf pup on the Apache-Sitgreaves National Forest, Arizona in April 2019. The Mexican wolf pup, born in captivity at the Endangered Wolf Center, was placed in the Prime Canyon den alongside seven wild born pups. Five of the pups were documented surviving to the end of the year. The alpha male of the Prime Canyon pack, AM1471 has been an integral part of improving the genetic diversity of the wild population, with his own genetic contribution and with raising additional cross-foster pups in the wild. Photograph: © Interagency Field Team.

of Game and Fish, US Department of Agriculture–Animal and Plant Health Inspection Service–Wildlife Services, US Forest Service, US Fish and Wildlife Service, and White Mountain Apache Tribe have been critical to the successful establishment of a population of Mexican wolves in the United States. The minimum population count indicates at least 196 wolves in New Mexico and Arizona at the end of 2021 (Figure 11.1). Given the ability of the agencies and the wolves to adapt to challenges, the recovery goals adopted in the 2017 recovery plan (an average of 320 wolves over 8 years in the United States and an average of 170 wolves in Mexico over eight years) appear to be achievable, ensuring that Mexican wolves persist in the Southwest and in New Mexico for the foreseeable future.

ACKNOWLEDGMENTS

The reintroduction and the recovery of species that are extirpated from the wild constitute an extremely difficult task. We thank all (too numerous to list) of the dedicated biologists, volunteers, and interns who have been involved in Mexican wolf recovery and reintroduction efforts and data collection. We thank J. W. Cain III, J. K. Frey, and J.-L. Cartron for their helpful comments and review of drafts of this publication. The opinions herein represent the views of the authors and not necessarily those of the US Fish and Wildlife Service or the US Department of Agriculture-Wildlife Services.

LITERATURE CITED

Adams, L. G., R. O. Stephenson, B. W. Dale, R. T. Ahgook, and D. J. Demma. 2008. Population dynamics and harvest characteristics of wolves in the Central Brooks Range, Alaska. *Wildlife Monographs* 170:1–25.

Ballard, W. B., L. Carbyn, and D. W. Smith. 2003. Wolf interactions with non-prey. In *Wolves: behavior, ecology, and conservation*, ed. L. D. Mech and L. Boitani, 259–71. Chicago: Chicago University Press.

Ballard, W. B., R. Farnell, and R. O. Stephenson. 1983. Long distance movement by gray wolves (*Canis lupus*). *Canadian Field-Naturalist* 97:333.

Ballard, W. B., D. W. Lutz, T. Keegan, L. H. Carpenter, and J. C. deVos Jr. 2001. Deer-predator relationships: a review of recent North American studies with emphasis on mule and black-tailed deer. *Wildlife Society Bulletin* 29:99–115.

Ballard, W. B., H. A. Whitlaw, B. F. Wakeling, R. L. Brown, J. C. deVos Jr., and M. C. Wallace. 2000. Survival of female elk in northern Arizona. *Journal of Wildlife Management* 64:500–504.

Bangs, E. E., J. A. Fontaine, M. D. Jimenez, E. H. Bradley, C. C. Niemeyer, D. W. Smith, C. M. Mack, V. Asher, and J. K. Oakleaf. 2005. Managing wolf-human conflict in the northwestern United States. In *People and wildlife: conflict or coexistence?*, ed. R. Woodroffe, S. Thirgood, and A. Rabinowitz, 340–56. Cambridge: Cambridge University Press.

Bangs, E. E., S. H. Fritts, J. A. Fontaine, D. W. Smith, K. M. Murphy, C. Mack, and C. C. Niemeyer. 1998. Status of gray wolf restoration in Montana, Idaho, and Wyoming. *Wildlife Society Bulletin* 26:785–98.

Bednarz, J. A. 1988. *The Mexican wolf: biology, history, and prospects for reestablishment in New Mexico*. Endangered species report 18. Albuquerque, NM: US Fish and Wildlife Service.

Berger, K. M., and E. M. Gese. 2007. Does interference competition with wolves limit the distribution and abundance of coyotes? *Journal of Animal Ecology* 76:1075–85.

Beschta, R. L., C. Eisenberg, J. W. Laundré, W. J. Ripple, and T. P. Rooney. 2014. Predation risk, elk, and aspen: comment. *Ecology* 95:2669–71.

Beschta, R. L., and W. J. Ripple. 2010. Mexican wolves, elk, and aspen in Arizona: is there a trophic cascade? *Forest Ecology and Management* 260:915–22.

Bjorge, R. R., and J. R. Gunson. 1985. Evaluation of wolf control to reduce cattle predation in Alberta. *Journal of Range Management* 38:483–487.

Boertje, R. D., M. A. Keech, and T. Paragi. 2010. Science and values influencing predator control for Alaska moose management. *Journal of Wildlife Management* 74:917–28.

Boertje, R. D., M. A. Keech, D. D. Young, K. A. Kellie, and C. T. Seaton. 2009. Managing for elevated yield of moose in interior Alaska. *Journal of Wildlife Management* 73:314–27.

Boertje, R. D., P. Valkenburg, and M. E. McNay. 1996. Increase in moose, caribou, and wolves following wolf control in Alaska. *Journal of Wildlife Management* 60:474–89.

Bogan, M. A., and P. Melhop. 1983. Systemic relationships of gray wolves (*Canis lupus*) in southwestern North America. *Occasional Papers of the Museum of Southwestern Biology* 1:1–21.

Boitani, L. 2003. Wolf conservation and recovery. In *Wolves: behavior, ecology, and conservation*, ed. L. D. Mech and L. Boitani, 317–40. Chicago: Chicago University Press.

Boitani, L. and P. Ciucci. 2009. Wolf management across Europe: species conservation without boundaries. In *A new era for wolves and people— wolf recovery, human attitudes, and policy*, ed. M. Musiani, L. Boitani, and P. C. Paquet, 15–39. Calgary: University of Calgary Press.

Boyd, D. K., P. C. Paquet, S. Donelon, R. R. Ream, D. H. Pletscher, and C. White. 1995. Transboundary movements of a recolonizing wolf population in the Rocky Mountains. In *Ecology and conservation of wolves in a changing world*, ed. L. Carbyn, S. H. Fritts, and D. Seip, 135–40. Edmonton: Canadian Circumpolar Institute.

Breck, S. W., B. M. Kluever, M. Panasci, J. Oakleaf, T. Johnson, W. Ballard, L. Howery, and D. L. Bergman. 2011. Domestic calf mortality and producer detection rates in the Mexican wolf recovery area: implications for livestock management and carnivore compensation schemes. *Biological Conservation* 144:930–36.

Brodie, J., H. Johnson, M. Mitchell, P. Zager, K. Proffitt, M. Hebblewhite, M. Kauffman, B. Johnson, J. Bissonnette, C. Bishop, J. Gude, J. Herbert, K. Hersey, M. Hurley, P. M. Lukacs, S. McCorquodale, E. McIntire, J. Nowak, H. Sawyer, D. Smith, and P. J. Whites. 2013. Relative influence of human harvest, carnivores, and weather on adult female elk survival across Western North America. *Journal of Applied Ecology* 50: 295–305.

Brown, D, ed. 1983. *The wolf in the Southwest: the making of an endangered species*. Tucson: University of Arizona Press.

Cariappa, C. A., J. K. Oakleaf, W. B. Ballard, and S. W. Breck. 2011. A reappraisal of the evidence for regulation of wolf populations. *Journal of Wildlife Management* 75:726–730.

Carrera, R., W. Ballard, P. Gipson, B. T. Kelly, P. R. Krausman, M. C. Wallace, C. Villalobos, and D. B. Wester. 2008. Comparison of Mexican wolf and coyote diets in Arizona and New Mexico. *Journal of Wildlife Management* 72:376–81.

Carroll, C., R. J. Fredrickson, and R. C. Lacy. 2014. Developing metapopulation connectivity criteria from genetic and habitat data to recover the endangered Mexican wolf. *Conservation Biology* 28:76–86.

Carroll, C., M. K. Phillips, C. A. Lopez-Gonzalez, and N. H. Schumaker. 2006. Defining recovery goals and strategies for endangered species: the wolf as a case study. *BioScience* 56:25–37.

Carroll, C., M. K. Phillips, N. H. Schumaker, and D. W. Smith. 2003. Impacts of landscape change on wolf restoration success: planning a reintroduction program based on static and dynamic spatial models. *Conservation Biology* 17:536–48.

Chambers, S., S. R. Fain, B. Fazio, and M. Amaral. 2012. An account of the taxonomy of North American wolves from morpological and genetic analyses. *North American Fauna* 77:1–67.

Crabtree, R. L., and J. Sheldon. 1999. Coyotes and canid coexistence. In *Carnivores in ecosystems: The Yellowstone experience*, ed. T. W. Clark et al, 127–63. New Haven, CT: Yale University Press.

Creel, S. 2005. Dominance, aggression, and glucocorticoid levels in social carnivores. *Journal of Mammalogy* 86:255–64.

Creel, S., D. Christianson, and J. A. Winnie Jr. 2011. A survey of the effects of wolf predation risk on pregnancy rates and calf recruitment in elk. *Ecological Applications* 21:2847–53.

Dekker, D., W. Bradford, and J. R. Gunson. 1995. Elk and wolves in Jasper National Park, Alberta from historical times to 1992. In *Ecology and conservation of wolves in a changing world*, ed. L. Carbyn, S. H. Fritts, and D. Seip, 85–94. Edmonton: Canadian Circumpolar Institute.

Eberhardt, L. L., P. J. White, R. A. Garrott, and D. B. Houston. 2007. A seventy-year history of trends in Yellowstone's northern elk herd. *Journal of Wildlife Management* 71:594–602.

Fan, Z., P. Silva, I. Gronav, S. Wang, A. S. Armero, R. M. Schweizer, O. Ramirez, J. Pollinger, M. Galaverni, D. O. Del-Vecchyo, L. Du, W. Zhang, Z. Zhang, J. Xing, C. Vilà, T. Marques-Bonet, R. Godinho, B. Yue, and R. K. Wayne. 2016. Worldwide patterns of genomic variation and admixture in gray wolves. *Genome Research* 26:163–73.

Findley, J. S., A. H. Harris, D. E. Wilson, and C. Jones. 1975. *Mammals of New Mexico*. Albuquerque: University of New Mexico Press.

Fischer, H. 1989. Restoring the wolf: Defenders launches a compensation fund. *Defenders* 66:35–39.

Fredrickson, R. J., P. W. Hedrick, R. K. Wayne, B. M. vonHoldt, and M. K. Phillips. 2015. Mexican Wolves Are a Valid Subspecies and an Appropriate Conservation Target. *Journal of Heredity* 106: 415–416.

Fritts, S. H. 1983. Record dispersal by a wolf from Minnesota. *Journal of Mammalogy* 64:166–67.

Fritts, S. H., and L. Carbyn. 1995. Population viability, nature reserves, and the outlook for gray wolf conservation in North America. *Restoration Ecology* 3:26–33.

Fritts, S. H., C. Mack, D. W. Smith, K. M. Murphy, M. K. Phillips, M. D. Jimenez, E. E. Bangs, J. A. Fontaine, C. C. Niemeyer, W. G. Brewster, and T. Kaminski. 2001. Outcomes of hard and soft releases of reintroduced wolves in central Idaho and the Greater Yellowstone Area. In *Large Mammal Restoration: Ecological and Sociological Challenges in the 21st Century*, ed. D. Maehr, R. F. Noss, and J. Larkin, 125–47. Washington, DC: Island Press.

Fritts, S. H., and L. D. Mech. 1981. Dynamics, movements, and feeding ecology of a newly protected wolf population in northwestern Minnesota. *Wildlife Monographs* 80:1–79.

Fuller, T. K. 1989. Population dynamics of wolves in north-central Minnesota. *Wildlife Monographs* 105:3–41.

———. 1990. Dynamics of a declining white-tailed deer population in north-central Minnesota. *Wildlife Monographs* 110:3–37.

Fuller, T. K., W. E. Berg, G. L. Radde, M. S. Lenarz, and G. B. Joselyn. 1992. A history and current estimate of wolf distribution and numbers in Minnesota. *Wildlife Society Bulletin* 20:42–55.

Fuller, T. K., L. D. Mech, and J. F. Cochrane. 2003. Wolf population dynamics. In *Wolves: behavior, ecology, and conservation*, ed. L. D. Mech and L. Boitani, 161–91. Chicago: Chicago University Press.

Gasaway, W. C., R. O. Stephenson, J. L. Davis, P. E. K. Shepherd, and O. E. Burris. 1983. Interrelationship of wolves, prey, and man in interior Alaska. *Wildlife Monographs* 84:1–50.

Gipson, P. S., W. B. Ballard, and R. M. Nowak. 1998. Famous North American wolves and the credibility of early wildlife literature. *Wildlife Society Bulletin* 26:808–16.

Gipson, P. S., E. E. Bangs, T. N. Bailey, D. K. Boyd, H. D. Cluff, D. W. Smith, and M. D. Jimenez. 2002. Color patterns among wolves in western North America. *Wildlife Society Bulletin* 30:821–30.

Goljani Amerkhiz, R., J. K. Frey, J. W. Cain III, S. W. Breck, and D. L. Bergman. 2018. Predicting spatial factors associated with cattle depredations by the Mexican wolf (*Canis lupus baileyi*) with recommendations for depredation risk modeling. *Biological Conservation* 224:327–35.

Hailer, F., and J. A. Leonard. 2008. Hybridization among three native North American *Canis* species in a region of natural sympatry. PLOS One 3. https://journals.plos.org/plosone/article?id=10.1371/journal.pone.0003333.

Hall, R. 1981. *Mammals of North America*. 2nd ed. New York: Wiley.

Hamlin, K. L., and J. A. Cunningham. 2009. *Monitoring and assessment of wolf-ungulate interactions and population trends within the Greater Yellowstone Area, southwestern Montana, and Montana statewide: final report*. Helena, MO.

Harrington, F. H., and C. S. Asa. 2003. Wolf communication. In *Wolves: behavior, ecology, and conservation*, ed. L. D. Mech and L. Boitani, 66–103. Chicago: Chicago University Press.

Harrington, F. H., and L. D. Mech. 1978. Wolf vocalization. In *Wolf and man: evolution in parallel*, ed. R. L. Hall and H. S. Sharp, 109–32. New York: Academic Press.

———. 1982. An analysis of howling response parameters useful for wolf pack censusing. *Journal of Wildlife Management* 46:686–93.

Hatlauf, J., K. Bayer, A. Trouwborst, and K. Hacklander. 2021. New rules or old concepts? The golden jackal (*Canis aureus*) and its legal status in Central Europe. *European Journal of Wildlife Research* 67:25. https://doi.org/10.1007/s10344-020-01454-2.

Hebblewhite, M. 2013. Consequences of ratio-dependent predation by wolves for elk population dynamics. *Population Ecology* 55:511–22.

Hebblewhite, M., and D. Smith. 2010. Wolf community ecology: ecosystem effects of recovering wolves in Banff and Yellowstone National Parks. In *The world of wolves: new perspectives on ecology, behavior, and policy*, ed. M. Musiani, L. Boitani, and P. Paquet, 1–52. Calgary: University of Calgary Press.

Hedrick, P. W. 2013. Adaptive introgression in animals: examples and comparison to new mutation and standing variation as sources of adaptive variation. *Molecular Ecology* 22:4606–18.

Hedrick, P. W., P. S. Miller, E. Geffen, and R. Wayne. 1997. Genetic evaluation of three captive Mexican wolf lineages. *Zoo Biology* 16:47–69.

Heffelfinger, J., R. M. Nowak, and D. Paetkau. 2017a. Clarifying historical range to aid recovery of the Mexican wolf. *Journal of Wildlife Management* 81:766–777.

———. 2017b. Revisiting revising Mexican wolf historical range: a reply to Hendricks et al. *Journal of Wildlife Management* 81:1334–37.

Hendricks, S., P. S. Clee, R. J. Harrigan, J. P. Pollinger, A. H. Freedman, R. Callas, P. J. Figura, and R. K. Wayne. 2016. Re-defining historical geographic range in species with sparse records: implications for the Mexican wolf reintroduction program. *Biological Conservation* 194:48–57.

Hendricks, S. A., S. Koblmüller, R. J. Harrigan, J. A. Leonard, R. M. Schweizer, B. M. von Holdt, R. Kays, and R. K. Wayne. 2017. Defense of an expanded historical range for the Mexican wolf: a comment on Heffelfinger et al. *Journal of Wildlife Management* 81:1331–33.

Hendricks, S. A., R. M. Schweizer, and R. K. Wayne. 2019. Conservation genomics illuminates the adaptive uniqueness of North American gray wolves. *Conservation Genetics* 20:29–43.

Husseman, J., D. Murray, G. Power, C. Mack, C. R. Wenger, and H. Quigley. 2003. Assessing differential prey selection patterns between two sympatric carnivores. *Oikos* 101:591–601.

Jensen, W. F., T. K. Fuller, and W. L. Robinson. 1986. Wolf (*Canis lupus*) distribution on the Ontario-Michigan border near Sault Ste. Marie. *Canadian Field-Naturalist* 100:363–66.

Jimenez, M. D., E. E. Bangs, D. K. Boyd, D. W. Smith, S. A. Becker, D. E. Ausband, S. P. Woodruff, E. H. Bradley, J. Hoylan, and K. Laudon. 2017. Wolf dispersal in the Rocky Mountains, western United States 1993–2008. *Journal of Wildlife Management* 81:581–592.

Kauffman, M. J., J. F. Brodie, E. S. Jules, and S. Url. 2010. Are wolves saving Yellowstone's aspen? A landscape-level test of a behaviorally mediated trophic cascade. *Ecology* 91:2742–2755.

Klein, D. R. 1995. The introduction, increase, and demise of wolves on Coronation Island, Alaska. In *Ecology and conservation of wolves in a changing world*, ed. L. Carbyn, S. H. Fritts, and D. Seip, 275–80. Edmonton: Canadian Circumpolar Institute.

Koblmüller, S., C. Vilà, B. Lorente-Galdos, M. Dabad, O. Ramirez, T. Marques-Bonet, R. K. Wayne, and J. A. Leonard. 2016. Whole mitochondrial genomes illuminate ancient intercontinental dispersals of grey wolves (*Canis lupus*). *Journal of Biogeography* 43:1728–38.

Kortello, A. D., T. E. Hurd, and D. L. Murray. 2007. Interactions between cougars (*Puma concolor*) and gray wolves (*Canis lupus*) in Banff National Park. *Ecoscience* 14:214–222.

Krofel, M., J. Hatlauf, W. Bogdanowicz, L. A. D. Campbell, R. Godinho, V. Yadvendiadev, V. Jhala, A. C. Kitchener, K. P. Koepfli, P. Moehlman, H. Senn, and C. Sillero-Zubiri. 2021. Towards resolving taxonomic uncetainties in wolf, dog and jackal lineages of Africa, Eurasia and Australasia. *Journal of Zoology* 316:155–68.

Larsen, T., and W. J. Ripple. 2006. Modeling gray wolf (*Canis lupus*) habitat in the Pacific Northwest, U.S.A. *Journal of Conservation Planning* 2:30–61.

Leonard, J. A., C. Vilà, and R. K. Wayne. 2005. Legacy

lost: Genetic variability and population size of extirpated US grey wolves (*Canis lupus*). *Molecular Ecology* 14:9–17.

Leonard, J. A., and R. K. Wayne. 2008. Native Great Lakes wolves were not restored. *Biology Letters* 4:95–98.

Lopez, B. H. 1978. *Of wolves and men*. New York: Charles Scribner.

Lucherini, M., and E. M. L. Vidal. 2008. *Lycalopex gymnocercus* (Carnivora: Canidae). *Mammalian Species* 820:1–9.

Macnab, J. 1985. Carrying capacity and related slippery shibboleths. *Wildlife Society Bulletin* 13:403–10.

MacNulty, D. R., D. R. Stahler, and D. W. Smith. 2020. Limits of wolf predatory performance. In *Yellowstone wolves*, ed. D. W. Smith, D. R. Stahler, and D. R. MacNulty, 149–156. Chicago: University of Chicago Press.

Martinez-Meyer, E., A. Gonzalez-Bernal, J. A. Velasco, T. L. Swetnam, Z. Y. Gonzalez-Saucedo, J. Servin, C. A. Lopez-Gonzalez, J. K. Oakleaf, S. Liley, and J. R. Heffelfinger. 2021. Rangewide habitat suitability analysis for the Mexican wolf (*Canis lupus baileyi*) to identify recovery areas in its historical distribution. *Diversity and Distributions* 27:643–654.

McBride, R. T. 1980. *The Mexican wolf (Canis lupus baileyi): a historical review and observations on its status and distribution*. US Fish and Wildlife Service, Region 2, Albuquerque, NM.

McIntyre, R., J. B. Theberge, M. T. Theberge, and D. W. Smith. 2017. Behavioral and ecological implications of seasonal variation in the frequency of daytime howling by Yellowstone wolves. *Journal of Mammalogy* 98:827–34.

McLeod, S. R. 1997. Is the concept of carrying capacity useful in variable environments. *Oikos* 79:529–42.

McRoberts, R. E., and L. D. Mech. 2014. Wolf population regulation revisited—again. *Journal of Wildlife Management* 78:963–67.

Mech, L. D. 1970. *The wolf: the ecology and behavior of an endangered species*. New York: Doubleday.

———. 2006a. Prediction failure of a wolf landscape model. *Wildlife Society Bulletin* 34:874–877.

——— 2006b. Mladenoff et al. rebut lacks supportive data. *Wildlife Society Bulletin* 34:882–83.

———. 2012. Is science in danger of sanctifying the wolf? *Biological Conservation* 150:143–49.

Mech, L. D., and L. Boitani. 2003. Wolf social ecology. In *Wolves: behavior, ecology, and conservation*, ed. L. D. Mech and L. Boitani, 1–34. Chicago: Chicago University Press.

Mech, L. D., S. H. Fritts, G. L. Radde, and W. J. Paul. 1988. Wolf distribution and road density in Minnesota. *Wildlife Society Bulletin* 16:85–87.

Mech, L. D., and P. D. Karns. 1977. *Role of the wolf in a deer decline in the Superior National Forest*. St. Paul, MN: USDA Forest Service Research Paper NC-148.

Mech, L. D., T. J. Meier, J. W. Burch, and L. G. Adams. 1995. Patterns of prey selection by wolves in Denali National Park, Alaska. In *Ecology and conservation of wolves in a changing world*, ed. L. Carbyn, S. H. Fritts, and D. Seip, 231–43. Edmonton: Canadian Circumpolar Institute.

Mech, L. D., and R. O. Peterson. 2003. Wolf-prey relations. In *Wolves: behavior, ecology, and conservation*, ed. L. D. Mech and L. Boitani, 131–60. Chicago: Chicago University Press.

Merkle, J. A., P. R. Krausman, D. W. Stark, J. K. Oakleaf, and W. B. Ballard. 2009. Summer diet of the Mexican gray wolf (*Canis lupus baileyi*). *Southwestern Naturalist* 54:480–85.

Mexican Wolf Interagency Field Team and Mexican Wolf Adaptive Management Oversight Committee. 2005. *Mexican wolf Blue Range reintroduction project 5-year review: technical component*. Albuquerque, NM: US Fish and Wildlife Service.

Mladenoff, D. J., M. K. Clayton, T. A. Sickley, and A. P. Wydeven. 2006. L. D. Mech critique of our work lacks scientific validity. *Wildlife Society Bulletin* 34:878–881.

Mladenoff, D. J., T. Sickley, R. Haight, and A. Wydeven. 1995. A regional landscape analysis and prediction of favorable gray wolf habitat in the northern Great Lakes region. *Conservation Biology* 9:279–94.

Mladenoff, D. J., T. A. Sickley, and A. P. Wydeven. 1999. Predicting gray wolf landscape recolonization: logistic regression models vs. new field data. *Ecological Applications* 9:37–44.

Musiani, M., L. Boitani, and P. C. Paquet. 2009. Introduction—newly recovering wolf populations produce new trends in human attitudes and policy. In *A new era for wolves and people—wolf recovery, human attitudes, and policy*, ed. M. Musiani, L. Boitani, and P. C. Paquet, 1–12. Calgary: University of Calgary Press.

National Academies of Sciences, Engineering, and Medicine. 2019. *Evaluating the taxonomic status of the Mexican gray wolf and the red wolf.* Washington, DC: National Academies Press. https://doi.org/10.17226/25351

National Research Council. 1997. *Wolves, Bears and Their Prey in Alaska: Biological and Social Challenges in Wildlife Management.* Washington, DC: National Academy Press

Nelson, E. W., and E. A. Goldman. 1929. A new wolf from Mexico. *Journal of Mammalogy* 10:165–66.

Nowak, R. M. 1978. Evolution and taxonomy of coyotes and related *Canis.* In *Coyotes: biology, behavior, and management*, ed. M. Bekoff, 3–16. New York: Academic Press

———. 1979. *North American Quaternary Canis.* Monographs of Museum of Natural History, University of Kansas 6.

———. 1995. Another look at wolf taxonomy. In *Ecology and conservation of wolves in a changing world*, ed. L. Carbyn, S. H. Fritts, and D. Seip, 375–97. Edmonton: Canadian Circumpolar Institute.

———. 2003. Wolf evolution and taxonomy. In *Wolves: behavior, ecology, and conservation*, ed. L. D. Mech and L. Boitani, 239–58. Chicago: Chicago University Press.

Oakleaf, J. K., C. Mack, and D. L. Murray. 2003. Effects of wolves on livestock calf survival and movements in central Idaho. *Journal of Wildlife Management* 67:299–306.

Oakleaf, J. K., D. L. Murray, J. R. Oakleaf, E. E. Bangs, C. M. Mack, D. W. Smith, J. A. Fontaine, M. D. Jimenez, T. J. Meier, and C. C. Niemeyer. 2006. Habitat selection by recolonizing wolves in the northern Rocky Mountains of the United States. *Journal of Wildlife Management* 70:554–63.

Packard, J. M. 2003. Wolf behavior: reproductive, social, and intelligent. In *Wolves: behavior, ecology, and conservation*, ed. L. D. Mech and L. Boitani, 35–65. Chicago: Chicago University Press.

Parsons, D. R. 1996. Case study: the Mexican wolf. *New Mexico Journal of Science* 36:101–23.

———. 1998. Green fire returns to the Southwest: reintroduction of the Mexican wolf. *Wildlife Society Bulletin* 26:799–809.

Perri, A. R., K. J. Mitchell, A. Mouton, S. Álvarez-Carretero, A. Hulme-Beaman, J. Haile, A. Jamieson, J. Meachen, A. T. Lin, B. W. Schubert, et al. 2021. Dire wolves were the last of an ancient New World canid lineage. *Nature* 591:87–91.

Peterson, R. O., and P. Ciucci. 2003. The wolf as a carnivore. In *Wolves: behavior, ecology, and conservation*, ed. L. D. Mech and L. Boitani, 104–30. Chicago: Chicago University Press.

Prassack, K. A., and L. C. Walkup. 2022. Maybe so, maybe not: *Canis lepophagus* at Hagerman Fossil Beds National Monument, Idaho, USA. *Journal of Mammalian Evolution* 29:313–33. https://link.springer.com/article/10.1007/s10914-021-09591-4.

Ramler, J. P., M. Hebblewhite, D. Kellenberg, and C. Sime. 2014. Crying wolf a spatial analysis of wolf location and depredations on calf weight. *American Journal of Agricultural Economics* 96:631–656.

Reed, J. E., W. B. Ballard, P. S. Gipson, B. T. Kelly, P. R. Krausman, M. C. Wallace, and D. B. Wester. 2006. Diets of free-ranging Mexican gray wolves in Arizona and New Mexico. *Wildlife Society Bulletin* 34:1127–33.

Ripple, W. J., R. L. Beschta, J. K. Fortin, and C. T. Robbins. 2014a. Trophic cascades from wolves to grizzly bears in Yellowstone. *Journal of Animal Ecology* 83:223–33.

Ripple, W. J., J. A. Estes, R. L. Beschta, C. C. Wilmers, E. G. Ritchie, M. Hebblewhite, J. Berger, B. Elmhagen, M. Letnic, M. P. Nelson, O. J. Schmitz, D. W. Smith, A. D. Wallach, and A. J. Wirsing. 2014b. Status and ecological effects of the world's largest carnivores. *Science* 343, article 1241484. doi:10.1126/science.1241484.

Robinson, M. J. 2005. *Predatory bureaucracy: the extermination of wolves and the transformation of the West.* Boulder: University Press of Colorado.

Rothman, R. J., and L. D. Mech. 1979. Scent-marking in lone wolves and newly formed pairs. *Animal Behaviour* 27:750–52.

Sacks, B. N., K. J. Mitchell, C. B. Quinn, L. M. Hennelly, M-H. S. Sinding, M. J. Statham, S. Preckler-Quisquater, S. R. Fain, S. L. Vanderzwan, J. A. Meachen, E. A. Ostrander, and L. A. F. Frantz. 2021. Pleistocene origins, western ghost lineages, and the emerging phylogeographic history of the red wolf and coyote. *Molecular Ecology* 30:4292–94.

Scudday, J. 1972. Two recent records of Gray Wolves in west Texas. *Journal of Mammalogy* 53:598.

Shivik, J. A. 2006. Tools for the edge: what's new for conserving carnivores. *BioScience* 56:253–59.

Sinding, M-H. S., S. Gopalakrishan, F. G. Vieira, J. A. Samaniego Castruita, K. Raundrup, M. P. Heide-Jørgensen, M. Meldgaard, B. Petersen, T. Sicheritz-Ponten, J. B. Mikkelsen, U. Marquard-Petersen, R. Dietz, C. Sonne, L. Dalén, L. Bachmann, Ø. Wiig, A. J. Hansen, and M. T. P. Gilbert. 2018. Population genomics of grey wolves and wolf-like canids in North America. *PLOS Genetics* 14(11): e1007745. https:// doi.org/10.1371/journal.pgen.1007745.

Smith, D. W., E. E. Bangs, J. K. Oakleaf, C. Mack, J. A. Fontaine, D. K. Boyd, M. D. Jimenez, D. H. Pletscher, C. C. Niemeyer, T. J. Meier, D. R. Stahler, J. Holyan, V. J. Asher, and D. L. Murray. 2010. Survival of Colonizing Wolves in the northern Rocky Mountains of the United States, 1982–2004. *Journal of Wildlife Management* 74:620–34.

Smith, D. W., T. D. Drummer, K. M. Murphy, D. S. Guernsey, and S. B. Evans. 2004. Winter prey selection and estimation of wolf kill rates in Yellowstone National Park, 1995–2000. *Journal of Wildlife Management* 68:153–66.

Sommers, A. P., C. C. Price, C. D. Urbigkit, and E. M. Peterson. 2010. Quantifying economic impacts of large-carnivore depredation on bovine calves. *Journal of Wildlife Management* 74:1425–34.

Stahler, D. R., D. W. Smith, and D. S. Guernsey. 2006. Foraging and feeding ecology of the gray wolf (*Canis lupus*): lessons from Yellowstone National Park, Wyoming, U.S.A. *Journal of Nutrition* 136:1923S–26S.

Stahler, D. R., D. Smith, and R. Landis. 2002. The acceptance of a new breeding male into a wild wolf pack. *Canadian Journal of Zoology* 80:360–65.

Strong, D. R. 1992. Are trophic cascades all wet? *Ecology* 73:747–54.

Tedford, R. H., X. Wang, and B. E. Taylor. 2009. Phylogenetic systematics of the North American fossil Caninae (Carnivora: Canidae). *Bulletin of the American Museum of Natural History* 325:1–218.

Thiel, R. P. 1985. Relationship between road densities and wolf habitat suitability in Wisconsin. *American Midland Naturalist* 113:404–7.

Truett, J. 1996. Bison and elk in the American Southwest: in search of pristine. *Environmental Management* 20:195–206.

US Fish and Wildlife Service (USFWS). 1976a. *Determination that two species of butterfly are threatened species and two species of mammals are endangered species*. 41:177736–40. Washington, DC: Office of the Federal Register.

———. 1976b. *Endangered status for 159 taxa of animals*. 41:24062–67.

———. 1982. *Mexican Wolf Recovery Plan*. Albuquerque, NM: US Fish and Wildlife Service.

———. 1998a. *Final rule—establishment of a nonessential experimental population of the Mexican gray wolf in Arizona and New Mexico*. Vol. 63. Washington, DC: Office of the Federal Register.

———. 1998b. *Mexican wolf recovery program, progress report # 1, reporting period: January 1–December 31, 1998*. Albuquerque, NM.

———. 1999. *Mexican wolf recovery program, progress report # 2, reporting period: January 1–December 31, 1999*. Albuquerque, NM.

———. 2000a. *Mexican wolf recovery program, progress report # 3, reporting period: January 1–December 31, 2000*. Albuquerque, NM.

———. 2000b. *Environmental assessment for the translocation of Mexican wolves throughout the Blue Range wolf recovery area in Arizona and New Mexico*. Albuquerque, NM: US Fish and Wilidlife Service.

———. 2003. *Mexican wolf recovery program, progress report # 6, reporting period: January 1–December 31, 2003*. Albuquerque, NM.

———. 2013. *Mexican wolf recovery program, progress report # 16, reporting period: January 1–December 31, 2013*. Albuquerque, NM.

———. 2014a. *Final environmental impact statement for the proposed revision to the regulations for the nonessential experimental population of the Mexican wolf (Canis lupus baileyi)*. USFWS, Region 2, Albuquerque, NM.

———. 2014b. *Mexican wolf recovery program, progress report # 17, reporting period: January 1–December 31, 2014*. Albuquerque, NM.

———. 2015a. *Revision to the regulations for the*

nonessential experimental population of the Mexican wolf. Washington, DC: Office of the Federal Register.

———. 2015b. *Mexican wolf recovery program, progress report # 18, reporting period: January 1–December 31, 2015.* Albuquerque, NM.

———. 2017. *Mexican wolf recovery plan.* USFWS, Region 2, Albuquerque, NM.

———. 2018. *Mexican wolf recovery program, progress report # 21, reporting period: January 1–December 31, 2018.* Albuquerque, NM.

Vilà, C., I. Amorim, J. Leonard, D. Posada, J. Castroviejo, F. Petrucci-Fonseca, K. Crandall, H. Ellegren, and R. Wayne. 1999. Mitochondrial DNA phylogeography and population history of the grey wolf *Canis lupus. Molecular Ecology* 8:2089–103.

VonHoldt, B. M., J. A. Cahill, Z. Fan, I. Gronau, J. Robinson, J. P. Pollinger, B. Shapiro, J. Wall, and R. K. Wayne. 2016. Whole-genome sequence analysis shows that two endemic species of North American wolf are admixtures of coyote and the gray wolf. *Science Advances* 2:e1501714.

VonHoldt, B. M., J. P. Pollinger, D. A. Earl, J. C. Knowles, A. R. Boyko, H. Parker, E. Geffen, M. Pilot, W. Jedrzejewski, B. Jedrzejewska, V. Sidorovich, C. Greco, E. Randi, M. Musiani, R. Kays, C. D. Bustamante, E. A. Ostrander, J. Novembre, and R. K. Wayne. 2011. A genome-wide perspective on the evolutionary history of enigmatic wolf-like canids. *Genome Research* 21:1294–305.

Vucetich, J. A., M. Hebblewhite, D. W. Smith, and R. O. Peterson. 2011. Predicting prey population dynamics from kill rate, predation rate and predator-prey ratios in three wolf-ungulate systems. *Journal of Animal Ecology* 80:1236–45.

Vucetich, J. A., and R. O. Peterson. 2004. The influence of top-down, bottom-up and abiotic factors on the moose (*Alces alces*) population of Isle Royale. *Proceedings of the Royal Society of London. Series B: Biological Sciences* 271:183–89.

Vucetich, J. A., D. W. Smith, and D. R. Stahler. 2005. Influence of harvest, climate and wolf predation on Yellowstone elk, 1961–2004. *Oikos* 111:259–70.

Wang, X. 1994. Phylogenetic systematics of the Hesperocyoninae (Carnivora: Canidae). *Bulletin of the American Museum of Natural History* 221:1–207.

Wang, X., and R. H. Tedford. 2007. Evolutionary history of Canids. In *The Behavioural Biology of Dogs,*

ed. P. Jensen, 3–20. Wallingford, UK: Centre for Agriculture and Bioscience International.

———. 2008. How dogs came to run the world. *Natural History* 117:18–23.

Wang, X., R. H. Tedford, and M. Antón. 2008. *Dogs: their fossil relatives and evolutionary history.* New York: Columbia University Press.

Wang, X.-M., R. H. Tedford, B. Van Valkenburgh, and R.K. Wayne. 2004. Ancestry: evolutionary history, molecular systematics, and evolutionary ecology of Canidae. In *The biology and conservation of wild canids,* ed. D. W. Macdonald and C. Sillero-Zubiri, 39–54. New York: Oxford University Press.

Wayne, R. K., and S. M. Jenks. 1991. Mitochondrial DNA analysis implying extensive hybridization of the endangered red wolf *Canis rufus.* Nature 351(6327):565–68.

Wheeldon, T. J., B. R. Patterson, and B. N. White. 2010. Sympatric wolf and coyote populations of the western Great Lakes region are reproductively isolated. *Molecular Ecology* 19:4428–40.

Wheeldon, T. J., and B. N. White. 2009. Genetic analysis of historic western Great Lakes region wolf samples reveals early *Canis lupus/lycaon* hybridization. *Biology Letters* 5:101–4.

White, P. J., and R. A. Garrott. 2005. Northern Yellowstone elk after wolf restoration. *Wildlife Society Bulletin* 33:942–55.

Wilmers, C. C., R. L. Crabtree, D. W. Smith, K. M. Murphy, and W. M. Getz. 2003. Facilitation by introduced trophic top predators: grey wolf in Yellowstone National Park subsidies to scavengers. *Journal of Animal Ecology* 72:909–16.

Wilson, P. J., S. Grewal, I. D. Lawford, J. N. M. Heal, A. G. Granacki, D. Pennock, J. B. Theberge, M. T. Theberge, D. R. Voigt, W. Waddell, R. E. Chambers, P. C. Paquet, G. Goulet, D. Cluff, and B. N. White. 2000. DNA profiles of the eastern Canadian wolf and the red wolf provide evidence for a common evolutionary history independent of the gray wolf. *Canadian Journal of Zoology* 78:2156–66.

Winnie, J. A., Jr. 2012. Predation risk, elk, and aspen: test of a behaviorally mediated trophic cascade in the Greater Yellowstone Ecosystem. *Ecology* 93:2600–14.

———. 2014. Predation risk, elk, and aspen: reply. *Ecology* 95:2671–74.

Wozencraft, W. C. 2005. Order Carnivora. In *Mammal species of the world: a taxonomic and geographic reference*, Vol. 1, ed. D. E. Wilson and D. M. Reeder, 532–628. 3rd ed. Baltimore: Johns Hopkins University Press.

Wright, G. J., R. O. Peterson, D. W. Smith, and T. O. Lemke. 2006. Selection of northern Yellowstone elk by gray wolves and hunters. *Journal of Wildlife Management* 70:1070–78.

Wydeven, A. P., R. L. Jurewicz, T. R. Van Deelen, J. Erb, J. H. Hammill, D. E. Beyer Jr., B. Roell, J. E. Wiedenhorft, and D. Weitz. 2009. Gray wolf conservation in the Great Lakes Region of the United States. In *A new era for wolves and people—wolf recovery, human attitudes, and policy*, ed. M. Musiani, L. Boitani, and P. C. Paquet, 69–93. Calgary: University of Calgary Press.

Young, S. P., and E. A. Goldman. 1944. *The wolves of North America*. Pt. 1, *Their history, life habits, economic Status, and control*. Washington, DC: American Wildlife Institute.

GRAY FOX (*UROCYON CINEREOARGENTEUS*)

Matthew J. Gould, Jean-Luc E. Cartron, and Gary W. Roemer

Naturally shy and often wary of humans, gray foxes (*Urocyon cinereoargenteus*; also called northern gray foxes) are nonetheless curious creatures, inquisitive and often mellow once in hand. They are the second largest fox native to New Mexico. Twenty-three male gray foxes live-captured across New Mexico and western Texas, from the Guadalupe Mountains and Carlsbad Caverns National Parks west to the Mogollon Mountains and north to the Manzano Mountains, weighed an average of 3.5 kg (SD = 0.7) (7.7 lbs ± 1.5) and were slightly heavier (~11%) than 20 females, which averaged 3.2 kg (SD = 0.4) (7.0 lbs ± 0.9) (G. Roemer, unpubl. data). Only the North American red fox (*Vulpes fulva*) is larger, ranging from 3.4–8.7 kg (7.5–19.2 lbs), whereas both the kit fox (*V. macrotis*) and swift fox (*V. velox*) are smaller, averaging only about 2 kg (4.4 lbs) (see Chapters 13–15).

Although gray foxes have a pelage often resembling that of both kit and swift foxes, they are easily recognized by the black stripe running along the dorsal length of their tail and ending in a black tip (Photo 12.1). In contrast, both kit and swift foxes have a black-tipped tail but lack the black tail stripe. Belying their name, North American red foxes have a pelage that can be variably colored, but they sport a white-tipped tail. Like the red fox, the gray fox shows rufous coloration on various parts of its body. That rufous coloration is mostly restricted to the ventral and lateral surfaces and found between, and often

Photo 12.1. Gray fox (*Urocyon cinereoargenteus*) in New Mexico's Guadalupe Mountains on 15 June 2007. Note the peppery gray dorsal pelage and the black stripe along the dorsal side of the tail. Gray fur extends laterally down the legs to where it is replaced by rufous. Along the inside surface of the legs, the gray and rufous grade into white. Photograph: © Jared A. Grummer.

blending with, the peppery gray dorsal pelage and white ventral region. In contrast, when exhibiting the most common red color variant, red foxes have a red dorsal and lateral pelage with darker, often black, legs. The difference in pelage color along with the distinctive tail markings, easily distinguishes the two species.

The gray fox's head is also distinctive, gray with black patches on the lateral sides

(*opposite page*) Photograph: © John McClure.

Photo 12.2 (*top*). Gray fox head close-up. The head is mainly gray with distinctive black, white, and rufous markings. The long, pointed ears are typically rufous at the base. Gray fox photographed in Hidalgo County on 18 May 2006. Photograph: © Robert Shantz.

Photo 12.3 (*bottom*). Gray fox head close-up (right lateral view). Note the small black patches on the lateral sides of the muzzle and along the lips, white fur just below and lateral to the tip of the snout and extending behind onto the cheeks and downward to at least some of the throat. Photograph: © Pat (Ranger) Ward.

of the muzzle in the vicinity of the vibrissae, with black outlining the lips of both jaws. Small white patches are usually present just below and lateral to the tip of the snout, or rhinarium (Photos 12.2 and 12.3). White fur on the muzzle extends both behind the lateral black patches to the cheek and downward to at least some of the throat. The ventral surface of the neck can be all white down to the chest but typically joins with rufous below the

throat and laterally along the neck. The dorsal surface of the ears is usually rufous at the base with the hair becoming gray distally with a white rim demarcating the ear and a fringe of longish white hair covering the inside of the pinnae. Variable degrees of white and rufous color are also found on the chest and extend throughout the belly where they blend to gray and white caudally. The body and tail are mostly gray, with the latter having the aforementioned conspicuous black stripe and black tip. The gray of the body extends partially down the legs giving way to mostly rufous both in the middle and toward the rear, whereas the ventral or inside surfaces of the legs can be white. The pelage is relatively short (2 to 4 cm [0.8–1.6 in] deep) with a single molt resulting in a thin summer coat and a dense winter coat.

The skull of the gray fox resembles that of other fox species except that the temporal ridges broaden posteriorly and form a distinct "violin-shape," also termed lyrate in imitation of the musical instrument lyre (Photo 12.4). In the swift fox the temporal ridges are also lyrate, but they are narrower and less conspicuous than in the gray fox. In *V. fulva* and *V. macrotis* the temporal ridges instead converge toward the occipital ridge, resulting in a narrower structure posteriorly. The lower and posterior border of the mandibles of the gray fox also has a conspicuous notch (Fuller and Cypher 2004; Photo 12.5). The gray fox's dental formula is the same as that of the kit, swift, and red foxes, or 3/3, 1/1, 4/4, 2/3 = 42.

One unique aspect of gray foxes is their ability to climb trees, adding a third dimension to their foraging (see Grinnell et al. 1937; Gunderson 1961), unlike what is typically seen in other canid species. This ability is a consequence of unique limb and muscle anatomy (Feeney 1999). The limbs of canids are typically long, adapted for cursorial movement, but they are shorter in the gray fox. The radius can also be rotated around the ulna to a greater degree in the gray fox compared to most canids, thus allowing the

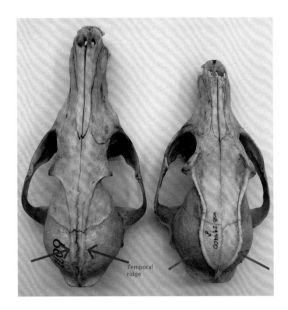

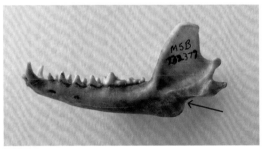

Photo 12.4 (*top*). Superior view of red fox (*Vulpes fulva*, left) and gray fox (*Urocyon cinereoargenteus*; right) skulls from New Mexico. Note the larger size of the red fox's skull and the gray fox skull's widely separated temporal ridges outlining a distinct violin shape or lyrate. Photograph: © Jon Dunnum.

Photo 12.5 (*bottom*). Gray fox mandible showing the conspicuous notch at the posterior end of the horizontal ramus. Photograph: © Jon Dunnum.

Photo 12.6 (*top*). Gray fox in a desert willow (*Chilopsis linearis*) in Luna County on 23 November 2013. Photograph: © John McClure.

Photo 12.7 (*bottom*). Gray fox in a tree in Colfax County on 31 March 2001. Photograph: © Clint Henson.

forelimb to be medially rotated rather than being restricted to the sagittal plane. In addition, the hind limbs can be abducted to a greater degree, possibly allowing for enhanced purchase against a climbing surface such as a tree. And finally, the claws are semi-retractable so they can remain sharp (Feeney 1999). Findley et al. (1975) reported observations of gray foxes in Emory oaks (*Quercus emoryi*) in Hidalgo County, in addition to junipers (*Juniperus* spp.) and cottonwoods (*Populus* spp.). Although we have never ourselves detected a gray fox in a tree, one of us (G. W. Roemer) once observed a semi-domesticated island fox (*U. littoralis*), a congener of the gray fox, "playing" with a hunting dog on Santa Cruz Island, California. The fox repeatedly ran between, and scaled, two adjacent trees as quickly and as adeptly as a cat to elude the frustrated dog.

The extant Canidae evolved in North America from a common ancestor during the Oligocene (32 Ma) and later spread via both the Beringian (6 Ma) and Panamanian (3 Ma) land bridges to Asia and South America, respectively (Wang and Tedford 2008). The most likely progenitor of the modern-day gray fox was the slightly larger, so-called progressive gray fox (*U. progressus*), which was found in North America as early as the Hemphillian and through the Blancan, but fossil remains are scarce (Kurtén and Anderson 1980). The Hemphillian is a North American faunal stage extending from 10.3 Ma to 4.9 Ma, and it was succeeded by the Blancan faunal stage that extended from ~4.75 to 1.8 Ma, and thus ended in the early Pleistocene. Fossil specimens associated with the modern-day gray fox are widespread across North America, having been found in nearly 40 localities within the present-day states of Arizona, Arkansas, California, Florida, Georgia, Kansas, Kentucky, Missouri, New Mexico, Pennsylvania, Tennessee, and Texas; fossil specimens date back to the Irvingtonian (1.9 to 0.25 Ma) (Kurtén and Anderson 1980). In New Mexico, gray fox fossil remains exist from cave deposits primarily in the southern part of the state (e.g., U-Bar and Pendejo caves) and date back to the mid- and late Wisconsin (Harris 1987, 2003; Chapter 1).

A consensus tree involving >2,000 base pairs of three protein-coding regions within the mtDNA genome and 57 morphological characters, revealed that the genus *Urocyon* is the most basal genus within the Canidae and that it split from the genera *Nyctereutes* (raccoon dog, indigenous to eastern Asia) and *Otocyon* (bat-eared fox, found in Africa), in addition to a "red fox-like" clade consisting of the genus *Vulpes*, approximately 10 Ma (Wang et al. 2004). Gray foxes are evolutionarily distinct from the vulpine foxes, and there are only two extant, currently recognized species within the genus *Urocyon*, the other being the island fox (Wayne et al. 1997). The island fox is a diminutive species restricted to the California Channel Islands, and it is a direct descendant of coastal California gray foxes (Collins 1993; Goldstein et al. 1999; Roemer et al. 2004).

Multiple studies that explored variation both within the mtDNA genome and at the cytochrome *b* gene within the mtDNA genome have found cryptic divergence within gray foxes yielding two North American groups, eastern and western gray foxes. The two lineages likely diverged ~500,000–800,000 years ago with secondary contact between the two clades occurring on the southern Great Plains (Goddard et al. 2015; Reding et al. 2021). This pattern coincides with other carnivore and forest dependent taxa that were isolated into refugia during the frequent glaciation events of the Pleistocene but subsequently expanded their geographic range during the Holocene (e.g., Aubry et al. 2009; Reding et al. 2012; see also Chapter 7). The long period since divergence, however, suggests that the two clades were isolated through multiple glaciation events such as the Last Glacial Maximum (~23,000 years ago) and the Penultimate Glacial Maximum (~140,000 years ago; Reding et al. 2021). Reding et al. (2021) identified both lower genetic diversity in the northeastern United States and a pattern indicative of rapid colonization of the area following the recession of glacial ice. They also did not identify any additional cryptic lineages within the east-west gray fox lineages, putting the validity of subspecific designations into question. They did note, however, that the east-west split at the Great Plains corresponds with well-known phenotypic differences between eastern and western gray foxes. Gray foxes in the east tend to be darker, duller, and less silvery while those in the west have a slenderer body, longer tail, and more pointed ears (Reding et al. 2021). Given the incongruence between the distribution of gray fox subspecies and mtDNA haplotypes, the authors suggest that a taxonomic revision based on a range-wide analysis using phenotypic, nuclear, and mitochondrial data is necessary (Reding et al. 2021). In that same study, the island fox was found to represent a monophyletic group nested within the western lineage, a pattern similar to that involving the western spotted skunk (*Spilogale gracilis*) and the island spotted skunk (*S. gracilis amphialus*), another California Channel Islands restricted species (Reding et al. 2021). Interestingly, the island fox is considered a separate species from the gray fox while the island spotted skunk is considered a subspecies of the western spotted skunk (Reding et al. 2021).

Across its range, the gray fox has several common names including tree fox and, in Spanish, zorro (fox), zorro gris (gray fox), and zorro plateado (silver fox; Fuller and Cypher 2004). Gato de monte (cat of the woods) is another name used in some regions of Latin America due to the cat-like appearance of the gray fox's face. However, gato de monte is also a common name given to different species depending on the country (e.g., margay, ocelot, and jaguarundi; G. Palomo, pers. comm.). There are as many as 16 recognized subspecies (Fritzell and Haroldson 1982). Only *U. c. scottii* is found in New Mexico.

DISTRIBUTION

Presently, gray foxes are found in northern South America, in both Venezuela and Columbia, north throughout Central America—though curiously absent from some Caribbean watersheds along the Atlantic coast—and Mexico, including the Baja Peninsula, and across the United States from Florida to Oregon, but absent in some parts of the Great Plains and the northern Rocky Mountains (Fuller and Cypher 2004). At the northern limit of their distribution, finally, they occur in small numbers in southern Canada from Manitoba east to New Brunswick, with breeding populations confirmed in Ontario and suspected in Quebec (Fuller and Cypher 2004; Environment and Climate Change Canada 2017). Gray foxes have been documented on two islands off the coast of Mexico: Tiburon Island in the Gulf of California (Collins 1993) and Cozumel Island off the

Yucatan Peninsula (Gompper et al. 2006). Since the 1930s and 1940s, the gray fox has expanded its range and/or recolonized areas it formerly occupied at or near the northern and northeastern edges of its distribution, possibly in relation to climate change (Fritzell and Haroldson 1982; McAlpine et al. 2008).

The gray fox occurs throughout most of New Mexico, primarily at middle elevations, where it is common around foothills and in areas of rugged topography (Bailey 1931; Findley 1987; Map 12.1). The species has been documented at numerous locations such as Chaco Culture National Historical Park and Blanco in the northwest (Bailey 1931; Bogan et al. 2007); El Malpais National Monument and surrounding lava fields near Grants in Cibola County (Hooper 1941; Bogan et al. 2007); the Animas Mountains (Cook 1986) in the Bootheel region; Carlsbad Caverns National Park (Geluso and Geluso 2004), the San Andres National Wildlife Refuge (Halloran 1946), and White Sands Missile Range (Hobert et al. 2016) in south-central New Mexico; the Llano Estacado in the southeast (Aday and Gennaro 1973; Stuart and Anderson 1993; Choate 1997; Frey 2003); the Middle Rio Grande Valley, the Sandia Mountains, Salinas Pueblo Missions National Monument, and Petroglyph National Monument in the central part of the state (Ivey 1957; Bogan et al. 2007); Bandelier National Monument near Los Alamos in Sandoval and Los Alamos counties; and Sugarite Canyon State Park in Colfax County (Frey and Schwenke 2012). Bailey (1931:301) referred to gray foxes as "particularly numerous" in the Mogollon, Sacramento, Guadalupe, Jemez, Zuni, and Chuska mountains as well as in the Black Range. Halloran (1946) described the gray fox as the most common fox on the San Andres National Wildlife Refuge, while Geluso and Geluso (2004) reported as many as 82 observations of the species at Carlsbad Caverns National Parks during surveys from 1973 to 2000.

Bailey (1931) described the species as most

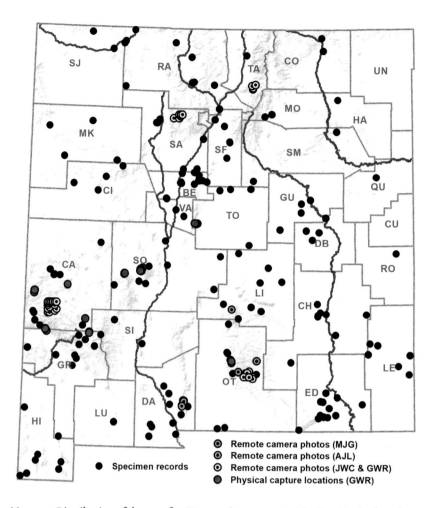

Map 12.1. Distribution of the gray fox (*Urocyon cinereoargenteus*) in New Mexico based on both museum specimen records and research by the authors and collaborators (remote camera photos and trapping locations; MJG = Matthew J. Gould; AJL = Andrew J. Lawrence; JWC = James W. Cain III; and GWR = Gary W. Roemer).

abundant in the Upper Sonoran Zone, which lies at elevations ranging generally from ~1,370 to 2,130 m (4,500–7,000 ft) and constitutes about three-fourths of the state's area, and as present but far less common in the Transition Zone (~2,130–2,590 m [7,000–8,500 ft]). As parts of various research projects, two of us (G. W. Roemer and M. J. Gould) and collaborators have captured or photographed gray foxes in southern New Mexico in the Guadalupe and Sacramento mountains, the uplands adjacent to the Gila River Valley, the Black Range,

and the Mogollon Mountains; and farther north in the San Mateo, Manzano, Jemez, and Taos mountains, mostly at elevations encompassing the upper Sonoran and Transition zones. During a recent cougar (*Puma concolor*) research project, gray foxes were incidentally photographed on multiple occasions in the Transition Zone in the Jemez, Taos, and Sacramento mountains, even as high as 2,959 m (9,708 ft) and 2,890 m (9,480 ft) on the Gila National Forest (J. W. Cain III and G. W. Roemer, unpubl. data). Veals (2018) similarly recorded the presence of gray foxes in the Pinaleño and White mountains of southeastern Arizona at elevations of 1,530 to 2,985 m (~5020–9795 ft), or the study's entire elevational range. In the northern half of New Mexico, gray foxes reach at least the Transition Zone, as observed in the Sandia Mountains (Ivey 1957) or above Lake Maloya at Sugarite Canyon State Park, at an elevation of approximately 2,290 m (7,515 ft) (J.-L. E. Cartron, pers. obs.). The highest-elevation record remains a specimen collected at 3,230 m (10,600 ft) on 19 July 1964 between Mount Taylor and Mosca Peak (MSB:Mamm:21176). At the highest elevations of New Mexico in the Sangre de Cristo and San Juan mountains, the gray fox does appear to be absent, replaced instead by the North American red fox, better adapted to heavy snow cover (see Chapter 15).

In Bailey's (1931) time, gray foxes appeared to be only occasional visitors and/or were limited in their distribution to areas of suitable topography both in the lowest valleys (Lower Sonoran Zone) of New Mexico—all in the south and southeast—and on the state's eastern plains. With human land use practices, however, gray foxes might have expanded their distribution and/or increased in numbers in those areas of the state. This appears to be the case on the Llano Estacado, the very large mesa spanning both sides of the New Mexico–Texas state line, where gray foxes have been found in recent decades in proximity to towns such as Portales in Roosevelt County and Hobbs and Lovington in Lea County (Stuart and Anderson 1993; Frey 2003; see under "Habitat Associations").

HABITAT ASSOCIATIONS

Although gray foxes are habitat generalists associated with a variety of land-cover types throughout their range, they occur most predictably in forests and woodlands and utilize brushy vegetation and rocky areas (Fritzell and Haroldson 1982). In Central and South America, gray foxes inhabit subtropical and tropical

Photo 12.8 (*top*). Gray fox photographed in July 2014 at an elevation of ~2,660 m (8,730 ft), within the Agua Chiquita Creek watershed in the Sacramento Mountains. The dominant land-cover type was classified as mixed timber, or a combination of Douglas-fir (*Pseudotsuga menziesii*) and white fir (*Abies concolor*). Photograph: © Matthew J. Gould.

Photo 12.9 (*bottom*). Gray fox photographed on 24 October 2015 by a remote camera in riparian woodland along the Gila River in Grant County. Photograph: © Keith Geluso.

dry and moist forests (e.g., Pineda-Guerrero et al. 2015), though in Belize, they prefer a mixed landscape of well-vegetated savannah, pine forest, and agricultural lands, avoiding the interior of the tropical moist broadleaf forest, using instead only its edges (Harmsen et al. 2019). In eastern North America, gray foxes are strongly

Photo 12.10. Gray fox visiting a wildlife drinker on 11 May 2011 in Goat Draw on the Sevilleta National Wildlife Refuge. Goat Draw is located in an intermontane valley in the Los Pinos Mountains, in the northeastern corner of the refuge. The dominant vegetation is juniper savanna. Photograph: © Sevilleta National Wildlife Refuge, US Fish and Wildlife Service.

Photo 12.11. Gray fox leaping on a small rodent (possibly a woodrat [*Neotoma* sp.]) across a pond in a Los Alamos residential area on 23 September 2012. Photograph: © Hari Viswanathan.

Photo 12.12 (*top*). Gray fox on a wooden fence in Santa Fe, Sante Fe County on 27 October 2020. Photograph © Ella Meyer.

Photo 12.13 (*bottom*). Gray Fox on a wooden headframe covering an underground mine shaft southwest of Lordsburg in Hidalgo County, New Mexico (18 May 2006). Gray fox use of reclaimed mines has been documented in parts of the species' range. Photograph: © Robert Shantz.

associated with deciduous hardwood forests and mosaics of agricultural fields and woodlands (Wood et al. 1958; Hall 1981) whereas in the western United States, they also often use brushy and rocky areas (Grinnell et al. 1937; Hardy 1945; Johnson et al. 1948; Leopold 1959). In New Mexico, gray foxes appear most common in pinyon-juniper woodlands, but they are also found in ponderosa pine forests with their occurrence in grasslands conditioned by the local presence of rocky outcrops or encroaching juniper woodland (Findley et al. 1975). In apparent contradiction of Findley et al. (1975), the recent use of remote cameras in mountain ranges around the state shows that gray foxes frequently reach mixed-conifer forests (Photo 12.8). The species is also known to use riparian woodlands such as those along the Rio Grande (Hink and Ohmart 1984) and the Gila River (Geluso 2016).

In our own experience, gray foxes are also

found in New Mexico in desert scrub vegetation types common to the foothill regions or bajadas of desert mountain ranges. Farther away from foothills, however, they appear to be scarce or replaced by kit foxes. During intensive remote camera surveys over a four-year period, no gray foxes were documented in White Sands National Park in Otero and Doña Ana counties (Robinson et al. 2014; G. W. Roemer and F. Abadi, unpubl. data), but a few (n = 10 photographs) were detected with remote cameras on White Sands Missile Range, New Mexico within the rockier bajadas nearer to the San Andres Mountains and adjacent to lava flows (Hobert et al. 2016). Gray foxes are known to occur in sage scrub and chaparral habitat in southern California (Farías González et al. 2012). In Pecos County, Texas, on the northern edge of the Chihuahuan Desert, gray foxes were found on lower slopes with creosote, *Opuntia*, and agave shrublands and in mesquite-greasewood shrublands in valley bottoms (Atwood et al. 2011).

The water requirements of the gray fox are not well understood, but the species lacks some of the adaptations of kit foxes for heat radiation and water conservation (see Cypher 2003). Gray fox visits to wildlife drinkers have been documented in some of New Mexico's more arid locations (e.g., Photo 12.10). In western Texas, visits by gray foxes to artificial water sources suggested that they select sources adjacent to rugged escape terrain, perhaps to minimize the risk of agonistic interactions with coyotes (*Canis latrans*; Atwood et al. 2011; see also under "Interspecific Interactions"). Thus, in desert environments where water is scarce and rocky escape cover is lacking, the lack of access to water may be compounded by the increased risk of conflict with a larger competitor.

Gray fox use of reclaimed mines has been recorded in parts of the species' range including in New Mexico (Yearsley and Samuel 1980; Photos 12.13 and 12.14). In New Mexico as elsewhere, gray foxes also frequently reside within urban and

Photos 12.14a and b (*top to bottom*). (a) Gray fox on 13 December 2018 at the bottom of a pit-like opening descending into the main working body of an abandoned mine in Grant County, New Mexico. (b) Main cavern within that mine on 16 July 2019, when possibly the same gray fox was observed with a desert cottontail (*Sylvilagus audubonii*) it had just caught. (a) Photograph: © Dillon Metcalfe; (b) Photograph: © Jean-Luc Cartron.

exurban areas (Harrison 1997). Their presence has been documented in residential neighborhoods in many cities and towns around the state, in particular Las Cruces, where G. W. Roemer has photographed the species with remote cameras. Based on observations of gray foxes inhabiting the communities of Tijeras, Cedar Crest, and Sandia Park in Bernalillo County, Harrison (1997) suggested that the species may only be tolerant of urban development if the density of housing

Photo 12.15. Gray fox with rodent prey in Los Alamos County on 15 May 2021. Photograph: © Hari Viswanathan.

is less than 50–125 residences/km². Studies in California, North Carolina, and Texas, however, have found gray foxes living in varying degrees of urbanization, including areas with housing density two to three times greater (236–347 buildings/km²) as long as there was access to forested areas nearby (Riley 2006; Kapfer and Kirk 2012; Lombardi et al. 2017).

LIFE HISTORY

Diet and Foraging

The diversity of land-cover types occupied by the gray fox has, unsurprisingly, led to a catholic and, arguably, the most diverse diet among North American canids (Hockman and Chapman 1983; Cypher 1993, 2003; Neale and Sacks 2001; Cunningham et al. 2006). Recorded dietary items commonly include rodents, lagomorphs, birds (including largely passerines but also some larger species such as ducks and pheasants [*Phasianus colchicus*]), insects, reptiles, scavenged ungulates, and soft and hard mast (Cypher 1993, 2003; Neale and Sacks 2001). In New Mexico, all information on the diet of gray foxes is anecdotal, with Bailey (1931) mentioning the importance of juniper berries, cactus fruit, insects (grasshoppers, crickets, beetles) and small rodents. To the west in Arizona's southern Mazatzal Mountains northeast of

Phoenix, Cunningham et al. (2006) studied the diet of gray foxes in vegetation consisting primarily of interior chaparral with Madrean evergreen woodland and ponderosa pine forest also present. The study was conducted after a 237 km² wildfire in the Mazatzal Mountains, but the diet of gray foxes was found to be similar in burned and unburned areas. The primary food source (n = 609 scats) was soft mast (53% and 54% of all scats in burned and unburned areas, respectively) with the most common fruits being manzanita (*Arctostaphylos pungens*), one-seed juniper (*Juniperus monosperma*), and prickly pear cactus (*Opuntia engelmannii*) followed by small mammals (burned = 39% and unburned = 44%), birds, insects, and reptiles (burned = 26% and unburned = 20%), and herbaceous material (burned = 11% and unburned = 7%; Cunningham et al. 2006). In Maryland, plants were again the main food source for gray foxes, with persimmon fruits (*Diospyros virginiana*) the most frequently detected item and corn (*Zea mays*) accounting for the greatest percentage in terms of weight (Hockman and Chapman 1983). Gray foxes in California and Illinois displayed similar diversity in their diet with fruits being the dominant food type for part of the year, followed by a shift toward a more meat-based diet in the winter when the availability of fruits was limited (Cypher 1993; Neale and Sacks 2001). Although fruits may comprise a substantial portion of the diet of gray foxes, it is unlikely that foxes could meet their energy requirements and persist on a completely frugivorous diet. Gray foxes experimentally fed mice versus Himalaya berries (*Rubus procerus*) lost weight on the latter diet (Ball and Golightly 1992).

Gray foxes are more active at night, at which time they conduct most of their foraging (Fuller and Cypher 2004). The foraging behavior of gray foxes has received little attention and we can only speculate that the tactics employed (e.g., hunting, foraging, and scavenging) are as diverse as their diet. A rabies vaccination study on free-ranging gray foxes in central Texas found no preference by foxes regarding the type, location, or combination of artificial baits and odor attractants (e.g., baits: wax-lard cakes with marshmallow icing and dog food–based and fish meal–based polymer cubes; odor attractants: beef lard, valeric acid, butyric acid, and granulated sugar) suggesting the foraging behavior of the gray fox may indeed be as opportunistic as its diet is general (Steelman et al. 2000).

Reproduction and Social Behavior

In New Mexico, there is scarce information on the home range size, social behavior, and reproductive biology of the gray fox. Instead, it is necessary that we glean insight from research spanning the species' distribution and with comparison to the island fox, a congener. The size of gray fox home ranges is highly variable. In a mixture of residential and undeveloped areas in Bernalillo and Socorro counties, New Mexico, the home range of 12 individuals averaged 734 ha (SD = 244) (1,814 acres) based on a 95% adaptive kernel (AK) and 481 ha (SD = 179) (1,189 acres) using a 95% minimum convex polygon (MCP; Harrison 2002). Average home range size in the Pinaleño and White Mountains of Arizona, where vegetation types consisted of chaparral, oak woodlands, pinyon-juniper, ponderosa pine, and mixed conifers, was 378 ha (SD = 274) (934 acres) and 515 ha (SD = 3.91) (1,273 acres) for two males and a female using a 95% kernel density estimator and MCP, respectively (Veals 2018). In California, there was no statistically significant difference found among home range sizes for males and females in urban and rural zones with home range size ranging from 78.3 to 100.3 ha (SD = 31.3–51.8) (193 to 248 acres) using a 95% AK and ranging from 59.2 to 87.9 ha (SD = 14.5–61.9) (146 to 217 acres) using a 95% MCP (Riley 2006). Home range size also did not differ statistically between male and female gray foxes in Georgia, but did so among seasons with home range size smallest in

Photo 12.16 (*top*). Gray fox in an abandoned house, Doña Ana County, 3 February 2005. The fox apparently used the house as its den. Photograph: © James E. Zabriskie.

Photo 12.17 (*bottom*). Orphaned gray fox pup brought to the Wildlife Center in Española, 19 May 2009. Photograph: © Alissa D. Mundt.

the spring at 160.8 ha (SD = 31.7; n = 23) (397 acres) and largest in the winter at 217.0 ha (SD = 53.8; n = 12) (536 acres) using a 95% AK (Deuel et al. 2017). Research thus shows that males and females tend to have similarly sized home ranges, likely due to a lack of pronounced sexual dimorphism in the species and possibly a consequence of a monogamous mating system, except during the spring when females are whelping pups and their movements are more restricted (Chamberlain and Leopold 2000, 2002; Veals 2018).

What we know to be ubiquitous for the species is that gray foxes are territorial but also exhibit social monogamy with the formation of pair-bonds between mates (Deuel et al. 2017). Evidence for both territoriality and mated pair-bonds stems from research that has evaluated the spatial overlap of home ranges among neighboring foxes (Nicholson et al. 1985; Tucker et al. 1993; Chamberlain and Leopold 2000; Riley 2006; Deuel et al. 2017). In California and Georgia, the proportion of spatial overlap for home ranges of mated-pairs averaged 0.82–0.94 with mates often observed resting together during daylight hours (Riley 2006; Deuel et al. 2017). Meanwhile, spatial overlap among male-male, female-female, and non-mated-pair male-female home ranges were 1.7–4.0 times less than the overlap observed for mated-pairs (overlap: 0.19–0.37; Riley 2006) with individuals exhibiting the classic territorial behavior commonly observed among canid species.

While uncommon in the class Mammalia, social monogamy is a common trait in the family Canidae (Moehlman 1989; Deuel et al. 2017). The spatial behavior of mated pairs in Mississippi and Georgia suggests that pair-bonds occur year-round, rather than solely throughout the breeding season, as the home ranges of mated pairs overlapped more so than wandering ventures among non-paired neighbors across all seasons (Chamberlain and Leopold 2000; Deuel et al. 2017). While socially monogamous, extra-pair copulations do occur and have resulted in multiple paternity in gray fox litters that was confirmed through genetic analysis (Weston Glenn et al. 2009). Extra-pair copulations between neighboring males and females are believed to be a mating strategy employed by larger and behaviorally dominant males with the benefit of enhancing their direct fitness (Deuel et al. 2017).

Island foxes exhibit similar home range overlap among socially monogamous mated pairs and extra-pair copulations between neighboring males and females (Roemer et al. 2001). Given that island foxes are direct descendants of gray foxes, that both have similar social systems exhibiting territoriality and high overlap among mated pair home ranges, and show evidence of extra-pair copulations, it is possible that gray foxes have a similar reproductive physiology. Extra-pair fertilizations, as discussed in the previous paragraph, are likely facilitated by induced estrus, possibly followed by induced ovulation, a reproductive physiology characterizing the island fox (Asa et al. 2007). Induced estrous occurs when females are in the appropriate period of their menstrual cycle but need a male present to cause or induce estrus; ovulation then follows and the female is receptive and capable of fertilization for only a short period, perhaps as short as 40 hours (Ralls et al. 2013). This results in all females breeding during the same season, but conception is asynchronous. Males guard females to ensure their reproductive success and prevent cuckoldry, but once their mate has been inseminated and is no longer receptive, males can be free to seek extra-pair copulations (Ralls et al. 2013).

Gray foxes reach sexual maturity at 9–10 months of age (Fritzell and Haroldson 1982). Courtship between males and females occurs during late fall and into winter with mature male sex cells, the spermatozoa, forming in October and November (Aldridge 2008). The breeding season spans from late winter through early spring with populations in more southerly latitudes breeding earlier than those in northern latitudes and breeding opportunities occurring when females enter into their monestrous state (i.e., enter into heat once during a breeding cycle) for one to six days (Ferguson and Larivière 2002; Aldridge 2008). For example, in New York, the breeding season occurs from mid-January to May with the highest percentage of foxes mating in early March (Sheldon 1949), whereas peak-breeding in Florida occurred in early February (Wood 1958). Most gray fox litters in Alabama were found to occur in early April suggesting that most successful breeding opportunities occurred sometime in February (Nicholson et al. 1985).

Gray foxes appear to be flexible in where they rear their pups. Hollow logs, rock piles, earthen holes, brush piles, root systems of fallen trees, and buildings have all been documented as den sites (Nicholson et al. 1985; Aldridge 2008). Sullivan (1956), however, noted that gray fox dens in Alabama were commonly found near permanent water sources suggesting some dependency in den site selection. Gestation has been variously estimated to last 50–63 days with females giving birth to, on average, three to four pups (range: 1–10) that weigh 86–100 grams (~ 3–3.5 oz) per pup, or roughly the weight of a deck of playing cards (Nowak 1999; Ferguson and Larivière 2002). Paternal care has been documented in the species as both males and females constrict the core area of their home range around their den and concurrently engage in pup-rearing duties (Nicholson et al. 1985; Deuel et al. 2017). In the unfortunate circumstance that a mate dies before the pups become independent, either mate will continue to raise the litter until the pups disperse. Success for the male, however, is only possible if the pups have already been weaned prior to the loss of the female (Chamberlain and Leopold 2002). Pups leave the den at ~10–12 weeks old and will forage with their parents, ultimately becoming independent by the end of summer or early fall (Nicholson et al. 1985; Nowak 1999). Dispersal often

occurs by the end of the calendar year. Males disperse farther than females, who often display natal-biased philopatry by setting up their territory relatively close to the home ranges of their parents (Nicholson et al. 1985).

Interspecific Interactions

The broadly distributed gray fox lives sympatrically with the full complement of New Mexico's carnivores except for the state's transient and high-elevation resident, the Canada lynx (*Lynx canadensis*; see Chapter 6). The ubiquity of the gray fox implies a wide breadth of interspecific interactions with research outside the state indicating to us that most interactions likely involve coyotes, bobcats (*Lynx rufus*), and striped skunks (*Mephitis mephitis*), as well as white-backed hognosed skunks (*Conepatus leuconotus*), hooded skunks (*Mephitis macroura*), and North American red foxes in some areas (Chamberlain and Leopold 2005; Farías et al. 2005; Parsons et al. 2019). Such interactions are likely predicated on all these mesocarnivores being habitat generalists, leading to considerable overlap in diets and habitat use. Despite those shared similarities, coyotes, bobcats, and striped skunks in particular elicit different behavioral responses from gray foxes, providing a unique insight into multiple ecological principles (e.g., niche partitioning and intraguild predation).

The dietary overlap among gray foxes, coyotes, bobcats, and striped skunks is centered around small mammals (particularly, rodents [Rodentia spp.] and hares and rabbits [Lagomorpha spp.]) with prey size positively correlated with predator size (Fedriani et al. 2000; Neale and Sacks 2001). Investigation into resource partitioning has revealed a more nuanced picture of food competition. Diet studies in California found high dietary overlap between gray foxes and coyotes during summer and fall when fruits were prevalent on the landscape (Fedriani et al. 2000; Neale and Sacks 2001), but fruits seemingly represented a larger

Photo 12.18. A North American red fox (*Vulpes fulva*) and a gray fox interact aggressively with each other on the San Joaquin River National Wildlife Refuge in California (5 March 2013). Photograph: © US Fish and Wildlife Service Pacific Southwest Region.

energy source for foxes as their foraging activity was focused on fruits during abundant mast crops (Neale and Sacks 2001). Conversely, gray fox and coyote diets were dissimilar during the winter and spring when the diet composition of coyotes shifted toward ungulates (Neale and Sacks 2001). Gray foxes and bobcats showed an inverse pattern as their winter and spring diets were similar, both species consuming a large proportion of rodents and foxes being unable to forage for fruits in that season (Neale and Sacks 2001). Dissimilarity arose between the two in the summer and fall as bobcats consumed a larger proportion of lagomorphs and foxes shifted their diet toward fruits (Neale and Sacks 2001). Outside of New Mexico, striped skunks show similar diets as gray foxes, with the exception that invertebrates constitute a larger

portion of the striped skunk's diet (Greenwood et al. 1999; Azevedo et al. 2006; and see Chapter 28).

A similar story to the dietary overlap among gray foxes, coyotes, bobcats, and striped skunks can be recited regarding the habitat use of gray foxes as all four mesocarnivores use the same land-cover types (Neale and Sacks 2001; Veals 2018). These findings, however, also hold nuance as patterns of space use by gray foxes are contingent on the coexisting species of mesocarnivore. Research in the Pinaleño and White Mountains of Arizona showed that the probability of occupancy of an area by gray foxes was reduced—though it still remained relatively high—when bobcats and coyotes were present, whereas occupancy rates did not change in the presence of striped skunks (Veals 2018). Conversely, the probability of occupancy of an area by gray foxes was not reduced by coyotes in urban Nacogdoches, Texas (Lombardi et al. 2017). Neale and Sacks (2001) also concluded that gray foxes in California did not avoid areas inhabited by coyotes and bobcats, but their inferences were based on the presence of scats resulting in the inability to evaluate potential temporal segregation by gray foxes. An examination of water use in the Chihuahuan Desert of western Texas revealed that both bobcats and coyotes influenced timing of visitation by gray foxes, with foxes visiting water sources primarily at night and in particular between 00:00 and 04:00 hrs (Atwood et al. 2011). Coyotes and bobcats also visited water sources more often at night, but were more catholic in their visitation rates and frequently visited water during the day as well. Farías González et al. (2012) and Fedriani et al. (2000) found no difference in the land-cover types used among gray foxes, coyotes, and bobcats in California; however, camera and foothold traps, along with radio-telemetry collars, allowed them to determine that gray foxes used brushier land-cover more frequently than coyotes or bobcats, which they postulated gray foxes used as escape cover to avoid predation by the latter two species.

Intraguild predation (IGP) is an extreme form of interference competition where a subordinate species (IG prey) may be killed by a dominant species (IG predator) and excluded from a particular land-cover type (e.g., Chapters 7 and 9). IGP theory predicts three equilibria as it relates to resource availability and the presence and absence of the competing species: 1) In resource-scarce areas, the IG predator cannot "invade" the system because there are not enough resources to support it, and the IG prey persists due to being a superior exploitative competitor; 2) Intermediate resource levels allow a stable coexistence because though present, the level of predation on the IG prey caused by the IG predator is not sufficient enough to exclude the IG prey; 3) Resource rich areas promote a higher abundance of the IG predator resulting in an increased predation rate that can ultimately exclude the IG prey (Holt and Polis 1997). Intraguild predation has been well documented in canid-canid interactions especially in species that differ in size (e.g., Robinson et al. 2014). Nevertheless, observing such theoretical equilibria can be a difficult task, but research has documented an inverse relationship in relative abundance between gray foxes and coyotes, perpetuated by high mortality rates of the IG prey (i.e., gray fox; Fedriani et al. 2000). Interestingly, this dynamic is likely influenced by other heterospecifics (i.e., other species besides just the IG predator and prey). For example, preliminary research findings in New Mexico show evidence that as cougar activity increases, the composition of the mesocarnivore community changes, with the gray fox increasing and coyote and bobcat decreasing in relative abundance (T. Perry and J. Evens, Furman University, pers. comm.; and see Chapters 7, 9, and 10). In the eastern United States, cameras set at sites baited with carcasses to attract eagles were used to address a potential effect of coyote abundance on gray fox occupancy (Egan et al. 2021). Although gray foxes were just as likely to be found at sites where coyotes were present as where they were absent, the most supported occupancy models suggested that estimates of relative coyote abundance negatively influenced gray fox occupancy. Egan et al. (2021) argued that coyote expansion into the eastern United States was driving gray fox declines, but studies have shown that relative abundance indices can be misleading (Sollmann et al. 2013) and the probability of occupancy is not necessarily a good indicator of abundance (Dibner et al. 2017). Further, estimates of gray fox occupancy were not available prior to coyote expansion (nor the precolonial presence of wolves [Canis lupus]), another reason to view Egan et al.'s (2021) claim with caution. In southern Illinois, the range of both North American red foxes and gray foxes appears to be contracting, and though the latter species also appeared to exhibit spatial partitioning with respect to coyotes, habitat played a greater role in predicting the presence of six different carnivore species, including the gray fox (Lesmeister et al. 2015). It was also observed that gray foxes were more likely to occur with coyotes at sites in hardwood forests, possibly because the climbing ability of the gray fox facilitates coexistence by enabling it to avoid intraguild predation (Lesmeister et al. 2015). Further exploration of the intraguild relations between gray fox and other carnivores is certainly warranted.

How much red and gray foxes interact in New Mexico remains unknown. Hockman and Chapman (1983) studied the diet of the two species in Maryland and concluded that competition between the red fox, which is primarily a predator of small mammals, and the more omnivorous gray fox is unlikely except in human-altered landscapes and/or in the presence of anthropogenic food sources. Though smaller, gray foxes are believed to be the more aggressive, dominant competitor, capable of displacing red foxes in some areas, based in part on observed interactions between the two species in traps or in cages and local fluctuations in bounty records (Richmond 1952; Rue 1969; Carey 1982). Parsons et al. (2019) used a multi-species occupancy model to explore interactions among four competing predators (gray fox, North American red fox, coyote, and bobcat) along an urban gradient around Washington, DC and Raleigh, North Carolina. Contrary to a priori expectations (the clumped distribution of limited food resources in urban areas should lead to increased interspecific competition), they found negative interactions between gray and red foxes all along the housing density gradient, with the probability of occupancy by either species to be significantly lower in the presence of the other at both low (<0.5 house/km²) and high levels of urbanization (Parsons et al. 2019). Thus, the potential for competition between red and gray foxes in New Mexico

might exist not just in urban areas but wherever the two species share the same habitats.

STATUS AND MANAGEMENT

The International Union for Conservation of Nature and Natural Resources lists the gray fox as a Species of Least Concern and considers it to be stable across its range (Cypher et al. 2008), except in Canada, where the species is considered threatened owing, in part, to suspected historical declines and a limited geographic range (Environment and Climate Change Canada 2017). Other studies, however, have suggested that gray foxes may be declining in certain areas in the eastern and midwestern regions of the United States (e.g., Lesmeister et al. 2015; Eagan et al. 2021). The New Mexico Department of Game and Fish (NMDGF) has historically monitored the gray fox population using scent-post transects for small carnivores as an index of population size, tracking the trend in visitation rates and occurrence as a metric of the distribution of the species, but it is now also assessing the efficacy of camera traps to monitor gray foxes across New Mexico (N. Forman and R. Winslow, NMDGF, pers. comm.). The abundance and density of gray foxes in New Mexico is currently unknown as data have not been collected and used to estimate either demographic parameter. To the east in Nacogdoches, Texas, gray fox density was estimated to be 0.05 gray foxes/km² during the fall (Lombardi et al. 2017).

The gray fox is designated as a protected furbearer species in New Mexico, and a license is required to take or trap them. As of 2019, there was no harvest limit for the gray fox with the take season occurring from 1 November–15 March. Between 1980 and 2019, annual reported gray fox harvest in New Mexico has varied from a minimum of 324 to a maximum of 6,234 (data provided by the NMDGF). Since 2008, when mandatory reporting was implemented, an average of 2,447 (SD = 777) gray foxes have been harvested

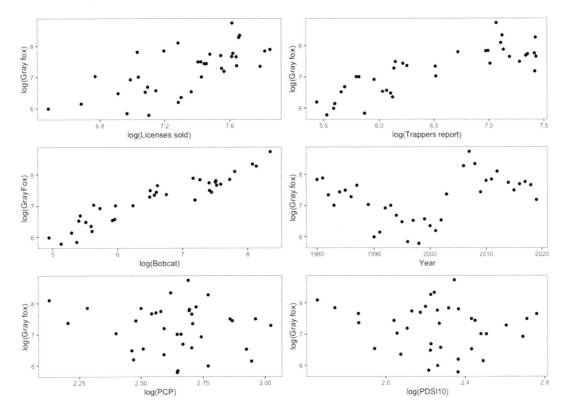

Figure 12.1. Harvest of gray fox (*Urocyon cinereoargenteus*) in New Mexico from 1980 to 2019 as a function of the number of licenses sold, the number of trappers reporting, bobcat (*Lynx rufus*) harvest, year, and two environmental variables: precipitation (PCP) and a modification of the Palmer Drought Severity Index (PDSI10).

annually by 1,368 (SD = 287) trappers representing an average of 79% (SD = 9.8%) of all trapper licenses sold. The gray fox is the most trapped fur-bearer species in New Mexico and comprises, on average, 44% of that total annual harvest (NMDGF Harvest Records; www.wildlife.state.nm.us). Nationally, gray fox furs on average sold between $10.65–$40.13 between 2002 to 2018, but did not appear to be a highly sought-after fur product as most furs remained unsold during the 2016–2018 auctions (Fur Harvesters Auction Inc. 2021).

From the 1970s to the 1980s, the gray fox harvest in North America grew from 182,000 to 301,000 (Obbard et al. 1999). Data on harvest rates are often difficult to amass, can be obfuscated by differing recording practices and a lack of mandatory reporting, and may include a composite harvest rate (e.g., all fox species recorded together); fluctuations in fox harvest can vary with environmental and economic factors and trapper attitudes, including resource conservation perspectives. The gray fox harvest in New Mexico was correlated with the number of licenses sold (r = 0.68) and the number of trappers actually reporting (r = 0.82), which was simply the number of licenses sold multiplied by reporting rate, suggesting that trapper effort naturally influences harvest rate (Figure 12.1).

Gray fox harvest was also positively correlated with the number of bobcats harvested (r = 0.94) and year (r = 0.3), though the latter relationship is non-linear (Figure 12.2). Finally, gray fox harvest was negatively correlated with precipitation (r = -0.14) and paradoxically, negatively correlated

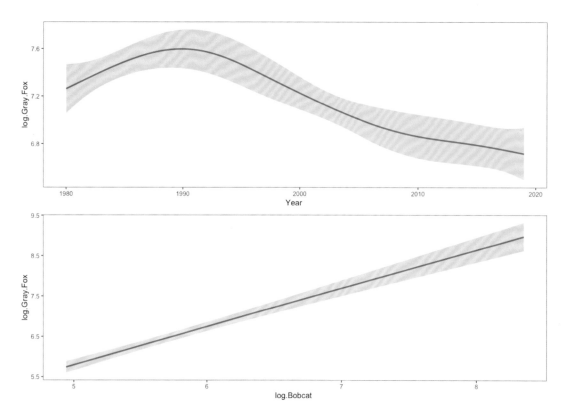

Figure 12.2. The relationship between gray fox (*Urocyon cinereoargenteus*) harvest and both year (*top*) and bobcat (*Lynx rufus*) harvest (*bottom*).

with the Palmer Drought Severity Index (r = -0.15) (Figure 12.1); these correlations decreased to near zero when a one-year lag was introduced. We are not sure why harvest rates of foxes increased when conditions were more arid, but we speculate that it could be related to an increase in the probability of capturing gray foxes as a result of lower resource productivity owing to drought conditions (i.e., there is less prey available to foxes, so they are easier to capture).

We used a General Additive Model (GAM) to examine how these and other factors predicted gray fox harvest. The most supported model included a linear effect of bobcat harvest with a non-linear effect of year (Figure 12.2). The reason for the declining trend in gray fox harvest across time is not clear. Trappers may be targeting bobcats and incidentally capturing gray foxes,

because bobcat pelts are worth more than gray fox pelts. Between 2002 to 2018, gray fox pelts averaged $22.77, whereas bobcat pelts averaged $346.82 (Fur Harvesters Auction Inc. 2021). Consequently, bobcat harvest might be a proxy for trapping effort, and as bobcat harvest increases, so might gray fox harvest. Bobcat harvest is highly correlated with the number of trappers reporting (r = 0.89), and the number of trappers reporting had a positive effect on gray fox harvest rate (~ß = 0.25) when alone or coupled with precipitation and PDSI, but it was only significant when coupled with year (ß = 0.58, SE = 0.24, adj. r^2 = 0.875, p = 0.04) and when coupled with bobcat harvest its effect size dropped to near zero. Thus, the correlation between bobcat harvest and gray fox harvest is most likely tied to trapping effort. The effect of year on gray fox harvest is not clear.

Finally, we attempted to determine whether pelt prices influenced the gray fox harvest. We surmised that if trappers were pursuing bobcats and incidentally taking gray foxes, that bobcat pelt prices may predict gray fox harvest rates. We found limited data, from 2002 to 2018, to support an analysis exploring the effect of pelt prices. Bobcat pelt price did not predict gray fox harvest ($ß = 0.17$, $SE = 0.20$, adj. $r^2 = -0.02$, $p = 0.41$), but gray fox pelt price did ($ß = 1.48$, $SE = 0.19$, adj. $r^2 = 0.83$, $p \ll 0.001$). Because this analysis involved a limited dataset and because harvest of both species fluctuates over time, a more comprehensive analysis, perhaps over a longer timeframe and/or by county, is warranted. Additional factors including environmental variables, those expected to influence furbearer populations and those that might influence trapper success, together with the attitudes of trappers toward furbearer management and resource conservation, should also be explored.

Despite their designation as a Species of Least Concern, gray foxes garner considerable attention from state and federal health and wildlife agencies tasked with monitoring, controlling, and attempting to eradicate pathogens residing in wildlife populations. The intermingling of domestic and wild animal populations can exacerbate control efforts as the larger effective host population size can lead to a higher prevalence of the disease and can even threaten populations of conservation concern (Van de Bildt et al. 2002). For example, evidence of canine parvovirus, canine adenovirus, canine heartworm, and *Leptospira interrogans* was detected in gray foxes in California with the seroprevalence rate (the presence of pathogens in a population measured by antibody presence) for canine parvovirus being higher in the urban than rural areas (Riley et al. 2004). Gray foxes also possess a number of parasites such as nematodes and fleas. A study in El Cimatario National Park in the state of Queretaro in central Mexico recorded nine species of nematodes through analysis of gray fox scats (Hernández-Camacho et al. 2011). Meanwhile, Patrick and Harrison (1995) and Harrison et al. (2003) documented six species of fleas on gray foxes in New Mexico with four of the six flea species being known carriers of plague. The latter fact led to the authors' recommendation that every gray fox in New Mexico should be considered a potential carrier of the bacterial disease (Harrison et al. 2003).

Gray foxes are also a primary reservoir host for rabies in the Southwest. The species harbors the Arizona gray fox variant and the Texas gray fox variant, two unique variants of the rabies lyssavirus (Velasco-Villa et al. 2017). The former circulates in gray fox populations in Arizona, New Mexico, and northwestern Mexico, while the latter is restricted to west-central Texas (Velasco-Villa et al. 2017). Epizootic events (epidemics in animals) in Texas and Arizona last occurred in 1988 and 2008, respectively (Dyer et al. 2014; Velasco-Villa et al. 2017; Veals 2018). Texas initiated an oral rabies vaccination program in 1996, whereby over 53 million doses of an oral rabies vaccine were dropped via aircraft along the leading edge of the epizootic. This effort intended to create a zone of vaccinated gray foxes in an attempt at halting the spread of the virus (Texas Department of State Health Services, Zoonosis Control Branch 2020). The program was highly successful as the number of rabies cases reported dropped from 244 gray foxes in 1995 to zero by 2009 (Texas Department of State Health Services, Zoonosis Control Branch 2020). No new cases were reported until 2013 when a single-rabid domestic cow (*Bos taurus*) tested positive for the Texas gray fox variant, highlighting the difficult task of eliminating the virus from a wild population (Dyer et al. 2014). To date, there have been no additional cases of the Texas gray fox variant of rabies reported In Texas (Texas Department of State Health Services, Zoonosis Control Branch 2020).

Photo 12.19. Gray fox captured on 18 March 2007 as part of a study exploring the genetic connectivity of mesocarnivores between Carlsbad Caverns National Park and Guadalupe Mountains National Park. Captured mesocarnivores were PIT-tagged for future identification and blood was drawn for genetic and disease surveys. McKittrick Canyon, Guadalupe Mountains National Park, Texas. Photograph: © Pat (Ranger) Ward.

Arizona also implemented an oral rabies vaccination program in response to the 2008 epizootic event with 130,000 oral rabies vaccination baits dropped via aircraft with distribution centered on Coconino County (USDA, APHIS 2009). The program was in response to 277 gray foxes testing positive for rabies (Veals 2018). In New Mexico, only the Arizona gray fox rabies variant has been detected with the first case of an infected gray fox detected in Catron County, near Glenwood, in 2007 (Albuquerque Journal News Staff 2010). Rabies monitoring programs are currently on-going in both states; however, eradicating the virus from the landscape is likely to be a difficult, if not impossible, task. DeYoung et al. (2009) assessed the genetic structure and dispersal of gray foxes in west-central Texas in an attempt to assist rabies

management efforts by identifying the spatial extent of gray fox populations in the oral rabies vaccination region. Their research detected no genetic structuring or population boundaries with genetic similarity among individuals over distances >30 km, suggesting long distance movements and dispersal are common in that area. Therefore, management plans attempting to control the virus likely need to be implemented over large landscape scales to ensure oral rabies vaccination programs are effective (DeYoung et al. 2009).

Although a rare event, rabid gray foxes have bitten humans including a 19-year-old woman in 2008 in Catron County, and, in Lincoln County, a 78-year-old woman in 2015 and a 29-year-old man in 2019 (NMDGF 2008; Scutti 2015; Stallings 2019). The gray fox that attacked the 78-year-old victim tested positive for a previously undetected strain of rabies, the first found in the United States in several years. State officials hypothesize that the rabid gray fox contracted the virus through contact with an infected bat based on similarities between the novel strain and a known bat strain (NMDOH 2015).

Research needs for the gray fox in New Mexico include regional estimates of population abundance along with estimating survival and fecundity to model population dynamics, which could prove valuable for understanding how environmental characteristics and anthropogenic factors (e.g., increasing urbanization and human-induced harvest) influence spatial and temporal trends in gray fox populations. An assessment of phylogeographic structure (spatial scale at which populations may be connected) and dispersal patterns would assist in understanding landscape connectivity, which could help explain the emergence and dynamics of infectious disease, including rabies (Biek and Real 2010). Finally, explorations of how top-down interspecific interactions (e.g., intraguild predation, exploitative competition, and mesopredator release), driven by sympatric predator populations (i.e., coyotes and cougars), and how bottom-up trophic interactions, driven by fluctuating prey populations that also interact with this suite of predators (e.g., through apparent competition), influence gray fox populations could contribute to our understanding of the landscape community ecology of New Mexico's mammals. Clearly, much remains to be learned about the gray fox, but even though it is the furbearer most often trapped in New Mexico, the species likely ranks among the Carnivores least at risk owing, in part, to its statewide distribution, its broad ecological niche, and because it is not perceived as a threat to livestock.

LITERATURE CITED

Aday, B. J., Jr., and A. L. Gennaro. 1973. Mammals (excluding bats) of the New Mexican Llano Estacado and its adjoining river valleys. *Studies in Natural Sciences* 1:1–33.

Albuquerque Journal News Staff. 2010. Raccoon, fox from Sierra County test positive for rabies. *Albuquerque Journal*. Monday, 22 February 2010.

Aldridge, B. M. 2008. Gray fox (*Urocyon cinereoargenteus*). *Mammals of Mississippi* 9:1–7.

Asa, C. S., J. E. Bauman, T. J. Coonan, and M. M. Gray. 2007. Evidence for induced estrus or ovulation in a canid, the island fox (*Urocyon littoralis*). *Journal of Mammalogy* 88:436–40.

Atwood, T. C., T. L. Fry, and B. R. Leland. 2011. Partitioning of anthropogenic watering sites by desert carnivores. *Journal of Wildlife Management* 75:1609–15.

Aubry, K. B., M. J. Statham, B. N. Sacks, J. D. Perrine, and S. M. Wisely. 2009. Phylogeography of the North American red fox: vicariance in Pleistocene forest refugia. *Molecular Ecology* 18:2668–86.

Azevedo, F. C. C., V. Lester, W. Gorsuch, S. Larivière, A. J. Wirsing, and D. L. Murray. 2006. Dietary breadth and overlap among five sympatric prairie carnivores. *Journal of Zoology* 268:127–35.

Bailey, V. 1931 (=1932). *Mammals of New Mexico*. North American Fauna 53. Washington, DC: US Department of Agriculture, Bureau of Biological Survey.

Ball, L. C., and R. T. Golightly Jr. 1992. Energy and nutrient assimilation by gray foxes on a diet of mice and Himalaya berries. *Journal of Mammalogy* 73:840–46.

Biek, R., and L. A. Real. 2010. The landscape genetics of infectious disease emergence and spread. *Molecular Ecology* 19:3515–31.

Bogan, M. A., K. Geluso, S. Haymond, and E. W. Valdez. 2007. Mammal inventories for eight National Parks in the Southern Colorado Plateau Network. Fort Collins, CO: Department of the Interior, National Park Service Natural Resource Technical Report NPS/SCPN/NRTR-2007/054.

Carey, A. B. 1982. The ecology of red foxes, gray foxes, and rabies in the eastern United States. *Wildlife Society Bulletin* 10:18–26.

Chamberlain, M. J., and B. D. Leopold. 2000. Spatial use patterns, seasonal habitat selection, and interactions among adult gray foxes in Mississippi. *Journal of Wildlife Management* 64:742–51.

———. 2002. Movements and space use of gray foxes (*Urocyon cinereoargenteus*) following mate loss. *American Midland Naturalist* 142:409–12.

———. 2005. Overlap in space use among bobcats (*Lynx rufus*), coyotes (*Canis latrans*), and gray foxes (*Urocyon cinereoargenteus*). *American Midland Naturalist* 153:171–79.

Choate, L. L. 1997. The mammals of the Llano Estacado. *Special Publications, Museum of Texas Tech University* 40:1–240.

Clifford, D. L., R. Woodroffe, D. K. Garcelon, S. F. Timm, and J. A. K. Mazet. 2007. Using pregnancy rates and perinatal mortality to evaluate the success of recovery strategies for endangered island foxes. *Animal Conservation* 10:442–51.

Collins, P. W. 1993. Taxonomic and biogeographic relationships of the island fox (*Urocyon littoralis*) and gray fox (*U. cinereoargenteus*) from Western North America. In *Third California Islands Symposium*, ed. F. G. Hochberg, 351–90. Santa Barbara, CA: Santa Barbara Museum of Natural History.

Cook, J. A. 1986. The mammals of the Animas Mountains and adjacent areas, Hidalgo County, New Mexico. *Occasional Papers of the Museum of Southwestern Biology* 4:1–45.

Cunningham, S. C., L. Kirkendall, and W. Ballard. 2006. Gray fox and coyote abundance and diet responses after a wildfire in central Arizona. *Western North American Naturalist* 66:169–80.

Cypher, B. L. 1993. Food item use by three sympatric canids in southern Illinois. *Transactions of the Illinois State Academy of Science* 86:139–44.

———. 2003. Foxes (*Vulpes* species, *Urocyon* species, and *Alopex lagopus*). In *Wild mammals of North America: biology, management, and economics*, 2nd ed., ed. G. A. Feldhamer, B. C. Thompson, and J. A. Chapman, 511–46. Baltimore: Johns Hopkins University Press.

Cypher, B. L., T. K. Fuller, and R. List. 2008. Gray fox. *Urocyon cinereoargenteus*. The IUCN Red List of Threatened Species 2008:e. T22780A9385878. https://www.iucnredlist.org/species/22780/46178068

Deuel, N. R., L. M. Conner, K. V. Miller, M. J. Chamberlain, M. H. Cherry, and L. V. Tannenbaum. 2017. Gray fox home range, spatial overlap, mated pair interactions and extra-territorial forays in southwestern Georgia, USA. *Wildlife Biology* 4:1–10.

DeYoung, R. W., A. Zamorano, B. T. Mesenbrink, T. A. Campbell, B. R. Leland, G. M. Moore, R. L. Honeycutt, and J. J. Root. 2009. Landscape-genetic analysis of population structure in the Texas gray fox oral rabies vaccination zone. *Journal of Wildlife Management* 72:1292–99.

Dibner. R. R., D. F. Doak, and M. Murphy. 2017. Discrepancies in occupancy and abundance approaches to identifying and protecting habitat for an at-risk species. *Ecology and Evolution* 7:5692–702.

Dyer, J. L., P. Yager, L. Orciari, L. Greenberg, R. Wallace, C. A. Hanlon, and J. D. Blanton. 2014. Rabies surveillance in the United States during 2013. *Journal of the American Veterinary Medical Association* 245:1111–23.

Egan, M. E., C. C. Day, T. E. Katzner, and P. A. Zollner. 2021. Relative abundance of coyotes (*Canis latrans*) influences gray fox (*Urocyon cinereoargenteus*) occupancy across the eastern United States. *Canadian Journal of Zoology* 99:63–72.

Environment and Climate Change Canada. 2017. *Recovery Strategy for the Grey Fox (Urocyon cinereoargenteus) in Canada*. Species at Risk Act Recovery Strategy Series. Ottawa: Environment and Climate Change Canada.

Farías, V., T. K. Fuller, R. K. Wayne, and R. M. Sauvajot. 2005. Survival and cause-specific mortality of gray foxes (*Urocyon cinereoargenteus*) in southern California. *Journal of Zoology* 266:246–54.

Farías González, V., T. K. Fuller, and R. M. Sauvajot. 2012. Activity and distribution of gray foxes (*Urogyon cinereoargenteus*) in southern California. *Southwestern Naturalist* 57:146–81.

Fedriani, J. M., T. K. Fuller, R. M. Sauvajot, and E. C. York. 2000. Competition and intraguild predation among three sympatric carnivores. *Oecologia* 125:258–70.

Feeney, S. 1999. Comparative osteology, myology, and locomotor specializations of the fore and hind limbs of the North American foxes *Vulpes vulpes* and *Urocyon cinereoargenteus*. PhD dissertation, University of Massachusetts, Amherst.

Ferguson, S. H., and S. Larivière. 2002. Can comparing life histories help conserve carnivores? *Animal Conservation* 5:1–12.

Findley, J. S. 1987. *The natural history of New Mexican mammals*. Albuquerque: University of New Mexico Press.

Findley, J. S., A. H. Harris, D. E. Wilson, and C. Jones. 1975. *Mammals of New Mexico*. Albuquerque: University of New Mexico Press.

Frey, J. K. 2003. Distributional records and natural history notes for uncommon mammals on the Llano Estacado of eastern New Mexico. *New Mexico Journal of Science* 43:1–24.

Frey, J. K., and Z. J. Schwenke. 2012. Mammals of Sugarite Canyon State Park, Colfax County, New Mexico. *Occasional Papers, Museum of Texas Tech University* 311:1–24.

Fritzell, E. K., and K. J. Haroldson. 1982. *Urocyon cinereoargenteus*. *Mammalian Species* 189:1–8.

Fuller, T. K., and B. L. Cypher. 2004. Gray fox *Urocyon cinereoargenteus*. In *Canids: foxes, wolves, jackals,* and dogs: status survey and conservation action plan, ed. C. Sillero-Zubiri, M. Hoffmann, and D. W. Macdonald, 92–97. 2nd ed. Gland, Switzerland, and Cambridge, UK: IUCN/SSC Canid Specialist Group.

Fur Harvesters Auction Inc. 2021. Auction results. http://furharvesters.com/auctionresults.html. Accessed 15 February 2021.

Geluso, K. 2016. *Mammals of the active floodplains and surrounding areas along the Gila and Mimbres Rivers, New Mexico*. Final Report submitted to the New Mexico Department of Game and Fish.

Geluso, K. N., and K. Geluso. 2004. Mammals of Carlsbad Caverns National Park, New Mexico. *Bulletin of the University of Nebraska State Museum* 17:1–180.

Goddard, N. S., M. J. Statham, and B. N. Sacks. 2015. Mitochondrial analysis of the most basal canid reveals deep divergence between Eastern and Western North American gray foxes (*Urocyon* spp.) and ancient roots in Pleistocene California. *PLOS One* 10(8):e0136329. doi:10.1371/journal.pone.0136329.

Goldstein, D. B., G. W. Roemer, D. A. Smith, D. E. Reich, A. Bergman, and R. K. Wayne. 1999. The use of microsatellite variation to infer patterns of migration, population structure and demographic history: an evaluation of methods in a natural model system. *Genetics* 151:797–801.

Gompper, M. E., A. E. Petrites, and R. L. Lyman. 2006. Cozumel island fox (*Urocyon* sp.) dwarfism and possible divergence history based on subfossil bones. *Journal of Zoology* 270:72–77.

Greenwood, R. J., A. B. Sargeant, J. L. Piehl, D. A. Buhl, and B. A. Hanson. 1999. Foods and foraging of prairie striped skunks during the avian nesting season. *Wildlife Society Bulletin* 27:823–832.

Grinnell, J., J. Dixon, and J. M. Linsdale. 1937. *Fur-bearing mammals of California*. Vol. 2. Berkeley: University of California Press.

Gunderson, H. L. 1961. A self-trapped gray fox. *Journal of Mammalogy* 42:270.

Hall, E. R. 1981. *The mammals of North America*. 2nd ed. Vol. 2. New York: John Wiley.

Halloran, A. F. 1946. The carnivores of the San Andres Mountains, New Mexico. *Journal of Mammalogy* 27:154–61.

Hardy, R. 1945. The influence of types of soil upon the local distribution of some mammals in southwestern Utah. *Ecological Monographs* 15:71–108.

Harmsen, B. J., E. Sanchez, O. A. Figueroa, S. M. Gutierrez, C. P. Doncaster, and R. J. Foster. 2019. Ecology of a versatile canid in the Neotropics: Gray foxes (*Urocyon cinereoargenteus*) in Belize, Central America. *Mammal Research* 64:319–32.

Harris, A. H. 1987. Reconstruction of mid-Wisconsin environments in southern New Mexico. *National Geographic Research* 3:142–51.

———. 2003. *The Pleistocene vertebrate fauna from Pendejo Cave*. In *Pendejo Cave*, ed. R. S. MacNeish and J. G. Libby, 36–65. Albuquerque: University of New Mexico Press.

Harrison, R. L. 1997. A comparison of gray fox ecology between residential and undeveloped rural landscapes. *Journal of Wildlife Management* 61:112–22.

———. 2002. Estimating gray fox home-range size using half-night observation periods. *Wildlife Society Bulletin* 30:1273–1275.

Harrison, R. L., M. J. Patrick, and C. G. Schmitt. 2003. Foxes, fleas and plague in New Mexico. *Southwestern Naturalist* 48:720–22.

Hernández-Camacho, N., R. Pineda-López, C. A. López-González, and R. W. Jones. 2011. Nematodes parasites of the gray fox (*Urocyon cinereoargenteus* Schreber, 1775) in the seasonally dry tropical highlands of central Mexico. *Parasitology Research* 108:1425–29.

Hink, V. C., and R. D. Ohmart. 1984. *Middle Rio Grande Biological Survey*. Final Report. Army Corps of Engineers Contract No DACW47–81-C0015. Tempe, AZ: Arizona State University, Center for Environmental Studies.

Hobert, J., D. Burkett, M. Hartsough, R. Wu, and G. Villegas. 2016. *2015–2016 White Sands Missile Range Mesocarnivore Third Annual Report*. US Army, White Sands Missile Range.

Hockman, G. J., and J. A. Chapman. 1983. Comparative Feeding Habits of Red Foxes (*Vulpes vulpes*) and Gray Foxes (*Urocyon cinereoargenteus*) in Maryland. *American Midland Naturalist* 110:276–85.

Holt, R.D., and G.A. Polis. 1997. A theoretical framework for intraguild predation. *American Naturalist* 149:745–64.

Hooper, E. T. 1941. Mammals of the lava fields and adjoining areas in Valencia County, New Mexico.

Museum of Zoology, University of Michigan, Miscellaneous Publications 51:1–52.

Ivey, R. D. 1957. Ecological notes on the mammals of Bernalillo County, New Mexico. *Journal of Mammalogy* 38:490–502.

Johnson, D. H., M. D. Bryant, and H. H. Miller. 1948. Vertebrate animals of the Providence Mountains area of California. *University of California Publications in Zoology* 48:221–375.

Kapfer, J. M., and R. W. Kirk. 2012. Observations of gray foxes (*Urocyon cinereoargenteus*) in a suburban landscape in the piedmont of North Carolina. *Southeastern Naturalist* 11:511–516.

Kurtén, B., and E. Anderson. 1980. *Pleistocene mammals of North America*. New York: Columbia University Press.

Leopold, A. S. 1959. *Wildlife of Mexico: the game birds and mammals*. Berkeley: University of California Press.

Lesmeister, D. B., C. K. Nielsen, E. M. Schauber, and E. Hellgren. 2015. Spatial and temporal structure of a mesocarnivore guild in midwestern North America. *Wildlife Monographs* 191:1–61. doi:10.1002/wmon.1015

Lombardi, J. V., C. E. Comer, D. G. Scognamillo, and W. C. Conway. 2017. Coyote, fox, and bobcat response to anthropogenic and natural landscape features in a small urban area. *Urban Ecosystems* 20:1239–48.

McAlpine, D., J. D. Martin, and C. Libby. 2008. First occurrence of the grey fox, *Urocyon cinereoargenteus*, in New Brunswick: a climate-change mediated range expansion? *Canadian Field-Naturalist* 122:169–71.

Moehlman, P. D. 1989. Intraspecific variation in canid social systems. In *Carnivore Behavior, Ecology and Evolution*, ed. L. Gittleman, 143–63. Ithaca, NY: Cornell University Press.

Neale, J. C. C., and B. N. Sacks. 2001. Food habits and space use of gray foxes in relation to sympatric coyotes and bobcats. *Canadian Journal of Zoology* 79:1794–800.

New Mexico Department of Game and Fish (NMDGF). 2008. Another Silver City fox tests positive for rabies; catwalk fox bites woman. 20 March 2008. http://www.wildlife.state.nm.us/legacy/publications/press_releases/documents/2008/032008rabidfoxes2.html

New Mexico Department of Health (NMDOH). 2015. "Rabid Fox from Lincoln County has new rabies strain." 19 May 2015. https://www.nmhealth.org/news/disease/2015/5/?view=264

Nicholson, W. S., E. P. Hill, and D. Briggs. 1985. Denning, pup-rearing, and dispersal in the gray fox in East-Central Alabama. *Journal of Wildlife Management* 49:33–37.

Nowak, R. M. 1999. *Walker's mammals of the world*. Vol. 1. 6th ed. Baltimore: Johns Hopkins University Press.

Obbard, M. E., J. G. Jones, R. Newman, A. Booth, A. J. Satterthwaite, and G. Linscombe. 1999. Furbearer harvests in North America. *In Wild furbearer management and conservation in North America*, ed. M. Novak, J. A. Baker, M. E. Obbard, and B. Malloch, 1007–33. Ontario, Canada: Ministry of Natural Resources.

Parsons, A. W., C. T. Rota, T. Forrester, M. C. Baker-Whatton, W. J. McShea, S. G. Schuttler, J. J. Millspaugh, and R. Kays. 2019. Urbanization focuses carnivore activity in remaining natural habitats, increasing species interactions. *Journal of Applied Ecology* 56:1894–904.

Patrick, M. J., and R. L. Harrison. 1995. Fleas on gray foxes in New Mexico. *Entomological Society of America* 32:201–4.

Pineda-Guerrero, A., J. F. González-Maya, and J. Pérez-Torres. 2015. Conservation value of forest fragments for medium-sized carnivores in a silvopastoral system in Colombia. *Mammalia* 79:115–19.

Ralls, K., J. N. Sanchez, J. Savage, T. J. Coonan, B. R. Hudgens, and B. L. Cypher. 2013. Social relationships and reproductive behavior of island foxes inferred from proximity logger data. *Journal of Mammalogy* 94:1185–96.

Reding, D. M., A. M. Bronikowski, W. E. Johnson, and W. R. Clark. 2012. Pleistocene and ecological effects on continental-scale genetic differentiation in the bobcat (*Lynx rufus*). *Molecular Ecology* 21:3078–93.

Reding, D. M., S. Castañeda-Rico, S. Shirazi, C. A. Hofman, I. A. Cancellare, S. L. Lance, J. Beringer, W. R. Clark, and J. E. Maldonado. 2021. Mitochondrial genomes of the United States distribution of gray fox (*Urocyon cinereoargenteus*) reveal a major phylogeographic break at the

Great Plains Suture Zone. *Frontiers in Ecology and Evolution* 9:666800. https://www.frontiersin.org/articles/10.3389/fevo.2021.666800/full.

Richmond, N. D. 1952. Fluctuations in gray fox population in Pennsylvania and their relationship to precipitation. *Journal of Wildlife Management* 16:198–206.

Riley, S. P. D. 2006. Spatial ecology of bobcats and gray foxes in urban and rural zones of a national park. *Journal of Wildlife Management* 70:1425–35.

Riley, S. P. D., J. Foley, and B. Chomel. 2004. Exposure to feline and canine pathogens in bobcats and gray foxes in urban and rural zones of a national park in California. *Journal of Wildlife Diseases* 40:11–22.

Robinson, Q. H., D. Bustos, and G. W. Roemer. 2014. The application of occupancy modeling to evaluate intraguild predation in a model carnivore system. *Ecology* 95:3112–23.

Roemer, G. W., T. J. Coonan, L. Munson, and R. K. Wayne. 2004. The island fox. In *Canids: foxes, wolves, jackals, and dogs: status survey and conservation action plan.*, ed. C. Sillero-Zubiri, M. Hoffmann, and D. W. Macdonald, 97–105. 2nd ed. Gland, Switzerland, and Cambridge, UK: IUCN/SSC Canid Specialist Group.

Roemer, G. W., D. A. Smith, D. K. Garcelon, and R. K. Wayne. 2001. Behavioural ecology of the island fox. *Journal of Zoology* 255:1–14.

Rue, L. L., III. 1969. *The world of the red fox.* New York: J. B. Lippincott.

Scutti, S. 2015. New Mexico DOH reports new strain of Rabies after gray fox attacks and bites woman. *Medical Daily.* 20 May 2015. https://www.medicaldaily.com/new-mexico-doh-reports-new-strain-rabies-after-gray-fox-attacks-and-bites-woman-334216

Sheldon, W. G. 1949. Reproductive behavior of foxes in New York State. *Journal of Mammalogy* 30:236–46.

Sollmann, R., A. Mohamed, H. Samejima, and A. Wilting. 2013. Risky business or simple solution—relative abundance indices from camera-trapping. *Biological Conservation* 159:405–12.

Stallings, D. L. 2019. Capitan man bitten by rabid fox. *Ruidoso News.* 27 September 2019. https://www.ruidosonews.com/story/news/local/community/2019/09/27/

new-mexico-man-bitten-rabid-gray-fox-pet-animal-rabies-vaccine/3789543002/

Steelman, H. G., S. E. Henke, and G. M. Moore. 2000. Bait delivery for oral rabies vaccine to gray foxes. *Journal of Wildlife Diseases* 36:744–51.

Stuart, J. N., and R. E. Anderson. 1993. The gray fox, *Urocyon cinereoargenteus*, on the Llano Estacado of New Mexico. *Texas Journal of Science* 45:354–55.

Sullivan, E. G. 1956. Gray fox reproduction, denning, range, and weights in Alabama. *Journal of Mammalogy* 37:346–51.

Texas Department of State Health Services, Zoonosis Control Branch. 2020. Texas oral rabies vaccination program (ORVP) 1995–2020. https://www.dshs.texas.gov/IDCU/disease/rabies/orvp/information/Summary.doc. Accessed 26 February 2021.

Tucker, R. L., H. A. Jacobson, and M. R. Spencer. 1993. Territoriality and pair-bonding of gray foxes in Mississippi. *Proceedings of the Annual Conference of the Southeastern Association of Fish and Wildlife Agencies* 47:90–98.

United States Department of Agriculture, Animal and Plant Health Inspection Service (USDA, APHIS). 2009. ORV information by state. https://www.aphis.usda.gov/aphis/ourfocus/wildlifedamage/programs/nrmp/ORV-Information-by-State?st=AZ. Accessed 26 February 2021.

Van de Bildt, M. W. G., T. Kuiken, A. M. Visee, S. Lema, T. R. Fitzjohn, and A. D. M. E. Osterhaus. 2002. Distemper outbreak and its effect on African wild dog conservation. *Emerging Infectious Disease* 8:211–13.

Veals, A. M. 2018. Spatial ecology of the gray fox (*Urocyon cinereoargenteus*) in southeastern Arizona. MS thesis, University of Arizona, Tucson.

Velasco-Villa, A., L. E. Escobar, A. Sanchez, M. Shi, D. G. Streicker, N. F. Gallardo-Romero, F. Vargas-Pino, V. Gutierrez-Cedillo, I. Damon, and G. Emerson. 2017. Successful strategies implemented towards the elimination of canine rabies in the Western Hemisphere. *Antiviral Research* 143:1–12.

Wang, X., and R. H. Tedford. 2008. *Dogs: their fossil relatives and evolutionary history.* New York: Columbia University Press.

Wang, X., R. H. Tedford, B. Van Valkenburgh, and

R. K. Wayne. 2004. Ancestry. In *The biology and conservation of wild canids*, ed. D. W. Macdonald and C. Sillero-Zubiri, 39–54. New York: Oxford University Press.

Wayne, R. K., E. Geffen, D. J. Girman, K. P. Koepfli, L. M. Lau, and C. R. Marshall. 1997. Molecular systematics of the Canidae. *Systematic Biology* 46:622–53.

Weston Glenn, J. L., D. J. Civitello, and S. L. Lance. 2009. Multiple paternity and kinship in the gray fox (*Urocyon cinereoargenteus*). *Mammalian Biology* 74:394–402.

Wood, J. E. 1958. Age structure and productivity of a gray fox population. *Journal of Mammalogy* 39:74–86.

Wood, J. E., D. E. Davis, and E. V. Komarek. 1958. The distribution of fox populations in relation to vegetation in southern Georgia. *Ecology* 39:160–62.

Yearsley, E. F., and D. E. Samuel. 1980. Use of reclaimed surface mines by foxes in West Virginia. *Journal of Wildlife Management* 44:729–34.

KIT FOX (*VULPES MACROTIS*)

Jonathan L. Dunnum and Joseph A. Cook

The kit fox (*Vulpes macrotis*), the smallest of New Mexico's four fox species, is North America's most arid-adapted canid. Easily recognizable but rarely seen, it is the embodiment of desert adaptation, a suite of traits and behaviors allowing it to thrive in the arid and semi-arid deserts and shrublands of the western United States and northern Mexico. Within New Mexico, the kit fox occupies elevations below 2,300 m (~ 7,550 ft) from the northwestern corner through the southeastern portion of the state. Over this large range in New Mexico, observations are nonetheless rare due to a spotty distribution and nocturnal activity patterns. Accounts by those fortunate enough to have seen this little fox describe a charismatic and fascinating animal. Vernon Bailey (1931:299) observed kit foxes at dusk and remarked: "Their motions were the consummation of grace and speed as they glide cautiously from place to place and dart with wonderful quickness across the mesas or from one concealing gulch to another." Respect and admiration for the kit fox is also evident in Native American cultures of the Southwest. Diné (Navajo) consider the kit fox to be a worthy animal, much respected for its speed and quickness and bringing good luck when crossing one's path (Kluckhohn 1944). The Hopi Water Coyote clan is associated with the kit fox and among O'odham tribes (Pima and Papago), kit foxes are never eaten and are considered friends of humankind (Native Languages of the Americas 2019).

While kit foxes exhibit the typical fox-like appearance, adaptative pressure amid desert conditions has led to the evolution of distinctive characteristics setting it apart from other New Mexican foxes. The smallest body size and longest relative ear length of any North American canid, coupled with long legs, a slender build, and a long black-tipped tail, promote relatively easy recognition of this fox in most areas. The head is narrow and the rostrum long, with distinctive, dark markings on the sides of the muzzle, lower lip, and posterior one-third of the upper lip (McGrew 1979). The large ears are tan or gray on the back, changing to buff or orange at the

Photo 13.1. The kit fox (*Vulpes macrotis*). Note the slender build, long legs, rusty yellowish-gray pelage, black-tipped tail, long ears, and dark markings on the sides of the muzzle. Photograph: © Robert Shantz.

(*opposite page*) Photograph: © Robert Shantz.

base. The pinnae have a thick border of white hairs on the forward inner edge and inner base (McGrew 1979).

Kit fox pelage is generally a rusty yellowish-gray, but coloration varies somewhat across the species' geographic distribution. The grizzled appearance is due to guard hairs that are typically black-tipped or composed of two bands of black separated by a band of white (McGrew 1979). Seasonally, the dorsal pelage varies from rusty-tan to buff-gray in the summer to silvery gray in the winter, but the belly remains whitish-yellow year-round. The shoulders and outside of the legs are brownish yellow with inner portions of the legs whitish-yellow. The fur is short except on the very long (~40% of total length) bushy tail. The skull lacks a pronounced sagittal crest and is narrow and delicate with a long, slender rostrum and inflated tympanic bullae (McGrew 1979).

The overall appearance of kit foxes matches their desert environment extremely well, making them difficult to spot in the wild. Bailey (1931:299) reported:

Photo 13.2: Small, young adult female with a uniquely reddish pelage at White Sands National Park. Photograph: © Jessica K. Buskirk.

> Out in the Jornado [sic] Valley, a couple of miles west of the little town of Engle, the writer noticed not far from the roadside a furry ball near a low mound of earth. Its color

Photos 13.3a, b, c, and d (*left to right*). Kit fox pelage. From left to right: winter dorsal (MSB:Mamm:214964), summer dorsal (MSB:Mamm:123103), and ventral (MSB:Mamm:140706) pelage, and black-tipped tail (MSB:Mamm:36138). Photographs: © Jon Dunnum.

Photo 13.4. Skull of MSB:Mamm:66086—*Vulpes macrotis*, 15 November 1990, Jornada Experimental Range, Doña Ana County, New Mexico. Photograph: © Jon Dunnum.

blended perfectly with the desert soil, and the object would have been overlooked except for the two sharp points [ears] that were conspicuously non earthlike.

Secondary sexual variation in size is inconspicuous, but males nonetheless tend to be heavier than females (McGrew 1979). Average measurements of specimens at the Museum of Southwestern Biology (MSB) of adult *V. m. neomexicana* from New Mexico reflect the pattern of slightly larger males [♀ (21) total length = 780 mm (30.7 in), tail length = 297 mm (11.7 in), hind foot length = 119 mm (4.7 in), ear length = 76 mm (3 in), weight = 2121 g (74.8 oz); ♂ (50) total length = 811 mm (31.9 in), tail length = 309 mm (12.2 in), hind foot length = 125 mm (4.9 in), ear length = 79 mm, (3.1 in), weight = 2297 g (81 oz)]. In a series of 10 males and nine females of San Joaquin kit fox (*V. m. mutica*) in California, females averaged only 3–4% smaller in all external measurements, but nearly 15% lighter in weight (Grinnell et al. 1937). Dragoo (1988) found no significant secondary sexual variation in cranial measurements between male and female foxes.

A variety of characters differentiate kit foxes from other New Mexican canids, including extremely small size and relatively much longer ears that easily distinguish them from wolves (*Canis lupus baileyi*) and adult coyotes (*Canis latrans*). North American red foxes (*Vulpes fulva*) average about 25% larger, and exhibit variable dorsal coloration but typically are a reddish chestnut. They usually have black legs and feet and a white-tipped tail. In contrast, kit foxes sport a black-tipped tail and their feet and legs are whitish or of similar color as the body. Gray foxes (*Urocyon cinereoargenteus*) are also larger and possess a band of black fur running down the dorsal portion of the tail that is not seen in kit foxes. Differentiating between the kit fox and the closely related swift fox (*Vulpes velox*) is certainly the most problematic. Kit foxes are similar to swift foxes in terms of behavior, abilities, ecological requirements, and prey base and thus also are quite similar phenotypically. However, the two species can be distinguished on the basis of morphology. The kit fox is smaller, with longer (typically >75mm [2.95 in]) and more closely set ears. It is broader between the eyes and possesses a narrower rostrum and larger (deeper) bullae than the swift fox (Thornton and Creel 1975; Dragoo et al. 1987; O'Farrell 1987).

The subspecies of kit fox occurring within New Mexico, *V. m. neomexicana*, was described by Merriam (1902:74) from the "San Andreas Range, New Mexico (about 50 miles north of El Paso)." The type specimen was an adult male collected in April 1899 by C. M. Barber (Original No. 2055x) and cataloged in the Biological Survey Collection at the United States National Museum, now the National Museum of Natural History (USNM 98646). Halloran (1945:93) subsequently established the type locality as "Baird's Ranch, eastern side of San Andres Mountains, Doña Ana County, New Mexico (about 50 mi N El Paso, Texas)."

There have been conflicting views in regard to the taxonomic status of the kit fox in relation to the swift fox. As already discussed, the two are not dramatically different morphologically, though

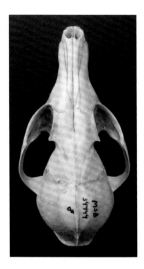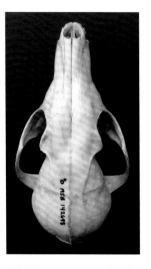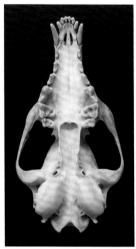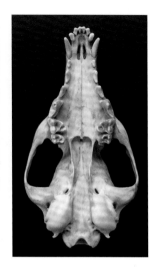

Photos 13.5 a, b, c, and d (*left to right*). Dorsal and ventral views of kit fox (photos a and c: MSB:Mamm:54444) and swift fox (photos b and d: MSB:Mamm:142698) skulls. Note the larger auditory bullae and narrower rostrum and interorbital breadth in the kit fox. Photographs: © Jon Dunnum.

consistent differences have been detected, as might be expected of closely related species or well-differentiated subspecies (Dragoo et al. 1990). In analyses of cranial morphology, Rohwer and Kilgore (1973) found morphologically intermediate animals in the Pecos River drainage in eastern New Mexico but suggested there was no evidence for gene flow extending more than ~30–50 km (20–30 miles) from the presumed area of contact. Thornton and Creel (1975), utilizing hemoglobin data, also detected a hybrid zone in the same Pecos River Valley area. Dragoo et al. (1990) found that *V. velox* and *V. m. neomexicana* were nearly identical based on allozyme protein data and suggested that these were subspecies that had differentiated morphologically but had not completed the process of speciation.

Subsequent research (Mercure et al. 1993) using different markers (mtDNA restriction site and cytochrome *b* sequence data) found kit and swift fox formed distinct clades and showed genetic divergence equal to that seen between either one of these species and the Arctic fox (*Vulpis lagopus*). Mercure et al. (1993) also identified individuals with both swift and kit fox

haplotypes and recognized kit and swift forms as distinct species, but noted that hybridization was likely occurring in the contact zone. Schwalm et al. (2014) also found that the swift fox south of the Canadian and Conchas rivers in eastern New Mexico had a unique mitochondrial haplotype more similar to kit fox haplotypes than to other swift foxes, providing further evidence of hybridization (i.e., mitochondrial capture). Recent compilations (MacDonald 2001; Feldhamer et al. 2003; Frey 2004; Wozencraft 2005; Malaney et al., 2022) have recognized two distinct species.

Regardless of the taxonomic names applied to the two forms, the narrow zone of contact between them remains a point of interest. The southern Rocky Mountains serve as a barrier separating the northern distributions of the kit and swift foxes. Farther south, hybridization takes place along the eastern edge of the Chihuahuan Desert, with a <100 km (~60 mile-) wide contact zone that falls along the Pecos River drainage in eastern New Mexico and western Texas. This region marks the eastern edge of the kit fox distribution and closely corresponds to the interface between the desert and high plains grasslands ecoregions. Further research on population dynamics in the contact zone is warranted, especially as future environmental conditions shift.

DISTRIBUTION

Kit foxes inhabit the Mojave, Sonoran, Chihuahuan, and Great Basin deserts and adjacent arid lands of western North America, typically at elevations ranging between 400 and 1900 m (1,310–6,235 ft). In the United States, the species occurs from southern California east to western Colorado and western Texas, north and northeast into southern Oregon and Idaho. In Mexico, it is found on the Baja California Peninsula, east across northern Sonora and Chihuahua to western Nuevo León, and south into northern Zacatecas (McGrew 1979; Hall 1981). The eastern edge of the species' distribution is the southern Rocky Mountains in Colorado and northern New Mexico, the Pecos River valley through central and southern New Mexico and west Texas, and the eastern Chihuahuan Desert in southwestern Texas and northern Mexico.

Although there are no data documenting continental-scale shifts in the distribution of the kit fox, there is evidence of a regional decline due to agricultural and industrial habitat conversion and urban development in the past century. In Colorado, Fitzgerald (1996) suggested that the species' range has contracted significantly. In California, the range of V. m. mutica is much reduced compared to historical accounts from the San Joaquin Valley (McGrew 1979). In that state, V. m. macrotis formerly occurred from Riverside County northwestward to Los Angeles County, but was extirpated from those areas by 1910 (Grinnell et al. 1937). There has been no assessment of distributional shifts within New Mexico.

There are only a few fossil records from the southwestern United States, thus, there is considerable uncertainty as to where V. macrotis occurred during full glacial advances of the Pleistocene (Harris 2016). Colbert (1950) reported kit

Photo 13.6. Kit fox in Tooele County, Utah, April 2014. The kit fox is distributed across the arid regions of the western United States from Oregon to Texas and into northern Mexico. Photograph: © Robert C. Lonsinger.

foxes from Ventana Cave (11,500 years before present [BP]) in southern Arizona. If that species identification is correct, kit foxes occupied the Sonoran Desert in the late Wisconsin glaciation. Jasso (2001) reported both kit (17) and swift (472) foxes from four fossil sites in New Mexico: Harris' Pocket, Animal Fair, and Isleta Caves No. 1 and 2. Kit fox specimens from Isleta Caves (NM: Bernalillo Co. Sec 31, T8N, R1E; ca. 1716 m.) are late Pleistocene/Holocene (11,000–17,000 BP) and within the current geographic range of V. macrotis (Harris 2016). On the other hand, the Harris' Pocket and Animal Fair fossil sites (NM: Eddy Co. 1,280 m (4,200 ft) are late Pleistocene (about 14,500–20,000 BP) in content. Re-evaluation of Jasso's (2001) material from these latter sites suggests that they may be V. velox and thus only a single V. macrotis specimen of Pleistocene age has been recovered from the NW Talus Slope site of New Mexico. The NW Talus Slope is older than the Animal Fair date of about 15,000 BP and may be late mid-Wisconsin. The presence of both species may indicate a confluence of range boundaries for these species at that time encompassing an area similar to that of today

(Harris 2016). There are several Pleistocene sites in California that have reported the kit fox, and in western Texas, the Lubbock Lake Site yielded *V. macrotis* from Clovis-age sediments of around 11,100 BP (Johnson 1986). The Lubbock Lake Site is well within the range of *V. velox* today but Johnson (1986) specifically implied that the skull and mandibles were not *V. velox*.

Faunal remains of the kit (or swift) fox have been excavated from two archaeological sites in the Mimbres watershed; Las Hermanas (1,372 m elev.), from the Late Pithouse/Classic Mimbres period (CE550–1000) and Bobcat Cave (1,509 m elev.) from the Archaic-Animas period (2000BCE–CE1450) (Schollmeyer and MacDonald 2020).

Within New Mexico, the species' present-day distribution is throughout the lower elevations (1,000–2,300 m) (3,281–7,546 ft) of roughly half the state (Map 13.1). Three records (MSB:Mamm:24744–24746) exist from "Ladron Peak" that would extend the elevational range up to 2,700 m (8,858 ft). But because the exact location of collection is unclear and the peak is difficult to access, it is likely the three specimens were from lower elevations within the Ladron Mountains. Kit foxes range from San Juan County in the northwest along the western edge of the southern Rocky Mountains south through Bernalillo and Torrance counties in central New Mexico, into Chavez and Eddy counties in the south. As stated, the Pecos River Valley serves as the eastern extent in the central and southern portions of New Mexico.

Two hundred and eleven kit fox specimens are archived in 20 US natural history museum collections, with the majority held in the Museum of Southwestern Biology and the US National Museum of Natural History (Appendix 13.1). The type was collected in 1899. Other historical specimens from the early 1900s were collected by James H. Gaut, Vernon Bailey, and Edward A. Goldman and are also archived at the USNM.

Later collections from the 1920s and 1930s by James Stokley Ligon, Edward Edgington, and Adrey E. Borell are housed in the Museum of Vertebrate Zoology, Yale Peabody Museum, and MSB. Prominent collectors from the 1950s, 1960s, and 1970s include Robert Dewitt Ivey, William J. Koster, Arthur H. Harris, and James L. Patton. More recent collections over the past few decades have been added during various carnivore related projects initiated by governmental agencies (i.e., New Mexico Department of Game and Fish, US Department of Agriculture, and US Department of Energy).

Museum specimen records exist from 21 of the 33 counties in New Mexico including most western and southern counties and along the eastern edge of the distribution from San Juan, Bernalillo, Torrance, Guadalupe, De Baca, Roosevelt, Chavez, Eddy, and Lea counties (Map 13.1; Appendix 13.1).

The overall distribution across the state is fairly well documented with museum records, though some areas still lack specimens. In the northwestern part of the state, there are records from San Juan and McKinley counties. Early accounts by Bailey (1931:299) noted the presence of kit foxes in San Juan County and suggested they were relatively abundant then:

> Near Liberty, in the San Juan Valley, Clarence Birdseye was told by a trapper that these desert foxes were quite common in the open valley, and in the sand near Fruitland and Shiprock small tracks were noticed that may have been made by them, but no specimens were obtained.

Since that time, specimens have been collected in northwestern New Mexico from the same areas as well as around Farmington and Aztec. Harris (1963) reported on specimens from the grasslands south of the San Juan River, a sight record in the Gallegos Canyon region, and relayed a rancher's

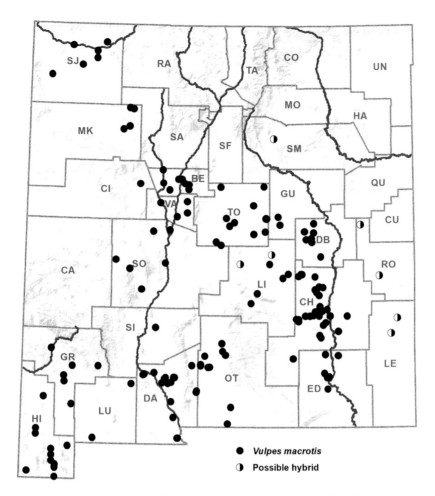

Map 13.1. Distribution of the kit fox (*Vulpes macrotis*) in New Mexico based on museum specimen records. Along the zone of contact between the kit fox and the swift fox (*V. velox*) in eastern New Mexico, the two species are known to hybridize.

remark that kit foxes were the most common carnivore in the area. Some recent records originate from near Aztec and just north of Newcomb on the Navajo Indian Reservation, while others from McKinley County include animals from the Estrella and Hospah areas.

Indicative of the uneven sampling across the state, no specimens are available for Sandoval County, though kit foxes certainly occur there, with specimens secured just across the county lines in Bernalillo and McKinley counties. The northeastern-most record within New Mexico is a skin collected in 1895 from Las Vegas in San

Miguel County (UMMZ 41714), which is currently assigned to *V. macrotis* but was not examined by us. Edward Edgington (US Biological Survey) collected a male and a female (MSB:Mamm 102111, 101289) in December 1928 near Albuquerque. These represent the oldest specimens from north-central New Mexico. No other specimens from the same area were taken until the 1950s, when Dewitt Ivey collected four individuals from the overgrazed mesas west and east of Albuquerque.

Records in the south-central part of the state include animals from the desert grasslands and

Ladrone Mountains of Socorro County, down through Lincoln County and the Capitan Mountains, the San Andres Mountains, White Sands, and the Jornada Experimental Range in Doña Ana County. Halloran (1946:158) commented on animals collected near the San Andres Mountains in the 1940s, claiming that "at no time did we record this fox in other than the Lower Sonoran [sic] plains and washes." Animals were collected in the White Sands area a few miles from the Baird Ranch (type locality), near Parker Lake, and on the Jornada Experimental Range.

From 1902 to 1903, J. H. Gaut collected five kit foxes (four males and one female) from localities in south-central New Mexico and deposited them in the USNM; two northwest of Loveless Lake in the Capitan Mountains; two east of Parker Lake in the Organ Mountains—in traps baited with prairie dogs; and another south of Tularosa (see Bailey 1931). In his annotated list of the mammals of the Tularosa Basin, Blair (1941) reported finding numerous tracks attributed to kit foxes on White Sands. Further specimens from White Sands were collected in the 1970s by Lesley Carraway and deposited at University of Texas, El Paso (UTEP). Recent camera trap surveys on the White Sands National Monument—now a national park—during the spring and early summer of 2011 and 2012 (Robinson et al. 2014) detected kit foxes on 263 separate instances. Specimens collected during the mid-1990s on Fort Bliss are held at Texas Tech University (TTU).

By comparison, specimens and observations from the west-central part of the state are rare. No records are available from Catron County, while only a single record exists for Cibola County and only a few are known for Grant and Sierra counties. A number of specimens have been taken in Socorro County, including two females and a male from northwest of Monica Spring in the Plains of San Agustin, in September 1909. At that same location, Edward A. Goldman reportedly (Bailey 1931) observed these foxes

sitting outside their burrow entrances and moving about restlessly. Before diving into the burrow, they had watched him intently until he was within 100 yards (Bailey 1931).

Only occasional records or accounts of kit foxes originate from the Gila or Mimbres river valley areas. Edward A. Goldman collected one (USNM 158889) 14.5 km (9 miles) north of Faywood in September 1908. Vernon Bailey (1931:299) collected two males (USNM 148292, 148293) from Cliff along the Gila River in August 1903 and November 1906.

> While out setting traps toward sunset on the mesa top, the writer [Bailey] saw two of these foxes playing about their burrows evidently just waking up for their night of hunting. There were several large burrows close below the crest of the ridge that evidently constituted a breeding den, as there were signs of long occupation with well worn trails and scattered remains of food.

Jones (2016) reported no photographs from Ladder Ranch camera-traps (2008–2012) and no sightings or specimens during 2014–2016 surveys of the greater Gila region. Over this same period, camera-trapping on the River Ranch in Luna and Grant counties and in the Cliff and Gila area recorded no kit foxes, while no specimens were secured and no observations were made despite many hours of night driving in the area (K. Geluso, pers. comm. 2016). Mammalogist and long-time Gila area resident, Steven O. MacDonald (pers. comm. 2016), reported only seeing kit foxes once or twice, "very briefly near the highway between Hurley and Deming," drawing a possible link between the paucity of recent sightings and the loss of prairie dogs and banner-tailed kangaroo rats in the area (S. MacDonald, pers. comm. 2016).

In New Mexico's "Bootheel," there are observations and specimens dating from the early 1900s through the mid-1980s. In 1910, H. Hotchkiss collected two males (USNM 167995, 167996) from the Cloverdale Ranch in the Animas Valley. Edward A. Goldman later found a den and observed a fox sitting at the entrance to its burrow at the southern end of the Animas Valley near the Mexican border (Bailey 1931). In 1971, L.S. Dillon secured four specimens from the Lordsburg area, which are deposited at the Field Museum of Natural History. In his account on the mammals of the Animas Mountains, Cook (1986:25) reported kit foxes from lower elevations than those occupied by the gray fox, and observed them in particular in "the level grasslands southwest of the main massif" at three localities just south of the Diamond A Ranch (formerly Gray Ranch) headquarters and near the junction of NM 79 and NM 338. Just a few years later (1984–85), Rex Jensen collected a large series of kit foxes nearby from the Antelope Wells area. Those specimens were archived at the Fort Hayes State Museum and at MSB.

The eastern edge of the distribution in New Mexico is well documented along the Pecos River Valley through Torrance, De Baca, Guadalupe, Chaves, and Eddy counties. Kit fox specimens have been collected from near Moriarty and Vaughn, through the Roswell area, the Bitter Lake National Wildlife Refuge, and into Carlsbad Caverns National Park. Kit foxes trapped in Chaves County by the US Department of Agriculture's (USDA) Animal Damage Control and subsequently archived by C. Gregory Schmitt (New Mexico Department of Game and Fish) make up the majority of these eastern specimens. In Carlsbad Caverns National Park, Geluso and Geluso (2004) reported sightings only from the reef area but suggested that kit foxes are also likely common on the seabed along the base of the escarpment.

As would be expected from within the contact zone, there are specimens exhibiting intermediate kit fox–swift fox morphological characteristics. Species identification thus remains provisional in some cases and the specimens discussed below

are labeled possible hybrids on our distribution map. The easternmost New Mexico record attributed by us to *macrotis* is a skull (TTU-M 1678) obtained 11.7 km (7.3 miles) west of Tatum, Lea County, which extends the range of *V. macrotis* within New Mexico approximately 120–130 km (75–80 miles) northeastward, and makes it overlap with that of *V. velox* (Packard and Bowers 1970). That skull is currently identified as *V. velox velox* in the Natural Science Research Laboratory (NSRL) at Texas Tech University, but our examination of the specimen indicates it is a kit fox. Another specimen (MSB:Mamm:142666) from close to that same locality (8 mi SW Tatum; Ingles Ranch) and previously identified as *velox* also exhibits *macrotis* morphological characteristics and has been reidentified as such by us (J. Dunnum, pers. obs.). The Tatum localities are farther east than other specimens from around Maljamar, Lea County, which morphologically appear to be swift fox. Two other specimens from Lea County had originally been assigned to *V. macrotis* but fell within Mercure et al.'s (1993) swift fox clade based on mtDNA data. Unfortunately, no museum catalog

Photo 13.7. Aerial view of the Pecos River north of Bitter Lake in Chaves County (the eastern bank is in the foreground). The eastern edge of the kit fox's distribution in New Mexico lies along the Pecos River Valley through Torrance, De Baca, Guadalupe, Chaves, and Eddy counties. Photograph: © Mark L. Watson.

numbers were reported, so re-examination or verification of exact locality is not possible.

Other interesting specimens from the eastern edge of the contact zone are two specimens (MSB:Mamm:142649, 142712) from Roosevelt County previously attributed to *V. velox*. Based on examination of New Mexico kit fox specimens for this work, we have reassigned these to *V. macrotis* based on cranial features and ear lengths that exceeded 76 mm. Finally, two specimens (MSB:Mamm:96188, 142703) from Lincoln County west of the purported contact zone exhibit small ears (70 mm [2.76 in] and 69 mm [2.72 in], respectively) and intermediate cranial morphology, suggesting they might represent hybrids.

In summary, the area within and around the contact zone in southeastern New Mexico contains both kit and swift fox forms as well as morphologically intermediate individuals. Mercure et al. (1993) found 41% of the foxes in that area had swift fox genotypes, while 59% had kit fox genotypes. A focused and comprehensive survey of the foxes across eastern New Mexico is required to properly delineate the distributional boundaries of the kit and swift foxes as well as understand the population dynamics of this contact zone.

HABITAT ASSOCIATIONS

Across its range, the kit fox inhabits arid and semi-arid regions characterized by desert scrub, chaparral, halophytic, and grassland communities (McGrew 1979; O'Farrell 1987). In New Mexico, it occurs across most of the lowland vegetation types of the southern, central, and northwestern areas of the state. All three studies on New Mexican populations of kit fox were conducted within the Tularosa Basin in the Chihuahuan Desert (Rodrick and Mathews 1999; Ewald 2009; Robinson et al. 2014). Habitat on the MacGregor Range of the Fort Bliss Military Reservation consisted of creosote (*Larrea tridentata*) and tarbush (*Flourensia cemua*) flats; grasslands predominantly characterized by burrograss (*Scleropogon brevifolius*) interspersed

with cactus (*Opuntia* spp.) and yucca (*Yucca* spp.); mesquite (*Prosopis glandulosa*) dunes, and four-wing saltbush (*Atriplex canescens*) dunes (Rodrick and Mathews 1999). Ewald's (2009) study was conducted on the Armendaris Ranch, between the Fra Cristobal Mountains and White Sands Missile Range, with vegetation consisting mainly of black grama (*Bouteloua eriopoda*), soaptree yucca (*Yucca elata*), Torrey ephedra (*Ephedra torreyana*), and mesquite (Ewald 2009). Robinson et al.'s (2014) research took place at White Sands National Monument, with white gypsum dune fields and a patchwork of dunes and grasslands that transition to four-wing saltbush or mesquite shrublands, together with patches of spiny allthorn (*Koeberlinia spinosa*).

Kit foxes live in open areas in burrows that they typically excavate themselves (Findley 1987), though use of modified American badger (*Taxidea taxus*), ground squirrel, prairie dog (*Cynomys* spp.), and kangaroo rat (*Dipodomys* spp.) burrows has also often been reported (Morrell 1972; List 1997; Koopman et al. 1998). Areas with loose, friable, mostly loamy soils are preferred, as those soils are more easily excavated and have enough cohesiveness to remain intact over multiple seasons but enough drainage to keep dens from flooding (Ewald 2009; Dempsey et al. 2015). Favorable denning habitat is also characterized by short, sparse vegetation, where foxes can likely detect predators more easily. On the Armendaris Ranch in south-central New Mexico, den sites were in areas with lower vegetation density compared with the surrounding landscape, but den entrances were located closer to shrubs than random points (Ewald 2009). Arjo et al. (2003) reported that areas around multi-use dens had sparser vegetation, but natal dens were often associated with thicker vegetation, perhaps providing both better cover for young foxes and higher invertebrate abundance. In the Tularosa Basin of south-central New Mexico, Rodrick and Mathews (1999) found dens mostly on slopes of

Photo 13.8 (*right*). A pair of kit foxes in desert scrub on the Jornada Experimental Range, Doña Ana County. Photograph: © David C. Lightfoot.

less than 5 degrees, oriented to the northwest and located in well-drained silty or sandy loam. Warrick and Cypher (1998) similarly reported that kit foxes generally avoid rugged terrain and prefer areas with slopes less than 5%. At the same time, however, foxes denning in lower-elevation sites often travel to more rugged locations to forage (Arjo et al. 2003; Kozlowski et al. 2008).

Occasionally, the species also inhabits agricultural lands, orchards, and urban environments (Morrell 1972; Newsome et al. 2010). Kozlowski et al. (2008) reported that foxes (referred to in their study as "city foxes") at Dugway Proving Ground in western Utah tended to use areas around human developments. Those foxes occupied and presumably found protection in flat lowland areas containing old vehicles, pipes, fences, buildings, and construction debris. At night, they traveled from those areas to forage at rugged, structurally complex sites in greasewood and stable dune environments. Road systems are not necessarily a major deterrent either. Kit fox scat is readily collected along roadways, kit foxes have been observed hunting and traveling along roads (i.e., the trans-mountain highway in El Paso, and on the Armendaris Ranch), and research shows their dens to often be located more closely to roads than

Photo 13.9. The gypsum dunefield at White Sands National Park represents another example of kit fox habitat in New Mexico. The kit fox vixen and her three pups are photographed by a remote camera next to the entrance to their den. Photograph: © Jessica K. Buskirk.

would be expected by chance alone (Ewald 2009; Dempsey et al. 2015; R. Lonsinger, pers. comm.).

LIFE HISTORY

Kit foxes have evolved to thrive in desert environments through a suite of behavioral, morphological, and physiological adaptations. Unlike many other fox species that den only when rearing pups, kit foxes den year-round. They are predominantly nocturnal, spending the days underground to avoid the heat of the day and thereby reducing heat intake and water loss. Other adaptations include a large surface area–to–body mass ratio and large ears, which both favor non-evaporative heat dissipation. Kit foxes also have the ability to vary the rate at which they pant (Klir and Heath 1992). They possess dense fur on their footpads and toes, which creates a broader surface or "sand shoe" (Miller and Stebbins 1964). Grinnell et al. (1937) suggested that the long hair on the feet improves traction on sandy surfaces, but it also likely reduces heat absorption.

Kit foxes are able to obtain all necessary water from their food, though to do so they must consume approximately 150% of their daily energy requirements (Golightly and Ohmart 1984). They need only 101 g (3.6 oz) and 175 g (6.2 oz) of wet prey mass to meet daily energy and water requirements, respectively (Golightly and Ohmart 1984), or about four kangaroo rats daily. This metabolic adaptation enables them to persist in areas far from any source of drinkable water.

While relatively well studied in some parts of its range, the kit fox has been the focus of limited research in New Mexico. Other than studies by Rodrick and Mathews (1999), Ewald (2009), and Robinson et al. (2014), much of what we know about kit fox population dynamics, natural history, ecology, and evolution comes from studies of populations elsewhere.

Diet and Foraging

Kit foxes are nocturnal, opportunistic predators. They typically emerge around dusk to forage during the night and then return to their den just before sunrise. They are most active during the hours just after sunset, often decreasing their activity in the middle of the night, and then becoming more active again in the early morning (Zoellick 1985; but see Kozlowski et al. 2008). Foraging foxes typically make straight-line movements, systematically hunting for prey, often going out to the edge of their home ranges and then moving back to the interior. Tree climbing is rare in canids, and kit foxes are indeed primarily terrestrial hunters. However, Murdoch et al. (2004) observed a female San Joaquin kit fox foraging for insects in the bark of a chaste tree (*Vitex agnus-castus*).

Kit foxes were found to travel maximum distances of 3.0 to 4.8 km (~1.9–3 miles) from their dens (Zoellick 1985; Schmidly 2004). The average total distance traveled by foxes in California was 11.5 km (7.1 mi) per night, about 18% less than foxes in Arizona, where prey biomass was substantially lower (Zoellick et al. 2002). Males and females traveled similar distances while foraging, more during the breeding season, followed by the pup-rearing stage, with the shortest distances being recorded during the pup-dispersal period (Zoellick et al. 2002).

Home range size also may vary greatly across the geographic distribution of kit foxes depending on both resource availability and levels of intraguild predators (especially the coyote). Southern New Mexico populations located in the Chihuahuan Desert had home ranges from about 6 km² (2.3 sq mi) to nearly 10 km² (~3.9 sq mi) with no significant differences between males and females (Ewald 2009). Zoellick (1985) also found no differences between sexes in western Arizona, but home ranges were larger (about 11.2 km² [4.3 sq mi]). Home ranges of kit foxes in Mexico decreased as the availability of prairie dog towns increased (Moehrenschlager et al. 2007).

Some California populations averaged between 4 and 5 km² (~1.5–1.9 sq mi) ranges (Spiegel and Bradbury 1992; Zoellick et al. 2002), whereas others were about 11 km² (4.2 sq mi). At the high extreme, home ranges in western Utah populations averaged over 20 km² (7.7 sq mi) (Dempsey 2013) in areas with large coyote populations. Similar patterns of inflated home ranges have been observed in other canid systems with smaller foxes and larger intraguild canids. For example, Kamler et al. (2013) reported increased home range sizes of

Photo 13.10 (*right*). Kit fox vixen carrying part of a snake, 16 May 2009, Hidalgo County, New Mexico. Photograph: © Robert Shantz.

Photos 13.11a, b, and c (*below, left to right*). Prominent in the kit fox's diet are species such as the banner-tailed kangaroo rat (*Dipodomys spectabilis*), black-tailed jackrabbit (*Lepus californicus*), and Gunnison's prairie dog (*Cynomys gunnisoni*). Photograph a: © David C. Lightfoot; Photograph b: © Robert Shantz; Photograph c: © James N. Stuart.

Photo 13.12. Kit fox holding four deer mice in Socorro County on 14 April 2019. Photograph: © WSMR Garrison Environmental Division and ECO Inc., Ecological Consultants.

Photo 13.13. Kit fox chasing a desert cottontail (*Sylvilagus audubonii*) in Otero County on 10 August 2015. Photograph: © WSMR Garrison Environmental Division and ECO Inc., Ecological Consultants.

both bat-eared foxes (*Otocyon megalotis*) and cape foxes (*Vulpes chama*), in the presence of a dominant intraguild predator, the jackal (*Canis aureus*).

The kit fox's diet is predominantly composed of rodents and lagomorphs, but the full suite of prey items is extensive and diet shifts can occur due to seasonal availability or major declines in preferred prey. In studies of the San Joaquin kit fox (*V. m. mutica*) from California, Morrell (1972)

observed only kangaroo rats being fed to pups by their parents. Of 52 scats investigated, more than 80% contained kangaroo rats (*Dipodomys*) and 52% rabbits (*Lepus, Sylvilagus*) remains. Similarly, Laughrin (1970) found 80–90% of approximately 600 scats consisting of *Dipodomys* remains across the range of *V. m. mutica*. But while kit foxes prey heavily on kangaroo rat populations, not all interactions between predator and prey end with a devoured kangaroo rat. Randall (2001) reported that giant kangaroo rats (*Dipodomys ingens*) foot-drum in the presence of a kit fox and documented six natural encounters in which kangaroo rats foot-drummed and the kit fox moved away. Foot-drumming is believed to serve in some cases as a signal to potential predators, letting them know that they have been detected (Randall and Matocq 1997).

In *V. m. nevadensis* from western Utah, Egoscue (1962) found that black-tailed jackrabbits (*Lepus californicus*) were the primary prey and made up over 94% of the diet of a family unit (two adults and five pups) over the pup rearing season, despite *Dipodomys* spp. being among the most common rodents in the area. Black-tailed jackrabbits weigh approximately the same as kit foxes (~2.5 kg [5.5 lbs]), and thus represent a highly profitable resource when available. However, likely due to a decline in jackrabbit populations in their study area in Utah, Arjo et al. (2007) reported that kit foxes appear to have switched prey over a 30 year period. Similarly, there is some evidence that during years of high abundance of some invertebrates (e.g., Jerusalem [*Stenopelmatus fuscus*] and Mormon crickets [*Anabrus simplex*]), kit foxes may shift their diet primarily to those less energy profitable, but also low-cost, prey items (R. Lonsinger, pers. comm.; and see below). At the same time, White et al. (1996) found that San Joaquin kit foxes relied on just a few small mammal species as their staple prey base and did not appear to shift their diet to non-mammalian prey in relation to short-term density declines in their preferred mammalian prey.

The overall breadth of prey taxa consumed by kit foxes is extensive and in addition to mammals, also includes arachnids, insects, reptiles, and birds. Kozwalski et al. (2008) documented seasonal changes in kit fox diet and showed that while kangaroo rats and other rodents were the predominant prey throughout the year, during the summer there was a significant increase in the consumption of reptiles, scorpions, and insects. During that season, insects represented the second most common prey item. Arachnids and insects included scorpions (*Centruroides* spp.), Mormon crickets, Jerusalem crickets, grasshoppers (*Orthoptera* spp.), beetles (*Tenebrionidae*), and various larvae. Among reptiles, gopher snake (*Pituophis melanoleucus*), sagebrush lizard (*Sceloporus graciosus*), side-blotched lizard (*Uta stansburiana*) and western whiptail (*Aspidoscelis tigris*) were all documented in the kit fox's prey base, while avian prey included western meadowlark (*Sturnella neglecta*) and horned lark (*Eremophilia alpestris*) (Kozwalski et al. 2008), in addition to burrowing owl (*Athene cunicularia*) (Egoscue 1956). Kozwalski et al. (2008) also documented rodents as the largest prey group and included the desert woodrat (*Neotoma lepida*), western harvest mouse (*Reithrodontomys megalotis*), deer mouse (*Peromyscus maniculatus*), pinyon mouse (*Peromyscus truei*), northern grasshopper mouse (*Onychomys leucogaster*), meadow vole (*Microtus pennsylvanicus*), plains pocket gopher (*Thomomys bottae*), Townsend's ground squirrel (*Urocitellus townsendii*), white-tailed antelope squirrel (*Ammospermophilus leucurus*), Ord's kangaroo rat (*Dipodomys ordii*), and chisel-toothed kangaroo rat (*D. microps*). Lagomorphs included mountain cottontail (*Sylvilagus nuttallii*) and black-tailed jackrabbit. Miscellaneous prey included badger, marmot (*Marmota flaviventris*), porcupine (*Erethizon dorsatum*) and long-tailed weasel (*Neogale frenata*). Two ungulate species, pronghorn (*Antilocapra americana*) and mule deer (*Odocoileus hemionus*) were also represented in kit fox scats (Kozlowski et al. 2008). For kit foxes occupying areas in and around prairie dog colonies, prairie dogs constituted the single most abundant item in kit fox diets (List 1997; List et al. 2003).

While kit foxes are primarily carnivorous, they also exploit other resources when needed. In periods of food scarcity, they have been reported to eat the fruit of cactus such as saguaro

(*Carnegiea gigantea*) (Egoscue 1956), juniper (*Juniperus* sp.), Russian olive (*Elaeagnus angustifolia*), various grasses, as well as ungulate (as carrion presumably) (Kozwalski et al. 2008). Anthropogenic food consumption by kit foxes subjected to urban encroachment on their habitat is also well documented. Human food packaging material was found in scat studies from populations in both California and Utah (Kozwalski et al. 2008; Newsome et al. 2010).

Despite the relatively wide array of prey species and subtleties of kit fox diet, the reliance on kangaroo rats appears to be strong enough that it has played an important role in shaping kit fox distribution. Benson (1938) noted that the eastern part of *V. macrotis*' range closely parallels that of the banner-tailed kangaroo rat (*Dipodomys spectabilis*), while Grinnell et al. (1937) similarly suggested that the kit fox's dependency on kangaroo rats accounts for its distribution in California. The large-bodied species of kangaroo rats (i.e., *D. spectabilis*, *D. deserti*, *D. ingens*) are likely critical, but six of the smaller (~35–45 g [1.23–1.59 oz] species (e.g., *D. merriami* and *D. ordii* in New Mexico) will meet the daily energy and water requirements of an adult kit fox, so those species too are likely important drivers of kit fox occurrence.

In New Mexico, *Dipodomys* spp. and *V. macrotis* distributions are quite congruent and though there have been no specific kit fox diet studies conducted in the state, previous authors reported anecdotal observations supporting the role of kangaroo rats in the fox's diet:

> At a den near Cliff there were scattered bits and bones of jackrabbits, cottontails, and part of the skeleton of a bird that had served as food. The foxes are readily caught in traps baited with carcasses of prairie dogs, rabbits, or the small animals that have been skinned for specimens, but their regular fare undoubtedly includes most of the small nocturnal rodents of the region. The observation that

their range corresponds closely to that of the large kangaroo rat would suggest this rodent is one of their favorite prey [Bailey 1931:299–300].

Swift and kit foxes are desert and grassland species, most common where soft soils support large populations of rodents, especially kangaroo rats, on which these little foxes prey [Findley et al. 1975:287].

Reproduction and Social Behavior

New Mexico populations have not been well studied with respect to reproduction and social behavior, but substantial data exist from other parts of the species' distribution. Kit foxes are primarily monogamous, but while some pairs apparently mate for life, there is extensive variation regarding the length of time that pair-bonds last. Ralls et al. (2007) found that pair-bonds dissolved for various reasons, most commonly due to the death of a pair mate. New pairs formed throughout the year, males and females remained together

Photo 13.14. Kit fox pair at White Sands National Park. The kit fox on the right is fitted with a GPS/VHF collar and marked on its sides with a unique dye pattern for easy recognition during research. Photograph: © Jessica K. Buskirk.

Photo 13.15. Kit fox vixen with pups, 16 May 2009, Hidalgo County, New Mexico. Kit fox litters average four pups but can range from one to nine. Photograph: © Robert Shantz.

Photo 13.16. Kit fox vixen nursing pup, 16 May 2009, Hidalgo County, New Mexico. Pups are usually independent by four to five months and disperse a few months thereafter. Photograph: © Robert Shantz.

Photos 13.17a, b, and c (*left to right*). Kit fox pups, 16 May 2009, Hidalgo County, New Mexico. Photographs: © Robert Shantz.

throughout the annual reproductive cycle, and mating partners who both survived to the next breeding season usually remained together.

Possible cases of polygamy (one male and two females) have been reported in California and Utah populations (Egoscue 1962; Ralls et al. 2001). In each of the three instances recorded in Utah (Egoscue 1962), one of the females appeared younger than the other and the litters differed in age. One of the observed family groups consisted of thirteen foxes, two adult females, one adult male, and two litters of five pups each, living together in a single den.

Based on information gathered outside New Mexico, females select a natal den in the fall (October and November) and, if not already paired, males join them at that time (Egoscue 1956). Kit foxes are monestrous (Asa and Valdespino 2003), and mating usually takes place from December through February. The length of the gestation period is estimated to range from 49 to 56 days, with litters born from January to March (Egoscue 1962; Zoellick et al. 1987). In the San Joaquin kit fox, typical litter size averages about four and ranges from 1–9 (Cypher 2003), while in New Mexico it is only known from anecdotal observations, including Cook's (1986) reported sighting of an adult female with her three kits in the Animas Mountain area in May and June 1981 (and see Photos 13.15–13.17). While nursing, the female rarely leaves the den, with the male then responsible for most of the food provisioning. After weaning, both parents bring food to the pups. Regurgitation has not been reported; instead, adults bring whole animals to the pups (Findley 1987).

Annual reproductive success is strongly influenced by food availability (Egoscue 1975; White and Garrott 1997) and at one location in California ranged from 20 to 100% over a 16-year period (Cypher et al. 2000). Average generation time is ~2.5 years. In the San Joaquin kit fox, young from previous years, usually females, may delay dispersal and assist with raising the current year's litter (Koopman et al. 2000; Ralls et al. 2001). Pups are usually independent by around 4–5 months and the observed mean age at dispersal was eight months (range = 4–32 months). Significantly more males (49.4%) than females (23.8%) dispersed. About 33% of juveniles dispersed from their natal territory, and most of those (65.2%) died within ten days of leaving. Predation by coyotes was the most common cause of death. Koopman et al. (2000) found that a greater proportion of males were likely to reproduce than females. Peak dispersal times has been shown to vary by site. California populations dispersed throughout the year but with a peak in July and August (Koopman et al. 2000), while in Utah the typical dispersal period was mid-August to December (Dempsey et al. 2014).

Year-round, kit foxes utilize large numbers of dens (see below), which can serve different purposes including daytime resting, predator escape, extreme temperature avoidance, moisture conservation, and birthing and rearing of young. They move frequently, abandoning dens and occupying new ones as needed (Egoscue 1956). Most dens have an average of two to seven entrances, and underground chambers average 2–5 m (6.6–16.4 ft) in length (O'Farrell 1987). Kit foxes often modify the dens or burrows of other animals, enlarging structures built by prairie dogs, kangaroo rats, and badgers to suit their needs (McGrew 1977; List 1997).

Seton (1925) suggested that a reason for year-long denning is the necessity of a refuge from coyotes. Contrary to an earlier school of

Photo 13.18. Kit fox emerging from its den, 30 April 2009, Hidalgo County, New Mexico. Kit foxes typically emerge shortly before sundown, spend the night foraging, then return to their dens just before sunrise. Photograph: © Robert Shantz.

Photo 13.19. The kit fox in this photograph and its mate had dug a den into an old banner-tailed kangaroo rat (*Dipodomys spectabilis*) mound on the Jornada Experimental Range, Doña Ana County, New Mexico. Photograph: © David C. Lightfoot.

thought, coyotes were not absent from the arid west prior to European colonization (see Chapter 10). The notion that they were restricted to the Great Plains (e.g., Moore and Park 1992; Gompper 2002) has now been discredited, and their range expansion has been shown to have occurred instead mostly to the north and east. Had the coyote been limited historically to prairie ecosystems, kit foxes would not have evolved with them, and it would have been more likely that denning was driven by the need to temper the harsh environmental conditions of desert living. Given what we now know about the early distribution of coyotes (see Chapter 10), it is impossible to rule out that, historically, escaping from coyotes represented the primary purpose of dens. On the Armendaris Ranch in south-central New Mexico, kit foxes avoided prairie dog colonies as denning habitat, a pattern also observed by List (1997) in the Chihuahuan Desert in northern Mexico, suggesting they were avoiding coyotes which were utilizing the colonies extensively (Ewald 2009). Rodrick and Mathews (1999) suggested that the distinctive "keyhole" shape (e.g., greater height than width) of their burrow entrances was also a predator defense adaptation, allowing for quick entrance by kit foxes but impeding predators such as coyotes and badgers.

Dens are distributed throughout home ranges, and an individual fox typically uses about 11 dens during a given year, though accounts of up to 50 also exist (Koopman et al. 1998; Ralls et al. 1990). Rodrick and Mathews (1999) showed that kit foxes in New Mexico's Chihuahuan Desert used more dens during the breeding and pup-rearing seasons. Conversely, Morrell (1972) and Scrivner et al. (1987) both found higher numbers of dens being utilized during dispersal than during the breeding and pup-rearing seasons in California, suggesting higher den occupancy was possibly a result of juvenile foxes beginning to expand their ranges and disperse. Rodrick and Mathews (1999) reported no differences between natal and non-natal dens and no seasonal variation in the frequency with which foxes change dens. Female foxes will often reuse the same natal dens year after year (O'Farrell 1987), and some foxes even return to damaged or destroyed dens and rebuild them (Golightly 1981). A further potential reason for increased numbers of dens during the summer is that flea populations also increase quickly at that time and may cause kit foxes to avoid or reduce ectoparasitic infestations (Koopman et al. 1998).

While kit foxes vocalize less extensively than other North American canids, they do possess a wide array of barks, growls, yelps, purrs, croaks, and snarls that serve to warn, intimidate, or signal alarm, distress, or fear (Egoscue 1962; Hoffmeister 1986; Murdoch et al. 2008). Murdoch et al. (2008) documented barking and a higher pitched "chittering" in San Joaquin kit foxes. The barks were similar in pitch, duration, and frequency to those of the arctic and swift foxes, but in contrast to those species, were very rare and concentrated around mating activities. The chittering vocalizations were observed in two different contexts; either emitted softly between closely related foxes during social activities or uttered loudly by individuals from different social groups and associated with aggressive behaviors at territorial boundaries (Murdoch et al. 2008). Calls were not used by solitary foxes to broadcast their presence in an area, suggesting that territories probably are maintained primarily through other means such as scent marking (Murdoch 2003).

Chemical communication through scent marking by kit foxes has not been much studied or well described. Some research found no strong evidence for regular scent marking of objects, home range boundaries, or trails with urine or feces (Egoscue 1962; O'Farrell 1987). However, latrine use has been documented. Ralls and Smith (2004) in particular observed many latrines with accumulations of both feces and urine.

Interspecific Interactions

Of principal importance to kit fox population dynamics are their interactions with their small mammal prey and with other sympatric, medium-sized carnivores. Changes in kangaroo rat and black-tailed jackrabbit densities have the potential to greatly influence kit fox populations and interspecific competition with co-occurring canids and felids. Kit foxes experience intense interference and exploitative competition with many other predatory species, including the coyote, North American red fox, American badger, bobcat (*Lynx rufus*), and even golden eagle (*Aquila chrysaetos*) (Ralls and White 1995; Cypher and Spencer 1998; Kozlowski et al. 2008; Clark 2009), but sympatric canids likely constitute the most influential kit fox competitors and predators.

In southern New Mexico, kit fox and coyote interactions were found to be complex and highly dependent upon prey density (Robinson et al. 2014; see also Chapter 10). Kit foxes, because they are a superior exploitative competitor, persisted in habitats of low resource abundance, which could not support coyotes. In adjacent areas of higher prey density where this spatial partitioning could not occur, the two species coexisted, with coyote occupancy strongly correlated with small mammal abundance. These findings indicate that kit foxes avoid coyotes, but that their adaptations to arid conditions enable them to exploit habitats unsuitable for coyotes. Consequently, the primary driver of this spatial separation is the inability of coyotes to use prey-poor habitats, rather than kit fox avoidance of coyotes (Robinson et al. 2014). Lonsinger et al. (2017) found similar results but also suggested kit foxes will utilize riskier habitats (e.g., coyote occupied areas) to secure sufficient resources.

In Utah, Kozlowski and colleagues (2008) found that the competitive pressures exerted on the kit fox population by sympatric coyotes forced kit foxes to seek areas of more abundant cover during foraging activities. Despite the spatial

Photos 13.20a and b (*top to bottom*). Kit fox-badger interaction at White Sands National Park on 14 November 2019. (a) A badger approaches the kit fox den; (b) the kit fox defends its den and chases the badger away. Photographs: © Jessica K. Buskirk.

partitioning of foraging habitat between the two species, incidents of interference competition were numerous and 56% of known kit fox deaths were attributed to predation by coyotes (Kozlowski et al. 2008). In contrast, White et al. (1994) found that San Joaquin kit foxes do not avoid coyotes and may be able to coexist by exploiting particular prey species better than coyotes. By maintaining numerous dens throughout their home ranges, they increase their chances of escaping predation. Even so, coyotes still proved to be the main cause of San Joaquin kit fox mortality (Cypher and Spencer 1998; Cypher et al. 2000).

Overall, while mortality due to coyotes may be substantial, in most cases coyotes do not appear to be consuming kit foxes, in what may be an extreme form of interference competition, or interspecific competitive killing (Lourenço et al. 2014). Kozlowski et al. (2008) and Byerly et al. (2018) found no evidence of kit fox remains in coyote scats (>1,900) in western Utah. Kit fox remains were found in only one of 1,088 coyote stomachs from New Mexico, and one of 569 stomachs from Texas (Sperry 1941). In cases where environmental conditions and resource availability necessitate it, coyotes will feed, however, on kit foxes. During a period of extreme drought in the Carrizo Plain Natural Area in southern California, coyotes not only killed kit foxes but also consumed them (Ralls and White 1995). Robinson et al. (2014) suggested that coyotes could be actively hunting kit foxes at White Sands National Monument in southern New Mexico, given the arid setting and considering that a single kit fox could meet both the daily energy and water requirements of a coyote.

Large differences in prey abundance likely explain why pronounced spatial partitioning has been documented between coyotes and kit/swift foxes in some studies (e.g., Arjo et al. 2007; Nelson et al. 2007; Kozlowski et al. 2008) and not others (e.g., Moehrenschlager et al. 2007), and why both strategies, spatial partitioning, and co-occurrence, were observed in yet another (Robinson et al. 2014). Although likely of minimal impact compared to coyotes, other sympatric carnivores may also have some bearing on local fox populations. For example, Benedict and Forbes (1979) reported on the remains of 12 kit foxes in an active bobcat den in Oregon. During Ewald's (2009) study on the Armendaris Ranch in New Mexico, two radio collars originally placed on kit foxes were recovered without animals, likely due to raptor predation (T. Waddell, Armendaris Ranch manager, pers. comm. 2005).

Communal den use by other members of the faunal community is another nexus of interspecific interaction. Denning sites near urban habitat may act as a focal point for interspecific interactions between kit foxes and skunks.

About 30% of dens in an urban area population on the California State University–Bakersfield campus were utilized by both kit foxes and striped skunks (Mephitis mephitis), though only occasionally at the same time. North American red foxes and feral cats (Felis catus) were also seen using kit fox dens in the same area (Harrison et al. 2011). Overlapping use of dens between kit foxes and skunks raises the possibility of pathogen exchange between these two species. Rabies of possible mephitid origin was recorded in kit foxes in San Luis Obispo County in California and may have contributed to the dramatic decline of a kit fox population in that state (White et al. 2000). Additionally, the use of latrines within kit fox den complexes by skunks could potentially expose kit foxes to a variety of other pathogens such as parvovirus, which can be transmitted via contact with contaminated fecal matter (Afonso et al. 2007).

Although interactions with pathogens are not well documented rangewide or in New Mexico, a few studies have been conducted on that topic. Eleven species of fleas were found on kit foxes in New Mexico and of those, nine potentially carry plague (Harrison et al. 2003). In an endoparasite survey of New Mexican kit foxes, Ubelaker et al. (2014) reported 88.7% were infected with cestodes or nematodes. Cestodes included *Dipylidium caninum*, *Mesocestoides variabilis* and *Taenia multiceps*, while nematodes consisted of *Ancylostoma caninum*, *Toxocara canis*, *Toxocara cati*, *Toxascaris leonina*, and *Physaloptera rara*.

STATUS AND MANAGEMENT

The kit fox is a species of conservation concern that has experienced rangewide population declines (Dempsey et al. 2014). Declines have been commonly attributed to broad-scale habitat conversion, decreased prey abundances, and competition with coyotes (White and Garrott 1997; White et al. 2000; Arjo et al. 2007; Moehrenschlager et al. 2007; Kozlowski et al. 2012).

Robust population trend data for most states are lacking, but studies show declines and threats to habitat are widespread across the range.

In western Utah, for example, populations have decreased significantly, and current kit fox densities (0.02/km²) are far lower than historical estimates (0.1–0.8/km²; Egoscue 1956, 1962, 1975), including lower estimates for the endangered San Joaquin kit fox populations in California (Arjo et al. 2007; Lonsinger et al. 2018). Those San Joaquin kit fox populations had estimated densities ranging from 0.15 to 0.24/km² over a three-year period on one study site (White et al. 1996) and from 0.2 to 1.7/km² over 15 years on another site (Cypher et al. 2000). In contrast, populations closely associated with prairie dog towns in Chihuahua (0.32–0.8/km²; List 1997), Coahuila, and Nuevo León (0.1/km²; Côtera 1996) maintained somewhat higher densities, but prairie dogs and kit foxes in those areas south of the border are also threatened by expansion of cropland agriculture (Ceballos et al. 2005).

In New Mexico, population status and trends are based solely on personal observational accounts, harvest data, and assumptions based on declines in habitat and prey base. Recent scent-post surveys by NMDGF personnel suggest that kit foxes may remain common in suitable habitat. In short grass and open desert areas, in particular, they can reliably be expected to visit scent posts (F. Winslow, pers. comm.). However, the reduction in semi-arid grasslands due to desertification and fragmentation caused by petroleum development has reduced habitat for major prey taxa such as prairie dogs and bannertailed kangaroo rats. Many of these critical prey species were severely impacted by the massive governmental poisoning campaigns that began in the early 1900s. Those campaigns continue today and have resulted in the extirpation of prairie dogs over large parts of their distribution and huge reductions in ground squirrels, jackrabbits, and kangaroo rats (Barko 1997; Oakes

2001). In 1908, Vernon Bailey (1931) reported a continuous prairie dog colony from Deming to Hachita through the Animas and Playas valleys and estimated the county contained 6,400,000 prairie dogs. According to Findley (1987), MSB mammalogists worked throughout that area from the 1950s through 1970s but never saw a single prairie dog. Like prairie dogs, foxes were frequently encountered in the Gila/Cliff area in the early 1900s but now are quite rare (S. MacDonald, pers. comm.). Most current prairie dog colonies in New Mexico are patchy and small, and they bear little resemblance to the huge contiguous colonies of the past.

Photo 13.21. Assisted by Lindy Gasta (right), graduate student Jessica Buskirk (left) is processing a kit fox on 4 March 2020 for her research at White Sands National Park. Each animal is weighed and measured, its physical condition assessed. Biological samples include hair, whiskers, ectoparasites, and blood (for both genetic and disease sampling). Those animals in good enough condition are not only marked and tagged with a PIT tag but also fitted with a GPS/VHF collar (weighing less than 5% of the fox's body weight). The unique dye mark drawn on both lateral sides are for visual resighting and remain until the animal's coat is shed. Kit foxes are common in areas of suitable habitat at White Sands National Park. Photograph: © Jessica K. Buskirk.

Economic Importance

Historically, kit foxes have never been of major economic importance in New Mexico. Their fur, which Bailey (1931:300) described as displaying "no marked character of beauty," has little market value (O'Farrell 1987). The commercial fur market for primarily overseas fashion clothing drives domestic trapping and is a price-driven enterprise. Low selling price and relatively low densities in the state have precluded the species from becoming a frequently sought target of trapping efforts, but kit foxes are nonetheless harvested in New Mexico. Opportunistic harvest and bycatch of predator control aimed at other species (i.e., USDA animal damage control efforts) are ongoing and certainly have occurred dating back to the late 19th century, as documented by Bailey (1931:299):

> At Albuquerque, in July 1889, . . . a number of skins were seen in the fur stores there, and they were said to be frequently brought in for sale. At that time they brought from 25 to 50 cents apiece.

From 1970 through 2019, nearly 22,000 kit foxes might have been harvested in New Mexico (NMDGF archives; Association of Fish and Wildlife Agencies 2021; note that through the 1999–2000 hunting season, swift and kit foxes were lumped together as kit foxes in harvest reports). Annual harvest numbers varied greatly over the 50-year period with the majority taken during the late 1970s to late 1980s. In most years, 2,000 to 3,000 kit foxes were taken annually, though in 1979 as many as 3,983 were harvested. Since the 1990s, New Mexico has seen a huge drop in the average reported annual take, which now averages around 100 per year (Fig. 13.1). While harvest reporting methods and accuracy vary greatly and trends should be interpreted with caution, kit fox harvests appear to have decreased by at least an order of magnitude since the high point in the 1970s and 80s. According to Rick Winslow of the New Mexico Department of Game and Fish, harvest of both kit and swift foxes has been low for about the last 20–25 years in New Mexico because there is essentially no market for them (and see Chapter 14).

From 1970 through 2018, nearly 94,000 kit foxes were harvested across the species' range within the United States (Association of Fish and Wildlife Agencies 2021). Because accurate population density estimates are lacking for the kit fox across most of the species' distribution, the impact of this level of harvest is unknown, though generally regarded as minimal. Hunting and trapping of kit foxes are prohibited in Idaho, Colorado, and Oregon. In New Mexico, Utah, Arizona, Nevada, and Texas, kit foxes are currently listed as a protected furbearer species and thus subjected to regulated harvests, though with few limits.

Threats/Protection

Despite the potential impact of fur harvesting, habitat loss and degradation due to agricultural, industrial, and urban development represent the main threats to the kit fox across its range. In California, Fiehler et al. (2017) documented the absence of sensitive taxa such as the short-nosed kangaroo rat (*Dipodomys nitratoides brevinasus*), Nelson's antelope squirrel (*Ammospermophilus nelson*), and San Joaquin kit fox in areas with high levels of oilfield development. Substantial fragmentation and degradation of habitat due to petroleum extraction activities in northwestern and southeastern New Mexico are likely major threats, compounded by habitat conversion for agricultural expansion (see Ceballos et al. 2005) and the issue of desertification. Also of considerable concern is the widespread addition or development of anthropogenic water sources. These are aimed at improving grazing conditions for livestock

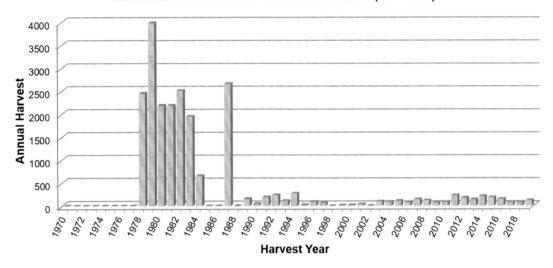

Kit fox harvest data for the State of New Mexico (1970-2019).

Figure 13.1. Annual harvest of kit foxes in New Mexico between 1970 and 2019 (based on furbearer harvest reports). Prior to and including the 1999–2000 hunting season, swift and kit foxes were lumped together as kit foxes in harvest reports.

(which incidentally means that livestock grazing may not be as compatible with kit foxes as suggested) and improving or expanding habitat for game (e.g., quail, mule deer) or endangered species (e.g., Sonoran pronghorn [*Antilocapra americana sonoriensis*]). The addition of water is expected to impact kit foxes by allowing larger-bodied carnivores to invade and exploit formerly inhospitable areas (e.g., coyotes in many desert systems) (Arjo et al. 2007; Kozlowski et al. 2012). As noted earlier, intraguild predation by coyotes is a major threat to kit foxes and likely affects survivorship, causes mesopredator suppression, and influences not only kit fox abundance but also the ability of the species to occupy suitable habitat, spread, or move around the landscape.

Not all habitat alteration is necessarily harmful to the kit fox, but at a minimum the anthropogenic disturbances mentioned above have not been sufficiently studied. Cypher et al. (2003) suggested that compatible land uses included livestock grazing, oil and gas development, and military training activities as long as sufficient food and denning sites remained available. Pesticides, road traffic, illegal hunting, and predator control programs have all had some detrimental effect but do not seem to be critical population-regulating factors. Disease does not appear to be a significant factor in most kit fox populations, although pathogens are poorly studied. Nonetheless, all these impacts may have cumulative effects and the limited estimates of population size available (e.g., commercial harvest records) suggest a significant decline in this species that warrants more intensive monitoring and investigation. Limited specimen availability and associated documentation for this species in New Mexico limits our ability to track many of these threats, either temporally or spatially (Cook et al., 2016; Malaney and Cook 2018; Cook and Light 2019).

There is a wide range of estimated survival rates across the range of kit foxes. Annual survival probabilities for adults ranged from 20 to 81% over 16 years with a mean of 44% for San

Joaquin kit fox (Cypher et al. 2000), while populations in Utah experienced survival rates of about 60% in adults and 30% for juveniles (Kluever 2015). Regrettably, no information is available on that topic from New Mexico.

In regard to global conservation status, the species does not meet the criteria for inclusion in the International Union for Conservation of Nature's (IUCN's) Red List threatened categories and is considered to be of "Least Concern" (Cypher and List 2014). All populations occurring in Mexico are considered "Vulnerable" (SEMARNAT 2010), but the conservation measures that this status grants for the species are not enforced and habitat conversion remains a major threat. The endangered Mexican subspecies *Vulpes macrotis zinseri* is restricted to the arid plains of the Sierra Madre Mountains but apparently is genetically very closely related to southwestern US populations (Maldonado et al. 1997). Efforts are underway to protect prairie dog towns in both eastern and western Mexico, which are strongholds for the kit fox, but specific actions focusing on the species have yet to be undertaken (Cypher and List 2014).

In the United States, kit fox populations have been granted various levels of state or federal protection. In Colorado and Oregon, where NatureServe considers the species to be "critically imperiled," the kit fox is state-listed as Endangered and Threatened, respectively. It is also listed at the state level in Idaho and Utah (state-sensitive in both cases). California's San Joaquin kit fox (*V. m. mutica*) benefits from both state and federal protection. It is listed as a threatened and "fully protected" species under California state laws, and designated as Endangered federally under the Endangered Species Act, with a federal recovery plan that emphasizes protection of essential habitat and demographic and ecological research in both natural and anthropogenically modified landscapes (Cypher and List 2014). In New Mexico and all other states within the kit fox distribution, the species has not been granted any special status, and it is listed by NatureService as "apparently secure." Since no reintroductions are occurring anywhere in the species' range, no captive breeding programs currently exist, though live foxes are kept at various institutions (i.e., the Arizona-Sonora Desert Museum in Tucson, AZ; and the California Living Museum in Bakersfield, CA) for display and educational purposes (Cypher and List 2014).

Because our knowledge of New Mexico's kit fox populations is derived from just a few studies (Rodrick and Mathews 1999; Ewald 2009; Robinson et al. 2014), we are left to assume that conservation threats here are similar to those across the species' distribution. Human-induced habitat conversion, reduction in prey species habitat, intraguild predation, and the impending effects of climate change are all challenges this southwestern desert fox faces. Mitigating these threats where possible will be critical to sustaining our kit fox populations into the future. The kit fox merits far greater attention, as there is much to learn about this charismatic, arid-adapted carnivore and its compelling story of desert survival.

LITERATURE CITED

Afonso, E., P. Thulllez, D. Pontier, and E. Gilot-Fromont. 2007. Toxoplasmosis in prey species and consequences for prevalence in feral cats: not all prey species are equal. *Parasitology* 134:1963–71.

Arjo, W. M., T. J. Bennett, and A. J. Kozlowski. 2003. Characteristics of current and historical kit fox (*Vulpes macrotis*) dens in the Great Basin Desert. *Canadian Journal of Zoology* 81:96–102.

Arjo, W. M., E. M. Gese, T. J. Bennett, and A. J. Kozlowski. 2007. Changes in kit fox–coyote–prey relationships in the Great Basin Desert, Utah. *Western North American Naturalist* 67:389–402.

Asa, C. S., and C. Valdespino. 2003. A review of small canid reproduction. In *The swift fox: ecology and conservation of kit foxes in a changing world*, ed. M. Sovada and L. Carbyn, 117–24. Regina: Canadian Plains Research Center, University of Regina .

Association of Fish and Wildlife Agencies. 2021. US Furbearer Harvest Statistics Database 1970–2018. http://furbearermanagement.com/furbearer-reports/. Accessed 17 September 2021.

Bailey, V. 1931 (=1932). *Mammals of New Mexico*. North American Fauna 53. Washington, DC: US Department of Agriculture, Bureau of Biological Survey.

Barko, V. A. 1997. History of policies concerning the black-tailed prairie dog: a review. *Proceedings of the Oklahoma Academy of Science* 77:27–33.

Benedict, E. M., and R. B. Forbes. 1979. Kit fox skulls in a southeastern Oregon cave. *Murrelet* 60:25–27.

Benson, S. B. 1938. Notes on kit foxes (*Vulpes macrotis*) from Mexico. *Proceedings of the Biological Society of Washington* 51:17–24.

Blair, W. F. 1941. Annotated list of mammals of the Tularosa Basin, New Mexico. *American Midland Naturalist* 26:218–29.

Byerly, P. A., R. C. Lonsinger, E. M. Gese, A. J. Kozlowski, and L. P. Waits. 2018. Resource partitioning between kit foxes (*Vulpes macrotis*) and coyotes (*Canis latrans*): a comparison of historical and contemporary dietary overlap. *Canadian Journal of Zoology* 96:497–504.

Ceballos, G., R. List, J. Pacheco, P. Manzano-Fischer, G. Santos, and M. Royo. 2005. Prairie dogs, cattle, and crops: diversity and conservation of the grassland–shrubland habitat mosaic in northwestern Chihuahua. In *Biodiversity, Ecosystems and Conservation in Northern Mexico*, ed. J.-L. E. Cartron, G. Ceballos, and R. S. Felger, 425–38. Oxford: Oxford University Press.

Clark, H. O. 2009. Species at risk: golden eagle predation on arid-land foxes. *Endangered Species Update: Science, Policy, and Emerging Issues* 26:10–14.

Colbert, E. H. 1950. The fossil vertebrates. In *The Stratigraphy and Archaeology of Ventana Cave*, ed. E. W. Haury, 126–48. Tucson and Albuquerque: University of Arizona Press and University of New Mexico Press.

Cook, J. A. 1986. The mammals of the Animas Mountains and adjacent areas, Hidalgo County, New Mexico. *Occasional Papers of the Museum of Southwestern Biology* 4:1–45.

Cook, J. A., S. Greiman, S. Agosta, R. P. Anderson, B. S. Arbogast, R. J. Baker, W. Boeger, R. D. Bradley, D. R. Brooks, R. Cole, J. R. Demboski, A. P. Dobson, J. L. Dunnum, R. P. Eckerlin, J. Esselstyn, K. Galbreath, J. Hawdon, H. Hoekstra, S. Kutz, J. Light, L. Olson, B. D. Patterson, J. L. Patton, A. J. Phillips, E. Rickart, D. S. Rogers, M. Siddall, V. Tkach, and E. P. Hoberg. 2016. Transformational principles for NEON sampling of mammalian parasites and pathogens: a response to Springer and colleagues. *BioScience* 66:917–19.

Cook, J. A., and J. E. Light. 2019. The emerging role of mammal collections in 21st Century mammalogy. *Journal of Mammalogy* 100:733–50. doi:10.1093/jmammal/gyy148

Côtera, M. 1996. Untersuchungen zur ökologischen anpassung des wüstenfuchses *Vulpes macrotis zinseri* B. in Nuevo León, Mexiko. PhD dissertation, Wildlife Institute, Faculty of Forestry, University of Munich.

Cypher, B. L. 2003. Foxes (*Vulpes* species, *Urocyon* species, and *Alopex lagopus*). In *Wild mammals of North America: biology, management, and economics*, 2nd ed., ed. G. A. Feldhamer, B. C. Thompson, and J. A. Chapman, 511–46. Baltimore: Johns Hopkins University Press.

Cypher, B. L., P. A. Kelly, and D. F. Williams. 2003. Factors influencing populations of endangered San Joaquin kit foxes: implications for conservation and recovery. In *The swift fox: ecology and conservation of kit foxes in a changing world*, ed. M. Sovada and L. Carbyn, 125–37. Regina: Canadian Plains Research Center, University of Regina.

Cypher, B., and R. List. 2014. Kit fox. *Vulpes macrotis*.

The IUCN Red List of Threatened Species 2014:e. T41587A62259374. http://dx.doi.org/10.2305/IUCN.UK.2014-3.RLTS.T41587A62259374.en. Downloaded on 26 April 2019.

Cypher, B. L., and K. A. Spencer. 1998. Competitive interactions between coyotes and San Joaquin kit foxes. *Journal of Mammalogy* 79:204–14.

Cypher, B. L., G. D. Warrick, M. R. Otten, T. P. O'Farrell, W. H. Berry, C. E. Harris, T. T. Kato, P. M. McCue, J. H. Scrivner, and B. W. Zoellick. 2000. Population dynamics of San Joaquin kit foxes at the Naval Petroleum Reserves in California. *Wildlife Monographs* 145:1–43.

Dempsey, S. J. 2013. Evaluation of survey methods and development of species distribution models for kit foxes in the Great Basin Desert. MS thesis, Utah State University.

Dempsey, S. J., E. M. Gese, and B. M. Kluever. 2014. Finding a fox: an evaluation of survey methods to estimate abundance of a small desert carnivore. *PLOS One* 9(8):e105873. doi:10.1371/journal.pone.0105873

Dempsey, S. J., E. M. Gese, B. M. Kluever, R. C. Lonsinger, and L. P. Waits. 2015. Evaluation of scat deposition transects versus radio telemetry for developing a species distribution model for a rare desert carnivore, the kit fox. *PLOS One* 10(10):e0138995. doi:10.1371/journal.pone.0138995

Dragoo, J. W. 1988. Systematic and evolutionary relationships among North American arid-land foxes. MS thesis, Fort Hays State University, Hays, KS.

Dragoo, J. W., J. R. Choate, and T. P. O'Farrell. 1987. Intrapopulational variation in two samples of arid-land foxes. *Texas Journal of Science* 39:223–32.

Dragoo J. W., J. R. Choate, T. L. Yates, and T. P. O'Farrell. 1990. Evolutionary and taxonomic relationships among North American arid-land foxes. *Journal of Mammalogy* 71:318–32.

Egoscue, H. J. 1956. Preliminary studies of the kit fox in Utah. *Journal of Mammalogy* 37:351–57.

———. 1962. Ecology and life history of the kit fox in Tooele County, Utah. *Ecology* 43:481–97.

———. 1975. Population dynamics of the kit fox in western Utah. *Bulletin of the Southern California Academy of Science* 74:122–27.

Ewald, M. L. 2009. Den site characteristics and home range sizes of kit foxes (*Vulpes macrotis*) in a Chihuahuan Desert grassland. MS thesis, New Mexico State University, Las Cruces.

Feldhamer, G. A., B. C. Thompson, and J. A. Chapman, eds. 2003. *Wild mammals of North America: biology, management, and economics,* 2nd ed. Baltimore: Johns Hopkins University Press.

Fiehler, C. M., B. L. Cypher, and L. R. Saslaw. 2017. Effects of oil and gas development on vertebrate community composition in the southern San Joaquin Valley, California. *Global Ecology and Conservation* 9:131–41.

Findley, J. S. 1987. *The natural history of New Mexican mammals.* Albuquerque: University of New Mexico Press.

Findley, J. S., A. H. Harris, D. E. Wilson, and C. Jones. 1975. *Mammals of New Mexico.* Albuquerque: University of New Mexico Press.

Fitzgerald, J. P. 1996. *Status and distribution of the kit fox (Vulpes macrotis) in western Colorado.* Final Report. Colorado Division of Wildlife Project No. W-153-R-7.

Frey, J. K. 2004. Taxonomy and distribution of the mammals of New Mexico: an annotated checklist. *Occasional Papers, Museum of Texas Tech University* 240:1–32.

Geluso, K. N., and K. Geluso. 2004. Mammals of Carlsbad Caverns National Park, New Mexico. *Bulletin of the University of Nebraska State Museum* 17:1–180.

Golightly, R. T., Jr. 1981. Comparative energetics of two desert canids: the coyote (*Canis latrans*) and the kit fox (*Vuples macrotis*). PhD dissertation, Arizona State University, Tempe.

Golightly, R. T., and R. D. Ohmart. 1984. Water economy of two desert canids: coyote and kit fox. *Journal of Mammalogy* 65:51–58.

Gompper, M. E. 2002. Top carnivores in the suburbs? Ecological and conservation issues raised by colonization of north-eastern North America by coyotes: the expansion of the coyote's geographical range may broadly influence community structure, and rising coyote densities in the suburbs may alter how the general public views wildlife. *Bioscience* 52:185–90.

Grinnell, J., J. S. Dixon, and J. M. Linsdale. 1937. *Fur-bearing mammals of California.* Vol. 2. University of California Press: Berkeley.

Hall, E. R., 1981. *The mammals of North America*. 2nd ed. Vol. 2. New York: John Wiley.

Halloran, A. F. 1945. The type locality of *Vulpes macrotis neomexicanus*. *Journal of Mammalogy* 26:92–93.

———. 1946. The carnivores of the San Andres Mountains, New Mexico. *Journal of Mammalogy* 27:154–61.

Harris, A. H., 1963. Ecological distribution of some vertebrates in the San Juan Basin, New Mexico. *Museum of New Mexico Papers in Anthropology* 8:1–63.

———. 2016. Pleistocene vertebrates of southwestern United States and northwestern Mexico. www.utep.edu/leb/pleistnm/taxaMamm/Vulpes.htm. Accessed 11 September 2016.

Harrison, R. L., M. J. Patrick, and C. G. Schmitt. 2003. Foxes, fleas, and plague in New Mexico. *Southwestern Naturalist* 48:720–22.

Harrison, S. W. R., B. L. Cypher, S. Bremner-Harrison, and C. L. V. H. Job. 2011. Resource use overlap between urban carnivores: implications for endangered San Joaquin kit foxes (*Vulpes macrotis mutica*). *Urban Ecosystems* 14:303–11.

Hoffmeister, D. F. 1986. *Mammals of Arizona*. Tucson and Phoenix: University of Arizona Press and Arizona Game and Fish Department.

Jasso, H. A. 2001. Systematics of Pleistocene *Vulpes* in New Mexico. MS thesis, University of Texas at El Paso.

Jones, A. K. 2016. Mammals of the Greater Gila Region. MS thesis, University of New Mexico, Albuquerque.

Johnson, E. 1986. Late Pleistocene and early Holocene vertebrates and paleoenvironments on the Southern High Plains, U.S.A. *Geographie Physique et Quaternaire* 40:249–61.

Kamler, J. F., U. Stenkewitz, and D. W. MacDonald. 2013. Lethal and sublethal effects of black-backed jackals on cape foxes and bat-eared foxes. *Journal of Mammalogy* 94:295–306.

Klir, J. J., and J. E. Heath. 1992. An infrared thermographic study of surface temperature in relation to external thermal stress in three species of foxes: the red fox (*Vulpes vulpes*), arctic fox (*Alopex lagopus*), and kit fox (*Vulpes macrotis*). *Physiological zoology* 65:1011–21.

Kluckhohn, C. 1944. Navaho Witchcraft. *Papers of the Peabody Museum of American Archaeology and Ethnology* 22:1–254.

Kluever, B. M. 2015. Relationships between water developments and select mammals on the US Army dugway proving ground, Utah. PhD dissertation, Utah State University, Logan.

Koopman, M. E., B. L. Cypher, and J. H. Scrivner. 2000. Dispersal patterns of San Joaquin kit foxes (*Vulpes macrotis mutica*). *Journal of Mammalogy* 81:213–22.

Koopman, M. E., J. H. Scrivner, and T. T. Kato. 1998. Patterns of den use by San Joaquin kit foxes. *Journal of Wildlife Management* 62:373–79.

Kozlowski, A. J., E. M. Gese, and W. M. Arjo. 2008. Niche overlap and resource partitioning between sympatric kit foxes and coyotes in the Great Basin Desert of western Utah. *American Midland Naturalist* 160:191–208.

———. 2012. Effects of intraguild predation: evaluating resource competition between two canid species with apparent niche separation. *International Journal of Ecology*, Volume 2012, Article ID 629246. doi:10.1155/2012/629246

Laughrin, L. 1970. *San Joaquin kit fox, its distribution and abundance*. Administrative report 70-2. California Department of Fish and Game.

List, R. 1997. Ecology of the kit fox (*Vulpes macrotis*) and coyote (*Canis latrans*) and the conservation of the prairie dog ecosystem in northern Mexico. PhD dissertation, University of Oxford.

List, R., and B. L. Cypher. 2004. Kit fox (*Vulpes macrotis*). In *Canids: foxes, wolves, jackals, and dogs: status survey and conservation action plan*, ed. C. Sillero-Zubiri, M. Hoffmann, and D. W. Macdonald, 105–9. 2nd ed. Gland, Switzerland, and Cambridge, UK: IUCN/SSC Canid Specialist Group.

List, R., P. Manzano-Fischer, and D. W. MacDonald. 2003. Coyote and kit fox diets in prairie dog towns and adjacent grasslands in Mexico. In *The swift fox: ecology and conservation of kit foxes in a changing world*, ed. M. Sovada and L. Carbyn, 183–88. Regina: Canadian Plains Research Center, University of Regina.

Lonsinger, R. C., E. M. Gese, L. L., Bailey, and L. P. Waits. 2017. The roles of habitat and intraguild predation by coyotes on the spatial dynamics of kit foxes. *Ecosphere* 8(3):e01749. doi:10.1002/ecs2.1749

Lonsinger, R. C., P. M. Lukacs, E. M. Gese, R. N. Knight, and L. P. Waits. 2018. Estimating densities for sympatric kit foxes (*Vulpes macrotis*) and coyotes (*Canis latrans*) using noninvasive genetic sampling. *Canadian Journal of Zoology* 96:1080–89.

Lourenço, R., V. Penteriani, J. E. Rabaça, and E. Korpimäki. 2014. Lethal interactions among vertebrate top predators: a review of concepts, assumptions and terminology. *Biological Reviews* 89:270–83.

Macdonald, D. W. 2001. Foxes. In *The Encyclopedia of Mammals*, ed. D. W. Macdonald, 54–61. London: Brown Reference Group.

Malaney, J., and J. A. Cook. 2018. A perfect storm for mammalogy: declining sample availability in a period of rapid environmental degradation. *Journal of Mammalogy* 99:773–88.

Malaney, J. L., J. L. Dunum, and J. A. Cook. 2022. Checklist of New Mexico mammals. *New Mexico Museum of Natural History Science Bulletin* 87:361–69.

Maldonado, J. E., M. Cotera, E. Geffen, and R. K. Wayne. 1997. Relationships of the endangered Mexican kit fox (*Vulpes macrotis zinseri*) to North American arid-land foxes based on mitochondrial DNA sequence data. *Southwestern Naturalist* 42:460–70.

McGrew, J. C. 1977. Distribution and habitat characteristics of the kit fox (*Vulpes macrotis*) in Utah. MS thesis, Utah State University, Logan.

———. 1979. *Vulpes macrotis. Mammalian Species* 123:1–6.

Mercure, A., K. Ralls, K. P. Koepfli, and R. K. Wayne. 1993. Genetic subdivisions among small canids: mitochondrial DNA differentiation of swift, kit, and arctic foxes. *Evolution* 47:1313–28.

Merriam, C. H. 1902. Three new foxes of the kit and desert fox groups. *Proceedings of the Biological Society of Washington* 15:73–74.

Miller, A. H., and R. C. Stebbins. 1964. *The lives of desert animals in Joshua Tree National Monument*. Berkeley: University of California Press.

Moehrenschlager, A., R. List, and D. W. MacDonald. 2007. Escaping intraguild predation: Mexican kit foxes survive while coyotes and golden eagles kill Canadian swift foxes. *Journal of Mammalogy* 88:1029–39.

Moore, G. C., and G. R. Parker. 1992. Colonization by the eastern coyote (*Canis latrans*). In *Ecology and management of the eastern coyote*, ed. A. H. Boer, 23–38. Fredericton: Wildlife Research Unit, University of New Brunswick.

Morrell, S. H. 1972. Life history of the San Joaquin kit fox. *California Fish and Game* 58:162–174.

Murdoch, J. D. 2003. Scent marking behavior of San Joaquin kit foxes. MS thesis, University of Denver, Denver, Colorado.

Murdoch, J. D., K. Ralls, and B.L. Cypher. 2004. Two observations of tree climbing by the San Joaquin kit fox. *Southwestern Naturalist* 49:522–23.

Murdoch, J. D., K. Ralls, B. L. Cypher, and R. P. Reading. 2008. Barking vocalizations in San Joaquin kit foxes (*Vulpes macrotis mutica*). *Southwestern Naturalist* 53:118–24.

Native Languages of the Americas. 2019. http://www.native-languages.org/kit-fox.htm. Accessed 21 February 2019.

Nelson, J. L., B. L. Cypher, C. D. Bjurlin, and S. Creel. 2007. Effects of habitat on competition between kit foxes and coyotes. *Journal of Wildlife Management* 71:1467–75.

Newsome, S. D., K. Ralls, C. V. H. Job, M. L. Fogel, and B. L. Cypher. 2010. Stable isotopes evaluate exploitation of anthropogenic foods by the endangered San Joaquin kit fox (*Vulpes macrotis mutica*). *Journal of Mammalogy* 91:1313–21.

Oakes, C. L. 2001. History and consequence of keystone mammal eradication in the desert grasslands: the Arizona black-tailed prairie dog (*Cynomys ludovicianus arizonensis*). PhD dissertation, University of Texas, Austin.

O'Farrell, T. P. 1987. Kit fox. In *Wild furbearer management and conservation in North America*, ed. M. Novak, J. A. Baker, M. E. Obbard, and B. Malloch, 422–31. Ontario, Canada: Ministry of Natural Resources.

Packard, R. L., and J. H. Bowers. 1970. Distributional notes on some foxes from western Texas and eastern New Mexico. *Southwestern Naturalist* 14:450–51.

Ralls, K., B. Cypher, and L. K. Spiegel. 2007. Social monogamy in kit foxes: formation, association, duration, and dissolution of mated pairs. *Journal of Mammalogy* 88:1439–46.

Ralls, K., K. L. Pilgrim, P. J. White, E. E. Paxinos, M. K. Schwartz, and R. C. Fleischer. 2001. Kinship, social relationships, and den sharing in kit foxes. *Journal of Mammalogy* 82:858–66.

Ralls, K., and D. A. Smith. 2004. Latrine use by San Joaquin kit foxes (*Vulpes macrotis mutica*) and coyotes (*Canis latrans*). *Western North American Naturalist* 64:544–47.

Ralls, K., and P. J. White. 1995. Predation on San Joaquin kit foxes by larger canids. *Journal of Mammalogy* 76:723–29.

Ralls, K., P. J. White, J. Cochran, and D. B. Siniff. 1990. *Kit fox-coyote relationships in the Carrizo Plain Natural Area*. Annual report to the U.S. Fish and Wildlife Service, October 31, 1990. Department of Zoological Research, National Zoological Park, Smithsonian Institution, Washington, DC.

Randall, J. A. 2001. Evolution and function of drumming as communication in mammals. *American Zoologist* 41:1143–56.

Randall, J. A., and M. D. Matocq. 1997. Why do kangaroo rats (*Dipodomys spectabilis*) footdrum at snakes? *Behavioral Ecology* 8:404–41.

Robinson, Q. H., D. Bustos, and G. W. Roemer. 2014. The application of occupancy modeling to evaluate intraguild predation in a model carnivore system. *Ecology* 95:3112–23.

Rodrick, P. J., and N.E. Mathews. 1999. Characteristics of natal and non-natal kit fox dens in the northern Chihuahuan Desert. *Great Basin Naturalist* 59:253–58.

Rohwer, S. A., and D. L. Kilgore Jr. 1973. Interbreeding in the arid-land foxes, *Vulpes velox* and *V. macrotis*. *Systematic Zoology* 22:157–65.

Schmidly, D. J., 2004. *The mammals of Texas*. Rev. ed. Austin: University of Texas Press.

Schollmeyer, K. G., and S. O. MacDonald. 2020. Faunal remains from archaeology sites in southwestern New Mexico. *Occasional Papers of the Museum of Southwestern Biology* 13:1–58.

Schwalm, D., L. P. Waits, and W. B. Ballard. 2014. Little fox on the prairie: genetic structure and diversity throughout the distribution of a grassland carnivore in the United States. *Conservation genetics* 15:1503–14.

Scrivner, J. H., T. P. O'Farrell, and T. T. Kato. 1987. *Dispersal of San Joaquin kit foxes, Vulpes macrotis mutica, on Naval Petroleum Reserve #1, Kern County, California*. U.S. Department of Energy Topical Report, EG&G/EM Santa Barbara Operations Report EGG 10280–2190.

SEMARNAT (Ministry of Environment and Natural Resources of Mexico). 2010. *Norma Oficial Mexicana NOM-059-SEMARNAT-2010, Protecciónambiental-Especies nativas de México de flora y fauna silvestres-Categorías de riesgo y especificaciones para su inclusión, exclusión o cambio-Lista de especies en riesgo*. Diario Oficial de la Federación (Mexico's federal register), 30 December 2010.

Seton, E. T. 1925. *Lives of game animals*. Vol. 1, Pt. 1, *Cats, wolves, and foxes*. Garden City, NJ: Doubleday and Doran.

Sperry, C. C., 1941. Food habits of the coyote. US Department of the Interior Fish and Wildlife Service. *Wildlife Research Bulletin* 4:1–75.

Spiegel, L. K., and M. Bradbury. 1992. Home range characteristics of the San Joaquin kit fox in western Kern County, California. *Transactions of the Western Section of the Wildlife Society* 28:83–92.

Thornton, W. A., and G. C. Creel. 1975. The taxonomic status of kit foxes. *Texas Journal of Science* 26:127–36.

Ubelaker, J. E., B. S. Griffin, G. M. Konicke, D. W. Duszynski, and R. L. Harrison. 2014. Helminth parasites from the Kit Fox, *Vulpes macrotis* (Carnivora: Canidae), from New Mexico. *Comparative Parasitology* 81:100–104.

Warrick, G. D., and B. L. Cypher. 1998. Factors affecting the spatial distribution of a kit fox population. *Journal of Wildlife Management* 62:707–17.

White, P. J., W. H. Berry, J. J. Eliason, and M. T. Hanson. 2000. Catastrophic decrease in an isolated population of kit foxes. *Southwestern Naturalist* 45:204–11.

White, P. J., and R. A. Garrott. 1997. Factors regulating kit fox populations. *Canadian Journal of Zoology* 75:1982–88.

White, P. J., K. Ralls, and R. A. Garrott. 1994. Coyote-kit fox interactions as revealed by telemetry. *Canadian Journal of Zoology* 72:1831–36.

White, P. J., C. A. Vanderbilt White, and K. Ralls. 1996. Functional and numerical responses of kit foxes to a short-term decline in mammalian prey. *Journal of Mammalogy* 77:370–76.

Wozencraft, W. C. 2005. Order Carnivora. In *Mammal species of the world: a taxonomic and geographic reference*, Vol. 1, ed. D. E. Wilson and D. M. Reeder,

532–628. 3rd ed. Baltimore: Johns Hopkins University Press.

Zoellick, B. W. 1985. Kit fox movements and home range use in western Arizona. MS thesis, University of Arizona, Tucson.

Zoellick, B. W., C. E. Harris, B. T. Kelly, T. P. O'Farrell, T. T. Kato, and M. E. Koopman. 2002. Movements and home ranges of San Joaquin kit foxes (*Vulpes macrotis mutica*) relative to oil-field development. *Western North American Naturalist* 62:151–59.

Zoellick, B. W., T. P. O'Farrell, P. M. McCue, C. E. Harris, and T. T. Kato. 1987. *Reproduction of the San Joaquin kit fox on Naval Petroleum Reserve No. 1, Elk Hills, California: 1980–1985 (No. EGG-10282–2144).* EG and G Energy Measurements Inc., Goleta, CA. Santa Barbara Operations.

SWIFT FOX (*VULPES VELOX*)

Robert L. Harrison, Jean-Luc E. Cartron, and Jonathan L. Dunnum

Swift foxes (*Vulpes velox*) are small, house-cat-sized members of the dog family, Canidae, and are found in the shortgrass and mixed-grass prairies of the Great Plains (Egoscue 1979). Also known as swift kit foxes or northern kit foxes, swift foxes received their name from the belief that they were the fastest mammal on the plains, even faster than coyotes (*Canis latrans*; Seton 1937). In reality, small animals frequently appear to be moving at higher speeds than they actually are. Swift foxes can attain the impressive speed of 60 km/hr (~37 mph; Moehrenschlager and Sovada 2004), but they are no match for coyotes, who can run as fast as 80 km/hr (50 mph) (Fisher 1975; Sooter 1943). To elude coyotes, their main predators (e.g., Rongstad et al. 1989; Carbyn 1998; Sovada et al. 1998), swift foxes must also rely upon their cryptic coloration and a system of dens in the open grasslands (McGee et al. 2006).

The coat colors of swift foxes match their grassland home well. New Mexico's short-grass prairie in particular is a yellowish brown or bronze for much of the year, and against such a background, the tan or buff (moderate orange-yellow) swift fox pelage on the sides and legs is inconspicuous. As with many mammals, countershading (gradation of color from a darker dorsum to a lighter ventrum to counter directional lighting from the sun; see Ruxton et al. 2004) likely also contributes to concealment. Ventrally, the pelage is light buff or white, whereas on the head, back, and tail, it is gray with black and white guard hairs that give

Photo 14.1. Robert Harrison holding an adult swift fox during research conducted in northeastern New Mexico from 1998 through 2001. Swift foxes are about the same size as house cats. Their pelage is a handsome mixture of gray, white, black, and tan. Note also the black-tipped tail. Photograph: © Robert L. Harrison.

(*opposite page*) Photograph: © Mike Dunn.

Photo 14.2. Swift fox showing typical grizzled gray dorsal pelage and countershading, or gradation of color from a darker dorsum to a lighter ventrum, likely for concealment. Photograph: © Mike Dunn.

Photos 14.3a, b, and c (*left to right*). Swift fox specimen MSB 1238 from Curry County, New Mexico, showing typical winter pelage with longer and paler fur ventrally. Photographs: © Jon Dunnum.

Photo 14.4. Leucistic, radio-collared juvenile female swift fox in 2000 near New Mexico in Dallam County, Texas. The pelage is white, with some black guard hairs on the tail, back, and muzzle. The eyes are blue. Photograph: © Jan F. Kamler.

Table 14.1. Body measurements of male vs. female swift foxes in northeastern New Mexico. Data from Harrison (2003). *Significant difference at $P \leq 0.05$.

	n	Mean	Minimum	Maximum
Weight (kg)*				
Males	18	2.24	2.0	2.5
Females	9	1.97	1.6	2.3
Total (body + tail) length (mm)				
Males	11	808.9	750	870
Females	10	781.9	725	830
Body length (mm)*				
Males	11	523.0	500	545
Females	10	503.5	475	540
Tail length (mm)				
Males	11	285.9	250	340
Females	10	278.4	250	302
Hind foot length (mm)*				
Males	11	121.5	115	127
Females	10	116.3	109	126
Ear length (mm)				
Males	10	63.8	59	68
Females	10	62.5	57	68

it a grizzled appearance. The tail is tipped with black, and there are black spots on either side of the snout. From the side, the tail may appear to have a black dorsal stripe, similar to that of gray foxes (*Urocyon cinereoargenteus*), but that is an illusion created by the guard hairs. A rare leucistic color phase was documented near New Mexico in Dallam County, Texas (Kamler and Ballard 2003; Photo 14.4). In the leucistic color phase, the pelage is white, while the eyes are light blue, instead of the normal yellow (Kamler and Ballard 2003). The leucistic color phase is distinct from albinism as the eyes of albinos are pink.

Sexual dimorphism exists in the swift fox with males slightly larger than females (e.g., Sheldon 1992; Harrison 2003; and see Chapter 13 for a similar pattern in the kit fox [*Vulpes macrotis*]).

In northeastern New Mexico, males were significantly larger than females in weight, body length, and hind foot length (Harrison 2003; Table 14.1).

Swift foxes are very similar to kit foxes. At a distance the two species are indistinguishable. At closer range, the ears of kit foxes appear to be larger than those of swift foxes relative to the size of their heads. Kit fox ears are indeed longer (>75 mm, ~3 in)—and more pointed—than those of swift foxes (<75 mm, ~3 in; Cypher 2003; Chapter 13). In the kit fox, the ears are also set closer together and the cranium is characterized by a narrower rostrum and larger (deeper) auditory bullae (Thornton and Creel 1975; Dragoo et al. 1987). Kit fox tails, finally, extend to 62% of body length, compared to 52% in swift foxes (Thornton and Creel 1975).

The ranges of the swift and kit foxes generally meet along the Pecos River in eastern New Mexico (see under "Distribution"; Chapter 13), with hybridization occurring in the contact zone (Rohwer and Kilgore 1973), a fact that has contributed to a debate over whether the two foxes should be lumped together in a single species (e.g., Hall 1981; Jones et al. 1982; Stromberg and Boyce 1986). Dragoo et al. (1990) reported that 1) cranial measurements distinguished between the kit fox of the Southwest and the swift fox of the Great Plains but not among nominal subspecies of the two foxes; and 2) genetically, two of the previously recognized subspecies of the kit fox were more similar to the swift fox than to two other previously recognized subspecies of the kit fox. In contrast, Mercure et al. (1993) found through DNA sequencing that the degree of divergence between kit and swift foxes is similar to that between these two taxa and the arctic fox (*Vulpes lagopus*). Given that the artic fox is a morphologically distinct species placed in a separate genus, Mercure et al. (1993) concluded that swift and kit foxes should be considered separate species. The taxonomic separation between swift and kit foxes might perhaps be best referred as clinal differentiation, involving gradual changes of characteristics from one species to the other over distance, rather than a clear break between the two foxes at the contact zone (Dragoo and Wayne 2003).

Some taxonomic confusion has also existed regarding the designation of subspecies within the swift fox. Merriam (1902) originally described two subspecies, *V. v. hebes* in the northern portion of the range and *V. v. velox* in the south (including New Mexico). Stromberg and Boyce (1986) examined 250 museum specimens and, despite considerable geographic variation, could find no justification for the subspecies designations. Schwalm et al. (2014) examined the nuclear and mitochondrial DNA of swift foxes south of the Canadian border and identified both rare haplotypes and (up to seven) genetic groups akin to Evolutionary Significant Units, rather than subspecies. Clusters of rare mtDNA haplotypes were detected, separated by natural and/or anthropogenic geographic barriers including the Conchas and Canadian rivers in northeastern New Mexico. Despite those barriers, genetic structuring showed gene flow occurring across the species' distribution and was compatible with the species' history of human-caused decline followed by a rapid range re-expansion (Schwalm et al. 2014; see under "Distribution" below). The Conchas and Canadian river basins acted as a particularly strong barrier limiting gene flow. To the south of it (in east-central and southeastern New Mexico), a unique mtDNA haplotype was more similar to the kit fox's, with the Conchas and Canadian river basins preventing or limiting the transfer of that haplotype to the swift fox genetic group farther north, in the extreme northeastern corner of the state (Schwalm et al. 2014). An area of rough topography in Baca County in southeastern Colorado represented a weaker geographic barrier separating the genetic group in extreme northeastern New Mexico from populations farther north (Schwalm et al. 2014; see under "Habitat Associations" for a discussion of topography).

Vegetation cover along New Mexico's Canadian River is dominated by thick stands of mesquite in which swift foxes are not found (Harrison 2003, Harrison et al. 2004). This vegetation type extends into Texas where there is extensive agricultural development, areas in which swift foxes are also not found. Before western Texas was developed for agriculture, New Mexico swift fox populations south of the Canadian River would have been connected more directly to swift foxes farther north. Now, however, swift foxes in east-central and southeastern New Mexico are notably isolated from swift foxes farther north, though not isolated from kit foxes to the west. Whether these swift foxes lose some of their genetic distinctiveness through hybridization with kit foxes, or follow their own evolutionary pathway, remains unclear.

The fact that swift and kit foxes hybridize does not in itself make them a single species, as other clearly distinct species can also hybridize. Examples are coyotes and wolves (*Canis lupus*), and harp seals (*Phoca groenlandica*) and hooded seals (*Cystophora cristata*; Wayne and Brown 2001). Swift and kit foxes also occupy different habitats, which suggests that over evolutionary time—with the possible exception noted above—they will continue to diverge as they respond to different selection pressures. The taxonomic and evolutionary relationship between kit and swift foxes remains perhaps somewhat unresolved, but today most authors (e.g., Frey 2004; Bradley et al. 2014; Burgin et al. 2018; Malaney et al. 2022) and most game management agencies including the New Mexico Department of Game and Fish treat the two foxes as separate species. The distinction is not entirely academic, as it has conservation implications (see under "Status and Management" below).

DISTRIBUTION

Historically, swift foxes ranged from eastern New Mexico and western Texas northward through the Great Plains shortgrass and mixed-grass prairies to Montana, North Dakota, Alberta, Saskatchewan, and Manitoba. However, by the early 1900s they were considered extirpated from Canada, Montana, North Dakota, Nebraska, Kansas, and Oklahoma and acutely depleted elsewhere in the United States (Sovada et al. 2009; see under "Status and Management"). Today the range of swift foxes is approximately 30% of the historical extent in the United States and Canada (calculated from Sovada et al. 2009), but there are significant areas of suitable yet unoccupied land-cover types, particularly shortgrass and mixed prairies, especially in Canada, North Dakota, Nebraska, and Texas.

Swift fox populations continue to occupy most of their historical habitat within Colorado,

Kansas, and Oklahoma (Dowd Stukel 2017). In Wyoming and Montana, populations are expanding (Harris 2020). North Dakota has a small but viable population in its southwestern corner (Tucker 2020). Swift foxes have been found in western South Dakota since 2000 (Dowd Stukel 2020), but may be declining in some areas (Nevison et al. 2017). Surveys in Texas and Nebraska have been very limited, and the status of the swift fox in those two states is uncertain (see Swift Fox Conservation Team annual reports 1995–2018, under Colorado Parks and Wildlife 2021a).

There have been a number of swift fox reintroductions in an effort to restore populations to unoccupied yet suitable areas within the species' historical range. The first such effort occurred in Canada near the Alberta/Saskatchewan/Montana border (Carbyn 1998; Herrero 2003). Occupancy of the reintroduced population has declined, but it is being closely monitored (Cullingham and Moehrenschlager 2019). Subsequent reintroductions are ongoing in Montana (Fort Peck Indian Reservation and Blackfeet Indian Reservation), South Dakota (Turner Bad River Ranch, Pine Ridge Indian Reservation, Badlands National Park, and Lower Brule Sioux Reservation) and Alberta (Kainai [Blood Tribe] Lands) (Swift Fox Conservation Team annual reports 1995–2018; Ausband and Foresman 2007; Dowd Stukel and Fecske 2007; Schroeder 2007; Sovada et al. 2009; Sasmal et al. 2013).

Within New Mexico, swift foxes occupy the prairie ecosystems east of the Pecos River valley, though a few records exist farther west (Map 14.1). Hubbard (1994) reviewed museum records and found only two specific records of swift foxes in New Mexico from 1850 to 1950 despite surveys by early naturalists in the state. After 1950, swift fox specimens began to appear in New Mexico museum collections, suggesting that the decline and rebound of swift fox populations elsewhere in the Great Plains may have occurred in New Mexico as well. In 1997, Harrison and Schmitt

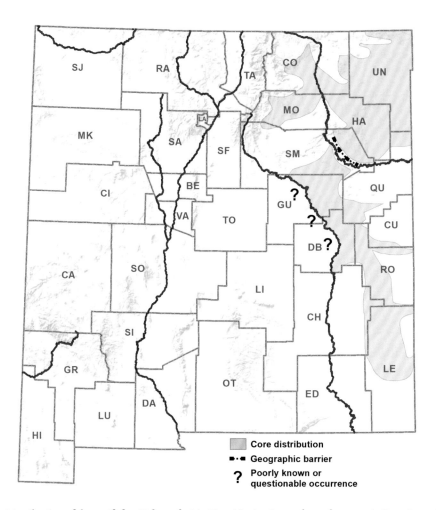

Map 14.1. Distribution of the swift fox (*Vulpes velox*) in New Mexico (range boundary map). Genetic studies indicate the current existence of two distinct populations in New Mexico. An area of rugged topography in San Miguel's Conchas and Canadian river basins may serve as a geographic barrier preventing gene flow between the two populations. A unique, rare mtDNA haplotype more similar to the kit fox (*V. macrotis*) has been found in the southern population. A zone of contact between swift and kit foxes exists along the Pecos River, with hybridization occurring between the two species. The extent of the swift fox's distribution west of the Pecos River in Guadalupe and De Baca counties is unclear (see Appendix 14.1 and Chapter 13). Swift foxes likely occupy only a portion of their historical range in New Mexico. They are absent in intensive agricultural areas of eastern Quay, Curry, Roosevelt, and Lea counties, as well as in dense shrublands in southeastern Quay County.

(2003) surveyed the thirteen easternmost counties of New Mexico and found evidence of swift foxes throughout their historical range, though not in the intensive agricultural areas of Quay, Curry, Roosevelt, and Lea counties, nor in dense shrublands in southeastern Quay County. Harrison and Schmitt (2003) used track stations (one-m² [~11-sq-ft] areas containing a scented lure and covered with sifted sand for identification of tracks), spotlighting, museum specimen records, New Mexico Department of Game and Fish fur harvest records, and US Department of Agriculture Wildlife Services records. By collecting scats (feces) in 2002, Harrison et al. (2004)

confirmed the distribution of swift foxes in New Mexico as described in Harrison and Schmitt (2003). Although Harrison et al. (2004) did not count foxes, they did find more scats north of the Canadian River, suggesting that swift foxes are more common in the wetter, cooler areas of northeastern New Mexico than in the drier, warmer plains of east-central and southeastern New Mexico. There are no large areas of unoccupied swift fox habitat in New Mexico.

HABITAT ASSOCIATIONS

Swift foxes are denizens of open short-grass and mixed-grass prairies (Egoscue 1979; Scott-Brown et al. 1987). They are most commonly found in rangeland areas of low slope and short vegetation,

Photo 14.5. Swift fox adult and pups in Weld County, Colorado. The swift fox is native to the Great Plains region of North America. Its distribution once extended from Montana, North Dakota, Alberta, Saskatchewan, and Manitoba south to eastern New Mexico and western Texas. Due largely to habitat loss, only about 30% of the species' historical distribution may be occupied today. Photograph: © Mike Dunn.

Photo 14.6. The northeastern plains of New Mexico are home to one of the state's two populations of swift foxes, both genetically distinct. Photograph: © Mark L. Watson.

Photo 14.7. Mills Canyon along the Canadian River (view to the north in March 2017) near Roy in Harding County, New Mexico. Out on the northeastern plains of New Mexico, the Canadian River has cut a canyon nearly 300 m (1,000 ft) deep. Areas of rugged topography are generally avoided by swift foxes and in some locations act as geographic barriers limiting gene flow. Photograph: © Daniel Kalal.

Photo 14.8 (*upper left*). Kiowa National Grassland near Roy in Harding County, New Mexico around 2000. Prime swift fox habitat is open, treeless, shortgrass prairie. Photograph: © Robert L. Harrison.

Photo 14.9 (*upper right*). Kiowa National Grassland near Roy in Harding County, New Mexico around 2000. Swift fox habitat in New Mexico is colored tan and bronze for most of the year. Photograph: © Robert L. Harrison.

Photo 14.10 (*bottom*). Three swift fox pups in grassland just outside their den in northeastern New Mexico. Swift foxes avoid areas with a high density of shrubs but may use grasslands with yuccas at least 5 m (16 ft) apart. Photograph: © Don MacCarter/New Mexico Department of Game and Fish.

Photo 14.11. New Mexico State Road 39 as it cuts through shortgrass prairie in northeastern New Mexico (view to the south; March 2017). Although swift foxes experience mortality from collisions with vehicle traffic, they tend to den closer to roads (or in areas with more roads) than expected by chance alone. Photograph: © Daniel Kalal.

Photo 14.12. Swift foxes on an unpaved road on the eastern plains of Colorado, 18 July 2017. The species appears to benefit from its association with roads as coyotes (*Canis latrans*), a top source of swift fox mortality, often avoid them. Note also the cattle guard (swift foxes often deposit scats where they encounter unusual features on the prairie, such as fence corners and cattle guards). Photograph: © Mike Dunn.

whereas they avoid areas of dense shrubs and rugged topography (Cameron 1984; Harrison and Schmitt 2003; Russell 2006). In South Dakota, Sasmal et al. (2011) found that swift foxes used grassland, sparse vegetation, and prairie dog towns in proportion to their availability, but used woodland, shrubland, pasture/cropland, and developed areas less than they were available.

Swift foxes may occur in what are considered atypical habitats, such as cropland and mixed shrub-grasslands (Cutter 1958; Wooley et al. 1995; Sovada et al. 1998, 2001; Olson 1999; Jackson and

Choate 2000; Matlack et al. 2000; Olson and Lindzey 2002), but generally not without consequences in terms of lower probability of area occupancy (and thus likely also lower density, provided that sampling meets the assumption of spatial independence), lower resource use, lower habitat connectivity, and perhaps also reduced fitness. On the eastern plains of Colorado, for example, Finley et al. (2005) live-trapped swift foxes on 72 31.2 km² (~12 sq mi) grids with varying proportions of shortgrass vs. (dry and irrigated) agricultural and other vegetation cover. As determined by Finley et al. (2005), the probability of swift fox occupancy ranged from 34% for grids with no shortgrass prairie habitat to 93% for grids consisting of 100% shortgrass prairie. In northeastern Montana, Butler et al. (2020) studied home range size and resource use in a swift fox population occupying a landscape mosaic of shortgrass and mixed-grass prairie, shrublands, and dry and irrigated agriculture. The relative probability of use of an area rose by 3.3% for every 1% increase in the proportion of grassland. Home range size was positively correlated with the proportion of cropland, an indication that swift foxes needed to travel over increasingly large distances to find denning sites and food as agricultural fields dominated more and more (Butler et al. 2020). In northern Texas, adult and juvenile swift foxes were strongly associated with shortgrass prairie (46% of the study area), used dry agricultural fields in lower proportion compared to their availability (25% of the study area), and completely avoided irrigated fields (Kamler et al. 2003). In Kansas, Sovada et al. (1998, 2003) also found that swift foxes did not use irrigated cropland areas. During the pup-rearing season in western South Dakota, female swift foxes used grasslands in relation to their availability but avoided shrublands, crop fields, and pastures (Sasmal et al. 2011). In contrast to the studies above, Matlack et al. (2000) reported no differences in swift fox relative abundance and mortality between shortgrass prairie

and cropland in western Kansas (see also Sovada et al. 1998 for similar results regarding mortality), but swift foxes in shortgrass prairie were larger and in better condition. Based on an analysis of genetic structuring, finally, Schwalm et al. (2014) stated that agricultural development appeared to limit gene flow on a local scale. Approximately 41% of the historical range of swift foxes in the United States and Canada has been converted to cropland (calculated from Sovada et al. 2009). Red foxes (*Vulpes fulva*) also significantly declined on the Great Plains with European settlement, but they are also able to exploit some agricultural areas (see Chapter 15). Intraguild competition and predation are common among canids, but it is unknown if red foxes, which are larger than swift foxes, limit the distribution or abundance of swift foxes in croplands.

In New Mexico, Harrison and Schmitt (2003) and Harrison et al. (2004) did not find evidence of swift foxes in the agricultural areas of Curry and Roosevelt counties. Harrison and Schmitt (2003) reported that swift foxes in New Mexico prefer gramma (*Bouteloua*) rangeland free of shrubs with grass lengths <30 cm (12 in). *Bouteloua* rangeland was preferred over little bluestem (*Schizachyrium*) or three-awn (*Aristida*) rangeland. Harrison and Schmitt (2003) suggested that swift fox habitat use is unaffected by shrub density as long as the average nearest neighbor distance between shrubs is greater than 15 m (49 ft). Swift fox occupancy declines as the spacing between shrubs (as measured by nearest neighbor distance) decreases below 15 m (49 ft), and ceases altogether at distances less than 5 m (16 ft). Swift foxes do not occur in dense stands of mesquite (*Prosopis*), sagebrush (*Artemesia*), or shin oak (*Quercus havardii*), but do use stands of *Yucca* with individual plants between 5 and 15 m (16 and 49 ft) apart. Thompson and Gese (2007) also reported that swift fox density decreased as shrub density increased in southeastern Colorado. In this case, the underlying mechanism was an increase of prey density in areas of dense shrubs, which led to higher use by coyotes.

Swift—and kit—foxes are unusual among canids because they are dependent upon dens and use them every day throughout the year (e.g., Scott-Brown et al. 1987; Cypher 2003; R. L. Harrison, pers. obs.). There is little cover in their short-grass prairie homes, so dens are essential for safety from predators such as coyotes. Most dens have one opening and may be spotted at a distance by the large apron of excavated soil extending about 1 m (~3.3 ft) from the opening (Cutter 1958). Typically, an accumulation of scats is present. (R. L. Harrison, pers. obs). Natal dens are usually larger and may have several openings. Inside, dens consist of several tunnels extending up to several meters in length and over one meter in depth below the surface (Seton 1937; Cutter 1958; Kilgore 1969). Swift foxes may dig their own dens or modify those of other species such as prairie dogs (*Cynomys* spp.) or American badgers (*Taxidea taxus*) (Cypher 2003).

Rangewide, the majority of observed dens have been found on shallow slopes or flat terrain (Harrison and Hoagland 2003). For example, in Wyoming, Olson (2000) found dens on slopes of <3% more than expected and dens on slopes of 3–6% less than expected. Of 41 dens observed by Cameron (1984) in Colorado, 31 (76%) were located on flat terrain or at the base of a hill. In a study on the Kiowa National Grassland and private lands near Roy in Harding County, New Mexico, Harrison (2003) found that the average slope at swift fox dens was less than at random points in the study area. The need for swift foxes to have a long line of sight and better detect coyotes (and other potential predators) likely explains the negative association between swift foxes and rugged terrain and/or higher slopes at a landscape scale (Cameron 1984) and for traveling between foraging patches within the home range (Butler et al. 2019). The same need seemingly applies to the selection of a den site, including in northeastern

New Mexico where, in two separate studies, dens occurred on slightly higher ground compared to their immediate surroundings. Near Roy in Harding County, Harrison (2003) found that dens were located preferentially near the tops of hillsides, while according to Kintigh and Andersen (2005; study area mostly around Clayton in Union County) the den elevation index (percentage of the ground within 500 m of the den that occurred at lower elevation) was significantly higher than found at randomly selected points.

In Wyoming, Olson (2000) reported that swift foxes select for loamy soils and against clayey soils. In northeastern New Mexico, however, den sites were found significantly more in association with not only sandy-loam and loam but also clay soils (Harrison 2003). Swift foxes are much better diggers than humans. One of us (R. L. Harrison) excavated two swift fox dens to recover radio-collared foxes that had died. Despite having the advantage of a shovel, the task took several hours as the ground seemed as hard as concrete.

Numerous studies have also established that throughout the species' distribution, swift fox dens tend to be located near roads or closer to roads than expected by chance alone (Hines and Case 1991; Pruss 1999; Olson 2000; Harrison 2003; Russell 2006; Butler et al. 2019; but see Mitchell 2018). In northeastern New Mexico, Harrison (2003) found that swift foxes selected den sites closer to roads compared to random points, while Kintigh and Andersen (2005) indicated a positive association between swift fox dens and higher road densities (rather than proximity to roads). Proposed explanations for the association between swift foxes and roads—despite vehicle collisions representing an important or leading cause of mortality (e.g., Kamler et al. 2003)—include 1) greater availability of carrion and small mammals along roadsides (Hines and Case 1991; Klausz 1997); 2) avoidance and/or enhanced detection of coyotes (Russell 2006; Butler et al. 2019); and 3) use of roads as travel corridors (Hines and

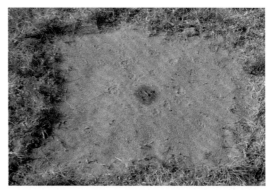

Photo 14.13 (*above*). Swift fox research by one of us (R. L. Harrison) involved capturing and radio-collaring 36 swift foxes (18 males and 18 females) in Harding and Colfax counties in northeastern New Mexico. The use of radio collars allowed tracking of swift foxes to their dens, which could then be characterized in terms of number of entrances and surrounding vegetation. Note the black dye on the fox's fur, which is also wearing a radio collar. Dying fur in unique patterns is a useful but temporary means to identify individual foxes for research. Photograph: © Robert L. Harrison.

Photo 14.14 (*top*). Scent station on the Kiowa National Grassland in Harding County, New Mexico. The scent station is a 1-m² [~11-sq-ft] area containing a scented lure and covered with sifted sand for identification of tracks. Swift foxes are readily attracted to scent stations, leaving many tracks. Photograph: © Robert L. Harrison.

Photo 14.15 (*bottom*). During Harrison and Schmitt's (2003) research in northeastern New Mexico, radio-collared and fur-dyed swift foxes frequently provided additional genetic information by depositing scats on scent stations. Photograph: © Robert L. Harrison.

Photo 14.16 (*left*). During the daytime, swift foxes stay in or very close to their dens, which provide them with essential protection from coyotes. Photograph: © Mike Dunn.

Case 1991; Pruss 1999). Although based on fine-scale movement of foxes rather than den site location, Butler et al.'s (2019) study in Montana provided support for the explanation centered on avoidance and/or detection of coyotes across all types of activity patterns (resting, foraging, and traveling) during the winter.

Swift foxes likely can survive without free-standing water. Flaherty and Plakke (1986) found that swift foxes can remain in water balance with food alone. Swift fox habitat is generally semi-arid, and free-standing water is often undependable. Water sources exist in many areas but are typically artificial (Rosenstock et al. 1999).

If not harassed, swift foxes can be quite tolerant of low-impact nearby human activity such as water well maintenance (Cutter 1958; Kilgore 1969). However, more intense development, such as quarries, gravel pits, and mixed urban/built-up land will likely be avoided (Sasmal et al. 2011).

LIFE HISTORY

Diet and Foraging

Mammals, birds, and invertebrates make up the bulk of the swift fox's diet (Harrison 2001). Mammals are taken consistently throughout the year, while birds and invertebrates are consumed seasonally as available. The relative importance of prey groups in the diet varies considerably between studies. Among mammals, rabbits and hares have often been reported as the main prey (Cutter 1958; Kilgore 1969; Zumbaugh and Choate 1985), but swift foxes may also rely primarily on rodents instead (Hines and Case 1991; Zimmerman 1998; see review in Harrison 2001). Reptiles and amphibians represent only occasional prey items (Harrison 2001). Carrion also forms part of the swift fox's diet, and has been reported to occur in up to 38% of scats (Hines and Case 1991). Plant remains in scats have usually been found in trace amounts, but vegetation that could not be consumed incidentally or occurred in large volumes in scat indicate that swift foxes deliberately consume plant material. Sovada et al. (2001) reported that swift foxes readily consume commercially grown sunflower seeds.

In northeastern New Mexico, invertebrates, especially grasshoppers (Orthoptera), were found in over 80% of scats collected by Harrison (2003), and they were the most frequent prey on an annual basis. Mammals and birds were second and third most frequent at 43% and 17%, respectively. Next to kangaroo rats (*Dipodomys* spp.), lagomorphs were the most often detected mammals in scats. Other mammalian prey consisted primarily of a variety of other rodents besides kangaroo rats (Cricetidae, Geomyidae, Heteromyidae, Muridae, and Sciuridae). Vegetation, reptiles, and carrion (pronghorn and cattle) were consumed less often (Harrison 2003).

Because the swift fox consumes a wide variety of prey, it tends to be considered an opportunistic predator (Olson 2000). Opportunistic predators consume prey in proportion to their abundance, and this indeed appears to be true of the swift fox. The frequencies of birds, insects, and vegetation in scats generally follow seasonal trends of availability. Kilgore (1969) found that the relative frequencies of rodent and shrew remains in scats were very similar to the relative capture rates of those two taxa in snap traps. In Wyoming, the occurrence of pronghorn (*Antilocapra americana*) in scats peaked in fall (i.e., the hunting season) and winter, at which times pronghorn mortality translated into a greater availability of carrion (Olson 2000; hunters do not always harvest animals in their entirety as they often leave the internal organs in the field, and pronghorn that are not immediately killed may escape and die without being recovered). Kamler et al. (2007) studied swift fox diet in two separate areas, one of continuous native prairie and the other of fragmented prairie interspersed with agricultural fields. Insects were eaten more frequently on continuous prairie, whereas mammals, birds,

Photo 14.17 (*top*). Adult female swift foxes may give birth to as many as eight pups per litter. The pup-rearing season is a demanding time for swift fox mothers. Photograph: © Mike Dunn.

Photo 14.18 (*middle*). Pups are weaned at the age of six to seven weeks. Photograph: © Mike Dunn.

Photo 14.19 (*bottom*). Pups grow up with their siblings and often play together on long summer evenings, but will have gone separate ways by the end of their first winter. They must learn quickly to be alert for coyotes and other potential threats. Photograph: © Mike Dunn.

and crops represented a larger portion of the diet on the fragmented site. Leporids were more abundant on the fragmented site, and swift foxes responded to their greater availability by taking and consuming them more frequently. Diet diversity was greater on the fragmented site, showing swift foxes to be opportunistic as well as adaptable predators. The fact that certain foods have been found out of season in their diet only indicates that swift foxes cache items such as ground squirrels (Pruss 1994) and eggs (Sovada et al. 2001).

Swift foxes are nocturnal, leaving their dens to forage at sunset and re-entering them at dawn (Covell et al. 1996; Kitchen et al. 1999; Cypher 2003). They may be active during daylight, but at that time usually remain close to their dens. In the evenings they often lounge near the entrances of their dens, especially in the summer (R. L. Harrison, pers. obs.). Swift foxes usually hunt alone, traveling up to 1 km/hr (0.6 mph) during the night (Kitchen et al. 1999). They may cover an average of 13 km (8 mi) during winter, and 5.7 km (3.5 mi) during summer (Covell et al. 1996).

Reproduction and Social Behavior

Breeding occurs in winter (Kilgore 1969), followed by a gestation period of 51 days (Schroeder 1985). Typically, four to five pups (range one to eight) are born in spring (Kilgore 1969; Hillman and Sharps 1978). Pups open their eyes at 10–15 days and are weaned at six to seven weeks (Kilgore 1969). Food availability strongly influences reproductive success (Cypher 2003). Mortality is high among pups as they are vulnerable to many predators (Sovada et al. 1998). By fall pups have adult pelage and size and begin to disperse (Cypher 2003). Dispersal can occur as late as February (Mitchell 2018). Swift foxes can reproduce within their first year (Scott-Brown et al. 1987).

Swift foxes often den together, especially during early portions of the mating season (Covell 1992; Kitchen et al. 2005a). Families of swift foxes may be readily observed at their dens in early summer, when parents and pups play in the cool of the evening before the parents leave for nighttime foraging (R. L. Harrison, pers. obs.). Juvenile males are more likely to disperse than juvenile females. Juvenile females are likely to remain with their mothers as helpers or to reproduce within or adjacent to their natal home range (Kamler et al. 2004a). As a result, neighboring foxes, especially females, are more likely to be related than non-neighbors (Kitchen et al. 2005b).

Average home range size estimates range from 418 to 4,661 ha (1.6–18.0 mi²; Hines and Case 1991; Zimmerman 1998; Kitchen et al. 1999; Harrison 2003; Lebsock et al. 2012; Thompson and Gese 2012; Mitchell 2018; Butler et al. 2019, 2020). In northeastern New Mexico, Harrison (2003) found annual home range sizes to average 1,495 ha (8.5 mi²). Home ranges can shift spatially as swift foxes occasionally occupy new areas and abandon old ones if their neighbors die (Harrison 2003; R. L. Harrison, pers. obs.). Home range size estimates obtained by examining the distribution of dens can be different from those obtained with nighttime observations. During daytime there is relatively little overlap between the home ranges of adjacent swift fox families, but during nighttime overlap increases, especially during the dispersal season (Schauster et al. 2002; and see Butler et al. 2020 for use of GPS tracking collars to determine home range size in northeastern Montana). Accumulations of scats at specific locations (latrines) may aid in territorial defense. Darden et al. (2008) found that latrines were more common within cores of territories and in areas of overlap between family groups.

Although food availability and habitat quality are often cited as the factors influencing canid social systems, predation and predator control can also be important. In Texas, Kamler et al. (2004b) documented polygynous groups, communal denning, and non-breeding females in a high-density swift fox population where coyotes were controlled to protect livestock and thus experienced low survival. In a nearby area where coyotes were not deliberately controlled, coyote survival was higher and swift fox density was lower. In the latter area, only monogamous pairs of swift foxes were found (Kamler et al. 2004c).

The social organization of swift foxes is unusual among canids, as it is female-based rather than male-based (Kamler et al. 2004b). In most canid families, males provide food to females and young and maintain territories. However, because of the species' primarily insectivorous diet in summer when pups are being reared (Kitchen et al. 1999; Lemons 2001; Harrison 2003), the role of males is diminished as pups beginning to forage can readily acquire insects. Female swift foxes maintain territories: when a male dies, females remain in the same home range, but when females die, males emigrate (Kamler et al. 2004b).

Individual swift foxes may be identified by sonograms of their vocalizations (Darden et al. 2003). Identification of individuals opens the possibility of counting actual numbers of foxes for management rather than using relative indices. Also, Avery (1989) found that captive foxes will respond to tapes of the sounds of other foxes. While studying population survey methods, Harrison et al. (2002) attempted to stimulate free-ranging foxes to vocalize by playing tapes of swift fox vocalizations in their study area in New Mexico. However, the foxes did not respond as hoped, and the strange sounds carried further than expected over the open grasslands, disturbing local homeowners. Vocalizations of the swift fox include barks, growls, yaps, and whines (Avery 1989). A distinctively raspy "whar-whar-whar" sound is used to maintain connections between adults or between adults and kits (Cameron 1984).

Interspecific Interactions

In many areas the greatest single source of swift fox mortality today consists of predation by coyotes (e.g., Kitchen et al. 1999; Matlack et al. 2000; Olson and Lindzey 2002; Allardyce and Sovada 2003; Karki et al. 2007; Sovada et al. 2009; Gese and Thompson 2014; Mitchell 2018). The ecological relationship between coyotes and swift foxes is referred to as interference competition rather than predation, as the foxes are often not eaten when killed and the two species consume many of the same prey (Sovada et al. 1998; Kitchen et al. 1999). Coyotes hampered the reintroduction of swift foxes in Canada, with most fox deaths occurring within the first month after release (Carbyn et al. 1994). In southern Colorado, Kitchen et al. (1999) reported a high degree of spatial overlap between coyotes and swift foxes—in contrast North American red foxes tend to establish home ranges only where coyotes are absent (Voigt and Earle 1983; Harrison et al. 1989). All swift fox home ranges were partially or entirely overlapped by coyote home ranges, and foxes did not time their activities in order to avoid coyotes (Kitchen et al. 1999). Although coyotes consumed a greater proportion of larger prey such as jackrabbits, the two species also exhibited a large amount of dietary overlap. With coyotes being responsible for 48% of detected fox mortality events, Kitchen et al. (1999) concluded that swift foxes were able to coexist with coyotes largely as

a result of the year-round use of their dens. Still according to Kitchen et al. (1999), coyote attacks on swift foxes tended to be more successful away from fox dens. As noted earlier, the positive association between swift fox dens and proximity to roads may also help foxes subsist in association with coyotes (Russell 2006; Butler et al. 2019). In some areas, coyotes tend to avoid roads (seemingly perceived as threats), thus creating areas of lower predation risk for swift foxes (Butler et al. 2019).

Reduction of coyote numbers (i.e., control) can increase swift fox survival and density (Kamler et al. 2003; Karki et al. 2007). In northern Texas, Kamler et al. (2003) reported higher swift fox survival, density, and juvenile recruitment (measured as the ratio of juveniles to adults) following removal of coyotes. However, in southeastern Colorado, Karki et al. (2007) found that while swift fox survival increased with coyote control, reproduction and density remained unchanged. Karki et al. (2007) speculated that the swift fox population in their study area was saturated, having observed that density remained unchanged due to increased dispersal. How swift foxes respond to the risk of predation by coyotes likely varies seasonally. In winter, swift foxes are resource-stressed, and their behavior is governed by the need for food acquisition more than that of avoiding predation. During the rest of the year, however, their behavior is driven by predation avoidance (Thompson and Gese 2012).

North American red foxes may pose a threat to swift foxes (Sovada et al. 2009). Ralls and White (1995) reported that red foxes will kill kit foxes and exclude them from areas they would otherwise occupy. Red foxes have not been reported to kill swift foxes, but Mitchell (2018) found that swift fox presence correlated negatively with red fox occupancy in the Dakotas. Anecdotal evidence also suggests that red foxes may exclude swift foxes from agricultural areas that would otherwise be suitable for swift foxes. For example, swift foxes were not found in the extensive agricultural areas of western Texas, whereas red foxes were (Kamler et al. 2005; Schwalm et al. 2012). Conversely, swift foxes were documented in agricultural areas of Kansas, but red foxes were not (Sovada et al. 1998). Red foxes have been reported trapped in all New Mexico counties within the historical swift fox's range (NMDGF unpublished Hunter Harvest Report Program, Summary of Results—Furbearers; see also Kamler and Ballard 2002 and Chapter 15). A study of interactions between red and swift foxes would be very worthwhile.

Besides coyotes, other confirmed predators of swift foxes include humans, American badgers, golden eagles (*Aquila chrysaetos*), and bobcats (*Lynx rufus*; Thompson 2006). Human-caused mortality can be significant. Swift foxes may be killed by vehicles, agricultural machinery, poisoning for control of other species, and fur trappers (see under "Status and Management" below; Sovada et al. 1998, 2009; Thompson 2006). Death on roadways is often a significant cause of mortality (Sovada et al. 2009; Mitchell 2018). In northwestern Texas, near New Mexico, Kamler et al. (2003) monitored a total of 42 swift foxes, with the leading cause of mortality represented by vehicle collision (42%), ahead of coyote predation (33%).

Diseases and Pathogens

Several infectious diseases have been documented in swift foxes, including canine parvovirus, tularemia, canine distemper, and plague (e.g., Miller et al. 2000; Olson and Lindzey 2002; Gese et al. 2004; Mitchell 2018). A high prevalence of antibodies against canine parvovirus was found in swift fox populations in southeastern Colorado (Gese et al. 2004) and in northwestern South Dakota and southwestern North Dakota (Mitchell 2018). Both studies also revealed some exposure to canine distemper, as did research in Wyoming documenting two fatal cases of that disease in swift foxes (Olson

and Lindzey 2002). Other viral diseases, such as vesicular stomatitis and Cache Valley virus, have also been reported as affecting the swift fox (Miller et al. 2000; Cypher 2003).

Swift foxes carry a variety of ecto- and endoparasites, including fleas, tapeworms, and nematodes (Miller et al. 1998; Pybus and Williams 2003; Pence et al. 2004; Criffield et al. 2009). Fleas are especially common on swift foxes due to the species' daily use of dens. Many of the fleas found on swift foxes in New Mexico can carry bubonic plague (Harrison et al. 2003). The tapeworm *Echinococcus multilocularis* is of also particular concern as it too can infect humans, producing severe debilitation and possibly death. *E. multilocularis* has been found in North American red foxes within the distribution of the swift fox, but not yet in swift foxes (Pybus and Williams 2003). However, while 31 swift fox specimens from New Mexico tested negative for *E. multilocularis*, 74% of them were positive for one or several other endoparasites, including four nematode species (*Physaloptera rara, Physaloptera* sp., *Toxascaris leonina*, and *Toxocara canis*) and two cestodes (*Dipylidium caninum* and *Mesocestoides variabilis*) (Ubelaker et al. 2014).

STATUS AND MANAGEMENT

Many factors were implicated in the early decline of swift foxes, particularly poison control programs directed at wolves, fur trapping, the conversion of prairies to agriculture, the collapse of the prairie ecosystem following the extirpation of bison (*Bison bison*), the loss of large animal carcasses left over from wolf kills, and, inaccurately, the spread of red foxes (Scott-Brown et al. 1987; Allardyce and Sovada 2003; Sovada et al. 2009; and see Chapter 15). The use of poison to eliminate wolves may have been of special significance, as swift foxes readily take poisoned meat from carcasses (Young 1944). Elimination of poison control programs, decreases in fur prices, restrictions on trapping, and reductions in the numbers of farms and ranches coincided with the recovery of populations beginning in the 1950s (Allardyce and Sovada 2003). In particular, poisoning programs changed from broadcasting poison bait to the use of toxicants delivered at stations and designed more specifically for coyotes (e.g., M-44 cyanide capsule; see Robinson 1953).

Official concern over the decline of the swift fox in the northern portion of its range began in 1978 with the declaration of the swift fox as extirpated in Canada by the Committee on the Status of Endangered Wildlife in Canada (Sovada et al. 2009). The reintroduction program in Canada began at a small private wildlife refuge (Smeeton et al. 2003), but quickly expanded into a major program involving governmental, commercial, and non-profit organizations. The first releases began in 1983 (Carbyn et a1.1994). Success of the program led to declaration of the swift fox as endangered in Canada in 1999 for continued protection of the species (Sovada et al. 2009).

The US Fish and Wildlife Service (USFWS) listed the swift fox as a Category 2 candidate for endangered species listing in 1982 (USFWS 1982). This listing indicated that while some information suggested that the swift fox was in trouble, sufficient information was not available to reach a firm conclusion. Had the swift fox been listed federally as endangered, then the USFWS would have overseen recovery programs conducted by individual state agencies. The specter of federal oversight prompted wildlife management agencies within the ten US states in the historical range of the swift fox to form the Swift Fox Conservation Team (SFCT) in 1994 (Colorado Parks and Wildlife 2001a). Additional US federal and Canadian agencies, including the Bureau of Land Management, the National Park Service, the US Fish and Wildlife Service, and the Alberta Fish and Wildlife Division have since joined the SFCT (Colorado Parks and Wildlife 2021a; see Appendix 14.2 for a complete list of organizations involved). SFCT annual reports and other documents are available from Colorado Parks and Wildlife 2021a.

The SFCT developed a formal recovery plan for the swift fox, the *Conservation assessment and conservation strategy for swift fox in the United States* (Colorado Parks and Wildlife 2001a; Dowd Stukel 2011). Partially in response to the high degree of commitment shown by state agencies to protect the species, the US Fish and Wildlife Service concluded in 1995 that listing of the swift fox as endangered was "warranted but precluded by higher priority listings" (USFWS 1995). This decision meant that while the Fish and Wildlife Service had concluded that the swift fox qualified to be listed as endangered, the agency's limited resources had to be allocated to other, more critically endangered species. During that same period the taxonomy of swift and kit foxes was of special interest. Had swift and kit foxes been lumped together as a single species, some of the impetus for swift fox recovery and monitoring would have been lost. Most populations of kit foxes are not considered to be endangered (see Chapter 13), and kit foxes in general are thought to be more abundant than swift foxes. Recognition of the swift fox as its own distinct species, together with its unknown population status, stimulated a great deal of ecological and other research by both universities and government agencies. It was found that the range of the swift fox was greater and less fragmented than previously thought. In 2001, the US Fish and Wildlife Service removed the swift fox from its list of candidate species but recommended continued monitoring of populations (USFWS 2001).

In New Mexico, Hubbard (1994) analyzed New Mexico Department of Game and Fish fur harvest reports in response to a US Fish and Wildlife Service request for information on the status of the swift fox. Hubbard (1994) found no indication that trapping had negatively impacted swift fox populations during the period 1969–1993. Thus, he did not recommend that swift foxes be listed

Photo 14.20. Swift fox in Colorado on 28 May 2016. During the late 1800s and the early to mid-20th century, swift foxes declined throughout much of their range due to many factors including poison control programs directed at wolves, fur trapping, and habitat loss. Swift foxes have since been reintroduced in the northern part of the species' range. Farther south, some populations have rebounded naturally since about 1950, including possibly in New Mexico, but likely remain below historical levels. Today, swift fox conservation benefits from inter-jurisdictional, regional, and international cooperation. Photograph: © Mike Dunn.

as threatened or endangered by the state of New Mexico under its Wildlife Conservation Act. At that time, swift foxes were listed as "Protected Furbearers" (Schmitt 1995). In 2005, swift foxes remained protected furbearers, but were also listed as a Species of Greatest Conservation Need in the Comprehensive Wildlife Conservation Strategy (Stuart 2006). In 2016, swift foxes lost their status as a Species of Greatest Conservation Need while remaining a protected furbearer (Stuart and Murphy 2017). Killing of protected

furbearers by trapping or hunting is regulated by the New Mexico Department of Game and Fish. Unlimited numbers of swift foxes may be taken during the fur trapping season (1 November–15 March) by any legal means, including guns, traps, falconry, and snares (NMDGF 2020).

Elsewhere, the swift fox is legally classified as a furbearer and may be trapped or otherwise captured in four states: Colorado, Kansas, Montana, and Texas (Colorado Parks and Wildlife 2021b; Kansas Department of Wildlife and Parks 2023;

Montana Fish, Wildlife and Parks 2023; Texas Parks and Wildlife Department 2023). No trapping of swift foxes is allowed in Nebraska, North and South Dakota, Oklahoma, and Wyoming, though incidental taking is allowed in some circumstances (Nebraska Game and Parks Commission 2021; North Dakota Game and Fish Department 2021; South Dakota Department of Game, Fish and Parks 2021; Oklahoma Department of Wildlife Conservation 2023; Wyoming Game and Fish Department 2021).

One past conservation effort in New Mexico revolved around helping swift foxes avoid coyote predation. From 2002 to 2004, artificial swift fox escape dens were installed on the Kiowa National Grassland near Clayton in Union County—as well as on the adjacent Rita Blanca National Grassland in northwestern Texas (Stuart 2008; Laws

Photo 14.21. Artificial escape den for swift foxes on 29 April 2008 on the Kiowa National Grassland near Clayton in northeastern New Mexico. The escape den consists of a ~4.3 m (~14 ft) long PVC pipe ~20 cm (~8 in) in diameter and buried just underground with both ends remaining open. Swift fox survival is increased where artificial dens are provided for escape from coyotes, but the dens must be maintained for long-term use. Photograph: © James N. Stuart.

2017). The dens consisted of corrugated plastic sewer pipes 4 m (13 ft) long and 20 cm (8 in) in diameter and buried with only the two ends exposed (Photo 14.21). The openings of the dens were large enough for swift foxes to enter, but not coyotes. Holes were cut in the middle of the pipes to allow foxes to expand the dens below ground. An experimental trial in northwestern Texas found that swift fox survival, distribution, and abundance were higher in areas where the artificial dens had been installed, but no subsequent change in the recruitment of juveniles into the adult population was observed (McGee et al. 2006). Also, no change was found in home range size or the annual number of dens used, den sharing, or distance between dens (McGee et al. 2007). Foxes apparently used the artificial dens for escape, but during daytime used their regular dens (McGee et al. 2007). In 2012, Laws (2017) returned to the same experimental sites in northwestern Texas. He found that most (73%) of the artificial dens had become unusable due to infilling with soil, damage by cattle, and general degradation of the pipes. Periodic maintenance of the artificial dens is thus required for continued benefit to swift foxes.

Trapping of swift foxes for sale of their pelts to the garment industry remains a potential concern for the conservation of the species. Over 100,000 pelts were sold in London from 1853–1877 (Rand 1948, cited in Allardyce and Sovada 2003). Fur harvests decreased with the decline of swift foxes, and have not returned to previous levels. Today, swift fox pelts have little value, and are not traded by major fur dealers (Fur Harvesters Auction Inc. 2020). However, fashion trends are fickle, and the possibility that swift fox pelts may once again become desirable cannot be completely discounted.

In New Mexico, commercial and sport trappers usually catch swift foxes incidentally when trapping coyotes, as swift foxes are attracted to the same lures as coyotes. From 2010 to 2020, the reported annual fur harvest of swift foxes (plus likely misclassified kit foxes, see below) varied from 0 to 17, and averaged 8.4 (NMDGF unpublished fur harvest records). An average of only 82.3% of trappers reported their harvest during that period. Also, because of the low value of swift fox pelts, trappers may simply discard swift fox carcasses in the field without recording how many were trapped. These facts imply that the actual mortality was likely somewhat larger than the reported harvest. Reported rangewide trapping mortality in swift fox populations appears to be relatively small compared to mortality from coyotes and vehicles, but more information is needed to accurately assess the relative significance of these factors.

The SFCT continues to meet annually to discuss swift fox population monitoring and conservation strategy. Each agency pursues its own monitoring program chosen according to needs and conditions within a given state. Methods include systematic trapping (using non-lethal cage traps), track surveys, pelt tagging records, observations by agency employees and the public, track station surveys, automatic camera surveys, trapper reports, roadkill records, spotlighting, and scat surveys.

The New Mexico Department of Game and Fish has used three methods to survey and monitor swift fox populations: fur harvest reports, and scent-station and scat surveys. The agency began recording the numbers of all foxes reported killed by trappers by county in the early 1970s (NMDGF unpublished fur harvest records). At first, different fox species were not reported separately. Beginning with the 1976–1977 season, however, kit and swift foxes were reported together as kit foxes, but separated from red and gray (*Urocyon cinereoargenteus*) foxes. Today swift and kit foxes are reported separately, but kit and swift foxes are often confused in reports. The name of the county can help separate kit and swift fox harvest, but this method becomes unreliable in the general

zone of contact between the two species (Maps 13.1 and 14.1). Swift or kit foxes reported trapped in Guadalupe, De Baca, and Chaves counties in particular may belong to either species, or they could also be hybrids (see also Chapter 13). Reports from Chaves County are particularly questionable because the zone of contact might have shifted east there and most of the county now lies within kit fox range (Map 13.1). Harvest reports cannot be used to index population numbers, not just due to the species identification issues mentioned above, but also because the importance of trapper effort relative to swift fox population size is unknown. Furthermore, prior to 2010, an average of only 39.7% of trappers reported, rendering harvest reports from that period of little use for calculation of historical trapping mortality. Since 2010, reporting rates have ranged between 72.5 and 87.0% (NMDGF unpublished fur harvest records).

The first field swift fox survey in New Mexico was an extensive scent-station and spotlight survey in the winter of 1996–1997 (Harrison and Schmitt 2003; see under "Distribution," above). Subsequently, Harrison et al. (2002) compared trapping, track searches, scat searches, spotlighting, calling, and automatic cameras (which were film cameras at the time), and concluded that scat surveys were the most efficient. Scats are easy to find due to the habit of swift foxes of depositing scats at conspicuous locations within the open prairie, including cattle guards, road or fence intersections, and utility boxes. Track surveys are popular in states with wetter climates where soils take and hold tracks more clearly, but New Mexico soils were deemed too dry and variable for track surveys to be useful in the state (Harrison et al. 2002).

In 2002, 2005, and 2008, scat surveys were conducted on a predetermined set of 98 transects located along roads throughout eastern New Mexico's thirteen counties overlapping the historical range of the swift fox (Harrison et al. 2004;

Stuart 2008, 2011). During those three years, totals of 603, 200, and 894 scats were collected, respectively. The 2002 survey was successful at confirming the presence of swift foxes throughout their historical range in New Mexico, but problems of survey timing and scat degradation while in storage severely hampered DNA confirmation of scat species origin and limited the conclusions that could be drawn from the 2005 and 2008 surveys.

In 2020, the New Mexico Department of Game and Fish began deployment of automatic cameras in Mora, Union, Colfax, and Harding counties (N. Forman, NMDGF, pers. comm.). Currently, the agency is developing a spatial capture-recapture model using collected scats, and it proposes to collect tissue samples from trappers to determine the distribution of swift foxes in New Mexico and the extent of hybridization (J. Stuart, NMDGF, pers. comm.).

In New Mexico, the outlook for the swift fox remains uncertain. On the one hand, there is little direct trapping pressure on swift foxes and large areas of shortgrass prairie (and other likely suitable land-cover types) remain. Where needed, actions can be taken to aid swift foxes including coyote control and controlling shrub encroachment in grasslands. On the other hand, monitoring of populations by the New Mexico Department of Game and Fish has been sporadic and threats remain from incidental fur trapping and from road mortality. Solar and wind energy development is now occurring in the state, and prairie habitat in southeastern New Mexico has been heavily impacted by oil and gas development. Most looming, however, is perhaps the threat of climate change in New Mexico's eastern plains, given the east-west moisture gradient of the Great Plains and the anticipated pronounced rise in temperatures (Chapter 5).

In addition to the substantial ordinary challenges of finding food, raising pups, and surviving winters, swift foxes have persisted through loss of habitat, poisoning campaigns, fur trapping, and an increase in the numbers of coyotes. In the hands of an inquisitive biologist, swift foxes seem surprisingly small and vulnerable. That they have survived is testimony to their adaptable and resilient nature.

LITERATURE CITED

Allardyce, D., and M. A. Sovada. 2003. A review of the ecology, distribution, and status of swift foxes in the United States. In *The swift fox: ecology and conservation of kit foxes in a changing world*, ed. M. Sovada and L. Carbyn, 3–18. Regina: Canadian Plains Research Center, University of Regina.

Ausband D. E., and K. R. Foresman. 2007. Dispersal, survival, and reproduction of wild-born, yearling swift foxes in a reintroduced population. *Canadian Journal of Zoology* 85:185–89.

Avery, S. R. 1989. Vocalization and behavior of the swift fox (*Vulpes velox*). MA thesis. University of Northern Colorado, Greeley.

Bradley, R. D., L. K. Ammerman, R. J. Baker, L. C. Bradley, J. A. Cook, R. C. Dowler, C. Jones, D. J. Schmidly, F. B. Stangl, R. A. Van Den Bussche, and B. G. Würsig. 2014. Revised checklist of North American mammals north of Mexico, 2014. *Occasional Papers, Museum of Texas Tech University* 327:1–27.

Burgin, C. J., J. P. Colella, P. L. Kahn, and N. S. Upham. 2018. How many species of mammals are there? *Journal of Mammalogy* 99:1–14.

Butler, A. R., K. L. S. Bly, H. Harris, R. M. Inman, A. Moehrenschlager, D. Schwalm, and D. S. Jachowski. 2019. Winter movement behavior by swift foxes (*Vulpes velox*) at the northern edge of their range. *Canadian Journal of Zoology* 97:922–30.

———. 2020. Home range size and resource use by

swift foxes in northeastern Montana. *Journal of Mammalogy* 101:684–96. doi:10.1093/jmammal/gyaa030.

Cameron, M. W. 1984. The swift fox (*Vulpes velox*) on the Pawnee National Grassland: its food habits, population dynamics and ecology. MS thesis, University of Northern Colorado, Greeley.

Carbyn, L. N. 1998. *Updated COSEWIC status report on Swift fox (Vulpes velox)*. Committee on the Status of Endangered Wildlife in Canada. Ottawa: Canadian Wildlife Service, Environment Canada.

Carbyn, L. N., H. J. Armbruster, and C. Mamo. 1994. Swift fox reintroduction in Canada from 1983 to 1992. In *Restoration of endangered species*, ed. M. L. Bowles and C. J. Whelan, 247–271. Cambridge: Cambridge University Press.

Colorado Parks and Wildlife. [2021a]. Swift Fox Conservation Team. Under "Swift Fox Conservation Team reports, newsletters, and other documents," provides a link to a list of Swift Fox Conservation Team annual and biennial reports (1995–2018) (on the linked webpage, click on "View PDF" or "Download PDF," and then select year). https://cpw.state.co.us/learn/pages/SwiftFoxConservationTeam.aspx. Accessed 3 February 2021.

Colorado Parks and Wildlife. [2021b]. Furbearers and small game, except migratory birds. https://cpw.state.co.us/Documents/RulesRegs/regulations/Ch03.pdf. Accessed 3 February 2021.

Covell, D. F. 1992. Ecology of the swift fox (*Vulpes velox*) in southeastern Colorado. MS thesis. University of Wisconsin, Madison.

Covell, D. F., D. S. Miller, and W. H. Karasov. 1996. Cost of locomotion and daily energy expenditure by free-living swift foxes (*Vulpes velox*): a seasonal comparison. *Canadian journal of Zoology* 74:283–90.

Criffield, M. A., M. V. Reichard, E. C. Hellgren, D. M. Leslie Jr., and K. Freel. 2009. Parasites of swift foxes (*Vulpes velox*) in the Oklahoma panhandle. *Southwestern Naturalist* 54:492–498.

Cullingham, C. I., and A. Moehrenschlager. 2019. Genetics of a reintroduced swift fox population highlights the need for integrated conservation between neighboring countries. *Animal Conservation* 22:611–21.

Cutter, W. L. 1958. Denning of the swift fox in northern Texas. *Journal of Mammalogy* 39:70–74.

Cypher, B. L. 2003. Foxes (*Vulpes* species, *Urocyon* species, and *Alopex lagopus*). In *Wild mammals of North America: biology, management, and economics*, 2nd ed., ed. G. A. Feldhamer, B. C. Thompson, and J. A. Chapman, 511–46. Baltimore: Johns Hopkins University Press.

Darden, S. K., T. Dabelsteen, and S. B. Pedersen. 2003. A potential tool for swift fox (*Vulpes velox*) conservation: individuality of long-range barking sequences. *Journal of Mammalogy* 84:1417–27.

Darden, S. K., L. K. Steffensen, and T. Dabelsteen. 2008. Information transfer among widely spaced individuals: latrines as a basis for communication networks in the swift fox? *Animal Behaviour* 75:425–32.

Dowd Stukel, E., ed. 2011. *Conservation assessment and conservation strategy for swift fox in the United States—2011 update*. Pierre: South Dakota Department of Game, Fish and Parks.

———ed., 2017. *Swift fox conservation team: report for 2015–2016*. Wildlife Division Report No. 2017-04. Pierre: South Dakota Department of Game, Fish and Parks.

———. 2020. South Dakota Game, Fish and Parks update, 2017–2018. In *Swift Fox Conservation Team: report for 2017–2018*, ed. H. F. Harris, 30–31. Glasgow: Montana Fish, Wildlife and Parks.

Dowd Stukel, E., and D. M. Fecske. 2007. *Swift fox conservation team: report for 2005–2006*. Pierre, SD: Department of Game, Fish, and Parks, and Bismarck, ND, Game and Fish Department.

Dragoo, J. W., J. R. Choate, and T. P. O'Farrell. 1987. Intrapopulational variation in two samples of arid-land foxes. *Texas Journal of Science* 39:223–32.

Dragoo, J. W., J. R. Choate, T. L. Yates, and T. P. O'Farrell. 1990. Evolutionary and taxonomic relationships among North American arid-land foxes. *Journal of Mammalogy* 71:318–32.

Dragoo, J. W., and R. K. Wayne. 2003. Systematics and population genetics of swift and kit foxes. In *The swift fox: ecology and conservation of kit foxes in a changing world*, ed. M. Sovada and L. Carbyn, 207–21. Regina: Canadian Plains Research Center, University of Regina.

Egoscue, H. J. 1979. *Vulpes velox. Mammalian Species* 122:1–5.

Finley, D. J., G. C. White, and J. P. Fitzgerald. 2005. Estimation of swift fox population size and occupancy rates in eastern Colorado. *Journal of Wildlife Management* 69:861–73.

Fisher, J. 1975. The plains dog moves east. *National Wildlife* 13:1417.

Flaherty, M., and R. Plakke. 1986. Response of the swift fox, *V. Velox*, to water stress. *Journal of the Colorado-Wyoming Academy of Science* 18:51.

Frey, J. K. 2004. Taxonomy and distribution of the mammals of New Mexico: an annotated checklist. *Occasional Papers, Museum of Texas Tech University* 240:1–32.

Fur Harvesters Auction Inc. 2020. Auction Results. www.furharvesters.com/auctionresults.html. Accessed 3 February 2021.

Gese, E. M., S. M. Karki, M. L. Klavetter, E. R. Schauster, and A. M. Kitchen. 2004. Serologic survey for canine infectious diseases among sympatric swift foxes (*Vulpes velox*) and coyotes (*Canis latrans*) in southeastern Colorado. *Journal of Wildlife Diseases* 40:741–48.

Gese, E. M, and C. M. Thompson. 2014. Does habitat heterogeneity in a multi-use landscape influence survival rates and density of a native mesocarnivore? *PLOS One* 9(6):e100500. doi:10.1371/journal.pone.0100500

Hall, E. R. 1981. *The mammals of North America*. 2nd ed. Vol. 2. New York: John Wiley.

Harris, H. F., ed. 2020. *Swift fox conservation team: report for 2017–2018*. Glasgow: Montana Fish, Wildlife and Parks.

Harrison, D. J., J. A. Bissonette, and J. A. Sherburne. 1989. Spatial relationships between coyotes and red foxes in eastern Maine. *Journal of Wildlife Management* 53:181–185.

Harrison, R. L. 2001. A literature review of swift fox diet and prey density studies. In *Swift Fox Conservation Team 2000 annual report*, ed. C. G. Schmitt and B. Oakleaf, 95–103. Sante Fe: New Mexico Department of Game and Fish.

———. 2003. Swift fox demography, movements, denning, and diet in New Mexico. *Southwestern Naturalist* 48:261–273.

Harrison, R. L., D. J. Barr, and J. W. Dragoo. 2002. A comparison of population survey techniques for swift foxes (*Vulpes velox*) in New Mexico. *American Midland Naturalist* 148:320–37.

Harrison, R. L., P. S. Clarke, and C. M. Clarke. 2004. Indexing swift fox populations in New Mexico using scats. *American Midland Naturalist* 151:42–49.

Harrison, R. L., and J. W. Hoagland. 2003. A literature review of swift fox habitat and den-site selection. In *The swift fox: ecology and conservation of kit foxes in a changing world*, ed. M. Sovada and L. Carbyn, 78–79. Regina: Canadian Plains Research Center, University of Regina.

Harrison, R. L., M. J. Patrick, and C. G. Schmitt. 2003. Foxes, fleas, and plague in New Mexico. *Southwestern Naturalist* 48:720–22.

Harrison, R. L., and C. G. Schmitt. 2003. Current swift fox distribution and habitat selection within areas of historical occurrence in New Mexico. In *The swift fox: ecology and conservation of kit foxes in a changing world*, ed. M. Sovada and L. Carbyn, 71–77. Regina: Canadian Plains Research Center, University of Regina.

Herrero, S. 2003. Canada's experimental reintroduction of swift foxes into an altered ecosystem. In *The swift fox: ecology and conservation of kit foxes in a changing world*, ed. M. Sovada and L. Carbyn, 33–38. Regina: Canadian Plains Research Center, University of Regina.

Hillman, C. N., and J. C. Sharps. 1978. Return of swift fox to northern Great Plains. *Proceedings of the South Dakota Academy of Science* 57:154–62.

Hines, T. D., and R. M. Case. 1991. Diet, home range, movements, and activity periods of swift fox in Nebraska. *Prairie Naturalist* 23:131–38.

Hubbard, J. P. 1994. The status of the swift fox in New Mexico. Unpublished report to the New Mexico Department of Game and Fish.

Jackson, V. L., and J. R. Choate. 2000. Dens and den sites of the swift fox, *Vulpes velox*. *Southwestern Naturalist* 45:212–20.

Jones, J. K., Jr., D. C. Carter, H. H. Genoways, R. S. Hoffmann, and D. W. Rice. 1982. Revised checklist of North American mammals north of Mexico, 1982. *Occasional Papers, Museum of Texas Tech University* 107:1–22.

Kamler, J. F., and W. B. Ballard. 2002. A review of native and nonnative red foxes in North America. *Wildlife Society Bulletin* 30:370–79.

———. 2003. White color phase of the swift fox, *Vulpes velox*. *Canadian Field-Naturalist* 117:468–69.

Kamler, J. F., W. B. Ballard, E. M. Gese, R. L. Harrison, and S. M. Karki. 2004a. Dispersal characteristics of swift foxes. *Canadian Journal of Zoology* 82:1837–42.

Kamler, J. F., W. B. Ballard, E. M. Gese, R. L. Harrison, S. Karki, and K. Mote. 2004b. Adult male emigration and a female-based social organization in swift foxes, *Vulpes velox*. *Animal Behaviour* 67:699–702.

Kamler, J. F., W. B. Ballard, R. L. Gilliland, P. R. Lemons II, and K. Mote. 2003. Impacts of coyotes on swift foxes in northwestern Texas. *Journal of Wildlife Management* 67:317–23.

Kamler, J. F., W. B. Ballard, R. L. Harrison, and C. Gregory Schmitt. 2005. Range expansion of red foxes in northwestern Texas and northeastern New Mexico. Southwestern Naturalist 56(1):100–101.

Kamler, J. F., W. B. Ballard, P. R. Lemons, and K. Mote. 2004c. Variation in mating system and group structure in two populations of swift foxes, *Vulpes velox*. *Animal Behaviour* 68:83–88.

Kamler, J. F., W. B. Ballard, M. C. Wallace, and P. S. Gipson. 2007. Diets of swift foxes (*Vulpes velox*) in continuous and fragmented prairie in northwestern Texas. *Southwestern Naturalist* 52:504–10.

Kansas Department of Wildlife and Parks. [2023]. Pdf 115-25-11. Furbearers; open seasons and limits. https://ksoutdoors.com/Services/Law-Enforcement/Regulations. Accessed 14 February 2023.

Karki, S. M., E. M. Gese, and M. L. Klavetter. 2007. Effects of coyote population reduction on swift fox demographics in southeastern Colorado. *Journal of Wildlife Management* 71:2707–18.

Kilgore, D. L., Jr. 1969. An ecological study of the swift fox (*Vulpes velox*) in the Oklahoma panhandle. *American Midland Naturalist* 81:512–34.

Kintigh, K. M., and M. C. Andersen. 2005. A den-centered analysis of swift fox (*Vulpes velox*) habitat characteristics in northeastern New Mexico. *American Midland Naturalist* 154:229–39.

Kitchen, A. M., E. M. Gese, and E. R. Schauster. 1999. Resource partitioning between coyotes and swift foxes: space, time, and diet. *Canadian Journal of Zoology* 77:1645–56.

Kitchen, A. M., E. M. Gese, S. M. Karki, and E. R. Schauster. 2005a. Spatial ecology of swift fox social groups: from group formation to mate loss. *Journal of Mammalogy* 86:547–54.

Kitchen, A. M., E. M. Gese, L. P. Waits, S. M. Karki, and E. R. Schauster. 2005b. Genetic and spatial structure within a swift fox population. *Journal of Animal Ecology* 74:1173–81.

Klausz, E. E. 1997. Small mammal winter abundance and distribution in the Canadian mixed grass prairies and implications for the swift fox. MS thesis, University of Alberta, Edmonton.

Laws, C. D. 2017. Status and long-term use of artificial escape dens by swift foxes in northwest Texas. MS thesis, Texas Tech University, Lubbock.

Lebsock, A. A., C. L. Burdett, S. K. Darden, T. Dabelsteen, M. F. Antolin, and K. R. Crooks. 2012. Space use and territoriality in swift foxes (*Vulpes velox*) in northeastern Colorado. *Canadian Journal of Zoology* 90:337–44.

Lemons, P. R., II. 2001. Ecology of swift foxes (*Vulpes velox*) in northwest Texas: diets and den site activity. MS thesis, Texas Tech University, Lubbock.

Malaney, J. L., J. L. Dunnum, and J. A. Cook. 2022. Checklist of New Mexico Mammals. *New Mexico Museum of Natural History and Science Bulletin* 88:361–70.

Matlack, R. S., P. S. Gipson, and D. W. Kaufman. 2000. The swift fox in rangeland and cropland in western Kansas: relative abundance, mortality, and body size. *Southwestern Naturalist* 45:221–25.

McGee, B. K., W. B. Ballard, and K. L. Nicholson. 2007. Swift fox, *Vulpes velox*, den use patterns in northwestern Texas. *Canadian Field-Naturalist* 121:71–75.

McGee, B. K., W. B. Ballard, K. L. Nicholson, B. L. Cypher, P. R. Lemons II, and J. F. Kamler. 2006. Effects of artificial escape dens on swift fox populations in northwest Texas. *Wildlife Society Bulletin* 34:821–27.

Mercure, A., K. Ralls, K. P. Koepfli, and R. K. Wayne. 1993. Genetic subdivisions among small canids: mitochondrial DNA differentiation of swift, kit, and arctic foxes. *Evolution* 47:1313–28.

Merriam, C. H. 1902. Three new foxes of the kit and desert fox groups. *Proceedings of the Biological Society of Washington* 15:73–74.

Miller, D. S., B. G. Campbell, R. G. McLean, E. Campos, and D. F. Covell. 1998. Parasites of swift fox (*Vulpes velox*) from southeastern Colorado. *Southwestern Naturalist* 43:476–79.

Miller, D. S., D. F. Covell, R. G. McLean, W. J. Adrian, M. Niezgoda, J. M. Gustafson, O. J. Rongstad, R. D. Schultz, L. J. Kirk, and T. J. Quan. 2000. Serologic survey for selected infectious disease agents in swift and kit foxes from the western United States. *Journal of Wildlife Disease* 36:798–805.

Mitchell, E. L. 2018. Distribution, ecology, disease risk, and genetic diversity of swift fox (*Vulpes velox*) in the Dakotas. MS thesis, South Dakota State University, Brookings. https://openprairie.sdstate.edu/etd/2692

Moehrenschlager, A., and M. Sovada. 2004. Swift fox *Vulpes velox*. In *Canids: foxes, wolves, jackals, and dogs: status survey and conservation action plan*, ed. Sillero-Zubiri, C., M. Hoffmann, and D. W. Macdonald, 109–16. 2nd ed. Gland, Switzerland, and Cambridge, UK: IUCN/SSC Canid Specialist Group.

Montana Fish, Wildlife and Parks. [2023]. 2022 Wolf Furbearer Trapping. https://fwp.mt.gov/binaries/content/assets/fwp/hunt/regulations/2022/wolf-and-furbearer-final-for-web.pdf. Accessed 14 February 2023.

Nebraska Game and Parks Commission. [2021]. Furbearer Species. outdoornebraska.gov/FurbearerSpecies. Accessed 3 February 2021.

Nevison, S., J. J. Jenks, E. Childers, and J. Delger. 2017. Swift fox monitoring in southwestern South Dakota. In *Swift Fox Conservation Team: report for 2015–2016*, ed. E. Dowd Stukel, 67–68. Wildlife Division Report No. 2017-04. Pierre: South Dakota Department of Game, Fish and Parks.

New Mexico Department of Game and Fish (NMDGF). 2020. *2020–2021 New Mexico Furbearer Rules and Info*. Santa Fe, NM: New Mexico Department of Game and Fish.

North Dakota Game and Fish Department. [2021]. Species of Conservation Priority. https://gf.nd.gov/wildlife/scp. Accessed 3 February 2021.

Oklahoma Department of Wildlife Conservation. [2023]. Furbearer Regulations. https://www.wildlifedepartment.com//hunting/regs/furbearercoyote-regulations. Accessed 14 February 2023.

Olson, T. L. 1999. *Swift fox use of habitats, foods, and security cover in southeast Wyoming*. Progress report, Wyoming Cooperative Fish and Wildlife Research Unit, University of Wyoming, Laramie.

———. 2000. Population characteristics, habitat selection patterns, and diet of swift foxes in southeast Wyoming. MS thesis. University of Wyoming, Laramie.

Olson, T. L., and F. G. Lindzey. 2002. Swift fox survival and production in southeastern Wyoming. *Journal of Wildlife Management* 83:199–206.

Pence, D. B., J. F. Kamler, and W. B. Ballard. 2004. Ectoparasites of the swift fox in northwestern Texas. *Journal of Wildlife Diseases* 40:543–47.

Pruss, S. D. 1994. An observational natal den study of wild swift fox (*Vulpes velox*) on the Canadian Prairie. Master's thesis (Environmental Design), University of Calgary, Alberta.

———. 1999. Selection of natal dens by the swift fox (*Vulpes velox*) on the Canadian prairies. *Canadian Journal of Zoology* 77:646–52.

Pybus, M. J., and E. S. Williams. 2003. A review of parasites and diseases of wild swift fox. In *The swift fox: ecology and conservation of kit foxes in a changing world*, ed. M. Sovada and L. Carbyn, 231–36. Regina: Canadian Plains Research Center, University of Regina.

Ralls, K., and P. J. White. 1995. Predation on San Joaquin kit foxes by larger canids. *Journal of Mammalogy* 76:723–29.

Rand, A. L., 1948. *Mammals of the eastern Rockies and western plains of Canada*. National Museum of Canada bulletin 108, Biological Series No. 35.

Robinson, W. B. 1953. Population trends of predators and fur animals in 1080 station areas. *Journal of Mammalogy* 34:220–27.

Rohwer, S. A., and D. L. Kilgore Jr. 1973. Interbreeding in the arid-land foxes, *Vulpes velox* and *Vulpes macrotis*. *Systematic Zoology* 22:157–65.

Rongstad, O. J., T. R. Laurion, and D. E. Andersen. 1989. *Ecology of swift fox on the Pinon Canyon Maneuver Site, Colorado*. Final report, Directorate of Engineering and Housing, Fort Carson, CO.

Rosenstock, S. S., W. B. Ballard, and J. C. deVos Jr. 1999. Viewpoint: benefits and impacts of wildlife water developments. *Journal of Range Management* 52:302–11.

Russell, T. A. 2006. Habitat selection by swift foxes in Badlands National Park and the surrounding area in South Dakota. MS thesis, South Dakota State University, Brookings.

Ruxton G. D., M. P. Speed, and D. Kelly. 2004. What, if anything, is the adaptive function of counter-shading? *Animal Behaviour* 68:445–51.

Sasmal, I., J. A. Jenks, T. W. Grovenburg, S. Datta, G. M. Schroeder, R. W. Klaver, and K. M. Honness. 2011. Habitat selection by female swift foxes (*Vulpes velox*) during the pup-rearing season. *Prairie Naturalist* 43(1–2):29–37.

Sasmal, I., J. A. Jenks, L. P. Waits, M. G. Gonda, G. M. Schroeder, and S. Datta. 2013. Genetic diversity in a reintroduced swift fox population. *Conservation Genetics* 14:93–102.

Schauster, E. R., E. M. Gese, and A. M. Kitchen. 2002. Population ecology of swift foxes (*Vulpes velox*) in southeastern Colorado. *Canadian Journal of Zoology* 80:307–19.

Schmitt, C. G. 1995. Swift fox investigations in New Mexico, 1995. In *Report of the Swift Fox Conservation Team 1995*, ed. S. H. Allen, J. W. Hoagland, and E. Dowd Stukel, 27–332.

Schroeder, C. 1985. A preliminary management plan for securing swift fox, for reintroduction into Canada. MS thesis, University of Calgary, Calgary, Alberta.

Schroeder, G. M. 2007. Effects of coyotes and release site selection on survival and movement of trans-located swift foxes in the badlands ecosystem of South Dakota. MS thesis, South Dakota State University, Brookings.

Schwalm, D., L. P. Waits, and W. B. Ballard. 2014. Little fox on the prairie: genetic structure and diversity throughout the distribution of a grassland carnivore in the United States. *Conservation Genetics* 15:1503–14.

Schwalm, D., H. Whitlaw, W. Ballard, E. B, Fish, and H. A. Whitlaw. 2012. Distribution of the swift fox in Texas. *Southwestern Naturalist* 57:393–98.

Scott-Brown, J. M., S. Herrero, and J. Reynolds. 1987. Swift fox. In *Wild furbearer management and conservation in North America*, ed. M. Novak, J. A. Baker, M. E. Obbard, and B. Malloch, 433–41. Ontario, Canada: Ministry of Natural Resources.

Seton, E. T. 1937. *Lives of game animals. Vol. 1, Cats, wolves, and foxes.* New York: Literary Guild of America.

Sheldon, J. 1992. *Wild dogs: the natural history of the non-domestic Canidae.* San Diego: Academic Press.

Smeeton, C., K. Weagle, and S. S. Waters. 2003. Captive breeding of the swift fox at the Cochrane Ecological Institute, Alberta. In *The swift fox: ecology and conservation of kit foxes in a changing world*, ed. M. Sovada and L. Carbyn, 199–203. Regina: Canadian Plains Research Center, University of Regina.

Sooter, C. A. 1943. Speed of predator and prey. *Journal of Mammalogy* 24:102–3.

South Dakota Department of Game, Fish and Parks. [2021]. Trapping. https://gfp.sd.gov/trapping. Accessed 3 February 2021.

Sovada, M. A., C. C. Roy, J. B. Bright, and J. R. Gillis. 1998. Causes and rates of mortality of swift foxes in western Kansas. *Journal of Wildlife Management* 62:1300–306.

Sovada, M. A., C. C. Roy, and D. J. Telesco. 2001. Seasonal food habits of swift fox (*Vulpes velox*) in cropland and rangeland landscapes in western Kansas. *American Midland Naturalist* 145:101–11.

Sovada, M. A., C. C. Slivinski, R. O. Woodward, and M. L. Phillips. 2003. Home range, habitat use, litter size, and pup dispersal of swift foxes in two distinct landscapes of western Kansas. In *The swift fox: ecology and conservation of kit foxes in a changing world*, ed. M. Sovada and L. Carbyn, 149–60. Regina: Canadian Plains Research Center, University of Regina.

Sovada, M. A., R. O. Woodward, and L. D. Igl. 2009. Historical range, current distribution, and conservation status of swift fox, *Vulpes velox*, in North America. *Canadian Field-Naturalist* 123:346–67.

Stromberg, M. R., and M. S. Boyce. 1986. Systematics and conservation of the swift fox, *Vulpes velox*, in North America. *Biological Conservation* 35:97–110.

Stuart, J. N. 2006. Swift fox research in New Mexico: 2004 update. In *Swift Fox Conservation Team: annual report for 2004.*, ed. J. N. Stuart and S. Wilson, 11–13. Sante Fe: New Mexico Department of Game and Fish, and Lincoln: Nebraska Game and Parks Commission.

———. 2008. Swift fox surveys and other activities in New Mexico, 2006–2007. In *Swift Fox Conservation Team: report for 2007*, ed. B. Krueger and M.

Ewald, 7–12. Laramie: Wyoming Game and Fish Department.

———. 2011. Swift fox conservation activities in New Mexico, 2009–2010. In *Swift Fox Conservation Team: report for 2009–2010*, ed. K. Bly, 19–22. Bozeman, MO: World Wildlife Fund, and Helena: Montana Department of Fish, Wildlife and Parks.

Stuart, J. N., and S. M. Murphy. 2017. Status of swift foxes in New Mexico: 2016 update. In *Swift Fox Conservation Team: report for 2015–2016*, ed. E. Dowd Stukel, 35–37. Wildlife Division Report No. 2017-04. Pierre: South Dakota Department of Game, Fish and Parks.

Swift Fox Conservation Team. 1997. *Conservation assessment and conservation strategy for swift fox in the United States* [ed. R. L. Kahn, L. Fox, P. Horner, B. Giddings, and C. Roy]. Pierre: South Dakota Department of Game, Fish, and Parks.

Texas Parks and Wildlife Department. [2023]. https://tpwd.texas.gov/regulations/outdoor-annual/hunting/nongame-and-other-species. Accessed 14 February 2023.

Thompson, C. M. 2006. Landscape-level influences on swift fox (*Vulpes velox*) demographics in southeastern Colorado. PhD dissertation, Utah State University, Logan.

Thompson, C. M., and E. M. Gese. 2007. Food webs and intraguild predation: community interactions of a native mesocarnivore. *Ecology* 88:334–46.

———. 2012. Swift foxes and ideal free distribution: relative influence of vegetation and rodent prey base on swift fox survival, density, and home range size. *International Scholarly Research Network Zoology*. Volume 2012, Article ID 197356. doi:10.5402/2012/197356.

Thornton, W. A., and G. C. Creel. 1975. The taxonomic status of kit foxes. *Texas Journal of Science* 26:127–136.

Tucker, S. 2020. 2017–2018 North Dakota report to the Swift Fox Conservation Team. In *Swift fox conservation team: report for 2017–2018*, ed. H. F. Harris, 25–29. Glasgow: Montana Fish, Wildlife and Parks.

Ubelaker, J. E., B. S. Griffin, K. M. Mendoza, D. W. Duszynski, and R. L. Harrison. 2014. Distributional records of helminths of the swift fox (*Vulpes velox*) from New Mexico. *Southwestern Naturalist* 59:129–32.

United States Fish and Wildlife Service (USFWS). 1982. Endangered and threatened wildlife and plants; Review of vertebrate wildlife for listing as endangered or threatened species. *Federal Register* 47:58454–59 (30 December 1982).

———. 1995. Endangered and threatened wildlife and plants: 12-month finding for a petition to list the swift fox as endangered. *Federal Register* 60:31663–66 (16 June 1995).

———. 2001. Endangered and threatened wildlife and plants: annual notice of findings on recycled petitions. *Federal Register* 66:1295–300 (8 January 2001).

Voigt, D. R., and B. D. Earle. 1983. Avoidance of coyotes by red fox families. *Journal of Wildlife Management* 47:852–57.

Wayne, R. K., and D. M. Brown. 2001. Hybridization and conservation of carnivores. In *Carnivore Conservation.*, ed. J. L. Gittleman, S. M. Funk, D. Macdonald, and R. K. Wayne, 145–62. Cambridge: Cambridge University Press.

Wooley, T. P., F. G. Lindzey, and R. Rothwell. 1995. Swift fox surveys in Wyoming—annual report. In *Report of the Swift Fox Conservation Team 1995*, ed. S. H. Allen, J. W. Hoagland, and E. Dowd Stukel, 60–80.

Wyoming Game and Fish Department. [2021]. Swift fox, *Vulpes velox*. https://wgfd.wyo.gov/wgfd/media/content/pdf/habitat/SWAP/Mammals/Swift-fox.pdf. Accessed 3 February 2021.

Young, S. P. 1944. *The wolves of North America*. Pt. 1, *Their history, life habits, economic Status, and control*. Washington, DC: American Wildlife Institute.

Zimmerman, A. L. 1998. Reestablishment of swift fox in north central Montana. MS thesis, Montana State University, Bozeman.

Zumbaugh, D. M. and J. R. Choate. 1985. Winter food habits of the swift fox on the central high plains. *Prairie Naturalist* 17:41–47.

NORTH AMERICAN RED FOX (*VULPES FULVA*)

Jennifer K. Frey, Matthew E. Gompper, and Jean-Luc E. Cartron

The North American red fox (*Vulpes fulva*) is almost certainly the most misunderstood of the carnivore species occurring in New Mexico. Until recently, much of the popular and scientific literature portrayed that some, or even all, red foxes in North America were not native (e.g., Presnall 1958; Kamler and Ballard 2002). The notion persisted that American colonists imported red foxes from England for fox hunting (see Frey 2013), and that these exotic foxes then spread westward across the continent as an invasive species upsetting natural ecosystems and requiring population control (GISD 2012; Kamler and Ballard 2002; Lewis et al. 1999). Based on genetic studies we now know that this widely held belief was almost totally incorrect and that red foxes in North America are indeed native (Statham et al. 2012, 2014). However, the faulty ideas and negative attitudes toward red foxes prevailed for many decades, and likely resulted in depressed scientific interest in the North American red fox. This in turn likely forestalled conservation measures for some populations that are threatened by habitat loss and other factors. Even today in New Mexico, the biology and conservation status of the North American red fox are mostly unknown. This chapter aims to reverse that.

While the red fox is fairly common across much of North America, it is less common to come across the species in the more arid southwestern United States. As a result, many fail to recognize its occurrence, perhaps mistaking it for the similar-sized and more widespread gray fox (*Urocyon cinaeroargenteus*) or the smaller but locally abundant kit fox (*V. macrotis*). This lack of recognition of southwestern red foxes is reflected in the scientific literature as well. While there is a large body of research on red foxes in North America (as well as on the closely related *V. vulpes* in Europe, Asia, and Australia), there has been almost no study of the ecology of the species in the southwestern United States. In part this may be because of the mistaken consideration that the species is an exotic or introduced denizen of the region (see Frey 2013). And in part this may be because of a belief that the species is so well studied elsewhere that there is little to gain by examining it in New Mexico. Such mindsets are unfortunate, as what little we know about the biology of the red fox in New Mexico suggests there is indeed much to learn. This includes how this animal makes a living in arid-land ecosystems, as well as whether New Mexico contains unique populations worthy of special management attention.

Originally, two species of red foxes were recognized. The Eurasian red fox was described in 1758 as *Canis vulpes* by the originator of the modern system of taxonomy, Carl Linnaeus (Linnaeus 1758). Early naturalists and explorers in

(*opposite page*) Photograph: © Roger Hogan.

North America found a similar, yet distinct type of red fox along the eastern seaboard of the United States, which was named *Canis fulvus* in 1820 (Desmarest 1820:203). Both species of red foxes were then placed in the genus *Vulpes*, the true foxes, to separate them from the dogs and wolves in the genus *Canis* (Bowditch 1821). Thus, the names became *V. vulpes* and *V. fulva*, with the later change due to the requirement that the gender of the species name match the genus name. Compared to red foxes from Europe (*V. v. crucigera* [Bechstein 1789]), *V. fulva* of southeastern North America is smaller with a shorter tail, the feet and legs are more blackish, the rusty color of the face is more pale and mixed with white hairs, and cranial and dental characters are distinct (Baird 1857; Merriam 1900; Churcher 1959). As Baird (1857:126) declared, "There is never any difficulty in separating skins of the two kinds by their external appearance when mixed with each other." Bangs (1897) further thought that the skulls of the two species could always be distinguished. The North American red fox has a more delicate and narrower skull with a more pointed muzzle (Baird 1857). Whereas the profile of the skull across the muzzle and head is straight in the Eurasian red fox, it is angled in the North American red fox (Bangs 1897). No museum specimens of the European red fox, *V. v. crucigera*, has ever been collected in North America (Frey 2013).

However, several things happened to shake this well-founded scientific view. Most importantly, there came into being a common belief that the American colonists imported red foxes from England for equestrian sport hunting with hounds. It was often presumed that the colonists' reasoning was that the gray fox was the only species of fox occurring in the colonies and that its manner of flight did not provide as much sport, both of which were arguably incorrect (Frey 2013). Second, people reported observations of red foxes occurring in places where they did not think they historically occurred, which suggested

the red fox was expanding its range (Churcher 1959). Meanwhile, beginning in the mid-19th century other forms of red foxes were discovered by explorers in the mountains of western North America, notably the Rocky Mountain red fox (*V. f. macroura*; Baird 1852). Because Rocky Mountain red foxes were discovered in locations that had not yet been colonized by Europeans, it followed that they could not have been imported and therefore they were presumed to be native (Kamler and Ballard 2002). In addition, red foxes have valuable pelts, which led to the development of captive farming of red foxes on Prince Edward Island in southeastern Canada in the 1890s, which then spread to many parts of North America and eventually other continents (Westwood 1989). Releases of animals from fur farms or translocations of red foxes as game species resulted in human-mediated movements of red foxes from eastern North America to locations further west (Statham et al. 2012). Taken together, these factors eventually led to the extreme idea that all red foxes in North America are descended from introduced European red foxes, except for relic populations of native foxes at high elevation in the mountains of western North America (Kamler and Ballard 2002). The expansion of these "exotic" European red foxes across North America was thought to be due to differences in morphology, behavior, and habitat that allowed them to be both successful and harmful to other wildlife (Kamler and Ballard 2002).

Fortunately, genetic data are revealing the true story about the origins of red foxes in North America, though the story is not yet complete and some findings vary among studies. The overarching pattern is that all red foxes originated in the Middle East. At or before the Illinoian glacial period, foxes from Asia colonized North America across the Bering Strait, giving rise to two evolutionary lineages that diverged when they became reproductively isolated, and today these correspond to the Eurasian red fox, *V. vulpes*, and

Photo 15.1. North American red fox near Chama, Rio Arriba County, in January 2010. The red fox is a medium-sized canid with adaptation for living in cold and snowy environments, including a thick pelage, dense fur around the toes, allowing the feet to serve as snowshoes, and a bushy tail that is used to cover the face and provide insulation when the animal is curled up and resting. Photograph: © Roger Hogan.

the North American red fox, *V. fulva* (Aubrey et al. 2009; Statham et al. 2014; Sacks et al. 2018). Populations of red foxes in Alaska and western Canada exhibit a minority of admixture and mitochondrial DNA haplotypes related to those found in East Asian red foxes. This indicates some historical transfer of genetic material from one species to the other through hybridization during the late Pleistocene (Sacks et al. 2018). However, this ancient hybridization does not obscure the overriding nuclear DNA (i.e., the main genetic material of an organism, which comprises the chromosomes inside the nucleus of cells) pattern of two distinct species, each of which is endemic to its own continent (Aubry et al. 2009; Statham

et al. 2014; Sacks et al. 2018). The continental division of *V. fulva* and *V. vulpes* is a pattern of diversification similar to that of other northern latitude trans-Beringian species groups (Waltari et al. 2007). However, not all authorities currently recognize both species and instead some refer to all red foxes, including the North American species, as *V. vulpes*.

Genetic studies have revealed maternally inherited European mtDNA in 17% of red foxes from localized areas in the eastern United States, mostly from urban areas east of the Appalachian Mountains in a small area of the mid-Atlantic states (New Jersey's eastern shore, Chesapeake Bay, and southern Maryland), as well as the

Photo 15.2. Skull of a female North American red fox (lateral view) collected on 18 December 1996 southeast of Gallup in McKinley County (Museum of Southwestern Biology catalog number 142679). Compared to other species of foxes, red foxes have relatively massive skulls that provide for a powerful bite force and also include a large brain cavity. Such skull features are typical of canids that specialize on hunting mammal prey, such as coyotes (*Canis latrans*) and wolves (*C. lupus*). The red fox is more carnivorous than other species of fox, and it specializes on hunting small and medium-sized mammals. Photograph: © Jon Dunnum.

Hudson lowlands of western New York (Kasprowicz et al. 2016). Previously, Frey (2013) reviewed the evidence for importation of red foxes from Europe into the United States and traced it all to two second-hand sources that identified early colonial introduction sites as New York and the eastern shore of Maryland. After importation, the foxes were reported to have dispersed to Virginia by crossing the Chesapeake Bay when it was frozen. Thus, the genetic data support the stories of importation of Eurasian red foxes brought from Europe and later successfully breeding with North American red foxes (Kasprowicz et al. 2016). Additional genetic data from the male-inherited Y-chromosome indicate that introgression of alleles from European to North American red foxes was limited and that male offspring produced through hybridization between the two species may be infertile (Kuo et al. 2019). Thus, the genetic data, while acknowledging that hybridization did historically occur in localized

areas of the eastern United States, also have demonstrated that red foxes in North America are a distinct species that has not been displaced or subjected to extensive hybridization with European red foxes. Put simply, North American red foxes are native and represent a distinct species, *V. fulva*.

DESCRIPTION OF THE SPECIES

The North American red fox is a large species of fox, second in size only to some forms of the Eurasian red fox (Castelló 2018). Measurements representative of the two subspecies reported from New Mexico (*V. f. fulva* and *V. f. macroura*) are 56–72 cm (22–28 in) for body length; 31–40 cm (12–16 in) for tail length, 38–41 cm (11–16 in) for height at the shoulder, and 3–7 kg (6.6–15.4 lbs) for weight (Castelló 2018). Males are slightly larger than females (Bailey 1931). The species has an overall body plan typical of the Canidae, or dog family, which evolved in open grasslands and have adaptations for cursorial (i.e., running) locomotion. It is associated with a digitigrade posture (i.e., North American red foxes walk on their toes), with relatively long limbs and blunt non-retractable claws. The limbs are restricted in motion to mainly the sagittal plane (forward and backward). The tail is moderately long, which serves as a counterbalance during rapid changes in direction. The senses are keen and the ears are conspicuous and upright. The skull is robust with prominent ridges for the insertion of powerful jaw muscles that generate a large bite force (Photo 15.2). The muzzle is elongated and harbors 42 teeth with diversified function, including large, sharp canine teeth (often called fangs), which are used to kill and hold prey, and the carnassials, which are a special pair of large, powerful teeth consisting of the fourth upper premolar and first lower molar that together function as scissors to shear chunks of meat.

The pelage of the North American red fox is among the most beautiful and highly desired of pelts. The species has adaptations for cold, snowy environments and consequently the hair is relatively long, silky, and lustrous, and typically appears fluffy. The feet are densely furred, causing the near obliteration of the toe pads, which serve as snowshoes (Grinnell et al. 1937). The pelage is especially well developed in the winter. For instance, there is usually a ruff of long silky hairs around the top of the eyes and cheeks, and this is especially prominent in winter. The tail has exceptionally long hairs and is proportionately bushier than in other species of canids. The tail has the greatest thermoregulatory effectiveness and is used when the animal is curled at rest to protect other parts of the body that have relatively short hair and lose the greatest amount of heat in the cold, such as the face, lower limbs, and paws (Klir and Heath 1992).

Three principal pelage color patterns are recognized in the North American red fox, though additional color phases exist, together with individual variation within phases. In fact, variation in color can be so extreme that the fur industry sometimes refers to a wild red fox as a "colored fox" (Voigt 1999). The Latin word *fulva* refers to deep yellow, reddish yellow, gold, or tawny (i.e., yellowish brown). Indeed, the name *V. fulva* describes well aspects of the typical "red phase" color pattern of the North American red fox. The red phase is the most common pelage color phase, and it is the only one found in some warmer regions, such as much of the eastern United States. This pelage phase is colorful, bright, and varied. The overall impression is that the animal is reddish. However, yellowish hues are typically present on the hind-end, tail, and especially the ruff around the forehead. The shoulders are usually a bright rich orange. The hind parts typically appear tinged with gray, which is usually darkest on the tail, sometimes appearing a deep charcoal or almost black. The lower face and the neck and chest are usually whitish, and this may extend to the inner legs and underparts. The outer legs are

Photo 15.3. North American red fox near Questa in Taos County on 3 June 2020. The scientific name of the North American red fox, *Vulpes fulva*, refers to the yellowish coloration of its silky pelage, especially in the ruff of fur around the eyes. In contrast, the Eurasian red fox (*V. vulpes*) typically has pelage that is relatively dull and is more uniformly coppery or brownish orange. Photograph: © J. T. Radcliff and Jerry Hardy.

usually blackish, but this is highly variable and many North American red foxes in the American Southwest have very reduced, or are almost entirely lacking, black on the legs. The back of the ears is black (though this may be reduced in the Southwest), and the tail is usually tipped with white (though the white tip is absent in some cases).

The other two main color phases of the North American red fox are the "silver phase" and the "cross phase." Originally these two phases were thought to be different species, C. (*Vulpes*) *argentatus* and C. (*Vulpes*) *decussatus*, respectively (Bangs 1897), until it was discovered that all three phases could occur in a single litter (Cross 1941). It is now known that the three phases are an inherited trait determined by two genes: the typical red phase is the result of all dominant alleles being present (homozygous dominant); the silver phase has all recessive alleles for at least one of the genes (homozygous recessive); and the cross

Photos 15.4a and b (*left to right*). Two juvenile, littermate red foxes in Gallup, McKinley County, on 10 May 2015. North American red foxes exhibit considerable color variation, including three primary phases. Both cubs (a and b) belong to the red phase, but are relatively pale. (a) This cub displays classic North American red fox characteristics of reddish pelage, black legs, black behind the ears, and a distinctly white-tipped tail. (b) This cub is paler than its littermate and has greatly reduced black on the legs and a grayish tail, as often observed in New Mexico. Photographs: © David Mikesic.

Photo 15.5. Silver phase North American red fox captured on 29 October 2008 in a pigeon coop in Raton, Colfax County, by the New Mexico Department of Game and Fish and later released unharmed. The pelage of the silver phase North American red fox is commercially the most valuable and was the focus in the development of fur-farming. The North American red fox primarily consumes medium and small mammal prey, but it is an opportunistic predator and will take birds when available. Photograph: © Clint Henson.

Photo 15.6. A cross phase Rocky Mountain red fox (*V. f. macroura*) near Platoro in Conejos County, Colorado on 6 July 2021. The cross phase color variant is blackish across most of the body except the ears, side of the face and neck, lower side of the body, and back of the upper thighs. Cross phase North American red foxes are the result of mating between red and silver phase parents (in such a case, all three phases may be represented in a litter). Photograph: © Mark L. Watson.

Photo 15.7. First record of a cross phase North American red fox in New Mexico; captured by remote camera on 30 May 2017 near Williams Lake at ~3,360 m (11,040 ft) elevation in the Sangre De Cristo Mountains. Note the black underside of the neck and chest. Photograph: © Brian Jay Long.

phase is in between, typically having a dominant and recessive allele (heterozygous) for both genes (Voigt 1999). The silver phase is black with the exception of the white tip to the tail and light frosting (i.e., silver) on the tips of some or all of the guard hairs of the ruff around the face, the body, and sometimes the tail (the muzzle, ears, legs, and belly are usually black). A "black fox" is a silver phase that lacks the light tips to the guard hairs. The cross phase has characteristics of both the red and silver phase. Its face and tail are generally black, and the midline of the back and shoulders are blackish or gray in a cross-like pattern (Obbard 1999).

The dark phases associated with the recessive alleles are more commonly associated with boreal zones, and hence are more common at higher latitudes and at high elevations in the various mountain chains extending southward into the United States. For instance, in the extreme northern and northwestern parts of the range of the North American red fox, populations consist of 60% red, 35% cross, and 5% silver; while in the southern low-elevation portions of the range 99% of individuals are red (Obbard 1999). The silver phase is the rarest and most valuable pelage. Because of its natural rarity and commercial value, this phase was the primary focus for development of fox fur-farming (Johansson 1947; Voigt 1999). Obbard (1999) thought that wild cross phase foxes only occurred at northern latitudes and that any wild-caught cross foxes from the lower 48 states would be escapees from fur farms. However, that view is erroneous, as the cross phase can naturally occur from a mating between red and silver phase foxes at any location (Obbard 1999). The silver and cross phases are known to occur naturally in the mountains of the western United States (e.g., Cowan 1938; Armstrong 1972). In New Mexico, most red foxes are red phase, though silver and cross foxes have been documented (Photos 15.5 and 15.7).

Besides the three main color phases, there are additional variants. In the arid Great Basin desertlands of northwestern New Mexico and northeastern Arizona, there also exists a blond color phase that is associated with a very pale pelage, often greatly suffused with gray, and with greatly reduced black on the legs and ears (Mikesic and LaRue 2003). In addition to the main color phases and their variants found in wild red foxes, selective breeding in fur farms has resulted in a variety of additional variation and expression of mutant genes that are not found in the wild populations, such as white (complete or partial albinism) and piebald (i.e., white spots; Obbard 1999; Ratliff 2011). Red foxes have been domesticated (Trut 1999), and these unusual pelage colors are typical of domesticated red foxes sold in the pet trade (note that the ownership of red foxes is not legal in New Mexico).

COMPARISON WITH OTHER SPECIES
Because of the extreme variation in pelage color patterns, identifying red foxes in New Mexico is not always straightforward and specimens

Photo 15.8. North American red fox in Rio Arriba County on 3 May 2009. In New Mexico, pale red foxes with greatly reduced black on the ears and legs can resemble young coyotes (*Canis latrans*), particularly when there is no size reference. However, the red fox has a longer tail that reaches the ground when standing, the legs are usually darker, instead of lighter, compared to the rest of the body, and the tail is usually tipped in white rather than black (but note that coyotes with a white-tipped tail have been reported and some red foxes lack the white tip). Photograph: © Roger Hogan.

should be closely compared with other species of foxes as well as with young coyotes (*C. latrans*). Many reports of "red fox" are misidentifications of gray foxes, because both species can exhibit extensive areas of reddish pelage (see also Chapter 12). However, in the gray fox the reddish pelage is always limited to the head, neck, and undersides (sometimes including the legs and tail), while in the red fox the reddish pelage is more extensive on the body (except in the silver and some cross phases). The back of the ears of a red fox is blackish, while the ears are reddish in the gray fox. In the red fox, the legs are usually (but not always) gray or black and are darker than the body, while in the gray fox the legs are usually reddish or grayish and are paler or the same tone as the body. Further, in the red fox, the tail usually ends in a white or pale tip, while in the gray fox the tail usually has a thin black dorsal stripe and black tip. The red fox is also larger than a gray fox, has relatively longer limbs and shorter tail, and has a more dog-like face.

Young and pale-colored red foxes can appear similar to the swift fox (*V. velox*) and kit fox (Chapters 13 and 14). The swift fox and kit fox are smaller than the red fox. Although both are similar to the red fox in that they show extensive reddish and grayish pelage, the swift and kit foxes have pale reddish legs and black tips on the tail. The kit fox usually appears to have proportionally longer ears. The skull of the red fox is obviously larger and more massive than those of the swift fox or kit fox (Creel and Thornton 1971). Based on morphological evidence, Thornton et al. (1971)

described putative hybrids between the red fox and the kit fox in western Texas (Thornton et al. 1971). These reported hybrids are similar to the kit fox in general appearance, but have pelage characteristics similar to the red fox's body coloration of predominantly red with gray tinge (the opposite of the kit fox); the tail resembles that of a kit fox and lacks or has a greatly reduced white tip; the backs of the ears and feet are blackish as in the red fox (Thornton et al. 1971). Cases of putative hybrids between the red fox and the arctic fox (*V. lagopus*) have also been described, but further genetic analyses have shown these cases to be erroneous (Yannic et al. 2017). Thus, further studies may be required to confirm the presence of red fox–kit fox hybrids.

It may be possible to confuse a pale blonde red fox, which also tends to have reduced melanism in the ears and feet, as a juvenile coyote. Red foxes have relatively longer tails, which can touch the ground while the animal is standing, while the tails of coyotes and wolves are proportionately shorter, usually extending only to the hock when standing. In addition, the legs of coyotes are as pale, or paler, than the body and may have a reddish tinge. Further, for coyotes, the tip of the tail is usually black, though coyotes with a white-tipped tail have been reported (see Chapter 10), and some red foxes lack the white-tipped tail.

SUBSPECIES

The small number of red fox museum specimens that have been collected in New Mexico has hindered the ability to evaluate subspecies within the state. In fact, there has been no comprehensive evaluation of subspecies in New Mexico based on an examination of specimens. However, recent genetic studies have provided valuable new information about the degree of relationship among populations as well as the taxonomic status of red foxes in New Mexico (Aubry et al. 2009; Sacks et al. 2010; Statham et al. 2012, 2014; Merson et al. 2017; Sacks et al. 2018). Evidence

suggests that two or three subspecies of North American red foxes occur in New Mexico.

Aubry et al. (2009) first elucidated the evolutionary history of red foxes in North America. Ancestral red foxes originally colonized North America from Eurasia 130,000–300,000 years ago across the Bering land bridge during the Illinoian glacial period (Nearctic clade). During the subsequent Wisconsin glacial period 10,000–100,000 years ago (the most recent glacial period), red foxes stemming from the first colonization were isolated into two different regions: one in the Southwest (Mountain subclade) and one in the Southeast (Eastern subclade). More recent genetic analyses indicate that during the same time, red foxes in Alaska had secondary contact with the Eurasian red fox (Holarctic clade) when Alaska was separated from the rest of North America by ice sheets, but connected to Asia by the Bering land bridge (Staham et al. 2014; Sacks et al. 2018). Since melting of the Wisconsin glaciers and warming of the climate during our current interglacial, these subclades have shifted distributions tracking suitable climates. Those in the south tracked their preferred cool climate northward or higher in elevation and those in the north expanded their range southward to occupy habitat exposed by the melting glaciers. Thus, the Mountain subclade retreated to the tops of mountains in the western United States, the Eastern subclade expanded northward to occupy the eastern United States and Canada, and the Alaskan red foxes spread southward throughout Alaska and into western Canada.

The Mountain subclade of the North American red fox persists primarily as relict populations in high-elevation alpine and subalpine habitats in the western United States, and these have differentiated into three distinct subspecies: *V. f. macroura* in the Rocky Mountains, *V. f. cascadensis* in the Cascade Range, and *V. f. necator* in the Sierra Nevada (Perrine et al. 2007; Volkmann et al. 2015; Castelló 2018; Quinn et al. 2022). Until relatively recently, it was thought that all low-elevation populations of red foxes in the western United States were introduced (Kamler and Ballard 2002). However, additional genetic studies have revealed that the population of red foxes in California's Sacramento Valley, which exists in atypical semi-desert habitat, also belongs to the Mountain subclade and was described as a fourth subspecies in this subclade, *V. f. patwin* (Sacks et al. 2010). Thus, indigenous North American red foxes are known from both high and low elevations in the western United States.

The eastern American red fox, *V. f. fulva*, is contained within the Eastern subclade (Statham et al. 2012; Castelló 2018). This is the red fox of the eastern United States that the early literature thought was a product of importation of Eurasian red foxes from Europe (Frey 2013). It should again be emphasized, however, that genetic data show that *V. f. fulva* is actually native to eastern North America and not a product of importation (with the exception of historical hybridization in a localized area near the site of an introduction; Statham et al. 2012; Kuo et al. 2019). This subspecies seems to be adaptable to a wide variety of habitats, including those that have been modified by humans. Consequently, evidence suggests that its range may have naturally expanded both southward and westward—with changes that occurred post-European colonization—and that it now occurs as far west as Texas, or even possibly New Mexico (Frey 2013; Castelló 2018; and see below). Across North America, some of this expansion might have been due to translocations for fox-chasing sporting events (Hatcher and Wigtil 1985; Lee et al. 1993). In addition, red fox fur-farming was developed in the 1890s on Prince Edward Island on the eastern coast of Canada. The farmed animals were primarily silver foxes of the subspecies *V. f. rubricosa* that originated on or near Prince Edward Island and hence are also members of the Eastern subclade (Statham et al. 2011, 2012). The captive rearing of red foxes was a profitable business and spread across many areas of the United States

in the early 1900s. Releases or escapes from fur farms resulted in the establishment of non-native populations of eastern subclade red foxes in the western United States. These generally occurred at low elevations in urban and highly human-altered environments (Lewis et al.1999; Sacks et al. 2016; Merson et al. 2017). Populations of red foxes in the western United States that are thought to be recently established may be derived from either native local animals (western subclade) or translocated fur-farm foxes (eastern subclade). For instance, recently established populations of red foxes in western Washington, southern California, and the front range of Colorado are from the eastern subclade and were likely derived from translocated fur-farm animals, whereas populations thought to be recently established in the Great Basin and western Oregon are primarily derived from native red foxes (Sacks et al. 2010; Statham et al. 2012; Merson et al. 2017; Black et al. 2018). It is possible that indigenous red foxes occurring in the Four Corners area represent a unique genetic type, possibly referable to a new subspecies (Mikesic and LaRue 2003).

Frey (2004) recognized two subspecies of the North American red fox in New Mexico, including the Rocky Mountain red fox (*V. f. macroura*) and the eastern American red fox (*V. f. fulva*), though extensive genetic testing has not been conducted. The Rocky Mountain red fox, *V. f. macroura*, is part of the Mountain subclade and was described from the Wasatch Mountains, Utah. Its historical distribution extends across the Rocky Mountains from southern Canada south into Arizona and New Mexico (Hall 1981; Volkmann et al. 2015). Bailey (1931) and Hall (1981) considered all red foxes in New Mexico to belong to this subspecies. *V. f. macroura* is larger and has a paler and more yellow pelage than the eastern American red fox (*V. f. fulva*; Barnes 1927; Bailey 1931). The legs are often grayish rather than black, and the dark color is often greatly reduced in extent, particularly on the hind legs where it might only extend upward to the hock. Measurements of a male and female from Liberty and the Taos mountains, respectively, are (in mm) 1080, 992 (42.5, 39 in) for total length; 422, 388 (17, 15 in) for tail length; and 175, 173 (7, 7 in) for hind foot length. However, given that these measurements are from a single male and a single female, readers should be aware of the potential for these values to be a poor reflection of "average" sizes. The general color of the red phase of this subspecies is "straw yellow" becoming more reddish and darker over the shoulders (Bailey 1931:296). The cross phase is overall yellowish with a dusky band across the shoulders. The silver phase, including the black form, is uncommon (Bailey 1931). It is generally assumed that the eastern American red fox (*V. f. fulva*) did not historically occur in New Mexico, but, if now truly present, arrived through natural expansion of the subspecies' range in response to changes in the environment, or facilitated by introductions and expansion of animals translocated for fur farms or hunting (but see under "Distribution"). Compared to *V. f. macroura*, *V. f. fulva* has a shorter tail, smaller hind feet, a skull with much smaller auditory bullae and a wider basioccipital, and a dentition characterized by smaller carnassials and smaller first upper molars (Merriam 1900). In typical *V. f. fulva*, the legs are black and the black color nearly reaches the elbow of the front legs and extends upward above the hock and knee and onto the thigh on the hind legs.

Vulpes is a Latin noun that simply means "a fox" (Gotch 1979). Interestingly, the word *vulpes* also pertains to cunning or craftiness, which is an obvious reference to the cleverness of this animal. When foxes were placed in the genus *Vulpes* (which is a feminine Latin noun), the International Rules of Biological Nomenclature required that the species name change to *fulva*, the feminine form of the adjective *fulvus*. Thus, the proper name for the North American red fox is *V. fulva*, not *V. fulvus* as is frequently published (e.g., Wozencraft 2005 as *V. v. fulvus*).

Photo 15.9. North American red fox photographed on 2 November 2021 in the valley of the San Juan River at Waterflow, San Juan County at an elevation of ~1580 m (5,200 ft). The red fox's distribution in New Mexico includes the San Juan River watershed in the northwestern portion of the state. Photograph: © Marty Peale, Brian Long, and Jon Klingel. 2021 New Mexico Department of Game and Fish, Share with Wildlife, Mink Survey, northern New Mexico. US Fish and Wildlife Service Wildlife Restoration Grant W-208-R-1.

DISTRIBUTION

The red foxes (forms traditionally referred to as *V. vulpes* and *V. fulva*) have the largest geographic range of any carnivoran (Larivière and Pasitschniak-Arts 1996). The distribution of this group is Holarctic, extending across most of the boreal and temperate regions of Eurasia and including portions of northern Africa, in addition to most of North America north of Mexico (exclusive of portions of the Florida peninsula and some desert regions). The North American red fox was introduced to some Arctic islands in the northern Pacific through the commercial fur trade (Castelló 2018). The Eurasian red fox was introduced to Australia for sport hunting and expanded its distribution to inhabit most of that continent (Abbot 2011), in addition to recent establishment on nearby Tasmania (Sarre et al. 2012). Unlike in Australia, the introductions of

Photos 15.10a and b (*left to right*). Recent records of the North American red fox from the Jemez Mountains in northern New Mexico. a) Months of camera monitoring for cougars (*Puma concolor*) by the New Mexico Department of Game and Fish resulted in a single photograph of a red fox on 5 October 2018, between Cuba and San Pedro Parks; Photograph: © New Mexico Department of Game and Fish; b) National Park Service record on 19 February 2021 from near San Juan Canyon north of Ponderosa, also in the Jemez Mountains; Photograph: © Mark A. Peyton.

Photo 15.11. North American red fox photographed on 20 March 2017 at a wildlife road-crossing structure in Tijeras Canyon, which separates the Sandia Mountains to the north and the Manzanita and Manzano mountains to the south in central New Mexico. This photograph verifies the occurrence of the North American red fox in the Sandia-Manzano mountain chain. Photograph: © New Mexico Department of Transportation and Arizona Game and Fish Department.

Photo 15.12. North American red fox photographed along NM 288 about 24 km (15 mi) north of Clovis in Curry County on 6 May 2014. The area is part of the Llano Estacado, a large tableland region of the Southern Great Plains. Records indicate that red foxes historically occurred in many areas of the Great Plains but that they declined along with the loss of other species such as the swift fox (*V. velox*), gray wolf (*Canis lupus*), and American bison (*Bos bison*). Today, red foxes are known to occur in scattered locations across the eastern plains of New Mexico, and are not uncommon in the vicinity of some towns. This red fox displays the deep reddish orange pelage color consistent with the eastern American red fox (*V. f. fulva*), though genetic studies have not yet been conducted to confirm the presence of this subspecies in New Mexico. Photograph: © Nancy S. Cox.

European red foxes to North American have left behind only minor genetic signatures in localized areas, likely due to the existence of the native North American red fox that prevented expansion of the introduced fox (Kasprowicz et al. 2015). The success of introductions of red foxes to areas where they did not historically occur (or were presumed to have been absent, as in the case of North America), along with the ecological harm caused by an exotic predator on native biodiversity, led to the red fox being classified as one of the 100 most invasive species in the world by the Invasive Species Initiative (Lowe et al. 2000). Nonetheless, the designation of the North America red fox as invasive is not appropriate, except perhaps in localized contexts such as the case of non-native red foxes (i.e., Eastern subclade) in southern California (Sacks et al. 2016).

There is uncertainty about the pre-Columbian distribution of the North American red fox stemming in large part from the ongoing confusion regarding the species' origins, putative introductions, and subsequent movements. Much of the literature regarding the distribution of the North American red fox is based on scant information, and it often regurgitates inaccurate or speculative prior information (Frey 2013). Some scenarios posited in the literature are clearly inaccurate based on available data. For instance, an often-cited map of the historical (ca. 1600) distribution of red foxes in North America (Voigt 1999:382) shows that red foxes were absent from most areas in North America south of ca. 41° N latitude, which is the approximate latitude of the northern border of Colorado. Such maps helped feed the false notion that contemporary populations of red foxes represent recent invaders. In contrast, genetic data clearly demonstrate that the ancient indigenous Nearctic clade of the North American red fox occurs in many areas of the western United States south of 41° latitude, including the southern Rocky Mountains (Colorado and New Mexico), the Great Basin (Utah, Nevada), and

the Sierra Nevada (California; Aubry et al. 2009). Likewise, Voigt's map excluded the entire United States east of the Mississippi and Ohio rivers from the natural range of the red fox. Genetic and other lines of evidence indicate that red foxes were present within the eastern United States at the time of European settlement (Frey 2013).

In New Mexico, records of red foxes are uncommon, particularly prior to the recent widespread use of cell phone cameras, remote game cameras, and social media. Word of mouth and trapping records might be unreliable for this species because people sometimes call gray foxes with reddish pelage "red fox" (Clothier 1957). For instance, 27 records of "red fox" taken by Wildlife Services between 1991 and 2003 incidental to the take of coyotes included two animals from Hidalgo County, where red foxes do not occur. Such incorrect information calls into question the identification of purported records from other counties (Chaves [2 individuals], Colfax [4], Lea [6], Lincoln [3], McKinley [5], Quay [1], Roosevelt [2], and Torrance [2]). More broadly, such sparsity of records makes it difficult to reach firm conclusions about the origins, distribution, habitat, and ecological relations of red foxes in the state. In general, red foxes are reported most commonly from the northern half of the state. However, the species does occur elsewhere in New Mexico with their primary distribution in three regions: the northern mountains, the San Juan River Basin, and the eastern plains (Map 15.1 and Appendix 15.1).

In north-central New Mexico, most records of red foxes are from the higher elevations of the San Juan and Sangre de Cristo mountains, which likely reflects both higher habitat suitability and more extensive, early mammalogy fieldwork in those mountains. Recently, however, remote cameras have also documented the occurrence of red foxes in the Jemez Mountains (N. Forman, pers. comm.; M. Peyton, pers. comm.; Photos 15.10 a and b). The elevational distribution of the

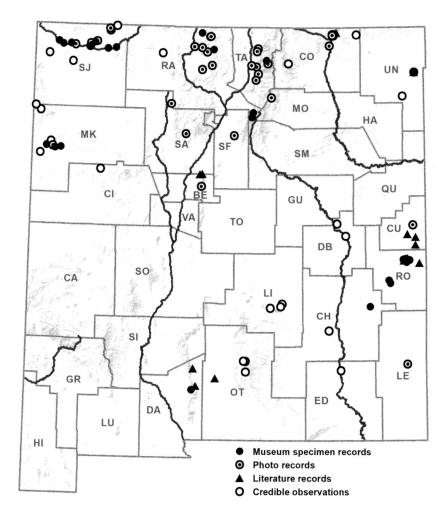

● Museum specimen records
◉ Photo records
▲ Literature records
○ Credible observations

Map 15.1. Current distribution of the North American red fox in New Mexico based on verified and credible occurrence records. North American red fox records published in the literature are those documented by Halloran (1946, 1964), Clothier (1957), Frey (2003, 2004), Kamler et al. (2005), Julyan and Stuever (2005), and Frey and Schwenke (2012). Observations deemed credible are those by qualified biologists. See also Appendix 15.1 for a list of all North American red fox records in New Mexico.

red fox extends upward to the crest of Wheeler Peak in the Sangre de Cristo Mountains, the highest point in New Mexico (3,409 m [13,167 ft]; Bailey 1931). However, red foxes also have been verified at lower elevations in this region such as in the Chama River valley (western Rio Arriba County), and on the Raton Mesa (eastern Colfax County) (Frey and Schwenke 2012).

Red foxes are occasionally observed in the Sandia-Manzano-Capitan-Sacramento mountain chain in central and south-central New Mexico. Clothier (1957) obtained reports of red foxes in the Sandia Mountains during the first rigorous study of mammals of the range. Recent photographs and observations document that they continue to exist in the Sandia and Manzano mountains (Polechla 2005). For instance, the New Mexico Department of Transportation used a remote camera to document a red fox using a wildlife crosswalk on New Mexico Highway 333 in Tijeras Canyon (Photo

15.11). Biologists have occasionally observed red foxes in the Capitan and Sacramento mountains, primarily at higher elevations, though thus far there are no photographs or specimens from these mountains (H. Walker, pers. comm; R. Stewart, pers. comm.).

Red foxes occur broadly within the San Juan River watershed, south of the Colorado border and west of the San Juan Mountains (Map 15.1 and Appendix 15.1). Farther to the west, the distribution of the red fox extends into northeastern Arizona, where the species is reasonably common on the Navajo Nation within the watersheds of the San Juan and Little Colorado rivers (Mikesic and LaRue 2003). Trapping records maintained by the Navajo Nation from 1994 to 2019 indicate that red foxes are regularly taken throughout most of the New Mexico portion of the Navajo Nation: four were taken from Unit 2 (northwest of Crystal), 83 from Unit 3 (north of NM134 and west of US491), 96 from Unit 4 (south of NM134 to near Vanderwagen and west of US491 and NM602), 156 from Unit 13 (east of NM371 and west of US491), and 45 from Unit 14 (southwest of US550 and east of NM371)—none were taken from Unit 15 (outlying areas of Ramah, Alamo, and To'hajiilee).

In the 1990s, the New Mexico Department of Game and Fish approved the importation of 100 live-trapped red foxes from California's San Francisco Bay area to the Zuni Reservation. This was proposed as an alternative to lethal removal of the foxes, which were imported eastern American red foxes (*V. f. fulva*) and not native to California nor New Mexico. The importation of the 100 foxes to New Mexico required that they be spayed/neutered and vaccinated. Fortunately, the translocation never took place (C. Nagano, pers. comm.). While the San Juan River watershed, set within the arid Great Basin Desert region, seems like an odd place for red foxes to occur, the species' presence was indeed reported in the first biological explorations that occurred in that part of the state (Bailey 1931; Eaton 1937; Halloran 1962,

1964, 1965; Harris 1963). It seems unlikely that their presence in the region is due to introductions, especially given the high frequency of individuals with unique pale coloration, suggesting the evolution of in situ adaptations (Hoffmeister 1986; Mikesic and LaRue 2003).

It is often claimed that red foxes are not native to the Great Plains. However, red foxes were reported from numerous sites in the Great Plains prior to European settlement (Zumbaugh and Choate 1985; Frey 2013). For instance, the red fox was reported from a number of locations in western Kansas during the 1800s, though it was far less common in that region than the sympatric swift fox (Carter 1939). Population declines of both species occurred, and by 1900 the swift fox was rare, while the red fox was no longer reported in western Kansas by 1910 (Carter 1939). These population declines coincided with declines in large game due to overexploitation, declines in other carnivores such as wolves due to control efforts, and declines in white-tailed jackrabbits (*Lepus townsendii*) due to a combination of factors (Brown et al. 2020). It seems possible that the decline of red foxes on the Great Plains was mediated by direct predator control as well as the extirpation of wolves that allowed for increases in coyotes, a potential competitor of the red fox. Today, red foxes once again appear to occur sparingly across the Great Plains, including eastern New Mexico. However, while the swift fox is now a conservation priority, virtually nothing has been done to understand the ecology or conservation status of red foxes on the plains.

Records of red foxes are scattered across the eastern plains region of New Mexico, from near the Colorado border south to the Llano Estacado, and from the Texas border west to the Sandia-Manzano-Sacramento ranges (Frey 2003, 2004; Map 15.1 and Appendix 15.1). Kamler et al. (2005) speculated that red foxes did not historically occur in the eastern plains region of New Mexico, but that they recently spread westward to occupy the region from Texas. Evidence for this assertion included a lack of records in eastern New Mexico until 1958 and the frequently repeated assertion that red foxes are not native to Texas but rather are the result of introductions to eastern and central Texas beginning in the late 1890s, from which foxes multiplied and spread west. Given the overreliance on arguments that red foxes were introduced, it is worth considering alternative arguments. First, the mammals of eastern New Mexico have been poorly studied, with even many large charismatic species not documented until the publication of the first comprehensive survey work by Aday and Gennaro in 1973, or for many areas of the Llano Estacado of eastern New Mexico, until the 2000s (Frey 2003). Thus, recent documentation of red foxes from the eastern plains does not equate with them being a recent arrival. Second, information about the introduction of red foxes to Texas was provided by Allen (1896), who reproduced a letter from the secretary of the Texas Fox Hunters' Association of Waco Texas that described the introduction of ca. 100 red foxes from "the East" to the region around the Brazos River and further east around Marshall near the Louisiana border. However, given the central and eastern Texas regions referred to in the letter, these introductions do not discount the possibility that red foxes were already present and native to portions of Texas, including western Texas. Indeed, Marcy (1854) reported red foxes from the Red River, at a time when human population densities were low throughout much of Texas. Regardless, the native distribution of red foxes in Texas might have little to do with the overall distribution of red foxes in New Mexico, given that the species is known to be native to other regions of New Mexico besides the eastern plains.

Several records of red foxes in New Mexico require special mention. While camping at Willow Creek in the Mogollon Mountains, Catron County, Bailey (1931) reported smelling a red fox.

We do not know that the smell of a red fox is sufficiently distinct from that of a gray fox (which is known to occur on Willow Creek) to allow for reliable documentation, and hence we discount this report. More generally, there are no valid records of red foxes from Catron County or anywhere within the Gila National Forest. Jones et al. (2021) reported specimens of red foxes from near Dusty (Socorro County), but an examination of the specimens demonstrates these were misidentified gray foxes. This underscores the importance of preserving specimens of putative red foxes to allow for verification of identification.

The most unusual records of red foxes in New Mexico are from Doña Ana County in the Chihuahuan Desert region of the south-central part of the state (Map 15.1 and Appendix 15.1). One of these records was an animal trapped in 1942 on the San Andres National Wildlife Refuge and deposited in the US National Museum (Halloran 1942, 1946). This remote and rugged refuge was established in the southern portion of the San Andres Mountains in 1941 to protect desert bighorn sheep (*Ovis canadensis nelsoni*). Predator trapping was then initiated as a management action in portions of the refuge. Arthur Halloran was the first refuge manager and served from 1941 to 1944. Halloran (1946) noted that in contrast to many other desert mountain ranges where livestock grazing had altered the flora and fauna, the San Andres Mountains were relatively lightly utilized and retained their original biodiversity. In addition to the specimen, Halloran (1946) also reported observations of red foxes from several other locations in the San Andres Mountains (Ash Canyon, Lostman Canyon) and adjacent Tularosa Valley (Point of Sands). Halloran (1942, 1946) considered that the records represented an isolated population, similar to the endemic *V. f. patwin* in the upper Sacramento Valley of California, given that local people had long been aware of red foxes in the region and there were no known introductions. If there once was an endemic population of red foxes in the Tularosa Basin, a lack of subsequent records suggests it is probably no longer extant. Alternatively, it is possible that robust predator control activities that targeted coyotes in the early and mid-1900s could have allowed an "ecological release" on red foxes allowing them to temporarily expand their distribution to places vacated by the larger competitor. However, red foxes also have been documented more recently in similar environments within the Chihuahuan Desert region of Trans-Pecos, Texas (Swepston 1981). Townsend (1893:315) reported "red and silver foxes (*Vulpes*)" from the Organ Mountains, but he did not mention the kit fox or gray fox, both of which are relatively common in the vicinity of the Organ Mountains. Thus, Townsend's "red and silver foxes" likely referred to those common species of the Organ Mountains instead of *V. fulva*.

Newly discovered or expanding red fox populations are frequently attributed to escapees from fur farms. Fur-farming began in the 1890s with silver phase red foxes derived primarily from Prince Edward Island, Canada (Statham et al. 2011). The industry focused on these silver phase red foxes, which had the most valuable pelts, and these were traded to fur farms in other regions reaching a peak in the 1940s (see also Chapter 24). Fur-farming appears to have been an uncommon enterprise in New Mexico, and the 1940 US census reported only 11 fox farms in the state (US Department of Commerce, Bureau of the Census 1942). Given the great value of the stock, it is unlikely that these foxes would be intentionally released. Further, survival of pen-reared animals in the wild is low. For instance, four pen-reared red foxes with functioning radio collars were all dead within five weeks, due to either humans or dogs (Pledger 1975). Similarly, five (two pen-reared, three wild-caught) radio-collared red foxes released into southeastern Oklahoma were all dead within two months (Hatcher and Wigtil 1985). Researchers observed that the pen-reared animals seemed unaware of dangers posed by humans and their activities (e.g., vehicles and roads). Further, genetic studies have suggested that native foxes may strongly exclude introduced foxes in some settings, hindering introgression of putatively non-native red fox alleles into native red fox populations (Sacks et al. 2011; Cross et al. 2018). Thus, given that there are no records of any introductions of red foxes into New Mexico, and given the evidence of low survival and possible exclusion of introduced foxes in landscapes where native red fox populations occur, it seems entirely conjectural to suggest that fur farms or translocations are responsible for any of the populations of red foxes in the state.

Red fox fossils dating back to the Pleistocene and remains discovered at pre-Columbian archaeological sites also deserve to be mentioned as they constitute additional evidence of the species occurring in North America long before the arrival of the Europeans. In New Mexico, red fox remains have been identified both at Pleistocene sites (Harris 2013) in the north-central, central, eastern, and southeastern parts of the state and at several ancestral pueblo sites (Allen 1954; Lang and Harris 1984; Map 15.2).

Specimens of red foxes in New Mexico have been referred to two subspecies (Frey 2004) though there has not been any comprehensive analysis of geographic variation, genetics, or taxonomy of red foxes in the state. Specimens from the Llano Estacado of Texas and New Mexico and the Oklahoma panhandle near the New Mexico border were referred to *V. f. fulva*, the eastern American red fox (Dalquest et al. 1990; Choate 1997; Frey 2003). Frey (2003) noted that two specimens from the Llano Estacado of New Mexico had a bright red pelage that resembled typical *V. f. fulva* from the eastern United States. Overall, however, the taxonomic designation of specimens from the southern Great Plains as *V. f. fulva* appears influenced by the perceptions that red foxes were not native to this region (e.g., Choate 1997; also distribution map of Hall 1981), despite

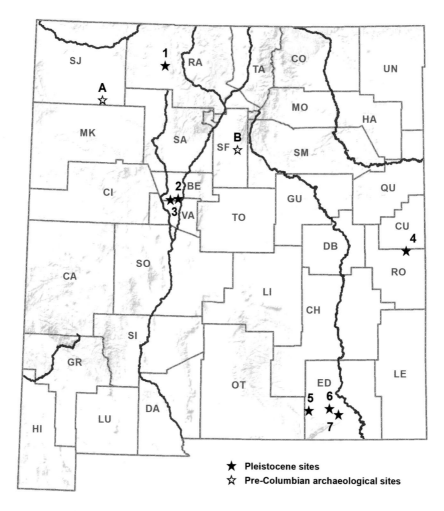

Map 15.2. New Mexico Pleistocene (Harris 2013) and pre-Columbian (Holocene) archaeological (Allen 1954; Lang and Harris 1984) sites with North American red fox remains. Pleistocene sites: 1 = San Antonio (SAM) Cave; 2 = Isleta Cave No. 1; 3 = Isleta Cave No. 2; 4 = Blackwater Loc. No. 1; 5 = Burnet Cave; 6 = Dry Cave, Camel Room; 7 = Dark Canyon Cave. Pre-Columbian archaeological sites: A = Pueblo Bonito (Chaco Canyon); B = Arroyo Hondo Pueblo. Together with other lines of evidence, the existence of North American red fox remains at multiple Pleistocene and pre-Columbian archaeological sites further underscores that the species is native and not the result of importation by European settlers.

evidence from Kansas suggesting that red foxes were common on the plains of that state prior to European colonization (Zumbaugh and Choate 1985).

Specimens of red foxes from the northern mountains of New Mexico and San Juan River watershed have been referred to *V. f. macroura*, the Rocky Mountain red fox (Bailey 1931; Hall 1981). In the Rocky Mountains, recent genetic analyses have demonstrated that indigenous populations of *V. f. macroura* persist at high elevations (Volkmann et al. 2015; Merson et al. 2017; Cross et al. 2018). However, the genetics of red foxes at lower elevations, especially in human-altered landscapes in the Southern Rocky Mountains such as near Denver, were admixed (i.e., hybridized) with "non-native" North American red foxes likely originating from fur-farm stock and originally derived from eastern Canada and Alaska (Merson et al. 2017). Unlike in New Mexico (11 farms) and Arizona (four farms), fur-farming was a much more extensive practice in Colorado with 139 silver fox farms reported in the 1940 census (US Department of Commerce, Bureau of the Census 1942) and an epicenter of fur production around Denver (https://coloradoencyclopedia. org/article/colorado%E2%80%99s-second-fur-trade). Genetic analyses are needed to determine the extent to which red foxes in northern New Mexico represent genetically intact *V. f. macroura*. In addition, it is possible that animals from the San Juan Basin of Arizona and New Mexico represent a distinctive form of the red fox, given its unusual low-elevation desert habitats and distinctive pale pelage. Genetic analyses have described a similarly unusual low-elevation population of red foxes from the Sacramento Valley of California as a distinctive subspecies (*V. f. patwin*). This population is isolated from the native high-elevation form in the Sierra Nevada (*V. f. necator*), but threatened by the presence of non-native red foxes in adjacent low-elevation areas (Sacks et al. 2010). Similar patterns could be operating in the Southwest.

HABITAT ASSOCIATIONS

Red foxes (*V. vulpes* and *V. fulva*) are usually considered quintessential habitat generalists that can use a variety of ecosystems, adapt to habitat changes, and exploit novel environments (Larivière and Pasitschniak-Arts 1996). However, the wide-ranging red fox group is phylogenetically diverse, with many evolutionary lineages that are restricted to particular geographic areas (e.g., Kutschera et al. 2013; Statham et al. 2014; Leite et al. 2015). Thus, while as a whole the group may

Photo 15.13. Rocky Mountain red fox (*V. f. macroura*) with typical pelage pattern of this subspecies (pale overall, yellowish ruff around face, grayish hind end and tail), photographed on 2 June 2022 at 3,215 m (10,544 ft) elevation at Brazos Pass, Rio Arriba County, in the San Juan Mountains. Most records of red foxes in New Mexico are from the north-central mountains region where prime high elevation habitat is in open subalpine forests and forest edges, subalpine meadows and grasslands, and tundra. Photograph: © Brian Jay Long.

Photo 15.14. North American red fox at Des Montes, Taos County, on 18 January 2022. The surrounding vegetation consists mainly of big sagebrush (*Artemisia tridentata*) and scattered junipers (*Juniperus* spp.) on the edge of irrigated farmland, mostly alfalfa fields, pastures, and orchards. The species is regarded as a habitat generalist because it occupies many ecosystem types across its broad range, though within them it tends to select open areas. Photograph: © Brian Jay Long.

Photo 15.15. Ash Canyon in the San Andres Mountains, Doña Ana County, which is within the Chihuahuan Desert region. In 1942 a North American red fox was captured on a ridge at 1,676 m (5,500 ft) elevation in the San Andres National Wildlife Refuge in Upper Sonoran grassland vegetation consisting of sideoats grama (*Bouteloua curtipendula*), hairy grama (*B. hirsuta*), resinweed (*Viguiera* sp.), sotol (*Dasylirion wheeleri*), yucca (*Yucca* sp.), and prickly pear (*Opuntia* sp.). Other red foxes were reportedly observed in Ash Canyon and nearby locations in the 1940s, though none have been documented in Doña Ana County since. Photograph: © US Fish and Wildlife Service.

Photo 15.16. North American red fox in Colorado on 30 May 2020. Research suggests that the distribution and abundance of the North American red fox is influenced by competition with other larger canids, particularly the coyote (*Canis latrans*). Red foxes might be better able to persist in places that either provide protection from coyotes or have minimal coyote presence, such as urban areas. Photograph: © Ulli Limpitlaw.

appear to be highly adaptable, the regional taxonomic units, lineages, or populations may exhibit specializations. Unfortunately, because of the false notion that the North American red fox was an exotic invasive, there has been relatively little interest in scientific study of its habitat relations and large knowledge gaps persist. To prevent overgeneralizing, information in this section is derived from only the most relevant literature.

Like their larger cousins the wolf and the coyote, red foxes are pursuit predators with cursorial (i.e., running) adaptations. As a result, they tend to select relatively open vegetation structures, even within areas considered "forested." This tendency to select relatively open habitats provides a common thread that links the seemingly disparate vegetation types used by red foxes—from alpine tundra to the edges of subalpine forests to suburbs. It also can explain changes in distribution. For instance, in eastern North America the cutting of previously dense forests for settlement and agriculture during colonial times likely allowed red foxes to expand their distribution (Frey 2013). Second, red foxes are perhaps the most carnivorous of all foxes, with a diet dominated by rabbits and small mammals such as voles (Larivière and Pasitschniak-Arts 1996; see under "Diet and Foraging"). Therefore, habitat selection is also driven by prey availability (Larivière and Pasitschniak-Arts 1996). Lastly, intraguild competition with larger canids, especially the coyote, can potentially influence the distribution and habitat selection of red foxes (Gese et al. 1996). Some environments might be usable by red foxes in the absence or low density of coyotes, while other areas might provide a refugium for red foxes if coyotes are reluctant to use them (e.g., urban areas). Thus, there might be considerable variation in habitat selection among different populations.

Few studies have evaluated habitat use by the Rocky Mountain red fox (V. f. macroura), considered a high-mountain specialist. In the Yellowstone Ecosystem, red foxes occurring at high elevation (>1600 m [>5,250 ft]) are relict indigenous populations of V. f. macroura, while those at low elevation have mtDNA signatures admixed with foxes indigenous to the Great Plains and derived from fur farms (Swanson et al. 2005; Cross et al. 2018). Van Etten et al. (2007) used distance-based analyses to investigate habitat selection by 16 radio-collared red foxes (likely representing V. f. macroura) in the Yellowstone Ecosystem. Their analyses suggested that red foxes selected home ranges near Douglas-fir (Pseudotsuga menziesii) and far from mesic environments; they also selected areas within home ranges near Douglas-fir and far from ecotones. However, a subsequent study reported that distance-based analyses provided extensive misclassification and recommended they not be used for habitat selection (Bingham et al. 2010). In contrast, Fuhrmann (1998) compared points collected along tracks made by red foxes in the snow to available points drawn from within 500 m of the tracks. He found that red foxes traveled in areas with low topographic roughness, near escape cover, and near edges—defined by a major change in vertical vegetation. They also selected mesic meadow and spruce-fir forest vegetation, but avoided xeric non-forested areas, recently burned forest, whitebark pine (Pinus albicaulis), regenerating spruce-fir, and aspen (Populus tremuloides). When foraging, red foxes mainly used areas near the edge between spruce-fir forests and mesic meadows at high elevation and sagebrush communities at low elevation (Fuhrman 1998).

Similar habitat has been described for the other two endemic high-elevation red foxes in the Cascade Mountains (V. f. cascadensis) and Sierra Nevada (V. f. necator), which select subalpine forests, parklands, and alpine tundra and meadows near the treeline but avoid dense forest (Aubry 1983; Quinn et al. 2018; USFWS 2018). In New Mexico, Bailey (1931) reported red foxes as common in the Canadian zone (i.e., mixed coniferous forests and meadows) and above the timberline in the Sangre de Cristo Mountains. In very high-elevation environments, red foxes may only use alpine tundra during the summer and shift to conifer forests during the winter, due to absence of key food (such as lagomorphs or ungulate carrion) in the tundra during winter (Cagnacci et al. 2004; Perrine 2005). The avoidance of dense forest, whether through behavioral or physiological reasons, is thought to restrict the endemic mountain red foxes to high elevations and result in barriers that have prevented them from reaching some potential habitat, while also preventing non-native red foxes (i.e., fur-farm derived) from infiltrating the high elevations occupied by the three endemic montane forms (Aubry 1984).

Virtually nothing is known about red foxes in the southern Great Plains. In southwestern Kansas and the Oklahoma Panhandle, most records of red foxes are from areas near rivers (Zumbaugh and Choate 1985; Dalquest et al. 1990), all of which have headwaters in New Mexico. On the Llano Estacado in Texas, records of red foxes are primarily from the eastern portion of the plain but details as to habitat are unknown (Choate 1997). In the Texas Panhandle, some of these records originate from near towns and cities, including one red fox captured on an airport runway in Plainview (Packard and Bowers 1970) and several others observed in Lubbock (Choate 1997). Red foxes frequently exploit urban areas (e.g., Mueller et al. 2018). For instance, in New Mexico they have been documented from Portales, Clovis, Santa Fe, Gallup, and Farmington, among other cities (Map 15.1 and Appendix 15.1).

The San Juan Basin of Arizona and New Mexico harbors a variety of ecosystems across a wide range of elevations generally from 1,410 m (~4,640 ft; but down to 1,110 m [~3,640 ft] at Lake Powell) to over 2,740 m (~9,000 ft) in the mountains. Vegetation types are highly diverse, ranging from highly productive irrigated farmland in the valley of the San Juan River and its principal tributaries to Great Basin Desert scrub and arid grasslands on the extensive lower elevation plains, pinyon-juniper woodland on mid-elevation mesas, and coniferous forests, montane meadows, and grasslands in the Chuska Mountains and other ranges. Much of this region is contained within the Navajo Nation. On the Navajo Nation in Arizona and southern Utah, the median elevation for red fox records was 1,793 m (5,883 ft), and most (68%) records were from ~1,650–2,010 m (5,400–6,600 ft), though records extended from 1207–2378 m (3,960–7,800 ft; Mikesic and LaRue 2003). Although sampling was not uniform across the study area, these elevations suggest that red foxes in this region are not specializing on high-elevation environments as they do in the Rocky Mountain region. Further, of 101 records for which vegetation type was determined, 75% were in desert scrub or desert grassland (Great Basin Desert scrub 42%, Great Basin grassland 33%), with 26% in Great Basin conifer woodland (i.e., pinyon-juniper woodland; Mikesic and LaRue 2003). This region is one of the most arid portions of the United States, with annual precipitation ranging from ca. 15–30 cm (6–12 in), a seemingly unlikely place for a red fox to exist. However, there are other endemic arid-adapted red foxes at the species' southern range limits. In North America, the Sacramento Valley red fox (*V. f. patwin*) is endemic to arid grasslands of the northern portion of the Central Valley of California, though that region is considerably more mild than the San Juan Basin as average precipitation in that region reaches 45 cm (17.7 in) and there are wetlands in the valley (Black et al. 2019). Examples of endemic Eurasian red foxes (*V. vulpes*) that exist in such arid environments include *V. v. arabica* from the Arabian Peninsula, *V. v. pusilla* found in deserts of the Middle East and western India, and *V. v. aegyptiacus* found in wadis (dry drainages) and along the edges of deserts in Egypt (Castelló 2018).

LIFE HISTORY

Diet and Foraging

Although red foxes are commonly regarded as omnivorous, this designation is something of a misnomer as omnivory and carnivory exist on a continuum and even species generally considered carnivorous such as wolves sometimes eat plants and invertebrates. Thus, simple laundry lists of food items recorded as consumed by a species may be lengthy and diverse, but can mask the actual underlying diet selection. Relative to other species of foxes, the red foxes (*V. fulva* and *V. vulpes*) have morphological adaptations more typical of carnivorous canids that hunt prey, including relatively long legs, a large bite force, and a large brain (Damasceno et al. 2013). In terms of diet, the life history of red foxes is more similar to that of larger canid cousins than to other foxes. For instance, in Kentucky the diets of the North American red fox and the coyote were highly similar with both specializing on mammalian prey, though red foxes consumed proportionately more small mammals (76% versus 57%), slightly fewer rabbits (18% vs. 22%), and fewer large mammals (8% vs. 28%) than coyotes (Crossett and Elliott 1991). In Maryland, a comparison of the diets of the North American red fox and gray fox revealed that the red fox primarily consumed meadow voles (*Microtus pennsylvancus*) and cottontails (*Sylvilagus floridanus*), while the gray fox primarily consumed plants (Hockman and Chapman 1983). Indeed, a recent global review found that small (e.g., rodents) and medium-sized (e.g., rabbits) mammals were associated with the highest percent frequency of occurrence in the diet of the North American red fox (Castañeda et al. 2022). In contrast, birds, fruit, and insects were of moderate occurrence, while vegetation, large mammals (i.e., carrion), garbage, and reptiles were of relatively minor occurrence (Castañeda et al. 2022).

Overall, red foxes appear to be generalist predators that focus on the most abundant and easily

Photo 15.17. North American red fox after a successful hunt in a prairie dog colony near Tierra Amarilla in Rio Arriba County on 29 July 2011. The diet of the North American red fox has not been studied in New Mexico, but likely is dominated by small and medium-sized mammals that are locally abundant. In the species' shorter summer pelage, the long legs of the red fox are apparent. Relatively long legs are adaptations for cursorial hunting and are a feature shared by the North American red fox's cousins, the coyote (*Canis latrans*) and wolf (*C. lupus*), more so than its close relatives the other species of foxes, all of which are more omnivorous in their diet. The rarity of red foxes in New Mexico may be the result of a reduced small mammal prey base, as well as carnivore communities dominated by coyotes. Photograph: © Sally King.

obtainable prey, resulting in geographic variation in diet selection depending on local environments (Kidawa and Kowalczyk 2011; Castañeda et al. 2022). In addition, there is variation in diet among demographic groups. Studies on the Eurasian red fox suggest that the adult female has a narrower diet, composed more exclusively of key mammalian prey, in comparison with the adult males and juveniles. For instance, in Australia both adult male and juvenile Eurasian red foxes consumed relatively large amounts of carrion, while the adult females were more specialized on small mammals and invertebrates (Forbes-Harper et al. 2017). Similarly, in Poland Eurasian red foxes specialized on voles (*Microtus* spp.), but adult males and juveniles had larger food niche breadths (Kidawa and Kowalczyk 2011).

Unfortunately, there is little information on the diet of red foxes in New Mexico, but inferences can be made based on studies from other areas. For instance, in the Cascade red fox, one of the endemic subspecies of mountain red foxes, most of the items found in scats across the year were mammals (Aubry 1983). During winter, the scats contained mostly mammals, including snowshoe hares (*Lepus americanus*), voles (*Myodes, Phenacomys*), pocket gophers (*Thomomys*), and occasional birds and garbage. In summer, mammals were still the dominant diet items, mainly pocket gophers and voles, but the scats also contained moderate amounts of plants (mainly fruit) and insects (Aubry 1983). Similarly, for the Sierra Nevada red fox, the diet was dominated by mammals in all seasons, with the primary food items consisting of pocket gophers and mule deer (*Odocoileus hemionus*), the later likely consumed as carrion (Perrine 2005). In winter, the Sierra Nevada red foxes also consumed small rodents (murid mice and jumping mice; Perrine 2005). Pocket gophers and voles are common in meadows and grasslands at mid- to high elevations in New Mexico and they are capable of achieving high population densities that can support diverse carnivore communities (Frey 2018). It is likely that these rodents form a key part of the diet of red foxes in the mountainous areas of New Mexico.

While prey items for red foxes tend to be smaller-bodied (e.g., cottontail-size and smaller), red foxes do occasionally prey on species that are nearly their own size (e.g. wild turkeys [*Meleagris gallopavo*], juvenile ungulates). Further, red foxes will scavenge large-bodied species when the opportunity arises. For example, reports exist of red foxes scavenging stranded marine mammals, deer, bison, and even a polar bear (*Ursus maritimus*) carcass (Haelters et al. 2016; Jung et al. 2020). While those reports are of limited relevance to the feeding ecology of the species in New Mexico, reports of red foxes scavenging on bison, domestic sheep, and deer (Hewson 1984; Young et al. 2015; Selva et al. 2003) suggest that such events may also occur in the Southwest. Indeed, in urban settings where food refuse is available, or in rural areas where hunting results in discarded body parts (e.g., gut piles), red foxes may gain a significant portion of their diet by scavenging (Saunders et al. 1993; Contesse et al. 2004; Carrasco-Garcia et al. 2018).

Reproduction and Social Behavior

Reproduction has not been studied in New Mexico's red foxes, and Bailey's (1931) notes on this topic are very limited, mentioning only that litters of four have been reported, with six likely representing the maximum number of young based on the number of mammae. As documented through research outside New Mexico, red foxes are capable of breeding in their first year (Storm et al. 1976; Allen 1984), and females have a single estrus per year that lasts only one to six days (Asdell 1964). Breeding in North America may occur from December through April, with the peak typically in January and early February, but with later dates recorded in the north compared to the south (Sheldon 1949; Richards and Hine 1953; Hoffman and Kirkpatrick 1954; Layne and McKeon 1956; Storm et al. 1976). The natal den is excavated underground or modified from those of other burrowing mammals (Samuel and Nelson 1982). As reported by Sheldon (1949) for red foxes in the state of New York, the gestation period lasts 51–54 days. The young are born in March to May in North America in general (Voigt 1999), and specifically March or April in Texas (Schmidly and Bradley 2016). Based on counts of offspring inside dens (instead of numbers of corpora lutea, placental scars, or embryos), litter size in Iowa and Illinois averaged 3.5 and 3.8, respectively (Storm et al. 1976), while in California's Sacramento Valley, most litters ranged in

Photo 15.18. Three red fox cubs at the entrance to their den in Weld County, Colorado, on 21 April 2011. Red fox reproduction has not been studied in New Mexico, but in North America young are usually born in March to May. Bailey (1931) mentions litters of four having been documented in New Mexico. Photograph: © Mike Dunn.

Photo 15.19. Red fox cub just outside its den in Weld County, Colorado, 28 April 2011. Red foxes create dens by modifying another animal's burrow or digging their own one. Parental care is performed by both sexes, though the diet of juveniles more closely resembles that of the adult male instead of the adult female's. Photograph: © Mike Dunn.

Photo 15.20. Red foxes (here photographed in Rio Arriba County in 2009 or 2010) are usually monogamous. However, in some populations of Eurasian and North American red foxes, males may mate with several females (polygyny), and females may mate with several males (polyandry), thereby producing litters of mixed paternity. Photograph: © Roger Hogan.

size from one to four (Converse 2012). Larger numbers of offspring have been documented throughout the range of red foxes, but they are the product of multiple litters occurring together in the same den (e.g., Storm et al. 1976; Converse 2012; see below). The dispersal of young occurs in the fall (Storm et al. 1976).

The norm in the Canidae is social monogamy (though not exclusive mating) with pair-bonds lasting at least one breeding season and parental care by both sexes (Malcolm 1985; Asa and Valdespino 1998; Kleiman 2011; Macdonald et al. 2019). Both the European and North American red foxes, however, display flexible social systems, often deviating from social monogamy in response to changing environmental conditions (Cypher 2003). Genetic studies on European red foxes further show that mating strategies and parentage patterns can be highly complex, especially in populations characterized by high densities and abundant food resources (Baker et al. 2004; Iossa et al. 2009). In such populations, red foxes form social groups with plural breeding (aggregation of breeding females) and litters of mixed paternity, together reflecting a high degree of polygyny (males mate with multiple females), polyandry (females mate with multiple males), and polygynandry (both males and females have multiple mating partners). In a high-density European red fox population in the UK, Baker et al. (2004) found that 38–69% of all litters were of mixed paternity, each having been sired by two to seven different males. Dominant males mated with subordinate females within the same social group. Both dominant and subordinate males traveled outside their territories and sired cubs with dominant and subordinate females from neighboring groups. Dominant females did not mate with subordinate males within their social group but engaged in extra-pair copulations with both dominant and subordinate males from other groups (Baker et al. 2004). In California, Converse (2012) reported similar findings for

V. f. patwin in the Sacramento Valley, with genetic monogamy detected in only 23 (47%) of 49 social groups. Polygyny, polyandry, and polygynandry were found in 11 (22%), nine (18%), and six (12%) of all the social groups, respectively, with 31% of all litters exhibiting mixed paternity. Converse (2012) also observed that seven (14%) of 51 dens were shared by at least two family groups (communal denning).

The complexities of mating and social systems and of patterns of relatedness have yet to be fully explained in red foxes, but they likely relate to the ability of males to monopolize reproduction and the extent to which females are able to choose amongst potential mates. Further considerations include the net cost of tolerating subordinates where resources are abundant, the benefit and cost of philopatry relative to dispersal, and the likelihood of inheriting a territory and dominance status (see Creel and Macdonald 1995). Research in the UK showed red fox social groups to be mainly matrilineal, consisting in most cases of the dominant mated pair in addition to a subordinate female that is related to the dominant female but not the dominant male (Iossa et al. 2009; but see Converse 2012). Like for most canids, dispersal appeared to be typically male-biased, with females of the year instead remaining in their natal territory or simply moving to a neighboring area. Low relatedness between dominant males and subordinate females was the result of a high interannual turnover of dominant males (especially at low fox densities) and the occurrence of polyandry (at high densities) (Iossa et al. 2009). In New Mexico, there is no information on the social ecology and mating strategies of North American red foxes, nor how these could relate to environmental conditions, population densities, survivorship, and dispersal.

Interspecific Interactions

One of the fascinating aspects of the ecology of carnivores is how these species interact with one another. The ecological dynamics of mammalian carnivores are often characterized by extreme aggression between species, often culminating in members of one species killing those of another. This interaction has come to be known as *intraguild predation*, a type of interference competition; larger carnivores kill smaller carnivores for reasons that are not directly related to nutritional needs so much as due to a hypothesized need to reduce the numbers of potential competitors. These interactions are particularly well documented for canids and felids, and as such red fox numbers could be limited by the impacts of other carnivores. For instance, gray wolves are known to kill coyotes, and coyotes are known to kill multiple fox species including red foxes (Chapter 10, Chapter 14). Bobcats (*Lynx rufus*) may also play a role in limiting red foxes (Chapter 7), though unlike in the case of coyotes, the bobcat-red fox interaction is less clear-cut. In any case, while red foxes are not a prey species of these larger predators (foxes that are killed are typically not eaten), such predation risk might influence habitat availability for foxes (Moll et al. 2018). Indeed, it is easy to conjecture that the abundance of coyotes in New Mexico might limit the distribution and population sizes of red foxes across the state, much like coyotes limit where kit foxes may be found or where they may be abundant (Robinson et al. 2014). Coyotes are known to kill red foxes (e.g., Wooding 1984; Sargeant and Allen 1989) and in montane regions may limit the distribution of red foxes to higher elevations (Dekker 1989; Fuller and Harrison 2006). In Illinois, Gosselink et al. (2003) also found that red foxes used abandoned farms and urban landscapes as apparent refugia from coyotes as the latter avoided human-dominated areas. In an urban area (Madison, Wisconsin), coyotes and red foxes selected different habitats that facilitated coexistence; coyotes selected natural areas and avoided heavily developed areas, while red foxes

Photos 15.21a, b, and c (*top to bottom*). Within just a few days in May 2017, a remote camera located in an open subalpine forest near Williams Lake in New Mexico's Sangre de Cristo Mountains captured three species of carnivores that research suggests might interact: (a) Pacific marten (*Martes caurina*), (b) coyote (*Canis latrans*), and (c) North American red fox. All three species have overlapping diets and thus likely compete with one another. Interference competition and/or predation might occur between species pairs: numerous studies indicate that the larger coyote might kill or negatively interact with the smaller red fox, while research on the Eurasian red fox (*Vulpes vulpes*) and European pine marten (*Martes martes*) suggests the potential for North American red foxes to kill Pacific martens. Photographs: © Brian Jay Long.

Photo 15.22. Single image frame extracted from a video recorded 29 September 2021 showing a North American red fox–striped skunk (*Mephitis mephitis*) interaction in a low-density residential area north of Questa, Taos County. In the video, the skunk is seen repeatedly charging the red fox to claim or defend cat food left on the ground to attract wildlife. Photograph: © Dan Frank Kuehn.

avoided natural areas and selected open areas (Mueller et al. 2018). Similar patterns occur in parts of New Mexico, with red foxes more frequently observed in towns (e.g., Portales), and coyotes more frequently observed away from towns.

The flip side of the intraguild predation coin is how red foxes might interact with New Mexico's smaller carnivore species or even similar-sized species. For instance, in other settings red foxes are known to kill martens (*Martes* spp.) and feral cats (*Felis catus*). In New Mexico's northern mountains, North American red foxes overlap in their distribution with Pacific martens (*Martes caurina*; Chapter 20; Photos 15.21a and c), with possible local competition and/or predator-prey interactions between the two species. Interactions between the larger red foxes and either Pacific or American martens (*M. americana*) have not been studied in North America, though Thompson (1994) reported one marten in Ontario being killed by a red fox. In Scandinavia, red foxes are known to prey on European pine martens (*M. martes*), occasionally tracking them into their underground resting dens during the day

(Lindström et al. 1995). The two species share some of the same habitat and diet (e.g., Storch et al. 1990), with evidence suggesting that high red fox population densities can result in lower marten numbers (Lindström et al. 1995; Overskaug 2000; but see Kurki et al. 1998). In northeastern Poland, however, Brzeziński et al. (2014) documented a case of "reversed predation" by *M. martes* on red fox cubs. If true predation might be the underlying mechanism generally responsible for red foxes killing martens (see Storch et al. 1990), Brzeziński et al. (2014) argued that instances of "reversed predation" by martens could be the result of interference competition.

How might red foxes in New Mexico interact with species such as the ringtail (*Bassariscus astutus*), weasels (*Mustela and Neogale* spp.), the kit fox, and the swift fox? And perhaps more intriguing given their similar sizes, how might sympatric red and gray foxes interact (e.g., Rich et al. 2018)? In Scandinavia, Eurasian red foxes may negatively impact arctic foxes (Frafjord et al. 1989). Interactions between North American red foxes and gray foxes are still poorly documented in many regions (Chapter 12).

Disease Ecology

We know a good deal about the pathogens (including microparasites such as viruses and bacteria, macroparasites such as gastrointestinal helminths, and ectoparasites such as ticks and fleas) of red foxes, perhaps because many Carnivora share a similar pathogen community. As such, domestic dogs (*Canis familiaris*) and cats are also susceptible to many of the same pathogens as are red foxes, and thus a fair bit of veterinary parasitology that has focused on domestic carnivores is also relevant to red foxes. Further, while the harvest of New Mexican red foxes for the fur trade is of relatively minor economic importance, red fox pelt harvest elsewhere in its range is quite important, and thus there is a need to understand the mechanism by which pathogens cause disease and thereby population fluctuations. And finally, red foxes are potential vectors of the causative agents of rabies and echinococcosis, which represents a risk to humans. Thus, the pathogens of red foxes have historically drawn the attention of veterinary scientists, public health specialists, as well as wildlife managers.

Pathogens cause disease, and a non-exclusive list of diseases of particular importance to red foxes includes rabies, canine distemper virus (CDV), canine parvovirus (CPV), canine adenovirus, canine herpesvirus, canine parainfluenza virus (the agent of kennel cough in domestic dogs), mange (caused by the ectoparasitic mite *Sarcoptes scabiei*), echinococcosis (caused by the tapeworm *Echinococcus multilocularis*), and heartworm (*Dirofilaria immitis*). Such a list should not be viewed as all encompassing; there are a multitude of other parasitic species that have been identified from red foxes (Davidson et al. 1992; Larivière and Pasitschniak-Arts 1996; Campbell et al. 2020; Medkour et al. 2020). But this is no different from any other wildlife species—they all have the potential to harbor a diverse parasitic community. Readers should recognize, however, that the occurrence of any particular parasite in a host population varies dramatically from place to place, and even when occurring in a particular location, the proportion of individuals infected or the impacts on the host population as a whole might be quite low (e.g., Gortázar et al. 1998).

In the late 1990s, the New Mexico Department of Game and Fish collected eight red foxes, and the ectoparasites and gastrointestinal parasites from these animals were reported by Harrison et al. (2003) and Ubelaker et al. (2013), respectively. Fleas identified on red foxes included *Cediopsylla inaequalis* (McKinley County; this species may be *Spilopsyllus inaequalis*; Ford et al. 2004), *Euhoplopsyllus glacialis* (Roosevelt County), and *Pulex simulans* (Roosevelt County; Ford et al. 2004 also list this species as occurring on the red fox in San Juan County). Internal parasites identified were

Mesocestoides variabilis and *Ancylostoma caninum* from foxes collected in McKinley County, and *Spirocerca lupi* from a red fox from Roosevelt County. Given that phylogenetically close host species tend to share many of the same parasites (Huang et al. 2014), in the absence of further data on the ecto- and endoparasites of red foxes, one might tentatively assume that the species is also able to harbor the parasitic species that have previously been recorded for swift and kit foxes in regions where their ranges are geographically close and diets are similar. Geographically relevant surveys of the parasites of swift and kit foxes include Miller et al. (1998), Mayberry et al. (2000), Pence et al. (2004), Criffield et al. (2009), and Ubelaker et al. (2014a, b).

With regard to pathogens of public health concern, we are aware of no reports of rabies diagnosed in New Mexico red foxes, though this may simply be a function of relatively low population densities. Lyssavirus, the causative agent of rabies, occurs in several strains that are adapted for transmission within their hosts, and for foxes there are several strains, including arctic fox and gray fox strains. The fox strains of rabies virus do occur in New Mexico, but are apparently principally observed in gray foxes (Monroe et al. 2016; Ma et al. 2018). Nonetheless, because of the potential risk to humans, all foxes should be considered as potential carriers of rabies virus and extreme care should be used when handling them. Behavior alone is a poor mechanism for diagnosing rabies in foxes. The "furious" form of rabies does not occur in all cases, and infection by many pathogens may result in behavior that is easily misdiagnosed as rabies. For example, infection by CDV and CPV can result in such symptoms as ataxia, ocular and nasal discharge, and animals being out and about at abnormal times such as midday.

Similarly, the role of red foxes in the transmission of *E. multilocularis* to people in New Mexico and the broader southwestern United States is

Photo 15.23. A road-killed red fox cub on 29 May 2022 along US 64 north of Taos, New Mexico. The red fox is identified as a species of concern in the 2022 New Mexico Wildlife Corridors Action Plan, which identifies and ranks road corridors in need of wildlife road-crossing structures in areas with the highest number of wildlife-vehicle collisions and/or important habitat linkages. Red fox roadkills have been found in particular in Taos County including along US 64 and NM 68. Photograph: © Brian Jay Long.

Photo 15.24. Red fox on 28 May 2017 at a wildlife road-crossing underpass near Aztec, San Juan County. Based on research in the western United States, red foxes are known to cross roads safely using overpasses, culverts, and bridge underpasses. Note the very pale tail of this Four Corners red fox. The pale Four Corners red fox could be a unique form and is primarily associated with arid environments at low to mid-elevations. Photograph: © New Mexico Department of Transportation.

likely relatively minor, though it should be noted that there have been few studies of the parasite in southwestern wildlife (Cerda et al. 2018). This is in contrast to the importance of foxes and the parasite in regions of Europe (Oksanen et al. 2016). While there have been cases of echinococcosis in people in New Mexico (Schantz et al. 1977), these cases most likely derive from domestic dogs, and even further north in the Rocky Mountains, *Echinococcus* spp. have not been reported from red foxes (e.g., Pipas et al. 2021).

STATUS AND MANAGEMENT

The red fox is a protected furbearer in New Mexico (New Mexico Statute 17-5-2). Consequently, a furbearer license is required to take or possess the species and fur dealers must have a license to buy or sell skins. It is not legal to possess live red foxes in New Mexico, either as ranched animals or as exotic pets. During the 2022–2023 trapping season, the open season for the red fox extended from 1 November to 15 March, with no bag limit. Based on furbearer harvest records across the three seasons from 2018–2021, on average 118 red foxes (454 total) were taken each season, making it the fourth, fifth, and seventh most frequently taken species each year. Most (57%) of the red foxes were taken from a single county, San Juan County, in the northwestern corner of the state. The remainder were taken from 14 other counties, mostly in the northern and eastern parts of the state. Reports of red foxes taken in Grant County likely represent misidentified gray or kit foxes. In recent years red fox pelt prices have been relatively low, with prime pelts from northern North America averaging $15 and those from other areas averaging $5-$16 (https://www.trappingtoday.com/2022-fur-prices-fur-harvesters-march-auction-results).

There is considerable concern about the conservation status of endemic forms of red foxes in the western United States. The Sierra Nevada red fox,

V. f. necator, historically occupied high elevations of the southern Cascade Mountains of Oregon and northern California and the Sierra Nevada in California (USFWS 2018). In the Sierra Nevada it is only known to persist in one location with a population of ca. 10–50 adults (USFWS 2018); this distinct population segment has been listed as Endangered under the Endangered Species Act (USFWS 2021). In Oregon, the Sierra Nevada red fox is considered a Sensitive species by the US Forest Service, and research efforts are ongoing to better describe the status of this population, with the Oregon Cascade Mountains possibly serving as a refugium from more severe threats to habitat at more southern latitudes. Farther north in Washington, the Cascade red fox (*V. f. cascadensis*) is considered critically imperiled with a small effective population size (N_e=16; Akins et al. 2018), but little is known about its distribution and threats to its persistence, all necessary information for conservation actions. The Cascade red fox is currently being considered for Endangered status by Washington (Lewis et al. 2022). Similarly, the Sacramento Valley red fox (*V. f. patwin*) was only recently discovered following genetic research (Sacks et al. 2010), but its current distribution and population size are equivalent to those of the federally Endangered San Joaquin kit fox (*V. m. mutica*), and therefore likely require conservation attention (Black et al. 2019).

Research on the Rocky Mountain red fox (*V. f. macroura*) is lagging behind that for other endemic forms, particularly in the Southern Rocky Mountains, and it is largely assumed (though unproven) that populations of this subspecies are widespread and stable. However, Rocky Mountain red foxes are facing similar threats and challenges as the other endemic western red fox forms and recent data indicate non-native fur-farm genes have proliferated in some areas of Colorado and infiltrated into other areas (Merson et al. 2017).

The three endemic montane subspecies of the North American red fox may be at risk from climate change. All three utilize cold, high-elevation alpine tundra, subalpine forest, parkland, meadows, and upper montane forests. Increased temperature and aridity can increase the risk of wildfire and promote forest insect outbreaks that kill trees (see Chapter 5). Reduced snowpack can allow coyotes (a competitor and predator of red foxes) access to high-elevation environments year-round and can encourage tree encroachment into parklands and meadows causing loss of habitat (https://wdfw.wa.gov/species-habitats/species/vulpes-vulpes-cascadensis#conservation). It is also possible that the endemic high-elevation red foxes have physiological or behavioral adaptations to cold environments that may be disrupted by warming. Where low and high-elevation forms of red foxes are currently reproductively isolated from one another, climate change represents a threat through yet another mechanism. It could facilitate the introgression of non-native genes into high-elevation fox populations, by wildfires destroying dense forests currently acting as the separation between low- and high-elevation forms.

The endemic populations of red foxes found in the western United States are at risk of genetic introgression from populations of red foxes that are admixed with non-native (e.g., fur-farm) DNA (Quinn et al. 2022). In some areas, such as the Cascades of Washington, there has been little or no introgression of non-native genes into the endemic populations, possibly due to dense forests that separate the endemic populations at high elevations from the admixed populations at low elevations (Aubry 1984, Akins et al. 2018). In other areas, such as the Greater Yellowstone Ecosystem, there also has been no genetic introgression of non-native genes from lower elevation, but in this case the reason is because low-elevation populations appear to be the indigenous Great Plains form (Cross et al. 2018). However, in other areas such as Colorado where some low-elevation populations are admixed with fur-farm genes, barriers may be more diffuse allowing introgression of non-native genes into some high-elevation populations (Merson et al. 2017). It is likely that habitat alterations, such as the development of large wildfire scars, could lead to heightened disruption of endemic gene complexes.

Although there has been a considerable amount of recent genetic research in the western United States that has shown endemic red foxes at high elevations and admixed red foxes with non-native genes at low elevation, it is important to bear in mind that some endemic populations of red foxes occur at low elevations and that not all of these are admixed with non-native genes. Therefore, one should not assume that all high-elevation foxes are genetically pure endemic forms, nor should one assume that all low-elevation foxes are derived from non-native sources. Indeed, the recent discovery of the Sacramento Valley red fox, which had been assumed to be non-native because it occurred at low elevation, actually was a distinct endemic form worthy of taxonomic recognition (*V. f. patwin*), which emphasizes that judgements about the value of a population should not be made without appropriate data. It was only through rigorous scientific research that its status as a rare low-elevation endemic form requiring conservation actions became known (Sacks et al. 2010; Black et al. 2019). In addition, new research (Quinn et al. 2022) suggests that most low-elevation populations in the Intermontane West, including many assumed to be expanding non-native populations, had an endemic western genetic makeup. Regardless, to safeguard native populations of red foxes in New Mexico, it is imperative that research be conducted on the genetics of historical and contemporary populations, taxonomy of the San Juan Basin form, as well as distribution, habitat selection, and demography.

The North American red fox has been undervalued due to misconceptions about its origins, distribution, and ecology. Research has demonstrated that this species has a complex evolutionary history that has resulted in numerous endemic forms. New Mexico is at a crossroads between the Great Plains, Rocky Mountains, and Great Basin, and therefore may harbor a high diversity of red fox forms, some of which may require conservation efforts. We urge for more research that can reveal this diversity, both evolutionarily and ecologically. There still is much to learn about this most misunderstood part of New Mexico's natural heritage.

ACKNOWLEDGMENTS

We thank David Mikesic and the Navajo Nation for providing data on the red fox. Brian Long provided photographic records and locational information from several areas of New Mexico.

LITERATURE CITED

Abbott, I. 2011. The importation, release, establishment, spread, and early impact on prey animals of the red fox *Vulpes vulpes* in Victoria and adjoining parts of south-eastern Australia. *Zoologist* 35:463–533.

Aday, B. J., and A. L. Gennaro. 1973. Mammals (excluding bats) of the New Mexican Llano Estacado and adjacent river valleys. *Studies in Natural Science, Eastern New Mexico University* 1:1–33.

Akins, J. R., K. B. Aubry, and B. N. Sacks. 2018. Genetic integrity, diversity, and population structure of the Cascade red fox. *Conservation Genetics* 19:969–80.

Allen, G. M. 1954. Canid remains from Pueblo Bonito and Pueblo del Arroyo. *Smithsonian Miscellaneous Collections* 124:385–89.

Allen, J. A. 1896. On mammals collected in Bexar County and vicinity, Texas, by Mr. H.P. Attwater, with field notes by the collector. *Bulletin of the American Museum of Natural History* 8:47–80.

Allen, S. H. 1984. Some aspects of reproductive

performance in female red fox in North Dakota. *Journal of Mammalogy* 65:246–55.

Armstrong, D. M. 1972. Distribution of mammals in Colorado. *Monograph of the Museum of Natural History, University of Kansas* 3:1–415.

Asa, C. S., and C. Valdespino. 1998. Canid reproductive biology: an integration of proximate mechanisms and ultimate causes. *American Zoologist* 38:251–59. doi:10.1093/icb/38.1.251.

Asdell, S. A. 1964. *Patterns of Mammalian Reproduction.* 2nd ed. Ithaca, NY: Cornell University Press.

Aubry, K. B. 1983. The Cascade red fox: distribution, morphology, zoogeography, and ecology. PhD Dissertation, University of Washington.

———. 1984. The recent history and present distribution of the red fox in Washington. *Northwest Science* 58:69–79.

Aubry, K. B., M. J. Statham, B. N. Sacks, J. D. Perrine, and S. M. Wisely. 2009. Phylogeography of the North American red fox: vicariance in Pleistocene forest refugia. *Molecular Ecology* 18:2668–86.

Bailey, V. 1931 (=1932). *Mammals of New Mexico.* North American Fauna 53. Washington, DC: US Department of Agriculture, Bureau of Biological Survey.

Baird, S. F. 1852. Mammals. In *Exploration and survey of the valley of the Great Salt Lake of Utah including a reconnaissance of a new route through the Rocky Mountains,* by H. Stansbury, 309–13. USA Senate Executive document no. 3. Philadelphia: Lippincott, Grambo.

———. 1857. Mammals. In *Reports of explorations and surveys, to ascertain the most practicable and economical route for a railroad from the Mississippi River to the Pacific Ocean,* Vol. 8, Pt. 1, 1–757. Washington, DC: House of Representatives executive document no. 91.

Baker, P. J., S. M. Funk, M. W. Bruford, and S. Harris. 2004. Polygynandry in a red fox population: implications for the evolution of group living in canids? *Behavioural Ecology* 15:766–78.

Bangs, O. 1897. Description of a new red fox from Nova Scotia. *Proceedings of the Biological Society of Washington* 11:53–55.

Barnes, C. T. 1927. Utah mammals. *Bulletin of University of Utah* 17:1–183.

Bingham, R. L., L. A. Brennan, and B. M. Ballard. 2010. Discrepancies between euclidean distance and compositional analysis of resource selection data with known parameters. *Journal of Wildlife Management* 74:582–87.

Black, K. L., S. K. Petty, V. C. Radeloff, and J. N. Pauli. 2018. The Great Lakes region is a melting pot for vicariant red fox (*Vulpes vulpes*) populations. *Journal of Mammalogy* 99:1229–36.

Black, K. M., S. Preckler-Quisquater, T. J. Batter, S. Anderson, and B. N. Sacks. 2019. Occupancy, habitat, and abundance of the Sacramento Valley red fox. *Journal of Wildlife Management* 83:158–66.

Brown, D. E., A. T. Smith, J. K. Frey, and B. R. Schweiger. 2020. A review of the ongoing decline of the white-tailed jackrabbit. *Journal of Fish and Wildlife Management* 11:341–52.

Brzeziński, M., L. Rodak, and A. Zalewski. 2014. "Reversed" intraguild predation: red fox cubs killed by pine marten. *Acta Theriologica* 59:473–77.

Cagnacci, F., A. Meriggi, and S. Lovari. 2004. Habitat selection by the red fox *Vulpes vulpes* (L. 1758) in an alpine area. *Ethology, Ecology and Evolution* 16:103–16.

Campbell, S. J., W. Ashley, M. Gil-Fernandez, T. M. Newsome, F. Di Giallonardo, A. S. Ortiz-Baez, J. E. Mahar, A. L. Towerton, M. Gillings, E. C. Holmes, and A. J. Carthey. 2020. Red fox viromes in urban and rural landscapes. *Virus Evolution* 6(2):veaa065. https://pubmed.ncbi.nlm.nih.gov/33365150/.

Carrasco-Garcia, R., P. Barroso, J. Perez-Olivares, V. Montoro, and J. Vicente. 2018. Consumption of big game remains by scavengers: a potential risk as regards disease transmission in central Spain. *Frontiers in Veterinary Science* 5. doi:10.3389/fvets.2018.00004.

Carter, F. L. 1939. A history of the changes in population of certain mammals in western Kansas. MS thesis, Fort Hays Kansas State College, Hays.

Castañeda, I., T. S. Doherty, P. A. Fleming, A. M. Stobo-Wilson, J. C. Z. Woinarski, and T. M. Newsome. 2022. Variation in red fox *Vulpes vulpes* diet in five continents. *Mammal Review.* 52:328–42. https://doi.org/10.1111/mam.12292.

Castelló, J. R. 2018. *Canids of the world: wolves, wild dogs, foxes, jackals, coyotes, and their relatives.* Princeton, NJ: Princeton University Press.

Cerda, J. R., D. E. Buttke, and L. R. Ballweber. 2018. *Echinococcus* spp. tapeworms in North America. *Emerging Infectious Diseases* 24:230–35.

Choate, L. L. 1997. The mammals of the Llano Estacado. *Special Publications, Museum of Texas Tech University* 40:1–240.

Churcher, C. S. 1959. The species status of the New World red fox. *Journal of Mammalogy* 40:513–20.

Clothier, R. 1957. Distribution of the mammals of the Sandia and Manzano mountains, New Mexico. PhD dissertation, University of New Mexico, Albuquerque

Contesse, P., D. Hegglin, S. Gloor, F. Bontadina, and P. Deplazes. 2004. The diet of urban foxes (*Vulpes vulpes*) and the availability of anthropogenic food in the city of Zurich, Switzerland. *Mammalian Biology* 69:81–95.

Converse, K. E. 2012. Genetic mating system and territory inheritance in the Sacramento Valley red fox. MS thesis, California State University, Sacramento.

Cowan, I. M. 1938. Geographic distribution of color phases of the red fox and black bear in the Pacific Northwest. *Journal of Mammalogy* 19:202–6.

Creel, S., and D. W. Macdonald. 1995. Sociality, group size, and reproductive suppression among carnivores. *Advances in the Study of Behavior* 24:203–57.

Creel, G. C., and W. A. Thornton. 1971. A note on the distribution and specific status of the fox genus *Vulpes* in west Texas. *Southwestern Naturalist* 15:402–4.

Criffield, M. A., M. V. Reichard, E. C. Hellgren, D. M. Leslie, and K. Freel. 2009. Parasites of swift foxes (*Vulpes velox*) in the Oklahoma Panhandle. *Southwestern Naturalist* 54:492–98.

Cross, E. C. 1941. Colour phases of the red fox (*Vulpes fulva*) in Ontario. *Journal of Mammalogy* 22:25–39.

Cross, P. R., B. N. Sacks, G. Luikart, M. K. Schwartz, K. W. Van Etten, and R. L. Crabtree. 2018. Red fox ancestry and connectivity assessments reveal minimal fur farm introgression in Greater Yellowstone Ecosystem. *Journal of Fish and Wildlife Management* 9:519–30.

Crossett, R. L., II., and C. L. Elliot. 1991. Winter food habits of red foxes and coyotes in central Kentucky. *Proceedings of the Annual Conference of the Southeastern Association of Fish and Wildlife Agencies* 45:97–103.

Cypher, B. L. 2003. Foxes (*Vulpes* species, *Urocyon* species, and *Alopex lagopus*). In *Wild mammals of North America: biology, management, and economics*,

2nd ed., ed. G. A. Feldhamer, B. C. Thompson, and J. A. Chapman, 511–46. Baltimore: Johns Hopkins University Press.

Dalquest, W. W., F. B. Stangl Jr., and J. K. Jones Jr. 1990. Mammalian zoogeography of a Rocky Mountain-Great Plains interface in New Mexico, Oklahoma, and Texas. *Special Publications, Museum of Texas Tech University* 34:1–78.

Damasceno, E. M., E. Hingst-Zaher, and D. Astua. 2013. Bite force and encephalization in the Canidae (Mammalia: Carnivora). *Journal of Zoology* 290:246–54.

Davidson, W. R., M. J. Appel, G. L. Doster, O. E. Baker, and J. F. Brown. 1992. Diseases and parasites of red foxes, gray foxes, and coyotes from commercial sources selling to fox-chasing enclosures. *Journal of Wildlife Diseases* 28:581–89.

Dekker, D. 1989. Population fluctuations and spatial relationships among wolves, *Canis lupus*, coyotes, *Canis latrans*, and red foxes, *Vulpes vulpes*, in Jasper National Park, Alberta. *Canadian Field-Naturalist* 103:261–64.

Desmarest, M. A. 1820. *Mammalogie ou description des espèces de mammifères*. Paris: Mme Veuve Agasse.

Eaton, T. H. 1937. *Mammals of the Navajo country*. Berkeley, CA: National Youth Administration.

Elmhagen, B., and S. P. Rushton. 2007. Trophic control of mesopredators in terrestrial ecosystems: top-down or bottom-up. *Ecology Letters* 10:197–206.

Forbes-Harper, J. L., H. M. Crawford, S. J. Dundas, N. M. Warburton, P. J. Adams, P. W. Bateman, M. C. Calver, and P. A. Fleming. 2017. Diet and bite force in red foxes: ontogenetic and sex differences in an invasive carnivore. *Journal of Zoology* 303:54–63.

Ford, P. L., R. A. Fagerland, D. W. Duszynski, and P. J. Polechla. 2004. *Fleas and lice of mammals in New Mexico*. US Department of Agriculture, Forest Service, Rocky Mountain Research Station. Gen. Tech. Rep. RMRS-GTR-123.

Frafjord, K., D. Becker, and A. Angerbjörn. 1989. Interactions between arctic and red foxes in Scandinavia—predation and aggression. *Arctic Institute of North America* 42:354–56.

Frey, J. K. 2003. Distributional records and natural history notes for uncommon mammals on the

Llano Estacado of eastern New Mexico. *New Mexico Journal of Science* 43:1–24.

———. 2004. Taxonomy and distribution of the mammals of New Mexico: an annotated checklist. *Occasional Papers, Museum of Texas Tech University* 240:1–32.

———. 2013. Re-evaluation of the evidence for the importation of red foxes from Europe to colonial America: origins of the southeastern red fox (*Vulpes v. fulva*). *Biological Conservation* 158:74–79.

———. 2018. Beavers, livestock, and riparian synergies: bringing small mammals into the picture. In *Riparian research and management: past, present, and future*, Vol. 1, ed. R. R. Johnson, S. W. Carothers, D. M. Finch, K. J. Kingsley, and J. T. Stanley, 85–101. Fort Collins, CO: US Forest Service, Rocky Mountain Research Station, Gen. Tech. Rep. RMRS-GTR-377.

Frey, J. K., and Z. J. Schwenke. 2012. Mammals of Sugarite Canyon State Park, Colfax County, New Mexico. *Museum of Texas Tech University, Occasional Papers* 311:1–24.

Fuhrmann, R. T. 1998. Distribution, morphology, and habitat use of the red fox in the northern Yellowstone ecosystem. MS thesis, Montana State University, Bozeman.

Fuller, A. K., and D. J. Harrison. 2006. *Ecology of red foxes and niche relationships with coyotes on Mount Desert Island, Maine*. Final contract report to Acadia National Park and USDI National Park Service Northeast Region. Orono, ME: Department of Wildlife Ecology, University of Maine.

Gese, E. M., T. E. Stotts, and S. Grothe. 1996. Interactions between coyotes and red foxes in Yellowstone National Park, Wyoming. *Journal of Mammalogy* 77:377–82.

Global Invasive Species Database (GISD). 2012. Managed by the Invasive Species Specialist Group, IUCN Species Survival Commission. http://www.iucngisd.org/gisd/. Accessed 8 June 2021.

Gortázar, C., R. Villafuerte, J. C. Blanco, and D. Fernandez-de-Luco. 1998. Enzootic sarcoptic mange in red foxes in Spain. *Zietschrift für Jagdwissenschaft* 44:251–56.

Gosselink, T. E., T. R. Van Deelen, R. E. Warner, and M. G. Joselyn. 2003. Habitat partitioning and spatial use of coyotes and red foxes in east-central Illinois. *Journal of Wildlife Management* 67:90–103.

Gotch, A. F. 1979. *Mammals—their Latin names explained*. Poole, UK: Blandford Press.

Grinnell, J., J. J. Dixon, and J. M. Linsdale. 1937. *Fur-bearing mammals of California*. Vol. 2. Berkeley: University of California Press.

Haelters, J., E. Everaarts, P. Bunskoek, L. Begeman, J. W. Hinrichs, and L. IJsseldijk. 2016. A suspected scavenging event by red foxes (*Vulpes vulpes*) on a live, stranded harbour porpoise (*Phocoena phocoena*). *Aquatic Mammals* 42:227–32.

Hall, E. R. 1981. *The mammals of North America*. 2nd ed. Vol. 2. New York: John Wiley.

Halloran, A. F. 1942. The western red fox in southern New Mexico. *Journal of Mammalogy* 23:223.

———. 1946. The carnivores of the San Andres Mountains, New Mexico. *Journal of Mammalogy* 27:154–61.

———. 1962. An Arizona specimen of the red fox. *Journal of Mammalogy* 43:432.

———. 1964. *The mammals of Navajoland*. Navajo Tribal Museum, Navajoland Publications, 4.

———. 1965. Carnivore notes from the Navajo Indian Reservation. *Southwestern Naturalist* 10:139–40.

Harris, A. H. 1963. Ecological distribution of some vertebrates in the San Juan Basin, New Mexico. *University of New Mexico Press, Papers in Anthropology* 8:1–63.

———. 2013. *Pleistocene vertebrates of Arizona, New Mexico, and Trans-Pecos Texas*. https://www.utep.edu/leb/PleistNM2.pdf.

Harrison, R. L., M. J. Patrick, and C. G. Schmitt. 2003. Foxes, fleas, and plague in New Mexico. *Southwestern Naturalist* 48:720–22.

Hatcher, R. T., and G. W. Wigtil. 1985. Fate of red foxes released into southeastern Oklahoma. *Proceedings of the annual conference SEAFWA*. 39:321–25.

Hewson, R. 1984. Changes in the numbers of foxes (*Vulpes vulpes*) in Scotland. *Journal of Zoology* 203:561–69.

Hockman, J. G., and J. A. Chapman. 1983. Comparative feeding habits of red foxes (*Vulpes vulpes*) and gray foxes (*Urocyon cinereoargenteus*) in Maryland. *American Midland Naturalist* 110:276–85.

Hoffman, R. A., and C. M. Kirkpatrick. 1954. Red fox

weights and reproduction in Tippecanoe County, Indiana. *Journal of Mammalogy* 35:504–9.

Hoffmeister, D. F. 1986. *Mammals of Arizona*. Tucson: University of Arizona Press and Arizona Game and Fish Department.

Huang, S., O. R. P. Bininda-Emonds, P. R. Stephens, J. L. Gittleman, and S. Altizer. 2014. Phylogenetically related and ecologically similar carnivores harbour similar parasite assemblages. *Journal of Animal Ecology* 83:671–80.

Iossa, G., C. D. Soulsbury, P. J. Baker, K. J. Edwards, and S. Harris. 2009. Behavioral changes associated with a population density decline in the facultatively social red fox. *Behavioral Ecology* 20:385–95. https://doi.org/10.1093/beheco/arn149.

Johansson, I. 1947. The inheritance of the platinum and white face characters in the fox. *Hereditas* 33:152–74.

Jones, A. K., S. W. Liphardt, J. L. Dunnum, T. W. Perry, J. Malaney, and J. A. Cook. 2021. An overview of the mammals of the Gila region, New Mexico. *Therya* 12:213–36.

Jung, T. S., M. J. Suitor, S. Barykuk, J. Nuyaviak, D. C. Gordon, D. Gordon Jr, and E. Pokiak. 2020. Red Fox (*Vulpes vulpes*) scavenging on the spring sea ice: potential implications for Arctic food webs. *Canadian Field-Naturalist* 134:144–46.

Kamler, J. F., and W. B. Ballard. 2002. A review of native and nonnative red foxes in North America. *Wildlife Society Bulletin* 30:370–79.

Kasprowicz, A. E., M. J. Statham, and B. N. Sacks. 2016. Fate of the other redcoat: remnants of colonial British foxes in the eastern United States. *Journal of Mammalogy* 97:298–309.

Kidawa, D., and R. Kowalczyk. 2011. The effects of sex, age, season and habitat on diet of the red fox *Vulpes vulpes* in northeastern Poland. *Acta Theriologica* 56:209–18.

Kleiman, D. G. 2011. Canid mating systems, social behavior, parental care and ontogeny: are they flexible? *Behavioral Genetics* 41:803–9.

Klir, J. J. and J. E. Heath. 1992. An infrared thermographic study of the surface temperature in relation to external thermal stress in three species of foxes: the red fox (*Vulpes vulpes*), arctic fox (*Alopex lagopus*), and kit fox (*Vulpes macrotis*). *Physiological Zoology* 65:1011–21.

Kuo, Y. H., S. L. Vanderzwann, A. E. Kasprowicz, and B. N. Sacks. 2019. Using ancestry-informative SNPs to quantify introgression of European alleles into North American red foxes. *Journal of Heredity* 110:782–92. https://doi.org/10.1093/jhered/esz053.

Kurki, S., A. Nikula, P. Helle, and H. Lindén. 1998. Abundances of red fox and pine marten in relation to the composition of boreal forest landscapes. *Journal of Animal Ecology* 67:874–86.

Kutschera, V. E., N. Lecomte, A. Janke, N. Selva, A. A. Sokolov, T. Haun, K. Steyet, C. Nowak, and F. Hailer. 2013. A range-wide synthesis and timeline for phylogeographic events in the red fox (*Vulpes vulpes*). *BMC Evolutionary Biology* 13:114. http://www.biomedcentral.com/1471-2148/13/114.

Lang, R. W., and A. H. Harris. 1984. *The faunal remains from Arroyo Hondo Pueblo, New Mexico: a study in short-term subsistence change*. Arroyo Hondo Archaeological Series 5. Santa Fe, NM: School of American Research Press.

Larivière, S., and M. Pasitschniak-Arts. 1996. *Vulpes vulpes. Mammalian Species* 537:1–11.

Layne, J. N., and W. H. McKeon. 1956. Some aspects of red fox and gray fox reproduction in New York. *New York Fish and Game Journal* 3:44–74.

Lee, G. W., K. A. Lee, and W. R. Davidson. 1993. Evaluation of fox-chasing enclosures as sites of potential introduction and establishment of *Echinococcus multilocularis*. *Journal of Wildlife Diseases* 29:498–501.

Lewis, J. C., J. R. Akins, and T. Chestnut. 2022. State of Washington draft status report for the Cascade red fox. Washington Department of Fish and Wildlife, Wildlife Program.

Lewis, J. C., K. L. Sallee, and R. T. Golightly Jr. 1999. Introduction and range expansion of nonnative red foxes (*Vulpes vulpes*) in California. *American Midland Naturalist* 142:372–81.

Lindström, E. R., S. M. Brainerd, J. O. Helldin, and K. Overskaug. 1995. Pine marten—red fox interactions: a case of intraguild predation? *Annales Zoologici Fennici* 32:123–130.

Lowe, S., M. Browne, S. Boudjelas, and M. De Poorter. 2000. *100 of the world's worst invasive alien species: a selection from the global invasive species database*. Invasive Species Specialist Group, Species Survival Commission (SSC) of the World Conservation Union (IUCN).

Ma, X., B. P. Monroe, J. M. Cleaton, L. A. Orciari, Y. Li, J. D. Kirby, R. B. Chipman, B. W. Petersen, R. M. Wallace, and J. D. Blanton. 2018. Rabies surveillance in the United States during 2017. *Journal of the American Veterinary Medical Association* 253:1555–68.

Macdonald, D. W., L. A. D. Campbell, J. F. Kamler, J. Marino, G. Werhahn, and C. Sillero-Zubiri. 2019. Monogamy: cause, consequence, or corollary of success in wild canids? *Frontiers in Ecology and Evolution* 7:341. doi.org/10.3389/fevo.2019.00341.

Malcolm, J. R. 1985. Paternal care in canids. *American Zoologist* 25:853–56. doi:10.1093/icb/25.3.853.

Marcy, R. B. 1854. Exploration of the Red River of Louisiana, in the year 1852. House of Representatives executive document, A. O. P. Nicholson, Washington.

Mayberry, L. F., A. G. Canaris, J. R. Bristol, and S. L. Gardner. 2000. Bibliography of parasites and vertebrate hosts in Arizona, New Mexico and Texas (1893–1984). *Faculty Publications from the Harold W. Manter Laboratory of Parasitology* 2:1–100.

Medkour, H., Y. Laidoudi, J. L. Marié, F. Fenollar, B. Davoust, and O. Mediannikov. 2020. Molecular investigation of vector-borne pathogens in red foxes (*Vulpes vulpes*) from southern France. *Journal of Wildlife Diseases* 56:837–50.

Merriam, C. H. 1900. Preliminary revision of the North American red foxes. *Proceedings of the Washington Academy of Sciences* 2:661–76.

Merson C., M. J. Statham J. E. Janecka, R. R. Lopez, N. J. Silvy, and B. N. Sacks. 2017. Distribution of native and nonnative ancestry in red foxes along an elevational gradient in central Colorado. *Journal of Mammalogy* 98:365–77.

Mikesic, D. G., and C. T. LaRue. 2003. Recent status and distribution of red foxes (*Vulpes vulpes*) in northeastern Arizona and southeastern Utah. *Southwestern Naturalist* 48:624–34.

Miller, D. S., B. G. Campbell, R. G. McLean, E. Campos, and D. F. Covell. 1998. Parasites of swift fox (*Vulpes velox*) from southeastern Colorado. *Southwestern Naturalist* 43:476–79.

Moll, R. J., J. D. Cepek, P. D. Lorch, P. M. Dennis, T. Robison, J. J. Millspaugh, and R. A. Montgomery. 2018. Humans and urban development mediate the sympatry of competing carnivores. *Urban Ecosystems* 21:765–78.

Monroe, B. P., P. Yager, J. Blanton, M. G. Birhane, A. Wadhwa, L. Orciari, B. Petersen, and R. Wallace. 2016. Rabies surveillance in the United States during 2014. *Journal of the American Veterinary Medical Association* 248:777–88.

Mueller, M. A., D. Drake, and M. L. Allen. 2018. Coexistence of coyotes (*Canis latrans*) and red foxes (*Vulpes vulpes*) in an urban landscape. PLOS One 13(1):e0190971. https://pubmed.ncbi.nlm.nih.gov/29364916/.

Obbard, M. E. 1999. Fur grading and pelt identification. *In Wild furbearer management and conservation in North America*, ed. M. Novak, J. A. Baker, M. E. Obbard, and B. Malloch, 717–826. Ontario: Ontario Fur Managers Federation. CD-ROM.

Oksanen, A., M. Siles-Lucas, J. Karamon, A. Possenti, F. J. Conraths, T. Romig, P. Wysocki, A. Mannocci, D. Mipatrini, G. La Torre, and B. Boufana. 2016. The geographical distribution and prevalence of *Echinococcus multilocularis* in animals in the European Union and adjacent countries: a systematic review and meta-analysis. *Parasites and Vectors* 9:1–23.

Overskaug, K. 2000. Pine marten *Martes martes* versus red fox *Vulpes vulpes* in Norway: an inter-specific relationship? *Lutra* 43:215–21.

Packard, R. L., and J. H. Bowers. 1970. Distributional notes on some foxes from western Texas and eastern New Mexico. *Southwestern Naturalist* 14:450–51.

Pence, D. B., J. F. Kamler, and W. B. Ballard. 2004. Ectoparasites of the swift fox in northwestern Texas. *Journal of Wildlife Diseases* 40:543–47.

Perrine, J. D. 2005. Ecology of red fox (*Vulpes vulpes*) in the Lassen Peak region of California, USA. PhD Dissertation, University of California, Berkeley.

Perrine, J. D., J. P. Pollinger, B. N. Sacks, R. H. Barrett, and R. K. Wayne. 2007. Genetic evidence for the persistence of the critically endangered Sierra Nevada red fox in California. *Conservation Genetics* 8:1083–95.

Pipas, M. J., D. R. Fowler, K. D. Bardsley, and B. Bangoura. 2021. Survey of coyotes, red foxes and wolves from Wyoming, USA, for *Echinococcus granulosus* s. l. *Parasitology Research* 120:1335–40.

Pledger, M. 1975. In pursuit of the declining red fox. *Arkansas Game and Fish* 7(4):14–16.

Polechla, P. J. 2005. Mammals. In *Field Guide to the Sandia Mountains* ed. R. Julyan and M. Stuever, 169–95. Albuquerque: University of New Mexico Press.

Presnall, C. C. 1958. The present status of exotic mammals in the United States. *Journal of Wildlife Management* 22:49–50.

Quinn, C. B., J. R. Akins, T. L. Hiller, and B. N. Sacks. 2018. Predicting the potential distribution of the Sierra Nevada red fox in the Oregon Cascades. *Journal of Fish and Wildlife Management* 9:351–66.

Quinn, C. B., S. Preckler-Quisquater, J. R. Akins, P. R. Cross, P. B. Alden, S. L. Vanderzwan, J. A. Stephenson, P. J. Figura, G. A. Green, T. L. Hiller, and B. N. Sacks. 2022. Contrasting genetic trajectories of endangered and expanding red fox populations in the western US. *Heredity* 129:123–36. https://doi.org/10.1038/s41437-022-00522-4.

Ratliff, E. 2011. Taming the wild. *National Geographic* 219(3):34–59.

Rich, M., C. Thompson, S. Prange, and V. D. Popescu. 2018. Relative importance of habitat characteristics and interspecific relations in determining terrestrial carnivore occurrence. *Frontiers in Ecology and Evolution*, 6: doi.org/10.3389/fevo.2018.00078.

Richards, S. H., and R. L. Hine. 1953. *Wisconsin fox populations.* Wisconsin Conservation Department Technical Wildlife Bulletin No. 6.

Ritchie, E. G., and C. N. Johnson. 2009. Predator interactions, mesopredator release and biodiversity conservation. *Ecology Letters* 12:982–98.

Robinson, Q. H., D. Bustos, and G. W. Roemer. 2014. The application of occupancy modeling to evaluate intraguild predation in a model carnivore system. *Ecology* 95:3112–23.

Sacks, B. N., J. L. Brazeal, and J. Lewis. 2016. Landscape genetics of nonnative red fox of California. *Ecology and Evolution* 6:4775–91.

Sacks, B. N., Z. T. Lounsberry, and M. J. Statham. 2018. Nuclear genetic analysis of the red fox across its trans-Pacific range. *Journal of Heredity* 109:573–84.

Sacks, B. N., M. Moore, M. J. Statham, and H. U. Wittmer. 2011. A restricted hybrid zone between native and introduced red fox (*Vulpes vulpes*) populations suggests reproductive barriers and competitive exclusion. *Molecular Ecology* 20:326–41.

Sacks, B. N., M. J. Statham, J. D. Perrine, S. M. Wisely, and K. B. Aubry. 2010. North American montane red foxes: expansion, fragmentation, and the origins of the Sacramento Valley red fox. *Conservation Genetics* 11:1523–39.

Samuel, D. E., and B. B. Nelson 1982. Foxes. In *Wild mammals of North America: Biology, management, and economics*, ed. J. A. Chapman and G. A. Feldhamer, 475–90. Baltimore: Johns Hopkins University Press.

Sargeant, A. B., and S. H. Allen. 1989. Observed interactions between coyotes and red foxes. *Journal of Mammalogy* 70:631–33.

Sarre, S. D., A. J. MacDonald, C. Barclay, G. R. Saunders, and D. S. L. Ramsey. 2012. Foxes are now widespread in Tasmania: DNA detection defines the distribution of this rare but invasive carnivore. *Journal of Applied Ecology* 50:459–68.

Saunders, G., P. C. L. White, S. Harris, and J. M. V. Rayner 1993. Urban foxes (*Vulpes vulpes*): food acquisition, time and energy budgeting of a generalized predator. *Symposia of the Zoological Society of London* 65(4):215–34.

Schantz, P. M., C. F. Von Reyn, T. Welty, F. L. Andersen, M. G. Schultz. and I. G. Kagan. 1977. Epidemiologic investigation of echinococcosis in American Indians living in Arizona and New Mexico. *American Journal of Tropical Medicine and Hygiene* 26:121–26.

Schmidly, D. J., and R. D. Bradley. 2016. *The mammals of Texas.* 7th ed. Austin: University of Texas Press.

Selva, N., B. Jedrzejewska, W. Jedrzejewski, and A. Wajrak. 2003. Scavenging on European bison carcasses in Bialowieza primeval forest (eastern Poland). *Ecoscience* 10(3):303–11.

Sheldon, W. G. 1949. Reproductive behavior of foxes in New York State. *Journal of Mammalogy* 30:236–46.

Smith, C. H. 1840. *The natural history of dogs. Canidae or Genus Canis of authors including also the genera Hyaena and Proteles.* Vol. 2. Edinburgh: W. H. Lizars.

Statham, M. J., J. Murdoch, J. Janecka, K. B. Aubry, C. J. Edwards, C. D. Soulsbury, O. Berry, Z. Wang,

D. Harrison, M. Pearch, L. Tomsett, J. Chupasko, and B. N. Sacks. 2014. Range-wide multilocus phylogeography of the red fox reveals ancient continental divergence, minimal genomic exchange and distinct demographic histories. *Molecular Ecology* 23:4813–30.

Statham, M. J., B. N. Sacks, K. B. Aubry, J. D. Perrine, and S. M. Wisely. 2012. The origins of recently established red fox populations in the United States: translocations or natural range expansions? *Journal of Mammalogy* 93:52–65.

Statham, M. J., L. N. Trut, B. N. Sacks, A. V. Kharloamova, I. N. Oskina, R. G. Gulevich, J. L. Johnson, S. V. Temnykh, G. M. Acland, and A. V. Kukekova. 2011. On the origin of the domesticated species: identifying the parent population of Russian silver foxes (*Vulpes vulpes*). *Biological Journal of the Linnean Society* 103:168–75.

Storch, I., E. Lindström, and J. de Jounge. 1990. Habitat selection and food habits of the pine marten in relation to competition with the red fox. *Acta Theriologica* 35:311–20.

Storm, G. L., R. D. Andrews, R. L. Phillips, R. A. Bishop, D. B. Siniff, and J. R. Tester. 1976. Morphology, reproduction, dispersal, and mortality of midwestern red fox populations. *Wildlife Monographs* 49:3–82.

Swanson, B. J., R. T. Furmann, and R. L. Crabtree. 2005. Elevational isolation of red fox populations in the Greater Yellowstone Ecosystem. *Conservation Genetics* 6:123–31.

Swepston, D. A. 1981. New records of the red fox (*Vulpes vulpes*) in west Texas. *Southwestern Naturalist* 25:565.

Thompson, I. D. 1994. Marten population in uncut and logged boreal forest in Ontario. *Journal of Wildlife Management* 58:272–80.

Thornton, W. A., G. C. Creel, and R. E. Trimble. 1971. Hybridization in the fox genus *Vulpes* in West Texas. *Southwestern Naturalist* 15:473–84.

Townsend, C. H. T. 1893. On the life zones of the Organ Mountains and adjacent region in southern New Mexico, with notes on the fauna of the range. Science 22:313–15.

Trut, L. N. 1999. Early canid domestication: the farm-fox experiment. *American Scientist* 87:160–69.

Ubelaker, J. E., B. S. Griffin, D. W. Duszynski, and R. L. Harrison. 2013. Distribution records for helminths of the red fox *Vulpes vulpes* from New Mexico. *Southwestern Naturalist* 58:111–12.

Ubelaker, J. E., B. S. Griffin, G. M. Konicke, D. W. Duszynski, and R. L. Harrison. 2014a. Helminth parasites from the kit fox, *Vulpes macrotis* (Carnivora: Canidae), from New Mexico. *Comparative Parasitology* 81:100–104.

Ubelaker, J. E., B. S. Griffin, K. M. Mendoza, D. W. Duszynski, and R. L. Harrison. 2014b. Distributional records of helminths of the swift fox (*Vulpes velox*) from New Mexico. *Southwestern Naturalist* 59:129–32.

United States Department of Commerce, Bureau of the Census. 1942. *Sixteenth census of the United States: 1940. Agriculture*. Vol. 1, *First and second series state reports*. Pt. 6, *Statistics for counties: farms and farm property, with related information for farms and farm operators, livestock and livestock products, and crops*. Washington, DC: United States Printing Office.

United States Fish and Wildlife Service (USFWS). 2018. *Species status assessment report for the Sierra Nevada distinct population segment of the Sierra Nevada red fox*. Sacramento, CA.

———. 2021. Endangered species status for the Sierra Nevada distinct population segment of the Sierra Nevada red fox. *Federal Register* 86(146):41743–58 (3 August 2021).

Van Etten, K. W., K. R. Wilson, and R. L. Crabtree. 2007. Habitat use of red foxes in Yellowstone National Park based on snow tracking and telemetry. *Journal of Mammalogy* 88:1498–507.

Voigt, D. R. 1999. Red fox. In *Wild furbearer management and conservation in North America*, ed. M. Novak, J. A. Baker, M. E. Obbard, and B. Malloch, 379–92. Ontario: Ontario Fur Managers Federation. CD-ROM.

Volkmann, L. A., M. J. Statham, A. Ø. Mooers, and B. N. Sacks. 2015. Genetic distinctiveness of red foxes in the Intermountain West as revealed through expanded mitochondrial sequencing. *Journal of Mammalogy* 96:297–307.

Waltari, E., E. P. Hoberg, E. P. Lessa, and J. A. Cook. 2007. Eastward ho: phylogeographical perspectives on colonization of hosts and parasites across the Beringian nexus. *Journal of Biogeography* 34:561–74.

Westwood, R. C. 1989. Early fur farming in Utah. *Utah Historical Quarterly* 57:320–39.

Wooding, J. B. 1984. Coyote food habits and the spatial relationship of coyotes and foxes in Mississippi and Alabama. MS thesis, Mississippi State University.

Wozencraft, W. C. 2005. Order Carnivora. In *Mammal species of the world: a taxonomic and geographic reference*, Vol. 1, ed. D. E. Wilson and D. M. Reeder, 532–628. 3rd ed. Baltimore: Johns Hopkins University Press.

Yannic, G., M. J. Statham, L. Denoyelle, G. Szor, G. Q. Qulaut, B. N. Sacks, and N. Lecomte. 2017. Investigating the ancestry of putative hybrids: are arctic fox and red fox hybridizing? *Polar Biology* 40:2055–62.

Young, A., N. Márquez-Grant, R. Stillman, M. J. Smith, and A. H. Korstjens. 2015. An investigation of red fox (*Vulpes vulpes*) and Eurasian badger (*Meles meles*) scavenging, scattering, and removal of deer remains: forensic implications and applications. *Journal of Forensic Sciences* 60:S39–S55.

Zumbaugh, D. M., and J. R. Choate.1985. Historical biogeography of foxes in Kansas. *Transactions of the Kansas Academy of Science* 88(1–2): 1–13.

BLACK BEAR (*URSUS AMERICANUS*)

Cecily M. Costello

As familiar as many people are with popular images of the black bear—the cuddly teddy bear, the zoo attraction, the frightening movie menace, or the pesky campground raider—misconceptions abound about the real animal and how it lives in the wild. Less known still is how it lives in the arid Southwest, an area renowned for its scaly reptiles and prickly cacti, more than its furry mammals. No less compelling than many widely held notions, the true ecology of this fascinating animal, in New Mexico and beyond, is the subject of this chapter. Much of the information about New Mexico black bears is derived from a comprehensive eight-year field study conducted by the author and colleagues during 1992–2000. This study took place in the Sangre de Cristo and Mogollon mountains, and involved over 500 uniquely identified individuals, radio-telemetry monitoring of over 300 individuals, and other extensive field work. Additional information is gleaned from other studies in New Mexico, surrounding states, and Mexico, including more recent investigations.

The American black bear is the smallest, most abundant, and most widely distributed member of the Ursidae family in North America. The Ursidae family is of recent origin, believed to have diverged from the Canidae family approximately 20–25 million years ago (McLellan and Reiner 1994). Black bears are one of eight ursid species worldwide. At least 2 million years ago, after radiating to North America from Asia, a small forest-adapted ancestor (probably *Ursus abstrusus*) gave rise to the modern American black bear (Stirling and Derocher 1989). Despite climatic changes and competition with various species, the black bear was sufficiently adaptable to survive to the present day virtually unchanged from 1 million years ago (Stirling and Derocher 1989).

The New Mexico black bear (*U. a. amblyceps* Baird 1859) is one of 16 subspecies currently recognized (Larivière 2001). It is found in Arizona, New Mexico, southern Colorado, southern Utah, northern Mexico, and western Texas (see under "Distribution"). Recent genetic analyses generally support a unique southwestern lineage (Puckett et al. 2015), though evidence points to some western Texas and northern Mexico populations belonging instead to an eastern lineage (Onorato et al. 2004). Fossils of black bears in New Mexico are known at least as far back as 40,000 years ago (Harris 1993; see Chapter 1). Within their evolutionary history, black bears have coexisted with several other ursid species, including the now-extinct short-faced bear (*Arctodus simus*) and the North American spectacled bear (*Tremarctos floridanus*). The brown bear (*Ursus arctos*; also known as grizzly bear), which coexists with black bears in northwestern North

(*opposite page*) Photograph: © Geraint Smith.

Photo 16.1 (*above*). Black bear at Bandelier National Monument on 24 July 2007. Black bears are characterized by a thickset body, short powerful limbs, and a short tail. They have large heads with small round eyes and rounded, erect ears. The claws on their feet are non-retractable. Black bears at Bandelier National Monument on the Pajarito Plateau are part of a larger population found in a region encompassing the San Juan and Jemez mountains in northern New Mexico. Photograph: © Sally King.

Photos 16.2a and b (*top to bottom*). Black bears are capable swimmers, as illustrated here by an adult female swimming with her cub across Navajo Lake in northwestern New Mexico on 4 August 2011. (a) The two bears were first spotted in Navajo Lake where the distance from shore to shore is about 1.6 km (1 mi); (b) the cub was clinging to the mother's back as she swam and reached the eastern shore of the lake. Photographs: © Mark H. Meier.

Photo 16.3 (*left*). Armed with strong musculature and stout curved claws, black bears excel at climbing trees, even those with trunks as large as this ponderosa pine at Bandelier National Monument in north-central New Mexico. Photograph: © Sally King.

America today, radiated into North America only about 75,000–100,000 years ago, and probably reached the Southwest about 10,000–12,000 years ago. From that time on and until recently, black and grizzly bears both inhabited New Mexico and probably shared similar distributions. However, grizzly bears were extirpated from New Mexico by the late 1930s (Brown 1985; Chapter 17).

Black bears are characterized by a thickset body, short powerful limbs, a short tail, and a large head with small round eyes and rounded erect ears. Their feet have five toes tipped with curved, non-retractable claws approximately 4 cm (1.5 in) long. Their locomotion is plantigrade, meaning their heels are flat on the ground. They walk with a pigeon-toed gait and may appear slow and lumbering. Black bears are, however, quite agile and athletic. They are capable of running 40 km/hr (25 mph), have been known to swim distances in excess of 3 km (1.9 mi; Photos 16.2a and b), and are excellent tree climbers, even as adults (Photo 16.3). They are known to stand, and even walk, on their hind feet, usually to see over obstacles like tall grass (Brown 2009).

Black bears have a thick, long fur. Although the typical pelage is black throughout most of the species' distribution, color can and does vary among individuals (e.g., Lewis et al. 2020). Four color phases or morphs are generally recognized, reflecting both polygenic and environmental influences and, in one case, a recessive mutation (Rogers 1980; Rounds 1987; Lewis et al. 2020). In eastern and far northern populations, from Louisiana to the maritime provinces of Canada to Alaska, nearly all black bears exhibit the black morph. Brown-morph individuals exist in significant proportions in the Southwest, the Rocky Mountains all the way to the Yukon, and in regions surrounding the Great Basin (Rounds 1987). Brown fur has been explained as an adaptation to heat stress and more open habitats (Rogers 1980; Rounds 1987). A unique white or cream-colored morph, often referred to as the Kermode bear or spirit bear, occurs along the southern coastal regions of British Columbia (Hornaday 1905), and a bluish-gray color morph, known as the glacier bear, is found along the northern coast of British Columbia and the southern coast of Alaska (Anderson 1945; Lewis et al. 2020). Most dark-colored bears are marked with light brown fur around their muzzle, and many have light patches on their brows. Some bears also have a patch or chevron of lighter fur below their neck and between their front legs, known as a chest blaze.

In New Mexico, a minority of black bears are black morphs. Most are brown morphs, with hues ranging from blonde to cinnamon to chocolate to liver (Photos 16.4, 16.5, 16.6, 16.7). In our sample, we found that black bears in the Sangre de Cristo Mountains were 86% brown morph and 14% black morph, while bears in the Mogollon Mountains were 62% brown morph and 38% black morph (Costello et al. 2001). Due to bleaching and shedding, the hue of brown-morph individuals often changed over time. Even black-morph bears experienced some bleaching in their coat. Bears typically molted between May and August. During mid-summer, we often saw bears with short, newly grown, dark brown fur on their legs and lower body, crowned by a line of the previous year's bleached fur down their head and spine (Photo 16.8). As noted by Bailey (1931), we found that color morph frequently varied between mother and offspring or between littermates. Most cubs observed during winter den visits had a white chest blaze, ranging from a few white hairs to a sizable patch, but the majority of them had lost this marking by the time they matured (Photos 16.9, 16.10). Among hundreds of bears captured for our study, only one adult male featured a noticeably large chevron on his chest.

Members of the Ursidae have the smallest young at birth of any placental mammal, as compared to the size of their mother (Ramsay and Dunbrack 1986). Black bear cubs weigh

Photo 16.4. A minority of New Mexico black bears are black. This black-morph adult male, captured for research on 20 May 1998 using a foot snare, received an injection of immobilizing drugs from Bob Inman in the Mogollon Mountains in Catron County. Photograph: © Cecily Costello.

Photo 16.5. Hues of brown-morph bears can vary. One common color is "chocolate," as seen in this adult male photographed on 8 August 2010 at Bandelier National Monument in north-central New Mexico. Photograph: © Sally King.

Photo 16.6. An adult female and two yearling "cinnamon" bears on 30 April 2011 at Bandelier National Monument in north-central New Mexico. Brown fur has been explained as an adaptation to heat stress and more open habitats in parts of the species' geographic range. Photograph: © Sally King.

Photo 16.7. Prevailing sunny conditions in New Mexico often cause bleaching in brown-morph bears, making them appear blonde, like this adult female study bear photographed on 2 June 1994 in the Sangre de Cristo Mountains in Colfax County. Photograph: © Cecily Costello.

approximately 440 g (or 16 oz) at birth (Alt 1989; Mesa-Cruz et al. 2020), but gain weight about ninefold by 14 weeks. After this modest start, black bears can continue to grow, even beyond adulthood, to attain a body mass exceeding 300 kg (or 660 lbs) in some regions of North America (Pelton 2003). Black bears display considerable sexual size dimorphism, with adult males weighing as much as 2.2 times more than adult females (Alt 1980; Noyce and Garshelis 1994). In the Sangre de Cristo and Mogollon populations, body mass of bears began to diverge between sexes before sexual maturity and the gap between males and females subsequently increased with age (Fig. 16.1; Costello et al. 2001). From spring to early fall, most adult male bears weighed about

90–125 kg (200–275 lbs), while most adult females weighed about 55–70 kg (120–155 lbs). Among males, the maximum body mass was 159 kg (350 lbs), recorded twice, and as that body weight was in both cases measured during spring, it is likely the two individuals may have exceeded 180 kg (400 lbs) before denning the previous fall. Among females, the maximum body mass was 113 kg (250 lbs) during fall, likely representing a year-round maximum. Among both sexes, body mass began to decline at older ages, especially above 15 years. Body mass fluctuated seasonally and across years, due to variation in food abundance and, in adult females, the energetic costs of lactation. Adult female body mass was usually lowest in the late summer during years cubs were produced.

Black bears are popular symbols in New Mexico. The Zuni people consider the black bear to be guardian and master of the West (one of six regions of the world) and often symbolize it in carvings known as fetishes (Cushing 1994). This traditional depiction of a bear, often marked with an arrow-shaped heart line, is now a popular symbol in contemporary Native American art evocative of the

Photo 16.8. The color of brown-morph individuals can change considerably due to bleaching and shedding. This black bear, photographed at Bandelier National Monument in north-central New Mexico and likely a young male, is growing in a new coat of dark brown fur, but the remnants of last year's bleached fur persists on his head and spine. Photograph: © Sally King.

Photo 16.9. During our study, most black bear cubs observed during winter den visits had a white chest blaze, ranging from a few white hairs to a sizable patch of white fur. In the Sangre de Cristo Mountains in Colfax County, Mark Haroldson holds an eight-week-old male cub with a distinct chevron-shaped chest blaze on 25 March 1997. Photo: © Cecily Costello.

Photo 16.10. Black bear cub, about eight months old, at Bandelier National Monument in north-central New Mexico on 15 August 2010. Note the obvious chest blaze, which is a patch of lighter fur located below the neck and between the front legs. This chest marking often fades by adulthood. Photograph: © Sally King.

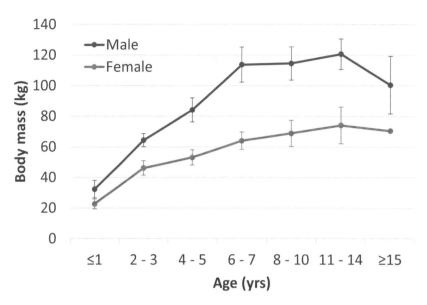

Figure 16.1. Mean body mass (kg) of male and female black bears as a function of age in the Sangre de Cristo and Mogollon mountains, New Mexico, 1992–2000 (Costello et al. 2001).

Southwest. The black bear was selected as New Mexico's official state animal in 1963 by the State Legislature, and the head of a black bear appears as the symbol of the New Mexico Department of Game and Fish (Fig. 16.2). In 1950, the story of an injured, orphaned black bear cub rescued after a forest fire in the Lincoln National Forest near Capitan, Lincoln County, became national news.

Fig. 16.2. The black bear was selected as New Mexico's official state animal in 1963 by the New Mexico Legislature. The head of a black bear appears as the symbol of the New Mexico Department of Game and Fish.

Later, this brown-morph cub was transferred to the National Zoo in Washington, DC, where he became the living embodiment of "Smokey Bear," a wildfire-prevention mascot first created by the US Forest Service in 1944. Upon his death in 1976, Smokey's remains were brought back to New Mexico and buried at the newly created Smokey Bear Historical Park in Capitan (Young 1984).

DISTRIBUTION

Throughout the species' evolutionary history, the distribution of black bears has been largely defined by the extent of forest in North America (Herrero 1972; Stirling and Derocher 1989). Black bears have inhabited eastern deciduous forests from Florida to Maine, boreal forests from Newfoundland to Alaska, and montane forests from Mexico to Alberta. Fossil evidence suggests black bears were never common in open vegetation communities such as the Great Plains, the Great Basin, or the arctic tundra, possibly due to competition with larger ursids such as short-faced bears and brown bears (Stirling and Derocher 1989).

After European settlement of North America,

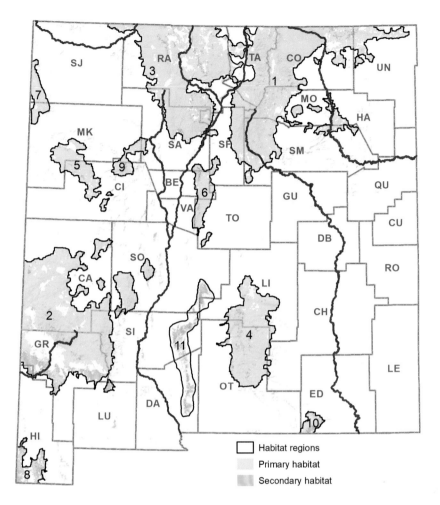

Map 16.1. Predicted black bear habitat in New Mexico (Costello et al. 2001). Primary habitats included closed-canopy forest and closed-canopy woodland cover types, along with 500-m edges of open-canopy woodlands or grasslands. Secondary habitats included shrubland cover types adjacent to primary habitat. Distinct tracts are numbered from largest to smallest by total area of primary and secondary habitats.

black bear numbers and distribution were significantly reduced due to habitat loss, unlimited hunting, and use of poisons. Beginning in the 19th century, black bears were eliminated or greatly reduced in several US states, including Illinois, Ohio, Kentucky, Florida, Alabama, Louisiana, Arkansas, Missouri, and Texas (Servheen 1989). However, during the last century, reforestation and legal limits on hunting have allowed populations to recover in many regions. The black bear populations in Arkansas, and subsequently Missouri,

Louisiana, Mississippi, and Oklahoma, were also augmented with bears transplanted from Minnesota in the 1950s. Today, black bear distribution is expanding and is known to include 41 US states, 12 Canadian provinces or territories, and 6 Mexican states (Garshelis et al. 2016). Throughout their current distribution, bears are variously protected as game animals or under the designation of threatened or endangered species.

As elsewhere, the distribution of black bears in New Mexico is correlated with that of forests and

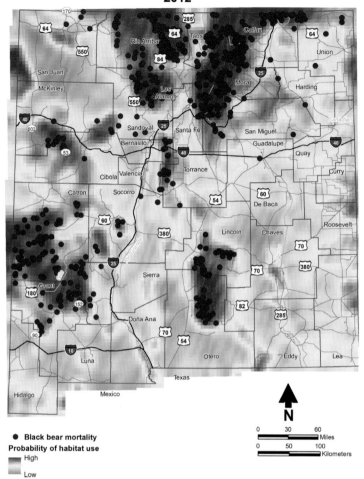

Black bear mortality
Probability of habitat use
High
Low

N

0   30   60
Miles
0   50   100
Kilometers

woodlands, primarily within mountain ranges with peaks above 2,500 m (8,200 ft; Costello et al. 2001). As such, bears are found in various distinct areas of the state, separated by large expanses of lower-elevation grassland or desert habitat incompatible with continuous occupation. To predict the total distribution of black bears in New Mexico, we developed a map of black bear habitat based on land-cover characteristics and biological information from our field studies, and we verified its accuracy with strong corroboration from locations of hunter-killed bears (Costello et al. 2001; Map 16.1). On the basis of our map, black bears currently occupy at least some portion of 25 of 33 New Mexico counties.

Two of the three largest tracts of bear habitat in New Mexico are located in the southern Rocky Mountains and are contiguous with bear habitat in Colorado. These include the Sangre de Cristo Mountains (Map 16.1 [Habitat Region 1]) and the region encompassing the San Juan and Jemez mountains and upper elevations of the Colorado Plateau (Map 16.1 [Habitat Region 3]). Reoccupation of black bears in the western panhandle of Oklahoma during the 1980s and 1990s has been attributed to emigration from the Sangre de Cristo region (Kamler et al. 2003). The other largest tract is contiguous with black bear habitat in Arizona and includes the Mogollon Mountains, the Black Range, the San Francisco Mountains, and other,

2014

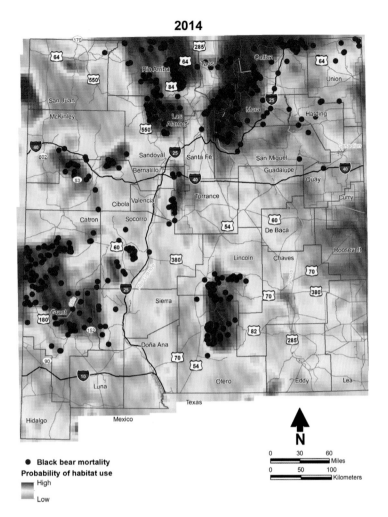

Maps 16.2a and b (*opposite page and above*). Predicted probability of habitat use by black bears in New Mexico overlaid with black bear mortality locations, based on occupancy modeling of detections at baited hair snares in the Sangre de Cristo Mountains during 2012 and the Sacramento Mountains during 2014 (adapted from Gould et al. 2019). Models did not account for cover type and therefore overpredicted use of open vegetation, such as in eastern New Mexico.

smaller mountain ranges in west-central New Mexico (Map 16.1 [Habitat Region 2]). The Sacramento and Capitan mountains form a relatively large tract in the south-central part of the state (Map 16.1 [4]). Other smaller, isolated tracts of habitat include the Zuni Mountains (Map 16.1 [5]), Sandia and Manzano mountains (Map 16.1 [6]), Chuska Mountains (Map 16.1 [7]), and San Mateo Mountains (i.e., Mount Taylor area; Map 16.1 [9]). The small tract of black bear habitat in the Bootheel region includes the Peloncillo and Animas

mountains (Map 16.1 [8]) and is part of the sky island complex of Arizona, New Mexico, Sonora, and Chihuahua. The small tract in the Guadalupe Mountains (Map 16.1 [10]) in southeastern New Mexico is shared with Texas. The small tract of secondary habitat in south-central New Mexico includes the Oscura, San Andres, and Organ mountains (Map 16.1 [11]); black bears in this tract are primarily associated with the closed-canopy pinyon (*Pinus edulis*) woodlands in the Oscura Mountains (J. Frey, pers. comm.). These separate

Photo 16.11. Mountainous regions of New Mexico, generally with peaks above 2,500 m (8,200 ft), form the core of black bear habitat in New Mexico. This June 1994 photograph of Baldy Mountain in Colfax County illustrates the variety of cover types available in the foothills and on the slopes of mountains, including low-elevation woodlands, mid-elevation forests, and high-elevation alpine communities. Photograph: © Don Jones.

or isolated habitat tracts essentially correspond to distinct black bear populations in New Mexico. However, some individuals may move between areas (see under "Movements, Home Range, and Social Behavior").

Gould et al. (2019) used occupancy modeling to infer black bear habitat use in the Sangre de Cristo Mountains during 2012 and 2013 and the Sacramento Mountains during 2014, based on bear detections at baited hair-snare sites in forest and woodland cover types (Costello et al. 2001). Habitat use was positively associated with primary productivity and negatively associated with road density. Applied statewide (Maps 16.2a and b), the models were spatially correlated with bear mortality locations during the same year and further corroborated the tracts of suitable habitat identified during our study (Map 16.1). However, Gould et al. (2019) recognized that the

model overpredicted habitat use in open cover types (especially irrigated areas in eastern New Mexico), because it was generated using data collected within closed canopy cover types and cover was not accounted for in the model.

HABITAT ASSOCIATIONS

In New Mexico, black bears are highly selective of closed-canopy forest and woodland communities that provide visual and thermal cover, and a variety of plant and animal foods (Costello et al. 2001). Mid- to upper elevation conifer forests represent the core of bear habitat in New Mexico (Costello et al. 2001; Guntly 2016; Bard and Cain 2020; Photo 16.11). Ponderosa pine (*Pinus ponderosa*) and mixed conifer forests dominated by Douglas-fir (*Pseudotsuga menziesii*) and white fir (*Abies concolor*) are interspersed with stands of quaking aspen (*Populus tremuloides*). At higher elevations, especially in the northern part of the state, subalpine forests consisting of Englemann spruce (*Picea engelmannii*), subalpine fir (*Abies lasiocarpa*), bristlecone pine (*Pinus aristata*), and limber pine (*Pinus flexilis*) are also important habitats. Typically staying within 500 m (1600 ft) of a forest edge, bears will also use interspersed meadows of mixed grasses and forbs (Photo 16.12), as well as alpine tundra communities surrounded by scree and talus slopes at the highest elevations. The herbaceous plants, grasses, and mast-producing shrubs and trees associated with these various forest communities provide food during all seasons. Common mast-producing species include chokecherry (*Prunus virginiana*), gooseberry or currant (*Ribes* spp.), serviceberry (*Amalanchier* spp.) and kinikinnick (*Arctostaphylos uva-ursi*). In addition, the plentiful woody debris provides a diversity of colonial insects for consumption. The acorns of Gambel oak (*Quercus gambelii*), a common understory species at mid elevations, are a major food for bears in most regions of New Mexico (Costello et al. 2001; see under "Diet").

Photo 16.12 (*top left*). Although closed-canopy forests and woodlands constitute preferred habitat, black bears will use interspersed meadows of mixed grasses and forbs. Bears typically stay within about 500 m (550 yards) of cover, like this young, radio-marked male photographed on 23 June 1993 in the Sangre de Cristo Mountains in Colfax County. Photograph: © Cecily Costello.

Photo 16.13 (*bottom left*). Bears seek many important foods, especially hard and soft mast species, in lower-elevation woodlands, like this individual photographed on 13 August 1997 in the foothills of the Sangre de Cristo Mountains in Colfax County. Photograph: © Cecily Costello.

Photo 16.14 (*right*). Black bears utilize riparian habitats for the abundant foods associated with the moist conditions. Wet drainage bottoms, like Devils Creek, Catron County (photographed in July 1993), also provide welcome thermal cover, water, and travel corridors, which are especially important in the more arid regions of New Mexico. Photograph: © Cecily Costello.

At the lower elevations, where they are connected or interspersed with the conifer forests mentioned above, woodland habitats are also important (Photo 16.13). These woodlands, along with associated shrublands, provide important foods, especially mast. In the cooler and moister regions of the state, they are typically dominated by pinyon (*Pinus edulis*), Rocky Mountain juniper (*Juniperus scopulorum*), and one-seed juniper (*J. monosperma*). Associated shrublands might include such mast-producing species as Gambel oak, wavyleaf oak (*Q. undulata*), squawbush (*Rhus trilobata*), and hawthorn (*Crataegus* spp.) (see also Chapter 2). Woodlands in the warmer and drier regions are often dominated by pinyon, alligator juniper (*J. deppeana*), and Utah juniper (*J. osteosperma*). The diversity of mast-producing shrubs tends to be higher and, in addition to the species already mentioned, might include gray oak (*Q. grisea*), emory oak (*Q. emoryi*), Arizona white oak (*Q. arizonica*), silverleaf oak (*Q. hypoleucoides*), netleaf oak (*Q. rugosa*), manzanita (*Artostaphylos* spp.), buckthorn (*Rhamnus* spp.) and silktassel (*Garrya* spp.). The abundance of prickly pear cactus (*Opuntia* spp.) also makes these important habitats for foraging (Costello et al. 2001).

Riparian areas found at various elevations provide important grasses, sedges, herbaceous plants, and mast-producing trees, including chokecherry, wild plum (*Prunus americana*), and Arizona walnut (*Juglans major*; Photo 16.14). They also provide thermal cover, water, and travel corridors which are especially important in the more arid regions of the state. Riparian corridors enhance the ability of bears to move between isolated tracts of bear habitat (Atwood et al. 2011; Gantchoff and Belant 2017).

Bears make their winter dens in all the same habitats they use during the active season. Use of mixed conifer and pinyon-juniper habitats was most common in the Sangre de Cristo and Mogollon mountains (Costello et al. 2001). Most den sites were located on the mid-to-upper portion of slopes, while by comparison few were located on ridge-tops or in bottoms. The wide variation in observed den locations suggests bears are not highly selective for any particular elevation, slope, or aspect when choosing a den site. However, Bard and Cain (2020) found that black bears in the Jemez Mountains were more likely to select den and bedding sites with lower horizontal visibility and higher basal area providing hiding cover. Mollohan (1987) reported similar results for adult female bears in northern Arizona and found that 74% of summer beds were located on the uphill side of a tree that offspring could climb for security or escape.

LIFE HISTORY
Diet and Foraging
Although taxonomically included in the Carnivora, black bears are, in fact, omnivores. Throughout North America, diets of black bears are dominated by plant matter (Hatler 1972; Beeman and Pelton 1980; Graber and White 1983; MacHutchon 1989; Raine and Kansas 1990). Compared with other families in the order Carnivora, ursids have evolved various traits associated with increased herbivory, including reduced cheek teeth, expanded molar chewing surfaces, and longer claws for digging (Herrero 1978; Bunnell and Tait 1981). At the same time, they have maintained the unspecialized gastrointestinal tract of carnivores, only lengthened (Davis 1964). They lack the morphological or physiological traits of true herbivores that would enable them to digest cellulose (Bunnell and Hamilton 1983; Pritchard and Robbins 1990). These constraints impose limits on the nutritional value of some plant foods, but still allow them to digest proteins as efficiently as most other carnivores (Bunnell and Hamilton 1983).

Bears are highly opportunistic feeders in New Mexico as elsewhere. Their diets vary annually in association with fluctuations in food availability,

Photo 16.15. Black bears feed on a variety of soft and hard mast species in New Mexico. Chokecherries (*Prunus virginiana*), most often growing near streams, are a common summer food for bears in many regions of the state. Photograph: © Cecily Costello.

Photo 16.16. In the more arid regions of the state, the fruit of prickly pears (*Opuntia* spp.) can be an important food source, especially during years of abundant fruit production. Photograph: © Bob Inman.

Photo 16.17 (*near right*). Bears will dig up and consume the stems and fruit of squawroot (*Conopholis mexicana*), a parasitic, non-photosynthesizing plant. Squawroot appears to be a particularly important food resource for black bears in the mountains of central and south-central New Mexico (Sandia, Manzano, and Gallinas mountains, and the ranges on White Sands Missile Range) (R. Winslow, pers. comm.). Photograph: © Bob Inman.

Photo 16.18 (*far right*). Gambel oak (*Quercus gambelii*) and other oak acorns are key fall foods. During our field study, occasional failures of acorn crops were associated with skipped reproductive opportunities and measurable declines in natality rates and cub survival. Photograph: © Cecily Costello.

especially mast crops and insect populations (Costello et al. 2001). It is not unusual for a particular food to dominate a seasonal diet in one year, only to be virtually absent the following year, a pattern that we observed in response to the fluctuating availability of prickly pear fruit and yellowjackets (*Vespula* spp.) in the Mogollon Mountains. Bear diets also vary seasonally in accordance with annual plant phenology and developmental changes in prey vulnerability. Importantly, bears engage in hyperphagia during

fall, when they spend almost all their waking hours finding and consuming food to build up fat stores necessary for winter hibernation. In New Mexico, bears generally forage heavily on vegetation and insects during spring and early summer; switch to soft mast (berries, pomes, and drupes) when it becomes available in late summer; and concentrate on hard mast (acorns and other nuts) during fall (Costello et al. 2001; Table 16.1; Fig. 16.3). While conforming to this general pattern, the specific species consumed by New

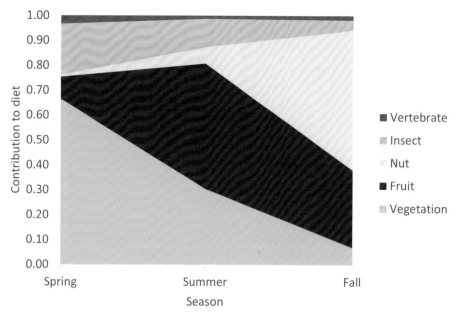

Figure 16.3. Observed proportional contribution of various food types in the diet of black bears in the Sangre de Cristo and Mogollon mountains, New Mexico, 1992–1995 (Costello et al. 2001). Contribution to diet was visually estimated as percent volume of each food item in the analysis of scats, following methods described by Hatler (1972).

Mexico black bears vary across the state, due to differences in habitat composition.

Vegetative matter is a major component of bear diets during all seasons, but is most dominant during spring. Digestible protein content of vegetation, especially forbs, is highest during the early growing season (Noyce and Garshelis 1998; Rode et al. 2001), and bears concentrate on these foods prior to onset of the fruiting season. Important grass species include bluegrass (*Poa*), fescue (*Festuca*), and muhly (*Muhlenbergia*), while frequently consumed forbs include vetch (*Vicia* spp.), peavine (*Lathyrus* spp.), and golden pea (*Thermopsis rhombifolia*). Bailey (1931:354) reported black bears stripping bark from trees "in search of the borers that work underneath." However, the "bear scratches showing on the naked trunks" were likely mixed with tooth marks and indicate that cambium feeding was the primary activity. Spring season cambium feeding, particularly on conifers, has been observed in many studies of black bears in the west, but is especially

common in the Pacific Northwest (Raine and Kansas 1990; Noble and Meslow 1998).

New Mexico bears feed on a wide array of fleshy fruits, but only a few species are distributed widely enough to become a major food for a regional population (Photo 16.15). Some of the more commonly consumed fruits statewide are chokecherry and gooseberry in the moister forests of northern New Mexico, and prickly pear, manzanita, and squawroot (*Conopholis mexicana*) in the more arid regions of the state (Photos 16.16, 16.17). Where alligator juniper exists, mostly in the Mogollon and Bootheel regions, bears consume the fresh berries during fall, as well as over-wintered berries during spring and summer. Although bears also feed on the fruits of one-seed, Utah, and Rocky Mountain junipers, these appear less preferred, possibly due to their smaller size. Consumption of juniper berries appears to be more extensive when hard mast production fails (Costello et al. 2001).

Table 16.1. Known and probable foods of black bears in New Mexico identified during bear studies conducted in the state (Costello et al. 2001; Guntly 2016; Kindschuh et al. 2016), as well as research in surrounding states and Mexico (Zager and Beecham 1982; McClinton et al. 1992; Hellgren 1993; Baldwin and Bender 2009; and López-González et al. 2009).

Type	Scientific Name	Common Name
Vegetation (leaves, roots, flowers)	*Agave* spp.	Agave
	Allium spp.	Wild onion
	Carex spp.	Sedge
	Cirsium spp.	Thistle
	Conopholis mexicana	Squawroot
	Dasylirion spp.	Sotol
	Equisetum spp.	Horsetail
	Festuca spp.	Fescue grass
	Fragaria spp.	Strawberry
	Heracleum lanatum	Cowparsnip
	Lathyrus spp.	Peavine
	Ligusticum porteri	Licorice-root
	Lomatium spp.	Biscuitroot
	Muhlenbergia spp.	Muhly grass
	Osmorhiza spp.	Sweet cicely
	Piptochaetium spp.	Speargrass
	Poa spp.	Bluegrass
	Prosopis glandulosa	Honey mesquite
	Ranunculus spp.	Buttercup
	Robinia neomexicana	New Mexico locust
	Rorippa spp.	Yellowcress
	Smilacina racemosa	False Soloman's seal
	Swertia spp.	Felwort
	Taraxicum spp.	Dandelion
	Thermopsis spp.	Golden pea
	Trifolium spp.	Clover
	Vicia spp.	Vetch
	Yucca spp.	Yucca
Soft mast	*Actea arguta*	Red baneberry
	Amelanchier spp.	Serviceberry
	Arbutus spp.	Madrone
	Arctostaphylos spp.	Manzanita
	Arctostaphylos uva-ursi	Kinnikinnick
	Berberis trifoliolata	Agarito

continued on next page

Type continued	Scientific Name	Common Name
	Conopholis mexicana	Squawroot
	Cornus stolonifera	Red-osier dogwood
	Crataegus sp.	Hawthorn
	Echinocereus spp.	Hedgehog cactus
	Garrya wrightii	Wright silktassel
	Juniperus spp.	Juniper
	Lonicera spp.	Honeysuckle
	Opuntia spp.	Prickly pear cactus
	Prunus americana	Wild plum
	Prunus virginiana	Chokecherry
	Rhamnus sp.	Buckthorn
	Rhus trilobata	Skunkbush sumac
	Ribes spp.	Gooseberry or currant
	Rosa spp.	Wild rose
	Rubus stigosus	Raspberry
	Sambucus spp.	Elderberry
Hard mast	*Juglans major*	Arizona walnut
	Pinus edulis	Pinyon
	Quercus spp.	Oak
Invertebrates	Coleoptera	Beetles
	Euxoa auxiliaris	Army cutworm moth
	Formicidae	Ants
	Orthoptera	Locusts
	Diplocentrus peloncillensis	Scorpion
	Vespula spp.	Yellowjackets
Vertebrates	*Bos taurus*	Cattle
	Cervus canadensis	Elk
	Lagomorpha	Rabbits
	Neotoma cinerea	Bushy-tailed woodrat
	Odocoileus hemionus	Mule deer
	Odocoileus virginianus	White-tailed deer
	Passeriformes	Passerines
	Rodentia	Rodents
	Sciuridae	Squirrels
	Sylvilagus nuttallii	Mountain cottontail
	Urocyon cinereoargenteus	Gray fox
	Ursus americanus	Black bear

In New Mexico, acorns represent both the primary hard mast and a major fall food (Photo 16.18; Costello et al. 2001; Guntly 2016). Bears will also feed on over-wintered acorns during spring following abundant crops. With its wide geographic and elevational distribution, Gambel oak is likely the most important oak species. The nuts of pinyon trees are also consumed, and they represent the only other widely distributed hard mast available to bears in New Mexico. However, pinyons produce plentiful crops only periodically. In local areas, the interval between good nut crops often varies from 2 to 5 years, and may in some cases even exceed 10 years (Jeffers 1995). Consequently, though bears may feed heavily on pine nuts during bumper years, it cannot be considered a staple of the black bear diet to the same degree as acorns. For example, we observed moderate-to-excellent mast productivity of Gambel and/or wavyleaf oaks during 4 of 8 years in the Sangre de Cristo Mountains, and moderate-to-good productivity of Gambel and/or gray oaks during 4 of 8 years in the Mogollon Mountains. In contrast, moderate seed productivity of pinyon occurred during only 1 of 8 years in either study area, and no good or excellent crops were observed at all (Costello et al. 2001).

Regularity of mast crops differs between soft and hard mast producers, due to their differing methods of seed dispersal. When hard mast is consumed, the seed is destroyed. Therefore, hard mast producers tend to synchronize reproduction and produce sporadic crops that are large enough to satiate seed consumers, ensuring that some seeds are left to germinate. In contrast, when soft mast is consumed, seeds are scarified in the guts of consumers and then dispersed through scat deposition. Soft mast crops are generally more consistent between years (Inman and Pelton 2002).

Coniferous forests, like those in New Mexico and in most of western North America, possess lower diversity and abundance of hard and soft mast, compared to deciduous forests in eastern North America (Clark et al. 2020). Consistency in hard mast availability is enhanced in Eastern forests because multiple genera often coexist, such as oak, hickory (Carya spp.), beech (Fagus grandifolia), and hazel (Corylus spp.). Additionally, soft mast functions as a surrogate when hard mast production is low (Inman and Pelton 2002). In contrast, with oak being the single predominant hard mast genera and many soft mast species occurring with only patchy distribution, black bears in New Mexico and the Southwest are faced with spatially variable and temporally unpredictable mast abundance. The resulting "boom-and-bust" food economy likely influenced local adaptation and continues to influence black bear behavior and population dynamics (see under "Reproduction and Social Behavior").

The animal matter in New Mexico black bear diets is largely insects (Costello et al. 2001; Guntly 2016), which are valuable sources of protein and fat in both the larval and adult stages (Noyce et al. 1997; Swenson et al. 1999). Ants (Formicidae) constitute a significant portion of the spring and summer diets. Bears also consume other insects, such as yellowjackets and locusts (Orthoptera), especially during years when they are particularly abundant. In the northern regions of the state, some bears forage on aggregations of army cutworm moths (Euxoa auxiliaries) on the high-elevation talus slopes of the Jemez Mountains (Coop et al. 2005) and we also observed this feeding activity in the Sangre de Cristo Mountains. This is a geographically limited food source, however, and it is therefore valuable to only a few bears residing near the timber line. Another invertebrate, the scorpion Diplocentrus peloncillensis, was identified as a food item in the Sierra de San Luis in Sonora, Mexico, just south of the New Mexico border (López-González et al. 2009).

Much of the vertebrate meat in black bear diets is obtained through scavenging, but black bears are also predators. Among ungulate species present in New Mexico, bears have been identified as predators of elk (*Cervus canadensis*: e.g. Griffin et al. 2011), white-tailed deer (*Odocoileus virginianus*: e.g. Carstensen et al. 2009), mule deer (*O. hemionus*: e.g. Monteith et al. 2014), moose (*Alces alces*: e.g. Schwartz and Franzmann 1991; moose are accidental in occurence in New Mexico), and, less frequently, pronghorn (*Antilocapra americana*: Barnowe-Meyer et al. 2009) and bighorn sheep (*Ovis canadensis*: Sawyer and Lindzey 2002). Bears primarily prey on neonates, especially during the first few weeks of life when young are relatively immobile and rely on hiding cover for protection. Predation on adult ungulates has also been observed (elk: Banmore and Stradley 1971; white-tailed deer: Svoboda et al. 2011; moose: Austin et al. 1994). Predation on ungulates is probably a learned behavior, and the propensity for predation can vary among individuals and populations (Zager and Beecham 2006). As opportunistic feeders, bears will prey on various other vertebrates when they are vulnerable. Black bears have been documented to prey or attempt to prey on a number of other species present in New Mexico, including beavers (*Castor canadensis*; Smith et al. 1994a), porcupines (*Erithizon dorsatum*; Brown and Babb 2009), yellow-bellied marmots (*Marmota flaviventris*; Van Vuren 2001), great blue herons (*Ardea herodias*; Foss 1980), northern flickers (*Colaptes auratus*; Deweese and Pillmore 1972; Walters and Miller 2001), red-naped sapsuckers (*Sphyrapicus nuchalis*; Walters and Miller 2001), bank swallows (*Riparia riparia*; Morlan 1972), bald eagles (*Haliaeetus leucocephalus*; McKelvey and Smith 1979), and dusky grouse (*Dendragapus obscurus*; Sullivan 1979).

During our study, we found evidence of limited neonate ungulate consumption from analyses of scats (Costello et al. 2001). Outside of Raton in Colfax County, Guntly (2016) observed a modest spike in the presence of mule deer in scats during the fawning season but documented no bear-caused mortality among the sample of fawns monitored for survival. In contrast, during a study on the Valle Vidal Unit of the Carson National Forest and adjacent Vermejo Park Ranch in northern New Mexico during 2009–2012, Tatman et al. (2018) found that black bear predation was the leading cause of mortality (39%) for elk during their first year of life. The average age of bear-killed calves was 16 days (range 2–103). Two-thirds of all documented black bear predation events took place within the calves' first three weeks of life, and 98% occurred within the first 3 months of life. Elk calf survival rates were greater during years of higher, experimental harvest of black bears, predominantly males. Overall survival and risk of predation were also influenced by birth mass, and therefore Tatman et al. (2018) concluded that bear predation was not entirely additive. Kindschuh et al. (2016) performed ground investigations of GPS locations for 24 black bears in the Jemez Mountains during 2012–2014 and found predation or scavenging events involving 56 ungulates (elk calves, adult mule deer, adult elk, and cattle [*Bos taurus*]) and 17 smaller vertebrates (mountain cottontails [*Sylvilagus nuttallii*], bushy-tailed woodrats [*Neotoma cinerea*], and passerines).

Hibernation

The dormancy of plants, often coupled with snow cover, makes winter a time of food shortage for black bears, considering their largely herbivorous diet and inability to digest cellulose. It is believed bear hibernation evolved primarily as a response to this seasonal scarcity of food. Although quite different from hibernation among smaller mammals, such as ground squirrels, the physiological state attained by bears is generally considered true hibernation (Folk et al. 1976; Hellgren 1998). Nelson (1980) argued it is the most refined response to starvation of any mammal. For periods of up to seven months, a hibernating bear does not eat, drink, defecate, or urinate (Folk et al. 1976; Nelson 1980; Hellgren 1998). In all hibernators, metabolic activity is generated from energy stored in the form of fat, but most other hibernators must arouse periodically to feed. Bears are capable of recycling the waste products of fat metabolism into lean body mass, while other hibernators must arouse and eliminate wastes through urination or suffer toxemia (Nelson et al. 1973; Hellgren 1998). Bears achieve energy savings by reducing their metabolic rate to 25% of normal and heart rate to 16% of normal (Tøien et al. 2011). Concurrent with these metabolic reductions, black bear body temperature drops modestly from 38° to 31–35° C (from 100.4° to 87.8–95° F). By comparison, most other hibernators decrease metabolic rates and body temperatures to <10% of normal. For example, the metabolic rate of hibernating ground squirrels (*Urocitellus* spp.) has been measured as low as 1% of normal (Yensen and Sherman 2003), with body temperature dropping to 3–5° C and sometimes below freezing (Nelson and Robbins 2015). Bears can achieve energy savings without dramatic changes in body temperature, largely because of their lower surface-area to volume ratio. This maintenance of near normal body temperature allows bears to quickly arouse and respond to disturbance. Remarkably, female black bears give birth in winter dens, and in addition to their own metabolic requirements must fulfill the energetic demands of gestation and lactation during the hibernating phase. Black bears have been observed to lose 14–34% of their body weight during the denning period, nursing females an additional 9% (Hock 1960; Erickson and Youatt 1961; Tietje and Ruff 1980).

The timing and duration of denning can vary based on a number of factors, including latitude, elevation, annual food availability, sex, and reproductive status. It is generally thought that bears forage until they encounter a decreasing or negative energy return per unit of search effort (Lindzey and Meslow 1976; Johnson and Pelton 1980). Bears at northern latitudes and higher elevations tend to enter dens earlier, remain denned longer, and emerge later than bears in southern latitudes and at lower elevations (Smith et al. 1994b). Typically, female bears enter dens earlier and emerge from dens later than male bears, and the denning period is longest for females giving birth in the den (e.g., Tietje and Ruff 1980). Finally, bears often remain active later during the fall when food is most abundant (e.g., Johnson and Pelton 1980). The onset of hibernation is probably not controlled by changes in weather. Rather, inclement weather such as snowfall typically coincides with decreased food availability, and tends to compound the negative energy return by increasing the effort required to obtain food (Schwartz et al. 1987; Schooley et al. 1994).

In New Mexico, the vast majority of black bears probably enter dens sometime between mid-October and mid-November and exit dens during April (Fig. 16.4), putting the average denning duration at about five to six months. A latitudinal difference in denning chronology was observed within the state (Inman et al. 2007). On average, bears in the Sangre de Cristo Mountains entered dens one to two weeks earlier and emerged from dens one to two weeks later than bears in the Mogollon Mountains. Among bears in the Sangre de Cristo Mountains, the typical pattern of earlier entry and later emergence by reproductive females was quite conspicuous. However, no differences in denning chronology were observed between male and female bears in the Mogollon Mountains (Inman et al. 2007). In the Mogollon Mountains, bears were simply observed to enter dens later when oak production was particularly high (Costello et al. 2001).

Some black bears, residing in regions with mild conditions and year-round food availability, forgo denning entirely. Absence of denning, by males and non-pregnant females, has been observed in Mexico (Doan-Crider and Hellgren 1996) and Texas (Mitchell at al. 2005), where habitats are comparable to some areas of southern New Mexico. Doan-Crider and Hellgren (1996)

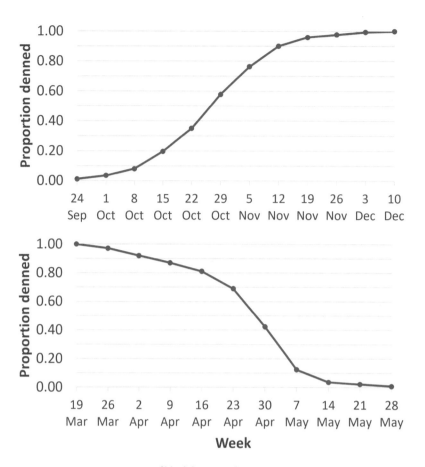

Figure 16.4. Approximate proportion of black bears in dens during the fall den entry period (top) and the spring den emergence period (bottom), by week, averaged for the Sangre de Cristo and Mogollon mountain populations, 1992–2000 (Costello et al. 2001).

noted that winter-active black bears in Mexico fed on madrone berries (*Arbutus* spp.) as well as sotol (*Dasylirion* spp.) and yucca (*Yucca* spp.) leaves throughout January, but signs of feeding decreased during February–March. Bailey (1931:355) reported that "even in the milder sections of the State [black bears] become extremely fat in autumn, and with the first heavy snowstorms and cold weather in the mountains they seek their winter quarters in some hollow bank, cave, or scooped-out hollow under a big log or [boulder], where they pass the winter in a state of inactive lethargy." Our study results corroborate this assertion. Among hundreds of radio-marked

bears in the Sangre de Cristo and Mogollon mountains, no evidence of individuals forgoing denning was ever observed (Costello et al. 2001). Nevertheless, two females in the Mogollon Mountains postponed den entry until January–February during an especially mild winter when acorns were abundant.

New Mexico black bears use a wide variety of structures for their dens, and there is no indication that suitable den sites are limited (Costello et al. 2001). Bears commonly excavate their dens by digging under boulders, trees, stumps, or downfall (Photo 16.19). Sometimes, they simply dig into the ground without any supporting structure.

Photo 16.20. Bears frequently make use of natural cavities within or amongst rocks, like this den photographed on 1 March 2000 and occupied by a lone radio-marked female black bear in the Sangre de Cristo Mountains in Colfax County. As long-persisting features of the landscape, some natural rock cavities may have seen use by black or grizzly bears (*Ursus arctos*) for thousands of years. Photograph: © Cecily Costello.

Photo 16. 19. Black bears in New Mexico frequently excavate their dens by digging under boulders, trees, stumps, or downfall. This den, photographed on 27 January 1999 and occupied by a radio-marked subadult male in the Mogollon Mountains, is a typical example of an excavated rock den. The berm (i.e., mound of excavated material) shows significant weathering, which indicates previous use of this den, possibly by the same individual or another bear. Photograph: © Bob Inman.

Photo 16.21. Female black bears are more likely than males to use hollow tree cavities for dens. Here, Larry Temple peaks into the ground-level entrance of a live tree den occupied by a radio-marked female and her two neonate cubs in the Sangre de Cristo Mountains in Colfax County, 16 March 1995. Photograph: © Cecily Costello.

Photo 16.22. Viewed from the entrance about 12 m (40 ft) up, a radio-marked adult female black bear hibernates at the bottom of a hollow, dead tree in the Sangre de Cristo Mountains in Colfax County, 13 March 1998. Unable to reach the bears, we verified the presence of two cubs by listening to their bawling vocalizations. Photograph: © Cecily Costello.

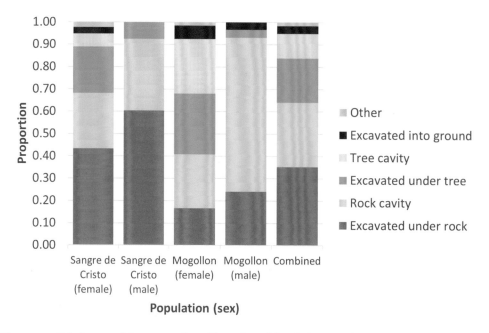

Figure 16.5. Relative use of den types selected by male and female bears in the Sangre de Cristo and Mogollon mountains, 1992–2000.

Bears also utilize natural rock cavities, hollows inside live or dead trees, and on rare occasions, man-made mine openings (Photos 16.20, 16.21, 16.22). Finally, a few bears will simply construct ground nests made with conifer boughs or other plant materials. Relative use of the various den types varies geographically and by sex, as revealed by a comparison of 390 dens used by black bears in the Sangre de Cristo and Mogollon mountains (Fig. 16.5; Costello et al. 2001). Females used dens associated with trees, especially tree cavities, more often than males. The reason for this difference is not known, but it may be that tree dens provide superior thermal insulation, important to reproductive females (Johnson et al. 1978). Or, perhaps, trees large enough to accommodate adult males are less common. Females with offspring may also be more selective for den types or locations that provide security from predators, disturbance, or infanticidal males (Johnson and Pelton 1981; Weaver and Pelton 1994, Gantchoff et al. 2019). All in all, certain den types tend to

be more closely associated with specific habitats, so this likely accounts for some of the variation in den selection. Most adult bears denned within their established home ranges; however, some bears, especially males, denned in distant areas where they had just foraged during the fall (Costello 2008; Noyce and Garshelis 2011).

Repeated use of particular dens appears to be quite common, especially those dens associated with rocks or boulders (Costello et al. 2001). In our study, for example, 15% of dens excavated under boulders were known to have been used in previous years, either by the same bear or by a different individual. Another 35% of dens under boulders were suspected of previous use, because of weathering or plant growth on the berm (or mound of excavated material). Reuse of natural rock cavities, hollow trees, and dens excavated under trees was also fairly common. Several study bears were observed to use the same den 2–5 times during the 8-year study period. Due to their durable nature, some of the natural rock cavities we visited, I imagine, may have seen use by black or even grizzly bears for thousands of years.

In the Sangre de Cristo and Mogollon mountains, the majority of observed dens had only one entrance (Costello et al. 2001). Bears are not often credited with being nimble, but the remarkably small den openings demonstrate that they are quite flexible. I am a slender person (more so in the 1990s when we conducted our research), but even I got stuck—temporarily—in den openings on a few occasions. Believe me, the prospect of being stuck inside a den with a live (albeit drugged) bear is something that will get your heart racing! Contrary to the popular image of bears hibernating inside roomy caves, most observed dens were just large enough for a bear to rest comfortably. Bears often lined their dens with bedding materials, hauled in from the surrounding landscape (Photo 16.23). Common bedding materials included pine needles, twigs, leaves, grasses, and conifer boughs. Dens lacking a bed were characterized by an earthen floor, sometimes strewn with stones. Although bedding materials were sometimes used to block the entrance, more often the den was open to the outside surroundings. The minimal number of openings, snug size, and use of bedding materials all probably help in the retention of body heat inside the den. At higher elevations, and especially on north-facing slopes, dens might also be covered by an insulating blanket of snow, as was the case in our study (Photo 16.24). Even when snow is over 1 meter (~3.3 ft) deep, there is usually a small hole in the snow kept open by body heat generated from within the den.

Reproduction, Survival, and Longevity

Like most other members of the order Carnivora, black bears can be classified as a K-selected species, characterized by slow maturation, low reproductive output, and long life spans (Caughley 1977). Female black bears take several years to reach reproductive maturity. Across their entire

Photo 16.23. When preparing their dens, black bears often line them with bedding materials hauled in from the surrounding landscape, like the radio-marked mother of this two-month-old cub in the Sangre de Cristo Mountains, Colfax County, 17 March 2000. While pregnant in the fall, she raked in leaves, pine needles, twigs, and duff, which then provided a soft bed and insulation for her litter of two cubs born in mid-winter. Photograph: © Cecily Costello.

Photo 16.24. At higher elevations dens are often covered by an insulating blanket of snow, like this black bear den in the Mogollon Mountains in Catron County, 12 February 1998. Despite modest declines in body temperature during hibernation, the heat emanating from a bear is often sufficient to melt the snow above the den chamber. Photograph: © Bob Inman.

Photo 16.25. In the Sangre de Cristo Mountains, Colfax County, Don Jones holds a litter of 2 cubs on 25 March 2000. Most of the black bear litters we observed during our study consisted of two cubs. At roughly 10 weeks old and equipped with long, sharp claws, these cubs are nearly ready to emerge from the den and are capable of climbing trees to escape potential threats. Photograph: © Cecily Costello.

range, females generally produce their first litter at ages between 3 and 8 years (Larivière 2001; Pelton 2003). In our study, we found that about 75% of females produced their first litter either at age 5 or 6 years, evenly divided between the two ages. Fewer than 10% of females reproduced as 4-year-olds, while roughly 15% first reproduced as 7-year-olds. We observed one female that did not produce her first litter until the age of 9 years (Costello et al. 2003). This female was captured and translocated twice (due to minor conflicts with humans) and made her way back to her own home range each time. The physical demands of this travel, as well as the reduced opportunity for breeding, may have contributed to the unusual delay.

Breeding and parturition are separated by as many as 8 months in female black bears, as the result of a process known as delayed implantation (Wimsatt 1963). Following fertilization during a spring-summer mating season, eggs divide until the blastocyst stage (about 300 cells) and remain within the fallopian tubes for several months. In late fall, the blastocyst migrates down the fallopian tubes and implants in the uterine wall, at which time gestation begins (Wimsatt 1963). The actual gestation length is approximately 60 days (Hellgren et al. 1991) and cubs are generally born during January in the den, though dates in late December and early February have been documented (Alt 1983; Bridges et al. 2011). Delayed implantation may be adaptive in bears for two primary reasons. First, it allows breeding to occur early in the active season when it would not interfere with the prolonged fall foraging necessary to build up fat stores for hibernation (Ewer 1973). Secondly, even after mating occurs, it is postulated that delayed implantation may allow those females with fat stores insufficient for gestation and lactation to forgo implantation. Lack of litter production has been linked to food shortages and poor nutritional status in many studies (e.g., Noyce and Garshelis 1994), including our own in New Mexico (Costello et al. 2003).

Black bear litter sizes are known to range from one to five in wild populations (e.g., Alt 1989; Kolenosky 1990), but litter sizes are generally smaller in western versus eastern North America (Clark et al. 2020). We documented a range of only one to three cubs among 115 litters observed during our study. Litters of two cubs were most common, accounting for 71% of the total (Photo 16.25). One-cub litters accounted for about 24%, and were more common among first-time mothers. Only 5% of litters consisted of three cubs, and these were typically produced by older, larger females. Four-cub litters have been reported anecdotally in New Mexico but were never observed during our research. In our study the observed sex ratio of 187 cubs of known sex did not differ from parity: 51% female to 49% male.

After their birth in the den, cubs remain with their mothers for approximately 16–18 months, denning with them during their second winter.

Following den emergence in the spring, yearling bears generally become independent between May and July, at which time the female is usually receptive to mating again. Bears have been observed to give birth to newborn cubs after fall separation from the previous year's cubs (LeCount 1983) or even in the presence of yearlings (Alt 1989). Nonetheless, these extreme cases appear to be extremely rare, and no evidence of their occurrence has been noted in New Mexico.

The prolonged dependence of offspring on their mother sets the minimum successful birth interval at 2 years. When a female suffers the loss of her entire litter, she may reproduce immediately the following year if the loss occurred during the breeding season, setting the minimum unsuccessful breeding interval at 1 year. It is not uncommon for an adult female to skip 1, or even 2 reproductive years, usually as a result of food shortages and poor physical condition (Larivière 2001; Pelton 2003). In New Mexico, we observed 69 litter intervals ranging from 1 to 3 years. Successful intervals were 2–3 years with a mean of 2.1, while unsuccessful intervals were 1–3 years with a mean of 1.3. Over two-thirds of skipped reproductive years coincided with exceptionally low counts of acorn production (i.e., failures), indicating that a lack of food, and poor physical condition, were the causative agents. Oak failures had significant impacts on bear reproduction. Natality rates (number of cubs/female ≥ 4 years old/year) decreased by more than 60% following years of oak failure, and recruitment rates (number of yearlings/female ≥5 years old/year) decreased by more than 70% two years after oak failure. Overall, based on the Sangre de Cristo and Mogollon populations, adult female black bears produced an average of 0.8 cubs per female per year and 0.4 yearlings per female per year. (Costello et al. 2003)

Beston (2011) compared published black bear population vital rates across North America and reported that natality was lowest in the Southwest, owing to smaller average litter sizes and longer average birth intervals. Additionally, cub survival rates were lowest in the Southwest (Beston 2011). We found that mean litter size was not associated with annual variation in acorn production in New Mexico, but cub survival and litter loss were associated with acorn failures (Costello et al. 2003). This suggests that low food availability does not act as a proximate factor explaining smaller litter sizes. Instead, I suspect that the unpredictable, boom-and-bust food economy in the Southwest has acted as an ultimate factor optimizing a smaller average litter size, by minimizing energetic investment in cubs that may be lost and maximizing energy reserves for future reproductive effort.

The multi-year birth interval means that only about half of all the adult females are receptive for mating during any year. Asynchrony in their estrous cycles further limits the number of females eligible for mating at any given time during a breeding season. This means that adult males far outnumber receptive females and breeding competition is extremely intense. Several males may compete for the opportunity to mate with an estrous female, and their interactions may include vocal and physical displays, as well as fighting. Under these circumstances, larger males generally gain access to females and it is not unusual for a small number of males to dominate reproduction (Kovach and Powell 2003). Our DNA analyses of paternity in the Sangre de Cristo and Mogollon populations support this concept. Breeding age of males ranged from 3 to 21 years, but we found that males of intermediate age were the most successful breeders, with a peak at about 10 years of age. We observed many fighting wounds on adult males, especially on the head and neck, including superficial scratches from claws, deep gashes from canines, and even bitten-off ears. Frequency of fresh wounds peaked at approximately 13 years, while body size peaked at about 11–12 years. The concurrence of these peak ages of body size, wounding, and reproductive success suggests that mating opportunities are enhanced by the expression of dominance during physical confrontations. Success of young males (7 years old) was negatively associated with density of mature males. It may be that at low densities, smaller, subordinate males may find unattended females more often than at high male densities, and this may provide them with mating opportunities (Costello et al. 2009). Judging from signs of female estrus, observations of male-female pairs, and evidence of male fighting, it appears that the mating season in New Mexico ranges from May to September, with a peak in June (Costello et al. 2001, 2009), similar to other populations (Bridges et al. 2011). DNA analyses indicated both male and female black bears often mate with more than one partner, even during the same year. Approximately 28% of multi-cub litters had multiple paternity, and males were observed to sire cubs with up to 3 different females during the same year (Costello et al. 2009).

Annual survival of black bears is lowest among cubs and usually increases with age (Lee and Vaughan 2005). Across the black bear's range, cub survival rates vary widely among populations, usually falling within the range of 40% to 90% (Beston 2011). Estimates of cub survival were 48–81% among populations studied in Arizona (LeCount 1982, 1987b), Colorado (Beck 1991; Baldwin and Bender 2009), and Mexico (Doan-Crider and Hellgren 1996). Observed cub survival in our study was 55% (n = 148), or at the mid- to lower end of that spectrum. Survival was lower among litters of first-time mothers (38%) compared to those of multiparous mothers (62%). The cause of death was rarely documented, only because neonates were far too small to be fitted with radio transmitters. However, we did document cub deaths resulting from intraspecific predation and automobile collision. Other studies suggest that most cub mortality is from natural causes,

including intra- and interspecific predation, disease, starvation, abandonment, accidents, and flooding of natal dens. Human-caused mortalities are overall less common among cubs, but they include collisions with vehicles, poaching, and, rarely, hunter harvest—despite customary prohibitions against taking bears younger than 1 year (Alt 1984; LeCount 1987a; Elowe and Dodge 1989; Garrison et al. 2007).

In New Mexico, observed yearling survival among 72 radio-marked bears was 86% (Costello et al. 2001). Among yearlings, 44% of documented mortalities were from natural causes, 33% were due to humans, and 22% were from unknown causes. Natural causes included intraspecific predation, predation by cougar (*Puma concolor*), and starvation. Human-caused mortalities included legal hunting, and depredation and illegal kill (Costello et al. 2001). Beck (1991) documented a yearling survival rate of 94% in Colorado.

In New Mexico, observed annual survival rates for adult and subadult females were above 90%, and rates for adult and subadult males were above 80% (n = 591 bear-years). Unlike cub and yearling mortality, all observed adult male mortality was anthropogenic, as was two-thirds of observed female mortality. Legal hunting was the leading cause of death, accounting for about half of all documented mortality among both sexes. Illegal and depredation killing, as well as collisions with road traffic, represented other important mortality sources (Map 16.3). Natural mortality, in the form of intraspecific predation, was observed among adult females, and accounted for about 9% of deaths. The remaining female mortalities were from unknown causes, some of which were likely natural. A 16-year-old female was observed with a debilitating dermatitis, which we believe was a form of mange (see under "Diseases and Parasites"; Costello et al. 2006). Survival rates of independent bears observed during our study were similar to many other hunted populations and fell well within the range observed for

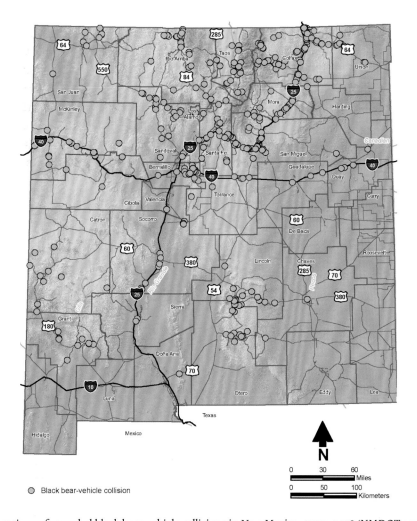

Map 16.3. Locations of recorded black bear-vehicle collisions in New Mexico, 2002–2018 (NMDOT, unpubl. data).

southwestern populations (LeCount 1990; Beck 1991; Doan-Crider and Hellgren 1996; Baldwin and Bender 2009).

Black bears have been known to live to 39 years of age in the wild (K. Noyce, Minnesota Department of Natural Resources, pers. comm), but bears older than 25 years are rarely observed. Across North America, the primary method for estimating the age of research-captured and hunter-killed bears is the cementum annuli method (Willey 1974). This involves extracting a small, vestigial premolar tooth, which can be safely removed from live bears under anesthesia.

When a thin cross-sectional segment of the tooth is viewed, annual layers of cementum can be counted to estimate age, similar to growth rings on a tree. These estimates are relatively reliable, but their accuracy decreases with age (Costello et al. 2004). We captured or monitored 21 different individuals who reached an estimated age of at least 15 years. Combining data from both of our study populations, these older cohorts represented approximately 11% of the adult populations (Costello et al. 2001). Among them, we only witnessed the natural demise of one bear—the female with suspected mange described above.

Photo 16.26. Movements of black bears across harsh, xeric environments once appeared anomalous, but evidence suggests they may be an integral part of the ecology of bears in the Southwest. They allow individuals to gain access to better conditions in response to drought, wildfire, or other perturbations. This photograph shows a bear that crossed grasslands south of Raton, Colfax County on 25 July 2011, possibly after being displaced by a wildfire that burned about 11,330 ha (28,000 acres) earlier that year. Photograph: © Jean-Luc Cartron.

Photo 16.27. A remote camera captured this photo of a large male black bear on 6 September 2017 in open grasslands of the Mockingbird Mountains, Sierra County, roughly 50 km (~30 mi) from the closest primary habitat in the Sacramento Mountains, though only a small distance from secondary habitat in the Oscura Mountains. Photograph: © White Sands Missile Range Garrison Environmental Division and ECO Inc., Ecological Consultants.

Photo 16.28. Near the bottom of this photograph, a bear can be seen traveling down an arroyo on the San Andres National Wildlife Refuge, Doña Ana County, within a tract of secondary habitat but roughly 60 km (~35 mi) from the nearest primary habitat in the Sacramento Mountains. Judging from the inset photo of this individual, the bear appears to be a subadult male—the sex-age class most likely to undertake long-distance dispersal and provide genetic exchange between isolated populations. Photographs: © Mara E. Weisenberger.

All the oldest individuals we observed were still alive when we removed their radio collars at the end of the study, including a 23-year-old male from the Sangre de Cristo Mountains population and a 27-year-old female from the Mogollon Mountains population. We believe this particular female, as well as another 24-year-old female from the Mogollon population, had reached reproductive senescence, having produced their last observed litters at the ages of 24 and 22 years, respectively.

Use of a bear population model with observed reproductive and survival rates indicated that New Mexico populations were stable or slightly increasing by no more than 2–4% per year on average (Costello et al. 2001).

Movements, Home Range, and Social Behavior

Black bears are solitary animals (e.g., Pelton 2003). Except for the long association of mothers and their dependent offspring, most social interactions are temporary and brief. Mating pairs usually unite for only a few days to two weeks. At long-lasting concentrated food sources, such as salmon streams or garbage dumps, groups of bears may congregate and form social hierarchies, largely governed by body size and sex (Rogers 1987; Chi 1999). Although most populations are spatially dispersed, black bears still exist within a social community of familial and non-familial neighbors. Social interactions are likely influenced by kinship, competition, and dominance. Black bears are known to rub and chew on many objects, including trees, signs, fence posts, power poles, and buildings. They possess scent glands, and it is believed marking behavior is a form of chemical and visual communication (Kendall and Stetz 2010; Taylor et al. 2015).

At any given time, a local black bear population is composed primarily of resident bears with established home ranges, but it also includes transient, non-resident bears undergoing natal dispersal (i.e., movements away from the birth site to the site of first reproduction; see Costello 2010). Natal dispersal is known to be highly male-biased in black bears. While nearly all young males disperse away from their natal range, most females remain in the vicinity of their birth site, a characteristic known as natal philopatry (Rogers 1987; Elowe and Dodge 1989; Beck 1991; Schwartz and Franzmann 1992; Lee and Vaughn 2004). Dispersal patterns were quite apparent in our study populations in New Mexico, as evidenced by movement data from radio-marked bears whose natal range was known or surmised from genetic identification of their mother (Costello 2008, 2010). Nearly all males in the sample emigrated from their natal range, usually between the ages of 1 and 3 years. Some young males remained transient for several years, moving their center of activity about 15–70 km (~9.5–43.5 mi) with each successive year, for an annual average of about 40 km (~25 mi). These movements sometimes took bears into the neighboring states of Colorado and Arizona. We observed that most dispersing males settled into permanent home ranges by the age of 4 years, and all settled by the age of 7 years. In contrast, nearly all females in our sample established their home range such that it overlapped that of their mother. We observed few exceptions (based on DNA evidence, rather than field observation) from these patterns in either sex (see below).

During our study, we found that once bears established home ranges, they displayed a high fidelity to them (Costello 2010). By the age of 7 years, no bears of either sex were observed to permanently disperse from their previous year's home range, although one Mogollon Mountain female was found to reside within two distinct home ranges separated by 33 km (21 mi). One range was situated near Bearwallow Mountain in Catron County, whereas the other was found within the Gila Wilderness to the south. Both ranges were used during all years and various seasons, indicating the split home range did not simply represent seasonal movements. Interestingly, we were able to monitor the movements of two of her female offspring as independent yearlings, and we found that one settled in her mother's northern home range, whereas the other settled in her mother's southern home range. Perhaps the mother's use of two distinct home ranges allowed for reduced competition among her philopatric daughters.

Established home ranges of male black bears are typically 2–4 times the size of female home ranges (e.g., Lindzey and Meslow 1977; Koehler and Pierce 2003). Among both sexes, the size and extent of the home range can vary significantly across seasons and years. Black bears generally concentrate their activity within a core area of multi-annual use during the spring and summer, but often travel widely during fall hyperphagia (e.g., Lindzey and Melsow 1977; Garshelis and Pelton 1981; Hellgren and Vaughan 1990). During our study, using a sample of resident, radio-marked bears with adequate radio-telemetry locations, we delineated multi-annual core and total home ranges utilizing 50% and 95% fixed kernel analyses, respectively (Costello 2008). Among 17 males, mean core range size was 87 km^2 (~ 33.5 mi^2) (range 65–118 km^2 [~25–45.5 mi^2]) and mean total range size was 463 km^2 (~179 mi^2) (range 291–844 km^2 [~112.5–326 mi^2]). Among 60 females, mean core range size was 18 km^2 (~7 mi^2) (range 13–42 km^2 [~5–16 mi^2]) and mean total range size was 89 km^2 (~34.5 mi^2) (range 48–239 km^2 [~18.5–92.5 mi^2]). Core and total home range sizes were slightly larger for females in the Mogollon Mountains than for females in the Sangre de Cristo Mountains. Guntly (2016) similarly estimated 50% core and 95% total home range sizes for bears captured on the National Rifle Association Whittington Center near Raton in Colfax County during 2014–2016. Among four males, mean core range size was 43 km^2 (16 mi^2) and total range size 389 km^2 (150 mi^2). For one

female monitored, core range size was 18 km^2 (7 mi^2) and total range size was 103 km^2 (40 mi^2).

Overlap of neighboring home ranges can be extensive (e.g., Garshelis and Pelton 1981; Horner and Powell 1990; Samson and Huot 2001). In our populations, we found that within-sex overlap of the core home range was comparable between males and females (Fig. 16.6). Overlap of a female by a neighboring female and of a male by another male reached 97% and 78%, respectively. Overlap of a male by a female and of a female by a male were as high as 51% and 100%, respectively (Costello 2008). Some authors have argued that females are more territorial than males, but share space with relatives (e.g., Rogers 1987). Our spatial genetic analyses did not support this idea. We found that half of females that overlapped another female by more than 50% were likely unrelated to that neighbor. In other words, females appeared equally likely to tolerate a non-relative as a relative within their core home range (Costello 2008). Schenk et al. (1998) also found low relatedness among females with overlapping home ranges in Ontario.

Still, as a result of natal philopatry, female bears often form "matrilinear assemblages," where home ranges of related females (i.e., mothers, daughters, and sisters) are geographically clustered (Støen et al. 2005). In contrast, as a result of natal dispersal, male home ranges are generally unstructured with respect to male relatives, and largely separated from those of related females. Inbreeding avoidance is a likely cause for these behaviors. Spatial genetic analyses within our study populations indicated that roughly 85% of male-female pairs that lived close enough to mate with one another were, indeed, unrelated. Only about 5% of pairs consisted of close relatives and most of these were father-daughter relationships, which would be expected given female philopatry and adult male home range fidelity. At the same time, the observed presence of a few mature brother-sister and mother-son pairs in proximity to one another suggests the male-bias in dispersal is not complete. Furthermore, the sex bias in dispersal also appeared less pronounced in the lower-density Mogollon Mountains population. This may suggest that intraspecific competition, for mates and/or for other resources, may play a role in the rate of, or distance traveled during, male dispersal (Costello et al. 2008).

The movements of black bears in New Mexico varied according to the seasonal needs of mating, hyperphagia, and hibernation (Costello 2008). We examined monthly movement patterns, as related to the estimated center and boundary of the core home range. The annual pattern began with most bears denning close to their home range center within their core home range. Post-denning movements were generally small, but increased during the mating season, especially for males. During fall hyperphagia, when bears were building up fat stores for hibernation, bears often ranged farthest from their core range. In the days prior to denning, bears typically returned to their core range and reduced their movement rates (Fig. 16.7).

During the spring and summer, we found that males and females show divergent tendencies in their movement patterns, even beyond a simple difference in magnitude (Fig. 16.7). As much of this period is the mating season, these differences likely relate to the diverse strategies males and females use to maximize their fitness. During the mating season, when females may come into estrus at any time, males appear to maximize their chances of encountering receptive females by exhibiting high mobility but concentrating their movements within or near their core home range. The dispersed distribution of females, along with the fact that few of them are estrous at any given time, necessitates a substantial search effort for receptive females by males. Although males spent significantly more time outside of their core range during June and July compared to females, the mean distance from home range center was

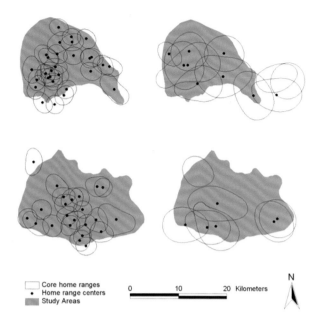

Figure 16.6. Overlap of core home ranges and proximity between home range centers for black bears: (top) Sangre de Cristo Mountain females (left) and males (right); (bottom) Mogollon Mountain females (left) and males (right). Maps depict the maximum number of ranges that were occupied simultaneously and overlapped the study areas.

significantly less than that observed during the mast season (Costello 2008). The distance between successive recorded locations was higher during the mating season than during the mast season, indicating that males were repeatedly traversing their core range, similar to other studies which showed movement rates were highest during the breeding season (Alt et al. 1980; Young and Ruff 1982; Hirsch et al. 1999). This canvassing behavior probably maximizes a male's chance of encountering familiar females at the time that they come into estrus (Costello 2008).

In contrast, females spend most of the spring and summer within—rather than simply near—their core home range (Fig. 16.7). The benefit of remaining within a familiar area is likely the efficiency with which females can exploit local food resources, important for short- and long-term

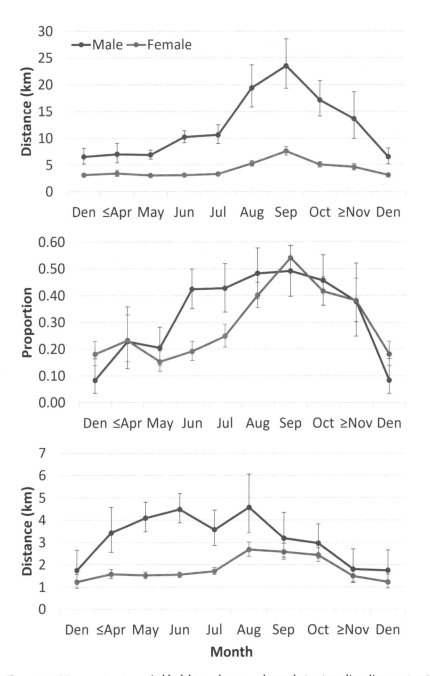

Figure 16.7. Movement patterns in black bears, by sex and month: (top) median distance (±95% CI) between each location and the home range center; (center) proportion of locations (±95% CI) occurring outside of core home range contours; and (bottom) median distance (±95% CI) between successive locations obtained 7–21 days apart. Distance between successive locations during the den period pertains to individuals that moved from one den to another.

reproductive success. Among females, the size and breadth of movements were lowest for females with young cubs (Costello 2008). Both median distance from the home range center and median distance between successive locations were roughly 25% smaller for new mothers than for other females. This might be explained by the overall lower mobility of small cubs, coupled with the need to take part in the stationary behavior of suckling. Females with cubs may also concentrate their activity near the central, most familiar part of their home range to avoid threatening encounters with other bears, especially males. Intraspecific predation of cubs has been observed in black bears (LeCount 1987a; Garrison et al. 2007) and brown bears (Troyer and Hensel 1962; Mattson et al. 1992; Olsen 1993; Swenson et al. 2001), and some argue that infanticide in ursids is primarily perpetrated by males in an effort to increase their reproductive success by mating with the mother (Bellemain et al. 2006). Some evidence from our study supports this idea. We analyzed dynamic interactions, or the tendency for pairs of individuals to be close together at the same time within seasons. Among all types of pairs, the only pattern of avoidance we detected was between males and those females accompanied by cubs (Costello 2008). In addition, we observed two events of likely predation on adult females and their cubs, both of which occurred during the spring. In the first instance, evidence pointed to the mother having been killed in a struggle with another bear. In the second case, evidence was not conclusive, but bear sign in the area suggested the predator may have been a bear (Costello et al. 2001). No cub carcasses were found, suggesting they may have been consumed.

During fall, the characteristics of male and female bear movements were much more similar than in the spring and summer, as both sexes spent most of their time foraging in preparation for hibernation (Fig. 16.7). Distance from the home range center was highest during fall for both sexes, and bears often spent 40–50% of their time outside of their core home range. These long-range movements were associated with concentrations of important mast species, particularly oaks. Bears often moved toward lower elevations to feed on acorns, and these movements were affected by acorn production. Median distance from home range center was about 25% higher for both sexes during years of oak failure than during other years, indicating bears had to search a wider area to find food (Costello 2008).

Using camera traps, Lara-Díaz et al. (2018) found that black bears in northeastern Mexico concentrated their activities during the day and crepuscular hours, similar to black bears in other regions (e.g., Garshelis and Pelton 1980). During our research, highest activity was observed at dusk. Spatial analyses indicated that the proportion of diurnal activity at each site was negatively associated with annual temperature, human density, and road density, whereas it was positively associated with annual rainfall.

Limited genetic exchange likely occurs among populations in the distinct tracts of bear habitat in New Mexico (Map 16.1) through dispersal, mostly of males. Onorato et al. (2004) observed high levels of genetic partitioning among the isolated bear populations in western Texas, northern Mexico, and southern New Mexico, due to low rates of male-mediated gene flow, and even lower rates of female-mediated gene flow. Onorato et al. (2007) found that gene flow was negatively correlated with distance between populations. Expanses of grassland or desert likely impede movements of black bears between distinct tracts of habitat in the Southwest, however various examples of long-distance movements between isolated mountain ranges have been documented, either by radio-telemetry or by estimation of the shortest-distance to occupied range or genetically determined population of origin. Observed or estimated distances traversed included 135 km (84 mi) by an adult male from the Mogollon population in New Mexico (Costello et al. 2001); 29–128 km (18–79 mi) by females and 80–135 km (50–84 mi) by males in the Texas, Chihuahua, and Coahuila regions (Hellgren et al. 2005); 37–87 km (23–53 mi) by males and 193 km (120 mi) for an unknown-sex bear in Utah (Auger et al. 2005); 250 km (155 mi) by a male in Durango, Mexico (Camargo-Aguilera et al. 2017); 182 km (113 mi) by a male in Queretaro, Mexico (López González et al. 2019); and 135 km (84 mi) by an unknown subadult in Texas (Yancey and Lockwood 2019). Movements like these across harsh, xeric environments in the southwestern United States and northern Mexico once appeared anomalous, but evidence suggests they may be an integral part of the ecology of bears in this region, allowing individuals to gain access to better conditions in response to drought, wildfire, or other perturbations (Garshelis et al. 2016; Photos 16.26, 16.27, 16.28; and see Chapter 5). Urbanization, highways, and the border fence or wall between the United States and Mexico may act as barriers to such movements and reduce connectivity among populations in northern Mexico, Arizona, and New Mexico (Atwood et al. 2011; Lara-Díaz et al. 2013). An examination of locations of black bear vehicle collisions during 2002–2018 in New Mexico shows highways represent not only a significant source of mortality within tracts of suitable habitat, but also act as barriers to dispersal and long-range movements. In particular, vehicle collisions on US interstates 25 and 40 may be reducing connectivity among the San Juan, Sangre de Cristo, and Sandia mountains; among the Chuska, Zuni, and San Mateo mountains; and between the Gila region and Sacramento Mountains (Map 16.3).

Liley and Walker (2015) documented another unusual movement of an adult male black bear that traveled from Vermejo Park Ranch (northeastern New Mexico) to Castle Rock (central Colorado) and back over 304 days during July 2011–May 2012. This journey was primarily through suitable bear habitat. Because the bear was equipped with a GPS collar, Liley and Walker (2015) were able to show that the travel path distance was at least 1,483 km (921 mi) and the straight-line distance was 282 km (175 mi), making it one of the most extreme examples of movement ever documented for a non-dispersing black bear. During this movement, the bear crossed Interstate 25 twice as well as several major rivers, including the Arkansas River, approximately 100 m (~330 ft) wide at the crossing (Liley and Walker 2015).

Interspecific Interactions

Black bears interact with other vertebrates as predators but also as prey and as competitors. Black bears, especially juveniles, are preyed on by a variety of species. Documented predators of black bear cubs and yearlings include golden eagles (*Aquila chrysaetos*; Nelson 1957), wolves (*Canis lupus*; Rogers and Mech 1981; Rogers 1987), cougars (LeCount 1987a; Costello et al. 2003), and bobcats (*Lynx rufus*; LeCount 1987a). Although now extirpated, grizzly bears were likely once a significant threat to both juvenile and adult black bears in New Mexico (Mattson et al. 1992; Bertram and Vivion 2002; Gunther et al. 2002; and see Chapter 17).

With the absence of grizzlies from New Mexico, black bears overall likely exhibit low levels of competition with the state's other larger carnivores, such as cougars, coyotes, bobcats, and wolves, due to differences in their diets, diel activity patterns, and reproductive cycles (Maehr 1997). Nonetheless, one likely form of interference competition is kleptoparasitism (i.e., food stealing). Kleptoparasitism can represent a nutritional windfall for the usurper and a significant nutritional loss to the victim. For example, black bears have been observed to usurp kills from cougars. In Glacier and Yellowstone national parks, Murphy et al. (1998) estimated that the biomass of cougar prey stolen by grizzly and black bears represented a gain of 71–113% of the daily energy needs of the bears, but a loss of 17–26% of the daily energy requirements of the cougars. Due to their large size, black bears are likely the winner in most contests with cougars, coyotes, and bobcats. Contests between black bears and lone wolves have been observed to go both ways (Gehring 1993; Fremmerlid and Latham 2009), but wolf packs have the advantage of numbers over lone bears. On rare occasions, interactions at kills, as well as other encounters (wolves seeking black bears in their dens, for example), might also result in interspecific predation. Although black bears have occasionally been observed to kill wolves, wolves dominate these interactions more frequently (Rogers and Mech 1981; Horejsi et al. 1984; Paquet and Carbyn 1986).

Diseases and Parasites

In general, disease is not considered a significant source of mortality in black bear populations (Pelton 2003). However, the prevalence of pathogens and their true impact on black bear morbidity and mortality have not been well studied in the Southwest, including in New Mexico. Bard and Cain (2019) collected 36 blood samples from 12 female and 17 male black bears in 2016–2017 in the Jemez Mountains. Seroprevalence was highest for plague (*Yersinia pestis*; 55%) and *Toxoplasma gondii* (37%). Some tested bears also had antibodies against West Nile virus (11%), *Francisella tularensis* (10%), and canine parvovirus (3%). No antibodies to canine distemper, *Borrelia burgdorferi*, *Rickettsia* spp., or *Babesia* spp. were detected. No mortality related to exposure to pathogens was recorded (Bard and Cain 2019). An undiagnosed dermatitis, characterized by alopecia and edema around the eyes and muzzle, has been observed in black bears in the Southwest, including the Sangre de Cristo and Mogollon mountains (Costello et al. 2006), the Jemez Mountains (Bard and Cain 2019), and Colorado (Beck 1991). The usually mild symptoms resemble mange and manifest temporarily during hibernation: we observed them during 23% of den

handlings but during <1% of active season captures during 1992–2000. The condition appears common but rarely life-threatening. We observed only one adult female with symptoms that worsened over several years, including extensive anterior alopecia and severe lichenification extending from the ears to the muzzle. Although her death resulted from our den handling, we suspect she would not have survived through the following spring given that she was severely emaciated and essentially blind from callousing over her eyes (Costello et al. 2006).

STATUS AND MANAGEMENT

The history of the black bear in New Mexico, and more generally in the Southwest, is a wildlife management success story. After decades of intense persecution by humans, the black bear was once reduced to very low numbers. Thanks to timely legal protections, not to mention the quality habitat available in New Mexico and the resilient nature of the species, black bears did not disappear like the Southwestern grizzly bear population and Merriam's elk (*C. c. merriami*). In fact, today black bears are flourishing in New Mexico, so much so that some of the same public animosity that led to their near demise may be emerging again.

Early reports of the black bear's distribution in New Mexico are limited, but it is reasonable to conclude that black bears inhabited all the forested areas of the Southwest prior to European settlement. Reporting on their exploration during 1871–1874, Coues and Yarrow (1875:66) indicated that black bears were "quite numerous" in the territories of Arizona, Colorado, and New Mexico. In his review of historical records of black bears from 1825 to 1915, Bailey (1931) included records within all the habitat tracts occupied by black bears today (Map 16.1). But, by the early 1900s, evidence indicates that black bear populations were already reduced, due to unlimited hunting and use of poisons (Bailey

1931; Brown 1985). Much of the mortality was the result of government-sponsored anti-predator programs aimed at eliminating loss of livestock to grizzly and black bears, wolves, and other carnivores (Brown 1985). In describing the Sangre de Cristo Mountains in 1913, Bailey (1913:56) reported that "white-tailed and mule deer are present, although becoming scarce, coyotes and black bears are fairly common, and there are still a few grizzlies or silvertips, gray wolves, and red foxes." In describing the Sacramento Mountains, Bailey (1913:72) stated, "Only a few years ago [the area] was famous for its variety and abundance of game, especially elk, mule deer, white-tailed deer, antelope, bighorn, black and silver-tip bears, and wild turkeys. The elk are now exterminated and other game birds and animals are becoming scarce, but it is hoped that they can be protected so that present numbers at least shall be maintained." In 1924, the US Forest Service estimated that only 1,500 black bears remained in the national forests of New Mexico, Arizona, southern Colorado, and southern Utah combined (Brown 1985). In 1925, the New Mexico population estimate was only 660 black bears (NMDGF 1926). A 1927 map of the occupied range in New Mexico still showed the species' presence within all the tracts occupied today, except for the Sandia and Manzano mountains, suggesting black bears were extirpated or extremely scarce in these two ranges (Ligon 1927). Four years later, Bailey (1931:350) stated that "black bears are still more or less common in most of the mountain ranges in New Mexico, where there is timber or chaparral to afford them cover and food. In the more open areas they are becoming scarce, and in some of the heavily timbered areas have been trapped and hunted until no longer abundant."

Responding to public and legislative support for protection of bears, the New Mexico Department of Game and Fish (NMDGF) classified the black bear as a big game species in 1927. The protection that this decision conferred had significant results, and the state's bear population appeared to have somewhat rebounded by the 1940s. In 1941, more than 3,500 bears were estimated to reside in the national forests of the Southwest (Brown 1985). By 1967, the New Mexico black bear population was estimated at 3,000 and stable (Lee 1967). In 2001, two independently derived population estimates, one from state-wide population modeling and the other from extrapolation of our field study data, put the New Mexico statewide bear population at roughly 5,200–6,000 bears ≥1 year old (Costello et al. 2001). During 2012–2014, more refined population density estimates for the Sangre de Cristo, Sandia, and Sacramento mountains were obtained through spatially explicit capture-recapture analyses of DNA samples collected by hair snagging (Gould et al. 2018). Updated statewide estimates, incorporating Gould et al.'s (2018) study results, put the current statewide population at about 8,000 bears (NMDGF 2020). It is worth noting that a timeline connecting these various population estimates is consistent with annual growth rates of approximately 2–4% per year, matching the annual growth rates considered reasonable based on population modeling with observed vital rates (Costello et al. 2001).

The first hunting regulations were enacted in 1927, setting a bag limit of one bear per season. Black bear, deer, and turkey (*Meleagris gallopavo*) were included in a single big game license, and this regulation remained in place until 1981. In 1971, a regulation was adopted prohibiting the harvest of bears less than 1 year of age or females accompanied by young. In 1978, a mandatory hide-tagging program was instituted and two further requirements were added in 1985—proof of sex and collection of premolar teeth for cementum aging. In 1982, facilitated in this by the separate black bear hunting license, the New Mexico Department of Game and Fish initiated a survey of randomly selected license

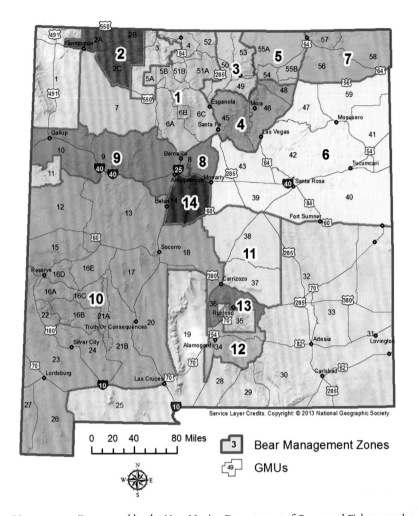

Map 16.4. Bear Management Zones used by the New Mexico Department of Game and Fish to regulate the harvest of black bears since 2004. Total harvest is limited to 8–12% of estimated population size within a management zone while the female limit is calculated as 40% of the total limit. GMUs are the New Mexico Department Game and Fish's Game Management Units.

holders. In 2004, largely in response to findings from the bear study, the agency launched a zone management program, with harvest limits set within each of 14 Bear Management Zones (Map 16.4). Currently, the limit for the total number of harvested bears is calculated as 8–12% of population size (depending on the zone), and the female limit is calculated as 40% of the total limit. Hunting within Bear Management Zones closes when harvest reaches 90% of the total limit, 90% of the female limit, or when the season ends, whichever comes first. Documented bear deaths occurring

due to depredation kills and other causes (e.g., roadkill, accidents) count toward harvest limits (NMDGF 2020; and see differences with cougar management in Chapter 9).

In 1927, the first bear hunting season was set for 10–31 October. Since that time, the timing and duration of hunting seasons have varied. By the late 1970s, bear seasons encompassed 7–8 months of the year, including parts of April, May, June, August, September, October, November, December, and January. In 1992, due to concerns about potential over-harvest, the New Mexico

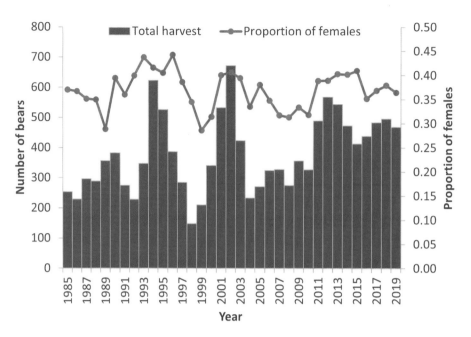

Figure 16.8. Number of black bears harvested annually and proportion of females in kill, as recorded through the mandatory hide-tagging program of the New Mexico Department of Game and Fish, 1985–2019.

Department of Game and Fish eliminated the spring bear season and reduced the fall season to 1 September–31 October. Since 1998, the exact timing of hunting seasons has continued to vary, with start dates ranging from mid-August to early September and end dates ranging from mid-November to mid-December.

Since 1985, when recording the sex and age of harvested bears first became mandatory, state-wide harvests have varied from 148 to 672 bears taken per year, with an average of 380 bears (Fig. 16.8). The overall trend is slightly increasing, at the rate of roughly 5 more bears taken per year. The proportion of females in the harvest has ranged from 29–44%, with a consistent average of about 37%. Many of the higher harvests observed in the mid-1990s and the early 2000s were associated with drought conditions and poor mast crops. As bears are known to increase fall movements when food is more limited, this increased seasonal vulnerability to harvest, and

other human-caused mortality, is not surprising. Higher bear harvest levels have been associated with shortages of natural foods in other states (McDonald et al. 1994; Noyce and Garshelis 1997). The higher harvests beginning in 2011 were associated with the New Mexico Department of Game and Fish's decision to increase harvest limits, motivated, at least in part, by a goal of reducing bear-human conflicts.

Dealing with human-bear conflict is as much a part of bear management as regulating harvest. The opportunistic feeding habits and curious nature of black bears make them especially prone to conflict with humans. Sometimes bears are attracted to domestic fruit in human residential areas (Photo 16.29). But the root of most bear-human conflict is the availability of anthropogenic foods—everything from unsecured garbage to hummingbird feeders. Incidents were relatively uncommon until the late 1980s, when a combination of factors, including drought,

Photo 16.29. Anthropogenic foods can attract black bears to campgrounds and residential areas. This bear seeks the fruit from an apple tree in a residential neighborhood in Raton, Colfax County on 11 September 2006. Photograph: © Clint Henson.

scarcity of natural foods, and growth in human residential development increased encounters between humans and bears. For example, during the summer of 1989, bear encounters escalated in Albuquerque. The problems were brought to nationwide attention when a young female bear, later nicknamed "Sparky," was injured climbing a power pole. Over the last few decades, bear complaints and public demands for control of the black bear population have been commonplace. Efforts to educate the public about bears and reduce availability of anthropogenic foods have also increased.

As a forest-adapted species possessing a life-long ability to climb trees, black bears evolved to respond to threats from other species or conspecifics by climbing a tree or fleeing into understory vegetation. This contrasts with grizzly bears, who developed behaviors allowing them to defend themselves on open ground, having evolved in both treed and non-treed habitats (e.g.,

tundra, prairie, and coastal flats). Herrero (1972) attributed the less aggressive nature of black bears, including a lower inclination to attack humans, to these divergent selective pressures. Comparing bear-human encounters for the two species in Alaska during 1880–2015, Smith and Herrero (2018) reported that black bears, who far outnumber grizzly bears in the state, inflicted only 9% of injuries and 10% of fatalities. Black bears generally avoid humans and, undoubtedly, countless interactions between people and black bears occur every year without incident (Herrero 2018). Very rarely, black bears do injure or kill humans, and the number of incidents has increased over the last few decades (Herrero 2018). The increase is likely best explained by human population growth and increasing numbers of people living, working, and recreating in bear habitats (Herrero et al. 2011; Smith and Herrero 2018). Most injuries caused by black bears are non-life-threatening, but Herrero et al. (2011)

reported 54 incidents during which people were killed by non-captive black bears in North America during 1960–2009. When detailed, reports suggested that food or garbage attracted the bear in 38% of cases and that the bear acted as a predator 88% of the time (Herrero et al. 2011). Since 2000, the New Mexico Department of Game and Fish has kept records of bear-human incidents (R. Winslow, NMDGF, pers. comm.). One human fatality (the only known fatality in the state) and 27 events of minor or severe injury have been recorded. Most incidents occurred at campgrounds or residences (78%) and many of these involved bears attracted by scents or anthropogenic foods. Other incidents involved people on foot (22%) and likely occurred because of surprise encounters and bears defending themselves or their cubs. Under either scenario, minor injuries generally resulted when the bear was startled by people, reacted by biting or scratching, and then fled or ended the contact. Less frequently, bears inflicted serious injury during more sustained attacks. In these cases, bears may have been motivated by predation or predatory instincts may have been stimulated by the interaction. Most incidents occurred during the summer months, and they were more frequent during drought years (R. Winslow, NMDGF, pers. comm.).

Increases in bear-human conflict often coincide with failures in the production of natural foods, such as nuts and berries (Howe et al. 2010; Obbard et al. 2014). Using radio-telemetry monitoring of black bears in Aspen, Colorado, Baruch-Mordo et al. (2014) observed that bears frequented human-developed areas and became more nocturnal in poor food years, but the pattern was reversible. The same individuals generally confined their movements to wildland areas in subsequent good food years. Evidence from our Sangre de Cristo population also supports the notion that most bears avoid anthropogenic foods, even when they might be readily available (Beckmann et al. 2008). The Sangre de Cristo Mountains study population showed considerable potential for conflict, given that it resided close to three towns and Philmont Scout Ranch, a recreation property that hosted about 20,000 visitors per year and harbored many campsites. Despite this high potential for conflict, most bears did not engage in nuisance or depredation activities. Although 70% of females and 84% of males overlapped areas or sites of potential conflict, only about 16% and 26% were known or suspected of nuisance activity, respectively (Beckmann et al. 2008). In addition, where anthropogenic food was made less available, fewer bears got into trouble. At least 52 bears had home ranges partly or entirely within the boundaries of Philmont Scout Ranch. However, due to the ranch implementing comprehensive bear-human management measures, all designed to ensure that bears did not obtain anthropogenic foods (Ricklefs 2005), only 6 (12%) of these bears were involved in any conflict (Beckmann et al. 2008). In contrast, bears were far more likely to engage in nuisance activity where garbage was unsecured and other attractants existed. Of 26 bears with home ranges that overlapped towns, 18 (69%) were involved in conflict. Roughly speaking, these bears were 13 times more likely to engage in conflict than bears that overlapped only Philmont Scout Ranch, and 55 times more likely than bears that did not overlap any sources of potential conflict (Beckmann et al. 2008). Clearly, efforts to reduce the availability of anthropogenic food can be quite effective at reducing conflicts, even amid higher-density bear populations, like the Sangre de Cristo study population. Ricklefs (2005) reported that during 1992–2002, approximately 2,000 groups of visitors hiked and camped along Philmont Scout Ranch's trails each summer and 40% of them reported seeing a bear. Although most bears appeared human-habituated, few were food-conditioned, because of the program to secure attractants.

In our statewide habitat analysis in 2001, we

Photo 16.30. Campground in Colfax County south of Raton. Continued use of open dumpsters in campgrounds and residential areas contributes to black bear–human conflict in New Mexico. Photograph: © Cecily Costello.

found that 32% of statewide suitable bear habitat was under private ownership and 17% was within 5 km (~3 mi) of human-populated areas. This juxtaposition of humans and black bears represents a significant potential for conflict. The percent of bear habitat within 5 km of humans differed regionally, with the highest proportions observed in the Sandia-Manzano Mountains (51%), the Sacramento Mountains (28%), and the Sangre de Cristo Mountains (23%)—the very same regions with the highest human population growth during 1990–2000 (US Census Bureau 2003). Human growth rates in the Sandia-Manzano Mountains and Sacramento Mountains regions were more than double the estimated 2% annual growth rate for the black bear population, suggesting that human population growth was possibly a better explanation for the increasing incidence of bear-human conflict during the 1990s than was bear population growth. Human population growth in New Mexico since 2000 has been modest compared to most neighboring states, which

rank among the fastest-growing populations in the nation (US Census Bureau 2010, 2020). Nonetheless, management of bear-human conflict is and will continue to be a challenge in New Mexico, especially given that the use of open dumpsters, often scattered throughout residential areas, is still quite common (Photo 16.30). Unless some effort is made to restrict bear access to garbage, such as community-wide adoption of bear-resistant containers, some bears will continue to seek anthropogenic foods, and thus continue to pose a threat to property and human safety. Like the success on Philmont Scout Ranch, sanitation and bear-proofing measures are quite effective in residential settings. Johnson et al. (2018) conducted a before-after-control-impact experiment to evaluate the effectiveness of bear-proofing measures to reduce conflict and document their influence on human attitudes in Durango, Colorado. After bear-resistant containers were deployed, trash-related conflicts were 60% lower, and compliance of residents with local wildlife ordinances was 39%

higher in treatment areas compared to control areas. Additionally, support for bear-proofing ordinances increased and perceived risk of a black bear breaking into homes decreased among residents in the treatment area. Similar success could be achieved in New Mexico communities, but a desire for coexistence and a willingness to invest in bear-proofing equipment are needed.

Although they share a similar life history with their cousins throughout North America, the black bears of New Mexico and the Southwest are unique. As such they are all the more worthy of our respect and admiration, and they should remain an important conservation priority. Over millennia, they have adapted to an arid, naturally fragmented landscape and a boom-and-bust food economy in various ways, such as developing light-colored coats, adopting smaller litter sizes, and fostering a propensity for long-range movements. During the last century, protected as a game species, they have succeeded in repopulating all the special places where they historically lived. And during the last few decades, they have proved adaptable to human development along the urban-wildland interface, often resisting the temptations we humans offer. While the large expanses of remote, public lands in New Mexico are likely to ensure that black bears persist in New Mexico well into the future, we should not forget that the human footprint is increasingly interfering with natural processes (Chapter 5) and actions taken today can provide benefits for wildlife and natural landscapes for years to come. My hope is for heightened accommodation for the needs of black bears on our shared landscape. In particular, I hope that more communities will be inspired to implement sanitation programs to reduce bear-human conflict and that the public will be motivated to invest in highway and infrastructure mitigation to safeguard travel corridors. These actions would be the true testament that New Mexicans value this fascinating species.

ACKNOWLEDGMENTS

I wish to acknowledge my fellow biologists, coauthors, and administrators of the New Mexico Black Bear Study from Hornocker Wildlife Institute, New Mexico State University, New Mexico Department of Game and Fish, Wildlife Conservation Society, and Montana State University. I am especially grateful to Don Jones for being my treasured field partner for eight wonderful years. I wish to thank my husband, Mark Haroldson, for sharing many memorable days in the field and for his unwavering support throughout completion of the study and my doctoral program. I am especially indebted to Bob Inman, Kris Inman, Margaret Kirkeminde, Deborah Perkins, Carrie Hunt, Larry Temple, Nick Smith, Bob Ricklefs, Maurice Hornocker, Howard Quigley, and Bruce Thompson.

LITERATURE CITED

Alt, G. L. 1980. Rate of growth and size of Pennsylvania black bears. *Pennsylvania Game News* 51:7–17.

———. 1983. Timing of parturition of black bears (*Ursus americanus*) in northeastern Pennsylvania. *Journal of Mammalogy* 64:305–7.

———. 1984. Black bear cub mortality due to flooding of natal dens. *Journal of Wildlife Management* 48:1432–34.

———. 1989. Reproductive biology of female black bears and early growth and development of cubs in northeastern Pennsylvania. PhD dissertation, University of West Virginia, Morgantown.

Anderson, R. M. 1945. Summary of Canadian black bears with description of two new northwestern species. *Provancher Society of Natural History of Canada* 1944:17–33.

Atwood, T. C., J. K. Young, J. P. Beckmann, S. W. Breck, J. Fike, O. E. Rhodes Jr., and K. D. Bristow. 2011. Modeling connectivity of black bears in a desert sky island archipelago. *Biological Conservation* 144:2851–62.

Auger, J., J. D. Heward, H. L. Black, and G. Wallace. 2005. Movements of Utah black bears: implications for management and conservation. *Western Black Bear Workshop* 8:72–80.

Austin, M. A., M. E. Obbard, and G. B. Kolenosky. 1994. Evidence of a black bear, *Ursus americanus*, killing an adult moose, *Alces alces*. *Canadian Field-Naturalist* 108:236–38.

Bailey, V. 1913. *Life zones and crop zones of New Mexico*. North American Fauna 35. US Department of Agriculture, Bureau of Biological Survey, Washington, D.C.

———. 1931 (=1932). *Mammals of New Mexico*. North American Fauna 53. Washington, DC: US Department of Agriculture, Bureau of Biological Survey.

Baldwin, R. A., and L. C. Bender. 2009. Foods and nutritional components of diets of black bear in Rocky Mountain National Park, Colorado. *Canadian Journal of Zoology* 87:1000–1008.

Banmore, W., and D. Stradley. 1971. Predation by black bear on mature male elk. *Journal of Mammalogy* 52:199–202.

Bard, S. M., and J. W. Cain III. 2019. Pathogen prevalence in American black bears (*Ursus American amblyceps*) of the Jemez Mountains, New Mexico, USA. *Journal of Wildlife Diseases* 55:745–54.

———. 2020. Investigation of bed and den site selection by American black bears (*Ursus americanus*) in a landscape impacted by forest restoration treatments and wildfires. *Forest Ecology and Management* 460:117904.

Barnowe-Meyer, K. K., P. J. White, T. L. Davis, and J. A. Byers. 2009. Predator-specific mortality of pronghorn on Yellowstone's northern range. *Western North American Naturalist* 69:186–94.

Baruch-Mordo, S., K. R. Wilson, D. L. Lewis, J. Broderick, J. S. Mao, and S. W. Breck. 2014. Stochasticity in natural forage production affects use of urban areas by black bears: implications to management of human-bear conflicts. *PLOS One* 9:e85122. doi:10.1371/journal.pone.0085122.

Beck, T. D. I. 1991. *Black bears of west-central Colorado*. Technical Publication No. 39. Denver: Colorado Division of Wildlife.

Beckmann, J. P., L. Karasin, C. M. Costello, S. Matthews, and Z. Smith. 2008. Coexisting with black bears: perspectives from four case studies across North America. WCS Working Paper 33, Wildlife Conservation Society, Bronx, NY.

Beeman, L. E., and M. R. Pelton. 1980. Seasonal foods and feeding ecology of the black bears in the Smoky Mountains. *International Conference on Bear Research and Management* 4:141–47.

Bellemain, E., J. E. Swenson, and P. Taberlet. 2006. Mating strategies in relation to sexually selected infanticide in a non-social carnivore: the brown bear. *Ethology* 112:230–37.

Bertram, M. R., and M. T. Vivion. 2002. Black bear monitoring in eastern interior Alaska. *Ursus* 13:69–77.

Beston, J. A. 2011. Variation in life history and demography of the American black bear. *Journal of Wildlife Management* 75:1588–96.

Bridges, A. S., M. R. Vaughan, and J. A. Fox. 2011. American black bear estrus and parturition in the Alleghany Mountains of Virginia. *Ursus* 22:1–8.

Brown, D. E. 1985. *The grizzly in the Southwest*. Norman: University of Oklahoma Press.

Brown, D. E., and R. D. Babb. 2009. Status of the porcupine (*Erithizon dorsatum*) in Arizona, 2000–2007. *Journal of the Arizona-Nevada Academy of Science* 41:36–41.

Brown, G. 2009. The bear almanac: a comprehensive

guide to the bears of the world. Guilford, CT: Lyons Press.

Bunnell, F. L., and T. Hamilton. 1983. Forage digestibility and fitness in grizzly bears. *International Conference on Bear Research and Management* 5:179–85.

Bunnell, F. L. and D. E. N. Tait. 1981. Population dynamics of bears—implications. In *Dynamics of Large Mammal Populations*, ed. C. W. Fowler and T. D. Smith, 75–98. New York: John Wiley.

Camargo-Aguilera, M. G., N. E. Lara-Díaz, H. Coronel-Arellano, and C. A. López-González. 2017. One black bear (*Ursus americanus*) connects the great sierras: genetic evidence. *Therya* 8: 277–82.

Carstensen, M., G. D. Delgiudice, B. A. Sampson, and D. W. Kuehn. 2009. Survival, birth characteristics, and cause-specific mortality of white-tailed deer neonates. *Journal of Wildlife Management* 73:175–83.

Caughley, G. 1977. *Analysis of vertebrate populations*. London: John Wiley.

Chi, D. K. 1999. The effects of salmon availability, social dynamics, and people on black bear (*Ursus americanus*) fishing behavior on an Alaskan salmon stream. PhD dissertation, Utah State University, Logan.

Clark, J. D., J. P. Beckmann, M. S. Boyce, B. D. Leopold, A. E. Loosen, and M. R. Pelton. 2020. American black bear (*Ursus americanus*). In *Bears of the world: ecology, conservation and management*, ed. V. Penteriani and M. Melletti, 122–38. Cambridge: Cambridge University Press.

Coop, J. D., C. D. Hibner, A. J. Miller, and G. H. Clark. 2005. Black bears forage on army cutworm moth aggregations in the Jemez Mountains, New Mexico. *Southwestern Naturalist* 50:278–81.

Costello, C. M. 2008. The spatial ecology and mating system of black bears (*Ursus americanus*) in New Mexico. PhD dissertation. Montana State University, Bozeman.

———. 2010. Estimates of dispersal and home range fidelity in American black bears. *Journal of Mammalogy* 91:116–21.

Costello, C. M., S. R. Creel, S. T. Kalinowski, N. V. Vu, and H. B. Quigley. 2008. Sex-biased natal dispersal and inbreeding avoidance in American black bears as revealed by spatial genetic analyses. *Molecular Ecology* 17:4713–23.

———. 2009. Determinants of male reproductive success in American black bears. *Behavioral Ecology and Sociobiology* 64:125–34.

Costello, C. M., K. H. Inman, D. E. Jones, R. M. Inman, B. C. Thompson, and H. B. Quigley. 2004. Reliability of the cementum annuli technique for estimating age of black bears in New Mexico. *Wildlife Society Bulletin* 32:169–76.

Costello, C. M., D. E. Jones, K. A. Green Hammond, R. M. Inman, K. H. Inman, B. C. Thompson, R. A. Deitner, and H. B. Quigley. 2001. *A study of black bear ecology in New Mexico with models for population dynamics and habitat suitability*. Final Report, Federal Aid in Wildlife Restoration Project W-131-R. Santa Fe: New Mexico Department of Game and Fish.

Costello, C. M., D. E. Jones, R. M. Inman, K. L. Inman, B. C. Thompson, and H. B. Quigley. 2003. Relationship of variable mast production to American black bear reproductive parameters in New Mexico. *Ursus* 14:1–16.

Costello, C. M., K. S. Quigley, D. E. Jones, R. M. Inman, and K. H. Inman. 2006. Observations of denning-related dermatitis in American black bears. *Ursus* 17:186–90.

Coues, E., and H. C. Yarrow. 1875. Report upon the collections of mammals made in portions of Nevada, Utah, California, Colorado, New Mexico, and Arizona during the years 1871, 1872, 1873, and 1874. In *Annual report upon the geographical surveys west of the one-hundredth meridian in the states and territories of California, Oregon, Nevada, Texas, Arizona, Colorado, Idaho, Montana, New Mexico, Utah, and Wyoming*, Vol. 5, 35–129. Washington, DC: Government Printing Office.

Cushing, F. H. 1994. *Zuni fetishes*. Reprint of the Second Annual Report of the Bureau of Ethnology, 1883. Las Vegas: KC Publications.

Davis, D. D. 1964. The giant panda: a morphological study of evolutionary mechanisms. *Fieldiana Zoology Memoirs* 3:1–339.

Deweese, L. R., and R. E. Pillmore. 1972. Bird nests in an aspen tree robbed by black bear. *Condor* 74:488.

Doan-Crider, D. L., and E. C. Hellgren. 1996. Population characteristics and winter ecology of black bears in Coahuila, Mexico. *Journal of Wildlife Management* 60:398–407.

Elowe, K. D, and W. E. Dodge. 1989. Factors affecting black bear reproductive success and cub survival. *Journal of Wildlife Management* 53:962–68.

Erickson, A. W., and W. G. Youatt. 1961. Seasonal variations in the hematology and physiology of black bears. *Journal of Mammalogy* 42:198–203.

Ewer, R. F. 1973. *The carnivores.* Ithaca, NY: Cornell University Press.

Folk, G. E., Jr., A. Larson, and M. A. Folk. 1976. Physiology of hibernating bears. *International Conference on Bear Research and Management* 3:373–80.

Foss, E. 1980. A black bear in a great blue heron colony. *Murrelet* 61:113.

Fremmerlid, M., and A. D. M. Latham. 2009. Lone wolf, *Canis lupus*, displaced from a kill by an adult black bear, *Ursus americanus*, in northeastern Alberta. *Canadian Field-Naturalist* 123:266–67.

Gantchoff, M. G., and J. L. Belant. 2017. Regional connectivity for recolonizing American black bears (*Ursus americanus*) in southcentral USA. *Biological Conservation* 214:66–75.

Gantchoff, M. G., D. Beyer, and J. L. Belant. 2019. Reproductive class influences risk tolerance during denning and spring for American black bears (*Ursus americanus*). *Ecosphere* 10:e02705. 10.1002/ecs2.2705.

Garrison, E. P., J. W. McCown, and M. K. Oli. 2007. Reproductive ecology and cub survival of Florida black bears. *Journal of Wildlife Management* 71:720–27.

Garshelis, D. L., and M. R. Pelton. 1980. Activity of black bears in the Great Smoky Mountains National Park. *Journal of Mammalogy* 61:8–19.

———. 1981. Movements of black bears in the Great Smoky Mountains National Park. *Journal of Wildlife Management* 45:912–25.

Garshelis, D. L., B. K. Scheick, D. L. Crider, D. L. Beecham, and M. E. Obbard. 2016. American black bear. *Ursus americanus*. The IUCN Red List of Threatened Species 2016:e.T41687A45034604, http://dx.doi.org/10.2305/IUCN.UK.2016-3. RLTS.T41687A45034604.en.

Gehring, T. M. 1993. Adult black bear, *Ursus americanus*, displaced from a kill by a wolf, *Canis lupus*, pack. *Canadian Field-Naturalist* 107:373–74.

Gould, M. J., J. W. Cain III, G. W. Roemer, W. R. Gould, and S. G. Liley. 2018. Density of American

black bears in New Mexico. *Journal of Wildlife Management* 82:775–88.

Gould, M. J., W. R. Gould, J. W. Cain III, and G. W. Roemer. 2019. Validating the performance of occupancy models for estimating habitat use and predicting the distribution of highly mobile species: a case study using the American black bear. *Biological Conservation* 234:28–36.

Graber, D. M., and M. White. 1983. Black bear food habits in Yosemite National Park. *International Conference on Bear Research and Management* 5:1–10.

Griffin, K. A, M. Hebblewhite, H. S. Robinson, P. Zager, S. M. Barber-Meyer, D. Christianson, S. Creel, N. C. Harris, M. A. Hurley, D. H. Jackson, B. K. Johnson, W. L. Myers, J. D. Raithel, M. Schlegel, B. L. Smith, C. White, and P. J. White. 2011. Neonatal mortality of elk driven by climate, predator phenology and predator community composition. *Journal of Animal Ecology* 80:1246–57.

Gunther, K. A., M. J. Biel, N. Anderson, and L. Waits. 2002. Probable grizzly bear predation on an American black bear in Yellowstone National Park. *Ursus* 13:372–74.

Guntly, K. M. 2016. Black bear (*Ursus americanus*) movements, diet, and impacts on ungulate neonate survival on the National Rifle Association Whittington Center, New Mexico. MS thesis, Texas Tech University, Lubbock.

Harris, A. H. 1993. Quaternary vertebrates of New Mexico. *Vertebrate Paleontology in New Mexico, New Mexico Museum of Natural History Bulletin* 2:179–97.

Hatler, D. F. 1972. Food habits of black bears in interior Alaska. *Canadian Field-Naturalist* 86:17–31.

Hellgren, E. C. 1993. Status, distribution, and summer food habits of black bears in Big Bend National Park. *Southwestern Naturalist* 38:77–80.

———. 1998. Physiology of hibernation in bears. *Ursus* 10:467–77.

Hellgren, E. C., D. P. Onorato, and J. R. Skiles. 2005. Dynamics of a black bear population within a desert metapopulation. *Biological Conservation* 122:131–40.

Hellgren, E. C., and M. R. Vaughan. 1990. Range dynamics of black bears in Great Dismal Swamp, Virginia-North Carolina. *Proceedings of the Annual Conference of Southeastern Fish and Wildlife Agencies* 44:268–78.

Hellgren, E. C., M. R. Vaughan, F. C. Gwazdauskas,

B. Williams, P. F. Scanlon, and R. L. Kirkpatrick. 1991. Endocrine and electrophoretic profiles during pregnancy and nonpregnancy in captive female black bears. *Canadian Journal of Zoology* 69:892–98.

Herrero, S. 1972. Aspects of evolution and adaptation in American black bears (*Ursus americanus* Pallas) and brown and grizzly bears (*U. arctos* Linné) of North America. In *Bears—their biology and management*, ed. S. Herrero, 221–31. International Union for the Conservation of Nature Publication, n.s. 23.

———. 1978. A comparison of some features of the evolution, ecology and behavior of black and grizzly/brown bears. *Carnivore* 1:7–17.

———. 2018. Bear attacks: their causes and avoidance. 3rd ed. New York: Lyons and Burford.

Herrero, S., A. Higgins, J. E. Cardoza, L. I. Hajduk, and T. S. Smith. 2011. Fatal attacks by American black bear on people: 1900–2009. *Journal of Wildlife Management* 75:596–603.

Hirsch, J. G., L. C. Bender, and J. B. Haufler. 1999. Black bear, *Ursus americanus*, movements and home ranges on Drummond Island, Michigan. *Canadian Field-Naturalist* 113:221–25.

Hock, R. J. 1960. Seasonal variation in physiological functions of arctic ground squirrels and black bears. In *Mammalian Hibernation*, ed. C. P. Lyman and A. R. Dawe, 155–77. Cambridge, MA: Harvard University Press.

Horejsi, B. L., G. E. Hornbeck, and R. M. Raine. 1984. Wolves, *Canis lupus*, kill female black bear, *Ursus americanus*, in Alberta. *Canadian Field-Naturalist* 98:368–69.

Hornaday, W. T. 1905. A new white bear, from British Columbia. *Annual Report of the New York Zoological Society* 9:81–86.

Horner, M. A., and R. A. Powell. 1990. Internal structure of home ranges of black bears and analyses of home-range overlap. *Journal of Mammalogy* 71:402–10.

Howe, E. J., M. E. Obbard, R. Black, and L. L. Wall. 2010. Do public complaints reflect trends in human–bear conflict? *Ursus* 21:131–42.

Inman, R. M., C. M. Costello, D. E. Jones, K. H. Inman, B. C. Thompson, and H. B. Quigley. 2007. Denning chronology, hunt season timing, and design of bear management units. *Journal of Wildlife Management* 71:1476–83.

Inman, R. M., and M. R. Pelton. 2002. Energetic production by soft and hard mast foods of American black bears in the Smoky Mountains. *Ursus* 13:57–68.

Jeffers, R. M. 1995. Pinon pine seed production, collection, and storage. In *Desired future conditions for pinyon-juniper ecosystems.*, ed. D. W. Shaw, E.F. Aldon, and C. LoSapio, 191–97. General Technical Report RM-258. Fort Collins, CO: US Department of Agriculture, Forest Service, Rocky Mountain Research Station.

Johnson, H. E., D. L. Lewis, S. A. Lischka, and S. W. Breck. 2018. Assessing ecological and social outcomes of a bear-proofing experiment. *Journal of Wildlife Management* 82:1102–14.

Johnson, K. G., D. O. Johnson, and M. R. Pelton. 1978. Simulation of winter heat loss for a black bear in a closed tree den. *Proceedings of the Eastern Workshop on Black Bear Management and Research* 4:155–66.

Johnson, K. G., and M. R. Pelton. 1980. Environmental relationships and the denning period of black bears in Tennessee. *Journal of Mammalogy* 61:653–60.

———. 1981. Selection and availability of dens for black bears in Tennessee. *Journal of Wildlife Management* 45:111–19.

Kamler, J. F., L. A. Green, and W. B. Ballard. 2003. Recent occurrence of black bears in the southwestern great plains. *Southwestern Naturalist* 48:303–6.

Kendall, K., and J. Stetz. 2010. Grizzly and black bear marking behavior. *Intermountain Journal of Sciences* 16:126.

Kindschuh, S. R., J. W. Cain III, D. Daniel, and M. A. Peyton. 2016. Efficacy of GPS cluster analysis for predicting carnivory sites of a wide-ranging omnivore: the American black bear. *Ecosphere* 7:e01513. https://doi.org/10.1002/ecs2.1513.

Koehler, G. M., and D. J. Pierce. 2003. Black bear home-range sizes in Washington: climatic, vegetative, and social influences. *Journal of Mammalogy* 84:81–91.

Kolenosky, G. B. 1990. Reproductive biology of black bears in east-central Ontario. *International Conference on Bear Research and Management* 8:385–92.

Kovach, A. I., and R. A. Powell. 2003. Effects of body size on male mating tactics and paternity in black bears, *Ursus americanus*. *Canadian Journal of Zoology* 81:1257–268.

Lara-Díaz, N. E., H. Coronel-Arellano, C. A. López-González, G. Sánchez-Rojas, and J. E. Martínez-Gómez. 2018. Activity and resource selection of a threatened carnivore: the case of black bears in northwestern Mexico. *Ecosphere* 9:e01923. 10.1002/ecs2.1923.

Lara-Díaz, N. E., C. A. López-González, H. Coronel-Arellano, and A. González-Bernal. 2013. Black bear connectivity in the Sky Islands of Mexico and the United States. In *Merging science and management in a rapidly changing world: biodiversity and management of the Madrean Archipelago III*, ed. G. J. Gottfried, P. F. Ffolliott, B. S. Gebow, L. G. Eskew, and L. C. Collins, 263–68. RMRS-P-67. Fort Collins, CO: US Department of Agriculture, Forest Service, Rocky Mountain Research Station.

Larivière, S. 2001. *Ursus americanus. Mammalian Species* 647:1–11.

LeCount, A. L. 1982. Characteristics of a central Arizona black bear population. *Journal of Wildlife Management* 46:861–68.

———. 1983. Evidence of wild black bears breeding while raising cubs. *Journal of Wildlife Management* 47:264–68.

———. 1987a. Causes of black bear cub mortality. *International Conference on Bear Research and Management* 7:75–82.

———. 1987b. *Characteristics of a northern Arizona black bear population—a final report*. Phoenix: Arizona Game and Fish Department.

———. 1990. *Characteristics of an east-central Arizona black bear population*. Phoenix: Arizona Game and Fish Department Technical Report No. 2.

Lee, D. J., and M. R. Vaughan. 2004. Black bear family breakup in Western Virginia. *Northeastern Naturalist* 11:111–22.

———. 2005. Yearling and subadult black bear survival in a hunted Virginia population. *Journal of Wildlife Management* 69 :1641–51.

Lee, L. 1967. *Bears in New Mexico wildlife management*. Sante Fe: New Mexico Department of Game and Fish.

Lewis, T., G. Roffler, A. Crupi, R. Maraj, and N.

Barten. 2020. Unraveling the mystery of the glacier bear: Genetic population structure of black bears (*Ursus americanus*) within the range of a rare pelage type. *Ecology and Evolution* 10:7654–68. doi:10.1002/ece3.6490.

Ligon, J. S. 1927. *Wild life of New Mexico: its conservation and management*. Santa Fe, NM: State Game Commission, Department of Game and Fish.

Liley, S. G., and R. N. Walker. 2015. Extreme movement by an American black bear in New Mexico and Colorado. *Ursus* 26:1–6.

Lindzey, F. G., and E. C. Meslow. 1976. Winter dormancy in black bears in southwestern Washington. *Journal of Wildlife Management* 40:408–15.

———. 1977. Home range and habitat use by black bears in southwestern Washington. *Journal of Wildlife Management* 41:413–25.

López González, C. A., M. G. Camargo-Aguilera, K. U. Saucedo, and N. E. Lara Díaz. 2019. A wandering black bear (*Ursus americanus*, Pallas 1780) in the Sierra Gorda Biosphere Reserve, Queretaro. *American Midland Naturalist* 182:252–59.

López-González, C. A., R. W. Jones, C. Silva-Hurtado, and I. A. Sáyago-Vázquez. 2009. Scorpions are a food item of American black bears in Sonora, Mexico. *Western North American Naturalist* 69:131–33.

MacHutchon, A. G. 1989. Spring and summer food habits of black bears in the Pelly River Valley, Yukon. *Northwest Science* 63:116–18.

Maehr, D. S. 1997. The comparative ecology of bobcat, black bear, and Florida panther in south Florida. *Bulletin of the Florida Museum of Natural History* 40:1–176.

Mattson, D. J., R. R. Knight., and B. M. Blanchard. 1992. Cannibalism and predation on black bears by grizzly bears in the Yellowstone Ecosystem, 1975–1990. *Journal of Mammalogy* 73:422–25.

McClinton S. F., P. L. McClinton, and J. V. Richerson. 1992. Food habits of black bears in Big Bend National Park. *Southwestern Naturalist* 37:433–35.

McDonald, J. E., D. P. Fuller, J. E. Cardoza, and T. K. Fuller. 1994. Influence of hunter effort and food abundance on Massachusetts black bear harvest. *Eastern Workshop on Black Bear Research and Management* 12:159.

McKelvey, R. W., and D. W. Smith. 1979. A black bear in a bald eagle nest. *Murrelet* 60:106–7.

McLellan, B., and D. C. Reiner. 1994. A review of bear evolution. *International Conference on Bear Research and Management* 9:85–96.

Mesa-Cruz, J. B., C. Olfenbuttel, M. R. Vaughan, J. L. Sajecki, and M. J. Kelly. 2020. Litter size and cub age influence weight gain and development in American black bears (*Ursus americanus*). *Journal of Mammalogy* 101:564–73.

Mitchell, F. S., D. P. Onorato, E. C. Hellgren, J. R. Skiles Jr., and L. A. Harveson. 2005. Winter ecology of American black bears in a desert montane island. *Wildlife Society Bulletin* 33:164–71.

Mollohan, C. M. 1987. Characteristics of adult female black bear daybeds in northern Arizona. *International Conference on Bear Research and Management* 7:145–49.

Monteith, K. L, V. C Bleich, T. R. Stephenson, B. M. Pierce, M. M. Connor, J. G. Kie, and R. T. Bowyer. 2014. Life-history characteristics of mule deer: effects of nutrition in a variable environment. *Wildlife Monographs* 186:1–62.

Morlan, R. E. 1972. Predation at a northern Yukon bank swallow territory. *Canadian Field-Naturalist* 86:376.

Murphy, K. M., G. S. Felzien, M. G. Hornocker, and T. K. Ruth. 1998. Encounter competition between bears and cougars: some ecological implications. *Ursus* 10:55–60.

Nelson, J. N. 1957. Bear cub taken by an eagle. *Victoria Naturalist* 14:62–63.

Nelson, O. L., and C. T. Robbins. 2015. Cardiovascular function in large to small hibernators: bears to ground squirrels. *Journal of Comparative Physiology B* 185:265–79.

Nelson, R. A. 1980. Protein and fat metabolism in hibernating bears. *Federation Proceedings* 39:2955–58.

Nelson, R. A., H. W. Wahner, J. D. Jones, R. D. Ellefson, and P. E. Zollman. 1973. Metabolism of bears before, during, and after winter sleep. *American Journal of Physiology* 224:491–96.

New Mexico Department of Game and Fish (NMDGF). 1926. *Report for the fiscal years 1925–1926*. Santa Fe: New Mexico Department of Game and Fish.

———. 2020. New Mexico hunting rules and information. Santa Fe: New Mexico Department of Game and Fish.

Noble, W. O., and E. C. Meslow. 1998. Spring foraging and forest damage by black bears in the central coastal ranges of Oregon. *Ursus* 10:293–98.

Noyce, K. V., and D. L. Garshelis. 1994. Body size and blood characteristics as indicators of condition and reproductive performance in black bears. *International Conference on Bear Research and Management* 9:481–96.

———. 1997. Influence of natural food abundance on black bear harvest in Minnesota. *Journal of Wildlife Management* 61:1067–74.

———. 1998. Spring weight changes in black bears in northcentral Minnesota: the negative foraging period revisited. *Ursus* 10:521–31.

———. 2011. Seasonal migrations of black bears (*Ursus americanus*): causes and consequences. *Behavioral Ecology and Sociobiology* 65:823–35.

Noyce, K. V., P. B. Kannowski, and M. R. Riggs. 1997. Black bears as ant-eaters: seasonal associations between bear myrmecophagy and ant ecology in north-central Minnesota. *Canadian Journal of Zoology* 75:1671–86.

Obbard, M. E., E. J. Howe, L. L. Wall, B. Allison, R. Black, P. Davis, L. Dix-Gibson, M. Gatt, and M. N. Hall. 2014. Relationships among food availability, harvest, and human–bear conflict at landscape scales in Ontario, Canada. *Ursus* 25:98–110.

Olsen, T. L. 1993. Infanticide in brown bears, *Ursus arctos*, at Brooks River, Alaska. *Canadian Field-Naturalist* 107:92–94.

Onorato, D. P., E. C. Hellgren, R. A. Van Den Bussche, and D. L. Doan-Crider. 2004. Phylogeographic patterns within a metapopulation of black bears (*Ursus americanus*) in the American Southwest. *Journal of Mammalogy* 85:140–47.

Onorato, D. P., E. C. Hellgren, R. A. Van Den Bussche, D. L. Doan-Crider, and J. R. Skiles Jr. 2007. Genetic structure of American black bears in the Desert Southwest of North America: conservation implications for recolonization. *Conservation Genetics* 8:565–76.

Paquet, P. C., and L. N. Carbyn. 1986. Wolves, *Canis lupus*, killing denning black bears, *Ursus americanus*, in the Riding Mountain National Park Area. *Canadian Field-Naturalist* 100:371–72.

Pelton, M. R. 2003. Black bear. In *Wild mammals of North America: biology, management, and economics*, 2nd ed., ed. G. A. Feldhamer, B. C. Thompson, and J. A. Chapman, 547–55. Baltimore: Johns Hopkins University Press.

Pritchard, G. T., and C. T. Robbins. 1990. Digestive and metabolic efficiencies of grizzly and black bears. *Canadian Journal of Zoology* 68:1645–51.

Puckett, E. E., P. D. Etter, E. A. Johnson, and L. S. Eggert. 2015. Phylogeographic analyses of American black bears (*Ursus americanus*) suggest four glacial refugia and complex patterns of postglacial admixture. *Molecular Biology and Evolution* 32:2338–50.

Raine, R. M., and J. L. Kansas. 1990. Black bear food habits and distribution by elevation in Banff National Park, Alberta. *International Conference on Bear Research and Management* 8:297–304.

Ramsay, M. A., and R. L. Dunbrack. 1986. Physiological constraints on life history phenomena: the example of small bear cubs at birth. *American Naturalist* 127:735–43.

Ricklefs, B. 2005. Bear-human management on Philmont Scout Ranch. *Western Black Bear Workshop* 8:84–90.

Rode, K. D., C. T. Robbins, and L. A. Shipley. 2001. Constraints on herbivory by grizzly bears. *Oecologia* 128:62–71.

Rogers, L. L. 1980. Inheritance of coat color and changes in pelage coloration in black bears in northeastern Minnesota. *Journal of Mammalogy* 61:324–27.

———. 1987. Effect of food supply and kinship on social behavior, movements, and population growth of black bears in northeastern Minnesota. *Wildlife Monographs* 97:1–72.

Rogers, L. L., and L. D. Mech. 1981. Interactions of wolves and black bears in northeastern Minnesota. *Journal of Mammalogy* 62:434–36.

Rounds, R. C. 1987. Distribution and analysis of colour morphs of the black bear (*Ursus americanus*). *Journal of Biogeography* 14:521–38.

Samson, C., and J. Huot. 2001. Spatial and temporal interactions between female American black bears in mixed forest of eastern Canada. *Canadian Journal of Zoology* 79:633–41.

Sawyer, H., and F. Lindzey. 2002. *A review of predation on bighorn sheep (Ovis canadensis)*. Laramie: Wyoming Cooperative Fish and Wildlife Research Unit.

Schenk, A., M. E. Obbard, and K. M. Kovacs. 1998. Genetic relatedness and home-range overlap among female black bears (*Ursus americanus*) in northern Ontario, Canada. *Canadian Journal of Zoology* 76:1511–19.

Schooley, R. L., C. R. McLaughlin, G. J. Matula Jr., and W. B. Krohn. 1994. Denning chronology of female black bears: effects of food, weather, and reproduction. *Journal of Mammalogy* 75:466–77.

Schwartz, C. C., and A. W. Franzmann. 1991. Interrelationships of black bears to moose and forest succession in the northern coniferous forest. *Wildlife Monographs* 113:1–58.

———. 1992. Dispersal and survival of subadult black bears from the Kenai Peninsula, Alaska. *Journal of Wildlife Management* 56:426–31.

Schwartz, C. C., S. D. Miller, and W. W. Franzmann. 1987. Denning ecology of three black bear populations in Alaska. *International Conference on Bear Research and Management* 7:281–91.

Servheen, C. 1989. The status and conservation of bears of the world. *International Conference on Bear Research and Management* 2:1–32.

Smith, D. W., D. R. Trauba, R. K. Anderson, and R. O. Peterson. 1994a. Black bear predation on beavers on an island in Lake Superior. *American Midland Naturalist* 132:248–55.

Smith, M. E., J. L. Hechtel, and E. H. Follmann. 1994b. Black bear denning ecology in interior Alaska. *International Conference on Bear Research and Management* 9:513–22.

Smith, T. S., and S. Herrero. 2018. Human-bear conflict in Alaska: 1880–2015. *Wildlife Society Bulletin* 42:254–63.

Stirling, I., and A. E. Derocher. 1989. Factors affecting the evolution and behavioral ecology of the modern bears. *International Conference on Bear Research and Management* 8:189–204.

Støen, O., E. Bellemain, S. Sæbø, and J. E. Swenson. 2005. Kin-related spatial structure in brown bears *Ursus arctos*. *Behavioral Ecology and Sociobiology* 59:191–97.

Sullivan, M. G. 1979. Blue grouse hen–black bear confrontation. *Canadian Field-Naturalist* 93:200.

Svoboda, N. J., J. L. Belant, D. E. Beyer, J. F. Duquette, H. K. Stricker, and C. A. Albright. 2011. American black bear predation of an adult white-tailed deer. *Ursus* 22:91–94.

Swenson, J. E., A. Jansson, R. Riig, and F. Sandegren. 1999. Bears and ants: myrmecophagy by brown bears in central Scandinavia. *Canadian Journal of Zoology* 77:551–61.

Swenson, J. E., F. Sandegren, S. Brunberg, and P. Segerstrom. 2001. Factors associated with loss of brown bear cubs in Sweden. *Ursus* 12:69–80.

Tatman, N. M., S. G. Liley, J. W. Cain III, and J. W. Pitman. 2018. Effects of calf predation and nutrition on elk vital rates. *Journal of Wildlife Management* 82:1417–28.

Taylor, A. P., M. L. Allen, and M. S. Gunther. 2015. Black bear marking behavior at rub trees during the breeding season in northern California. *Behavior* 152:1097–111.

Tietje, W. D., and R. L. Ruff. 1980. Denning behavior of black bears in boreal forest of Alberta. *Journal of Wildlife Management* 44:858–70.

Tøien, Ø., J. Blake, D. M. Edgar, D. A. Grahn, H. C. Heller, and B. M. Barnes. 2011. Hibernation in black bears: independence of metabolic suppression from body temperature. *Science* 331:906–9.

Troyer, W. A., and R. J. Hensel. 1962. Cannibalism in brown bear. *Animal Behavior* 10: 231.

United States Census Bureau. 2003. *2000 census of population and housing, population and housing unit counts PHC-3-33, New Mexico*. Washington, D.C.

———. 2010. *Annual estimates of the resident population for the United States, regions, states, and Puerto Rico: April 1, 2000 to July 1, 2010 (NST-EST2010–01)*. Washington, DC: US Census Bureau, Population Division.

———. 2020. *Annual estimates of the resident population for the United States, regions, states, and the District of Columbia: April 1, 2010 to July 1, 2020 (NST-EST2020)*. Washington, DC: US Census Bureau, Population Division.

Van Vuren, D. H. 2001. Predation on yellow-bellied marmots (*Marmota flaviventris*). *American Midland Naturalist* 145:94–100.

Walters, E. L., and E. H. Miller. 2001. Predation on nesting woodpeckers in British Columbia. *Canadian Field-Naturalist* 115:413–19.

Weaver, K. M., and M. R. Pelton. 1994. Denning ecology of black bears in the Tensas River Basin of Louisiana. *International Conference on Bear Research and Management* 9:427–33.

Willey, C. H. 1974. Aging black bears from first premolar tooth sections. *Journal of Wildlife Management* 38:97–100.

Wimsatt, W. A. 1963. Delayed implantation in the Ursidae, with particular reference to the black bear, *Ursus americanus*. In *Delayed implantation*, ed. A. C. Enders, 69–77. Chicago: University of Chicago Press.

Yancey, F. D., II, and M. W. Lockwood. 2019. First record of the American black bear (*Ursus americanus*) from the Chinati Mountains of western Texas. *Southwestern Naturalist* 63:133–36.

Yensen, E., and P. W. Sherman. 2003. Ground squirrels (*Spermophilis* and *Ammospermophilus* species). In *Wild mammals of North America: biology, management, and economics*, 2nd ed., ed. G. A. Feldhamer, B. C. Thompson, and J. A. Chapman, 211–31. Baltimore: Johns Hopkins University Press.

Young, B. F., and R. L. Ruff. 1982. Population dynamics and movements of black bears in east central Alberta. *Journal of Wildlife Management* 46:845–60.

Young, J. V. 1984. *The state parks of New Mexico*. Albuquerque: University of New Mexico Press.

Zager, P., and J. Beecham. 1982. *New Mexico black bear studies, report for contract number 516-7-28*. Santa Fe: New Mexico Department of Game and Fish.

———. 2006. The role of American black bears and brown bears as predators on ungulates in North America. *Ursus* 17:95–108.

BROWN BEAR (*URSUS ARCTOS*)

David E. Brown, Matthew J. Gould, and Jean-Luc E. Cartron

No life history study has ever been conducted on New Mexico's brown bears (historically and taxonomically referred to as grizzly bears; see below), nor is one now possible. The grizzly, while formerly a well-known and relatively common resident of New Mexico, has been extirpated from the "Land of Enchantment" since sometime during the 1930s (Lee 1967). New Mexico's last grizzly might have been a large male taken near Cliff in Grant County during the spring of 1931, though the last museum specimen, now housed at the US National Museum, is a skull dated 1935 (but quite possibly older) from the Magdalena Mountains (Brown 1996). Nor is there much likelihood of a reintroduction in the future, there being no serious proposal to return the grizzly to any of its former haunts in New Mexico. In large measure, wildlife management agencies are unlikely to even consider a reintroduction in the face of expected opposition by local ranching communities and for fear of legal liability in the event of physical or financial harm caused by a released animal. This is perhaps unfortunate in that much of the bear's former habitat remains intact.

Evolutionarily, the brown bear is a relative newcomer to the bear family, Ursidae, diverging from the American black bear lineage only 940,000 years ago (Hailer et al. 2014). Originating in Asia, brown bears radiated throughout the world to inhabit every continent except South America and Antarctica. A short ~300,000 years later during the mid-Pleistocene, a northern brown bear population was isolated in the arctic coastal regions and adapted to a highly carnivorous diet and marine lifestyle that ultimately led to the speciation of the brown bear's closest relative, the polar bear (*Ursus maritimus*; Kutschera et al. 2014; Hailer and Welch 2016). After crossing the Bering land bridge from Asia into North America less than 250,000 years ago, the brown bear entered Alaska approximately 100,000 years before present (ybp) and there it remained until glacial retreat opened a southerly path 13,000 ybp (McLellan and Reiner 1994). Kurtén and Anderson (1980) make no specific references to *Ursus arctos* in New Mexico, pointing out that all the subglacial records for this species in North America are during late Pleistocene-Holocene times, and postdate the Sangamonian Stage, or last interglacial period. This species was therefore absent in New Mexico prior to the late Wisconsin ca. 15,000 ybp and arrived only recently, comparatively speaking.

After decades of taxonomic confusion, grizzly and brown bears are now known as one and the same species, *Ursus arctos* (e.g., Rausch 1963; Kurtén 1973; Schwartz et al. 2003). At times, the two names have been used interchangeably, but "grizzly bear" is reserved for the North American

(*opposite page*) Photograph: © Pat Gaines.

subspecies of brown bear, *Ursus arctos horribilis*—except for the Kodiak Archipelago off the coast of Alaska (see below). Size is the main characteristic used anywhere to distinguish the brown bear from its smaller congener the black bear (*Ursus americanus*). Exceptionally, a male grizzly in New Mexico might have reached ~2 m (~6 ft 6 in) in total length with a standing height upward of ~2.4 m (8 ft) (Brown 1996). More typical of a large grizzly in New Mexico, however, would have been an adult 1.5 m (5 ft) in total length. Although grizzly weights have been reported from the Southwest, most of them are only estimates, and probably too high (e.g., ~360 to 455 kg [800 to 1,000 lbs]) as many were based on squared feet of hide, an unreliable method (Brown 1996). Accurate weights of New Mexico grizzlies are hard to come by for obvious reasons, but Starker Leopold's (1959) range of ~135–230 kg (300 to 700 lbs) for adults seems not too far off the mark, given recent studies of Rocky Mountain grizzlies. Measurements of grizzlies in the Greater Yellowstone Ecosystem show that females range from ~90–220 kg (200–485 lbs) and males ~120–325 kg (265–720 lbs) (van Manen and Haroldson 2017). Thus, even a small adult grizzly would still be expected to exceed, if not greatly exceed, a black bear in size. Southwestern male black bears uncommonly exceed ~135 kg (300 lbs), and females are usually less than ~73 lbs (160 lbs) (LeCount 1977; Waddell and Brown 1984; see Chapter 16).

Other traits used to distinguish grizzlies from black bears include the grizzly's dished face (due to a depression in the frontonasal region of the skull; see Merriam 1918), a distinctive shoulder hump, short and round ears, and the long (~5–10 cm [2–4 in]), slightly curved claws on the forefeet that help grizzlies with digging but limit adult bears' ability to climb trees (Brown 1996; see Chapter 16). By comparison, the black bear tends to have a straight brow line, giving the face less of a dished appearance. It also lacks a prominent

shoulder hump and has longer and more pointed ears, and its claws on all four feet are both shorter (<5 cm [2 in]) and more tightly curved. Pelage color would not have been a reliable trait to distinguish between grizzly and black bears in the Southwest, but the hairs of grizzly bears are often white- or silver-tipped, giving it a grizzled appearance and thus its nickname "silvertip" (Brown 1996). The hairs of grizzly bears also have their own, distinctive texture, causing them to be longer, silkier, and wavier than those of black bears, which are more uniform and shorter (Brown 1996). The skulls of grizzlies and black bears are structurally very similar, and even one experienced with both bears should not assign species before taking measurements and taking into consideration the animal's relative age. With those caveats in mind, however, rostrum length in adult grizzlies normally exceeds 17.8 cm (7 in), while in black bears it is less than 16.5 cm (6 1/2 in) (Hoffmeister 1986). The dental formula of the grizzly bear is identical to that of the black bear, or I3/3, C1/1, PM 2–4/2–4, M 2/3 (or a total of 42 teeth). In the grizzly bear, however, the second molar (M2) is broadest at its anterior end, whereas in the black bear M2 is broadest approximately halfway between the anterior and posterior margins, and the last upper molar is >31 mm (1.2 in) for the grizzly bear and <31 mm for the black bear (Foresman 2001).

In early-day New Mexico (including before statehood), body size would not have been as diagnostic as it is now, for one simple reason: observers just did not know how many bear species were present. Making species identification even more complicated, both brown and black bears also exhibit much variation in size, pelage, and behavior, and both are sexually dimorphic as well, with males larger than females. At the time that grizzlies still roamed New Mexico, individual and sex-based variation tended to muddle species identification. As a result, earlier investigators did not always differentiate between grizzlies and black bears, and

Photo 17.1 (*above*). Adult grizzly bear from Alaska. Brown bears (*Ursus arctos*), both grizzlies (*U. a. horribilis*) and Kodiak brown bears (*U. a. middendorffi*), are identified by their large size, prominent shoulder hump, dished face (from a depression in the frontonasal region of the skull), and short and round ears. The grizzly bear no longer occurs in New Mexico. Based on verified records, it was extirpated from the state during the 1930s. Photograph: © Hari Viswanathan.

Photo 17.2 (*right*). Yearling in spring (18 May 2012) in Yellowstone National Park, with a pelage that looks over-wintered and has not started to shed. Four age classes are recognized in grizzly bears: 1) cubs of the year; 2) yearlings (one year old); 3) subadults (two to three years old); and 4) adults (≥ 4 years old). Compared to subadults, yearlings still have fairly short snouts. Dispersal of this bear and other yearlings would happen the following year. Photograph: © Pat Gaines.

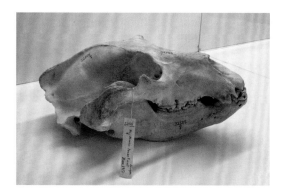

Photo 17.3. Skull of a female grizzly bear (lateral view) collected in 1917 by J. Stokley Ligon in New Mexico's Pecos Wilderness. Skull measurements led Merriam (1918) and Bailey (1931) to erroneously conclude that New Mexico was home to as many as seven species of grizzly bears. Differences in skull size are now known to be the result of sexual dimorphism (males are much larger than females), growth past sexual maturity, and individual morphological variation within and between populations. Photograph: © Jon Dunnum.

government reports and scientific communications often referred to one or both species using generic names such as "big bears" or brown bears (e.g., Bailey 1931; Brown 1996). In Mexico, the grizzly was given such diverse names as *oso blanco* (white bear), *oso plateado* (silver bear), *oso gris* (gray bear), *oso grande* (big bear), or simply *el oso* (the bear; Sheldon 1925; Leopold 1959). In the United States, grizzly bears were referred to as variegated bears (for the variation in pelt tones), white bears, gray bears, red bears, cinnamon bears, brown bears, silver bears, and, most often, silvertips (Brown 1996), sometimes resulting in confusion with the cinnamon phase of the black bear (e.g., Coues and Yarrow 1875:65–66; Duffen 1960). That grizzlies show great individual variation in coat color was illustrated by the pelage of mixed various shades on a cub taken in Taos County on 22 July 1918 (Bailey 1931). Some grizzlies undoubtedly could change colors when they shed their coats, as documented also in black bears (see, e.g., Waddell and Brown 1984;

see Chapter 16). The hide of a large male killed in 1930 in New Mexico's Black Range was still variegated and dark in color when one of us (D. E. Brown) viewed it 50 years later. An adult female from the Blue Range along the New Mexico–Arizona border was much lighter and closely resembled a female described by Bailey (1931) from Arizona's White Mountains and a grizzly taken in the Sangre de Cristo Mountains by George Alfred "Skipper" Viles, the latter now on display at the New Mexico Game and Fish office in Santa Fe.

Merriam (1918), on the basis of skull measurements, postulated that there were 96 subspecies of the brown bear in North America. He and

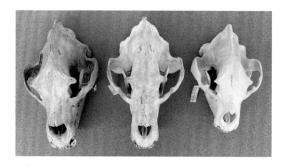

Photos 17.4a and b. a) (*top*) Dorsal view of the skulls of an adult grizzly bear from Alaska (MSB 14250; left), an adult female grizzly bear from New Mexico (MSB 32505, center), and a wild-caught, adult male black bear (*Ursus americanus*) (MSB 64927, right) from the Roswell Zoo. b) (*bottom*) Adult Kodiak brown bear (*Ursus arctos middendorffi*) from the Kodiak Archipelago (MSB 23004, left), and the same grizzly bear skulls from Alaska (center) and New Mexico (right) as in Photo 17.4a. The sexes of the Alaska grizzly and the Kodiak brown bear are unknown. Photographs: © Jon Dunnum.

Vernon Bailey (1931) recognized seven species of grizzlies in New Mexico, going as far as describing more than one from the same mountain range! Modern mammalogists now follow Rausch (1963), who documented geographic clinal variation (a gradual change in a feature across the range of a species that is often correlated with an environmental transition) in condylobasal length (i.e., distance between the rearmost aspect of the occipital cusps and the anterior margin of the intermaxillary bones) and on that basis ascribed all grizzly populations to *Ursus arctos horribilis*. As pointed out by Rausch (1963), much of the past taxonomic confusion stemmed from a lack of attention to the influence of sexual size dimorphism, growth past sexual maturity, and the considerable amount of individual morphological variation within populations. Genetic evidence supports this assertion and posits that there are two subspecies of brown bear in North America, the grizzly bear and the broader-skulled Kodiak bear (*U. a. middendorffi*) of the Kodiak Archipelago (Paetkau et al. 1998; Miller et al. 2006; and see Photo 17.4b). Paetkau et al. (1998) also note that ecological differences, rather than genetic differences, are often the best explanation for differences in body size among grizzlies as some groups are limited to a more vegetation-based diet while others have access to a more protein-dominated diet such as salmon (*Salmo* spp.) runs. In retrospect, given all the potential for morphological variation in the species, the range in size of six adult male New Mexican grizzly skulls from different localities is not that great, with lengths from 284 to 338 mm (11.2 to 13.3 in) (mean = 315.5 mm [12.4 in]) and widths of 199 to 234 mm (7.8 to 9.2 in) (mean = 213 mm [8.4 in]) (Bailey 1931), possibly with larger specimens originating from the higher, more northern portions of the state and one smaller specimen from the Mexican border area, or simply reflecting differences in age, sex, or both.

Grizzly and black bear tracks are similar, with subtle differences (Photos 17.5a and b). The toes in a grizzly's footprint are closer together and less curved compared with a black bear's. The longer claws of a grizzly's front feet also register farther (≥ 5 cm [2 in]) in front of the toes. The imprint left by the front foot is squarer: a line can be drawn straight across the track in the space between the pad and the toes. Furthermore, the back pad of the front paw for a grizzly is square-shaped rather than the kidney shape exhibited by black bears. A hind footprint 18 cm or longer (≥ 7 in) belongs to a grizzly. Any print 12.7 cm (5 in) or greater in width is also considered evidence of a grizzly. Note that the medium in which a paw print occurs can distort the size or shape of these characteristics. Identifying grizzlies from black bears based on scat morphology is unreliable, with more reliable methods used by researchers now including specially trained dogs (*Canis familiaris*) and genetic analysis.

DISTRIBUTION AND ABUNDANCE

Before European settlement of North America, the grizzly bear was widely distributed from northern Canada and Alaska south to central Mexico and from the Pacific coast east nearly to the Mississippi River (Hall and Kelson 1959). Early travelers in New Mexico reported relatively more encounters with grizzlies than black bears (Pattie 1833; Davis 1982) because grizzlies were more frequently associated with river bottoms and other lower-elevation environments. Elsewhere, in the foothills and mountains, the reverse might have been true, with black bears more common given their clear preference for closed canopy woodlands and forests (see Chapter 16). Near the Santa Rita copper mines at the southern end of the Mogollon Mountains, J. H. Clark collected four black bears but only one grizzly in 1855 (Brown 1996).

Reports dating back to the years after the Civil War no longer mentioned grizzlies from riparian thickets along New Mexico's larger rivers; instead, black bears now seemed more common compared

Photos 17.5a and b. Grizzly bear tracks in mud. a) (*left*) Front tracks are wider; rear tracks are longer (the small round heel pad of the front foot rarely registers). Each foot has five toes; the large toe is on the outside. Note also the imprint of the claws. b) (*right*) In grizzlies the imprint left by the front foot is squarer than in black bears, and a line can be drawn straight across the track in the space between the pad and the toes. Photographs 17.5 a: NPS/ Eric Johnston; 17.5 b: Jim Peaco.

Photo 17.6. The distribution of the grizzly bear once extended from the Pacific Coast east nearly to the Mississippi River and from northern Alaska and Canada to central Mexico. Roughly between 1850 and 1950, grizzly bears were extirpated from nearly all areas south of the Canadian border. Their current distribution is largely restricted to Alaska (including Katmai National Park and Preserve on the Alaska Peninsula; this photo) and Canada (British Columbia, Alberta, and the territories) and only covers about 6% of their historical range in the lower 48 states. Photograph: © Hari Viswanathan.

to grizzlies in New Mexico (Brown 1996). Those grizzlies living in the accessible terrain had apparently been extirpated by settlement, leaving only those that lived at a higher elevation. Both New Mexico and Arizona had a sizeable Native American population ranging through the lowlands prior to either Spanish or Anglo settlement. Additionally, New Mexico had a large Spanish presence before American explorers began to arrive. The extirpation of grizzlies in the lowlands during the exploration and settlement phase might have been linked to superior firepower brought by the Americans, the use of hounds in hunting, and perhaps also different attitudes toward bears. More powerful rifles and ultimately repeating rifles made tackling a grizzly a much safer proposition. Reports of grizzlies in the riparian thickets along the larger watercourses decreased as these areas became the focal points of travel and settlement. A typical report of the time is that of Farrar (1968:147), for the period 1871 to 1875 in the Raton Mountains along the New Mexico-Colorado border, about 80 km (50 mi) east of Trinidad: "We saw black bears every few days and occasionally a grizzly came along."

By the next decade, the grizzly was mainly limited to elevations higher than those occupied by black bears. No longer were grizzlies found in river valleys or even in the open woodlands of the foothills. Stevens (1943:17), who hunted grizzlies in the 1880s and 1890s in the mountains of east-central New Mexico and knew bears well, stated that grizzlies were usually above the black bears: "At about eighty-five hundred feet elevation, the timber changes to spruce, balsam [Douglas-fir] and quaking aspen thickets, and here the grizzly makes his home." By the turn of the century Bailey (1931:367) also considered the grizzly an animal of New Mexico's high mountains: "In this region the grizzly bears seem to be almost entirely mountain-forest animals, keeping in the dense growth of timber, especially in the deep canyons and almost impenetrable jungles of windfall and second growth, which follow the burning of forest areas."

These accounts do not mean that grizzlies in New Mexico were always confined to high mountain peaks. Then, as now, the animals ranged widely and used a great variety of habitats. A bear in a montane meadow one day could be in a canyon bottom the next, and in alpine tundra that afternoon. Grizzlies followed food sources, and because they were naturally more open country animals than their black bear relatives, they often frequented open grassland, meadows, or rockslides. When startled, however, they would invariably head upslope to dense aspens and other mountain-forest retreats (e.g., Stevens 1943).

Locality Records

A number of grizzly specimens and literature records are available for New Mexico—particularly from the mountains in west-central portions of the state (Map 17.1). These include the Mogollon, Tularosa, San Francisco, Saliz, Gallo, and Pinos Altos mountains, the Datil and Gallinas mountains, the Magdalena Mountains, the San Mateo Mountains and the Black Range (Bailey 1931; Findley et al. 1975). The National Museum possesses the skulls of two males collected on 10 October 1911 and 20 June 1918 from the Mogollon Mountain complex adjacent to the New Mexico–Arizona border and two females collected in 1893 in the Pinos Altos Mountains near Silver City. Also present at this museum are a male sent in by Vernon Bailey and Victor Culbertson in 1895; an animal of unknown sex sent in by Bailey and H. H. Hotchkiss on 14 June 1906; a female and cubs taken on 16 December 1918; a female collected on 6 May 1920; and the skull of a male taken by Dub Evans in April 1930—all from the Black Range (also known as the Mimbres Mountains). Other grizzlies in the museum include a skull collected at Kid Springs in the Datil Range on 13 October 1905 and the already mentioned male skull from Magdalena Baldy dated 1935.

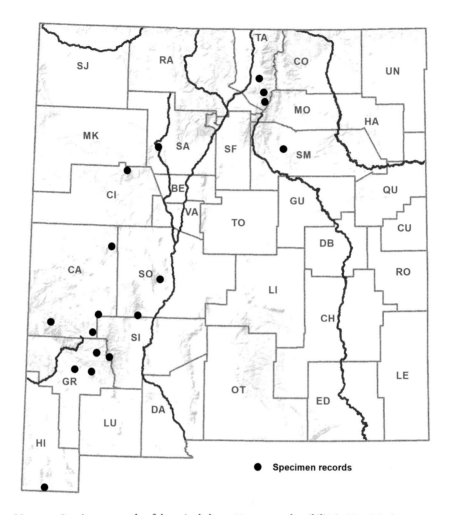

Map 17.1. Specimen records of the grizzly bear (*Ursus arctos horribilis*) in New Mexico.

Grizzlies from other New Mexican ranges in the National Museum include a female and two cubs taken by J. T. McMullin in the Sangre de Cristo Range, and there are good literature records from this range as well as from such other northern New Mexico ranges as the Chuska, Gallinas, Jemez, San Pedro, San Juan, and Zuni mountains. Less well documented, but reported to be grizzly range, were the Guadalupe, Sacramento, and El Capitan mountains in the southeastern portion of the state (Bailey 1931).

The occurrence of grizzlies in the Peloncillo, San Luis, and Animas mountains in New Mexico's Bootheel region is also well documented, but the status of the species in the Manzano, Sandia, and San Andres mountains is less certain (Bailey 1931). Although Pattie (1833) reported a number of grizzly bear encounters along the Rio Grande and other major waterways, almost all of the succeeding references are to montane habitats. We are unaware of any evidence of a "plains grizzly" occurring east of the Pecos in eastern New Mexico.

HABITAT ASSOCIATIONS

As elsewhere in the Southwest, New Mexico's grizzlies required a wide range of habitats (Brown

Photos 17.7. Undated photos of a grizzly in a leg-hold trap, reputed to show the last "great bear" taken in the Black Range—sometime after the big male grizzly killed by G. W. ("Dub") Evans in April 1930 above Hillsboro. Unfortunately, the name of the trapper remains unknown. The photos, mounted as a threesome, were in the possession of a local rancher, Jimmy Bason (now deceased), without information regarding its source. Note that the vegetation in the background appears to be mixed conifer forest, suggesting that the bear might have been trapped near the top of the mountain. Photographs: courtesy of Harley Shaw.

1996). An individual bear could either use a relatively small, diverse area, such as a mountain mass with its numerous canyons and foothills, or roam over a larger, more homogeneous range with local areas of seasonal importance. Large mountains with steep relief and high elevations generally possess a greater array of habitats than similar-sized areas having less topographic relief (Brown 1996). Either way, it took a large amount of land and a variety of habitats to support a grizzly bear population. The movements of a band of grizzly bears—perhaps as many as 13—are described in Barker's (1953:18–19) book *Beatty's Cabin* as witnessed by Lewis Lindsay Dyche, who followed the trails of overturned logs and stones seemingly in every direction one day, "through the woods . . . through a number of grassy parks, down a small stream, up another, and then over a mountain" in the Pecos Wilderness.

While widely used by grizzlies elsewhere in the Rocky Mountains, specific records of grizzlies collected in New Mexico's alpine tundra (found in the San Juan, Sangre de Cristo and Sacramento

mountains above ~3,350–3,510 m [11,000–11,500 ft]) are lacking. Nonetheless, grizzlies were present in all of New Mexico's mountains possessing alpine tundra, and as suggested in Barker's (1953) account above, grizzlies undoubtedly ranged through these habitats in midsummer as they do farther north (Aune and Stivers 1982; Craighead et al. 1982). Bailey (1931:367) noted that grizzlies in the mountains of northern New Mexico ranged from timberline to the lowest timber," but that this distribution, ca. 1904 to 1906 may have been relictual because of vigorous, long-term persecution by hunters.

Numerous references place New Mexican grizzlies in boreal forest environments (Subalpine Conifer Forests and Meadows) from ~2,440–2,590 to 3,510 m (8,000–8,500 to 11,500 ft) elevation, and subalpine conifers provided important habitat for these bears as did adjacent meadows and aspen stands (Brown 1996). Similarly to Stevens (1943), Bailey (1931:367) referred to grizzlies commonly occurring in spruce (*Picea* spp.) and other subalpine vegetation and called this zone "the breeding zone of the bears." As the grizzlies came out of hibernation, the riparian bottoms of the high country provided both food and cover, and the bears fed on early green marsh plants, such as sedges (*Carex* spp.) and rushes (*Scirpus* spp., *Juncus* spp., *Eleocharis* spp., etc.). Beavers (*Castor canadensis*), too, were likely taken. In spring and early summer, the subalpine meadows supplied green herbaceous plants, as well as roots, ants, gophers (family Geomyidae), yellow-bellied marmots (*Marmota flaviventris*) and other small game. The subalpine conifer forest with its deadfalls, blowdowns, and old burns represented excellent escape cover and provided some of the most reliable foods such as currants and berries; quaking aspen (*Populus tremuloides*) thickets were said to be of special importance due to their herbaceous understories (Brown 1994).

At lower elevations, below the subalpine forest, grizzlies used the dense, mixed conifer forests of Douglas-fir (*Pseudotsuga menziesi*), white fir (*Abies concolor*), aspen and ponderosa pine (*Pinus ponderosa*) as bedding sites (Brown 1996). In the more open areas, where grasses and forbs flourished, the forest was of at least seasonal importance for foraging. Monotypic stands of ponderosa pines, however, were likely poor habitats for bears unless Gambel oak (*Quercus gambelli*) was also present. Key habitats within the 1,980 to 2,590 m (6,500 to 8,500-ft) montane zone consisted of meadows and aspen patches—not only in valley bottoms but also those on steep slopes and in rugged canyons. It was also here, in the montane meadows, that the bears found the sedges, horsetails (*Equisetum* spp.), and other herbaceous plants that were so important to their survival in the early spring. Later, as the season progressed, the grizzlies worked the meadow edges and canyons for grass, currants (*Ribes* spp.), raspberries (*Rubus strigosus*), and thimbleberries (*R. parviflorus*). Gambel oak, both as a tree and in shrubform thickets, supplied acorns, shade, and, if the occasion required, escape cover (Brown 1996).

Woodlands dominated by evergreen oaks (*Quercus* spp.), junipers (*Juniperus* spp.), or Mexican piñons (*Pinus cembroides*) also provided seasonal habitat for grizzlies (Brown 1996). The periodic presence of mast from these trees, combined with the proximity of grasslands and chaparral made these *encinals* (oak groves) especially attractive in late summer and early fall. Perhaps because of the open nature of many of these vegetation types, there are few reports of grizzlies in the more northern and drier pinyon-juniper communities and in the oak woodlands south of the Mogollon Rim. But even here, in the Animas, San Luis, and Peloncillo mountains, there are a few reports of grizzlies after 1890 (Brown 1996).

Although limited in extent in New Mexico, there is frequent mention of grizzlies in the higher (above 1,680 m [5,500 ft]) chaparral communities—especially after 1880 (French 1965a and b; Bailey 1931). Reference to grizzlies in manzanita thickets are especially common, the bears using these communities both as escape cover and to feed on manzanita berries and other mast. It is speculated that with the deterioration of riparian and grassland habitats from livestock grazing after 1880, the grizzly tended to disproportionately use chaparral as an alternative habitat. By then the abundance of livestock provided a ready source of protein that allowed the bears to remain in heavy cover and thereby gain some protection here from humans and their dogs (Brown 1996).

Most references to grizzlies in open grasslands (plains and semi-desert grasslands) are in late summer and prior to 1880 (Pattie 1831; Baird 1859; French 1965a and b). Even in these early accounts the grizzlies are usually near woodlands and riparian communities, and it is suspected that the attraction of these open environments was the new growth of summer growing grasses, roots, and, later in the year, cactus fruit. Baird (1859) makes mention of a root particularly sought by grizzlies—perhaps biscuit-root, which is a favorite late-summer food of grizzlies in the Rocky Mountains and also found in certain grasslands along the US-Mexican border (Kearney and Peebles 1960).

Many accounts place grizzlies in riparian deciduous forests and other streamside communities (e.g., Pattie 1833; Duffen 1960). Indeed, these and other wetlands were key habitats for New Mexico grizzlies, as they are today for their more northern brethren (Brown 1996). In early spring, streamside vegetation, marshy *ciénegas*, and cottonwood (*Populus* spp.) stands supported an understory of green plants not yet available on the uplands. The willow and cottonwood-timbered bottoms also furnished a source of cover, and, as the summer progressed, furnished cool retreats—an important consideration for these Holarctic animals. Here too, the grizzlies could find the occasional tidbit of opportunity—a deer fawn, a lodge of young beavers, or a floundered cow. Unfortunately for the bears, these were also the habitats most impacted by Anglo settlement (Brown 1996).

LIFE HISTORY

The following life history information was gleaned from early reports and/or extrapolated from recent studies of grizzlies elsewhere in the Rocky Mountains.

Diet

Rangewide, the grizzly bear has a very broad diet that includes vegetative matter, in the form of graminoids, forbs, roots, bulbs, fruits, and nuts, as well as animal foods, most notably insects, birds, mammals, and fish (Schwartz et al. 2003). Grizzlies may graze herbaceous vegetation, harvest fruit from shrubs, dig for roots, bulbs, and corms, overturn rocks looking for insects, and either prey on mammals as large as moose (*Alces alces*) and elk (*Cervus canadensis*) or scavenge their carcasses. They are opportunistic in their feeding habits, as in Alaska when they congregate along rivers to feed on runs of salmon swimming upstream (Schwartz et al. 2003). Important differences in diet among populations (i.e., from largely carnivorous to largely herbivorous) have led some authors to describe the grizzly as a flexible omnivore (Robbins et al. 2004). In late summer, grizzlies enter a state of hyperphagia, during which they eat voraciously to dramatically increase their body fat reserves (López-Alfaro et al. 2015).

In the summer in New Mexico, grizzlies roamed throughout their home range feeding on carrion, yampa roots (*Perideridia gairdneri*), onions (*Allium* spp.), green herbaceous vegetation and ants (Brown 1996). Canyons and riparian bottoms were especially favored as feeding sites (Sheldon 1925), but generally the bears were now moving upslope. By midsummer the bears often frequented aspen stands, making heavy use of high-elevation meadows, rockslides, and outcrops. There they grazed on vegetation green from the summer rains, dug for tubers and small

Photo 17.8. Grizzly bear on 11 July 2021 at Brooks Falls, Alaska. Although grizzlies consume much vegetative matter, they also rely on animal food (more so than black bears [*Ursus americanus*]), including mammals, birds, and fish, obtained through predation or scavenging. Grizzly bears are efficient predators where their prey base is concentrated and vulnerable, as is the case with sockeye salmon (*Oncorhynchus nerka*) runs. Photograph: © Hari Viswanathan.

mammals, tore open logs in search of grubs, and fed on early New Mexico raspberries (*Rubus neomexicanus*), American red raspberries (*R. strigosus*), currants, and other fruits (Brown 1996). With the approach of fall, the grizzlies concentrated their feeding in areas where mast was available. Historical accounts have New Mexico grizzlies feeding on piñon nuts (*Pinus edulis, P. cembroides*), acorns (*Quercus* spp.), manzanita berries (*Arctostaphylos* spp.), juniper berries (e.g., *Juniperus deppeana*), and prickly-pear fruits (*Opuntia* spp.), as well as carrion (Bailey 1931; Duffen 1960; French 1965a and b).

Hibernation

As there are no extant populations of grizzly bears in New Mexico, we can only hypothesize what the denning habits of New Mexico grizzlies might have been from historical records and from life history information from present-day conspecifics to the north and heterospecifics within New Mexico. Several accounts have suggested that Southwest grizzlies may have hibernated irregularly (e.g., Bailey 1931:359). This was emphatically refuted by Sheldon (1925:160), who found grizzlies in Mexico to have much the same denning habits as conspecifics farther north in the Rocky Mountains: "from November to April they hibernated far within the ranges; they did not leave the interior recesses of the mountains until near the first of May." Montague Stevens (1943) likewise stated that New Mexico grizzlies began hibernating about 20 November and did not leave their dens until the middle of March. All Southwest grizzlies of record were taken between mid-April and November except for a female grizzly and three yearling cubs taken by government hunter T. T. Loveless on 18 December, and a big male noted by bear hunter Ben Lilly on 11 March (Brown 1996).

Contemporary research on grizzlies shows that males typically enter and exit their dens later in the fall and earlier in the spring, respectively, than all other sex and age classes, while females with cubs enter and exit their dens earlier and later, respectively, to provide care to their highly dependent cubs, who also have limited mobility (González-Bernardo et al. 2020). Globally, the denning chronology of brown bears follows the general pattern of populations in more southern latitudes denning later in the year and for shorter periods of time than populations in northern latitudes—the pattern may not be followed by all age classes but holds true in particular for adult males and potentially also for adult females (González-Bernardo et al. 2020). In North America, this general trend has been observed through various studies. On the east front of the Rocky Mountains in Montana, Aune and Stivers (1982) found that grizzlies denned between 2 and 23 November and emerged between 10 March and 29 April, moving away from the den site between 13 and 28 May. In the Greater Yellowstone Ecosystem from 1975 to 1999, the mean week of den entry for females with cubs was the first week of November while females without cubs and males entered in mid-November. Come spring, females with and without cubs emerged the third and fourth week of April, respectively, a full month later than males (Haroldson et al. 2002). The pattern of earlier entry and later emergence continues north with grizzly bears in British Columbia, Canada, entering and emerging from their dens approximately a week earlier and later, respectively, than those in the Greater Yellowstone Ecosystem (Pigeon et al. 2016b). The hibernation period extends even further for grizzly bears in Canada's central arctic with den entry beginning in mid-October and emergence in the first week of May, highlighting the adaptive significance of hibernation in response to limited or non-existent food resources during the winter months (McLoughlin et al. 2002).

Environmental factors, specifically, snow, temperature, and food availability, have been identified as factors that influence the denning chronology of grizzly bears. The first major snowfall may serve as a cue for grizzlies to enter their dens while greater snow depths during fall may prolong hibernation bouts (González-Bernardo et al. 2020). As average daily temperature begins to drop, so, too, does the body temperature, and thus the heart rate, of grizzly bears, as they physiologically prepare to hibernate (González-Bernardo et al. 2020). Warm fall temperatures have been shown to delay den entry while colder temperatures have resulted in early entry (González-Bernardo et al. 2020). With warming spring temperatures, comes the slow physiological process of bears waking up from their slumber. Food availability is posited as the primary factor affecting den entry and exit with abundant food and mild winters resulting in bears spending the winter outside their dens (González-Bernardo et al. 2020).

Black bears in New Mexico exhibit the same latitudinal denning chronology pattern: in the Mogollon Mountains, they enter dens later and emerge from dens one to two weeks earlier than those in the Sangre de Cristo Mountains to the north (see Chapter 16). Research in Mexico and Texas noted that some males and non-pregnant female black bears forgo denning when winters are mild and abundant food resources are present (Doan-Crider and Hellgren 1996; Mitchell et al. 2005; see Chapter 16). Given the latitudinal shift in denning chronology in both grizzlies and black bears, it seems logical that the denning dates for New Mexico grizzlies would follow a similar pattern with a longer mean hibernation period than populations to the south and a shorter mean hibernation period compared to populations to the north.

Latitude also appears to play a role in the elevation at which grizzlies select den sites. For example, den sites in the Greater Yellowstone Ecosystem were found at an average of 2,636 m (8,648 ft; Podruzny et al. 2002). Traveling north, the mean elevation of den sites drops to 2,124 m (6,968 ft)

in west-central Montana (Servheen and Klaver 1983), 2,280 m (7,480 ft) in Banff National Park, Alberta, Canada (Vroom et al. 1980), 1,250 m (4,101 ft) in southwestern Yukon, Canada (Pearson 1975), 1,316 m (4,318 ft) in Denali National Park, Alaska (Libal et al. 2011), and 1,040 m (3,412 ft) in the Brooks Range of northeastern Alaska (Reynolds et al. 1976). Elevation is thought to be a proxy for snow depth, with deeper insulating layers being posited as an important component to survival by decreasing the energy cost to an individual during hibernation, requiring populations at lower latitudes to travel up in elevation to obtain snow depths sufficient for thermal insulation (Vroom et al. 1980). Compared to males, females have been shown to hibernate at higher elevations with the sexual segregation of den sites hypothesized as a mechanism to reduce to the probability of infanticide (Libal et al. 2011).

Dens are typically excavated into steeper (20–40°) slopes that help provide structural integrity and sufficient drainage properties while allowing snow to accumulate without entering the den chamber (Vroom et al. 1980; Servheen and Klaver 1983; Libal et al. 2011). Thick timbered areas, where dens are often found, provide better protection from the elements and better security from disturbance events (González-Bernardo et al. 2020). Dens are frequently found on south-facing slopes and in areas where spring foods at den emergence are more abundant (Pigeon et al. 2016a; González-Bernardo et al. 2020). Den emergence is a critical time; grizzly bears must find foods high in protein to compensate for the weight lost during hibernation. These needs are supplied by carrion and green herbaceous plants such as spring beauty (*Claytonia lanceolata* = *C. rosea*), cow parsnip (*Heracleum lanatum*), sedges, horsetails, dandelions (*Taraxacum* spp.), and the new growth of cool-season grasses (e.g., *Melica porteri*). The search for food sources draws the grizzlies downhill, often along drainages. As a result, riparian habitats are much used in the quest for green vegetation and other early food supplies (C. Jonkel, pers. comm.).

Reproduction, Survival, and Longevity

Because nothing in the literature indicates that the breeding habits of southwestern grizzlies differed from conspecifics elsewhere, we assume that the breeding season in New Mexico likely occurred sometime during the late spring (earlier than for populations to the north) and lasted for approximately two months (García-Rodríguez et al. 2020). Latitude has been found to be a significant predictor of the onset of the breeding season with the variable being a proxy for photoperiod, the latter having a well-known effect on the endocrine rhythms and thus timing of reproduction in many mammals (García-Rodríguez et al. 2020). During the breeding period, male grizzlies come in contact with, and then follow, receptive females for days or weeks until the female comes into estrus. The strategy is aimed at increasing the male's probability of breeding with the female, with larger, more dominant males being more successful. Even after copulation, both the male and female will continue mating with other individuals resulting in multiple-paternity litters.

Female grizzlies, like female black bears, delay pregnancy until hibernation. Known as seasonal delayed implantation, eggs that are fertilized in the spring and are early in their development (i.e., blastocysts) do not implant into the uterine wall until the body has assessed whether the female possesses adequate fat levels necessary for reproduction and lactation (Spady et al. 2007). Females must meet a minimum threshold of 19% body fat when entering the den to initiate reproduction—up to 33% depending on the number of cubs and the duration of lactation during denning (López-Alfaro et al. 2013). After a short gestation period (~60 days), an average of two cubs weighing 450 g (1 lb) and about 20 cm (8 in) in length are born in late January and early February (van Manen and Haroldson 2017). The age at which grizzlies reach

sexual maturity varies drastically geographically ranging from 3–10 years (Schwartz et al. 2003; van Manen and Haroldson 2017). Females typically breed once every three years although that interval can be shorter depending on the age at which offspring are weaned or if the loss of a litter occurs. For example, male bears may engage in infanticide, whereby they kill non-sired cubs in order to bring a female into estrus and generate another mating opportunity (van Manen and Haroldson 2017). Females usually do not come into estrus while nursing cubs; however, mixed-age litters, though rare, have been observed and are likely due to cubs and yearlings being temporarily separated from their mother, which then allows females to breed and produce cubs in the following year (van Manen and Haroldson 2017). Offspring are usually weaned at two and a half years of age, with age at weaning heavily influenced by food availability such as in the resource-poor North Slope of Alaska, where offspring are not weaned until four and a half years of age (van Manen and Haroldson 2017).

Grizzly bear survival rates are lowest among cubs-of-the-year and increase with each subsequent age class (yearling, 1 year old; subadult, 2–3 years old; adult, ≥4 years old). Cub survival is also the most variable with estimated survival rates ranging from 13 to 87% depending in part on the estimation method (Schwartz et al. 2003). Yearling survival is also highly variable with survival ranging from ~30% in areas approaching carrying capacity to 90% in other regions (Schwartz et al. 2003; van Manen et al. 2016). Male and female adult survival rates are often greater than 80% and 90%, respectively, with lower survival rates for adult males in areas with legal hunting (Schwartz et al. 2003; Garshelis et al. 2012; van Manen et al. 2014). Female survival is the most important metric when monitoring the demographic health of a grizzly population as adult females have the largest effect on population growth.

Bears are aged by counting the annuli of the

Photo 17.9. Female grizzly bear and cubs in Yellowstone National Park on 8 July 2015. Grizzly bears have one of the lowest reproductive rates among terrestrial mammals, with adult females typically breeding only once every three years and producing an average of just two cubs. Photograph: NPS/ Eric Johnston.

cementum of sectioned teeth by a technique similar to counting tree rings. Each annulus represents a year's growth. Although this is considered a modern aging method, the layered deposition of cementum was first noted by the legendary Grizzly Adams, who kept a stable of tamed California grizzlies (Hittel 1911). Wild grizzlies have been known to live for more than 25 years, but bears older than 20 years invariably show dental problems that likely affect survival (Craighead et al. 1974). Female grizzly bears are able to breed for the majority of their lifespan, but because they typically reach sexual maturity at four to six years old and produce a litter only every three years, they may only produce three to six litters in total. As a result, grizzly bears have a low recruitment rate relative to other mammals. It means that grizzly bear populations may

decline even under small declines in female adult survival.

Home Range, Movements, and Social Behavior

Grizzlies do not defend established territories; rather, they have overlapping home ranges, the size of which varies based on sex, age, reproductive status, availability of food, and season (Schwartz et al. 2003). In the conterminous United States, the average home range size for adult males (475–2,162 km² [183–835 mi²]) is generally 2–6.5 times greater than the average home range size for adult females (130–358 km² [50–138 mi²]) (USFWS 2021a). Home range size also varies greatly with bear density (smaller home ranges are observed at higher bear densities) and habitat productivity (larger home ranges are associated with resource-poor habitat). In general, home range size increases with age and the absence of dependent young (USFWS 2021a). As observed also in black bears (Chapter 16), subadult male grizzly bears disperse farther from their natal range than subadult females. Females, meanwhile, will often establish their home range within or overlapping the home range of their mother. Known as female-biased philopatry, sex-biased dispersal is thought to help reduce the likelihood of breeding among closely related individuals, allows female bears to remain in a familiar territory, and reduces intra-specific conflict as neighbors tend to be closely related (McLellan and Hovey 2001; Schwartz et al. 2003). The average dispersal distance for male grizzly bears from the center of their mother's home range is 30–42 km (19–26 mi), compared to just 10–14 km (6–9 mi) for female bears (McLellan and Hovey 2001; Proctor et al. 2004). The average home range for males in the northern Rocky Mountains generally exceeds 415 km² (180 mi²), while females have ranges of less than 260 km² (100 mi²) (sources cited in Brown 1982). Densities in the northern Rocky Mountains during the 1970s and 1980s generally ranged from one grizzly per 21 to 155 km² (8–60 mi²).

Historical accounts of grizzlies in New Mexico indicate that in general individual home range size fluctuated over the course of the active season due to seasonal mating and foraging behaviors (Brown 1996). During the breeding season, males enlarged their home ranges as they searched for receptive females. Home range size remained fairly constant over the remainder of the summer as grizzly bears exploited spatially and temporally patchy food resources within their home range. With fall drawing near, the grizzlies entered hyperphagia, resulting in expanded home ranges as they searched for food resources. Later in the autumn, when the bears had sufficiently fattened, they would seek out den sites.

The grizzly bear is not a social animal (Brown 1996). An adult generally treats any other bear as an interloper and will kill or drive away any smaller bear that is not a potential mate. Small cubs will be killed by male bears if not treed or protected by their mother. This solitary behavior appears to be an adaptation to limited food supplies and is modified in situations where food is plentiful. Hence, communal feeding behavior has been observed at garbage dumps, at ranch bone-yards, and in areas where fish or dead animals are abundant (e.g., Craighead 1979).

The grizzly is highly intelligent, secretive, and, when hunted, largely nocturnal. Aune and Stivers (1982) found that grizzlies on the east slope of the Rockies did not become fully active until between 9:00 p.m. and midnight, and retired between 6:00 a.m. and 9:00 a.m. Thus, the period when they were feeding and traveling was less than nine hours a day. Bears that had less contact with people (in remote areas) were active in the early evening and early morning hours (Aune and Stivers 1982).

Like black bears, grizzlies make well-defined pathways to and from characteristic bedding sites in heavy cover where they avoid the heat of midday. These bedding sites are interconnected to favorite foraging areas and escape retreats by a system of well-worn trails that facilitate rapid movement with a minimum of noise. Trails and typical bedding sites in New Mexico were described by Dyche in Edwords (1893:69):

The [grizzly's] trail led into a deep fir forest and it was almost dusk under the trees. The pines interlaced at the top and the ground was covered with a thick bed of needles, shredded fir-cones that had been opened by squirrels looking for the seeds and leaves, which formed a carpet in some places three feet thick. In this mass of debris were found many bear beds, where the animals had scooped out great hollows and made comfortable sleeping places.

Often in dense vegetation, such as in aspen thickets or willow-lined stream bottoms, these trail-ways took on the appearance of tunnels and were virtually invisible to a man on horseback (Stevens 1943). Once discovered, however, these paths could be used to trap the bears in "trail sets."

Interspecific Interactions

Before their extirpation from New Mexico, interspecific interactions between grizzlies and the remaining carnivores of the state would have been dominated by grizzly bears. The impetus driving these conflicts was likely kleptoparasitism, i.e., the stealing of resources by one animal from another, whereby, a grizzly bear pushes and claims the kill made by a heterospecific competitor. Grizzlies have been documented to usurp wolves (*Canis lupus*) (Jimenez et al. 2008) and cougars (*Puma concolor*) (Murphy et al. 1998) from their kills garnering a caloric bonanza needed for hibernation with the loser, as in the case of cougars, losing 17–26% of their daily energy requirements (Murphy et al. 1998). Subordinate heterospecifics have good reason to concede a kill if it means avoiding further confrontation with a grizzly bear as grizzlies have been known to kill wolves and black bears (Gunther et al. 2002; Jimenez et al. 2008).

In New Mexico, the black bear was likely the biggest competitor to the grizzly bear with the two species having similar niches as opportunistic omnivores with vegetation often as the primary food source in preparation for hibernation. Rangewide, the larger and more aggressive

Photo 17.10. A grizzly feeds on a carcass at Yellowstone National Park on 16 April 2013. Also present on the scene, and waiting for their share of the carcass, are a wolf (*Canis lupus*, left), two coyotes (*C. latrans*), and several ravens (*Corvus corax*). Before their extirpation from New Mexico, grizzlies would have been apex predators and dominant competitors to black bears (*Ursus americanus*), cougars (*Puma concolor*), and wolves (other than those occurring in larger packs). Photograph: Jim Peaco.

grizzly dominates interspecific interactions at the individual level through interference competition with the dominant bear reigning over concentrated food sources such as carcasses and spawning salmon while being a source of intraguild predation (Gunther et al. 2002; Schwartz et al. 2010). In response, black bears shift their daily activity patterns and are more active during the middle of the day, in contrast to their crepuscular competitor, to reduce temporal overlap (Schwartz et al. 2010). At the population level, however, competition between the two ursids may favor black bears as higher densities and smaller home ranges allow them to better use dispersed and patchily distributed food sources such as acorns and berries (Stetz et al. 2018). Stetz et al. (2018) found that primary productivity of vegetation, quantified using remotely sensed photosynthetic activity via satellite imagery, was predictive of variation in seasonal density patterns in black bears. Grizzly bear density, however, was not explained by the distribution of food resources or other hypothesized abiotic factors, but instead by the density of black bears, resulting in lower densities of grizzlies in areas with higher black bear density (Stetz et al. 2018). These results highlight the ability of a subordinate competitor to influence density patterns in the behaviorally dominant competitor through exploitative competition (Stetz et al. 2018).

STATUS AND MANAGEMENT

Recovery Zones

Grizzly bears once spanned across much of western North America from Alaska to central Mexico and from the West Coast to the Great Plains. An estimated 50,000 grizzlies once inhabited the conterminous United States, before habitat loss, hunting, and human and livestock related conflicts resulted in fewer than 1,000 individuals remaining by the early 1970s (Servheen 1990, 1999). By then occupying only ~2% of their original range, grizzly bears in the conterminous

United States were listed as Threatened under the Endangered Species Act on 28 July 1975. Despite the listing and the measure of protection it afforded, the hunting of grizzlies continued to be allowed in northwestern Montana until 1991. Two years later, in 1993, the US Fish and Wildlife Service updated the 1982 *Grizzly bear recovery plan* (the guiding document for grizzly bear recovery and management) by identifying six ecosystems suitable for grizzly bear recovery. These recovery zones included the North Cascades Ecosystem of north central Washington (25,000 km² [9,500 mi²]), the Selkirk Ecosystem of northern Idaho, northeastern Washington, and southeastern British Columbia (5,700 km² [2,200 mi²]), the Cabinet-Yaak Ecosystem of northwest Montana and northern Idaho (6,700 km² [2,600 mi²]), the Northern Continental Divide Ecosystem of northwest Montana (25,000 km² [9,600 mi²]), the Bitterroot Ecosystem of central Idaho and western Montana (15,000 km² [5,800 mi²]), and the Greater Yellowstone Ecosystem of northwestern Wyoming, eastern Idaho, and southwestern Montana (25,000 km² [9,500 mi²]; USFWS 2004). The recovery plan also recommended the evaluation of a seventh recovery zone in the San Juan Mountains of southern Colorado, where grizzlies were thought to have been extirpated by 1951 until a hunting guide in 1979 killed an adult female grizzly bear in self-defense (DMNS 2021; USFWS 2021a). The female weighed ~160 to 180 kg (350 to 400 lbs) and was 16 to 20 years old, a genetic analysis further determining that the bear carried a genetic signature unique to grizzlies of the San Juan Mountains of southern Colorado and northern New Mexico (DMNS 2021). No evidence of grizzly bear occurrence has been found in the San Juan Mountains since.

As of 2021, grizzly bear populations have grown and expanded into ~6% of their historical range in the conterminous United States, occupying in particular four of the six recovery zones (Selkirk, Cabinet-Yaak, Northern Continental

Photo 17.11. Grizzly bear in Yellowstone National Park on 4 July 2006. The grizzly bear in the conterminous United States is listed as Threatened under the Endangered Species Act. Six recovery zones have been established including the Greater Yellowstone Ecosystem of northwestern Wyoming, eastern Idaho, and southwestern Montana. Photograph: © Pat Gaines.

Divide, and Greater Yellowstone ecosystems; USFWS 2021a). Plans to repopulate the two unoccupied recovery zones (Bitterroot and North Cascade ecosystems) have been proposed but have not been fruitful. In 2000, the US Fish and Wildlife Service proposed releasing 25 grizzlies into the Bitterroot Ecosystem but was met with staunch opposition by state governments and the public. A year later the US Fish and Wildlife Service abandoned the proposed reintroduction. The National Park Service in 2014 announced the first steps toward reintroducing grizzlies to the North Cascades by starting a three-year environmental analysis. That analysis, however, was not finished within the expected time frame and in 2020 US Secretary of the Interior David Bernhardt announced his agency would not move forward with reintroduction efforts despite support for reintroduction by former Secretary of the Interior Ryan Zinke in 2018 (Johnson 2018; Geranios 2020). Today, the movement of grizzly bears from one recovery zone to another is limited to the Cabinet-Yaak Ecosystem augmentation program in which a few bears per year are relocated from the nearby Northern Continental Divide Ecosystem to the small, isolated population to the west. Elsewhere, grizzly bear recovery is focused on identifying and conserving landscapes that maintain structural connectivity (i.e., characteristics and distribution of habitat) with the hope that these habitat linkages and movement corridors promote functional connectivity (i.e., exchanging of individuals and their genes) among the recovery zones as grizzly bear populations grow and expand (Peck et al. 2017; Sieracki and Bader 2020).

The two recovery zones with the largest grizzly populations, and therefore those most likely to help the species expand its distribution into smaller or unoccupied recovery zones, are the Northern Continental Divide and Greater Yellowstone ecosystems. The Northern Continental Divide Ecosystem has an estimated 1,068 grizzlies (95% confidence interval = 906–1,243), up from the estimated 300 individuals at the time of the listing under the Endangered Species Act in 1975, and it has been growing on average 2.3% annually since 2004 (Costello and Roberts 2020; USFWS 2021b). As of December 2021, the population was listed as Threatened under the Endangered Species Act, although the state of Montana had just petitioned the US Fish and Wildlife Service to delist grizzlies in the Northern Continental Divide Ecosystem (Governor's Office 2021). The Greater Yellowstone Ecosystem is slightly smaller with an estimated 921 individuals (95% confidence interval = 889–1121), or roughly three times the number (220–320 grizzlies) estimated at the time of listing (Burnham and Mott 2021; Gould et al. 2023). Based on multiple metrics, population growth appears to be slowing down and it is believed that grizzlies in the Greater Yellowstone Ecosystem are approaching carrying capacity (van Manen et al. 2016). Although that population is still considered Threatened, the listing status for the Greater Yellowstone population has been tumultuous. The US Fish and Wildlife Service first proposed but ultimately removed endangered species protections in 2005 and 2007, respectively (Burnham and Mott 2021). Lawsuits quickly followed the delisting, and in 2009, Greater Yellowstone grizzly bears were again back on the endangered species list with the District Court citing concerns over the decline of a key food source, whitebark pine (*Pinus albicaulis*) seeds, due to climate change. The US Fish and Wildlife Service and others appealed the relisting decision in August 2011, but the US Ninth Circuit Court of Appeals, three months later, sided with the District Court, keeping federal protections for grizzlies intact (Burnham and Mott 2021). The US Fish and Wildlife Service again announced it was delisting grizzlies in 2017, citing increases in abundance and occupied range, with groups opposed to the delisting again quickly filing lawsuits. Idaho and Wyoming each proposed a hunting season for the grizzly bear, but those plans were short-lived when the District Court halted the hunts and returned federal protection to Greater Yellowstone grizzlies in the fall of 2018. That decision may be short-lived as the state of Wyoming has petitioned the US Fish and Wildlife Service to delist the Greater Yellowstone Ecosystem population from the Endangered Species List (WGFD 2022). The Selkirk and Cabinet-Yaak ecosystems have an estimated 50–60 grizzlies in each recovery zone, with an annual growth rate of 2.5% for the former and 0.9% for the latter from 1983 to 2019 (USFWS 2021a, 2021b). Both populations are still listed as Threatened under the Endangered Species Act (USFWS 2021a).

The US Fish and Wildlife Service recently conducted a Species Status Assessment for grizzly bears in the conterminous United States (USFWS 2021a). The assessment evaluates the current and future status of a species based on the resiliency, redundancy, and representation of the species. Known as the "three Rs," these principles describe the ability of grizzly bears to withstand stochastic (resiliency) and catastrophic (redundancy) events and their ability to adapt to long-term changes in their environment (representation). In that document, the US Fish and Wildlife Service conducted an analysis that identified "secure core" areas (lands managed by federal, tribal, or state entities that are at least 1,011 ha [2,500 acres] in size and > 500 m [0.31 mi] away from roads) across the grizzly bear's historical range in the conterminous United States. Theoretically, these criteria identify areas that could hold a self-sustaining grizzly bear population. Outside of the six recovery zones, the Sierra Nevada Mountain Range in California was the largest area of secure core. Next, the US Fish and Wildlife Service analyzed habitat security for the Sierra Nevada and Colorado's San Juan Mountains (based on the recommendation of the 1993 *Grizzly bear recovery plan*) by identifying the amount of secure core and secure habitat (lands managed by federal, tribal, or state entities that are at least 4 ha [10 acres] in size and >500 m [0.31 mi] away from roads) in each area.

The Species Status Assessment identified 43% (2,262,701 ha [5,591,258 acres]) and 47% (2,482,520 ha [6,134,441 acres]) of the Sierra Nevada Mountain Range and 52% (1,395,148 ha [3,447,488 acres]) and 56% (1,488,389 ha [3,677,890 acres]) of the San Juan Mountains as secure core and secure habitat, respectively (USFWS 2021a). The percentages identified in the Sierra Nevada Mountain Range and San Juan Mountains are 1.46–1.81 times less than the secure core and secure habitat identified in the Northern Continental Divide and

Greater Yellowstone ecosystems (USFWS 2021a). Based on research conducted in the Northern Continental Divide Ecosystem, 68% secure core is required for successful female grizzly bear reproduction, meaning that it is unlikely that those two areas could support a self-sustaining population without immigration from outside recovery zones. The Sierra Nevada and San Juan Mountains, however, are extremely isolated from the six recovery zones as they lie >880 km (550 mi) and >480 km (300 mi) away, respectively, from the nearest extant grizzly population. Thus, it is unlikely that male bears would recolonize or supplement reintroduced populations in the Sierra Nevada and the San Juan Mountains as the maximum observed dispersal distance for male bears ranges from 67–176 km (42–109 mi) in the Northern Continental Divide and Greater Yellowstone ecosystems (USFWS 2021a). To maintain the demographic and genetic health of hypothetical populations in the two mountain ranges would require a perpetual augmentation program using the other recovery zones (USFWS 2021a).

In New Mexico, the Mogollon Mountains (southwestern corner of the state just north of Silver City) were identified as an area that contained a substantial amount of secure habitat. This area, however, contains smaller amounts of secure core than the Sierra Nevada Mountain Range and San Juan Mountains, with the intervening matrix between the Mogollons and extant grizzly populations to the north containing high human population densities and anthropogenic infrastructure. Therefore, the Species Status Assessment did not further analyze the Mogollon Mountains to assess their ability to support a self-sustaining grizzly bear population (USFWS 2021a). Since it is unlikely that the Sierra Nevada Mountain Range and the San Juan Mountains would support self-sustaining grizzly bear populations, we postulate that grizzly bears are not likely to recolonize or be reintroduced in New Mexico without a reduction in human–grizzly bear conflicts to a level acceptable by a majority of the public.

Early Grizzly Bear: Human Conflicts in New Mexico

The extirpation of the grizzly bear from New Mexico by 1935 is detailed further on based on Brown (1996) and citations therein. Grizzlies—and black bears, for that matter—were justifiably accused of despoiling cornfields; raiding beehives; killing sheep, hogs, and cattle; and breaking into cabins, stores, caches, and trap sets. Their attacks on cattle and sheep earned them the enmity of New Mexican livestock producers. Opinions about the severity of their depredations varied. Some competent ranchers and observers considered the grizzly's reputation as a stock killer undeserved, particularly on the larger ranches before overgrazing became widespread. Consider French's (1965a:233) opinion about grizzlies and cattle on the WS Ranch in the middle 1880s:

> In those days encounters with bears were not infrequent. They really are most interesting animals, and despite what a great many people maintain I never could find that they were destructive to cattle. My personal experience is that they never molest them until they grow old and the procuring of other food becomes difficult, and cattle are not much afraid of them.

This "benign" opinion largely ended with the overstocking of livestock on the ranges in New Mexico after 1885. Even before then, the grizzly's penchant for carrion made him a likely suspect whenever his tracks were found around the carcass of a cow that might have died from any of a number of causes. Nonetheless, its need for ready protein when it came out of hibernation made every grizzly a potential stock killer. This was especially so where grass was scarce and cattle were plentiful.

The attitudes of later ranchers were summed up by Vic Culberson, owner of the GOS Ranch, who told Vernon Bailey (1931:360): "The destruction of these grizzlies was absolutely necessary before the stock business of the region could be maintained on a profitable basis . . . The grizzlies were at one time almost as bad as the big wolves in their depredation on the range cattle."

By 1900 all grizzlies were considered stock killers. Barker (1966:x) echoed the opinion of most ranchers: "I feel very strongly that the assumption that grizzly bears originally fed on carrion and rarely attacked or killed cattle themselves is entirely wrong. My personal experience and the experiences of others in New Mexico will not bear out any such theory."

Most ranchers could not imagine a time when the big bears would not take livestock (Brown 1996). Mired animals, calves, cows with calf, newly weaned calves, and unfolded sheep were all susceptible to attacks by grizzlies. Unlike cougars and wolves, grizzlies would select a large cow if this was the easiest animal to obtain. Cattle were most often taken in spring and fall, but depredations were noted at all times of the year that the bears were active. On the other hand, oftentimes grizzlies were near ranching operations for years without incident (Brown 1996).

The descriptions of grizzlies slapping cattle across the head and breaking their necks, while colorful, are mostly imaginative. Grizzlies generally grab and pull their victims toward them, biting them on the backs and neck. The massive jaws often broke back and neck vertebrae, hence the descriptions of broken necks (Stevens 1943; Leopold 1959). Carcasses fed on by bears are readily identifiable, for the hide is peeled back in a characteristic manner, with little hide or hair consumed. Oftentimes too, cattle killed by grizzlies were found close to water as if they had been ambushed while drinking (Stevens 1943). Unlike black bears, grizzlies may drag a kill to a chosen site and partly bury it (Roy and Dorrance 1976).

Sheep losses in New Mexico, in contrast to

Photo 17.12. Grizzly bear taken near Chloride, New Mexico by Christian Olson in the 1880s. Photograph: Henry A. Schmidt. Negative No. 12241 from Museum of New Mexico, Santa Fe.

Photo 17.13. Sheep grazing in 1903 on the Santa Fe National Forest in New Mexico. More than a century of overgrazing beginning in the mid to late 19th century had severe long-lasting impacts on New Mexico's rangelands and set the stage for an inevitable conflict between livestock producers and land managers on one side and the grizzly on the other. Sheep numbers skyrocketed to 4 million sheep in New Mexico in 1880 (Denevan 1967), greatly contributing to rangelands being denuded of their grass cover. Photograph: Walter J. Lubken.

Photo 17.14. Cattle on Beattie's Cabin Park (Pecos Wilderness) grazing on hillside at the edge of timber in 1924. Overgrazing from the overstocking of New Mexico's ranges likely contributed to grizzlies turning to livestock as an alternative, easily obtainable source of food. Note that prior to 1900, the same area was also grazed but almost exclusively by sheep. Photograph: E. S. Shipp.

Photo 17.15. Effects of livestock grazing on the Carson National Forest in New Mexico. In this photograph taken on 26 August 1953, there is a striking contrast in density and vigor of grass inside the fence enclosure and outside where it is grazed by livestock. The enclosure was fenced in 1925 and enlarged in 1940. By the 1950s, too late for the grizzly, the impact of livestock grazing had been well documented. Photograph: W. L. Hansen.

cattle losses, were greatest from midsummer to the end of the grazing season (Brown 1996). A sheep-killing grizzly might take one to three sheep over a period of several days, though as many as two dozen sheep might be killed at a time (Johnson and Griffel 1982). Like attacks on cattle, most sheep depredations took place at night or in the very early hours of the morning.

The grizzly is an opportunistic feeder and readily takes to cattle, sheep, or garbage if any of these items are easily obtainable (Murie 1948; Johnson and Griffel 1982). Once an individual bear learns how to take livestock, he or she becomes a stock killer. Thus habituated, such grizzlies were more easily trapped, for they almost always returned to feed on their kills. The greater the number of stock killed, the greater the vulnerability of the grizzly to the trap (Johnson and Griffel 1982).

To Kill a Grizzly

With the advent of ranching, former mountain men and commercial hunters readily turned to bounty hunting as a means of making a living at what they knew best. As stockmen may not have the time or expertise to rid their range of predators, the issuance of bounties became increasingly popular. A hunter could shop around for an area where predators were abundant or where local ranchers were offering lucrative bounties. Although fraud might be widespread, and predators usually reduced only temporarily, government bounties became a fixture in the Southwest, New Mexico included. Such monies were irregularly appropriated, however, and oftentimes individual cattlemen or livestock associations found it necessary to offer additional bounties for specific areas, particular species, or even individual animals. Grizzlies were sometimes the target of the latter, and amounts of up to $300 might be paid for individual bears (Young and Beyers 1980). That was good money at the turn of the century and attracted professional bear hunters from a wide area.

Hunting Southwest grizzlies was initially mostly a trailing process, the hunter following up on "sign" and hoping to be able to stalk and get a shot at the bear with a large-caliber rifle. The hope was to catch the bear in an opening; otherwise the hunter might come on his adversary in heavy cover and get off a quick shot before the bear disappeared or charged. The latter alternative could be dangerous as well as generally less productive, and a more common practice was to make a stand over bait. Using this method, the hunter either staked out a carcass that showed signs of having been recently fed on by a bear or established his own bait station in hopes of attracting a bear. After a wait of several hours or several nights, and if the wind was right, one might get the reward of a shot. Both stalking and hunting over bait became less productive as grizzlies became scarce and more wary. Experienced grizzlies rarely exposed themselves and foraged almost entirely at night, thus requiring more sophisticated techniques.

Professional hunters had long known that a pack of good dogs was an essential bear hunting tool. Well-trained hounds, followed by one or more men on horseback, could trail a bear whenever a fresh track was found. There was no set breed of bear dog. Most were lion hounds—blueticks, with bloodhounds and other trailing breeds mixed in (see, e.g., Stevens 1943). It was not unusual to include some "fighting dogs" to "worry" a corned bear and take the heat off the trailing dogs.

Once the dogs hit a fresh trail, their barking and baying led the hunters toward the bear. The grizzly's response was usually to head uphill to try and shake his pursuers in heavy cover or rough terrain (Stevens 1943; Ellison 1968). Because a grown grizzly does not climb trees, a tired bear would often make his stand at the base of a tree or under a ledge of rock. According to Stevens (1943), cornered grizzlies tended to attack individual dogs, while the smaller black bears rushed at the collective pack.

both the sportsmen and the ranchers agreed upon, and they thought that the best way to accomplish such a program was through the federal government. If the US Forest Service collected the fees to graze the forests that provided the grizzly bears' habitat, then it was the government's job to remove the predators. Although the US Forest Service was already controlling predators (Bailey 1907), the agency had other programs to administer. Would not a full-time predator control agency do an even better job? Besides, the western states were paying out more than a million dollars a year in bounties. That cost could be eliminated once and for all if the job was done right.

Such sentiments were compelling and strongly sold (Brown 1996). On 30 June 1914, the US Congress authorized the Predatory Animal and Rodent Control (PARC) branch of the Biological Survey within the US Department of Agriculture, and made it responsible for experiments and demonstrations in destroying wolves, prairie dogs (*Cynomys* spp.), and other animals injurious to agriculture and animal husbandry. A sum of $125,000 was appropriated for the 1914–1915 fiscal year to employ 300 hunters. By the close of the next fiscal year, 1915–1916, the PARC had been organized into control districts and staffed. The inspector for the New Mexico–Arizona District was J. Stokley Ligon.

Ligon soon proved himself a professional field biologist and predator hunter, who was also a competent administrator (Brown 1996). He hired the best hunters, among them Eddy Anderson, Ben Lilly, C. C. Wood, Eddie Ligon, and T. T. Loveless. In the first year of operation, 1915–1916, Ligon employed 32 men and concentrated his efforts in New Mexico. Only seven "bears" were taken by the force that year, one more than the number of cougars taken. Many more bears were killed by ranchers and bounty hunters, but the program was just getting under way (Ligon 1916).

Twelve bears were taken in the 1916–1917 fiscal year, and in his annual report Ligon (1917:2) had this to say about bears:

> Damage by bear has also been greatly reduced during the year. Several bear were killed in Eastern Arizona, that were destructive to livestock, and the big male grizzly and female grizzly that were killed on Mt. Taylor and the head of the Pecos, in New Mexico, have proven to be of great importance, since their destruction has put an end to the great loss of cattle in the two districts. There are still a few large bear in some of the higher ranges of the District—the Pecos Forest, Black Range and Mogollons of New Mexico, and in the White Mountains of and a few other points in Arizona, but the wolf and lion men can easily pick those among them that are a nuisance and kill them when they are located. It is of importance to note here that on July 29, 1916, Ed Steele and Bart Burnam, two Government hunters, shot and wounded a large bear in Western New Mexico, about 20 miles North-West of Reserve. On recent visits to that locality we are informed that the great damage done by a certain bear ceased with the wounding of this animal, and the ranchers are quite sure now that the wound was a fatal one.

One other bear was not reported in the tally, since neither the skin nor the skull had been received in the district office (Brown 1996). Thus it appears that while the records did not differentiate between species the difference was appreciated, and grizzlies were usually referred to as the "big bears" or "large bears." Bears were not considered game animals in New Mexico and considered generally as second-class predators or nuisances (see, e.g., Bailey 1931).

In 1915 the Great War was on in Europe, and American commodities were in great demand (Brown 1996). Prices for beef and mutton rose rapidly. Good economic times also kindled a demand for access to the remaining open ranges in the most inaccessible parts of the national forests. New settlers, cow outfits large and small, loggers, miners, and even some old market hunters went to their Congressman and demanded the abolition of the national forests. The least that America could do for the war effort was to sacrifice its natural resources—at a profit to its citizens. The response was to relax the grazing restrictions on the national forests.

When America entered the war in Europe, the war on predators intensified (Brown 1996). At a conference in Albuquerque on 1 November 1917, the president of the New Mexico Cattle Growers Association, the president of the New Mexico College of Agriculture and Mechanical Arts, and other interested parties arrived at the following estimate of damage sustained by the livestock industry in New Mexico from wolves, cougars, "big bears," coyotes, bobcats (*Lynx rufus*) and "wild dogs" during the preceding year: cattle, 24,350 head lost at a value of $1,374,000; sheep, 165,000 head lost at a value of $1,320,000; and horses, 850 head lost at a value of $21,250; adding up to a total loss of $2,715,250 (Gish 1977, in Brown 1983).

These figures were repeated often and widely, and they were used to recruit support for Ligon and his PARC. Thus the states of New Mexico and Arizona began to cooperate with PARC, contributing funds and personnel to the cause. Ligon (1918:5–6) states the following in his report:

Good progress has also been made in big bear and lion work. We have a close check on the number of big bear in the district as well as being familiar with their exact range. From the fact that those that remain occupy the roughest and most heavily forested mountains makes their capture rather difficult . . . As the wolves are brought under control attention will be directed to bears and lions.

Six "big bears" had been killed during 1917–1918 by Ligon's PARC force, four in July 1917, one in May 1918, and one in June 1918. Four of those were taken in New Mexico, all by federal hunters (Brown 1996). The six grizzlies brought to 25 the total bears killed by the district. Damage attributed to predatory bears by Ligon during 1917–1918 was estimated at $10,000.

Ligon (1918:18–19) went on to discuss depredations by bears in detail, and cautioned against the political consequences of the failure to eliminate them.

As the wolves and lions are killed out in certain districts, much light is thrown on the case against bears as predatory animals. Guilt is now being placed on them, where in years gone by it was generally supposed that bears did little killing of domestic stock. They are becoming more destructive to cattle in recent years. The dry seasons have probably added to their killing, since the shortage of feed has created a demand for range everywhere, even in the highest and most heavily forested regions—the home of big bears—thus throwing the helpless stock into the very haunts of the animals.

The damage from bears has been greatest in the Taos and Pecos Mountains, and in the Black Range and Mogollon Mountains in New Mexico . . . West of Chloride, New Mexico, along the crest of the Black Range, many cattle were killed in the early spring. This was probably the work of two or three grizzlies. The heaviest losses, however, were sustained in the Taos and Pecos Mountains about the head of the Pecos River, where the number of cattle killed will be more than a hundred head. We are working after these cattle-killers and expect to have the guilty animals destroyed by the end of the year.

While cattle are the animals usually killed by grizzlies and the larger brown and black bears, it develops, beyond any doubt, that all kill domestic stock under certain conditions . . .

I feel sure that the losses in cattle alone in New Mexico during the present spring and summer will aggregate more than two hundred head. This is rather a serious matter when we consider the fact that big bears generally kill big animals—often cows carrying calves.

To fail to listen to the requests from ranchmen for protection against bears would have a serious weakening effect on our organization. Destructive animals, of whatever species, should be controlled. There is no danger of bears being exterminated so long as we have our parks and wild northern woods. Even our reluctance in killing the smaller bears creates discord between our methods and the interest of the ranchmen.

I am preparing a detailed report on the bear as a predatory animal, in which I hope to show clearly the real status of bears in the district.

Ligon was an astute observer and knew the reasons for the increase in bear depredations, but he was reluctant to state them lest it lower his political support among the livestock producers (Brown 1996). Later he would, but it would be too late for the few grizzlies remaining in New Mexico.

In the meantime, the grizzly was to face a new threat. Ligon (1918:6) reported, "We are making extensive preparations to carry out a poison campaign the coming winter—from November to March." Although grizzlies would be mostly in hibernation during this time, the procedures for using this method were not always strictly adhered to.

By the end of World War I, it was increasingly apparent that New Mexico's grizzlies were nearing extirpation (Brown 1996). In 1917, Ligon estimated that only 48 grizzlies remained in the state—in the Sangre de Cristo, San Juan, Jemez, and San Mateo mountains; Mount Taylor; the Black Range; and the Mogollon Mountains complex (Bailey 1931; Map 17.2).

The Last Holdouts

Ligon (1919; as quoted in Brown [1996]) noted the decline in grizzlies during the war years and considered the PARC at least partially responsible: "The constant trapping and hunting for wolves, lions and big bears in the forests of the state during the last three years, have greatly reduced these animals." Ligon went on to say that "the stock killing bears have been reduced in numbers, but those that remain occupy practically the same districts." Ligon (1919; as quoted in Brown [1996]) also felt some concern about the growing sentiment among sportsmen that bears were game animals in need of protection. An astute politician, Ligon now presented the PARC as the agency best able to judge the guilt or innocence of bears:

The last four years of predatory animal operations have demonstrated conclusively the folly of letting everyone destroy bears with the hope of getting the destructive individuals. None but experienced and equipped hunters continuously succeed in getting the cattle killers. Three-fourths of the bears killed during the last fiscal year were harmless as regards destructiveness to livestock, and no bear molests game to any extent. The result is that most of the smaller bears are wasted on account of being killed in summer, when neither meat nor hides are of value.

Hunters and others, in their desire to kill bears, report them as being killers of livestock, while the kind and size of the animal disproves their contentions.

Ligon (1919; as quoted in Brown [1996]) also

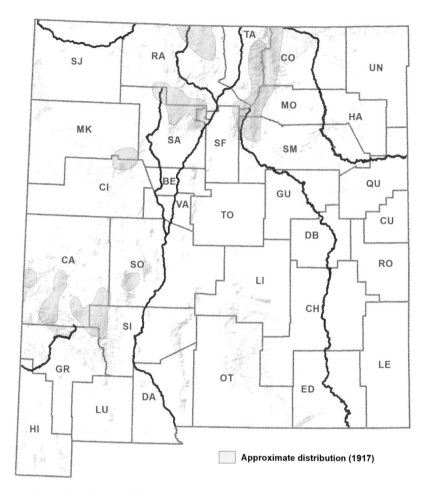

Map 17.2. Approximate distribution of the 48 grizzly bears thought to be remaining in New Mexico in 1917, from a map prepared by J. Stokley Ligon (adapted from Bailey 1931). Note that Ligon's mapped distribution does not include the Magdalena Mountains—where a skull dated 1935 was obtained by the US National Museum.

appeared to be changing his opinion on the need to kill bears:

> The smaller bears need protection the same as deer and turkey, leaving the matter of control of individuals in the hands of Government forces that will operate under permit or agreement.
>
> During the last few months destruction of small bears by citizens of the mountainous districts has been greater than usual, and in but few cases have the hides been worth taking from the animals.
>
> It appears that uniform state laws for the protection of bears, with provisions for the care of destructive animals, be made. Such laws would, no doubt, meet with the approval of both ranchmen and sportsmen of the country . . .
>
> By recent investigation I found that many cattle, for which bears are credited as having killed, die from effects of poison weeds. This, however, does not remove the guilt from the larger animals, that we know are very destructive to cattle.

Nonetheless, 28 bears were taken in New Mexico during the 1918–1919 fiscal year by PARC and state hunters, eight of them by the famous houndsman Ben Lilly (Ligon 1919; Photo 17.18). At least some of these animals were said to be grizzlies.

Then a lack of funds and manpower caused the PARC to let up on its program against bears during 1919–1920 (Brown 1996). Ligon (1920; as quoted in Brown [1996]), again reversing himself, reported:

> The bear situation is one that has not been met. No concentrated effort has been made to get the depredators; but the time is at hand for our service to add up the heavy losses from this source and this is a more difficult and expensive problem than that of getting other species. Hunting big bear is the most dangerous of predatory work, and requires much experience to make it successful. During the last four months bear have killed no less than 800 head of cattle in the state.

Only 10 bears had been taken by PARC hunters in Arizona and New Mexico in 1919–1920 (Brown 1996). Another 19 were killed by state hunters in New Mexico (Ligon 1920). Although there is no way to tell from the report, none of these bears appears to have been grizzlies. Grizzlies and other "big bears" were still on Ligon's (1920; as quoted in Brown [1996]) mind, however, and he planned to get them:

> Big bear have caused an unusual amount of disturbance in several districts during the last three months. They have not only held their own in some high districts, but their numbers have increased since none of the large grizzlies have been killed by Bureau men or others. There are as many as twenty big grizzlies in the state distributed over the highest portions of the Mogollons, Black Range, San Mateo, and Sangre De Cristo mountain systems; all cattle killers. Definite and expensive methods will have to be followed to get these particular depredators.

Photo 17.18. US Biological Survey employees in their hunting camp, West Turkey Creek, Gila Wilderness, August 1920. Left to right: Ben Lilly, Walt Hotchkiss, and J. Stokley Ligon. The Predatory Animal and Rodent Control branch of the US Biological Survey was authorized by Congress in 1914. As Predatory Animal Inspector for the New Mexico–Arizona District, J. Stokley Ligon hired some of the best professional hunters. Ben Lilly in particular was a notorious "hound dog man," who specialized in hunting bears and cougars (*Puma concolor*). Predators such as grizzly bears were considered "varmints" that posed a threat to livestock (Sweet 2002). Photograph: J. E. Hawley.

Ligon made good on his promise. In July and August 1920, he personally supervised a few chosen hunters in an organized hunt in the Black Range and Mogollon Mountain regions against wolves, cougars, and bears. "Many . . . predatory bears were treed and killed" (Bliss 1921:2). The reference to treeing indicates that these were mostly black bears, and indeed there were no longer that many grizzlies left to be taken. Still, the Black Range and the Mogollons were grizzly strongholds, and an unknown number of grizzlies were taken. The PARC report for 1920–21 is brief and, as usual, vague about the species of bear. All in all, nine bears were taken that year by the PARC force in New Mexico, while state cooperators bagged another eight (Bliss 1921).

That there was to be no let up on the war on grizzlies was shown by new State Supervisor Bliss' (1922:3) annual report for New Mexico in 1921–1922: "Few brown and black bears are predatory under normal field food conditions, but those of the harmful class along with grizzlies must be dealt with accordingly."

Bliss (1922:5) estimated that only six grizzlies remained in New Mexico and predicted their extinction: "The few grizzly bears remaining will eventually fall victims to still hunting [following up grizzlies from carcasses or bait] maneuvers. Even a well trained pack of dogs is confronted by a difficult task to bag a powerful grizzly in flight."

Nonetheless, the beginning of a softened policy on the taking of bears is evident (Brown 1996). Only two bears were recorded as having been taken by PARC hunters in New Mexico in 1921–1922 with state hunters taking another 14 (Bliss 1922). Such a respite, if indeed this is what it was, was too little too late, however.

Thirty bears were taken by PARC hunters and their state cooperators in Arizona and New Mexico in 1922–1923 (Pineau 1923; Musgrave 1923). Musgrave's monthly newsletters to his field men attest that bears were being downplayed as PARC targets with hunters encouraged to make sure that any bears taken were in fact stock killers—a fine point too often ignored. Bears were rated at 10 points toward the 15 points needed to get on the monthly "honor roll." A government employee who failed to produce was dropped from employment. At least one bear was a well-known grizzly, and in the summer of 1923 the last Pecos grizzly of record was killed. Elliot Barker (1953:171) described how this bear met his end:

That night I [Barker] phoned Skipper Viles to be on the lookout for the grizzled old stock killer for he was headed back his way. Skipper reported that the bear had come back to one of his previous kills the night before and gotten caught in a trap waiting for his big foot, and

that he had shot him that morning. That, as far as I know, was the last grizzly bear of the Pecos high country. From time to time, there have been reports of tracks of a grizzly being seen but certainly none lives there, and it could be only one occasionally passing through.

At the time I was a bit jealous of Skipper because he, not I, had killed the bear. Now, since it proved to be the last one. I am mighty glad I didn't kill it. Mrs. Viles still has the rug and it is really a nice one.

Also in 1923, 103,000 poison baits were distributed to cooperating ranchers in New Mexico—a program that would be greatly expanded in the coming years.

Despite a reported population increase in the number of bears and a more enlightened PARC policy, 1923–1924 was a hard year for southwestern bears (Brown 1996). At least 45 bears were taken by the PARC in Arizona and New Mexico—a new record (Ligon 1924; Musgrave 1924). Some of these animals were known to have been poisoned and there were probably others (Musgrave 1924). Ligon had returned from service with the state of New Mexico and was again New Mexico District supervisor for the PARC. As usual, he filed a voluminous report, noting that private houndsmen were taking over the control of "stock-killing bears," a role welcomed by Ligon and the PARC.

As for the declining status of grizzlies, Ligon (1924:15) was not ready to let up:

One or more big cattle-killing bears is still at large in the Sangre de Cristo Range, ranging from the head of the Pecos River northward and a few cattle-killers remain in the Black Range, Mogollon, San Mateo and Zuni Mountains. Even in the mountain ranges mentioned there has not been serious complaint during the past year, although the remaining grizzlies may be expected to kill if cattle are available, no matter where they range.

Ligon (1924:16) went on to seemingly contradict his earlier statements regarding the efficiency of hunters, noting that, while the grizzlies were becoming scarce, "the smaller bears are rather numerous in all mountainous sections of the state, and unless there is much more effort put forth by sportsmen to kill them than has been the case during the past few years, they will no doubt continue to hold their own."

As suggested earlier, the PARC's expanded use of poison may have been a factor in the demise of the far-ranging grizzly. Ligon (1924:2) considered poison an indispensable tool for the killing of predatory animals, especially coyotes, and enthusiastically reported that "poison [strychnine] has come to stay." Not only was the PARC placing poison and setting out poison bait stations, but "approximately 155,000 poison baits [were] distributed to cooperators [stockmen and state predator control agents] free of charge" (Ligon 1924: Table 3). That more bears were not killed was due to the reluctance of US Forest Service administrators to condone poisoning in the national forests, and objections of houndsmen, who were prevented from working cougars and bears in poison areas. This was a real concern to Ligon (1924:21–22):

> In actual experience, with but few exceptions, poison has proved to be detrimental on the National Forests in New Mexico. Commercial poison has been used by the Manager of the VT Ranch in the Black Range, Datil National Forest, during the last three or four years. The result has been that no Biological Survey wolf nor mountain lion hunters, on account of their valuable dogs (well trained dogs are valued at $100 to $500) could be induced to hunt this range or the country closely adjoining it.

The spring and fall poison campaigns and the large amount of unsupervised poisoning by ranchers undoubtedly killed many bears. Many

Photo 17.19. Elliott Speer Barker (1886–1988) was born in Texas but grew up in northern New Mexico. Before World War I, he worked for the US Forest Service in the Jemez, Pecos and Carson national forests, rising through the ranks to become the forest supervisor of the Coconino National Forest in Arizona. After World War I, he owned a ranch near Las Vegas, New Mexico and also worked as a guide before being appointed state game warden and director of the New Mexico Department of Game and Fish in 1931. He is remembered as a conservationist and an author. As a forest ranger in New Mexico, however, he took several grizzlies, and in one of his books wrote that he felt "very strongly that the assumption that grizzly bears originally fed on carrion and rarely attacked or killed cattle themselves is entirely wrong." Photograph: © New Mexico Department of Game and Fish.

discrepancies also existed between where poisoning operations were planned and where they actually took place (Brown 1996).

In 1924 the US Forest Service began keeping estimates of the numbers of big-game species in the national forests (Brown 1996). New Mexico's rangers, true to the tradition of Ligon and others, did not initially separate bears by species, a distinction that would be made in subsequent years.

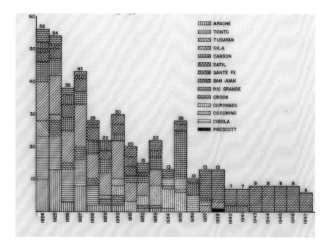

Figure 17.1. Annual grizzly bear numbers on national forests of the Southwest (including the San Juan and Rio Grande national forests in Colorado), as estimated by the US Forest Service beginning in 1924. Annual estimates were based on both numbers of livestock depredation complaints and reports by Forest Service rangers. No grizzly bear was thought to be remaining on any southwestern forest in 1948.

These figures were merely educated guesses and were not based on any census techniques other than stock-killing complaints and reports from Forest Service rangers. They were, in retrospect, optimistic and represent only an idea of the various numbers of grizzlies still extant in the forests. Fifty-four grizzlies were estimated to roam the national forests of the Southwest in 1925 (Fig. 17.1).

Almost 162,000 poison baits were issued to cooperators in the 1924–1925 fiscal year, and approximately 195,000 km² (75,000 mi²) of New Mexico were treated (Pope 1925). Sixteen bears were reported taken in the state by federal and state hunters. That the policy of taking only known stock killers was neither popular nor always obeyed is illustrated in PARC hunter Ed Steele's narrative report to District Supervisor E. F. Pope (1925:10–11):

A few days ago I had four bear up at once. There was an old bear that was fat and I knew she had

to be eating beef to be in that shape this time of year so I killed her. She had beef in her stomach and the meat, I don't think if a fried steak was given to anyone to eat, that they could detect it from beef steak. She was by far the fattest bear I ever saw this time of year. The other bear were in good shape, but not as fat and I really had no evidence on them so I just let them go unharmed, but it was about all I could do to leave them for it sure hurts your dogs to tree bear and let them go and my dogs have kept away from the small bear and have only run two or three bad bear and got whipped off till they need some encouragement. To do good work on the large bear they should be allowed to work on some small bear and I am inclined to think that the three I let go will go ahead and kill but I don't know of course.

I am going to trap the bear at Eagle Peak for I know about that biggest grizzly for I run him all one day last year and there has been some better trained bear dogs than mine after him and they don't have any luck with him. He don't care for dogs, he just keeps out of the way of the hunters and fights the dogs off, so I am going to try and hang a trap on him.

The next year the US Forest Service estimated that only 38 grizzlies remained in the Southwest's forests. Twenty-nine bears, at least four of them cubs, were killed in New Mexico by the PARC and cooperating New Mexico state hunters (Pope 1926). Eight of these were killed by Ed Steele in localities where bears had been reported killing livestock. Pope (1926:9) went on to elaborate:

During the spring months of the past year many claims were advanced that bear were killing cattle. Several of the claims were substantiated and a number of bears were killed by our hunters. The main cause for so many bear becoming predatory was attributed to the scarcity, or almost complete absence, of succulent vegetation on the mountain ranges due to lack of moisture. During the present spring with these conditions almost exactly reversed, it will be interesting to note the actions of bear in their relation to livestock as compared with their behavior last spring. Thus far, during the present season, only one report of depredations by bear has been received. This was the case of a large grizzly bear which was killed by a local hunter near Raton, just after the bear had killed a steer.

The above was the last New Mexico grizzly reported from east of the Rio Grande and the range of the grizzly was becoming more restricted each year. Pope (1926:8) understood the relationship between available forage and stock killing and blamed the dry spring of 1925 for a rash of bear-depredation complaints:

It is believed that an unusual lack of moisture and consequent scarcity of succulent vegetation during the early spring months of the calendar year 1925 was largely responsible for so many bears acquiring predatory habits . . .

During the latter half of the fiscal year almost no complaint of depredations of bears was received, and only one bear was killed by a salaried hunter during the period from January 1 to June 30. There was no evidence that the bear killed was a predatory animal and the hunter who killed the animal contrary to instructions was promptly dismissed from the service.

Changing Attitudes

Bears were now gaining public support as game animals, and there was a movement among sportsmen to afford them some protection. The PARC, always attuned to changes in the political wind, reported a total of only 10 bears by government hunters in Arizona and New Mexico during the 1926–1927 fiscal year (Musgrave 1927; Gray 1927).

Ligon, now working again for the state of New Mexico and no longer with the PARC, stated a case for the bears—even grizzlies. A good observer and a competent field biologist, Ligon (1927:95) had known for some time the true reasons behind bear depredations:

As a group, bears are not livestock killers but individuals, especially among the grizzlies, occasionally vary from traditional habits and become serious destroyers of sheep and cattle. Poverty stricken ranges, result of excessive range utilization, and drought often render their usual food so scanty that out of need bears become killers; hence, as respects losses from bears, forage conservation would result in increased saving of cattle and sheep. Since bears feed on carrion, they are frequently the victims of circumstantial evidence. Guilt as to having been the killer of livestock is often erroneously charged to them.

These habitants of the mountains and forests impart a touch of the primitive and a flavor of romance that nothing else can replace. Their absence would take away one of the most interesting phases of the original wild life that all would regret.

Ligon (1927:95) went on to describe the grizzly's status in New Mexico, and expressed the hope that a remnant population might be preserved:

Unfortunately, the more predatory habits of the grizzlies have resulted in their being almost totally exterminated in the State. Revised policies as affects grazing in the higher portions of the mountains, where these bears make their home, may bring about a situation which will permit a few grizzlies to survive. All true sportsmen of the State hope that such a situation will develop.

In 1927 the New Mexico legislature placed all

bears on the list of protected big-game animals, to be taken only during authorized open seasons. Provision was made in Game Code, however, for stockmen and government hunters to take individual bears that were known to be stock killers. The legislation was enacted with almost no opposition (Ligon 1927).

In 1928, the US Forest Service estimated that only 16 grizzlies remained in the national forests of New Mexico (Fig.17.1). Not included of course were any transient grizzlies that might come in from adjacent Colorado or Arizona. The hope was that enough animals remained for the species to survive.

Only one bear was reported taken by government hunters in New Mexico in the fiscal year 1928, and that was by a state cooperator (Gray 1928). Several stock-killing bears were reported, however, and permits were issued by the New Mexico Game and Fish Department for a bear killed in the Mogollon Mountains, another in the San Mateo Mountains, and still another that was killed in "self-defense" (Gray 1928). Also of concern was a general uproar over the continued widespread use of strychnine poison (now in "2-D" capsule form), drop baits and baited carcasses by both stockmen and PARC hunters (Gray 1928).

In 1928–1929 only one "stock-killing bear" was taken in New Mexico—by a New Mexico state hunter. That sheep continued to be raised in bear country was a problem, however, causing Gray (1929:2) to report: "The policy relating to the distribution of 2-D poison has been modified so that members of the Cattle Growers and Wool Growers can secure sufficient quantity upon written application approved by the Secretary of these associations."

For its part, the US Forest Service was gaining control over the grazing problem on which it had worked so long and which had been set back during the war years. By now, however, it was probably too late for the grizzly given the continued attitude of stock-raisers and the freedom given to them to take matters in their own hands.

New Mexico District Supervisor J. C. Gatlin (1930:15) put it this way: "In most of these [depredation] cases the ranch-men themselves obtained a permit from the State Game Warden and killed the culprits. We feel that whenever it is possible this method is preferable to having our field force do this work. Adverse criticism is also avoided."

No bears were reported taken by the PARC in 1930 (Gatlin 1930), and the Forest Service raised its estimate of Southwest grizzlies to 30. In fact, though, the great bear was now just about gone from New Mexico.

No mention was made in the New Mexico report of the big male grizzly killed by G. W. ("Dub") Evans in April 1930, even though the skull was turned in to the US Biological Survey. This grizzly hunt in the Black Range above Hillsboro was detailed in the *El Paso Times* and in Evans's (1951) *The Slash Ranch hounds*. The story does little to explain the stock-killing aspects of the slain grizzly, but illustrates the stockmen's attitude toward grizzly hunting as a duty mixed with sport. The PARC may have changed its policies, but the stockmen had not changed theirs.

Despite only two cubs being taken in 1930–1931 in New Mexico by the PARC—both by accident in coyote traps—the ranchmen continued as before (Gatlin 1931). An article in the 17 April 1931 issue of the *Silver City Enterprise* read:

To Carl and Blue Rice of Cliff goes the credit for killing one of the largest grizzly bears ever seen in this section. They were riding their range and came upon a dead cow on Rain Creek, and finding the tracks of a big grizzly bear around the carcass, they went to the nearest phone and called up Supervisor James A. Scott and asked him to secure from the state game department a special permit for killing the bear out of season.

As far as we know this animal from the Mogollon Mountains is the last written account of the taking of a grizzly in New Mexico. Nevertheless, there may have been some later records. The grizzly skull in the US National Museum cataloged as collected by G. W. Evans and coming from Magdalena Baldy, New Mexico is dated 1935. No other accounts of this specimen have been found, and in a conversation with one of us (D. E. Brown) at Rancho Montosa in 1982, G. W. Evans Jr. said that he had never heard of his father killing a grizzly on Magdalena Baldy but that he well remembered the grizzly on the wall taken in 1930 on Hillsboro Peak in the Black Range. The possibility therefore exists that the 1935 skull may have been found or given to the US Biological Survey several years after it was taken. The younger Evans also recalled another grizzly killed before 1935, near Alma in Catron County, New Mexico. For his part, Norman Woolsey (per. comm., 1982) remembered a grizzly killed about 1933 near Sacaton, which is near Alma, and which was shown to the schoolchildren at Cliff. These bears, however, are thought to be the same animal reported in the Enterprise account.

Ghosts from the Past

By 1940 the only grizzlies left in the Southwest were the few the Forest Service estimated to be in southern Colorado and some rumored to be in Chihuahua, Mexico (Brown 1996). Then, in the *Annual wildlife report* of the US Forest Service's Region 3, there came a report of grizzlies in New Mexico: "Several grizzly bears are reported on the Jemez District of the Santa Fe Forest. On several occasions, bear hunters have had their dogs badly torn up by bears that will not tree. Tracks and feeding habits of these bears along with hair from one animal that one dog got hold of prove the existence of grizzly in this area" (Johnson and Gee 1941, as quoted in Brown 1996).

Whether these were accounts of one or more grizzlies that had wandered down from Colorado is not known. Next year's *Annual wildlife report* stated: "The report of a grizzly bear on the Santa Fe Forest made last year caused considerable interest. We have no information concerning this report this year."

Reports of grizzlies in southern Colorado continued, however, in the San Juan Mountains, in the upper Rio Grande wilderness of the Rio Grande and San Juan national forests, and on the Tierra Amarilla Land Grant that straddles the border with New Mexico.

In August 1951, a sheepherder killed a young grizzly near Plataro, on the headwaters of the Navajo River in the San Juan National Forest. More sheep were lost to bears, and the next month Ernie Wilkinson, a predator trapper with the US Fish and Wildlife Service, trapped and killed a two-year-old grizzly near Starvation Gulch in the upper Rio Grande wilderness, west of Creede, Colorado. Another grizzly was also taken just to the north, in Saguache County, in 1954 (T. Rausch, Colorado Division of Wildlife, 1979, quoted in Bissell 1980). The US Forest Service revised its figures for the number of grizzlies on the San Juan National Forest upward through the 1950s. Could there also have been a few holdouts southward in New Mexico? Whatever the case, no grizzlies were reportedly taken, officially or unofficially, during the next 25 years from either state.

On the evening of 23 September 1979, Ed Wiseman, of Crestone, Colorado, was seriously mauled by a grizzly near the head of the Navajo River just northwest of Blue Lake, in the San Juan National Forest within approximately 110 km (70 mi) of the New Mexico line. The particulars of the attack were described by Bissell (1980), and the incident was investigated by the Colorado Division of Wildlife in that the bear had been killed during the attack. On the day after Wiseman's rescue, 25 September, Dick Weldon, district

THE LAST GRIZZLY OF THE PECOS

This grizzly bear was killed by George (Skipper) Viles near Beatty's Cabin in the spring of 1923. In memory of her husband, Mrs. Viles presented the hide to the people of New Mexico.

The bear was taken because it had killed 11 head of cattle in as many days. When natural foods, such as deer, are scarce, a predator must seek substitutes. Domestic livestock are also easier to catch than wild deer, particularly for an old bear with worn teeth and claws.

Under such circumstances, man is probably justified in eliminating the offending animal. Large-scale predator control, however, is unnecessary and inadvisable. Today, we recognize the value of maintaining a rich diversity of wildlife species, both prey and predator, in our wild areas.

The New Mexico Division of the Izaak Walton League of America made this display possible. The League regrets the untimely extinction of any species and works to preserve endangered species.

We do not know the exact dates of the final disappearance of grizzlies from New Mexico. In late April of 1930, a cattle-killing grizzly, said to be more than 50 years old, was taken in the Black Range of southwestern New Mexico. In the early 1960's, a grizzly was reported seen near Raton. The report, like several others from 1930 to 1960, was not verified.

Grizzlies can now live only in wildernesses large enough to give the grizzly relative freedom from conflict with man's interests. They still exist in Alaska and a few northern states and Canadian provinces. Some persons, however, agree with the feeling of Aldo Leopold, one of America's greatest conservationists:

Relegating grizzlies to Alaska is about like relegating happiness to heaven; one may never get there.

Photos 17.20a and b. Last grizzly of record from the Pecos Wilderness (and the northern part of the state), killed in 1923 by George Alfred "Skipper" Viles. a) (*top*) Made into a rug, the pelt is now on display in the New Mexico Department of Game and Fish's main office. b) (*bottom*) A commemorative sign is placed next to it. In his book *Beattie's cabin*, Elliott Barker (1953:171) writes, "At the time, I was a bit jealous of Skipper because he, not I, had killed the bear. Now, since it proved to be the last one, I am mighty glad I didn't kill it." Photographs: © Mark L. Watson.

wildlife manager of the Colorado Division of Wildlife, and two other men visited the site in a helicopter. What they found was a full-grown female grizzly. Unfortunately, the helicopter had been damaged on landing, requiring Weldon and the pilot to walk out while the others skinned the bear.

The next day a leased helicopter took out the skull and hide, leaving the carcass and its all-important reproductive organs behind. Had these body parts been retrieved, it could have been determined whether the bear had bred and if other grizzlies were likely to be present.

On 29 September, days after the attack, another helicopter was dispatched to retrieve the carcass and repair the damaged helicopter. After securing the bear to the sling, the helicopter pilot took off, but his craft crashed on leaving Plataro en route for Alamosa. No one was hurt and the carcass was placed in the freezer at the La Jara Fish Hatchery.

A necropsy showed that the bear had probably bled to death and that she was more than 16 years old according to the annual rings in a sectioned tooth. Her weight was estimated at between ~160–180 kg (350–400 lbs), and she was considered to have been "in remarkably good condition." Her reproductive history remained unknown; the uterus had spoiled and could not be examined for placental scars. The hide, the skull, and part of the skeleton were placed in the Denver Museum of Natural History (Bissell 1980).

Numerous attempts to find additional grizzlies in the San Juan Mountains, some of them much publicized, proved negative or inconclusive. More than 40 years have passed, and the chance that a small, breeding population of grizzlies remains tucked away in the mountains along the Colorado–New Mexico border is almost non-existent. Until such time as grizzlies are reintroduced to the area, if ever, it must be assumed that this bizarre incident is the last chapter in the Southwest's grizzly's passage into history.

NOTE AND ACKNOWLEDGMENTS

Some of the text in this chapter was written by the first author (D. E. Brown) for his earlier book, *The grizzly in the Southwest* (University of Oklahoma Press, 1996), and reproduced here with only slight changes. The authors wish to thank Sara Szakaly (National Wildlife Research Center) for copies of annual predatory animal control reports from 1924 to 1931, Harley Shaw for his insightful and thorough review of the chapter and for copies of historical grizzly photos; Jon Dunnum for the grizzly skull photos; and Hari Viswanathan and Pat Gaines for grizzly bear images from outside the Southwest.

LITERATURE CITED

Aune, K., and T. Stivers. 1982. *Rocky Mountain Front grizzly bear monitoring and investigation. Annual Report*. Bureau of Land Management; Montana Department Fish, Wildlife and Parks; US Forest Service; Williams Exploration Company and Sun Exploration Company.

Bailey, V. 1907. Wolves in relation to stock, game and the National Forest reserves. *US Department of Agriculture Forest Bulletin* 72:1–31.

———1931 (=1932). *Mammals of New Mexico*. North American Fauna 53. Washington, DC: US Department of Agriculture, Bureau of Biological Survey.

Baird, S. F. 1859. Mammals of the boundary. In *Report on the U.S. Mexican boundary survey, under order of Lieut. Col. W. H. Emory, Major First Cavalry, and United States Commissioner*, Vol. 2, Pt. 2, 1–62. Washington, DC: Department of Interior.

Barker, E. S. 1953. *Beatty's cabin*. Albuquerque: University of New Mexico Press.

———. 1966. [1963] Letter to Levon Lee, Chief of Game Management, New Mexico Game and Fish Department. In *The grizzly bear: portraits from life*, ed. B. D. Haynes and E. Haynes, x–xii. Norman: University of Oklahoma Press.

Bissell, S. J. 1980. Grizzly bear incident. September 1979 summary report. Colorado Division of Wildlife compendium of reports, correspondence and newspaper columns. Colorado Division of Wildlife non-game files.

Bliss, C. F. 1921. *Predatory animal control*. New Mexico District Annual Report. US Department of Agriculture, Bureau of Biological Survey.

———. 1922. *Predatory animal control*. New Mexico District Annual Report. US Department of Agriculture, Bureau of Biological Survey.

Brown, D. E., ed. 1983. *The wolf in the Southwest: the making of an endangered species*. Tucson: University of Arizona Press.

———, ed. 1994. *Biotic communities: southwestern United States and northwestern Mexico*. 2nd ed. Salt Lake City: University of Utah Press.

———. 1996. *The grizzly in the Southwest*. Norman: University of Oklahoma Press.

Brown, D. L. 1982. *Grizzly bear recovery plan*. US Fish and Wildlife Service, in cooperation with the Montana Department of Fish, Wildlife and Parks.

Burnham, J., and N. Mott. 2021. Timeline: a history of grizzly bear recovery in the lower 48 states. Montana Public Radio. 2 April 2021, updated 2 November 2021. https://www.mtpr.org/montana-news/2021-04-02/timeline-a-history-of-grizzly-bear-recovery-in-the-lower-48-states.

Costello, C. M., and L. L. Roberts. 2021. *Northern Continental Divide Ecosystem Grizzly Bear Monitoring Team Annual Report, 2020*. Kalispell: Montana Fish, Wildlife & Parks.

Coues, E., and H. C. Yarrow. 1875. Report upon the collections of mammals made in portions of Nevada, Utah, California, Colorado, New Mexico, and Arizona during the years 1871, 1872, 1873, and 1874. In *Annual report upon U.S. geographical and geological explorations and surveys west of the one hundredth meridian, in charge of 1st Lieutenant B. M. Wheeler*, Vol. 5, 35–129, 960–79. Washington, DC: US Government Printing Office.

Craighead, F. C., Jr. 1979. *Track of the grizzly*. San Francisco: Sierra Club Books.

Craighead, J. J., J. S. Sumner, and G. B. Scaggs. 1982. A definitive system for analysis of grizzly bear habitat and other wilderness resources. University of Montana Foundation. *Wildlife-Wildlands Monograph* 1:1–251.

Craighead, J. J., J. Varney, and F. C. Craighead Jr. 1974. A population analysis of the Yellowstone grizzly-bears. *Bulletin of the Montana Forest and Conservation Experiment Station School of Forestry, University of Montana, Missoula* 40:1–20.

Davis, G. P., Jr. 1982. *Man and wildlife in Arizona: the American exploration period 1825–1865*. Phoenix: Arizona Game and Fish Department.

Denver Museum of Nature and Science (DMNS). [2021]. Colorado's Last Grizzly Bear. https://www.dmns.org/science/featured-collections/zoology/colorados-last-grizzly-bear/. Accessed 24 December 2021.

Doan-Crider, D. L., and E. C. Hellgren. 1996. Population characteristics and winter ecology of black bears in Coahuila, Mexico. *Journal of Wildlife Management* 60:398–407.

Duffen, W. A., ed. 1960. Overland via jackass mail in 1858: the diary of Phocian R. Way. *Arizona and the West* 2:35–43, 147–64, 279–92, 353–70.

Edwords, C. E. 1893. Camp fires of a naturalist: the story of fourteen expeditions after North American mammals from the field notes of Louis Lindsay Dyche. New York: D. Appleton.

Ellison, G. R. ("Slim"). 1968. *Cowboys under the Mogollon Rim*. Tucson: University of Arizona Press.

Evans, G. W. 1951. *Slash Ranch hounds*. Albuquerque: University of New Mexico Press.

Farrar, H. R. 1968. Tales of New Mexico Territory, 1868–1876. *New Mexico Historical Review* 43:147–49.

Findley, J. S., A. H. Harris, D. E. Wilson, and C. Jones. 1975. *Mammals of New Mexico*. Albuquerque: University of New Mexico Press.

Flader, S. L. 1974. *Thinking like a mountain*. Columbia: University of Missouri Press.

Foresman, K. R. 2001. *Key to the mammals of Montana*. Lawrence, KS: Allen Press.

French, W. J. 1965a. *Some recollections of a western ranchman: New Mexico, 1883–1899*. Vol. 1. New York: Argosy-Antiquarian.

———.1965b. *Further recollections of a western ranchman: New Mexico, 1883–1899*. Vol. 2. New York: Argosy-Antiquarian.

García-Rodríguez, A., R. Rigg, I. Elguero-Claramunt, K. Bojarska, M. Krofel, J. Parchizadeh, T. Pataky, I. Seryodkin, M. Skuban, P. Wabakken, F. Zięba, T. Zwijacz-Kozica, and N. Selva. 2020. Phenology of brown bear breeding season and related geographical cues. *European Zoological Journal* 87:552–58.

Garshelis, D. L., M. L. Gibeau, and S. Herrero. 2012. Grizzly bear demographics in and around Banff National Park and Kananaskis Country, Alberta. *Journal of Wildlife Management* 69:277–97.

Gatlin, J. C. 1930. *Predatory animal control*. New Mexico District Annual Report. US Department of Agriculture, Bureau of Biological Survey.

———. 1931. *Predatory animal control*. New Mexico District Annual Report. US Department of Agriculture, Bureau of Biological Survey.

Geranios, N. K. 2020. Federal government gives up grizzly reintroduction in North Cascades. Oregon Public Broadcasting. 8 July 2020. https://www.opb.org/news/article/grizzly-bear-reintroduction-north-cascades-give-up/

Gish, D. M. 1977. An historical look at the Mexican gray wolf (*Canis lupus baileyi*) in early Arizona Territory and since statehood. Unpublished report. Albuquerque: US Fish and Wildlife Service.

González-Bernardo, E., L. F. Russo, E. Valderrábano, Á. Fernández, and V. Penteriani. 2020. Denning in brown bears. *Ecology and Evolution* 10:6844–62.

Gould, M. J., F. R van Manen, M. A. Haroldson, J. J. Clapp, J. A. Dellinger, D. Thompson, and C. M. Costello. 2023. Vital rates, population size, population growth rate, and percent mortality. In *Yellowstone grizzly bear investigations: annual report of the Interagency Grizzly Bear Study Team, 2022*, ed. F. T. van Manen, M. A. Haroldson, and B. E. Karabensh, 12–13. Bozeman, MT: US Geological Survey.

Governor's Office. 2021. Gov. Gianforte: Montana petitioning federal government to delist NCDE grizzly bears. montana.gov. State of Montana Newsroom. 6 December 2021. https://news.mt.gov/Governors-Office/

Gov-Gianforte-Montana-Petitioning-Federal-Government-to-Delist-NCDE-Grizzly-Bears.

Gray, A. E. 1927. *Predatory animal control*. New Mexico District Annual Report. US Department of Agriculture, Bureau of Biological Survey.

———. 1928. *Predatory animal control*. New Mexico District Annual Report. US Department of Agriculture, Bureau of Biological Survey.

———. 1929. *Predatory animal control*. New Mexico District Annual Report. US Department of Agriculture, Bureau of Biological Survey.

Gunther, K. A., M. J. Biel, N. Anderson, and L. Waits. 2002. Probable grizzly bear predation on an American black bear in Yellowstone National Park. *Ursus* 13:372–74.

Hailer, F., and A. J. Welch. 2016. Evolutionary history of brown and polar bears. In *Encyclopedia of Life Sciences*, 1–8. Chichester: John Wiley.

Hall, E. R., and K. R. Kelson. 1959. *The mammals of North America*. Vol. 2. New York: Ronald Press.

Haroldson, M. A., M. A. Ternent, K. A. Gunther, and Charles C. Schwartz. 2002. Grizzly bear denning chronology and movements in the Greater Yellowstone Ecosystem. *Ursus* 13:29–37.

Hittell, T. H. 1911. *The adventures of James Capen Adams, mountaineer and grizzly bear hunter of California*. New York: Charles Scribner's Sons.

Hoffmeister, D. F. 1986. *Mammals of Arizona*. Tucson and Phoenix: University of Arizona Press and Arizona Game and Fish Department.

Housholder, B. 1971. *The grizzly bear in Arizona*. Phoenix: Bob Housholder.

Jimenez, M. D., V. J. Asher, C. Bergman, E. E. Bangs, and S. P. Woodruff. 2008. Gray wolves, *Canis lupus*, killed by cougars, *Puma concolor*, and a grizzly bear, *Ursus arctos*, in Montana, Alberta, and Wyoming. *Canadian Field-Naturalist* 122:76–78.

Johnson, F. W., and M. A. Gee. 1941. *Report on wildlife within National Forests of Arizona and New Mexico, 1940*. US Department of Agriculture, Forest Service, Region 3.

Johnson, G. 2018. Zinke supports restoration of grizzlies in North Cascades. *Washington Times*. 23 March 2018. https://www.washingtontimes.com/news/2018/mar/23/zinke-supports-restoration-of-grizzlies-in-north-c/.

Johnson, S. J., and D. E. Griffel. 1982. Sheep losses on grizzly bear range. *Journal of Wildlife Management* 46:786–90.

Kearny, T. H., and R. H. Peebles 1960. *Arizona flora*. Berkeley: University of California Press.

Kurtén, B. 1973. Transberingian relationships of *Ursus arctos* Linn, (brown and grizzly bears). *Commentationes biologicae* 65:1–10.

Kurtén, B., and E. Anderson. 1980. *Pleistocene mammals of North America*. New York: Columbia University Press.

Kutschera, V. E., T. Bidon, F. Hailer, J. L. Rodi, S. R. Fain, and A. Janke. 2014. Bears in a forest of gene trees: phylogenetic inference is complicated by incomplete lineage sorting and gene flow. *Molecular Biology and Evolution* 31:2004–17.

LeCount, A. 1977. Using chest circumference to determine bear weight. *Arizona Game and Fish Department Wildlife Digest* 11:1–2.

Lee, L. 1967. Bear. *In New Mexico wildlife management*, 83–87. Santa Fe: New Mexico Department of Game and Fish.

Leopold, A. 1959. *Wildlife of Mexico: the game birds and mammals*. Berkeley: University of California Press.

———. 1966. *A Sand County almanac*. Enl. ed. New York: Oxford University Press.

Libal, N. S., J. L. Belant, B. D. Leopold, G. Wang, and P. A. Owen. 2011. Despotism and risk of infanticide influence grizzly bear den-site selection. *PLOS One* 6(9):e24133. doi:10.1371/journal.pone.0024133

Ligon, J. S. 1916. *Predatory animal control*. New Mexico District Annual Report. Albuquerque: US Department of Agriculture, Bureau of Biological Survey.

———. 1917. *Predatory animal control*. New Mexico District Annual Report. Albuquerque: US Department of Agriculture, Bureau of Biological Survey.

———. 1918. *Predatory animal control*. New Mexico District Annual Report. Albuquerque: US Department of Agriculture, Bureau of Biological Survey.

———. 1919. *Predatory animal control*. New Mexico District Annual Report. Albuquerque: US Department of Agriculture, Bureau of Biological Survey.

———. 1920. *Predatory animal control*. New Mexico District Annual Report. Albuquerque: US Department of Agriculture, Bureau of Biological Survey.

———. 1924. *Predatory animal control*. New Mexico

District Annual Report. Albuquerque: US Department of Agriculture, Bureau of Biological Survey.

———. 1927. *Wild life of New Mexico: its conservation and management*. Santa Fe: State Game Commission, Department of Game and Fish.

López-Alfaro, C., S. C. P. Coogan, C. T. Robbins, J. K. Fortin, and S. E. Nielsen. 2015. Assessing nutritional parameters of brown bear diets among ecosystems gives insight into differences among populations. *PLOS One* 10(6):e0128088. doi:10.1371/journal.pone.0128088.

López-Alfaro, C., C. T. Robbins, A. Zedrosser, S. E. Nielsena. 2013. Energetics of hibernation and reproductive trade-offs in brown bears. *Ecological Modeling* 270:1–10.

McLellan, B. N., and F. W. Hovey. 2001. Habitats selected by grizzly bears in multiple use landscapes. *Journal of Wildlife Management* 65:92–99.

McLellan, B., and D. C. Reiner. 1994. A review of bear evolution. *International Conference on Bear Research and Management* 9:85–96.

McLoughlin, P. D., H. D. Cluff, and F. Messier. 2002. Denning ecology of barren-ground grizzly bears in the central Arctic. *Journal of Mammalogy* 83:188–98.

Mealy, S. P. 1980. The natural food habits of grizzly bears in Yellowstone National Park, 1973–74. In *Bears—their biology and management*, ed. C. J. Martinka and K. L. McArthur, 281–92. Bear Biology Association Conference Series 3.

Merriam, C. H. 1918. Review of the grizzly and big brown bears of North America. *North American Fauna* 4:1–136.

Miller, C. R., L. P. Waits, and P. Joyce. 2006. Phylogeography and mitochondrial diversity of extirpated brown bear (*Ursus arctos*) populations in the contiguous United States and Mexico. *Molecular Ecology* 15:4477–85.

Mitchell, F. S., D. P. Onorato, E. C. Hellgren, J. R. Skiles Jr., and L. A. Harveson. 2005. Winter ecology of American black bears in a desert montane island. *Wildlife Society Bulletin* 33:164–71.

Murie, A. 1948. Cattle on grizzly bear range. *Journal of Wildlife Management* 12:57–72.

Murphy, K. M., G. S. Felzien, M. G. Hornocker, and T. K. Ruth. 1998. Encounter competition between bears and cougars: some ecological implications. *Ursus* 10:55–60.

Musgrave, M. E. 1923. *Predatory animal control*. Arizona District Annual Report. Phoenix: US Department of Agriculture, Biological Survey.

———. 1924. *Predatory animal control*. Arizona District Annual Report. Phoenix: US Department of Agriculture, Biological Survey.

———. 1927. *Predatory animal control*. Arizona District Annual Report. Phoenix: US Department of Agriculture, Biological Survey.

Paetkau, D., G. F. Shields, and C. Strobeck. 1998. Gene flow between insular, coastal and interior populations of brown bears in Alaska. *Molecular Ecology* 7:1283–92.

Pattie, J. O. 1833. *The personal narrative of James Ohio Pattie of Kentucky*. Ed. John H. Wood. Cincinnati, OH: T. Flint.

Pearson, A. M. 1975. *The northern interior grizzly bear Ursus arctos L*. Canadian Wildlife Service Report Series 34, Ottawa.

Peck, C. P., F. T. van Manen, C. M. Costello, M. A. Haroldson, L. A. Landenburger, L. L. Roberts, D. D. Bjornlie, and R. D. Mace. 2017. Potential paths for male-mediated gene flow to and from an isolated grizzly bear population. *Ecosphere* 8(10):e01969. 10.1002/ecs2.1969.

Pigeon, K. E., S. D. Côté, and G. B. Stenhouse. 2016a. Assessing den selection and den characteristics of grizzly bears. *Journal of Wildlife Management* 80:884–93.

Pigeon, K. E., G. Stenhouse, and S. D. Côté. 2016b. Drivers of hibernation: linking food and weather to denning behaviour of grizzly bears. *Behavioral Ecology and Sociobiology* 70:1745–54.

Pineau, E. L. 1923. *Predatory animal control*. New Mexico District Annual Report. Albuquerque: US Department of Agriculture, Bureau of Biological Survey.

Podruzny, S. R., S. Cherry, C. C. Schwartz, and L. A. Landenburger. 2002. Grizzly bear denning and potential conflict areas in the Greater Yellowstone Ecosystem. *Ursus* 13:19–28.

Pope, E. F. 1925. *Predatory animal control*. New Mexico District Annual Report. Albuquerque: US Department of Agriculture, Bureau of Biological Survey.

———. 1926. *Predatory animal control*. New Mexico District Annual Report. Albuquerque: US Department of Agriculture, Bureau of Biological Survey.

Proctor, M. F., B. N. McLellan, C. Strobeck, and R. M. R. Barclay. 2004. Gender-specific dispersal distances of grizzly bears estimated by genetic analysis. *Canadian Journal of Zoology* 82:1108–18.

Rausch, R. L. 1963. Geographic variation in size of North American brown bears, *Ursus arctos* L., as indicated by condylobasal length. *Canadian Journal of Zoology* 41:33–45.

Reynolds, H. V., J. A. Curatolo, and R. Quimby. 1976. Denning ecology of grizzly bears in northeastern Alaska. *International Conference on Bear Research and Management* 3:403–9.

Robbins, C. T., C. C. Schwartz, and L. A. Felicetti. 2004. Nutritional ecology of ursids: a review of newer methods and management implications. *Ursus* 15:161–71.

Roy, L. D., and M. J. Dorrance. 1976. *Methods of investigating predation of domestic livestock: a manual for investigating officers.* Edmonton: Alberta Agricultural Plant Industry Laboratory.

Schwartz C. C., S. L. Cain, S. Podruzny, S. Cherry, and L. Frattaroli. 2010. Contrasting activity patterns of sympatric and allopatric black and grizzly bears. *Journal of Wildlife Management* 74:1628–38.

Schwartz, C. C., S. D. Miller, and M. A Haroldson. 2003. Grizzly bear. In *Wild mammals of North America: biology, management, and economics*, 2nd ed., ed. G. A. Feldhamer, B. C. Thompson, and J. A. Chapman, 556–86. Baltimore: Johns Hopkins University Press.

Servheen, C. 1990. *The status and conservation of the bears of the world.* Eighth International Conference on Bear Research and Management Monograph Series Number 2.

———. 1999. Status and management of the grizzly bear in the lower 48 United States. In *Bears: status survey and conservation action plan*, ed. C. Servheen, S. Herrero, and B. Peyton, 50–54. Gland, Switzerland: IUCN/SSC Bear and Polar Bear Specialist Groups.

Servheen, C., and R. Klaver. 1983. Grizzly bear dens and denning activity in the Mission and Rattlesnake mountains, Montana. *International Conference on Bear Research and Management* 5:201–207.

Shaw, H. G., and M. E. Weisenberger. 2011. Twelve hundred miles by horse and burro: J. Stokley Ligon and New Mexico's first breeding bird survey. Tucson: University of Arizona Press.

Sheldon, C. 1925. Big game of Chihuahua, Mexico. In *Hunting and conservation*, ed. C. Sheldon and G. G. Grinnell. New Haven, CT: Boone and Crockett Club and Yale University Press.

Sieracki, P., and M. Bader. 2020. Analysis of road density and grizzly bears in the Ninemile Demographic Connectivity Area Montana. Flathead-Lolo-Bitterroot Citizen Task Force Technical Report 01-20.

Spady, T. J., S. G. Lindburg, and B. S. Durrant. 2007. Evolution of reproductive seasonality in bears. *Mammal Review* 37:21–53.

Stetz, J. B., M. S. Mitchell, and K. C. Kendall. 2018. Using spatially-explicit capture recapture models to explain variation in seasonal density patterns of sympatric ursids. *Ecography* 41:1–12.

Stevens, M. H. 1943. *Meet Mr. Grizzly: a saga on the passing of the grizzly.* Albuquerque: University of New Mexico Press.

Sweet, J. R. 2002. Men and varmints in the Gila Wilderness, 1909–1936: the wilderness ethics and attitudes of Aldo Leopold, Ben Lilly, J. Stokley Ligon, and Albert Pickens towards predators. *New Mexico Historical Review* 77(4):369–97. https://digitalrepository.unm.edu/nmhr/vol77/iss4/2

United States Fish and Wildlife Service. 2004. *Species assessment and listing priority assignment form.* 30 June 2004.

———. 2021a. *Biological report for the grizzly bear (Ursus arctos horribilis) in the lower-48 states.* Version 1.1, 31 January 2021. Missoula, MO.

———. 2021b. *Grizzly bear in the lower-48 states (Ursus arctos horribilis) 5-year status review: summary and evaluation.* March 2021. Denver, CO.

van Manen, F. T., M. R. Ebinger, M. A. Haroldson, R. B. Harris, M. D. Higgs, S. Cherry, G. C. White, and C. C. Schwartz. 2014. Re-Evaluation of Yellowstone grizzly bear population dynamics not supported by empirical data: response to Doak and Cutler. *Conservation Letters* 7:323–31.

van Manen, F. T., and M. A. Haroldson. 2017. Reproduction, survival, and population growth. In *Yellowstone grizzly bears: ecology and conservation of an icon of wilderness*, ed. P. J. White, K. A. Gunther, and F. T. van Manen, 29–45. Yellowstone Forever, Yellowstone National Park, WY.

van Manen, F. T., M. A. Haroldson, D. D. Bjornlie, M. R. Ebinger, D. J. Thompson, C. M. Costello, and

G. C. White. 2016. Density dependence, white-bark pine, and vital rates of grizzly bears. *Journal of Wildlife Management* 80:300–313.

Vroom, G. W., S. Herrero, and R. T. Ogilvie. 1980. The ecology of winter den sites of grizzly bears in Banff National Park, Alberta. *International Conference on Bear Research and Management* 4:321–30.

Waddell, T. L., and D. E. Brown. 1984. Black bear population characteristics in an isolated Southwest mountain range. *Journal of Mammalogy* 65:350–51.

Wyoming Game and Fish Department (WGFD). 2022. Governor Gordon submits petition to remove Greater-Yellowstone grizzlies from the Endangered Species List. 1 November 2022. https://wgfd.wyo.gov/News/Governor-Gordon-Submits-Petition-To-Remove-Greater.

Young, F. M., and C. Beyers. 1980. *Man meets grizzly.* Boston: Houghton Mifflin.

Young. S. P., and E. A. Goldman. 1944. *The wolves of North America.* Washington, DC: American Wildlife Institute.

WOLVERINE (*GULO GULO*)

Jennifer K. Frey

Even in places where the wolverine is well established, it is a mere ghost, rarely observed by humans. To someone unacquainted with the species, a wolverine might resemble a small, short-eared bear with a long bushy tail. In fact, wolverines are not related to bears at all, but rather are a super-sized member of the weasel family (Mustelidae). On the rare occasion a wolverine is actually observed by a person, it usually is from a distance and often with the animal on an open substrate, such as snow or rock, that provides little size reference, all of which makes it difficult to judge how large the animal really is. In actuality, and despite their ferocious reputation, wolverines are much smaller than bears—about the same size as a beagle dog. They stand about 30–45 cm (12–18 in) tall at the shoulders, and they are 65–107 cm (26–42 in) in length, which includes the 17–26 cm (7–10 inch) tail. Like other weasels, there is pronounced sexual dimorphism with females 10% smaller in measurements and 30% smaller in weight compared to males (Hall 1981). Typical weights range from 11–18 kg (24–40 lbs) for males and 6–12 kg (13–26.5 lbs) for females (Banci 1994).

Wolverines have a thick body with medium-length, heavily muscled legs; a broad, bear-like face with small eyes and a prominent muzzle; short rounded ears situated on the sides of the head; and a droopy, bushy tail that extends to the hock (i.e., the tail is shorter than the hind leg). The feet, which serve as snowshoes, are large and broad with soles densely haired surrounding six small naked pads. The feet have five toes, each tipped with a long, curved, semi-retractable claw (Pasitschniak-Arts and Larivière 1995). The pelage is adapted for cold, snowy environments. It is dense and shaggy, particularly in the winter. The shiny hair is rich in oils that help shed frost and prevent the hair from filling with snow, making it a popular trim for parka hoods (Banci 1994). Though there is considerable variation in the color pattern of the pelage, the general pattern is dark brown to blackish with a pale buffy stripe along the sides that extends from the front shoulder (sometimes starting at the head) posteriorly across the hips and then meeting over the base of the tail. The pale lateral stripe often sharply contrasts with the legs and distal end of the tail, both usually blackish. The typically pale claws similarly stand out sharply next to the dark color of the feet and legs. The muzzle and area around the eyes are blackish, but the forehead and ruff around the face are often pale buff, sometimes giving a prominent pale eyebrow. The underside of the neck and chest usually has pale yellowish or whitish marks, which can be used to identify individual animals (Magoun et al. 2011). When newborn, wolverine kits are white (Chadwick 2010).

Wolverines are primarily terrestrial and are usually considered to have plantigrade (flat-footed like a bear) foot posture and locomotion (Carrano 1997), which facilitates travel through soft snow. However, while the front foot is

(*opposite page*) Photograph: © Kalon Baughan.

Photo 18.1. Wolverine photographed at Alaska's McNeil River State Wildlife Sanctuary. The wolverine resembles a small, short-eared bear but is actually a member of the weasel family, the Mustelidae. It is characterized by a stocky body with a thick coat of fur, a bushy tail, and medium-length, heavily muscled legs. The wolverine has a distinctive yellowish lateral stripe that extends along the body to the tail, but it is difficult to see on this particular animal, which appears wet. The bear-like face is broad with small eyes, yellowish eyebrows, and a prominent muzzle. Wolverines walk on the soles of their feet. Despite their reputation for being ferocious, they are no larger than medium-sized dogs. Photograph: Drew Hamilton; © Alaska Department of Fish and Game.

invariably plantigrade, during locomotion the rear parts of the hind feet often do not touch the ground. The wolverine's gait is a lumbering, loping gallop. Wolverines are also scansorial—their long claws and maneuverable limbs allow them to easily climb rocks and cliffs, and like their cousins the martens, they easily climb trees. They can also dig burrows and swim across rivers (USFWS 2018).

The wolverine shares many of the same skull and dental characteristics as found in the weasels (*Mustela* and *Neogale*) and martens (*Martes*; see Hall 1981). Like the martens, each side of the jaw has three upper and lower incisors, one upper and lower canine, four upper and lower premolars, one upper molar and two bottom molars, for a total of 38 teeth (Coues 1877). However, to

a large degree the cranial similarities end there. While the skulls of weasels and martens are delicate, smooth, and armed with sharp pointed teeth for catching and slicing small rodents, the skull of the wolverine is massive, with prominent ridges for the insertion of powerful muscles that work a jaw armed with large, robust teeth that can crush large bones of ungulates. The carnassial teeth (i.e., specialized shearing teeth in carnivores formed by the fourth upper premolar and first lower molar) are especially well developed for shearing, which allows wolverines to eat frozen carcasses and tough hides of long dead ungulates (Pasitschniak-Arts and Larivière 1995).

Besides bears, the species most likely to be confused with a wolverine in New Mexico include the Pacific marten (*Martes caurina*), American badger

(*Taxidea taxus*), and yellow-bellied marmot (*Marmota flaviventris*), all of which may be found in some of the same habitats used by wolverines. The marten is smaller and more weasel-like in overall appearance, with a long tubular body and relatively short legs (see Chapter 20). The marten has a more pointed face, more prominent cat-like ears, and it lacks the pale lateral stripe. The badger is slightly smaller than the wolverine, and it has a robust short-legged appearance (see Chapter 25). However, the badger has much shorter legs such that the body is usually held near the ground. The overall coloration of the badger is gray, and it has a distinct black and white color pattern on the face including a white stripe that extends down the middle of the face from the nose to the back of the neck or beyond. The marmot, also known as "rock chuck" or "whistle pig," is actually a type of ground squirrel. The yellow-bellied marmot is smaller than a wolverine (47—68 cm [18.5—27 in] long), and it has shorter legs with the body held near the ground. The marmot has a blunt arched nose and squirrel-like face (no muzzle) with tiny ears. In most yellow-bellied marmots, the pelage color on the body grades uniformly and lacks distinct lateral stripes (other than occasional molt patterns), and there usually are white markings above the nose and around the mouth (this same area of the face is black on a wolverine).

The etymology of *wolverine* is not precisely known. One interpretation is that the name is derived from the Old English *wulf(a)-ryne* (=wolvering), meaning a diminutive wolf (Arngart 1979). Other older common names used primarily before the 1900s include skunkbear and quickhatch (Seton 1909). The name *skunk bear* is a reference to both the striped pelage pattern and the species' pungent anal gland secretions (Seton 1909). The name *quickhatch* was used primarily in Canada's Hudson Bay region and is thought to have been derived from Cree or Algonquin (Billings 1856). Canadian French refer to the wolverine as *carcajou*, which is derived from an Algonquin

word for the animal (Platt 1901; Bishop 2006). According to Coues (1877) the scientific name *Gulo* is a Latin translation of the Scandinavian and Russian names for the animal. It means a glutton, and apparently refers to its voracity. The French and English followed suit also referring to it as "glouton" and "glutton," respectively. Coues (1877) disputed the common belief that the German name for the wolverine, *Vielfrass*, similarly meaning glutton, is a misnomer that was supposedly a misinterpretation of the Swedish or Norwegian word *fjall*, meaning rock or cliff. Rather, he provided evidence that the name was intentionally meant to connote the animal's voracity. To make matters even more confusing, Kvam et al. (1988) stated that the German *Vielfrass* was the same as the old Norwegian *fellfross*, which means mountain cat or mountain bear; this is an often repeated but uncited Wiki claim (sometimes as a mistranslation of the Old Swedish *Fjellfräs*, meaning mountain cat).

Gulo gulo is the sister taxon to the martens (genus *Martes*), the two having diverged from each other 6.3 million years ago (Li et al. 2014). At one time wolverines in Eurasia and North America were considered separate species, *G. gulo* and *G. luscus*, respectively (Pasitschniak-Arts and Larivière 1995). However, today these two forms are regarded as subspecies of a single wide-ranging Holarctic species (Wilson and Reeder 2005).

DISTRIBUTION

The wolverine is an iconic denizen of the cold "North Country." It has a circumpolar distribution that encompasses much of the arctic and subarctic portions of Eurasia and North America (Pasitschniak-Arts and Larivière 1995). The wolverine's southern distributional limits roughly correspond with the occurrence of snow cover during the spring maternal denning period (Aubry et al. 2007; Copeland et al. 2010). Thus, in western North American its historical southern range limits were confined to the alpine zones of major mountain

Photo 18.2. Eurasian wolverine (*Gulo gulo* ssp. *gulo*). At one time wolverines in Eurasia and North America were considered separate species, *G. gulo* and *G. luscus*, respectively. However, today these two forms are treated as subspecies of a single, wide-ranging species. Photograph: © Marc Baldwin—wildlifeonline.me.uk.

Photo 18.3. Wolverine photographed at McNeil River State Wildlife Sanctuary in Alaska, at a latitude of about 59° N. The species has a circumpolar distribution that corresponds mainly to the boreal and tundra biomes of North America, Europe, and Asia. Southward, its range extends only about as far as the presence of snow cover persisting through the spring maternal denning period. Photograph: Drew Hamilton; © Alaska Department of Fish and Game.

ranges, including the Sierra Nevada in California, as well as two major southern branches of the Rocky Mountains—the Wasatch Range and other peaks in south-central Utah and the Southern Rocky Mountains, which encompass southeastern Wyoming, Colorado, and northern New Mexico (Aubry et al. 2007). However, genetic data indicate that the wolverine was nearly or completely extirpated from the contiguous United States in the early 1900s (McKelvey et al. 2014). Current populations in this region are thought to be descended from animals dispersing from Canada into the United States (McKelvey et al. 2014). The process of natural dispersal and restoration of populations in the United States appears to be continuing, at least for historical habitat that is not too isolated (see below).

Determining the distribution and reproductive status of wolverine populations is difficult given the species' rarity and association with remote habitats. However, use of modern methods such as arrays of remote cameras and the analysis of DNA from specially designed hair snares to determine if females are present in a population are shedding light on the species' status (Lukacs et al. 2020). Currently, resident breeding populations of wolverines in the western contiguous United States are only known from the Northern Rocky Mountains (Montana, Wyoming, Idaho) and the Cascade Range (Washington; Lukacs et al. 2020). Recently, individual males, which are more likely to engage in ultra-long-distance dispersal than are females, also have been documented in the Wallowa Mountains in northeastern Oregon (Magoun and Valkenburg 2012), the northern Sierra Nevada in California (Moriarty et al. 2009), the Wind River Range in Wyoming (Lukacs et al. 2020), northern Utah (USFWS 2018), and the Southern Rocky Mountains in Colorado (Inman et al. 2009; Packila et al. 2017). The species is considered "functionally extinct" (i.e., if present, the population is so small that it is not viable) in Oregon, California, Nevada, Utah, Colorado, and New Mexico (Aubry et al. 2007).

Evidence for the historical occurrence of wolverines in New Mexico is scant (Frey 2006), and it was not until recently that its former occurrence in the state was considered verified (Aubry et al. 2007; Map 18.1). This is not surprising given that in addition to wolverines being naturally rare and occupying remote habitats, New Mexico is at the extreme southern limits of the species' range. As an example of the paucity of information concerning

its distribution, there are fewer than 100 museum specimens of wolverines from across the entire contiguous United States (Aubry et al. 2007). In New Mexico, Findley et al. (1975) did not mention wolverines in their monograph on the mammals of the state. Similarly, Frey and Yates (1996) did not include the wolverine on a list of mammals

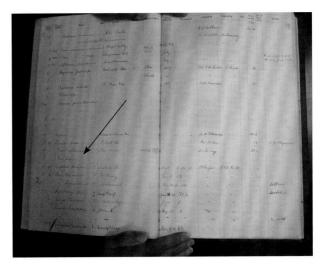

Photo 18.4. Today, Ernest Thompson Seton (1860–1946) is remembered as an influential writer, artist, and naturalist. Having emigrated as a young child from the UK to Canada, he lived in Manitoba from 1882 until 1930, then moved to Santa Fe, New Mexico where he remained until his death. In a short communication published in 1931 in the *Journal of Mammalogy*, Seton recounts being visited in Santa Fe by Flaming Arrow, a Native American from Acoma Pueblo. Flaming Arrow's ancestors evidently encountered wolverines in the mountains of northern New Mexico. Flaming Arrow even produced a carving which Seton immediately recognized as the head of a wolverine. From his years in Canada, Seton was quite familiar with the species and had already published *The life histories of northern animals: an account of the mammals of Manitoba*, which includes a chapter on the wolverine. Photograph available from the United States Library of Congress's Prints and Photographs Division (George Grantham Bain collection).

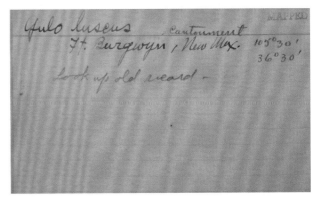

Photos 18.5a and b (*top to bottom*). a) US National Museum skin catalog entry for USNM 3767 *Gulo luscus* from New Mexico, collected by Dr. John Strong Newberry and cataloged on 21 February 1860. b) US National Museum life history file card for *Gulo luscus* from "Ft. Burgwyn, New Mex." Data likely added subsequent to the main entry include "cantonment" (i.e., Cantonment Burgwyn was the original name of the fort), "105 30' 36 30'" (i.e., approximate location of the fort), and "Look up old record," and the card is stamped "MAPPED." Photographs: © Suzanne Peurach.

Photo 18.6. Located 16 km (10 mi) south of Taos in Taos County, Fort Burgwin was a US military cantonment that remained in operation from 1852 to 1860. The wolverine specimen secured by John Strong Newberry ca. 1860 (perhaps 1859) apparently originated from the vicinity of Fort Burgwin in the Sangre de Cristo Mountains. The elevation at Fort Burgwin is only about 2,255 m (7,400 ft), but many of New Mexico's tallest peaks occur nearby in the Taos Mountains, part of the Sangre de Cristo mountain range. Today, the site of the old cantonment is owned by Southern Methodist University. Photograph: © Geraint Smith.

Photo 18.7. M56 (a male wolverine first caught by biologists near Grand Teton National Park in northwestern Wyoming in December 2008) photographed on 25 May 2015 near the Fourth of July Trail head west of Nederland in Colorado. The wolverine darted in front of the two photographers' car, leaped over an adjacent creek bed, and remained perched on a tuft of snow in utter stillness as the photographers scrambled for their cameras. From Wyoming, M56 traveled south more than 900 km (~560 miles). Photograph: © Brad Taylor and John Gallagher.

verified as having occurred in New Mexico, but did include it as a species that was likely historically present but was extirpated prior to being verified. Previously, I (Frey 2004) included the wolverine as a member of the mammal fauna of New Mexico based, in part, on a report by Elliot Coues in his monograph on the weasel family in North America. Coues (1877:49–50) stated: "It is probable that its extreme limit is even somewhat further than this [Great Salt Lake], reaching in the mountains to the borders of Arizona and New Mexico and corresponding latitudes in California. Of this, I was assured by hunters whose statements I had no reason to doubt, and who were evidently acquainted with the species." Bailey (1931:322) thought that Coues' statements were "based on verbal reports that Coues had from the residents of northern New Mexico, where he came more or less in contact with the local hunters and trappers in 1864." Bailey (1931) subsequently concluded that wolverines historically occurred in New Mexico based on Coues' report and his own perception that the Sangre de Cristo Mountains contained appropriate environment for the species.

Unbeknownst to many researchers, however, additional evidence for the historical occurrence of wolverines in the state is available. The wolverine was recognized by local Native American culture in New Mexico. Famed naturalist Ernest Thompson Seton (1931:166), writing from his ranch near Santa Fe, stated:

On the evening of the same day [17 December 1930], an Acoma Indian named Flaming Arrow was visiting me. He told me his people had a hunting song in which they extolled the three great hunters—first, the Mountain Lion; second, the Eagle; and third, the Ho-Hö-an or Ko-Kö-an. He did not know the white man's name for the last; but said it was as big as a dog, somewhat like a small bear, but it had a bushy tail. Its coat was rough, its feet black,

and its back nearly white. It was terribly fierce.

Then, from his medicine bag, he produced a carving, a small effigy of the creature's head. It was an excellent likeness, and there can be no doubt that the Ko-Ko-an is the wolverine. The Indian said it was formerly found in all these mountains, but had disappeared. None of the present generation had seen one.

Acoma Pueblo lies in northern New Mexico, west of Albuquerque. Santa Fe is located in northern New Mexico at the southwestern base of the Sangre de Cristo Mountains. It seems likely that when Seton referred to "all these mountains" he meant the Sangre de Cristo and other adjacent mountains in northern New Mexico.

More recently, Aubry et al. (2007) conducted an exhaustive review of the distribution of the wolverine in the contiguous United States based on records that were verifiable (i.e., associated with physical evidence) or documented (i.e., accounts of animals killed or captured but lacking physical evidence); no anecdotal records (i.e., observations) were used. They found 35 historical (pre-1960) records of wolverines from the Southern Rocky Mountains, primarily from Colorado. In addition, one of the records they considered verifiable was from New Mexico, and it represented the first specific proof of wolverines having occurred in the state (Aubry et al. 2007:2150):

Archival records at the National Museum of Natural History indicate that a wolverine skin was obtained in the vicinity of Ft. Burgwin in the Sangre de Cristo Mountains near present-day Taos, New Mexico by J. S. Newberry in 1860. (National Museum of Natural History, Skin Catalog 3768)

Further investigation of this record requires that some details be corrected. The skin catalog number for the record of a *Gulo luscus* from

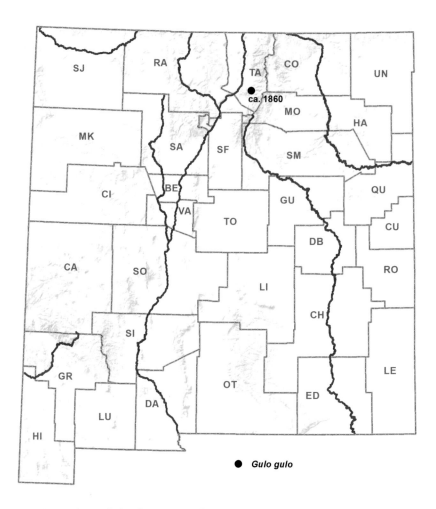

Map 18.1. New Mexico's only verified wolverine record, a specimen (National Museum of Natural History, Skin Catalog 3767) collected by J. S. Newberry ca. 1860 (perhaps in 1859) near Fort Burgwin in the Sangre de Cristo Mountains.

New Mexico is 3767 (3768 is a bat from Wyoming; Photo 18.5a). The record did not indicate a collection date; rather, the record was cataloged on 21 February 1860, and hence the specimen likely was collected sometime prior to 1860. Other entries on the same catalog page have collection dates from 1859 (Photo 18.5a). Newberry served as naturalist and geologist for scientific expeditions of the West in the mid-1800s. For instance, he accompanied the Ives Expedition to determine the navigability of the Colorado River for supplying military forts in New Mexico and Utah in 1857, then in 1859 the Macomb Expedition to

parts of New Mexico, Colorado, and Utah (Newberry 1876). I have not been able to find a specific record that Newberry visited Fort Burgwin, but it certainly is possible as there were few American outposts in northern New Mexico at the time. Lastly, the record does not have a specific location associated with it, other than "New Mexico" (Photo 18.5a). However, a card in the museum life history files has an entry for "*Gulo luscus* Ft. Burgwyn, New Mex." Data likely added to the card subsequent to the main entry are "cantonment" (i.e., Cantonment Burgwyn was the original name of Fort Burgwin); "105 30' 36 30'" (i.e., this is the approximate latitude and longitude of Fort Burgwin); "Look up old record"; and the card is stamped "MAPPED" (Photo 18.5b). Indeed, the museum archives include an old, unpublished map of *Gulo* specimen records that includes Fort Burgwin (R. Fisher pers. comm.). Fort Burgwin (= Cantonment Burgwin) was located ca. 10 km (6 mi) south of Ranchos de Taos, Taos County. It was established shortly after the formation of the New Mexico Territory in 1850 and was in operation from 1852 to 1860. Other important natural history specimens also were collected near Fort Burgwin at this time, including the first specimens of snowshoe hare (*Lepus americanus*) and the first specimen of the endangered New Mexico jumping mouse (*Zapus luteus*).

On basis of the Fort Burgwin record, Aubry et al. (2007) concluded that the southern distributional limit of the wolverine may have been in New Mexico. Previously, although unaware of the Fort Burgwin record, I (Frey 2006) also argued that New Mexico should be included within the natural range of the wolverine. I based that conclusion on several lines of evidence: 1) wolverines were well documented in adjacent areas of Colorado (e.g., Cary 1911; Warren 1942; Armstrong 1972; Nead et al. 1985); 2) there was plausible reason for paucity of records in New Mexico; 3) continuous alpine and subalpine communities extend from the verified record localities in Colorado to areas in New Mexico, and 4) there are no biogeographic breaks in the distributions of other species between the verified records in Colorado and areas in New Mexico. Thus, I concluded that the historical natural distribution of the wolverine included the San Juan and Sangre de Cristo mountains in New Mexico (Frey 2006). The Fort Burgwin record provides strong support for that conclusion. Subsequent sophisticated habitat models have also predicted that wolverine habitat exists in the San Juan and Sangre de Cristo mountains of New Mexico (Inman et al. 2013; Carroll et al. 2020), and the US Fish and Wildlife Service concurred that the historical range of the wolverine included the mountains of northern New Mexico (USFWS 2018).

The current status of the wolverine in the Southern Rocky Mountains is not clear. Aubry et al. (2007) reported that the last documented or verifiable wolverine record from Colorado dated back to 1919, presumably that reported by Grinnell (1926). Consequently, they (Aubry et al. 2007) concluded that the wolverine had been extirpated from the Southern Rocky Mountains. However, Nead et al. (1985) reported 57 observations of wolverines in Colorado and concluded that the wolverine may still exist in the state. Aubry et al. (2007) did not cite Nead et al. (1985), and so it is unclear if they considered those 57 records in their analysis. The Colorado Division of Wildlife (CDOW) has made considerable efforts to confirm sightings of wolverines in that state, including 12 surveys involving 9,388 km (5833.5 miles) of snow tracking, 62 hair-snag locations, 110 remote camera locations, and 686 trap-nights of snares (CDOW 2010). Although these and other surveys found probable tracks of wolverines, the CDOW concluded that if any wolverines occurred in Colorado, "their numbers were so small that they did not represent a viable population" (CDOW 2010:5). However, the US Fish and Wildlife Service (2018) reported two records from Colorado in the period 2004–2017. In New Mexico there have also been a few unverified reports (Frey 2006; USFWS 2018), but no additional conclusive evidence. Thus, it seems likely that the wolverine is extirpated or functionally extinct (i.e., its population is so small as to not be self-sustaining) in New Mexico and elsewhere within the Southern Rocky Mountains.

If the wolverine is functionally extinct in the Southern Rocky Mountains, what accounts for the occasional unverified observations of wolverines in both Colorado and New Mexico? For instance, McDonald (1985) provided a detailed description of a first-hand observation of a wolverine in alpine tundra on Latir Peak, in the Sangre de Cristo Mountains of New Mexico. One possibility is that the observations are misidentifications of marmots, black bears (*Ursus americanus*), badgers, or other animals. On the other hand, it is possible that people really have seen wolverines, despite their "extirpated" status. How could that be? If we consider that at least some of the observations were actually wolverines, then the most likely explanation is that wolverines, especially males, are tremendous travelers that are capable of very long-distance dispersal. Such rare events used to be almost impossible for wildlife biologists to document. This is no longer the case, thanks to the advent of new, more sophisticated technology, such as remote cameras and GPS-satellite collars. A recent example, with important connotations for New Mexico, illustrates this point.

In June 2009, a wild wolverine was documented in Colorado, the first verified record in the state or anywhere else in the Southern Rocky Mountains since ca. 1919 (Packila et al. 2017). This was not an accidental discovery. The subadult male was a research subject, known as M56, that was part of an investigation on the wolverine in the Greater Yellowstone Ecosystem (Packila et al. 2017). Male M56 was first caught by biologists near Grand Teton National Park in northwestern Wyoming in December 2008 (Inman et al. 2009). The wolverine was implanted with a radio transmitter and fitted with a GPS collar (Packila et al. 2017). The GPS collar failed, but M56's travels between northwestern Wyoming and Colorado are known in rather exacting detail based on radio signals from the implant. The realm between the Northern Rocky Mountains and the Southern Rocky Mountains in southern Wyoming is known as the Red Desert, and it is a mostly low-elevation, flat, arid expanse—wholly unlike the alpine environments usually occupied by wolverines in the western United States. But cross the Red Desert he did, a journey totaling

more than 900 km (559 mi). Ironically, one of the first places M56 arrived at in Colorado was Rocky Mountain National Park, which during the 1990s featured as a prominent player in plans to potentially restore the wolverine to the state (Seidel 1998b). But, apparently not finding a mate to give him reason to establish a home range in the park, M56 continued to move southward to the vicinity of Leadville near the center of the state, and only about 241 km (150 mi) from the New Mexico border (CDOW 2010). Coincidentally, just prior to the arrival of M56 in Colorado, Schwartz et al. (2009) produced a least-cost corridor map for wolverines in the Rocky Mountains based on a correlation between persistent spring snow and genetic similarity. They concluded that natural recolonization of wolverines to Colorado from Yellowstone was possible but challenging. Interestingly, their most likely dispersal route was similar to the actual path used by M56, and this route eventually continues southward into the San Juan Mountains, which span the Colorado-New Mexico border (Schwartz et al. 2009). M56 was known to inhabit Colorado until 2012. Unfortunately, he evidently continued to wander and eventually headed back north and wound up in North Dakota (a straight-line distance of 826 km [513 mi]), where he was shot and killed in 2016 (Packila et al. 2017). There is no evidence that M56 ever made it as far south as New Mexico. However, long-distance dispersals may not be as rare as we once thought (Packila et al. 2017) and M56's journey demonstrates that wolverines have the capacity to disperse from their current core range in the Northern Rocky Mountains to New Mexico. Perhaps other wolverines have made such a trek.

There have been other similar examples of long-distance dispersals by wolverines in other regions. For instance, wolverines were last documented in California in the 1920s and are considered extirpated in the Sierra Nevada (Aubry et al. 2007). However, in 2008 a wolverine was photographed by a remote camera that had been deployed by a graduate student studying Pacific martens. Further study of this animal, including investigations of its genetics and of the stable carbon (δ^{13}C) and nitrogen (δ^{15}N) isotope composition of hair samples, revealed that it was a male that had likely dispersed from the Northern Rocky Mountains, possibly from the Sawtooth Mountain region of southwestern Idaho, corresponding to a straight-line distance of more than 644 km (400 mi; Moriarty et al. 2009). Traditional data and conventional scientific wisdom suggest that it is only males that make such epic journeys. If true, then such wanderings would not be able to restore the wolverine to regions where they have been extirpated because females would still be needed to form a viable reproducing population. However, a study in the North Cascades documented a young female that traveled more than 483 km (300 mi) before being killed by a cougar (*Puma concolor*) (Aubry et al. 2012). Her journey did not take her across flat deserts or other locations that are atypical for a wolverine, and it is unknown to what extent females will cross low-quality habitats. Nonetheless, the possibility exists, however remote, that one day wolverines could naturally reoccupy the Southern Rocky Mountains—that is, if we do not give them a helping hand to reoccupy the area first (see under "Status and Management").

Population densities of wolverines are always naturally low, even less than those of wolves (*Canis lupus*), even in ideal habitats (Pasitschniak-Arts and Larivière 1995; Ruggiero et al. 2007). In North America, though densities range up to 14 wolverines per 1000 km^2 (386 sq mi), they are usually much lower even in high-quality habitat (Inman et al. 2012a). In British Columbia, wolverine densities were estimated at one per 161 km^2 in high-quality habitat, one per 244 km^2 in moderate-quality habitat, and one per 3,333 km^2 in low-quality habitat (Lofroth and Krebs 2007). In the contiguous United States, density estimates were highest in northwestern Montana (15.4/1000 km^2) and lowest in the Greater Yellowstone Ecosystem (3.5/1000 km^2; Inman et al. 2012a). The current total population of wolverines in the contiguous United States has been estimated at 250–300 individuals (USFWS 2010). However, a genetic study of the Northern Rocky Mountain population, which is the core of the western US population, found that the number of reproducing individuals was only 35 (Schwartz et al. 2009). A habitat suitability model for wolverines in the Rocky Mountain region of the contiguous United States reveals island-like habitat patches corresponding to disjunct mountain ranges and peaks (Inman et al. 2013). The Southern Rocky Mountains (including northern New Mexico) represent one of the largest areas of potential habitat for wolverines in the western United States (Inman et al. 2013). It was estimated that the Southern Rocky Mountains could support 137 wolverines, or 21.3% of the total possible population size for the western United States (Inman et al. 2013). In New Mexico, high-quality wolverine habitat was predicted to occur in the northern San Juan Mountains and the higher-elevation portions of the Sangre de Cristo Mountains (Inman et al. 2013; Carroll et al. 2020).

HABITAT ASSOCIATIONS

Like many large carnivores, wolverines do not necessarily select specific vegetation types. Rather, other factors such as prey availability may determine where they occur. In the case of the wolverine, recent research has established that its geographic range is closely associated with regions that have cool summer temperatures (i.e., maximum average August temperature <22° C [= 72° F] and a high probability of snow cover during the spring maternal denning period (Aubry et al. 2007; Copeland et al. 2010). Spring snow also predicts dispersal routes, which can lead to gene flow among populations or recolonization of historical range (Schwartz et

Photo 18.8. Aerial view of Rocky Mountain National Park in Colorado. Rocky Mountain National Park is located within wolverine's modeled habitat based on persistent spring snow cover. It also corresponds to the location where M56 was first photographed in Colorado. Photograph: © Eric Odell / Colorado Parks and Wildlife.

Photo 18.9. Northern New Mexico's Wheeler Peak Wilderness as viewed from the Moreno Valley on 3 February 2013. The Wheeler Peak Wilderness in the Sangre de Cristo Mountains contains the highest elevation point in New Mexico, Wheeler Peak (4,013.3 m [13,161 ft]). Depending on annual snowfall, there may be snow on the north slopes of the Wheeler Peak Wilderness throughout most of the year. Persistent snow cover through the end of the reproductive denning period (late April to mid-May at more northern latitudes) is a key habitat requirement for wolverines. Photograph: © Geraint Smith.

al. 2009). Thus, the relationship of the wolverine with persistent spring snow cover has been considered obligatory (Copeland et al. 2010). However, other factors are also important to wolverines. Carroll et al. (2020) developed a resource selection model across the wolverine's historical range in the western United States, based on six variables, each of which represents a specific biological function. Wolverines were positively associated with higher elevations (adjusted for latitude), which provide low temperatures, alpine meadows for prey, and cool environments for long-term caching of food. They were positively associated with ridge and peak landforms, which indicate low temperatures, prey, and steep terrain. They were positively associated with areas of high snow, which provides wolverines with a competitive advantage over other species and enhances predator avoidance. They were positively associated with alpine meadows and grasslands for hunting and rocky areas for denning and caching. They were negatively associated with human-modified environments and distance to familiar features. This model predicted that core habitat for wolverines exists in the San Juan and Sangre de Cristo mountains in northern New Mexico (Carroll et al. 2020; and see Maps 18.2 and 18.3).

Given the wolverine's association with spring snowpack and cool summer temperatures, there is a trend for the species to occur at higher elevations at lower latitudes in western North America. In the Greater Yellowstone Ecosystem, which harbors the one, viable population nearest to New Mexico, wolverines selected high elevations at or above treeline where freezing temperatures may occur year-round and where there is deep snow in winter (Inman et al. 2012b). Selected habitat also was on steep terrain with a mixture of coniferous forest, alpine meadow, boulders (e.g., talus), and avalanche chutes (Inman et al. 2012b). There may be subtle seasonal variation in habitats used by wolverines (Hornocker and Hash 1981; Copeland et al. 2007). For instance, in the Greater Yellowstone Ecosystem, wolverines selected elevations at and above treeline in summer, but used a slightly lower elevation at treeline during winter (Inman et al. 2011). It is likely that wolverines historically occupied similar alpine and subalpine environments in the Southern Rocky Mountains (i.e., southeastern Wyoming, Colorado, and northern New Mexico; Aubry et al. 2007). For instance, the above habitat description matches available environments along much of the crest of the Sangre de Cristo Mountains in New Mexico. Thus, historical habitats of the wolverine in New Mexico were likely: alpine tundra, which is located above the treeline generally >3,500 m (11,500 ft); subalpine coniferous forest, which is usually dominated by Engelmann spruce (*Picea engelmannii*) and subalpine fir (*Abies lasiocarpa*) and generally occurs above 3,000–3,650 m (9,500–12,000 ft); and subalpine meadows, which are found within forest openings (Dick-Peddie 1993; see Chapter 2).

Given the tightrope that wolverines walk with respect to securing enough food in a cold, unproductive environment versus the need to fuel reproduction and the growth and development of young, den site selection is particularly important for this species. Wolverine kits are born in February through mid-March, during the midst of winter (Inman et al. 2012b). Reproductive dens that are used from birth to weaning, are nearly always located under snow (Magoun and Copeland 1998; Magoun et al. 2017). Typically, these dens are long, complex tunnels built into deep hard-packed snowdrifts or shorter snow tunnels that lead to spaces under boulders or fallen trees covered in snow. In snowdrifts, often there is just a single entry hole ca. 35 cm (14 in) wide showing on the outside. Although in some regions wolverines are known to den in locations that lack snow (Jokinen et al. 2019), the presence of deep snow from February to May is usually considered crucial because of

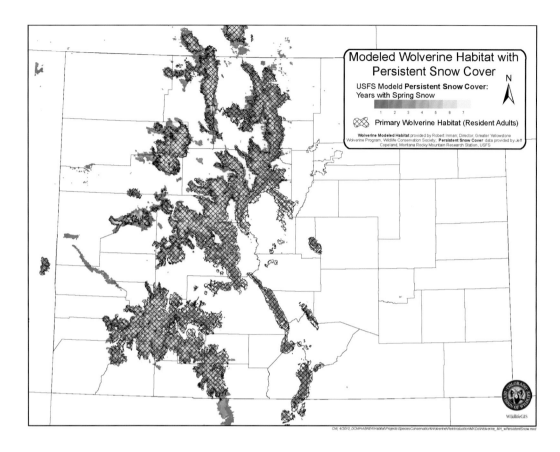

Map 18.2. Wolverine habitat suitability model for Colorado developed based on predicted persistent spring snow cover. The importance of persistent spring snow cover is based on Copeland et al. (2010), which found that all 562 reproductive dens from Fennoscandia and North America occurred at sites with persistent spring snow cover through the end of the wolverine's reproductive denning period (24 April to 15 May). Note that predicted habitat extends into northern New Mexico. Reproduced with permission from Colorado Parks and Wildlife.

the insulation it provides to the kits (Magoun and Copeland 1998). Favored den sites may be used repeatedly (Magoun and Copeland 1998).

Wolverines are usually viewed as requiring remote undisturbed wilderness and avoiding human disturbance (Pasitschniak-Arts and Larivière 1995). However, this might be a spurious pattern resulting from wolverines selecting habitats that tend to be inhospitable to humans (Copeland et al. 2007). For instance, in Scandinavia wolverine home ranges avoided human development, but that result could have been because most of the human infrastructure was in forested valley bottoms, which are naturally avoided by wolverines (May et al. 2006). Within home ranges, wolverines did not avoid human structures such as houses and cabins (May et al. 2006). Similarly, Copeland et al. (2007) found that wolverines were less likely to use areas near roads, but they speculated that pattern might have been because most of the roads were in low-elevation, peripheral regions of the study area. They also found that wolverines did not avoid human trails and that they used unmaintained winter roads used for snowmobile access to trapping sites (Copeland et al. 2007). Recent research, however, suggests

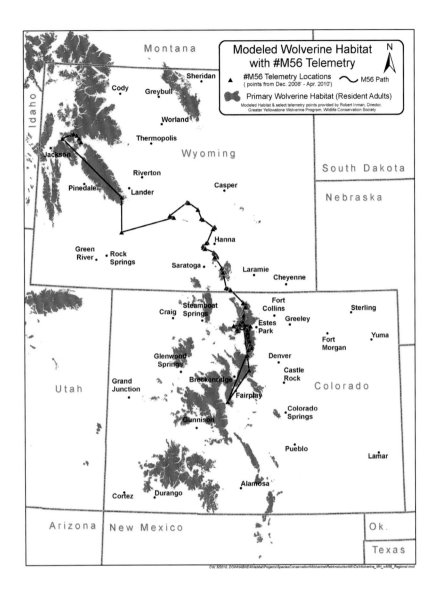

Map 18.3. In late May 2009, a radio-tagged subadult male wolverine (M56) from Wyoming near Grand Teton National Park travelled more than 900 km (~560 mi) south to Colorado. Telemetry data collected on M56 demonstrated that modeled habitat matched his habitat use. The wolverine remained in Colorado for over two years, and his travels took him to within 241 km (150 mi) of New Mexico. Reproduced with permission from Colorado Parks and Wildlife.

that wolverines do indeed avoid many types of human disturbance. Scrafford et al. (2018) confirmed that not only did wolverines avoid industrial roads, they also changed their behavior (i.e., increased their movement speed) near roads, especially as traffic volume increased. Stewart et al. (2016) also found that wolverines behaved differently in heavily human-modified environments in that they spent less time at some sites, possibly because they perceived them as less safe. Wolverines avoid areas where humans engage in backcountry winter recreation, especially off-road recreation, and for females, especially off-road motorized recreation (Krebs et al. 2007;

Heinemeyer et al. 2019). In the Rocky Mountains of Alberta, wolverines were less likely to occur at sites where the forest had burnt, or where timber harvesting or petroleum exploration had occurred (Fisher et al. 2013). Even during dispersal, wolverines avoid human disturbance, which is exemplified by the path taken by M56 as he made his way from near Grand Teton National Park, Wyoming, to Rocky Mountain National Park, Colorado. Traffic on Wyoming Highway 130 caused M56 to stop and detour along an alternate route to cross elsewhere. His travels took him to the edge of a small housing development, which he immediate skirted by 0.8 km (0.5 mi) and then continued on his way south (Inman et al. 2009).

Information about dispersal habitat for wolverines in the western United States is important, given the species' natural rarity and propensity for long-distance travel, and the fragmented distribution of core habitat (Carroll et al. 2020; Lukacs et al. 2020). Dispersal is essential not only for restoring populations that were extirpated, but also for maintaining the genetic health of populations. Carroll et al. (2020) used circuit theory to understand dispersal habitat across the historical range of the wolverine in the western United States. They found that while wolverines strongly select specific habitat features for establishing home ranges, they also use a broader array of habitats for dispersal. Yet, during dispersal wolverines do not move indiscriminately across landscapes, but rather use pathways of lower habitat resistance (Carroll et al. 2020). Gene flow among populations of wolverines in the Northern Rocky Mountains region was associated with high snow depth, high terrain ruggedness, and low housing density (Balkenhol et al. 2020). Of these three habitat linkage attributes, snow depth was the most important overall, as well as at small scales within habitat complexes (Balkenhol et al. 2020). Across broad scales, which are used for long-distance dispersal in environments dissimilar to core habitat, low housing density and terrain ruggedness were the most influential factors on genetic structure (Balkenhol et al. 2020).

LIFE HISTORY
Diet and Foraging

The wolverine's morphology, behavior, and life history are uniquely adapted to take advantage of cold regions with low productivity (Inman et al. 2012a). Like other mustelids, the wolverine is mostly carnivorous, and it is opportunistic in its food selection. Thus, there is a long "laundry list" of food items that may be consumed by wolverines, including carrion, especially that of ungulates; ungulates that are young, injured, or mired in deep powdery snow, such as deer (*Odocoileus*), elk (*Cervus canadensis*), and bighorn sheep (*Ovis canadensis*); medium-sized mammals such as marmots (*Marmota*), beaver (*Castor*), muskrat (*Ondatra zibethicus*), porcupine (*Erethizon dorsatum*), snowshoe hare, and North American red fox (*Vulpes fulva*); small mammals such as ground squirrels (e.g., *Callospermophilus*, *Neotamias*), voles (*Microtus* and *Myodes*), and shrews (*Sorex*); birds such as geese (*Chen*), magpies (*Pica*) and ptarmigans (*Lagopus*); fish; insects; and berries (e.g., *Vaccinium*; *Arctostaphylos*) (Banci 1994; Pasitschniak-Arts and Larivière 1995)—all of which are available in New Mexico! However, such laundry lists belie the unique specializations of the wolverine's diet and how the species' food habits are intimately tied to its reproduction, demography, habitat selection, and distribution.

In the wolverine's cold environment, food availability may vary radically by season. Overall, large ungulate carrion is the primary food of wolverines (Dalerum et al. 2009). However, ungulates may only be present in wolverine habitat during the brief growing season, and then they migrate to winter ranges below the wolverine's elevational zone. A key adaptation to contend with this variation in ungulate presence is scavenging. Ernest Thompson Seton even referred to the wolverine as the hyena of the North, due to its role as a scavenger of large ungulate carcasses, similar to the role of hyenas scavenging ungulates killed by lions (*Panthera leo*) in Africa. The wolverine's massive skull and cheek teeth enable it to crush and eat bones of adult ungulates, while the well-developed carnassial teeth allow it to shear off and consume frozen tissue or old, leathery skin (Banci 1994; Pasitschniak-Arts and Larivière 1995). So specialized are wolverines at eking out a living in a desolate environment that at times their only food might be old bones (Banci 1994). Another key adaptation for survival is that wolverines cache food in cool protected places such as among boulders or under tree boughs in dense conifer forest (Samelius et al. 2002; Wright and Ernst 2004). Cool protected cache sites help to protect the food from competitors, including both decay organisms and other carnivores (Inman et al. 2012a). Caching appears to be an innate behavior and one wolverines engage in regardless of the season (Inman et al. 2012a). In a study in Scandinavia, wolverines cached perishable food, both carrion and prey, year-round and at multiple locations throughout their home ranges, relying on it both in the short-term and long-term (Van der Veen et al. 2020).

Successful reproduction requires vastly more energy than sheer existence. During the brief growing season in the climates wolverines favor, limited food availability can hamper reproduction and even cause starvation (Inman et al. 2012a). Thus, food availability is tied to wolverine demographic and other life history parameters including reproductive rates, territoriality, home range size, and densities. Inman et al. (2012a) proposed that lactation, which is the most energetically costly period (February–April), is fueled by food caches amassed over the winter. In fact, the need to refrigerate food caches for long-term storage may be the ultimate reason wolverines are associated with areas of persistent spring snow cover, which is known as the "refrigeration-zone" hypothesis (Inman et al. 2012a).

Photo 18.10. The yellow-bellied marmot (*Marmota flaviventris*; photographed at Williams Lake near Wheeler Peak, Taos County on 26 July 2011) is a large alpine ground squirrel that serves as important prey for wolverines where these occur. Photograph: © James N. Stuart.

Although wolverines are scavenging specialists, they can hunt and kill animals as prey. In general, most of the scavenging in non-coastal regions involves large ungulates and occurs in the winter, while most of the predation is on birds, small mammals, and neonatal ungulates, and it occurs in the summer (Inman and Packila 2015). It is thought that the critical post-weaning growth period, when females are provisioning the young, is dependent on such hunting (Inman et al. 2012a). It is during the brief summer growing season that voles, ground squirrels, and marmots are especially susceptible to predation since they are either exposed due to lack of snow cover (voles) or out of hibernation (ground squirrels and marmots). Reliance on important prey species such as snowshoe hares, porcupines, beavers, and marmots appears to depend on their respective abundance in any given region (Lofroth et al. 2007). For instance, in Alberta wolverines used large food items (ungulate carcasses or killed beavers) mainly in the spring (Scrafford and Boyce 2018). Marmots may be a particularly important prey for denning females (Lafroth et al. 2007), but wolverines also might dig marmots out of hibernation burrows (Hornocker and Hash 1981). In Scandinavia, wolverines preyed on reindeer in the summer when calves were abundant, but they scavenged more in the winter when carrion was more readily available, particularly in the presence of other predators that provided carcasses (Mattisson et al. 2016). Thus, in areas where larger predators, such as wolves or cougars, are present, the summer diet may also continue to include scavenged, large ungulates.

Wolverines have been recorded as successfully killing larger ungulates, though the vast majority of such records have involved the Eurasian wolverine (*G. g. gulo*), which is larger than the North American wolverine (*G. g. luscus*). Wolverine predation on larger ungulates likely occurs mainly in deep snow. However, Magoun et al. (2018) reported that a North American wolverine was able to successfully prey on caribou (*Rangifer tarandus*) by pursuing them for long distances (4–62 km [2.5–38.5 mi]) to exhaustion. Nearly all reports of wolverines killing livestock (i.e., domestic sheep and domesticated reindeer) have been from Eurasia (Banci 1994; Landa et al. 1999). I am aware of only one report of wolverine killing livestock (i.e., sheep) in North America (CDOW 2010).

Reproduction and Social Behavior

As with other mustelids, wolverines are polygamous (one male mates with several females) and mating usually occurs in summer. However, the fertilized egg does not directly develop into a fetus. Rather, it ceases development at the eight-cell stage until it implants into the uterus in early winter and then continues to develop (i.e., a process called delayed implantation). Wolverines usually give birth in late February (average February through mid-March), which is the earliest birth date for any non-hibernating northern carnivore (Inman et al. 2012b). Litter sizes are usually one to four (Pasitschniak-Arts and Larivière 1995). Reproductive dens are built under snow and hence they must later be abandoned, usually in late April or early May, due to snowmelt (Banci 1994; Magoun et al. 2017). As a result, while the kits are still too young to travel when the mother is hunting, they are left behind at rendezvous sites. Once old enough to travel, they stay with the mother during the growing season, but in November, at an age of only about 6 months old, they disperse and start living independently (Pasitschniak-Arts and Larivière 1995; Chadwick 2010). Females usually can produce their first litter when two years old, while males do not gain sexual maturity until at least two years of age and some do not produce sperm until more than two years old (Banci 1994; Pasitschniak-Arts and Larivière 1995). Wolverines usually disperse away from the mother's home range by the age of two years old with the onset of sexual maturity (Chadwick 2010). Although most mature females might become pregnant each year, litter loss can be high. Reproductive success is probably mostly a function of winter food availability (Banci 1994; Persson 2005).

To a large extent the social life (or lack thereof) of wolverines reflects adaptations to living in an environment with limited food resources. Wolverines have exceptionally large home ranges, especially when considered relative to their body size; in the Greater Yellowstone Ecosystem, male home ranges averaged 797 km² (308 mi²) while females averaged 303 km² (117 mi²) (Inman et al. 2012b). Boundaries of territories are marked with urine, excrement, and scent from anal glands (Koehler et al. 1980). Home ranges generally do not overlap among same-sex individuals, though a large male home range usually overlaps the home ranges of several females with which it breeds. Wolverines have high movement rates, and they regularly patrol their entire home range—they can cover the distance equivalent to the perimeter of their home range in less than a week (Inman et al. 2012b).

Photo 18.11. An adult female wolverine and her cubs foraging together on 15 July 2007 in the Greater Yellowstone Ecosystem in southwestern Montana. This type of talus slope is typical summer habitat for wolverines hunting yellow-bellied marmots (*Marmota flaviventris*). Photograph: © Mark Packila.

Traditionally, wolverines have been considered solitary animals. However, recent studies have suggested more social interactions among family members than once realized. For instance, yearling females have been observed to use their mother's home range as they explore the region, typically setting up their own home range near the mother's (Chadwick 2010). This implies that not only the mother but also the resident male (likely the father) tolerate the presence of the yearling. This helps to ensure that the yearling can gain experience and independence without having to risk contending with intraspecific aggressions with older resident animals. More surprising is the observation that young wolverines of either sex occasionally meet and travel with either their mother or father. The observation of young animals that have already separated from their mothers and visiting the father is almost unheard of in solitary, territorial animals (Chadwick 2010).

Interspecific Interactions

Wolverines scavenge on large ungulates killed by more proficient predators such as wolves and cougars. It is thought that the extirpation of wolves, in particular, has had a negative impact on wolverines (Van Dijk et al. 2008a). By the same token, restoration of wolf populations in some regions may have led to recolonization by wolverines (Van Dijk et al. 2008a). Wolves benefit wolverines by providing them with a source of carrion and allowing them to shift their diet away from smaller animals, which they otherwise typically must hunt (Van Dijk et al. 2008b). This may be especially important for reproductive females since the availability of large ungulate carrion negates the need to expend energy hunting (Van Dijk et al. 2008b). However, there is a trade-off because wolves also represent the wolverine's main predator (Pasitschniak-Arts and Larivière 1995). Wolverines appear to avoid direct confrontations with wolves by largely restricting themselves to higher elevations and avoiding wolf trails; rather, they make only short excursions to lower elevations to find wolf kills and thus exhibit temporal (as well as spatial) separation from wolves (Van Dijk et al. 2008a). This may explain why wolverine cache sites usually are located in places with a good field of view, rugged terrain, or near a large "climbing tree" that is used for escape (Wright and Ernst 2004; Van der Veen et al. 2020). Wolverines also cache the carcasses of animals killed by other predators closer to the kill site, which may allow the wolverine to make multiple trips to cache the food while the other carnivore is away. Wolverines may follow red fox (*Vulpes* spp.) trails to locate carcasses of animals killed by wolves (Van Dijk et al. 2008a), which suggests that red foxes do not pose a threat to wolverines. A synthesis of mortality in wolverines during 12 radio-telemetry studies in North America found that 11 (18%) of 62 deaths were due to predation by wolves, cougars, and other wolverines (Krebs et al. 2004). Part of the reason wolverines might select rugged high elevations is to lessen the risk of predation (Magoun and Copeland 1998).

STATUS AND MANAGEMENT

Since 1800 the distribution of the wolverine has significantly contracted, particularly along its southern range margins (Aubry et al. 2007; Ruggiero et al. 2007). Evidence suggests that the wolverine was likely extirpated from the entire contiguous United States by the early 1900s with subsequent dispersal from source populations in Canada allowing recolonization of some historical habitat (McKelvey et al. 2014). However, that process of natural recolonization is not complete and is stymied by the fragmented nature of core habitat and long distances over inferior habitat that must be crossed to occupy some historical core habitat such as the Sierra Nevada and Southern Rocky Mountains. The primary cause of the decline of wolverine populations was killing by humans, including overexploitation through

hunting and trapping as well as predator control activities (Banci 1994; Frey 2006; Aubry et al. 2007). New Mexico does not have any laws that specifically cover the wolverine because it was only recently recognized as having historically occurred in the state, not to mention that it is now considered extirpated.

Since the mid-1990s there have been several attempts to list wolverines in the contiguous United Stated under the federal Endangered Species Act (ESA), most of which failed due either to a lack of information or technical disputes, namely whether the contiguous United States represents a "distinct population segment" or a significant portion of the wolverines' range (USFWS 2010). In December 2010, based on new information and a re-examination of the evidence, the US Fish and Wildlife Service found that the wolverine in the contiguous United States did in fact represent a distinct population segment and that it did warrant being added to the Endangered Species Act, though the listing was precluded by higher priority actions (USFWS 2010). Within the petition finding, New Mexico was regarded as a region of historical occurrence where "wolverines . . . existed as reproducing and potentially self-sustaining populations prior to human-induced extirpation, and where re-establishment of those populations is possible given current habitat condition and management" (USFWS 2010:78036). In 2013, the Fish and Wildlife Service proposed listing the distinct population segment of the wolverine in the contiguous United States as threatened and proposed establishment of a non-essential experimental population in Colorado, Wyoming, and New Mexico to allow for federal reintroduction efforts of wolverines into the Southern Rocky Mountains (USFWS 2013a,b). However, in 2014 the Fish and Wildlife Service withdrew the proposal to list the wolverine as threatened, along with the proposed non-essential experimental population (USFWS 2014). In 2016, a court overturned the decision to

Photo 18.12. Male wolverine on snow in Lewis and Clark County, Montana (6 April 2016). Southwestern Canada and Montana serve as the source population from which wolverines can disperse to reoccupy historical habitat in the western United States. Photograph: © Kalon Baughan.

withdraw the 2014 listing proposal and remanded the decision back to the Fish and Wildlife Service for further consideration. Most recently, in October 2020, the Fish and Wildlife Service withdrew the 2013 listing proposal based on its analysis indicating that wolverine populations in the northwestern part of the contiguous United States are stable and that wolverines move across the Canadian border (USFWS 2018, 2020).

The future for wolverines in New Mexico will likely depend on what happens in Colorado. Colorado occupies the majority of the Southern Rocky Mountains (the remainder is in southern Wyoming and northern New Mexico), and it has been estimated that the Southern Rocky Mountains could support approximately 137 wolverines, a significant number given that the current estimated population size in the western United States is only 318 (Inman et al. 2013). Colorado has had a long-standing interest in restoring the wolverine within its boundaries. In 1997 a team of

interested biologists from various state and federal agencies formed CLAWS, the Colorado Lynx and Wolverine Strategy, a group that worked to develop strategies for the restoration of the two forest carnivores in Colorado (Seidel 1998a). In 1998 a decision was made by the CLAWS team to proceed with introductions of both species, and in November of that year the Colorado Wildlife Commission formally approved the plan. In part, the Commission was influenced by its desire to maintain some state control over the Canada lynx (*Lynx canadensis*), which at the time was under consideration for listing as a federally endangered species (Seidel 1998c; Chapter 6). The Colorado Cattlemen's Association opposed the reintroduction of both species due to worries that the Fish and Wildlife Service could reduce grazing allotments (Seidel 1998c). Four areas were evaluated as potential wolverine release sites. The area selected for the initial release of wolverines was the San Juan and Rio Grande national forests in the San Juan Mountains, which abut the New Mexico border in southwestern Colorado. It was chosen due to its large size, the large extent of wilderness and roadless areas within its boundaries, moderate levels of human impact, the presence of elk, and abundant potential den sites (Siedel 1998b). The release of wolverines, however, was to be delayed until 1999–2000 so that immediate efforts could be concentrated on restoring lynx, both because source populations of lynx happened to be peaking at that time and due to limited funding and manpower (Seidel 1998bc). The planned reintroduction of wolverines never occurred.

In July 2010, and perhaps inspired by M56, Colorado Parks and Wildlife was granted permission to once again explore the potential for restoring wolverines to Colorado with stakeholders, and a draft plan for wolverine reintroduction was produced in November of that year (CDOW 2010). The draft plan was developed as a source of information to aid in planning and decision-making. The stated goals of the reintroduction were to "establish a viable population of wolverine in Colorado which would contribute to the rangewide conservation of the species, serve as a continental-level refuge for the species in response to potential effects of climate change, which thus could offset the need for protections under the Endangered Species Act" (CDOW 2010:2). It called for the reintroduction of up to 36 wolverines beginning in 2012 (CDOW 2010). A primary opponent to the plan was the ski industry due to concerns that existing operations or future expansions onto federal land might be restricted due to presence of wolverines, especially if they were to become listed under the federal Endangered Species Act (Finley 2010). The wolverine restoration plan was put on hold pending the outcome of the federal listing process (Finley 2012). Most recently, in October 2020, following the announcement that the Fish and Wildlife Service was withdrawing its listing proposal, the Director of Colorado Parks

and Wildlife stated that the agency was once again interested in restoring wolverines to the state of Colorado (Boster 2020).

Within the contiguous United States, trapping is considered the primary source of additive mortality for wolverines, and as such it can reduce a population's density (Krebs et al. 2004; Squires et al. 2007). The largest population of wolverines in the contiguous United States is found in Montana due to its nearness to populations in Canada (Cegelski et al. 2003), and hence it is the only state that allowed trapping for wolverines in recent decades (Banci 1994; USFWS 2018). Since 2012, however, Montana has not allowed trapping for wolverines. This reversal follows a petition that was filed to end wolverine trapping in Montana until the Fish and Wildlife Service determined whether the wolverine would be listed under the Endangered Species Act (Bishop and Hernandez 2012). As of 2021, the trapping season in Montana (and elsewhere in the contiguous United States) remained closed. However, regulated trapping for wolverines does occur in Alaska and Canada. Recent research found that wolverine trapping in southern British Columbia and Alberta was not sustainable, and it was recommended that harvest be reduced by at least 50% to allow populations to recover (Mowat et al. 2019). Such actions are relevant to New Mexico because southwestern Canada and Montana serve as the source population from which wolverines can disperse to reoccupy historical habitat in the western United States (Cegelski et al. 2003; Inman et al. 2013).

Given that persistent spring snow cover is critical to maintaining viable populations of wolverines, climate change has been recognized as an important threat to the species (Copeland et al. 2010; McKelvey et al. 2011a; Inman et al. 2012b; and see Chapter 5). Wolverines use snow for reproductive denning, food acquisition and preservation, avoidance of predators and competitors, and traveling, the last providing connectivity among populations, genetic cohesion, and the potential for colonization of historical habitat (Schwartz et al. 2009; Copeland et al. 2010; Inman et al. 2012a,b; Carroll et al. 2020). A warming climate that reduces the availability of snow, particularly persistent spring snow, is therefore a threat to the species' persistence. If carbon dioxide emissions persist at the medium-low to high levels, it has been predicted that there may be little or no spring snow cover by 2050 in the Northern Rocky Mountains where wolverines currently occur (Peacock 2011). Further, maximum daily temperatures are predicted to exceed 32° C (90° F) during most of August, but wolverine habitat is associated with mean August temperatures less than 22° C (72° F; Peacock 2011; see Chapter 5). Consequently, it has been predicted that the wolverine's range will be reduced by 23% over the next 30 years and by 75% over the next 75 years (USFWS 2023). Thus, by 2045 it is projected that maintaining current populations in the contiguous United States would require humans to move animals among subpopulations to offset the greater inability for natural dispersal (USFWS 2023). Brodie and Post (2010, 2011) postulated that a recent decline in Canadian wolverines was due to reduced average snowfall, though this conclusion has been disputed (DeVink et al. 2011, McKelvey et al. 2011b). The Southern Rocky Mountains may represent a key to the long-term viability of wolverines in the western United States. Although this region is at the southern edge of the wolverine's distribution, it also has the highest average elevation of any area south of the Canadian border. Thus, the high peaks of the Southern Rockies may provide an important refugium from a warming climate (Peacock

2011). For example, models for another alpine mammal that co-occurs with the wolverine, the American pika (*Ochotona princeps*), predicted the smallest proportion of habitat loss in the southern Rocky Mountains and Sierra Nevada and the greatest habitat loss in the Northern Rocky Mountains (Calkins et al. 2012).

As far as documented by science, wolverines have been absent from New Mexico for more than 160 years. However, with their progressive recovery in the contiguous United States, and the recent attention given to the species by natural resource agencies, scientists, and the public, the prospects seem hopeful that the wolverine may in the future once again roam the high mountains of northern New Mexico. Of course, that likelihood will be increased if a restoration program is initiated in Colorado. But even if not, M56 taught a lesson that the seemingly impossible can become possible as a result of the wolverine's tremendous capacity for long-distance travel. Therefore, naturalists in New Mexico should continue to remain astute to the possible rare presence of wolverines in the state. What a treat it will be to the lucky observer who is able to see this most unique carnivore—keep your camera at the ready.

ACKNOWLEDGMENTS

I thank Octaviano Lucero for assistance compiling literature, Jim Stuart for providing information on the legal status of wolverines in New Mexico, Robert Fisher and Suzy Puerach for details on the Fort Burgwin record, and Meghan Riley for helpful comments on an earlier version of the chapter.

LITERATURE CITED

Armstrong, D. M. 1972. *Distribution of mammals in Colorado*. Monograph of the Museum of Natural History, University of Kansas Number 3.

Arngart, O. 1979. The word "wolverine." *Notes and Queries* 26:494–95.

Aubry, K. B., K. S. McKelvey, and J. P. Copeland. 2007. Distribution and broadscale habitat relations of the wolverine in the contiguous United States. *Journal of Wildlife Management* 71:2147–58.

Aubry, K. B., J. Rohrer, C. M. Raley, R. D. Weir, and S. Fitkin. 2012. *Wolverine distribution and ecology in the North Cascades Ecosystem*. 2012 Annual Report 21 November 2012. http://wolverinefoundation. org/wp-content/uploads/2011/02/ncws_2012annual_rept.pdf

Bailey, V. 1931 (=1932). *Mammals of New Mexico*. North American Fauna 53. Washington, DC: US Department of Agriculture, Bureau of Biological Survey.

Balkenhol, N., M. K. Schwartz, R. M. Inman, J. P. Copeland, J. S. Squires, N. J. Anderson, and L. P. Waits. 2020. Landscape genetics of wolverines (*Gulo gulo*): scale-dependent effects of bioclimatic, topographic, and anthropogenic variables. *Journal of Mammalogy* 101:790–803.

Banci, V. 1994. Wolverine. In *The scientific basis for conserving forest carnivores: American marten, fisher, lynx and wolverine in the western United States*, ed. L. F. Ruggiero, K. K. Aubry, S. W. Buskirk, L. J. Lyon, and W. J. Zielinski, 99–123. Fort Collins, CO: US Forest Service, Gen. Tech. Rep. RM-254.

Billings, E. 1856. Natural history of the wolverine or caracajou (*Gulo luscus*). *Canadian Naturalist and Geologist* 1(4):241–46.

Bishop, J. E. 2006. Comment dit-on *tchistchimanisi8* en francais? The translation of Montagnais ecological knowledge in Antoine Silvy's *Dictionnaire montagnais-francais* (ca. 1678–1684). MA essay, Memorial University of Newfoundland, St. John's.

Bishop, M., and S. Hernandez. 2012. Petition to end the trapping of wolverine (*Gulo gulo*) in Montana. Submitted to Montana Fish Wildlife & Parks commission, Montana Department of Fish, Wildlife and Parks, and Joe Mauirer, Director of Montana Department of Fish, Wildlife and Parks by the Western Environmental Law Center, 31 July 2012.

Boster, S. 2020. On the heels of wolf debate in Colorado, could wolverines be next? *Gazette*, 7 December 2020 (updated 11 January 2022). https://gazette.com/life/on-heels-of-wolf-debate-in-colorado-could-wolverines-be-next/article_f5261168-2dd7-11eb-8956-47c054c2b455.html

Brodie, J. F., and E. Post. 2010. Nonlinear responses of wolverine populations to declining winter snowpack. *Population Ecology* 52:279–87.

———. 2011. Wolverines and declining snowpack: response to comments. *Population Ecology* 53:267–69.

Calkins, M. T., E. A. Beever, K. G. Boykin, J. K. Frey, and M. C. Andersen. 2012. Not-so-splendid isolation: modeling climate-mediated range collapse of a montane mammal *Ochotona princeps* across numerous ecoregions. *Ecography* 35:780–91.

Carrano, M. T. 1997. Morphological indicators of foot posture in mammals: a statistical and biomechanical analysis. *Zoological Journal of the Linnean Society* 121:77–104.

Carroll, K. A., A. J. Hansen, R. M. Inman, R. L. Lawrence, and A. B. Hoegh. 2020. Testing landscape resistance layers and modeling connectivity for wolverines in the western United States. *Global Ecology and Conservation* 23:e01125.

Cary, M. 1911. *A biological survey of Colorado*. North American Fauna 33. Washington, DC: US Department of Agriculture, Bureau of Biological Survey.

Cegelski, C. C., L. P. Waits, and N. J. Anderson. 2003. Assessing population structure and gene flow in Montana wolverines (*Gulo gulo*) using assignment-based approaches. *Molecular Ecology* 12:2907–18.

Chadwick, D. H. 2010. *The wolverine way*. Ventura, CA: Patagonia Books.

Colorado Division of Wildlife (CDOW). 2010. Draft plan to reintroduce wolverine to Colorado. Unpublished white paper, draft 11/4/2010, 28 pp.

Copeland, J. P., K. S. McKelvey, K. B. Aubry, A. Landa, J. Persson, R. M. Inman, J. Krebs, E. Lofroth, H. Golden, J. R. Squires, A. Magoun, M. K. Schwartz, J. Wilmot, C. L. Copeland, R. E. Yates, I. Kojola, and R. May. 2010. The bioclimatic envelope of the wolverine (*Gulo gulo*): do climatic constraints limit its geographic distribution? *Canadian Journal of Zoology* 88:233–46.

Copeland, J. P., J. M. Peek, C. R. Groves, W. E. Melquist, K. S. McKelvey, G. W. McDaniel, C. D. Long, and C.-E. Harris. 2007. Seasonal habitat associations of the wolverine in central Idaho. *Journal of Wildlife Management* 71:2201–12.

Coues, E. 1877. Fur-bearing animals: a monograph of North American Mustelidae. *U.S. Geologic and Geographical Survey of the Territories, Miscellaneous Publications* 8:1–348.

Dalerum, F., K. Kunkel, A Angerbjorn, and B. S. Shults. 2009. Diet of wolverines (*Gulo gulo*) in the western Brooks Range, Alaska. *Polar Research* 28:246–53.

DeVink, J., D. Berezanski, and D. Imrie. 2011. Comments on Brodie and Post: harvest effort: the missing covariate in analyses of furbearer harvest data. *Population Ecology* 53:261–62.

Dick-Peddie, W. A, ed. 1993. *New Mexico vegetation: past, present, and future*. Albuquerque: University of New Mexico Press.

Findley, J. S., A. H. Harris, D. E. Wilson, and C. Jones. 1975. *Mammals of New Mexico*. Albuquerque: University of New Mexico Press.

Finley, B. 2010. Colorado ski group wary of possible reintroduction of wolverines with protected status. *Denver Post*, 21 December 2010. http://www.denverpost.com/news/ci_16907832.

———. 2012. Wolverine M56 goes solo in Colorado as feds mull endangered status. *Denver Post*, 11 November 2012. http://www.denverpost.com/environment/ci_21973904/wolverine-m56-goes-solo-colorado-feds-mull-endangered.

Fisher, J. T., S. Bradbury, B. Anholt, L. Nolan, L. Roy, J. P. Volpe, and M. Wheatley. 2013. Wolverines (*Gulo gulo luscus*) on the Rocky Mountain slope: natural heterogeneity and landscape alteration as predictors of distribution. *Canadian Journal of Zoology* 91:706–16.

Frey, J. K. 2004. Taxonomy and distribution of the mammals of New Mexico: an annotated checklist. *Occasional Papers, Museum of Texas Tech University* 240:1–32.

———. 2006. Inferring species distributions in the absence of occurrence records: an example considering wolverine (*Gulo gulo*) and Canada lynx (*Lynx canadensis*) in New Mexico. *Biological Conservation* 130:16–24.

Frey, J. K., and T. L. Yates. 1996. Mammalian diversity in New Mexico. *New Mexico Journal of Science* 36:4–37.

Grinnell, G. B. 1926. Some habits of the wolverine. *Journal of Mammalogy* 7:30–34.

Hall, E.R. 1981. *The mammals of North America*. 2nd ed. Vol. 2. New York: John Wiley.

Heinemeyer, K., J. Squires, M. Hebblewhite, J. J. O'Keefe, J. D. Holbrook, and J. Copeland. 2019. Wolverines in winter: indirect habitat loss and functional response to backcountry recreation. *Ecosphere* 10:e02611.

Hornocker, M. G., and H. S. Hash. 1981. Ecology of the wolverine in northwestern Montana. *Canadian Journal of Zoology* 59:1286–301.

Inman, R. M., B. L. Brock, K. H. Inman, S. S. Sartorius, B. C. Aber, B. Giddings, S. L. Cain, M. L. Orme, J. A. Fredrick, B. J. Oakleaf, K. L. Alt, E. Odell, and G. Chapron. 2013. Developing priorities for metapopulation conservation at the landscape scale: wolverines in the western United States. *Biological Conservation* 166:276–86.

Inman, R. M., A. J. Magoun, J. Persson, and J. Mattisson. 2012a. The wolverine's niche: linking reproductive chronology, caching, competition, and climate. *Journal of Mammalogy* 93:634–44.

Inman, R., and M. Packila. 2015. Wolverine (*Gulo gulo*) food habits in Greater Yellowstone. *American Midland Naturalist* 173:156–61.

Inman, R., M. Packila, K. Inman, B. Aber, R. Spence, D. McCauley. 2009. Greater Yellowstone Wolverine Program. Progress report, December 2009. Wildlife Conservation Society.

Inman, R. M., M. L. Packila, K. H. Inman, A. J. McCue, G. C. White, J. Persson, B. C. Aber, M. L. Orme, K. L. Alt, S. L. Cain, J. A. Fredrick, B. J. Oakleaf, and S. S. Sartorius. 2012b. Spatial ecology of wolverines at the southern periphery of distribution. *Journal of Wildlife Management* 76:778–92.

Jokinen, M. E., S. M. Webb, D. L. Manzer, and R. B. Anderson. 2019. Characteristics of wolverine (*Gulo gulo*) dens in the lowland boreal forest of north-central Alberta. *Canadian Field-Naturalist* 133:1–15.

Koehler, G. M., M. G. Hornocker and H. S. Hash. 1980. Wolverine marking behavior. *Canadian Field-Naturalist* 94:339–41.

Krebs, J., E. Lofroth, J. Copeland, V. Banci, D. Cooley, H. Golden, A. Magoun, R. Mulders, and B. Shults. 2004. Synthesis of survival rates and causes of mortality in North American wolverines. *Journal of Wildlife Management* 68:493–502.

Krebs, J., E. C. Lofroth, and I. Parfitt. 2007. Multiscale

habitat use by wolverines in British Columbia, Canada. *Journal of Wildlife Management* 71:2180–92.

Kvam, T., K. Overskaug, and O. J. Sorensen. 1988. The wolverine *Gulo gulo* in Norway. *Lutra* 31:7–20.

Landa, A., K. Gudvangen, J. E. Swenson, and E. Roskaft. 1999. Factors associated with wolverine *Gulo gulo* predation on domestic sheep. *Journal of Applied Ecology* 36:963–73.

Li, B., M. Wolsan, D. Wu, W. Zhang, Y. Xu, and Z. Zeng. 2014. Mitochondrial genomes reveal the pattern and timing of marten (*Martes*), wolverine (*Gulo*), and fisher (*Pekania*) diversification. *Molecular Phylogenetics and Evolution* 80:156–64.

Lofroth, E. C., and J. Krebs. 2007. The abundance and distribution of wolverines in British Columbia, Canada. *Journal of Wildlife Management* 71:2159–69.

Lofroth, E. C., J. A. Krebs, W. L. Harrower, and D. Lewis. 2007. Food habits of wolverine *Gulo gulo* in montane ecosystems of British Columbia, Canada. *Wildlife Biology* 2:31–37.

Lukacs, P. M., D. E. Mack, R. Inman, J. A. Gude, J. S. Ivan, R. P. Lanka, J. C. Lewis, R. A. Long, R. Sallabanks, Z. Walker, S. Courville, S. Jackson, R. Kahn, M. K. Schwartz, S. C. Torbit, J. S. Waller, and K. Carroll. 2020. Wolverine occupancy, spatial distribution, and monitoring design. *Journal of Wildlife Management* 84:841–51.

Magoun, A. J., and J. P. Copeland. 1998. Characteristics of wolverine reproductive den sites. *Journal of Wildlife Management* 62:1313–20.

Magoun, A. J., C. R. Laird, M. A. Keech, P. Valkenburg, L. S. Parrett, and M. D. Robards. 2018. Predation on caribou (*Rangifer tarandus*) by wolverines (*Gulo gulo*) after long pursuits. *Canadian Field-Naturalist* 132:382–85.

Magoun, A. J., C. D. Long, M. K. Schwartz, K. L. Pilgrim, R. E. Lowell, and P. Valkenburg. 2011. Integrating motion-detection cameras and hair snags for wolverine identification. *Journal of Wildlife Management* 75:731–39.

Magoun, A. J., M. D. Robards, M. L. Packila, and T. W. Glass. 2017. Detecting snow at the den-site scale in wolverine denning habitat. *Wildlife Society Bulletin* 41:381–87.

Magoun, A. J., and P. Valkenburg. 2012. Wallowa wolverine project: 2011–2012—April progress report. 2 May 2012, 6 pp.

Mattisson, J., G. R. Rauset, J. Odden, H. Andren, J. D. C . Linnell, and J. Persson. 2016. Predation or scavenging? Prey body condition influences decision-making in a facultative predator, the wolverine. *Ecosphere* 7:e01407.10.1002/ecs2.1407.

May, R., A. Landa, J. van Dijk, and R. Andersen. 2006. Impact of infrastructure on habitat selection of wolverines. *Wildlife Biology* 12:285–95.

McDonald, C. 1985. *Wilderness: a New Mexico legacy*. Sante Fe: Sunstone Press.

McKelvey, K. S., K. B. Aubry, N. J. Anderson, A. P. Clevenger, J. P. Copeland, K. S. Heinemeyer, R. M. Iman, J. R. Squires, J. S. Waller, K. L. Pilgrim, and M. K. Schwartz. 2014. Recovery of wolverines in the western United States: recent extirpation and recolonization or range retraction and expansion? *Journal of Wildlife Management* 78:325–34.

McKelvey, K. S., J. P. Copeland, M. K. Schwartz, J. S. Littell, K. B. Aubry, J. R. Squires, S. A. Parks, M. M. Elsner, and G. S. Mauger. 2011a. Climate change predicted to shift wolverine distributions, connectivity, and dispersal corridors. *Ecological Applications* 21:2882–97.

McKelvey, K. S., E. C. Lofroth, J. P. Copeland, K. B. Aubry, and A. J. Magoun. 2011b. Comments on Brodie and Post: Climate-driven declines in wolverine populations: causal connection or spurious correlation? *Population Ecology* 53:263–66.

Moriarty, K. M., W. J. Zielinski, A. G. Gonzales, T. E. Dawson, K. M. Boatner, C. A. Wilson, F. V. Schlexer, K. L. Pilgrim, J. P. Copeland, and M. K. Schwartz. 2009. Wolverine confirmation in California after nearly a century: native or long-distance immigrant? *Northwest Science* 83:154–62.

Mowat, G., A. P. Clevenger, A. D. Kortello, D. Hausleitner, M. Barrueto, L. Smit, C. Lamb, B. Dorsey, and P. L. Ott. 2019. The sustainability of wolverine trapping mortality in southern Canada. *Journal of Wildlife Management* 84:213–26.

Nead, D. M., J. C. Halfpenny, and S. Bissell. 1985. The status of wolverines in Colorado. *Northwest Science* 8:286–89.

Newberry, J. S. 1876. *Report of the exploring expedition from Santa Fe, New Mexico, to the junction of the Grand and Green Rivers of the great Colorado of the West, in 1859: under the command of Capt. J. N. Macomb, Corps of topographical engineers*. Washington, DC: Government Printing Office.

Packila, M. L., M. D. Riley, R. S. Spencer, and R. M. Inman. 2017. Long-distance wolverine dispersal from Wyoming to historic range in Colorado. *Northwest Science* 9:399–407.

Pasitschniak-Arts, M., and S. Larivière. 1995. *Gulo gulo*. *Mammalian Species* 499:1–10.

Peacock, S. 2011. Projected 21st century climate change for wolverine habitats within the contiguous United States. *Environmental Research Letters* 6:1–9.

Persson, J. 2005. Female wolverine reproduction: reproductive costs and winter food availability. *Canadian Journal of Zoology* 83:1453–59.

Platt, J. 1901. "Kinkajou." *Notes and Queries* 176:386.

Ruggiero, L., K. McKelvey, K. Aubry, J. Copeland, D. Pletscher, and M. Hornocker. 2007. Wolverine conservation and management. *Journal of Wildlife Management* 71:2145–46.

Samelius, G., R. T. Alisauskas, S. Larivière, C. Bergman, C. J. Hendrickson, K. Phipps, and C. Wood. 2002. Foraging behaviours of wolverines at a large arctic goose colony. *Arctic* 55:148–50.

Schwartz, M. K., J. P. Copeland, N. J. Anderson, J. R. Squires, R. M. Inman, K. S. McKelvey, K. L. Pilgrim, L. P. Waits and S. A. Cushman. 2009. Wolverine gene flow across a narrow climatic niche. *Ecology* 90:3222–32.

Scrafford, M. A., T. Avgar, R. Heeres, and M. S. Boyce. 2018. Roads elicit negative movement and habitat-selection responses by wolverines. *Behavioral Ecology* 29:534–42.

Scrafford, M. A., and M. S. Boyce. 2018. Temporal patterns of wolverine (*Gulo gulo luscus*) foraging in the boreal forest. *Journal of Mammalogy* 99:693–701.

Seidel, J., ed. 1998a. *Clawmark: a newsletter about the Colorado lynx and wolverine strategy for re-establishment of these forest carnivores* 1(1). Colorado Division of Wildlife.

———, ed. 1998b. *Clawmark* 1(2). Colorado Division of Wildlife

———, ed. 1998c. *Clawmark* 1(3). Colorado Division of Wildlife

Seton, E. T. 1909. *The life histories of northern animals: an account of the mammals of Manitoba*. New York: Scribner.

———. 1931. Two records for New Mexico. *Journal of Mammalogy* 12:166.

Squires, J. S., J. P. Copeland, T. J. Ulizio, M. K. Schwartz, and L. F. Ruggiero. 2007. Sources and patterns of wolverine mortality in western Montana. *Journal of Wildlife Management* 71:2213–20.

Stewart, F. E. C., N. A Heim, A. P. Clevenger, J. Paczkowski, J. P. Volpe, and J. T. Fisher. 2016. Wolverine behavior varies spatially with anthropogenic footprint: implications for conservation and inferences about declines. *Ecology and Evolution* 6:1493–503.

United States Fish and Wildlife Service (USFWS). 2010. Notice of 12-month finding on a petition to list the North American wolverine as endangered or threatened; proposed rule. *Federal Register* 75(239):78030–61 (14 December 2010).

———. 2013a. Threatened status for the distinct population segment of the North American wolverine occurring in the contiguous United States; proposed rule. *Federal Register* 78(23):7864–90 (4 February 2013).

———. 2013b. Establishment of a nonessential experimental population of the North American wolverine in Colorado, Wyoming, and New Mexico; proposed rule. *Federal Register* 78(23):7890–904 (4 February 2013).

———. 2014. Proposed Rules withdrawal; Threatened status for the distinct population segment of the North American wolverine occurring in the contiguous United States; Establishment of a nonessential experimental population of the North American wolverine in Colorado, Wyoming, and New Mexico. *Federal Register* 79(156):47522–545 (13 August 2014).

———. 2018. Species status assessment report for the North American wolverine (*Gulo gulo luscus*). Version 1.2. March 2018. US Fish and Wildlife Service, Mountain-Prairie Region, Lakewood, CO.

———. 2020. U.S. wolverines are healthy. US Fish and Wildlife Service. https://www.fws.gov/press-release/2020-10/us-wolverines-are-healthy.

———. [2023.] North American wolverine (*Gulo gulo luscus*). ECOS Environmental Conservation Online System. https://ecos.fws.gov/ecp/species/5123.

Van der Veen, B., J. Mattisson, B. Zimmermann, J. Oden, and J. Persson. 2020. Refrigeration or anti-theft? Food-caching behavior of wolverines (*Gulo gulo*) in Scandinavia. *Behavioral Ecology and Sociobiology* 74, article52. https://link.springer.com/article/10.1007/s00265-020-2823-4.

Van Dijk, J., T. Andersen, R. may, R. Andersen, R. Andersen, and A. Landa. 2008a. Foraging strategies of wolverines within a predator guild. *Canadian Journal of Zoology* 86:966–75.

Van Dijk, J., L. Gustavsen, A. Mysterud, R. May, Ø. Flagstad, H. Brøseth, R. Andersen, R. Andersen, H. Steen, and A. Landa. 2008b. Diet shift of a facultative scavenger, the wolverine, following the recolonization of wolves. *Journal of Animal Ecology* 77:1183–90.

Warren, E. R. 1942. *The mammals of Colorado: their habits and distribution*. 2nd ed. Norman: University of Oklahoma Press.

Wilson, D. E., and D. M. Reeder, eds. 2005. *Mammal species of the world: a taxonomic and geographic reference*. 3rd ed. Baltimore: Johns Hopkins University Press.

Wright, J. D., and J. Ernst. 2004. Wolverine, *Gulo gulo luscus*, resting sites and caching behavior in the boreal forest. *Canadian Field-Naturalist* 118:61–64.

NORTHERN RIVER OTTER (*LONTRA CANADENSIS*)

Melissa Savage and James N. Stuart

The northern river otter (*Lontra canadensis*), also called the North American river otter, Nearctic otter, or simply river otter, is a semi-aquatic member of the Mustelidae, or weasel family. The species formerly occurred in at least some perennial rivers in the state (Findley et al. 1975). Later considered to be extirpated in New Mexico, the river otter is now making a comeback through a combination of reintroduction efforts and natural dispersal from restored populations in adjacent states (Long 2010; Converse et al. 2014; Savage and Klingel 2015). Already, this lively carnivore can once again be encountered in some parts of northern New Mexico.

The river otter is a mainly piscivorous predator with a streamlined body plan (Melquist and Dronkert 1987). The body is long and cylindrical, or torpedo-like; the abdomen is the widest part of the body, and the neck is not distinctly narrower than the head. The long, stout tail tapers to a point and makes up about one-third of the animal's total length. The dense, short fur is typically a rich brown or blackish-brown, with a paler brown underside and the lower part of the face and the throat a pale gray or light brown. The head is somewhat flattened with a short muzzle and small ears, the eyes are large and dark colored, and the nose is black and broad. The legs of river otters are powerful and short in relation to the long, sleek body. The feet are five-toed and fully webbed with tufts of hair between the toes; the hind feet have four distinct callous pads which provide traction on slick surfaces. Males have a small scent gland between the pads of their hind feet that presumably is used to mark territories (Melquist et al. 2003).

The river otter is a medium-sized carnivore but fairly large for a member of the weasel family. Both female and male river otters average around 1.25 m (4 ft) in total length, from snout to tail tip. River otters have an average body mass ranging from about 5 to 14 kg (~11–31 lbs), with males weighing an average 5% to 17% more than females (Jackson 1961; Larivière and Walton 1998; Melquist and Dronkert 1987—body measurements exist for only one New Mexico specimen mentioned below). Males and females are otherwise identical in general appearance, and, because river otters are not easily handled, the gender of an individual usually cannot be determined unless the animal is immobilized, thereby allowing examination of the genitalia. The river otter propels itself swiftly in the water by undulating its muscular body and using its tail as a rudder as well as for propulsion. Physiological adaptations for its aquatic lifestyle include oxygen conservation measures such as bradycardia (slowing of the heartbeat), which allows a river otter to remain underwater for up to four minutes (Melquist and Dronkert 1987). On land, the

(*opposite page*) Photograph: © Dennis Buchner.

Photo 19.1. Northern river otter on a boulder at the Rio Grande del Norte National Monument in Taos County, 19 October 2019. The river otter is a streamlined semi-aquatic carnivore that consumes primarily fish and crustaceans. Photograph: © Britt Runyon.

Photo 19.2. Two northern river otters, in the Rio Grande in the Orilla Verde Recreation Area, Taos County on 14 March 2019. River otters are strong swimmers that propel themselves swiftly underwater by undulating their muscular body and tail. Photograph: © Surasit Khamsamran.

river otter can efficiently cover ground by walking or bounding, and by sliding across snow and ice (Melquist and Hornocker 1983).

The river otter can potentially be confused with other mammal species that occur or occurred until recently in New Mexico. It resembles the American mink (*Neogale vison*; Chapter 24), a species of mustelid that also frequents rivers and streams, though the mink can be distinguished by its much smaller size, darker brown fur, a longer and more tapered snout, and, often, a distinctive white patch of fur on its chin or chest. River otters are probably most often confused with two rodent species that have similar aquatic habits, the American beaver (*Castor canadensis*) and the muskrat (*Ondatra zibethicus*). The beaver

Photo 19.3. Three northern river otters running on the ice at the Orilla Verde Recreation Area in Taos County, 23 February 2021. While most adept in water, river otters move easily on land with a rocking gait and are able to disperse over some watershed boundaries. Photograph: © Gak Stonn.

is also brownish in coloration, but is bulkier and has a flattened, hairless, paddle-like tail, unlike the long, tapered, furred tail of the river otter. Muskrats are also brownish in overall color but are much smaller than either river otters or beavers and have a long, hairless, rat-like tail that is flattened from side to side. Both beavers and muskrats are often seen swimming at the water surface in rivers and streams where river otters also occur, sometimes leading to misidentification, especially when viewed from a distance. However, river otters usually appear more active and agile when swimming compared to the other two species.

The river otter is classified in the genus *Lontra* along with other New World species of river otters (Melquist et al. 2003). Its closest relative is the Neotropical otter (*L. longicaudis*), very similar in appearance and with a distribution that extends from northern Mexico to South America (Hall 1981). Up until about the 1980s, the New World species of river otters were classified in the genus *Lutra*, together with the Eurasian otter (*Lutra lutra*), but they are presently considered part of a lineage distinct from Old World otters (van Zyll de Jong 1972).

The taxonomy of geographic races within *L. canadensis* is poorly studied and needs to be genetically assessed (Melquist et al. 2003). Although seven subspecies were recognized by Hall (1981) in the United States and Canada, the validity and geographic limits of these various subspecies are debatable (van Zyll de Jong 1972; Serfass et al. 1998). Most native populations of the river otters that formerly occurred in New Mexico were assigned to the subspecies *L. c. sonora*, the southwestern or Arizona river otter, which was originally described by Rhoads (1898) from specimens collected in 1886 at Montezuma Well in the Verde River watershed of Arizona (Mearns 1891). This form, which apparently occurred throughout portions of Arizona, California, Colorado, Nevada, New Mexico, and

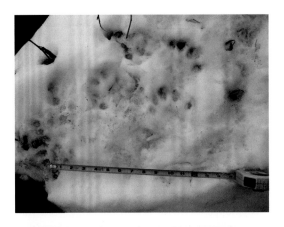

Photos 19.4a and b (*top to bottom*). Northern river otter tracks in the snow in the upper Rio Grande watershed in northern New Mexico. Tracks are about 7.5 cm (3 in) wide and 10 cm (4 in) long, and often show heel pad and claws. Webbing between toes is rarely visible. Photographs: © Brian Jay Long.

Utah, is now considered extinct (Frey and Yates 1996; Compton 2000; NMDGF 2006a; but see Gallo-Reynoso et al. 2019). In the late 1980s, when *L. c. sonora* was considered a candidate for federal protection under the Endangered Species Act, Bates (1988) suggested there was insufficient evidence to support the recognition of a southwestern subspecies. Bailey (1931) had earlier noted the difficulty in assessing geographic variation in river otters from the Southwest, given the paucity of specimens for study, but suggested that more than one species (or perhaps subspecies) occurred in New Mexico. Since few examples of *L. c. sonora* were preserved prior to its apparent disappearance, the issue of its taxonomic distinctiveness remains unresolved. Researchers in Mexico (Ceballos and Carillo-Rubio 2017; Gallo-Reynoso et al. 2019) have discovered what may be a relict population of the northern river otter in the upper Rio Conchos basin in the state of Chihuahua. The Rio Conchos is a tributary of the lower Rio Grande, thus raising the possibility that *L. c. sonora* might not be extinct as was previously thought. A genetic analysis is needed to confirm findings by Ceballos and Carillo-Rubio (2017) and Gallo-Reynoso et al. (2019).

The recognition of subspecies in the river otter is further complicated by reintroductions throughout the species' historical range. In the Southwest, the states of Arizona, Colorado, Utah, and, most recently, New Mexico have all pursued river otter release programs since the 1970s using animals obtained from widely divergent geographical locations in North America (Bricker et al. 2022; see under "Status and Management" below). Arizona reintroduced river otters from Louisiana (*L. c. lataxina*) into the Verde River (Britt et al. 1984). Colorado used animals from California, Oregon, Minnesota, Michigan, Wisconsin, Washington, Alaska, and Newfoundland, Canada (all *L. c. canadensis*) and Louisiana (*L. c. lataxina*) (CDW 2003). Utah translocated river otters from Alaska (*L. c. pacifica* or possibly *L. c. kodiacensis*) and Nevada (Raesly 2001; but see corrections to Raesly 2001 in Bricker et al. 2022). Initial reintroductions in the Rio Grande basin of New Mexico were of *L. c. pacifica* from Washington State (Stuart 2014), whereas supplemental releases in the same drainage beginning in 2021 were of *L. c. lataxina* from Louisiana (see under "Reintroduction and Monitoring").

Given the diverse geographic origins of the populations that now occur within the historical range of *L. c. sonora*, and the likelihood that interbreeding has occurred among at least some of them, the question of whether there was a southwestern subspecies may hinge on results of further research on the upper Conchas basin otter population in Mexico.

DISTRIBUTION

The historical distribution of *Lontra canadensis* included virtually all of Canada except the arctic tundra, the continental United States except for the southwestern deserts, and a few areas in northern Mexico, making it one of the most widely distributed of any North American mammal (Toweill and Tabor 1982; Melquist et al. 2003; and see Ceballos and Carillo-Rubio 2017 and Gallo-Reynoso et al. 2019). Over-harvesting for the fur trade combined with degradation of river habitat, water pollution, and urbanization led to a precipitous decline in populations of the species across much of its range by the early 20th century (Toweill and Tabor 1982; Route and Peterson 1988; and see under "Status and Management" below). By the mid-1900s, the species was extirpated from many locations in the contiguous United States (Toweill and Tabor 1982; Melquist and Dronkert 1987). However, reintroduction programs in 21 states and one Canadian province during the late 20th century, and in New Mexico from 2008 to 2010, together with reduced trapping pressure and improvements in habitat and water quality, set the stage for a broadly successful comeback across the species' historical range (Bricker et al. 2022). Northern river otters are now considered to have a wide distribution and to be locally abundant, especially along the Gulf Coast, in northern states, including Alaska, and throughout much of Canada.

In New Mexico, verified historical records of river otters are scarce (Findley et al. 1975). Before European settlement, river otters were known to some of the Native American peoples in the state (Polechla 2000; Polechla and Carrillo-Rubio 2009). Bailey (1931:324) noted that the "Taos Indians are familiar with them, and bits of river otter fur were seen on their clothing and ornaments as well as on those of the Jicarilla Apaches farther west." Although the name *nutria* ("otter" in Spanish) intriguingly appears in numerous New Mexico place names, it is likely that the term was applied to the beaver by the early Spanish-speaking colonists of the region (Julyan 1996). Among the first Europeans likely to have encountered river otters in the West were French or French-Canadian trappers, who arrived in the 18th and early 19th centuries (Scurlock 1998:119). In response to Zebulon Pike's 1810 account of his expedition into what at the time was still New Spain, Anglo-Americans arrived to trap and trade for furs (Scurlock 1998:119). The Old Santa Fe Trail was an early and important trade route for buffalo hides, mules, and the pelts of the American beaver and other furbearing animals that were destined for markets in the east. During the 19th century, the town of Taos, New Mexico became an important base for these trappers who were the very image of mountain men, living roughly off the backcountry (Beck 1962:107). By the 1920s and 1930s, river otters were very nearly trapped out of most watersheds, mostly as incidental catch in the beaver trade (deBuys 1985).

The earliest mention in the literature of the river otter in New Mexico seems to date to 1825 when the Pattie brothers took one in a beaver trap on the Gila River near the present-day Arizona–New Mexico border (Bailey 1931; Pattie 2001). For the starving trapping party, this animal "served for breakfast and supper" (Pattie 2001:95). Bailey (1931:328) wrote that "in 1906, the writer was told that there were still a few Arizona river otters along the headwaters of the Gila River in southwestern New Mexico." He also noted second-hand reports of river otters from "the upper Rio Grande and Canadian Rivers in the northeastern part of the state, where the species is probably different [than in the Gila]" (Bailey 1931:324). In 1846, a river otter was apparently killed (but not preserved) at the confluence of the Canadian and Mora rivers in San Miguel County, whereas on the upper Rio Grande the species was reported from "near Española, Rinconada, and Cienequilla [= Cieneguilla; probably a reference to present-day Pilar, Taos County]" (Bailey 1931:324).

The only native river otter ever verified in New Mexico based on a specimen was an adult male taken in a beaver trap in 1953 by a Department

Photo 19.5. Taxidermy mount of a native northern river otter trapped in 1953 on the Gila River south of Cliff in Grant County. No other verified, historical record of the northern river otter exists from New Mexico. As a result of intensive trapping throughout the 19th century, northern river otters were already considered very rare by the time New Mexico achieved statehood in 1912. The mounted specimen is an adult male preserved in the Museum of Southwestern Biology at the University of New Mexico. Photograph: © Jennifer Miyashiro.

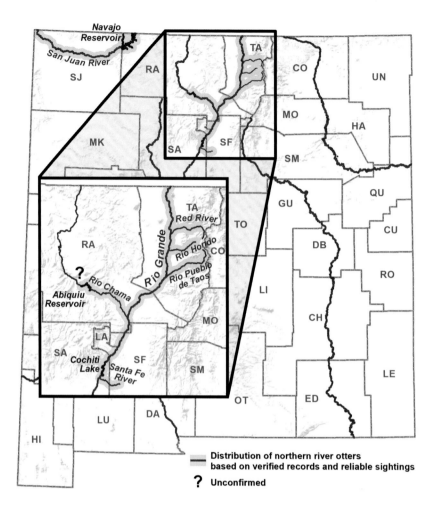

Map 19.1. Distribution of the reintroduced northern river otter in New Mexico. Following successful reintroductions, the species occurs in the San Juan and upper Rio Grande basins in the northern third of the state.

of Game and Fish trapper on the Gila River, 1.6 km (1 mi) south of the town of Cliff, Grant County (McClellan 1954). This animal, an adult male that weighed 11.3 kg (25 lbs) and measured 123.5 cm (48 5/8 in) total length, is now preserved as a taxidermy mount in the Museum of Southwestern Biology at the University of New Mexico (Photo 19.5). Despite various survey efforts and a number of eyewitness reports of varying credibility from the Rio Grande and Pecos River basins (reviewed by Stahlecker 1986; Polechla et al. 1994; Polechla 2000; NMDGF 2006a; Polechla and Carrillo-Rubio 2009), no additional verified

records were obtained from New Mexico during the latter half of the 20th century.

At present, the river otter is known to occur in two river systems in New Mexico, the upper Rio Grande, in the north-central part of the state, and the San Juan River in the northwestern corner of the state. River otters in the upper Rio Grande were reintroduced in recent years as part of a restoration program in New Mexico (see under "Status and Management" below), whereas the San Juan river otters are probably the result of migration from reintroduction sites in southwestern Colorado and/or Utah.

During the 1970s and 1980s, Colorado carried out releases of river otters in the Los Pinos and Piedra rivers, tributaries to the San Juan River that drain into Navajo Reservoir in New Mexico. Following these releases, river otters were observed occasionally in the Navajo Reservoir area on the Colorado–New Mexico border (CDW 2003) and confirmed there in the early 2000s based on genetic analysis of collected scat (i.e., feces) found on the reservoir shore (Polechla et al. 2004; Polechla and Walker 2008). River otters also have been recently verified on the Colorado River in Lake Powell, Arizona, downstream from its confluence with the San Juan River, not far below the New Mexico border (C. Day, pers. comm.). Since then, several reliable observations and photographs of river otters have been obtained from the San Juan River in New Mexico from the tailwater below Navajo Dam to as far downriver as the Shiprock area (NMDGF, unpubl. data; Map 19.1). Although there has been speculation that river otters in the San Juan River could represent a remnant native population (Polechla et al. 2004; Polechla and Walker 2008), the most logical explanation is that these animals are derived from the reintroductions into the Piedra River in southwestern Colorado, releases in the Escalante River of Utah in 2005 (Utah Division of Wildlife Resources 2010), or both. In the future, river otters may expand their range into more of the upper Colorado River watershed as a result of these releases.

The state of Colorado did not include the Rio Grande basin in its reintroduction program (CDW 2003), so there was little opportunity for Colorado river otters to disperse into the Rio Grande in northern New Mexico. However, beginning in 2008, New Mexico launched its own reintroduction program in the Rio Grande basin on the Rio Pueblo de Taos, on Taos Pueblo land, Taos County (see under "Status and Management" below). Dispersing individuals from these reintroductions were later detected more than

Photo 19.6. Northern river otter in the Rio Grande in the Orilla Verde Recreation Area, Taos County, 5 December 2020. The upper Rio Grande, from Cochiti Dam to the Colorado border, together with some tributaries, such as the Rio Pueblo de Taos and the Red River, is currently home to a restored river otter population. River otters in the San Juan River likely originated from releases by the Colorado Division of Wildlife into the Los Pinos and Piedra rivers. Photograph: © Trey Flynt.

120 river km (75 river mi) from the release site (J. Stuart, unpubl. data). As of 2016, river otters or their tracks and scat had been detected in the upper Rio Grande of New Mexico from the Colorado border downriver to Cochiti Dam (including the tailwater below the dam), and in tributaries such as the lower Rio Chama, the lower Red River, the lower Santa Fe River (upstream of Cochiti Reservoir), and in the Rio Pueblo de Taos and its own tributary streams (e.g., the Rio Grande del Rancho and the Rio Lucero) (Long 2010; NMDGF, unpubl. data; Map 19.1). In British Columbia, Canada, Crowley et al. (2017) showed that river otter latrine sites (communal "restrooms") can be used as accurate predictors of the species' distribution except at times of ice cover (Photo 19.8). Through scat surveys, twenty active river

Photo 19.7. Northern river otter tracks on a frozen Rio Grande at the Colorado state line, 2 January 2018. River otters running in snow occasionally drop to their bellies and slide to conserve energy. Photograph: © Daniel Boyes.

Photo 19.8. Northern river otter scat (i.e., feces) found at the confluence of the Red River and the Rio Grande in Taos County, 29 January 2018. Latrine sites, or collections of scat, are used by river otters to share information through scent and to identify the territories of other individuals. Photograph: © Daniel Boyes.

otter latrine sites were discovered or confirmed in 2018 along the upper Rio Grande and tributaries in Taos County (Wolf-Gonzalez 2020). The northernmost latrine sites were found on the Colorado-New Mexico border. Two latrine sites occurred along the Red River near its confluence with the Rio Grande, two others along the lower Rio Pueblo de Taos southwest of the town of Taos (Wolf-Gonzalez 2020). In 2011, a river otter was photographed on the Rio Grande near Creede in Mineral County, Colorado, or more than 230 river km (~147 river mi) upstream from the New Mexico state line (B. Woodward, Colorado Division of Wildlife, pers. comm.). Whether that river otter truly originated from the New Mexico release site is unknown, as the possibility of movements between the headwaters of different river basins cannot be ruled out (see under "Status and Management"). Further dispersal of river otters in the Rio Grande drainage system is possible, though

establishment downriver from Cochiti Dam might be limited by occasional low river flows in this reach. While it is still too early to determine the overall success of the New Mexico reintroduction program (see Cox and Murphy 2019), the species appears to have regained a foothold in its historical range in the Rio Grande basin north of Mexico.

HABITAT ASSOCIATIONS

The river otter is a highly adaptable species that can use a variety of aquatic environments throughout its range. An adequate prey base, clean water, and suitable structures within and near the water to provide shelter, den sites, and foraging areas are the essential requirements (Melquist and Dronkert 1987). Rivers and large streams, estuaries, lakes, and marshes are all suitable for this species. The river otter is primarily a freshwater species but also readily enters

Photo 19.9. Northern river otter habitat at the confluence of the Rio Pueblo de Taos and the Rio Grande in Taos County. Photograph: © James N. Stuart.

brackish and even salt water in coastal areas. Deep pools in streams that have been impounded by the American beaver provide excellent habitat, and the presence of this dam-building rodent is often key to providing a mosaic of lotic and lentic environments that the river otter and its prey rely on (Melquist et al. 2003).

River otters are also capable of using isolated ponds that are within foraging distance of occupied streams and rivers, and this ability has likely facilitated the dispersal of individuals between drainage systems in more mesic areas of North America. River otters also will seasonally shift their use of aquatic environments in response to changes in water quantity or prey availability. For example, some shallow streams and ponds may be used in summer but abandoned in the winter when they freeze over. Diverse and suitable riparian vegetation adjacent to rivers and wetlands is also an important component of otter habitat as

Photo 19.10. Northern river otter habitat on the upper Rio Grande near Buckman, Santa Fe County: looking upstream near the upper end of White Rock Canyon, 24 February 2009. Photograph: © James N. Stuart.

Photo 19.11 (*top*). A mated pair of northern river otters preparing to line their den with grass among boulders on the banks of the Rio Grande in the Orilla Verde Recreation Area, Taos County on 30 November 2020. Photograph: © Surasit Khamsamran.

Photo 19.12 (*bottom*). Northern river otter on the Rio Grande at Buckman in Santa Fe County, about 5 miles south of San Ildefonso Pueblo, 30 March 2013. Photograph: © Jonathan Batkin.

it provides den and escape sites and helps maintain water quality in the adjacent aquatic environment (Melquist et al. 2003).

River otters do not construct their own dens for shelter or for rearing young. Thus, the availability of den sites can be a limiting factor in determining occupancy of a site. Nevertheless, river otters are behaviorally flexible and opportunistic in their choice of den sites, and will readily use bankside tree root systems, crevices in boulder

piles, hollow logs, logjams and other large collections of woody debris, natural recesses in banks, and abandoned dens of other species, especially bank burrows and lodges constructed by beavers (Lauhachinda 1978; Toweill and Tabor 1982; Melquist and Hornocker 1983; Mason and MacDonald 1986; Bradley 1994).

River otters also require protected resting sites for a variety of activities, including rolling in grass or sand to dry themselves, and grooming their dense fur to maintain its insulative properties. Resting sites are often associated with latrine sites on land that individuals and groups of river otters use for scent marking and for communal deposition of feces and urine (Melquist and Hornocker 1983). The choice of denning and resting sites is influenced by their proximity to good foraging areas and safety as much as by structural characteristics (Larsen 1983).

In the arid environment of New Mexico, river otter habitat is limited to perennial rivers and major streams, including backwaters and nearby ponds, and, to a lesser extent, reservoirs. Because reservoirs have fluctuating water levels and sloping shorelines that usually offer little vegetative cover or potential den sites, they typically are not used as regularly (Reid et al. 1994). However, recent observations of river otters in Cochiti Reservoir in New Mexico indicate at least occasional use of this water body. Similarly, many New Mexico rivers can be highly dynamic or "flashy," with flooding occurring during some seasons, and low or no flows in others, making them unsuitable for year-round occupation by river otters. Compared to water bodies used elsewhere in North America, the few perennial waters that are suitable for this species in the Southwest have little or no surface water beyond the river itself (NMDGF 2006a).

Historically, river otters inhabited a number of rivers in New Mexico that may no longer be suitable for the species. An analysis by the New Mexico Department of Game and Fish (2006a) found that three rivers, the upper Rio Grande (including its main tributary, the Rio Chama), the Gila River, and the San Juan River, still provide environments deemed suitable for river otters. The Rio Grande, New Mexico's longest river, bisects the state as it flows from north to south. In its northernmost reach, between the Colorado border and Cochiti Reservoir, the river is largely contained within a series of canyons and gorges separated by broad alluvial valleys with cottonwood and willow bosques (riparian forests). Although the river valleys in this reach have been largely developed, the canyon portions of the river are largely undisturbed. River flows in the upper Rio Grande are perennial and of greater volume than any other river in the state (NMDGF 2006a). The San Juan River below Navajo Reservoir in northwestern New Mexico is a large river with abundant prey, well-developed riparian vegetation, and low human population. However, most of the land the river flows through is private and human impacts such as overgrazing and energy extraction are common. The Gila River, in the southwestern corner of the state, originates in the mountains of the Gila and Aldo Leopold wilderness areas and flows through New Mexico into Arizona. Much of the river runs through designated wilderness or public lands, and though water levels are lower than either the Rio Grande or the San Juan River, year-round flows are adequate and reliable for most of the river's length (NMDGF 2006a).

The aquatic habitat where river otters occur in New Mexico today can be characterized as highly linear and restricted to narrow river channels within broad valleys and canyons. The associated riparian areas along these rivers are vegetated by narrow and intermittent bosques composed of dense thickets of willow (*Salix* spp.) and stands of cottonwood (*Populus* spp.), box elder (*Acer negundo*), and juniper (*Juniperus* spp.). Although den sites in New Mexico have not been described, they are likely located beneath partially exposed root systems of trees, jumbles of boulders and log piles, or dens constructed by beavers, either in riverbanks or the traditional stick lodges. Preferred resting sites are likely located amid vegetation thickets or deep grasses. Much of the riverbank in the upper Rio Grande canyons, including the Taos Gorge and White Rock Canyon, consists of sandy beaches interspersed with large aggregations of basaltic boulders and a narrow ribbon of riparian vegetation. Unlike many other places where the river otter occurs in North America, the presence of large aggregations of rocks adjacent to the river may be important in providing shelter and den sites. For example, in western Colorado, in arid and semi-arid landscapes similar to New Mexico's, river otters prefer latrine locations with overhanging cliffs and rocks rather than trees, and large rocks for scent marking (Depue and Ben-David 2010). On the upper Rio Grande in New Mexico, Converse et al. (2014) observed a family group of river otters using a jumbled pile of basaltic boulders on the riverbank as cover. Most of the twenty latrine sites included in Wolf-Gonzalez's (2020) study were characterized by 1) an accessible riverbank; 2) vegetation cover; and/or 3) a large boulder or rock formation overlooking the river.

LIFE HISTORY

The life history of the river otter remains largely unstudied in New Mexico and most of what is known about the biology, ecology, and behavior of this species is derived from studies of populations outside the state. Other than Wolf-Gonzalez's (2020) diet research in Taos County, the detailed ecological study of river otters in the Payette River basin of Idaho by Melquist and Hornocker (1983) perhaps provides the information best representative of the ecology of river otters that once inhabited New Mexico. Information from reintroduced populations in Colorado and Arizona (e.g., Taylor et al. 2003; Polechla 2010; Armstrong et al. 2011) is also useful.

Photo 19.13. In New Mexico, the invasive crayfish (*Faxonius virilis*) can form the majority of the river otter diet, especially in the warmer months. Photograph: © Jill Terese Wussow.

Photo 19.14. White sucker (*Catostomus commersonii*) from the Rio Grande just downstream of Cochiti Lake in Sandoval County. Fish (mainly Salmonidae and Catastomidae) and crayfish dominate the diet of northern river otters in New Mexico. The white sucker, which is a common non-native catastomid in the upper Rio Grande, likely contributes significantly to the diet. Photograph: © Jean-Luc Cartron.

Diet and Foraging

River otters do not hibernate and are active year-round. Like other members of the Mustelidae, their metabolic rate is high and they must feed on a regular basis, including in extremely cold weather, so much of their caloric intake must be used for thermoregulation (Larivière and Walton 1998). They will typically eat 15 to 20% of their body weight in food daily (Kruuk 2006:216), and therefore must forage in relatively productive waters with sufficient available prey (Serfass and Rymon 1985).

Although river otters are opportunistic predators, their diet is primarily composed of fish, often largely supplemented by crayfish (Cambaridae). Where and when crayfish are especially abundant, they may even comprise a dominant portion of the diet (e.g., Taylor et al. 2003). Other aquatic or terrestrial prey that are taken opportunistically include reptiles, frogs, aquatic invertebrates other than crayfish, and the occasional small mammal or bird (Melquist and Hornocker 1983). River otters preferentially feed on fish in inverse proportion to their prey's swimming speed, to minimize hunting effort (Ryder 1955), and larger (e.g., greater than 25 cm [10 in]) and slower fish hide and maneuver less easily in a chase than smaller fish (Melquist and Hornocker 1983). The relative importance of crayfish in a river otter's diet tends to increase during warmer months, when their availability is higher at the same time that fish are quicker and therefore more difficult to catch (Stearns and Serfass 2011; Barding and Lacki 2012; Cosby 2013; Feltrop et al. 2016). River otters do not typically scavenge or eat carrion (Melquist and Dronkert 1987).

Until recently, information on the food habits of river otters in southwestern river systems was limited and derived only from research in Arizona. River otters were reintroduced into the Verde River of Arizona in 1981 (Britt et al. 1984) and subsequent prey studies revealed that virile crayfish (Faxonius virilis) were found in most sampled river otter scat, along with a variety of fish species and lesser numbers of frog and mammal remains (Taylor et al. 2003; Polechla 2010). In a survey of potential river otter prey in the upper Rio Grande of New Mexico, Caldwell (2008) found that the fish community was dominated by non-native fish such as white sucker (Catostomus commersonii; 73%), brown trout (Salmo trutta; 11%), smallmouth bass (Micropterus dolomieu; 6%), and common carp (Cyprinus carpio; 3%). Wolf-Gonzalez (2020) found that 66.2% and 61.8% of river otter scat in the upper Rio Grande watershed contained crayfish and fish, respectively, with mollusks, birds, reptiles, and mammals less frequently identified (Wolf-Gonzalez 2020). Among fish, Salmonidae (39.6%) and Catostomidae (37.1%) dominated fish families in terms of frequency of detection; remains of fish belonging to Esocidae (likely northern pike,

Esox lusicus, 14.3%), Cyprinidae (Rio Grande chub, Gila pandora; longnose dace, Rhinichthys cataractae; common carp; flathead chub, Platygobio gracilis; fathead minnow, Pimephales promelas; red shiner, Cyprinella lutrensis; 12.4%), and Centrarchidae (smallmouth bass; largemouth bass, Micropterus salmoides; bluegill, Lepomis macrochirus; green sunfish, L. cyanellus; 3.2%) were also detected. During the wet (and warm) season (June and September samples), catastomid fish represented the most frequently consumed prey. Given the higher relative abundance of white sucker (Catostomidae) and brown trout (Salmonidae) reported by Caldwell (2008) in the upper Rio Grande, river otters thus seem to catch fish based largely on their availability. The probability of a scat containing fish was more than three times higher during the dry (and cold) season (February–April and December samples; 0.74) than during the wet season (0.23). Crayfish (not identified to species or even genus but perhaps virile crayfish) were observed in scats more frequently during the wet season (94.5%)—at which time they were the most frequently consumed prey—compared to the dry season (61.3%). The mollusks observed in river otter scats were all identified as the non-native Asiatic clam (Corbicula fluminea) (Wolf-Gonzalez 2020).

River otters forage in water features where their aquatic prey tend to congregate, such as in logjams or in undercut banks in shallow streams, in deep pools in larger streams, or along shorelines in lakes and reservoirs (Melquist and Hornocker 1983). Long, sensitive vibrissae, or facial whiskers, aid river otters in foraging in turbid waters. Individuals use a strategy of actively foraging in an area until prey becomes scarce before moving to another location. River otters are efficient predators, as suggested by a recorded average dive time of 21 seconds despite their ability to stay underwater much longer (Ben-David et al. 2000).

Reproduction, Social Behaviors, and Movements

Both male and female river otters reach sexual maturity at around two years of age and although females may breed as early as that, males often do not until they are a few years older (Toweill and Tabor 1982). Notably, the reproductive cycle of river otters is discontinuous. Copulation takes place in late winter or early spring, but the early-stage fertilized embryo does not immediately implant in the uterine wall. The phenomenon, known as "delayed implantation," can postpone development of a fertilized zygote for up to ten months (Melquist and Hornocker 1983). The actual period of true gestation after implantation is 60 to 63 days, such that the entire reproductive cycle from mating to parturition can be as long as 12 months (Larivière and Walton 1998). In Idaho, Melquist and Hornocker (1983) reported that estrus in females begins immediately after parturition, and copulation was observed in one pair in late April, about 3–4 weeks after the female had birthed (Liers 1951). This reproductive strategy is presumably beneficial as it allows both mating and birthing to coincide with times when the weather is mild, food is ample, and the prior year's offspring are independent of the mother.

River otter cubs are typically born in late winter or early spring, but exact parturition dates can vary considerably depending on geographic location (Harris 1968). In Colorado, young are probably born in March and April (Armstrong et al. 2011). Litter size usually ranges from one to three cubs (Larivière and Walton 1998), but is most commonly two. Males do not usually participate in raising the young. The cubs are altricial at birth and develop slowly, remaining in the den for as long as two months and are weaned at three months (Larivière and Walton 1998). Even then, cubs need to be taught by their mother to swim and capture prey. Young river otters leave their family group as early as their first fall, or in the following spring, when the mother is preparing to birth her next litter (Liers 1951; Melquist and Hornocker 1983). Dispersing young have a high rate of mortality as they seek a suitable place of their own (Tabor and Wright 1977). Small litters, coupled with a long dependence of the young on the mother, result in a low rate of reproduction. Although river otters can be relatively long-lived, up to 14 years in the wild (Stevenson 1977; Melquist and Dronkert 1987), most live much shorter lives (Kruuk 2006:209).

Little information on reproduction is available from the reintroduced population in New Mexico. In July 2009, an adult female river otter collected as a roadkill near Pilar, Taos County appeared to have been recently lactating, suggesting that she had produced a litter. On 30 June 2012, a group of five river otters was observed on the Rio Grande upriver of the Rio Pueblo de Taos confluence; one animal that appeared to be leading the group was larger than the others, suggesting a female and her young (E. Rominger, pers. comm.). Converse et al. (2014) documented a family group of river otters consisting of an adult and at least three juveniles from the same general area in June 2013, providing the first definitive evidence of reproduction in the reintroduced population on the Rio Grande. Cox and Murphy's (2019) estimates of current river otter population size and annual population growth rate in the upper Rio Grande constitute positive evidence of continued reproduction (see under "Status and Management").

River otters are relatively social for mustelids (Beckel-Kratz 1977). Family groups of a mother and her offspring, sometimes including grown offspring, are the most common social group (Melquist and Dronkert 1987). So-called bachelor groups, which can consist of several gregarious and unrelated males, may stay together for much of the year, sharing dens, hunting, and playing together (Shannon 1989). Groups of several river otters may cooperatively forage together (Serfass 1995) and use the same dens, latrines, and resting sites (Larivière and Walton 1998). Mated pairs of

Photo 19.15. Five river otters together along the upper Rio Grande on 19 October 2019. A family group usually consists of a mother and cubs, and occasionally a youngster from a prior litter. Males do not participate in raising the young, but sometimes form bachelor groups. © Britt Runyon

river otters appear to travel together only briefly, and, more rarely, young males may travel with females (Melquist and Hornocker 1983). In general, both female and male river otters have core activity areas, with significant levels of overlap (Kruuk 2006:10). Lone adults, usually females, may join family groups for some periods of time (Melquist and Hornocker 1983). The dynamics of these shifting and ephemeral social groups appear to be influenced by many factors, such as the abundance of suitable habitat and prey, breeding season, and population structure.

River otters are known for their liveliness. Sliding across snow field or ice may be an efficient mode of movement in winter for river otters, but repeatedly sliding down snowy slopes appears to be just for amusement (Stevens and Serfass 2005). Play behavior, such as wrestling and chasing, seems to be most typical for juveniles (Melquist and Hornocker 1983).

As for many carnivores, olfactory communication among river otters is important. Individuals will repeatedly scent-mark over time via scat, urine, and secretions from an anal sac deposited on boulders, logs, or sandy beaches (Olson et al. 2008). These latrines are used repeatedly by resident river otters. The scat of the river otter, which typically contains undigested bones and scales of fish or crayfish parts, is loose in consistency, and musky or fishy in odor when fresh. What information is being conveyed among individuals by scent marking is not known (Kruuk 2006:86). It appears likely that olfactory messages may function to space out individuals in a population and thereby reduce conflict, while not serving strictly to mark territories (Melquist and Hornocker 1983). Home ranges appear to be delineated by familiarity and security, and are not strongly defended, though the presence of an adult female with young may influence use of an area by other otters (Melquist and Hornocker 1983). In addition, river otters use a variety of vocal signals such as chirping, growling, and whistling to communicate with one another (Beckel-Kratz 1977).

As energetic predators with high food demands, linear home ranges may overlap extensively among individuals in riverine environments (Gorman et al. 2006). Melquist and Hornocker (1983) reported high levels of variability in Idaho home ranges, from 8 km (5 mi) for a juvenile female river otter in fall to 78 km (48 mi) for a yearling male in summer. Reid et al. (1994) found even larger home ranges along rivers in Canada, with averages of 58 km (36 mi) for females and 182 km (113 mi) for males (Spinola et al. 2008). Typically, home ranges for males are much larger than for females, especially when the latter are constrained in their movements by caring for young (Melquist and Hornocker 1983). Wide variation in home range may depend on many factors, such as topography, season, prey availability, and weather. Kruuk (2006:63) suggests the term "group ranges" for river otters since a given reach of river often serves as core habitat for family groups or other somewhat fluid "packs" or associations of individuals. River otters do not appear to have a strong territorial response to the movement of other otters through their home range (Melquist and Hornocker 1983).

River otter densities are also variable depending on the suitability of the habitat, which overall seem to be substantially higher in coastal bays compared to inland riverine systems (Melquist and Hornocker 1983; Testa et al. 1994; Bowyer et al. 2003; Mowry et al. 2011; Brzeski et al. 2013). In Idaho, Melquist and Hornocker (1983) estimated an average of one river otter per 2.4 km (1.5 mi) of river. In the reintroduced population of the Verde River of Arizona, there is an estimated one river otter per 1 to 1.5 km (~0.6–1 mi) (P. Barber, Arizona Game and Fish Department, pers. comm.). Cox and Murphy (2019) used a spatial capture-recapture model to estimate densities of 0.23–0.28 river otter per kilometer (~0.6 mi) in the upper Rio Grande. In general, it is difficult to estimate the abundance of river otters due to their elusive nature, mobility, and sparse distribution, in addition to the difficulty of extrapolating abundance from otter sign such as scat or tracks. Most population studies of river otters stress presence/absence and trends in distribution, but the most precise estimates of territory size and abundance rely on tracking individual otters using radio-telemetry devices and/or genetic analysis of scats (Melquist and Hornocker 1983; Spinola et al. 2008; Cox and Murphy 2019; see under "Status and Management" below).

The vast majority of river otter travel occurs in water, up and downstream along rivers, since otters are highly competent swimmers, but they will also walk or run parallel to rivers to cover ground, take shortcuts across land at river bends, or cross frozen sections of streams and lakes to access open water during winter. River otters also commonly move on land between nearby water bodies or drainage systems. Overland travel may be most common when the family group breaks up and young river otters seek their own territories (Melquist and Hornocker 1983).

Everyday movements within watersheds vary by season, gender, and individual predilection, but the movement of 10 river otters tracked in Idaho ranged from 7 to 36 km (~4 to 22 mi) in 24 hours (Melquist and Hornocker 1983). Dispersing yearlings traveled farthest; the longest single day trek of a dispersing yearling was recorded at 42 km (26 mi) (Melquist and Hornocker 1983). River otters may even cross ridges separating one watershed from another (Melquist and Hornocker 1983), though the distance otters are willing to travel overland and across obstacles is unstudied.

Interspecific Interactions

The river otter has a commensal relationship with beavers in North America. Beaver-dammed ponds on streams provide both foraging opportunities and escape cover for beavers and river otters alike, especially during periods of low flow (Tumlison et al. 1982). Ruiz-Olmo et al. (2001) have shown higher levels of mortality in Eurasian otters during extreme drought. In arid environments such as New Mexico, drought is a significant concern for river otter populations where beaver populations have been greatly reduced, particularly on many smaller-order streams in northern New Mexico (Small et al. 2016). River otters commonly use old beaver lodges for denning; up to 38% of otter dens are in beaver lodges in Idaho (Melquist and Hornocker 1983). On many larger perennial rivers in New Mexico, such as the Rio Grande and San Juan River, beavers may forgo dam construction and instead use bank burrows as den sites, which could potentially be appropriated by river otters. Rarely, river otters have been reported to prey upon beaver (Greer 1955).

Throughout most of their range in North America, river otters have no natural predators while they are in the water. However, coyotes (*Canis latrans*), wolves (*Canis lupus*), bobcats (*Lynx rufus*), black bear (*Ursus americanus*), and cougar (*Puma concolor*) may opportunistically prey upon river otters on land, particularly when they travel between water bodies (Route and Peterson 1991). Still, natural predation represents an insignificant source of mortality in most river otter populations (Toweill and Tabor 1982).

Diseases and Pathogens
A number of disease-causing bacteria and viruses have been reported in river otters (Melquist et al. 2003). River otters are known to contract rabies (Serfass et al. 1995), canine distemper (Melquist et al. 2003), respiratory tract and urinary infections (Hoover et al. 1984; Route and Peterson 1991), and hepatitis and pneumonia (Harris 1968), among other diseases. They are also susceptible to endo- and ectoparasites (Serfass et al. 1995; Melquist et al. 2003). As a precautionary measure, river otters obtained in Washington State for the New Mexico reintroduction program were given a medical evaluation, vaccinated for rabies and distemper, and treated with Ivermectin, an anti-parasitic, prior to their importation to the state (Stuart 2014). In general, however, disease and parasitism are not known to be a significant cause of mortality in wild river otter populations (Harris 1968; Melquist et al. 2003).

STATUS AND MANAGEMENT
Conservation Status
Long valued primarily for its lustrous pelt, the river otter is increasingly viewed by biologists and wildlife managers as an important component in managing healthy aquatic systems. Carnivore species, many of which are now imperiled, have been recently more fully recognized as crucial players in the maintenance of biodiversity and ecosystem function (e.g., Ripple et al. 2014). The river otter, as a top predator in aquatic environments, has the ability to greatly influence food webs and community structure in freshwater ecosystems. At the same time, river otters are sensitive to adverse changes in their environments and for that reason have been considered a "sentinel species" for assessing the health of aquatic ecosystems (Bowyer et al. 2003).

The status and management of the river otter throughout its range has long been driven by the species' economic importance as a furbearer. River otters are still widely trapped today for their pelt. As of 2013, 37 of the 49 states in the United States that have otter populations—Hawaii has none—allowed some level of commercial trapping of this species (Bricker et al. 2022). Although multiple factors were involved in river otter declines in North America during the 1800s and 1900s, unregulated trapping was likely the single most important. Fur trapping in North America by Europeans began as early as the 1600s (Route and Peterson 1988), and accelerated considerably in the early 1800s, driven in large part by the high demand for beaver pelts (Weber 1971). In the 1700s and 1800s, fur trapping in the Southwest was an important industry. River otters, which were less common than beavers in southwestern rivers and streams, were likely taken most often as unintended, but valuable, "bycatch" in beaver traps. So heavily were these two mammals trapped, in what was then the northernmost outpost of Mexico, that even government authorities took notice. In 1838 the departmental junta of Chihuahua acted to protect both the river otter and beaver along the Rio Grande by declaring a six-year moratorium on trapping of these two species (Weber 1971; deBuys 1985). deBuys (1985:98) calls this act "one of the first conservation measures in the North American West."

The ongoing demand for pelts resulted in intensive trapping in New Mexico throughout the 19th century. Huey (1956:1) noted that this trapping had "resulted in the near extinction of beavers in most of New Mexico late in the century." Little mention is made of the river otter in New Mexico during this time period, and it is likely that the less abundant otter was even more completely eradicated from the state by trapping than the beaver. By the time that New Mexico achieved statehood and the Department of Game and Fish was created, both in 1912, the river otter was too rare to be of much consideration for wildlife managers. Bailey (1931:324) concluded that river otters were "so rare as to be of little economic importance" in New Mexico.

The river otter is classified as a protected furbearer under New Mexico statute, though the trapping season is currently closed. This protection, originally applied to native river otters, also extends to the reintroduced animals present today. During the early 1900s, there was no closed season on the river otter or most other "fur animals." After the passage of a state law in 1939 to regulate the harvest of furbearers, the New Mexico Game Commission established trapping rules for most furbearer species. In 1953, the season on beavers was reopened (with restrictions)

after having been closed for 56 years due to past over-harvesting (Huey 1956). But the river otter was not mentioned as a harvestable furbearer in any year after passage of the 1939 law, presumably due to its likely extirpation from the state. In some states where river otters have been reintroduced in recent decades, such as Illinois and Missouri, trapping seasons have been implemented once again. However, given the river otter's limited habitat in New Mexico, a trapping season seems unlikely.

The native subspecies of the river otter (*L. c. sonora*) was listed as Endangered under the New Mexico Wildlife Conservation Act in 1975, only to be delisted in 1985 based on the assumption that no otters persisted in the state (Frey and Yates 1996; Jones and Schmitt 1997; Compton 2000). The subspecies was formerly a candidate for listing under the US Endangered Species Act (Spicer 1987), and it is still informally considered a Species of Concern by the US Fish and Wildlife Service though there is no evidence that it persists anywhere in its former range except possibly in Mexico's upper Rio Conchas basin (see under Distribution). The river otter was classified in 2006 as a Species of Greatest Conservation Need in the New Mexico Wildlife Action Plan (NMDGF 2006b) in anticipation of its reintroduction to the state, though this designation does not provide any additional protection.

At the global level, the river otter is listed in Appendix II of the Convention on International Trade in Endangered Species of Wild Fauna and Flora (CITES) (Wozencraft 2005), and classified as a Species of Least Concern by the International Union for the Conservation of Nature (IUCN; Serfass and Polechla 2008), suggesting that populations in North America are considered to be stable at present.

Reintroduction and Monitoring

Following the river otter's decline in the 20th century, efforts to re-establish the species in

Photo 19.16. USDA Wildlife Services biologist Darren Bruning helps load an otter carrier on board a Cessna 185 Skywagon in the state of Washington on 19 October 2009. In the fall and winter of 2008, 2009, and 2010, Wildlife Services live-trapped 33 otters in Washington and transported them to New Mexico for release along the Rio Pueblo de Taos, a tributary of the Rio Grande in Taos County. Darren Bruning accompanied the otters during the eight-hour flights from Olympia, Washington to Taos. Photograph: © Darren L. Bruning

Photo 19.17. Northern river otter holding pens at the reintroduction site along the partly iced over Rio Pueblo de Taos on 20 December 2008. Hay bales were brought to the site to provide an insulation layer on top of the two occupied pods as well as bedding inside for the otters. Photograph: © James N. Stuart.

Photo 19.18. Darren Bruning releasing an otter from its pen along the Rio Pueblo de Taos by removing the carrier, 6 November 2009. After two to three days of monitoring and feeding, the otters were released to the nearby river. The otters were released on Taos Pueblo land, with permission and assistance from Taos Pueblo. Photograph: © James N. Stuart.

Photo 19.19. Darren Bruning and Bureau of Land Management biologist Valerie Williams stand in the river by one of several fish pens at the reintroduction site on 21 November 2008. Fish trapped in the pens were fed as live prey to the river otters in their pods. In this particular instance, no fish were found in the pen, which had possibly been raided by a recently released otter. Photograph: © James N. Stuart.

Photo 19.20 (*above*). Members of the New Mexico River Otter Working Group, who assisted in the restoration effort, watching the first otter releases, above the Rio Pueblo de Taos, Taos County on 22 November 2008. Photograph: © James N. Stuart.

Photo 19.21 (*top right*). River otter leaving its pod at the release site along the Rio Pueblo de Taos, 23 October 2009. The otter transport carrier is on the left, the holding pen on the right. Photograph: © James N. Stuart.

Photo 19.22 (*bottom right*). Newly released northern river otter on the bank of the Rio Pueblo de Taos, 14 October 2008. Thirty-three river otters were introduced to the upper Rio Grande watershed in several release events from 2008 to 2010. Photograph: © James N. Stuart.

much of the southwestern United States have been largely successful. In the US Southwest, the species was reintroduced into five watersheds in Colorado from the 1970s to the 1990s (Armstrong et al. 2011), into the Verde River of Arizona in the 1980s (Britt et al. 1984; Hoffmeister 1986), and, in Utah, into the Green River in the 1980s and 1990s and the Escalante River in 2005 (B. Bates, pers. comm.). By the late 1990s, river otters inhabited at least parts of their historical range in each of those states, but not in New Mexico. Wildlife agencies in Colorado and Utah currently consider their reintroduced populations to be stable or growing; in Arizona, however, the population

status of the reintroduced river otter is now considered uncertain (Roberts et al. 2020).

New Mexico was a latecomer to river otter reintroduction in comparison with other states. Although discussions with the New Mexico Department of Game and Fish to reintroduce the species began in the early 1980s following the releases in Arizona (M. Savage, pers. obs.), these plans remained stalled for 20 years due to the concerns of wildlife managers and other biologists over a possible remnant native population in New Mexico. However, with still no evidence of a persisting native population, the thinking began to shift toward reintroduction as the only means to restore the species in the state. In 2000, the New Mexico River Otter Working Group, a coalition of non-profit organizations, state and federal agencies, Taos Pueblo (a Native American Reservation), and members of the public, was formed to work with the New Mexico Department of Game and Fish in pursuing the goal of restoring river otters. After several years of investigation and planning, the New Mexico Department of Game and Fish (2006a) issued a Feasibility Study that examined the suitability of various rivers in the state for river otter reintroduction. Among the variables that were considered were prey availability, water quantity and quality, length of contiguous stream reaches, stream structure, potential interactions with other species, human activity and land management, and the potential for connectivity with existing river otter populations in adjacent states.

The Feasibility Study, which was accepted by the New Mexico Game Commission, identified the upper Rio Grande and the Gila River as the most suitable rivers for otter reintroduction. Several factors made the Rio Grande in Taos County the best choice for initial releases, including the support of the US Bureau of Land Management, Taos Pueblo, and the general public. The releases were conducted on Taos Pueblo land, with the assistance of tribal authorities. The Gila River was also considered a suitable location for river otter releases, though the decision to allow reintroductions in this river system remains complicated by concerns over increasing predation pressure on several endangered aquatic species in the drainage basin, particularly fish species such as Gila trout (*Oncorhynchus gilae*) and spikedace (*Meda fulgida*) (Propst 1999). Although solutions to these issues are currently being investigated, the proposed reintroduction of the river otter to the Gila River is suspended at this time.

In 2007, a source of river otters in the Puget Sound area south of Seattle was identified by the USDA Wildlife Services in Washington State, where animal damage control personnel were engaged in the removal of "nuisance" otters that annoyed homeowners by frequenting boat docks and waterfront homes (D. Bruning, pers. comm.). Rather than euthanizing these animals, Wildlife Services agreed to give them to New Mexico. During fall and winter in the years 2008–2010, Wildlife Services live-trapped 33 river otters and transported them to New Mexico by truck and airplane (Stuart 2014; Photo 19.16). Once in New Mexico, animals were temporarily held in pens on the bank of the Rio Pueblo de Taos (a tributary of the Rio Grande) on Taos Pueblo land (Photo 19.17). After two to three days of monitoring and feeding, the river otters were released and allowed to disperse to the nearby river (Photo 19.18). This "soft release" method proved to be highly successful as all animals were healthy at the time of release (D. Bruning, pers. comm.; Stuart 2014).

Released river otters dispersed rapidly in the Rio Pueblo de Taos drainage system and downstream into the Rio Grande and the Orilla Verde Recreation Area. From there, some animals moved north through the Taos Gorge as far as the Colorado state line and south into White Rock Canyon to Cochiti Reservoir (Long 2010; Savage and Klingel 2015; Cox and Murphy 2019). Many informal reports of sightings, including those accompanied by clear photographs of groups of river otters, provided documentation that the animals were persisting and reproduction was occurring (Converse et al. 2014; NMDGF, unpubl. data).

The reintroduction program did not proceed without setbacks. Four of the released river otters were killed, including three that were trapped in body-gripping traps legally set for beavers in the Rio Pueblo de Taos drainage system, and a fourth that was struck by a vehicle near the town of Pilar (NMDGF, unpubl. data). An initial survey that relied on visual detection of river otters or their scat and tracks (Long 2010) documented the rapid dispersal of newly released animals throughout the upper Rio Grande watershed. Dispersal from the reintroduction site was to be expected based on the propensity of river otters to travel long distances (Spinola et al. 2008), and appears to reflect the presence of sufficient habitat.

A key component of managing the reintroduced populations is surveying and monitoring, which can be challenging for a species that is elusive and relatively uncommon. A variety of techniques, both non-invasive and those requiring capture and handling of animals (Melquist et al. 2003; Boyle 2006; Serfass et al. 1996), have been used for providing river otter presence or absence data in a given watershed. Reports from trappers are useful where the species is harvested but lacking for fully protected populations such as those in New Mexico. In many states, including New Mexico, shoreline surveys of rivers and other water bodies for latrine sites and tracks has been used to document the presence of river otters (e.g., Long 2010) and, in the case of latrines, can also provide estimates of relative abundance (Mowry et al. 2011). In New Mexico, reports from the public have been valuable in defining the distribution of river otters, particularly when accompanied by photographs of the animal or its tracks or by detailed descriptions of the animal and its behavior (Savage and Klingel 2015). The use of camera traps (wildlife cameras) has also been successfully used in New Mexico to document the presence of river otters in some parts of the Rio Grande (Long 2010; R. Conn, pers. comm.).

Opportunistic citizen-science monitoring has resulted in >150 reported sightings of individual river otters, their tracks, or scats since the original introduction (Savage and Klingel 2015). Most of those sightings occurred along the Rio Grande, but additional confirmations along the Chama River, Red River, Rio Hondo, and Rio Pueblo de Taos, as well as near the headwaters of the Rio Grande in southern Colorado (Long 2010; Savage and Klingel 2015; CPW 2018), suggested that population growth and range expansion had occurred.

With recent advances in methods for DNA collection and analysis, genetic studies of river otter populations have become more feasible, not only to determine the presence of the species in particular watersheds but also to identify genetic diversity within populations and numbers of individuals present (Bricker et al. 2022). Genetic analysis of scat samples was used in New Mexico to verify the presence of river otters in the San Juan River drainage (Polechla et al. 2004). In 2018, a non-invasive genetic sampling of fecal deposits, or scats, in a capture-recapture survey was conducted along the Rio Grande River, Red River, Rio Hondo, and Rio Pueblo de Taos (Cox and Murphy 2019). The survey led to estimates of population growth of the river otter population, while also assessing genetic diversity and divergence from the source population. This methodology of scat analysis has proven reliable for documenting river otter population size and genetic profile elsewhere (Mowry et al. 2011; Brzeski et al. 2013). River otter population size, across 359 km of river, from La Mesilla in Rio Arriba County, New Mexico to Alamosa National Wildlife Refuge, Colorado, was estimated at 83–100 animals, with an annual population growth rate of 1.12–1.15/ year (Cox and Murphy 2019). Model precision was hampered by sample analysis issues and model capabilities, resulting in a wide confidence interval around population estimates. The study concluded that there may be as few as 47 or as many as 176 river otters in the upper Rio Grande river network studied (Cox and Murphy 2019).

Cox and Murphy (2019) also characterized the genetic diversity of the upper Rio Grande otter population. The relatively small number of animals released from 2008 to 2010 (33 river otters in total) had raised concern that a founder effect, combined with genetic drift, would result in limited genetic diversity in a population isolated from any other otter genetic pool. The estimated size of the current river otter population in the upper Rio Grande was shown to be well below the minimum necessary to prevent inbreeding depression (Cox and Murphy 2019). In addition, the study found a strong bottlenecking founder effect in the New Mexico population when compared to the genetic profile of trapped river otters from the source population in Washington State. Cox and Murphy (2019) concluded that this reintroduced river otter population has elevated vulnerability to inbreeding and loss of fitness. The New Mexico Department of Game and Fish, based on these findings, began a supplemental release program in the upper Rio Grande during 2021, using river otters obtained by a commercial trapper in southern Louisiana. These additional river otters (nine released as of September 2021) are expected to enhance the genetic diversity of the existing population in the Rio Grande basin.

Habitat and Species Management

River otters still face threats to their habitat in New Mexico, but some environmental conditions have improved. Beaver populations in New Mexico have increased in some river systems, through better management and improved stream conditions, though the species is still far less abundant or widespread than it was prior to the days of unregulated trapping (Small et al. 2016). Nonetheless, the presence of beavers on the same rivers occupied by river otters presumably benefits the latter species through habitat modification. Although there is no legal harvest of river otters in New Mexico, beaver trapping continues in the state, albeit much less intensively than in the past. Areas closed to beaver trapping include the Wild Rivers and Orilla Verde recreation areas on the Rio Grande del Norte National Monument that encompasses part of the Rio Grande basin currently occupied by river otters. However, where the two species do co-occur and beaver trapping is legal, river otters are vulnerable to accidental killing in beaver trap sets.

The greatest concern for maintaining river otters in New Mexico is the threat to both the continued availability of perennial water in occupied river basins and the species' aquatic prey base. Water is a resource in high demand in a dry state like New Mexico, where diversion and de-watering of streams and rivers for agriculture and urban use is not uncommon. Dams do not by themselves necessarily pose a threat to river otter populations. For example, individuals have been observed to use the tailwater below Navajo Dam (J. Tensfield, pers. comm.), and the presence of a river otter foraging immediately below Cochiti Dam suggests that these obstacles can be circumvented, at least occasionally (T. Beauchene, pers. comm.; New Mexico Department of Game Fish, unpubl. data). However, if long stretches of river become ephemeral or de-watered from diversion dams, they will be unable to serve as river otter habitat. In addition, many reservoirs are not ideal habitats, due to water level fluctuations and a lack of structural diversity along the shore which would provide escape cover and habitat for prey (Melquist and Hornocker 1983; Melquist et al. 2003).

Rivers and riparian areas are among the most threatened ecosystems in New Mexico and the most vulnerable to climate change impacts (NMDGF 2006b; see Chapter 5). The threat to instream water resources is enhanced during times of periodic drought, which appear to be increasingly common in the region. Warmer temperatures predicted for the near-term future in the Southwest (Seager et al. 2007; Chapter 5) will exacerbate the reduction and fragmentation of free-flowing rivers and result in reduced quality of instream and riparian habitat for a variety of wildlife species.

Mining and other industrial impacts to water quality have declined in the state in recent decades. The river otter is at the top trophic level of the aquatic food web, and therefore susceptible to the accumulation in its prey of toxins such as polychlorinated biphenyls (PCBs). The threat of bioaccumulation of toxins has been reduced, as many mines and other industries in and around New Mexico have now closed or reduced their pollution of waterways due to stricter regulation. However, the need to monitor and improve water quality in New Mexico rivers continues. A preliminary analysis of fish bodies from the upper Rio Grande suggests that concentrations of PCBs, DDT, and heavy metals currently represent a low to moderate risk to river otter health (Caldwell 2008). The San Juan River, which supports not only a prized trout fishery but endangered species of fishes, also warrants close monitoring to ensure that healthy aquatic conditions persist. The Gold King Mine spill in the Animas River (a tributary to the San Juan River) in 2015 is a recent example of how historical mining continues to pose a threat to water quality in New Mexico's rivers. PCBs, DDT, and heavy metals do not represent the only threat to water quality as plastic waste has become ubiquitous in aquatic environments. Based on visual observations alone, microplastics (plastics <5 mm) were found in approximately 5% of river otter scat samples in Wolf-Gonzalez's (2020) study in Taos County.

Upland erosion from logging in the 1800s and 1900s has also been reduced as forests have grown back, reducing the sediment input to rivers. But a more recent threat to rivers in the state is an increased frequency of catastrophic wildfires (Chapter 5). Frequent, low-severity fires were historically the norm in many forested ecosystems in New Mexico, and such fires do not result in high levels of erosion. However, there has been a rapid increase in recent years in the number of severe wildfires in watershed forests. Extreme post-wildfire floods carry silt and ash into streams and may result in substantial declines in fish biomass, as documented in the Gila River basin (Whitney et al. 2015). A major wildfire within a drainage system occupied by river otters could pose a significant threat to the persistence of the population if the prey base is severely reduced.

Intensive cattle grazing in riparian areas throughout the 1800s and much of the 1900s reduced riparian vegetation that affords escape cover and resting sites for river otters. Unrestricted access of cattle to riparian areas has diminished from intensive past use, but where cattle still graze along rivers, there may be adverse effects to habitat used by river otters. Small et al. (2016) found that intense livestock grazing along streams in northern New Mexico adversely affected the mutualistic relationship between streamside willows and beavers, thus greatly reducing the abundance of beaver dams. Because river otters often use beaver-created ponds and den sites, the reduction in beaver numbers in such watersheds may limit their suitability for river otters.

Inbreeding depression, or the reduction of biological fitness due to breeding among closely related individuals, is a concern for small and isolated populations of river otters, especially if the founding individuals were few in number and from the same location (Cox and Murphy 2019). For example, some restrictions in gene flow have been detected among reintroduced river otter populations in the Gunnison River of Colorado due to a mere 15 km (9.3 mi) gap in habitat. In New Mexico, the introduction of additional river otters from southern Louisiana in 2021 will likely benefit the species in the state by contributing to its long-term genetic diversity.

Because river otters are piscivores and are adept at finding concentrations of their favored prey, they occasionally come into conflict with humans (Bricker et al. 2022). Elsewhere in their range, river otters occasionally have caused damage to fish populations in hatcheries and private fishing ponds and are sometimes viewed with antipathy by anglers (Melquist et al. 2003; Bricker et al. 2022; J. Stuart, pers. obs.). One fish hatchery in the Verde River drainage of Arizona experienced problems with "nuisance" river otters that became resident near rearing ponds and consumed captive fish; these animals were periodically live-trapped and relocated (D. Billingly, Arizona Game and Fish Department, pers. comm.). Similarly, river otters have been observed to take rainbow trout (*Oncorhynchus mykiss*) from overflow ponds at the Red River Fish Hatchery in Taos County, New Mexico, but have not affected fish-rearing at the facility itself (J. Stuart, pers. obs.). Incidents involving reintroduced river otters and captive fish populations have been rare so far in New Mexico. Despite the potential for river otters to affect fish-rearing facilities or their reliance on game fish as a large part of their diet in the upper Rio Grande (Wolf-Gonzalez 2020), their ability to damage wild populations of valuable gamefish such as trout is often exaggerated (Melquist et al. 2003; Bricker et al. 2022). Overall, river otters also have proved to be popular in New Mexico, and conflicts with the public are rare. Since reintroductions began, river otter watching has become a source of pleasure to the lucky hiker, angler, kayaker, and rafter who happens to encounter a family in the upper Rio Grande or San Juan River.

River otters have only recently reoccupied portions of their historical distribution in New Mexico, and much remains unknown about their life history. However, based on their rapid dispersal in the Rio Grande following recent releases, and preliminary evidence of reproduction, it appears that this river system is a suitable home for this flexible, resilient mammal. The re-establishment of river otters in the San Juan River of New Mexico has not been monitored but also seems likely, thanks to successful reintroductions in Colorado and Utah. In some rivers of the state, such as the Gila River, the species might contribute to the health of rivers they occupy through predation on non-native fishes and crayfish, thereby benefiting native aquatic species. However, even if restoration of river otter populations is eventually considered successful, limited aquatic and riparian habitat will likely constrain the species to small and potentially vulnerable populations in New Mexico. Careful monitoring and protection will be needed to ensure that river otters remain a part of our state's fauna for the foreseeable future.

ACKNOWLEDGMENTS

We thank Jon Klingel, Darren Bruning, Merav Ben-David, John Cox, and Gabriela Wolf-Gonzalez for information and assistance. We also thank Tom Serfass, Jennifer K. Frey, Jean-Luc E. Cartron, and an anonymous reviewer for comments on earlier versions of the manuscript.

LITERATURE CITED

Armstrong, D. M., J. P. Fitzgerald, and C. A. Meaney. 2011. *Mammals of Colorado*. 2nd ed. Boulder: University Press of Colorado.

Bailey, V. 1931 (=1932). *Mammals of New Mexico*. North American Fauna 53. Washington, DC: US Department of Agriculture, Bureau of Biological Survey.

Barding, E. E., and M. J. Lacki. 2012. Winter diet of river otters in Kentucky. *Northeastern Naturalist* 19:157–64.

Bates, B. 1988. Reintroduction of river otter in Utah. Utah Division of Wildlife Resources. Unpublished report.

Beck, W. A. 1962. *New Mexico: a history of four centuries*. Norman: University of Oklahoma Press.

Beckel-Kratz, A. 1977. Preliminary observations on the social behavior of the North American otter (*Lutra canadensis*). *Otters: The Journal of the Otter Trust* (Annual Report):28–32.

Ben-David, M., T. M. Williams, and O. A. Ormseth. 2000. Effects of oiling on exercise physiology and diving behavior of river otters: a captive study. *Canadian Journal of Zoology* 78:1380–90.

Bowyer, R. T, G. M. Blundell, M. Ben-David, S. C. Jewett, T. A Dean, and L. K. Duffy. 2003. Effects of the *Exxon Valdez* oil spill on river otters: injury and recovery of a sentinel species. *Wildlife Monographs* 153:1–53.

Boyle, S. 2006. North American river otter (*Lontra canadensis*): a technical conservation assessment. Unpublished report. USDA Forest Service, Rocky Mountain region, Species Conservation Project.

Bradley, P. V. 1994. Otter limits. *Natural History* 103:36–45.

Bricker, E. A., T. L. Serfass, Z. L. Hanley, S. S. Stevens, K. J. Pearce, and J. A. Bohrman. 2022. Conservation status of the North American river otter in the United States and Canada: assessing management practices and public perceptions of the species. In *Small carnivores: evolution, ecology, behaviour and conservation*, ed. E. Do Linh San, J. J. Sato, J. L. Belant, and M. J. Somers, 509–35. Oxford: Wiley-Blackwell.

Britt, T. L., R. Gerhart, and J. S. Phelps. 1984. Wildlife surveys and investigations: river-otter stocking. Arizona Project No. W-53-R-34, 125–34. Phoenix: Arizona Game and Fish Department.

Brzeski, K. E., M. S. Gunther, and J. M. Black. 2013. Evaluating river otter demography using noninvasive genetic methods. *Journal of Wildlife Management* 77:1523–31.

Caldwell, C. 2008. Otter prey study: contaminant assessment of their potential prey base. Unpublished report. New Mexico Department of Game and Fish, Santa Fe.

Ceballos, G., and E. Carrillo-Rubio. 2017. Redescubrimiento y estado de conservación de la nutria de río del norte (*Lontra canadensis*) en México. *Revista Mexicana de Mastozoología Nueva Época* 7:1–12.

Colorado Division of Wildlife (CDW). 2003. State of Colorado river otter recovery plan: revision of 1980, 1984, and 1988 draft plans. Denver: Colorado Division of Wildlife.

Colorado Parks and Wildlife (CPW). 2018. *Confirmed river otter observations from sightings database (1976–2017)*. Denver: Colorado Parks and Wildlife.

Compton, L. A. 2000. Status of southwest river otters, *Lontra canadensis sonora*, in the Colorado River through Grand Canyon National Park. *River Otter Journal* 9:10–11.

Converse, R. L., M. Baron-Deutsch, A. Gjuillin, and A. J. Rowe. 2014. Reproduction of reintroduced North American river otter (*Lontra canadensis*)

confirmed in New Mexico. *IUCN Otter Specialist Group Bulletin* 31:35–39.

Cosby, H. A. 2013. Variation in diet and activity of river otters (*Lontra canadensis*) by season and aquatic community. MS thesis, Humboldt State University, Arcata, California.

Cox, J. J., and S. M. Murphy. 2019. Demographic and genetic status of a reintroduced river otter population in north-central New Mexico. Final report to the New Mexico Department of Game and Fish Share with Wildlife Program. Project # 171012. Unpublished report to New Mexico Department of Game and Fish, Santa Fe.

Crowley, S. M., C. J. Johnson, and D. P. Hodder. 2017. Are latrine sites an accurate predictor of seasonal habitat selection by river otters (*Lontra canadensis*) in freshwater systems? *Mammal Research* 62:37–45.

deBuys, W. E. 1985. *Enchantment and exploitation: the life and hard times of a New Mexico mountain range*. Albuquerque: University of New Mexico Press.

Depue, J. E., and M. Ben-David. 2007. Hair sampling techniques for river otters. *Journal of Wildlife Management* 71:671–74.

———. 2010. River otter latrine site selection in arid habitats of western Colorado, USA. *Journal of Wildlife Management* 74:1763–67.

Feltrop, P. D., C. K. Nielsen, and E. M. Schauber. 2016. Asian Carp in the Diet of River Otters in Illinois. *American Midland Naturalist* 176:298–305.

Findley, J. S., A. H. Harris, D. E. Wilson, and C. Jones. 1975. *Mammals of New Mexico*. Albuquerque: University of New Mexico Press.

Frey, J. K., and T. L. Yates. 1996. Mammalian diversity in New Mexico. *New Mexico. Journal of Science* 36:4–37.

Gallo-Reynoso, J. P., S. Macías-Sánchez, V. A. Nuñez-Ramos, A. Loya-Jaquez, I. D. Barba-Acuña, L. del Carmen Armenta-Méndez, J. J. Guerrero-Flores, G. Ponce-García, and A. A. Gardea-Bejar. 2019. Identity and distribution of the Nearctic otter (*Lontra canadensis*) at the Río Conchos Basin, Chihuahua, Mexico. *Therya* 10(3):243–53. DOI: 10.12933/therya-19-894 ISSN 2007-3364.

Gorman, T. A., J. D. Erb, B. R. McMillan, and D. J. Martin. 2006. Space use and sociality of river otters (*Lontra canadensis*) in Minnesota. *Journal of Mammalogy* 87:740–47.

Greer, K. R. 1955. Yearly food habits of the river otter in the Thompson Lakes region, northeastern Montana, as indicated by scat analysis. *American Midland Naturalist* 54:299–313.

Hall, E. R. 1981. *The mammals of North America*. 2nd ed. New York: John Wiley.

Harris, C. J. 1968. *A study of the recent Lutrinae*. London: Weidenfeld and Nicolson.

Hoffmeister, D. F. 1986. *Mammals of Arizona*. Tucson and Phoenix: University of Arizona Press and Arizona Game and Fish Department.

Hoover, J. P., C. R. Root, and M. A. Zimmer. 1984. Clinical evaluation of American river otters in a reintroduction study. *Journal of the American Veterinary Medical Association* 185:1321.

Huey, W. S. 1956. New Mexico Beaver Management. *New Mexico Department of Game and Fish Bulletin* 4:1–49.

Jackson, H. H. T. 1961. *Mammals of Wisconsin*. Madison: University of Wisconsin Press.

Jones, C., and C. G. Schmitt. 1997. Mammal species of concern in New Mexico. In *Life among the muses: papers in honor of James S. Findley*, ed. T. L. Yates, W. I. Gannon, and D. E. Wilson, 179–205. Albuquerque: University of New Mexico, Museum of Southwestern Biology. *Special Publication of the Museum of Southwestern Biology, University of New Mexico* 3:1–290.

Julyan, R. 1996. *The place names of New Mexico*. Albuquerque: University of New Mexico Press.

Kruuk, H. 2006. *Otters: ecology, behaviour and conservation*. Oxford: Oxford University Press.

Larivière, S., and L. R. Walton. 1998. *Lontra canadensis*. *Mammalian Species* 587:1–8.

Larsen, D. N. 1983. Habits, movements and foods of river otters in coastal southeastern Alaska. MS thesis, University of Alaska, Fairbanks.

Lauhachinda, V. 1978. Life history of the river otter in Alabama with emphasis on food habits. PhD dissertation, University of Auburn, Alabama.

Liers, E. E. 1951. Notes on the River Otter (*Lutra canadensis*). *Journal of Mammalogy* 32:1–9.

Long, B. J. 2010. River otter monitoring in the upper Rio Grande watershed in northern New Mexico, October 14, 2008 through January 21, 2010. Unpublished report to New Mexico Department of Game and Fish, Santa Fe.

Mason, C. F., and S. M. MacDonald. 1986. *Otters: ecology and conservation*. Cambridge: Cambridge University Press.

McClellan, J. 1954. An otter catch on the Gila River in southwestern New Mexico. *Journal of Mammalogy* 35:443–444.

Mearns, E. A. 1891. Notes on the otter (*Lutra canadensis*) and skunks (genera *Spilogale* and *Mephitis*) of Arizona. *Bulletin of the American Museum of Natural History* 3:252–62.

Melquist, W. E., and A. E. Dronkert. 1987. River otters. In *Wild furbearer management and conservation in North America*, ed. M. Novak, J. A. Baker, M.E. Obbard, and B. Malloch, 626–41. Ontario, Canada: Ministry of Natural Resources.

Melquist, W. E., and M. G. Hornocker. 1983. Ecology of river otters in west central Idaho. *Wildlife Monographs* 83:1–60.

Melquist, W. E., P. J. Polechla Jr., and D. Toweill. 2003. River otter (*Lontra canadensis*). In *Wild mammals of North America: biology, management, and economics*, 2nd ed., ed. G. A. Feldhamer, B. C. Thompson, and J. A. Chapman, 708–34. Baltimore: Johns Hopkins University Press.

Mowry, R. A., M. E. Gompper, J. Beringer, and L. S. Eggert. 2011. River otter population size estimation using noninvasive latrine surveys. *Journal of Wildlife Management* 75:1625–36.

New Mexico Department of Game and Fish (NMDGF). 2006a. Feasibility study: potential for restoration of river otters in New Mexico. Unpublished report to the New Mexico Game Commission from New Mexico Department of Game and Fish, Santa Fe.

———. 2006b. *Comprehensive Wildlife Conservation Strategy for New Mexico*. Santa Fe: New Mexico Department of Game and Fish.

Olson, Z. H., T. L. Serfass, and O. E. Rhodes. 2008. Seasonal variation in latrine site visitation and scent marking by Nearctic River Otters (*Lontra canadensis*). *IUCN Otter Specialist Group Bulletin* 25:108–20.

Pattie, J. O. 2001. *The personal narrative of James O. Pattie*. Santa Barbara, CA: Narrative Press.

Polechla, P. J. 2000. Ecology of the river otter and other wetland furbearers in the upper Rio Grande. Unpublished report. US Bureau of Land Management, Taos, NM.

———. 2010. River otter (*Lontra canadensis*) dietary

study in the Verde River, Arizona: otter-crayfish ecological interactions. Unpublished report. The Nature Conservancy, Arizona Field Office.

Polechla, P. J., A. G. Burns, S. Rist, K. A. Moore, and J. W. Dragoo. 2004. First physical evidence of the Nearctic river otter (*Lontra canadensis*) collected in New Mexico, USA, since 1953. *IUCN Otter Specialist Group Bulletin* 21:70–74.

Polechla, P. J., and E. Carrillo-Rubio. 2009. Historic and current distributions of river otters (*Lontra canadensis* and *Lontra longicaudis*) in the Rio Grande or Rio Bravo del Norte drainage of Colorado and New Mexico, USA and of Chihuahua, Mexico and adjacent areas. *IUCN Otter Specialist Group Bulletin* 26:82–96.

Polechla, P. J., S. DesGeorges, R. Gardiner, B. Hayes, G. Long, and D. Storch. 1994. Field reconnaissance for river otters of the upper Rio Grande, Cow Paddy to Lee's Trail, Taos County, New Mexico. Unpublished report. US Bureau of Land Management, Taos, NM.

Polechla, P. J., and S. Walker. 2008. Range extension and a case for a persistent population of river otters (*Lontra canadensis*) in New Mexico. *IUCN Otter Specialist Group Bulletin* 25:13–22.

Propst, D. L. 1999. *Threatened and endangered fishes of New Mexico*. Technical Report Number 1. Santa Fe: New Mexico Department of Game and Fish.

Raesly, E. J. 2001. Progress and status of river otter reintroduction projects in the United States. *Wildlife Society Bulletin* 29:856–62.

Reid, D. G., T. E. Code, A. C. H. Reid, and S. M. Herrero. 1994. Food habits of the river otter in a boreal ecosystem. *Canadian Journal of Zoology* 72:1314–24.

Rhoads, S. N. 1898. Contributions to a revision of the North American beavers, otters and fisher. *Transactions of the American Philosophical Society* n.s. 19:417–39.

Ripple, W. J., J. A. Estes, R. L. Beschta, C. C. Wilmers, E. G. Ritchie, M. Hebblewhite, J. Berger, B. Elmhagen, M. Letnic, M. P. Nelson, O. J. Schmitz, D. W. Smith, A. D. Wallach, and A. J. Wirsing. 2014. Status and ecological effects of the world's largest carnivores. *Science* 343, article 1241484. doi:10.1126/science.1241484

Roberts, N. M., M. J. Lovallo, and S. M. Crimmins. 2020. River otter status, management, and distribution in the United States: evidence of large-scale population increase and range expansion. *Journal of Fish and Wildlife Management* 11:279–86.

Route, W. T., and R. O. Peterson. 1988. *Distribution and abundance of river otter in Voyageurs National Park, Minnesota*. Washington, DC: National Park Service Research/Resources Management Report MWR-10.

———. 1991. An incident of wolf, *Canis lupus*, predation on a river otter, *Lutra canadensis*, in Minnesota. *Canadian Field-Naturalist* 105:567–68.

Ruiz-Olmo, J., J. M. Lopez-Martin, and S. Palazon. 2001. The influence of fish abundance on the otter (*Lutra lutra*) populations in Iberian Mediterranean habitat. *Journal of Zoology* 254:325–36.

Ryder, R. A. 1955. Fish predation by the otter in Michigan. *Journal of Wildlife Management* 19:497–98.

Savage, M., and J. Klingel. 2015. Citizen monitoring after an otter restoration (*Lontra canadensis*) in New Mexico. *IUCN Otter Specialist Group Bulletin* 32:21–24.

Scurlock, D. 1998. *From the rio to the sierra: an environmental history of the middle Rio Grande basin*. General Technical Report RMRS-GTR-5. Fort Collins, CO: US Department of Agriculture, Forest Service, Rocky Mountain Research Station.

Seager, R., M. Ting, I. Held, Y. Kushnir, J. Lu, G. Vecchi, H-P. Huang, N. Harnik, A. Leetmaa, N. C. Lau, C. Li, J. Velez, and N. Naik. 2007. Model projections of an imminent transition to a more arid climate in southwestern North America. *Science* 316:1181–84.

Serfass, T. L. 1995. Cooperative foraging by North American river otters, *Lutra canadensis*. *Canadian Field-Naturalist* 109:458–59.

Serfass, T. L., R. P. Brooks, J. M. Novak, P. E. Johns, and O. E. Rhodes. 1998. Genetic variation among populations of river otter in North America: considerations for reintroduction projects. *Journal of Mammalogy* 79:736–46.

Serfass, T. L., R. P. Brooks, T. J. Swimley, L. M. Rymon, and A. H. Hayden. 1996. Considerations for capturing, handling, and translocating river otters. *Wildlife Society Bulletin* 24:25–31.

Serfass, T. L., and P. J. Polechla. 2008. North American river otter. *Lontra canadensis*. The IUCN Red

List of Threatened Species. IUCN 2013, www. iucnredlist.org.

Serfass, T. L., and L. M. Rymon. 1985. Success of river otters reintroduced into Pine Creek Drainage in northcentral Pennsylvania. *Transactions of the Northeast Section of the Wildlife Society* 41:138–49.

Serfass, T. L., M. T. Whary, R. L. Peper, R. P. Brooks, T. J. Swimley, W. R. Lawrence, and C. E. Rupprecht. 1995. Rabies in a river otter (*Lutra canadensis*) intended for reintroduction. *Journal of Zoo and Wildlife Medicine* 26:311–14.

Shannon, J. S. 1989. Social organization and behavioral ontogeny of otters (*Lutra canadensis*) in a coastal habitat in northern California. *IUCN Otter Specialist Group Bulletin* 4:8–13.

Small, B.A., J. K. Frey, and C. C. Gard. 2016. Livestock grazing limits beaver restoration in northern New Mexico. *Restoration Ecology* 24(5):646–55.10.1111/rec.12364.

Spicer, R. B. 1987. Status of the Arizona river otter, *Lutra canadensis sonora* (Rhoads), along the Colorado River in Arizona. Unpublished report. US Fish and Wildlife Service, Office of Endangered Species, Albuquerque, NM.

Spinola, R. M., T. Serfass, and R. P. Brooks. 2008. Survival and post-release movements of river otters translocated to Western New York. *Northeastern Naturalist* 15:13–24.

Stahlecker, D. W. 1986. A survey for the river otter (*Lutra canadensis*) in Taos and Colfax counties, New Mexico. Unpublished report. New Mexico Department of Game and Fish, Santa Fe.

Stearns, C. R., and T. L. Serfass. 2011. Food habits and fish prey size selection of a newly colonizing population of river otters (*Lontra canadensis*) in eastern North Dakota. *American Midland Naturalist* 165:169–184.

Stevens, S. S., and T. L. Serfass. 2005. Sliding behavior in Nearctic river otters: locomotion or play? *Northeastern Naturalist* 12:241–44.

Stevenson, A. B. 1977. Age determination and morphological variation in Ontario otters. *Canadian Journal of Zoology* 55:1577–83.

Stuart, J. N. 2014. Restoration and monitoring of river otter in New Mexico. Unpublished report to US Fish and Wildlife Service. New Mexico Department of Game and Fish, Santa Fe.

Tabor, J. E., and H. H. Wright. 1977. Population status

of river otter in Western Oregon. *Journal of Wildlife Management* 41:692–99.

Taylor, M., J. E. Rettig, and G. R. Smith. 2003. Diet of re-introduced river otters, *Lontra canadensis*, in north-central Arizona. *Journal of Freshwater Ecology* 18:337–38.

Testa, J. W., D. F. Hollerman, R. T. Bowyer, and J. B. Faro. 1994. Estimating populations of marine river otters in Prince William Sound, Alaska, using radiotracer implants. *Journal of Mammalogy* 75:1021–32.

Toweill, D. E., and J. E. Tabor. 1982. River otter. In *Wild mammals of North America: biology, management, and economics*, ed. J. A. Chapman and G. A. Feldhamer, 688–703. Baltimore: Johns Hopkins University Press.

Tumlison, R., M. Karnes, and A. W. King. 1982. The river otter in Arkansas: II. Indications of a beaver-facilitated commensal relationship. *Proceedings of the Arkansas Academy of Science* 36:73–75.

Utah Division of Wildlife Resources. 2010. *Northern river otter management plan V. 2.0.* Utah Division of Wildlife Resources Publication No. 10-22.

Van Zyll de Jong, C. G. 1972. *A systematic review of the Nearctic and Neotropical river otters.* Life Science Contributions No. 80. Ontario: Royal Ontario Museum.

Weber, D. J. 1971. *The Taos trappers: the fur trade in the far Southwest, 1540–1846.* Norman: University of Oklahoma Press.

Whitney, J. E., K. B. Gido, T. J. Pilger, D. L. Propst, and T. F. Turner. 2015. Consecutive wildfires affect stream biota in cold- and warmwater dryland river networks. *Freshwater Science* 34:1510–26.

Wolf-Gonzalez, G. A. 2020. Diet of a recently reintroduced river otter (*Lontra canadensis*) population in Taos County, New Mexico. MS thesis, University of Kentucky, Lexington.

Wozencraft, W. C. 2005. Order Carnivora. In *Mammal species of the world: a taxonomic and geographic reference*, Vol. 1, ed. D. E. Wilson and D. M. Reeder, 532–628. 3rd ed. Baltimore: Johns Hopkins University Press.

PACIFIC MARTEN (*MARTES CAURINA*)

James N. Stuart and Brian J. Long

The Pacific marten (*Martes caurina*), also called pine marten or simply marten, is a rarely encountered member of the weasel family that occurs in the subalpine and upper montane coniferous forests of northern New Mexico. The species has a dense, soft, medium-length pelage that is typically tan to chocolate brown in color, often darker on the tail tip and limbs. The head may be grayish, contrasting with the rest of the body. The fur of the throat and upper chest is pale cream or straw-colored to bright yellow or orange, forming a distinct bib-like patch, and this yellowish or orangish coloration can extend on the venter as far back as the abdomen and onto the sides of the body (Merriam 1890). The Pacific marten has a long body, slender torso, rather short legs, and a bushy tail that is about one-third of the animal's total length. The moderately wide face is pointed and rather fox-like in profile, and the erect cat-like ears are rounded at the tips and broad at the base. The feet are five-toed, though the first toe on the front foot does not always register in snow tracks (Elbroch 2003). Martens are well adapted for an arboreal lifestyle; the hind limbs can be rotated in a manner similar to that seen in tree squirrels, which allows the marten to climb rapidly and also descend a tree trunk headfirst (Armstrong et al. 2011). The light-colored claws on each foot can be partially retracted, presumably an adaptation for both climbing trees and foraging on the ground (reviewed by Buskirk and Ruggiero 1994; Clark 1999).

Photo 20.1. Pacific marten in the upper Rio Hondo watershed near Taos in the Sangre de Cristo Mountains on 22 April 2003. Martens are semi-arboreal carnivores characterized by a long body, short legs, and a bushy tail. The triangular face is reminiscent of a fox. The cat-like ears are erect and rounded. Photograph: © Brian Jay Long.

The Pacific marten is a medium-sized mustelid that is somewhat smaller than a small housecat. Little information is available on body size in martens in New Mexico, but in Colorado the species is reported to have a total length of 460–750 mm (~18–30 in), of which the tail constitutes 170–250 mm (~7–10 in), and the body weight is 0.5–1.2 kg (1.1–2.6 lbs) (Armstrong et al. 2011). Adult males are 20–40% larger in body size than adult females.

The marten superficially resembles several other species of mammals that also occur or

(*opposite page*) Photograph: © Sole Marittimi.

Photos 20.2a and b (*left to right*). The Pacific marten belongs to the weasel family and is somewhat smaller than a domestic cat. The medium-length, dense pelage is typically tan to chocolate brown in color, often darker on the tail tip and the limbs. The head may be a contrasting gray, the throat and upper chest a pale cream or straw colored to bright yellow or orange that can extend to the abdomen and sides. The marten in the photographs is fitted with a radio collar after it was captured near Taos Ski Valley. Photographs: © Brian Jay Long.

Photo 20.3. Pacific marten on a tree trunk in the upper Rio Hondo watershed near Taos in the Sangre de Cristo Mountains in March 2003. Martens are excellent climbers and can descend tree trunks head-first, largely as a result of the hind limbs being able to rotate at the hips. Photograph: © Brian Jay Long.

formerly occurred in northern New Mexico. It can best be distinguished from other, potentially sympatric species of weasels by its greater body size, fox-like head profile, and erect, cat-like ears. The long-tailed weasel (*Neogale frenata*), American ermine (*Mustela richardsonii*), and black-footed ferret (*Mustela nigripes*) are all smaller than martens, more elongated in body shape, and they lack the dense, rich brownish pelage of the marten while having a more rounded, blunter snout. The mink (*Neogale vison*) is also somewhat smaller than the marten, and though it is typically dark-colored (usually blackish) with a dense pelage, it is more weasel-like in overall appearance, with smaller ears and a blunter, more rounded snout. The mink also often has a white chest patch, unlike the distinct yellow or orange bib seen in the marten. Another mustelid, the wolverine (*Gulo gulo*; considered extirpated in New Mexico) is larger than the marten and, with its heavy build and particularly broad face, resembles a short bear rather than a weasel; it also sports a broad, yellowish stripe extending from the shoulders along each side of the body and usually a white blaze on the chest (see Chapter 18). Because of similar arboreal habits, the marten potentially could be confused with various tree squirrels, such as the red squirrel (*Tamiasciurus* spp.), but its broad face with forward-facing eyes, distinct snout, and greater size set it apart. The yellow-bellied marmot (*Marmota flaviventris*), a large terrestrial squirrel often found in similar habitats as the marten, also has an overall dark brownish pelage and a longish bushy tail but has a rounded head, small ears, distinct white facial markings, and is much stockier in overall build.

The taxonomy of martens in North America has been a subject of debate for many years. During the first half of the 20th century, it was generally recognized that two species of martens occur in North America: the American marten (*Martes americana*) and Pacific marten (*M. caurina*) (e.g., Rhoads 1902; Bailey 1931). In his study of skull morphology and pelage in North American martens, Wright (1953) proposed that only one species, *M. americana*, occurs in Canada and the United States, and relegated the species *M. caurina* (described by C.

Hart Merriam in 1890 from Washington State) to subspecies status. Wright's argument was based largely on the presence of apparent intergrades between these two forms in western Montana. Hagmeier (1961) subsequently recognized intraspecific, evolutionary divergence of two distinct lineages: an "americana" group found in most of Canada and Alaska and in the contiguous United States east of the Rocky Mountains, and a "caurina" group found in the Pacific coast states and provinces and the central and southern Rocky Mountains. This single-species taxonomic arrangement was followed for the next few decades until genetic evidence provided new and further support for recognizing the "americana" and "caurina" groups as separate and distinct entities (Clark et al. 1987; Carr and Hicks 1997; McGowan et al. 1999; Wozencraft 2005) and eventually, once again, as full species (MacDonald and Cook 2007; Pauli et al. 2011, 2012). Although genetic evidence of hybridization between the two clades possibly complicates any taxonomic conclusions (Small et al. 2003), we nonetheless follow Dawson and Cook (2012) and

  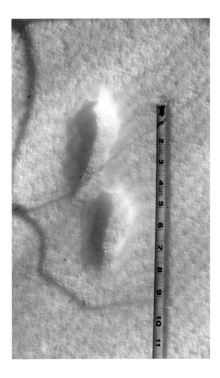

Photos 20.4a, b, and c (*left to right*). In snow the most commonly seen marten tracks are the slightly offset paired tracks of the easy running gait (the 2x2 gait; Photo 20.4a). The hind foot registers on top of the front foot track. The distance between the pairs of tracks can vary from ~23–117 cm (9–46 in), and when walking tracks of all four feet are seen. The actual length of the print is approximately 5 cm (2 in) (track plate; Photo 20.4b) but in loose snow prints can measure ~10–13 cm (4–5 in; Photo 20.4c). The straddle of a pair of tracks is roughly 10 cm (4 in) in snow. It can be difficult to distinguish the tracks of a small marten from a large long-tailed weasel (*Neogale frenata*), but weasel tracks tend to be more erratic as they zigzag back and forth. Photographs: © Brian Jay Long.

Photo 20.5. In New Mexico's Sangre de Cristo Mountains, Pacific martens have been found dependably in the subalpine coniferous forest surrounding Williams Lake below Wheeler Peak in Taos County. Photograph: © Geraint Smith.

Photo 20.6. Pacific marten on 8 April 2020 near Serpent Lake in the Sangre de Cristo Mountains. The New Mexico Department of Game and Fish conducted a survey for martens from the fall of 2019 through the spring of 2020. Motion-activated remote cameras were deployed at 64 monitoring sites in the Sangre de Cristo Mountains. Each site was monitored at all times by a primary camera. Most sites were also monitored by a secondary camera installed 500 m away from the primary camera for about two to three weeks in the fall. The secondary cameras were rotated through most of the monitoring sites for additional coverage. The detections occurred in the Taos Ski Valley area on 15 September 2019 and on 15 January and 8 April 2020; and in the Serpent Lake area on 8 April 2020. Photograph: © Nick Forman/New Mexico Department of Game and Fish.

Photo 20.7. Marten on a tree trunk in Columbine Canyon (near Red River) in the Taos Mountains on 30 December 2018. Earlier, one of the authors (B. J. Long) had observed marten tracks in the area and was able to confirm the presence of the species using a remote camera. The camera is pointed at a tree where meat or fish scraps were placed in a mesh bag. Columbine Canyon is within the known range of the marten in the Taos Mountains. Photograph: © Brian Jay Long.

Photo 20.8. Pacific marten (center of the photo) on the slopes just above Lake Katherine in Santa Fe County on 29 July 2015. The species' distribution reaches the southern end of the Sangre de Cristo Mountains in northeastern Santa Fe County and, based on two 1884 specimen records, also in northwestern San Miguel County. Photograph: © Diego Romero

Photo 20.9. Marten in the Brazos Cliffs area of the San Juan Mountains east of Tierra Amarilla, Rio Arriba County on 24 May 2018. Compared to the Sangre de Cristo Mountains, there are fewer reliable records of martens from the San Juan Mountains. Photograph: © Mark L. Watson.

Bradley et al. (2014), among others, in recognizing *M. caurina* as a separate species. Because almost all the life history and ecological studies on martens during the late 20th and early 21st centuries recognized *M. americana* as the only marten species in North America (thus obscuring the differences between these two distinct forms), we have focused our review of the scientific literature mainly on data from populations of the western ("caurina") group, now called *M. caurina*, which includes the martens found in New Mexico.

Although 14 subspecies *of Martes americana* (including *M. caurina*) were recognized by Hall (1981), other researchers have suggested that far fewer geographic races warrant recognition (Hagmeier 1961; Clark et al. 1987). Populations in the Southern Rocky Mountains were described by Rhoads (1902) as *Martes caurina origenes* (the Rocky Mountain marten) based on a specimen from Garfield County, Colorado (Bailey 1931). However, this geographic race has more recently been considered by some as indistinguishable from, and therefore synonymous with, *M. caurina caurina* of the Pacific Northwest (e.g., Clark et al. 1987). We

consider the taxonomic status of the subspecies in the Southern Rockies as unresolved.

DISTRIBUTION

The American and Pacific martens are associated with boreal and subalpine coniferous forest extending from Alaska across most of Canada to Newfoundland and (formerly) southward to Ohio in the eastern United States (Hall 1981). In western North America, the Pacific marten is distributed in the states and provinces along the Pacific coast from southeastern Alaska to central California and in the central and southern Rocky Mountains from Montana and Idaho to northern New Mexico (Clark et al. 1987; Powell et al. 2003). Although the range of both marten species north of the contiguous United States remains essentially intact, some reduction in populations of the Pacific marten in parts of the Pacific Northwest and Rocky Mountains has been documented (Clark et al. 1987; Buskirk and Ruggiero 1994). Clark et al. (1987) suggested that some range retraction may have occurred in New Mexico during historical times; however, there is

Photos 20.10a and b. a) (*left*) Lagunitas Creek east of Chama, Rio Arriba County, in the San Juan/Tusas Mountains; b) (*right*) Pacific marten caught on camera along Lagunitas Creek on 15 June 2021. Photograph 20.10a: © Marty Peale; Photograph 20.10b: © Brian Long, Marty Peale, and Jon Klingel. 2021 New Mexico Department of Game and Fish, Share with Wildlife, Mink Survey, northern New Mexico. US Fish and Wildlife Service Wildlife Restoration Grant W-208-R-1.

little evidence for a significant change in distribution in the state due to the rather small number of confirmed locations that have been documented even going as far back as the late 1800s. The range of the Pacific marten in the Southern Rocky Mountains (southern Wyoming to northern New Mexico) appears to be disjunct from that of more northerly populations in the Rockies (Powell et al. 2003).

Within its limited range in New Mexico, the Pacific marten has been documented in the Sangre de Cristo Mountains (a complex of several named mountain ranges) and the San Juan Mountains in the northern part of the state in parts of Colfax, Mora, Rio Arriba, San Miguel, Santa Fe, and Taos counties (Map 20.1). Known locations range in elevation from 2,560 m (8,400 ft) up to the alpine zone at 3,350–3,960 m (11,000–13,000 ft) (Long 1999, 2001a). Major areas of occurrence in the Sangre de Cristo and San Juan ranges are likely continuous with populations in adjacent southern Colorado (Findley et al. 1975; Armstrong et al. 2011), but are separated from each other by the Rio Grande Gorge and intermountain basins in northern New Mexico (Map 20.1). The marten populations at the southern end of the Sangre de Cristo Mountains (in southern Colorado and

northern New Mexico) are now possibly isolated from more northern populations as a result of the large 2018 Spring Creek Fire at La Veta Pass on the boundary between Costilla and Huerfano counties, Colorado. Although subalpine coniferous forests appear continuous between the Taos and Pecos marten populations, there has not been any verified reports of martens from this gap area. In particular, limited surveys including snow tracking, track plates, and baited camera stations have failed to document any marten, and though there are two east-west highways across this north-south montane corridor, no roadkills have been reported either. On the west side of the Rio Grande there is very little subalpine coniferous forest between the San Juan population and the Jemez Mountains, where the species might also occur. The lower elevations of the Chama River valley might preclude any dispersal between these ranges. However, the capability of martens to disperse through low-quality habitat is not fully understood.

Very few museum specimens of the marten have been secured in New Mexico since the first was obtained in the 1870s. Findley et al. (1975) listed only four specimens preserved in natural history collections from three locations: Chama, Rio Arriba

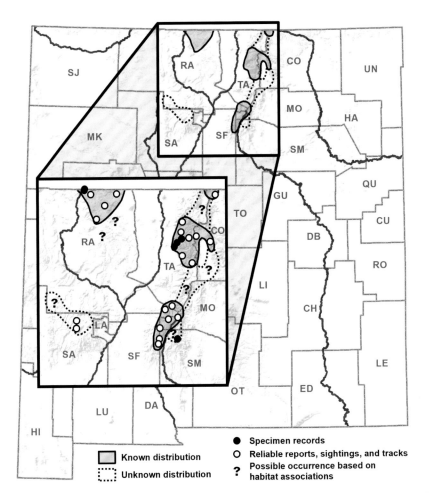

Map 20.1. Distribution of the Pacific marten based on specimen records and reliable reports, sightings, and tracks. Two specimen records (KU KUM 1466 and KU KUM 1467; shown as one location) serve as the basis for extending the marten's distribution into northwestern San Miguel County. The two specimens, two male martens, were collected on 15 August and 15 September 1884, respectively, by Lewis L. Dyche in New Mexico's "Las Vegas Mountains." According to Hoffmeister (1945:41), Dyche collected specimens for several seasons in the vicinity of the Harvey Ranch, "about twenty-five miles by trail up the Gallinas River (northwest) from Las Vegas, San Miguel County." On their distribution map for the species, Findley et al. (1975) did plot the two marten records northwest of Las Vegas in San Miguel County in the Sangre de Cristo Mountains. A marten confiscated at a hunter checkpoint at Raton Pass in Colfax County likely originated in Colorado and is therefore excluded from the distribution.

County; the "Las Vegas Mountains" (the southeastern portion of the Sangre de Cristo Mountains), San Miguel County; and Taos (presumably the nearby Taos Mountains), Taos County. And to Findley et al.'s (1975) tally of four, only three more New Mexico specimens have been added more recently. The specimen listed by Findley et al. (1975) from the Chama area was taken in 1893 (Bailey 1931), while more recently another was collected as a roadkill in 1995 (Long 1999; MSB:Mamm:281916). The Taos specimen (a skin) known to Findley et al. (1975) provided what is probably the earliest record from the state and was secured by Henry Crécy Yarrow in August 1874. The Taos locality was at the time "the southernmost on record thus far for the species" (Coues and Yarrow 1875:61). Long (1999) later identified two additional specimen records obtained from near Taos Ski Valley, Taos County.

Another specimen of the Pacific marten, now preserved in the Museum of Southwestern Biology, was confiscated by New Mexico Department of Game and Fish officers at a hunter checkpoint near the Colorado state line at Raton Pass, Colfax County in 1988. Although martens are known from ca. 50 km (30 mi) to the west in the Costilla Range (where one set of tracks was collected on a track plate by B. J. Long), it is likely the Raton Pass marten was actually taken in Colorado (NMDGF, unpubl. data), though it is still cataloged as a specimen from New Mexico.

Other records of martens not based on preserved specimens are available from photographs (including remote camera images); detection via snow-track and track-plate surveys; and observation reports in the scientific literature or compiled from various individuals, the latter of varying reliability (Bailey 1931; Long 1999, 2001a; NMDGF, unpubl. data; Map 20.1). Most of these records are from the Sangre de Cristo Mountains and fall mainly within two "core areas" of occurrence in the Taos and Pecos mountains (Long 2001a), primarily within the Wheeler Peak and Pecos wilderness areas on the Carson and Santa Fe national forests. Bailey (1931) summarized the available visual reports from the Sangre de Cristo Mountains including a sighting in 1881 near Truchas Peaks, in what is now Mora or Rio Arriba county (Edwords 1893) and from Twining in the upper Rio Hondo drainage (near the present-day village of Taos Ski Valley), Taos County. Since the mid-1900s, the area around Taos Ski Valley and nearby Williams Lake in the Wheeler Peak Wilderness has provided a number of reliable reports (Berghofer 1967; Long 1999, 2001a, 2003a) and this might be the best area in the state to find martens today. In 2012, a probable sighting was made of a marten above timberline on Lobo Peak, west of Taos Ski Valley (M. East, pers. comm.), and the species was documented via remote cameras in the Taos Ski Valley area (Gavilan Canyon and near Williams Lake) as recently as 2014–2015 (Long et

al. 2015). Snow tracks were observed and photographed near Capulin Peak, adjacent to Taos Pueblo land, in 2003 (Long 2003b). Two older sightings mentioned in the literature are considered reliable. One of them is from Lost Lake, south of Red River, in Taos County, dating back to 1961 (Berghofer 1967), the other from the Rio de La Casa basin (Rincon Bonito area), west of Cleveland, Mora County, perhaps from the 1940s (Barker 1953). The headwaters of the Rio Santa Barbara and Rio San Leonardo (north of Truchas Peak) on the northwestern side of the Pecos Wilderness in Rio Arriba County may represent other good locations for martens based on the number of reliable reports we have received (Long 2001a; NMDGF, unpubl. data). Two other records from the Pecos Wilderness are based on photographs of individuals: one of a marten in a tree near Hidden Lake below Jicarilla Peak in July 2004, and another of a marten crossing a talus field at Lake Katherine in July 2015 (Photo 20.8). In 1996, three black bear researchers observed a marten at 3,414 m (11,200 ft) on Touch-Me-Not Mountain in the Cimarron Range on the Colin Neblett Wildlife Area, Colfax County (C. Costello, in litt.). In 2001, marten tracks were collected on a track plate in Long Canyon in the Costilla Range, northwestern Colfax County, during a track-plate survey (Long 2001b). Other reports based on sightings considered to be reliable are from the Heart Lake area near Latir Peak in the Latir Peak Wilderness; the Rio Valdez, northeast of Cowles in the Pecos Wilderness; and the Elk Mountain area, including the Bull Creek drainage south of that peak (Long 1999, 2001a; NMDGF, unpubl. data).

Martens are less often reported from the San Juan Mountains in Rio Arriba County. Long (2001a) collected marten tracks on a track plate near the head of Apache Canyon, northeast of Chama, in 2001. A reliable sighting obtained in 2005 is from the Brazos Meadows west of the Cruces Basin Wilderness on the Carson National Forest (NMDGF, unpubl. data). Another reliable

Photo 20.11. Pacific marten in subalpine coniferous forest in the Rio Hondo watershed near Taos, Taos County on 22 April 2003. Dominant trees consist of Englemann spruce (*Picea engelmannii*) and subalpine fir (*Abies lasiocarpa*). Photograph: © Brian Jay Long.

Photo 20.12. Pacific marten habitat between Williams Lake and Twining along the Lake Fork trail in Taos County on 4 September 2008. Note the extensive ground cover of logs and other coarse woody debris, which serves to break up snow cover in winter. Martens use the subnivean zone thus created between ground and snow both as a shelter and for hunting prey. Photograph: © James N. Stuart.

sighting dating back to around 2000 was on private land along the East Fork of Wolf Creek in the upper Rio Chama drainage (NMDGF, unpubl. data). The southernmost record we have from the San Juan Mountains, obtained by a remote camera in 2018, is from east of Tierra Amarilla in the Brazos Cliffs area (Photo 20.9; NMDGF, unpubl. data). The relative paucity of reports from the San Juan Mountains compared to the Sangre de Cristo Mountains might be partly attributable to less public access, partly to a lesser amount of subalpine coniferous forest in that range.

A few intriguing reports of the Pacific marten have been obtained from the Jemez Mountains in Sandoval and Los Alamos counties. In August 1990, two wildlife biologists observed an animal they identified as a marten cross a road near Redondo Peak on the Valles Caldera National Preserve (VCNP; Jones and Schmitt 1997). Several other unverified reports from the Jemez Mountains, a few of which had enough detail to be considered probable, are available from the 1950s

Photo 20.13. A stump and a downed log have created a natural opening under the snow cover for a Pacific marten rest site in the upper Rio Hondo watershed near Taos, Taos County (22 April 2003). Elevation, topographic ruggedness, prey availability, forest canopy height, and winter snow depth may all be important features of Pacific marten habitat in New Mexico. Photograph: © Brian Jay Long.

Photo 20.14. A Pacific marten peers out of a hole in the snow in Taos County on 22 April 2003. Pacific martens depend on being able to access the subnivean zone during periods of deep snow cover in winter. Photograph: © Brian Jay Long.

Photo 20.15. Pacific marten habitat along the trail from the Taos Ski Basin to Williams Lake in Taos County. Although martens avoid venturing far from tree cover, they also use talus slopes and boulder fields. Photograph: © Mark L. Watson.

Photos 20.16. Use of talus slopes by Pacific martens at Lake Katherine in the Pecos Wilderness has been documented more than once. Photographs: © Brianne Kenny.

Photo 20.17. Red squirrel (*Tamiasciurus* sp.) midden in the Urraca Wildlife Management Area north of Questa in the Sangre de Cristo Mountains. Red squirrel middens are both food caches and debris piles containing the shredded cones already eaten by squirrels. They are a typical component of marten habitat, with research indicating that not only are red squirrels important prey but also their middens serve as winter resting sites and summertime maternal dens for martens. Photograph: © Mark L. Watson.

through 2012 with most from the general vicinity of the VCNP (Long 2002; NMDGF, unpubl. data). No martens were detected during survey efforts in the Jemez using remote cameras and track plates in 2002 at the VCNP, or in 2009 and 2014 in the San Pedro Parks Wilderness on the west side of the mountain range (Long 2009; Long et al. 2015). One observation report that we consider possibly valid was of a marten seen on Mesa Alta just north of the Jemez Mountains and south of the Rio Chama in Rio Arriba County (Long 2002). Long (2001a) suggested that the Tusas Range and Brazos Canyon area in the southern San Juan Mountains could provide a dispersal corridor for martens to range as far south as the Jemez Mountains. He also noted that the San Pedro Parks Wilderness, which has an extensive area of sub-alpine coniferous forest, warrants further survey effort. Frey et al. (2007) accepted the occurrence of the species in the Jemez Mountains based on

observation reports. However, in the absence of a voucher specimen or conclusive photograph, we consider the presence of the Pacific marten in this mountain range to be unverified.

Almost no information is available on the prehistoric occurrence and distribution of the Pacific marten in New Mexico. Lang and Harris (1984) reported possible marten remains among faunal skeletal material collected at the Arroyo Hondo Pueblo archeological site southeast of Santa Fe, Santa Fe County. Presumably any martens that might have been obtained by Native Americans who formerly occupied this area would have originated in the nearby Sangre de Cristo Mountains.

HABITAT ASSOCIATIONS

Both the American and Pacific martens are among North America's most specialized carnivores in terms of habitat requirements (reviewed by Buskirk and Powell 1994), and as such have been considered useful indicators of ecosystem health in boreal and old-growth forests (Wiebe et al. 2013). Martens are animals of either high latitudes or high elevations in North America where they are primarily associated with boreal or subalpine coniferous forest, respectively; within these forests they select mature and old-growth stands dominated by spruce (*Picea* spp.), fir (*Abies* spp.) and pine (*Pinus* spp.). Optimal habitat has at least a 30% canopy cover and complex physical structure at, or near, ground level such as large amounts of downed logs (e.g., wind-felled trees), stumps, tree snags, and sometimes rocky outcroppings or taluses (Koehler and Hornocker 1977; Buskirk and Ruggiero 1994). Regenerating stands of trees are generally less preferred than old-growth stands where both are available in a given area (Thompson et al. 2012). Forests that have been impacted by large wildfires or intensive logging are generally not suitable for martens, though at least in Alaska, partially burned or early successional forests in which downed logs are left in place may be used (see Buskirk and Ruggiero 1994). In Washington State, Pacific martens were found to frequent old-growth (at least 82-years-old) spruce-fir forests but infrequently used less mature (43-years-old or younger) stands (Koehler et al. 1990). Fragmented forests that contain openings less than 100 m (328 ft) apart were associated with lower marten capture rates in Utah (Hargis et al. 1999). Pacific martens may also forage in riparian zones along streams in mountain canyons and meadows within forests where herbaceous plants and low shrubs predominate. Broadleaf (deciduous) forests seem not to be favored habitat anywhere in the inclusive range of American and Pacific martens (see Buskirk and Ruggiero 1994). Martens that do not have an established territory, such as young and inexperienced individuals that are dispersing from their natal area, may be found in forest types that are of low habitat quality, such as immature or less-dense stands (Thompson et al. 2012).

In New Mexico, the Pacific marten is primarily associated with the high-elevation Hudsonian life zone, along with adjacent lower elevation areas where it grades into the Canadian zone; these zones correspond to subalpine coniferous forest and upper montane coniferous forest, respectively (Dick-Peddie 1993; see also Chapter 2). Dominant tree species within these forests include Engelmann spruce (*Picea engelmannii*) and subalpine fir (*Abies lasiocarpa*) in the subalpine coniferous forests (i.e., spruce-fir forests), and blue spruce (*Picea pungens*), white fir (*Abies concolor*), Douglas-fir (*Pseudotsuga menziesi*), limber pine (*Pinus flexilis*), and quaking aspen (*Populus tremuloides*) in the upper montane coniferous forests. Understory plants from habitats in New Mexico include serviceberry (*Amelanchier utahensis*), baneberry (*Actaea rubra*), elderberry (*Sambucus racemosa*), cliffbush (*Jamesia americana*), kinnikinnick (*Arctostaphylos uva-ursi*), common juniper (*Juniperus communis*), and whortleberry (*Vaccinium myrtillus*) (Long 2001a; C. Costello, in litt.). Lower-elevation forests, below about 2,600 m (8,500 ft) and dominated by ponderosa pine (*Pinus ponderosa*), are apparently not used or used infrequently (see Buskirk and Ruggiero 1994). In Colorado, martens were observed only infrequently in ponderosa pine or aspen stands that were located just below lodgepole pine (*Pinus contorta*) stands (Yeager and Remington 1956). Long (2001a) suggested that the overall ruggedness of the Sangre de Cristo Mountains, combined with a greater abundance of talus fields, cirques, streams, and lakes, might make this range more suitable for martens than the San Juan Mountains in New Mexico. Despite some important limitations and potential biases, a GIS model developed for New Mexico based on known marten localities indicated that elevation, topographic ruggedness, the availability of important rodent prey, and forest canopy height might be the most important variables in predicting suitable marten habitat, especially when combined with winter snow depth (Menke and Perry 2009). In Colorado, Pacific martens primarily used older-aged Engelmann spruce-subalpine fir and lodgepole pine, at least in summer (Baldwin and Bender 2008). Notably, the species shares many of the same habitat requirements as the snowshoe hare (*Lepus americanus*), which has a distribution similar to that of the Pacific marten in New Mexico (Frey and Malaney 2006; see also Chapter 6).

Various researchers have noted the tendency for the Pacific marten to remain close to tree cover and avoid open areas (Streeter and Braun 1968; reviewed by Powell et al. 2003), and in Colorado, wetlands and uplands dominated by herbaceous vegetation were avoided (Baldwin and Bender 2008). However, martens also use talus and boulder fields within forests and even venture above the treeline where perhaps the spaces among large rocks provide adequate escape cover and thermally suitable resting sites while also harboring prey animals (see Buskirk and Ruggiero 1994). Pacific martens have been observed in the tree-less alpine zone above about 3,658 m (12,000 ft) in both New Mexico and Colorado, including within boulder fields and on open tundra (Streeter and Braun 1968; Long 2001a). Even meadows and burned areas with cover in the form of downed timber or rock piles may be used (Koehler and Hornocker 1977). Although these non-forested areas can provide foraging habitat in summer, they are likely avoided in winter when snowpack can limit the marten's access to escape cover and prey (see Buskirk and Ruggiero 1994).

A crucial component of the Pacific marten's life during winter is the availability of subnivean ("underneath the snow") environment that can be accessed during periods of deep snowpack (Sherburne and Bissonette 1994). The subnivian environment used by martens is created by coarse woody debris on or near the ground such as wind-felled trees, accumulations of logs, and tree stumps (see also Chapter 21 for an extensive discussion on the role of the subnivian space in providing rodent prey). In high-quality marten habitat, woody debris is often so extensive as to make human travel on foot extremely difficult. Coarse woody debris serves to intercept snowfall and break up deep snow cover, allowing martens to access shelter and forage for prey at ground level beneath snow (Sherburne and Bissonette 1994).

Another common feature of forests used by Pacific martens is the abundance of feeding middens (large piles of the discarded remains of conifer cones found around conifer trees) belonging to red squirrels (*Tamiasciurus hudsonicus* and *T. fremontii*). Red squirrels use such middens as sites to cache conifer cones for food during winter and maintain access tunnels to them during periods of snowfall. Middens not only signify the presence of an important prey species for martens but also may be used by these carnivores as winter resting sites and summertime maternal dens (Sherburne and Bissonette 1993; Ruggiero et al. 1998). The southwestern red squirrel (*T. fremontii*), although more widespread and less specialized in its habitat requirements in New Mexico than the Pacific marten, is common in areas occupied by martens in the state.

Microclimatic conditions are an important determinant of where martens may occur, and a diversity of structures such as tree snags, logs, and rocks are necessary to provide sites for denning or sheltering (see Buskirk and Ruggiero 1994). During warmer months, martens may be less selective in where they rest. However, in winter when the energetic costs associated with freezing temperatures are much higher, the selection of well-insulated resting sites can be critical to survival. During winter in southeastern Wyoming, adult Pacific martens rested more often in spruce-fir forests than in other stand types where they made extensive use of partially decomposed, coarse woody debris that seemed to provide the thermal properties required to prevent heat loss (Buskirk et al. 1989). During spring and summer in Wyoming, females were found to use rock crevices, hollows in large tree snags and logs, and red squirrel middens as natal dens

Photos 20.18a and b. Trail camera photos showing the same tree being visited multiple times in March 2003 in the upper Rio Hondo watershed in Taos County. One of the authors (B. J. Long) used elk or deer meat secured to tree trunks to attract martens to sites monitored by remote cameras. Pacific martens are active at all times of year, both during the day and at night. Photographs: © Brian Jay Long.

(parturition sites) and maternal dens (sites where kits are raised but not whelped) (Ruggiero et al. 1998).

A natal den was documented during a radio telemetry study at Taos Ski Valley in 2003. A collared female was tracked to a standing snag about 30 cm (12 in) in diameter with a woodpecker hole about 6 m (20 ft) above ground. A large pile of woody debris covered the snow around the base of the tree, apparently excavated by the marten to enlarge the cavity (B. J. Long, pers. obs.).

LIFE HISTORY

Although some research on the life history of the Pacific marten has been conducted in New Mexico (e.g., Long 2003a), most of the published information is from elsewhere in the species' range. Good reviews of the biology of both the American and Pacific martens include those by Strickland et al. (1982), Strickland and Douglas (1987), Clark et al. (1987), Buskirk and Ruggiero (1994), Powell et al. (2003), and Aubry et al. (2012). In the following discussion, we rely primarily on information obtained from Pacific marten populations outside of New Mexico.

Diet and Foraging

Martens do not hibernate and instead are active year-round. Because they lack the ability to store large fat reserves or withstand prolonged fasting, martens must forage frequently (see Powell et al. 2003), and their activity can be both diurnal and nocturnal. In one study in California, most nocturnal activity by Pacific martens apparently occurred during winter, whereas daytime activity was observed primarily in summer, the opposite of what might be expected if martens were adjusting their activity based on ambient temperatures (Zielinski et al. 1983). Patterns of activity are likely influenced instead by whether their main prey are nocturnal or diurnal, and so may vary depending on the season and what prey are available in a given area. Although some foraging is conducted in trees, most prey is likely sought on the ground. In winter, martens will tunnel beneath the snow to find prey in subnivean cavities. Like most weasels, they are active, curious, and quick-moving predators, often standing on their hind legs to survey their surroundings when on the ground (Strickland and Douglas 1987). Although Clark (1999) reported that martens may occasionally enter water, such as mountain streams, in pursuit of prey and are capable of swimming and diving, this behavior apparently has not been documented by other researchers.

Martens feed primarily on animal prey and are adept hunters. The diet includes a wide range of

Photos 20.19a, b, and c (*left to right*). The diet of Pacific martens has not been studied in New Mexico, but based on research elsewhere it likely consists primarily of small mammals. The least chipmunk (*Neotamias minimus*; photographed between Twining and Williams Lake in Taos County on 15 September 2008 in Photo 20.19a), golden-mantled ground squirrel (*Callospermophilus lateralis*, Photo 20.19b), and red squirrel (*Tamiasciurus* sp.; Photo 20.19c) occur within the range of the marten in New Mexico, and all have been documented as prey in other western states. Photographs: © James N. Stuart.

prey species though small mammals (especially cricetid rodents) make up the bulk of their food (Weckwerth and Hawley 1962; Koehler and Hornocker 1977; see also Buskirk and Ruggiero 1994). A study of stomach and intestinal contents in Pacific martens collected in northern Colorado during fall and winter indicated that voles (*Microtus* spp.) were the most important prey items (present in ~80% of the samples), followed by shrews (*Sorex* spp.; ~40%) with other items such as red squirrels, snowshoe hares, birds, beavers (*Castor canadensis*), fish, insects, and vegetative matter present in lesser frequencies and in some cases presumably scavenged (Gordon 1986). Food habits have not been studied in New Mexico, but many of the same prey species reported by researchers in other western states are also found within the range of martens in New Mexico (Findley et al. 1975). These species include Gapper's red-backed vole (*Myodes gapperi*), heather vole (*Phenacomys intermedius*), montane vole (*Microtus montanus*), long-tailed vole (*M. longicaudus*), red squirrel, least chipmunk (*Neotamias minimus*), golden-mantled ground squirrel (*Callospermophilus lateralis*), northern pocket gopher (*Thomomys talpoides*), American pika (*Ochotona princeps*), snowshoe hare, and several species of shrews. The greatest diversity of foods is likely consumed during summer. In winter, when prey diversity is reduced, voles (found in the subnivean

Photo 20.20. American pika (*Ochotona princeps*) on 8 October 2016 in a boulder field in Taos County near Williams Lake. Pacific martens likely prey on pikas where the two species co-occur in the Taos Ski Basin and surrounding area. Photograph: © Jean-Luc Cartron.

zone), red squirrels, and hares are the staple food items in some populations (Zielinski et al. 1983; see also Buskirk and Ruggiero 1994). In Wyoming, Pacific martens may be found more often in proximity to red squirrel middens when snow is present, suggesting both the importance of red squirrels in the marten's winter diet and the midden itself, which provides shelter and access to foraging areas below the snow (Sherburne and Bissonette 1993).

Birds, not just nestlings and eggs, may be

taken during spring and summer months when they are presumably more vulnerable at nest sites (Weckwerth and Hawley 1962; Clark 1999). In Idaho, martens have been suspected of taking eggs and fledglings from the cavity nests of the boreal owl (*Aegolius funereus*), a species that also co-occurs with the Pacific marten in northern New Mexico (Stahlecker 2010).

Insects and vegetable matter are consumed but to a lesser extent than vertebrate prey. In Montana, Weckwerth and Hawley (1962) reported remains of social species of Hymenoptera (hornets, yellow jackets, and ants) and smaller quantities of beetles in marten scats. Streeter and Braun (1968) reported an observation of a Pacific marten feeding on moths above timberline in Colorado. Vegetable matter such as the soft mast and seeds of shrubs and trees may be consumed in late summer or fall, including the fruits of blueberry and huckleberry (*Vaccinium* spp.) and raspberry (*Rubus* spp.) (see Buskirk and Ruggiero 1994).

Martens are also known to rely on carrion when available, and they may consume portions of the carcasses of ungulates and perhaps other animal species too large to be taken as prey (Weckwerth and Hawley 1962). Lapinski (2006) described an observation from Idaho of a Pacific marten feeding on a portion of an elk (*Cervus canadensis*) carcass that had been hung in camp by a hunting party. The marten apparently had made multiple visits to one of the hind quarters and was holed up in a large cavity that it had "excavated" in the meat! In Colorado, a marten was observed by a hunting party feeding on the discarded viscera of an elk that had been killed (Yeager and Remington 1956). Incidentally, odoriferous meat, such as canned fish or chicken, evidently attracts martens. Even large pieces of elk or deer meat, secured to a tree trunk with wire mesh, have been used effectively by one of us (B. J. Long) in New Mexico as a lure to attract martens to survey sites equipped with remote cameras (see under "Status and Management").

Although Pacific martens do most of their hunting in association with tree or shrub cover or along streams, they are also known to leave forested areas to forage in nearby talus fields in pursuit of American pikas and possibly other animals such as chipmunks and ground squirrels. A marten was observed and photographed foraging in a talus field near Lake Katherine in 2015 where it captured an unidentified prey (D. Romero, pers. comm.; see Photo 20.8). Presumably hunting in talus is conducted mostly during summer months when prey are active above ground. Bailey (1931:322), citing Edwords (1893), stated that "in 1881, at his camp near Truchas Peaks, Dyche saw some martens catching conies [pikas] in the rocks, and succeeded in getting two of them." It is unknown to what extent American pikas are preyed upon by martens in New Mexico, although the two species co-occur in several places in the Sangre de Cristo Mountains, including at Taos Ski Valley.

Reproduction and Social Behavior

Both male and female martens reach sexual maturity in about one year but may not be able to successfully breed for another year, perhaps less in the case of females (see Powell et al. 2003). The sex ratio in populations where trapping is not allowed is likely close to 1:1, but may be female-biased in harvested populations as males tend to be taken in traps more frequently (see Buskirk and Ruggiero 1994). Males are polygynous (i.e., they mate with more than one female) and females are likely polyandrous (i.e., they mate with more than one male). Females enter estrus during summer, and it is during this period that most social activity occurs among individuals. Yeager and Remington (1956) suggested that the mating period in Colorado lasted from about mid-June to mid-August and this likely also corresponds to the breeding season in New Mexico. Courtship and copulation were described by Ruggiero and Henry (1993) at a study site in

Wyoming. A courting pair often engages in much running and tree-climbing over a period of one to three hours and may cover distances of 0.5 to 1.3 km (0.3 to 0.8 mi). Males will frequently scent-mark and follow scent trails during the courtship period. Copulation can take place in trees or on the ground and may occur at least three times within the span of 1.5 hours; each copulation can last from two to 47 minutes, during which the male may grasp the female by the neck with his teeth. Often these bouts of mating take place in the presence of the female's kits (Ruggiero and Henry 1993). At his study site at Taos Ski Valley, Long (2003a) observed a radio-collared female marten being pulled from her den by her neck by another marten, presumably a male, on July 24; a juvenile had been observed at that same location two days earlier, suggesting that the local female was still being accompanied by young. It is likely that copulation induces the female to ovulate, thereby increasing the odds that her eggs will be fertilized (Strickland and Douglas 1987).

Following insemination, a fertilized zygote in a female marten develops into a blastocyst and then becomes inactive in a process called delayed implantation. The blastocyst remains in this suspended state for 7.0–8.5 months through the fall and winter. During late winter or early spring, possibly in response to increasing daylight, the blastocyst implants in the wall of the uterus and development of the fetus proceeds. Gestation lasts about 27 days and the young martens, called kits, are usually born in late March or April but sometimes as late as May, based on research in Montana (Jonkel and Weckwerth 1963). A female produces only one litter per year, and the number of kits ranges from one to five but is typically three or four (see Powell et al. 2003). Kits are born blind and nearly hairless but develop rapidly; they begin taking solid food at about 40 days of age and are usually weaned at about two months old. Yeager and Remington (1956) reported an observation in Colorado of an adult

Photo 20.21. Pacific marten in the upper Rio Hondo watershed in Taos County on 22 April 2003. Home ranges were estimated by one of us (B. J. Long) for three martens captured and fitted with radio collars in the Taos Ski Basin in 2003. The three radio-collared martens were tracked during the late winter and spring. The estimated home ranges of the two radio-collared females were 1.2 km² (0.47 mi²) and 2.9 km² (1.12 mi²), respectively, while the male had an estimated home range of 2.5 km² (0.97 mi²). Photograph: © Brian Jay Long.

marten traveling on the ground accompanied by four kits on July 7. Young martens leave the company of the mother in late summer but typically do not disperse until the following year (see Buskirk and Ruggiero 1994). In Oregon, young Pacific martens were at least 14 months old when they began to disperse from their natal area during June-August, and by 16 months were considered adults (Bull and Heater 2001a). In a Wyoming study, the age at which a marten was considered to be an adult was about one year (Buskirk et al. 1989). The presence of kits or subadults that have not yet separated from their mother may be used by males to locate females during the summer mating season, when the female is typically still engaged in rearing young (Ruggiero and Henry 1993). Although the male takes no part in the care of his offspring, his continued presence in the area may exclude other males and thereby enhance foraging opportunities for his mate and their kits (Strickland and Douglas 1987).

Although martens can live up to 13 years (Powell et al. 2003), this is likely exceptional. In populations that are subjected to trapping, the average lifespan is much shorter, perhaps only a few years (see Buskirk and Ruggiero 1994).

Martens are generally considered to be shy animals, and they live mostly solitary lives, avoiding contact with other members of their species except during the breeding season. They are rarely seen by humans, even those who live in contact with them, and yet some martens can be inquisitive and will even approach people (Buskirk and Ruggiero 1994).

Like all carnivores, olfactory communication is important. Martens possess scent glands, which are not as odoriferous (at least to humans) as those of some other weasels (Buskirk and Ruggiero 1994). Both sexes have an abdominal gland covered with short bristly hairs that produces an oily fluid with the distinctive musky odor characteristic of the species, in addition to paired anal scent glands and glands on the plantar feet. Both males and females scent-mark by dragging their bellies over logs or branches (Clark et al. 1987). One of us (B. J. Long), while conducting a summertime survey along a trail at Taos Ski Valley, watched a marten scent marking by dragging its abdomen on rocks, then noted the strong musky scent originating from those same rocks.

Martens make a variety of agonistic and distress calls when in live traps including vocalizations that sound like huffs, pants, growls, chuckles, screams, and whines (Belan et al. 1978), and they may hiss, growl, or make chuckling sounds when approached by a human in the wild, either while on the ground or from a perch in a tree (Yeager and Remington 1956). During mating encounters, martens may utter cries and growls that resemble those of housecats fighting (Ruggiero and Henry 1993).

Adult martens of the same sex compete for and defend exclusive territories. The territory of a male is often two to three times that of a female and typically overlaps that of one or more females. Presumably, the location and size of these territories determines to a large extent which individuals in a population will mate. Juvenile martens may be more transient and not have established territories; 47 of 85 martens captured during a study in Montana were considered transients, indicating that the fluctuations in the composition of a population can be substantial (Hawley and Newby 1957). One juvenile male marten in this study traveled approximately 40 km (25 miles) between August and October. In Oregon, juveniles were found to disperse 28–43.2 km (17.4–26.8 miles) (Bull and Heater 1995, 2001b). Territorial aggression among male Pacific martens can be severe and in Oregon has been shown to result in mortalities (Bull and Heater 1995, 2001a).

The home range of a marten may be larger than a defended territory and includes the total area in which an individual may travel to obtain prey or other resources, including lower-quality habitat. In a short-term study in New Mexico, Long (2003a) reported home range sizes for three radio-collared individuals from Taos Ski Valley that were tracked during late winter and spring in 2003. The two females had estimated home ranges of 1.2 km² (0.47 mi²) and 2.9 km² (1.12 mi²), respectively, and the male had a home range of 2.5 km² (0.97 mi²). The male used an equipment shed in the ski area as a rest site on numerous occasions and was a frequent visitor to a nearby restaurant, where he was recaptured twice. Long-term studies of home range in other states yield somewhat different estimates, presumably due to differences in habitat quality or study methodology. In Montana, mean home range sizes were 2.4 km² (0.92 mi²) for males and 0.70 km² (0.27 mi²) for females (Hawley and Newby 1957). In a study in

Photo 20.22. Multiple sets of tracks belonging to a marten, an American ermine (*Mustela richardsonii*), and possibly a red squirrel (*Tamiasciurus fremonti*). Underneath the snow is a fallen tree, which is a typical activity center for multiple species, with its cavities and subnivean passageways to the forest floor where voles (*Microtus* spp.) and other prey species live. In areas where martens are found in New Mexico, the tracks of other predators include the long-tailed weasel (*Neogale frenata*), ermine (generally in forest openings and meadows), and bobcat (*Lynx rufus*), all species that potentially interact with martens. Photograph: © Brian Jay Long.

northeastern Oregon, home ranges were substantially greater and averaged 27.2 km² (10.5 m²) for 10 males and 14.2 km² (5.5 mi²) for nine females (Bull and Heater 2001b). The authors suggested that these larger home range sizes could be attributed to several factors including prey availability, intraspecific behavior, predation, habitat characteristics, and density of martens (Bull and Heater 2001b).

Interspecific Interactions and Disease

In addition to being harvested by humans for their fur, martens are subject to predation by several other species of mammals and birds including the coyote (*Canis latrans*), North American red fox (*Vulpes fulva*), bobcat (*Lynx rufus*), Canada lynx (*Lynx canadensis*), cougar (*Puma concolor*), great horned owl (*Bubo virginianus*), northern goshawk (*Accipiter gentilis*), and golden eagle (*Aquila chrysaetos*) (reviewed by Buskirk and Ruggiero 1994; Bull and Heater 2001a; Armstrong et al. 2011). Research in Oregon indicates that most predation on martens occurs during the non-winter months (Bull and Heater 2001a).

Disease transmission to martens from other species is not well understood. Several Pacific martens in California were found to be seropositive for sylvatic plague (*Yersinia pestis*) antibodies during early winter, following a period when these animals were preying on chipmunks and ground squirrels, the apparent source of plague-carrying fleas found on the martens (Zielinski 1984).

Photo 20.23. Pacific marten on 22 April 2003 in the upper Rio Hondo watershed in Taos County. Most of the marten habitat in New Mexico is found on the Carson and Santa Fe national forests, and the majority of occurrence records are from localities within or adjacent to designated wilderness areas where human disturbance of habitat is minimal. Threats to the species in the state include climate-driven impacts affecting snow cover and an increased incidence of high-severity fires. Photograph: © Brian Jay Long.

Sylvatic plague, though widespread in many other mammal species in our state, has not been reported in New Mexico martens.

STATUS AND MANAGEMENT

The Pacific marten is an important furbearer in Canada and the Pacific Northwest. Both the American and Pacific martens are relatively easily trapped due to their inquisitive nature, and in some regions occupied by the American marten (e.g., eastern Canada and the north-central United States) trapping has contributed to population declines (reviewed by Buskirk and Ruggiero 1994).

The Pacific marten (as "pine marten") is classified as a protected furbearer in New Mexico under state law (see Chapter 4), though trapping of this species is no longer allowed by regulation. In the early 1900s, the species was considered an unprotected fur animal with no closed season. Following the passage of a state law in 1939 to regulate the harvest of furbearers, the New Mexico Game Commission established an open season on martens and several other furbearers for licensed trappers and set an annual harvest season extending from either 1 October to 1 March or 1 November to 31 March. The season on martens was closed in 1947 and has remained so up to the present with few exceptions. During 1963–1964, the season was reopened in the state for "research purposes" though no martens were harvested during that trapping year (Berghofer 1967:191). One-month trapping seasons were also opened in December during both 1968 and 1969, but whether any martens were taken at those times remains unknown.

Due to the species' limited distribution in New Mexico and its reliance on old-growth forests, the Pacific marten (as American marten) was listed in 1975 as Threatened under the New Mexico Wildlife Conservation Act of 1974, one of the first mammals to receive some protection under that state law (Jones and Schmitt 1997). The US Forest Service considers the marten a Sensitive Species on both the Carson and Santa Fe national forests in New Mexico. Local threats to martens in New Mexico include human development in the Taos Ski Valley area, including on the Carson National Forest where the expansion of recreational facilities and trails all involve the cutting of old-growth forest in prime marten habitat.

Based on the accounts of several 20th-century wildlife managers, the marten was likely never common in New Mexico. Ligon (1927:187) noted that it is a "valuable fur-bearer which is found sparingly in New Mexico. It is confined to the forests of the high altitudes of the Sangre de Cristo Range, from the head of the Pecos northwards. Its scanty numbers prevent it from doing much damage to small game in this State." Bailey (1931) commented on the high market value of marten fur and, given the species' scarcity in the state, the desirability of more effective laws governing its take. (He also noted that captive rearing of martens for the fur trade, though "tempting," had not been effective.) Because the marten was never trapped in large numbers in New Mexico—Barker (1953:148), for instance, considered it "exceedingly rare"—it therefore has always been of little economic importance in our state. Lee (1967:2) noted that the species had a "limited season" in New Mexico and that "high prices may be obtained for pine marten skins, but these animals are so rare and their habitat so difficult to access in the wintertime when the pelts are prime, that they are rarely taken and infrequently show up on the fur sale reports."

In adjacent Colorado, which has more extensive subalpine forests and a more robust and widespread population of the Pacific marten than New Mexico, the species is presently a harvestable furbearer that can be taken from 1 November through the end of February. During the period 1975–1994, more than 4,100 martens were legally harvested in Colorado, or an average of about 200 per year. Colorado banned the use of lethal and leg-hold traps in 1996 and the current legal methods of take are by hunting or use of cage or box-type live traps. A closed season was implemented in Colorado in 1995 due to concerns about the sustainability of the harvest rates in the state. The season was reopened in 2006, and as of 2020 the annual reported harvest of martens (not available for every year) was quite variable (Colorado Parks and Wildlife, unpubl. data). Beginning in the April 2021 license year, Colorado implemented a licensing change for furbearer harvesters that will help secure more precise harvest estimates for the marten in the future (M. Vieira, pers. comm.).

In states where martens are harvested as furbearers, data from trappers can provide important information on the distribution and status of the species which in turn can inform management decisions. In New Mexico, where the species is not harvested, such information is more difficult to obtain. A variety of non-invasive survey methods have been devised for martens including track surveys, either in snow or through the use of track plates; the use of hair snares at bait stations; scat surveys, usually accompanied by DNA analysis to determine species, or the use of trained scat-detection dogs; and remote cameras, also called game cameras or camera traps, which are typically placed at bait stations or scent stations (Zielinski and Kucera 1995; Long and McKay 2012). In New Mexico, snow-track, track-plate, and remote camera surveys have all been used with varying degrees of success to better define the distribution and habitat requirements of martens (e.g., Long 1999, 2001a, 2001b, 2002, 2009; Long et al. 2015). With recent improvements in remote camera technology, it is likely that in the future, this technique will prevail as the most effective and cost-efficient method to survey for and monitor this species in New Mexico.

Because direct mortality of the Pacific marten by humans is believed to be infrequent in New Mexico, the main focus for conservation is protection of its limited and specialized habitat. Fragmentation of marten habitat even at a small spatial scale from forest destruction or alteration can result in abandonment of an area by martens; at a larger scale, habitat fragmentation can isolate small populations which then become more prone to extirpation (Hargis et al. 1999). The majority of marten habitat in New Mexico is contained within the Carson and Santa Fe national forests, and most locality records known to us indicate that the state's most robust populations occur within or adjacent to designated wilderness areas (i.e., the Wheeler Peak, Pecos, Latir Peak, and possibly also Cruces Basin wildernesses), where subalpine forests are expansive and human disturbance of habitat is minimal. The removal of old-growth conifer stands and the resulting fragmentation of forests through commercial logging were historically the most important threat to marten habitat in New Mexico, but this activity has waned as forest management practices have changed. Since the decline of commercial logging in the 1990s, most tree removal in New Mexico's montane forests has been conducted selectively to reduce fire risk in the vicinity of human dwellings or for small-bole timber or fuel wood. Today and for the foreseeable future, catastrophic wildfire and forest die-offs due to drought, disease, and outbreaks of insect pests pose the greatest threats to marten habitat. In particular, Engelmann spruce-subalpine fir forests are not fire adapted and large-scale wildfires can eliminate large swaths of this forest type important for martens (Koehler et al. 1975; Koehler and Hornocker 1977; see also Chapters 2 and 5). For example, much of the subalpine forest in the Jemez Mountains—where the Pacific marten possibly occurs—has been devastated by several major wildfires since the 1970s.

The long-term effects of climate change on the Pacific marten are a cause for concern (Chapter 5). Changes in precipitation patterns and increased temperatures in the Southwest are

Photo 20.24. Unleashed dog (*Canis familiaris*) in marten habitat near Arroyo Seco. Human development and the increasing presence of dogs are a concern in some areas of the marten's distribution in New Mexico. Dogs can prey on local wildlife, and their scent and droppings may also interfere with the olfactory communication of many native animals (see Chapter 33). Photograph: © Brian Jay Long.

likely to result in the retraction of subalpine forests to higher elevations and latitudes, thus resulting in a curtailment of marten habitat in New Mexico (Lawler et al. 2012; Chapter 5). The shorter-term possible effects of climate change include a reduction in seasonal snowpack, soil moisture, and available prey for martens, as well as an increase in the frequency of catastrophic forest fires and tree loss due to insect infestations as the climate in the Southern Rockies becomes warmer and drier.

A reduction in understory shrubs and herbaceous vegetation in coniferous forests can potentially impact the prey base of martens, particularly small rodents. In particular, understory vegetation can be reduced from overgrazing by livestock. The reliance of martens on the red squirrel—both as prey and as producers of middens—is not yet fully understood. Red squirrel populations in the Southern Rockies and elsewhere in the Southwest are limited to high-elevation conifer forests and are often disjunct and isolated from each other, thus potentially more prone to local extirpation. Maintenance of these squirrel populations within the range of the

Pacific marten in New Mexico is therefore likely necessary for the persistence of marten populations in the state.

The encroachment of roads and other development into marten habitat can be the source of additional impacts on the species. At least three of the few museum specimens obtained in New Mexico were collected as roadkills (Long 1999). Although road traffic is probably not a significant source of mortality in New Mexico at present, an increase in housing, recreational developments, and associated traffic is to be expected in the future, at least in portions of the species' range. A study of off-highway vehicle use in Pacific marten habitat in California did not indicate that martens were impacted by motorized recreation, though additional research on this increasingly popular activity is warranted (Zielienski et al. 2008). It should be noted that martens have shown some level of accommodation with human development (Koehler et al. 1975), as in the semi-developed Taos Ski Valley area, where the species has persisted and been found to use homes, sheds, and other man-made structures as shelters (Long 2003a). We do not know if the operation of ski recreation areas within marten range in New Mexico significantly affects marten use of adjacent forests, though this would be useful to investigate. Another factor that has not been studied is the effect of dogs that increasingly are brought into marten habitat by hikers in the Taos Ski Valley area and elsewhere (see Photo 20.24).

Although much is known about the biology of the Pacific marten, many unanswered questions remain that are likely important to the conservation and management of the species in North America (e.g., reviews by Buskirk and Ruggiero 1994; Aubrey et al. 2012). Additional research in New Mexico, at the southern limit of the species' range and where it has been little studied, is especially warranted.

ACKNOWLEDGMENTS

We thank J. Apker, M. East, E. Goldstein, J. Klingel, E. Nelson, D. Romero, S. Romero, N. Forman, and M. L. Watson for assistance and information, and J.-L. E. Cartron, J. K. Frey, and an anonymous reviewer for constructive comments on the manuscript.

LITERATURE CITED

Armstrong, D. M., J. P. Fitzgerald, and C. A. Meaney. 2011. *Mammals of Colorado*. 2nd ed. Boulder: University Press of Colorado.

Aubry, K. B., W. J. Zielinski, M. G. Raphael, G. Proulx, and S. W. Buskirk, eds. 2012. *Biology and conservation of martens, sables, and fishers: a new synthesis*. Ithaca, NY: Comstock Publishing Associates.

Bailey, V. 1931 (=1932). *Mammals of New Mexico*. North American Fauna 53. Washington, DC: US Department of Agriculture, Bureau of Biological Survey.

Baldwin, R. A., and L. C. Bender. 2008. Distribution, occupancy, and habitat correlates of American martens (*Martes americana*) in Rocky Mountain National Park, Colorado. *Journal of Mammalogy* 89:419–27.

Barker, E. S. 1953. *Beatty's cabin*. Albuquerque: University of New Mexico Press.

Belan, I., P. N. Lehner, and T. Clark. 1978. Vocalizations of the American pine marten, *Martes americana*. *Journal of Mammalogy* 59:871–74.

Berghofer, C. B. 1967. Protected furbearers in New Mexico. In *New Mexico Wildlife Management*, 187–194. Santa Fe: New Mexico Department of Game and Fish.

Bradley, R. D., L. E. Ammerman, R. J. Baker, L. C. Bradley, J. A. Cook, R. C. Dowler, C. Jones, D. J. Schmidly, F. B. Stangl Jr., R. A. Van Den Bussche, and B. Würsig. 2014. Revised checklist of North American mammals north of Mexico, 2014. *Occasional Papers, Museum of Texas Tech University* 327:1–27.

Bull, E. L., and T. W. Heater. 1995. Intraspecific predation on American marten. *Northwestern Naturalist* 76:132–34.

———. 2001a. Survival, causes of mortality, and reproduction in the American marten in northeastern Oregon. *Northwestern Naturalist* 82:1–6.

———. 2001b. Home range and dispersal of the American marten in northeastern Oregon. *Northwestern Naturalist* 82:7–11.

Buskirk, S. W., S. C. Forrest, M. G. Raphael, and H. J. Harlow. 1989. Winter resting site ecology of marten in the central Rocky Mountains. *Journal of Wildlife Management* 53:191–96.

Buskirk, S. W., and R. A. Powell. 1994. Habitat ecology of fishers and American martens. In *Martens, sables, and fishers: biology and conservation*, ed. S. W. Buskirk, A. S. Harestad, M. G. Raphael, and R. A. Powell, 283–96. Ithaca, NY: Cornell University Press.

Buskirk, S. W., and L. F. Ruggiero. 1994. American marten. In *The scientific basis for conserving forest carnivores: American marten, fisher, lynx, and wolverine in the western United States*, ed. L. F. Ruggiero, K. B. Aubry, S. W. Buskirk, L. J. Lyon, and W. J. Zielinski, 7–37. USDA Forest Service. General Technical Report RM-254.

Carr, S. M., and S. A. Hicks. 1997. Are there two species of pine marten in North America? Genetic and evolutionary relationships within *Martes*. In *Martes: Taxonomy, Ecology, Techniques, and Management*, ed. G. Proulx, R. Goddard, and H. Bryant, 15–28. Edmonton: Provincial Museum of Alberta.

Clark, T. W. 1999. American marten, *Martes americana*. In *The Smithsonian Book of North American Mammals*, ed. D. E. Wilson and S. Ruff, 165–66. Washington, DC: Smithsonian Institution Press.

Clark, T. W., E. Anderson, C. Douglas, and M. Strickland. 1987. *Martes americana*. *Mammalian Species* 289:1–8.

Coues, E., and H. C. Yarrow. 1875. Report upon the collections of mammals made in portions of Nevada, Utah, California, Colorado, New Mexico, and Arizona, during the years 1871, 1872, 1873, and 1874. In *Annual report upon U.S. geographical and geological explorations and surveys west of the one *hundredth meridian, in charge of 1st Lieutenant B. M. Wheeler*, Vol. 5, 65–66, 960–79. Washington, DC: US Government Printing Office.

Dawson, N. G., and J. A. Cook. 2012. Behind the genes: diversification of North American martens (*Martes americana* and *M. caurina*). In *Biology and conservation of martens, sables, and fishers: a new synthesis*, ed. K. B. Aubry, W. J. Zielinski, M. G. Raphael, G. Proulx, and S. W. Buskirk, 23–38. Ithaca, NY: Comstock Publishing Associates.

Dick-Peddie, W. A, ed. 1993. *New Mexico vegetation: past, present, and future*. Albuquerque: University of New Mexico Press.

Edwords, C. E. 1893. *Camp-fires of a naturalist: the story of fourteen expeditions after North American mammals, from the field notes of Lewis Lindsay Dyche. . . .* New York: D. Appleton.

Elbroch, M. 2003. *Mammal tracks and sign: a guide to North American species*. Mechanicsburg, PA: Stackpole Books.

Findley, J. S., A. H. Harris, D. E. Wilson, and C. Jones. 1975. *Mammals of New Mexico*. Albuquerque: University of New Mexico Press.

Frey, J. K., M. A. Bogan, and T. L. Yates. 2007. Mountaintop island age determines species richness of boreal mammals in the American Southwest. *Ecography* 30:231–40.

Frey, J. K., and J. L. Malaney. 2006. Snowshoe hare (*Lepus americanus*) and mountain cottontail (*Sylvilagus nuttallii*) biogeography at their southern range limit. *Journal of Mammalogy* 87:1175–82.

Gordon, C. C. 1986. Winter food habits of the pine marten in Colorado. *Great Basin Naturalist* 46:166–68.

Hagmeier, E. 1961. Variation and relationships in North American marten. *Canadian Field-Naturalist* 75:122–38.

Hall, E. R. 1981. *The mammals of North America*. 2nd ed. Vol. 2. New York: John Wiley.

Hargis, C. D., J. A. Bissonette, and D. L. Turner. 1999. The influence of forest fragmentation and landscape pattern on American marten. *Journal of Applied Ecology* 36:157–72.

Hawley, V. D., and F. E. Newby. 1957. Marten home ranges and population fluctuations. *Journal of Mammalogy* 38:174–83.

Hoffmeister, D. F. 1945. Snow bunting in New Mexico. *Condor* 47:41.

Jones, C., and C. G. Schmitt. 1997. Mammal species of concern in New Mexico. In *Life among the muses: papers in honor of James S. Findley*, ed. T. L. Yates, W. I. Gannon, and D. E. Wilson, 179–205. Albuquerque: University of New Mexico, Museum of Southwestern Biology. *Special Publication of the Museum of Southwestern Biology, University of New Mexico* 3:1–290.

Jonkel, C. J., and R. P. Weckwerth. 1963. Sexual maturity and implantation of blastocysts in the wild pine marten. *Journal of Wildlife Management* 27:93–98.

Koehler, G. M., J. A. Blakesley, and T. W. Koehler. 1990. Marten use of successional forest stages during winter in north-central Washington. *Northwestern Naturalist* 71:1–4.

Koehler, G. M., and M. G. Hornocker. 1977. Fire effects on marten habitat in the Selway-Bitterroot Wilderness. *Journal of Wildlife Management* 41:500–505.

Koehler, G. M., W. R. Moore, and A. R. Taylor. 1975. Preserving the pine marten: management guidelines for western forests. *Western Wildlands* 2:31–36.

Lang, R. W., and A. H. Harris. 1984. *The faunal remains from Arroyo Hondo Pueblo, New Mexico: a study in short-term subsistence change*. Arroyo Hondo Archaeological Series 5. Santa Fe, NM: School of American Research Press.

Lapinski, M. 2006. *Wilderness predators of the Rockies: the bond between predator and prey*. Guilford, CT: Globe Pequot Press.

Lawler, J. J., H. D. Stafford, and E. H. Girvetz. 2012. Martens and fishers in a changing climate. In *Biology and conservation of martens, sables, and fishers: a new synthesis*, ed. K. B. Aubry, W. J. Zielinski, M. G. Raphael, G. Proulx, and S. W. Buskirk, 371–97. Ithaca, NY: Comstock Publishing Associates.

Lee, L. 1967. *Furbearers and predators of New Mexico*. Santa Fe: New Mexico Department of Game and Fish.

Ligon J. S. 1927. *Wild life of New Mexico: its conservation and management*. Santa Fe, NM: State Game Commission, Department of Game and Fish.

Long, B. J. 1999. Detection surveys for American

marten (*Martes americana*) in the Rio Hondo drainage, Taos Co., N.M. Unpublished report. T & E Inc., Las Cruces, NM and Share with Wildlife Program, New Mexico Department of Game and Fish, Santa Fe.

———. 2001a. The distribution of American marten in northern New Mexico: the combined result of track-plate and snow-tracking surveys, June 1997 through June 2001. Unpublished report. Share With wildlife Program, New Mexico Department of Game and Fish, Santa Fe and Turner Foundation, Atlanta, GA.

———. 2001b. American marten surveys on Vermejo Park Ranch and adjacent lands. Unpublished report. Turner Foundation Inc., Atlanta, GA.

———. 2002. Detection surveys for American marten (*Martes americana*) on the Valles Caldera National Preserve, Jemez Mountains, New Mexico. Unpublished report. Valles Caldera National Preserve, Jemez Springs, NM.

———. 2003a. Winter and spring home ranges of American marten in northern New Mexico, along with observations on rest site and den site characteristics. Unpublished report. Share with Wildlife Program, New Mexico Department of Game and Fish, Santa Fe; T & E Inc., Cortaro, AZ; and US Forest Service, Carson National Forest, Questa, NM.

———. 2003b. Detection surveys for American marten (*Martes americana*) in Taos Canyon, Taos Co., N.M. Unpublished report. Carson National Forest, Camino Real District, Taos, NM.

———. 2009. American marten survey report. Unpublished report. Share with Wildlife Program, New Mexico Department of Game and Fish, Santa Fe.

Long, B., M. East, and J. Klingel. 2015. Snow-tracking surveys and camera trapping for American marten in the Pecos Wilderness and San Pedro Parks Wilderness areas, in north central New Mexico, USA. Unpublished report. Share with Wildlife Program, New Mexico Department of Game and Fish, Santa Fe.

Long, R. A., and P. McKay. 2012. Noninvasive methods for surveying martens, sables, and fishers. In *Biology and conservation of martens, sables, and fishers: a new synthesis*, ed. K. B. Aubry, W. J. Zielinski, M. G. Raphael, G. Proulx, and S. W.

Buskirk, 320–42. Ithaca, NY: Comstock Publishing Associates.

MacDonald, S. O., and J. A. Cook. 2007. *Mammals and amphibians of southeast Alaska*. University of New Mexico, Museum of Southwestern Biology, Special Publication No. 8.

McGowan, C., W. S. Davidson, and L. A. Howes, 1999. Genetic analysis of an endangered pine marten (*Martes americana*) population from Newfoundland using randomly amplified polymorphic DNA markers. *Canadian Journal of Zoology* 77:661–66.

Menke, K. A., and T. W. Perry. 2009. Determining suitable habitat for American Marten (*Martes americana*) using a Geographic Information System (GIS). Unpublished report. Share with Wildlife Program, New Mexico Department of Game and Fish, Santa Fe.

Merriam, C. H. 1890. Description of a new marten (*Mustela caurina*) from the northwest coast region of the United States. *North American Fauna* 4:27–29.

Pauli, J. N., W. P. Smith, and M. Ben-David. 2012. Quantifying dispersal rates and distances in North American martens: a test of enriched isotope labeling. *Journal of Mammalogy* 93:390–98.

Pauli, J. N., J. P. Whiteman, B. G. Marcot, T. M. McClean and M. Ben-David. 2011. A DNA-based approach to age martens. *Journal of Mammalogy* 92:500–510.

Powell, R. A., S. W. Buskirk, and W. J. Zielinski. 2003. Fisher and marten (*Martes pennanti* and *Martes americana*). In *Wild mammals of North America: biology, management, and conservation*, 2nd ed., ed. G. A. Feldhamer, B. C. Thompson, and J. A. Chapman, 635–49. Baltimore: Johns Hopkins University Press.

Rhoads, S. N. 1902. Synopsis of the American martens. *Proceedings of the Academy of Natural Sciences, Philadelphia* 54:443–60.

Ruggiero, L. F., and S. E. Henry. 1993. Courtship and copulatory behavior of *Martes americana*. *Northwestern Naturalist* 74:18–22.

Ruggiero, L. F., D. E. Pearson, and S. E. Henry. 1998. Characteristics of American marten den sites in Wyoming. *Journal of Wildlife Management* 62:663–73.

Sherburne, S. S., and J. A. Bissonette. 1993. Squirrel middens influence marten (*Martes americana*) use of subnivean access points. *American Midland Naturalist* 129:204–7.

———. 1994. Marten subnivean access point use: response to subnivean prey levels. *Journal of Wildlife Management* 58:400–405.

Small, M. P., K. D. Stone, and J. A. Cook. 2003. American marten (*Martes americana*) in the Pacific Northwest: population differentiation across a landscape fragmented in time and space. *Molecular Ecology* 12:89–103.

Stahlecker, D. W. 2010. Boreal owl (*Aegolius funereus*). In *Raptors of New Mexico*, ed. J.-L. E. Cartron, 649–57. Albuquerque: University of New Mexico Press.

Streeter, R. G., and C. E. Braun. 1968. Occurrence of pine marten, *Martes americana*, (Carnivora: Mustelidae) in Colorado alpine areas. *Southwestern Naturalist* 13:449–51.

Strickland, M. A., and C. W. Douglas. 1987. Marten. In *Wild furbearer management and conservation in North America*, ed. M. Novak, J. A. Baker, M. E. Obbard, and B. Malloch, 531–46. Ontario, Canada: Ministry of Natural Resources.

Strickland, M. A., C. W. Douglas, M. Novak, and N. P. Hunziger. 1982. Marten, *Martes americana*. In *Wild mammals of North America: biology, management, and* economics, ed. J. A. Chapman and G. A. Feldhamer, 599–612. Baltimore: Johns Hopkins University Press.

Thompson, I. D., J. Fryxell, and D. J. Harrison. 2012. Improved insights into use of habitat by American martens. In *Biology and conservation of martens, sables, and fishers: a new synthesis*, ed. K. B. Aubry, W. J. Zielinski, M. G. Raphael, G. Proulx, and S. W. Buskirk, 209–30. Ithaca, NY: Comstock Publishing Associates.

Weckwerth, R. P., and V. D. Hawley. 1962. Marten food habits and population fluctuations in Montana. *Journal of Wildlife Management* 26:55–74.

Wiebe, P. A., J. M. Fryxell, I. D. Thompson, L. Borger, and J. A. Baker. 2013. Do trappers understand marten habitat? *Journal of Wildlife Management* 77:379–91.

Wozencraft, W. C. 2005. Order Carnivora. In *Mammal species of the world: a taxonomic and geographic reference*, Vol. 1, ed. D. E. Wilson and D. M. Reeder, 532–628. 3rd ed. Baltimore: Johns Hopkins University Press.

Wright, P. L. 1953. Intergradation between *Martes americana* and *Martes caurina* in western Montana. *Journal of Mammalogy* 34:74–86.

Yeager, L. E., and J. D. Remington. 1956. Sight observations of Colorado martens, 1950–1955. *Journal of Mammalogy* 37:521–24.

Zielinski, W. J. 1984. Plague in pine martens and the fleas associated with its occurrence. *Great Basin Naturalist* 44:170–75.

Zielinski, W. J., and T. E. Kucera, eds. 1995. American marten, fisher, lynx, and wolverine: survey methods for their detection. Albany, CA: US Department of Agriculture, Forest Service, Pacific Southwest Research Station, Gen. Tech. Rep. PSW-GTR-157.

Zielinski, W. J., K. M. Slauson, and A. E. Bowles. 2008. Effects of off-highway vehicle use on the American marten. *Journal of Wildlife Management* 72:1558–71.

Zielinski, W. J., W. D. Spencer, and R. H. Barrett. 1983. Relationship between food habits and activity patterns of pine martens. *Journal of Mammalogy* 64:387–96.

AMERICAN ERMINE (*MUSTELA RICHARDSONII*)

Jennifer K. Frey

New Mexico's smallest carnivore, the American ermine is a tiny weasel with a short, black-tipped tail. It has a typical weasel body plan with short legs, a long body and neck, and a small wedge-shaped triangular head with short round ears. The dorsum is uniformly brown (white in winter), the venter is usually nearly pure white (occasionally very pale yellow); the face is all one color. Only the eyes and the tip of the tail are black. The southern border of the species' range is northern New Mexico. New Mexico's ermines belong to the subspecies *Mustela richardsonii muricus*, which ranges throughout much of the Mountain West in the contiguous United States from southwestern Montana, southeastern Washington, and Oregon south into the Sierra Nevada Mountains, Great Basin, and Southern Rocky Mountains including New Mexico (Hall 1951, 1981). *M. r. muricus* is the smallest subspecies of the American ermine. Hall (1951) thought that all *M. r. muricus* were about the same size, with those from New Mexico no smaller than those from the northern part of the subspecies' distribution. However, Hall had few specimens to work with, and based on the larger series of ermine specimens now available from New Mexico, it appears that New Mexico ermines are likely the smallest form of the species (Table 21.1). Measurements of New Mexico ermines tend to be smaller than the range of variation reported for American ermines in published sources (e.g., Hall 1951; King 1983; Fagerstone 1999). This holds true even for other populations of *M. r. muricus*. For instance, while male ermines in the Sierra Nevada are about the same size as males in New Mexico, reported mass for female ermines in the Sierra Nevada (44.7 g [1.58 oz]; Fitzgerald 1977) averaged 28% higher compared to New Mexico females (see below). In addition, most sources proclaim the least weasel (*M. nivalis* [= *M. rixosa*]) as the smallest mustelid and the smallest living carnivore in the world (e.g., Fagerstone 1999). However, ermines have a proportionately longer tail than least weasels (e.g., ~25% of total length in New Mexico ermine vs. ~15% in North American least weasels). Consequently, the body size and mass of New Mexico ermines are smaller than those of the least weasel (data on least weasels taken from Hall [1951] and Fagerstone [1999]). Thus, New Mexico ermines are not only the smallest known mustelid but also the smallest known carnivore!

Weasels exhibit significant sexual dimorphism in size (Hall 1951; Fagerstone 1999), and ermines in New Mexico are no exception (Table 21.1). To put the size of New Mexico ermines into perspective, males (the larger gender) average 59 g (2.1 oz) in body weight and 229 mm (9.0 in) in total length, about the same as a chipmunk (*Neotamias*; Photo 21.3). Female ermines in New Mexico average 35 g (1.2 oz) in body weight and 201 mm (7.9 in) in total length, which is approximately equivalent to the weight of a montane vole (*Microtus montanus*) and the length of a brush deermouse (*Peromyscus boylii*; Frey 2007). Either gender is small enough

(*opposite page*) Photograph: © Evan Kipp.

Table 21.1. Descriptive statistics for four standard external measurements (mm) and mass (g) of the American ermine (*Mustela richardsonii*) collected in New Mexico.

	Total Length	Tail Length	Hindfoot Length	Ear Length	Mass (g)
Female					
Mean	201.3	49.9	23.8	12.8	34.6
Minimum	161	32	21	6	20.7
Maximum	220	60	26	17	41
SD	13.82	6.58	1.69	2.85	5.80
n	13	13	13	13	11
Male					
Mean	228.7	61.9	29.9	16.0	58.9
Minimum	201	48	26	12	45
Maximum	272	85	36	19	70
SD	17.02	7.74	2.67	2.10	8.62
n	15	15	15	15	11

to fit comfortably in the palm of a human hand, and most people seeing one will think they have seen some kind of a mouse rather than a weasel.

In New Mexico, identification of *M. richardsonii* should be relatively straightforward—provided a specimen is available to allow for close inspection and accurate measurements. However, I am aware of at least two ermine specimens in the University of New Mexico's Museum of Southwestern Biology that were originally misidentified as the long-tailed weasel (*Neogale frenata*), and so caution is warranted in identifying putative specimens. In New Mexico, the American ermine only requires a careful comparison with the long-tailed weasel, which has a distribution overlapping the ermine's in the northern part of the state. Long-tailed weasels are larger than ermines (Figure 23.1 in Chapter 23). For females, ermines average 62% of the total length and 37% of the mass of the long-tailed weasel. The difference is even more striking for males, wherein ermines average only 50% of the total length, and a mere 17% of the mass, of long-tailed weasels. There is no overlap in size between the two species in New Mexico, even between male

ermines and female long-tailed weasels (Figure 23.1 in Chapter 23). In addition, the ermine has a relatively shorter tail than the long-tailed weasel (see photos in Chapter 23 for comparison). In New Mexico, the tail of an ermine is less than half of the body length (i.e., not including tail), while in the long-tailed weasel it is more than half the body length (not including tail; Table 21.2). Another metric of this tail proportion is to compare it with hind foot length: in the ermine, the tail length is about twice the hind foot length, while in the long-tailed weasel the tail is about four times the length of the hind foot.

Ermines can be readily identified in comparison with the low-elevation form of the long-tailed weasel, *N. f. neomexicana*, which occurs broadly throughout the state (Chapter 23). This is a very large subspecies of long-tailed weasel that has a distinct, white facial mask (and hence sometimes is called "bridled weasel"), reddish brown dorsum, and yellow or orangish belly, including in winter (see Chapter 23). More difficult are comparisons of the ermine with montane forms of the long-tailed weasel, both *N. f. nevadaensis*, which

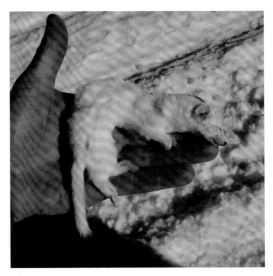

Photo 21.1. American ermine in summer pelage on 15 September 2020 between Mitchell Lake and Blue Lake in the Indian Peaks Wilderness, Colorado. Note the typical weasel body plan with the elongated body and neck, the short legs, and the small wedge-shaped triangular head with short round ears. The summer pelage is characterized by a white or occasionally very pale yellow venter and a uniformly chocolate brown dorsum. The eyes and the tip of the tail are black in both the summer and winter pelages. Although the American ermine and the much larger long-tailed weasel (*Neogale frenata*) may be difficult to distinguish at a distance or on photographs, the length of the tail relative to the length of the body is diagnostic. Tail length is less than half of body length in the ermine, while in the long-tailed weasel the tail is more than half the body length. Photograph: © Leif Saul.

Photo 21.2. American ermine in winter pelage found dead along snowmobile tracks in the West Elk Mountains of west-central Colorado. A member of the weasel family (Mustelidae), the American ermine is so small it might be mistaken for an odd-looking mouse. Males are larger than females, but individuals of either gender fit comfortably into the palm of a human hand. The length of the tail is about twice that of the hindfoot (measured from heel to claw), rather than four times as in the long-tailed weasel (*Neogale frenata*). Photograph: © Tim Shortell

Photo 21.3. Adult male (*top*) and female (*bottom*) American ermines (*Mustela richardsonii muricus*) collected in New Mexico, compared for size with a Colorado chipmunk (*Neotamias quadrivittatus*; *center*). The subspecies *M. r. muricus* in New Mexico may represent the smallest extant carnivore known anywhere in the world. Note also that based on measurements, New Mexico's adult male American ermines average 13% longer (total length) than adult females (see Table 21.1). Photograph: © Jennifer K. Frey.

Table 21.2. Comparison of morphological features of the American ermine (*Mustela richardsonii*) and the long-tailed weasel (*Neogale frenata*) in New Mexico.

	M. richardsonii	N. frenata
Total length	<275 mm	>310 mm
Tail length	≤85 mm	≥115 mm
Tail length: body length	≤45%	≥60%
Hind foot length	female <27 mm; male <37 mm	female >31 mm; male >43 mm
Tail length: hind foot length	mean 2.1 (range 1.7–2.6)	mean 3.9 (range 3.2–4.7)
Mass	<71 g	>75 g
Face and head pelage color	uniform	uniform or with contrasting pattern
Dorsal body pelage	dark brown in summer and white in winter	light reddish brown all year or dark brown in summer and white in winter
Ventral body pelage	white or yellow	yellow (white in winter) or orange
Tail pelage	black tip	black tip
Postglenoid length of skull/condylobasal length[1]	>46 %	<46%

1. From Hall 1951.

co-occurs with the ermine in northern New Mexico, and *N. f. arizonensis*, which is a small "ermine-like" long-tail weasel that occurs in some isolated mountains south beyond the range of the ermine in the Southwest. The ermine and the montane subspecies of long-tailed weasels have similar pelage color patterns. Montane long-tailed weasels entirely lack or have greatly reduced white markings on the face, similar to the ermine (Photo 21.4). Thus, presence of a mask can confirm a specimen as long-tailed weasel, but the absence of a mask is not a character that can be used to distinguish the species. In summer, both ermines and montane long-tailed weasels have a grayish brown dorsum, whitish or yellowish venter, brown feet and legs, and a black-tipped tail (Photos 21.4 and 21.5). Some references have suggested that the ermine differs from the long-tailed weasel in having the pale ventral coloration extend down the inside of the legs (Fagerstone 1999). However, that character is not diagnostic in New Mexico, where some long-tailed weasels

also exhibit it. Findley et al. (1975) and Findley (1987) stated that ermines usually possess white venters while long-tailed weasels usually have a venter with an "orange" wash. However, there is considerable variation and overlap in color of the venter such that identification cannot be made on the basis of venter color alone. Both species exhibit venter colors ranging from whitish to yellowish, though ermines tend to be more whitish (sometimes with a yellowish wash) and montane long-tailed weasels tend to be more yellowish (Photo 21.5).

A source of confusion is that some weasels turn white in winter and others do not, and any weasel (regardless of species) with a white pelage is commonly referred to as "ermine" (Hall 1951; Fagerstone 1999). In fact, to avoid confusion, some authors have adopted calling *M. richardsonii* "short-tailed weasel" or "American stoat" instead of American ermine. Some authors (e.g., King 1983; Findley 1987) have reported that "southern" populations of ermines in North America do not turn white in winter but become a paler brown. However, this pale brown winter pelage is probably restricted to low-elevation Pacific coastal populations (Merriam 1896; Hall 1951; Fagerstone 1999). One winter specimen of *M. richardsonii* from New Mexico (MSB 60780), collected on 11 February, is white all over save for the black-tipped tail (Photo 21.6). An animal found dead in the Jemez Mountains also showed a white pelage (R. Parmenter, pers. comm.). Swickard et al. (1971) mentioned that an animal captured in the Jemez Mountains on 5 November exhibited a pelage mottled brown and white. Fitzgerald et al. (1994) reported that ermines turn white in winter in Colorado (and see Photo 21.2). Thus, it is presumed that all ermines in New Mexico similarly turn white in winter. A specimen (MSB 143855) with brown pelage has a museum tag with a date 9 March 2004. However, this specimen had been first taken to a wildlife rehabilitation center, with the date of collection recorded as "9/3/2004." Hence, the date on the specimen tag is likely an error and the actual collection date should be 3 September 2004. Montane populations of long-tailed weasels in northern New Mexico also turn white in winter (see Chapter 23 for further discussion). The presence of a white winter pelage is yet another trait that cannot be used to discriminate between the species, though a brown winter pelage would indicate long-tailed weasel.

The ermine is distinguished from the black-footed ferret (*Mustela nigripes*) by its much smaller size (black-footed ferret total length = 457–552 mm [18–21.7 in]; mass = 530–1,300 g [18.7–45.9 oz]; Fitzgerald et al. 1994; Chapter 22), a white or yellow belly that contrasts with a brown dorsum (black-footed ferrets are tannish all around the body), the lack of black feet, and the absence of a black facial mask across the eyes. *M. richardsonii* is distinguished from the Pacific marten (*Martes caurina*), again by its much smaller size (marten total length = 460–750 mm [18.1–29.5 in]; mass = 500–1,200 g [17.6–42.3 oz]; Fitzgerald et al. 1994),

Photo 21.4. Comparison of size and pelage of the foreparts of an "ermine-like" long-tailed weasel (*Neogale frenata arizonensis*; top) and an American ermine (*Mustela richardsonii*; bottom). Note subtle differences in the coloration of the underparts, pure white in the *M. richardsonii* specimen, but only pure white under the head and neck and yellowish posteriorly in the *N. frenata* specimen. Photograph: © Jennifer K. Frey.

Photo 21.5. Comparison of ventral pelage color in representative specimens of the American ermine (*Mustela erminea muricus*; bottom four specimens, where bottom two are females, upper two are males) and the long-tailed weasel (*Neogale frenata arizonensis*; top three). Note also the difference in body size and tail length, both between the two taxa and between male and female ermine specimens. Photograph: © Jennifer K. Frey.

Photo 21.6. Comparison of winter (top) and summer (bottom) pelage of American ermines (*Mustela richardsonii*) collected in New Mexico. Photograph: © Jon Dunnum.

and also by its proportionately shorter tail, a white or yellow belly that contrasts with a brown dorsum (martens are all brown with a yellow chest patch), and the lack of a bushy tail (Chapter 20). *M. richardsonii*, finally, is distinguished from the American mink (*Neogale vison*) by its much smaller size (mink total length = 491–720 mm [19.3–28.3 in]; mass = 525–780 g [18.5–27.5 oz]; Fitzgerald et al. 1994), a proportionately shorter tail, and a white or yellow belly that contrasts with the brown dorsum (minks are all dark brown or blackish with occasional white patches on the venter; Chapter 24).

Until recently, ermines were considered to belong to a single species, *M. erminea*, that had an expansive Holarctic distribution across which it exhibited considerable geographic variation in morphology, with a total of 34 subspecies recognized worldwide (Colella et al. 2021) and 20 subspecies recognized in North America alone (Hall 1981; King 1983). A general trend was that ermines were much larger in Eurasia and at more northern latitudes (Hall 1951; Fagerstone 1999). Recent molecular genetic studies have revealed that *Mustela erminea* was actually comprised of four major clades (e.g., evolutionary lineages) that evolved in different glacial refugia (Dawson et al. 2014; Colella et al. 2021). Three of these four clades have now been recognized as distinct species. The deepest divergence is between the clade representing Eurasia, Alaska, and some islands of the Pacific Northwest, and a clade representing the rest of North America. Ermines from Eurasia and Alaska are referred to as *M. erminea* (Beringian ermine), those from Haida Gwaii Islands of the Pacific coast of Canada as *M. haidarum* (Haida ermine), and those from the remainder of North America as *M. richardsonii* (American ermine; Colella et al. 2021). Within *M. richardsonii*, there are two major clades (Fleming and Cook 2002; Dawson et al. 2014; Colella et al. 2021): a "western clade" that includes New Mexico north through British Columbia and a "continental clade" from much of Canada and the eastern United States. Genetic signatures suggest that corresponding populations contracted into remnant habitat patches following the end of the last glacial period (Dawson et al. 2014). Similar patterns have been found for other western montane species, often revealing endemic diagnosable taxa (e.g., Aubry et al. 2009; Galbreath et al. 2010; Statham et al. 2012; Hope et al. 2014, 2016). The analyses of Colella et al. (2021) found support for splitting the western clade and the continental clade into separate species at >90% posterior probability, but the authors noted that morphological differences in size among North American ermines are primarily driven by local adaptation to climate rather than evolutionary origins (Eger 1990). However, many subspecies representing the western clade of *M. richardsonii* are characterized by a very small body size in comparison with the larger *M. richardsonii* of the continental clade found in the eastern United States and much of Canada, and in this chapter are referred to as small ermines. As of yet no taxonomic recommendations have been made for recognizing the western clade, which includes New Mexico, as a distinct species, though future analyses could. Further, *M. r. muricus*, the subspecies that occurs in New Mexico, was an outlier in a plot of mtDNA haplotypes (genetic variants) among subspecies in the western clade (Colella et al. 2021), suggesting it may represent unique genetic diversity.

As currently recognized, the range of *M. r. muricus* extends from southwestern Montana, southeastern Washington, and Oregon south into the Sierra Nevada, Great Basin, and Southern Rocky Mountains including northern New Mexico (Hall 1981). A subspecies of ermine, *M. r. leptus*, was originally described by Merriam (1903) from Silverton, San Juan County in southwestern Colorado, with contiguous high-elevation environments extending to the south through La Plata County and across the state line into New Mexico. It may be regrettable that, without explaining his decision, Hall (1945) synonymized "*leptus*" within *M. r. muricus* (= *M. r. murica*), which was described from the Sierra Nevada in El Dorado County, California. Later, Hall (1951:163) noted that a comparison of specimens from the Southern Rocky Mountains in Colorado with specimens from the Sierra Nevada, California, gave "no basis for recognizing more than one subspecies" and that specimens from northern New Mexico were as large as specimens from the northern portion of the subspecies' range. The conclusion Hall (1951) reached might have been the direct result of small sample sizes. Several montane mammals have distinct subspecies in the Southern Rocky Mountains (Findley and Anderson 1956), and no study has evaluated the phylogenetic patterns within these populations to determine if *leptus* might represent a distinct Southern Rocky Mountains form.

There is considerable confusion in the use of common names for different species of weasels around the world (King and Powell 2007). The Beringian ermine is commonly called "stoat" in Europe. Beringian ermines are much larger than American ermines—even larger than our long-tailed weasels—and hence much about their respective life histories differ in subtle but important ways. For that reason, in this chapter I use information about American ermines wherever possible. Some treatments use the name "short-tailed weasel" to refer to *M. richardsonii* and *M. erminea* (e.g., Fagerstone 1999). One expressed reason is that it prevents the confusion surrounding the word "ermine," which is used by trappers to describe the pelt of any species of weasel in white winter pelage. While I understand the merits of using the term "short-tailed weasel," I prefer to continue using the more traditional name "ermine" for this species because the term "short-tailed weasel" can also describe the least weasel (which also has a short tail), while the common name "ermine" both was used in the seminal taxonomic works on the species (Hall 1945, 1951) and conforms to important mammal references (e.g., King 1983; Baker et al. 2003; Wozencraft 2005; American Society of Mammalogists Biodiversity Database). In New Mexico and elsewhere in the western United States, the use of the term "ermine" should provide little confusion for amateur or professional students of mammalogy, though caution is warranted in popular or non-technical references where the scientific name is not provided. The subspecies name *muricus* was given to mean "a mouser" (Bangs 1899), which is an excellent description of the food habits and way of life of this tiny weasel.

DISTRIBUTION

Ermine species are mainly associated with arctic

Photo 21.7. A female American ermine (*Mustela richardsonii muricus*) collected in July 1963 in the Taos Ski Valley, Taos County, New Mexico (TTU 1866). New Mexico lies at the southern edge of the American ermine's distribution. All of New Mexico's ermines belong to the subspecies *M. r. muricus*. Photograph: © Jennifer K. Frey.

and boreal zones (King 1983), and their distribution closely matches global surface temperature regions characterized by cool to cold winters (i.e., below 10° C [50° F]) and mild to cold summers (i.e., below 20° C [68° F]; Parkins 1984). The Beringian ermine, *M. erminea*, from Europe was introduced to New Zealand, where it is considered invasive (King 1983). As in other boreal animals that rely on cold climates, the distribution of the American ermine, *M. richardsonii*, exhibits a trend toward occurrence at higher elevations at lower latitudes. Thus, although the bulk of the natural geographic range of *M. richardsonii* lies north of 45° N latitude, its distribution also extends southward in mountainous regions that maintain cold climate conditions, which in the western contiguous United States includes the Sierra Nevada, mountains of the Great Basin, and the Rocky Mountains (Hall 1981). The southernmost occurrence of *M. richardsonii* in North America is in New Mexico.

Little is known about *M. richardsonii* in New Mexico or elsewhere in the western contiguous United States (but see Fitzgerald 1977; Cain et al. 2003, 2006). Findley et al. (1975) reported only eight specimens and three literature records of the ermine in their monograph on the mammals of New Mexico. Consequently, Michael Calkins and I undertook a study to compile and synthesize all available information on the American ermine in New Mexico in order to better understand its conservation status (see Frey and Calkins 2010, 2014). We obtained data from museum specimens, field notes, catalogs, and other associated museum data. We found a total of 60 records of *M. richardsonii* from ca. 45 locations in New Mexico, while two additional specimens have since been deposited in a museum, for a current total of 62 records (Map 21.1; Appendix 21.1). Nearly all the localities were restricted to the major mountain ranges of northern New Mexico, including the San Juan Mountains, Sangre de Cristo Mountains, and Jemez Mountains.

However, there also was an unusual record of an American ermine captured by Zachary Schwenke and me at Sugarite Canyon State Park on the Raton Mesa Group in northeastern Colfax County (Frey and Schwenke 2007, 2012). The Raton Mesa Group is located ~70 km (45 mi) east of the crest of the Sangre de Cristo Mountains along the Trinidad Escarpment, which is a relatively high-elevation (>7,500 ft) eastern extension of the Sangre de Cristo Mountains. However, based on intervening land-cover types, the population of *M. richardsonii* on the Raton Mesa Group, which reaches elevations of 2,962 m (9,718 ft), is likely isolated from populations in the Sangre de Cristo Mountains.

The elevation of 42 unique locations of ermines in New Mexico averaged 2,773 m (9,095 ft) and ranged from 2,274 to 3,542 m (SD = 316; 7,459—11,618 ft; Frey and Calkins 2010). This is a higher range of elevations than reported for Colorado, where American ermines have been reported from several locations <2,130 m (7,000 ft; Armstrong 1972). Possibly the lowest elevation record for ermines in New Mexico is a specimen from "Teseque [sic]," Santa Fe County (MSB 60780). This specimen is in white winter pelage, and the tag indicates it was killed by a domestic cat (*Felis catus*) on 11 February 1977 and collected by W. (Bill) Isaacs (Photo 21.6). The specimen bears a New Mexico Department of Game and Fish (NMDGF) Endangered Species Program number (ESP-326) and a preparatory number MCC-155 (the specimen was prepared by Marshal Conway, who was at the time the endangered species mammalogist for the NMDGF). Unfortunately, the NMDGF has no record of this specimen (J. N. Stuart, pers. comm.). Bill Isaacs was a well-known botanist and mycologist in the Santa Fe area. Marshal Conway (pers. comm.) confirmed that the specimen was collected in 1977 by Bill Isaacs, and the collection location reported by Isaacs was Tesuque. However, town names are sometimes used for collections made anywhere

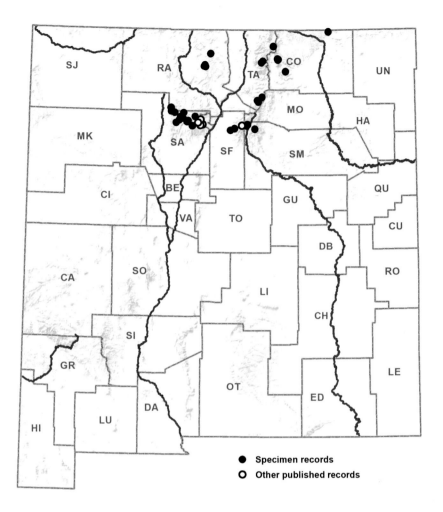

Map 21.1. Distribution of the American ermine (*Mustela richardsonii*) in New Mexico based on museum specimen and other published records.

in the vicinity (e.g., Frey and Malaney 2006) and so the precise collection location of this specimen is not known with certainty.

The town of Tesuque is located at an elevation of 2,063 m (6,767 ft) in an irrigated agricultural valley along the Rio Tesuque at the western edge of the Sangre de Cristo Mountains. However, residential streets extend upward into the foothills of the mountains to elevations of ca. 2,255 m (7,400 ft), which is approximately the lower-elevational range limit of the species in New Mexico. Elevations of 2,255 m (7,400 ft) are within 3 km (1.9 mi) of the town center of Tesuque in the valley. Thus,

not knowing specifically where the specimen was collected means that the low-elevation limits in New Mexico remain unknown. A second specimen, (MSB 143855) reported as from "Tesuque," was an animal brought to a wildlife rehabilitation center. Based on rehabilitation center records, "Tesuque" was the donor's address, while the specimen was actually found in "Pacheco Canyon," which is a canyon through which the Rio Chupadero flows between 2,316 and 2,804 m (7,600–9,200 ft) on the western edge of the Sangre de Cristo Mountains northwest of the town of Tesuque.

Armstrong (1972) reported three records of *M. richardsonii* in Colorado from <2,134 m (7,000 ft). One of these was a white animal captured 3 April 1932 in Denver, Denver County, at an elevation of ca. 1,615 m (5,300 ft; Seton 1933; Miller 1933; Armstrong 1972). Given that the Tesuque specimen similarly exhibited a winter pelage, it remains a possibility that the distribution of *M. richardsonii* extends to lower elevations during the winter. Seasonal shifts in distribution and habitat use also have been postulated for other "boreal" species in the Southern Rocky Mountains (e.g., Malaney and Frey 2006). Beringian ermines have been recorded dispersing up to 22 mi (35 km), but that species is relatively large-bodied and likely can disperse over much larger distances (Burns 1964); dispersal distances in *M. r. muricus* are unknown.

American ermines are not known to occur on any of the small isolated mountains south of the Jemez and Sangre de Cristo mountains. Other taxa with a similar distribution pattern include the cinereus shrew (*Sorex cinereus*), bushy-tailed woodrat (*Neotoma cinerea orolestes*) and American pika (*Ochotona princeps*). Based on the interpretation of fossil records and nested patterns of montane mammal communities, this group of species is presumed to be more prone to extinction than are species with broader distributions (Patterson and Atmar 1986; Harris 1990; Patterson 1995; Frey et al. 2007). Reduction in habitat due to a warmer, post-Pleistocene climate is considered the primary driver of these extinction events, which ultimately has been responsible for determining the species' current distribution patterns (Patterson 1980, 1984; Grayson 1981, 1987). For example, a fossil ermine dated to ca. 12,000 years before present was found in southeastern New Mexico (Eddy County), ca. 400 km (250 mi) south of current populations of ermines in the Sangre de Cristo Mountains (Harris 1993). In contrast, more extinction resistant species, such as the western water shrew (*Sorex navigator*), Fremont's

red squirrel (*Tamiasciurus fremonti*), southern red-backed vole (*Myodes gapperi*), and long-tailed vole (*Microtus longicaudus*), have been able to persist in many mountain ranges in the Southwest, as has the long-tailed weasel, which is less of a habitat and diet specialist than is the ermine. As such, the ermine is highly prone to extinction, and its modern distribution is a product of historical extinction events that eliminated it from areas where it previously occurred. Future reductions in habitat, whether due to climate change or other anthropogenic causes, thereby pose a particular risk to this species (see under "Status and Management").

Discredited records

Ivey (1957) reported observing a weasel in the Sandia Mountains, Bernalillo County at 2,895 m (9,500 ft) elevation in mixed coniferous forest that he concluded was an ermine. However, there have been no subsequent observations (or specimens) of ermines from this mountain range despite proximity to the University of New Mexico, which has a strong program in mammalogy. Ivey observed the front portion of the animal's body, except the feet, for about one minute from a distance of about 3.3 m (10 ft). He described it as "small size, chocolate brown back and white underparts," "larger than the least weasel," and "clearly not the large, pale, bridled weasel of the valley" (Ivey 1957:500). That is a very good description of a high-elevation montane long-tailed weasel in New Mexico (see Chapter 23). First, he was able to discern it as a weasel from ca. 3.3 m away with only the forepart of the body showing. The size comparison, "larger than the least weasel," suggests that the animal was a long-tailed weasel since New Mexico ermines are at least as small, if not smaller, than the least weasel. Long-tailed weasels that occur at high elevations outside of the range of the ermine in the Southwest are ermine-like in appearance—they are relatively small and the pelage is chocolate brown above lacking a mask on the face, and whitish below, often pure white under the chin. This is in stark contrast to the low-elevation long-tailed weasel found throughout the state (*N. f. neomexicana*), which is large and pale reddish brown with a distinct facial mask (Photos 23.1 and 23.11 in Chapter 23). Consequently, it seems likely that what Ivey observed was a montane ermine-like long-tailed weasel.

Together with colleagues Michael Bogan and Terry Yates (Frey et al. 2007), I previously provided additional independent lines of evidence that Ivey's Sandia Mountains weasel was not an ermine. Based on the size and isolated nature of mountaintop habitat, it is possible to predict the number of species that occur in a mountain range, and our analysis found that the mammal fauna of the Sandia Mountains had more species than was predicted. The unusual pattern was resolved when the number of species on this mountaintop island was reduced by one. Similar adjustments on other mountain ranges produced no similar effects. Thus, we concluded that the likely cause of the unusual community composition of the Sandia Mountains was due to the inclusion of Ivey's (1957) ermine record. As additional support that this ermine record was erroneous, we pointed out that all other species of montane mammals known to occur in the Sandia Mountains also had been documented in other interconnected mountain ranges such as the Manzano Mountains, which was not the case for the ermine (Frey et al. 2007).

The ermine and long-tailed weasel are similar in external appearance. Unless a specimen is in hand or there are photographs that clearly show size and tail length, reports of ermines should be treated as probably inaccurate. It is likely that most observations reported as "ermine" in New Mexico were actually long-tailed weasels. *N. f. nevadensis*, which is the form of long-tailed weasel that co-occurs with the ermine in northern New Mexico, is a relatively small long-tailed

weasel with relatively dark summer pelage, and white pelage in winter (see Chapter 23). These are characteristics usually associated with the ermine. Further, most observations reported as "ermine" are of animals recognizable as weasels. It is unlikely that casual observation of a New Mexico ermine would evoke the idea of a weasel. Rather, an observer might think they had seen a mouse or some other unknown tiny animal. For example, Dixon (1931) reported an ermine chasing a young pika in Colorado, but he originally thought the animal chasing the pika was a chipmunk. Because of the high likelihood for misidentification, all observational records of ermines maintained in files of the NMDGF Conservation Services Division were discredited with the exception of one possible observation by a former state furbearer biologist. For similar reasons, furbearer records maintained by the NMDGF do not contain any reliable data about the ermine. State-maintained furbearer records are based on trapper reported harvests. Since trappers commonly consider any white weasel an "ermine," any white weasel trapped is likely incorrectly reported as an ermine (F. Winslow, pers. comm.). For example, between 2006 and 2009 and 2016 and 2019 the New Mexico Hunter Harvest Report summaries for furbearers recorded four long-tailed weasels and six ermines taken in the state (NMDGF, unpubl. data). Four of the six reported "ermines" came from counties where the species does not occur. Further, the tiny size of these weasels would make them unlikely targets as furbearer quarry and unlikely to be captured in conventional furbearer traps. Thus, most, if not all, reports of ermines in the furbearer records are likely long-tailed weasels.

Because of the difficulty studying small-bodied weasels, there are no available data on densities of ermines in New Mexico or elsewhere in the western contiguous United States, while estimates are rare elsewhere. In Ontario, the density of American ermines was 5.97/km² and 10.53/km² in preferred habitat (Simms 1979a). However, the density of ermines in New Mexico is likely to be much lower due to the specialized and limited habitat that they select in the state. The ermine is likely to be rare and declining in New Mexico (see under "Status and Management").

HABITAT ASSOCIATIONS

Using the compilation of ermine records in New Mexico, Mike Calkins and I developed habitat suitability models for the species in New Mexico based on suites of bioclimatic and biophysical variables (Frey and Calkins 2010, 2014). For climatic variables, we found that the ermine was associated with regions that received at least 150 mm (5.9 in) of precipitation, which corresponds to roughly 1.5 m (5 ft) of snow, during January, February, and March (Frey and Calkins 2014). We consider winter snow cover a key habitat requirement for the species. Under snow cover, water vapor migrates upward from the warmer ground producing a "subnivian [i.e., below snow] space" of fragile and loose snow crystals at ground level (Korslund and Steen 2006). During winter, small mammals are closely associated with the subnivean space, which they use for thermal insulation, access to plant foods, and protection from most predators (small specialist weasels such as the ermine excluded). For instance, in California's high Sierra Nevada, under the cover of snow during winter, the ground does not freeze and some plants remain green, providing food for voles (Fitzgerald 1977). During this time, montane voles make nests on the surface of the ground within the subnivean space, though they switch to underground burrows for nests during summer once the snow has melted (Fitzgerald 1977). Voles are the key prey for ermines (see under "Diet and Foraging"). Adequate and appropriate snow cover is considered a requirement for high population densities of voles, which in turn drive higher densities of ermines (King 1983; Hornfeldt et al. 2005). Snow cover also may directly reduce predation on ermines by other predators, avian and mammalian (Powell 1973). Indeed, most mortality to *M. richardsonii* is due to predation by birds in winter (Linnell et al. 2017).

Based on studies conducted in eastern North America, Simms (1979b) hypothesized that snow cover at more northern latitudes conferred an advantage to small weasels because it restricted access to subnivean foraging spaces for larger species. The results of her field study led her to conclude that female *M. r. cicognannii* (the subspecies of American ermine in the northeastern United States) were optimally sized for foraging in subnivean environments whereas long-tailed weasels (particularly males) were too large. Given that both female and male *M. r. muricus* in New Mexico are smaller than female *M. r. cicognannii* (comparative data from Hall 1951), all ermines in New Mexico are likely ideal in body size for subnivean foraging. Conversely, Simms (1979b) hypothesized that competitive interference interactions with the larger long-tailed weasel limited distribution of the ermine at more southern (warmer) locations. This same process might also be applicable to elevational gradients in the western United States, such that ermines exploit the ideal snow conditions of the high mountains, while long-tailed weasels, which are not as ideally suited for subnivean foraging, also occur in low-elevation areas with limited snowfall, potentially competing with ermines and limiting their distribution.

The best predictors of ermine occurrence in New Mexico based on biophysical variables were, in order of decreasing importance: 1) proximity to perennial streams; 2) elevation; 3) land cover; and 4) proximity to springs (Frey and Calkins 2014). The variable with the greatest contribution to the model (46.7%) was proximity to perennial streams. This result was supported by anecdotal information on habitats and small mammal communities associated with New Mexico ermine records, all indicating that ermines are indeed strongly tied to riparian areas (see below).

Streams and springs support lush, diverse, and productive early successional vegetation that provides ideal habitat for important ermine prey species (e.g., voles and shrews) and cover from predators. American ermines in Ontario also were found to prefer early successional stages dominated by graminoids or a mixture of graminoids and shrubs (Simms 1979b).

Elevation was our predictive model's second highest contributing variable (33.2%), reflecting the ermine's association with high-elevation montane areas in New Mexico. Our third predictor variable, land cover (e.g., vegetation communities) did not contribute strongly to the model (14.1%). One reason it was not a good predictor is because the resolution of map data is coarse and hence small patches of riparian vegetation within a map pixel get coded as the surrounding vegetation type. For instance, the four ermines I have caught in New Mexico were all from riparian areas, but when their locations were added onto the land-cover map, they were classified into the dominant forest vegetation type surrounding the capture site. The most frequent land-cover types associated with ermines in New Mexico were spruce-fir forest, mixed conifer forest, ponderosa pine forest, and montane-subalpine grassland (Frey and Calkins 2014; Appendix 21.2). These land-cover types correspond with the subalpine, upper montane, and lower montane forest zones. Our data should not be construed to mean that the ermine prefers these forest types. Rather, ermines are primarily associated with riparian zones nestled within these forests.

To complement the results of the habitat suitability models, I searched museum records for first-hand information about the habitat where ermine specimens were taken in New Mexico and found habitat data for 22 specimen records (Appendix 21.2). Of the 22 specimens, more than half (i.e., 54%) were described as taken along streams (this does not mean that the other half of the specimens were not also collected along streams, but rather that a stream was not specifically mentioned in the museum data). Most (75%) of the collection sites associated with streams were described as being located in meadows or associated with sedges; the presence of sedges implies a low-gradient area with low tree cover such as a wet meadow. Based on mapped locations it is likely that the remaining 25% of specimens reported as taken along streams were also within meadows. The next most frequently reported vegetation type at ermine capture locations was "meadow," which accounted for 27% of the 22 specimens. In New Mexico, "meadow" usually refers to a complex herbaceous system found in mountain valley bottoms with low gradients. Typically, meadows have wetland areas around seeps or a stream on the valley floor. The riparian wetland then grades up the valley slopes through a drier upland area dominated by grasses to the forest edge (in contrast, "grassland" refers to dry upland herbaceous systems on exposed slopes, hills, and rolling terrain). Thus, reference to a "meadow" in the specimen data likely indicates the presence of a riparian wetland. Altogether, 82% of New Mexico ermine specimens with habitat data were described as having been collected from a stream and/or meadow. The most frequently mentioned plant type in the habitat descriptions was "sedge" (41% of specimens). Sedges are found in wetland areas with deep saturated soils and where there is an open tree canopy such as found along streams and seeps in montane meadows. Lastly, the presence of beaver (*Castor canadensis*) activity was noted at six (27%) of the locations, which is a high frequency given the relatively sparse occurrence of beavers along high-elevation streams in northern New Mexico (Small et al. 2016).

Besides riparian areas, there also were a few records of ermines taken in other vegetation types. Two ermine specimens were taken on the slopes above the valley floor. One of these locations was on the hillside adjoining a valley

Photo 21.8. Price Lakes, Archuleta County, Colorado, near the New Mexico border. Zachary Schwenke and the author (Jennifer Frey) captured American ermines in the tall sedges surrounding the lakes while surveying for New Mexico jumping mice (*Zapus luteus*). Photograph: © Jennifer K. Frey.

bottom that had a stream in a meadow where other ermine specimens were collected; vegetation on the hillside was not documented. Another location was in an un-logged forest above the valley bottom; the valley bottom at this location has a weakly developed meadow system and intermittent stream (J. K. Frey, pers. obs.). One putative record (an observation) was from a grove of aspen and spruce with abundant herbaceous ground cover that was surrounded by a complex of wet meadows with talus and spruce-fir forest; many seeps were present. Finally, one record was from a rock slide (i.e., probably talus) on a north-facing mountain slope where proximity to water is unknown. Thus, while New Mexico ermines appear to be primarily selecting riparian areas in montane meadows, they can use other areas such as forest and talus. However, it remains unknown

to what extent and for what purposes ermines use different available environments.

Other records of *M. r. muricus* from arid regions at the species' southern range boundary (i.e., Utah, Arizona) also were taken in riparian situations (e.g., Anderson 1955; Egoscue 1957; Stock 1970; Berna 1991). For instance, in the Sierra Nevada ermines were associated with willows in wet meadows (Cain et al. 2006). One of four specimens reported by Durrant (1952:417) from Utah was taken "in a moist mountain meadow, in long grass (knee deep), in heavy stands of willow (*Salix*)." In southeastern Idaho, *M. r. muricus* was most abundant on sites saturated with water versus sites that were damp or dry (Austin and Pyle 2004). In the Cascade Range, Oregon, *M. r. streatori* was captured significantly more frequently in riparian versus upland sites and the percent of females in breeding condition was greatest in riparian areas (Doyle 1990). Two of the few specimens of ermines taken in Colorado with habitat information were from "a moist hay meadow interspersed with willow stands in which long-tailed weasels were never captured" (Fitzgerald et al. 1994:337). Four ermines captured by me in New Mexico were all taken on saturated soil.

The distribution and abundance of ermine species is tightly linked to the distribution and abundance of prey species (King 1983; King and Powell 2007). In the mountains of northern New Mexico, the highest diversity and abundance of

Photos 21.9a–d. Examples of trap locations where the author (Jennifer Frey) and Zachary Schwenke captured American ermines in New Mexico while surveying for the New Mexico jumping mouse (*Zapus luteus*). In each photograph a–d, Zach is holding a Robel pole (a device used to measure the height of the cover provided by the vegetation) at the capture location. a) (*top left*) A small tributary to Soda Pocket Creek in Sugarite Canyon State Park, Colfax County, New Mexico. Zach is standing in the tiny creek where the vegetation is dominated by sedges and water hemlock (*Cicuta* sp.). b) (*top right*) A wet meadow dominated by beaked sedge (*Carex rostrata*) on a tributary to Holman Creek, Taos County, New Mexico. The bottom of the valley floor is covered with shallow water (Zach is standing in water), and the cover provided by the sedges is about 76 cm (30 in) tall. c) (*bottom left*) A wet sedge meadow created by a beaver (*Castor canadensis*) along the Rio Pueblo, Taos County, New Mexico. Zach is standing in a water-filled channel created by the beaver, and the herbaceous cover is about 84 cm (23 in) tall. d) (*bottom right*) Diverse wet sedge meadow along the Rio Cebolla, Sandoval County, New Mexico. Zach is standing in a water-filled beaver channel, and the herbaceous cover is about 66 cm (26 in) tall. Photographs: © Jennifer K. Frey.

small mammals are in riparian zones (Frey 2003, 2018). This is particularly true for voles, shrews, and jumping mice, which are important prey to ermines (see below; King 1983). Riparian zones produce luxuriant vegetation growth, which provides diverse and abundant food for small mammals and cover from predators (Frey 2018). Riparian zones and meadows are ideal for most species of voles, and under ideal conditions vole populations can become exceptionally dense, which in turn can fuel higher densities of ermines (King 1983). The primary sources of mortality for ermines are predation and low prey availability. As small mammals themselves, ermines likely rely on vegetation cover for concealment from predators during non-snow periods. Raptors are major predators of ermines, and tall vegetation may reduce the ability of raptors to detect weasels (Grimm and Yahner 1987). Most predation on ermines is due to raptors during winter, which is likely facilitated by reduced plant cover.

A final habitat suitability model for the ermine in New Mexico—a combination of the models based on bioclimatic and biophysical variables (Figure 21.1)—identified suitable areas in the San Juan Mountains, Jemez Mountains, and Sangre de Cristo Mountains as well as Raton Mesa (Colfax County) and Capulin Mesa (Rio Arriba County) (Frey and Calkins 2014). No habitat for ermines was predicted to occur in the Sandia Mountains, which further supports the conclusion that the observation of a weasel by Ivey (1957) was in fact a long-tail weasel (see under "Distribution"). Importantly, the model also predicted that ermine habitat is fragmented within each mountain range. This may result in the isolation of subpopulations, a risk to their long-term persistence due to small population size and lack of immigrants, though there are no demographic data to either support or refute this hypothesis. It should be emphasized that areas predicted to be suitable for ermines do not necessarily support ermine populations. In essence, the model depicts the ideal maximum distributional capacity of the species in a broad-scale context. However, this is likely to be a gross overestimation of actual habitat. The occurrence of ermine populations in New Mexico likely requires an abundant and diverse riparian and wet-meadow small-mammal prey community. If local conditions do not support abundant and diverse small mammal prey, such as where livestock grazing reduces herbaceous cover and willows (e.g., Small et al. 2018), for example, then ermines will not be present. Recent studies aimed at surveying riparian areas and small mammal communities in northern New Mexico have revealed substantial declines in the distribution of quality riparian areas and declines in small mammal communities, including the widespread extirpation of at least one species, the New Mexico jumping mouse (*Zapus luteus* [= *Z. hudsonius luteus*]), which utilizes habitat similar to that of the American ermine (Frey 2006; Frey and Malaney 2009).

LIFE HISTORY
Diet and Foraging
Ermine species are considered specialist predators of microtine rodents (i.e., lemmings and voles), in much the same way that the Canada lynx (*Lynx canadensis*) is considered a specialist predator on snowshoe hares (*Lepus americanus*; Sims 1979; Fagerstone 1999; King and Powell 2007; see Chapter 6). Lemmings are mostly restricted to Arctic regions, while voles occur primarily throughout the boreal and montane zones of the Northern Hemisphere. Voles are small (adult mass ~25–60 g [~0.9–2.1 oz] in New Mexico) terrestrial rodents that typically are associated with dense herbaceous plants such as grasses and sedges, which they use for food and cover. Voles are relatively unique among small mammals in that they eat the green leafy parts and stems of herbaceous plants, which is an essentially unlimited food supply. In contrast, other small mammals usually consume seeds and insects, which may be more limited in

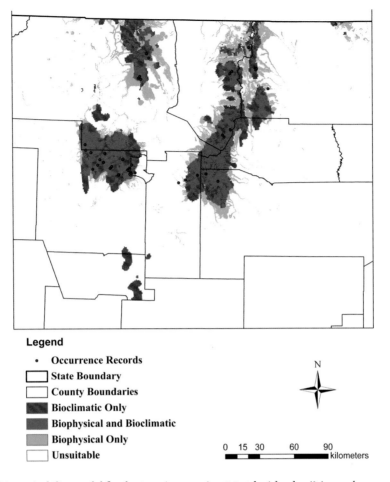

Legend

- **Occurrence Records**
- **State Boundary**
- **County Boundaries**
- **Bioclimatic Only**
- **Biophysical and Bioclimatic**
- **Biophysical Only**
- **Unsuitable**

N

0 15 30 60 90 kilometers

Figure 21.1. Habitat suitability model for the American ermine (*Mustela richardsonii*) in north-central New Mexico based on bioclimatic and biophysical variables. The small dots represent occurrence records. Predicted areas of ermine occurrence are in blue. Regions in red were predicted to be associated with a suitable climate but not suitable biophysical conditions. Regions in green were predicted to have suitable biophysical conditions but not a suitable climate. From Frey and Calkins (2010).

supply. Thus, voles can become quite abundant in places where there is tall dense cover to conceal them from the wealth of avian, mammalian, and reptilian predators that feed on them (Frey 2018). Most voles make distinctive bare "footpaths" and tunnels through the vegetation, which lead to underground burrows used for nesting. During winter, they rely on the subnivean spaces beneath the snow for both nesting and access to food plants, many of which remain green under the snow. Predation by ermines is thought to be a major controller of vole population cycles (Fagerstone 1999).

Mustea richardsonii is exquisitely designed as a predator of voles and other mice (Merriam 1896). For instance, it has a narrow skull with a reduction in bony protuberances and angles and a long braincase that holds a relatively large brain but maintains a narrow circumference. These modifications allow an ermine's head to pass through small openings, which is an important factor for a predator that dashes down burrows after a meal. Likewise, with respect to the body, Merriam continues with a colorful description (1896:6):

The body as a whole has undergone parallel modification, presenting the extreme degree of slenderness known among the mammalia. This type of weasel seems to have been developed for the express purpose of preying upon field mice or voles, its narrow skull and cylindrical body enabling it to enter and follow their runways and subterranean galleries. The extreme development of the type is presented in [the least weasel and smallest ermines], whose exceedingly small size and almost serpentine form make it possible for them to traverse the burrows of even the smaller mice.

The specialized body plan of the American ermine is supremely adapted for hunting voles where they live (in burrows and under snow), but not without a tremendous cost. The narrow, tube-shaped body results in a high ratio of skin surface to body volume. In addition, the tiny body and active predatory lifestyle mean the fur must be short. After all, a ~20-cm-long (8-in-long) ermine with ~8-cm-long (3-in-long) hair would look like a fluff-ball, which is not the sort of animal that could actively hunt prey. Because of both of these factors, small weasels lose body heat very rapidly. To compensate, ermines use well-insulated habitats in winter (e.g., the subnivean environment) and have a high metabolism, which requires a relatively steady intake of large amounts of food fuel. Thus, the ermine must be an active, roving vole-killing machine.

Ermines hunt energetically by using their senses of hearing, sight, and especially smell, to actively search for prey as they explore every nook and cranny within every patch of cover (King and Powell 2007). While smell is primarily used to find prey, the pursuit is sparked by the sight of moving prey, regardless of whether the ermine is hungry (King and Powell 2007). Being a hunter is a dangerous activity because of the risk of being harmed by the prey. Fortunately for

Photo 21.10. American ermine with a western deermouse (*Peromyscus sonoriensis*, formerly *P. maniculatus*) on 8 August 2015 near Summit Lake on the trail to Mt. Evans, Colorado. Photograph: © Jamie Simo.

Photo 21.11. American ermine on 8 June 2021 with its rodent prey near the western shore of Colorado's Grand Lake just outside Rocky Mountain National Park, at an elevation of 2,530 m (8,300 ft). The rodent appears to be a western meadow vole (*Microtus drummondii*; formerly *M. pennsylvanicus*). Note the large size of the vole compared to the ermine. Photograph: © Stella F. Uiterwaal.

ermines, hunting voles might be a little analogous to hunting a dairy cow—the voles are slow and just do not put up a whole lot of resistance. Ermines kill their rodent prey with a precisely delivered bite to the back of the neck or base of the skull such that the long upper canines skewer the back of the vertebral column (or skull), meeting the lower canines from below. The death of the rodent is nearly instantaneous. To initially subdue the animal and insure an expertly placed kill-bite, the ermine may first grab the rodent with its forelegs and wrap its long body around it, continuing to grip until the prey is dead. Ermines sometimes start a meal by licking blood from their prey, which may have led to the false idea that ermines suck blood from the neck of their victims like mythical vampires (e.g., Newhouse 1914). Ermines usually begin their meal by eating the brain, often leaving the rostrum and teeth behind as waste. The brain must be their favorite part because sometimes it will be all they eat if ample food is available. They then work backward, eating the rest of the body, but leaving the less palatable bony parts (feet and tail) and digestive tract for the end. If there is plenty of meat, they will avoid eating skin, hair, and large bones (King and Powell 2007). Even though the digestive tract is usually left for the end (especially the stomach), consumption of the small intestines is thought to be important because it supplies necessary carbohydrates in the form of partially digested plant material (King and Powell 2007).

Alas, though an ermine killing a vole may be a safer prospect for the predator than an African lion (*Panthera leo*) killing a giraffe (*Giraffa* sp.), the real problem for an ermine is finding enough food at regular intervals to meet the high energy demands of its long, narrow, active body. Based on data collected on *M. nivalis*, small weasels need to consume 24–42% of their body weight in prey every day (Fitzgerald 1977; Gillingham 1984). From the perspective of a predator, there are three key variables that determine survivorship and reproductive success: 1) the energy required to capture a prey item including search time and handling, 2) the nutritional value of the prey, and 3) the number of prey available. Controlled studies have shown that ermines have to expend less energy hunting voles as compared with other small mammals such as shrews and mice (e.g., Raymond et al. 1990). Further, most small mammals, whether shrews, voles, or mice, have about the same nutritional value when corrected for weight, which ranges from 5.0 to 5.5 calories per dry gram (Raymond et al. 1990). However, voles are larger than mice and hence a 41.5 g (~1.5 oz) vole is worth 54 calories while a 17.3 g (~0.6 oz) deermouse is worth only 28 calories (Raymond et al. 1990). Thus, besides voles being easy prey, ermines might specialize on them due to their higher calorie content. An even more important reason is probably that voles are often abundant and they can go through population explosions, at which time they become even more plentiful. For instance, in montane areas of the Southwest, the highest concentrations of small mammals are almost always those of voles in riparian areas (Frey 2018). However, the same species that tend to be plentiful may also experience population crashes. These dramatic shifts in abundance of the key prey mean that ermines experience periods of both bounty and starvation. Unlike in other carnivores that switch to other species when their main prey declines, Fitzgerald (1977) found that ermines continue to prey on voles even as their numbers decline. Thus, ermine populations are likely to be much more closely tied to the dynamics of their prey than are other carnivores.

Although voles are usually the most abundant small mammal prey, ermines can still run the risk of not finding enough of them to live or successfully reproduce. The stomachs of small weasels are simply not large enough to hold big meals, and weasels cannot store large amounts of fat because an obese weasel would not be able to hunt efficiently in small burrows. In addition, ermines have short digestive tracts, which, in conjunction with a high metabolism, means that a meal passes completely through in just 2.3–4.5 hours (Short 1961). Consequently, ermines must eat frequently. Captive North American least weasels, which have larger bodies than New Mexico ermines, had to eat meals every 2.5 to 3 hours, totaling 5–10 meals per day (Gillingham 1984). Small weasels are not likely to be able to survive more than 24 hours without eating (Gillingham 1984). Because ermines can eat only so much during each meal and because their prey tends to be at least as large or larger than themselves, ermines cache their excess prey. Caching also accounts for the fact that weasels will reflexively kill many prey animals at one time if given the opportunity. Caching provides an important safety net with respect to the uncertainty of when the ermine will obtain its next meal.

POTENTIAL DIET OF AMERICAN ERMINES IN NEW MEXICO

Although specialized on voles, to some extent the prey choice of ermines is influenced by availability (abundance), as well as opportunity (encounters). Consequently, a complete list of foods eaten by ermines could be quite long and includes not only small mammals but also juvenile lagomorphs (i.e., rabbits and pikas), birds, reptiles, amphibians, and invertebrates (Fagerston 1999). To evaluate potentially important prey for ermines in New Mexico, I compiled information on other small mammals collected together with ermines in the state (see Table 21.3). These data only include species that are readily captured by the same methods (e.g., Sherman traps) used to secure specimens of ermines in New Mexico. Thus, species such as gophers, rabbits, pikas, tree squirrels, and other larger mammals are not included in this evaluation, though they also are unlikely to be regular prey of ermines due to their larger size. Data were available for 18 ermine sampling events based on either specific information on trap lines as recorded in field notes or specimens collected during the same field trip when ermines were captured as recorded in museum catalogs. The field trip data based on museum catalog records can be biased because they reflect only those animals preserved as specimens. Thus, they might not accurately portray all species captured or actual numbers or proportions of species captured.

A total of 17 species of terrestrial small mammals were captured together with ermines in New Mexico (Table 21.3). Observed small mammal communities tended to be diverse, averaging seven species but with up to 12 species documented. The most frequent and numerous mammals captured along with ermines were voles (*Microtus*; 120 individuals recorded at 17 sites) and shrews (*Sorex*; 114 individuals recorded at 14 sites). These two taxa are primarily associated with riparian areas, especially wet meadows. The two most common voles captured with ermines were the long-tailed vole and montane vole. The distribution of the long-tailed vole includes all areas of the state where the ermine occurs, and it is primarily associated with riparian areas dominated by diverse herbaceous plants (grasses, sedges, and forbs) interspersed with riparian shrubs such as willows (*Salix* spp.) and alders (*Alnus* spp.). In contrast, the montane vole is limited to the San Juan and Jemez mountains and is primarily associated with riparian areas dominated by graminoid vegetation (grasses and sedges), though it also commonly occurs in other mesic grass-dominated areas (e.g., drier parts of montane meadows). The meadow vole (*M. pennsylvanicus*) has similar habitat associations as *M. montanus*. However, it was less frequently documented with ermines because its distribution is restricted to the Sangre de Cristo Mountains region and it tends to occur at lower elevations than the other two voles. The large-bodied western water shrew is restricted in distribution to riparian areas along perennial streams. The smaller-bodied montane shrew (*S. monticolus*) and cinereus shrew are most abundant in riparian zones, though both may be caught in other moist situations.

Following voles and shrews, the North American deermouse (*Peromyscus maniculatus;* now reclassified as *P. sonoriensis*) was the next most common mammal documented with ermines in New Mexico. The deermouse is a generalist that uses a wide range of vegetation types. Chipmunks (*Neotamias*) formed the fourth most frequent and numerous taxonomic group. Most of these records were of the least chipmunk (*N. minimus*), which is typically associated with relatively open vegetation such as meadows. In contrast, the Colorado chipmunk (*N. quadrivittatus*) is more closely associated with mixed coniferous forest. The fifth most frequent and numerous group consisted of jumping mice (*Zapus*). Four of the New Mexico ermine specimens were captured during surveys for the endangered New Mexico jumping mouse. This jumping mouse requires tall, dense herbaceous riparian vegetation on saturated soil, particularly dominated by sedges and forbs (Frey and Malaney 2009; Frey 2017). The western jumping mouse (*Zapus princeps*) also is primarily associated with riparian areas, though it uses a broader range of conditions (Frey 2006). The remaining six species, which were least frequently encountered and least numerous, are primarily associated with upland areas. The golden-mantled ground squirrel (*Callospermophilus lateralis*) is primarily associated with dry grassy montane meadows and other forest openings. The southern red-backed vole is primarily associated with mature spruce-fir forest. Woodrats, including the bushy-tailed woodrat (*Neotoma cinerea*) and Mexican woodrat (*N. mexicana*) as well as other deer mice, including the brush deermouse and rock deermouse (*P. nasutus*), are primarily associated with rocky areas in coniferous forests. Some of these species (bushy-tailed woodrat and brush deermouse) were never caught on the same trap line as an ermine, which indicates that different microhabitats were sampled in the field

trip data. Thus, the small mammal communities associated with ermines in New Mexico were primarily associated with riparian areas, which provides an independent line of evidence that ermines specialize on riparian zones in New Mexico.

It must be emphasized that the list of small mammals captured together with the ermine represents its potential prey base. Ermines may never, or only rarely, utilize some of these species as food (e.g., *P. boylii, N. cinerea*). Until a food habits study is conducted in New Mexico or southern Colorado, we can only assume that voles make up the majority of the diet. In addition, there are other species not included in Table 21.3 that might provide occasional or locally abundant food for ermines in New Mexico including other long-tailed shrews (*Sorex*), small pikas (*Ochotona*

Photo 21.12. Comparison of body size among small, co-occurring mammals in New Mexico (from left to right): western deermouse (*Peromyscus sonoriensis*, formerly *P. maniculatus*), montane vole (*Microtus montanus*), western meadow vole (*M. drummondii*), Colorado chipmunk (*Neotamias quadrivittatus*), female American ermine (*Mustela richardsonii*), male American ermine, Mexican woodrat (*Neotoma mexicana*), and mountain cottontail (*Sylvilagus nuttallii*). Photograph: © Jennifer K. Frey.

Table 21.3. Records of other terrestrial small mammals captured with the ermine (*Mustela richardsonii*) in New Mexico. Only species readily captured in traps devised for small terrestrial mammals are included (i.e., exludes gophers, tree squirrels, rabbits, etc). Mountain ranges: JZ: Jemez; SDC: Sangre de Cristo; SJ: San Juan.

Location	Mountain Range	County	Locality	Source	Known Inclusive Dates at Site	Scope	Cinereus Shrew	Montane Shrew	Water Shrew	Unidentified Shrew	Least Chipmunk
2	SJ	Rio Arriba	24.1 mi W, 3.8 mi N of Tres Piedras	Catalog	22–23 Sept 1973	Location	3	3	4		10
3	SJ	Rio Arriba	Upper Canjilon Lake	Catalog	7 Sept 1974	Location	2	9	5		2
10	SDC	Colfax	Philmont Scout Ranch, 17 mi NW of Cimarron, French Henry Camp	Catalog	18–29 June 1968	Location	2	2	2		
12	SDC	Taos	Twining, 10,700 ft	Catalog	6–8 August 1904	Location	5		1		3
15	SDC	Taos	Rio La Junta , 17 mi S, 6 mi E of Taos, 9,400 ft.	Catalog	3–5 Sept 1976	Location	1	2	5		
20	SDC	San Miguel	1.5 km N of Elk Mtn., Santa Fe Mtns., 3,447 m	Catalog	20–21 Sept 2008	Location	4	3			1
26	JZ	Sandoval	16 mi N of Jemez Springs	Catalog	7–9 Sept 1979	Location		28	6	1	1
26	JZ	Sandoval	2 mi E, 15 mi N of Jemez Springs	Catalog	5–7 Sept 1980	Location		2	1	1	2
29	JZ	Sandoval	Fenton Lake, Jemez Mtns.	Catalog	4–5 August 1979	Location		1			
31	JZ	Sandoval	7.5 mi N, 6 mi E of Jemez Springs	Catalog	21–22 July 1981	Location					2
33	JZ	Sandoval	6.5 mi. N, 5 mi E of Jemez Springs	Catalog	23–24 July 1981	Location					
42	JZ	Sandoval	3.25 mi N, 10 mi E of Jemez Springs	Catalog	20–22 Sept 1996	Location		2	2		4
5	SJ	Rio Arriba	4.25 mi N, 6.5 mi E of Canjilon, Canjilon Creek Campground	Field notes	25–26 July 1980	Trap line	1	4	4		
6	SDC	Colfax	Sugarite Canyon State Park, small tributary to Soda Pocket Creek	Field data	11–12 July 2006	Trap line					
7	SDC	Taos	Southern tributary to Holman Creek, along Forest Road 1950	Field data	24–26 July 2006	Trap line					
17	SDC	Taos	Rio Pueblo, beaver ponds just below junction Agua Sarca Canyon	Field data	9–10 July 2006	Trap line		1			
28	JZ	Sandoval	Jemez Mtns., W of Slope Barley Canyon	Field notes	21 August 1992	Trap line					
30	JZ	Sandoval	Rio Cebolla, 0.6 mi SW of FR 376 bridge over Rio Cebolla	Field data	14–15 August 2006	Trap line		1	2		
43	JZ	Sandoval	0.2 mi E of Las Conchas Campground on NM Rt. 4	Field notes	18–20 Sept 1998	Trap line		2	1	1	2
						Number	18	60	33	3	27
						Sites	7	13	11	3	9

Other Small Terrestrial Mammals Captured (Number of Specimins at Each Location)

Colorado Chipmunk	Unidentified Chipmunk	Golden-mantled Ground Squirrel	Western Jumping Mouse	New Mexico Jumping Mouse	Long-tailed Vole	Montane Vole	New Mexico Meadow Vole	Unidentified Vole	Southern Red-backed Vole	Bushy-tailed Woodrat	Mexican Woodrat	Western Deermouse	Brush Deermouse	Rock Deermouse	Unidentified Deermouse	American Ermine	Species Richness	Number
		1	1		6	9			1	1		2				1	12	42
						1		3								1	7	23
1		2			1					1		1				1	9	13
					1				8							1	6	19
			3		1					1		1				1	8	15
		1							7							1	6	17
					7	8		1				2				1	9	55
		1			3	4					1	6	1		1	3	12	26
				1	1	5										1	5	9
					2								1			2	4	7
					1											1	2	2
		2			2	2				1	3	13				1	10	32
	4							9							17	3	7	42
				2			2					7				1	4	12
1												3				1	3	5
		3	5													1	4	10
	1							1	1							1	4	4
				3	2	4										1	6	13
1					7	20					3	26		1		1	11	65
3	5	10	9	6	34	53	2	14	17	4	7	61	2	1	18	24	7	22
3	2	6	3	3	12	8	1	4	4	4	3	9	2	1	2	19		

princeps), juvenile rabbits (*Lepus americanus, Sylvilagus nuttallii*), western harvest mice (*Reithrodontomys megalotis*), western heather voles (*Phenacomys intermedius*), birds, amphibians, snakes, fishes, and invertebrates (e.g., Fagerstone 1999). Male ermines are more likely to deviate from a vole diet than are females (Raymond et al. 1990). Weasels of different body size exploit prey of different sizes, and this pattern pertains both to comparisons between sexes as well as among different species of weasel. Photo 21.12 shows the size of New Mexico's American ermine compared with representative small mammals with which it co-occurs in the state. This illustrates how mice and voles are appropriately sized prey for New Mexico ermines, but larger species such as cottontails (*Sylvilagus* spp.) likely are not.

Reproduction and Social Behavior

Based on data collected elsewhere in North America (reviewed by Fagerstone 1999), ermine species have only a single litter each year, and it is born in spring (usually April). However, mating occurs during the previous spring or summer. Adult females usually mate in spring, just after giving birth. Mating typically occurs in summer for females in their first year, though some are mated as early as spring. Ermines are rather unusual in that newborn females, while unweaned and their eyes still closed, become reproductively mature. The mother also comes into estrus soon after birth. Thus, as surprising as it might sound, an adult male may mate with both the mother and the infant females in the nest (King and Powell 2007). It is the act of copulation that induces the female to ovulate (i.e., release eggs from ovary so that they are available to be fertilized by the sperm). Thus, by the end of summer most female ermines are pregnant. However, the embryonic development ceases at the blastocyst stage (a hollow ball of 100–200 cells) before implanting in the uterus, and the embryo over-winters in this state, which is a reproductive strategy called delayed implantation. In spring, the increasingly longer day length stimulates the blastocyst to implant into the uterus with a subsequent gestation period of about four weeks (King and Powell 2007). Because the spring molt from the winter white to summer brown pelage is also stimulated by day length, birth can be predicted to occur about 3.5 weeks after the first brown hairs appear on the mother's nose (King and Powell 2007). Birth dates at the ermine's southern range limits in the western United States are expected to be late March or early April (King and Powell 2007). In males, testes start to become active in February and first mating occurs in the spring after birth.

It is thought that ermine maternal dens are in burrows with narrow entrances to provide protection from other weasels (King and Powell 2007). However, during winter, *M. r. muricus* in the Sierra Nevada were found to nest in subnivean non-breeding nests of voles (Fitzgerald 1977). These non-breeding vole nests are located on the surface of the ground (under the snow) and constructed of grasses with a distinct inner chamber lined with finely shredded grass. Vole nests used by ermines were easily identifiable because the entrances were littered with the uneaten parts of voles, there were nearby latrines of weasel scat, and the ermines modified the inner nest chamber by lining it with a thick layer of vole hair (also including bone fragments and teeth). Given that snow cover in the high mountains at the southern edge of the ermine's range persists until April or May (Fitzgerald 1977), perhaps modified vole nests are similarly used by pregnant or nursing ermines. Once a suitable den is found, the mother stores up caches of food and awaits the birth of her young.

Newborn ermines cannot maintain their own body temperature and hence must rely on the mother for warmth. When the mother must leave the nest, say, to hunt, the young huddle together to try to stay warm. However, if the mother is away for any extended period, their body temperatures will fall until they enter a torpid state in which their heart rate, breathing, and metabolism slow and their bodies become cold and stiff (Segal 1975). Because growth is slowed during these torpors, young weasels born to mothers experiencing food shortages (meaning longer and more frequent periods away from the nest hunting) will be small and may die (Powell and King 1977). Ermines are solitary animals with no formation of pair-bonds, and the male takes no part in caring for the young. Observed cases of male ermines bringing dead rodents to a female's nest are thought to represent "presents" to the female to entice her to mate, as females come into estrus again soon after birth (Powell and King 2007).

In general, weasels have very unstable populations. Although most females might become pregnant, there often may be low survival of the embryos, to the point of essentially no successful reproduction in some years (King and Powell 2007). Reproductive success is tied to the abundance of small rodent prey (King and Powell 2007). Weasels are relatively short-lived animals, and life span is even shorter for the smaller forms of weasels. There are no good data on the life span of *M. richardsonii* in western North America, but it is expected to be less than that of *M. erminea*, which probably live a year or less on average with only a small proportion surviving as long as three years (King and Powell 2007). Thus, because there is only one reproductive period per year, and in some years little or no reproduction, small ermines such as *M. r. muricus* might routinely exist at the brink of local extinction. However, during population highs of voles, the reproductive success of ermines should likewise be high. With an average successful litter size of perhaps six and upward to as much as nine or 10 (King and Powell 2007), the summer population may mostly be made up of young-of-the-year animals.

Little is known about reproduction in New Mexico ermines. Sex-ratios from field studies are usually 1:1 (Fagerstone 1999), and of the New Mexico specimens examined by Frey and Calkins (2010) with reported gender, 23 were male and 23 were female. Seven female specimens had recorded reproductive data; none had embryos or was lactating (one was collected in March and six in July). Of the male specimens, 12 had recorded reproductive data. Two reported as scrotal (i.e., the testes were descended into the scrotum indicating sexual readiness) were captured on 10 and 24 July (one with testes = 6 mm [0.24 in]). Three non-scrotal animals were collected on 4 August (testes = 8 mm [0.31 in]), 15 August (testes = 4 mm [0.16 in]; epididymis not swollen) and 19 September. Testes lengths were available for 10 specimens (including those already mentioned) in June, 10 mm (0.39 in); July, 6, 8 mm; August 4, 4, 6, 8 mm; September, 4, 6 mm; and October, 5 mm (0.2 in).

Interspecific Interactions

Long-tailed weasels occur throughout the range of the American ermine in New Mexico. Although long-tailed weasels are more habitat generalists and tend to select larger prey than ermines, they also prey on the species included in the ermine's narrower diet (i.e., voles, mice, and shrews). Thus, the two species of weasel exhibit significant overlap in diet (Simms 1979b). In addition, long-tailed weasel habitat may also overlap with that of the more specialized American ermine (e.g., Cain et al. 2006). Perhaps more importantly, long-tailed weasels have been known to kill and/or displace smaller weasels (e.g., Polderboer 1941; Simms 1979b; St-Pierre et al. 2006). Thus, there is not only indirect competition for food and space, but also the possibility of direct interference competition between the two species. Competition between the species creates unstable communities in which it is likely that one species will go extinct unless prey populations are high (Powell and Zielinski 1983). Even though the ermine is a more efficient predator of usually the most abundant prey (voles), it is still more likely to be on the losing side of competitive interactions because long-tailed weasels 1) are physically dominant and hence can displace ermines; 2) take a greater variety of prey, which is an advantage when vole populations are low; and 3) can utilize a wider range of environments, which is an advantage if riparian areas are degraded and, simply, because there are more areas available for occupancy.

Ermines in New Mexico are tiny and hence they are likely just as susceptible to predation as are other small mammals. The only record of predation on ermines in New Mexico involved a domestic cat, but that is a consequence of the lack of studies. In Finland, red foxes (*Vulpes vulpes*) were observed to eat weasels (Beringian ermine and least weasel), especially in winter and when vole populations were high (Dell'Arte et al. 2007). However, predation by other mammalian predators appears to be uncommon, perhaps because weasels have musk glands similar to skunks, which can release a powerful stench when the weasel is frightened (King and Powell 2007). In other regions of North America, reported predators of ermines have included the great horned owl (*Bubo virginianus*), northern goshawk (*Accipiter gentilis*), rough-legged hawk (*Buteo lagopus*), and golden eagle (*Aquila chrysaetos*; Seidensticker 1968; Springer 1975; Olendorff 1976; Squires 2000). Although weasels are usually a negligible component (e.g., <1%) of raptor diets, one study in Michigan estimated that birds of prey took ca. 70% of the spring weasel population (Craighead and Craighead 1956) and, consequently, population models predicted that raptors can limit weasel populations (Powell 1973). In fact, the white pelage, black-tipped tail and erratic movements of the ermine are considered defenses against raptor predation during winter when herbaceous cover lies under a blanket of snow (i.e., the raptor focuses on the black tip rather than the body; Powell 1982). In contrast, during the growing season ermines likely depend on tall herbaceous vegetation as concealment from predators. This may be another reason ermines in the arid West prefer riparian zones. For instance, the abundance of American ermines was depressed in mowed versus unmowed roadsides in Minnesota, which was not associated with an abundance of voles (Grimm and Yahner 1987). The authors concluded that the ermine's association with tall herbaceous vegetation was for protection from raptor predation.

Ermines are prone to harbor a roundworm parasite known as *Skrjabingylus nasicola* (reviewed by Powell and King 2007). The worms lodge in the nasal sinuses behind the eyes, creating swelling in the skull. Heavy infestations can cause reductions in survival and reproduction. The parasite larvae live in a snail or slug intermediate host. The snail or slug is then eaten by a shrew (or rodent), which is then consumed by the ermine. The prevalence of infection by the parasite varies geographically. Dougherty and Hall (1955) found this parasite in three of 18 (17%) specimens of *M. r. muricus*. However, it is unknown if any of those samples were from New Mexico or elsewhere in the Southern Rocky Mountains, and they mapped this region as having insufficient data to determine the prevalence of the parasite.

Methods for Capturing American Ermines for Research

To study an animal, one must either capture it or observe it through non-invasive techniques. One reason for the paucity of information about *M. r. muricus* is because it is both rare and difficult to capture or observe. All specimens of American ermines taken in New Mexico appear to have been incidental captures during small mammal surveys, except for one caught by a cat and two found dead on the ground. Specimens in other states also have been taken from domestic animals or found dead (e.g., Seton 1933; Hayward 1949; Vazquez 1956). Animals found dead are presumed to have been killed by predators and then dropped (e.g., Hayward 1949).

Small mammal surveys are usually conducted by setting large numbers of snap traps, pitfall traps, or small live traps. There are occasional records of ermines being taken in mouse snap traps, but at least some of these instances apparently were accidental as the ermine passed by rather than was attracted to the trap (e.g., Bailey 1931; Durrant 1952; Egoscue 1957). Likewise, small-bodied ermines and least weasels are only incidentally captured in pitfall traps (i.e., Berna 1991; Handley and Kalko 1993; Bellows et al. 1999; Murphy et al. 2007). Pitfall traps are not likely to be efficient for trapping weasels unless the buckets are very deep or are partially filled with fluid to prevent the animal from jumping out, but which causes them to drown (see Sikes et al. 2016 for more information about this lethal capture method). However, the problem with this method is that it does not allow for studies that require live animals and it can result in mortality of non-target species. The only known ermine specimen taken in Arizona was an incidental capture in a pitfall trap during a study of salamanders (order Urodela) (Berna 1991). Similarly, commercially available wire-cage live traps are not effective in capturing the smallest weasels. Wilson and Carey (1996) found that *M. r. olympica* in Washington was able to crawl in and out of Tomahawk traps through the space around the doors—the ermines entered (and then left) closed traps to feed on captured rodents.

Steel traps (i.e., "leg-hold" traps), which are a standard method of furbearer harvest, can be used to capture long-tailed weasels and large forms of ermine such as *M. erminea* and *M. richardsonii* from northern and eastern North America (e.g., *M. r. richardsonii, M. r. bangsi, M. r. cicognanii*; e.g., Egoscue 1957; McDonald and Case 1990; Fagerstone 1999). To improve capture rates, such traps are often deployed inside a baited wooden box, and this method has been recommended for animal damage control purposes (https://icwdm.org/species/carnivores/weasels/weasel-damage-prevention-and-control-methods/). However, some trappers use large rat snap traps or body-grip traps placed in wooden boxes to harvest larger weasels. In 2009, the Colorado Trappers Association reported online that captures of "ermine" were mostly incidental to trapping for martens and minks. I am unaware of any verified reports of *M. r. muricus* taken in furbearer sets in Arizona, Utah, Colorado, or New Mexico. Given the tiny size of *M. r. muricus*, it is unlikely that these ermines would be taken in conventional furbearer trap sets.

A most unusual and fanciful method for trapping ermines was found in a 1903 article in an Ellensburgh, Washington newspaper (Anonymous 1903), which evokes images of the boy with his tongue stuck to a metal pole in the classic movie *A Christmas Story*. In order not to mar the pelt by using a snare, the article instructs that the trapper place his hunting knife, which has been smeared with grease, on the trail of a young ermine. The knife becomes frosted with ice, which the ermine then licks. "But, alas, for the resemblance between ice and steel! Ice turns to water under the warm tongue; steel turns to fire that blisters and holds the foolish little stoat by his inquisitive tongue, a hopeless prisoner, until the trapper comes" (Anonymous 1903). Fortunately, better methods have since been devised for the mammalogist or wildlife biologist to study ermines!

Of 11 New Mexico ermine specimens for which the method of capture is known, all but one were taken in a Sherman live trap; the other was secured in a mouse kill-trap (probably a snap trap) baited with squirrel meat (Bailey 1931). Sherman traps are long rectangular boxes that may resemble a rodent burrow when the trap door is open; they also typically smell like rodents from prior captures. Sherman traps are probably ideal traps for capturing small ermines, such as *M. r. muricus*. In New Mexico, ermines are within the size and weight range of other small mammals (e.g., large shrews, mice, voles, or chipmunks) that are effectively and efficiently sampled with these traps. In addition, Sherman traps are probably especially effective traps for small-bodied ermines, whose natural behavior is to investigate holes and crevices for prey (King 1983) and several studies have reported ermines captured in Sherman traps (e.g., Hayward 1949; Doyle 1990; McDonald and Case 1990; Wilson and Carey 1996). Sherman traps became commercially available in 1955 (https://www.shermantraps.com/) and were widely used by the 1970s (based on the number of scientific papers published per decade and containing the

Photo 21.13. The subspecies of American ermine occurring in New Mexico and adjacent areas (*Mustela richardsonii muricus*) is very small and can be successfully captured in Sherman live traps. This large adult male ermine held by Zachary Schwenke was one of several the author and Zach captured at Price Lakes on the south slope of Navajo Mountain in Archuleta County, Colorado, near the New Mexico border. Ermines are difficult to handle and so they were processed inside a zippered tent to prevent escapes. Zach is holding the Sherman trap that caught the ermine in his left hand. Photograph: © Jennifer K. Frey.

words "Sherman live trap" found in the digital library of journals and books in JSTOR). Prior to 1955, specimens of small ermine subspecies were particularly rare, and Anderson (1955) thought that the apparent rarity of ermines was due to ineffectual trapping methods. All but two New Mexico specimens of ermine were taken after 1960, which was likely due to the increased use of Sherman traps during mammal survey work. Small ermines caught in Sherman and other live traps are often found dead, presumably due to their high metabolic rate and large surface-to-volume ratio (J. K. Frey, pers. obs.; Hayward 1949; Fitzgerald 1977). Consequently, studies requiring live animals will need to modify methods to prevent trap mortalities.

Searches for tracks and scat failed to document ermines during a six-year study of weasels in Gunnison County, Colorado (Quick 1951), where one had previously been captured with a Sherman trap (Hayward 1949). In contrast, Fitzgerald (1977) studied *M. r. muricus* in the Sierra Nevada by winter snow-tracking and spring searches for winter nests. Hair traps and tracking tunnels have proven effective for studying the larger Beringian ermines and least weasels in Europe, and these methods could potentially be applied to small-bodied ermines (Mowat and Poole 2005; King et al. 2007; Garcia and Mateos 2009). Besides Sherman traps, the only other method proven effective for studying small ermines was the use of track plates to determine areas of activity and then use of automatic cameras to evaluate predation attempts on songbird nests (Cain et al. 2003, 2006). Remote cameras have been used to survey for larger forms of weasels, but they often have low detection rates and are unlikely to be effective for the diminutive *M. r. muricus*. However, remote cameras can be paired with a "Mostela" box (custom-made wooden box for capturing mustelids) specially designed for small weasels such as *M. r. muricus* (Mos and Hofmeester 2020).

STATUS AND MANAGEMENT

Legally, the American ermine is considered a furbearer in New Mexico with an open sustainable harvest limit estimated to be 9,727–14,591 (NMDGF 2006). There are two problems with this situation. First, ermines in New Mexico are so tiny that they are not captured via conventional furbearer trapping techniques and the pelts have essentially no value. Records from the New Mexico Department of Game and Fish indicate that there is little or no harvest of this species. Thus, having a season on ermines in New Mexico makes little sense. Second, if ermines were to become targets of furbearer harvest, the estimated sustainable harvest is certainly much too high given what is now known about the species' distribution and habitat. The above sustainable harvest limit represents 13% of a New Mexico ermine population estimate that relies on an estimated density of four to six ermines per square kilometer in the state and an estimated occupied range of 18,707 km² (7,220 sq mi) (Thompson et al. 1996). Such a density estimate is typical for Beringian ermines in Europe, but not for the American ermine in western North America, where it should be much lower (i.e., <1/ km²). Second, Thompson and colleagues' estimate of the occupied range was a gross overestimation because they included any area mapped as subalpine coniferous forest, montane coniferous forest, montane scrub, or montane grassland (rather than riparian). Two counties outside of the actual known range of the species (i.e., Bernalillo and San Juan counties) were also included. In contrast, the more sophisticated habitat suitability models described above predicted that the ermine's entire geographic range in New Mexico was <7,000 km² (<2,700 mi²; Frey and Calkins 2011, 2014). However, even that is a gross overestimate of the actual area likely to be occupied by ermines in New Mexico because the species is restricted to areas with very specific microhabitat conditions (i.e., tall herbaceous riparian vegetation). There

are no data on habitat occupancy. Although there is essentially no harvest of this species in New Mexico, closing the season would serve to highlight the species' status as one of conservation concern (e.g., similar to the American mink and the white-nosed coati [*Nasua narica*]). The state of Colorado recently changed the status of the ermine from harvestable furbearer to protected non-game species. However, in New Mexico non-game species benefit from few protections and hence such a change of status would be counterproductive in the state. Further, despite the warning by Frey and Calkins (2010, 2014) that the American ermine is likely facing conservation threats and should be considered in conservation planning, the species was not included as a Species of Greatest Conservation Need in the most recent State Wildlife Action Plan for New Mexico (NMDGF 2016).

Photo 21.14. An American ermine (*Mustela richardsonii*) in summer pelage captured by the author (Jennifer Frey) at Price Lakes, Archuleta County, Colorado, near the New Mexico border. The American ermine is listed as a protected furbearer in New Mexico, with an estimated sustainable harvest limit that is almost certainly much too high. The species is very rare in New Mexico, its distribution is highly patchy and mainly restricted to sedge wetlands still harboring beavers (*Castor canadensis*). Photograph: © Jennifer K. Frey.

The American ermine can be considered extremely rare in New Mexico based on a restricted distribution, isolated populations, the species' patchy local distribution, and low abundance. Ermines also appear to be rare in neighboring states. In Arizona, there is but a single record of the American ermine (Berna 1991). A search for Utah specimens using the Mammal Networked Information System revealed only 13 specimens. In Colorado, Armstrong (1972:275) considered the ermine "nowhere common," and Fitzgerald et al. (1994) noted that only two ermines were captured in 14 years of research on small mammals in South Park.

A recent study on the status of ermines (combined *M. erminea* and *M. richardsonii*), long-tailed weasels, and least weasels in North America based on a review of historical harvest data, museum records, iNaturalist records, and camera survey results concluded that, unlike declines observed for long-tailed and least weasels, ermines had not experienced any notable change in their distribution across the continent (Jachowski et al. 2021). However, that study likely provided an incomplete picture of the status of ermines in New Mexico due to the inclusion of fossils in the historical museum records, misidentifications in the iNaturalist database, and lack of a harvest record or camera surveys within the ermine's range in the state. In contrast, other evidence indicates that American ermines are almost certainly declining in New Mexico. Captures of ermines have happened only during a fraction of all small mammal surveys in the San Juan, Jemez, and Sangre de Cristo mountains. Most surveys of small mammals in this region have not documented ermines. The peak in ermine records was in the 1970s and 1980s, and all multiple captures of ermines during field trips (suggestive of ideal habitat conditions) occurred during or prior to that time. In addition, evidence suggests that ermines are likely to be sensitive to habitat changes, both directly and through impacts to small mammal prey populations. For example, between 2003–2012, I surveyed small mammals at 60 riparian locations in the San Juan, Jemez, and Sangre de Cristo mountains, for a total of 10,898 Sherman trap-nights (representing the cumulative sampling effort), but captured only four ermines (Frey 2003, 2006, 2007, 2012a,b; Frey and Schwenke 2007; Frey and Malaney 2009). All captures took place in remnant patches of sedge wetland that were not grazed by domestic livestock, two of them in beaver modified areas (Frey 2006). Beaver dams create and maintain wetlands that can support high densities of small mammals (Frey 2018). However, beaver distribution and abundance in New Mexico have been greatly reduced (Huey 1956; Small et al. 2018). The presence of beavers and associated well-developed riparian areas is rare and has declined since the 1980s (Frey and Malaney 2009; Small et al. 2018). The major cause of loss of beavers and declines in diversity and abundance of small mammal communities in montane riparian zones was determined to be the loss of herbaceous cover due to livestock grazing (e.g., Frey and Malaney 2009; Small et al. 2018). Other studies also have linked declines in voles and other riparian small mammals (e.g., jumping mice, shrews) to overgrazing (Fagerstone and Ramey 1996). Ermine populations can only be maintained in high-quality riparian areas that produce large populations of voles. Thus, livestock grazing that reduces plant cover or changes vegetation composition in riparian zones and montane meadows may be a key threat to the American ermine.

Climate change is likely to be an increasingly important threat to the American ermine at its southern distributional limits (see Chapter 5). The American ermine in western North America is extinction-prone, and its current occurrence in the Southern Rocky Mountains is a relic of a broader distribution during the cooler periods of the Pleistocene (see under "Distribution" above). Post-Pleistocene warming resulted in the extinction of ermines from low elevations and from smaller, more isolated mountain ranges. Accelerated contemporary climate warming is predicted to profoundly impact montane species. Higher temperatures are likely to cause an upward displacement of montane vegetation zones, reduction in riparian zones, and increased wildfire, all of which will increase habitat fragmentation and population decline for the ermine and other montane riparian species.

The American ermine may be especially vulnerable to climate change compared to many other boreo-montane species, because adequate snow cover with subnivean spaces is a key habitat requirement for the species, in terms of increased foraging efficiency, reduced competition with other predators of voles, increased winter survival of prey, and cover from predators. Mote et al. (2005) demonstrated significant reductions in snowpack throughout much of the western United States, and these trends are expected to continue with additional climate warming (see also Chapter 5). Copeland et al. (2010) found that spring snow cover was a good predictor of the breeding distribution of another boreal mustelid, the wolverine (*Gulo gulo*), and cautioned that reductions in spring snow cover associated with climate warming would pose a threat by reducing wolverine habitat and connectivity (see Chapter 18). These threats also are relevant to the American ermine in the mountainous regions at the southern edge of the species' range in North America. Mild and wet winters have been linked to recent declines in vole populations and several of their key predators (Hornfeldt et al. 2005). One particular climate threat to American ermines is increased ice formation in the subnivean space, which may result from changes in temperature and humidity to milder and wetter winter conditions (Kausrud et al. 2008). Such changes are anticipated to not only decrease prey populations but also have cascading effects on 1) relationships with, and

Photo 21.15. In New Mexico, American ermines are closely associated with wet meadows and riparian zones dominated by tall sedges. This habitat produces large abundance of voles (*Microtus*), which is the ermine's main prey. Beaver (*Castor canadensis*) activity increases the amount and quality of this habitat, but it can be degraded by livestock grazing which reduces the necessary plant cover. This ermine capture location (where the red flagging and Robel pole are placed) is on a beaver channel along the Rio Cebolla, Sandoval County. Management to restore and protect riparian vegetation can only benefit ermines. Photograph: © Jennifer K. Frey.

between, specialist (small weasels such as the ermine) and generalist predators (e.g., long-tailed weasels, foxes, raptors); and 2) changes in plant communities as a response to altered rodent grazing (Kausrud et al. 2008). Lastly, the reduced duration of snow cover due to climate change could be a particular threat to American ermines, which are evolutionarily programed to turn white in winter. A white animal is an easy target for predators when there is no snow on the ground, as has been recently found for other species that turn white in winter such as the snowshoe hare (Zimova et al. 2014) and mountain hare (*Lepus timidus*; Pedersen et al. 2017).

Other intrinsic biological characteristics contribute to concern about the conservation of the American ermine in New Mexico. Our broad-scale habitat suitability models indicated that ermines have both a naturally restricted and fragmented distribution in the state. However, ermines are habitat specialists mostly restricted to high-quality herbaceous riparian areas with high densities of voles, which are now patchy due to land-use practices such as nearly ubiquitous livestock grazing. Thus, the ermine's actual distribution is likely much more restricted and fragmented than indicated by the models. For species that have patchy distributions, it is critically important to maintain habitat connectivity among patches in order to maintain immigration. Second, where the species is present, the abundance of American ermines is generally very low and the species is the rarest one found in the local terrestrial small mammal community. This is due to its trophic status and behavior (e.g., territoriality) as a carnivore. Third, the fact that ermines can only produce a single litter each year reduces the ability of populations to rebound after crashes. Further, complete reproductive failure during some years is expected. Fourth, habitat fragmentation and low densities can make it difficult for individuals to find mates (i.e., Allee effect), and small isolated populations can be characterized by lowered genetic variation. Fifth, competition with long-tailed weasels creates unstable communities in which it is likely that ermine populations will go extinct unless prey populations are high (Powell and Zielinski 1983). Last, incomplete knowledge about the American ermine in western North America hampers its conservation and management.

Much information is presented in this chapter, but the ermine remains one of the most poorly understood mammals in the state and elsewhere throughout the range of the subspecies. Additional information is needed concerning its distribution and status, habitat associations, diet, response to human land-use practices, competitive relationships with other species, metapopulation dynamics, and genetics. Surveys are needed to better define the current distribution of the ermine in New Mexico and to evaluate population losses. Besides these important lines of research, Frey and Calkins (2010) also recommended that the species should be included in conservation and management plans at both the local and state levels and that the trapping season should be closed. Management to benefit American ermines in New Mexico includes restoring riparian zones and managing livestock grazing to foster lush vegetation and the presence of beavers.

ACKNOWLEDGMENTS

I thank J. Dunnum, W. Gannon, and J. Cook (Museum of Southwestern Biology) and A. Harris (University of Texas, El Paso) for useful information and for allowing me to examine specimens and other museum records. Special thanks to J. Dunnum and C. Ramotnik (MSB) for providing information on field notes in MSB. I thank the following for additional helpful information: N. Gilmore (Academy of Natural Sciences of Philadelphia), H. Garner and R. Baker (The Museum of Texas Tech University), P. Tebbel (The Wildlife Center, Española, New Mexico), R. Parmenter (Valles Caldera National Preserve), and M. Conway, J. Stuart, C. Painter, and F. Winslow (New Mexico Department of Game and Fish). Thanks finally to J. Cain for a helpful review of a previous version of this chapter.

LITERATURE CITED

Anderson, S. 1955. An additional record of *Mustela erminea* from Utah. *Journal of Mammalogy* 36:568.

Anonymous. 1903. Trapping ermine: different methods for the full grown and baby animals. *Ellensburgh Capital*, 21 March 1903.

Armstrong, D. M. 1972. *Distribution of mammals in Colorado*. Monograph of the Museum of Natural History, University of Kansas Number 3.

Aubry, K. B., M. J. Statham, B. N. Sacks, J. D. Perrine, and S. M. Wisely. 2009. Phylogeography of the North American red fox: vicariance in Pleistocene forest refugia. *Molecular Ecology* 18:2668–86.

Austin, J. E., and W. H. Pyle. 2004. Small mammals in montane wet meadow habitat at Grays Lake, Idaho. *Northwest Science* 78:225–33.

Bailey, V. 1931 (=1932). *Mammals of New Mexico*. North American Fauna 53. Washington, DC: US Department of Agriculture, Bureau of Biological Survey.

Baker, R. J., L. C. Bradley, R. D. Bradley, J. W. Dragoo, M. D. Engstrom, R. S. Hoffman, C. A. Jones, F. Reid, D. W. Rice, and C. Jones. Revised checklist of North American mammals north of Mexico, 2003. *Occasional Papers, Museum of Texas Tech University* 229:1–23.

Bangs, O. 1899. Descriptions of some new mammals from western North America. *Proceedings of the New England Zoological Club* 1:65–72.

Bellows, 1999. First record of the least weasel, *Mustela nivalis* (Carnivora: Mustelidae), from the coastal plain of Virginia. *Northeastern Naturalist* 6:238–40.

Berna, H. J. 1991. First record of the ermine (*Mustela erminea*) in Arizona. *Southwestern Naturalist* 36:245–46.

Burns, J. J. 1964. Movement of a tagged weasel in Alaska. *Murrelet* 45:10.

Cain, J. W., III, M. M. Morrison, and H. L. Bombay. 2003. Predator activity and nest success of willow flycatchers and yellow warblers. *Journal of Wildlife Management* 67:600–10.

Cain, J. W., III, K. S. Smallwood, M. M. Morrison, and H. L. Loffland. 2006. Influence of mammal activity on nesting success of passerines. *Journal of Wildlife Management* 70:522–31.

Colella, J. P., L. M. Frederick, S. L. Talbot, and J. A. Cook. 2021. Extrinsically reinforced hybrid speciation within Holarctic ermine (*Mustela* spp.) produces an insular endemic. *Diversity and Distributions* 27:747–62.

Copeland, J. P., K. S. McKelvey, K. B. Aubry, A. Landa, J. Persson, R. M. Inman, J. Krebs, E. Lofroth, H. Golden, J. R. Squires, A. Magoun, M. K. Schwartz, J. Wilmot, C. L. Copeland, R. E. Yates, I. Kojola, and R. May. 2010. The bioclimatic envelope of the wolverine (*Gulo gulo*): do climatic constraints limit its geographic distribution? *Canadian Journal of Zoology* 88:233–46.

Craighead, J. J., and F. C. Craighead. 1956. *Hawks, owls and wildlife*. Harrisburg, PA and Washington DC: Stackpole Company and Wildlife Management Institute.

Daly, C. 2006. Guidelines for assessing the suitability of spatial climate data sets. *International Journal of Climatology* 26:707–21.

———. 2008. Physiographically sensitive mapping of climatological temperature and precipitation across the conterminous United States. *International Journal of Climatology* 28:2031–64.

Dawson, N. G., A. G. Hope, S. L. Talbot, and J. A. Cook. 2014. A multilocus evaluation of ermine (*Mustela erminea*) across the Holarctic, testing hypotheses of Pleistocene diversification in response to climate change. *Journal of Biogeography* 41:464–75.

Dell'Arte, G. L., T. Laaksonen, K. Norrdahl, and E. Korpimaki. 2007. Variation in the diet composition of a generalist predator, the red fox, in relation to season and density of main prey. *Acta Oecologica* 31:276–81.

Dixon, J. 1931. Pikas versus weasel. *Journal of Mammalogy* 12:72.

Dougherty, E. C., and E. R. Hall. 1955. The biological relationships between American weasels (genus *Mustela*) and nematodes of the genus *Skrjabingylus* Petrov, 1927 (Nemotoda: Metastrongylidae), the causative organisms of certain lesions in weasel skulls. *Revista Ibérica de Parasitología*, Tomo Extraordinario, 531–76.

Doyle, A. T. 1990. Use of riparian and upland habitats by small mammals. *Journal of Mammalogy* 71:14–23.

Durrant, S. D. 1952. Mammals of Utah: taxonomy and distribution. *University of Kansas Publications, Museum of Natural History* 6:1–549.

Edwards, M. A., and G. J. Forbes. 2003. Food habits of ermine, *Mustela erminea*, in a forested landscape. *Canadian Field-Naturalist* 117:245–48.

Eger, J. L. (1990). Patterns of geographic variation in the skull of Nearctic Ermine (*Mustela erminea*). *Canadian Journal of Zoology* 68:1241–49.

Egoscue, H. J. 1957. Notes on Utah weasels. *Journal of Mammalogy* 38:411–12.

Elith, J., and C. H. Graham. 2009. Do they? How do they? WHY do they differ? On finding reasons for differing performances of species distribution models. *Ecography* 32:66–77.

Fagerstone, K. A. 1999. Black-footed ferret, long-tailed weasel, short-tailed weasel, and least weasel. *In Wild furbearer management and conservation in North America*, ed. M. Novak, J. A. Baker, M. E. Obbard, and B. Malloch, 549–73. Ontario, Canada: Ministry of Natural Resources.

Fagerstone, K. A., and C. A. Ramey. 1996. Rodents and lagomorphs. In *Rangeland wildlife*, ed. P. R. Krausman, 83–132. Denver, CO: Society for Range Management.

Findley, J. S. 1987. *The natural history of New Mexico mammals*. Albuquerque: University of New Mexico Press.

Findley, J. S., and S. Anderson. 1956. Zoogeography of the montane mammals of Colorado. *Journal of Mammalogy* 37:80–82.

Findley, J. S., A. H. Harris, D. E. Wilson, and C. Jones. 1975. *Mammals of New Mexico*. Albuquerque: University of New Mexico Press.

Fitzgerald, B. M. (1977). Weasel predation on a cyclic population of the montane vole (*Microtus montanus*) in California. *Journal of Animal Ecology* 46:367–97.

Fitzgerald, J. P., C. A. Meaney, and D. M. Armstrong. 1994. *Mammals of Colorado*. Niwot: Denver Museum of Natural History and University Press of Colorado.

Fleming, M. A., and J. A. Cook. 2002. Phylogeography of endemic ermine (*Mustela erminea*) in southeast Alaska. *Molecular Ecology* 11:795–807.

Freeman, E. A., and G. G. Moisen. 2008. A comparison of the performance of threshold criteria for binary classification in terms of predicted prevalence and kappa. *Ecological Modeling* 217:48–58.

Frey, J. K. 2003. Baseline inventory of small mammal prey-base communities on Carson National Forest, New Mexico. Final report submitted to Carson National Forest, Taos, New Mexico, 31 December 2003, 48 pp.

———. 2006. Status of the New Mexico meadow jumping mouse (*Zapus hudsonius luteus*) in the Sangre de Cristo Mountains, New Mexico. Final report submitted to New Mexico Department of Game and Fish, Santa Fe, 14 December 2006, 78 pp.

———. 2007. Key to the rodents of New Mexico. Final report submitted to New Mexico Department of Game and Fish, Santa Fe, New Mexico, 24 June 2007, 120 pp.

———. 2012a. Survey for the New Mexico meadow jumping mouse (*Zapus hudsonius luteus*) at Coyote Creek, Mora County, New Mexico. Final Report submitted to Marron and Associates, 31 August 2012, 10 pp.

———. 2012b. Survey for the New Mexico meadow jumping mouse (*Zapus hudsonius luteus*) on Carson National Forest, New Mexico. Final Report submitted to Carson National Forest, 5 December 2012, 71 pp.

———. 2017. Landscape scale and microhabitat of the endangered New Mexico meadow jumping mouse in the White Mountains, Arizona. *Journal of Fish and Wildlife Management* 8:39–58.

———. 2018. Beavers, livestock, and riparian synergies: bringing small mammals into the picture. In *Riparian research and management: past, present, and future*, Vol. 1, ed. R. R. Johnson, S. W. Carothers, D. M. Finch, K. J. Kingsley, and J. T. Stanley, 85–101. Fort Collins, CO: US Forest Service, Rocky Mountain Research Station, Gen. Tech. Rep. RMRS-GTR-377.

Frey, J. K., M. A. Bogan, and T. L. Yates. 2007. Mountaintop island age determines species richness of boreal mammals in the American Southwest. *Ecography* 30:231–40.

Frey, J. K., and M. T. Calkins. 2010. Status of the ermine (*Mustela erminea*) at its southern range limits, New Mexico. Final Report submitted to Share with Wildlife Program, New Mexico Department of Game and Fish, Santa Fe, 8 June 2010, 48 pp.

———. 2014. Snow cover and riparian habitat determine the distribution of the short-tailed weasel

(*Mustela erminea*) at its southern range limits in arid western North America. *Mammalia* 78:45–56.

Frey, J. K., and J. L. Malaney. 2006. Snowshoe hare (*Lepus americanus*) and mountain cottontail (*Sylvilagus nuttallii*) biogeography at their southern range limit. *Journal of Mammalogy* 87:1175–82.

———. 2009. Decline of the meadow jumping mouse (*Zapus hudsonius luteus*) in two mountain ranges in New Mexico. *Southwestern Naturalist* 54:31–44.

Frey, J. K., and Z. J. Schwenke. 2007. Mammals of Sugarite Canyon State Park, Colfax County, New Mexico. Final report submitted to New Mexico State Parks, Santa Fe, New Mexico, 16 November 2007, 49 pp.

———. 2012. Mammals of Sugarite Canyon State Park, Colfax County, New Mexico. *Occasional Papers, Museum of Texas Tech University* 311:1–24.

Galbreath, K. E., D. J. Hafner, K. R. Zamudio, and K. Agnew. 2010. Isolation and introgression in the Intermountain West: contrasting gene genealogies reveal the complex biogeographic history of the American pika (*Ochotona princeps*). *Journal of Biogeography* 37:344–362.

Garcia, P., and I. Mateos. 2009. Evaluation of three indirect methods for surveying the distribution of the least weasel *Mustela nivalis* in a Mediterranean area. *Small Carnivore Conservation* 40:22–26.

Gillingham, B. J. 1984. Meal size and feeding rate in the least weasel (*Mustela nivalis*). *Journal of Mammalogy* 65:517–19.

Graham, C. H., J. Elith, R. J. Hijmans, A. Guisan, A. T. Peterson, B. A. Loiselle, and The NCEAS Predicting Species Distributions Working Group. 2008. The influence of spatial errors in species occurrence data used in distribution models. *Journal of Applied Ecology* 45:239–47.

Grayson, D. K. 1987. The bieogeographic history of small mammals in the Great Basin: observations on the last 20,000 years. *Journal of Biogeography* 68:359–75.

Grimm, J. W., and R. H. Yahner. 1987. Small mammal responses to roadside habitat management in south central Minnesota. *Journal of the Minnesota Academy of Science* 53:16–21.

Guthrie, D. A., and N. Large. 1980. Mammals of Bandelier National Monument, New Mexico. Claremont Colleges, Claremont, California, Unpublished report to National Park Service PX7029-7-0807, 22 pp.

Hall, E. R. 1945. A revised classification of the American ermines with description of a new subspecies from the western Great Lakes region. *Journal of Mammalogy* 26:175–82.

———. 1951. American weasels. *University of Kansas Publications, Museum of Natural History* 4:1–466.

———. 1981. *The mammals of North America*. 2nd ed. Vol. 2. New York: John Wiley.

Handley, C. O., Jr., and E. K. V. Kalko. 1993. A short history of pitfall trapping in America, with a review of methods currently used for small mammals. *Virginia Journal of Science*, 44:19–26.

Harris, A. H. 1990. Fossil evidence bearing on southwestern mammalian biogeography. *Journal of Mammalogy* 71:219–29.

———. 1993. A late-Pleistocene occurrence of ermine (*Mustela erminea*) in southeastern New Mexico. *Southwestern Naturalist* 38:279–80.

Hayward, C. L. 1949. The short-tailed weasel in Utah and Colorado. *Journal of Mammalogy* 30:436–37.

Hoffmeister, D. F. 1986. *Mammals of Arizona*. Tucson and Phoenix: University of Arizona Press and Arizona Game and Fish Department.

Hope, A. G., J. L. Malaney, K. C. Bell, F. Salazar-Miralles, A. S. Chavez, B. R. Barber, and J. A. Cook. 2016. Revision of widespread red squirrels (genus: *Tamiasciurus*) highlights the complexity of speciation within North American forests. *Molecular Phylogenetics and Evolution* 100:170–82.

Hope, A. G., N. Panter, J. A. Cook. S. L. Talbot, and D. W. Nagorsen. 2014. Multilocus phylogeography and systematic revision of North American water shrews (genus: *Sorex*). *Journal of Mammalogy* 95:722–38.

Hornfeldt, B., T. Hipkiss, and U. Eklund. 2005. Fading out of vole and predator cycles? *Proceedings of the Royal Society B* 272:2045–49.

Huey, W. S. 1956. New Mexico beaver management. *New Mexico Department of Game and Fish Bulletin* 4:1–49.

Ivey, R. D. 1957. Ecological notes on the mammals of Bernalillo County, New Mexico. *Journal of Mammalogy* 38:490–502.

Jachowski, D., R. Kays, A. Butler, A. M. Hoylman, and M. Gompper. 2021. Tracking the decline of weasels in North America. *PLOS One* 16(7):e0254387. https://doi.org/10.1371/journal.pone.0254387.

Kausrud, K. L., A. Mysterud, H. Steen, J. O. Vik, E. Ostbuye, B. Cazelles, E. Framstad, A. M. Eikeset, I. Mysterud, T. Solhoy, and N. C. Stenseth. 2008. Linking climate change to lemming cycles. *Nature* 456:93–98.

King, C. M. 1983. *Mustela erminea. Mammalian Species* 195:1–8.

King, C. M., R. M. McDonald, R. D. Martin, G. W. Tempero, and S. J. Holmes. 2007. Long-term automated monitoring of the distribution of small carnivores. *Wildlife Research* 34:140–48.

King, C. M., and R. A. Powell. 2007. *The natural history of weasels and stoats: ecology, behavior, and management.* New York: Oxford University Press.

Korslund, L., and H. Steen. 2006. Small rodent inter survival: snow conditions limit access to food resources. *Journal of Animal Ecology* 75:156–66.

Linnell, M. A., C. W. Epps, E. D. Forsman, and W. J. Zielinski. 2017. Survival and predation of weasels (*Mustela erminea, Mustela frenata*) in North America. *Northwest Science* 91:15–26.

Liu, C., P. M. Berry, T. P. Dawson, and R. G. Pearson. Selecting thresholds of occurrence in the prediction of species distributions. *Ecography* 28:385–93.

Lozier, J. D., P. Aniello, and M. J. Hickerson. 2009. Predicting the distribution of Sasquatch in western North America: anything goes with ecological niche modeling. *Journal of Biogeography* 36:1623–27.

Malaney, J. A., and. J. K. Frey. 2006. Summer habitat use by snowshoe hare and mountain cottontail at their southern zone of sympatry. *Journal of Wildlife Management* 70:877–83.

McDonald, K. P., and D. S Case. 1990. The status of ermine (*Mustela erminea*) in Ohio. *Ohio Journal of Science* 90:46–47.

Merriam, C. H. 1896. Synopsis of the weasels of North America. *North American Fauna* 11:1–45.

———. 1903. Eight new mammals from the United States. *Proceedings of the Biological Society of Washington* 16:73–78.

Miller, F. W. 1933. Concerning a recent record of

Mustela rixosa in Colorado. *Journal of Mammalogy* 14:368.

Mos, J., and T. R. Hofmeester. 2020. The *Mostela*: an adjusted camera trapping device as a promising non-invasive tool to study and monitor small mustelids. *Mammal Research* 65:843–53.

Mote, P. W., A. F. Hamlet, M. P. Clark, and D. P. Lettenmaier. 2005. Declining mountain snowpack in western North America. *Bulletin of the American Meteorological Society* 86:39–49.

Mowat, G., and K. G. Poole. 2005. Habitat associations of short-tailed weasels in winter. *Northwest Science* 79:28–36.

Murphy, R. K., R. A. Sweitzer, and J. D. Albertson. 2007. Occurrence of small mammal species in mixed-grass prairie in northwestern North Dakota. *Prairie Naturalist* 39:91–95.

New Mexico Department of Game and Fish (NMDGF). 2006. Furbearer population assessment and harvest management matrix. Adopted 9-29-06. Unpublished white paper, 3 pp.

———. 2016. *State Wildlife Action Plan for New Mexico.* Santa Fe: New Mexico Department of Game and Fish.

Newhouse, S. 1914. *The Newhouse trapper's guide.* 12th ed. Oneida, NY: Oneida Community.

Olendorff, R. R. 1976. The food habits of North American golden eagles. *American Midland Naturalist* 95:231–36.

Parkins, A. E. 1984. Surface temperature regions. In *Goode's World Atlas*, ed. E. B. Espenshade Jr., 10. 16th ed. Chicago: Rand McNally.

Patterson, B. D. 1980. Montane mammalian biogeography in New Mexico. *Southwestern Naturalist* 25:33–40.

———. 1984. Mammalian extinction and biogeography in the southern Rocky Mountains. In *Extinctions*, ed. M. H. Nitecki, 247–94. Chicago: University of Chicago Press.

———. 1995. Local extinctions and the biogeographic dynamics of boreal mammals in the Southwest. In *Storm over a mountain island: conservation biology and the Mt. Graham affair*, ed. C. A. Istock and R. S. Hoffmann, 151–76. Tucson: University of Arizona Press.

Patterson, B. D., and W. Atmar. 1986. Nested subsets and the structure of insular mammalian faunas and archipelagos. *Biological Journal of the Linnean Society* 28:65–82.

Pearson, R. G., C. J. Raxworthy, M. Nakamura, and A. T. Peterson. 2007. Predicting species distributions from small numbers of occurrence records: a test case using cryptic geckos in Madagascar. *Journal of Biogeography* 34:102–17.

Pedersen S., M. Odden, H. C. Pedersen. 2017. Climate change induced molting mismatch? Mountain hare abundance reduced by duration of snow cover and predator abundance. *Ecosphere* 8(3):e01722. https://doi.org/10.1002/ecs2.1722

Polderboer, E. B., L. W. Kuhn, and G. O. Hendrickson. 1941. Winter and spring habits of weasels in central Iowa. *Journal of Wildlife Management* 5:115–19.

Powell, R. A. 1973. A model for raptor predation on weasels. *Journal of Mammalogy* 54:259–63.

———. 1982. Evolution of black-tipped tails in weasels: predator confusion. *American Naturalist* 119:126–31.

Powell, R. A., and C. M. King. 1977. Variation in body size, sexual dimorphism and age-specific survival in stoats, *Mustela erminea* (Mammalia: Carnivora), with fluctuating food supplies. *Biological Journal of the Linnean Society* 62:165–94.

Powell, R. A., and W. J. Zielinski. 1983. Competition and coexistence in mustelid communities. *Acta Zoologica Fennica* 174:223–27.

Quick, H. F. 1951. Notes on the ecology of weasels in Gunnison County, Colorado. *Journal of Mammalogy* 32:281–90.

Raymond, M., J. Robitaille, P. Lauzon, and R. Vaudry. 1990. Prey-dependent profitability of foraging behaviour of male and female ermine, *Mustela erminea*. *Oikos* 58:323–28.

Segal, A. N. 1975. Postnatal growth, metabolism and thermoregulation in the stoat. *Soviet Journal of Ecology* 6:28–32.

Seidensticker, J. C., IV. 1968. Notes of the food habits of the great horned owl in Montana. *Murrelet* 49:1–3.

Seton, E. T. 1933. Occurrence of the least weasel near Denver. *Journal of Mammalogy* 14:70.

Short, H. L. 1961. Food habits of a captive least weasel. *Journal of Mammalogy* 42:273–74.

Sikes, R. S., and the Animal Care and Use Committee of the American Society of Mammalogists.

2016. 2016 Guidelines of the American Society of Mammalogists for the use of wild mammals in research and education. *Journal of Mammalogy* 97:663–88.

Simms, D. A. 1979a. Studies of an ermine population in southern Ontario. *Canadian Journal of Zoology* 57:824–32.

———. 1979b. North American weasels: resource utilization and distribution. *Canadian Journal of Zoology* 57:504–20.

Small, B. A., J. K. Frey, and C. Gard. 2016. Livestock grazing limits beaver restoration in northern New Mexico. *Restoration Ecology* 24:646–55.

Springer, A. M. 1975. Observations of the summer diet of rough-legged hawks from Alaska. *Condor* 77:338–39.

Squires, J. R. 2000. Food habits of northern goshawks nesting in south central Wyoming. *Wilson Bulletin* 112:536–39.

Statham, M. J., B. N. Sacks, K. B. Aubry, J. D. Perrine, and S. M. Wisely. 2012. The origins of recently established red fox populations in the United States: translocations or natural range expansions? *Journal of Mammalogy* 93:52–65.

Stock, A. D. 1970. Notes on mammals of southwestern Utah. *Journal of Mammalogy* 51:429–33.

St-Pierre, C., J.-P. Ouellet, and M. Crete. 2006. Do competitive intraguild interactions affect space and habitat use by small carnivores in a forested landscape? *Ecography* 29:487–96.

Swickard, M., G. E. Hass, and R. P. Martin. 1971. Notes on small mammals infrequently recorded from the Jemez Mountains, New Mexico. *Bulletin of the New Mexico Academy of Science* 12:10–14.

Thompson, B. C., D. F. Miller, T. A. Doumitt, T. R. Jacobson, and M. L. Munson-McGee. 1996. *An ecological framework for monitoring sustainable management of wildlife: a New Mexico furbearer example.* US Department of the Interior, National Biological Service, Information and Technology Report 5.

Vazquez, A. W. 1956. A new southern record for *Mustela erminea cicognanii*. *Journal of Mammalogy* 37:113–14.

Wilson, T. M., and A. B. Carey. 1996. Observations of weasels in second-growth Douglas-fir forests in the Puget Trough, Washington. *Northwestern Naturalist* 77:35–39.

Wintle, B. A., J. Elith, and J. M. Potts. 2005. Fauna habitat modeling and mapping: A review and case study in the Lower Hunter Central Coast region of NSW. *Austral Ecology* 30:719–38.

Wozencraft, W. C. 2005. Order Carnivora. In *Mammal species of the world: a taxonomic and geographic reference*, Vol. 1, ed. D. E. Wilson and D. M. Reeder, 532–628. 3rd ed. Baltimore: Johns Hopkins University Press.

Zimova M., L. S. Mills, P. M. Lukacs, and M. S. Mitchell. 2014. Snowshoe hares display limited phenotypic plasticity to mismatch in seasonal camouflage. *Proceedings of the Royal Society B* 281:20140029. http://dx.doi.org/10.1098/rspb.2014.0029

BLACK-FOOTED FERRET (*MUSTELA NIGRIPES*)

Dustin H. Long and James N. Stuart

Among the rarest of New Mexico's carnivores, the black-footed ferret (*Mustela nigripes*) is at the center of one of the most compelling wildlife conservation stories in our state and elsewhere. Infrequently observed even before rangewide declines throughout the 20th century and eventual extirpation in the wild in 1987, the species was twice presumed extinct by wildlife biologists (see under "Status and Management"). In 1979, it appeared to have been lost after what was then thought to be the last surviving black-footed ferret—a female retrieved from the wild in Mellette County, South Dakota for a captive breeding program—died that year. Two years later, in 1981, another surviving, and indeed the last known, wild population of black-footed ferrets was discovered near Meeteetse, Wyoming. This population has since disappeared, but not before 18 individuals were captured and placed into a captive breeding program. Today, thanks to the dedicated work of numerous wildlife biologists, husbandry experts, and conservation advocates, the species is making a slow and difficult comeback in several western states through the implementation of captive propagation and reintroduction efforts. In 2012, New Mexico became the latest state to attempt the repatriation of the black-footed ferret to its former range, though it remains to be seen whether the species' return to the "Land of Enchantment" succeeds. As later told in this chapter, important challenges continue to confront reintroduction efforts.

The family Mustelidae is comprised of some of the most diverse and specialized carnivores, and within that family the black-footed ferret is more specialized than most (Anderson 1989). A species native to the North American continent, it is a long-bodied, short-legged, and relatively small carnivore but is larger than both the American ermine (*M. richardsonii*; Chapter 21) and the long-tailed weasel (*Neogale frenata*; Chapter 23). Adults are 480–610 mm (19–24 in) in total length with a 110–140 mm (4–6 in) tail, and weigh 600–1,400 g (1.3–3 lbs); females average 68% of male body weight and 93% of male body length (Anderson et al. 1986). The short, sleek fur is yellowish buff (occasionally whitish) on the upper parts of the body, becoming darker on the middle of the back, and grading to white on the belly and throat, and on much of the face. There are no color morphs. The pelage may become somewhat longer during the cold months of the year, but individuals do not develop the white pelage exhibited during winter by either the American ermine or montane populations of the long-tailed weasel (Biggins 2000; see Chapters 21 and 23). A blackish mask extends between and encircles the eyes and is most defined in young individuals. The feet, legs, and tip of the tail are black (Clark 1999; Svendsen 2003; Photo 22.1)

(*opposite page*) Photograph: © David A. Eads.

Photo 22.1. Captive black-footed ferret. Note the long body and short legs, the black feet, legs, and tip of the tail, and the black mask extending between and encircling the eyes, Photograph: © Kimberly Fraser/ US Fish and Wildlife Service.

Photo 22.2. Black-footed ferrets have distinctive black face masks that contrast with surrounding whitish fur on the sides of the head, the muzzle, and the throat. Photograph: © Kimberly Fraser/US Fish and Wildlife Service.

Among other mustelids in North America, the black-footed ferret most closely resembles the long-tailed weasel, which is widespread and fairly common in New Mexico (Chapter 23). Throughout much of the state, the long-tailed weasel, particularly the "bridled" (masked) morph, has a striking black-and-white face mask and, often, a tawny body color similar to that of the black-footed ferret, but lacks the black feet of its close relative (Photo 22.3). Weasels are

also considerably smaller (300–350 mm [12–14 in] in total length) and besides grasslands and prairie dog towns are also commonly found in rocky or wooded areas, atypical habitat for black-footed ferrets. Nonetheless, the two species have remarkably convergent color patterns and can be easily mistaken for one another, especially when observed only briefly. This similarity of appearance accounts for the many reports of black-footed ferrets in New Mexico that actually turn out to be sightings of long-tailed weasels.

The black-footed ferret also resembles the domestic ferret, the non-native and popular domesticated form of the European polecat (*Mustela putorius*). Now widely distributed via captive breeding and the pet trade, the domestic ferret occasionally escapes or is released from captivity and, though no breeding population is known to be established anywhere in New Mexico, it has been found on rare occasions in the wild in the state (NMDGF, unpubl. data). The European polecat is comparable in size to the black-footed ferret but usually lacks its distinctive face mask and color pattern. As a result of selective breeding, however, the domestic ferret can exhibit a variety of color patterns and could potentially be mistaken for our native ferret species (Photo 22.4).

Although similar in appearance to the long-tailed weasel, the black-footed ferret is only distantly related to other mustelids in North America and has been assigned to the subgenus *Putorius*, allying it instead with the Old World polecats (Abramov 2000; Wozencraft 2005; and see Chapter 24 for a discussion of recent taxonomic changes in the genus *Mustela*). Many of the 17 recognized subspecies of the Eurasian steppe polecat (*M. eversmanni*) bear a striking resemblance to the black-footed ferret, and in fact *M. eversmanni* is the black-footed ferret's closest relative, with fossil records suggesting the two species diverged between 0.5 and 2 million years ago (Wisely 2005). Hoffmann and Pattie (1968) considered the steppe polecat and the black-footed ferret to be "ecological equivalents" in that they serve similar functions in separate biological communities. Biggins et al. (2011) somewhat disagreed by stating that the steppe polecat could not be substituted for the black-footed ferret. Nonetheless, the similarity between the two species led to the use of steppe polecats and steppe polecat–black-footed ferret hybrids as surrogates for *M. nigripes* in early research on the North American species (Biggins et al. 2011).

Despite the species' formerly extensive range, morphological variation might have been lacking among historical populations of *M. nigripes*, to the point that no subspecies have been described (Hillman and Clark 1980; Wozencraft 2005). Genetic differentiation existed between historical Great Plains and Wyoming metapopulations, with restricted gene flow even before the black-footed ferret's decline in the 20th century (Wisely 2005). The loss of the Great Plains core population and population bottlenecks elsewhere were later reflected in a remarkable lack of genetic diversity in the Meeteetse, Wyoming population and may account for the failure of the first captive breeding efforts from 1976–1978 using animals from the Mellette County, South Dakota population (Biggins and Schroeder 1988; O'Brien et al. 1989). All black-footed ferrets that are known to exist today are descended from seven of the 18 individuals captured at Meeteetse during 1985–1987 (Garell et al. 1998). Consequently, both captive and reintroduced populations of black-footed ferret exhibit significantly less genetic variation than existed a mere hundred years ago. Nonetheless, the loss of approximately 90% of the species' genetic diversity has not up to this point manifested itself in reduced fecundity or physiological abnormalities common in inbred populations of other mammals (Wisely 2005). Some morphological differences exist between captive and wild-born black-footed ferrets (e.g., ferrets raised in captivity are smaller and shaped differently than wild ferrets), but reintroduced populations

Photo 22.3. Long-tailed weasel (*Neogale frenata*) at the Rio Grande Nature Center in Albuquerque on 30 March 2008. All long-tailed weasels—not just the "bridled" (masked) form represented in this photograph—can readily be mistaken for black-footed ferrets. Photograph: © James N. Stuart.

Photo 22.4. The domestic ferret (*Mustela putorius furo*) is the domesticated form of the European polecat (*Mustela putorius*), a species found in much of the western Palaearctic. Various pelage colors are recognized in the domestic ferret, including sable. Sable domestic ferrets that escape into the wild can be mistaken for black-footed ferrets. Photograph: © Travis Livieri/ Prairie Wildlife Research.

acquire the morphology of ancestral populations, suggesting that these differences are due to environmental rather than genetic factors (Wisely et al. 2002, 2005).

DISTRIBUTION
The black-footed ferret is exclusively a North American species. Its range once extended from

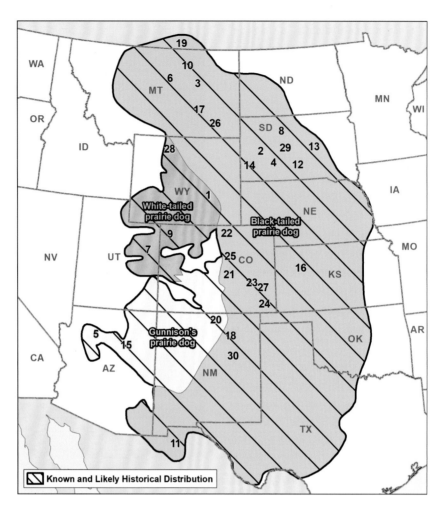

Map 22.1. Black-footed ferret (*Mustela nigripes*) historical distribution (known and likely) as the cumulative distribution of the white-tailed prairie dog (*Cynomys leucurus*), Gunnison's prairie dog (*C. gunnisoni*), and black-tailed prairie dog (*C. ludovicianus*). Reintroduction sites (with year of first release; * = site still active in fall 2018): (1) Shirley Basin, WY (1991)*; (2) Badland NP, SD (1994)*; (3) UL Bend NWR, MT (1994)*; (4) Conata Basin, SD (1996)*; (5) Aubrey Valley, AZ (1996)*; (6) Ft. Belknap Reservation, MT (1997)*; (7) Coyote Basin, UT (1999)*; (8) Cheyenne River Reservation, SD (2000); (9) Wolf Creek, CO (2001); (10) BLM 40-Complex, MT (2001); (11) Janos, Mexico (2001); (12) Rosebud Reservation, SD (2004); (13) Lower Brule Reservation, SD (2006)*; (14) Wind Cave NP, SD (2007)*; (15) Espee Ranch, AZ (2007); (16) Logan County, KS (2007)*; (17) Northern Cheyenne Reservation, MT (2008); (18) Vermejo Park Ranch #1, NM (2008); (19) Grasslands NP, SK, Canada (2009); (20) Vermejo Park Ranch #2, NM (2012); (21) Walker Ranch, CO (2013); (22) Soapstone Complex, CO (2014)*; (23) North Holly Complex, CO (2014); (24) Liberty Complex, CO (2014); (25) Rocky Mountain Arsenal NWR, CO (2015)*; (26) Crow Indian Reservation, MT (2015)*; (27) South Holly Complex (CO (2015); (28) Meeteetse, WY (2016)*; (29) Bad River Ranch, SD (2017); (30) Wagon Mound (Moore) Ranch, NM (2018)*. Adapted from Roelle et al. 2006, with updates provided by J. Hughes (Black-footed Ferret Reintroduction Program).

southern Saskatchewan and Alberta in Canada southward through the Great Plains and Intermontane Basins of the west-central United States to Trans-Pecos Texas (Clark 1989; Map 22.1). The species may also have occurred as far south as northern Chihuahua, Mexico (Miller et al. 1996), though evidence for this is lacking (Findley et al. 1975; Wozencraft 2005). *M. nigripes* appears to be a fairly recent species, the result of immigration of an ancestral form from Asia sometime between 1 and 2 million years ago, followed by diversification in North America (Wisely 2005). The earliest confirmed record of a black-footed ferret in North America, found in Cathedral Cave, Nevada is approximately 750,000–850,000 years old (Owen et al. 2000).

The historical distribution of the black-footed ferret coincides spatially with the combined geographic ranges of the black-tailed (*Cynomys ludovicianus*), Gunnison's (*C. gunnisoni*), and white-tailed (*C. leucurus*) prairie dogs, semi-fossorial colonial ground squirrels that have also experienced significant declines in abundance and distribution over the last century (Anderson et al. 1986; Biggins 2006; Map 22.1; see under "Habitat Associations," below). Prairie dogs are keystone species with which many other taxa have a close ecological relationship, including the black-footed ferret (Kotliar et al. 2006). They are also often viewed by humans as agricultural pests that destroy and consume vegetation; consequently, prairie dogs are frequently killed for the sake of livestock grazing in particular. Carr (1986:7) describes the attempted eradication of the prairie dog as "one of the most diligent vertebrate pest control exercises in history."

No natural populations of black-footed ferrets are known to persist within the species' historical range, and most populations in the wild were likely extirpated by the 1950s (Lockhart et al. 2006). The proximal causes for the demise and near extinction of the species can be directly attributed to the historical and ongoing widespread eradication of prairie dogs through poisoning; the conversion of native grasslands to cropland; and sylvatic plague (*Yersinia pestis*), a disease threat to both ferrets and prairie dogs (USFWS 2013a). Today, sylvatic plague and limited distribution and abundance of prairie dogs are considered to be the primary obstacles to the recovery of the black-footed ferret (Jachowski and Lockhart 2009).

In New Mexico, the black-footed ferret was formerly found within the combined geographic ranges of black-tailed and Gunnison's prairie dogs, which encompassed most of the state (Hubbard and Schmitt 1984; Map 22.2). Prehistoric evidence of the species' occurrence in the state during the late Pleistocene has been found in Eddy and Bernalillo counties (see Chapter 1), while the charred remains of black-footed ferrets dating back to 2,000–3,000 BP were discovered in Atlatl Cave in San Juan County (Hubbard and Schmitt 1984). According to Hubbard and Schmitt's (1984) review of the species in New Mexico, those are the only confirmed records of occurrence predating the 20th century in the state (but see below). Specimens and reliable observations from New Mexico nonetheless suggest that the species formerly ranged statewide except for the highest elevations and possibly the southwestern corner south of the Mogollon Plateau (Schmitt 1982; Hubbard and Schmitt 1984; Map 22.2 and Table 22.1). Findley et al. (1975) listed six specimens that had been preserved from New Mexico, most of them from McKinley and Cibola (formerly western Valencia) counties in the western part of the state, and one each from Santa Fe and Chaves counties. Perhaps not surprisingly, most ferret specimens and reliable observations in New Mexico were obtained during the 1920s when a federally sponsored program to eradicate prairie dogs and predators was underway (Hubbard and Schmitt 1984; Anderson et al. 1986). A specimen obtained in 1929 from near Agua Fria in Colfax County, at about 2,400 m

(8,000 feet) in elevation, was secured in a Gunnison's prairie dog colony that was being eradicated with carbon bisulphide gas (Aldous 1940; Hooper 1941). Hubbard and Schmitt's (1984) review identified both verified records based on preserved museum specimens and numerous observation records of varying reliability. Nine preserved specimens (three more than listed by Findley et al. [1975]) are available from 1915 to 1934 whereas no verified specimens have been collected since that time (see Map 22.2 and Table 22.1). Anderson et al. (1986) listed ten extant specimens collected in New Mexico that included both the nine mentioned by Hubbard and Schmitt (1984) and an additional individual identified through a mandible collected in Roswell, Chaves County by Vernon Bailey in 1899. The most recent observations that Hubbard and Schmitt (1984) considered

Table 22.1. Confirmed and highly probable black-footed records from New Mexico, in chronological order (adapted from Hubbard and Schmitt 1984 and Anderson et al. 1986).

Record (Number)	Date	Sex	Location	Collector
Confirmed (1)	June 1899	Unknown	Chaves County; Roswell	Vernon Bailey
Confirmed (1)	18 March 1915	Male	Catron County; Centerfire Basin	J. Stokley Ligon
Confirmed (1)	1 May 1918	Male	McKinley County; 10 mi NE of Mt. Taylor	J. Stokley Ligon
Confirmed (1)	15 October 1918	Male	Cibola County; 2 mi N. of Bluewater	C. P. Musgrave
Confirmed (1)	22 November 1918	Female	Catron County; Garcia Ranch, 75 mi SW of Magdalena	J. S. Felkner
Confirmed (1)	14 November 1925	Male	Bernalillo County; Albuquerque, 12th St.	J. Stokley Ligon
Highly probable (1)	Summer 1928		De Baca County; Ben Hall Ranch, SE of Ft. Sumner	Homer Pickens
Confirmed (1)	7 April 1929	Female	Lincoln County; 3 mi S of Picacho	Wharton Huber
Confirmed (1)	10 July 1929; killed 28 December 1929	Female	Colfax County, Moreno Valley, Agua Fria area	Shaler E. Aldous
Confirmed (1)	31 August 1930	Male	Santa Fe County; 8 mi SW of Santa Fe, near Arroyo Hondo	Theodore E. White
Highly probable (2)	Autumn(?) 1930	Unknown	Colfax County; Vermejo Park Ranch near Castle Rock	Elliot Barker
Highly probable (2)	1931	Unknown	Catron County; 10 mi S., 1.5 mi W. of Quemado	T. J. Lyon
Confirmed (1)	30 October 1934	Unknown	McKinley County; Gallup	M. E. Musgrave
Highly probable	1934	Unknown	Catron County; Snow Lake area	Arnold Bayne
Highly probable (6–8)	1934	Unknown	Chaves [or Roosevelt?] and Lea counties; Milnesand-Caprock area	Charles Walter
Highly probable (1)	Autumn(?) 1940	Unknown	Santa Fe County; E. side of Hwy 285 between Lamy and Hwy 85	Howard Campbell
Highly probable (1)	1940	Unknown	McKinley County; between Window Rock, AZ, and Mexican Springs	William E. Fair
Highly probable	1941	Unknown	Cibola County; Ramah area	Arnold Bayne
Highly probable	1942	Unknown	McKinley County; Gallup area	Arnold Bayne
Highly probable	1966	Unknown	Curry County; western boundary of Cannon Air Force Base	James R. Vaught

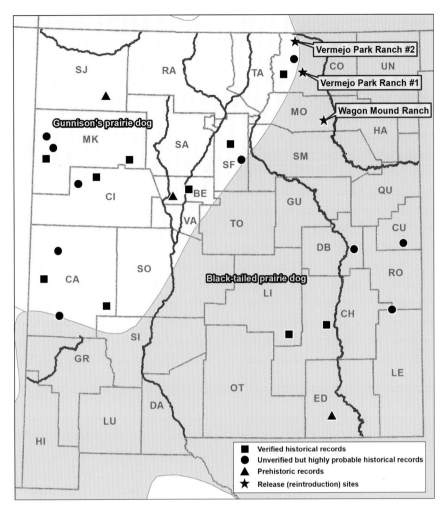

Map 22.2. Black-footed ferret historical (verified and highly probable) records and reintroduction sites in New Mexico, relative to the distribution of Gunnison's prairie dogs (*Cynonys gunnisoni*) and black-tailed prairie dogs (*C. ludovicianus*). Highly probable records are those of Hubbard and Schmitt (1984). The last highly probable record, from just outside Cannon Air Force Base in Curry County, dates back to 1966.

"probable" were of single animals sighted during daytime at Valle Grande, Sandoval County in May 1970 and another near Angel Fire, Colfax County in September 1981; both of these observations were in areas occupied by Gunnison's prairie dogs. No reliable reports of black-footed ferrets have been obtained in New Mexico in recent decades and, despite Hubbard and Schmitt's (1984) optimistic contention that the "species still likely occurs here," it seems more likely that the ferret had been extirpated from the state by the time of their 1984 review.

Although no natural black-footed ferret populations apparently persist today in New Mexico, experimental reintroductions took place at Vermejo Park Ranch in Colfax County and, most recently, also at Wagon Mound Ranch in Mora County. The details of these reintroductions are discussed below under "Status and Management."

HABITAT ASSOCIATIONS

Historically, populations of the black-footed ferret were coexistent with and dependent upon colonies of black-tailed, Gunnison's, and white-tailed prairie dogs (Anderson et al. 1986; Biggins 2006). Two other prairie dog species occur in North America, the Utah (*C. parvidens*) and Mexican (*C. mexicanus*) prairie dogs, but black-footed ferrets were in contrast never documented anywhere within their ranges (Lockhart et al. 2006). Such is the ecological dependency on black-tailed, Gunnison's, and white-tailed prairie dogs that though dispersing black-footed ferrets are reported in other habitats, no breeding populations have ever been documented anywhere but on colonies of one of those three sciurid rodents (Linder et al. 1972; Forrest et al. 1985). Therefore, much of our understanding of habitat use by *M. nigripes* can be deduced or inferred from historical and current habitat use by prairie dogs.

The available 20th-century distribution records from New Mexico suggest that the black-footed ferret used diverse biotic communities (prairie grasslands, semi-desert grasslands and shrublands, and montane meadows), provided that one of the species of prairie dogs indigenous to the state was present to provide both food and burrow systems for shelter (Hubbard and Schmitt 1984). The most widespread of the *Cynomys* species in North America, the black-tailed prairie dog, occupies both shortgrass prairie and desert grasslands of the xeric Southwest and the more mesic mixed-grass prairie of the Great Plains (Map 22.1). A few populations of black-tailed prairie dogs have also been reported in tall-grass prairie, where vegetation height can impair the ability of prairie dogs to scan for predators or communicate visually. Thus, occupied tall-grass prairie sites require substantial vegetation height reduction (e.g., heavy grazing) for prairie dogs and ferrets to persist. The US Fish and Wildlife Service (2009) estimated that black-tailed prairie dogs once occupied 32–42 million ha (80–104 million acres) in North America, encompassing 83% of the black-footed ferret rangewide locality records (Anderson et al. 1986);

and that their distribution in North America had been reduced by 2009 to 404,685–809,371 ha (1–2 million acres). In New Mexico, the black-tailed prairie dog was formerly abundant on the plains of the eastern and southern portions of the state (Hubbard and Schmidt 1984; Oakes 2000; Map 22.2) and occupied 2,687,000–3,622,000 ha (6–8.95 million acres) (Bailey 1931; USFWS 2009). In comparison with estimates of the black-tailed prairie dog's historical distribution in our state, the New Mexico Department of Game and Fish calculated in 2010 that the species now occupies only about 16,592 ha (41,000 acres) in New Mexico, mostly as small colonies on the eastern plains. Only 20% (two of ten) of the verified black-footed ferret records in New Mexico and 30% (three of ten) of the highly probable records as defined by Hubbard and Schmitt (1984) originated on black-tailed prairie dog colonies (see Map 22.2 and Table 22.1). The preponderance of records from Gunnison's prairie dog colonies likely stems from the fact that in New Mexico, prairie dog control occurred earlier and more effectively in the historical range of the black-tailed prairie dog (J. Frey, pers. comm.).

The Gunnison's prairie dog largely replaces the black-tailed prairie dog in distribution from central New Mexico westward into central and northern Arizona, and northward into southeastern Utah and southwestern Colorado (Map 22.1). *C. gunnisoni* occupies more diverse vegetation types than the black-tailed prairie dog, ranging from alluvial river valleys and shrub-dominated plains to plateau grasslands and mountain meadows, up to around 3,660 meters (12,000 feet) in elevation (see Maps 22.1 and 22.2). Compared to black-tailed prairie dogs, Gunnison's prairie dogs are more tolerant of not being able to scan their surroundings for predators, as evidenced by the fact that they often occur in shrub-steppe vegetation communities (Hoogland 1995). Gunnison's prairie dogs historically occupied approximately 9.7 million ha (24 million acres) (USFWS 2008), with colonies of this prairie dog species associated with 5.8% of all rangewide ferret locality records (Anderson et al. 1986). The US Fish and Wildlife Service (2008) estimated rangewide historical occupancy of the black-footed ferret on Gunnison's prairie dog colonies to be between 136,000–200,000 ha (340,000–500,000 acres). Approximately 4.1–7.3 million ha (10–18 million acres) of suitable vegetation types for Gunnison's prairie dogs occur in New Mexico, based on two predictive models (Seglund et al. 2006; Neville and Johnson 2007). However, based on survey data from 2010, the actual occupied acreage in New Mexico is probably substantially less, perhaps only 720,340–1,294,994 ha (1.78–3.2 million acres) (NMDGF, unpubl. data). The difference between occupied and suitable land cover may be due in part to the predictive models overestimating the amount of habitat in the state, or else it may reflect the history of regional decline of Gunnison's prairie dog colonies.

The white-tailed prairie dog, which has habitat requirements similar to those of Gunnison's prairie dogs, does not occur in New Mexico. This species of prairie dog, which was associated with the last known natural population of black-footed ferrets in Wyoming, is found in northwestern Colorado, northeastern Utah, central and western Wyoming, and southern Montana (see Map 22.1). White-tailed prairie dogs historically occupied 17–20 million ha (43–51 million acres) (Pauli et al. 2006) with colonies of this prairie dog species associated with 11.2% of rangewide locality records for black-footed ferrets (Anderson et al. 1986). The current rangewide distribution of the white-tailed prairie dog is only approximately 340,000 ha (840,158 acres) (Pauli et al. 2006).

All species of prairie dog dig extensive burrow systems and occur in colonies of varying size and density. Merriam (1902) and Bailey (1905) both mentioned a colony of black-tailed prairie dogs in the mixed-grass prairie of Texas estimated to be 400 km (248 miles) long and 160–240 km (100–150 miles) wide, thus perhaps covering as many as 9.6 million ha (23.7 million acres). In general, black-tailed prairie dogs are more social and tend to live at higher densities than the other two species (Hoogland 1995). Colonies of black-tailed prairie dogs also tend to be more conspicuous in the open prairie landscapes where they occur, as they have the unique habit of clipping vegetation on and around the colony and maintaining a distinct mound around many burrow openings. As expected, wild-born black-footed ferrets show a strong preference for large prairie dog colonies and portions of prairie dog colonies that persist at high population and burrow densities (Biggins et al. 1985, 1993; Eads 2009). However, a study of captive-born ferrets released into a black-tailed prairie dog colony in northeastern New Mexico (see under "Status and Management") indicated no measurable preference for areas with high burrow densities, perhaps because the ferrets were born in captivity and, consequently, were relatively naive (Chipault et al. 2012).

While many present-day populations of prairie dogs in North America appear to be extensive enough to support self-sustaining populations of black-footed ferrets, most of them are routinely decimated by sylvatic plague (see under "Life History").

LIFE HISTORY

Until recently, *M. nigripes* was one of the least-studied carnivores in North America. The species was known to many Native American tribes, which used ferret parts during rituals, and the species was mentioned by early fur trappers (Henderson et al. 1969; Clark 1975). However, it was not formally described and named, by John James Audubon and John Bachman, until 1851 (Clark 1986), and the ferret was rarely mentioned again in the scientific literature for the remainder of the century (Casey et al. 1986). It actually took the discovery of the last wild black-footed ferret population in 1981 and the ferret's designation as an Endangered species to prompt the initiation of numerous life history studies, which have been conducted in association with captive breeding and reintroduction programs. As a result, the life history of the black-footed ferret is now better understood in many respects than that of many other, more common mustelids. Several research volumes have been published that focus on various aspects of the species' biology and review the extensive literature, including those by Wood (1986), Clark (1989), Oldemeyer et al. (1993), Miller et al. (1996), Roelle et al. (2006), and Blake (2011). Relatively little research

Photo 22.5. An adult female black-footed ferret (*Mustela nigripes*) carrying her prey, a juvenile black-tailed prairie dog (*Cynomys ludovicianus*), in the Conata Basin, South Dakota, July 2009. Photograph: © David Eads.

Photo 22.6. Captive black-footed ferrets photographed learning to hunt live prey at the National Black-footed Ferret Conservation Center in Colorado. Prairie dogs account for about 90% of the black-footed ferret's diet. Photograph: © Mike Lockhart/US Fish and Wildlife Service.

on black-footed ferrets has been conducted in New Mexico, though, given that habitat requirements appear to be consistent rangewide, much of what has been learned elsewhere is likely relevant to New Mexico.

Diet and Foraging

M. nigripes is an apparent obligate associate of prairie dogs (Richardson et al. 1987; Biggins et al. 1993), as these colonial rodents and their burrows provide the ferret with its main prey and den sites, respectively (Sheets et al. 1972; Richardson et al. 1987). At Meeteetse, Wyoming and Mallette County, South Dakota, prairie dogs accounted for about 90% of the black-footed ferret's diet (Sheets et al. 1972; Campbell et al. 1987). And Bailey (1931:326), one of the first biologists to report on black-footed ferrets in New Mexico, offered these comments—somewhat disparagingly—about the species:

> These big weasels are almost invariably associated with prairie-dog towns, where they live among the burrows and feed on the prairie dogs, going down the burrows and capturing the occupants at will. Had they not been very scarce they would long since have exhausted their favorite food supply. High living on easily obtained fat prairie dogs seems to be the only explanation of their scarcity, as they are vicious little animals with few enemies.

Although black-footed ferrets today are highly dependent on prairie dogs, Owen et al. (2000) noted that 42% of fossil remains of the species are not associated with prairie dog remains and suggested that the close relationship between these animals may be a secondary effect of the colonization of the North American grasslands by black-footed ferrets within the last 800,000 years.

As is true for many other mustelids, a black-footed ferret's energy requirements are relatively high, but comparable to the predicted energy needs of mammals of similar body mass (Harrington et al. 2003). A ferret's metabolic rate requires 1,200 kJ or less per day (Harrington et al. 2005), and one black-tailed prairie dog alone can provide 4,000–5,000 kJ of energy (Powell et al. 1985). On the other hand, the ferret is not a hibernator, nor does it store large quantities of fat, so it must forage year-round. Biggins et al. (1993) estimated that a population of 763 prairie dogs are needed for one "ferret family" (i.e., one adult female, her young, and 0.5 male; male home territories typically encompass that of 2 female home territories) to survive one year, and that on average a single black-footed ferret kills and consumes 109 prairie dogs every year.

Biggins et al. (2011) and Eads et al. (2011) found that black-footed ferrets move the most during brightly moonlit nights and were most active during the hours after midnight. This nocturnal activity pattern differs from that of coyotes (*Canis latrans*) and might be a means to avoid contact with this and other predators, including diurnal raptors. In most instances, ferret predation on prairie dogs occurs in burrows at night and, consequently, is rarely observed. However, one of us (D. Long) and Vargas and Anderson (1998) have had the opportunity to witness and document multiple predation events in preconditioning pens (enclosures used to prepare captive-born ferrets for release to the wild; see under "Status and Management"), which seem to follow a similar pattern. Ferrets held in the preconditioning pens in New Mexico were observed attacking prairie dogs both above ground and in burrows. If the attack was initiated above ground, the ferret always attempted to drag the prairie dog into a burrow or artificial burrow tube to complete the kill. Never in the hundreds of predatory encounters observed in the New Mexico pens was a ferret observed to kill an adult prairie dog above ground (D. Long, pers. obs.). In most aboveground attacks that were successful, an experienced ferret would

attack and bite onto the head or neck of the prairie dog, which would result in the prairie dog "balling up," thus making the task of dragging, or sometimes carrying, it into the burrow easier for the ferret. Occasionally, the initial attack resulted in the ferret biting onto a part of the prairie dog other than the head or neck. Such cases often resulted in the prairie dog then defending itself and biting the ferret, which likely explains why ferrets were frequently observed aborting a predatory effort if unable to secure a firm bite on the head or neck in the opening seconds of an encounter. Once a prairie dog was pulled inside the burrow, however, the ferret would often readjust its bite (if necessary) onto the head or neck and then quickly moved its grip to the throat where it "chewed" at the area around the esophagus, resulting in hemorrhaging and visible contusions. During this final phase of the attack, a ferret would also frequently wrap itself around the prairie dog and brace its body against the side of the burrow, thus limiting its victim's ability to continue the struggle and presumably reducing the potential for injury to the ferret. Most of these predatory encounters resulted in the death of the prairie dog within a couple of minutes, though attacks sometimes lasted for four or more minutes, depending in part on the amount of prior hunting experience the ferret had. Such is the physical exertion involved that ferrets have been observed panting for prolonged periods after attacks (Vargas and Anderson 1998), and large prairie dogs, particularly adult males, are very difficult, even often impossible, for many ferrets to kill (D. Long, pers. obs.). And more energy may need to be spent in the wild, some female ferrets having been observed to carry a dead prairie dog from a burrow where the kill occurred to her maternal den site to feed her young (Hillman 1968).

Unlike captive individuals, wild ferrets will only rarely attack prairie dogs during daytime when they are above ground and away from their burrows (Clark et al. 1986). Undoubtedly, such attacks can be hazardous for the ferret as the exposure and commotion associated with capturing a prairie dog on the surface increases the ferret's own risk of predation, such as by raptors. In addition, it is common for black-tailed prairie dogs to "mob" and chase away a black-footed ferret that ventures away from its burrow during the daytime (Livieri et al. 2013).

The black-footed ferret is also an opportunistic carnivore and has been observed attacking, capturing, and consuming small mammal and bird species associated with prairie dog colonies and, occasionally, feeding on carrion (Linder et al. 1972; Clark et al. 1986; Eads 2012; D. Long, pers. obs.). The capture and killing of small rodent species (e.g., Ord's Kangaroo Rat [*Dipodomys ordii*.]) appears to require less skill and effort than that needed to kill a prairie dog. When capturing small rodents, the ferret generally bites the prey on the dorsal part of the neck or back and crushes the animal in its jaws (D. Long, pers. obs.). A ferret captured in New Mexico in 1929 was fed a variety of meats, milk, and bread while in captivity and reportedly was fond of fish (Aldous 1940), though these items certainly would not be found in the diet of wild ferrets.

Reproduction and Social Behavior

Black-footed ferrets become reproductively mature in their first year (Clark 1999). Changes in photoperiod in late winter and spring trigger reproductive activity, and breeding in the wild occurs in March–April (Anderson et al. 1986). In captive black-footed ferrets, the photoperiod is sometimes manipulated to meet management needs and to induce breeding at other times of the year (Branvold et al. 2003). Black-footed ferrets are polygynous (Miller et al. 1988), monoestrous, and induced ovulators (Williams et al. 1992); their breeding and whelping seasons roughly parallel those of prairie dogs. Unlike many mustelids, black-footed ferrets do not exhibit delayed implantation of the fertilized ova, and therefore the gestation period of about 42 days is relatively short for a member of that family (Carpenter and Hillman 1978). Most litters range in size from one to five kits with 3.3 kits and 3.4 kits being the average in the last two wild populations studied (Linder et al. 1972; Forrest et al. 1988), and 3.1 kits being the average in a reintroduced black-footed ferret population at Buffalo Gap National Grasslands, South Dakota (US Forest Service 2000). A group of seven kits was observed with a female ferret (dam) in the Conata Basin, South Dakota (D. A. Eads, Colorado State University, pers.

Photo 22.7. One-week-old black-footed ferrets. Most litters in the wild consist of one to five kits, whereas in captivity litter size can reach eight or nine. Photograph: © Robyn Bortner/US Fish and Wildlife Service.

Photo 22.8. Black-footed ferret newborn kit. Photograph: © Kimberly Fraser/US Fish and Wildlife Service.

comm.) and in captivity, litters of eight or nine kits have been documented (Branvold et al. 2003). Males (sires) do not appear to assist in raising young (Forrest et al. 1985). Dams are not committed to one den site and will routinely move kits between burrows, either by carrying them or, when they are more mature, leading them in a single-file "train" to a new burrow (Hillman 1968; Clark et al. 1986; Paunovich and Forrest 1987). Kits first appear above ground at about 45 days of age (in June–July) (Clark et al. 1986) and begin to participate in hunting forays in August. By September, as the kits are now transitioning from what Biggins et al. (1985) call social and dependent juveniles to solitary and independent individuals, they start to disperse from their natal area (Forrest et al. 1988). Data collected from the Meeteetse, Wyoming population indicated that juvenile mortality is high and the average life span is probably less than one year (Biggins et al. 2006). An individual in the wild rarely lives more than three years (Forrest et al. 1988), though older individuals have been detected, including a six-year-old wild-born adult that was monitored in South Dakota (Eads 2012).

The black-footed ferret is generally a solitary animal except during the reproductive season. Studies by Livieri and Anderson (2012), involving a reintroduced black-footed ferret population on a black-tailed prairie dog colony in South Dakota, indicate that males occupy mean home ranges of 131.8 ha (325 acres) whereas female home ranges average 64.7 ha (160 acres). Females have been reported to successfully raise a litter of kits on colonies as small as 10 ha (25 acres) (Hillman 1979), and it has been suggested that females may even successfully raise litters on colonies as small as 5 ha (12 acres) (Biggins et al. 2006). Overall, both male and female home ranges are strongly influenced by the species of prairie dog present, colony size, and prairie dog densities (Forrest et al. 1985; Biggins et al. 2006; Jachowski et al. 2010; Livieri and Anderson 2012).

The population density of female ferrets may be less than the predicted carrying capacity due to intrasexual (female vs. female) territoriality leading to reduced overlap in home ranges (Livieri and Anderson 2012). The territory of a male ferret typically overlaps the territories of two or more females and may not include any additional area beyond that occupied by the females. The sex ratio in most populations with established home ranges is one male for every two females (Forrest et al. 1988; Livieri and Anderson 2012).

Predation, Interspecific Interactions, and Disease

Based on the similar size of black-footed ferrets and prairie dogs and the strong association that many predators form with prairie dog colonies, it seems reasonable to assume that predators that take prairie dogs are also capable of preying on black-footed ferrets (Biggins 2000). The best information on predators of ferrets is derived from research on the Meeteetse, Wyoming population and more recent studies of the reintroduced Conata Basin, South Dakota population (Miller et al. 1996; Breck et al. 2006). Predation can account for up to 95% of the documented mortalities of newly released, captive-reared, and relatively naïve ferrets at reintroduction sites (Breck et al. 2006). In the last wild population at Meeteetse, Wyoming, an estimated 57% of known ferret mortality could be attributed to predation (Forrest et al. 1988), though the actual percentage was likely much higher (Breck et al. 2006).

Coyotes appear to be the most important predator of black-footed ferrets, accounting for about 60% of documented predation events and up to 95% of the mortality of captive-bred, reintroduced individuals (Biggins et al. 2006). American badgers (*Taxidea taxus*) and raptors including great horned owls (*Bubo virginianus*) account for a smaller portion of ferret mortality, though Breck et al. (2006) noted that great horned owls in particular can develop a "search image" for

Photo 22.9. Ring-reader in place at prairie dog burrow. The reader is used to detect the individual passive integrated transponder (PIT)-tag number of a reintroduced ferret after it has been observed at a burrow during nighttime spotlighting surveys. This monitoring technique allows researchers to track survivorship of ferrets on a reintroduction site. Each PIT tag is a small radio transponder that contains its own specific code. Photograph: © James N. Stuart.

black-footed ferrets on reintroduction sites, resulting in substantial losses of newly released animals. If unconfirmed predation events are included, coyotes may nonetheless account for 80–90% of predation events (Breck et al. 2006). A coyote is far less likely to immediately consume a ferret that it has killed and instead may bury it (Miller et al. 1996; Biggins 2000). Predation by coyotes on other carnivore species is common and presumably benefits the coyote by removing competition (Biggins et al. 2011).

Known causes of human related black-footed ferret mortality include vehicle impacts, shooting, trapping, and poisoning (Cahalane 1954; Hanebury and Biggins 2005). In New Mexico, Hubbard and Schmitt (1984) documented instances of human-caused mortalities of ferrets through trapping and prairie dog poisoning. The black-footed ferret is also vulnerable to bacterial and viral diseases. Around 1900, the invasive disease

commonly referred to as plague—sylvatic plague when it occurs in wild animals—made its entry into North America, likely aboard a trading ship traveling from Asia and docking in San Francisco (Biggins and Kosoy 2001). The disease, which is caused by the bacterium *Yersinia pestis* and relies on fleas for transmission, quickly spread among the native mammal communities on the West Coast and began to progress eastward. By the 1940s, the plague had been reported in New Mexico and 15 other western states (Barnes 1993). For unknown reasons, but possibly environmental conditions related to humidity and temperature, the plague made a stop at the 101st Meridian for 70 years and only in 2008 did it push eastward beyond what was formerly referred to as the "plague line." With this last intrusion, the plague now largely encompasses the entire historical range of prairie dogs and black-footed ferrets. In New Mexico and elsewhere in the West, important reservoir species for sylvatic plague may include mammals with high resistance to the disease such as certain species of voles (*Microtus* spp.), deer mice (*Peromyscus maniculatus*), kangaroo rats (*Dipodomys* spp.), and possibly many carnivores (Antolin et al. 2002).

Prairie dogs are highly vulnerable to plague, and mortality within a colony is typically very high during epizootic (outbreak) events (Antolin et al. 2002). However, a small number (less than 1%) of prairie dogs in a colony will sometimes survive the disease and are capable of repopulating the site over time (Cully 1997; D. Long, pers. obs.). As for the black-footed ferrets, they are often unable to survive major plague events since those individuals that do not succumb to the disease either starve or are forced to abandon the stricken colony (Cully 1993). Even for ferrets that do survive epizootic events, the disease can continue to pose a significant threat during so-called enzootic periods, when plague remains present in the environment without any noticeable prairie dog die-off. During enzootic periods, black-footed ferret survival is indeed significantly reduced (Matchett et al. 2010). The disease has been and remains a major impediment in the re-establishment of black-footed ferrets in the wild (Lockhart et al. 2006), and its ubiquity throughout the historical range of prairie dogs and black-footed ferrets alike often requires "heavy-handed" management to protect populations of both animals (see under "Status and Management").

Canine distemper, another highly virulent disease found in black-footed ferrets, was in part responsible for the loss of the last known wild populations in South Dakota and Wyoming (Carpenter et al. 1976; Forrest et al. 1988). Of the nine ferrets brought into captivity from the South Dakota population, four were inadvertently killed after vaccination with a modified live canine distemper virus (the vaccine had proven safe in European polecats but was lethal in black-footed ferrets). The first six black-footed ferrets removed from the wild at Meeteetse, Wyoming apparently were already infected with the virus and succumbed to the disease while in captivity (Lockhart et al. 2006).

STATUS AND MANAGEMENT
Historical Population Status

Not only are black-footed ferrets fossorial, nocturnal, and secretive by nature, they were never of economic importance as a furbearer. Consequently, very few records of occurrence exist, making it difficult to assess the historical abundance of the species throughout its range. Based on pre-settlement accounts of large prairie dog colonies measured in miles (see below), population estimates from the last two recorded wild ferret populations, and the availability of ferret specimens from throughout its historical range, the species might have once been common, if not abundant, in at least portions of its distribution. However, there is no consensus on this. Some authors have suggested that ferrets were probably common (Linder et al. 1972; Choate et al. 1982; Hubbard and Schmitt 1984; Anderson et al. 1986), whereas others have argued they were likely scarce (Bailey 1931; Cahalane 1954). Paleontological evidence, historical records, and studies of the last two wild populations before extirpation suggest to us that black-footed ferret densities were not uniform throughout the species' range. Perhaps ferrets were locally common, but, like many carnivores, they may also have persisted at relatively low densities throughout much of the species' historical range, due to territoriality and, in some cases, variable abundance of prey.

Estimates of historical and contemporary rangewide prairie dog occupation and, consequently, black-footed ferret populations are imprecise and oftentimes controversial (see Hubbard and Schmitt 1984). Early explorers and naturalists frequently failed to report on what we know today were extensive prairie dog complexes (Knowles et al. 2002), while other reports wildly exaggerated the extent of colonies (Virchow and Hygnstrom 2002; see Hubbard and Schmitt 1984). Records from the poisoning campaigns of the 1920s and 1930s provided acreage estimates of some areas being treated but frequently failed to identify the species being targeted, often describing them simply as "rodents." Prairie dog colonies also regularly fluctuate in size due to plague or drought, and, unless such impacts are monitored and quantified, estimates of total occupied hectares can be flawed, even today. In short, the information available to us is imprecise and oftentimes conflicting. The formerly extensive distribution of ferrets, along with the current distribution of other species that likely co-evolved with prairie dogs, such as the burrowing owl (*Athene cunicularia*) and the mountain plover (*Charadrius montanus*), provide strong evidence that *Cynomys* species were not only once widespread but also abundant in interior North America (Knowles et al. 2002).

At the beginning of the 20th century, the three species of prairie dogs on which ferrets are dependent occupied an estimated 41 million ha

(101,313,206 acres) of the grasslands and shrublands of western North America (Nelson 1919; Anderson et al. 1986). Federally sponsored campaigns to eradicate prairie dogs, conversion of grasslands for agricultural purposes, and sylvatic plague all reduced the extent of their range by 97% to approximately 1.2 million ha (2,965,264 acres), with two-thirds of the remaining colonies being small and isolated (Miller and Reading 2012). For the black-footed ferret, a species inextricably linked to large, healthy populations of prairie dogs, the loss of so much habitat and the fragmentation of what little remained resulted in a precipitous population decline throughout the early and mid-20th century, culminating in the near extinction of the species by the 1980s (USFWS 2013a).

In New Mexico, Shriver (1965) estimated that prairie dog colonies covered 4,836,398 ha (11,951,000 acres) in the state in 1919 (equivalent to about 15% of New Mexico's surface area). Hubbard and Schmitt (1984) estimated that federally sponsored prairie dog poisoning campaigns treated 4,370,604 ha (10,800,000 acres) of prairie dog colonies from 1931 to 1981. Recent estimates suggest there may be 736,932–1,311,586 ha (1,820,998–3,240,999 million acres) of prairie dog colonies remaining in New Mexico (NMDGF, unpubl. data). As is the case throughout most of North America, New Mexico now supports only a fraction of its historical prairie dog population.

By 1987, the known population of the black-footed ferret had been reduced to 18 individuals, all of which were in captivity, making the species one of the rarest mammals on Earth. Today the number of free-ranging, reintroduced ferrets varies from year to year, and survivorship is mainly affected by the presence or absence of plague epizootics. In 2012, the USFWS (2013a) estimated the wild ferret population to be 364 animals, which was down from an estimated 500–1,000 animals just a few years before. As part of the federal recovery effort, there were approximately 250–300 black-footed ferrets being maintained in captive breeding facilities throughout the United States. In 2018, after more reintroductions rangewide, the wild ferret population likely did not exceed 340 animals, or fewer than six years earlier (J. Hughes, pers. comm.).

Legal Status

The black-footed ferret was listed as Endangered throughout its range in 1967 under the Endangered Species Preservation Act, an anemic set of rules (which did little to prevent the decline and eventual extirpation of the Mellette County, South Dakota population), and later under the more robust Endangered Species Act (ESA) of 1973. In 2013 the US Fish and Wildlife Service (2013a) finalized a revised recovery plan for the black-footed ferret and outlined the criteria by which the species could be considered "recovered" in the wild. Downlisting from Endangered to Threatened status would require, among other criteria, a total of at least 1,500 free-ranging ferret adults in 10 or more populations with no fewer than 30 breeding adults in any one population (USFWS 2013a). The criteria for delisting (removal from the ESA list) included reaching a total of at least 3,000 free-ranging ferrets in 30 or more populations with no fewer than 30 breeding adults in any one population (USFWS 2013a). In the revised recovery plan, the US Fish and Wildlife Service (2013a) adopted a model developed by Ernst et al. (2004), described as a "technique to allocate hypothetical black-footed ferret recovery goals in an equitable fashion," based on the historical rangewide distribution of prairie dogs. Based on this allocation model, New Mexico would be responsible for 220 of the 1,500 wild free-ranging black-footed ferrets necessary to downlist and 440 of the 3,000 needed to delist.

All existing reintroduced populations of the black-footed ferret in the United States are currently categorized by the US Fish and Wildlife Service as experimental populations under either Section 10(j) or 10(a)(1)(A) of the ESA.

Both categories permit "incidental take" (i.e., unintentional harming or killing) of individual black-footed ferrets and therefore allow less protection than what would otherwise be provided by a "fully Endangered" designation where any "take" of the species is illegal. Despite the ferret's close ecological relationship with prairie dogs and their colonies, no critical habitat (i.e., areas of habitat believed to be essential to the species' conservation) has been proposed thus far by the US Fish and Wildlife Service.

In New Mexico, the black-footed ferret is categorized by state law as a protected furbearer, though no legal harvest has been allowed since at least the 1960s, at which time wildlife managers considered the species to be rare (Berghofer 1967). The black-footed ferret was listed as an Endangered species under the New Mexico Wildlife Conservation Act in 1975, only to be delisted in 1988 after survey efforts indicated the species was likely extirpated in the state (Jones and Schmitt 1997). Following the rediscovery of the black-footed ferret in Wyoming in 1981, the New Mexico Department of Game and Fish and US Bureau of Land Management initiated a publicity campaign to "ferret out"(!) reports of the species by New Mexico residents via the dissemination of advertisements, posters, and popular articles (Hubbard and Schmitt 1984; Photo 22.10). Although these efforts, combined with intensive surveys of many prairie dog colonies, failed to identify any remaining ferret populations in the state—a sighting from west of Angel Fire in Colfax County in 1981 was deemed a probable record (the last one) by Hubbard and Schmitt (1984)—they did serve to raise public awareness about the species. This increased awareness is reflected in the many reports from the public of black-footed ferret sightings that are still received today by the New Mexico Department of Game and Fish and other resource management agencies. Unfortunately, when photographs or details of observations are available, all such reports turn out to be

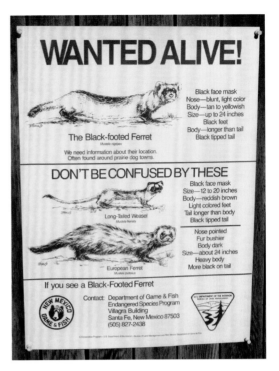

Photo 22.10. 1982 poster produced jointly by the New Mexico Department of Game and Fish and the US Bureau of Land Management to obtain any possible black-footed ferret observation records from the public in New Mexico. The species was suspected of being extinct throughout its North American range when a small population was discovered in Wyoming in 1981. All known surviving ferrets today are descended from that population. No verifiable reports were obtained from New Mexico in the 1980s, and the species was quite possibly extirpated in the state by the time the poster was made. Photograph: © James N. Stuart.

sightings of long-tailed weasel (J. Stuart, pers. obs.).

Throughout the species' range, the conservation and recovery of the black-footed ferret is inextricably tied to the conservation of prairie dogs (Miller and Reading 2012). All three species of prairie dog within the former range of the ferret have previously been considered for listing under the ESA. The US Fish and Wildlife Service determined that the status of the black-tailed prairie dog and of montane populations of

Gunnison's prairie dog in northern New Mexico and Colorado, both former federal Candidate species, did not warrant listing (USFWS 2009, 2013b). In New Mexico, both species of prairie dogs are considered Species of Greatest Conservation Need (SGCN) under the State Wildlife Action Plan (NMDGF 2016), but like all non-game species in the state otherwise do not receive any formal protection, except by those public land managers, tribal governments, or private property owners who seek to conserve the species on their lands. As of 2016, the black-footed ferret is also classified as a SGCN due to current efforts to reintroduce the species.

Recovery Efforts

With the demise of most wild populations of the black-footed ferret by the 1950s, conservation biologists eventually turned to what was believed to be the only option to save the species: captive propagation. The ferret population in Mellette County, South Dakota—believed to be the last at the time—had been studied since the mid-1960s, and in 1971 the first attempt was made to capture a sample of these animals for a captive breeding program at the Patuxent Wildlife Research Center in Maryland. Unfortunately, the program was stymied by both mortalities of adult captive ferrets caused by vaccine-induced canine distemper and poor survivorship of offspring (Lockhart et al. 2006). With the extirpation of the South Dakota wild population in 1974 and the death of the last captive ferret at Patuxent in 1979, it appeared the species had been lost.

Dr. James Carpenter (pers comm.), leader and research veterinarian at the Endangered Species Propagation Program at Patuxent at the time, relayed the events leading up to the black-footed ferret's presumed extinction. His account bears witness to the many challenges of developing a captive breeding program and the dedication and investment required for the conservation of a species.

Of the five black-footed ferrets used in the breeding program from 1976–1978, two were females. Only one of the females was ever receptive to the males. In 1976 this female produced the first litter of black-footed ferrets ever born in captivity. Unfortunately, four of the five kits were stillborn, and the remaining kit died because its mother did not provide it with adequate care.

Since earlier studies had shown the European polecat females readily accepted young from other polecats, as well as from other mustelids, our research team decided to remove the black-footed ferret kits immediately after the second birth in 1977 and to place them with a lactating European polecat. The black-footed ferret produced four stillborn kits and one weak kit. Although the surviving kit was readily accepted by the lactating European polecat and also received intermittent medical care, it died two days later.

In view of the female black-footed ferret's advancing age and her history of stillborn litters, in 1978 it was decided to take the young by caesarian section, hoping that they might be saved. The female black-footed ferret was "bred" successfully, and her abdomen became distended over the following 6 weeks. At 42 days the female was taken to a veterinary hospital and, although the surgery was successful, the female did not contain any young—she had a false pregnancy, probably a reflection of her age. Once again, producing black-footed ferrets in captivity had eluded us.

The species received another reprieve from extinction in September 1981, when another, and indeed the last, wild population of black-footed ferrets was discovered by ranchers John and Lucille Hogg near Meeteetse, Wyoming. Their dog, Shep, "got in a tangle" (Gustkey 1985) with and killed a black-footed ferret one night. John Hogg found the ferret carcass on his doorstep the following morning and, not knowing what it was, "threw it over the fence" (K. Frasier, pers.com.) (It bears mentioning that the Hogg family had been ranching in the area for generations and had never seen a black-footed ferret, illustrating just how secretive the species can be.) Lucille decided she wanted to have the unusual animal mounted, so she and her husband retrieved the carcass and delivered it to a local taxidermist. The taxidermist immediately recognized the animal, called the authorities, and informed the Hoggs that the carcass was being confiscated (Gustkey 1985). Thus, the black-footed ferret recovery community forever owes a debt of gratitude to the Hoggs and their dog Shep, for without that chance encounter between a ferret (probably a "wayfaring" juvenile; Carr 1986:4) and Shep that September night it is almost certain that the species would now be extinct.

Immediately following the fortuitous discovery, research began on the Meeteetse population and much of what we know today about the behavior, habitat, and ecology of the black-footed ferret was learned during those early studies. The white-tailed prairie dog colony complex that sustained the last black-tailed ferret population consisted of 37 colonies covering 2,995 ha (7,400 acres) (Clark 1986). During surveys in 1982, 1983, and 1984 the ferret population at Meeteetse was estimated at 61, 88, and 129 individuals, respectively (Clark 1986). In June 1985 sylvatic plague was detected in the prairie dog population at Meeteetse and ferret numbers declined sharply over the summer; by August the population had been reduced to an estimated 58 individuals (Lockhart et al. 2006). While researchers knew prairie dogs were susceptible to plague, it was thought that black-footed ferrets were immune, since Siberian and European polecats and other mustelids had demonstrated resistance to the disease (Lockhart et al. 2006). Due to declining

numbers in the wild in September 1985, the decision was made to capture six black-footed ferrets for captive breeding. Concurrently, it was discovered that the ferret population at Meeteetse was not only being impacted by plague but was also in the midst of a canine distemper outbreak. All six individuals brought into captivity died of distemper (Lockhart et al. 2006). Before the confirmation of canine distemper in the wild ferret population, field researchers were also beginning to question whether black-footed ferrets were indeed immune to the plague; however, the confirmation of canine distemper seemed to explain the observed declines at Meeteetse (D. Biggins, pers. comm.). Not until the mid-1990s, and the loss of 27 ferrets in captivity, was susceptibility to plague again considered—and subsequently confirmed as—a direct threat to ferrets (Godbey et al. 2004). After the loss of the first six ferrets brought into captivity and the confirmation of canine distemper in the wild population, the decision was made to try to capture all remaining ferrets in the wild and bring them into captivity; thus ended the last naturally occurring black-footed ferret population (Lockhart et al. 2006).

Black-footed ferrets captured at Meeteetse during 1985–1987 provided the captive breeding stock for all the ferrets alive today. Since 1991, captive-bred ferrets have been released in eight states (Arizona, Colorado, Kansas, Montana, New Mexico, South Dakota, Utah, and Wyoming), southern Saskatchewan in Canada, and northwestern Chihuahua in Mexico. However, even after more than 30 years of effort, the survival of the black-footed ferret still hangs in the balance. Reintroductions have so far failed at half of the release sites, including in northern Mexico and Saskatchewan. The wild ferret population reintroduced at the Conata Basin/Badlands, South Dakota site had reached 355 individuals in 2007, but by the fall of 2018 it had been reduced to just 119 (USFWS, unpubl. data).

At present, black-footed ferrets destined for

Photos 22.11a and b (*left to right*). "Triple-shooter" dispenser mounted on an ATV, used to distribute sylvatic plague vaccine (SPV) baits on a prairie dog colony in New Mexico. SPV baits are peanut-butter flavored pellets that contain the vaccine against plague. They are distributed on a prairie dog colony to inoculate a large percentage of the rodents against plague and are one of the more recent tools to manage the disease on reintroduction sites for the black-footed ferret. Photographs: © James N. Stuart.

release are produced from the captive stock at several breeding facilities located throughout the United States, the largest of which is managed by the US Fish and Wildlife Service at Carr, Colorado. Since 1986, and as of the 2019 breeding season, approximately 9,600 kits had been produced in captivity, many of which had been reintroduced to the wild (Black-footed Ferret Connections 2015; R. Bortner, pers. comm.). In most years, about 200 preconditioned black-footed ferrets have been available for release at reintroduction sites (J. Hughes, USFWS, pers. comm.). "Wild preconditioning" is a key component of captive-rearing of ferrets and involves exposing young captive ferrets to live prairie dogs, thus allowing them to develop hunting skills before their release. The allocation of captive-born ferrets to new or existing reintroduction sites is based on a ranking system that considers such factors as habitat quality; ongoing site management; disease presence and monitoring; ferret survival and population monitoring (for older

reintroduction sites); local reintroduction program management; and local research programs. The captive breeding program has been so successful that availability of animals is no longer the limiting factor in re-establishing the species; rather, the availability of suitable habitat and release sites is the primary obstacle to recovery (Lockhart et al. 2006), together with disease.

Disease Management in the Wild

For many years, there were only two possible main methods for mitigating the effects of the plague at black-footed ferret reintroduction sites: the application of a pulicide (an insecticide that kills fleas) into prairie dog burrows (Seery et al. 2003) and the vaccination of ferrets (Rocke et al. 2006). The most common and effective pulicide used at black-footed ferret release sites today is deltamethrin, which is applied to prairie dog burrows in the form of a fine powder (Seery et al. 2003). Without prophylactic pulicide treatment, most prairie dog complexes sufficient in size to support black-footed ferrets fall into a plague cycle characterized by population build-ups followed by precipitous declines (Oldemeyer et al. 1993). Such tremendous fluctuation in prairie dog populations renders most untreated sites unsuitable for ferrets due not only to the black-footed ferret's own susceptibility to the disease but also the loss of its primary food source during plague epizootics (Matchett et al. 2010). Deltamethrin may lead to an 88.5% vector reduction (Roth 2019), but the application of this and other pulicides has drawbacks, including the fact that it is labor-intensive. It may also fail to halt outbreaks if "dusting" is used too late (Abbott et al. 2012).

The US Geological Survey also developed a recombinant, injectable vaccine for black-footed ferrets that has proven effective in inducing an antibody response, thus reducing their susceptibility to plague (Rocke et al., 2006). About 69% of vaccinated ferrets exposed to high levels of the plague bacterium survived whereas similarly exposed unvaccinated ferrets all died (Rocke et al. 2006). Immunization requires two doses, preferably administered at a two-week interval—a task which is often problematic when working with animals as secretive and difficult to capture as black-footed ferrets. An obvious shortcoming of relying on the recombinant vaccine is that it also does not protect prairie dogs against plague epizootics (Abbott et al. 2012).

In 2012 the US Fish and Wildlife Service, the US Geological Survey, and several state, tribal, and private entities began field trials on an oral plague vaccine (also called sylvatic plague vaccine or SPV), which is delivered to prairie dogs through peanut butter–flavored bait. The goal of this vaccine program is to significantly reduce the impact of plague on prairie dog colonies, thereby benefitting both prairie dogs and ferrets. Lab results and preliminary field trials proved encouraging, but it remains to be determined just how effective the vaccine is in the field,

whether other species might be affected, and if the vaccine can be produced at a reasonable cost (Abbott et al. 2012).

Canine distemper, which is not as devastating to black-footed ferret populations as plague, is treated with a recombinant vaccine that can be administered to both captive animals and captured wild ferrets. The disease does not affect prairie dogs. Despite its impact to wild and captive ferrets in the 1970s and 1980s, at present the disease is more easily managed than plague and therefore is of secondary concern at most release sites (USFWS 2013a).

Reintroduction Efforts in New Mexico

New Mexico is a relative latecomer to the reintroduction efforts for the species. In 1998, the Turner Endangered Species Fund (TESF) established a captive breeding and preconditioning facility for black-footed ferrets at the Vermejo Park Ranch, a 238,280 ha (588,800 acres) property in Colfax County. From 1999 to 2005, the black-footed ferrets produced in captivity at the ranch were sent to reintroduction programs elsewhere in the United States as well as Canada and Mexico. In 2005, however, TESF shifted its focus to "wild preconditioning" of captive-born ferrets on black-tailed prairie dog colonies on the ranch. Between 2005 and 2007, 75 black-footed ferrets were released onto black-tailed prairie dog colonies, where they were allowed to hunt and interact in a natural state for a period of several weeks to several months before being recaptured. Initially, in 2005, all wild preconditioned ferrets were held temporarily in a protected 405-ha (1,000-acre) prairie dog colony encircled by electrified predator-exclusion netting. During 2006–2007, black-footed ferrets were again released into the same, protected prairie dog colony but were later recaptured and moved into unprotected colonies. Forty of the 75 thus-released animals were recaptured and, as planned, relocated to reintroduction sites outside of New Mexico for

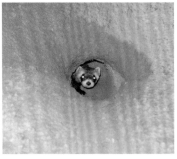

Photo 22.12. Captive-born black-footed ferret about to be released at Vermejo Park Ranch, New Mexico as part of a "wild preconditioning" experiment. This female black-footed ferret was subsequently recaptured and transferred to a permanent release site in Arizona. Photograph: © Dustin Long.

Photo 22.13. First documented wild-born black-footed ferret in New Mexico in perhaps 75 years; the ferret was born in 2009 on a black-tailed prairie dog colony. Photograph: © Dustin Long.

Photo 22.14. Anesthetized wild-born black-footed ferret at Vermejo Park Ranch, New Mexico. Wild-born black-footed ferrets were captured, anesthetized, and implanted with a small transponder chip to assist in subsequent identification. Photograph: © Dustin Long.

Photos 22.15a and b (*top to bottom*). Black-footed ferrets were released in the fall of 2012 on a Gunnison's prairie dog colony at Vermejo Park Ranch. Once all ferrets had been released, they were given time to settle into their new home. Staff returned to the release site after dark to check on them using spotlights. Photographs: © Della Garelle.

Photo 22.16. Captive-born black-footed ferret being released onto a Gunnison's prairie dog colony at Vermejo Park Ranch, New Mexico. Photograph: © Vermejo Park Ranch.

Photos 22.17a–d. Black-footed ferret reintroduction on Greg Moore's 25,000-acre ranch near Wagon Mound, Mora County on 26 September 2018. From many years of observation, rancher Greg Moore concluded that the black-tailed prairie dogs on his ranch could contribute to grassland health if their numbers were kept in check. To that end, he petitioned the US Fish and Wildlife Service to reintroduce black-footed ferrets on his property. A year later, with the help of New Mexico Department of Game and Fish, eight ferrets finally arrived and were released into existing prairie dog holes on the ranch. Photographs: © Scott Wilber / New Mexico Land Conservancy.

Photo 22.18. One of the eight black-footed ferrets released on 26 September 2018 on Greg Moore's ranch in Mora County. The carriers are opened at randomly selected prairie dog burrows and tilted towards the burrow opening. Most ferrets will not go willingly and instead typically hide in the short length of corrugated plastic tube that is provided for them inside the carrier. The easiest way to get the ferret in the burrow is to move the tube into the burrow opening using a stick. The ferret is then allowed to escape from the tube into the burrow, which might take up to a couple of minutes. Photograph: © Robert Muller.

permanent release (D. Long, unpubl. data). The fate of 35 animals released but not recaptured remained unknown, but those were presumed to have perished due to predation or, in at least one case, starvation.

During the same time period of 2005–2007, TESF was actively managing its Vermejo Park Ranch black-tailed prairie dog colonies, and by 2008 the prairie dog complex was deemed to be of sufficient size (2,790 ha; 6,900 acres) to serve as a ferret reintroduction site. In fall 2008, TESF began releasing captive-born black-footed ferrets in a study to determine whether a viable population could be established on the ranch. On September 16, 2009, the first wild-born black-footed ferret kit in New Mexico in perhaps 75 years was captured at the ranch (D.

Long, pers. obs.; Photo 22.13), an encouraging sign that the species could be re-established in the state. Later, multiple litters of wild-born kits were detected on the Vermejo Park Ranch, but despite that success, in late 2012 TESF discontinued black-footed ferret releases due to poor survival. The failure of the ferret restoration project on the black-tailed prairie dog colonies during the 2008–2012 period appeared to be closely linked to prairie dog pup production, which itself is positively correlated with spring and summer precipitation patterns (D. Long, unpubl. data). In short, drought years resulted in the production of few prairie dog pups, which in turn resulted in poor black-footed ferret survival and reproduction. Over the five-year course of the black-footed ferret releases onto black-tailed prairie dog colonies at the ranch, the black-footed ferret population peaked in the spring of 2011 with 19 individuals identified; by the fall of that year the population had been reduced to three individuals—all males. In 2012, black-footed ferret habitat conditions continued to deteriorate due to the intensifying drought and the decision was made to withdraw from further releases until conditions improved.

In September 2012, 20 black-footed ferrets were released, this time, onto the Gunnison's prairie dog colony at Castle Rock, the same Vermejo Park Ranch site where former New Mexico Game and Fish director Elliot Barker had trapped one ferret and seen another in 1930 (Hubbard and Schmitt 1984; see Table 22.1). Data collected by TESF from 2012 to 2015 and historical records suggested that black-footed ferrets might fare better in areas occupied by Gunnison's prairie dogs at Vermejo Park Ranch, provided that sylvatic plague could be managed. Unfortunately, the Castle Rock Gunnison's prairie dog colony was decimated by plague in 2015, resulting in the death of all wild ferrets at that site.

Most recently, in late September 2018, eight black-footed ferrets were released on a ranch just east of Wagon Mound in Mora County. The ferrets arrived from northern Colorado, where they had been raised in captivity. The ranch, owned by Greg Moore, harbors a relatively small (approximately 182-ha; 450-acre) black-tailed prairie dog colony that was treated prophylactically with the oral sylvatic plague vaccine in late 2018. Of the eight ferrets, only three appeared to have survived the first few months of their release. Four more ferrets were released in September 2019 on the same ranch, and another three in October 2020, again with captive-reared animals from the breeding facility in Colorado. The most recent nocturnal spotlighting survey and trapping, in September 2021, led to the detection of at least six animals including males, females, and young of the year. The New Mexico Department of Game and Fish has documented reproduction in the small, reintroduced Mora County population since the first summer (2019) following the initial release.

The reintroduction site in Mora County will continue to be monitored for the persistence of ferrets and continued reproduction, though as of 2022, it appears that the local prairie dog colony has drastically declined due to plague and/or drought, and the site is unlikely to serve as a ferret reintroduction site in the near future. More releases are possible on Vermejo Park Ranch, pending ongoing research on the effectiveness of the oral sylvatic plague vaccine in field conditions, or the development of more cost-effective techniques. In Montana, the US Fish and Wildlife Service is currently experimenting with baits that contain the insecticide Fipronil (a phenylpyrazole) instead of SPV, for a much-reduced cost.

The twin impacts of sylvatic plague and prairie dog poisoning have had, and continue to have, severe consequences for prairie dogs, both in the extirpation of these rodents from many places in the state and the fragmentation of their remaining range into populations too small and isolated to support black-footed ferrets. Although neither species of prairie dog is likely to become extinct in New Mexico in the foreseeable future, much of New Mexico where prairie dogs still exist is likely no longer suitable for a small carnivore that depends on abundant, robust, plague-free populations of prey.

Many challenges remain in re-establishing the black-footed ferret in New Mexico and elsewhere, primarily disease management, but also conservation of prey species and habitat protection. Although the ferret likely will never again be as widespread in the state as it was historically, opportunities may exist to establish other black-footed ferret populations on private or public lands in New Mexico where the establishment and protection of large prairie dog populations is feasible. As noted by others—and as is true of an increasing number of organisms worldwide—the black-footed ferret will likely persist as a species only through the intervention of conservation biologists and long-term, careful management of its habitat.

For the time being, we have narrowly avoided the prediction by naturalist E.T. Seton (1929:573) regarding the prairie dog and black-footed ferret: "Now that the big Demon of Commerce has declared war on the Prairie-dog, that merry little simpleton of the Plains must go . . . and with the passing of the Prairie-dog, the Ferret, too, will pass."

ACKNOWLEDGMENTS
We thank John Hughes (US Fish and Wildlife Service, Black-footed Ferret Recovery Program) for providing us with an update on the status of reintroductions rangewide; Robyn Bortner (US Fish and Wildlife Service, National Black-footed Ferret Conservation Center) for the updated number of ferrets produced in captivity; and David Eads for constructive comments on an earlier version of the manuscript.

LITERATURE CITED

Abbott, R. C., J. E. Osorio, C. M. Bunck, and T. E. Rocke. 2012. Sylvatic plague vaccine: a new tool for conservation of threatened and endangered species? *Ecohealth* 9:243–50.

Abramov, A. V. 2000. A taxonomic review of the genus *Mustela* (Mammalia, Carnivora). *Zoosystematica Rossica* 8:357–64.

Aldous, S. E. 1940. Notes on a black-footed ferret raised in captivity. *Journal of Mammalogy* 21:23–26.

Anderson, E. 1989. The phylogeny of Mustelids and the systematics of ferrets. In *Conservation biology and the black-footed ferret*, ed. U. S. Seal, E. T. Thorne, M. A. Bogan, and S. H. Anderson, 10–20. New Haven, CT: Yale University Press.

Anderson, E., S. C. Forrest, T. W. Clark, and L. Richardson. 1986. Paleobiology, biogeography, and systematics of the black-footed ferret, *Mustela nigripes* (Audubon and Bachman), 1851. *Great Basin Naturalist Memoirs* 8:11–62.

Antolin, M. F., P. Gober, B. Luce, D. E. Biggins, W. E. van Pelt, D. B. Seery, M. Lockhart, and M. Ball. 2002. The influence of sylvatic plague on North American wildlife at the landscape level, with special emphasis on black-footed ferret and prairie dog conservation. In *Transactions of the sixty-seventh North American wildlife and natural resources conference*, ed. J. Rahm, 104–127. Washington DC: Wildlife Management Institute.

Bailey, V. 1905. *Biological survey of Texas*. North American Fauna 25. Washington, DC: US Department of Agriculture, Bureau of Biological Survey.

———. 1931 (=1932). *Mammals of New Mexico*. North American Fauna 53. Washington, DC: US Department of Agriculture, Bureau of Biological Survey.

Barnes, A. M. 1993. A review of plague and its relevance to prairie dog populations and the black-footed ferret. In *Proceedings of the symposium on the management of prairie dog complexes for reintroduction of the black-footed ferret*, ed. J. L. Oldemeyer, D. E. Biggins, B. J. Miller, and R. Crete, 28–37. Washington, DC: US Fish and Wildlife Service.

Berghofer, C. B. 1967. Protected furbearers in New Mexico. In *New Mexico wildlife management*, 187–94. Santa Fe: New Mexico Department of Game and Fish.

Biggins, D. E. 2000. Predation on black-footed ferrets (*Mustela nigripes*) and Siberian polecats (*M. eversmanii*): conservation and evolutionary implications. PhD Dissertation, Colorado State University, Fort Collins.

———. 2006. The symposium in context. In *Recovery of the black-footed ferret: progress and continuing challenges*, ed. J. E. Roelle, B. J. Miller, J. L. Godbey, and D. E. Biggins, 3–5. United States Geological Survey Scientific Investigations Report 2005-5293, Reston, VA.

Biggins, D. E., L. R. Hanebury, B. J. Miller, and R. A. Powell. 2011. Black-footed ferret and Siberian polecat as ecological surrogates and ecological equivalents. *Journal of Mammalogy* 92:710–20.

Biggins, D. E., and M. Y. Kosoy. 2001. Influences of introduced plague on North American mammals—implications from ecology of plague in Asia. *Journal of Mammalogy* 82:906–16.

Biggins, D. E., J. M. Lockhart, and J. L. Godbey. 2006. Evaluating habitat for black-footed ferrets: revisions of an existing model. In *Recovery of the black-footed ferret: progress and continuing challenges*, ed. J. E. Roelle, B. J. Miller, J. L. Godbey, and D. E. Biggins, 143–50. United States Geological Survey Scientific Investigations Report 2005-5293, Reston, VA.

Biggins, D. E., B. J. Miller, L. R. Hanebury, B. Oakleaf, A. H. Farmer, R. Crete, and A. Dood. 1993. A technique for evaluating black-footed ferret habitat. In *Proceedings of the symposium on the management of prairie dog complexes for reintroduction of the black-footed ferret*, ed. J. L. Oldemeyer, D. E. Biggins, B. J. Miller, and R. Crete, 73–88. Washington, DC: US Fish and Wildlife Service.

Biggins, D. E., and M. H. Schroeder. 1988. Historical and present status of the black-footed ferret. In *Eighth Great Plains wildlife damage control workshop proceedings*, ed. D. W. Uresk, G. L. Schenbeck, and R. Cefkin, 93–97. Fort Collins, CO: USDA Forest Service, General Technical Report RM-154.

Biggins, D. E., M. Schroeder, S. Forrest, and L. Richardson. 1985. Movements and habitat relationships of radio-tagged black-footed ferrets. In *Black-footed ferret workshop proceedings*, ed. S. H. Anderson and D. B. Inkley, 11.1–11.17. Cheyenne: Wyoming Game and Fish Department.

Black-footed Ferret Connections. 2015. [Homepage].

http://blackfootedferret.org. Accessed 10 December 2015.

Blake, B. H., ed. 2011. Special feature [black-footed ferret]. *Journal of Mammalogy* 92:699–770.

Branvold, H. A., D. E. Biggins, and J. H. Wimsatt. 2003. Photoperiod manipulation to increase the productivity of black-footed ferrets (*Mustela nigripes*) and Siberian polecats (*M. eversmanii*). *Zoo Biology* 22:1–14.

Breck, S. W., D. E. Biggins, T. M. Livieri, M. R. Matchett, and V. Kopcso. 2006. Does predator management enhance survival of black-footed ferrets? In *Recovery of the black-footed ferret: progress and continuing challenges*, ed. J. E. Roelle, B. J. Miller, J. L. Godbey, and D. E. Biggins, 203–9. United States Geological Survey Scientific Investigations Report 2005-5293, Reston, VA.

Cahalane, V. H. 1954. Status of the black-footed ferret. *Journal of Mammalogy* 35:418–24.

Campbell, T. M., T. W. Clark III, L. Richardson, S. C. Forrest, and B. R. Houston. 1987. Food habits of Wyoming black-footed ferrets. *American Midland Naturalist* 117:208–10.

Carpenter, J. W., M. J. Appel, R. C. Erickson, and M. N. Novilla. 1976. Fatal vaccine-induced canine distemper infection of black-footed ferrets. *Journal of American Veterinary Medical Association* 169:961–64.

Carpenter, J. W., and C. N. Hillman. 1978. Husbandry, reproduction and veterinary care of captive ferrets. *American Association Zoo Veterinarians Annual Proceedings*, 36–47. Knoxville, TN.

Carr, A. 1986. Introduction: The black-footed ferret. *Great Basin Naturalist Memoirs* 8:1–7.

Casey, D. E, J. DuWaldt, and T. W. Clark. 1986. Annotated bibliography of the black-footed ferret. *Great Basin Naturalist Memoirs* 8:185–208.

Chipault, J. G., D. E. Biggins, J. K. Detling, D. H. Long, and R. M. Reich. 2012. Fine-scale habitat use of reintroduced black-footed ferrets on prairie dog colonies in New Mexico. *Western North American Naturalist* 72:216–27.

Choate, J. R., E. K. Boggess, and F. R. Hendersen. 1982. History of the black-footed ferret in Kansas. *Transactions of the Kansas Academy of Science* 85:121–32.

Clark, T. W. 1975. Some relationships between prairie dogs, black-footed ferrets, paleo-Indians, and ethnographically known tribes. *Plains Anthropologist* 20:71–74.

———. 1986. Technical introduction. *Great Basin Naturalist Memoirs* 8:8–10.

———. 1989. Conservation biology of the black-footed ferret, *Mustela nigripes*. *Wildlife Preservation Trust, Special Scientific Report* 3:1–175.

———. 1999. Black-footed ferret, *Mustela nigripes*. In *The Smithsonian book of North American mammals*, ed. D. E. Wilson and S. Ruff, 172–73. Washington, DC: Smithsonian Institution Press.

Clark, T. W., L. Richardson, S. C. Forrest, D. E. Casey, and T. M. Campbell III. 1986. Descriptive ethology and activity patterns of black-footed ferrets. *Great Basin Naturalist Memoirs* 8:115–34.

Cully, J. F. 1993. Plague, prairie dogs, and black-footed ferrets. In *Management of prairie dog complexes for the reintroduction of the black-footed ferret*, ed. J. L. Oldemeyer, D. E. Biggins and B. J. Miller, 38–48. Biological Report 13. Washington, DC: US Fish and Wildlife Service.

———. 1997. Growth and life-history changes in Gunnison's prairie dogs after a plague epizootic. *Journal of Mammalogy* 78:146–57

Eads, D. A. 2009. Evaluation and development of black-footed ferret resource selection models. MS thesis, University of Missouri, Columbia.

———. 2012. Notes on black-footed ferrets: Conata Basin, South Dakota, 2007–2009. *Western North American Naturalist* 72:191–95.

Eads, D. A., D. S. Jachowski, J. J. Millspaugh, and D. E. Biggins. 2011. Importance of lunar and temporal conditions for spotlight surveys of adult black-footed ferrets. *Western North American Naturalist* 72:179–90.

Ernst, A. E., A. L. Clark, and D. R. Gober. 2004. A habitat-based technique to allocate black-footed ferret recovery among jurisdictional entities. In *Recovery of the black-footed ferret: progress and continuing challenges*, ed. J. E. Roelle, B. J. Miller, J. L. Godbey, and D. E. Biggins, 89–95. United States Geological Survey Scientific Investigations Report 2005-5293, Reston, VA.

Findley, J. S., A. H. Harris, D. E. Wilson, and C. Jones. 1975. *Mammals of New Mexico*. Albuquerque: University of New Mexico Press.

Forrest, S. C., D. E. Biggins, L. Richardson, T. W. Clark, T. M. Campbell III, K. A. Fagerstone, and

E. T. Thorne. 1988. Population attributes for the black-footed ferret (*Mustela nigripes*) at Meeteetse, Wyoming, 1981–1985. *Journal of Mammalogy* 69:261–73.

Forrest, S. C., T. W. Clark, L. Richardson, and T. M. Campbell III. 1985. Black-footed ferret habitat: some management and reintroduction considerations. *Wyoming Bureau of Land Management Wildlife Technical Bulletin No. 2.*

Garell, D., A. Vargas, and P. Marinari. 1998. *Black-footed ferret species survival plan—annual report 1997.* Bethesda, MD: American Zoo and Aquarium Association.

Godbey, J. L., D. E. Biggins, and D. Garelle. 2004. Exposure of captive black-footed ferrets to plague and implications for species recovery. In *Recovery of the black-footed ferret: progress and continuing challenges*, ed. J. E. Roelle, B. J. Miller, J. L. Godbey, and D. E. Biggins, 233–237. United States Geological Survey Scientific Investigations Report 2005-5293, Reston, VA.

Gustkey, E. 1985. Black-footed animal, once believed extinct, has surfaced in Wyoming; curiosity-seekers head for Meeteetse: mysterious ferret has 'em guessing. *L.A. Times.* 17 March 1985.

Hanebury, L. R., and D. E. Biggins. 2005. History of searches for black-footed ferrets. In *Recovery of the black-footed ferret: progress and continuing challenges*, ed. J. E. Roelle, B. J. Miller, J. L. Godbey, and D. E. Biggins, 47–65. United States Geological Survey Scientific Investigations Report 2005-5293, Reston, VA.

Harrington, L. A., D. E. Biggins, and A. W. Alldredge. 2003. Basal metabolism of the black-footed ferret (*Mustela nigripes*) and Siberian polecat (*Mustela eversmannii*). *Journal of Mammalogy* 85:497–504.

———. 2005. Modeling black-footed ferret energetics: are southern release sites better? In *Recovery of the black-footed ferret: progress and continuing challenges*, ed. J. E. Roelle, B. J. Miller, J. L. Godbey, and D. E. Biggins, 286–88. United States Geological Survey Scientific Investigations Report 2005-5293, Reston, VA.

Henderson, F. R., P. F. Springer, and R. Adrian. 1969. The black-footed ferret in South Dakota. *South Dakota Department of Game, Fish, Parks Technical Bulletin* 4:1–37.

Hillman, C. N. 1968. Life history and ecology of the black-footed ferret in the wild. MS thesis, South Dakota State University, Brookings.

———. 1979. Prairie dog distribution in areas inhabited by black-footed ferrets. *American Midland Naturalist* 102:185–87.

Hillman, C. N., and T. W. Clark. 1980. *Mustela nigripes. Mammalian Species* 126:1–3.

Hoffmann, R. S., and D. L. Pattie. 1968. *A guide to Montana mammals.* Missoula: University of Montana Press.

Hoogland, J. L. 1995. *The black-tailed prairie dog: social life of a burrowing mammal.* Chicago: University of Chicago Press.

Hooper, E. T. 1941. Mammals of the lava fields and adjoining areas in Valencia County, New Mexico. *Miscellaneous Publications, Museum of Zoology, University of Michigan* 51:1–47.

Hubbard, J. P., and C. G. Schmitt. 1984. *The black-footed ferret in New Mexico.* Final report to Bureau of Land Management, Santa Fe, NM, Contract No. NM-910-CT1–7. Santa Fe: New Mexico Department of Game and Fish, Project FW-17-R.

Jachowski, D. J., and J. M. Lockhart. 2009. Reintroducing the black-footed ferret (*Mustela nigripes*) to the Great Plains of North America. *Small Carnivore Conservation* 41:58–64.

Jachowski, D. J., J. J. Millsbaugh, D. E. Biggins, T. M. Livieri, and M. R. Matchett. 2010. Home-range size and spatial organization of black-footed ferrets *Mustela nigripes* in South Dakota, USA. *Wildlife Biology* 16:1–11.

Jones, C., and C. G. Schmitt. 1997. Mammal species of concern in New Mexico. In *Life among the muses: papers in honor of James S. Findley*, ed. T. L. Yates, W.L. Gannon, and D. E. Wilson, 179–205. Albuquerque: University of New Mexico, Museum of Southwestern Biology.

Knowles C. J., J. D. Proctor, and S. C. Forrest. 2002. Black-tailed prairie dog abundance and distribution in the Great Plains based on historic and contemporary information. *Great Plains Research* 12:219–54.

Kotliar, N. B., B. J. Miller, R. P. Reading, and T. W. Clark. 2006. The prairie dog as a keystone species. In *Conservation of the black-tailed prairie dog, saving North America's western grasslands*, ed. J. L Hoogland, 53–64. Washington, DC: Island Press.

Linder, R. L., R. B. Dahlgren, and C. N. Hillman. 1972.

Black-footed ferret-prairie dog interrelationships. In *Symposium on rare and endangered wildlife of the southwestern United States*, 22–37. Santa Fe: New Mexico Department of Game and Fish.

Livieri, T. M, and E. M. Anderson. 2012. Black-footed ferret home ranges in Conata Basin, South Dakota. *Western North American Naturalist* 72:196–205.

Livieri, T. M., D. S. Licht, B. J. Moynahan, and P. D. Mcmillan. 2013. Prairie dog aboveground aggressive behavior towards black-footed ferrets. *American Midland Naturalist* 169:422–25.

Lockhart, J. M., E. T. Thorne, and D. R. Gober. 2006. A historical perspective on recovery of the black-footed ferret and the biological and political challenges affecting its future. In *Recovery of the black-footed ferret: progress and continuing challenges*, ed. J. E. Roelle, B. J. Miller, J. L. Godbey, and D. E. Biggins, 6–19. United States Geological Survey Scientific Investigations Report 2005-5293, Reston, VA.

Matchett, M. R., D. E. Biggins, V. Carlson, B. Powell, and T. Rockie. 2010. Enzootic plague reduces black-footed ferret (*Mustela nigripes*) survival in Montana. *Vector-Borne and Zoonotic Diseases* 10:27–35.

Merriam, C. H. 1902. The prairie dog of the Great Plains. In *USDA yearbook 1901*, 257–70. Washington, DC: US Government Printing Office.

Miller, B. J., S. H. Anderson, H. Stanley, M. W. DonCarlos, W. Michael, and T. E. Thorne. 1988. Biology of the endangered black-footed ferret and the role of captive propagation in its conservation. *Canadian Journal of Zoology* 66:765–73.

Miller, B., and R. P. Reading. 2012. Challenges to black-footed ferret recovery: protecting prairie dogs. *Western North American Naturalist* 72:228–40.

Miller, B., R. P. Reading, and S. Forrest. 1996. *Prairie night: black-footed ferrets and the recovery of endangered species*. Washington, DC: Smithsonian Institution Press.

Nelson, E. W. 1919. Annual report of Chief of Bureau of Biological Survey. In *USDA Annual Report, Department of Agriculture for year ended June 1919*, 275–98.

Neville, T. B., and K. Johnson. 2007. *Gunnison's prairie dog predictive range model and survey site selection for New Mexico*. Report to New Mexico Department of Game and Fish from Natural Heritage New Mexico, Publication. No. 07-GTR-320.

New Mexico Department of Game and Fish (NMDGF). 2016. *State Wildlife Action Plan for New Mexico*. Santa Fe: New Mexico Department of Game and Fish.

Oakes, C. L. 2000. History and consequence of keystone mammal eradication in the desert grasslands: the Arizona black-tailed prairie dog (*Cynomys ludovicianus arizonensis*). PhD dissertation, University of Texas at Austin.

O'Brien, S. J., J. S. Martenson, M. A. Eichelberger, E. T. Thorne, and F. Wright. 1989. Biochemical genetic variation and molecular systematics of the black-footed ferret. In *Conservation biology and the black-footed ferret*, ed. U. S. Seal, E. T. Thorne, M. A. Bogan, and S. H. Anderson, 21–48. New Haven, CT: Yale University Press.

Oldemeyer, J. L., D. E. Biggins, B. J. Miller, and R. Crete. 1993. *Management of prairie dog complexes for the reintroduction of the black-footed ferret*. Biological Report 13. Washington, DC: US Fish and Wildlife Service.

Owen, P. R., C. J. Bell, and E. M. Mead. 2000. Fossils, diet, and conservation of black-footed ferrets (*Mustela nigripes*). *Journal of Mammalogy* 81:422–33.

Pauli, J. N., R. M. Stephens, and S. H. Anderson. 2006. *White-tailed prairie dog (Cynomys leucurus): a technical conservation assessment*. USDA Forest Service, Rocky Mountain Region.

Paunovich, R., and S. C. Forrest. 1987. Activity of a wild black-footed ferret litter. *Prairie Naturalist* 19:159–62.

Powell, R. A., T. W. Clark, L. Richardson, and S. C. Forrest. 1985. Black-footed ferret (*Mustela nigripes*) energy expenditure and prey requirements. *Biological Conservation* 34:1–15.

Richardson, L., T. W. Clark, S. C. Forrest, and T. M. Campbell III. 1987. Winter ecology of black-footed ferrets (*Mustela nigripes*) at Meeteetsee, Wyoming. *American Midland Naturalist* 117:225–39.

Rocke, T. E., P. Nol, P. E. Marinari, J. S. Kreeger, S. R. Smith, G. P. Andrews, and A. W. Friedlander. 2006. Vaccination as a potential means to prevent plague in black-footed ferrets. In *Recovery of the black-footed ferret: progress and* continuing challenges, ed. J. E. Roelle, B. J. Miller, J. L. Godbey, and D. E. Biggins, 243–47. United States Geological Survey Scientific Investigations Report 2005-5293, Reston, VA.

Roelle, J. E., B. J. Miller, J. L. Godbey, and D. E. Biggins, eds. 2006. *Recovery of the black-footed ferret: progress and continuing challenges*. United States Geological Survey Scientific Investigations Report 2005-5293, Reston, VA.

Roth, J. D. 2019. Sylvatic plague management and prairie dogs—a meta-analysis. *Journal of Vector Ecology* 44:1–10.

Schmitt, G. 1982. Black-footed ferrets. *New Mexico Wildlife* 27(3):16–17.

Seery, D. B., D. E. Biggins, J. A. Montenieri, R. E. Enscore, D. T. Tanda, and K. L. Gage. 2003. Treatment of black-tailed prairie dog burrows with Deltamethrin to control fleas (Insecta: Siphonaptera) and plague. *Journal of Medical Entomology* 40:718–22.

Seglund, A. E., A. E. Ernst, and D. M. O'Neill. 2006. Gunnison's prairie dog conservation assessment. Unpublished final report to Western Association of Fish and Wildlife Agencies, Laramie, WY.

Seton E. T. 1929. *Lives of game animals. Vol. 2, Pt. 2*. Garden City, NJ: Doubleday and Doran.

Sheets, R. G., R. L. Linder, and R. B. Dahlgreen. 1972. Food habits of two litters of black-footed ferrets in South Dakota. *American Midland Naturalist* 87:249–51.

Shriver, R. V. 1965. Annual report, 1965 fiscal year, New Mexico district narrative. Albuquerque, NM: US Fish and Wildlife Service.

Svendsen, G. E. 2003. Weasels and black-footed ferret (*Mustela* species). In *Wild mammals of North America: biology, management, and conservation*, 2nd ed., ed. G. A. Feldhamer, B. C. Thompson, and J. A. Chapman, 650–61. Baltimore: Johns Hopkins University Press.

United States Fish and Wildlife Service (USFWS). 2008. Endangered and threatened wildlife and plants; 12-month finding on a petition to list the Gunnison's prairie dog as threatened or endangered. *Federal Register* 73(24):6660–84 (5 February 2008).

———. 2009. Endangered and threatened wildlife and plants; 12-month finding on a petition to list the black-tailed prairie dog as threatened or endangered. *Federal Register* 74(231):63343–66 (3 December 2009).

———. 2013a. Recovery plan for the black-footed ferret (*Mustelan igripes*). Denver: US Fish and Wildlife Service.

———. 2013b. Endangered and threatened wildlife and plants; 12 month finding on a petition to list the Gunnison's prairie dog as an endangered species. Federal Register 78(220):68660–85 (14 November 2013).

United States Forest Service. 2000. Summary of the black-footed ferret program: Conata Basin, Buffalo Gap National Grasslands, South Dakota. Interior, South Dakota: USDA, Forest Service.

Vargas, A., and S. H. Anderson. 1998. Ontogeny of black-footed ferret predatory behavior towards prairie dogs. *Canadian Journal of Zoology* 76:1696–704.

Virchow, D., and S. E. Hygnstrom. 2002. Distribution and abundance of black-tailed prairie dogs in the Great Plains: a historical perspective. *Great Plains Research* 12:197–218.

Williams, E. S., E. T. Thorne, D. R. Kwiatkowski, K. Lutz, and S. L. Anderson. 1992. Comparative vaginal cytology of the estrous cycle of black-footed ferrets (*Mustela nigripes*), Siberian polecats (*M. eversmanni*), and domestic ferrets (*M. putorius furo*). *Journal of Veterinary Diagnostic Investigation* 4:38–44.

Wisely, S. M. 2005. The genetic legacy of the black-footed ferret: past, present, and future. In *Recovery of the black-footed ferret: progress and continuing challenges*, ed. J. E. Roelle, B. J. Miller, J. L. Godbey, and D. F. Biggins, 37–43. United States Geological Survey Scientific Investigations Report 2005-5293, Reston, VA.

Wisely, S. M., J. J. Ososky, and S. W. Buskirk. 2002. Morphological changes to black-footed ferrets (*Mustela nigripes*) resulting from captivity. *Canadian Journal of Zoology* 80:1562–68.

Wisely, S. M., R. M. Santymire, T. M. Livieri, P. E. Marinari, J. S. Kreeger, D. E. Wildt, and J.

Howard. 2005. Environment influences morphology and development for in situ and ex situ populations of the black-footed ferret (*Mustela nigripes*). *Animal Conservation* 8:321–28.

Wood, S. L. 1986. The black-footed ferret. *Great Basin Naturalist Memoirs* 8:1–208.

Wozencraft, W. C. 2005. Order Carnivora. In *Mammal species of the world: a taxonomic and geographic reference*, Vol. 1, ed. D. E. Wilson and D. M. Reeder, 532–628. 3rd ed. Baltimore: Johns Hopkins University Press.

LONG-TAILED WEASEL (*NEOGALE FRENATA*)

Jennifer K. Frey

The long-tailed weasel (*Neogale frenata*) is New Mexico's most familiar and widely distributed member of the weasel family. It is North America's quintessential "weasel." However, for such a recognizable and widespread species, there has been little research on its natural history, including in our state, where diet and the species' association with prairie dog (*Cynomys* spp.) colonies warrant further studies. The long-tailed weasel has a typical weasel body plan consisting of a long, narrow neck and body, short legs, and a long tail. The head is relatively small, narrow, and triangular-shaped. Small rounded ears are situated far back on the sides of the head, which enhances the overall triangular appearance of the head. The tail is long in comparison with that in most other weasels, typically averaging just over half of the length of the head and body (53–70%, mostly around 60%) and about four times the length of the hind foot (Hall 1951). Further, though the tail is densely furred, it is not bushy and so appears rather thin. The tail always bears a black tip, which is thought to confuse avian predators as the weasel flees twisting and turning (Powell 1982). Males are substantially larger than females, though the degree of sexual dimorphism varies by subspecies, with the larger subspecies displaying greater differences between the sexes (Hall 1951). Males may weigh up to twice as much as females, and the skull of the male may be more than 80% heavier than that of the female. Much of the difference in the mass

of the skull is because the skull in females is relatively narrower (Hall 1951). The overall smaller body size and narrower skull in females is an adaptation that allows them to hunt rodents more efficiently in their burrows and hiding places (Hall 1951).

In New Mexico, the long-tailed weasel exhibits considerable geographic variation in both body size and pelage color patterns. Much of this variation is exemplified in the three named subspecies of long-tailed weasels that occur in New Mexico: *N. f. nevadensis* from the northern mountains and adjacent areas, *N. f. arizonensis* from the Mogollon Mountains, and *N. f. neomexicana* from suitable environments throughout the remainder of the state, mostly at lower elevations. Hall's (1951) distribution map included a portion of northeastern New Mexico within the range of the Great Plains subspecies, *N. f. longicauda*. However, no specimens of *N. f. longicauda* have been reported from New Mexico, and Armstrong (1972) regarded specimens of long-tailed weasels from Baca County in southeastern Colorado as *N. f. neomexicana*. Thus, I do not consider *N. f. longicauda* as occurring in New Mexico (see Frey 2004).

The most widely distributed subspecies of the long-tailed weasel in New Mexico, *N. f. neomexicana*, is also the most distinctive in appearance. In essence, the "New

(*opposite page*) Photograph: © Sally King.

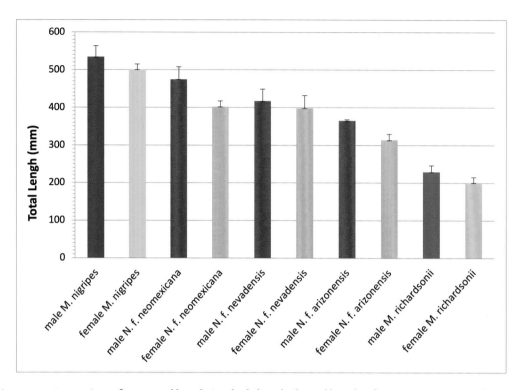

Figure 23.1. Comparison of mean total length (i.e., body length plus tail length) of museum specimens of some of New Mexico's weasels, the black-footed ferret (*Mustela nigripes*), New Mexico's three long-tailed weasel subspecies (*Neogale frenata neomexicana*, *N. f. nevadensis*, and *N. f. arizonensis*), and the New Mexico ermine (*M. richardsonii muricus*). Results are expressed as mean total length values ± standard deviations. Note that within a species males are larger than females, but that the size range of males of one species may overlap the size range of females of the next larger form.

Photo 23.1. An alert long-tailed weasel, *Neogale frenata neomexicana*, in winter pelage on 17 November 2011 in the valley of the Pecos River at Bitter Lake National Wildlife Refuge, Chaves County. Note the extensive white facial markings and very pale underparts, which seem to be typical of most *N. f. neomexicana* from the Pecos River region. Photograph: © Clifford G. Powell.

Mexico" subspecies is large, pale, and has distinct white facial markings, while the other two subspecies are relatively small, dark, and have reduced or absent white facial markings. Total length of male *N. f. neomexicana* averages 474 mm (18.7 in), while females average 402 mm (15.8 in). The dorsal body pelage color is light yellowish-brown and the undersides are paler yellowish or orangish. Animals from the valley of the Pecos River may be especially pale with the undersides near whitish. The pale ventral coloration extends over the tops of the forefeet and over the tops of the front portion of the hind feet. The white markings of *N. f. neomexicana*, which are confluent across the face, are more extensive than in any other subspecies found in the United States. (Photo 23.1). The white extends from the base of the ears, over the eyes, and onto the bridge of the nose, with at least some connection of white across this entire area. The white pattern contrasts sharply with a dark brown to blackish pelage (darker than the rest of the dorsal coloration) found around the eyes, over the nose, and on the top of the head. Because of this contrasting light and dark facial pattern, weasels with this pelage pattern are sometimes referred to as "masked weasels" or "bridled weasels" (Bailey 1931). Masking is a character that has been used to define subspecies of long-tailed weasels, with masked forms occurring in Florida and southwestern North America south to Nicaragua; unmasked forms occur elsewhere in North America, and in Central and South America (Hall 1951). The white facial markings of *N. f. neomexicana* seem to be most extensive in individuals from the Pecos River Valley (Photo 23.1), while in individuals from the Rio Grande Valley and the central part of the state the white pattern usually appears as three blocks or patches, two over the eyes and one on the nose, slightly separated or barely touching. *N. f. neomexicana* does not turn white in the winter.

In comparison with *N. f. neomexicana*, the two montane subspecies, *N. f. nevadensis* and *N. f. arizonensis*, are relatively small, dark, and have reduced or absent white facial markings (i.e., they appear "unmasked"). Of the two, the "Nevada" subspecies is larger and it overlaps the size range of female *N. f. neomexicana* (Fig. 23.1). In comparison with *N. f. nevadensis*, the skull of *N. f. arizonensis* is smaller, has less prominent ridges, and has more inflated auditory bullae (Hall 1946; Hoffmeister 1986). *N. f. arizonensis* may be half the mass as *N. f. nevadensis* (Mearns 1891), and it is as small as some forms of the American ermine (*M. richardsonii*; Hall 1951). Due to similarity in overall body morphology and life history, I have referred to *N. f. arizonensis* as "ermine-like" (Frey and Calkins 2010). Hall (1951) thought that intergradation occurred among subspecies resulting in difficulty identifying individuals in areas of overlapping distributions. However, Findley et al. (1975)

Photos 23.2a and b. a) (*top*) Comparison of pelage patterns of larger-bodied long-tailed weasels (*Neogale frenata*) in New Mexico. The top specimen is a male unmasked *N. f. nevadensis* in summer pelage from ~2,200 m (7,200 ft) elevation near Ocate, Mora County. The bottom specimen is a female masked *N. f. neomexicana* in winter pelage from ~2,160 m (7,100 ft) elevation in Sandia Park, Bernalillo County. b) (*bottom*) Close-up of the facial pattern. Photographs: © Jennifer Frey.

disagreed with that assertion, and I concur that the specimens and available photographs of *N. frenata* from New Mexico seem to allow a rather clean segregation of subspecies.

Although *N. f. nevadensis* and *N. f. arizonensis* differ in body size and skull characteristics, they are similar in pelage color patterns. During summer the dorsal body color is a moderate brown that lacks the yellowish tone of *N. f. neomexicana*. As in *N. f. neomexicana*, the undersides of summer *N. f. nevadensis* and *N. f. arizonensis* are yellowish, buffy, or orangish. Typically, the summer pelage lacks any white spots on the face. However, some individuals, particularly those nearer the edges of the subspecies boundaries, may have a white

Photos 23.3a and b. a) *(top)* A typical diminutive "ermine-like" long-tailed weasel, *Neogale frenata arizonensis*, hunting among rocks at Crescent Lake, in the White Mountains in Apache County, Arizona, on 17 July 2009. b) *(bottom)* Same weasel in side view. Photographs: © Robert Shantz.

"frontal" spot located on the bridge of the nose between the eyes. Less commonly, some individuals also may have a pair of white spots on either side of the head and situated between the eyes and ears. However, unlike in *N. f. neomexicana*, these white patches are always isolated (i.e., they never merge together). During winter, *N. f. nevadensis* molts into a pelage that is all white, except for the black tip of the tail (Hall 1946, 1951). Until recently it was unknown if *N. f. arizonensis* turns white in winter. However, Bill van Pelt of the Arizona Game and Fish Department (pers. comm.) reported seeing a white *N. f. arizonensis* at Sunrise Ski Area on Mount Baldy in the White Mountains of Apache County, Arizona, less than 50 km (30 mi) west of the New Mexico border. In addition, Van Pelt provided information about a white *N. f. arizonensis* seen on 18 January 2005 at Arizona Snowbowl ski area on San Francisco Peaks near Flagstaff, Coconino County, Arizona. Thus, at least some individuals of *N. f. arizonensis* turn white in winter as does *N. f. nevadensis*.

Across the range of the long-tailed weasel, northern populations have white winter pelage, while southern populations do not. The term "ermine" is sometimes used by trappers and other lay people to refer to any white weasel, which has caused confusion because "ermine" is also the accepted common name of *M. richardsonii* (Wilson and Reeder 2005; see Chapter 21) prompting some researchers to refer to *M. richardsonii* and related forms such as *M. erminea*, as short-tailed weasel or stoat (King and Powell 2007). Molt into a white winter pelage is not a character that has been used to define subspecies in *N. frenata* (Hall 1951). Between the distribution of winter-white and winter-brown populations, Hall (1951) described an intermediate "zone" where both brown and white long-tailed weasels may be found during winter. Hall (1951) thought this zone was primarily defined by pelage pattern in males, but he noted that there also appeared to be a narrower zone embedded

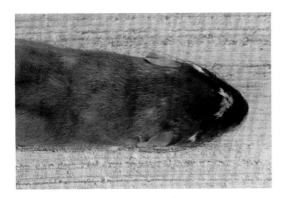

Photo 23.4 (*above*). Facial pelage pattern of a long-tailed weasel, *Neogale frenata nevadensis*, found dead on a road just north of Pecos, San Miguel County on 11 September 2008. This specimen of *N. f. nevadensis* is unusual because it has both a frontal spot and paired spots in front of the ears. Overall, however, the reduced white facial markings are consistent with *N. f. nevadensis*. Photograph: © James N. Stuart.

Photo 23.5 (*right*). A long-tailed weasel, *Neogale frenata nevadensis*, on 30 November 2013 in full white winter pelage from 6.6 km (4.1 mi) N, 2.7 km (1.7 mi) E of La Jara in Sandoval County, New Mexico. The elevation was ~2300 m (7600 ft). Photograph: © Rick Gonzales.

within the broader male zone, in which females also were either brown or white during winter. In the American Southwest, Hall (1951) mapped the boundary between winter-white and winter-brown pelage as along the northern borders of Arizona and New Mexico, with a small zone of overlap in northeastern New Mexico. Hall (1951) cited no specimens to justify the placement of the boundary line along those state borders. Rather, it appears the boundary line simply followed his interpretation of the southern range boundary of *N. f. nevadensis*. The implications of this map are that all long-tailed weasels in New Mexico should have a brown winter pelage, with perhaps the exception of some individuals along the Colorado state border and in northeastern New Mexico. However, specimen records disprove Hall's hypothesis that all long-tailed weasels south of the Colorado–New Mexico border

are brown in winter. For instance, Bailey (1931) reported a white *N. f. nevadensis* caught in Farmington, San Juan County. A winter specimen from Mount Taylor, Cibola County (=Valencia County on tag) that I refer to as *N. f. nevadensis*, has white pelage. In addition, white *N. f. arizonensis* have been observed, at least on the higher peaks in Arizona. In seeming conflict, Hoffmeister (1971) reported a brown long-tailed weasel (presumably *N. f. arizonensis*) found dead on 30 January 1935 in the (Harvey) mule barn at Grand Canyon Village within pinyon-juniper woodland zone on the South Rim, Coconino County, Arizona. However, examination of the specimen (Grand Canyon National Park catalog number 1468) shows that the mummified specimen was actually collected on 30 June 1935. Thus, there is no evidence for brown *N. f. arizonensis* or *N. f. nevadensis* in winter, though winter

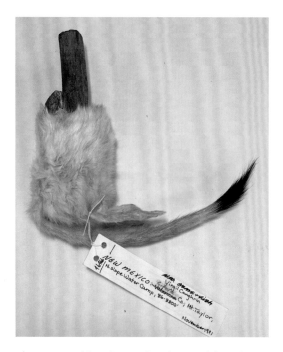

Photo 23.6. White winter pelage in a partial specimen (MSB:Mamm:46466) of long-tailed weasel (*Neogale frenata nevadensis*) collected in November 1981 on Mount Taylor, Cibola County, New Mexico. Photograph: © Jon Dunnum.

Powell 2007). In fall, the molt from brown to white begins on the belly and progresses upward along the sides. The white then appears all over the back in a speckled pattern before turning pure white (Hall 1951). The spring molt from white winter into brown summer pelage occurs in reverse, with the brown pelage first appearing along the midline of the back and then progressing down the sides, usually resulting in a sharp molt line (Hall 1951). Timing of molt is not precisely known for New Mexico weasels. A *N. f. nevadensis* caught near Twining (=Taos Ski Valley), Taos County, on 19 November was changing from brown to white (Bailey 1931). A *N. f. nevadensis*, photographed in Tesuque, Santa Fe County, on 9 April was beginning to molt into the summer pelage.

Photos 23.7a and b. a) (*top*) A long-tailed weasel (*Neogale frenata nevadensis*) from Tesuque, Santa Fe County on 9 April 2006. This weasel was starting to molt from its white winter pelage into its brown summer pelage; in spring, the brown color first starts to appear along the mid-dorsal line and then proceeds down the sides. The change in pelage color helps the weasel camouflage itself against would-be predators such as hawks. b) (*bottom*) Close-up view of the same weasel. Photographs: © Harrison Frazier.

records of these subspecies in Arizona and New Mexico are very sparse. Additional specimens are needed to more fully understand pelage patterns in long-tailed weasels in the Southwest. At this time, all that can be said for certain is that white winter weasels occur in at least some of the larger and higher mountain ranges south of the dividing line posited by Hall (1951) in the American Southwest. No winter specimens of *N. f. neomexicana* have showed any indication of turning white, and it seems likely that the white winter pelage is restricted to unmasked forms (*N. f. nevadensis* and *N. f. arizonensis*) in the Southwest.

Long-tailed weasels fall prey to other larger predators. The change in pelage color from brown to white in winter is thought to be an adaptation to help camouflage the weasels from predators such as hawks in snowy locations (King and

COMPARISONS WITH OTHER SPECIES

N. f. neomexicana may be confused with the black-footed ferret (*Mustela nigripes*; Figure 23.2; see Chapter 22). The two taxa are morphologically similar, and males of *N. f. neomexicana* are about the same size as black-footed ferrets (black-footed ferret total length = 457–552 mm [18.0–21.7 in]; mass = 530–1,300 g [18–46 oz]; Fitzgerald et al. 1994; Figure 23.1). To make matters worse, both taxa may be associated with prairie dog (*Cynomys*) towns, and both have prominent light and dark facial patterns and black-tipped tails. Upon closer observations, however, the two taxa can be distinguished. In *N. f. neomexicana* the facial pattern is one of white patches above the eyes on an otherwise dark face including the rostrum, while in the black-footed ferret there is a raccoon-like black mask across the eyes on an otherwise pale face including the rostrum (Figure 23.2). *N. f. neomexicana* also has a pale belly that sharply contrasts with a brown dorsum, while black-footed ferrets are about the same color of tan around the entire body. *N. f. neomexicana* has brown legs and pale toes, while black-footed ferrets have entirely black legs and feet.

In New Mexico, *N. f. nevadensis* may be confused with three other sympatric species mustelids, the American ermine, Pacific marten (*Martes caurina*), and American mink (*Neogale vison*). Of these, *N. f. nevadensis* is most likely to be confused with the ermine as both have similar overall morphology and pelage color patterns (see Chapter 21 for a more detailed discussion of the topic). To summarize, *N. f. nevadensis* is much larger than the ermine, averaging over 364 mm (14.3 in) compared to the majority of New Mexico ermines which are <246 mm (9.7 in). In addition, the tail is proportionally longer in *N. f. nevadensis* averaging >60% of the head and body length, in comparison to <45% of the head and body length in the ermine. *N. frenata* has a tail about four times the length of the hind foot, while *M. richardsonii* has a tail about two times the length of the hind

Figure 23.2. Comparison of the facial pattern in the "bridled" or "masked" long-tailed weasel (*Neogale frenata*; top) and the black-footed ferret (*Mustela nigripes*; bottom). Figure reproduced from Merriam (1896).

foot. Finally, the summer pelage of *N. f. nevadensis* tends to have a yellowish or buffy wash to the venter, while in the ermine the venter is usually (but not always) purer white. *N. f. nevadensis* turns white (at least some individuals do) in winter while the marten and mink do not. Large *N. f. nevadensis* may approach the size of martens and minks, but most martens and minks will nonetheless be larger (marten: total length =460–750 mm [18 to 29 in], mass =500–1,200 g [18 to 42 oz]; mink: total length =491–720 mm [19 to 28 in], mass = 525–780 g [18.5 to 27.5 oz]; Fitzgerald et al. 1994). Further, martens differ from *N. f.*

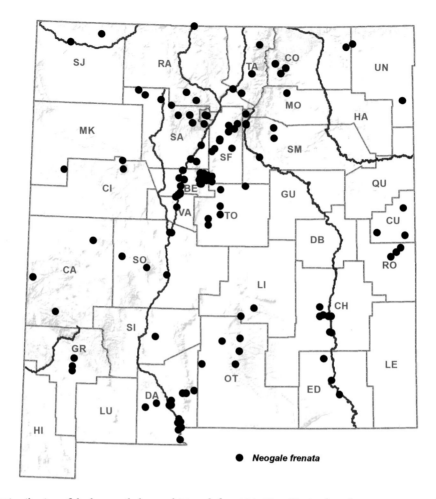

Map 23.1. Distribution of the long-tailed weasel (*Neogale frenata*) in New Mexico based on museum specimen records.

nevadensis in having a bushy tail, in being brown all over with a yellowish throat and chest patch (as opposed to having a whitish or yellowish belly that contrasts with a brown dorsum as in *N. f. nevadensis*), and in having larger prominent cat-like ears (as opposed to having smaller rounded ears as in *N. f. nevadensis*). Minks differ from *N. f. nevadensis* in being nearly uniformly dark brown or blackish, occasionally with small white patches on the throat or belly (as opposed to having a whitish or yellowish belly that contrasts with a brown dorsum as in *N. f. nevadensis*). In addition, minks have small, inconspicuous ears

that are nearly concealed by pelage. In New Mexico, the range of *N. f. arizonensis* does not overlap with any other species of mustelids that might pose identification problems.

SYSTEMATICS

Traditionally, it was thought that the long-tailed weasel was closely related to North America's other two common weasels, the American ermine (*M. richardsonii*) and the least weasel (*M. nivalis*), in addition to the mountain weasel (*M. altaica*) of Asia (King and Powell 2007). Recent genetic studies reveal instead that the long-tailed weasel is

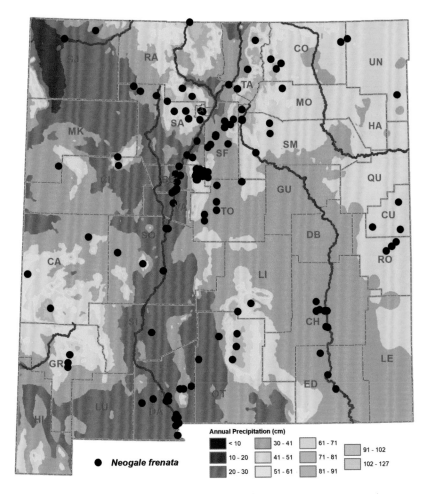

Map 23.2. Long-tailed weasel specimen records and mean annual precipitation in New Mexico, 1981–2010. Note the absence of records from many of the most arid regions (e.g., <250 mm [=10 in] per year), except in the riparian zone along some large rivers. Adapted from PRISM Climate Group 2019.

actually part of an endemic New World lineage, and that its closest relatives are the American mink and two poorly known species occurring in South America, the tropical weasel (*N. africana*) and the Columbian weasel (*N. felipei*; Harding and Smith 2009). The New World weasels were subsequently referred to the genus *Neogale* (Patterson et al. 2021). Harding and Dragoo (2012) posited that long-tailed weasels have their origin in tropical Mexico and have been present in temperate North America for ca. 2 million years. In contrast, ermines and least weasels, belonging to the genus *Mustela*, represent relatively recent

arrivals (ca. <0.7 million years), having colonized North America from Eurasia following one or more crossings of the Bering land bridge during glacial periods (e.g., Dawson et al. 2014).

DISTRIBUTION

The long-tailed weasel is purported to have the largest geographic range of any weasel in the Western Hemisphere (Sheffield and Thomas 1997). That range extends from the southern edge of the boreal forest zone in southern Canada southward across North and Central America and into northern and western South America to

Photo 23.8. A *Neogale frenata nevadensis* on 3 April 2007 in summer pelage from Maxwell National Wildlife Refuge in Colfax County. Although *N. f. nevadensis* is usually found in the mountains, this record verifies that its distribution extends out onto the adjacent plains in at least some areas. Photograph: © Patty Hoban.

northern Bolivia (Sheffield and Thomas 1997). In North America, however, aridity appears to limit the range of the long-tailed weasel in the West and hence occurrence records are lacking from the majority of the Baja California peninsula (but see Ralston and Clark 1971 and Ruiz-Campos et al. 2009) and the Mojave, Sonoran, and much of the Great Basin deserts, including southeastern California, much of Arizona, and portions of northwestern New Mexico (Huey 1960; Hall 1981).

Broadly speaking, the long-tailed weasel appears to occur, at least sporadically, throughout much of New Mexico (Map 23.1). It is found across all elevations within the state, from valley bottoms of major rivers to alpine environments above the treeline (Bailey 1913; Quick 1951; Findley et al. 1975). In New Mexico's most arid regions, however, it may only occur in association with riparian zones or well-developed grasslands where plant productivity is high enough to produce abundant prey (Map 23.2). For instance, in the Great Basin Desert region in the northwestern part of the state, there are no records of *N. frenata* from the arid desert scrub flats of the San Luis Valley of northwestern Taos County and adjacent Colorado, while in the San Juan Basin (San Juan and northern McKinley counties) the only records of *N. frenata* are from near major rivers including the San Juan and Animas rivers. Similarly, in the most arid portions of the Chihuahuan Desert region, such as Doña Ana County, the majority of records of *N. frenata* are from either the Rio Grande Valley or higher-elevation grasslands in the mountains.

The distributions of the three subspecies of long-tailed weasels in the state are not precisely determined. *N. f. nevadensis* has a broad distribution throughout the Great Basin, Sierra Nevada, and much of the Rocky Mountains. Its range extends into northern New Mexico, including the Sangre de Cristo Mountains, San Luis Valley, San Juan Mountains, Chuska Mountains, and the floodplain of the San Juan River and its tributaries (Hall 1951). I also regard specimens from the Jemez Mountains and Mount Taylor as referable to *N. f. nevadensis*. In the Sangre de Cristo Mountains, *N. f. nevadensis* is known from as far south as Ribera, San Miguel County, which is within the pinyon-juniper woodland zone along the Pecos River (Bailey 1931; Hall 1951). It also occurs on the eastern plains adjacent to the Sangre de Cristo Mountains, though range limits with *N. f. neomexicana* are uncertain (Photo 23.8).

N. f. arizonensis is endemic to the American Southwest, where it is known to occur in Arizona across the Mogollon Plateau from the Kaibab Plateau on the North Rim of the Grand Canyon and southeastward through the White Mountains (Hoffmeister 1986). Hall (1951, 1981) referred a specimen from ~2,600 m (8,500 ft) elevation on Willow Creek near the northern edge of the Gila

Wilderness Area in the Mogollon Mountains, Catron County, to *N. f. arizonensis*, and I concur with that designation. Thus, although the subspecies' exact range remains uncertain, it likely includes the adjacent outlying high peaks of the Mogollon Plateau such as the Black Range and Tularosa Mountains. However, whether specimens from the high peaks of the Magdalena and San Mateo mountains can also be assigned to *N. f. arizonensis* has yet to be determined.

Lastly, *N. f. neomexicana* is known from the lower elevations of northern New Mexico and southeastern Kansas southward through western Texas, New Mexico, and southeastern Arizona to southern Durango, Mexico. It occurs throughout New Mexico exclusive of the ranges of the other two subspecies and exclusive of regions that are too arid to support long-tailed weasels. In Arizona, Hoffmeister (1986) mapped a significant gap in the distribution of long-tailed weasels between the Mogollon Plateau, which is occupied by *N. f. arizonensis*, and the Chuska Mountains, which is occupied by *N. f. nevadensis*. The only exception was a single specimen from ~10 km (6 mi) west of Oraibi in Navajo County, Arizona (Hoffmeister 1966). In the original description of the specimen, Hoffmeister (1966) noted that characteristics of the pelage were inconsistent with both *N. f. nevadensis* and *N. f. arizonensis*. It is interesting that Hoffmeister (1986) later referred that same specimen to *N. f. arizonensis* with no justification. I consider the specimen from near Oraibi as referable to *N. f. neomexicana* on the basis of possessing pronounced white facial markings that are confluent across the face, in addition to orangish underparts, a short black tip on the tail, and pale tops of forefeet. Thus, the range of *N. f. neomexicana* extends westward from the Rio Grande Valley between the ranges of *N. f. nevadensis* to the north and *N. f. arizonensis* to the south into the watershed of the Little Colorado River (i.e., in New Mexico this would include Cibola and northern Catron counties). The range of *N. f. neomexicana* also wraps south of the range

Photo 23.9. A juvenile long-tailed weasel, *Neogale frenata neomexicana*, in winter pelage. It was raised by The Wildlife Center located in Española after having been caught by a dog on 27 October 2010 in Eldorado at Santa Fe, Santa Fe County. At the time of its rescue, the juvenile weasel weighed a mere 68 g (2.4 oz), but it then grew to 189 g (~7 oz) before its release in early December. As consistent with the *neomexicana* subspecies, the pelage showed no sign of turning white. Photograph: © Alissa Diane Mundt.

of *N. f. arizonensis* through the Bootheel region of New Mexico and into the Chihuahuan Desert grasslands and Madrean mountains of southeastern Arizona (Hoffmeister 1986). The range of *N. f. neomexicana* also extends northward to the plains between the Jemez and Sangre de Cristo mountains, as revealed by one individual found at Eldorado at Santa Fe, Santa Fe County, which is located in juniper savanna at the southwestern base of the Sangre de Cristo Mountains (Photo 23.9).

A lingering question concerns the taxonomic identity of long-tailed weasels occurring at high elevations in some of the larger mountain ranges south of the Jemez and Sangre de Cristo mountains and north and east of the Mogollon Mountains. Ivey (1957) described a small, dark, unmasked weasel that he considered to be an ermine in a forest dominated by white fir (*Abies concolor*) and Douglas-fir (*Pseudotsuga menziesii*) at ~2,900 m (9,500 ft) elevation in the Sandia Mountains, Bernalillo County. Based on several lines of evidence, I consider this record to represent an "ermine-like" long-tailed weasel resembling *N. f. nevadensis* (Frey et al. 2007; Frey and Calkins 2010, 2014). Thus, it is possible that other larger and higher mountain ranges, such as the Sacramento Mountains, also harbor ermine-like *N. frenata*. The ecological separation between these high-elevation forms and *N. f. neomexicana* is not known, though it is probably near the lower edge of the ponderosa pine forest zone. Specimens clearly referable to the larger, paler masked *N. f. neomexicana* have been found in pinyon-juniper woodland in the foothills, such as at Sandia Park. Additional specimens from isolated mountain ranges, particularly from high elevations, are needed for taxonomic evaluation.

Based wholly on anecdotal information, Hall (1951) speculated that the distribution of long-tailed weasels is limited where free drinking water is absent. Unfortunately, this has been taken as fact by subsequent researchers, which may have led to unfounded conclusions. As evidence of his theory, Hall cited two facts: 1) All four individuals trapped by him in Berkeley, California, were caught within 3 m (10 ft) from water. And 2) in the arid Pahranagat Valley in southern Nevada, long-tailed weasels only preyed on riparian rodents, primarily voles, but did not prey on kangaroo rats and other rodents found in adjacent desert scrub. Alternative explanations for these findings include 1) association of weasels with small mammal prey, which may be most abundant in riparian areas (Frey 2018); 2) avoidance of desert scrub because of heightened predation risk due to lack of concealment cover; and 3) saltatory locomotion of kangaroo rats foils a weasel's ability to capture and kill these rodents. Weasels are subject to intensive predation pressure by raptors (King and Powell 2007; see also under "Interspecific Interactions," below). I don't dispute that long-tailed weasels will drink water when available and that in the most arid regions long-tailed weasels are usually associated with riparian areas. However, I do not think that this is because of the free water, but rather due to the dense herbaceous vegetation found in such places and producing both abundant prey and cover (Frey 2018). In New Mexico, many specimens of long-tailed weasel have come from places far from any water, such as prairie dog towns on the Llano Estacado or White Sands. Long-tailed weasels in these areas may be able to derive enough water from their prey as do black-footed ferrets (Vargas and Anderson 1998; see Chapter 22) and many other arid-adapted carnivores. Chew (1965) estimated water intake acquired from prey in long-tailed weasels as 0.23 ml per gram per day, which vastly exceeded the amount estimated for both domestic cats (*Felis catus*; 0.08 ml/g/day) and ringtails (*Bassariscus astutus*; 0.07 ml/g/day). They considered the values for all three species as adequate for maintaining water balance. Thus, I do not consider free drinking water a limiting factor in the distribution of long-tailed weasels.

HABITAT ASSOCIATIONS

Habitat associations of the long-tailed weasel have not been studied in the American Southwest. Long-tailed weasels are usually considered habitat generalists in the sense that they occur in a wide range of different biotic communities (Simms 1979; King and Powell 2007). In New Mexico, *N. frenata* occurs virtually statewide and in such different environments as low-elevation semi-desert grasslands, marshes, and talus near the peaks of the highest mountains. However, actual habitat selection within this wide range of land cover is narrower. As with many other carnivores, an understanding of long-tailed weasel habitat associations really has more to do with recognizing the needs of the species' prey. Mustelids are strict carnivores and have relatively high metabolisms as a result of their slender body plan and active hunting style (Brown and Lasiewski 1972). Consequently, one of the most important factors governing the distribution

Photo 23.10 (*right*). Long-tailed weasel (*Neogale frenata nevadensis*) in coniferous forest at Trout Lakes in Rio Arriba County. Photograph: © James N. Stuart.

Photo 23.11 (*below*). A long-tailed weasel, *Neogale frenata neomexicana*, in the Chihuahuan Desert in Hembrillo Canyon in the San Andres Mountains, Sierra County, in August 2007. Photograph: © Greg Silsby.

and abundance of long-tailed weasels is an abundance of suitable prey (Hajduk 2011). For instance, a study in Indiana found that long-tailed weasels were more dependent on the presence of small and medium-sized prey than were other carnivores (Gehring and Swihart 2003). Thus, while the overall geographic range of the long-tailed weasel is essentially statewide, the local distribution of weasels may be spotty and limited to relatively productive habitats that afford concealment cover and food for abundant populations of small mammal prey. Based on assumed densities across broad vegetation types, the New Mexico Department of Game and Fish estimated there were 65,070–108,450 long-tailed weasels in the state (Thompson et al. 1992, 1996), but these numbers are likely grossly overestimated given the narrower habitat requirements of weasels within broader vegetation types.

Further, because long-tailed weasels are small enough to become prey themselves to other predators such as hawks, owls, and canids (see under "Interspecific Interactions," below), preferred habitats also provide adequate concealment cover for weasels from other predators (Powell 1973; Hajduk 2011). In Illinois, for example, long-tailed weasels selected home ranges with high ground cover, and core use areas with high vertical cover (Richter 2005). In Indiana, Gehring and Swihart (2003) found stronger habitat associations at the microhabitat scale, and reported that long-tailed weasels were associated with complex habitats (patches and corridors) typified by high ground cover and low canopy cover. In New Mexico, tall, dense ground cover in long-tailed weasel habitat usually consists of herbaceous plants, but also includes rodent burrows, log piles, and jumbled rocks (Photos 23.10 and 23.11). For instance, Bailey (1931) reported *N. f. neomexicana* living in pocket gopher burrows in irrigated alfalfa fields in the valleys of the Rio Grande and Gila River. In the White Mountains in adjacent Arizona, *N. f. arizonensis* is not uncommonly caught in the tall riparian sedges and grasses within the broad meadows along streams (pers. obs.). In the high Sangre de Cristo Mountains, *N. f. nevadensis* was often observed running through rockslides causing panic among the collared pikas (*Ochotona princeps*; Bailey 1931).

Due to differences in distribution and body size (and hence prey selection), habitat selection differs between the large masked *N. f. neomexicana* and the smaller non-masked *N. f. nevadensis* and *N. f. arizonensis*. Small body size in weasels is thought to be an adaptation that allows them to specialize on hunting voles (*Microtus*), which are often pursued in their burrows. Voles reach their highest abundance in riparian areas where tall, dense herbaceous ground cover occurs (Fagerstone and Ramey 1996; Frey 2018). Thus, it is not surprising that studies about habitat selection by smaller non-masked weasels in other regions of the Mountain West have confirmed the importance of riparian vegetation communities, especially in arid areas (e.g., Bailey 1931; Hall 1946; Egoscue 1957). For example, during a study comparing small mammal use of riparian

versus upland plant communities in the Cascade Mountains, Oregon, *N. frenata* was only captured in riparian areas, though captures were infrequent (Doyle 1990). At Jackson Hole, Wyoming, *N. f. nevadensis* was most common in a wetland community consisting of grasses, sedges, and willows (Negus and Findley 1959). In the high Sierra Nevada, *N. f. nevadensis* was less frequently documented in meadows than were ermines, which are smaller and hence more highly specialized on voles (Cain et al. 2003). At the same time, *N. f. nevadensis* was distributed throughout those meadows, whereas ermines were restricted to areas with willows near streams (Cain et al. 2006). While surveying for jumping mice (*Zapus luteus*) in the White Mountains, Arizona, I captured several *N. f. arizonensis* in herbaceous riparian areas that closely resembled those where I caught ermines in Colorado and northern New Mexico. However, other non-riparian vegetation types also are used by montane long-tailed weasels, especially those that provide ample cover afforded by herbaceous vegetation or rocks. For instance, at a study site located at ~2,900 m (9,500 ft) in elevation in the Rocky Mountains in Gunnison County, Colorado, *N. f. nevadensis* was most common on the edges of a Thurber's fescue (*Festuca thurberi*) meadow where chipmunks (*Neotamias*) and ground squirrels (*Callospermophilus*) were common (Findley and Negus 1953). At Jackson Hole, Wyoming, a *N. f. nevadensis* was observed in a sage-grass community, and one was taken in spruce-fir forest (Negus and Findley 1959). In New Mexico, *N. f. nevadensis* has frequently been reported from rockslides (e.g., Bailey 1931; Hill 1942; Findley et al. 1975).

The large, pale, masked *N. f. neomexicana* is primarily a denizen of grasslands and low-elevation riparian zones. Bailey (1931) considered *N. f. neomexicana* to be tightly linked to fertile river valley bottoms including those along the Rio Grande and Gila River, a habitat association he attributed to the local abundance of small rodents and lush vegetation cover. I know of no records of *N. f. neomexicana* from desert scrub vegetation types; all those from desert regions appear to be from embedded grasslands, woodlands, or riparian zones instead. In the Chihuahuan Desert region, for instance, specimens of *N. f. neomexicana* are available from extensive semi-desert grasslands of the Jornada Experimental Range, San Andres Pass, and White Sands National Park (Doña Ana and eastern Otero counties). On the Llano Estacado of eastern New Mexico, Best (1971) reported a specimen of *N. f. neomexicana* taken from a pasture dominated by grama grass (*Bouteloua* sp.), mesquite (*Prosopis* sp.) and yucca (*Yucca* sp.). Choate (1997) thought that *N. f. neomexicana* was most abundant on the Llano Estacado in Castro and Lamb counties, Texas, where there is a high concentration of playas that he thought might contribute to an abundance of prey in the form of small mammals and waterfowl. However, on the New Mexico portion of the Llano Estacado, I found the subspecies to be relatively common in open grasslands near prairie dog colonies. Castro and Lamb counties, Texas, also have relatively high densities of prairie dogs (Singhurst et al. 2010), and hence perhaps the abundance of long-tailed weasels in those counties was a result of the prairie dogs rather than (or in addition to) the playas.

LIFE HISTORY

Diet and Foraging

Few studies have attempted to quantify the diet of long-tailed weasels. Most of the available information is anecdotal or from other regions (e.g., Errington 1936). However, it is still possible to infer feeding habits for particular populations based on subspecies and habitats. Long-tailed weasels are strict carnivores. But, compared to other members of the weasel family that occur in New Mexico, the long-tailed weasel has a more generalized diet. After all, the species exhibits pronounced variation in morphology and it can

occupy a large range of land-cover types in the state ranging from arid grasslands to wetlands to cold talus slopes. Each vegetation type has its own suite of potential prey. Weasels are active at any time, and hence they can exploit either diurnal or nocturnal mammals (Brown and Lasiewski 1972). However, body size in weasels dictates the kind of prey that they are able to successfully attack and kill. Traditionally, it was thought that competition was a key driver of body size in weasels, though recent research points to the available prey base instead (Meiri et al. 2011). Furthermore, the act of predation is partially a learned behavior through experience with a particular type of prey, such that predators become more efficient with prey that is relatively more abundant. Small weasels tend to be adroit at pursuing and killing small mammals such as voles and mice within their tiny burrows and hiding places, while larger weasels are better suited to exploiting larger prey such as grounds squirrels (e.g., *Ictidomys*) and cottontails (*Sylvilagus*).

Very little is known about the food habits of the larger-bodied *N. f. neomexicana*. Since it occurs primarily in grasslands across most of the lower elevations of New Mexico, its diet likely consists mainly of upland small mammals. It seems likely that the diet primarily includes young cottontails (*Sylvilagus*), pocket gophers (Geomyidae), young prairie dogs (*Cynomys*), ground squirrels (especially *Ictidomys* and *Xerospermophilus*), cotton rats (*Sigmodon*), woodrats (*Neotoma*), and various mice (e.g., *Peromyscus*, *Mus*). For instance, the type specimen of *N. f. neomexicana* was collected in the valley of the Rio Grande, Doña Ana County, and had what appeared to be a northern grasshopper mouse (*Onychomys leucogaster*) in its stomach (Barber and Cockerell 1898). In a sandy bottom near El Paso, Texas, Bailey (1905) reported observing tracks of a weasel winding in and out of the burrows of kangaroo rats (*Dipodomys*). In Trans-Pecos Texas (Culberson County), a *N. f. neomexicana* collected from a rocky ridge was

Photo 23.12. A long-tailed weasel, *Neogale frenata nevadensis*, peering from the burrow of a Gunnison's prairie dog (*Cynomys gunnisoni*) near Taos in Taos County, New Mexico. Photograph: © Albert Ortega.

eating a white-throated woodrat (*Neotoma albigula*; Davis and Robertson 1944).

Long-tailed weasels occur in prairie dog colonies (e.g., Cary 1911; Findley 1987). On the Llano Estacado in eastern New Mexico, *N. f. neomexicana* seemed most common near colonies of black-tailed prairie dogs (*Cynomys ludovicianus*; pers. obs.). In the Janos-Nuevo Casas Grandes region of Chihuahua just south of the border with New Mexico, *N. f. neomexicana* were found in grasslands with prairie dogs, but were absent in grasslands that lacked prairie dogs (Ceballos et al. 1999). At Meeteetse, Wyoming, long-tailed weasels co-occur in prairie dog colonies with black-footed ferrets (*Mustela nigripes*), which are the quintessential predators of prairie dogs (Clark 1989). Both black-tailed prairie dogs and Utah prairie dogs (*C. parvidens*) are known to plug entrances to burrows that contain long-tailed weasels (Hoogland 2003). This implies that prairie dogs view long-tailed weasels as potential predators because they are known to plug burrows to prevent predation by black-footed ferrets

and snakes (Hoogland 2003). Given the adaptability of predation by long-tailed weasels, it seems likely that larger long-tailed weasels prey on prairie dogs, at least smaller individuals. This may be especially true for *N. f. neomexicana*, as the males of this subspecies overlap the size range of female black-footed ferrets, though Hoogland (2001) also reported that *N. f. nevadensis* killed >40 suckling Utah prairie dogs (*C. parvidens*). Long-tailed weasels, especially the smaller forms (e.g., *N. f. arizonensis*, female *N. f. nevadensis*), also may prey on the suite of other small mammals that also utilize prairie dog burrows, such as mice, voles, ground squirrels, and rabbits (Clark 1989). Further study on the relationship between long-tailed weasels, prairie dogs, and black-footed ferrets is needed (and see under "Interspecific Interactions," below).

In New Mexico, Bailey (1931) reported that the remains of chipmunks, pocket gophers, and ground squirrels were occasionally found in the stomachs of *N. f. nevadensis*. In meadows in the high Sierra Nevada, California, both *N. f. nevadensis* and ermines specialize on voles (primarily *Microtus montanus*) with population densities of both weasels tracking those of voles (Fitzgerald 1977). However, while ermines were tightly linked to riparian zones in the bottoms of meadows, long-tailed weasels ranged over larger areas of the meadows (Fitzgerald 1977). Yet prey remains in winter dens of long-tailed weasels still contained 98.0% voles; the only other prey remains consisted of two pocket gophers (*Thomomys monticola*), one deer mouse (*Peromyscus maniculatus*), and one shrew (*Sorex vagrans*; Fitzgerald 1977). Similar results were found in an analysis of hair in scats, 72% of which contained voles, 22% gophers, and 6% deer mice (Fitzgerald 1977). Thus, at least during winter, small-bodied long-tailed weasels may be as specialized and dependent on vole populations as are ermines, which are regarded a quintessential vole hunter (Frey and Calkins 2014; Chapter 21). The relatively small body size of long-tailed weasels that occur at high elevations (e.g., N. f. nevadensis and N. f. arizonensis) may be an adaptation for exploiting voles as prey. When hunting voles in the snow, long-tailed weasels often tunneled below the surface of the snow for several meters or yards. In contrast, ermines rarely dug tunnels (and for only short distances); rather, they accessed the subnivean (i.e., below the snow) runways of voles via natural holes formed around the branches of partially buried trees, perhaps due to their tiny body size (Fitzgerald 1977; see Chapter 21). As a result, long-tailed weasels were able to hunt voles in the middle of the meadow while ermines were more confined to locations with shrubs and trees (Fitzgerald 1977).

During warmer seasons, the diet of smaller long-tailed weasels broadens to include other species (Polderboer et al. 1941), such as chipmunks that hibernate in the winter. In the Rocky Mountains in Gunnison County, Colorado, Quick (1951) examined prey remains in scats of N. f. nevadensis collected in all seasons and in all vegetation zones above ~2,300 m (7,500 ft) elevation. Again, voles (Microtus) were the most frequently (52% of scats) detected food item (Quick 1951). The next most frequent food items were deer mice (Peromyscus maniculatus [now classified as P. sonoriensis]; 19.5 % of scats) and chipmunks (Neotamias; 18.2% of scats). Other frequent food items included northern pocket gopher (Thomomys talpoides), bushy-tailed woodrats (Neotoma cinerea), and adult paper wasps (Vespula). Rare identified food items (<5% of scats) included seeds of currants (Ribes), seeds and fruits of Vaccinium, larvae of moths (Lepidoptera), larvae of caddisflies (Trichoptera), grasshoppers (Tettigoniidae), an adult fly (Tabanidae), a flea (Siphonaptera), a teal (Anas), a brown creeper (Certhia), a bat (Chiroptera), mountain cottontail (Sylvilagus nuttallii), collared pika, Gunnison's prairie dog (Cynomys gunnisoni), golden-mantled ground squirrel (Callospermophilus lateralis), and a western jumping mouse (Zapus princeps). In the case of the teal, the author speculated that the bird might have been incapable of flight due to a late summer molt (Quick 1951).

In a study in Alberta, Canada, N. f. longicauda, which are only slightly larger than New Mexico N. f. nevadensis, exhibited a 10-year population cycle that was synchronized with changes in density of snowshoe hares (Lepus americanus; Keith and Cary 1991). The authors of the study concluded that the abundance of snowshoe hares had a direct impact on the reproduction and/or survival of weasels, though they were not able to discount the possibility that weasel populations declined instead as a result of predation by larger predators (e.g., red fox [Vulpes fulva]) when the hare population declined. Snowshoe hares and several species of cottontails occur in New Mexico (Frey 2004; Frey and Malaney 2006). However, long-tailed weasels may be less dependent on leporids in the Southern Rocky Mountains region as compared to the Northern Rocky Mountains, where the prey base is less diverse. Although seemingly rather timid, rabbits are actually rather formidable prey for long-tailed weasels, especially for smaller individuals. Quick (1951) placed a young cottontail in a cage with a N. f. nevadensis. It took seven minutes of intensive effort for the weasel to subdue the rabbit, raising doubt as to whether it would have been successful in the wild (Quick 1951). In another caged trial, four adult N. f. noveboracensis, which are about the same size as N. f. nevadensis, were paired with adult cottontails (Allen 1938). Two of the weasels were reticent to attack the rabbits. Ultimately, each weasel was killed by powerful kicks to the abdomen by the cottontail's hind feet when the weasel finally was stimulated to attack. The other two weasels killed the cottontails after 10–15 minutes of struggling after severing the neck muscles from the skull (Allen 1938). Allen (1938) concluded that an adult cottontail in the open would be able to defend itself from attack by a long-tailed weasel, though adults in their burrows and young may be more susceptible to predation. It seems unlikely that the small-bodied *N. f. arizonensis* could successfully prey on any leporid other than the very young.

Pocket gophers (*Geomyidae*) may be especially important prey to long-tailed weasels in western North America, where the ranges of the two nearly coincide (Proulx and Drescher 1993). Pocket gophers are moderately large rodents that live in subterranean burrows. Their larger body size makes them an energetically more profitable prey animal than more common species such as voles. Gophers eat the roots of herbaceous plants in their burrows, and hence they may be quite abundant in meadows and grasslands. Many accounts report long-tailed weasels living in or being trapped from pocket gopher burrows or being driven out of burrows by irrigation waters (e.g., Criddle 1930; Bailey 1931; Hall 1941; Hoffmeister 1971, 1986; Proulax 2005). For example, Hall (1946) reported that J. R. Alcorn caught 22 *N. f. nevadensis* in burrows of Botta's pocket gopher (*Thomomys bottae*) in Nevada. Vaughan (1961) reported that while collecting northern pocket gophers during a study in Colorado, he captured *N. f. nevadensis* at a rate of one weasel per 35 gophers during a year when gopher population levels were high, to one per 120 gophers when gopher numbers were lower, and one per 300 gophers when gopher numbers were lowest. Moreover, long-tailed weasels have been caught in the burrows of *Geomys* (Vaughan 1961), and likely also use the burrows of *Cratogeomys*, both of which occur on the plains of eastern, central, and southern New Mexico. Long-tailed weasels have been observed to use pocket gopher burrows for retreats, foraging routes, and as maternal nest sites (Hall 1946; Vaughan 1961). In addition, the burrows made by gophers are used by a host of other small vertebrates that could provide prey for long-tailed weasels. However, the actual predator-prey relationship between long-tailed weasels and gophers has been unknown until rather recently. Of nine scats of long-tailed weasels collected from mounds of northern pocket gophers in Alberta, six consisted of the remains of gophers, while three contained the remains of deer mice (*Peromyscus maniculatus*; Proulx and Cole 1998). Proulx (2005) followed long-tailed weasels during winter in fields inhabited by northern pocket gophers and concluded that the weasels may have a "cognitive map" of the distribution of gophers in their home range and actively prey on them.

Long-tailed weasels hunt small mammals by actively pursuing them. Using a loping gait, they cover a large area centered on the runways and hiding spots of small mammals. Prey is usually first detected via scent and hearing, but long-tailed weasels also rely on their excellent ability to see moving prey, which might stimulate an attack (Fagerstone 1999). Once a potential prey has been detected, they may slow to a walk and extend their head while searching and honing (Quick 1951). Their slender heads, torsos, and short legs are adapted for entering rodent burrows and other hiding places. When long-tailed weasels surprise their prey, they rush it and grasp it with mouth and paws, often wrapping their body around it (Allen 1938; Hall 1951). A bite to the back of the skull or a bite to the neck that severs the neck muscle usually succeeds in killing the prey (Allen 1938; Hall 1951). Often the weasel will at first only eat the eyes and brains of its kill, whereupon it may curl up to sleep on top of the carcass (Quick 1951)! Later, it will consume the remainder of the carcass, sometimes only leaving the incisors behind (Fitzgerald 1977). In some cases, especially for larger prey such as a rabbit, the weasel will first make an incision and lap the blood from the freshly killed prey—an activity that led to the misconception that weasels suck blood from their prey vampire-style (Allen 1938). Long-tailed weasels may kill excess prey, which they cache in burrows or shrubs for later use (Weeks 1991).

Photo 23.13. A long-tailed weasel, *Neogale frenata arizonensis*, hunting among jumbled rocks on 24 July 2008 along the East Fork of the Black River in Apache County, Arizona. Note that this individual has a few white spots on the face. Photograph: © Robert Shantz.

Photo 23.14. Long-tailed weasel in winter pelage carrying its Mexican woodrat (*Neotoma mexicana*) prey in Cimarron Canyon (Clear Creek, Colin Neblett State Wildlife Area in Colfax County) on 25 February 2021. Photograph: © Marty Peale, Brian Long and Jon Klingel. 2021 New Mexico Department of Game and Fish, Share with Wildlife, Mink Survey, northern New Mexico. US Fish and Wildlife Service Wildlife Restoration Grant W-208-R-1.

Although small mammals make up the bulk of the diet, long-tailed weasels also may make use of other foods. The most unusual reported food item was probably a ~13 cm (5 in) long larval giant salamander (*Dicamptodon*; Sturges 1955). More commonly, long-tailed weasels also prey on birds and their eggs. For instance, Smith et al. (2003) recorded predation attempts on nests of ruffed grouse (*Bonasa umbellus*) by *N. f. noveboracensis*, which is similar in size to *N. f. nevadensis*. On three occasions a long-tailed weasel attacked adult females sitting on nests. In each instance the weasel jumped down toward the bird's head from a higher perch (downed log or root-wad), in one instance landing on the bird's back. In each case the bird flushed and avoided predation. In two of those instances the weasel then investigated the eggs but left them unharmed. However, on a third occasion, the weasel removed

Photo 23.15. Long-tailed weasel hunting a bushy-tailed woodrat (*Neotoma cinerea*) on a vertical cliff face in Yellowstone National Park on 23 February 2013. The life and death struggle went on for about 20 minutes before the woodrat finally lost. Photograph: © Meg Sommers.

Photo 23.16. Long-tailed weasel kit photographed at the New Mexico Wildlife Center in Española after it was rescued from a cat in May 2010. Kits are born blind and with only a light covering of fur. Photograph: © Alisson Diane Mundt.

eggs from the nest and subsequently consumed or cached them nearby under roots and leaves. Following several unsuccessful attempts at consuming one egg, the weasel eventually succeeded by curling tightly into a ball while lying on its side and holding the egg with all four feet. It then maneuvered the egg until it could access the apex of the egg, which it punctured. Through the hole, the weasel then lapped the contents out of the egg. Similar predation by long-tailed weasels on eggs has been reported for the blue-winged teal (*Spatula discors*; Teer 1964) and the northern bobwhite (*Colinus virginianus*; Stoddard 1931). Consumed eggs of blue-winged teals exhibited paired tooth punctures, holes in the shell, and missing embryos (Teer 1964).

Reproduction and Social Behavior

There are scant data on reproduction in long-tailed weasels. Males are thought to become reproductively active in their second year of life and remain sexually active throughout the summer

(Hall 1951). In contrast, females may breed during their first summer, with most breeding occurring in mid-summer (Hall 1951). Ovulation is induced by coitus, but females remain in heat for several weeks if no breeding occurs. The fertilized eggs (blastocyst) are not implanted into the uterus until ca. 27 days before parturition, resulting in a very long gestation period averaging 279 days (Wright 1963). This common phenomenon in carnivorans is known as delayed implantation. It allows breeding and birth to be disconnected so that birth may occur during the optimal time of year. Implantation is thought to occur in February or March with most births occurring in April (Hoffmeister 1986). There is but one litter per year with litter sizes averaging four to five but ranging from one to nine (Sheffield and Thomas 1997). Usually the female cares for the young alone, though there have been some reports of males consorting with females tending young (Hall 1951). The family breaks up when the young are three to four months old and capable of fending for themselves (King and Powell 2007). A female *N. f. neomexicana* taken on the Llano Estacado was lactating on 14 May (Choate 1997) and so was a female *N. f. nevadensis* from Ribera, San Miguel County, on 2 July (Bailey 1931). Hall (1951) defined an "adult" as being more than 10 months of age and exhibiting no visible sutures between the nasals and maxillae of the skull.

Long-tailed weasels are usually solitary, except when mating or, in the case of adult females, when they have young. On 13 July 2006, while conducting mammal surveys at Sugarite Canyon State Park, Colfax County, Zachary Schwenke and I were disturbed by a sudden commotion in the canopy of a large Gambel oak (*Quercus gambelii*) near our camp. At first we thought that some squirrels were engaged in a mating chase. Rather, we found an adult and two nearly full grown *N. f. nevadensis* dashing and chasing each other around the branches and trunk of the tree (Frey and Schwenke 2012). Eventually the trio continued their play on

Photo 23.17. A long-tailed weasel, *Neogale frenata neomexicana*, crouched on the branch of an elm tree (*Ulmus* sp.) near Mountainair, Torrance County. Long-tailed weasels are primarily ground-dwellers, but they also climb trees. Note the gap that isolates the central white facial mark in this individual. Photograph: © Ken Steiner.

the ground beneath the tree where they chased each other and tumbled about in the grass. They continued unabashed and without regards to our curiosity for several minutes, even occasionally digging into gopher burrows. Others have also noted that long-tailed weasels are adroit at climbing trees (e.g., De Vos 1960) and often indifferent to the presence of humans (Armitage 1961).

Long-tailed weasels may have large home ranges relative to their body size, perhaps as a result of being (nearly) strict carnivores. For instance, in Indiana, long-tailed weasels moved over larger distances than omnivorous species like opossums (*Didelphis virginiana*) and raccoons (*Procyon lotor*) (Gehring and Swihart 2004). Further, home range size in long-tailed weasels was found to be larger for males but also for both sexes in areas with lower prey abundance (Gehring and Swihart 2004). Quick (1944) thought that long-tailed weasels in Michigan traveled an average

of 3.2 km (2 mi) per day while foraging. As well, during the breeding season, males may expand their home ranges or even become nomadic in order to find females, but females maintain relatively stable home ranges at all times (Gehring and Swihart 2004; Hajduk 2011).

Population sizes of long-tailed weasels tend to be more stable over time than observed for other smaller species of weasels such as the American ermine, probably due to their more diverse food habits (King and Powell 2007). However, populations of long-tailed weasels do fluctuate, and local populations may become extinct when prey numbers are low (King and Powell 2007). In such instances, re-establishment of the population requires recolonization. Although specific data are lacking for long-tailed weasels, weasels in general are capable of dispersing rather long distances (>20 km [~12 mi]), at least in high-quality habitats (King and Powell 2007). Most of these long-distance dispersals are made by young males during years with high prey populations resulting in births of many weasels; the young males are thought to be repeatedly driven out of areas by established adult males (King and Powell 2007). In contrast, young female weasels tend to stay within 5 to 6 km (~3–4 mi) of their mothers as adult males appear to be more tolerant of their presence (King and Powell 2007). Hence, the re-establishment of populations that go extinct is likely limited primarily by the ability of females to disperse, rather than males.

Interspecific Interactions

N. f. nevadensis and ermines co-occur in the mountains of northern New Mexico. Because both of these weasels are often dependent on voles, strong interspecific competition may occur in areas of overlap, though long-tailed weasels tend to be rarer at a given location due to their larger home ranges (e.g., Fitzgerald 1977). St-Pierre et al. (2006) studied competitive interactions between ermines and long-tailed weasels in Canada. When co-occurring with ermines, long-tailed weasels responded by traveling over larger distances and increasing the size of their home ranges, which resulted in higher mortality from predation but ultimately reduced overall competition with ermines. In contrast, ermines reduced use of their preferred habitat when it was used by long-tailed weasels. In these ways, both species of weasels were able to coexist during the course of the study. Although mathematical models have not been able to predict long-term coexistence of two species of weasels, temporary coexistence may occur as a result of higher reproductive rates in ermines due to lower energy requirements, different predation abilities (ermines are dependent on voles in all seasons whereas long-tailed weasels can utilize other, larger prey species), fluctuation of prey populations, and predation on weasels by other predators (Powell and Zielinski 1983). Fluctuations in prey populations can cause one of the species to go locally extinct, so persistence of both species may only happen when prey populations are high (Powell and Zielinski 1983). Over the long term, maintaining both species requires that individuals are able to recolonize the area if the population becomes temporarily extinct (Powell and Zielinski 1983). Thus, high-quality habitat that supports abundant and diverse small mammal populations, especially voles, is a requisite to maintain persistence of both species of weasel (Frey 2018).

Long-tailed weasels are subject to predation by other mammalian carnivores, large raptors, and even rattlesnakes (*Crotalus*; Hall 1981). Examples of confirmed predators of *N. frenata* that occur in New Mexico include the coyote (*Canis latrans*; Bailey 1931), red fox (*Vulpes fulva*; Errington 1935; Latham 1952), gray fox (*Urocyon cinereoargenteus*; Latham 1952), domestic dog (*Canis familiaris*; Barber and Cockerell 1898), domesticated cat (*Felis catus*; Hamilton 1933; Latham 1952), and raptors including the bald eagle (*Haliaeetus leucocephalus*; J.-L. Cartron, pers. comm.), northern goshawk (*Accipiter gentilis*; Smithers et al. 2005), red-tailed hawk (*Buteo jamaicensis*; Gatto et al. 2005), ferruginous hawk (*B. regalis*; Cartron et al. 2004, 2010), golden eagle (*Aquila chrysaetos*; Olendorff 1976), and great horned owl (*Bubo virginianus*; Knight and Jackman 1984). Foxes seem to have a predisposition for killing, but not eating, long-tailed weasels and thence depositing their carcasses about their dens (Errington 1935; Latham 1952). Latham (1952) expounded on this behavior and implicated foxes in limiting populations of long-tailed weasels in Pennsylvania. Like other mustelids, long-tailed weasels have anal scent glands that produce a pungent odor. When seriously alarmed, long-tailed weasels utter a chirping screech that may be accompanied by a "stink bomb" to thwart attackers (King and Powell 2007). Perhaps this strong scent accounts for the seeming unwillingness of foxes to eat weasels. During monitoring of raptor populations around

New Mexico, J.-L. Cartron (Cartron et al. 2004; pers. comm.) found the remains of two long-tailed weasels, one at a bald eagle nest site in Colfax County, the other in a ferruginous hawk nest in Torrance County. Both nests were situated near prairie dog colonies, and in both cases the resident pair seemed to be relying extensively on prairie dogs as part of their diet. It is unknown whether these two anecdotal findings are examples of incidental predation, defined as "the fortuitous capture of an unexpected prey item, the consumption of which does not change the predator's foraging behavior" (Vickery et al. 1992:281). They might also illustrate the risk to a long-tailed weasel foraging in a prey-rich, but also predator-rich, environment perhaps lacking in dense ground cover.

Unlike black-footed ferrets and prairie dogs, which are both highly susceptible to plague, long-tailed weasels appear to be resistant to the disease, though they can be infected by the bacterium (Abbot and Rocke 2012). Thus, this disease has not contributed to declines of the weasel. Rather, the broader diet and plague resistance of long-tailed weasels may allow them to better exploit prairie dog colonies today versus historically, prior to the introduction of plague and loss of black-footed ferrets from most of their historical range.

STATUS AND MANAGEMENT

The long-tailed weasel is legally recognized as a protected furbearer in New Mexico (New Mexico Statute 17-5-2). Consequently, a furbearer license is required to take or possess the species and fur dealers must have a license to buy or sell skins. During the 2021–2022 trapping season, the open season for long-tailed weasels was 1 November 2021 to 15 March 2022, with no bag limit. Despite the fact that there is a legal season for trapping or hunting long-tailed weasels, there is virtually no take. For instance, during the four trapping seasons from 2017–2021,

trappers reported taking a single long-tailed weasel and a single "ermine." Pelts of long-tailed weasels have little value in the fur industry (Svendsen 2003). In May 2019 the average price for an "ermine" pelt (i.e., any white weasel) was $2.04 (Fur Harvesters Auction Inc. 2019).

Populations of long-tailed weasels have long been thought to be declining in many regions of the United States and Canada (e.g., Proulx and Drescher 1993; Fagerstone 1999; Landholt and Genoways 2000). Lowe (1964) thought that long-tailed weasels were probably extinct at lower elevations (<4,000 ft) in Arizona. Fagerstone (1999) suggested that declines in Canada were due to agricultural conversion of native vegetation (especially to monocultures), loss of riparian and wetland areas, and control of ground squirrels and pocket gophers with pesticides. Landholt and Genoways (2000) speculated that the decline of long-tailed weasels in Nebraska may be the result of environmental contaminants such as mercury and halogenated hydrocarbons. Fagerstone (1999) noted that the species' low reproductive rate barely compensates for natural mortality, limiting the ability of long-tailed weasel populations to withstand direct and indirect human impacts. She (Fagerstone 1999) urged for efforts to protect the species' habitat in areas where it was likely declining, particularly riparian corridors. Most recently, Jachowski et al. (2021) evaluated rangewide population trends of *N. frenata*, *M. richardsonii*, and *M. nivalis*, which suggested that weasel populations have substantially declined over the last century and that reductions appear to be most severe in southern sites in the eastern United States. For *N. frenata* Jachowski et al. (2021) suggested that detections since 2000 have been average across Arizona and New Mexico, with better than average detection in the northern mountains (Jachowski et al. 2021). However, a benchmark of the year 2000 is likely too recent to capture substantial changes in the distribution and abundance of weasels in New Mexico as

Photo 23.18. A long-tailed weasel, *Neogale frenata neomexicana*, in winter pelage on 30 March 2008 along the Rio Grande at the Rio Grande Nature Center State Park in Albuquerque, Bernalillo County. Note the arrangement of white facial markings as three blocks that just touch. Long-tailed weasels have likely been impacted by habitat loss and degradation including the conversion of flood-plains to agriculture. Photograph: © James N. Stuart.

most dramatic habitat alterations occurred many decades prior to 2000, such as those leading to the possible extirpation of low-elevation populations in Arizona and perhaps also some regions of New Mexico.

In New Mexico, long-tailed weasels were likely never abundant. For instance, during four summers of field research at Bandelier National Monument, which would seem to have excellent habitat for *N. f. nevadensis*, Guthrie and Large (1980) reported seeing only two individuals. Although specific information about population trends is unknown, it seems likely that long-tailed weasels have declined in distribution and abundance in New Mexico through changes in habitat. In particular, large tracks of former semi-desert grassland in the southern and northwestern corner of the state have converted to unproductive desert scrub (Dick-Peddie 1993). These and other grasslands were once covered with the burrows of prairie dogs, banner-tailed kangaroo rats (*Dipodomys spectabilis*), and other rodents, but systematic poisoning campaigns caused reduction or elimination of these species from broad swaths of

former habitat (Oakes 2000). Broad river valleys have been converted to agriculture or have grown over into wastelands of exotic plants such as salt-cedar (*Tamarix*) that are comparatively lifeless (Dick-Peddie 1993). In the absence of abundant herbaceous vegetation, small mammal communities in these woody riparian communities are usually dominated by deer mice (*Peromyscus* spp.; Ellis et al. 1997), which do not constitute a major prey item for long-tailed weasels, due to their small body size. In the mountains, excessive livestock grazing has altered grassland, riparian, and other vegetation types and reduced cover vital for voles and other small mammals, to the extent that some of these species are now critically imperiled (Fagerstone and Ramey 1996; Frey and Malaney 2009; Frey 2017, 2018). Cut over coniferous forest have now grown so dense that they harbor few small mammals for want of ground cover and food that normally would be provided by a healthy herbaceous layer (Wampler et al. 2008). Climate change is exacerbating these losses of habitat (see Chapter 5). These vegetation changes are so vast and often so substantial that only a wholesale change in attitude toward the land with concerted restoration efforts might make a substantial difference to the distribution and abundance of long-tailed weasels and the other species with which they interact.

LITERATURE CITED

Abbott, R. C., and T. E. Rocke. 2012. *Plague*. US Geological Survey Circular 1372.

Allen, D. L. 1938. Notes on the killing technique of the New York weasel. *Journal of Mammalogy* 19:225–29.

Armitage, K. B. 1961. Curiosity behavior in some mustelids. *Journal of Mammalogy* 42:276–77.

Armstrong, D. M. 1972. *Distribution of mammals in Colorado*. Monograph of the Museum of Natural History, University of Kansas No. 3.

Bailey, V. 1905. *Biological survey of Texas*. North American Fauna 25. Washington, DC: US Department of Agriculture, Bureau of Biological Survey.

———. 1913. *Life zones and crop zones of New Mexico*. North American Fauna 35. US Department of Agriculture, Bureau of Biological Survey, Washington, DC.

———. 1931 (=1932). *Mammals of New Mexico*. North American Fauna 53. Washington, DC: US Department of Agriculture, Bureau of Biological Survey.

Barber, C. M., and T. D. A. Cockerell. 1898. A new weasel from New Mexico. *Proceedings of the Academy of Natural Sciences of Philadelphia* 50:188–89.

Best, T. L. 1971. Notes on the distribution and ecology of five eastern New Mexico mammals. *Southwestern Naturalist* 16:210–11.

Brown, J. H., and R. C. Lasiewski. 1972. Metabolism of weasels: the cost of being long and thin. *Ecology* 53:939–43.

Cain, J. W., III, M. M. Morrison, and H. L. Bombay. 2003. Predator activity and nest success of willow flycatchers and yellow warblers. *Journal of Wildlife Management* 67:600–10.

Cain, J. W., III, K. S. Smallwood, M. M. Morrison, and H. L. Loffland. 2006. Influence of mammal activity on nesting success of passerines. *Journal of Wildlife Management* 70:522–31.

Cartron, J.-L. E., P. J. Polechla Jr., and R. R. Cook. 2004. Prey of nesting ferruginous hawk in New Mexico. *Southwestern Naturalist* 49:270–76.

Cartron, J.-L. E., J. M. Ramakka, and H. R. Schwartz. 2010. Ferruginous hawk (*Buteo regalis*). In *Raptors of New Mexico*, ed. J.-L. E. Cartron, 337–57. Albuquerque: University of New Mexico Press.

Cary, M. 1911. *A biological survey of Colorado*. North American Fauna 33. Washington, DC: US Department of Agriculture, Bureau of Biological Survey.

Ceballos, G., J. Pacheco, and R. List. 1999. Influence of prairie dogs (*Cynomys ludovicianus*) on habitat heterogeneity and mammalian diversity in Mexico. *Journal of Arid Environments* 41:161–72.

Chew, R. M., 1965. Water metabolism of mammals. In *Physiological Mammalogy*, Vol. 2, ed. W. W. Mayer and R. G. Van Gelder, 43–178. New York: Academic Press.

Choate, L. L. 1997. The mammals of the Llano Estacado. *Special Publication, Museum of Texas Tech University* 40:1–240.

Clark, T. W. 1989. Conservation biology of the black-footed ferret *Mustela nigripes*. *Wildlife Preservation Trust, Special Scientific Report* 3:1–175.

Criddle, S. 1930. The prairie pocket gopher, *Thomomys talpoides rufescens*. *Journal of Mammalogy* 11:265–80.

Davis, W. B., and J. L. Robertson Jr. 1944. The mammals of Culberson County, Texas. *Journal of Mammalogy* 25:254–73.

Dawson, N. G., A. G. Hope, S. L. Talbot, and J. A. Cook. 2014. A multilocus evaluation of ermine (*Mustela erminea*) across the Holarctic, testing hypotheses of Pleistocene diversification in response to climate change. *Journal of Biogeography* 41:464–75.

De Vos, A. 1960. Long-tailed weasel climbing trees. *Journal of Mammlaogy* 41:520.

Dick-Peddie, W. A. 1993. *New Mexico vegetation: past, present, and future*. Albuquerque: University of New Mexico Press.

Doyle, A. T. 1990. Use of riparian and upland habitats

by small mammals. *Journal of Mammalogy* 71:14–23.

Egoscue, H. J. 1957. Notes on Utah weasels. *Journal of Mammalogy* 38:411–12.

Ellis, L. M., C. S. Crawford, and M. C. Molles. 1997. Rodent communities in native and exotic riparian vegetation in the middle Rio Grande Valley of central New Mexico. *Southwestern Naturalist* 42:13–19.

Errington, P. L. 1935. Food habits of mid-west foxes. *Journal of Mammalogy* 16:192–200.

———. 1936. Food habits of a weasel family. *Journal of Mammalogy* 17:406–7.

Fagerstone, K. A. 1999. Black-footed ferret, long-tailed weasel, shot-tailed weasel, and least weasel. *In Wild furbearer management and conservation in North America*, ed. M. Novak, J. A. Baker, M. E. Obbard, and B. Malloch, 549–73. Ontario: Ontario Fur Managers Federation. CD-ROM.

Fagerstone, K. A., and C. A. Ramey. 1996. Rodents and lagomorphs. In *Rangeland wildlife*, ed. P. R. Krausman, 83–132. Denver, CO: Society for Range Management.

Findley, J. S. 1987. *The natural history of New Mexico mammals*. Albuquerque: University of New Mexico Press.

Findley, J. S., A. H. Harris, D. E Wilson, and C. Jones. 1975. *Mammals of New Mexico*. Albuquerque: University of New Mexico Press.

Findley, J. S., and N. C. Negus. 1953. Notes on the mammals of the Gothic region, Gunnison County, Colorado. *Journal of Mammalogy* 34:235–39.

Fitzgerald, B. M. 1977. Weasel predation on a cyclic population of the montane vole (*Microtus montanus*) in California. *Journal of Animal Ecology* 46:367–97.

Fitzgerald, J. P., C. A. Meaney, and D. M. Armstrong. 1994. *Mammals of Colorado*. Niwot: Denver Museum of Natural History and University Press of Colorado.

Frey, J. K. 2004. Taxonomy and distribution of the mammals of New Mexico: an annotated checklist. *Occasional Papers, Museum of Texas Tech University* 240:1–32.

———. 2017. Landscape scale and microhabitat of the endangered New Mexico meadow jumping mouse in the White Mountains, Arizona. *Journal of Fish and Wildlife Management* 8:39–58.

———. 2018. Beavers, livestock, and riparian synergies: bringing small mammals into the picture. In *Riparian research and management: past, present, and future*, Vol. 1, ed. R. R. Johnson, S. W. Carothers, D. M. Finch, K. J. Kingsley, and J. T. Stanley, 85–101. Fort Collins, CO: US Forest Service, Rocky Mountain Research Station, Gen. Tech. Rep. RMRS-GTR-377.

Frey, J. K., and M. T. Calkins. 2010. *Status of the ermine* (Mustela erminea) *at its southern range limits, New Mexico*. Final Report submitted to New Mexico Department of Game and Fish, Share with Wildlife Program.

———. 2014. Snow cover and riparian habitat determine the distribution of the short-tailed weasel (*Mustela erminea*) at its southern range limits in arid western North America. *Mammalia* 78:45–56.

Frey, J. K., and J. L. Malaney. 2006. Snowshoe hare (*Lepus americanus*) and mountain cottontail (*Sylvilagus nuttallii*) biogeography at their southern range limit. *Journal of Mammalogy* 87:1175–82.

———. 2009. Decline of the meadow jumping mouse (*Zapus hudsonius luteus*) in two mountain ranges in New Mexico. *Southwestern Naturalist* 54:31–44.

Frey, J. K., and Z. J. Schwenke. 2012. Mammals of Sugarite Canyon State Park, Colfax County, New Mexico. *Occasional Papers, Museum of Texas Tech University* 311:1–24.

Frey, J. K., T. L. Yates, and M. A. Bogan. 2007. Mountaintop island age determines species richness of boreal mammals in the American Southwest. *Ecography* 30:231–40.

Fur Harvesters Auction Inc. 2016. Auction Results. 24 and 25 May 2019. https://www.furharvesters.com/results/2019/May/may19us.pdf.

Gatto, A. E., T. G. Grubb, and C. L. Chambers. 2005. Red-tailed hawk dietary overlap with northern goshawks on the Kaibab Plateau, Arizona. *Journal of Raptor Research* 39:439–44.

Gehring, T. M., and R. K. Swihart. 2004. Home range and movements of long-tailed weasels in landscape fragmented by agriculture. *Journal of Mammalogy* 85:79–86.

Guthrie, D. A., and N. Large. 1980. Mammals of Bandelier National Monument, New Mexico.

Claremont Colleges, Claremont, California. Unpublished report to National Park Service PX7029-7-0807, 22 pp.

Hajduk, L. I. 2011. Space use and habitat selection of long-tailed weasels (*Mustela frenata*) in southern Illinois. MS thesis, Northern Illinois University.

Hall, E. R. 1946. *Mammals of Nevada*. Berkeley: University of California Press.

———. 1951. American weasels. *University of Kansas Publications, Museum of Natural History* 4:1–466.

———. 1981. *The mammals of North America*. 2nd ed. Vol. 2. New York: John Wiley.

Hamilton, W. J., Jr. 1933. The weasels of New York. *American Midland Naturalist* 14:289–337.

Harding, L. E., and J. W. Dragoo. 2012. Out of the tropics: a phylogenetic history of the long-tailed weasel, *Mustela frenata*. *Journal of Mammalogy* 93:1178–94.

Harding, L. E., and F. A. Smith. 2009. *Mustela* or *Vison*? Evidence for the taxonomic status of the American mink and a distinct biogeographic radiation of American weasels. *Molecular Phylogenetics and Evolution* 52:632–42.

Hill, J. E. 1942. Notes on mammals of northeastern New Mexico. *Journal of Mammalogy* 23:75–82.

Hoffmeister, D. F. 1966. Records of northern Arizona mammals. *Plateau* 39:90–93.

———. 1971. *Mammals of the Grand Canyon*. Urbana: University of Illinois Press.

———. 1986. *Mammals of Arizona*. Tucson and Phoenix: University of Arizona Press and Arizona Game and Fish Department.

Hoogland, J. L. 2001. Black-tailed, Gunnison's, and Utah prairie dogs reproduce slowly. *Journal of Mammalogy* 82:917–27.

———. 2003. Black-tailed prairie dog (*Cynomys ludovicianus* and allies). In *Wild mammals of North America: biology, management, and economics*, 2nd ed., ed. G. A. Feldhamer, B. C. Thompson, and J. A. Chapman, 232–47. Baltimore: Johns Hopkins University Press.

Huey, L. M. 1960. It is possible: a correction. *Journal of Mammalogy* 41:278–79.

Ivey, R. D. 1957. Ecological notes on the mammals of Bernalillo County, New Mexico. *Journal of Mammalogy* 38:490–502.

Jachowski, D., R. Kays, A. Butler, A. M. Hoylman, and M. Gompper. 2021. Tracking the decline of weasels in North America. *PLOS One* 16(7):e0254387. https://doi.org/10.1371/journal.pone.0254387.

Keith, L. B., and J. R. Cary. 1991. Mustelid, squirrel, and porcupine population trends during a snowshoe hare cycle. *Journal of Mammalogy* 72:373–78.

King, C. M., and R. A. Powell. 2007. *The natural history of weasels and stoats: ecology, behavior, and management*. New York: Oxford University Press.

Knight, R. L., and R. E. Jackman. 1984. Food-niche relationships between great horned owls and common barn-owls in eastern Washington. *Auk* 101:175–79.

Landholt, L. M., and H. H. Genoways. 2000. Population trends in furbearers in Nebraska. *Transactions of the Nebraska Academy of Sciences* 26:97–110.

Latham, R. M. 1952. The fox as a factor in the control of weasel populations. *Journal of Wildlife Management* 16:516–517.

Lowe, C. H. 1964. *The vertebrates of Arizona*. Tucson: University of Arizona Press.

Mearns, E. A. 1891. Descriptions of a new species of weasel, and a new subspecies of the gray fox, from Arizona. *Bulletin of the American Museum of Natural History* 3:234–38.

Meiri, S., D. Simberloff, and T. Dayan. 2011. Community-wide character displacement in the presence of clines: a test of Holarctic weasel guilds. *Journal of Animal Ecology* 80:824–34.

Merriam, C. H. 1896. Synopsis of the weasels of North America. *North American Fauna* 11:1–45.

Negus, N. C., and J. S. Findley. 1959. Mammals of Jackson Hole, Wyoming. *Journal of Mammalogy* 40:371–81.

Oakes, C. L. 2000. History and consequences of keystone mammal eradication in the desert grasslands: the Arizona black-tailed prairie dog (*Cynomys ludovicianus arizonensis*). PhD dissertation, University of Texas at Austin.

Olendorff, R. R. 1976. The food habits of North American golden eagles. *American Midland Naturalist* 95:231–36.

Patterson, B. D., H. E. Ramírez-Chaves, J. F. Vilela, A. E. R. Soares, and F. Grewe. 2021. On the nomenclature of the American clade of weasels

(Carnivora: Mustelidae). *Journal of Animal Diversity*. 3(2). doi:10.29252/JAD.2021.3.2.1.

Polderboer, E. B., L. W. Kuhn, and G. O. Hendrickson. 1941. Winter and spring habits of weasels in central Iowa. *Journal of Wildlife Management* 5:115–19.

Powell, R. A. 1973. A model for raptor predation on weasels. *Journal of Mammalogy* 54:259–63.

———. 1982. Evolution of black-tipped tails in weasels: predator confusion. *American Naturalist* 119:126–31.

Powell, R. A., and W. J. Zielinski. 1983. Competition and coexistence in mustelid communities. *Acta Zoologica Fennica* 174:223–27.

Proulx, G. 2005. Long-tailed weasel, *Mustela frenata*, movements and diggings in alfalfa fields inhabited by northern pocket gophers, *Thomomys talpoides*. *Canadian Field-Naturalist* 119:175–80.

Proulx, G., and P. J. Cole. 1998. Identification of northern pocket gopher, *Thomomys talpoides*, remains in long-tailed weasel, *Mustela frenata longicauda*, scats. *Canadian Field-Naturalist* 112:345–46.

Proulx, G., and R. K. Drescher. 1993. Distribution of the long-tailed weasel, *Mustela frenata longicauda*, in Alberta as determined by questionnaires and interviews. *Canadian Field-Naturalist* 107:186–91.

Quick, H. F. 1944. Habits and economics of the New York weasel in Michigan. *Journal of Wildlife Management* 8:71–78.

———. 1951. Notes on the ecology of weasels in Gunnison County, Colorado. *Journal of Mammalogy* 32:281–90.

Ralston, G. L., and W. H. Clark. 1971. Occurrence of *Mustela frenata* in northern Baja California, Mexico. *Southwestern Naturalist* 16:209.

Richter, S. M. 2005. Status and space use of the long-tailed weasel (*Mustela frenata*) in southern Illinois. PhD dissertation, Southern Illinois University, Carbondale.

Ruiz-Campos, G., R. Martínez-Gallardo, S. González-Guzmán, and J. Alaníz-García. 2009. The long-tailed weasel *Mustela frenata* (Mammalia: Mustelidae) in Baja California, México. *Texas Journal of Science* 61:229–32.

Sheffield, S. R., and H. H. Thomas. 1997. *Mustela frenata*. *Mammalian Species* 570:1–9.

Simms, D. A. 1979. North American weasels: resource utilization and distribution. *Canadian Journal of Zoology* 57:504–20.

Singhurst, J. R., J. H. Young, G. Kerouac, and H. A. Whitlaw. 2010. Estimating black-tailed prairie dog (*Cynomys ludovicianus*) distribution in Texas. *Texas Journal of Science* 62:243–62.

Smith, B. W., C. A. Dobony, J. W. Edwards, and W. M. Ford. 2003. Observations of long-tailed weasel, *Mustela frenata*, hunting behavior in central West Virginia. *Canadian Field-Naturalist* 117:313–15.

Smithers, B. L., C. W. Boal, and D. E. Ancersen. 2005. Northern goshawk diet in Minnesota: an analysis using video recording systems. *Journal of Raptor Research* 39:264–73.

St-Pierre, C., J.-P. Ouellet, and M. Crete. 2006. Do competitive intraguild interactions affect space and habitat use by small carnivores in a forested landscape? *Ecography* 29:487–96.

Stoddard, H. L. 1931. *The bobwhite quail: its habits, preservation and increase.* New York: Charles Scribner's Sons.

Sturges, F. W. 1955. Weasel preys on larval salamander. *Journal of Mammalogy* 36:567–68.

Svendsen, G. E. 2003. Weasels and black-footed ferret (*Mustela* species). In *Wild mammals of North America: biology, management, and economics*, 2nd ed., ed. G. A. Feldhamer, B. C. Thompson, and J. A. Chapman, 650–61. Baltimore: Johns Hopkins University Press.

Teer, J. G. 1964. Predation by long-tailed weasels on eggs of blue-winged teal. *Journal of Wildlife Management* 28:404–6.

Thompson, B. C., D. F. Miller, T. A. Doumitt, and T. R. Jacobson. 1992. *Ecologically-based management evaluation for sustainable harvest and use of New Mexico furbearer resources.* Report to New Mexico Department of Game and Fish, NMDGF/NMSU Joint Powers Agreement 516.04 and New Mexico Federal Aid Project W-129-R, Job 1, December 1992.

Thompson, B. C., D. F. Miller, T. A. Doumitt, T. R. Jacobson, and M. L. Munson-McGee. 1996. *An ecological framework for monitoring sustainable management of wildlife: a New Mexico furbearer example.* US Department of the Interior, National Biological Service, Information and Technology Report 5.

Vargas, A., and S. H. Anderson. 1998. Black-footed ferret (*Mustela nigripes*) behavioral development:

aboveground activity and juvenile play. *Journal of Ethology* 16:29–41.

Vaughan, T. A. 1961. Vertebrates inhabiting pocket gopher burrows in Colorado. *Journal of Mammalogy* 42:171–74.

Vickery, P. D., M. L. Hunter, and J. F. Wells. 1992. Evidence of incidental nest predation and its effect on nests of threatened grassland birds. *Oikos* 63:281–88.

Wampler, C. R., J. K. Frey, D. M. VanLeeuwen, J. C. Boren, and T. T. Baker. 2008. Mammals in mechanically thinned and non-thinned mixed-coniferous forest in the Sacramento Mountains, New Mexico. *Southwestern Naturalist* 53:431–43.

Weeks, H. P. 1991. Arboreal caching of prey by long-tailed weasels. *Prairie Naturalist* 25:39–42.

Wilson, D. E., and D. M. Reeder, eds. 2005. *Mammal species of the world: a taxonomic and geographic reference*. 3rd ed. Baltimore: Johns Hopkins University Press.

Wright, P. L. 1963. Variations in reproduction cycles in North American mustelids. In *Delayed implantation*, ed. A. C. Enders, 77–97. Chicago: University of Chicago Press.

AMERICAN MINK (*NEOGALE VISON*)

Jennifer K. Frey

The American mink (*Neogale vison*) is one of the least studied and most neglected mammals in New Mexico. It has been considered extinct in the state, and yet the first survey aimed specifically at the species did not occur until 2021 (Peale et al. 2022). Importantly, it is also an unheralded symbol of the abuse that has been inflicted upon our streams and rivers, setting off cascades of ecological changes and the decline of native aquatic and riparian species. Minks require healthy waterways, more so perhaps than northern river otters (*Lontra canadensis*), or even some of the fishes that make their homes within the water. The ecological systems that once supported minks are now largely forgotten in our collective memory. They have been replaced by the channelized rivers and streams we know today, with their dams, highly regulated flows, and altered floodplains (e.g., Scurlock 1998). In writing this chapter, I hope that it will stimulate interest in this bellwether species and that it will open the door to new research and new goals for our streams and rivers. The mink can serve as a poster child for the ecological restoration needed for riparian zones so that they may once again become highly productive places and ecological systems that produce abundant, clean water, and diverse and abundant native plants and animals.

The mink is amphibious and well adapted for living and hunting in water and the adjacent riparian zone. It is a member of the weasel family (Mustelidae), and thus, not surprisingly, it has a typical weasel body plan. The body is altogether long and thin, with a long neck and narrow head; the small inconspicuous rounded ears are situated far back and low on the sides of the head. At the same time, the body is more robust and less lithe than a comparable sized weasel (Coues 1877). The legs are short, and the feet have partial webbing between the base of the toes (Armstrong et al. 2011). The tail is moderate in length, averaging about 41–51% of the length of the head and body (Hall 1981). The tail is densely furred and tapering (not bushy or penicillate; Coues 1877). Females average about 10% smaller than males in total length, but their weight is up to 50% less (Armstrong et al. 2011). There are no data on the size of minks specific to New Mexico. External measurements of two males and one female from Colorado were as follows: total length—548 mm (21.6 in), 579 mm (22.8 in), and 451 mm (17.8 in); tail length—171 mm (6.7 in), 189 mm (7.4 in), and 150 mm (5.9 in), and hind foot length—58 mm (2.3 in), 69 mm (2.7 in), and 55 mm (2.2 in), respectively (Bailey 1931; Armstrong 1972). The pelage is one of the mink's most distinguishing characteristics. The fur is modified for aquatic life with long coarse guard hairs overlaying a very dense and soft underfur (Bailey 1931). The hairs have glands at their base that

(*opposite page*) Photograph: © Romain Baghi.

Photo 24.1. American mink on 16 February 2020 in Boulder County, Colorado. The mink has the typical weasel body plan (narrow head, long, slender body and neck, and short legs) adapted for chasing prey down into their burrows. Compared to a weasel, however, it has a somewhat heavier build and its tail is bushier. The fur is mostly brown or dark brown, grading to black in the tail. Photograph: © Leslie Sutton.

Photo 24.2. Some wild minks show white spots on the chin, breast, or belly. Boulder County, Colorado, 30 January 2020. Photograph: © Leslie Sutton.

secrete oils that repel water, preventing the fur from becoming waterlogged when swimming (Armstrong et al. 2011). The fur is glossy and dark brown all over, except that some individuals have white spots on the chin, breast, or belly (Bailey 1931; Armstrong et al. 2011). The coloration of wild minks can vary from chestnut brown to blackish, and the undersides may be slightly paler (Armstrong et al. 2011). In contrast, pelage color in farmed minks may be variations of white, blue, gray, brown, or black (Obbard 1999).

In New Mexico, the mink may co-occur with several other members of the weasel family including the river otter, Pacific marten (*Martes caurina*), long-tailed weasel (*Neogale frenata*), and American ermine (*Mustela richardsonii*). The overall dark brown coloration, together with the lack of white or yellow undersides, serves to distinguish the mink from the long-tailed weasel and the ermine. From the otter, the mink can be distinguished by its much smaller size and lack of fully webbed hind feet. The mink is about the same size as the Pacific marten (Armstrong et

al. 2011). However, the ears of the Pacific marten are prominent, whereas those of the mink are rather inconspicuous. Although the black-footed ferret (*M. nigripes*) is not likely to co-occur with the mink due to different habitat associations (i.e., black-footed ferrets are closely associated with prairie dog towns; see Chapter 22), the skeletal and dental remains of these two species are particularly difficult to distinguish (Anderson et al. 1986). Thus, interpretation of remains such as those collected at archeological sites requires diligent care.

In North America, there once were two species of mink. The second species was the sea mink (*N. macrodon*), which became extinct around 1860 but which formerly occurred along the northeastern Atlantic coast perhaps as far south as Massachusetts (Hall 1981; Mead et al. 2000; Sealfon 2007). There has been considerable confusion regarding the taxonomy of the weasel family, especially with respect to treatment of the Nearctic minks (i.e., American mink and sea mink). Both species of Nearctic minks were traditionally classified in

Photo 24.3. American mink in the Alder/Fingru Open Space in Boulder County, Colorado. Minks have small eyes and small, rounded ears located far back on the sides of the head. Note the white spot on the chin, as well as the darker brown fur coloration, which some wild minks exhibit. Photograph: © Tamar Krantz.

Photo 24.4. American mink on 5 February 2015 in Logan, Utah. Some free-ranging minks in Utah are black instead of brown as they are derived from escaped or released fur-farm stock. Black pelage is more desirable in the fur trade and is the result of intensive selection from the more common native brown color. Today, mink fur farms may raise animals with white, blue, gray, brown, or black pelage. Photograph: © Ryan P. O'Donnell.

the genus *Mustela*, which also included the true weasels as well as the European mink (*M. lutreola*). In that traditional arrangement, all minks usually were placed in the subgenus *Vison* on the basis of shared morphological adaptation to a semi-aquatic lifestyle (e.g., Youngman 1982). However, various morphologic, cytogenetic, and biochemical data suggested that the Nearctic minks were well diverged from most other *Mustela*, including the European mink (Harding and Smith 2009). Ultimately, those studies culminated in the placement of the two Nearctic mink species into a separate new genus, *Neovison* (Abramov 2000; Wozencraft 2005). Even more recent molecular genetic analyses based on more exhaustive sampling of the Mustelidae revised that arrangement. Harding and Smith (2009) found that *N. vison* was in a clade with other species of weasels that originated in the Western Hemisphere, including the long-tailed weasel and two South American species. Surprisingly, these "American weasels" were more closely related to weasels from southeastern Asia than to weasels with origins in Eurasia (including the European mink and species that migrated to North America over the Bering land bridge such as the ermine and black-footed ferret). Patterson et al. (2021) determined that the correct genus name for American weasels was *Neogale*, hence the American mink is now referred to as *Neogale vison*.

Traditionally, 15 subspecies of the American mink are recognized (Hall 1981). The subspecies in New Mexico, *N. v. energumenos*, is known as the Rocky Mountain mink because it occurs throughout most of the Rocky Mountain region northward to Alaska (Hollister 1913; Bailey 1931; Hall 1981; Frey 2004). It is possible that historical populations of minks from the Canadian River basin on the plains of northeastern New Mexico were referable to the Great Plains subspecies, *N. v. letifera* (Armstrong 1972).

DISTRIBUTION
The American mink has a broad natural distribution throughout most of North America. In

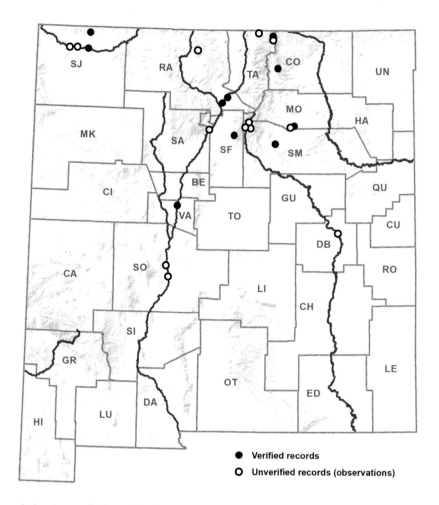

Map 24.1. Verified and unverified records of the American mink (*Neogale vison*) in New Mexico. The 1885 mink specimen obtained by Robert Shufeldt and bearing the location "Fort Wingate" in the western Zuni Mountains of present-day McKinley County is not shown on the map. Shufeldt was stationed at Fort Wingate during that time, but he conducted research on comparative osteology and collected specimens from around the entire Southwest. Although Schufeldt enjoyed hunting ducks on a small mountain stream in Milk Ranch Canyon near the fort (Shufeldt 1885), whether the 1885 specimen was indeed from the Fort Wingate area is unknown (see also Table 24.1).

the eastern half of the continent, minks occur from the southern edge of the Canadian tundra southward to the Gulf of Mexico (Hall 1981). In the western half of the continent, they are found from the southern edge of the Arctic slope in northern Alaska southward into the western United States, where their occurrence becomes increasingly limited by aridity. The species thus is absent from much of the Great Basin and the American Southwest, including southern California, all of Arizona, southern New Mexico, and most of western

Texas (Hall 1981). Its native range extends southward only as peninsulas along the major mountain ranges including the Sierra Nevada (California), Wasatch (Utah), and southern Rocky Mountains (New Mexico). The American mink is also now established as an exotic invasive species throughout large expanses of Eurasia and Patagonia, as well as on numerous islands due to introductions as fur animals or escapees from fur farms (Larivière 1999; Heptner and Sludskii 2002; Ibarra et al. 2009; Valenzuela et al. 2013).

New Mexico represents the southern periphery of the mink's range in the western half of North America. Historical records (Map 24.1; Table 24.1) suggest that minks formerly occurred along most perennial streams in the northern part of the state including in the following watersheds: 1) the San Juan River watershed, downstream probably at least through the Shiprock valley (San Juan County); 2) the Rio Grande watershed, downstream to at least Los Lunas (Valencia County) and potentially to the Bosque del Apache National Wildlife Refuge (Socorro County); 3) the Pecos River watershed, downstream to at least Fort Sumner (De Baca County); and 4) the Canadian River watershed, downstream for an unknown distance but at least eastward from the Sangre de Cristo Mountains onto the Upper Canadian Plateau, such as near Watrous (Mora County) and possibly historically to the borders of Oklahoma and Texas. The downstream limits of the mink in the Rio Grande watershed (including the Pecos River) are especially uncertain but conceivably could have extended southward beyond the Texas border. Merritt Cary provided a report of minks occurring near Fort Stockton, Pecos County, Texas, which is in the Pecos River watershed (Bailey 1905). Russell Arden Hill recounted his experience with a trapping party that took two dozen minks (in addition to 30 beavers [*Castor canadensis*] and 200 muskrats [*Ondatra zibethicus*]) on the Rio Grande of Texas between Indian Hot Springs (Hudspeth County) and Ruidosa (Presidio County) about the year 1928 (Abernethy 1990). Minks are indeed closely associated with both beavers and muskrats. The historical distribution of beavers and muskrats included the Rio Grande basin, downstream to the confluence of the Rio Grande and Pecos River in Val Verde County, Texas, with beavers occurring even farther downstream, to the mouth of the Rio Grande

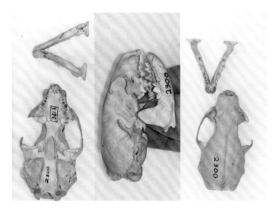

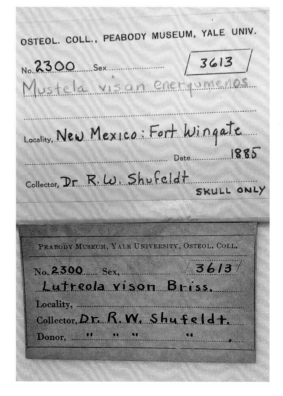

Photos 24.5a and b (*above and right*). American Mink specimen (YPM-MAM-007646; skull only) and associated labels at the Peabody Museum of Natural History at Yale University. The specimen is linked to the following accession record: "Received July 25, 1924. Gift from Dr. R. W. Shufeldt, Randall Mansions, 19th and Lamont Sts., Washington, DC. One box containing osteological collection." Fort Wingate, in present-day McKinley County, may or may not represent the location where the specimen was obtained. Photograph: © Kristof Zyskowski.

Table 24.1. Records of the American mink (*Neogale vison*) in New Mexico[1].

County	Locality	Date	Reference	Collector	Number of Animals	Nature
San Juan River Drainage						
San Juan	La Plata [La Plata River]	6 April 1892	AMNH MS-523/ MO-4114	C. P. Rowley	1	Specimen (skin, skull)
San Juan	Farmington	20 November 1908	USNM 158900	C. Birdseye	1	Specimen (skull)
San Juan	Animas River near Farmington	1908	Bailey (1931)	C. Birdseye	N/A	N/A
San Juan	San Juan River near Farmington	1908	Bailey (1931)	C. Birdseye	N/A	N/A
San Juan	Liberty [San Juan River]	October 1908	Bailey (1931)	V. Bailey party	N/A	N/A
San Juan	Fruitland [San Juan River]	October 1908	Bailey (1931)	V. Bailey party	N/A	N/A
San Juan	Vicinity of Chaco Canyon	<1937	Brand (1937)		N/A	N/A
McKinley	Fort Wingate	probably 1884–1888	YPM 7646	R. W. Shufeldt	1	Specimen (skull)
Rio Grande Drainage						
Taos	Costilla River, ~ 2,925 m (9,600 ft)	24 August 1904	USNM 133631	J. H. Gaut	1	Specimen (skin, skull)
Taos	Costilla River, 2,865 m (9,400 ft)	August 1904	Bailey (1931)	V. Bailey party	N/A	Tracks observed
Taos	Costilla River, ~ 2,440 m (8,000 ft), near Costilla	August 1904	Bailey (1931)	V. Bailey party	N/A	Tracks observed
Rio Arriba	Brazos River	Ca. 1892	Coape (1893)	A. P. F. Coape	1	Captured
Rio Arriba	Chama River		Berghofer (1967)		N/A	N/A
Rio Arriba	Velarde [Rio Grande]	25 December 1904	USNM 136273	C. C. Beattie	1	Specimen (skull)
Rio Arriba	Velarde [Rio Grande]	2 December 1904	USNM 136274	C. C. Beattie	1	Specimen (skull)
Rio Arriba	Velarde [Rio Grande]	6 March 1905	USNM 136276	C. C. Beattie	1	Specimen (skull)
Rio Arriba	Alcalde [Rio Grande]	25 February 1905	USNM 136275	C. C. Beattie	1	Specimen (skull)
Los Alamos/ Sandoval	Rio Grande near El Rito de los Frijoles	<1914	Henderson and Harrington (1914)	N. Dowell	N/A	N/A

Allen (1893:83): "One specimen, La Plata, 'N. Mex. Minks are quite plentiful on all the streams of this region, but I was only lucky enough to catch one, on the La Plata River" (Rowley, MS notes)."'

Bailey (1931:324–25): "In 1908 Clarence Birdseye reported minks as formerly common along the Animas and San Juan rivers near Farmington but at that time rather scarce. One trapper reported the diminishing numbers of 20, 12, and 10 taken there during the three preceding winters."

Bailey (1931:324–25): "In 1908 Clarence Birdseye reported minks as formerly common along the Animas and San Juan rivers near Farmington but at that time rather scarce. One trapper reported the diminishing numbers of 20, 12, and 10 taken there during the three preceding winters."

Bailey (1931:324–25): "In 1908 Clarence Birdseye reported minks as formerly common along the Animas and San Juan rivers near Farmington but at that time rather scarce. One trapper reported the diminishing numbers of 20, 12, and 10 taken there during the three preceding winters."

Bailey (1931:325): "A few were also reported at Liberty and Fruitland." Liberty was a settlement located 27 km (17 mi) west of Farmington, west of Waterflow, on the north bank of the San Juan River (Julyan 1998).

Bailey (1931:325): "A few were also reported at Liberty and Fruitland."

Brand (1937:48): "Among the carnivores occasionally seen in the Chaco area (and presumably more common in past time) are: . . . mink (*Lutreola vison energumenos*). . . . This report probably pertains to the nearby San Juan River."

Robert W. Shufeldt was stationed at Fort Wingate from 1884 to 1888, during which he collected specimens for his osteology collection in the Zuni Mountains and elsewhere in the region. All his specimens in the Yale Peabody Museum were labeled "Fort Wingate" rather than the specific location from which they came, including a *Lepus callotis*, which is not known to occur near Fort Wingate. Thus, it is possible the mink specimen did not come from McKinley County.

Bailey (1931:324): "On August 24, 1904 an immature male was collected by Gaut at 9,600 feet altitude on Costilla River in the Sangre de Cristo Mountains."

Bailey (1931:324): "At 9,400 feet [on the Costilla River] their tracks were seen in the mud around an old beaver house, and a family were evidently living in the beaver house, where they had well-worn entrances and trails." According to F. Bailey (1928) this location was 8 km (5 mi) below the mouth of Comanche Creek.

Bailey (1931:324): "Their tracks were common along this river [Costilla River] down to 8,000 feet near the town of Costilla in the Rio Grande Valley."

Coape (1893:261: "The next day I shot eight grouse, a mink, and a coyote, rather a mixed bag."

Berghofer (1967:189-190): "Through the 1940's the mink (*Mustela vison*), considered one of the most valuable furbearers, was common throughout the northern half of New Mexico and the Gila drainage, and plentiful in the Upper Rio Grande and Chama River drainage." Note: there is no evidence that the mink ever occurred in the Gila River drainage.

Bailey (1931:324): "The trappers along the Rio Grande above Santa Fe usually catch a few minks during the winter, and from one of these trappers during the winter of 1904–05 four skulls were obtained."

Bailey (1931:324): "The trappers along the Rio Grande above Santa Fe usually catch a few minks during the winter, and from one of these trappers during the winter of 1904–05 four skulls were obtained."

Bailey (1931:324): "The trappers along the Rio Grande above Santa Fe usually catch a few minks during the winter, and from one of these trappers during the winter of 1904–05 four skulls were obtained."

Bailey (1931:324): "The trappers along the Rio Grande above Santa Fe usually catch a few minks during the winter, and from one of these trappers during the winter of 1904–05 four skulls were obtained."

Henderson and Harrington (1914:23): "Mr. Dowell says mink occur along the Rio Grande near El Rito de los Frijoles."

continued on next page

Table 24.1. *continued*

County	Locality	Date	Reference	Collector	Number of Animals	Nature
Santa Fe	Santa Fe Canyon	22 February 1905	KU 1454	L. L. Dyche	1	Specimen (skin, skull)
Valencia	Rio Grande at Los Lunas; T7N, R2E	15 March 1919	MSB 15072	J. S. Ligon	1	Specimen (skull)
Socorro	Socorro [Rio Grande]	<1927	Ligon (1927)		N/A	N/A
Socorro	near San Antonio [Rio Grande]	September 1991–May 1992	Thompson et al. (1992)	D. Miller	1	Observation
	Pecos River Drainage					
San Miguel	Pecos River Mountains, ~ 2,440 m (8,000 ft)	Jul-Aug 1903	Bailey (1931)	V. Bailey party	N/A	Tracks observed
San Miguel	trout streams of the upper Pecos, ~ 2,620 m (8,600 ft)	Jul-Aug 1903	Bailey (1931)	V. Bailey party	N/A	N/A
San Miguel	Rio Mora [near junction Bear Creek, ca. 2,500 m (8,200 ft) elevation], Santa Fe National Forest; T18N, R12E, NE1/4 Sec. 14	summer 1995	Schmitt (2001)	Confidential	1	Observation
San Miguel	Las Vegas [Gallinas River watershed]	<1903	Stone and Rehn (1903)	M. Robbins	1	Specimen (skin)
De Baca	Pecos River, near Fort Sumner	Ca. 1866	Clifford (1871, 1877)	J. Clifford	1	Captured
	Canadian River Drainage					
Colfax	Cimarron Canyon, on US 64 about 0.4 km (0.25 mi) west of Clear Creek	1 April 1987	Hubbard (1987)	Confidential	2	Photographed
Colfax	Southwestern Colfax County	1938	Hill (1942)	J. E. Hill party	N/A	N/A
Mora	Mora River, Watrous	December 1966	Yarbrough and Studier (1968)	W. H. Bledsoe	5	Captured
Mora	Mora River, 5.6 km (3.5 mi) E of Watrous	January 1967	Yarbrough and Studier (1968)	W. H. Bledsoe	2	Captured
Mora	Mora River, 5.6 km (3.5 mi) E of Watrous	March 1967	MSB 24807, 24808	E. H. Studier (W. H. Bledsoe)	2	Specimen (skull)

1   Excluded from the table are two specimens in the Museum of Southwestern Biology (MSB 7623, 7844) that were cataloged as from Santa Fe (Santa Fe County), but actually were ranched minks donated to the museum by William Huey. Also excluded are pre-Columbian records: a partial dentary of Wisconsin/early Holocene age from Isleta Cave No. 2 (Bernalillo County) identified by Elaine Anderson (Harris 1993); two specimens from Arroyo Hondo Pueblo (Santa Fe County; located along a tributary of the Santa Fe River, ca. 7 km S of Santa Fe) that date to 1315–1330 and 1410–1425 (Lang and Harris 1984); and a ramus (AMNH M-249736) found in 1896 by B. T. Hyde at Pueblo Bonito Ruin in Chaco Canyon (San Juan County).

Ligon (1927:156): "Mink are not common in New Mexico, their habit being the Rio Grande Valley as far south as Socorro and along water-courses in the Sangre de Cristo Range."

Thompson et al. (1992:49): "One roadkill/observation survey respondent reported a mink sighting near San Antonio, Socorro county (Appendix L)." Thompson et al. (1992:Appendix L, p. 8; tabular entry): Species (total observed): "mink"; County: "Socorro"; Observer: "Miller, Damien NMSU"

Bailey (1931:324): "In the Pecos River Mountains a few mink tracks were seen along the river at 8,000 feet, and a skin was seen hanging in one of the houses at Willis [=Cowles] on the upper Pecos at 8,500 feet"

Bailey (1931:324): "In the Pecos River Mountains. . . . Others were reported at 8,600 feet along the trout streams of the upper Pecos, but they were by no means common at that time, July and August, 1903."

Schmitt (2001:1): "A single mink was observed running along the left bank (i.e., looking downstream of Rio Mora) of Rio Mora. [The observer] noticed the 'undulations' in its back as it was running along the bank of Rio Mora. Was seen ca. 10 seconds in the early afternoon ca. 1400 to 1600h from a distance of ca. 25ft. The specimen was ca. 18 inches in total length. Pelage coloration was described as dark brown."

Stone and Rhen (1903:25): "A skin of a mink collected at Las Vegas, San Miguel county, N.M., by Marshall Robbins, was submitted by Prof. Cockerell." The Academy of Natural Sciences of Philadelphia does not have a record of mink from New Mexico in their catalog or in the collection (N. Gilmore, pers. comm.).

Clifford (1871:66-67; 1877:134-135): "But my mink, Max, was a dear little pet. He was given me by a soldier at Fort Union, and had been captured on the Pecos River, near Fort Sumner. He was of a solid, dark-brown color, and the texture of his coat made it clear at once why a set of mink-furs is so highly prized by the ladies."

Hubbard (1987:1): "[A state park ranger] . . . found two recently-dead mink (*Mustela vison*) in Cimarron Canyon, Colfax County, on April 1, 1987. The exact locality was along US 64, about 1/4 mile west of Clear Creek—between Ute Park and Eagle Nest. The two were found side-by-side at about 9:30 a.m., probably having been killed by a vehicle the previous night. [The collector] noted the carcasses when crows flushed from beside the highway. There was some damage to the carcasses, but both animals were sexable as males. They are now fluid specimens at the Museum of Southwestern Biology, University of New Mexico, Albuquerque, NM." Note: The Museum of Southwestern Biology has no record of these specimens.

Hill (1942:78): "Mink occur in this area and have been trapped; but none was secured by the expedition."

Yarbrough and Studier (1968:105): "During the past winter (1966–1967), a commercial trapper, Walton H. Bledsoe, captured nine adult male mink, *Mustela vison*, near Watrous, N.M. The distribution of captures and localities are as follows—five in December, 1966, Mora River, Watrous, Mora Co., N.M.; two in January, 1967, 31/2 mi E of Watrous, Mora Co., N.M.; and, two in March, 1967, at the same locality. Of the nine mink, only the skulls of the last two were retrieved. These have been deposited in the Mus. of Southw. Biol. UNM (Nos. 24807, 24808)."

Yarbrough and Studier (1968: 105): "During the past winter (1966-1967), a commercial trapper, Walton H. Bledsoe, captured nine adult male mink, *Mustela vison*, near Watrous, N.M. The distribution of captures and localities are as follows--five in December, 1966, Mora River, Watrous, Mora Co., N.M.; two in January, 1967, 31/2 mi E of Watrous, Mora Co., N.M.; and, two in March, 1967, at the same locality. Of the nine mink, only the skulls of the last two were retrieved. These have been deposited in the Mus. of Southw. Biol. UNM (Nos. 24807, 24808)."

Yarbrough and Studier (1968: 105): "During the past winter (1966-1967), a commercial trapper, Walton H. Bledsoe, captured nine adult male mink, *Mustela vison*, near Watrous, N.M. The distribution of captures and localities are as follows--five in December, 1966, Mora River, Watrous, Mora Co., N.M.; two in January, 1967, 31/2 mi E of Watrous, Mora Co., N.M.; and, two in March, 1967, at the same locality. Of the nine mink, only the skulls of the last two were retrieved. These have been deposited in the Mus. of Southw. Biol. UNM (Nos. 24807, 24808)." MSB records reported collector as E.H. Studier, date 7 June 1967, and location "Mora River, 3.5 mi E Watrous."

Photo 24.6. The last two verified American mink records in the state were from Cimarron Canyon in northern New Mexico. Photograph: © Geraint Smith.

Photos 24.7a and b. The two American minks found dead on US Highway 64 in Cimarron Canyon in 1987 and representing the last verified occurrence of the species in New Mexico. The carcasses show the normal wild-type pelage color found in minks from New Mexico and Colorado. The two carcasses were reportedly deposited as specimens at the University of New Mexico Museum of Southwestern Biology, but seem to be missing today. The photographs taken in 1987 nonetheless help confirm that the two animals collected in Cimarron Canyon were probably native minks instead of fur farm escapees, which typically have black pelage. Photograph: © Rick Grothe.

(Hall 1981; Schmidly 2004). Temperature is likely not a limitation in this region as minks occur in southern Florida. Thus, there is no a priori reason to discredit Hill's account.

Most of the available New Mexico records of minks are from along larger rivers, such as the Rio Grande and the Animas, La Plata, San Juan, Chama, Mora, and Pecos rivers (Map 24.1; Table 24.1). However, minks also occurred along smaller, headwater streams. For instance, mink records are available from the Brazos River, a headwater tributary to the Chama River; the Rio Costilla, a headwater tributary to the Rio Grande; the Cimarron River, a headwater tributary to the Canadian River; and the Rio Mora, a headwater tributary to the Pecos River (Table 24.1). Such small streams were probably not incidental to the overall distribution of the mink in the state. This is exemplified by the account of A. P. F. Coape (1892), whose party trapped for beavers and minks on a small tributary to the Chama River on the north slope of the Jemez Mountains (i.e., opposite the Chama River from Arroyo del Cobre), probably Abiquiu Creek or Arroyo de los Frijoles, Rio Arriba County. A record of a mink from "above Santa Fe" attributed to Bailey

(1931:324) was incorrectly mapped by Findley et al. (1975) on the Santa Fe River. The actual location referenced by Bailey (1931:324) was "along the Rio Grande above Santa Fe" (see also Table 24.1). That said, a specimen in the University of Kansas Museum collected in 1905 confirms that minks historically occurred on the Santa Fe River above Santa Fe (Table 24.1).

In his review of economically important wildlife in New Mexico, Ligon (1927) made bare mention of minks, stating that they were not common in the state, presumably due to their restricted distribution. Bailey (1931), in contrast, described the mink as fairly common along streams in northern New Mexico. Berghofer (1967:189) thought the mink was "common" throughout its range in northern New Mexico and "plentiful" in the upper Rio Grande and Chama River watersheds through the 1940s, but since had been in steady decline for some unidentified reason. At the time of his writing, Berghofer (1967) thought there were very few minks to be found, mostly limited to the upper Rio Grande drainage. However, his comments must be examined with caution because he also stated that the mink occurred in the Gila River watershed; there is no evidence that minks ever occurred within this watershed.

The American mink is now regarded as extirpated from New Mexico (Jones and Schmitt 1997). The last wild-caught minks preserved as museum specimens were collected in 1967. In 1987, however, two minks were found dead on a road in Cimarron Canyon, Colfax County, and they were reportedly deposited as specimens at the University of New Mexico Museum of Southwestern Biology. I was unable to find the specimens or any record of them in the museum, but photos of the two minks exist, taken on the day they were discovered (Photo 24.7a, b). These photos not only confirm the authenticity of the original report cited in Hubbard (1987), but from the pelage of the two minks they seem to disprove the possibility that the two animals were farm-raised minks. Thus, I see no reason to think they were anything other than natural-born, wild minks. Fur farms were illegal in New Mexico at that time, and evidence indicates that mink farms in New Mexico were always quite rare when legal (see under "Status and Management," below). There also have been recent unverified reports of minks from the Pecos River near Lisboa Springs Hatchery and Pecos (San Miguel County), Pecos Baldy Lake (Mora County), and Vallecito River (Rio Arriba County; Schmitt 2001).

HABITAT ASSOCIATIONS

The mink is a quintessential wetland species. However, habitat requirements relate probably more to conditions that foster abundant prey than the vegetation per se (Loukmas and Halbrook 2001). Like other strictly carnivorous species with high metabolisms, minks must have continual access to high densities of suitable prey, or else they cannot thrive. The mink's diet primarily consists of riparian small mammals supplemented by locally abundant aquatic animals such as fishes, crayfish, and frogs (see under "Diet and Foraging"). In New Mexico, densities of riparian small mammal prey such as shrews, hispid cotton rats (*Sigmodon hispidus*), voles (*Microtus* spp.), muskrats, and jumping mice (*Zapus* spp.) are highest in wetlands and riparian zones with ample cover of herbaceous riparian plants (Frey 2018). These plants provide food and cover that sustain high rodent populations, especially voles in montane settings and cotton rats in lower-elevation locations. However, such plant communities also provide concealment cover for minks as protection from aerial predators such as hawks. Such vegetation types are best developed in association with abundant moist soil, such as that promoted by beaver activity and where livestock grazing is limited or absent (Frey 2018).

Due to both aridity and rugged topography, suitable riparian areas in New Mexico are usually confined to narrow linear strips (see Chapter 2).

Photo 24.8. Rio Cebolla above Seven Springs Hatchery in the Jemez Mountains, 26 August 2008. A beaver has recently built a dam (and lodge), creating a new marsh where a year earlier the stream was a narrow meandering channel. American minks depend on such wetlands with slow-moving waters and ample herbaceous vegetation. In New Mexico, water diversions and the construction of large dams eliminated much of the mink's historical habitat along the main rivers of the state. Along the smaller streams in the mountains, beaver populations were heavily impacted by the fur trade industry and by livestock grazing, resulting in further loss of mink habitat. Photograph: © James N. Stuart.

The most extensive wetlands suitable for mink populations would have occurred in the broad floodplains of major rivers. Historically, these river valleys offered a diversity of complex wetland habitats for wildlife, created by natural flooding, river meandering, and beaver activity. However, floodplains were the focus of early agricultural activities that eliminated wetlands in favor of irrigated cropland. Subsequently, major alteration of floodplains occurred in response to the construction of dams and canals, especially during the period from 1910 to the early 1970s (Scurlock 1998). However, even by the 1890s massive alterations had already occurred. For instance, on the principal river of the region—the Rio Grande—an estimated 925 ditches were by then drawing water from the river in Colorado alone, and there had been an estimated flow reduction of 200,000 acre-feet a year since 1880 that left the river dry by the time it reached Texas (Autobee 1994). Control of water flow allowed invasive saltcedar (*Tamarisk* spp.) and Russian olive (*Elaeagnus angustifolia*) to flourish and take over as monotypic stands in areas that were not farmed or otherwise developed. Today, floodplains with wetlands supporting herbaceous plant communities that fuel high densities of riparian rodents (i.e., mink prey) are largely confined to small isolated pockets in wildlife refuges and other protected locations.

As the larger rivers lost their associated wetlands, habitat for minks would have become restricted to the smaller rivers and headwater streams in the mountains. Historically, damming

Photo 24.9. Beaver-created wetland on Willow Creek at its confluence with the Pecos River near Tererro in San Miguel County on 14 May 2008 (the beaver dam has been partially breached). Wetlands formed by beaver activities meet the American mink habitat's requirements with herbaceous plants that provide cover for concealment against aerial predators and food for prey populations of rodents. Photograph: © James N. Stuart.

activities of beavers converted lower gradient reaches of these small streams into complex tiered wetlands, hosting a diversity of riparian vegetation communities that 1) supported diverse and abundant small mammal communities; and 2) maintained shallow, slow water areas required by minks for successful hunting. In the early 1800s, however, northern New Mexico became an epicenter for the Rocky Mountain fur trade industry that primarily focused on beavers (Weber 1971). The historical population of beavers in New Mexico was estimated to be between 62,500 and 281,400 (Wild 2011), but it declined to an unknown but drastically low number that resulted in the first legislative decision in 1897 to close beaver trapping in the state (Huey 1956). Although trapping was to resume, New Mexico began a restoration program in 1932 that recovered the beaver population to an estimated 7,954 animals by 1953 (Huey 1956). However, current populations of beavers on small streams are rare and likely much lower than reported by Huey in the 1950s (Small et al. 2016). With the drastic decline of beaver populations in the 1800s, their old dams breached over time, draining the wetlands that had been created and allowing floods to scour the valley bottoms (Fouty 2018). In addition, generations of livestock grazing eliminated the riparian shrubs needed by beavers as building material and food, thus preventing the species from reoccupying much of its historical range (Small et al. 2016). Excessive livestock grazing also keeps the vegetation along streams clipped low such that the cover required to support abundant small mammal populations is lacking (Frey 2018). Thus, with rare exceptions, the vegetation conditions that produce abundant mink food have been largely eliminated in the state.

Minks require clean water. Because they are near the top of the food chain, they are particularly susceptible to elevated environmental contaminants such as mercury and halogenated hydrocarbon compounds including DDT, PCBs, DDE, and dieldrin (see review in Landholt and Genoways 2000). For instance, it has been demonstrated that there is significant placental transfer of mercury to the fetus, and transfer of polychlorinated biphenyls (PCBs) to kits through their mother's milk (Wren et al. 1987a). Mink kits born to mothers exposed to PCBs had slower growth rates, while those born to mothers exposed to both PCBs and methylmercury had reduced survival (Wren et al. 1987b). In the Great Lakes region, research has suggested a causal link between lower mink population densities and higher levels of organochlorine chemicals (i.e., PCBs and dioxins; Wren 1991). Minks are sensitive bioindicators of environmental mercury levels (Wren et al. 1986), and deaths were attributed to a combination of exposure to cold and mercury (Wren et al. 1987a). Levels of PCBs and mercury are high enough in many of New Mexico's streams and lakes that the state issues human fish consumption advisories (https://www.env.nm.gov/surface-water-quality/fish-consumption-advisories/).

LIFE HISTORY
Diet and Foraging
The ecological niche of the American mink, in terms of both its habitat and food requirements, is intermediate between those of the American ermine and the northern river otter. Minks are exclusively carnivorous, but their diet is varied, typically reflecting what species of prey are locally abundant (Linscombe et al. 1982). Like the ermine, the mink retains a typical terrestrial

weasel body plan that is supremely adapted for hunting rodents in their burrows (see Chapter 21). In fact, small riparian mammals, especially muskrats, voles, cotton rats, shrews, and mice, are the most important component of the mink's diet during all seasons (Svihla 1931; Eagle and Whitman 1999). However, as with weasels, the tubular body plan and active lifestyle of the mink come at a tremendous metabolic cost (Iversen 1972)—one that requires a continual supply of abundant food energy. Consequently, minks can only exist where food supplies, especially riparian rodents, are abundant enough. Because food availability is a key limitation for them, they will kill surplus prey that they may cache for later use (Errington 1943; Larivière 2003). Most foraging occurs within 1–2 m (~3.3–6.6 ft) of the shoreline as they actively investigate any burrow, nook, cranny, log pile, or beaver structure for prey (Larivière 2003). They do not stalk or ambush but simply rush their prey, grasping it around the body, and killing it with a puncture of the canines to the head or neck in the same manner as other weasels (Poole and Dunstone 1976).

Unlike other terrestrial weasels, however, minks also have evolved specialized adaptations, such as partially webbed toes, for hunting animals in water. Heat loss is a major concern for minks especially if wet, but the thick underfur and oily guard hairs make the fur water-resistant, helping to prevent hypothermia (Larivière 1999, 2003). When submerged under water, the mink's heart rate is rapidly reduced to conserve oxygen (Gilbert and Gofton 1982). Because minks are incompletely adapted to aquatic hunting and have relatively poor dive endurance in comparison with better-adapted aquatic mammals, they hunt most efficiently in shallow water with slow currents (Dunstone 1983, 1998). Thus, they forage along shorelines and in small pools, focusing on areas where they have learned prey might occur (Dunstone 1998). However, minks have no special adaptations for seeing underwater (Sinclair et

Photo 24.10. Common muskrat (*Ondatra zibethicus*) at the Bosque del Apache National Wildlife Refuge on 2 March 2019. Muskrats are important in the diet of American minks. Photograph: © Mark L. Watson.

Photo 24.11. American mink with crayfish at Gross Reservoir in Boulder County, Colorado. Photograph: © Christopher Sichko.

al. 1974). Consequently, they find aquatic prey by peering into the water from above, unless surface reflection obscures the view, in which case they may stick their head underwater to seek moving prey (Poole and Dunstone 1976). Further, because the mink's swimming abilities are somewhat limited, it less often preys on larger (>20 cm) and faster-swimming fish such as salmonids (Burgess and Bider 1980; but see Lindstrom and Hubert 2004). When hunting aquatic animals, dives by minks usually last only 5–20 seconds (Poole and Dunstone 1976). When hunting small fish, minks simply grasp the body of the fish with their mouth, usually behind the head, and occasionally aided by the forefeet (Poole and Dunstone 1976). For large fish, a mink may grasp the fish's body with both its mouth and limbs. The mink then works its way up to the fish's head and grasps the front of the fish's face with its mouth and attempts to drag it ashore as the fish tires. Fish may be more prevalent in the diet of minks during winter when they are slower or during drought when they are caught in smaller pools (Gerell 1967; Dunstone 1978). Frogs and crayfish can be important foods during warmer months in locations where they are abundant (Korschgen 1958; Linscombe et al. 1982).

Minks and common muskrats (*Ondatra zibethicus*) form a special predator-prey relationship

Photo 24.12. American mink poised to pounce on a solitary sandpiper (*Tringa solitaria*) in the Running Deer Natural Area in Fort Collins, Colorado in 2009. Photograph: © Matthew M. Webb.

Photo 24.13. American mink scavenging on the carcass of a Canada goose (*Branta canadensis*) on 5 February 2015 in Logan, Utah. Photograph: © Ryan P. O'Donnell.

(Errington 1943; Shier and Boyce 2009). Muskrats are medium-sized, semi-aquatic, herbivorous rodents that are essentially large voles (Willner et al. 1980). Muskrats have a high reproductive rate, and, though they can outsize minks in weight, they are relatively meek and hence make an ideal prey for minks. Muskrats are associated with emergent herbaceous wetlands, and consequently they reach their highest abundance in regions with extensive marshes—not coincidentally these regions also have the highest densities of minks (Linscombe et al. 1982; Boutin and Birkenholz 1999). So dependent, in fact, are minks on muskrats in these regions that population cycles of the mink follow those of the muskrat, much the same way as between the Canada lynx (*Lynx canadensis*) and the snowshoe hare (*Lepus americanus*; e.g., Errington 1943; Viljugrenien et al. 2001; Shier and Boyce 2009; see Chapter 6). Where prey diversity is low, the strength of the relationship between the mink and the muskrat may be strongest (Erb et al. 2001; Shier and Boyce 2009). The distribution and abundance of muskrats in New Mexico have no doubt declined with the loss of wetlands. Particularly telling was the recent remote camera mink survey at 85 sites in northern New Mexico, during which only a single muskrat was detected (Peale et al. 2022).

Besides riparian rodents, fishes, and crayfish, minks may also take any encountered prey they can overwhelm and kill. If relatively abundant, such species can become secondarily or seasonally important food items. For instance, minks are agile tree climbers (Larivière 1996), which explains the regular presence of red squirrels (*Tamiasciurus hudsonius*) in their diet in the boreal forests of Canada (Shier and Boyce 2009). Examples of other locally or seasonally important foods include lagomorphs (cottontails and snowshoe hare), waterfowl and their eggs, and frogs (e.g., Korschgen 1958; Eberhardt 1973; Arnold and Fritzell 1987; Hoffman et al. 2009; Shier and Boyce 2009). Male minks, which are larger than

females, are more likely to hunt larger animals such as lagomorphs (the largest prey taken by minks) and adult muskrats, as well as to hunt in uplands (which they do to pursue lagomorphs); in contrast, females generally remain in the riparian zone and maintain a broader niche breadth, consuming relatively more small mammals, fishes, and crustaceans than males (Sealander 1943; Birks and Dunstone 1985). Food items taken incidentally include earthworms, insects, mollusks, salamanders, reptiles, songbirds, weasels, and carrion (Linscombe et al. 1982; Larivière 1999, 2003; Shier and Boyce 2009).

Reproduction and Social Behavior

Minks are generally solitary except during the mating season and except for females with young. Breeding is cued by photoperiodism and generally occurs in spring, during which time males may travel widely to find mates (Linscomb et al. 1982; Larivière 1999, 2003). Copulation induces ovulation of ova within 72 hours. Development of the fertilized eggs goes into stasis until these implant into the uterus during the next winter and then continue developing for another 28 to 30 days until the young are born, usually in April to June. Litter size averages four, and the young remain with the mother until fall (Larivière 1999, 2003). They become sexually mature at 10 months (Linscombe et al. 1982).

Interspecific Interactions

The original geographic ranges of the American beaver, common muskrat, and American mink tended to greatly overlap (Hall 1981). In places where natural wetlands are rare, such as in the arid West, beavers were a vital force in creating and maintaining wetlands (Frey 2018). Thus, the relationship between the beaver and the mink may have been stronger in this region. The wetlands created by beavers are ideal environments for the mink. The lush riparian growth promoted by beaver activities produces high diversities and

Photo 24.14. Adult female American mink and her four kits in Larimer County, Colorado. Litter size in American minks averages four. Photograph: © Lori Nixon.

densities of riparian rodents used as key mink prey (Frey 2018). Muskrats, in particular, benefit from beavers in a number of ways including through the creation of slow and deep waters, providing den locations in beaver lodges and access to felled tree leaves, and especially through stabilizing water levels and promoting sedges (*Carex* spp.) and other herbaceous wetland plants used for food (e.g., Leighton 1933; Proulx et al. 1987). Beaver lodges seem to be favored places for minks to den. For instance, Bailey (1931) reported well-worn entrances and trails made by minks in an old beaver lodge on the Costilla River, Taos County, at ~2,850 m (9,400 ft). Although minks may reside in beaver lodges, they apparently do not prey on beavers, including the kits (Brzezinski and Zurowski 1992).

Historically, the mink and the river otter, another species considered commensal with the beaver, would have co-occurred along many of the streams and rivers in New Mexico. Niche separation between minks and otters is thought

Photo 24.15. Bald eagle (*Haliaeetus leucocephalus*) holding an American mink in its talons in Jackson County, Texas. Photograph: © Rodney H. Cosper.

to occur through resource partitioning (Ben-David et al. 1996). In comparison with minks, river otters consume a higher proportion of fish and aquatic invertebrates and a lower proportion of mammals and birds (see Chapter 19). Thus, coexistence of the two species requires abundant terrestrial prey for the mink (e.g., Bonesi and Macdonald 2004a).

Humans are the primary predator of the American mink (Larivière 1999, 2003). However, a variety of other predators have been documented including red and gray foxes (*Vulpes fulva* and *Urocyon cinereoargenteus*), coyotes (*Canis latrans*), bobcats (*Lynx rufus*), river otters, and large raptors such as great horned owls (*Bubo virginianus*), bald eagles (*Haliaeetus leucocephalus*), and *Buteo* hawks (Larivière 1999, 2003; Photo 24.15).

STATUS AND MANAGEMENT

New Mexico Statutes Section 17-5-2 defines the mink as a furbearing species and hence all live animals and their pelts are the property of the state unless legally taken. However, there is currently no mechanism for the legal taking of minks since no open season has been declared for the species in nearly thirty years in New Mexico (see New Mexico Administrative Code 19.32.2.8). Since the 1951–1952 license year, the mink harvest season has in fact been closed except during 14 winters (1951–56, 1958–1960, and 1961–1970; Thompson et al. 1992). When open, the season was limited to the month of December except during 1953–1954 when it ran from 1 November to 31 March (Thompson et al. 1992). The New Mexico Department of Game and Fish reported the sale of only nine mink pelts between 1931 and 2008, one during the 1945–1946 season, six in 1950–1951, and two in 1969–70 (the report did not actually specify if these pelts originated from New Mexico or another state). Details of furbearer harvests were not reported in all years.

In 1975, the mink was listed as State Endangered, Group II (i.e., species whose prospects of survival or recruitment within the state are likely to become jeopardized in the foreseeable future; Jones and Schmitt 1997). However, in 1985, the species was delisted simply because it was presumed extirpated due to the absence of any verified record in the past decade (Jones and Schmitt

1997). In 1992, the mink was instead added to an informal list of extinct, extirpated, and vanishing species in New Mexico (NMDGF 1992). That list mentioned the mink's presumed extirpation, but also mentioned the two specimens collected in 1987 (misreported as 1988) and implied they could have escaped from fur farms (see under "Distribution" above). As noted already, the two minks collected in 1987 are more likely to have been naturally occurring animals.

Despite the change in the mink's legal designation in New Mexico, the species' current population status in the state remains uncertain. Extirpation has been presumed based simply on the lack of verified reports of the species. In the absence of formal scientific study, minks would be most likely documented by trappers, especially those trapping for muskrats. But, because there is no open season on the mink, it would be unlawful for someone to harvest that species, even accidentally. Consequently, any such take of a mink is likely to go unreported if it ever occurs. Based on the limited number of records, it seems nonetheless clear that minks have been at best rare in the state since the turn of the 20th century.

The decline of the mink population in New Mexico can be attributed primarily to the tremendous loss of wetlands in the state, though trapping during the 1800s and early 1900s also may have contributed to early losses in some areas. Bailey (1931) reported a trapper that took 42 minks near Farmington over three seasons in the early 1900s. In Nebraska, the introduction of agricultural pesticides in the 1940s and 1950s may have caused or accelerated declines in populations of the mink and other mustelids (Landholt and Genoways 2000). Similar impacts are likely in New Mexico, especially given that much of the best farmland in this arid state is located on river floodplains. But what about the status of the mink in New Mexico today? A review of the number of records of minks by decade reveals sporadic records since the 1960s. The last verified records were from 1987, but there

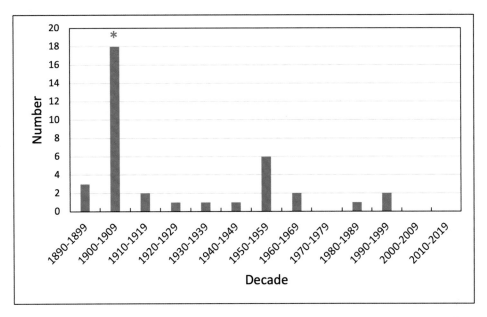

Figure 24.1. Number of records of the American mink (*Neogale vison*) in New Mexico by decade since 1890. The asterisk indicates a report of 42 minks trapped in 1905–1908 (Table 24.1).

were also two separate, compelling observations (one of a carcass) from the 1990s (Figure 24.1; Table 24.1). Most recently, in 2021, an extensive survey for minks in northern New Mexico using remote cameras at 85 sites in 11 watersheds did not yield any detections (Peale et al. 2022). However, remote cameras may have low detection rates for minks in comparison with other methods such as searching for scats and footprints (Bonesi and Macdonald 2004b) or environmental DNA (Robinson et al. 2017). Consequently, it remains a possibility that remnant populations of minks subsist in New Mexico. Additional field surveys of remaining healthy or healthier riparian areas are needed to determine if minks persist in the state. In addition, trappers should be allowed to report incidental capture of minks without penalty.

Mink Farming

Worldwide, the American mink is the animal most commonly farmed for its fur and it outranks all other farmed furs in terms of economic importance (United States International Trade

Commission 2004; Ward 2011). American minks are used in the commercial production of fur because their winter pelage is denser, longer, softer, and more luxuriant than in the European mink (Heptner and Sludskii 2002). Although some harvest of wild minks still occurs in some regions where the species remains abundant, mink farming now far exceeds it in importance. It has significant advantages to wild furskin trapping for production of pelts because it ensures controlled production of a large volume of high-quality fur and because pelage color can be controlled through selective breeding. Modern commercial fur-farming started in North American in the 1880s with the silver phase of the red fox, but mink farming was quickly added. Fur-farming reached a zenith in the United States during the 1920–1940s, with silver fox farming dominant prior to World War II and mink farming dominating after World War II, propelled by the development of mutant pelage colors starting in 1943 (International Fur Trade Federation, undated; Isto 2012). Since then, however, commercial fur-farming in the United

States has declined and, as of 2011, it accounted for only 5.9% of global farmed mink production (Ward 2011). Today the United States has only three million farmed minks (Maron 2021). China is currently the largest manufacturer and consumer of furs, and demand in southeastern Asia continues to increase with a growing middle class (Ward 2011).

There is evidence that mink farming occurred historically in New Mexico, but it seems to have been a rare enterprise. The New Mexico Department of Game and Fish apparently kept no records about historical fur-farming in the state, and it is uncertain when fur farms became illegal. In an unpublished manuscript, Elliot Barker mentioned that during the 1950–1951 federal fiscal year, somebody poached a deer in Cimarron Canyon, New Mexico in order to feed animals on a mink farm (M. W. Frentzel, pers. comm.). The US Census Bureau gathered data on fur-farming as an agricultural pursuit in 1940, at the height of the industry. The 1940 US census recorded that there was a single mink farm in New Mexico, but provided no information on location or production (US Department of Commerce, Bureau of the Census 1942). Kiker (1944) described the Taos Cañon Mink Ranch, located in Taos Canyon, Taos County, New Mexico. As of 1944, that mink farm was in its third year of operation. The source of the farm population was a pair of minks from Quebec that had black pelage, which was the original desirable color in captive-raised minks. The only other specific evidence I could find regarding historical mink farming in New Mexico concerned the Riedmont Fur Farm. This farm was located 5.7 km (3.5 mi) southeast (bearing 141°) of Tijeras, within what is today Bernalillo County's Sabino Canyon Open Space (www.bernco.gov/openspace). The farm was built by the Riedling Music Company family in the early 1920s, but it was abandoned in the late 1930s due to the Great Depression and concomitant crash of the fur markets. Apparently, the farm primarily raised

silver foxes, but it also held minks and rabbits (Ryan and Ausherman 2019). Several buildings on the farm are still standing.

Of recent concern is the worldwide pandemic due to the novel virus SARS-CoV-2, which causes the disease COVID-19 (coronavirus disease 2019) in humans. Farmed minks are susceptible to catching COVID-19 from human farm workers. The virus can then quickly spread through farmed mink populations causing mink mortality and possible routes of reinfection back to humans (Maron 2021). In some countries, millions of farmed minks have been euthanized to prevent transmission of the virus back to humans. It seems likely that wild mink populations would also be susceptible to contracting the virus, which could potentially impact small relict populations.

Future Prospects for the Mink in New Mexico

Is there hope for the mink in New Mexico? I hope the answer to that question is yes, but it will likely require a fundamentally different approach to how we manage riparian areas in the state. Minks cannot exist where excessive livestock grazing eliminates willows or reduces plant cover. On smaller streams, beavers must be allowed to exert their unique influence, creating the complex wetland ecosystems that foster both minks and their prey. Perhaps not all reaches of larger rivers must be so tightly controlled in terms of their flows, and perhaps some clever people can brainstorm how to re-tool our waterways so that they provide a fuller spectrum of ecosystem services. Perhaps the source of pollutants in our waterways can also be eliminated. Although there has been considerable alteration of our streams and rivers, there still exist a few places that have remained relatively unscathed. There is also renewed interest and effort in restoring impaired riparian and aquatic systems within the state, thus providing the potential for additional areas that can support minks in the future. Such changes can be achieved with vision and commitment. As a preeminent bellwether of riparian ecosystem health, the mink would benefit, but so too would a broad spectrum of other organisms, including us.

LITERATURE CITED

Abernethy, F. E. 1990. *The bounty of Texas*. Denton: University of North Texas Press.

Abramov, A. V. 2000. A taxonomic review of the genus *Mustela* (Mammalia, Carnivora). *Zoosystematica Rossica* 8:357–64.

Allen, J. A. 1893. List of mammals collected by Mr. Charles P. Rowley in the San Juan region of Colorado, New Mexico and Utah, with descriptions of new species. *Bulletin of the American Museum of Natural History* 5:69–84.

Anderson, E., S. C. Forrest, T. W. Clark, and L. Richardson. 1986. Paleobiology, biogeography, and systematics of the black-footed ferret, *Mustela nigripes* (Audubon and Bachman), 1851. *Great Basin Naturalist Memoirs* 8:11–62.

Armstrong, D. M. 1972. *Distribution of mammals in Colorado*. Monograph of the Museum of Natural History, University of Kansas No. 3.

Armstrong, D. M., J. P. Fitzgerald, and C. A. Meaney. 2011. *Mammals of Colorado*. 2nd edition. Boulder: University Press of Colorado.

Arnold, T. W., and E. K. Fritzell. 1987. Food habits of prairie mink during the waterfowl breeding season. *Canadian Journal of Zoology* 65:2322–24.

Autobee, R. 1994. Rio Grande project. Bureau of Reclamation. https://www.usbr.gov/projects/pdf.php?id=179.

Bailey, F. 1928. *Birds of New Mexico*. Santa Fe: New Mexico Department of Game and Fish.

Bailey, V. 1905. *Biological survey of Texas*. North American Fauna 25. Washington, DC: US Department of Agriculture, Bureau of Biological Survey.

———. 1931 (=1932). *Mammals of New Mexico*. North American Fauna 53. Washington, DC: US Department of Agriculture, Bureau of Biological Survey.

Ben-David, M., R. T. Bowyer, and J. B. Faro. 1996. Niche separation by mink and river otters: coexistence in a marine environment. *Oikos* 75:41–48.

Berghofer, C. B. 1967. Protected furbearers in New Mexico. In *New Mexico Wildlife Management*, 187–194. Santa Fe: New Mexico Department of Game and Fish.

Birks, J. D. S., and N. Dunstone. 1985. Sex-related differences in the diet of the mink *Mustela vison*. *Holarctic Ecology* 8:245–52.

Bonesi, L., and D. W. Macdonald. 2004a. Differential habitat use promotes sustainable coexistence between the specialist otter and the generalist mink. *Oikos* 106:509–19.

———. 2004b. Evaluation of sign surveys as a way to estimate relative abundance of American mink (*Mustela vison*). *Journal of Zoology* 262:65–72.

Boutin, S., and D. E. Birkenholz. 1999. Muskrat and round-tailed muskrat. *In Wild furbearer management and conservation in North America*, ed. M. Novak, J. A. Baker, M. E. Obbard, and B. Malloch, 315–24. Ontario: Ontario Fur Managers Federation. CD-ROM.

Brand, D. D. 1937. The report, part I, the natural landscape. In *Tseh so, a small house ruin, Chaco Canyon, New Mexico* (preliminary report), ed. D. D. Brand, F. M. Hawley, and F. C. Hibbens, 39–65. Albuquerque: University of New Mexico Bulletin.

Brzezinski, M., and W. Zurowski. 1992. Spring diet of the American mink *Mustela vison* in the Mazurian and Brodnica Lakelands, northern Poland. *Acta Theriologica* 37:193–98.

Burgess, S. A., and J. R. Bider. 1980. Effects of stream habitat improvements on invertebrates, trout populations, and mink activity. *Journal of Wildlife Management* 44:871–80.

Clifford, J. 1871. Something about my pets. *Overland Monthly* 6 (1 January):58–67.

———. 1877. *Overland tales*. San Francisco: A. L. Bancroft.

Coape, A. P. F. 1892. New Mexico bear hunting. *Forest and Stream: a weekly journal of the rod and gun*. 24 November 1892. 39(21):446.

———. 1893. New Mexico bears and bear dogs. *Forest and Stream*. 23 March 1893. 40(12):260–61.

Coues, E. 1877. Fur-bearing animals: a monograph of North American Mustelidae. *U.S. Geologic and Geographical Survey of the Territories, Miscellaneous Publications* 8:1–348.

De Reviziis, S. [2023]. The economic impact of the fur industry. welovefur.com. Accessed 9 March 2023. https://www.welovefur.com/the-economic-impact-of-the-fur-industry/.

Dunstone, N. 1978. The fishing strategy of the mink (*Mustela vison*); time-budgeting and hunting effort? *Behaviour* 67:157–77.

———. 1983. Underwater hunting behavior of the mink (*Mustela vison* Schreber): an analysis of constraints on foraging. *Acta Zoologica Fennica* 174:201–3.

———. 1998. Adaptations to the semi-aquatic habit and habitat. In *Behaviour and ecology of riparian mammals*, ed. N. Dunstone and M. L. Gorman, 1–16. Cambridge: Cambridge University Press.

Eagle, T. C., and J. S. Whitman. 1999. Mink. *In Wild furbearer management and conservation in North America*, ed. M. Novak, J. A. Baker, M. E. Obbard, and B. Malloch, 615–24. Ontario: Ontario Fur Managers Federation. CD-ROM.

Eberhardt, R. T. 1973. Some aspects of mink-waterfowl relations on prairie wetlands. *Prairie Naturalist* 5:17–19.

Erb, J., M. S Boyce, and N. C. Stenseth. 2001. Spatial variation in mink and muskrat interactions in Canada. *Oikos* 93:365–75.

Errington, P. L. 1943. An analysis of mink predation upon muskrats in north-central United States. *Research Bulletin of the Agricultural Experiment Station, Iowa State College of Agriculture and Mechanic Arts* 320:797–924.

Findley, J. S., A. H. Harris, D. E. Wilson, and C. Jones. 1975. *Mammals of New Mexico*. Albuquerque: University of New Mexico Press.

Fouty, S. C. 2018. Euro-American beaver trapping and long-term impact on drainage network form and function, water abundance, delivery, and system stability. In *Riparian research and management: past, present, and future*, Vol. 1, ed. R. R. Johnson, S. W. Carothers, D. M. Finch, K. J. Kingsley, and J. T. Stanley, 102–33. Fort Collins, CO: US Forest Service, Rocky Mountain Research Station, Gen. Tech. Rep. RMRS-GTR-377.

Frey, J. K. 2004. Taxonomy and distribution of the mammals of New Mexico: an annotated checklist. *Occasional Papers, Museum of Texas Tech University* 240:1–32.

———. 2018. Beavers, livestock, and riparian synergies: bringing small mammals into the picture. In *Riparian research and management: past, present, and future*, Vol. 1, ed. R. R. Johnson, S. W. Carothers, D. M. Finch, K. J. Kingsley, and J. T. Stanley, 85–101. Fort Collins, CO: US Forest Service, Rocky Mountain Research Station, Gen. Tech. Rep. RMRS-GTR-377.

Gerell, R. 1967. Food selection in relation to habitat in mink (*Mustela vison* Schreber) in Sweden. *Oikos* 18:233–46.

Gilbert, F. F., and N. Gofton. 1982. Heart rate values for beaver, mink and muskrat. *Comparative Biochemistry and Physiology* 73A:249–51.

Hall, E. R. 1981. *The mammals of North America*. 2nd ed. Vol. 2. New York: John Wiley.

Harding, L. E., and F. A. Smith. 2009. *Mustela* or *Vison*? Evidence for the taxonomic status of the American mink and a distinct biogeographic radiation of America weasels. *Molecular Phylogenetics and Evolution* 52:632–42.

Harris, A. H. 1993. Quaternary vertebrates of New Mexico. *Vertebrate Paleontology in New Mexico, New Mexico Museum of Natural History, Bulletin* 2:179–97.

Henderson, J., and J. P. Harrington. 1914. Ethnozoology of the Tewa Indians. *Bulletin of the Smithsonian Institution, Bureau of American Ethnology* 56:1–76.

Heptner, V. G., and A. A. Sludskii. 2002. *Mammals of the Soviet Union*. Vol. II, Pt. 1B, *Carnivores* (Mustelidae and Procyonidae). Washington, DC: Smithsonian Institution Libraries and National Science Foundation.

Hill, J. E. 1942. Notes on mammals of northeastern New Mexico. *Journal of Mammalogy* 23:75–82.

Hoffman, J. D., H. H. Genoways, and S. Wilson. 2009. Historical winter diets of mink (*Mustela vison*) in Nebraska. *Transactions of the Kansas Academy of Science* 112:119–22.

Hollister, N. 1913. A synopsis of the American minks. *Proceedings of the United States National Museum* 44:471–80.

Hubbard, J. P. 1987. Memorandum of the New Mexico Department of Game and Fish to Bruce Morrison regarding recent mink specimens from Cimarron Canyon, NM, 12 June 1987.

Huey, W. S. 1956. New Mexico beaver management.

New Mexico Department of Game and Fish Bulletin 4:1–49.

Ibarra, J. T., L. Fasola, D. W. Macdonald, R. Rozzi, and C. Bonacic. 2009. Invasive American mink *Mustela vison* in wetlands of the Cape Horn Biosphere Reserve, southern Chile: what are they eating? *Oryx* 43:87–90.

Isto, S. C. 2012. *The fur farms of Alaska: two centuries of history and a forgotten stampede.* Fairbanks: University of Alaska Press.

Iversen, J. A. 1972. Basal energy metabolism of mustelids. *Journal of Comparative Physiology* 81:341–44.

Jones, C., and C. G. Schmitt. 1997. Mammal species of concern in New Mexico. In *Life among the muses: papers in honor of James S. Findley,* ed. T. L. Yates, W. I. Gannon, and D. E. Wilson, 179–205. Albuquerque: University of New Mexico, Museum of Southwestern Biology. *Special Publication of the Museum of Southwestern Biology, University of New Mexico* 3:1–290.

Julyan, R. 1998. *The place names of New Mexico.* Albuquerque: University of New Mexico Press.

Kiker, V. 1944. Fur farm. *New Mexico Magazine.* March 1944:9, 33.

Korschgen, L. J. 1958. Food habits of mink in Missouri. *Journal of Mammalogy* 39:521–27.

Landholt, L. M., and H. H. Genoways. 2000. Population trends in furbearers in Nebraska. *Transactions of the Nebraska Academy of Sciences* 26: 97–110.

Lang, R. W., and A. H. Harris. 1984. *The faunal remains from Arroyo Hondo Pueblo, New Mexico: a study in short-term subsistence change.* Arroyo Hondo Archaeological Series 5. Santa Fe, NM: School of American Research Press.

Larivière, S. 1996. The American mink *Mustela vison* (Carnivora, Mustelidae) can climb trees. *Mammalia* 60:485–86.

———. 1999. *Mustela vison. Mammalian Species* 608:1–9.

———. 2003. Mink (*Mustela vison*). In *Wild mammals of North America: biology, management, and economics,* 2nd ed., ed. G. A. Feldhamer, B. C. Thompson, and J. A. Chapman, 662–71. Baltimore: Johns Hopkins University Press.

Leighton, A. H. 1933. Notes on the relations of beavers to one another and to the muskrat. *Journal of Mammalogy* 14:27–35.

Ligon, J. S. 1927. *Wildlife of New Mexico: its conservation and management.* Santa Fe: New Mexico State Game Commission.

Lindstrom, J. W., and W. A. Hubert. 2004. Mink predation on radio-tagged trout during winter in a low-gradient reach of a mountain stream, Wyoming. *Western North American Naturalist* 64:551–53.

Linscombe, G., N. Kinler, and R. J. Aulerich. 1982. Mink (*Mustela vison*). In *Wild mammals of North America: biology, management, and economics,* ed. J. A. Chapman and G. A. Feldhamer, 629–43. Baltimore: Johns Hopkins University Press.

Loukmas, J. J., and R. S. Halbrook. 2001. A test of the mink habitat suitability index model for riverine systems. *Journal of Wildlife Management* 29:821–26.

Maron, D. F. 2021. What the mink COVID-19 outbreaks taught us about pandemics. *National Geographic,* 24 February 2021. https://www.nationalgeographic.com/animals/article/what-the-mink-coronavirus-pandemic-has-taught-us

Mead, J. I., A. E. Spiess, and K. D. Sobolik. 2000. Skeleton of extinct North American sea mink (*Mustela macrodon*). *Quaternary Research,* 53:247–62.

New Mexico Department of Game and Fish (NMDGF). 1992. Checklist of the extinct, extirpated, and vanishing wildlife of New Mexico. Endangered Species Program, unpublished report, 12 February 1992, Santa Fe.

Obbard, M. E. 1999. Fur grading and pelt identification. *In Wild furbearer management and conservation in North America,* ed. M. Novak, J. A. Baker, M. E. Obbard, and B. Malloch, 717–826. Ontario: Ontario Fur Managers Federation. CD-ROM.

Patterson, B. D., H. E. Ramirez Chaves, J. F. Vilela, A. E. R. Soares, and F. Grewe. 2021. On the nomenclature of the American clade of weasels (Carnivora: Mustelidae). *Journal of Animal Diversity* 3(2). doi:10.29252/JAD.2021.3.2.1.

Peale, M., B. Long, and J. Klingel. 2022. *American mink (Neogale vison) habitat and population survey in northern New Mexico.* Final report submitted to New Mexico Department of Game and Fish Share with Wildlife Program, 12 January 2022.

Poole, T. B., and N. Dunstone. 1976. Underwater predatory behavior of the American mink (*Mustela vison*). *Journal of Zoology (London)* 178:395–412.

Proulx, G., J. A. McDonnell, and F. F. Gilbert. 1987.

The effect of water level fluctuation on muskrat, *Ondatra zibethicus*, predation by mink, *Mustela vison*. *Canadian Field-Naturalist* 101:89–92.

Robinson, L., S. A. Cushman, M. K. Lucid. 2017. Winter bait stations as a multispecies survey tool. *Ecology and Evolution* 7:6826–38.

Ryan, S., and S. Ausherman. 2019. *60 hikes within 60 miles: Albuquerque including Santa Fe, Mount Taylor, and San Lorenzo Canyon*. 3rd ed. Birmingham, AL: Menasha Ridge Press.

Schmidly, D. J. 2004. *The mammals of Texas*. Rev. ed. Austin: University of Texas Press.

Schmitt, C. G. 2001. Memorandum of the New Mexico Department of Game and Fish to ESP *Mustela vison* files, 26 July 2001.

Scurlock, D. 1998. *From the rio to the sierra: an environmental history of the middle Rio Grande basin*. General Technical Report RMRS-GTR-5. Fort Collins, CO: US Department of Agriculture, Forest Service, Rocky Mountain Research Station.

Sealander, J. A. 1943. Winter food habits of mink in southern Michigan. *Journal of Wildlife Management* 7:411–17.

Sealfon, R. A. 2007. Dental divergence supports species status of the extinct sea mink (Carnivora: Mustelidae: *Neovison macrodon*). *Journal of Mammalogy* 88:371–83.

Shier, C. J., and M. S. Boyce. 2009. Mink prey diversity correlates with mink-muskrat dynamics. *Journal of Mammalogy* 90:897–905.

Shufeldt, R. W. 1895. Some Fort Wingate reminiscenses—New Mexico. *The Nidiologist* 2:102-5.

Sinclair, W., N. Dunstone, and T. B. Poole. 1974. Aerial and underwater visual acuity in the mink *Mustela vison* Schreber. *Animal Behaviour* 22:965–74.

Small, B. A., J. K. Frey, and C. Gard. 2016. Livestock grazing limits beaver restoration in northern New Mexico. *Restoration Ecology* 24:646–55.

Stone, W., and J. A. G. Rehn. 1903. On the terrestrial vertebrates of portions of southern New Mexico and western Texas. *Proceedings of the Academy of Natural Sciences of Philadelphia* 55:16–34.

Svihla, A. 1931. Habits of the Louisiana mink (*Mustela vison vulgivagus*). *Journal of Mammalogy* 12:366–68.

Thompson, B. C., D. F. Miller, T. A. Doumitt, and T. R. Jacobson. 1992. *Ecologically-based management evaluation for sustainable harvest and use of New Mexico furbearer resources*. Santa Fe: New Mexico Department of Game and Fish.

United States Department of Commerce, Bureau of the Census. 1942. *Sixteenth census of the United States: 1940. Agriculture*. Vol. 1, *First and second series state reports*. Pt. 6, *Statistics for counties: farms and farm property, with related information for farms and farm operators, livestock and livestock products, and crops*. Washington, DC: United States Printing Office.

United States International Trade Commission. 2004. *Industry and trade summary: furskins*. USITC Publication 3666.

Valenzuela, A. E. J., A. R. Rey, L. Fasola, R. A. S. Samaniego, and A. Schiavini. 2013. Trophic ecology of a top predator colonizing the southern extreme of South America: feeding habits of invasive American mink (*Neovison vison*) in Tierra del Fuego. *Mammalian Biology* 78:104–10.

Viljugrenien, H., O. C. Lingjaerde, N. C. Stenseth, and M. S. Boyce. 2001. Spatio-temporal patterns of mink and muskrat in Canada during a quarter century. *Journal of Animal Ecology* 70:671–82.

Ward, S. 2011. US mink: state of the industry—2011. Fur Commission USA, 12 December 2011. Accessed 2 January 2013. https://faunalytics.org/wp-content/uploads/2015/05/Citation2185_USMinkStateOfIndustry2011.pdf.

Weber, D. J. 1971. *The Taos trappers: the fur trade in the far Southwest, 1540–1846*. Norman: University of Oklahoma Press.

Wild, C. 2011. *Beaver as a climate change adaptation tool: concepts and priority sites in New Mexico*. Santa Fe: Seventh Generation Institute.

Willner, G. R., G. A. Feldhamer, E. E. Zucker, and J. A. Chapman. 1980. *Ondatra zibethicus*. *Mammalian Species* 141:1–8.

Wozencraft, W. C. 2005. Order Carnivora. In *Mammal species of the world: a taxonomic and geographic reference*, Vol. 1, ed. D. E. Wilson and D. M. Reeder, 532–628. 3rd ed. Baltimore: Johns Hopkins University Press.

Wren, C. D. 1991. Cause-effect linkages between chemicals and populations of mink (*Mustela vison*) and otter (*Lutra canadensis*) in the Great Lakes basin. *Journal of Toxicology and Environmental Health* 33:549–85.

Wren, C. D., D. B. Hunter, J. F. Leatherland, and P.

M. Stokes. 1987a. The effects of polychlorinated biphenyls and mercury, singly and in combination, on mink. I: uptake and responses. *Archives of Environmental Contamination Toxicology* 16:441–47.

———. 1987b. The effects of polychlorinated biphenyls and mercury, singly and in combination, on mink. II: reproduction and kit development. *Archives of Environmental Contamination Toxicology* 16:449–54.

Wren, C. D., P. M. Stokes, and K. L. Fischer. 1986. Mercury levels in Ontario mink and otter relative to food levels and environmental acidification. *Canadian Journal of Zoology* 64:2854–59.

Yarbrough, J. W., and E. H. Studier. 1968. Mink in northeast New Mexico. *Southwestern Naturalist* 13:105.

Youngman, O. M. 1982. Distribution and systematics of the European mink *Mustela lutreola* Linnaeus 1761. *Acta Zoologica Fennica* 166:1–48.

25

AMERICAN BADGER (*TAXIDEA TAXUS*)

Robert L. Harrison and Jean-Luc E. Cartron

Appearances can indeed be deceiving, as in the case of the American badger (*Taxidea taxus*), a relatively large member of the Mustelidae (the weasel family). As many as 16 so-called badger species are known to occur worldwide (mainly in Asia), but they do not form a natural taxonomic assemblage or clade (i.e., all the descendants of one common ancestor) (e.g., Koepfli et al. 2008; Sato 2016). Instead, they are members of distinct lineages in the Mustelidae and the Mephitidae (the skunk family), previously lumped together based on ecomorphological similarities as they tend to fill the same ecological niche (Sato 2016). Within the badger ecomorph, the American badger alone occurs in North America, and it has no close, extant relatives. In his classification of the badgers, Sato (2016) assigns *Taxidea taxus* to its own subfamily Taxidiinae, the most basal lineage within the Mustelidae. Thus, the American badger may not be any more closely related to the superficially similar European badger (*Meles meles*) than it is to other mustelids with strikingly different morphologies and life habits including, in New Mexico, the semi-arboreal Pacific marten (*Martes caurina*; Chapter 20) and the semi-aquatic river otter (*Lontra canadensis*; Chapter 19). The intraspecific taxonomy of the American badger is less convoluted, with four subspecies widely recognized, of which only one, *T. t. berlandieri*, is found in New Mexico (Long 1972; Hall 1981; Long and Killingley 1983). *T. t. berlandieri* is occasionally

Photos 25.1a and b (*top to bottom*). Although the American badger (*Taxidea taxus*; Photo 25.1a) and the European badger (*Meles meles*; Photo 25.1b) show some striking similarities in general build and appearance, they are not close relatives. Badgers of the world form an ecomorph as they tend to fill the same ecological niche, but they constitute a non-monophyletic assemblage and are spread out over two separate families, the Mustelidae and the Mephitidae. In medieval England, both the striking head markings of European badgers and the conspicuous patterns on the shields of knights were referred to as "badges"; hence the name "badger" (Justice 2015). Photographs: © Sally King (a) and © Marc Baldwin (b).

(*opposite page*) Photograph: © Ed MacKerrow/In Light of Nature Photography.

referred to as the "Mexican" badger (e.g., Coues 1877; Bailey 1931; Long 1972; Schmidly 2002).

Long and Killingley (1983) describe badgers in general as "nature's supreme digging machine," with long front claws, short muscular forelimbs, partial webbing between toes, and squat bodies. Shared features of the forelimb structure include a robust humerus, large humeral epicondyles, and a long olecranon, but *Taxidea taxus* ranks among the more specialized badger species, with greater forelimb muscle mass, increased mechanical advantage of the elbow extensors, and the enhanced ability of the elbow extensor and carpal and digital flexor muscles to apply high out-force (Rose et al. 2014). Other shared or unique characteristics of the American badger include shovel-like hind claws, a conical head, small eyes that are protected from flying dirt particles by nictitating membranes, and short

Photo 25.2. An American badger ready to defend itself near Conchas Lake in San Miguel County on 12 August 2009, after seeking refuge within a pile of large concrete slabs just off a road. When threatened, American badgers vocalize by growling and hissing. The hissing of a badger in a trap is quite ominous. Combined with a tense posture and steady stare, a hissing badger presents the same prospect as a coiled rattlesnake, quite ready and willing to defend itself. Note the conical head, small eyes, and short ears, all representing adaptations to digging and burrowing. Photograph: © Christopher Newsom.

ears covered on both sides by long, stiff hairs (bristles) that similarly keep the ear canals clean during digging (Long and Killingley 1983). Upon close examination, a badger's skin appears to be too large for it, and the body underneath seems surprisingly small (R. L. Harrison, pers. obs.). Like other burrowing mammals, the American badger has very loose skin, which allows it to turn around in tight tunnels. All these specialized traits contribute to the American badger leading a semi-fossorial life, with digging and burrowing vital to its foraging strategy (Nowak 1999; Rose et al. 2014).

American badgers are not well known to the general public, but they are often perceived as irascible, even aggressive, animals, and instead of running will indeed often stand their ground against much larger opponents. The root of their pugnacity likely lies in their specialization for digging, which has led them to be somewhat slow runners. Over short distances, badgers are fast enough to capture rodents above ground (e.g., Eads et al. 2012) and outrun some humans (D. Eads, pers. comm.), but they are not built for speed like other carnivores. The tendency to fight is served by the very thick, loose skin, similar to that of Shar-pei dogs, allowing badgers to twist around and bite back even after being caught by a predator. The head markings of American badgers are also a striking combination of black and white that may serve as warning (aposematic) coloration, signaling to predators that they risk injury if they attack them, as a direct result of their aggressive defensive behavior (Newman et al. 2005).

The feet of an American Badger are blackish or black. Ventrally, the pelage varies from cream to buff. Dorsally, it has two layers. The outer layer is shaggy and consists of long black and white guard hairs, giving badgers their grizzled gray appearance. The inner layer (underfur) is denser, and its coloration varies over the badger's range, reflecting basic differences among subspecies,

Photos 25.3a, b, and c (*left to right*). Dorsal view of three UNM Museum of Southwestern Biology American badger specimens (MSB23259, MSB54854, and MSB85746) from New Mexico. The pelage is mainly gray with various amounts of white and brown. Note in particular the conspicuous, white mid-dorsal stripe that may extend only to the shoulders (a), or, alternately, run posteriorly to the rump as a continuous line (b) or a series of interrupted segments (c). Photographs: © Jon Dunnum.

Photo 25.4. Specimens of the American badger (dorsal views of head) showing differences in the length, thickness, and shape of the midline stripe. These differences in pattern can be used to identify individual badgers. Photographs: © Robert L. Harrison.

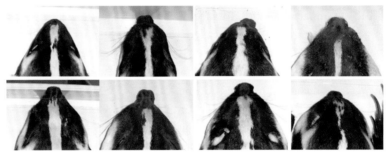

areas of intergradation including in New Mexico, and local variation (Long and Killingley 1983). In western portions of the species' distribution, the underfur has a rusty or reddish-brown coloration intermixed with white and silvery gray (Long and Killingley 1983). Proceeding eastward, including through New Mexico, the underfur has more grayish coloration mixed with yellowish-brown. Farther northeast, the underfur is darker, trending to dark brown and blackish mixtures with gray hair.

A white midline stripe extends from the nose over the head to the shoulders, or even runs the length of the body to the rump, but for some individuals, that dorsal stripe is a series of short, interrupted segments rather than a continuous line. The length and width of the stripe tend to show north-south clinal variation over the Great Plains, with longer stripes more prevalent in the south (Long 1972). No specimens west of Texas and Oklahoma were included in Long's (1972) analysis of clinal variation, and all that can be stated about New Mexico's American badgers is that some individuals in the same area may have stripes extending to the base of their tails, others only to the base of their heads (R. L. Harrison, pers. obs.). Research in New Mexico (Harrison 2016a; Gould and Harrison 2018) shows that variation in the pattern of the midline head stripe can be used to distinguish individuals in close-up photographs, just as pelage patterns of tigers (*Panthera tigris*) and bobcats (*Lynx* rufus) may be similarly used (Heilbrun et al. 2006; Karanth et al. 2006).

Reported lengths of adult badgers, including tails, vary from 600 to 890 mm (24–35 in, Nowak and Paradiso 1983; Messick 1987; Lindzey 2003). Tails (sometimes held erect while badgers walk) are bushy and vary from 98 to 174 mm (3.9–6.9 in). Weights of male and female badgers average approximately 8.2 and 6.2 kg (18 and 14 lbs), respectively (Lindzey 2003). As members of the more diminutive subspecies *T. t. berlandieri*, New Mexico badgers are likely smaller than those farther north (Long and Killingley 1983). Body measurements are available from the Museum of Southwestern Biology, which, it should be noted, tends to disproportionately represent collections from the central and northern parts of the state. Total body length of nine adult New Mexico male specimens averaged 677 mm (26.7 in) with a range of 610–750 mm (24–29.5 in). Total body length in 10 adult females was slightly smaller, averaging 658 mm (25.9 in) with a range of 608–704 mm (23.9–27.7 in). Weights (obtained from specimens in good condition when the animals were processed) averaged 6.7 kg (14.9 lbs) in 7 adult males but only 5.2 kg (11.5 lbs) in 10 adult females. One adult male specimen from Socorro County (MSB:Mamm:85725) weighed as much as 8.7 kg (19.2 lbs), one adult female specimen from Union County only 3.6 kg (8 lbs) (MSB:Mamm:294784). Overall, these measurements tend to confirm the smaller size of badgers in New Mexico compared to those farther north in the species' distribution. Whether the subspecies *berlandieri* exhibits any north-south clinal variation in body size or body weight is unknown. Of interest, for example, would be a comparison of body size between badgers in New Mexico and farther south.

There was no mention of badgers in the first compendium of New Mexico wildlife conservation and management written by naturalist J. Stokley Ligon (1927) with an exclusive focus on economically important game species. Vernon Bailey (1931), however, described the species in New Mexico, and his account of the badger's appearance, behavior, distribution, diet, and economic impact is arguably as complete and accurate as any modern summary. Afterward, distributional notes on the badger appeared in the scientific literature (e.g., Hooper 1941; Hill 1942; Halloran 1946), and a second compendium by the New Mexico Department of Game and Fish (1967) did include badgers, noting accurately that they occurred throughout the state. Since then, unfortunately, no modern scientific studies focusing primarily on badgers have been conducted in New Mexico, except for research by one of us (R. L. Harrison) on the Armendaris Ranch in Socorro and Sierra counties (Harrison 2015, 2016a, b; Gould and Harrison 2018; see below). At White Sands National Monument, Robinson et al. (2014) focused primarily on coyotes (*Canis latrans*) and kit foxes (*Vulpes macrotis*) in their occupancy modeling study but did obtain badger photographs from randomly distributed remote cameras. Although the number of photographs was insufficient to estimate badger occupancy, some data on habitat associations (G. Roemer et al., unpubl. data) were recorded and they are presented here. Despite additional research conducted in New Mexico on the species as an associate of prairie dog colonies (Goguen 2012; Davidson et al. 2018) and predator of prairie dogs (Kagel et al. 2020), most of our knowledge of American badgers and their ecology is still derived from studies conducted in the northern portion of their geographic range (see under "Distribution," below). Several potential differences between New Mexico badgers and those in more northern areas are nonetheless noted below.

DISTRIBUTION

The American badger has a wide distribution that extends from southwestern and central Canada south to the Baja California peninsula and central mainland Mexico, and from the Pacific coast east to the Great Lakes region and the Gulf Coast of Texas (Long 1972; Hall 1981). The Mexican subspecies, *T. t. berlandieri*, is found from southern California eastward through Arizona, New Mexico, and Texas—intergrading with *T. t. jeffersonii*

Photo 25.5. American badger at a wildlife drinker on the Armendaris Ranch in south-central New Mexico on 26 August 2012. Little research has been conducted on the American badger in the state, but one of the authors (R. L. Harrison) studied the species' habitat associations, population density, and patterns of activity on the Armendaris Ranch, located within the boundaries of the Chihuahuan Desert. Photograph: © Robert L. Harrison.

and *T. t. taxus* over a wide geographic area farther north—and southward to the southern end of the species' range in Mexico.

In New Mexico badgers occur throughout the state (Frey 2004; Map 25.1). Findley et al. (1975) stated that badgers are common in New Mexico, though rarely seen. In reality, the distribution and abundance of badgers in New Mexico have never been systematically studied. Reports from trappers compiled by the New Mexico Department of Game and Fish from 1980 to 2020, combined with museum specimen records, show that badgers can be found in every county of the state (Map 25.2; Table 25.1, Appendix 25.1). Trapping records are heavily influenced by the numbers of active trappers and traps set, both of which are not uniformly distributed throughout the state. Additionally, many trappers do not report their takings, and the numbers of animals killed and sold generally reflect the prices of pelts more than the population size of the target species. Similarly, in the case of badgers, museum records rely upon chance collections of specimens from roadkills and other haphazard sources rather

than systematic surveys. Thus, such records and actual population sizes are likely to be poorly correlated, though one would expect that in general more animals would be trapped and more specimens collected in areas with larger badger populations. Keeping such limitations in mind, in general these records suggest that badgers are relatively uniformly distributed throughout the state, with perhaps somewhat higher numbers in the central, southwestern, and northwestern counties (Table 25.1). Hill (1942) commented that near Cimarron in Colfax County, badgers seemed common on the high meadows along the eastern slopes of the Sangre de Cristo Mountains but rare farther east, on New Mexico's northeastern plains (see also under "Status and Management").

Currently, the New Mexico Department of Game and Fish is conducting camera trap surveys in selected areas of the state with emphasis on cougars (*Puma concolor*), bobcats, Pacific martens, and black bears (*Ursus americanus*). Observations of badgers were not reported, but individuals of that species did appear commonly in photographs except at elevations above ~2,750

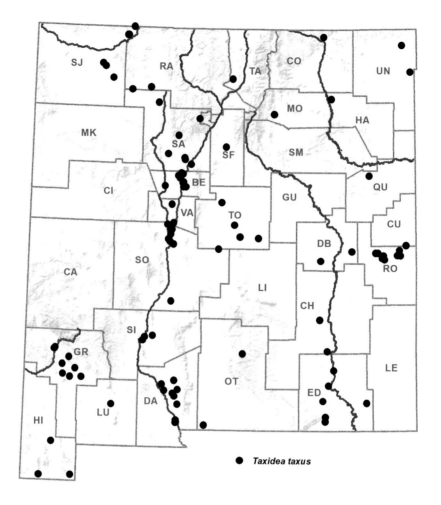

Map 25.1. Distribution of the American badger in New Mexico based on museum specimen records.

m (9,000 ft; N. Forman, NMDGF, pers. comm.). According to Long (1973), the upper limit of the American badger's elevational range is about 3,660 m (12,000 ft). In New Mexico, the species was reported as "often common near or above timber line," reaching at least in summer the alpine tundra biome (Bailey 1931:343; see also Chapter 2). More recent observations indicate that badgers, though perhaps in reduced numbers, have persisted at some of the higher elevations of the state including in Colfax County. On Vermejo Park Ranch, for example, biologist Dustin Long (pers. comm.) routinely found signs of badger activity at elevations above ~3,000 m

(10,000 ft) in association with Gunnison's prairie dog (*Cynomys gunnisoni*) colonies. In light of these old and more recent observations, the lack of specimen records from New Mexico's high mountains may seem surprising (see Appendix 25.1). One badger collected in July 1967 at only 2,700 m (9,000 ft) east of Cloudcroft in the Sacramento Mountains (UTEP:Mamm:1905) seemingly represents the highest specimen-based record for the state. The lack of specimen records from above 2,700 m (9,000 ft) clearly reflects a collection bias instead of the true upper elevational limit of the badger in New Mexico. In particular, the July 1938 mammalogical expedition in the Cimarron Mountains—on the Philmont

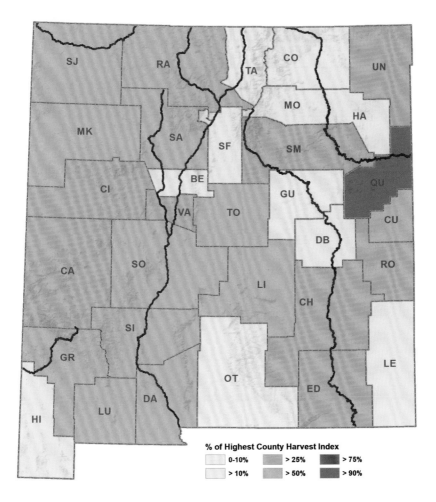

% of Highest County Harvest Index

0-10%	> 25%	> 75%
> 10%	> 50%	> 90%

Map 25.2. Harvest index by county, 2010–2020. All harvest indices are shown as percentages of the highest harvest index (in Quay County). See Table 25.1 for details on the calculation of harvest indices.

Ranch in Colfax County—attempted, but failed, to secure any specimens, despite traps being set near some of the highest peaks including in the vicinity of recent badger diggings (Hill 1942).

HABITAT ASSOCIATIONS

Choice of habitat for a badger depends upon several factors often operating at different spatial scales. Across its entire range, the species is most often associated with open grasslands, but it also occurs in many other vegetation types, including mixed, deciduous, redwood, and coniferous forests, sagebrush, creosote, and mesquite shrublands, marshes, brushy areas, deserts, alpine

tundra, parklands, and farms (Allen 1987; Lindzey 2003). In Idaho, for example, Todd (1980) captured badgers more often in annual grass, forb, and crested wheatgrass (*Agropyron cristatum*) communities but also recorded them in rabbitbrush (*Chrysothamnus* spp.) or sagebrush (*Artemisia tridentata*) shrubland. In coastal California, Quinn (2008) found that radio-collared badgers showed no habitat preferences when active, but located dens more frequently in native grassland and scrub vegetation than in riparian/wetland or urban areas, and more along mild slopes than on flat terrain. In eastern Washington, Paulson (2007) did not observe any habitat preferences

Table 25-1. Numbers of American badger museum specimens and reported number of American badgers trapped from 2010–2020 based upon New Mexico Department of Game and Fish records.* Harvest Index is Reported Harvest x 1000/ County Area.

County	Region	Museum Specimens**	Number Trapped Reported	County Area (km²)	Harvest Index	Regional Average Index
McKinley	Northwestern	1	86	14,113	6.09	7.04
San Juan	Northwestern	6	114	14,281	7.98	
Catron	West-central	0	59	17,943	3.29	3.35
Cibola	West-central	1	40	11,759	3.40	
Grant	Southwestern	9	58	10,272	5.65	4.71
Hidalgo	Southwestern	3	14	8,925	1.57	
Luna	Southwestern	2	53	7,679	6.90	
Los Alamos	North-Central	1	0	282	0.00	3.14
Rio Arriba	North-central	3	95	15,172	6.26	
Sandoval	North-central	7	68	9,609	7.08	
Santa Fe	North-central	2	9	4,944	1.82	
Taos	North-central	1	3	5,706	0.53	
Bernalillo	Central	11	2	3,020	0.66	5.16
Lincoln	Central	0	76	12,512	6.07	
Socorro	Central	21	98	17,216	5.69	
Torrance	Central	5	69	8,664	7.96	
Valencia	Central	2	15	2,766	5.42	
Doña Ana	South-central	15	58	9,860	5.88	4.29
Otero	South-central	2	12	17,164	0.70	
Sierra	South-central	4	68	10,826	6.28	
Colfax	Northeastern	1	24	9,731	2.47	2.81
Harding	Northeastern	1	11	5,506	2.00	
Mora	Northeastern	1	9	5,001	1.80	
San Miguel	Northeastern	0	37	12,217	3.03	
Union	Northeastern	3	47	9,920	4.74	
Curry	East-central	1	29	3,642	7.96	4.87
De Baca	East-entral	2	14	6,022	2.32	
Guadalupe	East-central	0	0	7,850	0.00	
Quay	East-central	1	83	7,394	11.23	
Roosevelt	East-central	11	18	6,343	2.84	
Chaves	Southeastern	1	53	15,724	3.37	3.42
Eddy	Southeastern	7	48	10,831	4.43	
Lea	Southeastern	0	28	11,378	2.46	

* No county-by-county trapping data were available for the 2011–2012 and 2014–2015 seasons.

** See Appendix 25.1 for a complete list of all American badger specimens from New Mexico.

Photos 25.6a and b (*left to right*). American badger mountain meadow habitat on 21 July 2013 at the Valles Caldera National Preserve. Note the presence of Gunnison's prairie dogs (*Cynomys gunnisoni*) in Photograph 25.6b. The association between American badgers and prairie dog colonies has been documented in many parts of the badger's distribution including in New Mexico. Photographs: © Mark L. Watson.

regarding land-cover type, soil type or depth, elevation, slope, or distance to roads or water when examined with large-scale GIS (Geographic Information System) databases. However, when habitat variables were measured directly within 5 m (16 ft) of individual burrows, badgers were found to select sites that 1) were near foraging areas with seemingly higher densities of their main prey, the northern pocket gopher (*Thomomys talpoides*); and 2) were associated with higher levels of horizontal cover (i.e., obscuring vegetation), where they could hide from predators. The availability of prey, in particular, is critical (see under "Life History" below), with numerous studies demonstrating that badgers are attracted to high concentrations of burrowing rodents (Lindzey 2003; Shaughnessy and Cifelli 2004; Paulson 2007). Also, because digging is so central to the ecology of badgers, soils must be friable (easily excavated) but firm enough to hold burrows, well drained, and not too rocky.

In New Mexico, badgers have been fairly well documented in grasslands, but they presumably occur in most other vegetation types. For example, the species was found in pinyon-juniper foothills, along arroyo bottoms in Bernalillo and adjacent counties (Ivey 1957). Around the lava fields of Valencia and southern McKinley counties, Hooper (1941) observed badgers in sandy plains wherever big sagebrush (*Artemisia tridentata*) was abundant. More recently, remote cameras deployed at White Sands National Monument recorded badgers mainly in honey mesquite (*Prosopis glandulosa*), pickleweed (*Salicornia* sp.), and fourwing saltbush (*Atriplex canescens*) shrublands and in gypsum interdune swale grasslands, but also, to a lesser extent, on vegetated dune outcrops and gypsum duneland—barren soils (G. Roemer, unpubl. data; see Robinson et al. 2014). Based upon studies from other states, museum specimens, road kills, and interviews, Thompson et al. (1992) determined

Photo 25.7. Three American badgers at or near the entrance to the family burrow on 10 July 2007 at Maxwell National Wildlife Refuge in Colfax County. Badger habitat at the refuge consists of shortgrass prairie. Photograph: © Patty Hoban.

Photo 25.8. American badger and surrounding habitat in Socorro County. Within the Chihuahuan Desert in New Mexico, the species occurs in both plains-mesa sand scrub and semi-desert grassland. Photograph: © WSMR Garrison Environmental Division and ECO Inc., Ecological Consultants.

primary habitats for badgers in New Mexico to be juniper savanna, plains-mesa grassland, desert grassland, Chihuahuan Desert scrub, Great Basin Desert scrub, plains-mesa sand scrub, and closed basin scrub, which together cover about two-thirds of New Mexico. All other vegetation types in New Mexico were considered to be secondary habitats, including alpine tundra, subalpine and mixed conifer forest, conifer and mixed woodlands, subalpine montane grassland, montane scrub, urban areas, farmland, lava beds, and sand dunes. However, Thompson et al. (1992) did not conduct ground-truthing surveys. In their final assessment the northeastern counties of New Mexico were considered to contain mostly primary habitat. Thus, those counties would be expected to contain average, if not above average, numbers of badgers, in contrast to results from trapper reports described above. Given the limitations of both trapper reports and Thompson et al. (1992), better approaches are needed to model badger habitat associations in New Mexico (see under "Status and Management").

A growing body of research, some of it already

mentioned, shows the existence of environmental drivers of badger habitat associations that are more important than vegetation type. Bailey (1931) described New Mexico's American badgers as primarily associated with burrowing rodents rather than specific vegetation communities, with rocky and timbered areas reportedly least favored because they have lower biomass of that type of prey. Bailey's (1931) assessment was based on observations alone—for example, the San Juan River Valley and surrounding higher country were mentioned as once harboring high numbers of badgers and an apparent abundance of prairie dogs (*Cynomys* spp.), ground squirrels (e.g., *Xerospermophilus spilosoma*), kangaroo rats (*Dipodomys* spp.), and pocket gophers (family Geomyidae). On Vermejo Park Ranch in Colfax County, Goguen (2012) surveyed birds and mammals along paired 600 m linear transects established 1) within Gunnison's prairie dog colonies; and 2) within associated shortgrass prairie without prairie dogs ("control" transects). Using distance sampling, relative abundance of badgers was estimated as 12 times higher in prairie dog colonies. With less than 20%

of detections along control transects, the American badger was one of six species showing a strong, positive association with Gunnison's prairie dog colonies. Similarly, on grasslands of the Sevilleta National Wildlife Refuge, a study using camera traps in treatment and control plots ranked the American badger among the four species most strongly associated with Gunnison's prairie dog colonies (Davidson et al. 2018).

Habitat associations of American badgers were examined by one of us (R. L. Harrison) on the Armendaris Ranch in the Chihuahuan Desert of southern New Mexico (see the discussion of density estimates under "Status and Management"). The Armendaris Ranch has an unusually large number of wildlife water sources (a.k.a drinkers or guzzlers), provided for its quail (*Callipepla squamata* and *C. gambelli*) hunting program (Rollins et al. 2006). With use of automatically triggered cameras to observe badgers at those anthropogenic water sources, the species was recorded most commonly in plains-mesa sand scrub (primarily in association with honey mesquite); it was also detected, but to a lesser extent, in semi-desert grassland, which on the Armendaris Ranch contains much less obscuring vegetation than sand scrub. However, soil depth, rather than vegetation type, was an important covariate explaining variation in badger density (Gould and Harrison 2018). Badgers drank only about half of the time when visiting water sources. Together with low rates of visitations, that finding suggested that even during summer months in a hot desert, water by itself plays a minimal role in the choice of habitat for a badger.

In addition to responding to environmental variations, badgers themselves modify habitats through their digging activities. Badgers create new holes and also enlarge old holes created by prey such as prairie dogs or kangaroo rats. Dens may be single tunnels when used for a short period. Natal and overwintering dens may be several meters long, with side tunnels and shallow pockets in the

Photo 25.9. American badger on 5 August 2008, just west of Alma in southwestern Catron County. Habitat consisted of rolling flats with a mix of junipers (*Juniperus* sp.) and overgrazed grasslands. Photograph: © Christopher Newsom.

walls; natal dens typically have a single entrance, a branching tunnel with dead-end, lateral tunnels, side pockets, and at least one chamber (Long and Killingley 1983). Many other species will occupy abandoned badger dens, including snakes, lizards, rabbits, foxes, rodents, and insects. When excavating dens, badgers create fan-shaped mounds of soil at the entrances. Mounds are usually approximately one meter (3 ft) in diameter, but may be larger. Where badgers are common, mounds may be significant features of the landscape. In Idaho, Eldridge (2004) reported that badger mounds on the Snake River Plain occupied 5–8% of the landscape, with an average mass of 26 metric tons per hectare (11.6 US tons/acre). Mound soils exhibited plant succession as they aged, significantly contributing to plant diversity in the area. In New Mexico, badger mounds appear to promote shrub maintenance in desertified grasslands (Eldridge et al. 2009).

LIFE HISTORY
Diet and Foraging
Over their entire range badgers consume a tremendous variety of small mammals and other

Photo 25.10. American badger pursing a prairie dog on 21 July 2013 at the Valles Caldera National Preserve. Although very broad, the badger's prey base consists largely of small, burrowing mammals. Photograph: © Mark L. Watson.

Photo 25.11. American badger after it killed a juvenile black-tailed prairie dog (*Cynomys ludovicianus*) at Vermejo Park Ranch, Colfax County on 4 August 2010. The badger was observed using tire-ruts in a two-track dirt road to covertly cross the prairie dog colony, occasionally stopping, lifting its head above the tire-rut, and scanning the colony from side to side. It chased the juvenile prairie dog over a distance of 22 m (72 ft) and captured it with a bite to the back, before shaking it from side to side for about 3 seconds. For more details, see Eads et al. 2012. Photograph: © David Eads.

prey, with seasonal variation and differences between juveniles and adults both documented (Messick and Hornocker 1981; Long and Killingley 1983). Lindzey (2003) lists 41 species of mammalian prey alone. According to Lampe (1982), about two-thirds of the small mammal species in the badger's prey base are either fossorial or frequently use burrows. Prairie dogs, ground squirrels, kangaroo rats, mice (*Peromyscus* spp.) and pocket gophers are commonly taken, but so are cottontails (*Sylvilagus* spp.) and jackrabbits (*Lepus* spp.). More rarely, skunks trapped in their dens may be eaten, and in at least one documented case, so were coyote cubs (Long and Killingley 1983). Insects and ground nesting birds are also part of the diet, though these may be preyed upon more commonly by young, dispersing badgers who have not developed the same foraging skills as adults and rely instead on chance encounters (Messick and Hornocker 1981). Lizards, amphibians, carrion, and, where available, fish are also consumed (Long and Killingley 1983). Plant material, including corn and peas, and mushrooms are occasionally eaten. Bailey (1931) described the diet of New Mexico badgers based upon stomach contents. His description is very similar to that reported from other areas, with primary reliance upon burrowing rodents such as prairie dogs, squirrels, and mice. Cottontails were occasionally taken, as were insects, especially cicadas. One snake was found in a stomach, and domestic chickens were consumed in a few instances.

Badgers are opportunistic predators, in the sense that they generally take prey in proportion to their abundance, as opposed to specializing in particular prey. For example, Messick and Hornocker (1981) observed that in southwestern Idaho, Piute ground squirrels (*Urocitellus mollis*)—formerly considered Townsend's ground squirrels (*Spermophilus townsendii*); (Yensen and Sherman 2003; Helgen et al. 2009)—were major prey items when population densities of these rodents were high. Piute ground squirrels contributed significantly less when their population numbers were lower, at which time badgers increased predation on other rodents and lagomorphs. At more northern latitudes, burrowing rodents are protected by frozen ground and deep snow during winter. As a result, badgers remain in dens for extended periods, and they may reduce their activity by over 90%, using fat reserves and even entering torpor (Harlow 1981; Messick 1987). New Mexico badgers on the Armendaris Ranch (elevation 1,600 m [~5,250 ft]) did not show periods of winter inactivity (R. L. Harrison, pers. obs.), but it is unknown if populations of *T. t. berlandieri* at higher elevations reduce their activity.

Badgers usually hunt alone. Excavation of burrow systems is the most common hunting technique (Michener 2004). When pursuing prey in burrows, badgers may block entrances by moving soil or sod from the mound of the victim or a nearby mound. Uprooted vegetation or snow may also be used. The intentional use of objects for a preconceived purpose constitutes tool using, and Michener (2004) observed a badger using wooden marker blocks and a rock, either alone or in conjunction with soil, to block entrances. Badgers are adept at locating and excavating hibernating and infant animals. On Michener's (2004) study site in Alberta, up to 60% of hibernating Richardson's ground squirrels (*Urocitellus richardsonii*) were taken, so many that badgers cached surplus prey in October and November.

Other hunting techniques are also used. Badgers can acquire prey above ground, sometimes using environmental features as cover before attacks (Michener 2004; Eads et al. 2012). Eads and Biggins (2008) suggested that badgers improved their chances of capturing black-tailed prairie dogs (*Cynomys ludovicianus*) by planning attacks using perceptions of speeds, angles, distances, and predicted escape responses of prey. A badger observed by Balph (1961) in Utah appeared to lie in wait just inside the entrance to an Uinta ground squirrel (*Urocitellus armatus*) burrow. The

Photo 25.12. American badger crossing a road in Colorado's Rocky Mountain National Park on 24 June 2011. The badger is holding a juvenile Wyoming ground squirrel (*Urocitellus elegans*) it just caught. Photograph: © Pat Gaines Photography.

badger had partially blocked the entrance, apparently for concealment. At that time of year (July) Uinta squirrels moved about rather freely and would dash into burrows when frightened. The level of intelligence demonstrated by badgers in these cases is quite remarkable (and see below the discussion of badger-coyote interactions).

Reproduction and Social Behavior

American badgers are promiscuous (polygynandrous), with males and females forming no pair bonds and instead mating with several members of the other sex (Lindzey 2003). Males are capable of breeding from late May through August or early September (Wright 1969; Long and Killingley 1983), at which time they expand their home range to overlap more those of females (Minta 1993; Collins et al. 2012; and see Hein and Andelt 1995 for a discussion of higher capture rates by humans during the badger's mating season as a result of increased movements). A small percentage of juvenile females breed (at four to five months old), but males do not exhibit full spermatogenesis until they are at least one year old (Messick and Hornocker 1981). Based on the

Photo 25.13. Badger family on 4 July 2009 at the Valles Caldera National Preserve. Adult females raise the young alone. Photograph: © Sally King.

histological examination of ovaries, conception typically occurs in late July and August (Wright 1966; Messick and Hornocker 1981), though mating has been observed as early as June in Idaho (Messick and Hornocker 1981) and as late as November in Nuevo León, Mexico (López Soto 1980).

American badgers in the northern portion of their distribution exhibit delayed implantation of the fertilized egg on the uterine wall until January or February (Hall 1946; Long and Killingley 1983; Lindzey 2003), and parturition thus occurs in March or early April, after a gestation period of only about five weeks (Long and Killingley 1983). Farther south, in the San Joaquin Valley of California, parturition may take place somewhat earlier, in February or March (Fry 1928), suggesting that implantation in some cases is delayed only until December. Young are born furred but blind and helpless (Long 1973; Long and Killingley 1983). Observations of a captive badger cub

reported by Mr. Tom Waddell, manager of the Armendaris Ranch in southern New Mexico, suggested that young may remain visually impaired for an extended period after their eyes open. Females raise the young alone, with no help from males (Davis 1946).

Based on research outside New Mexico, cubs remain mostly below ground for four to five weeks, and stay with their mothers for 10–12 weeks. Females lactate for five to six weeks, during which time they may also provide solid food to cubs (Messick and Hornocker 1981). By late May females begin to remain away from their litters for over 24 hours (Lampe and Sovada 1981). Families break up in June, July, or August. Badgers may live up to 14 years in the wild (Lindzey 2003), but most do not survive more than four years (Messick and Hornocker 1981).

Badgers are generally solitary, with the exceptions of mothers with young, breeding pairs, and occasional short-term sibling pairs (Lindzey

2003). Mutual avoidance appears important in territorial spacing and allows home ranges to overlap (Messick and Hornocker 1981), but fights between badgers do occur. Minta (1993) observed wounds on 73% of males and 50% of females. Badgers likely communicate with urine and feces, which are sometimes deposited in the entrances of dens where they are shaded and last longer. Badgers also have abdominal glands, which may be used in particular during the breeding season (López Soto 1980). During the litter-rearing period in Montana, a female badger known to have young traversed the edge of a black-tailed prairie dog colony and deposited scent marks at burrow openings surrounding the colony, suggesting the glands are also used outside of the breeding season (D. Eads, pers. comm.). Individuals observed with automatic cameras on the Armendaris Ranch occasionally examined water sources for several minutes, apparently probing smells. Vocal communications include hissing, growling, grunting, and purring. Hissing is the most common sound reported from captive badgers.

Interspecific Interactions

Badgers are themselves prey to larger predators, such as coyotes, cougars, wolves (*Canis lupus*), and bears (*Ursus* spp.) (Lindzey 2003), and predation by golden eagles (*Aquila chrysaetos*) has been documented in New Mexico (Stahlecker et al. 2010). In general, however, the badger's aggressive nature and fighting ability help protect it from natural enemies, though not from people (see under "Status and Management").

An unusual association between badgers and coyotes may be an example of cooperation advantageous to both, called mutualism (Minta et al. 1992; Thornton et al. 2018; see also Chapter 10). Native American folklore, early naturalists' reports, and anecdotal observations provide accounts of badgers and coyotes working together to secure rodents from burrows. Since coyotes are much larger than badgers (~10–12 kg

Photo 25.14. American badger and coyote (*Canis latrans*) in south-central New Mexico on 9 December 2015. Badger-coyote hunting associations have been documented in New Mexico and elsewhere. The association is beneficial to coyotes and likely also to badgers. When a burrowing rodent is located underground, the badger digs in pursuit of it. The coyote waits nearby in case the badger flushes out the prey aboveground. The rodent may remain underground to avoid the coyote, giving the badger a better chance to capture it. Photograph: © WSMR Garrison Environmental Division and ECO Inc., Ecological Consultants.

Photo 25.15. American badger harassed by a burrowing owl (*Athene cunicularia*) at the Sevilleta National Wildlife Refuge on 18 May 2015. Photograph: © Ana D. Davidson.

vs ~ 7 kg, 22–26 lb vs. 15 lb in New Mexico), it would seem more likely that coyotes would regard badgers as potential prey, or at least as competitors, and kill them as they do so many other species. However, Minta et al. (1992) were able to observe the association frequently on the National Elk Refuge in Wyoming in an area of high densities of coyotes, badgers, and Uinta ground squirrels. Coyotes and badgers were observed to move together, coyotes encouraging badgers to move by mock pursuit, leading, soliciting play, or scrambling about a specific site. In dense patches of squirrels, badgers and coyotes would simultaneously rush into the area, increasing chances to catch squirrels above ground or to trap them in shallow tunnels. If a squirrel was located underground, the coyote would wait nearby while the badger dug. Squirrels might be flushed out, giving the coyote a chance at capture. Or squirrels may remain within the burrows to avoid the coyote, giving the badger a better chance to capture it underground. Coyotes associated with badgers captured more squirrels than solitary coyotes. Since badgers caught and ate prey underground, it was not possible to directly observe predation by them. However, associated badgers spent less time above ground searching for other burrows, suggesting that prey had been obtained. Badgers usually ended the association by not re-emerging from a burrow. Coyotes were observed seizing prey from badgers on several occasions. The association appeared to be wholly beneficial to coyotes, but its overall benefit to badgers remains somewhat uncertain. It has not been documented in all parts of the badger's range. Minta et al. (1992) attributed the high frequency of the association in their study area to high densities of both predators and prey, and obscuring vegetation, which made hunting difficult. Using camera traps, Thornton et al. (2018) indeed found support for the existence of spatio-temporal variation in badger-coyote foraging associations with some of the same environmental drivers as

indicated by Minta et al. (1992). Coyote-badger associations have been reported from New Mexico (Photo 25.14; see also Chapter 10).

Near Ajo in southwestern Arizona, Devers et al. (2004) observed red-tailed hawks (*Buteo jamaicensis*) interacting with American badgers on six separate occasions all within a five-month period. Red-tailed hawks changed low perches to remain within a few meters of foraging badgers. In two instances, a red-tailed hawk dove down on a badger as though attempting to seize some escaping prey. No prey capture was observed, and the nature of the interaction remains unknown, whether interspecific competition or an ecological association beneficial to the hawk, while also not detrimental to the badgers (Devers et al. 2004).

STATUS AND MANAGEMENT

The American badger is considered a Species of Least Concern on the International Union for Conservation of Nature (IUCN) Red List due to being relatively common over a large range (IUCN 2012). The IUCN notes that badger populations have probably declined substantially due to conversion of native prairie to agriculture and reduction of prairie dogs and ground squirrels. Badgers remain threatened primarily by loss of habitat and prey, as well as by vehicle collisions and other factors (see below), but not yet at a level that would qualify them for a threat category. The size of the total American badger population is not known, but it is likely on the order of several hundred thousand (based on local population estimates and annual harvest; see below). The Canadian population is estimated to be around 30,000 (IUCN 2012). No estimates are available for the population in Mexico. Two subspecies are listed as endangered by the Committee on the Status of Endangered Wildlife in Canada (COSEWIC 2023). *T. t. jacksoni* occupies a small area in Ontario. The current population may be less than 200, and is threatened by development. *T. t. jeffersonii* occurs in British Columbia where its population is less than 600. The population of *T. t. jeffersonii* is fragmented as well as at risk from vehicle collisions and development. Canadian efforts to aid their badgers have included innovative research and translocation of badgers (e.g., Kinley and Newhouse 2008). In Mexico badgers are listed federally as endangered (Mexican Official Norm NOM-059-SEMARNAT-2010) and face severe conservation problems, especially hunting and poisoning to protect small domestic animals (Ceballos and Navarro 1991).

Humankind is without doubt the American badger's greatest known enemy, with human-related deaths occurring principally through vehicle collisions, conflicts with farmers and ranchers, and fur trapping. In Idaho, Messick and Hornocker (1981) reported that of 51 observed mortalities, 88% were due to people, either through roadkills, indiscriminate shooting, farmers destroying problem animals, interactions with farm machinery, or fur trappers. Badgers may harm agricultural interests in several ways, including damage to crops caused by digging for rodents within fields, damage to farm equipment or livestock from falling into excavations, damage to dams, water lines, levees, or dikes, occasional depredation on poultry or newborn lambs, and injuries to dogs who usually come out second best in a tangle with a badger (Long and Killingley 1983; Johnson 1983; Minta and Marsh 1988; Lindzey 2003). Minta and Marsh (1988) provide information on activity by the USDA Animal Damage Control (ADC)—renamed Wildlife Services, within the USDA Animal and Plant Health Inspection Service (APHIS)—in New Mexico for the years 1977, 1982, and 1987. During those years, ADC caught a total of 602 badgers (250 in 1977, 173 in 1982, and 179 in 1987), or about 3% of the agency's total carnivore/furbearer take for the state during those corresponding years. Even though most of the badger take (as high as 94% in 1987) was bycatch from predator control largely targeting coyotes, the vast majority of the badgers

Photo 25.16. American badger in Socorro County on 8 August 2008. The main threats to the American badger are anthropogenic and include collisions with road traffic and loss of habitat and prey. Also of concern are fur trapping and conflicts with farmers and ranchers. Photograph: © Douglas W. Burkett.

Photo 25.17. American badger roadkill along US Highway 285 near Artesia on 5 July 2011. The impact of roads and road traffic is unknown in New Mexico, but in some parts of the species' distribution, collisions with vehicles were documented as the leading source of mortality. Increased numbers of road-crossing structures for wildlife would likely benefit the species in New Mexico. Photograph: © Jean-Luc Cartron.

captured by ADC were killed rather than released (Minta and Marsh 1988). The benefits that badgers provide to agriculture by controlling rodents are unseen and usually unappreciated. An adult badger requires approximately two ground squirrels per day for body maintenance, and during breeding or raising cubs requires much more (Lindzey 2003). A study of the economic impact of rodent control by natural predators would be a valuable contribution to our appreciation of carnivores.

The effects of badger predation upon prey populations can be difficult to determine in the field because badgers are usually just one of several species of predators present. Coyotes, foxes, raccoons (Procyon lotor), bobcats, and raptors may prey upon the same species. The examination of prey remains can provide clues as to the identity of the predator. In the case of the American badger, however, many of the prey are small and may be consumed in one meal. Thus, there may not be any remains to examine, making any determination often impossible. Nonetheless, several studies of colonial ground squirrels have been able to provide information on the impact of badgers (Murie 1992; Michener 2004). Murie (1992) reported that badgers killed up to 56% of juvenile Columbian ground squirrels (Urocitellus columbianus) in a year, but higher survivorship of the remaining juveniles was able to maintain population size. In southern Alberta, densities of Richardson's ground squirrels remained high during a 15-year study, despite heavy predation by badgers documented in some years, both on hibernating individuals in the fall and litters in the spring (Michener 2004). Several badgers hunted at one time or another on the study site, but badger activity was limited or absent in some years (presumably as resident individuals died or dispersed to other locations). The absence (or relative absence) of badgers in those years seemingly offset high mortality rates in their prey when more hunted on the study site (Michener 2004). In the Sevilleta National Wildlife Refuge

in Socorro County, badgers severely hampered attempts to reintroduce Gunnison's prairie dogs in 2011. Reintroduced prairie dogs were very vulnerable during the first days after relocation to the refuge, when they were initially placed in artificial underground burrows. The prairie dogs almost immediately risked exposure to predators by leaving the artificial burrows in order to dig their own burrows (R. L. Harrison, pers. obs.). However, other studies show that the impact of badgers on reintroduced prairie dogs can be lessened by building enclosure fences (Truett et al. 2001). On a prairie dog reintroduction site on the Armendaris Ranch, Truett and Savage (1998) built four enclosures, each ~0.4 ha in size, of mesh chicken wire with 1) an interior apron buried as a U-shaped loop in a trench; and 2) electrical wires 10 cm above the ground, both inside (to prevent prairie dog dispersal) and outside (to repel badgers). During the one to two years the fences remained in place, badger diggings and tracks were observed immediately outside some of the fencing. Yet, no signs of badger intrusion were ever observed inside the enclosures (Truett and Savage 1998). To be successful, prairie dog reintroductions may thus initially require exclusion—or even relocation—of American badgers.

Badgers are hosts to numerous endoparasites (flatworms, ribbon worms, and tapeworms) and ectoparasites (fleas, lice, mites, and ticks; Lindzey 2003). They are often exposed to plague through their prey, particularly ground squirrels, and fleas in burrows. Among carnivores, sero-prevalence is highest in mustelids, probably reflecting degree of exposure from dietary habits (Salkeld and Stapp 2006). Messick et al. (1983) found that most of the badgers in their study area in Idaho had been exposed to plague, and that the percentage of badgers with antibodies indeed correlated with the percentage of ground squirrels in their diet. Many carnivores do not readily develop clinical symptoms of plague (e.g., Arjo et al. 2003; but see Chapter 22). However, they may

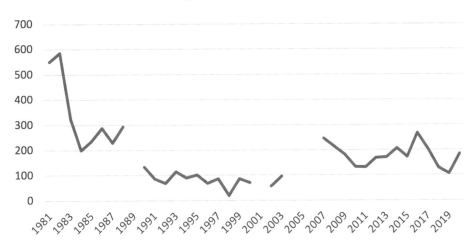

Number of Badgers Reported Trapped per Year

Figure 25.1. Numbers of American badgers reported trapped by fur trappers in New Mexico from 1918 to 2020 based upon New Mexico Department of Game and Fish fur harvest records. No data are available for the 1988–1989, 2000–2001, and 2003–2006 seasons. Reporting rates prior to 2006 were low and likely led to significant underestimation of numbers trapped.

spread it between colonies of rodents or even to other carnivores (Salkeld and Stapp 2006).

American badger fur was in the past used for paint and shaving brushes (Long and Killingley 1983), but is now primarily in demand for the garment industry. The demand for badger pelts has fluctuated with fashion. A notable peak occurred in the 1920s and led to a drastic population reduction in North America (Obbard et al. 1987). In the 1980s average annual harvest of badgers in the United States was about 33,800. From 1980 to 2020 an average of 179 badgers were reported killed annually by trappers in New Mexico, but additional unreported mortalities were likely (NMDGF, unpubl. records; Figure 25.1.).

The average price for a badger pelt at the Fur Harvesters Auction, one of the premier North American fur auctions, from 2003 through 2020 was $25.37, which is generally comparable to other species (Fur Harvesters Auction Inc. 2020). However, the number of badger pelts sold each year is low in comparison with significant year-to-year variation. In the 18 fur sale seasons at Fur Harvesters Auction between 2003 through 2020, an average of 408 pelts were sold during ten seasons, but in eight other seasons no badger pelts were sold. Most other furbearing species were sold each year and with numbers typically in the thousands.

In New Mexico, the American badger is listed as a "protected furbearer" (New Mexico Statute 17-2-2), meaning that any activities involving hunting or trapping of badgers are regulated by the state. Badgers can be trapped, snared, shot with guns or arrows, or hunted with dogs without limit from November 1 through March 15 (NMDGF 2020a). Managing sustainable harvest requires that population size trends be known, but just how many American badgers occur in New Mexico has not been determined. A non-technical publication by the New Mexico Department of Game and Fish in 1979 stated the population as 60,000 (Ames 1979), but no explanation of how this figure was obtained was presented. In 2006 the New Mexico Department of Game and Fish used population density estimates from studies in other states combined with area estimates for badger habitat in New Mexico to estimate the New Mexico population to be approximately 25,000. It is likely a gross overestimation, as shown by badger density estimates now available from research in New Mexico. Gould and Harrison (2018) used camera traps and a spatially explicit capture-recapture model to calculate American badger densities on the Armendaris Ranch, which is located in the south-central part of the state and characterized by Chihuahuan Desert vegetation. Their estimate of 0.1 badger/km² was significantly lower than estimates from previous studies farther north, with the exceptions of the endangered badger population in British Columbia (only 0.01 badger/km²; Klafki 2014) and an Illinois population in an agricultural landscape (0.15 badger/km²; Warner and Ver Steeg 1995). Gould and Harrison's (2018) estimates were 10 to 20, or even 50 times lower than those calculated in Idaho (1.6–5.1 badgers/km²; Messick and Hornocker 1981) and Wyoming (≥ 2 badgers/km²; Minta 1993). Given the wide variation in badger population densities among states and likely also among ecosystems as well as the high variability of demand for badger pelts, badger management by the New Mexico Department of Game and Fish should in our view be very conservative.

Badgers are relatively small and nocturnal, and they avoid people and spend their days hidden in dens. Directly counting them, as one might do with larger species like elk (*Cervus canadensis*) or deer (*Odocoileus* spp.), is not productive. Thus, wildlife biologists and game managers must sample populations, then estimate population parameters mathematically. In theory, this might be accomplished by four basic methods: interpretation of harvest data, trap and release programs for research purposes, remotely observing badgers without trapping them, and indirectly observing badgers by observing signs of their presence. On the Armendaris Ranch, one of us (R. L. Harrison) has studied the effectiveness of various survey methods, most of which prove useful within some limitations (Harrison 2015).

Interpretation of harvest data is the survey method used most commonly by game management agencies and is the method used by the New Mexico Department of Game and Fish. Since 2006 trappers in New Mexico have been required to report their catches, although only an average of 77% did (NMDGF 2020b). The level of activity of fur trappers varies with fashion and the prices of pelts, which makes year-to-year assessment of the population impact of trapping difficult to assess using only the total number of animals caught. A more useful index is the ratio of animals caught to number of licenses sold. Since 2006, the ratio of badgers reported caught has remained fairly steady at 8–15%, or 8–19% when adjusted for reporting rate (NMDGF 2020b). This indicates that at the state level, the badger population is not being depleted. However, with fewer than 500 badgers trapped per year spread over 33 counties in New Mexico, an average of at most 15 badgers are reported trapped per county each year, which is too small a number for useful statistics at the county level.

Intensive trapping programs for research purposes can be used to estimate badger densities in small areas, as in a capture-mark-recapture program (e.g., Robinson 1961). However, badgers are not easy to trap without injury. Badgers rarely will enter cage or box traps, and so are usually caught in foothold traps or snares. Foothold traps are the classic trap that seizes an animal's foot between two "jaws." Snares are loops of wire that catch either an animal's neck, torso, or foot. For greatest success, traps and snares must usually be located in the entrances of burrows. Trapping with foothold traps likely produces severe injuries to legs and feet and is labor-intensive (e.g., Onderka et al. 1990). Sedation may be necessary to attach a mark such as an ear tag or a radio collar, but sedation presents hazards as well (Kreeger 1996). Despite these limitations, Hein and Andelt (1995) used live-capture and marking to calculate a minimum relative density estimate (0.27/km²) for a 70-km² area in Colorado, with capture success enhanced by Carman's Canine Call.

Badgers may be observed without deliberate physical capture in several ways. Sightings of badgers by the public or game management employees may be compiled. Badgers may be spotted at night with spotlights or vehicle headlights. Road-killed badgers may be tallied. Automatic cameras may be installed at bait or lure (scented) stations, either randomly distributed across the landscape or at specific sites associated with a greater probability of detection (e.g., Harrison 2015; Gould and Harrison 2018; Davidson et al. 2018). The latter is referred to as "camera-trapping." Due to the low density and small size of badgers, roadkills, sightings, and spotlighting generally produce very limited data (but see Eads et al. 2013). However, automatic cameras can be left in the field continuously. The bait or lure serves to ensure that animals enter the field of view of the cameras (but note that it may introduce a risk of bias in some types of research). The information that can be gained from the use of cameras includes not just presence/absence but also times of activity, family group size, and season of mating (when adults are seen together). Automatic wildlife cameras have been installed at many locations in New Mexico, including at the Sevilleta National Wildlife Refuge, White Sands Missile Range, White Sands National Monument, and in areas around the state selected by the New Mexico Department of Game and Fish for specific species' studies (see under "Distribution" above). Because individuals may be identified by their head stripes (described in the introduction to this chapter), local population size and density may even be estimated using a capture-recapture model, as on the Armendaris

Ranch. Cameras can even be incorporated into the design of experimental studies to study associations of badgers with colonial rodents.

The presence of badgers may be indirectly inferred by searching for signs such as digging, burrows, scats, hair, or tracks (Harrison 2015; Proulx and Do Linh San 2016). Searches may be conducted on foot or by vehicle. Searching for burrows is the most commonly used method to determine badger presence (e.g., Lay 2008; Quinn 2008). In much of New Mexico kit foxes and swift foxes (*Vulpes velox*) make burrows of similar size. However, the entrance to badger dens is somewhat distinctive in that it is often elliptical instead of round as in other species. Badger burrows and diggings may also show horizontal claw marks on the sides, though these are not visible in all soil types. Collecting scats is not a useful survey method for badgers because they often defecate underground and their scats are difficult to distinguish from the scats of other similar-sized carnivores without expensive DNA analyses. Badger hair can be collected from burrows, either in the dirt thrown out while digging, or by inserting hair snares in the entrance (Ministry of Environment, Ecosystems Branch for the Resources Information Standards Committee 2007). If hair or fresh scat can be collected and DNA extracted from it, then a wealth of data may be obtained, including individual identity, which can then be used for population estimates and an assessment of gene flow between areas, describing how populations are connected (Frantz et al. 2004; Ethier et al. 2012).

In New Mexico, track surveys are rendered useless because many soils are dry and sandy and do not take or hold good track impressions. However, in the past biologists often observed tracks on prepared surfaces such as sifted sand or metal plates covered with chalk or soot. Animals were attracted to the sites by a scented lure and then identified by their tracks (e.g., Harrison and Schmitt 2003). Such sites are known as scent posts, scent stations, or track stations. In general, badgers are not drawn as readily to these stations as other species such as foxes or coyotes, and this method is no longer widely used to survey badgers. Track stations were used by the New Mexico Department of Game and Fish for carnivore surveys but obtained fewer than four observations of badgers per year from 1999 to 2009 (NMDGF, unpubl. data). Automatic cameras have now replaced the use of track stations.

The American badger has not been the focus of much research in New Mexico; nor is it usually considered a high-priority management or conservation species by the New Mexico Department of Game and Fish. Knowledge of badgers in New Mexico has thus not come very far since Bailey's (1931) account of the species. Despite that, we do know enough to confidently state that they make a very unique contribution to the diversity of New Mexico's wildlife. Even the authors of the 1967 New Mexico Department of Game and Fish compendium who rated badgers last in terms of economic and aesthetic value noted that "badgers are [otherwise] highly valuable since they control some of the worst rodent pests with which the state is infested. Their continued presence is a benefit to the state and should be encouraged" (NMDGF 1967:193). New Mexico is fortunate to have badgers in the state. They are entirely worthy of our respect and attention.

ACKNOWLEDGMENTS
The authors are grateful to Jon Dunnum for help obtaining body measurements for badger specimens housed at the Museum of Southwestern Biology, to Gary Roemer for information on badger remote camera records and habitat associations at White Sands National Monument, and to all those who contributed photographs illustrating this chapter.

LITERATURE CITED

Allen, A. W. 1987. The relationship between habitat and furbearers. In *Wild Furbearer Management and Conservation in North America*, ed. M. Novak, J. A. Baker, M. E. Obbard, and B. Malloch, 164–79. Ontario, Canada: Ministry of Natural Resources.

Ames, N. 1979. *Wildlife and People in New Mexico*. Pt. 1, *Mammals and Birds*. Santa Fe: New Mexico Department of Game and Fish.

Arjo, W. M., E. M. Gese, C. Bromley, A. Kozlowski, and E. S. Williams. 2003. Serologic survey for diseases in free-ranging coyotes (*Canis latrans*) from two ecologically distinct areas of Utah. *Journal of Wildlife Diseases* 39:449–55.

Bailey, V. 1931 (=1932). *Mammals of New Mexico*. North American Fauna 53. Washington, DC: US Department of Agriculture, Bureau of Biological Survey.

Balph, D. F. 1961. Underground concealment as a method of predation. *Journal of Mammalogy* 42:423–24.

Ceballos, G., and D. Navarro. 1991. Diversity and conservation of Mexican mammals. In *Latin American Mammalogy*, ed. M. A. Mares and D. J. Schmidly, 167–98. Norman: University of Oklahoma Press.

Collins, D. P., III, L. A. Harveson, and D. C. Ruthven, III. 2012. Spatial characteristics of American badgers (*Taxidea taxus*) in southern Texas. *Southwestern Naturalist* 57:473–78.

Committee on the Status of Endangered Wildlife in Canada (COSEWIC). [2023]. cacosewic.ca. Accessed 15 March 2023.

Coues, E. 1877. Fur-bearing animals: a monograph of North American Mustelidae. *U.S. Geologic and Geographical Survey of the Territories, Miscellaneous Publications* 8:1–348.

Davidson, A. D., E. A. Hunter, J. Erz, D. C. Lightfoot,

A. M. McCarthy, J. K. Mueller, and K. T. Shoe-maker. 2018. Reintroducing a keystone burrow-ing rodent to restore an arid North American grassland: challenges and successes. *Restoration Ecology* 26:909–20.

Davis, W. B. 1946. Further notes on badgers. *Journal of Mammalogy* 27:175.

Devers, P. K., K. Koenen, and P. R. Krausman. 2004. Interspecific interactions between badgers and red-tailed hawks in the Sonoran Desert, southwestern Arizona. *Southwestern Naturalist* 49:109–11.

Eads, D. A., and D. E. Biggins. 2008. Aboveground predation by an American badger (*Taxidea taxus*) on black-tailed prairie dogs (*Cynomys ludovicianus*). *Western North American Naturalist* 68:396–401.

Eads, D. A., D. E. Biggins, T. M. Livieri, and J. J. Millspaugh. 2013. American badgers selectively excavate burrows in areas used by black-footed ferrets: implications for predator avoidance. *Journal of Mammalogy* 94:1364–70.

Eads, D. A., M. T. J. Hague, and C. G. Zoubek. 2012. American badger (*Taxidea taxus*) uses covert reconnaissance to ambush a black-tailed prairie dog (*Cynomys ludovicianus*). *Southwestern Naturalist* 57:465–67.

Eldridge, D. J. 2004. Mounds of the American badger (*Taxidea taxus*): significant features of North American shrub-steppe ecosystems. *Journal of Mammalogy* 85:1060–67.

Eldridge, D. J., W. G. Whitford, and B. D. Duval. 2009. Animal disturbances promote shrub maintenance in a desertified grass-land. *Journal of Ecology* 97:1302–1310. doi: 10.1111/j.1365-2745.2009.01558.x.

Ethier, D. M., A. Laflèche, B. J. Swanson, J. J. Nocera, and C. J. Kyle. 2012. Population subdivision and peripheral isolation in American badgers (*Taxidea taxus*) and implications for conservation planning in Canada. *Canadian Journal of Zoology* 90:630–39.

Findley, J. S., A. H. Harris, D. E. Wilson, and C. Jones. 1975. *Mammals of New Mexico*. Albuquerque: University of New Mexico Press.

Frantz, A. C., M. Schaul, L. C. Pope, F. Fack, L. Schley, C. P. Muller, and T. J. Roper. 2004. Estimating population size by genotyping remotely plucked hair: the Eurasian badger. *Journal of Applied Ecology* 41:985–95.

Frey, J. K. 2004. Taxonomy and distribution of the mammals of New Mexico: an annotated checklist. *Occasional Papers, Museum of Texas Tech University* 240:1–32.

Fry, W. 1928. The California badger. *California Fish and Game* 14:204–8.

Fur Harvesters Auction Inc. 2020. Auction Results. www.furharvesters.com/auctionresults.html. Accessed 30 December 2020.

Goguen, C. B. 2012. Comparison of bird and mammal communities on black-tailed prairie dog (*Cynomys ludovicianus*) colonies and uncolonized shortgrass prairie in New Mexico. *Journal of Arid Environments* 80:27–34.

Gould, M. J., and R. L. Harrison. 2018. A novel approach to estimating density of American badgers (*Taxidea taxus*) using automatic cameras at water sources in the Chihuahuan Desert. *Journal of Mammalogy* 99:233–41.

Hall, E. R. 1946. *Mammals of Nevada*. Berkeley: University of California Press.

———. 1981. *The mammals of North America*. 2nd ed. Vol. 2. New York: John Wiley.

Halloran, A. F. 1946. The carnivores of the San Andres Mountains, New Mexico. *Journal of Mammalogy* 27:154–61.

Harlow, H. J. 1981. Torpor and other physiological adaptations of the badger (*Taxidea taxus*) to cold environments. *Physiological Zoology* 54:267–75.

Harrison, R. L. 2015. A comparison of sign searches, live-trapping, and camera-trapping for detection of American badgers (*Taxidea taxus*) in the Chihuahuan Desert. *Western North American Naturalist* 75:387–95

———. 2016a. Noninvasive identification of individual American badgers by features of their dorsal head stripes. *Western North American Naturalist* 76:259–61.

———. 2016b. Badger behavior at anthropogenic water sources in the Chihuahuan Desert. *Western North American Naturalist* 76:1–5.

Harrison, R. L., and C. G. Schmitt. 2003. Current swift fox distribution and habitat selection within areas of historical occurrence in New Mexico. In *The swift fox: ecology and conservation of kit foxes in a changing world*, ed. M. Sovada and L. Carbyn, 71–77. Regina: Canadian Plains Research Center, University of Regina.

Heilbrun, R. D., N. J. Silvy, M. J. Peterson, and M. E. Tewes. 2006. Estimating bobcat abundance using automatically triggered cameras. *Wildlife Society Bulletin* 34:69–74.

Hein, E. W., and W. F. Andelt. 1995. Evaluation of indices of abundance for an unexploited badger population. *Southwestern Naturalist* 40:288–92.

Helgen, K. M., F. R. Cole, L. E. Helgen, and D. E. Wilson. 2009. Generic revision in the Holarctic Ground squirrel genus *Spermophilus*. *Journal of Mammalogy* 90:270–305.

Hill, J. E. 1942. Notes on mammals of northeastern New Mexico. *Journal of Mammalogy* 23:75–82.

Hooper, E. T. 1941. *Mammals of the lava fields and adjoining areas in Valencia County, New Mexico*. Miscellaneous Publications of the Museum of Zoology, University of Michigan No. 51.

International Union for the Conservation of Nature (IUCN). 2012. The IUCN Red List of Threatened Species. www.iucnredlist.org. Accessed 3 October 2012.

Ivey, R. D. 1957. Ecological notes on the mammals of Bernalillo County, New Mexico. *Journal of Mammalogy* 38:490–502.

Johnson, N. C. 1983. Badgers. In *Prevention and Control of Wildlife Damage*, ed. R. M. Timm, C1–C3. Lincoln: Great Plains Agricultural Council Wildlife Resources Committee and Cooperative Extension Service, University of Nebraska.

Justice, D. H. 2015. *Badger*. London: Reaktion Books.

Kagel, S. M., R. S. Ziejka, L. M. L. Averilla, B. A. Minnig, and J. L. Hoogland. 2020. Relentless predation on Gunnison's prairie dogs (*Cynomys gunnisoni*) by a single American badger (*Taxidea Taxus*). *Western North American Naturalist* 80:345–50. https://doi.org/10.3398/064.080.0306.

Karanth, K. U., J. D. Nichols, N. S. Kumar, and J. E. Hines. 2006. Assessing tiger population dynamics using photographic capture-recapture sampling. *Ecology* 87:2925–37.

Kinley, T. A., and N. J. Newhouse. 2008. Ecology and translocation-aided recovery of an endangered badger population. *Journal of Wildlife Management* 72:113–22.

Klafki, R. W. 2014. Road ecology of a northern population of badgers (*Taxidea taxus*) in British Columbia, Canada. MS thesis, Thompson Rivers University, Kamloops, British Columbia, Canada.

Koepfli, K.-P., K. A. Deere, G. J. Slater, C. Begg, K. Begg, L. Grassman, M. Lucherini, G. Veron, and R. K. Wayne. 2008. Multigene phylogeny of the Mustelidae: resolving relationships, tempo and biogeographic history of a mammalian adaptive radiation. *BMC Biology* 6:10.

Kreeger, T. J. 1996. *Handbook of wildlife chemical immobilization*. Laramie, WY: International Wildlife Veterinary Services.

Lampe, R. R. 1982. Food habits of badgers in east central Minnesota. *Journal of Wildlife Management* 46:790–95.

Lampe, R. P., and M. A. Sovada. 1981. Seasonal variation in home range of a female badger (*Taxidea taxus*). *Prairie Naturalist* 13:55–58.

Lay, C. 2008. The status of the American badger in the San Francisco Bay area. MS thesis. San Jose State University, California.

Ligon, J. S. 1927. *Wild life of New Mexico: its conservation and management*. Santa Fe: State Game Commission, Department of Game and Fish.

Lindzey, F. G. 2003. Badger. In *Wild mammals of North America: biology, management, and economics*, 2nd ed., ed. G. A. Feldhamer, B. C. Thompson, and J. A. Chapman, 683–691. Baltimore: Johns Hopkins University Press.

Long, C. A. 1972. Taxonomic revision of the North American badger, *Taxidea taxus*. *Journal of Mammalogy* 53:725–59.

———. 1973. *Taxidea taxus*. American Society of Mammalogy, Mammalian Species 26:1–4.

Long, C. A., and C. A. Killingley. 1983. *The badgers of the world*. Springfield, IL: Charles C. Thomas.

Lopez Soto, J. H. 1980. Datos ecológicos del Tlalcoyote *Taxidea taxus berlandieri* Baird (1858), en el Ejido Tokio, Galeana, Nuevo León, México. MS tesis, Universidad Autónoma de Nuevo León, Monterrey, Mexico.

Messick, J. P. 1987. North American badger. In *Wild furbearer management and conservation in North America*, ed. M. Novak, J. A. Baker, M. E. Obbard, and B. Malloch, 587–97. Ontario, Canada: Ministry of Natural Resources.

Messick, J. P., and M. G. Hornocker. 1981. Ecology of

the badger in southwestern Idaho. *Wildlife Monographs* 76:1–53.

Messick, J. P., G. W. Smith, and A. M. Barnes. 1983. Serologic testing of badgers to monitor plague in southwestern Idaho. *Journal of Wildlife Diseases* 19:1–6.

Michener, G. R. 2004. Hunting techniques and tool use by North American badgers preying on Richardson's ground squirrels. *Journal of Mammalogy* 85:1019–27.

Ministry of Environment, Ecosystems Branch for the Resources Information Standards Committee. 2007. Inventory methods for medium-sized territorial carnivores: badger. *Standards for Components of British Columbia's Biodiversity No. 25A.* Victoria, BC: Resources Information Standards Committee.

Minta, S. C. 1993. Sexual differences in spatio-temporal interaction among badgers. *Oecologia* 96:402–9.

Minta, S. C., and R. E. Marsh. 1988. Badgers (*Taxidea taxus*) as occasional pests in agriculture. *Proceedings of the Vertebrate Pest Conference* 13:199–208.

Minta, S. C., K. A. Minta, and D. F. Lott. 1992. Hunting associations between badgers (*Taxidea taxus*) and coyotes (*Canis latrans*). *Journal of Mammalogy* 73:814–20.

Murie, J. O. 1992. Predation by badgers on Columbian ground squirrels. *Journal of Mammalogy* 73:385–94.

New Mexico Department of Game and Fish (NMDGF). 1967. *New Mexico Wildlife Management.* Santa Fe: New Mexico Department of Game and Fish.

———. 2020a. *2020–2021 New Mexico Furbearer Rules and Info.* Santa Fe: New Mexico Department of Game and Fish.

———. 2020b. Furbearers (harvest reports). www.wildlife.state.nm.us/hunting/information-by-animal/furbearers. Accessed 22 December 2020.

Newman, C., C. C. Buesching, and J. O. Wolff. 2005. The function of facial masks in "midguild" carnivores. *Oikos* 108:623–33.

Nowak, R. M. 1999. *Walker's mammals of the world.* Vol. 2. 6th ed. Baltimore: Johns Hopkins University Press.

Nowak, R. M., and J. L. Paradiso. 1983. *Walker's mammals of the world.* Vol. 1. 4th ed. Baltimore: Johns Hopkins University Press.

Obbard, M. E., J. G. Jones, R. Newman, A. Booth, A. J. Satterthwaite, and G. Linscombe. 1987. Furbearer harvests in North America. In *Wild furbearer management and conservation in North America,* ed. M. Novak, J. A. Baker, M. E. Obbard, and B. Malloch, 1007–33. Ontario, Canada: Ministry of Natural Resources.

Onderka, D. K., D. L. Skinner, and A. W. Todd. 1990. Injuries to coyotes and other species caused by four models of footholding devices. *Wildlife Society Bulletin* 18:175–82.

Paulson, N. J. 2007. Spatial and habitat ecology of North American badgers (*Taxidea taxus*) in a native shrubsteppe ecosystem of eastern Washington. MS thesis, Washington State University, Pullman.

Proulx, G., and E. Do Linh San. 2016. Non-invasive methods to study American and European badgers—a review. In *Badgers: systematics, biology, conservation, and research techniques,* ed. G. Proulx and E. Do Linh San, 311–38. Sherwood Park, AB: Alpha Wildlife Publications.

Quinn, J. H. 2008. The ecology of the American badger *Taxidea taxus* in California: assessing conservation needs on multiple scales. PhD dissertation. University of California, Davis.

Robinson, Q. H., D. Bustos, and G. W. Roemer. 2014. The application of occupancy modeling to evaluate intraguild predation in a model carnivore system. *Ecology* 95: 3112–23.

Robinson, W. B. 1961. Population changes of carnivores in some coyote-control areas. *Journal of Mammalogy* 42:510–15.

Rollins, D., B. D. Taylor, T. D. Sparks, T. E. Wadell, and G. Richards. 2009. Species visitation at quail feeders and guzzlers in southern New Mexico. In *Gamebird 2006: Quail VI and Perdix XII, 31 May–4 June 2006,* ed. S. B. Cederbaum, B. C. Faircloth, T. M. Terhune, J. J. Thompson, and J. P. Carroll, 210–19. Athens, GA: Warnell School of Forestry and Natural Resources.

Rose, J., A. Moore, A. Russell, and M. Butcher. 2014. Functional osteology of the forelimb digging apparatus of badgers. *Journal of Mammalogy* 95:543–58.

Salkeld, D. J., and P. Stapp. 2006. Seroprevalence rates and transmission of plague (*Yersinia pestis*) in mammalian carnivores. *Vector-borne and Zoonotic Diseases* 6:231–39.

Sato, J. J. 2016. The systematics and taxonomy of the world's badger species—a review. In *Badgers: systematics, biology, conservation, and research techniques*, ed. G. Proulx and E. Do Linh San, 1–30. Sherwood Park, AB: Alpha Wildlife Publications.

Schmidly, D. J. 2002. *Texas natural history: a century of change*. Lubbock: Texas Tech University Press.

Shaughnessy, M. J., Jr., and R. L. Cifelli. 2004. Influence of black-tailed prairie dog towns (*Cynomys ludovicianus*) on carnivore distributions in the Oklahoma Panhandle. *Western North American Naturalist* 64:184–92.

Stahlecker, D. W., J.-L. E. Cartron, and D. G. Mikesic. 2010. Golden eagle (*Aquila chrysaetos*). In *Raptors of New Mexico*, ed. J.-L. E. Cartron, 371–91. Albuquerque: University of New Mexico Press.

Thompson, B. C., D. F. Miller, T. A. Doumitt, and T. R. Jacobson. 1992. *Ecologically-based management evaluation for sustainable harvest and use of New Mexico furbearer resources*. Santa Fe: New Mexico Department of Game and Fish.

Thornton, D., A. Scully, T. King, S. Fisher, S. Fitkin, and J. Rohrer. 2018. Hunting associations of American badgers (*Taxidea taxus*) and coyotes (*Canis latrans*) revealed by camera trapping. *Canadian Journal of Zoology* 96:769–73.

Todd, M. C. 1980. Ecology of badgers in southcentral Idaho with additional notes on raptors. MS. University of Idaho, Moscow.

Truett, J. C., J. L. D. Dullum, M. R. Matchett, E. Owens, and D. Seery. 2001. Translocating prairie dogs: a review. *Wildlife Society Bulletin* 29:863–72.

Truett, J. C., and T. Savage. 1998. Reintroducing prairie dogs into desert grasslands. *Restoration and Management Notes* 16:189–95.

Warner, R. E., and B. Ver Steeg. 1995. *Illinois badger studies*. Final report. Federal Aid Project No. W-103-R-1-6. Division of Wildlife Resources, Illinois Department of Natural Resources, Springfield.

Wright, P. L. 1966. Observations on the reproductive cycle of the American badger (*Taxidea taxus*). *Symposium of the Zoological Society of London* 15:27–45.

———. 1969. The reproductive cycle of the male American badger (*Taxidea taxus*). *Journal of Reproductive Fertility (Supplement)* 6:435–45.

Yensen, E., and P. W. Sherman. 2003. Ground squirrels: *Spermophilus* spp. and *Ammospermophilus* spp. In *Wild mammals of North America: biology, management, and economics*, 2nd ed., ed. G. A. Feldhamer, B. C. Thompson, and J. A. Chapman, 211–31. Baltimore: Johns Hopkins University Press.

26

WHITE-BACKED HOG-NOSED SKUNK (*CONEPATUS LEUCONOTUS*)

Jerry W. Dragoo and Christine C. Hass

Nocturnal and solitary, the white-backed hog-nosed skunk (*Conepatus leuconotus*) tends to avoid contact with humans whenever possible. Much of its behavior was poorly known until recently, with research typically revolving only around how the species foraged. *Conepatus* is as large as, or larger than, the striped and hooded skunks (*Mephitis mephitis* and *M. macroura*), and almost twice as large as spotted skunks (*Spilogale leucoparia* and *S. gracilis*) in New Mexico (Dragoo 2009; see Chapters 27, 28, and 29). White-backed hog-nosed skunks can be distinguished readily from the other skunks in New Mexico by the color pattern of the dorsal pelage. They are the only skunks that lack a white spot or medial bar between the eyes and have primarily black body fur with a single white stripe along the back. The tail is shorter (<half the skunk's total length) in relation to body length than in other skunks. *C. leuconotus* generally is more muscular in the upper body (Van De Graaff 1969) and resembles badgers more so than other species of skunks in this respect (Patton 1974; see Chapter 25). The rectangular-shaped scapula, strong forearms, and shape of the humeri of *C. leuconotus* resemble those of badgers (Van De Graaff 1969). Their upper body strength allows them to be adept climbers (Brashear et al. 2010), though not as agile as spotted skunks of the genus *Spilogale* (Patton 1974; Dragoo 2009; see Chapter 29).

Typically, the white along the back of white-backed hog-nosed skunks starts as a single

Photo 26.1. The rarely seen and poorly known white-backed hog-nosed skunk (photographed here on 4 March 2015 in Helms Valley, Lincoln County) is characterized by a single, broad dorsal white stripe that extends from the top of the head to the base of the tail, the latter being almost entirely white. Relative to body length, the tail is also shorter than in other skunks. Photograph: © WSMR Garrison Environmental Division and ECO Inc., Ecological Consultants.

wedge-shaped stripe on the head that widens near the shoulders to approximately the width of the back. The stripe ranges from substantially reduced or absent on the rump to completely covering the entire back. Throughout the range of *C. leuconotus*, the thickness of the white stripe may be influenced by a variety of factors including the amount of canopy cover and temperature consistency in their habitats. More canopy cover and greater temperature stability correlate with hog-nosed skunks tending to have less white. In areas with less cover and more temperature variation, such as in New Mexico and Arizona, the species tends to have a wider white stripe (Ferguson 2014; Ferguson et al. 2022). The tail is white

(*opposite page*) Photograph: © Christopher H. Taylor.

Photo 26.2. Size and dorsal color pattern comparison between New Mexico's three larger skunks: striped skunk (*Mephitis mephitis*), top; hooded skunk (*M. macroura*), middle; and white-backed hog-nosed skunk, bottom. The hog-nosed skunk is as large or slightly larger than the other two skunks, but its tail is much shorter. Other characteristics of the hog-nosed skunk include the single, broad white stripe on the back and the absence of a white spot or medial bar between the eyes. Photograph: © Jon Dunnum.

large and truncated (Gray 1865; Photo 26.7). The auditory bullae are not inflated, and the palate ends behind upper molars (Merriam 1902). The carnassial teeth are not well developed, and they, as well as the large upper molar, provide an increased crushing surface (Kurtén and Anderson 1980). The dental formula for hog-nosed skunks is typically I 3/3, C 1/1, P 2/3, M 1/2 = 32 (Dragoo 2009), though it can vary as the P2 may be present (Van Gelder 1968). Our colleague Adam Ferguson (pers. comm.) has recorded several museum specimens with a single P2, both P2s, and no P2s. *Conepatus*, like *Mephitis mephitis* and possibly all skunks, are born with one set of teeth as they resorb the milk teeth prior to birth (Verts 1967; Slaughter et al. 1974). Females only have three pairs of mammae, two pectoral and one inguinal, suggesting they

along its total length dorsally, but ventrally it can be black or white at the base (Dragoo et al. 2003).

The snout of this species is relatively long, and the nose pad, which is naked, is about 20 mm (~0.8 in) wide by 26 mm (~1 in) long and resembles the nose of a small hog (Dragoo 2009; Photo 26.3). The nostrils are located ventrally and open downward; the sense of smell is acute and used in locating and capturing buried prey (Bailey 1905). *C. leuconotus* has small and rounded ears. Its legs are stocky, and the feet are plantigrade, meaning that the soles of the feet are in contact with the ground while walking (Photo 26.4). The front feet are broad and armed with long, powerful claws; the hind feet are longer, with shorter claws. The soles are naked along the length of the feet (Photo 26.5).

Hog-nosed skunks are built for digging and climbing, the result of a well-developed bone structure and musculature of the upper body and very long foreclaws (Photo 26.6; Van De Graaff 1969; Dragoo 2009). The skull is relatively deep (deepest in the temporal region), and nares are

Photo 26.3. Close-up of the head of a white-backed hog-nosed skunk from New Mexico. The elongated snout is reminiscent of that of a small hog, with a wide, naked nose pad and nostrils that are located ventrally and open downward, all adaptations for detecting buried prey by smell. Note also the wedge-shaped start of the white dorsal stripe on the head, the relatively small eyes, and the small, rounded ears. The hog-nosed skunk in the photograph, a male, lived to be 14 years old in captivity. Photograph: © Jerry W. Dragoo.

have fewer young (Bailey 1931) than other skunk species in New Mexico.

Ranges of morphological measurements (J. W. Dragoo, unpubl. data; from museum specimens used in Dragoo et al. 2003) for adult males and females, respectively, are total length, 39.7–93.4 cm (15.63–36.77 in), 45.2–74.0 cm (17.79–29.13 in); length of tail, 14.0–41.0 cm (5.51–16.14 in), 12.2–34.0 cm (4.8–13.39 in); length of hind foot 2.2–9.0 cm (0.87–3.54 in), 3.0–9.0 cm (1.18–3.54 in); and length of ear 0.8–3.6 cm (0.31–1.42 in), 0.8–3.3 cm (0.31–1.3 in).

The scent glands of this species, as in all skunks, are at the base of the tail on either side of the rectum. They are covered by a layer of smooth muscle that contracts to force secretions through ducts to nipples just inside the anal sphincter. The rectum is everted to expose the nipples, which can be aimed toward a target (Dragoo 1993). Two major volatile components [(E)-2-butene-1-thiol and (E)-S-2-butenyl thioacetate] and four minor components (phenylmethanethiol, 2-methylquinoline, 2-quinoline-methanethiol, and bis[(E)-2-butenyl] disulfide) are found in anal sac secretions from white-backed hog-nosed skunks (Wood et al. 1993). The scent gland secretions of hog-nosed skunks have fewer volatile components (6) than those of hooded skunks (14), striped skunks (8), and spotted skunks (9) according to Wood et al. (2002).

Earlier taxonomists placed skunks in the weasel family, Mustelidae. However, Dragoo and Honeycutt (1997) conducted a mtDNA and morphological analysis on all the skunk genera and many representatives of the various weasel subfamilies. Their conclusion that the skunks (including the stink badgers [Mydaus spp.]) should be relegated to their own family, the Mephitidae, was later verified using nuclear DNA (Eizirik et al. 2010). Skunks diverged from the remainder of the order-group taxon Mustelida (Musteloidea [Procyonidae, Mustelidae] and Ailuridae) about 32–40 million years ago (Eizirik et al. 2010). It is

Photo 26.4. Similar to other skunks, black bears (*Ursus americanus*) and raccoons (*Procyon lotor*), hog-nosed skunks are plantigrade animals, meaning that the full length of their feet is in contact with the ground during locomotion. Photograph: © Christine C. Hass.

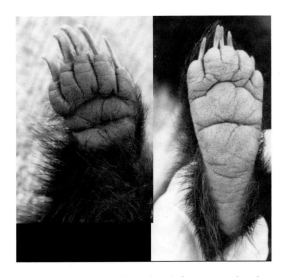

Photo 26.5. The broad front feet (left) are armed with five powerful claws that are about three times as long as those of the hind feet (right). Both front and hind feet have naked soles. The hind feet are notably longer than the front feet. Photograph: © Christine C. Hass.

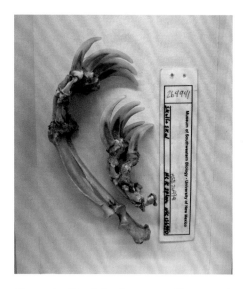

Photo 26.6. The skeletal anatomy of hog-nosed skunks shows strong forearms and powerful foreclaws adapted for digging. Photograph: © Jon Dunnum.

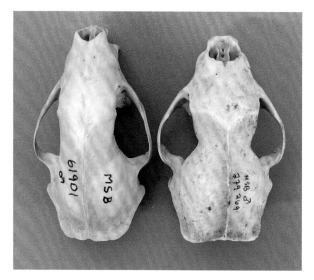

Photo 26.7. Comparison between the skulls of the striped skunk (*Mephitis mephitis*, left) and the white-backed hog-nosed skunk (right). Skunk skulls tend to be triangular in shape (similar to arrowheads) from the zygomatic arches forward. In the white-backed hog-nosed skunk, however, the skull is more massive and wider at the posterior margin, with larger nares and more widely spreading zygomatic arches. Photograph: © Jon Dunnum.

possible that the early "mustelids" actually may be more closely related to modern skunks than to modern weasels (Koepfli et al. 2017).

Currently, three species of hog-nosed skunks are recognized: *Conepatus chinga* (Molina's hog-nosed skunk), *C. leuconotus*, and *C. semistriatus* (striped hog-nosed skunk) (Dragoo et al. 2003; Wozencraft 2005; Schiaffini et al. 2013). *Conepatus chinga* (South America) and *C. semistriatus* (Central and South America) exhibit a striped color pattern like that seen in the striped skunk. Of these three hog-nosed skunks, the only one found in New Mexico is *C. leuconotus*.

Conepatus is derived from the Spanish *conepate* or *conepatl*, which means "skunk." *Conepatl* itself may be derived from *nepantla*, which in the Nahuatl language signified a subterranean dwelling (Coues 1877). The specific name *leuconotus* is of Greek origin meaning white (*leuco*) back (*nota* or *notum*). Before 2003, common names for

hog-nosed skunks in the United States and Mexico included badger skunk, rooter skunk, eastern and western hog-nosed skunks, Gulf Coast hog-nosed skunk, inland skunk, Texas skunk, and white-back skunk. Dragoo et al. (2003) referred to them as white-backed hog-nosed skunks (from the specific epithet) to distinguish them from the species of striped skunks of Central and South America, of which there were three at the time. Other names include American hog-nosed skunk, North American hog-nosed skunk (in South America [Schiaffini et al. 2013]), and common hog-nosed skunk (this name also applies to *C. chinga* in South America [Donadio et al. 2001; Arias et al. 2006]).

Hinrich Lichtenstein (1832) described two species of hog-nosed skunks in a set of two books. The first was a picture book that had colored drawings for representations of new or little-known mammals; the other had text explaining the illustrations. The pages of text were not

Photo 26.8. White-backed hog-nosed skunk photographed by a remote camera along the Gila River in Grant County on 14 September 2016. The Gila River Valley falls within that part of the species' distribution—in southern New Mexico—that has long been documented. In Bailey's (1931) time, the white-backed hog-nosed skunk's distribution was thought to extend only as far north as the Sandia Mountains. Because it is nocturnal and difficult to trap (traditional baits are ineffective for this largely insectivorous species), mapping the geographic range of the white-backed hog-nosed skunk has relied on locations of road-killed animals and, more recently, also the use of remote cameras. Photograph: © Keith Geluso.

Photos 26.9a and b. Two remote camera photo records from a) (top) Cat Mesa (2,520 m [8,260 ft]) and b) (bottom) San Juan Mesa (2,360 m [7,340 ft]), both in the Jemez Mountains in Sandoval County. The occurrence of hog-nosed skunks in the Jemez Mountains has been documented only recently, with the use of motion-activated, remote cameras. Photographs: © Mark Peyton / National Park Service.

Photo 26.10 (left). Road-killed white-backed hog-nosed skunk found on 2 October 2017 along I-25, just north of Raton, Colfax County. Together with another road-killed skunk seen just south of the Colorado border in 2003 in Colfax County at Raton Pass (Meaney et al. 2006), this record confirms the present occurrence of the species as far north as the Colorado state line. Photograph: © Mark L. Watson.

Photo 26.11 (above). Road-killed white-backed hog-nosed skunk found on 23 August 2017 along I-25, 2 mi east of Exit 330 to Bernal in San Miguel County, at an elevation of 1,945 m (6,383 ft). The distribution of the white-backed hog-nosed skunk extends as far east as Colfax, Harding, San Miguel, Guadalupe, De Baca, Chaves, and Eddy counties in New Mexico. Photograph: © Mark L. Watson.

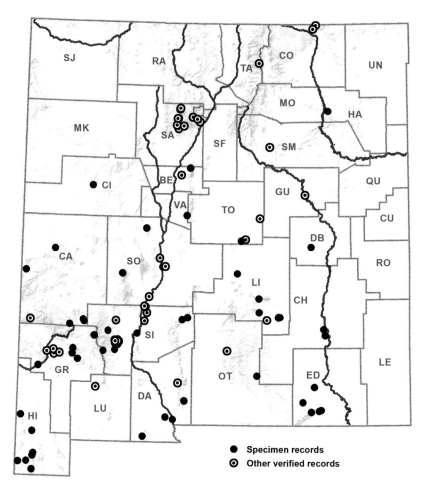

Map 26.1. Distribution of the white-backed hog-nosed skunk based on specimen and other verified records. Recent remote camera photos have documented the existence of a hog-nosed skunk population in the Jemez Mountains. Additional recent records include photographs of road-killed animals from as far north as the Colorado state line or the eastern third of New Mexico.

numbered, and the hog-nosed skunk illustrations were on Plate 40—Fig. 1 and Fig. 2 on the same page. Lichtenstein's (1832) first figure was of *Conepatus* (=*Mephitis*) *leuconotus*, and the second was *C. mesoleucus*. Later authors were under the impression that the two forms should be recognized as a single species (Coues 1877; Hall and Kelson 1952; Raun and Wilks 1961; Hall 1981).

A revision of white-backed hog-nosed skunks was performed in the late 1990s (Dragoo et al. 2003) and led to a better understanding of the taxonomy of *Conepatus*. Dragoo et al. (2003) examined specimens of the two species described

by Lichtenstein (1832) and the 12 subspecies subsumed under them. They concluded that there was not enough significant variation to recognize more than one species for all white-backed hog-nosed skunks. However, due to the lack of adequate specimens and the possibility of evolutionary significant units on the periphery of the species' range, to be conservative they recognized three subspecies, including one in east Texas (*C. l. telmalestes*) and one in southeastern Colorado and possibly northern New Mexico (*C. l. figginsi*). The final subspecies (*C. l. leuconotus*) occupies the rest of the known range of *C. leuconotus*. Dragoo et al. (2003) could not justify recognition of two unique subspecies in southern Colorado. Earlier, Armstrong (1972) had examined specimens from Colorado and concluded they represented only age and sex variation of the more widely distributed *C. l. leuconotus*. As we will be discussing only *Conepatus leuconotus* throughout the chapter, we will simply refer to them as "hog-nosed skunks" to distinguish them from other species of skunks in New Mexico.

DISTRIBUTION

Conepatus leuconotus ranges from the southwestern United States throughout most of Mexico, excluding the Baja and Yucatán peninsulas, reaching as far south as Nicaragua in Central America (Allen 1910; Hall 1981). Goodwin (1946) reported that the species possibly could be found as far south as northern Costa Rica, though no specimens or photos have been reported there yet. In the United States, it has been recorded as far north as Colorado (Warren 1921), Oklahoma (Caire et al. 1989), and Arizona (Hoffmeister 1986), including north of the Grand Canyon (Holton et al. 2021), as well as throughout much of Texas (Schmidly and Bradley 2016) and New Mexico (Hass and Dragoo 2017; Map 26.1).

According to Bailey (1931) and Findley (1987), in New Mexico hog-nosed skunks are primarily or exclusively restricted to the southern part of the state, with an old specimen record (MCZ 10513) collected in 1901 from as far north as the Sandia Mountains (Barber 1902) but with no subsequent records in that same area. Bailey (1931) reported that specimen as the northernmost record in the state. Leopold (1959:459) wrote, "Although hog-nosed skunks are not often seen in Mexico, and specimens are scarce in most collections, they actually may be more common than is generally supposed. Once I became familiar with their distinctive diggings and long-clawed tracks, I noted their presence in many localities from which no specimens

were obtained." Bailey (1905:202–3) held the same impression for *Conepatus* in Texas.

Because trapping the white-backed hog-nosed skunk is difficult, road surveys have often represented the most effective means of studying populations of this species (Dragoo 1993; Meaney et al. 2006). Opportunistic occurrence records have also been collected, some of them important. Working with Wildlife Rescue Inc. of New Mexico, J. W. Dragoo (unpubl. data) assisted with the relocation of a hog-nosed skunk found in a fast-food restaurant in the northeastern part of Albuquerque, Bernalillo County, just west of the Sandia Mountains. A road-killed animal was seen just south of the Colorado border in New Mexico in 2003 in Colfax County at Raton Pass (Meaney et al. 2006).

The distribution of the hog-nosed skunk in New Mexico has been poorly documented, except in the southern part of the state where it appears to be more abundant and where records exist from all counties east to the Pecos River (Hass and Dragoo 2007, 2009, 2017; Map 26.1). The more recent, widespread use of remote cameras, by scientists and lay people alike, has proven to be a valuable method for detecting hog-nosed skunks (Map 26.1). Notable is the remote camera photograph of a hog-nosed skunk in Mora County near Watrous (reported to and verified by the New Mexico Department of Game and Fish), as well as a published record in the next county over, Harding (Geluso 2002). Hog-nosed skunks have also been recently recorded by remote cameras at multiple locations in the Jemez Mountains and Pajarito Plateau including at Bandelier National Monument (S. Milligan, in litt.; M. Peyton, in litt.; Photos 26.9a and b). Hog-nosed skunks are now known to occur across most of New Mexico north to the Colorado border, except for the Great Basin desert region in the northwestern part of the state and much of the eastern tier of counties (Hass and Dragoo 2017; Map 26.1; see also under "Status and Management").

Photo 26.12. White-backed hog-nosed skunk along Willow Creek in the Mogollon Mountains, Catron County, at an elevation of approximately 2,500 m (8,200 ft). North facing slopes are characterized by upper mixed conifer transition with spruce-fir. The vegetation that grows on south facing slopes consists primarily of ponderosa pine (*Pinus ponderosa*) and Gambel oak (*Quercus gambelii*). The valley bottom is characterized by bluestem willow (*Salix irrorata*)/ mountain alder (*Alnus incana*) with a sedge (*Carex*)/ forb understory. In New Mexico, white-backed hog-nosed skunks were originally described as a species largely restricted to the "Lower Sonoran Valley and the hot canyons around the foothills of the desert mountains" (Bailey 1931:337). Photograph: © Jennifer K. Frey.

HABITAT ASSOCIATIONS

The elevational range of *C. leuconotus* is extensive. In Arizona, hog-nosed skunks occur up to ~2,740 m (9,000 ft) in the Pinaleño Mountains (Hoffmeister 1986). In Mexico, *C. leuconotus* ranges up to ~3,050 m (10,000 ft) (Cahalane 1961). Texas does not have any higher-elevation mountains, and therefore *Conepatus* is found at lower elevations all the way down to sea level in that state (Bailey 1905). The specimen collected by Barber (1902) in 1901 in a ponderosa pine forest in New Mexico's Sandia Mountains was found at about ~2,135 m (7,000

ft). Hog-nosed skunks have also been recorded by camera traps in New Mexico at elevations approaching or exceeding 2,590 m (8,500 ft) (J. K. Frey, in litt.; Photos 26.9a and 26.12).

White-backed hog-nosed skunks are generally found in scrub and woodland habitats, including thornscrub, riparian areas, and pine-oak forest. They occur in canyons, streambeds, and rocky terrain, from mountains to coastal plains, and they have also been found in cornfields surrounded by brushland, as well as cattle pastures when interspersed with mesquite shrubland, cactus, and native grassland (Dice 1937; Dalquest 1953; Hall and Dalquest 1963; Beasom 1974; Patton 1974; Schmidly and Hendricks 1984; Matson and Baker 1986; Schmidly 2002).

In New Mexico and Arizona, based on museum records, camera trap photos, roadkills, and track surveys, Hass and Dragoo (2007, 2009, 2017) showed that hog-nosed skunks were found 48 to 54% of the time in forests and woodlands, about 25% of the time in savanna and grasslands, approximately 13 to 14% in scrubland, and the least amount of time in urban and agricultural areas, 2.5 to 6.4%.

In southern New Mexico in particular, *Conepatus leuconotus* has been reported in vegetation types ranging from creosote desert to pine oak forests (Findley et al. 1975). Hog-nosed skunks also have been caught on cameras along a high-elevation creek adjacent to a north-facing slope with Engelmann spruce (*Picea engelmannii*), corkbark fir (*Abies lasiocarpa*), Douglas-fir (*Pseudotsuga menziesii*), ponderosa pine (*Pinus ponderosa*), aspen (*Populus*), and blue spruce (*Picea pungens*) (J. K. Frey, in litt.; Photo 26.12).

LIFE HISTORY
Diet and Foraging

No field studies of hog-nosed skunks have been conducted in New Mexico, so information on natural history is derived from studies outside of the state or of captive animals. Hog-nosed

Photo 26.13. While foraging, hog-nosed skunks dig up areas of soil measuring up to 12 m (40 ft) in diameter. Photographed along the Gila River near Cliff in Grant County, New Mexico, in September 2013. Photograph: © Christine C. Hass.

skunks are more insectivorous than other skunks and spend hours rooting or digging for grubs and larvae (Bailey 1905; Seton 1926). They are primarily nocturnal but may forage during the warm parts of the day in winter (Schmidly and Bradley 2016) and have even been observed feeding during the heat of the day in summer in Texas (Davis 1951).

Due to the species' primarily insectivorous diet, many naturalists have been unable to find bait suitable for trapping hog-nosed skunks and have had to capture individuals by hand instead (Bailey 1905; Dragoo 1993). The propensity for rooting by hog-nosed skunks is comparable to that of feral hogs (*Sus scrofa*; Patton 1974). While foraging, hog-nosed skunks dig up areas of soil

~12 m or 40 ft in diameter (Miller 1925; see Photo 26.13).

Insects may not always be plentiful, and *Conepatus leuconotus* is an opportunistic feeder that can also eat a variety of small vertebrates and fruits (Taylor 1953; J. W. Dragoo, pers. obs.). Hog-nosed skunks have been observed attacking and devouring small rodents (J. W. Dragoo, pers. obs.). Captive animals also consumed pears, raisins, zucchini, squash, green beans, radishes, green peppers, and a variety of other fruits and vegetables (Dragoo 2009). Many fruits have been used as baits to capture *Conepatus* (J. W. Dragoo, unpubl. data), although capturing by hand still is more efficient (Dragoo 1993). In New Mexico and Texas, *C. leuconotus* is rarely seen drinking and thus likely survives on water metabolized from the food it eats (Patton 1974; Dragoo 2009).

Denning

Normally during the day in Texas and Mexico (and presumably elsewhere), white-backed hog-nosed skunks retreat to underground burrows, brush piles, or rock crevices (Davis 1945; Leopold 1959). They den in hollows in the roots of trees or fallen trunks and in cavities under large rocks or in rock piles (Audubon and Bachman 1851; Bailey 1905). Dens also have been reported in caves, mine shafts, and woodrat nests (Hoffmeister 1986). Additionally, hog-nosed skunks use the abandoned burrows of other animals, and if all else fails, dig their own (Warren 1942). A captive pair cared for by one of us (J. W. Dragoo) would curl up in a half-buried hollow log while resting. The two skunks also excavated a tunnel at a branching point in the trunk of that log. The den angled downward ~0.6 m (2 ft) then turned back and up and ended in a larger opening at the end (J. W. Dragoo, pers. obs.).

Reproduction and Social Behavior

In Texas, mating occurs from late February through early March, and most adult females are pregnant by the end of March (Patton 1974). Captive males have exhibited breeding (mounting) behavior as early as late November, and females as early as mid-January (J. W. Dragoo, pers. obs.). Typically, gestation lasts about two months in Texas (Patton 1974). However, a captive female had a gestation period of at least 70 days (Dragoo 2009). In the other genera of Western Hemisphere skunks, variation in the length of the gestation period is attributed to delayed implantation (Mead 1968; Wade-Smith et al. 1980; Mead 1993). However, little is known about the reproductive biology of *Conepatus*, especially with respect to delayed implantation.

Parturition occurs in April and May. In western Texas, one female contained a single embryo

in late March and two others nursed young in April (Bailey 1905). Half-grown young have been observed in late July (Patton 1974) and in mid-August (Leopold 1959), and by late August young begin to disperse. Litter size for hog-nosed skunks ranges from one to five, though two to four is most common (Allen 1906; Davis 1945; Patton 1974; Dragoo 2009). In natural conditions, *Conepatus leuconotus* is not known to survive for more than three or four years (Patton 1974). However, J. W. Dragoo (unpubl. data) raised a litter of two (three were born, but one died the first night) in captivity for over 16 years.

White-backed hog-nosed skunks are generally solitary; however, they can tolerate conspecifics (Davis 1945; Dragoo 2009). Males and females stay together briefly during the breeding season, but females and young commonly are found together until the young disperse in late summer (Schmidly and Bradley 2016). Brashear et al. (2015) found that male home ranges were about three times larger than female home ranges (1.9 km² [0.7 mi²] for males and 0.64 km² [0.25 mi²] for females). They also found several instances of same-sex adults denning together or within 5 m (16 ft) of one another, and male/female denning together outside of the recognized mating season. The captive pair mentioned above shared a large enclosure with separate feeding and nesting areas. The male and female spent most of their time together and often shared the same nest box, especially in the colder months of winter (J. W. Dragoo, pers. obs.). Because these two animals were captive and neutered/spayed, however, their behavior might not have been typical of wild individuals in their native environments.

No information exists about hog-nosed skunk population densities in New Mexico. Population densities ranging from 0.6 ± 0.17 to 1.3 ± 0.26 individuals/km² have been reported from Mexico during various wet and dry seasons (Cervantes et al. 2002); the estimated density of hog-nosed skunks in west-central Texas was 2.6 individuals/km² (Brashear et al. 2015).

Interspecific Interactions

The white-backed hog-nosed skunk is sympatric with the striped skunk and two species of spotted skunks (*S. leucoparia* in the southern region and *S. gracilis* in the north) within much of its range in New Mexico and also overlaps with hooded skunks in the southwestern region of the state (Hass and Dragoo 2007, 2009, 2017). All species of skunks in New Mexico may even use the same den sites, though not at the same time; whereas use of other resources, such as food items, differs slightly among them (Patton 1974; Hass and Dragoo 2017). For example, whereas spotted skunks are more three dimensional in their foraging and striped skunks are more dietary generalists, hog-nosed skunks are more insectivorous. In detailed ecological niche modeling of hooded, striped, and hog-nosed skunks where they overlap in the southwestern United States and northern Mexico, all three species were found to occupy slightly different ecological niches (Hass and Dragoo 2017). However, niche modeling predicted that hog-nosed skunks would occupy more of the area of overlap among the three species. Based on average suitability maps, spatial niche overlaps were highest between hog-nosed and striped skunks (Hass and Dragoo 2017). According to Bailey (1903:203), "In general the habits of *Conepatus* and *Mephitis* are very similar even to a choice of the same brush patches and gulch bottoms for foraging ground. They must frequently meet, whether on friendly terms or otherwise."

Conepatus, like other skunks, is prey to canids such as coyotes (*Canis latrans*), dogs (*C. familiaris*), and foxes (*Vulpes fulva* and *Urocyon cinereoargenteus*); felids (*Puma concolor* and *Lynx rufus*); badgers (*Taxidea taxus*); and birds of prey such as great horned owls (*Bubo virginianus*) and eagles (*Aquila chrysaetos* and *Haliaeetus leucocephalus*) (Meaney et

al. 2006; Hass 2009). C. C. Hass (unpubl. data) also found a white-backed hog-nosed that had been fed on by a black bear (*Ursus americanus*). All those predators are non-discriminatory and opportunistically kill any of the skunk species in New Mexico and elsewhere (Dragoo 2009).

When threatened, the first response of a hog-nosed skunk is to flee and find cover. *C. leuconotus* has been observed to take refuge in prickly pear cactus when aggravated (Patton 1974), and Dragoo (1993) pulled one out from the middle of a prickly pear near Carlsbad Caverns National Park. According to Brashear et al. (2010), if prickly pear cacti are not readily available, hog-nosed skunks will climb trees to escape a potential predator. During flight a hog-nosed skunk may also turn toward its pursuer and, depending on the level of threat, stand on its hind legs and take two or three steps toward the pursuer. Next, it smacks the ground hard with its front paws while exhaling in a loud huff. Finally, it draws its paws under its body, kicking dirt backward. A defensive *C. leuconotus* crouches, stomps its front paws, raises its tail and holds it flat against its back, and bares its teeth. In this position it can bite and/or spray a predator (Dragoo 2009). *C. leuconotus* can squirt noxious liquid from its anal scent glands, either as a mist when the threat is not specifically located, or as a stream directed toward a specific threat. The mist can be emitted while the skunk is running (Dragoo 2009).

The white-backed hog-nosed skunk is host to numerous parasites, including fleas (*Pulex*) and ticks (*Ixodes texanus*), intestinal roundworms (*Psyalopteris maxillaris*), cestodes (*Oochoristica* and *Mesocestoides*), subcutaneous nematodes (*Filaria martis*), and *Skrjabingylus chitwoodorum*, the latter reported from the frontal sinuses region (Patton 1974). However, Hughes et al. (2018) suggest that the lack of frontal sinuses (see Van de Graaff [1969], who reports a homologous cavity located dorsal and posterior to the naso-turbinal) in the other species of skunks in Texas (and

Photo 26.14. Road-killed white-backed hog-nosed skunk near Vaughn, Guadalupe County on 4 June 2013. Collisions with road traffic are believed to be an important source of hog-nosed skunk mortality. Photograph: © Mark L. Watson.

New Mexico) make the finding of *S. chitwoodorum* suspect. Additionally, *C. leuconotus* in western Texas is infected with several helminths such as *Filaroides milksi*, *Filaria taxidaea*, *Gongylonema*, *Macracanthorhynchus ingens*, *Mathevotaenia mephitis*, *Oncicola canis*, *Pachysentis canicola*, *Physaloptera maxillaris*, and *P. rara* (Neiswenter et al. 2006).

Rabies has also been detected in *C. leuconotus* in Mexico (Aranda and Lopez-de Buen 1999) and Arizona (Hass 2003). It has not yet been reported from New Mexico (Dragoo et al. 2004), where specimens submitted for rabies testing to the New Mexico Department of Health Scientific Laboratory Division are still only reported as "skunks"; among those specimens, however, a few of them are actually hog-nosed skunks (J. W. Dragoo, pers. obs).

STATUS AND MANAGEMENT

C. leuconotus currently is not protected under the Endangered Species Act (Meaney et al. 2006). The International Union for the Conservation of Nature and Natural Resources (IUCN) has designated *C. leuconotus* as Lower Risk, least concern, but recognized that some populations may be

declining (Helgen 2016). White-backed hog-nosed skunk populations reportedly declined for many years throughout at least some portion of their historical range in the United States, with evidence that they were at some point undergoing a distribution-wide decline (Schmidly 2002; Dragoo et al. 2003; Meaney et al., 2006; Schmidly and Bradley 2016; see below).

The legal designation of *C. leuconotus* varies among US states. In Arizona, the white-backed hog-nosed skunk is classified as a "predator" and in Texas it is designated as a "furbearer." The species is legally harvested year-round in both states (Meaney et al. 2006). In Colorado, the species is classified by state wildlife agencies as a protected non-game species and in Oklahoma it is listed as a Category II Species of Concern, and there are no hunting/trapping seasons in those two states (Meaney et al. 2006). In New Mexico, hog-nosed skunks are regarded as unprotected furbearers by the New Mexico Department of Game and Fish. There are no closed hunting/trapping seasons or bag limits. The US Forest Service (USFS) considers *C. leuconotus* to be a Sensitive species in Colorado, New Mexico, Oklahoma, and Texas.

In Colorado, white-backed hog-nosed skunks were found historically in the southeastern corner of the state, where until recently the last reported specimens were those collected in the early 1930s (Miller 1933; Armstrong 1972). After decades of hiatus, *C. leuconotus* has again been recorded in Colorado. The recent records consist of a footprint in mud found in 1996 (we are not sure how this was documented) and two skulls (DMNS 9930 and 9989) discovered on the ground in 1997 and 2000 (Meaney et al. 2006). Hog-nosed skunks had not been detected in northern New Mexico either until more recently, except for the specimen mentioned by Geluso (2002), collected in Harding County in 1966 (ENMUNHM 651). It is possible they have made it back into peripheral areas of their historical range in the last 15–25 years, or else better detection methods (e.g., motion-activated remote cameras) for this species are simply showing their persistence where they were once believed to be declining or even no longer present (Meaney et al. 2006; Holbrook et al. 2012).

Threats to *C. leuconotus* include degradation, fragmentation, and loss of habitat; the alteration of natural fire regimes; interactions with feral hogs; road mortality; poison and pesticide control of predators and insect pests; disease; and livestock grazing (Schmidly 2002; Dragoo et al. 2003; Meaney et al. 2006). Because *C. leuconotus* is generally associated with rough, rocky areas and brushy areas, the conversion of native vegetation to row-crop agriculture may be partially responsible for any earlier decline of hog-nosed skunks in peripheral populations. According to Bailey (1931:338), "Most of their range, however, is over the desert valleys where no agriculture is possible and where they are of no great economic importance."

Robert F. Patton (1974:192) once described hog-nosed skunks as, "the antisocial, ill-tempered grouches of the skunk world. Every day is Monday morning to these slothlike animals." Hopefully, with the increased use of remote cameras for research and management purposes, we will find that hog-nosed skunks are not declining nor are they quite as rare as generally believed, and instead may just be trying to avoid us.

LITERATURE CITED

Allen, J. A. 1906. Mammals from the states of Sinaloa and Jalisco, Mexico, collected by J. H. Batty during 1904 and 1905. *Bulletin of the American Museum of Natural History* 22:191–262.

———. 1910. Additional mammals from Nicaragua. *Bulletin of the American Museum of Natural History* 28:87–115.

Aranda, M., and L. Lopez-de Buen. 1999. Rabies in skunks from Mexico. *Journal of Wildlife Diseases* 35:574–77.

Arias, S. M., M. J. Corriale, G. Porini, and R. F. Bo. 2006. *Proyecto de investigación y manejo del zorrino (Conepatus humboldtii y C. chinga) en la Provincia de Río Negro, Argentina*. Final Report. Secretaría de Ambiente y Desarrollo Sustentable, Buenos Aires, Argentina.

Armstrong, D. M. 1972. *Distribution of mammals in Colorado*. Monograph of the Museum of Natural History, University of Kansas Number 3.

Audubon, J. J., and J. Bachman. 1851. *The viviparous quadrupeds of North America (Genus Mephitis)*. New York: Arno Press.

Bailey, V. 1905. *Biological survey of Texas*. North American Fauna 25. Washington, DC: US Department of Agriculture, Bureau of Biological Survey.

———. 1931 (=1932). *Mammals of New Mexico*. North American Fauna 53. Washington, DC: US Department of Agriculture, Bureau of Biological Survey.

Barber, C. M. 1902. Note on little-known New Mexican mammals and species apparently not recorded from the territory. *Proceedings of the Biological Society of Washington* 15:191–93.

Beasom, S. L. 1974. Selectivity of predator control techniques in South Texas. *Journal of Wildlife Management* 38:837–44.

Brashear, W. B., R. C. Dowler, and G. Ceballos. 2010. Climbing as an escape behavior in the American hog-nosed skunk, *Conepatus leuconotus*. *Western North American Naturalist* 70:258–60.

Brashear, W. A., A. W. Ferguson, N. J. Negovetich, and R. C. Dowler. 2015. Spatial organization and home range patterns of the American hog-nosed skunk (*Conepatus leuconotus*). *American Midland Naturalist* 174:310–20.

Cahalane, V. H. 1961. *Mammals of North America*. New York: Macmillan.

Caire, W., J. D. Tyler, B. P. Glass, and M. A. Mares. 1989. *Mammals of Oklahoma*. Norman: University of Oklahoma Press.

Cervantes, F., J. Loredo, and J. Vargas. 2002. Abundance of sympatric skunks (Mustelidae: Carnivora) in Oaxaca, Mexico. *Journal of Tropical Ecology* 18:463–69.

Coues, E. 1877. *Furbearing animals*. Washington, DC: Government Printing Office.

Dalquest, W. W. 1953. *Mammals of the Mexican State of San Luis Potosi*. Baton Rouge: Louisiana State University Press.

Davis, W. B. 1945. Texas skunks. *Texas Game and Fish* (July):9–11, 25–26.

———. 1951. Texas skunks. *Texas Game and Fish* (March):19–21, 31.

Dice, L. R. 1937. Mammals of the San Carlos Mountains and vicinity: the geology and biology of the San Carlos Mountains, Tamaulipas, Mexico. *University of Michigan Studies, Science Series* 12:245–68.

Donadio, E., S. Di Martino, M. Aubone, and A. Novaro. 2001. Activity patterns, home-range, and habitat selection of the common hog-nosed skunk, *Conepatus chinga* (Mammmalia, Mustelidae), in northwestern Patagonia. *Mammalia* 65:49–53.

Dragoo, J. W. 1993. The evolutionary relationships of the skunks to each other and the rest of the weasels; with a note on behavioral idiosyncrasies. In *Eleventh Great Plains Wildlife Damage Control Workshop Proceedings, Kansas City, MO*, ed. R. A. Pierce and F. R. Henderson, 54–67. Kansas State University, Manhatten.

———. 2009. Family Mephitidae (skunks). In *Handbook of the mammals of the world*. Vol 1, *Carnivores*, ed. D. E. Wilson, and R. A. Mittermeier, 532–63. Barcelona: Lynx Edicions.

Dragoo, J. W., and R. L. Honeycutt. 1997. Systematics of mustelid-like carnivores. *Journal of Mammalogy* 78:426–43.

Dragoo, J. W., R. L. Honeycutt, and D. J. Schmidly. 2003. Taxonomic status of white-backed hog-nosed skunks, genus *Conepatus* (Carnivora: Mephitidae). *Journal of Mammalogy* 84:159–76.

Dragoo, J. W., D. K. Matthes, A. Aragon, C. C. Hass, and T. L. Yates. Identification of skunk species submitted for rabies testing in the Desert Southwest. *Journal of Wildlife Diseases* 40:371–76.

Eizirik, E., W. J. Murphy, K.-P. Koepfli, W. E. Johnson, J. W. Dragoo, R. K. Wayne, and S. J. O'Brien. 2010. Pattern and timing of diversification of the mammalian order Carnivora inferred from multiple nuclear gene sequences. *Molecular Phylogenetics and Evolution* 56: 49–63.

Ferguson, A. W. 2014. Evolution of skunks (Carnivora: Mephitidae) across the Mexican Transition Zone: understanding the influence of environmental variation on morphological and phylogeographic patterns. PhD dissertation, Texas Tech University, Lubbock.

Ferguson, A. W., R. E. Strauss, and R. C. Dowler 2022. Beyond black and white: Assessing color variation in the context of local environmental conditions in the aposematic American hog-nosed skunk *Conepatus leuconotus*. In *Small Carnivores: Evolution, Ecology, Behaviour and Conservation*, ed. E. Do Linh San, J. J. Sato, J. L. Belant, and M. J. Somers, 107–30. Oxford: Wiley-Blackwell.

Findley, J. S. 1987. *The natural history of New Mexican mammals*. Albuquerque: University of New Mexico Press.

Findley, J. S., A. H. Harris, D. E. Wilson, and C. Jones. 1975. *Mammals of New Mexico*. Albuquerque: University of New Mexico Press.

Geluso, K. 2002. Records of mammals from Harding County, New Mexico. *Southwestern Naturalist* 47:325–29.

Goodwin, G. G. 1946. Mammals of Costa Rica. *Bulletin of the American Museum of Natural History* 87:271–474.

Gray, J. E. 1865. Revision of the genera and species of Mustelidae contained in the British Museum. *Proceedings of the Zoological Society of London* 33:100–154.

Hall, E. R. 1981. *The mammals of North America*. 2nd ed. Vol. 2. New York: John Wiley.

Hall, E. R., and W. W. Dalquest. 1963. The mammals of Veracruz. *University of Kansas Publications, Museum of Natural History* 14:165–362.

Hall, E. R., and K. R. Kelson. 1952. Comments on the taxonomy and geographic distribution of some North American marsupials, insectivores and carnivores. *University of Kansas Publications, Museum of Natural History* 5:319–41.

Hass, C. C. 2003. Ecology of hooded and striped skunk in southeastern Arizona. Final report,

Arizona Game and Fish Department, Phoenix, Arizona.

———. 2009. Competition and coexistence in sympatric bobcats and pumas. *Journal of Zoology* 278:174–80.

Hass, C. C., and J. W. Dragoo. 2007. Distribution and habitat affiliations of 4 species of skunks (Mephitidae) in Arizona and New Mexico. 87th Annual Meeting of the American Society of Mammalogists, University of New Mexico, Albuquerque.

———. 2009. Distribution and habitat affiliations of 4 species of skunks (Mephitidae) in Arizona and New Mexico. 94th Annual Meeting of the Ecological Society of America, University of New Mexico, Albuquerque.

———. 2017. Competition and coexistence in sympatric skunks. In *Biology and conservation of the musteloids*, ed. D. W. Macdonald, C. Newman, and L. Harrington, 464–77. Oxford: Oxford University Press.

Helgen, K. 2016. American hog-nosed skunk. *Conepatus leuconotus*. The IUCN Red List of Threatened Species 2016:e.T41632A45210809. https://dx.doi.org/10.2305/IUCN.UK.2016-1.RLTS.T41632A45210809.en.

Hoffmeister, D. F. 1986. *Mammals of Arizona*. Tucson and Phoenix: University of Arizona Press and Arizona Game and Fish Department.

Holbrook, J. D., R. W. DeYoung, A. Caso, M. E. Tewes, and J. H. Young. 2012. Hog-nosed skunks (*Conepatus leuconotus*) along the Gulf of Mexico: population status and genetic diversity. *Southwestern Naturalist* 57:223–25.

Holton, B., T. Theimer, and K. Ironside. 2021. First records and possible range extension of the American hog-nosed skunk into Grand Canyon National Park, U.S.A. *Small Carnivore Conservation* 59:e59002. https://smallcarnivoreconservation.com/index.php/sccg/article/view/3457.

Hughes, M. R., N. J. Negovetich, B. C. Mayes, and R. C. Dowler. 2018. Prevalence and intensity of the sinus roundworm (*Skrjabingylus chitwoodorum*) in rabies-negative skunks of Texas, USA. *Journal of Wildlife Diseases* 54:85–94.

Koepfli, K.-P., J. W. Dragoo, and X. Wang. 2017. The evolutionary history and molecular systematics of the Musteloidea. In *Biology and conservation of the musteloids*, ed. D. W. Macdonald, C. Newman, and L. Harrington, 75–91. Oxford: Oxford University Press.

Kurtén, B., and E. Anderson. 1980. *Pleistocene mammals of North America*. New York: Columbia University Press.

Leopold, A. S. 1959. *Wildlife of Mexico: the game birds and mammals*. Berkeley: University of California Press.

Lichtenstein, H. 1832. *Darstellung neuer oder wenig bekannter Säugethiere in Abbildungen und Beschreibungen: von fünf und sechzig Arten auf funfzig colorirten Steindrucktafeln nach den Originalen des Zoologischen Museums der Universität zu Berlin*. Berlin: C. G. Lüderitz.

Matson, J. O., and R. H. Baker. 1986. Mammals of Zacatecas. *Special Publications of the Museum of Texas Tech University* 24:1–88.

Mead, R. A. 1968. Reproduction in western forms of the spotted skunk (genus *Spilogale*). *Journal of Mammalogy* 49:373–90.

———. 1993. Embryonic diapause in vertebrates. *Journal of Experimental Zoology* 266:629–41.

Meaney, C. A., A. K Ruggles, and G. P. Beauvais. 2006. American hog-nosed skunk (*Conepatus leuconotus*): a technical conservation assessment. Species Conservation Project. USDA Forest Service, Rocky Mountain Region.

Merriam, C. H. 1902. Six new skunks of the genus *Conepatus*. *Proceedings of the Biological Society of Washington* 15:161–65.

Miller, F. W. 1925. A new hog-nosed skunk. *Journal of Mammalogy* 6:50–52.

———. 1933. Two new Colorado mammals. *Proceedings of the Colorado Museum of Natural History* 22:1–3.

Neiswenter, S. A., D. B. Pence, and R. C. Dowler. 2006. Helminths of sympatric striped, hog-nosed, and spotted skunks in west-central Texas. *Journal of Wildlife Diseases* 42:511–17.

Patton, R. F. 1974. Ecological and behavioral relationships of the skunks of Trans Pecos Texas. PhD dissertation, Texas A&M University.

Raun G. G., and B. J. Wilks. 1961. Noteworthy records of the hog-nosed skunk (*Conepatus*) from Texas. *Texas Journal of Science* 13:204–5.

Schiaffini, M. I., M. Gabrielli, F. J. Prevosti, Y. P. Cardoso, D. Castillo, R. Bo, E. Casanave, and M.

Lizarralde. 2013. Taxonomic status of southern South American *Conepatus* (Carnivora: Mephitidae). *Zoological Journal of the Linnean Society* 167: 327–44.

Schmidly, D. J., 2002. *Texas natural history: a century of change*. Lubbock: Texas Tech University Press.

Schmidly, D. J., and R. D. Bradley. 2016. *The mammals of Texas*. 7th ed. Austin: University of Texas Press.

Schmidly, D. J., and F. S. Hendricks. 1984. Mammals of the San Carlos Mountains of Tamaulipas, Mexico. In *Contributions in Mammalogy in Honor of Robert L. Packard*, ed. R. E. Martin and B. R. Chapman, 15–69. Lubbock: Texas Tech University Press.

Seton, E. T. 1926. *Lives of game animals*. Vol. 2. Garden City, NJ: Doubleday and Doran.

Slaughter, B. H., R. H. Pine, and N. E. Pine. 1974. Erruption of cheek teeth in Insectivora and Carnivora. *Journal of Mammalogy* 55:115–25.

Taylor, W. P. 1953. Food habits of the hog-nosed skunk in Texas. In *Food habits of furbearers in relation to Texas game species*, 1–18. Final Project W-31-R-1. Austin: Texas Game and Fish Commission.

Van De Graaff, K. M. 1969. Comparative osteology of the skunks of the world. MS thesis, University of Utah.

Van Gelder, R. G. 1968. The genus *Conepatus* (Mammalia, Mustelidae): variation within a population. *American Museum Novitates* 2322:1–37.

Verts, B. J. 1967. *The biology of the striped skunk*. Urbana: University of Illinois Press.

Wade-Smith, J., M. E. Richmond, R. A. Mead, and H. Taylor. 1980. Hormonal and gestational evidence for delayed implantation in the striped skunk, *Mephitis mephitis*. *General and Comparative Endocrinology* 42:509–15.

Warren, E. R. 1921. The hog-nosed skunk (*Conepatus*) of Colorado. *Journal of Mammalogy* 2: 112.

———. 1942. *The mammals of Colorado: their habits and distribution*. 2nd ed. Norman: University of Oklahoma Press.

Wood, W. F., C. O. Fisher, and G. A. Graham. 1993. Volatile components in defensive spray of the hog-nosed skunk, *Conepatus mesoleucus*. *Journal of Chemical Ecology* 19:837–41.

Wood, W. F., B. G. Sollers, G. A. Dragoo, and J. W. Dragoo. 2002. Volatile components in defensive spray of the hooded skunk, *Mephitis macroura*. *Journal of Chemical Ecology* 28:1865–70.

Wozencraft, W. C. 2005. Order Carnivora. In *Mammal species of the world: a taxonomic and geographic reference*, Vol. 1, ed. D. E. Wilson and D. M. Reeder, 532–628. 3rd ed. Baltimore: Johns Hopkins University Press.

HOODED SKUNK (*MEPHITIS MACROURA*)

Christine C. Hass and Jerry W. Dragoo

Nearly a century ago, the early naturalist Vernon Bailey (1931) stated that little was known in New Mexico regarding the hooded skunk, *Mephitis macroura*. Not much has changed since then, as hooded skunks are probably among the most underreported carnivores in our state. There is, however, slightly more information about the hooded skunk from Texas and Arizona, so the bulk of data for this account will come from animals just outside of New Mexico. Habitat types used by hooded skunks in Texas and Arizona occur in New Mexico, so data from southern Arizona where we have studied them should apply to hooded skunks in our state.

The hooded skunk can be differentiated from the closely related striped skunk (*Mephitis mephitis*) by the long hairs on the nape of its neck and head (hence its common name), its relatively longer tail (≥50% of its total length), and its larger ears (see also Chapter 28). The skull is similar to that of a striped skunk, but the auditory bullae reportedly are more bulbous. This characteristic, however, is not easily distinguishable between the two species (C. C. Hass and J. W. Dragoo, pers. obs.; Hoffmeister 1986). Bailey (1931:336) wrote of them that "in habits [they] are rather more active and sprightly than are the heavier species of *Mephitis*, but they do not approach in weasellike motions the little [western] spotted skunks [(*Spilogale* spp.)]." In the eyes of mephitologist Robert F. Patton (1974:192), "Hooded skunks could be considered as the shy, regal ballerinas of

the circus world. They like to avoid trouble and do so usually by their very retiring manner." Seton (1929:370) described the species as a "long, slender weasel-like *Mephitis*, almost a *Spilogale*."

Color pattern, as in all skunks and more so in the genus *Mephitis*, is variable. Different researchers recognize different numbers of patterns among hooded skunks, with Patton (1974) and Schmidly and Bradley (2016) describing three in Texas; white-backed, black-backed (with side stripes), and a combination of the two. Hoffmeister described three color patterns in Arizona, white-backed, black-backed, and white-necked (the latter pattern referred to as "star" by the fur trade for striped skunks [Verts 1967; Hoffmeister 1986]). Based on our research in Arizona, which involved capturing 65 hooded skunks, we identified five pelage patterns along with several variants: 1) two stripes running down the side of the body from the shoulder to the mid-abdomen or hind quarters; 2) a single wide white band along the dorsum from the back of the head to the tail; 3) a combination of patterns one and two that may include additional stripes closer to the elbow; 4) a nearly all black pelage, which may have faint white stripes near the elbow, and 5) a "star" pattern corresponding to the hooded skunk being nearly black with a white cap on the back of the neck (Hass 2003; Photos 27.1 and 27.2). In both Texas and Arizona, the most common patterns among hooded skunks were two stripes along

(*opposite page*) Photograph: © Jerry W. Dragoo.

Photos 27.1 a, b, c, d, e, and f. Main black and white color patterns and variants in the hooded skunk: a) (*top left*) two stripes running down the side of the body from the shoulder to the mid-abdomen or hindquarters (referred to as pattern one in the text); b) (*top right*) a single wide white band along the dorsum from the back of the head to the tail (pattern two); c) (*middle left*) a combination of pattern one (a) and pattern two (b); d) (*middle right*) the nearly all-black pelage with possible faint white stripes near the elbow; e) (*bottom left*) the star pattern, consisting of a white cap on the back of the neck with the rest of the body being black; and f) (*bottom right*) one of several variants, showing the white cap and the white lateral stripe. Pattern one (a) and pattern two (b) represent the most common color patterns in both Texas and Arizona. Among the larger skunks, hooded skunks are the only ones that can show black on the back of the neck. Photographs: a, b, c, e, and f © Christine C. Hass; d © Travis Perry.

the side and a single wide band along the dorsum (patterns one and two; Patton 1974; Hass 2003). More than one of these patterns can be seen within a single litter (Photo 27.3).

Hooded skunks, like striped skunks, sport a thin white stripe between the eyes (Patton 1974; Hoffmeister 1986; Dragoo 2009; Chapter 28). Frequently, striped skunks with a lot of white down the back (two stripes merged into a single stripe) will be confused with and misidentified as hooded skunks. However, the single stripe found in hooded skunks is interspersed with black hair giving it a gray appearance, whereas, in striped skunks, the white is "pure." The pelage of the white-backed hooded skunk is often quite long, giving the animal a shaggy appearance. In addition, hooded skunks with patterns one and four are black on the dorsal area of their necks (Photos 27.1a and d); something not seen in striped skunks, which have white on the back of their necks (see Chapter 28). The hooded skunk's fur is long, light, soft, and fine compared to that of striped skunks. The all-black and some of the star morphs tend to have shorter hair, similar to striped skunks, yet still fine and soft. Although hooded skunks have proportionally larger tails than striped skunks, tail length can be difficult to determine from a distance due to the long hairs on the tip. In addition, the subspecies of striped skunk in eastern New Mexico and west Texas has a tail almost as long as that of a hooded skunk (Patton 1974; see Chapter 28). The white-backed form of the hooded skunk may also be confused with the white-backed hog-nosed skunk (*Conepatus leuconotus*), but the latter has a solid, dense white stripe along the back, with a relatively short white tail and no white stripe between the eyes (Chapter 26). Furthermore, hog-nosed skunks are larger and bulkier, with a large, blunt nose.

The hooded skunk was described and named by Hinrich Lichtenstein (1832) from specimens collected northwest of Mexico City. Currently, there are four subspecies recognized: *M. m.*

Photo 27.2. Variation in dorsal color pattern exhibited by three Museum of Southwestern Biology hooded skunk specimens from New Mexico. The first two specimens show the dorsal white stripe (pattern two) whereas the third is primarily black but exhibits the white cap on the back of the neck (the "star" pattern or pattern five). Regardless of color pattern, the species is characterized by long hairs on the nape of its neck and head forming a hood. Photograph: © Jon Dunnum.

Photo 27.3. Two juvenile hooded skunks in southern Arizona. Several color patterns may occur within the same litter. Photograph: © Christine C. Hass.

eximius (Hall and Dalquest 1950), *M. m. macroura* (Lichtenstein 1832), *M. m. milleri* (Mearns 1897), and *M. m richardsoni* (Goodwin 1957). Only one of those subspecies (*M. m. milleri*) occurs in New Mexico. Mearns (1897) tentatively described the northern hooded skunk as a distinct species,

Mephitis milleri, based on two specimens collected near Tucson, Arizona. The type specimen had only white stripes along the side of the body (pattern one). In contrast, the other specimen was similar to color pattern two with only white (gray) covering the back of the animal. The second specimen was similar to what Mearns (1897) considered *M. macroura*. He suggested color patterns in "*milleri*" were variable and indicated the possibility that this taxon, instead of being its own species, may instead represent a subspecies of the skunk described by Lichtenstein (1832). It was later relegated as such (Howell 1901).

Hooded skunks exhibit minimal sexual dimorphism. Male hooded skunks are slightly larger than females, in weight (males average just over 1 kg [2.2 lbs], whereas females average about 0.8 kg [1.8 lbs]), and length (males average from 687 to 750 mm [27–29 in] in different studies, whereas females average 662 to 750 mm [26–29 in]), but all studies show large overlaps (all data from Arizona and Texas; Patton 1974; Hoffmeister 1986; Hass 2003). Females have longer tails and bigger ears, but males have larger hind feet (Patton 1974; Hass 2003). Hooded skunks from the northern part of the species' distribution, including presumably in New Mexico, appear larger than in Mexico and Central America (Davis and Lukens 1958; Armstrong 1972; Janzen and Hallwachs 1982).

Hooded skunks share an important characteristic with the rest of the Mephitidae: enlarged scent glands at the base of the tail that are used to squirt a noxious fluid at a potential predator. The anal secretion of a hooded skunk in New Mexico was found to be composed of seven major components comprising 99% of the volatiles. These components include: (E)-2-butene-1-thiol, 3-methyl-1-butanethiol, S-(E)-2-butenyl thioacetate, S-3-methylbutenyl thioacetate, 2-phenylethanethiol, 2-methylquinoline, and 2-quinolinemethanethiol. There are several minor components as well, including phenylmethanethiol, S-phenylmethyl thioacetate, S-2-phenylethyl thioacetate,

bis[(E)-2-butenyl] disulfide, (E)-2-butenyl 3-methylbutyl disulfide, bis(3-methylbutyl) disulfide, and S-2-quinolinemethyl thioacetate (Wood et al. 2002). Two of these compounds (S-phenylmethyl thioacetate and S-2-phenylethyl thioacetate) are unique to hooded skunks (Wood et al. 2002).

As in all skunks, the anal glands of hooded skunks are used in self-defense (Dragoo 2009). One of us (J. W. Dragoo) observed (i.e., was the target of) an animal spraying nine times in 11 seconds and again three more times 90 minutes later. Based on our experience, the defensive behavior of *M. macroura* is similar to that of *M. mephitis*. The hooded skunk's first response, of course, is to run away. When cornered, however, a hooded skunk will turn to face a potential threat. It will raise its tail and arch its back, then stomp loudly with its front feet as it lunges forward. Often it will drag its front feet along the ground as it pulls its front legs to itself in preparation for another lunge. It will hiss as it lunges. If the predator does not retreat, it may then be sprayed in the face with a foul-smelling and temporarily blinding fluid secretion, an experience that it presumably will not forget soon. Hooded skunks appear to be a little more reticent to spray than striped skunks (J. W. Dragoo, pers. obs.).

The striking black and white pattern of the skunk's coat is believed to be aposematic, warning potential predators of the consequences of getting too close (Walton and Larivière 1994; Larivière and Messier 1996; Hunter 2009). The defensive displays of the skunks augment the black-and-white pattern of the coat to further deter predators (Medill et al. 2011). This warning coloration allows skunks to spend more time foraging and less time looking for predators (Speed et al. 2010; Fisher 2017). Studies on aposematic behavior have all been conducted on striped skunks. Further studies are needed to determine how the other skunks may differ (or not) in their anti-predator behavior.

DISTRIBUTION

Hooded skunks occur primarily in Central America and Mexico, and they reach their northern distribution in only three US states: Arizona, New Mexico, and Texas (Dragoo 2009). The southernmost subspecies, *M. m. richardsoni*, is found in Nicaragua and Costa Rica (Goodwin 1957; Janzen and Hallwachs 1982). Toward the middle of the hooded skunk's distribution, two other subspecies occur in Mexico, *M. m. eximius* in southern Veracruz (Hall and Dalquest 1950), and *M. m. macroura* from eastern Nayarit to Tamaulipas and south to Honduras. The northernmost subspecies is *M. m. milleri*, and it occurs from northern Mexico (Sonora and Sinaloa east to Coahuila) north into the southwestern United States (Dragoo 2009).

Audubon and Bachman (1854) reported hooded skunks, which they called large-tailed skunks, from between Houston and La Grange, Texas, nearly 1,300 km (~800 miles) northwest to Santa Fe, New Mexico. In doing so, however, they made what is now the classic mistake of misidentifying striped skunks with the broad-stripe pattern as hooded skunks (see Chapter 28). The description of the skunk and the drawing on the plate clearly demonstrate a striped skunk. Coues (1877) not only pointed out the error but also stated that the hooded skunk was not found in the United States. Given what we know now, was Coues incorrect with respect to that statement? At the time Audubon and Bachman first published their *Viviparous Quadrupeds of North America* (1845–1848), what is now New Mexico was still part of Old Mexico, and would not become part of the Union until 1850, with the current southern border not established until the Gadsden Purchase of 1853–1854 (Simmons 1988). Therefore, Coues may have been correct in recognizing that hooded skunks did not occur in the United States, though they did occur in territories owned by the United States. In describing the species, Mearns would also have

Photo 27.4. Hooded skunk from Sierra County just west of the Rio Grande. Photograph: © Travis Perry.

Photo 27.5. Hooded skunk photographed by a remote camera at the San Andres National Wildlife Refuge in October 2007 and representing a rare record of the species east of the Rio Grande in New Mexico. Photograph: © San Andres National Wildlife Refuge.

been the first one to establish that hooded skunks occurred in US territories (Mearns 1897).

In today's New Mexico, the hooded skunk is only found in the southwestern part of the state, where it has the smallest distribution of any of the skunks. Bailey, who reported specimens from the Animas Valley (Hidalgo County), suggested that

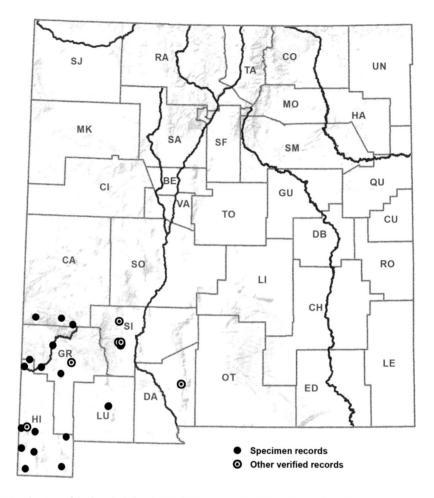

Map 27.1. Distribution of the hooded skunk (*Mephitis macroura*) in New Mexico based on museum specimen and other verified records.

the species was a Lower Sonoran life zone animal that occupied the valley country of southern Arizona and into southwestern New Mexico (Bailey 1931). Although the hooded skunk's distribution in New Mexico is more extensive than thought by Bailey (1931), most specimens and observations in New Mexico are west of the Rio Grande and south of the Gila River (Map 27.1), where the species is sympatric with the striped, hog-nosed, and western spotted skunks (Hass and Dragoo 2017). At the edge of the species' distribution in the state, a hooded skunk specimen was collected in open ponderosa forest on the Mogollon Plateau

in Catron County nearly 100 years ago (Hubbard 1972), and more recently, hooded skunks have been photographed by remote cameras on the San Andres National Wildlife Refuge in Doña Ana County (M. Weisenberger, US Fish and Wildlife Service, in litt.; Photo 27.5). Altogether, specimens have been collected or individuals observed in Catron, Doña Ana, Grant, Hidalgo, and Sierra counties (Hass and Dragoo 2009; T. Perry in litt.). Recent ecological niche modeling, based on climate and land-cover data, predicted habitat for hooded skunks in New Mexico below the Mogollon Mountains and eastward to the western boundary of the Tularosa Basin (Hass and Dragoo 2017).

The fossil record for hooded skunks in New Mexico is non-existent (Chapter 1), with the few fossil specimens from farther south in Sonora, Mexico based on limited material (Martinez-Lira et al. 2011). Given the nature of the similarity to striped skunks in skull morphology, it is nonetheless possible that many specimens may have been misidentified.

HABITAT ASSOCIATIONS

Hooded skunks occur from sea level to 3,110 m (~10,200 ft), in a wide variety of vegetation types across temperate and tropical regions (Davis and Russell 1954; Davis and Lukens 1958; Hubbard 1972; Findley et al. 1975; Hoffmeister 1986). They are most common in the arid lowlands (Davis and Russell 1954), but also occur in deciduous or ponderosa forest, along forest edges, and in pastures, rocky canyons, and riparian areas (Baker 1956; Findley et al. 1975; Janzen and Hallwachs 1982). In the southwestern United States, most specimen records and observations are from desert, closed and open shrubland, and grassland, up to montane woodland and forest communities (Cook 1986; Hass and Dragoo 2017). Dense, brushy streamside vegetation may be particularly important for hooded skunks in New Mexico, as "they are generally found along the stream valleys or in canyons, where they follow the trails and resort to rocky ledges or brushy bottoms for cover and concealment during the day" (Bailey 1931:336). In southeastern Arizona, hooded skunks use woodlands more than do striped and hog-nosed skunks (Hass 2003; Hass and Dragoo 2017), and our overall impression is that, compared to their relatives, they prefer denser vegetation. Hooded skunk specimens from New Mexico were collected in Chihuahuan desert scrub, closed basin desert scrub, desert grassland, montane scrub, coniferous and mixed woodland, and urban or agricultural areas (Hass and Dragoo 2017; vegetation

Photo 27.6. Hooded skunk in an arroyo in desert scrub above the Gila River floodplain. The species also occurs in riparian cottonwood woodland along the Gila River (Geluso 2016). Photograph: © Keith Geluso.

classification according to Dick-Peddie [1999] but see Chapter 2).

Ecological niche modeling of hooded, striped, and hog-nosed skunks along the US-Mexico border found that both hooded and hog-nosed skunks seemed to select habitats that were more "tropical" than did striped skunks. That is, they were more likely to be found in areas that were characterized by hot, dry springs, wet summers, and mild winters than were striped skunks (Hass and Dragoo 2017). Although the three species differed somewhat in habitat selection, we found no evidence of competitive exclusion. Hooded skunks appeared to have the narrowest habitat preferences; if habitat existed for hooded skunks, striped and hog-nosed skunks were likely present as well (Hass and Dragoo 2017).

Hooded skunks den in holes dug in the ground, in rock crevices, and in holes in logs. They have even been observed denning up to 2 m [~6.6 ft] above ground in trees (Hass 2003). They are more likely to den on rocky hillsides than sympatric striped skunks. In agricultural areas, they den along fencerows beneath irrigation canals and in heavily vegetated areas along streams (Patton 1974). Reed and Carr (1949) were surprised that a hooded skunk would use jumping cholla (*Cylindropuntia fulgida*) as escape cover. However, use of cacti is a common behavior for all of New Mexico's skunks (Dragoo 2009). Patton also found that hooded skunks were less likely than striped skunks to den around human dwellings in Texas (Patton 1974). However, radio-collared hooded skunks in Arizona were frequently located under buildings and sheds, and in culverts (Hass 2002). Unfortunately, the denning habits of hooded skunks in New Mexico have not been studied.

LIFE HISTORY
Diet and Foraging.
The diet of the hooded skunk consists primarily of insects (especially beetles and grasshoppers), fruits, small vertebrates, and bird eggs (Patton 1974; Hass and Dragoo 2017). In one study in Arizona, 94% of 49 scats contained beetles, whereas 16% contained orthopterans (grasshoppers and crickets; Hass 2003). Other common arthropods consumed included centipedes and caterpillars (>20% of scats); berries of alligator juniper (*Juniperus deppeana*) were found in 55% of scats. In Texas, insect material made up 74% of 10 stomach and intestinal contents, with vertebrates and plant material making up the rest (Patton 1974). When foraging, *M. macroura* moves slowly, sniffling among leaves and pouncing on grasshoppers and beetles (Dalquest 1953; Reid 1997). Vertebrates, such as rodents and reptiles, are taken opportunistically (J. Dragoo, pers. obs.; Hass 2003). Janzen and Hallwachs (1982:550–51) observed how one wild hooded skunk in Costa Rica opened an egg:

> When offered a raw and intact chicken egg, she held it against her chest with both front paws. Nearly simultaneously she arched her back, raised the front half of her body 10 to 20 cm above the ground, and threw the egg violently backwards between her legs. She then turned around and snuffled after it. The egg was thrown-rolled 50 to 150 cm; if the egg was unbroken, after she located it, she repeated the process in an apparently random direction. Eventually she hit something hard with the egg and it cracked or broke, whereupon she ate the contents.

Reproduction and Social Behavior
The reproductive biology of the hooded skunk is poorly known. Breeding likely takes place from mid-February to the end of March in Texas (Patton 1974) and during March in Arizona (Hass 2003). Like all carnivorans, male hooded skunks have a baculum, and, like all skunks, females have two pairs of inguinal, one pair of abdominal, and two pairs of pectoral mammae (Bailey 1931). Young hooded skunks are born in late May through early June (Patton 1974; Hass 2003). If a female loses a litter early or is not pregnant, she may stay reproductively receptive until August (Patton 1974; Hass 2003). Litter size ranges from three to eight (Bailey 1931; Patton 1974), and females may nurse the young through at least August (Patton 1974). Juvenile hooded skunks achieve full body size by one year of age (Hass 2003). No data are available on the gestation period of hooded skunks, so it is not known if they exhibit delayed implantation, as in striped and western spotted skunks (Wade-Smith and Verts 1982; Verts et al. 2001).

Hooded skunks remain active all year (Bailey 1931) and have been trapped during winter in New Mexico (J. Dragoo, pers. obs.) and Arizona (Hass 2003). Hooded skunks begin putting on weight during the summer and are heaviest by late summer and early fall. They can lose about 25% of their mass throughout the winter (Hass 2003).

Hooded skunks are typically nocturnal, but we have observed them also being active during daylight hours, especially during winter months. Although that is not atypical, diurnal activity could also be a sign of disease, such as rabies (see

Photo 27.7. Hooded skunk with at least four juveniles in Sierra County on 7 August 2015. Litter size ranges from three to eight, and in Arizona and Texas the young are born in late May–early June. Photograph: © Travis Perry.

below). Some of the individuals we found active during the day acted quite normally—out foraging in the morning or late afternoon during the winter, as though perhaps they did not obtain enough food while foraging the previous night. Others, including one confirmed with rabies (Hass and Dragoo 2006), appeared to be foraging (or otherwise moving) in a rather frantic manner.

Hooded skunks tend to be solitary except when females are raising young. Females normally do not den together during the winter months, as do striped skunks in colder regions (Verts 1967; Hwang et al. 2007). Males and females do not defend territories, and even though they are solitary, they are not necessarily antisocial with other hooded skunks or other species. Like other skunks, hooded skunks will often gather at particularly rich food sources, such as bird feeders and garbage dumps (Reid 1997).

Home ranges of hooded skunks can measure from 2.8–6 km² (692–1384 acres) (Ceballos and Miranda 1986; Hass 2003). Densities of 1.3–3.9 hooded skunks/km² [0.62 mi²] have been reported from southeastern Arizona (Hass 2003), with similar densities in southern Mexico (Cervantes et al. 2002). However, up to 25 hooded skunks/km² [0.62 mi²] have been reported from urban areas in central Mexico (Ceballos and Miranda 1986). The species is more common in urbanized areas than previously thought, though its numbers are underreported due to the difficulty in distinguishing them from striped skunks (Hass 2003).

Females, when not nursing, tend to stay at a den site longer than males (4.5 vs. 4 days; Hass 2003). Males will move greater distances from one den site to the next compared to females, and home ranges of hooded skunks tend to be smaller in

Photos 27.8a, b, and c (*top to bottom*). Hooded skunks (a) can occur in the same area as Rio Grande spotted skunks (b; *Spilogale leucoparia*), striped skunks (c; *Mephitis mephitis*), and white-backed hog-nosed skunks (*Conepatus leuconotus*; see Photo 26.8), as documented with these photographs in riparian woodland along the Gila River in Grant County. Photographs: © Keith Geluso.

urban areas (Hass 2003). Rheude (2008) showed that urban skunks shared mtDNA haplotypes with non-urban skunks, and that no unique haplotypes were found among urban skunks. This would suggest that gene flow occurs between urban and non-urban skunk populations.

Interspecific Interactions

During a two-year study in Arizona, survival rates of adult skunks were low (24–56%). Predation accounted for most of the mortality, but disease and parasites were additional contributing factors (Hass 2003). Great horned owls (*Bubo virginianus*), cougars (*Puma concolor*), bobcats (*Lynx rufus*), jaguars (*Panthera onca*), black bears (*Ursus americanus*), and coyotes (*Canis latrans*) may kill (or at least eat) hooded skunks (Leopold 1959; Hass 2003, 2009).

Hooded skunks have been trapped in areas where striped, hog-nosed, and western spotted skunks occur. Trap lines for hooded skunks have also captured white-nosed coatis (*Nasua narica*), gray foxes (*Urocyon cinereoargenteus*), and raccoons (*Procyon lotor*) (Hass 2003). In urban areas, hooded skunks resemble striped skunks in that they probably interact with dogs and cats, as well as with raccoons. During a study of skunk ecology in the Fort Huachuca Military Reservation and surrounding areas in southeastern Arizona, radio-collared hooded and striped skunks shared the same habitats and had similar den sites, but they did not share dens with any other skunks (whether same species or not), except for dens located under buildings in urban areas. There was almost complete overlap of denning ranges among sexes and species (Hass 2003; Hass and Dragoo 2017).

STATUS AND MANAGEMENT

Hooded skunks, like other skunks, probably do not live much beyond three to four years in the wild, although they may live at least 10 years in captivity (J. W. Dragoo, pers. obs.). Over a seven-year span in Arizona, one hooded skunk was recaptured almost 33 months after its initial capture; most of the others were recaptured <19 months after initial capture (Hass 2003). Besides predation, diseases contribute to an unknown extent to mortality. Roundworms (*Physaloptera maxillaries*), fleas (family: Pulicidae), and the nematode *Skrjabingylus chitwoodorum* all occur in hooded skunks in Trans-Pecos, Texas (Patton 1974). A new nematode, *Skrjabingylus santaceciliae*, has even been recently described from frontal sinus specimens in the hooded skunk (Carreno et al. 2005). The species is also susceptible to feline distemper (Patton 1974). Hooded skunks also have been found to harbor numerous fleas, ticks, lice, and mites (Hass 2003).

Rabies is a viral disease of the nervous system, capable of infecting most mammals. Bats and carnivores appear to be highly susceptible to the virus, and account for most of the human exposures to rabies in the United States (Birhane et al. 2017). Hooded skunks in Arizona were found to contract rabies, as well as respond to the rabies vaccine by producing antibodies (Hass and Dragoo 2006). Rabies, however, is rarely reported in hooded skunks (Ceballos and Miranda 1986; Aranda and López-de Buen 1999). There is no reason to suspect that hooded skunks do not contract the virus on a more regular basis, but rather they go unnoticed or misidentified as striped skunks as seen in a sample of skunks submitted for rabies testing in New Mexico (Dragoo et al. 2004). In Arizona, the number of reported cases of rabid skunks was related to precipitation one to two years before detection, indicating that the incidence of rabies was likely density dependent (more skunks equals more rabies; Verts 1967; Hass and Dragoo 2006).

The pelt of hooded skunks has never been considered to be of great economic value (see Bailey 1931); thus the hooded skunk has not been hunted to the extent that other skunk species (and furbearers in general) have. Its fur is very long, light, and, still today, not much sought after (Schmidly 2004). Its flesh has been served as food in some areas of the species' distribution, and the fat and scent glands have been used for medicinal purposes (Davis 1944; Dalquest 1953; Reid 1997). Indigenous populations in Mexico use not only fat but also meat of the hooded skunk to treat ailments such as coughs, stomachache, and asthma (Vazquez et al. 2006; Jacobo-Salcedo et al. 2011).

No hooded skunks were reportedly taken by fur trappers in New Mexico in most years between 2000 and 2018 (Association of Fish and Wildlife Agencies 2023). However, it is likely that some of the 1,000–2,000 "striped skunks" killed by private trappers and by APHIS-USDA Wildlife Services (USDA-APHIS Wildlife Services archives 2021 and previous years) each year in the southwestern corner of the state are actually hooded skunks. In New Mexico, skunks of all species are regarded as unprotected furbearers, and as a result may be taken year-round with no limits by state residents. Because no population estimates are available for the hooded skunk, it is unknown what impact trapping may have on the species on a local or more global scale. In addition, nuisance skunks may be trapped and euthanized by commercial pest control companies (translocation is typically now prohibited in New Mexico; see also below). Overall, however, we suspect that road traffic collisions constitute the greatest source of mortality for hooded skunks.

Because of their restricted range in New Mexico, hooded skunks are less likely to be considered nuisance animals than are striped skunks. In urban Tucson, Arizona, however, hooded skunks were found to be more numerous than striped skunks (Rheude 2008). Their attraction to urban and/or agricultural areas is for the same reasons in both species: the presence of garbage and pet food, water, and shelter. Limiting access to anthropogenic sources of food and to den sites will reduce the probability of skunks moving in where they are not wanted (Hass 2002).

Skunks kill an occasional chicken or dig holes

in a golf course looking for grubs and may be perceived as a nuisance. Although it may be tempting to capture and translocate them, a trap-vaccinate-relocate study on hooded skunks in Arizona revealed mixed results on the effectiveness and benefits of that management approach. Of three hooded skunks that were translocated from an urban site, one returned to its point of capture (approximately 10 km [6.2 mi] away) and later died from a human-induced cause (captured in a trap set for a squirrel); one died or was killed in the release area and fed upon by a great horned owl, and the last individual stayed in the release area and raised a litter (Hass 2002). Translocation also carries a risk of increasing the spread of diseases, including rabies (Conover 2002; Hass 2002). Currently, it is illegal or discouraged to translocate skunks in New Mexico; captured skunks must be euthanized (E. Goldstein pers. comm.).

Nuisance skunks are but one symptom of a larger problem; they are not the problem itself. A long-term solution to nuisance skunks should involve education (Hass 2002). Information should be provided on skunks focusing on why they are there, what will happen to a "problem" animal if it is removed, and potential benefits of leaving the animal in place. Hooded skunks play a beneficial role by eating insect pests and rodents in agricultural and urban areas (Ceballos and Miranda 1986; Hass 2003). As Bailey (1931:337) recognized in New Mexico, "the good that these skunks may do in destroying insects is probably of considerable importance locally and may overbalance their occasional mischief in destroying poultry or the eggs of game birds."

Even the skunks residing under buildings may be consuming unwanted insects and rodents, and plans can be made to exclude animals by covering potential den entrances, containing garbage, and feeding pets indoors. To avoid potential disease transmission (i.e., rabies), domestic pets should be vaccinated (Hass 2002).

Little is known about the overall status of the hooded skunk (Hoffmeister 1986), but it may be common in some parts of the species' range such as the southwestern United States, especially Arizona and possibly New Mexico. In Texas, however, hooded skunks have never been common. Bailey (1905) did not mention them in his survey of Texas—nor along the eastern half of New Mexico (i.e., east of the Rio Grande). Low hooded skunk population levels in Texas are a source of concern, Schmidly and Bradley (2016) going as far as suggesting that the species may no longer occur in that state, though hooded skunks have recently been photographed by automatic trail cameras in Big Bend National Park (Stevens 2017). In Arizona, at the northern end of the hooded skunk's range, populations appear to be cyclical in response to drought and disease (Hass and Dragoo 2006). In Mexico, hooded skunks are locally abundant (Cervantes et al. 2002), and they survive in human-altered environments such as cultivated fields, pastures, and suburban areas, though in that country little research has been conducted to further document their ecology (Ceballos and Miranda 1986). As one of the most overlooked animals in New Mexico, there is an obvious need for more research on hooded skunks in our state.

LITERATURE CITED

Aranda, M., and L. López-de Buen. 1999. Rabies in skunks from Mexico. *Journal of Wildlife Diseases* 35:574–77.

Armstrong, D. M. 1972. Mammals from the Mexican state of Sinaloa. III. Carnivora and Artiodactyla. *Journal of Mammalogy* 53:48–61.

Association of Fish and Wildlife Agencies. [2023]. Furbearer management. Under Surveys & reports, see US furbearer harvest statistics database 1970–2018 (a downloadable Excel file). https://www.fishwildlife.org/afwa-inspires/furbearer-management.

Audubon, J. J., and J. Bachman. 1854. *The quadrupeds of North America*. Vol. 3. New York: V. G. Audubon.

Bailey, V. 1905. *Biological survey of Texas*. North American Fauna 25. Washington, DC: US Department of Agriculture, Bureau of Biological Survey.

———. 1931 (=1932). *Mammals of New Mexico*. North American Fauna 53. Washington, DC: US Department of Agriculture, Bureau of Biological Survey.

Baker, R. H. 1956. Mammals of Coahila, Mexico. *University of Kansas Publications, Museum of Natural History* 9:125–35.

Birhane, M. G., J. M. Cleaton, B. P. Monroe, A. Wadhwa, L. A. Orciari, P. Yager, J. Blanton, A. Velasco-Villa, B. W. Petersen, and R. M. Wallace. 2017. Rabies surveillance in the United States during 2015. *Journal of the American Veterinary Medical Association* 250:1117–30.

Carreno, R. A., K. E. Reif, and S. A. Nadler. 2005. A new species of *Skrjabingylus petrov*, 1927 (Nematoda: Metastrongyloidea) from the frontal sinuses of the hooded skunk, *Mephitis macroura* (Mustelidae). *Journal of Parasitology* 91:102–7.

Ceballos, G., and A. Miranda. 1986. *Los mamíferos de Chamela, Jalisco: manual de campo*. Mexico City: Universidad Nacional Autónoma de México.

Cervantes, F. A., J. Loredo, and J. Vargas. 2002. Abundance of sympatric skunks (Mustelidae: Carnivora) in Oaxaca, Mexico. *Journal of Tropical Ecology* 18:463–69.

Conover, M. 2002. Wildlife translocation. In *Resolving human-wildlife conflicts: the science of wildlife damage management*, 211–28. Boca Raton, FL: CRC Press.

Cook, J. A. 1986. The mammals of the Animas Mountains and adjacent areas, Hidalgo County, New Mexico. *Occasional Papers of the Museum of Southwestern Biology* 4:1–45.

Coues, E. 1877. Fur-bearing animals: a monograph of North American Mustelidae. *U.S. Geologic and Geographical Survey of the Territories, Miscellaneous Publications* 8:1–348.

Dalquest, W. W. 1953. *Mammals of the Mexican state of San Luis Potosi*. Baton Rouge: Louisiana State University Press.

Davis, W. B. 1944. Notes on Mexican mammals. *Journal of Mammalogy* 25:370–403.

Davis, W. B., and P. W. Lukens Jr. 1958. Mammals of the Mexican state of Guerrero, exclusive of Chiroptera and Rodentia. *Journal of Mammalogy* 39:347–67.

Davis, W. B., and R. J. Russell. 1954. Mammals of the Mexican state of Morelos. *Journal of Mammalogy* 35:63–80.

Dick-Peddie, W. A, ed. 1999. *New Mexico vegetation: past, present, and future*. Albuquerque: University of New Mexico Press.

Dragoo, J. W. 2009. Family Mephitidae (skunks). In *Handbook of the mammals of the world*. Vol 1, *Carnivores*, ed. D. E. Wilson, and R. A. Mittermeier, 532–63. Barcelona: Lynx Edicions.

Dragoo, J. W., D. Matthes, A. Aragon, C. C. Hass, and T. L. Yates. 2004. Identification of skunk species submitted for rabies testing in the Desert Southwest. *Journal of Wildlife Diseases* 40:371–76.

Findley, J. S., A. H. Harris, D. E. Wilson, and C. Jones. 1975. *Mammals of New Mexico*. Albuquerque: University of New Mexico.

Fisher, K. A. 2017. Antipredator strategies of striped skunks in response to cues of aerial and terrestrial predators. PhD dissertation, California State University, Long Beach.

Geluso, K. 2016. *Mammals of the active floodplains and surrounding areas along the Gila and Mimbres rivers, New Mexico*. Final report, New Mexico Share with Wildlife contract.

Goodwin, G. G. 1957. A new kinkajou from México and a new hooded skunk from Central America. *American Museum Novitates* 1830:1–4.

Hall, E. R., and W. W. Dalquest. 1950. Geographic range of the hooded skunk, *Mephitis macroura*, with a description of a new subspecies from

Mexico. *University of Kansas Publications, Museum of Natural History* 1:575–80.

Hass, C. C. 2002. *Reduction of nuisance skunks in an urbanized area*. Final report to Arizona Game and Fish, Phoenix, AZ.

———. 2003. *Ecology of hooded and striped skunks in southeastern Arizona*. Final report to Arizona Game and Fish, Phoenix, AZ.

———. 2009. Competition and coexistence in sympatric bobcats and pumas. *Journal of Zoology* 278:174–80.

Hass, C. C., and J. W. Dragoo. 2006. Rabies in hooded and striped skunks in Arizona. *Journal of Wildlife Diseases* 42:825–29.

———. 2009. Distribution and habitat affiliations of 4 species of skunks (Mephitidae) in Arizona and New Mexico. 94th Annual Meeting of the Ecological Society of America. University of New Mexico, Albuquerque.

———. 2017. Competition and coexistence in sympatric skunks. In *Biology and conservation of the musteloids*, ed. D. W. Macdonald, C. Newman, and L. Harrington, 464–77. Oxford: Oxford University Press.

Hoffmeister, D. F. 1986. *Mammals of Arizona*. Tucson and Phoenix: University of Arizona Press and Arizona Game and Fish Department.

Howell, A. H. 1901. Revision of the skunks of the genus *Chincha*. *North American Fauna* 20:1–47.

Hubbard, J. P. 1972. Hooded Skunk on the Mogollon Plateau, New Mexico. *Southwestern Naturalist* 16:458.

Hunter, J. S. 2009. Familiarity breeds contempt: effects of striped skunk color, shape, and abundance on wild carnivore behavior. *Behavioral ecology* 20:1315–22.

Hwang, Y. T., S. Larivière, and F. Messier. 2007. Energetic consequences and ecological significance of heterothermy and social thermoregulation in striped skunks (*Mephitis mephitis*). *Physiological and Biochemical Zoology* 80:138–45.

Jacobo-Salcedo, M. del R., A. J. Alonso-Castro, and A. Zarate-Martinez. 2011. Folk medicinal use of fauna in Mapimi, Durango, México. *Journal of Ethnopharmacology* 133:902–6.

Janzen, D. H., and W. Hallwachs. 1982. The hooded skunk, *Mephitis macroura*, in lowland northwestern Costa Rica. *Brenesia* 19–20:549–52.

Larivière, S., and F. Messier. 1996. Aposematic behavior in the striped skunk, *Mephitis mephitis*. *Ethology* 102:986–92.

Leopold, A. S. 1959. *Wildlife of Mexico: the game birds and mammals*. Berkeley: University of California Press.

Lichtenstein, H. 1832. *Darstellung neuer oder wenig bekannter Säugethiere in Abbildungen und Beschreibungen: von fünf und sechzig Arten auf funfzig colorirten Steindrucktafeln nach den Originalen des Zoologischen Museums der Universität zu Berlin*. Berlin: C. G. Lüderitz.

Martinez-Lira, P., J. Arroyo-Cabrales, and J. P. Carpenter. 2011. Faunal remains and subsistence practices at the archaeological site La Playa, in Sonora, Mexico. *Kiva* 77:33–58.

Mearns, E. A. 1897. Preliminary diagnoses of new mammals of the genus *Mephitis, Dorcelaphus*, and *Dicotyles*, from the Mexican border of the United States. *Proceedings of the United States National Museum* 20:467–71.

Medill, S. A., A. Renard, and S. Larivière. 2011. Ontogeny of antipredator behaviour in striped skunks (*Mephitis mephitis*). *Ethology, Ecology and Evolution* 23:41–48.

Patton, R. F. 1974. Ecological and behavioral relationships of the skunks of Trans Pecos Texas, Texas. PhD dissertation, Texas A&M University.

Reed, C. A., and W. H. Carr. 1949. Use of cactus as protection by hooded skunk. *Journal of Mammalogy* 30:79–80.

Reid, F. A. 1997. *A field guide to the mammals of Central America and southeast Mexico*. New York: Oxford University Press.

Rheude, M. G. 2008. Movements of mesocarnivores in a fragmented desert environment. Master's thesis, University of Arizona, Tucson.

Schmidly, D. J. 2004. *The mammals of Texas*. Rev. ed. Austin: University of Texas Press.

Schmidly, D. J., and R. D. Bradley. 2016. *The mammals of Texas*. 7th ed. Austin: University of Texas Press.

Seton, E. T. 1929. *Lives of game animals*. Vol. 1, Pt. 1, *Cats, wolves, and foxes*. Garden City, NJ: Doubleday, Doran & Company.

Simmons, M. 1988. *New Mexico: an interpretive history*. Albuquerque: University of New Mexico Press.

Speed, M. P., M. A. Brockhurst, and G. D. Ruxton. 2010. The dual benefits of aposematism: predator avoidance and enhanced resource collection. *Evolution* 64:1622–33.

Stevens, S. 2017. Distribution and habitat selection of carnivores in Big Bend National Park, Texas. Master's thesis, Sul Ross State University, Alpine, Texas.

USDA-APHIS Wildlife Services archives. 2021. Program data reports: program data report G-2021. https://www.aphis.usda.gov/aphis/ourfocus/wildlifedamage/pdr/?file=PDR-G_Report&p=2021:INDEX:

Vazquez, P. E., R. M. Mendez, and O. G. Retana. 2006. Uso medicinal de la fauna silvestre en los altos de Chiapas, México. *Interciencia* 21:491–99.

Verts, B. J. 1967. *The biology of the striped skunk*. Urbana: University of Illinois Press.

Verts, B. J., L. N. Carraway, and A. Kinlaw. 2001. *Spilogale gracilis*. *Mammalian Species* 674:1–10.

Wade-Smith, J., and B. J. Verts. 1982. *Mephitis mephitis*. *Mammalian Species* 173.

Walton, L. R., and S. Larivière. 1994. A striped skunk, *Mephitis mephitis*, repels two coyotes, *Canis latrans*, without scenting. *Canadian Field-Naturalist* 108:492–93.

Wood, W. F., B. G. Sollers, G. A. Dragoo, and J. W. Dragoo. 2002. Volatile components in defensive spray of the hooded skunk, *Mephitis macroura*. *Journal of Chemical Ecology* 28:1865–70.

STRIPED SKUNK (*MEPHITIS MEPHITIS*)

Christine C. Hass and Jerry W. Dragoo

As one of the most recognizable animals in North America, the striped skunk, *Mephitis mephitis*, hardly needs an introduction. The size of a house cat, and adorned with a striking black and white coat, it is found in all the conterminous United States, southern Canada, and northern Mexico. Most studies of striped skunks have been in the northern parts of the species' range, with little research conducted in the Southwest, especially New Mexico. Indeed, the few studies of striped skunks in New Mexico have only focused on genetics and disease; there has been no ecological field research anywhere in the state. Striped skunks have been studied in Arizona and Texas, so we will have to extrapolate from those studies to further our understanding of striped skunks in New Mexico.

Striped skunks owe their name to the white stripe extending posteriorly from the nape and usually branching into a "V" at the shoulders, with two dorsal stripes, continuing to the rump and onto the tail (Walker 1964). However, striped skunks exhibit a lot of variation in the black and white pattern of their coat, and the species can be difficult to distinguish from the closely related hooded skunk (*Mephitis macroura*) in the field (see Chapter 27). Helpful for species identification is the fact that the striped skunk is slightly stockier, with shorter, coarser hair, and a shorter tail (< 50% of its total length: Hoffmeister 1986; Hwang and Larivière 2001; Hass 2003; Photo 28.2). The dorsal stripe is also pure

Photo 28.1. A typical black and white stripe pattern in a New Mexico striped skunk (*Mephitis mephitis*) from Moriarty, Torrance County (August 2011). The broad white stripe runs posteriorly from the nape before breaking into two dorsal stripes (the white "V" marking) at the shoulders. The stripes continue to the rump and onto the tail. Note also the thin white line extending from the nose to the top of the forehead, found in both striped skunks and their close relative the hooded skunk, *M. macroura*. The white V-shaped marking on the back of the body is not seen in all striped skunks, and instead the species exhibits much variation in the black and white stripe pattern. Photograph: © Jerry W. Dragoo.

white, as opposed to the white-backed version of the hooded skunk, which has black hairs mixed in. Stripe patterns in New Mexico (see Photos 28.3 and 28.4) have not been analyzed in detail, but Arizona striped skunks are usually of the broad-striped variety (Hass 2003),

(*opposite page*) Photograph: © Pat Gaines.

Photo 28.3. Striped skunk broad-stripe color variant southwest of Edgewood, Torrance County in September 2007. In some cases, the broad white stripe runs posteriorly from the nape and divides over the hips rather than at the shoulders, resulting in only a small area of black dorsally. Photograph: © Jerry W. Dragoo.

Photo 28.2. Dorsal view of two male skunk specimens from New Mexico: striped skunk (*Mephitis mephitis*; MSB:Mamm:64136; left) and hooded skunk (*Mephitis macroura*; MSB:Mamm:221896; right). Both species exhibit much variation in the black and white color pattern of their pelage, but the hooded skunk has a much longer tail. As is the case in the photo, some hooded skunks may present a white back, but black hairs are mixed in, in contrast to the stripes of the striped skunk, which are pure white. Photograph: © Jon Dunnum.

Photos 28.4a, b, and c (*left to right*). Variation in the black and white coat pattern of New Mexico striped skunks, as illustrated by three Museum of Southwestern Biology specimens. (a) MSB:Mamm:60673, a female collected on 25 December 1981 in Chaves County; (b) MSB:Mamm:156755, a male collected on 24 November 2006 in Bernalillo County; and (c) MSB:Mamm:160308, a female from Grant County (subspecies *M. m. estor*), collected on 13 March 1931. Photographs: © Jon Dunnum.

whereas Texas striped skunks show broad-stripe, narrow-stripe, and short-stripe patterns (Patton 1974; Schmidly and Bradley 2016). Hooded skunks are more variable in the width and number of white stripes, but all have longer hairs on the back of the neck (the "hood") and a much longer tail. Where striped and hooded skunks overlap (generally south of the Gila River and west of the Rio Grande), a skunk with a black nape is a hooded skunk, and a skunk with a small black patch near its tail, separating the dorsal stripes, is a striped skunk. Other patterns are more difficult to distinguish (Hoffmeister 1986; Hass 2003; and see Chapter 27 for a more detailed description of hooded skunks).

Both hooded and striped skunks have a thin white stripe between their eyes (Patton 1974; Hoffmeister 1986; Dragoo 2009a). This differentiates hooded and striped skunks from the other three species of skunks in New Mexico: white-backed hog-nosed skunks (*Conepatus leuconotus*) lack any white between their eyes, and western spotted skunks (*Spilogale gracilis* and *S. leucoparia*) have a large round or triangular white patch between their eyes, in addition to being much smaller than the other skunk species (Findley et al. 1975; Hoffmeister 1986).

Weights of adult striped skunks range from 1.4–8 kg (3–17.6 lbs) depending on age, gender, time of year, and latitude. In most populations, males are significantly larger than females (Dean 1965; Patton 1974; Schowalter and Gunson 1982; Fuller et al. 1985; Hansen et al. 2004). However, our research on Arizona striped skunks found no significant differences in body size or weight between males and females (Hass 2003), nor did a study in Tennessee (Bixler and Gittleman 2000). Southern skunks, especially those in southern Arizona and western Texas seldom exceed 2 kg (4.4 lbs), whereas weights of up to 8 kg (17.6 lbs) have been recorded in the northern United States and Canada (Allen 1939; Verts 1967; Patton 1974; Schowalter and Gunson 1982; Fuller and Kuehn

Photo 28.5. Striped skunk tracks in snow in Taos County, New Mexico (12 February 2022). The tracks of striped skunks resemble those of a bear except that they are much smaller. Front and rear imprints show five toes. Claws register as dots ahead of the toes. Photograph: © Brian Jay Long.

1985; Rosatte et al. 1991; Bixler and Gittleman 2000). Long-tailed Texas striped skunks (*M. m. varians*) had tails that made up 47–49% of their total length, compared to ≤40% in other populations (Verts 1967; Patton 1974; Fuller et al. 1985; Hass 2003).

What striped skunks and their relatives are best known for are the foul-smelling secretions produced from highly modified anal glands. The anal glands are modified apocrine glands, which produce a rich cocktail of thiols (formerly known as mercaptans) and thioacetates of these thiols, all sulfurous compounds capable of causing nausea and temporary blindness in sufficient doses (Verts 1967; Andersen and Bernstein 1975; Wood 1990; Dragoo 1993). These secretions are somewhat oily, adding substantial staying power to the odor, which can remain on surfaces for at least 30 days. The dried secretions take the form of acetates, which when hydrolyed, as after a rainstorm, break down to the more volatile thiols again producing the potent aroma (Wood 1999). Although tomato juice is commonly reported to remove skunk odor,

Photo 28.7. Striped skunk on a grassy slope on the Philmont Scout Ranch near Cimarron in Colfax County, New Mexico. Striped skunks occur in all parts of the state at both low and higher elevations. Photograph: © Travis W. Perry.

Photo 28.6 (*left*). Adult female striped skunk spraying one of the authors (J. W. Dragoo) while filming an episode of PBS Nature titled "Is That Skunk?" in July 2008. Note the nipples of the scent glands extruding from the anal sphincter. These nipples allow the skunk to direct the spray from the scent glands. A Photron Ultima APX high-speed HD industrial camera running at 1,000 frames per second was used to video document the release of the spray (see https://www.pbs.org/wnet/nature/is-that-skunk-the-answer-how-we-got-the-spray-shot/4598/). Courtesy of the Nature series, Thirteen/WNET Group (wnet.org), New York, NY. © 2000 WNET.ORG.

Photo 28.8. Striped skunk at an elevation of 2,280 m (7480 ft) west of Arroyo Seco in Taos County, New Mexico on 12 February 2022. Striped skunks have been reported as reaching ~2,680 m (8,800 ft) in the Chuska Mountains in the northwestern part of the state (Bailey 1931). They also have been recently photographed at higher elevations in several parts of the state including 2,500 m (8,200 ft) on Willow Creek in the Mogollon Mountains of Catron County (J. K. Frey, pers. comm.) and along the Rio Chama at 2,440 m (8,000 ft) and the Rio de los Pinos at 2,460 m (8,070 ft) in Taos County (B. Long, pers. comm.). Photograph: © Brian Jay Long.

a concoction developed by Paul Krebaum (Reese 1993; Wood 1999) and consisting of hydrogen peroxide, baking soda, and liquid detergent can neutralize the odor more effectively (Wood 1999; precise recipes can be found online).

Skunks can emit their anal gland secretions as either a fine mist or stream of droplets (Verts 1967; Dragoo 2009a; see Photo 28.6). They can precisely aim the ducts from which the spray is produced and can spray multiple times in succession (Dragoo 2009a). Skunks just a few days old are capable of spraying; however, fine-tuned control of the spray is only acquired with age. Young skunks lacking the speed and coordination to run and hide readily use spray as defense (Medill et al. 2011).

The black and white pelage, reinforced by antipredator displays, is believed to be aposematic, serving as a reminder of the consequences of coming too close (Walton and Larivière 1994;

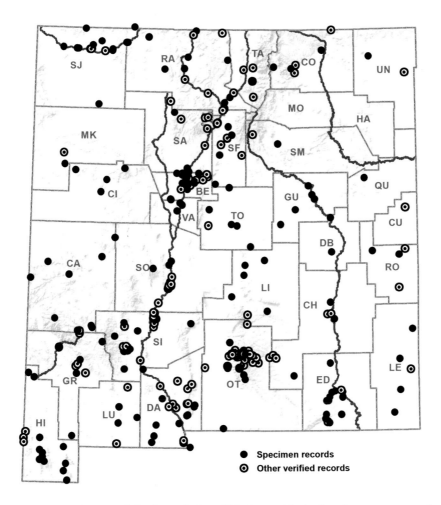

Map 28.1. Distribution of the striped skunk (*Mephitis mephitis*) in New Mexico based on museum specimen and other verified records.

Larivière and Messier 1996; Hunter 2009; see also Chapter 29). The warning coloration and potent weapon combined may allow skunks to spend more time foraging and less time looking out for potential predators (Speed et al. 2010; Fisher and Stankowich 2018; and see under "Interspecific Interactions," below).

There are 13 described subspecies (Hall 1981; Rue 1981). Bailey (1931) recognized three forms of striped skunks in New Mexico: *Mephitis mesomelas estor* (now *M. mephitis estor*), the smaller Arizona striped skunk; *M. mesomelas varians* (now *M. mephitis varians*), the long-tailed Texas striped skunk; and *M. hudsonica* (now *M. mephitis*

hudsonica), the large, dark northern plains striped skunk (Bailey 1931). Those three subspecies are still the ones recognized today in New Mexico (e.g., Frey 2004; and see under "Distribution").

DISTRIBUTION

Striped skunks occur throughout the conterminous United States and in northern Mexico and the southern Canadian provinces (Rosatte and Larivière 2003). In New Mexico, they are found statewide, or nearly so, as they have been reported from almost all counties (Findley et al. 1975; Map 28.1). Bailey (1931) reported that the species reached an elevation of at least 2,440

Photo 28.9. Striped skunk in riparian woodland at the Bosque del Apache National Wildlife Refuge on 25 May 2009. A habitat generalist, the striped skunk occurs in a wide variety of vegetation types. Camera trap project by Matt Farley, Jennifer Miyashiro, and J. N. Stuart.

m (8,000 ft) in both the Sangre de Cristo and Mogollon mountains, and 2,680 m (8,800 ft) in the Chuska Mountains.

The range limits of subspecies in New Mexico are poorly known (Frey 2004). The subspecies *M. m. hudsonica* occurs from the Sangre de Cristo and San Juan mountains in the north-central part of the state south to at least the Capitan Mountains (Bailey 1931; Armstrong 1972; Hall 1981; Frey 2004). *M. m. estor* is the subspecies found in the western part of the state from about the Continental Divide to the New Mexico/Arizona border (Frey 2004). Finally, *M. m. varians* occupies eastern New Mexico and the Tularosa Basin (Frey 2004; and see Dragoo 2009a). Intergrades between *estor* and *varians* may occur in the Rio Grande valley (Bailey 1931). Westward, *M. mephitis estor* occurs in all of Arizona, the only subspecies in that state (Hoffmeister 1986). East of New Mexico, *M. m. varians* is the subspecies found in western Texas, before being replaced by *M. m. mesomelas* east of the 100th meridian (Schmidly and Bradley 2016). No further studies have been conducted defining the exact ranges of all these putative subspecies.

HABITAT ASSOCIATIONS

Throughout their broad geographic range, striped skunks are found in all major vegetation types but tend to avoid alpine tundra and the driest deserts (Verts 1967; Rosatte 1987; Hass and Dragoo 2017). Striped skunks are often found in a mix of woodlands, brushy areas, and open fields, especially along edges (Wade-Smith and Verts 1982; Rosatte 1987; Bixler and Gittleman 2000). They are also very common in urban areas, where an abundance of food and fewer predators can lead to high densities (Rosatte et al. 1991). In much of the country, striped skunks are common in agricultural lands, which provide plentiful food and cover (Verts 1967; Rosatte 1987).

In the Southwest, striped skunks are less likely to use agricultural lands (perhaps due to the comparatively small amount of land devoted to agriculture) (Hass and Dragoo 2009, 2017). In a study of habitat affiliations of all the skunk species in Arizona and New Mexico, 30% of striped skunk observations were associated with forest and woodland, 28.4% with grassland and savannah, 19.5% with scrub, 17.1% with urban environments, and only 5.1 percent with agricultural land cover (Hass and Dragoo 2009). Striped skunks appear to use these vegetation types more evenly than other mephitids (Hass and Dragoo 2009), though Findley et al. (1975) suggested the species in New Mexico was most common in grassland and woodland. In Texas, striped skunks frequented agricultural fields, mesquite, deciduous woodlands, and pasture in accordance with availability (Patton 1974; Neiswenter and Dowler 2007).

LIFE HISTORY

The life history of striped skunks has not been studied in New Mexico. Thus, the information below is derived from research conducted in other parts of the species' distribution. As noted above, we emphasize findings from research in Arizona and Texas wherever possible.

Diet and Foraging

Skunks are primarily insectivorous, and most studies have found that insects comprise 80% or more of their diet, with beetles, moths, grasshoppers and crickets the most common insect foods (Verts 1967; Patton 1974; Greenwood et al. 1999; Hass 2003). As Patton (1974:92) stated, striped skunks "might also be considered opportunistic since they will eat anything palatable." "Anything palatable" includes fruits, crops, bees, small mammals, bird eggs, snails, crayfish, amphibians, reptiles, human refuse, and carrion (Verts 1967; Larivière 1998; Greenwood et al. 1999; Dragoo 2009a). Striped skunks are important nest predators on waterfowl, upland game, and ground nesting passerines (Chesness 1968; Duebbert 1974; Crabtree 1988; Greenwood et al. 1999; Byers et al. 2017).

Skunks have relatively limited eyesight (Johnson-Ulrich et al. 2017) and appear to detect most of their food using their sense of smell (Larivière 1998). They find most of their food by searching on the ground or digging, and are not as good climbers as spotted skunks. (Verts 1967; Dragoo 2009a). Prey is often captured by pouncing and pinning it with the front feet or digging it up with forefeet and claws (Verts 1967; Godin 1982; Dragoo 2009a). Striped skunks will use their front feet to roll caterpillars to remove hairs (setae), toads to remove skin toxins, and beetles to remove defensive sprays (Dragoo 2009a). A striped skunk, recorded by a trail camera, used a rock to break through the ice surface of a water bowl to get a drink, possibly a novel example of tool use (Pesendorfer et al. 2018).

Reproduction, Social Behavior, and Home Ranges

Mating takes place between February and April. Males come into breeding condition before the females and begin searching for mates as early as January (Seton 1913; Allen 1939; Hamilton 1963; Wade-Smith and Richmond 1978; Wade-Smith

Photos 28.10a and b. Striped skunks mating on 14 March 2022 along the Rio Grande in Rio Arriba County. No ecological research has been conducted on striped skunks in New Mexico, but the mating season is known to take place from February through April in other parts of the species' range. Photographs: © Sarah Nelson.

et al. 1980; Schowalter and Gunson 1982; Hansen et al. 2004). Older females come into estrus first. Female striped skunks are induced ovulators, with copulation serving as the external stimulus of ovulation. If mating is not successful, or the first litter is lost, the female may come into estrus again (Verts 1967; Wade-Smith and Richmond 1978). Juvenile females, born the previous spring, are capable of breeding before their first birthday, but often conceive slightly later and have smaller litters than older females (Patton 1974; Wade-Smith and Richmond 1978; Greenwood 1994). Gestation periods range from 59–77 days. Females that mate earlier in the season have longer gestations, with an obligate delay in implantation of one to two weeks (Wade-Smith et al. 1980). Male striped skunks are capable of breeding during their first year (Verts 1967).

Skunks are born in early May to mid-June,

Photo 28.11. Female striped skunk and her eight five-week kits (not all visible) in a man-made den south of Tijeras, New Mexico for filming the episode of PBS Nature "Is That Skunk?" in June 2008. Photograph: © Jerry W. Dragoo.

depending upon the population and the age of the mother. Females give birth to 2–12 (typically 5–7) blind and deaf kits in an insulated burrow (Verts 1967; Patton 1974). The kits open their eyes around 22–30 days, and they begin responding to sounds about the same time (Wade-Smith and Verts 1982). They start moving around the den about 40 days of age, and soon after start following their mother on feeding forays. They are usually on their own by 8–10 weeks of age (Verts 1967; Wade-Smith and Richmond 1978). In some areas it is common to see a sudden increase in the number of baby skunks killed on the road just as they start to disperse (C. C. Hass and J. W. Dragoo, pers. obs.). Young female skunks tend to stay near their natal home range, whereas males may disperse more than 100 km (~60 mi) away (Sargeant et al. 1982; Hansen et al. 2004).

Except for communal denning, breeding, and mothers with kits, striped skunks are solitary animals. They are not territorial but maintain overlapping home ranges of 0.2–5.1 km² (0.12–3.2 mi²), with males having larger home ranges than females (Rosatte and Gunson 1984; Greenwood et al. 1997; Larivière and Messier 1998; Bixler and Gittleman 2000; Hass 2003). Home ranges

are smaller in urban environments than in rural or natural areas (Rosatte et al. 1991; Hass 2003). Skunks appear to mutually avoid each other, even though they do not leave scent marks (Larivière and Messier 1998). Despite their relatively solitary nature, they produce a variety of sounds—raspy squeaks, hisses, growls, and grumbles—especially among the young in a litter. The young can be very communicative with each other as well as with their mother. The mother skunk will often "talk" back using a lower pitched guttural vocalization to get their attention (J. W. Dragoo, per. obs.).

Like all skunks, striped skunks are primarily nocturnal and retreat during the day to dens. Skunks rarely dig their own dens, but usually usurp those of other burrowers, or else squeeze their way into rock crevices, hollow logs, stumps, cacti, brush, or culverts, or they even use space

Photo 28.12. Two striped skunks together on the Philmont Scout Ranch in Colfax County on 20 November 2019. Associations between two or more striped skunks are rare other than in communal dens and during the breeding and kit-rearing seasons. Striped skunks are typically solitary animals. Photograph: © Travis W. Perry.

under human structures such as buildings and in piles of wood or debris (Verts 1967; Weller and Pelton 1987; Bixler and Gittleman 2000; Hass 2003; Doty and Dowler 2006). The structure of striped skunk dens varies throughout the year, with more insulating dens used during the winter, and more exposed dens (in logs or brush) used during the warmer months (Houseknecht and Tester 1978; Hass 2003; Doty and Dowler 2006). Dens may be shared with other species, including opossums (*Didelphis virginiana*), woodchucks (*Marmota monax*), cottontails (*Sylvilagus* spp.), and likely also woodrats (*Neotoma* spp.) (Shirer and Fitch 1970; Godin 1982; C. C. Hass, pers. obs.). Individuals have multiple dens throughout their home ranges, which are used repeatedly. Dens are often occupied for a few days before striped skunks move to a new area (Shirer and Fitch 1970; Houseknecht and Tester 1978; Weller and Pelton 1987; Larivière and Messier 1998; Hass 2003; Doty and Dowler 2006).

Wild and captive skunks will also modify den sites. A film crew built an "L"-shaped den of plywood, natural rocks, roots, and contoured colored concrete, with an opening at each end. In the center was a sky light which could be opened for filming (Photo 28.11). The constructed den was placed in an enclosure with a mother and 8 newly born kits. The mother proceeded to dig a hole in the ground at the junction of the "L," creating a burrow under the man-made structure (J. W. Dragoo, pers. obs.). Indoor captive skunks are adept at entering cabinets and drawers by using their forefeet. They will use clothing in drawers as bedding material (J. W. Dragoo, pers. obs.).

In southern Arizona and Texas, striped skunks are active all year (Patton 1974; Hass 2003; Doty and Dowler 2006; Cochran 2012). In colder parts of their range, they retreat to well-insulated dens for extended periods of time—up to 5 months (Allen 1939; Dean 1965; Verts 1967; Sunquist 1974; Mutch and Aleksuik 1977; Gehrt 2005; Hwang 2005). Communal dens usually include multiple females and occasionally one male, though most males den alone. Skunks have the remarkable ability to fatten and fast; that is, they put on large amounts of fat in the fall (as much as doubling their weight). In warmer areas, this extra fat is often enough to get them through the winter months, which are characterized by relative food scarcity (Hass 2003). In colder areas, however, surviving off nothing but their body fat also requires them to lower their metabolism (Mutch and Aleksuik 1977; Wade-Smith and Verts 1982). They do not hibernate, rather they enter daily torpor by lowering their body temperature (as much as 12° C [21.6° F] lower than normal—the most for any carnivore) for short periods of time (Mutch and Aleksuik 1977; Hwang 2005; Mustonen et al. 2013). This requires special metabolic processes to avoid the build up of toxic byproducts (Carey et al. 2003, Mustonen et al. 2013), as striped skunks may lose 20–60% of their body weight by the time they emerge in the spring (Allen 1939; Sunquist 1974; Schowalter and Gunson 1982; Fuller et al. 1985; Rosatte et al. 1991; Hwang 2005). "Social thermoregulation" (Hwang 2005) also appears to be an effective energy saving strategy: striped skunks sharing a den maintained warmer temperatures and were less likely to enter daily torpor than individuals alone in a den, and "huddlers" lost significantly less fat than solitary skunks (Hwang 2005). Several studies have documented that skunks maintain some activity during the winter but may also enter torpor for a few hours each day if outside conditions are extreme (Hwang 2005; Mustonen et al. 2013).

Aboveground activity by striped skunks during the winter is related to ambient temperature and breeding condition—males begin moving around in January and may share dens with different groups of females (Theimer et al. 2017), whereas females may not emerge from dens until March or April (Allen 1939; Dean 1965; Mutch and Aleksuik 1977; Schowalter and Gunson 1982; Hwang 2005). In Flagstaff, Arizona, female striped skunks were

Photo 28.13. Striped skunk and bobcat (*Lynx rufus*, shown by black arrow) photographed together at Mud Spring in the Sandia Mountains (at an elevation of ~2,160 m [7,100 ft]) on 20 September 2016. Despite their chemical defenses, striped skunks have several natural predators including the bobcat. Photograph: © Sandia Mountain Natural History Center—New Mexico Museum of Natural History and Science.

Photo 28.14. Striped skunk–cougar (*Puma concolor*) encounter at Paradise Spring in the Sandia Mountains on 9 June 2009. Using remote cameras, the Sandia Mountain Natural History Center has been monitoring Paradise Spring and other Sandia Mountain locations since 2008. After this photograph was taken, Sandia Mountain Natural History Center staff visited Paradise Spring to determine the outcome of the encounter and found skunk fur on the ground. A lingering skunk odor was also detected. Photograph: © Sandia Mountain Natural History Center—New Mexico Museum of Natural History and Science.

observed using communal dens between October and February; however, they never entered torpor and exhibited some aboveground activity each month. Male striped skunks occasionally shared dens with females, but remained active all winter (Theimer et al. 2016, 2017). Skunks in northern New Mexico probably exhibit similar behavior.

Population Densities and Survival

Densities of striped skunks range from 0.1 to 18.6 skunks per km² (0.62 mi²), with the highest densities reported in urban areas (Bjorge et al. 1981; Wade-Smith and Verts 1982; Rosatte et al. 1991; Hansen et al. 2004). Density estimates fluctuate widely among years due to outbreaks of disease (Verts 1967), possibly mediated by drought (Hass 2003) or severe winters (Rosatte et al. 1991). Annual survival rates of striped skunks range from 40 to 80% (Rosatte et al. 1991; Hass 2003; Hansen et al. 2004; Gehrt 2005), with few skunks

surviving to their third birthday. Captive striped skunks typically can live from 8–12 years (Rosatte et al. 2010), and we have heard reports of even longer life spans. Based on our experience, predation, disease, parasites, and interactions with people account for most of the mortality among wild striped skunks.

Interspecific Interactions

Seton (1913:106–107) noted, more than a century ago, "For few indeed there are in the land to-day that realize the gentleness and forbearance of this righteous little brother of ours, who, though armed with a weapon that will put the biggest and boldest to flight or disastrous defeat, yet refrains from using it until in absolute peril of his life, and then only after several warnings." Spraying is indeed a last-resort defense. A skunk's first choice is to run and hide. In some situations (see below), it goes through an elaborate series of behaviors to warn would-be predators to stay back (Medill et al. 2011). This begins with the skunk facing the threat, reflexively pointing its tail straight in the air, straightening its legs and arching its back, and fluffing out its fur to increase the appearance of size. It then jumps forward and stomps the ground with its forefeet, followed by quickly backing up while scraping its forefeet along the ground. A young striped skunk may repeat this "rush-and-stomp" several times, sometimes stomping so hard that its back feet come off the ground. An adult skunk is usually more stable on its feet and more threatening. A skunk may also hiss and growl during these displays. If the threat has not retreated yet, the skunk will then point the anal glands toward the offender, sometimes assuming a "U" shape, with both eyes and anus pointed at the threat. Lastly, it will take precise aim with its anal ducts and consciously let loose (Dragoo 2009a). Skunks that are suddenly and severely startled or feel cornered, may spray without going through the threat displays. A striped skunk's response to a threat varies with the type of predator; skunks are more likely to face a coyote (*Canis latrans*) defensively but tend to run and hide if they hear a great horned owl (*Bubo virginianus*) (Fisher and Stankowich 2018).

Despite their defensive secretions, striped skunks may not always escape predators. A remote camera set at the Sandia Mountain Natural History Center captured an initial encounter between a cougar (*Puma concolor*) and a striped skunk (Photo 28.14). The two animals, both in a defensive posture, were on opposite sides of a small watering hole. Staff at the center noticed a lingering skunk odor for several days after the photo was taken. They were curious as to the outcome of the encounter. Did the musk work? After further investigation of the area they found the remains (mostly hair) of a skunk (Anonymous 2009). In a study of bobcat (*Lynx rufus*) and cougar food habits in southeastern Arizona, about 6% of cougar diets consisted of striped, hooded,

and hog-nosed skunks (Hass 2009). Besides cougars, coyotes, and bobcats, great horned owls, eagles, badgers (*Taxidea taxus*), and other skunks (adult males on young of the year) can all prey on striped skunks (Sargeant et al. 1982; Wade-Smith and Verts 1982; Hass 2009). Thus, Leopold (1959:451) might have been overstating things a little when he wrote, "Armed with a particularly efficient weapon of defense, the striped skunk wanders over its range without apparent fear of man or beast."

Striped skunks are also host to a wide range of parasites, including fleas, mites, ticks, bot flies, tapeworms, and roundworms (Verts 1967; Lincoln 1973; Patton 1974; Stockdale 1974; Cawthorn 1976; Gehrt 2005; Brown et al. 2009). The nematode, *Skrjabingylus chitwoodorum*, is a common parasite of the frontal bones of striped skunks and may lead to significant skull deformation (Kirkland 1983; Maldonado and Kirkland 1986; Hughes et al. 2018). Skunks can carry heavy enough loads of internal and external parasites to jeopardize their survival during extremely cold winters or when resources are limited (Hass 2003; Gehrt 2005).

Striped skunks are hosts to many infectious diseases, including leptospirosis, listeriosis, tularemia, canine distemper, salmonella, Aleutian disease, West Nile virus, and the disease they are most commonly associated with, rabies (Wade-Smith and Verts 1982; Dragoo 2009b; LaDouecuer et al. 2014; Britton et al. 2015; Giannitti et al. 2017; see under "Status and Management").

STATUS AND MANAGEMENT

Striped skunks have been important in the fur trade for hundreds of years, not so much due to the value of their pelts but by the sheer number obtainable (Verts 1967; Schmidly and Bradley 2016). In New Mexico, striped (and other) skunks are unprotected and can be taken year-round with no limits (but note that non-residents still need a license). The most recent annual furbearer harvest reports released by the New Mexico Department

Photo 28.15. Two striped skunk babies admitted by Wildlife Rescue Inc. in Albuquerque in 2011. Many young skunks are brought to wildlife care centers and rehabilitators after their mothers have been killed or relocated. Although skunks are often considered nuisance animals, they are not aggressive and actually often benefit farmers and homeowners as they feed on insects and rodents. Relocated skunks typically experience high mortality in their new surroundings. Rather than relocation (no longer authorized in New Mexico), public education is critical to reduce conflicts between humans and skunks. Photograph: © Ruth Wheeler.

of Game and Fish have not included any information on the number of skunks trapped in the state. Earlier annual numbers were probably not reliable because, as unprotected furbearers, skunks harvested by trappers are seldom reported. Harvest of striped skunks may have declined, however, compared to the mid-to-late 1970s and early 1980s in particular (Association of Fish and Wildlife Agencies 2023). Reported harvest in the state was highest in 1979 (6,706 striped skunks), and though considerable variation can exist from year to year, the more recently reported numbers typically range in the low to mid-hundreds. To the reported harvest of striped skunks, one must also add mortality resulting from the animal damage control activities of the US Department of Agriculture's APHIS-Wildlife Services. In 2020, for example, USDA APHIS-Wildlife Services reported killing,

intentionally and unintentionally, ~420 striped skunks in New Mexico (USDA-APHIS Wildlife Services archives 2021; note that as no other species of skunks were listed, it is probable that all species of skunks are being lumped together as "striped skunks"). Overall, however, we believe numbers of striped skunks that are trapped for any reason every year may be far smaller than those of animals killed either on roads or inadvertently by pesticides used to poison rodents and insects that are later consumed by skunks. In addition, nuisance skunks were often relocated to different locations—which, in addition to potentially spreading disease, also carries a high mortality rate as animals try to find their way home or try to survive in unfamiliar areas (Conover 2002; Hass 2002).

Nuisance skunks are but one symptom of a larger problem; they are not the problem itself. Urban areas are attractive to skunks. Skunks generally have three basic needs (four during certain times of the year): food, water, and shelter. City recreational (parks, golf courses, etc.) and residential areas provide ample water as well as numerous den sites. Golf courses and lawns are also the source of plentiful food in the form of grubs. There are also plenty of food resources available from improperly contained garbage and pet food. Recognizing and removing the attractants will help eliminate the symptom. A long-term solution to nuisance skunks should involve education (Hass 2002; Dragoo 2009b). Information should be provided on skunks, focusing on why they are there, what will happen to a "problem" animal if it is removed, and potential benefits of leaving the animal in place. For example, skunks residing under buildings may be consuming less desirable insects and rodents. Plans can be made to exclude animals by blocking off potential den entrances, containing garbage, and feeding pets indoors. To avoid potential disease transmission (i.e., rabies), domestic pets should be vaccinated (Hass 2002; Hass and Dragoo 2006; Dragoo 2009b).

Rabies

Rabies is a viral disease of the nervous system, and capable of infecting most mammals. Bats and carnivores in particular are highly susceptible to the virus, and they account for most of the potential human exposures to rabies in the United States (Birhane et al. 2017). Striped skunks are one of the most common wildlife reservoirs of rabies virus in North America (Parker 1975; Birhane et al. 2017). Transmission typically occurs when an animal infected with rabies has the virus in its saliva and bites another animal of the same species. In very rare instances, the virus can be spread via aerosols in caves with high bat populations (Constantine 1962), or from the consumption of flesh of an infected animal (Verts 1967). The southwestern United States has several variants of rabies that maintain reservoirs in the skunk, gray fox (*Urocyon cinereoargenteus*), and populations of several species of bats (Birhane et al. 2017).

Rabies in the Southwest is considered enzootic—that is, it is always present at some level in one or more species. Mammals are most susceptible to their own specific strain of rabies, though some spillover does occur (Birhane et al. 2017). Spillover events typically do not last long; a notable exception occurred in Flagstaff, Arizona during 2001, when striped skunks were infected with the big brown bat (*Eptesicus fuscus*) strain of rabies. An extensive trap-vaccinate-release program reduced the incidence of rabid skunks in 2001; however, more appeared in 2004 (Leslie et al. 2006). Rabies surveillance, particularly in skunks, occurs regularly in southern Arizona, where animals are routinely trapped at several locations and their tissues submitted for testing. No such program is currently implemented in New Mexico, so far less is known about the prevalence of rabies in wildlife in the state. Skunks are not identified to species when submitted for rabies testing in New Mexico (Dragoo et al. 2004).

In many areas, rabies in skunks is also epizootic, with large numbers of rabid skunks seen at periodic intervals. In Arizona, the number of reported cases of rabid skunks was related to precipitation one to two years before detection, indicating that the incidence of rabies was likely density dependent (more skunks equals more rabies; Verts 1967; Hass and Dragoo 2006). Rabies is fatal to skunks, and rabies epizootics are followed by population crashes (Greenwood et al. 1997). In areas where striped skunks are active all year, rabies cases among skunks occur year-round but tend to peak during January-June (Pool and Hacker 1982; Hass and Dragoo 2006). In areas where skunks spend much of the year underground, cases of rabies peak in the second quarter of the year, with sometimes a second peak in the fall (Verts 1967; Gremillion-Smith and Woolf 1988). Most infections appear to be acquired during the mating season, when skunks are moving and fighting more, and during the fall when juveniles are dispersing (Verts 1967; Sargeant et al. 1982; Gremillion-Smith and Woolf 1988; Greenwood et al. 1997). In the Desert Southwest, rabies also may be spread during conflict over limited supplies of water during droughts (Pool and Hacker 1982; Hass and Dragoo 2006). Hwang and colleagues (2007) suggested that low body temperatures maintained by skunks that enter torpor during the winter would limit the spread of the virus within a skunk's body. Likewise, Mustonen et al. (2013). suggested that warming temperatures associated with climate change might increase skunk activity during the winter, perhaps increasing contact rates and disease transmission among skunks.

Skunks with rabies may exhibit increased agitation and hypersensitivity to light and sound (furious rabies); show lethargy and limited movement (paralytic rabies); or succumb with no outward sign of disease (Storm and Verts 1966; Verts 1967; Gough and Niemeyer 1975; Greenwood et al. 1997). They may have the virus in their saliva for a week or more before outward symptoms appear, but do not continuously shed the virus in their saliva (Verts 1967). In its terminal stage, rabies is an infection in the brain, and may cause "unusual" behaviors. This has led some to state that a skunk seen out during the day is almost certainly rabid (Parker 1975), but healthy skunks may be out during early morning and evening hours (C. C. Hass and J. W. Dragoo, pers. obs.).

Various strategies have been attempted to limit the prevalence and spread of rabies. Pybus (1988) identified 3 main approaches: 1) accept the presence of the rabies virus in wildlife populations and rely on an immune barrier by vaccinating pets and educating the public and public health officials; 2) vaccinate wildlife species against the rabies virus; and 3) reduce numbers of potential reservoirs. Alberta and Montana have adopted the third strategy, with extensive campaigns to eradicate skunks in areas where rabies is present, by trapping, shooting, and poisoning. This method seemed effective but requires constant maintenance with a long-term commitment (Rosatte et al. 1986; Pybus 1988), and has unknown ecological consequences. In addition, it was not possible to separate the effects of the eradication campaign from mortality caused by the virus itself; in other words, much of the population reduction may have been due to the lethal effects of the virus (MacInnes 1988). Overall, sustained population reduction of skunks and other mammalian reservoirs to eliminate rabies is not justified for ecological, economic, and ethical reasons (Rupprecht et al. 1995).

Vaccination of skunks, via injection or oral baits, has been shown to be effective at immunizing skunks against the virus (Rosatte et al. 2009). Oral baits are less effective in skunks compared to raccoons or foxes, but are likely the most cost-effective method for reducing the prevalence of rabies infections in a population (Rosatte et al. 1992, 2009; Hanlon et al. 2002; Brown et al. 2014).

Little is known about the overall status of the striped skunk in New Mexico. However, the species is highly adaptable and has a high reproductive rate. Despite constant pressure from predators, parasites, diseases, guns, traps, and speeding cars, it seems to be holding its own.

LITERATURE CITED

Allen, D. L. 1939. Winter habits of Michigan skunks. *Journal of Wildlife Management* 3:212–28.

Andersen, K. K., and D. T. Bernstein. 1975. Some chemical constituents of the scent of the striped skunk (*Mephitis mephitis*). *Journal of Chemical Ecology* 1:493–99.

Anonymous. 2009. In the spotlight. *Bear Claw: The Sandia Mountain Natural History Center Newsletter* 4(5):1.

Armstrong, D. M. 1972. Mammals from the Mexican state of Sinaloa. III. Carnivora and Artiodactyla. *Journal of Mammalogy* 53:48–61.

Association of Fish and Wildlife Agencies. [2023]. Furbear management. Under Surveys & reports, see US furbearer harvest statistics database 1970–2018 (a downloadable Excel file). https://www.fishwildlife.org/afwa-inspires/furbearer-management.

Bailey, V. 1931 (=1932). *Mammals of New Mexico*. North American Fauna 53. Washington, DC: US Department of Agriculture, Bureau of Biological Survey.

Birhane, M. G., J. M. Cleaton, B. P. Monroe, A. Wadhwa, L. A. Orciari, P. Yager, J. Blanton, A. Velasco-Villa, B. W. Petersen, and R. M. Wallace. 2017. Rabies surveillance in the United States during 2015. *Journal of the American Veterinary Medical Association* 250:1117–30.

Bixler, A., and J. L. Gittleman. 2000. Variation in home range and use of habitat in the striped skunk (*Mephitis mephitis*). *Journal of Zoology, London* 251:525–33.

Bjorge, R. R., J. R. Gunson, and W. M. Samuel. 1981. Population characteristics and movements of striped skunks (*Mephitis mephitis*) in central Alberta. *Canadian Field-Naturalist* 95:149–55.

Britton, A., T. Redford, J. J. Bidulka, A. P. Scouras, K. R. Sojonky, E. Zabek, H. Schwantje, and T. Joseph. 2015. Beyond rabies: are free-ranging skunks (*Mephitis mephitis*) in British Columbia reservoirs of emerging infection? *Transboundary and Emerging Diseases* 64: 603–12.

Brown, E. L., D. M. Roellig, M. E. Gompper, R. J. Monello, K. M. Wenning, M. W. Gabriel, and M. J. Yabsley. 2009. Seroprevalence of *Trypanosoma cruzi* among eleven potential reservoir species from six states across the southern United States. *Vector-Borne and Zoonotic Diseases* 10:757–63.

Brown, L. J., R. C. Rosatte, C. Fehlner-Gardiner, J. A. Ellison, F. R. Jackson, P. Bachmann, J. S. Taylor, R. Franka, and D. Donovan. 2014. Oral vaccination and protection of striped skunks (*Mephitis mephitis*) against rabies using ONRAB®. *Vaccine* 32:3675–79.

Byers, C., C. Ribic, D. Sample, J. Dadisman, and M. Guttery. 2017. Grassland bird productivity in warm season grass fields in southwest Wisconsin. *American Midland Naturalist* 178:47–63.

Carey, H. V., M. T. Andrews, and S. L. Martin. 2003. Mammalian hibernation: cellular and molecular responses to depressed metabolism and low temperature. *Physiological Reviews* 83:1153–81.

Cawthorn, R. J. 1976. Development of *Physaloptera maxillaris* (Nematoda: Physalopteroidea) in skunk (*Mephitis mephitis*) and the role of paratenic and other hosts in its life cycle. *Canadian Journal of Zoology* 54:313–23.

Chesness, R. A. 1968. The effect of predator removal on pheasant reproductive success. *Journal of Wildlife Management* 32:683–97.

Cochran, T. 2012. Circadian and seasonal activity patterns of sympatric hog-nosed (*Conepatus leuconotus*) and striped (*Mephitis mephitis*) skunks. Master's thesis, Angelo State University, San Angelo, Texas.

Conover, M. 2002. Wildlife translocation. In *Resolving human-wildlife conflicts: the science of wildlife damage management*, 211–28. Boca Raton, FL: CRC Press.

Constantine, D. G. 1962. Rabies transmission by non-bite route. *Public Health Reports* 77:287–89.

Crabtree, R. L. 1988. Effects of alternate prey on skunk predation of waterfowl nests. *Wildlife Society Bulletin* 16:163–69.

Dean, F. C. 1965. Winter and spring habits and density of Maine skunks. *Journal of Mammalogy* 46:673–675.

Doty, J., and R. Dowler. 2006. Denning ecology in sympatric populations of skunks (*Spilogale gracilis* and *Mephitis mephitis*) in west-central Texas. *Journal of Mammalogy* 87:131–38.

Dragoo, J. W. 1993. The evolutionary relationships of the skunks to each other and the rest of the weasels; with a note on behavioral idiosyncrasies. Eleventh Great Plains Wildlife Damage Control Workshop Proceedings, Kansas City, MO.

———. 2009a. Family Mephitidae (skunks). In *Handbook of the mammals of the world*. Vol 1, *Carnivores*, ed. D. E. Wilson, and R. A. Mittermeier, 532–63. Barcelona: Lynx Edicions.

———. 2009b. Nutrition and behavior of striped skunks. *Veterinary Clinics of North America Exotic Animal Practice* 12:313–26.

Dragoo, J. W., D. K. Mathes, A. Aragon, C. C. Hass, and T. L. Yates. 2004. Identification of skunk species submitted for rabies testing in the Desert Southwest. *Journal of Wildlife Diseases* 40:371–76.

Duebbert, H. F. 1974. Upland duck nesting related to land use and predator reduction. *Journal of Wildlife Management* 38:257–65.

Findley, J. S., A. H. Harris, D. E. Wilson, and C. Jones. 1975. *Mammals of New Mexico*. Albuquerque: University of New Mexico.

Fisher, K. A., and T. Stankowich. 2018. Antipredator strategies of striped skunks in response to cues of aerial and terrestrial predators. *Animal Behaviour* 143:25–34.

Frey, J. K., 2004. Taxonomy and distribution of the mammals of New Mexico: an annotated checklist. *Occasional Papers, Museum of Texas Tech University* 240:1–32.

Fuller, T. K., and D. W. Kuehn. 1985. Population characteristics of striped skunks in northcentral Minnesota. *Journal of Mammalogy* 66:813–15.

Fuller, T. K., D. W. Kuehn, P. L. Coy, and R. K. Markl. 1985. Physical characteristics of striped skunks in northern Minnesota. *Journal of Mammalogy* 66:371–74.

Gehrt, S. D. 2005. Seasonal survival and cause-specific mortality of urban and rural striped skunks in the absence of rabies. *Journal of Mammalogy* 86:1164–70.

Giannitti, F., M. Sadeghi, E. Delwart, M. Schwabenlander, and J. Foley. 2017. Aleutian disease virus-like virus (*Amdoparvovirus* sp.) infecting free-ranging striped skunks (*Mephitis mephitis*) in the midwestern United States. *Journal of Wildlife Diseases* 54:186–88.

Godin, A. J. 1982. Striped and hooded skunks. In *Wild mammals of North America: biology, management, and economics*, ed. J. A. Chapman and G. A. Feldhamer, 647–87. Baltimore: Johns Hopkins University Press.

Gough, P. M., and C. Niemeyer. 1975. A rabies epidemic in recently captured skunks. *Journal of Wildlife Diseases* 11:170–76.

Greenwood, R. J. 1994. Age-related reproduction in striped skunks (*Mephitis mephitis*) in the upper Midwest. *Journal of Mammalogy* 75:657–62.

Greenwood, R. J., W. E. Newton, G. L. Pearson, and G. J. Schamber. 1997. Population and movement characteristics of radio-collared striped skunks in North Dakota during an epizootic of rabies. *Journal of Wildlife Diseases* 33:226–41.

Greenwood, R. J., A. B. Sargeant, J. L. Piehl, D. A. Buhl, and B. A. Hanson. 1999. Foods and foraging of prairie striped skunks during the avian nesting season. *Wildlife Society Bulletin* 27:823–32.

Gremillion-Smith, C., and A. Woolf. 1988. Epizootiology of skunk rabies in North America. *Journal of Wildlife Diseases* 24:620–26.

Hall, E. R. 1981. *The mammals of North America*. 2nd ed. Vol. 2. New York: John Wiley.

Hamilton, W. J., Jr. 1963. Reproduction of the striped skunk in New York. *Journal of Mammalogy* 44:123–24.

Hanlon, C. A., M. Niezgoda, P. Morrill, and C. E. Rupprecht. 2002. Oral efficacy of an attenuated rabies virus vaccine in skunks and raccoons. *Journal of Wildlife Diseases* 38:420–27.

Hansen, L. A., N. E. Matthews, B. A. Vander Lee, and R. S. Lutz. 2004. Population characteristics, survival rates, and causes of mortality of striped skunks (*Mephitis mephitis*) on the southern high plains, Texas. *Southwestern Naturalist* 49:54–60.

Hass, C. C. 2002. *Reduction of nuisance skunks in an urbanized area*. Final report to Arizona Game and Fish, Phoenix, AZ.

———. 2003. *Ecology of hooded and striped skunks in southeastern Arizona*. Final report to Arizona Game and Fish, Phoenix, AZ.

———. 2009. Competition and coexistence in sympatric bobcats and pumas. *Journal of Zoology* 278:174–80.

Hass, C. C., and J. W. Dragoo. 2006. Rabies in hooded and striped skunks in Arizona. *Journal of Wildlife Diseases* 42:825–29.

———. 2009. Distribution and habitat affiliations of 4 species of skunks (Mephitidae) in Arizona and New Mexico. 94th Annual Meeting of the Ecological Society of America. University of New Mexico, Albuquerque.

———. 2017. Competition and coexistence in sympatric skunks. In *Biology and conservation of the musteloids*, ed. D. W. Macdonald, C. Newman, and L. Harrington, 464–77. Oxford: Oxford University Press.

Hoffmeister, D. F. 1986. *Mammals of Arizona*. Tucson and Phoenix: University of Arizona Press and Arizona Game and Fish Department.

Houseknecht, C. R., and J. R. Tester. 1978. Denning habits of striped skunks (*Mephitis mephitis*). *American Midland Naturalist* 100:424–30.

Hughes, M. R., N. J. Negovetich, B. C. Mayes, and R. C. Dowler. 2018. Prevalence and intensity of the sinus roundworm (*Skrjabingylus chitwoodorum*) in rabies-negative skunks of Texas, USA. *Journal of Wildlife Diseases* 54:85–94.

Hunter, J. S. 2009. Familiarity breeds contempt: effects of striped skunk color, shape, and abundance on wild carnivore behavior. *Behavioral ecology* 20:1315–22.

Hwang, Y. T. 2005. Physiological and ecological aspects of winter torpor in captive and free-ranging striped skunks. PhD dissertation, University of Saskatchewan.

Hwang, Y. T., and S. Larivière. 2001. *Mephitis macroura*. *Mammalian Species* 686:1–3.

Hwang, Y. T., S. Larivière, and F. Messier. 2007.

Energetic consequences and ecological significance of heterothermy and social thermoregulation in striped skunks (*Mephitis mephitis*). *Physiological and Biochemical Zoology* 80:138–45.

Johnson-Ulrich, Z., E. Hoffmaster, A. Robeson, and J. Vonk. 2017. Visual acuity in the striped skunk (*Mephitis mephitis*). *Journal of Comparative Psychology* 131:384–91.

Kirkland, G. L., Jr. 1983. Patterns of variation in cranial damage in skunks (Mustelidae: Mephitinae) presumably caused by nematodes of the genus *Skrjabingylus* Petrov 1927 (Metastrongyloidea). *Canadian Journal of Zoology* 61:2913–20.

LaDoueucer, E., M. Anderson, B. Ritchie, P. Ciembor, G. Rimoldi, M. Piazza, D. Pesti, D. Clifford, and F. Giannitti. 2014. Aleutian disease: an emerging disease in free-ranging striped skunks (*Mephitis mephitis*) from California. *Veterinary Pathology* 52:1250–53.

Larivière, S. 1998. The radiating mousing technique of the striped skunk. *Blue Jay* 56:218–20.

Larivière, S., and F. Messier. 1996. Aposematic behavior in the striped skunk, *Mephitis mephitis*. *Ethology* 102:986–92.

———. 1998. Denning ecology of the striped skunk in the Canadian prairies: implications for waterfowl nest predation. *Journal of Applied Ecology* 35:207–13.

Leopold, A. S. 1959. *Wildlife of Mexico: the game birds and mammals*. Berkeley: University of California Press.

Leslie, M. J., S. Messenger, R. E. Rohde, J. Smith, R. Cheshier, C. Hanlon, and C. E. Rupprecht. 2006. Bat-associated rabies virus in skunks. *Emerging Infectious Diseases* 12:1274–77.

Lincoln, R. C. 1973. The relationship of *Physaloptera maxillaris* (Nematoda: Physalopteroidea) to skunk (*Mephitis mephitis*). *Canadian Journal of Zoology* 51:437–41.

MacInnes, C. D. 1988. Control of wildlife rabies: the Americas. In *Rabies*, ed. J. B. Campbell, 381–405. Boston: Kluwer Academic Publishers.

Maldonado, J. E., and G. L. Kirkland. 1986. Relationship between cranial damage attributable to *Skrjabingylus* (Nematoda) and braincase capacity in the striped skunk (*Mephitis mephitis*). *Canadian Journal of Zoology* 64:2004–7.

Medill, S. A., A. Renard, and S. Larivière. 2011.

Ontogeny of antipredator behaviour in striped skunks (*Mephitis mephitis*). *Ethology, Ecology and Evolution* 23:41–48.

Mustonen, A.-M., J. Bowman, C. Sadowski, L. A. Nituch, L. Bruce, T. Halonen, K. Puukka, K. Rouvinen-Watt, J. Aho, and P. Nieminen. 2013. Physiological adaptations to prolonged fasting in the overwintering striped skunk (*Mephitis mephitis*). *Comparative Biochemistry and Physiology Part A: Molecular & Integrative Physiology* 166:555–63.

Mutch, G. R. P., and M. Aleksuik. 1977. Ecological aspects of winter dormancy in the striped skunk (*Mephitis mephitis*). *Canadian Journal of Zoology* 55:607–15.

Neiswenter, S. A., and R. C. Dowler. 2007. Habitat use of western spotted skunks and striped skunks in Texas. *Journal of Wildlife Management* 71:583–86.

Parker, R. L. 1975. Rabies in skunks. In *The natural history of rabies*, ed. G. M. Baer, 41–51. New York: Academic Press.

Patton, R. F. 1974. Ecological and behavioral relationships of the skunks of Trans Pecos Texas, Texas. PhD dissertation, Texas A&M University.

Pesendorfer, M. B., S. Dickerson, and J. W. Dragoo. 2018. Observation of tool use in striped skunks: how community science and social media help document rare natural phenomena. *Ecosphere* 9:1–5.

Pool, G. E., and C. S. Hacker. 1982. Geographic and seasonal distribution of rabies in skunks, foxes and bats in Texas. *Journal of Wildlife Diseases* 18:405–18.

Pybus, M. J. 1988. Rabies and rabies control in striped skunks (*Mephitis mephitis*) in three prairie regions of western North America. *Journal of Wildlife Diseases* 24:434–49.

Reese, K. M. 1993. Lab method deodorizes a skunk-afflicted pet. *Chemical and Engineering News* 10/18:90.

Rosatte, R. C. 1987. Striped, spotted, hooded, and hog-nosed skunk. In *Wild furbearer conservation and management in North America*, ed. M. Nowak, J. A. Baker, M. E. Obbard, and B. Malloch 599–613. Ontario, Canada: Ministry of Natural Resources.

Rosatte, R. C., D. Donovan, J. C. Davies, M. Allan, P. Bachmann, B. Stevenson, K. Sobey, L. Brown, A. Silver, K. Bennett, T. Buchanan, L. Bruce, M.

Gibson, A. Beresford, A. Beath, C. Fehlner-Gardiner, and K. Lawson. 2009. Aerial distribution of ONRAB® baits as a tactic to control rabies in raccoons and striped skunks in Ontario, Canada. *Journal of Wildlife Diseases* 45:363–74.

Rosatte, R. C., and J. R. Gunson. 1984. Dispersal and home range of striped skunks, *Mephitis mephitis*, in an area of population reduction in southern Alberta. *Canadian Field-Naturalist* 98:315–19.

Rosatte, R. C., and S. Larivière. 2003. Skunks. In *Wild mammals of North America: biology, management, and economics*, 2nd ed., ed. G. A. Feldhamer, B. C. Thompson, and J. A. Chapman, 692–707. Baltimore: Johns Hopkins University Press.

Rosatte, R. C., M. J. Power, and C. D. MacInnes. 1991. Ecology of urban skunks, raccoons, and foxes in metropolitan Toronto. In *Wildlife conservation in urban environments*, ed. L. W. Adams, 31–38. Columbia, MD: National Institute for Urban Wildlife.

Rosatte, R. C., M. J. Power, C. D. MacInnes, and J. B. Campbell. 1992. Trap-vaccinate-release and oral vaccination for rabies control in urban skunks, raccoons and foxes. *Journal of Wildlife Diseases* 28:562–71.

Rosatte, R. C., M. J. Pybus, and J. R. Gunson. 1986. Population reduction as a factor in the control of skunk rabies in Alberta. *Journal of Wildlife Diseases* 22:459–67.

Rosatte, R. C., K. Sobey, J. W. Dragoo, and S. D. Gehrt. 2010. Striped skunks and allies (*Mephitis* spp.). In *Urban Carnivores: ecology, conflict, and conservation*, ed. S. D. Gehrt, S. P. D. Riley, and B. L. Cypher, 97–106. Baltimore: Johns Hopkins University Press.

Rue, L. L. 1981. *Furbearing animals of North America*. New York: Crown Publishers.

Rupprecht, C. E., J. S. Smith, M. Fekadu, and J. E. Childs. 1995. The ascension of wildlife rabies: a cause for public health concern or intervention? *Emerging Infectious Diseases* 1:107–14.

Sargeant, A. B., R. J. Greenwood, J. L. Piehl, and W. B. Bicknell. 1982. Recurrence, mortality, and dispersal of prairie striped skunks, *Mephitis mephitis*, and implications to rabies epizootiology. *Canadian Field-Naturalist* 96:312–16.

Schmidly, D. J., and R. D. Bradley 2016. *The mammals of Texas*. 7th ed. Austin: University of Texas Press.

Schowalter, D. B., and J. R. Gunson. 1982. Parameters of population and seasonal activity of striped skunks, *Mephitis mephitis*, in Alberta and Saskatchewan. The *Canadian Field-Naturalist* 96:409–20.

Seton, E. T. 1913. *Wild animals at home*. New York: Doubleday, Page & Company.

Shirer, H. W., and H. S. Fitch. 1970. Comparison from radiotracking of movements and denning habits of the raccoon, striped skunk, and opossum in northeastern Kansas. *Journal of Mammalogy* 51:491–503.

Speed, M. P., M. A. Brockhurst, and G. D. Ruxton. 2010. The dual benefits of aposematism: predator avoidance and enhanced resource collection. *Evolution* 64:1622–33.

Stockdale, P. H. G. 1974. The development, route of migration, and pathogenesis of *Crenosoma mephitidis* in the skunk (*Mephitis mephitis*). *Canadian Journal of Zoology* 52:681–85.

Storm, G. L., and B. J. Verts. 1966. Movements of a striped skunk infected with rabies. *Journal of Mammalogy* 47:705–8.

Sunquist, M. E. 1974. Winter activity of striped skunks (*Mephitis mephitis*) in east-central Minnesota. *American Midland Naturalist* 92:434–446.

Theimer, T. C., J. M. Maestas, and D. L. Bergman. 2016. Social contacts and den sharing among suburban striped skunks during summer, autumn, and winter. *Journal of Mammalogy* 97:1272–81.

Theimer, T. C., C. T. Williams, S. R. Johnson, A. T. Gilbert, D. L. Bergman, and C. L. Buck. 2017. Den use and heterothermy during winter in free-living, suburban striped skunks. *Journal of Mammalogy* 98:867–73.

USDA-APHIS Wildlife Services archives. 2021. Program data reports: program data report G-2021. https://www.aphis.usda.gov/aphis/ourfocus/wildlifedamage/pdr/?file=PDR-G_Report&p=2021:INDEX:

Verts, B. J. 1967. *The biology of the striped skunk*. Urbana: University of Illinois Press.

Wade-Smith, J., and M. E. Richmond. 1978. Reproduction in captive striped skunks (*Mephitis mephitis*). *American Midland Naturalist* 100:452–55.

Wade-Smith, J., M. E. Richmond, R. A. Mead, and H. Taylor. 1980. Hormonal and gestational evidence for delayed implantation in the striped skunk, *Mephitis mephitis*. *General and Comparative Endocrinology* 42:509–15.

Wade-Smith, J., and B. J. Verts. 1982. *Mephitis mephitis*. *Mammalian Species* 173.

Walker, E. P. 1964. *Mammals of the world*. Baltimore: Johns Hopkins University Press.

Walton, L. R., and S. Larivière. 1994. A striped skunk, *Mephitis mephitis*, repels two coyotes, *Canis latrans*, without scenting. *Canadian Field-Naturalist* 108:492–93.

Weller, D. M. G., and M. R. Pelton. 1987. Denning characteristics of striped skunks in Great Smoky Mountain National Park. *Journal of Mammalogy* 68:177–79.

Wood, W. F. 1990. New components in defensive secretion of the striped skunk, *Mephitis mephitis*. *Journal of Chemical Ecology* 16:2057–65.

———. 1999. The history of skunk defensive secretion research. *Chemical Educator* 4:44–50.

WESTERN SPOTTED SKUNK (*SPILOGALE GRACILIS*) AND RIO GRANDE SPOTTED SKUNK (*S. LEUCOPARIA*)

Jerry W. Dragoo and Christine C. Hass

Descriptions of spotted skunks (genus *Spilogale*) made by natural historians allude to an active nature and to high intelligence. Bailey (1931:339) described New Mexico spotted skunks as "bright, quick, and active, with larger eyes and more intelligent expression than the other skunks." According to Patton (1974:191), spotted skunks "remind one of acrobatic clowns in a circus with their frolicsome play with each other when they are young and with their inquisitive and curious natures. They would have to be considered the pranksters of the skunks of Trans Pecos Texas." Seton (1929: 384) described them as "quick and alert, much more active than the skunks of the genus *Mephitis*; indeed, they are almost as active as weasels, and readily climb trees."

Spotted skunks are the smallest members of the Mephitidae and more weasel-like than any of the other skunks. Roughly the size of rock squirrels (*Otospermophilus variegatus*), male western forms of spotted skunks weigh about 500–600 g (1.1–1.3 lbs), with females about 20% smaller in Texas (Patton 1974) and California (Crooks 1994). Spotted skunks were captured incidentally in Arizona's Huachuca Mountains during studies of white-nosed coatis (*Nasua narica*; Hass 2021) and hooded and striped skunks (*Mephitis macroura* and *M. mephitis*; Hass 2003; Hass and Dragoo 2017). Of 18 spotted skunks captured, 17 were males. The 13 adult males (judging from tooth wear and scrotal testes) averaged 350 g (0.8 lb) in weight (range 100–600 g [0.2–1.3 lb]), with a mean total length

of 458 mm (18 in) (range: 440–505 mm [17–20 in]) and a mean tail length of 143 mm (6 in) (range: 123–157 mm [5–6 in]; C. C. Hass, unpubl. data). These numbers lie at the lower end of the spectrum of published measurements for western forms of spotted skunks (see Verts et al. 2001). Van Gelder (1959) provided external and cranial measurements for the then-recognized taxa of spotted skunks, documenting the extent of size variation between genders (sexual dimorphism) as well as among taxa.

Like the other skunks, spotted skunks have five toes and have plantigrade locomotion. The plantar pads on their feet are more subdivided (Photo 29.3) than in the other skunks, giving their tracks an almost squirrel-like appearance. Van Gelder (1959) provided a detailed forensic analysis of color patterns in spotted skunks. He suggested they were black animals with white markings in various complex configurations. He described these patterns as six white stripes along the back and sides of the skunk. Depending on visual perspective, a spotted skunk's color pattern can then look more spotted or more striped. Images of the color pattern of a single animal clearly indicate stripes from the side, but a frontal view of the same skunk in a handstand position reveals a more spotted pattern (Photo 29.4). Spotted skunks also have a large white spot between the eyes (Van Gelder 1959; Photo 29.4).

The bold black and white color pattern has been interpreted either as camouflage, where the

(*opposite page*) Photograph: © Douglas W. Burkett.

Photos 29.1a and b (*left and above*). A probable Rio Grande spotted skunk (*Spilogale leucoparia*) in Doña Ana County on 25 September 2007 (the same individual is also shown in the photograph introducing the chapter). Compared to New Mexico's three other skunks, *Spilogale leucoparia* and *S. gracilis* (the two western forms of spotted skunk recently elevated to species status) are smaller and more slender, with a weasel-like body; their eyes are larger. Spotted skunks are also characterized by their ability to climb trees. Photographs © Douglas W. Burkett

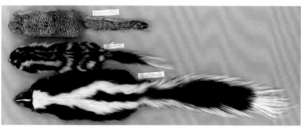

Photo 29.2. Western and Rio Grande spotted skunks (*Spilogale gracilis* and *S. leucoparia*) are closer in size to a rock squirrel (*Otospermophilus variegatus*) than to a striped skunk (*Mephitis mephitis*). Museum of Southwestern Biology specimens from New Mexico: *O. variegatus* (top; MSB:Mamm:269894), *Spilogale* cf. *leucoparia* from Grant County (middle; MSB:Mamm:160344), and *M. mephitis* (bottom; MSB:Mamm:140119). Photograph: © Jon Dunnum.

pattern may break up the body outline of spotted skunks on moonlit nights making them nearly invisible (Seton 1929), or as a warning signal to potential predators. Aposematic color patterns are highly visible and advertise prey's unprofitability, often linked to a high risk of injury to the would-be predator, whether from dangerous poison-glands or stings or simply the prey's aggressiveness and ability to fight back (Pocock 1908). In the case of a skunk, aposematism would advertise the foul-smelling spray from its anal glands (see below). In general, animals with aposematic colorations are notable for their slow, deliberate movements (Pocock 1908).

Caro et al. (2013) suggested the color pattern is both aposematism and camouflage. They photographed mounted spotted skunk, striped skunk, and bobcat (*Lynx rufus*) taxidermy specimens at close range against natural backgrounds. Specimens either had natural body color or were dyed to resemble striped skunks or bobcats. On the resulting photographs, Caro et al. (2013) measured and compared the mean hue, saturation,

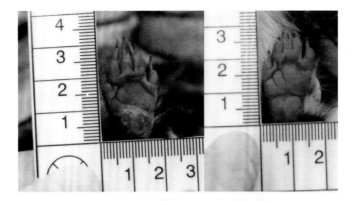

Photo 29.3. *Spilogale leucoparia* foot pads: front (left image) and back (right). Plantar pads on their feet are more subdivided than in the other skunks, with shorter claws. Photograph: © Christine C. Hass.

Photo 29.4. Two views of the same male animal, likely a western spotted skunk (*Spilogale gracilis*), from Bernalillo County, New Mexico, on 13 September 2009. This animal was rescued, rehabilitated in captivity, and later released into the wild. In the side view (left), the bold black-and-white color pattern appears more striped; in the front view (right), the white appears to form spots. Photographs: © Jerry W. Dragoo.

and brightness of mounts and backgrounds. Human observers were also used to compare the observability of mounts in the dim light of dusk and dawn. Overall, Caro et al.'s (2013) study suggests that when spotted skunks are observed from a distance, their color pattern makes them more cryptic to predators such as birds of prey (assuming birds of prey only use hue, saturation, and brightness to detect prey). In dim patchy light, a spotted skunk can disappear simply by holding still (Photo 29.5). However, up close,

the high-contrast black-and-white pattern becomes conspicuous and announces that it is a skunk. Part of the aposematism at close range is enhanced by the skunk's movement patterns, in particular the white flag of the tail (C. C. Hass and J. W. Dragoo, pers. obs).

As in other skunks, western forms of spotted skunks have two muscular musk glands at the base of the tail inside the anal sphincter. These glands can eject a noxious yellow fluid that is used in defense to deter a predator. The musk of western

Photo 29.5. In dim, patchy light, a spotted skunk can blend in seamlessly with its surrounding environment. From a distance the bold pattern of stripes and spots serves as camouflage, and the skunk can disappear simply by holding still. Photograph: © Christine C. Hass.

spotted skunks is composed of three major thiols—(E)-2-butene-1 -thiol, 3-methyl-1-butanethiol, and 2-phenylethanethiol—and several minor components such as phenylmethanethiol, 2-methyl-quinoline, 2-quinoline methanethiol, bis[(E)-2-butenyl] disulfide, (E)-2-butenyl 3-methylbutyl disulfide, and bis(3-methylbutyl) disulfide. Spotted skunks do not have the thiolacetates associated with the musk of species within the genera *Mephitis* and *Conepatus*, and while its spray is as pungent as in those other skunks it does not have the "staying power" without the thiolacetates. Thiolacetates have a higher molecular weight than thiols. When these are hydrolyzed, they release the more volatile thiols (Wood et al. 1991).

TAXONOMY

Spotted skunk taxonomy is, and always has been, in a state of flux. When we started writing this chapter there was one species (*Spilogale gracilis*) and two subspecies (*S. g. gracilis* and *S. g. leucoparia*) in New Mexico. Within the last few years several authors (see below) have conducted numerous phyologeographic, phylogenetic, and population genetic studies of spotted skunks leading us to conclude there are at least two species in the state, *S. gracilis*, the western spotted skunk (also called Rocky Mountain spotted skunk), and *S. leucoparia*, the Rio Grande spotted skunk (also called desert spotted skunk). In addition to these two species is a third form, "*arizonae*," currently classified with *S. leucoparia* but which might be distinct. Because no studies of ecology or life history have been conducted on the individual species in New Mexico, we consider them together as "western forms of spotted skunks." Frey (2004) listed the plains spotted skunk, *S. interrupta*, as possibly occurring in the eastern part of New Mexico based on records from the western panhandle and Llano Estacado of Texas.

Pre-Pleistocene divergence gave rise to eastern and western lineages of spotted skunks. These two lineages diverged approximately 1.36 Ma. Divergence of lineages within the eastern and western forms occurred more recently; approximately 0.25 to 0.12 Ma (Ferguson et al., 2017).

Merriam (1890a) described a new species of spotted skunk, *Spilogale gracilis*, which was based on two specimens collected in northern Arizona, and he compared it to the eastern *S. putorius*. Later, Merriam (1890b) provided the first taxonomic revision of spotted skunks and recognized two forms: eastern and western, based on the shape of the cranium. Skulls of the eastern form were narrower and more arched than those of the western form, which were flatter and wider.

Van Gelder (1959) provided a thorough revision of the spotted skunks based on color patterns and cranial variation (including age and gender) and concluded that, except for the pygmy spotted skunk (*Spilogale pygmaea*) of western Mexico, all other spotted skunks represented only a single species, *S. putorius*. Within New Mexico Van Gelder (1959) recognized two subspecies, *S. p. gracilis* in the northern section of the state, with specimens obtained from Colfax, Union, and McKinley counties, and *S. p. leucoparia* in the southern region, with specimens recorded from most of the southernmost counties (Eddy, Grant, Hidalgo, Luna, and Otero). *S. leucoparia* could be distinguished from *S. gracilis* based on longer and broader lateral stripes (Van Gelder 1959). Van Gelder (1959) suggested there was a wide swath of intergradation between the two subspecies in central New Mexico, but no specimens of intergrades were reported. Van Gelder (1959) remains the most comprehensive evaluation of morphological variation and taxon assignment in the state.

Mead later published two papers documenting reproduction in spotted skunks (Mead, 1968a, b). He indicated that the western form of the spotted skunk had a period of delayed implantation, whereas the eastern form did not. He concluded (Mead 1968b) that the two forms should be regarded as different species. Dragoo et al. (1993), based on a very limited molecular dataset, supported Mead's conclusion. A few studies published from the late 1960s through the mid-1990s examined chromosomal variation in spotted skunks. Owen (1996) suggested that the degree of chromosomal variation found in his study, as well as the previous chromosomal studies (Hsu and Mead 1969; Lee and Modi 1983), was indicative of differentiation of species within *Spilogale*.

Relying on those new data, Wozencraft (2005) recognized four species of spotted skunk: *S. putorius*, *S. gracilis*, *S. angustifrons*, and *S. pygmaea*, of which only *S. gracilis* occurred in New Mexico. Dragoo (2009) followed Wozencraft's taxonomy and included two subspecies, *S. g. gracilis* and *S. g. leucoparia*, in New Mexico. Dragoo (2009) also predicted that the variation in the number of chromosomes exhibited by *Spilogale* was such that more species were likely to be recognized through additional molecular research in addition to the morphological data.

Using mtDNA, Ferguson et al. (2017) produced a phylogeographic representation of the western forms of spotted skunks and found three distinct clades that had been formed as a result of Quaternary climatic changes. The "Western" clade was found broadly throughout the western United States and Baja Peninsula, with southeastern limits in central New Mexico (Socorro and Lincoln counties). The "East-central" clade was found in Texas and eastern Mexico, with northwestern limits in central New Mexico (Bernalillo, Lincoln, and Doña Ana counties). The "Arizona" clade was found in southern Arizona and northwestern Mexico. No specimens from New Mexico were included in this last clade, but likely would include skunks from the southwestern part of the state (see Howell 1906; Bailey 1931). No changes were suggested as the study primarily was focused on observations of phylogeographic variation.

A similar study (Shaffer et al. 2018) focusing on the eastern forms of spotted skunks found that the Florida spotted skunk (*S. p. ambarvalis*) was a monophyletic clade within the paraphyletic eastern spotted skunk (*S. p. putorius*) clade, and that the plains spotted skunk was a distinct clade. Shaffer et al. (2018) kept the existing taxonomy for all three clades. Bell (2020) was more assertive in his thesis and recommended the plains spotted skunk be elevated to species level, *S. interrupta*, based on over 400,000 single nucleotide polymorphisms (SNPs). He did not include all the taxa or populations of spotted skunks but did suggest the eastern spotted skunk and Florida spotted skunk may not be taxonomically distinct. He also concluded that *S. gracilis* and *S. leucoparia* should be recognized as distinct species, both of which occur in New Mexico.

Most recently, using ultra-conserved elements from the nuclear genome as well as mtDNA sequences, McDonough et al. (2022) performed a thorough evaluation of species limits and diversification of spotted skunks. Their analyses represent the most current understanding of *Spilogale* population genetics and taxonomy to date. McDonough et al. (2022) suggested a full taxonomic revision based on multiple sources of data (morphology, molecular, karyology, and ecology) is required. Until then, however, they made taxonomic recommendations assigning many of the above-mentioned clades (Ferguson et al. 2017; Shaffer et al. 2018; Bell 2020) as distinct species. This resulted in two species of spotted skunks in New Mexico; *Spilogale gracilis* (western spotted skunk) in the north, and *S. leucoparia* (Rio Grande spotted skunk) in the south.

Ferguson et al.'s (2017) "Arizona" clade was also identified by mtDNA by McDonough et al. (2022), who called it the "Sonora" clade instead. According to McDonough et al. (2022), that mtDNA clade is closely related to either *S. leucoparia* or *S. angustifrons* of southern Mexico. Calling it a subspecies at this time may not be valid as it would make whichever species it is listed under paraphyletic. As it is closely aligned to either species, it should itself be recognized as a species. However, the clade was detected using only mtDNA, not nuclear DNA, analyses (McDonough et al. 2022). Results of the nuclear DNA analyses indicated that the Arizona clade may be aligned with *S. leucoparia*. The Arizona clade recognized by Ferguson et al. (2017) was originally described as *Spilogale phenax arizonæ* from near Fort Verde, Arizona (Mearns 1891). Elliot (1903) elevated the taxon to species level. Howell (1906) listed a specimen of *S. arizonæ* from the Rio Mimbres in New Mexico. However, Bailey (1931) suggested that *S. arizonæ* occurred throughout the southwestern region of New Mexico. The populations of spotted skunks in southwestern New Mexico will require further analyses to ascertain taxonomic status. We tentatively consider all spotted skunks from southwestern New Mexico as *S. leucoparia*.

McDonough et al. (2022) only examined 208 specimens, none of which was from northern New Mexico. The taxonomic boundary lines recognized by Hall and Kelson (1959:930) are present on the map used by McDonough et al. (2022). It was an unfortunate use of an older map. The analysis of 208 specimens from Florida to California

and Oregon to Nicaragua does not leave a lot of room for discussion of species boundaries. Most researchers have used Van Gelder's (1959) map, based on museum specimens, for "*gracilis*" and "*leucoparia*." Hall (1981) reported *leucoparia* in the southern half of New Mexico and *gracilis* in the northern half. Verts et al. (2001) used Hall's (1981) map. McDonough et al. (2022) did not describe any new taxa, and their work, plus that of Ferguson et al. (2017), appears to support the species ranges described in Hall (1981) and Van Gelder (1959). The boundaries that existed among spotted skunk taxa are still recognized but may not be well defined. We are content saying that the species/population boundaries in New Mexico are not known. It represents the topic for a future graduate student project (and see Bailey 1931).

DISTRIBUTION

The current distribution of clades within the western lineage of spotted skunks is the result of Quaternary climatic changes. Three refugia formed during the last interglacial, leading to isolated populations in the Pacific northwest and central California, in central Sonora, Mexico, and in northeastern Mexico/southern Texas (Ferguson et al. 2017). The Arizona clade was likely isolated from other western forms by the Bouse Embayment (now the lower Colorado River). Other river systems, deserts (Mojave and Death Valley), and mountain ranges may also have played a role in isolating different populations (Ferguson et al. 2017). Today, spotted skunk populations of the southwestern United States and northern Mexico are found primarily in mountain ranges, with the surrounding deserts harboring much lower densities and serving to at least partially isolate populations (Van Gelder 1959).

Bailey (1931) recognized three species of spotted skunks in New Mexico, which likely correspond to the three mtDNA clades discussed above: 1) *S. arizonae*, the Arizona clade (here tentatively considered part of *S. leucoparia*); 2) *S.*

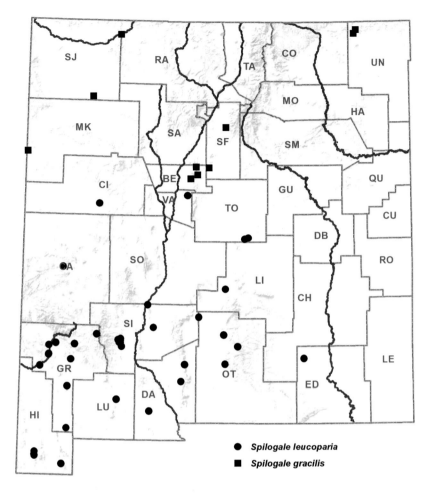

Map 29.1. Distribution of the Rio Grande spotted skunk (*Spilogale leucoparia*) and western spotted skunk (*Spilogale gracilis*) in New Mexico, based on museum specimen records, camera trap data, and sightings. The assignment of records to species in this map is tentative and generalized from distributional information in Van Gelder (1959), Findley et al. (1975), and Hall (1981). More research is needed to precisely assign each record to species.

leucoparia, the East-Central clade; and 3) *S. gracilis* (= *S. tenuis*), the Western clade. He described *S. arizonae* as occurring in the southwestern part of the state and perhaps reaching as far north as McKinley County southwest of Gallup. *S. arizonae* populations described by Bailey (1931) appear to belong to the Arizona clade of Ferguson et al. (2017) and the Sonoran clade of McDonough et al. (2022). Bailey (1931) reported them specifically from the Animas Mountains, Hachita, Deming, the Burro Mountains,

Redrock, and the head of the Mimbres River, and mentioned that they were common in particular in the Cliff and Glenwood areas along brushy bottoms of both the Gila and San Francisco river valleys. Bailey (1905, 1931) listed only one specimen of *S. leucoparia* from Tularosa in Otero County, but the East-Central clade is known from the deserts of Sonora and Chihuahua in Mexico, reaching into southwestern Texas and south-central and southeastern New Mexico (Ferguson et al. 2017). Ferguson et al.

Photo 29.6. Spotted skunk in a canyon bottom along the Mimbres River in Grant County, within the distribution of *Spilogale leucoparia*. The surrounding vegetation is oak, pinyon-juniper woodland. Photograph: © Keith Geluso.

(2017) suggested the range of this clade was possibly constrained by the Rio Grande but found a couple of specimens west of the river. In New Mexico, spotted skunk records (museum specimens, camera trap data, and sightings) potentially referring to *S. leucoparia* are available from the southeastern and south-central region of the state around Tularosa and from the Pecos Valley, and around Carrizozo in Lincoln County (Map 29.1). Finally, *S. gracilis* occurs throughout most of the western United States (Van Gelder 1959). Bailey (1931) showed that the species (he referred to it as *S. tenuis*) occurred in the northern third of New Mexico. Records of spotted skunks from Bernalillo, Colfax, San Juan, Torrance, and Valencia counties could be referable to *S. gracilis*. Based on the distribution of clades in Ferguson et al. (2017), it is possible that *S. gracilis* and *S. leucoparia* have overlapping distributions in central New Mexico or are hybridizing (Van Gelder 1959).

Additional field work will be required to determine the exact ranges of spotted skunks in New Mexico. Bailey (1931) called for local research to be conducted throughout the state to better understand the distribution and abundance of spotted skunks. Modern camera trap studies might accomplish that goal as they have proven an effective method for detecting spotted skunks (Hass and Dragoo 2009; J. K. Frey, pers. comm.). As no ecological work has been conducted on the individual spotted skunk taxa in New Mexico, we do not differentiate among them in the following sections.

HABITAT ASSOCIATIONS

Findley et al. (1975) described spotted skunks in New Mexico as occurring in rocky and brushy areas in woodland, grassland, and desert. Despite Van Gelder (1959) considering deserts as geographic barriers to movements among populations, Bailey (1931) indicated that spotted skunks occupied practically all the low country in the state, ranging in elevation through both the lower and upper Sonoran life zones (i.e., reaching pinyon-juniper or pinyon-juniper oak) (see Bailey [1913] for his description of the life zones of New Mexico). Within that general elevational range, Bailey (1931) indicated that spotted skunks were found in areas where protection can be afforded by vegetative cover. In the southwestern corner of the state, Bailey (1931) found spotted skunks to be mainly associated with rocky terrain, both canyons and cliffs, while also frequently seeking cover along brushy drainage bottoms and in rock piles within wooded foothills. Still according to Bailey (1931), spotted skunks may occasionally occur in desert regions along drainages, in weedy vegetation, or among cacti. Neiswenter and Dowler (2007) reported similar habitat preferences in western forms of spotted skunks in western Texas, particularly a strong association with mesquite (*Prosopis*), prickly pear (*Opuntia*), and underbrush for cover. Despite Neiswenter and Dowler's (2007) suggestion that western forms of spotted skunks were not found in agricultural areas, their occurrence in farm buildings and barns has been reported (Rosatte et al. 2010). Spotted skunks have also been found in urban areas, such as Albuquerque and Las Cruces in New Mexico (J. W. Dragoo, pers. obs.), further illustrating the wide variety of environments they occupy.

Based on museum specimens, remote camera photos, and sightings, 47 spotted skunk locations in New Mexico were found in the following International Geosphere–Biosphere Programme land-cover categories: 2% evergreen needleleaf forest, 9% woodland, 19% wooded grassland, 23% closed shrubland, 23% open shrubland, 15% grassland, 2% cropland, and 4% urban and built-up land (GBIF 2021; Hass and Dragoo 2017). A number of spotted skunk records are associated with springs and water drainages, including some of the above records as well as a specimen collected on 10 May 1966 south of Cliff along the Gila River (WNMU 6367) and another secured along the Los Pinos River on 27 August 1958 (MSB 5370). Overall, western forms of spotted skunks seem to prefer shrubby or wooded vegetation in association with complex physiography, often rocky areas.

LIFE HISTORY

Spotted skunks are secretive and primarily nocturnal, and Findley et al. (1975) suggested that this accounts for them not often being seen. Also notable is their ability to climb up and down trees squirrel-like.

Diet and Foraging

Taylor (1953) noted that spotted skunks were primarily omnivorous, but preferentially fed on insects, small mammals, and other vertebrates, in addition to plant products, primarily berries and larger fruits, when in season. In Texas, spotted skunks preyed upon beehives during inclement weather and were observed eating red spotted toads (*Bufo punctatus*) after a rainstorm (Patton 1974). Patton also reported that captive spotted skunks ate a variety of insects, along

Photo 29.7. Spotted skunk in southeastern Arizona on 4 October 2001. Rio Grande and western spotted skunks are nocturnal. Although diet and foraging information from New Mexico is largely anecdotal, studies of western spotted skunks in other states show them to be omnivorous, with insects, small mammals, reptiles, birds, and plant materials (fruits and berries) all consumed. Photograph: © Christine C. Hass

Photo 29.8. A female Rio Grande spotted skunk (*Spilogale* cf. *leucoparia*)—a rescued animal, rehabilitated and later released—from Doña Ana County, New Mexico climbing down a juniper tree (*Juniperus monosperma*) on 10 April 2004. The ability of spotted skunks to climb trees allows them to escape potential predators. It also likely provides additional foraging opportunities as spotted skunks in general exploit the vertical dimension more than other genera of skunks. Photograph: © Jerry W. Dragoo.

with small mammals (including kangaroo rats [*Dipodomys*] and pocket mice [*Chaetodipus*]) and fresh fish, but showed a preference for chicken eggs. Bailey (1936) included not just insects and small mammals in the diet of western spotted skunks in Oregon, but also reptiles, amphibians, small birds, and crayfish, adding that rarely poultry (young chicks) may be taken as well. A more recent study of spotted skunks on Santa Cruz Island again documented a diet composed of small mammals, reptiles, and birds, in addition to many insects (Jones et al. 2008).

Information on the diet of spotted skunks in New Mexico is only anecdotal. In the Animas Mountains in the southwestern part of the state, food of the spotted skunks consisted of grasshoppers and other insects, in addition to scorpions (order Scorpiones); also in southwestern New Mexico, spotted skunks caught small rodents, near and around cabins (Bailey 1931). A spotted skunk (likely *S. leucoparia*) in the Organ Mountains, Doña Ana County, was caught on camera with the remains of a mountain patchnosed snake (*Salvadora grahamiae*) in its mouth (Jacobson and Frey 2022). To trap spotted skunks, Bailey (1931) suggested that meat of any kind, fur, or feathers worked effectively as bait.

Working with Wildlife Rescue Inc. of New Mexico, one of us (J. W. Dragoo) fed captive spotted skunks a variety of steamed vegetables, such as broccoli, cauliflower, peas, corn, carrots, green beans, and various fruits such as apples, bananas, strawberries, tomatoes, grapes, and raisins. The diets were also supplemented with various grains such as rice, millet, oats, and barley, as well as cooked ground turkey and a high-quality commercial ferret food. When skunks were approaching a release date, insects such as mealworms, crickets, grasshoppers, and beetles were provided both to teach young spotted skunks to hunt and to provide protein enrichment (G. A. Dragoo, unpubl. data.). One unreleasable animal was maintained on this diet for over 12 years.

Reproduction and Social Behavior

As with many other aspects of their life history, the reproductive biology of spotted skunks is not well known in New Mexico but is assumed to be similar to what has been documented elsewhere. Mead (1968b) showed that adult male western forms of spotted skunks are ready to breed by June, and many juveniles (young of the year) by September. Mead (1968b) indicated that western forms of spotted skunks usually begin mating in September and both juvenile and adult females participate in breeding. Patton (1974), however, observed that some Rio Grande spotted skunks in the Trans-Pecos, Texas, may breed in July, and not all males will breed during their first fall season. The implantation of embryos is delayed for 200–280 days, with birth occurring about 30 days after implantation (Foresman and Mead 1973). Females usually give birth from March through early May. Mead (1967) suggested that *S. gracilis* is unlikely to give birth to a second litter within the same year given the long gestation period. However, though most young are born in the spring throughout the range of western forms of spotted skunks, there is some evidence of parturition occurring in a few females from southeastern Arizona in the fall (C. C. Hass, unpubl. data). Evidence from captive males in Oregon suggests that serum testosterone levels, testes size, and ejaculate volumes begin increasing in the fall as daylight decreases (Kaplan and Mead 1993).

Bailey (1931) reported that females had one pair of inguinal mammae, one pair of abdominal mammae, and two pairs of pectoral mammae. The average litter size of western forms of spotted skunks is about four (range 2–6; Mead 1968b), and the young are born blind and naked (J. W. Dragoo, pers. obs.). In January 2009, however, a wildlife rehabilitator in Idaho rescued a pregnant female spotted skunk, which gave birth to a litter of eight—one died shortly thereafter—on 6 March (S. England, pers. comm.). The eyes of young spotted skunks open after about 28 days.

The young are weaned after two months, and they begin to disperse after about three months of age (Dragoo 2009).

While playing, young, excited animals will let out a loud, ear-piercing, high-pitched screech (J. W. Dragoo, pers. obs.). Young males of the same litter will wrestle more aggressively than females (Dragoo 2009). Males will also tend to leave the maternal den sooner to look for mates (Patton 1974). During the late summer and early fall, male western forms of spotted skunks tend to be captured more than females. This could be due to increased movement of males during the breeding season, which corresponds to those months (Patton 1974; Doty and Dowler 2006).

Western forms of spotted skunks are solitary while foraging, and usually den alone but use multiple dens. Dens may be occupied by multiple animals but not necessarily at the same time. Crooks (1994) reported several female spotted skunks denning together on Santa Cruz Island; as there were no other burrowing animals on the island, skunks had to dig their own dens. Doty and Dowler (2006) performed a comprehensive study on the denning ecology of spotted skunks in western Texas. They found that *S. leucoparia* usually dens in areas that provide protective cover, such as under shrubs or prickly pear cactus. Dens may be placed above or below ground, as long as there is vegetative cover. In southeastern Arizona, a couple of spotted skunk dens were found in rocky outcrops (C. C. Hass, unpubl. data).

Few data are available regarding the life span of spotted skunks. It is likely similar to that of other skunks, or two to four years, a presumption shared by Van Gelder (1959), who suggested that a five-year-old, wild spotted skunk would be considered old. Captive animals have been reported to survive just under 10 years (Van Gelder 1959; Verts et al. 2001). However, one of us (J. W. Dragoo) rescued an animal in the fall of 1999; it died in July 2011, making it the longest lived (~12 years) spotted skunk on record.

Photo 29.9. Prickly pear (*Opuntia* spp.) is often found in denning areas as it provides protective cover for spotted skunks. A female Rio Grande spotted skunk (*Spilogale leucoparia*)—a rescued animal, rehabilitated and later released—from Doña Ana County, New Mexico uses the cactus as protection on 12 October 2006. Photograph: © Jerry W. Dragoo.

Interspecific Interactions

The activity period of western forms of spotted skunks may be influenced by interactions with, and avoidance of, potential predators and competitors. Island spotted skunks (*Spilogale gracilis amphiala*) are known to avoid foxes on California's Channel Islands (Santa Cruz and Santa Rosa islands; Crooks and Van Vuren 1995), and removal of foxes was one factor that may have led to an exponential increase in local spotted skunk numbers (Jones et al. 2008). Spotted skunks occur in sympatry with other species of skunks in New Mexico. In western Texas they were found to be more active after midnight, presumably to avoid competition with striped skunks, which were the dominant competitors and showed a peak in activity during the earlier part of the night (Neiswenter et al. 2010). Patton (1974) also demonstrated that spotted skunks were rarely captured before midnight for the same reason. He noted that the four species of skunks in western Texas generally try to avoid one another.

Photo 29.10. A male western spotted skunk (*Spilogale gracilis*)—a rescued animal, rehabilitated and later released—from Bernalillo County, New Mexico in handstand position on 13 September 2009. The handstand presumably makes spotted skunks appear larger, allowing them to display their aposematic coloration more effectively. Photograph: © Jerry W. Dragoo.

Spotted skunks are prey to foxes, coyotes (*Canis latrans*), bobcats (*Lynx rufus*), and golden eagles (*Aquila chrysaetos*) (Verts et al. 2001; Dragoo 2009; Hass 2009). Bailey (1905) also found the jaws of several western spotted skunks in regurgitated owl pellets in western Texas. The relationship between spotted skunks and rattlesnakes is interesting, as both animals are evidently capable of preying on each other. Spotted skunks will elicit a defensive posture and prolonged rattling from rattlesnakes (Cowles 1938), but also have been found in the stomachs of rattlesnakes (Klauber 1972). An occasional cougar (*Puma concolor*) will take a spotted skunk, but likely has to be very hungry. Otherwise, the little spotted skunk, while doing a handstand,

can be very intimidating to a cougar, as demonstrated in a video presented by Allen et al. (2013). Cougars likely do not perceive spotted skunks as a meal, but rather as an annoyance. However, bobcats are less intimidated and may get more of a belly full by eating spotted skunks. In a study of diets in bobcats and cougars in Arizona, a small percentage (3%) of the bobcat diet consisted of spotted skunks; in contrast, no spotted skunks were taken by cougars (Hass 2009).

When nervous, spotted skunks walk with heavy, deliberate steps (J. W. Dragoo, C. C. Hass, pers. obs.). When facing an enemy, spotted skunks will also smack the ground with their front paws and hiss as a warning. They also may rush forward, sometimes hopping, and then stand on their forepaws (front handstand) with their hind end elevated off the ground (Walker 1930; Photo 29.10). They can, but usually do not, spray in this position. They drop down on all four feet and face the target with both ends, then spray (J. W. Dragoo, pers. obs.). Spotted skunk are also adept climbers and can spray from an elevated perch (Verts et al. 2001). In general, releasing musk is a last resort.

Feral cats (*Felus catus*) and domestic dogs (*Canis domesticus*) sometimes kill spotted skunks. Humans are also a major source of spotted skunk mortality. Spotted skunks sometimes are shot as pests, trapped for fur, and poisoned during pest control operations, usually as non-target species (coyotes are often the intended target; Dragoo 2009). Findley et al. (1975) reported that spotted skunks in New Mexico are rarely found as roadkill. However, that does not mean they are not hit by cars, but rather that they may not be found afterward due to their diminutive size (J. W. Dragoo, unpubl. data).

STATUS AND MANAGEMENT

In urban areas, western forms of spotted skunks are rarely reported as nuisance animals. However, they have been collected in residential houses in Las Cruces and Albuquerque, as well as in rural settings such as the Fourth of July Campground in the Manzano Mountains. They occasionally enter houses through a pet door or establish a den in a garage. In such cases they are not always appreciated (J. W. Dragoo, unpubl. data).

Western forms of spotted skunks are not among the species listed under the Convention on International Trade in Endangered Species of Wild Fauna and Flora (CITES). They are also not listed by the International Union for the Conservation of Nature in its Red List of Threatened Species. Whether any of the western spotted skunk species are declining is unknown. In New Mexico, spotted skunks are not considered game animals and are not afforded any special protection (BISON-M; https://www.bison-m.org/booklet.aspx?SpeciesID=050747).

Spotted skunks are susceptible to various external and internal parasites such as bot flies, various fleas, lice, ticks, tapeworms, and roundworms, as well as other nematodes and coccidians (Patton 1974; Kirkland and Maldonado 1988; Neiswenter et al. 2006). One common name for spotted skunks is hydrophobia cat, suggesting that rabies is a problem in these animals. Coues (1877) reproduced two previously published articles (one by H. C. Hovey and the other by J. G. Janeway) dealing with skunks and rabies. Both focused on several cases of skunks causing "hydrophobia" (a clinical sign of rabies) in people. While his 1877 monograph was in press he received a correspondence from a colleague describing the spotted skunk's bite as "highly dangerous, causing a fatal disease like hydrophobia," but without giving any specific evidence in support of the claim (Coues 1877:227). Howell (1906) indicated that spotted skunks are known as "hydrophobia cats" throughout much of the western United States, and that despite most skunk rabies cases being linked to *Mephitis*, people fear them specifically for the risk of becoming infected.

Seton (1929:394) tells of houses in the Southwest being built with rooms separated by "an inconveniently high" barrier and an external door. He explains, "The reason given to me long ago for this inconvenient arrangement is that it is done to keep out the Phoby-cat." It was believed by the locals that the so-called hydrophobia cat (*Spilogale*) was able to infect anyone or any animal at any time with rabies. Seton (1929:396) goes on to explain that people living in the Southwest could cure rabies with the use of a gallstone

boiled in milk, and speedily laid on the wound. This must be held there until the stone turns green, then boiled again, and applied as before 10 or 15 times. If then, the stone no longer turns green, the poison has been sucked out and destroyed. These gallstones are found in the bladder of a deer and are esteemed priceless. A famous gallstone at Springer [in Colfax County, New Mexico], had a record of saving many lives in the early '[18]90s; and was in constant request during my residence in that region.

Seton (1929) did not put a lot of faith in the cure. For his part, he believed that rabies was more likely to be spread by striped skunks than by spotted skunks, and thus was unsure why the name hydrophobia-cat was applied to the little spotted skunk. Today, the consensus is that spotted skunks are rarely diagnosed with rabies (Krebs et al. 1995). Dragoo et al. (2004) examined 24 skunk specimens that were submitted for rabies testing in New Mexico and identified three of the five species occurring in the state among them. The only two not represented were the spotted skunks. However, this finding does not mean western forms of spotted skunks never contract rabies. Aranda and Lopez-de Buen (1999) reported seven spotted skunks with rabies in Mexico. The seven cases involved spotted skunks attacking humans in their sleep inside their bedrooms. More accurate data on rabies in spotted skunks is hampered by reports to state health departments in which submissions of animals for testing are often lumped together under the term "skunks" without differentiating species (Dragoo et al. 2004).

There is a current need for research on spotted skunks in New Mexico in several areas, including genetics, habitat use, distribution, and reproduction. Disease ecology and the ecological relationships between *Spilogale* and the other skunk species represent more areas ripe for research. There is much we still do not know about New Mexico's most attractive and playful little skunks.

LITERATURE CITED

Allen, M. L., L. M. Elbroch, and H. U. Wittmer. 2013. Encounter competition between a cougar, *Puma concolor*, and a western spotted skunk, *Spilogale gracilis*. *Canadian Field-Naturalist* 127:64–66.

Aranda, M., and L. Lopez-de Buen. 1999. Rabies in skunks from Mexico. *Journal of Wildlife Diseases* 35:574–77.

Bailey, V. 1905. *Biological survey of Texas*. North American Fauna 25. Washington, DC: US Department of Agriculture, Bureau of Biological Survey.

———. 1913. *Life zones and crop zones of New Mexico*. North American Fauna 35. US Department of Agriculture, Bureau of Biological Survey, Washington, DC.

———. 1931 (=1932). *Mammals of New Mexico*. North American Fauna 53. Washington, DC: US Department of Agriculture, Bureau of Biological Survey.

———. 1936. *The mammals and life zones of Oregon*. North American Fauna 55. Washington, DC: US Department of Agriculture, Bureau of Biological Survey.

Bell, Z. 2020. *Genomic markers recognition of at least four forms of spotted skunks in the United States*. Department of Zoology and Physiology. Laramie: University of Wyoming Master of Science Thesis.

Caro, T., T. Stankowich, C. Kiffner, and J. Hunter. 2013. Are spotted skunks conspicuous or cryptic? *Ethology, Ecology and Evolution* 25:144–60.

Coues, E. 1877. *Furbearing animals*. Washington, DC: Government Printing Office.

Cowles, R. B. 1938. Unusual defense postures assumed by rattlesnakes. *Copeia* 1938:13–16.

Crooks, K. R. 1994. Den-site selection in the island spotted skunk of Santa Cruz Island, California. *Southwestern Naturalist* 39:354–57.

Crooks, K. R., and D. van Vuren. 1995. Resource utilization by two insular endemic mammalian carnivores, the island fox and island spotted skunk. *Oecologia* 104:301–7.

Doty, J. B., and R. C. Dowler. 2006. Denning ecology in sympatric populations of skunks (*Spilogale gracilis* and *Mephitis mephitis*) in west-central Texas. *Journal of Mammalogy* 87:131–38.

Dragoo, J. W. 2009. Family Mephitidae (skunks). In *Handbook of the mammals of the world*. Vol 1, *Carnivores*, ed. D. E. Wilson and R. A. Mittermeier, 532–63. Barcelona: Lynx Edicions.

Dragoo, J. W., R. D. Bradley, R. L. Honeycutt, and J. W. Templeton. 1993. Phylogenetic relationships among the skunks: a molecular perspective. *Journal of Mammalian Evolution* 1:255–67.

Dragoo, J. W., D. K. Matthes, A. Aragon, C. C. Hass, and T. L. Yates. 2004. Identification of skunk species submitted for rabies testing in the desert southwest. *Journal of Wildlife Diseases* 40:371–376.

Elliot, D. G. 1903. Descriptions of apparently new species and subspecies of mammals from California, Oregon, the Kenai Peninsula, Alaska, and Lower California, Mexico. *Field Columbian Museum, Zoological Series* 3:153–73.

Ferguson, A. W., M. M. McDonough, G. I. Guerra, M. Rheude, J. W. Dragoo, L. K. Ammerman, and R. C. Dowler. 2017. Phylogeography of a widespread small carnivore, the western spotted skunk (*Spilogale gracilis*) reveals temporally variable signatures of isolation across western North America. *Ecology and Evolution* 7:4229–40.

Findley, J. S., A. H. Harris, D. E. Wilson, and C. Jones. 1975. *Mammals of New Mexico*. Albuquerque: University of New Mexico Press.

Foresman, K. R., and R. A. Mead. 1973. Duration of post-implantation in a western subspecies of the spotted skunk (*Spilogale putorius*). *Journal of Mammalogy* 54:521–23.

Global Biodiversity Information Facility (GBIF) 2021. [GBIF occurrence download]. https://doi.org/10.15468/dl.3bwn86.

Hall, E. R. 1981. *The mammals of North America*. 2nd ed. Vol. 2. New York: John Wiley.

Hall, E. R., and K. R. Kelson. 1952. Comments on the taxonomy and geographic distribution of some North American marsupials, insectivores and carnivores. *University of Kansas Publications, Museum of Natural History* 5:319–41.

———. 1959. *The mammals of North America*. Vol. 2. New York: Ronald Press.

Hass, C. C. 2003. *Ecology of hooded and striped skunks in southeastern Arizona*. Final Report to Arizona Game and Fish Department, Phoenix.

———. 2009. Competition and coexistence in sympatric bobcats and pumas. *Journal of Zoology* (London) 278:174–80.

———. 2021. *Nosey beast: natural history of the coatis*. Carson City, NV: Wild Mountain Echoes.

Hass, C. C., and J. W. Dragoo. 2017. Competition and coexistence in sympatric skunks. In *Biology and conservation of the musteloids*, ed. D. W. Macdonald, C. Newman, and L. Harrington, 464–77. Oxford: Oxford University Press.

Howell, A. H. 1906. Revision of the skunks of the genus *Spilogale*. *North American Fauna* 26:1–55.

Hsu, T. C., and R. A. Mead. 1969. Mechanisms of chromosomal changes in mammalian speciation. In *Comparative mammalian cytogenetics*, ed. K. Benirschke, 8–17. New York: Springer-Verlag.

Jacobson, H. N., and J. K. Frey. 2022. *Salvadora grahamiae grahamiae* (Mountain Patch-nosed Snake) predation. *Herpetological Review* 53:157–58.

Jones, K. L., D. H. van Vuren, and K. R. Crooks. 2008. Sudden increase in a rare endemic carnivore: ecology of the island spotted skunk. *Journal of Mammalogy* 89:75–86.

Kaplan, J. B., and R. A. Mead. 1993. Influence of season on seminal characteristics, testis size and serum testosterone in the western spotted skunk (*Spilogale gracilis*). *Journal of Reproduction and Fertility* 98:321–26.

Kirkland, G. L., and J. E. Maldonado. 1988. Patterns of variation in cranial damage attributable to *Skrjabingylus* sp. (Nematoda: Metastrongyloidea) in skunks (Mammalia: Mustelidae) from Mexico. *Southwestern Naturalist* 33:15–20.

Klauber, L. M. 1972. *Rattlesnakes: their habits, life histories, and influence on mankind*. Berkeley: University of California Press.

Krebs, J. W., M. L. Wilson, and J. E. Childs. 1995. Rabies, epidemiology, prevention, and future research. *Journal of Mammalogy* 76:681–94.

Lee, M. R., and W. S. Modi. 1983. Chromosomes of *Spilogale pygmaea* and *S. putorius leucoparia*. *Journal of Mammalogy* 64:493–95.

McDonough, M. M., A. W. Ferguson, R. C. Dowler, M. E. Gompper, and J. E. Maldonado. 2022. Phylogenomic systematics of the spotted skunks (Carnivora, Mephitidae, *Spilogale*): additional species diversity and Pleistocene climate change as a major driver of diversification. *Molecular Phylogenetics and Evolution* 167:107266. https://doi.org/10.1016/j.ympev.2021.107266.

Mead, R. A. 1967. Age determination in the spotted skunk. *Journal of Mammalogy* 48:606–16.

———. 1968a. Reproduction in eastern forms of the spotted skunk (genus *Spilogale*). *Journal of Zoology, London* 156:119–36.

———. 1968b. Reproduction in western forms of the spotted skunk (genus *Spilogale*). *Journal of Mammalogy* 49:373–90.

Mearns, E. A. 1891. Notes on the otter (*Lutra canadensis*) and skunks (genera *Spilogale* and *Mephitis*) of Arizona. *Bulletin of the American Museum of Natural History* 3:252–62.

———. 1890b. Contribution toward a revision of the little striped skunks of the genus *Spilogale*, with descriptions of seven new species. *North American Fauna* 4:1–15.

Merriam, C. H. 1890a. Results of a biological survey of the San Francisco mountain region and desert of the Little Colorado, Arizona. *North American Fauna* 3:1–101.

Neiswenter, S. A., and R. C. Dowler. 2007. Habitat use of western spotted skunks and striped skunks in Texas. *Journal of Wildlife Management* 71:583–86.

Neiswenter, S. A., R. C. Dowler, and J. H. Young. 2010. Activity patterns of two sympatric species of skunks (*Mephitis mephitis* and *Spilogale gracilis*) in Texas. *Southwestern Naturalist* 55:16–21.

Neiswenter, S. A., D. B. Pence, and R. C. Dowler. 2006. Helminths of sympatric striped, hog-nosed, and spotted skunks in west-central Texas. *Journal of Wildlife Diseases* 42:511–17.

Owen, J. G., R. J. Baker, and S. L. Williams. 1996. Karyotypic variation in spotted skunks (Carnivora, Mustelidae, *Spilogale*) from Texas, Mexico and El Salvador. *Texas Journal of Science* 48:119–22.

Patton, R. F. 1974. Ecological and behavioral relationships of the skunks of Trans Pecos Texas. PhD dissertation, Texas A&M University.

Pocock, R. I. 1908. Warning coloration in the musteline Carnivora. *Proceedings of the Zoological Society of London* 78:944–59.

Rosatte, R. C., K. Sobey, J. W. Dragoo, and S. D. Gehrt. 2010. Striped skunks and allies (*Mephitis* spp.). In *Urban Carnivores: ecology, conflict, and conservation*, ed. S. D. Gehrt, S. P. D. Riley, and B. L. Cypher, 97–106. Baltimore: Johns Hopkins University Press.

Seton, E. T. 1929. *Lives of game animals*. Vol. 2. Garden City, NJ: Doubleday and Doran.

Shaffer, A. A., R. C. Dowler, J. C. Perkins, A. W. Ferguson. M. M. McDonough, and L. K. Ammerman. 2018. Genetic variation in the eastern spotted skunk (*Spilogale putorius*) with emphasis on the plains spotted skunk (*S. p. interrupta*). *Journal of Mammalogy* 99:1237–48.

Taylor, W. P. 1953. *Food habits of the Rio Grande spotted skunk in central Texas*. Austin: Texas Game and Fish Commission.

Van Gelder, R. G. 1959. A taxonomic revision of the spotted skunks (genus *Spilogale*). *Bulletin of the American Museum of Natural History* 117:229–392.

Verts, B. J., L. N. Carraway, and A. Kinlaw. 2001. *Spilogale gracilis*. *Mammalian Species* 674:1–10.

Walker, A. 1930. The "hand-stand" and some other habits of the Oregon spotted skunk. *Journal of Mammalogy* 11:227–29.

Wood, W. F., C. G. Morgan, and A. Miller. 1991. Volatile components in defensive spray of the spotted skunk, *Spilogale putorius*. *Journal of Chemical Ecology* 17:1415–20.

Wozencraft, W. C. 2005. Order Carnivora. In *Mammal species of the world: a taxonomic and geographic reference*, Vol. 1, ed. D. E. Wilson and D. M. Reeder, 532–628. 3rd ed. Baltimore: Johns Hopkins University Press.

RINGTAIL (*BASSARISCUS ASTUTUS*)

Robert C. Lonsinger and Gary W. Roemer

*The beautiful little cacomistle has the face of a fox,
the body of a buffy-gray marten, and a long bushy
tail with a black tip and about seven black and seven
white bars or rings. . . . the broad black bands give it a
striking character shared by no other North American
animal except in a slight degree by the raccoon.*

—BAILEY 1931:346

The species described in such glowing terms by Bailey (1931) in his *Mammals of New Mexico* is known today as the ringtail (*Bassariscus astutus*) and not as the cacomistle (*B. sumichrasti*), the latter corresponding instead to the larger congener found farther south in Mexico and Central America. But with its conspicuous tail—which it does share with its tropical cousin—and its lithe and supple body, the ringtail is perhaps indeed one of the most handsome of North America's carnivores. Historically common in the rafters of miners' cabins and often described as an excellent mouser, it is nonetheless a little seen species that has not received as much attention as its larger carnivoran brethren. Ringtails are small, primarily nocturnal members of the Procyonidae, a New World carnivore family that includes two other species native to the United States, the white-nosed coati (*Nasua narica*; Chapter 31) and the raccoon (*Procyon lotor*; Chapter 32). Of the 13 known procyonids (see Wilson and Reeder 2005; and taxonomic augmentations and revisions by Helgen et al. 2009, 2013; and Louppe et al. 2020), the ringtail is one of the smallest. Only the recently

discovered olinguito (*Bassaricyon neblina*) is more diminutive in size (Helgen et al. 2013).

General Appearance

Ringtails are slight of build, with an elongated, cylindrical body and a nearly equally long tail, the latter ending in a black tip and showing a mostly white stripe on the underside as the black and white rings do not quite extend all the way around (Grinnell et al. 1937; Poglayen-Neuwall and Toweill 1988). Similar to raccoons, ringtails have a mask, but with white spectacles surrounding the eyes, reminiscent of eye rings. Proximally, the white spectacles bleed into a narrow,

Photo 30.1. Ringtail at Bandelier National Monument on 1 September 2009. Note the white facial markings around the eyes in the form of spectacles and the large, rounded ears. Photograph: © Sally King.

(*opposite page*) Photograph: © Jared A. Grummer.

Photo 30.2. Ringtail in Dog Canyon in the Guadalupe Mountains in Texas, just south of the New Mexico stateline. Ringtails have elongated bodies and long tails whose black and white rings are interrupted on the underside by a white, longitudinal stripe. The venter varies in color from fawn to yellow, especially along the lower sides; the abdomen may be slightly whiter. Photograph: © Robert Deans.

darkened muzzle, from which most of the vibrissae (tactile hairs including whiskers and bristles) emanate. The muzzle then turns white again adjacent to the rhinarium (or tip of the snout), which is itself usually a pink to black color. The ears of ringtails resemble those of domestic cats (*Felis catus*), being large and upright, but generally thinner and more rounded. Some ringtails also sport a "lightning bolt" of white fur on the top of their heads between the ears. The facial, moustacial, and mandibular vibrissae all provide tactile sensation to assist ringtails in navigating dark and narrow spaces (Poglayen-Neuwall and Toweill 1988; and see below). Ringtail pupils are round and the irises hazel or brown in color with a yellow to orange sheen in a spotlight at night. The dental formula is typical of most procyonids, or 3/3, 1/1, 3/4, 3/2, for a total of 40 teeth (for a more detailed description, see Poglayen-Neuwall and Toweill 1988).

A ringtail's body is typically uniformly gray to light brown to rust along the dorsal surface with the fur grizzled with black at the tips. Dorsal color has been described as varying with habitat, being darker at higher latitudes or altitudes and lighter in environments that are more arid (Dice and Blossom 1937; Poglayen-Neuwall and Toweill 1988), a pattern that we observed in particular along an elevational gradient in the Guadalupe Mountains of southern New Mexico and western Texas (R. C. Lonsinger and G. W. Roemer, unpubl. data; and see Photo 30.3). The underfur of the dorsal surface is dark gray to black. The venter is lighter, a fawn to yellow color, especially along the lower sides, with the abdomen sometimes becoming slightly whiter. Ringtails have additional long tactile vibrissae on the underside of the forearms, presumably to facilitate the capture of prey through "fishing," a method of capture that involves reaching into small openings to detect and hook prey (Toweill and Toweill 1978).

In part because of their cat-like ears, ringtails have often been referred to colloquially as ringtailed cats. Several other common names again evoke comparisons with a domestic cat or with a fox, including band-tailed cat, coon cat, coon fox, civet cat, and miner's cat. Ringtails are somewhat smaller, but with longer tails, compared to domestic cats, and they exhibit sexual dimorphism with males tending to be heavier and slightly longer. The average weight of ringtails from the Guadalupe Mountains was 1,211 g (SD ± 201) [~2.7 lbs ± 0.4] for 84 adult males and 971 g (± 141) [~2.1 lbs ± 0.3] for 60 adult females (R. C. Lonsinger and G. W. Roemer, unpubl. data). Average body and tail length for males was 41.8 cm (± 2.7) (~16.5 in ± 1.1) and 37.1 cm (± 2.7) (~14.6 in ± 1.1), respectively, whereas females were 39.3 cm (± 2.2) (~15.3 in ± 0.9) and 34.8 cm (± 3.2) (~13.7 in ± 1.3) in comparison. Eighteen ringtails (9 males and 9 females; all adults) captured on the Sevilleta National Wildlife Refuge in central New Mexico (Socorro County) showed similar patterns in weight and size (Harrison 2012). Males were heavier (1,000 g ± 100) (~2.2 lbs ± 0.2) than females (900 g ± 100) (~2 lbs ± 0.2), but were only slightly longer (body length: males = 39.4 cm ± 1.5 [~15.5 in ± 0.6], females = 37.8 cm ± 1.5 [~14.9 in ± 0.6]; tail length: males = 37.5 cm ± 2.1 [~14.8 in ± 0.8], females = 36.7 cm ± 1.7 [~14.4 in ± 0.7]). Three adult males captured on the Ladder Ranch (Sierra County) weighed 1,091.7 g (± 118.1) (~2.4 lbs ± 0.3) and four adult females weighed 957.5 g (± 69.9) (~2.1 lbs ± 0.2) (G. W. Roemer, unpubl. data).

Ringtails are known for their athleticism and grace, and we have seen them climb trees and vertical walls with ease. They can "ricochet" off canyon walls or small cliff outcrops, or "chimney stem" narrow crevices (Trapp 1972). In one instance we watched a released animal bounce from one foothold to another as it climbed a vertical cliff face until it rested on a small outcrop large enough only for its four feet to straddle. There it stopped and peered down at us from its safe refuge. Its climb happened so rapidly and we were so astounded that neither of us attempted a picture before it scaled the remainder of the cliff and disappeared. No doubt the long tail assists in maintaining balance (Taylor 1954), while semi-retractile claws and pad friction aid in grip and further facilitate the species' climbing ability (Trapp 1972). Ringtails are also capable of rotating their hind limbs externally by 180°, which helps them climb head first down trees similar to tree squirrels (Trapp 1972). This anatomical adaptation likely results from an extension of the ankle at the tibiotarsal joint coupled with inversions at the subastragalar and transverse tarsal joints (Liu et al. 2016). Rotations at the femoral-hip and intratarsal joints may also contribute to hind foot reversal (Trapp 1972), but this has not been verified (Liu et al. 2016).

Taxonomy

The family Procyonidae is composed of six extant genera (Wozencraft 2005). A cladistic analysis of morphological characters within that family seemingly identified two subfamilies or tribes,

Photos 30.3a and b (*left to right*). Variation in the dorsal coloration of ringtails. (a) Two specimens from Socorro County, New Mexico: MSB 87714 (foothills of the San Mateo Mountains at an elevation of ~1,680 m [~5,500 ft]; left) and MSB 282000 (Magdalena Mountains, elevation 3230 m [~10,600 ft]; right); the high elevation animal is somewhat grayer with less tan coloration; compared with (b): MSB 83892 (Mexico; Sonora, Isla Tiburon; left) with pale dorsal pelage and MSB 160391 (Texas; Llano County, within ~25 mi radius of Llano; right) with melanistic dorsal coloration. Photographs: © Jon Dunnum.

including one with *Bassariscus*, *Procyon*, and *Nasua* (Baskin, 1982, 1989, 1998, 2004; Decker and Wozencraft 1991). However, a more recent reassessment based on nine nuclear and two mitochondrial gene sequences revealed that the genus *Bassariscus* forms a monophyletic clade with *Procyon*, whereas *Nasua* belongs to another lineage with *Bassaricyon* (i.e., the olingos), with *Potos flavus*, the kinkajou, representing a sister taxon to both the *Bassariscus*/*Procyon* and *Nasua*/*Bassaricyon* clades (Koepfli et al. 2007). Results from the earlier analyses were interpreted as reflecting convergence among morphological traits, the consequence of similarities in diet and life habits among species occupying similar ecological niches (e.g., both raccoons and coatis primarily forage on the ground and use their forepaws to forage, whereas the olingos, kinkajou, and ringtails are more slender and lithe, and can be highly arboreal). Molecular-based and morphological phylogenies within the Procyonidae continue to be incongruent, with Ahrens (2012) finding evidence for homology, rather than adaptive convergence, in a set of 78 binary and multistate craniodental traits analyzed with high-resolution X-ray computed tomography.

Fossils of earlier procyonids discovered in western Europe date back to the late Eocene or late Oligocene (genus *Pseudobassaris*) (Koepfli et al. 2007). Fossil evidence for the presence of procyonids in North America is not found until the early and middle Miocene. Fossils of *Bassariscus* appear in the middle Miocene with this origination further supported by molecular data. *Bassariscus* is estimated to have diverged from *Procyon* during the end of the middle Miocene (11.4–12.8 million years ago [mya]), with the two sister species within the genus *Bassariscus* splitting in the late Miocene (9.1–10.3 mya) (Koepfli et al. 2007). This period of diversification within the family Procyonidae also coincides with the repeated formation of the Panamanian land bridge and the Great American Interchange of North American and South American faunas.

There are 14 recognized subspecies of *B. astutus*, of which two are present in New Mexico (Hall and Kelson 1959). Most of New Mexico is home to *B. a. flavus*, with *B. a. arizonensis* found in the southwestern and northwestern corners of the state. Based on morphometrics, Kortlucke (1984) suggested these 14 subspecies could be reduced to 3 subspecies, with all ringtails in New Mexico grouping with those in Texas and on the northern Mexican Plateau. There has been no comprehensive genetic re-evaluation of these subspecific designations.

DISTRIBUTION

Ringtails are distributed across most of Mexico, from the southern states of Guerrero, Oaxaca, and Veracruz north along the west coast of the United States to Oregon, eastward through much of the Southwest as far as Louisiana and Arkansas (Poglayen-Neuwall and Toweill 1988; Gehrt 2003). In northern Mexico, ringtail populations have been documented on three Gulf of California islands (i.e., Espíritu Santo, San José, and Tiburón) and on the Baja California Peninsula, while farther south they overlap with the northernmost distribution of the closely related cacomistle (Poglayen-Neuwall and Toweill 1988; Escobar-Flores et al. 2012). Within the United States, ringtails are widespread in the southwestern states of Arizona, New Mexico, and Texas (Gehrt 2003). They also occur in limited portions of California, Colorado, Kansas, Nevada, Oklahoma, Oregon, Utah, and Wyoming (Long and House 1961; Poglayen-Neuwall and Toweill 1988; Gehrt 2003). The northernmost extent of the ringtail distribution has typically been defined by populations in southwestern Oregon. Outlier records of the species from eastern Oregon (Bailey 1936) and southwestern Idaho (Larrison 1967) have been dismissed as released or escaped captive ringtails (see Poglayen-Neuwall and Toweill 1988). However, ringtails have been recently confirmed in south-central Idaho (Liebenthal 2014), representing a possible northern range expansion. As is true of most North American mesopredators, the species is believed to have expanded its historical

Photos 30.4 a (*above*) and b (*right*). (a) Three of four ringtails in a closed heading inside an underground mine in the Little Hatchet Mountains on 26 November 2010. (b) In the center of the photo is the headframe above the underground mine shaft where the ringtails were found. Photographs: © Robert Shantz.

distribution—though by only an estimated 2%–following the anthropogenic decline of top predator populations (Prugh et al. 2009).

Within New Mexico, museum records exist for 20 of New Mexico's 33 counties (Map 30.1; Appendix 30.1). Most specimen records are from rugged topography in the southern half of the state, including the southern and eastern edges of the Mogollon Plateau and outlying ranges, the sky island mountains and lava fields of the Bootheel and Deming Plain, and the San Andres-Organ and Capitan-Sacramento-Guadalupe mountain chains (and see Photos 30.4 and 30.5 for photographic records from the Little Hatchet and Oscura mountains). In central and northern New Mexico, specimens have similarly been collected in mountains and canyon country, particularly the Jemez, Cebolleta, and Zuni Mountains, the Sandia-Manzano-Los Pinos chain, and foothills and lower canyons surrounding the Sangre de Cristo Mountains (Map 30.1; Appendix 30.1). Ringtail specimen records exist from along some of the main rivers of the state including in alluvial valleys. Only one ringtail specimen was collected on the eastern plains of New Mexico east of the Pecos River, and none north of the Canadian River. Ringtails reach at least ~3,230 m (~10,600 ft)

Photo 30.5. Ringtail at Yates Spring in the Oscura Mountains, Socorro County, on 16 November 2012. Photograph: © WSMR Garrison Environmental Division and ECO Inc., Ecological Consultants.

Photo 30.6. Ringtail in Tijeras Canyon, Bernalillo County. Photograph: © New Mexico Department of Transportation in collaboration with the Arizona Game and Fish Department.

Photo 30.7. Ringtail photographed on 7 November 2015 by a remote camera in a canyon bottom along the Mimbres River in Grant County. Photograph: © Keith Geluso.

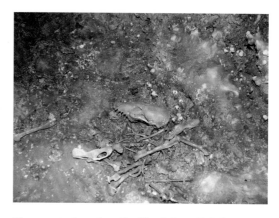

Photo 30.8. Complete, fossilized ringtail skeleton discovered in 2008 near the terminus of a cave at El Malpais National Monument in Cibola County. Photograph: © J. Judson Wynne.

in elevation in New Mexico, based on a specimen collected in the Magdalena Mountains (MSB:Mamm:282000; see Appendix 30.1).

Among some of the more notable ringtail records from the state are those from El Malpais National Monument (Bogan et al. 2007; Wynne 2013). Hooper (1941) first documented reports of ringtail occurrence from the lava fields and surrounding areas in Cibola County. Bogan et al. (2007) confirmed the presence of the species at El Malpais National Monument, and Wynne

(2013) discovered a complete ringtail skeleton, apparently >1,000 years old, near the terminus of one of the local caves (Photo 30.8). Wynne (2013:50) suggested that this "animal may have entered the cave to hunt bats, became disoriented, and, unable to find its way back to the entrance, died in the cave." Scats putatively belonging to raccoons, skunks, or ringtails were found during the ecological inventory of the caves but have not been identified to species (Wynne 2013).

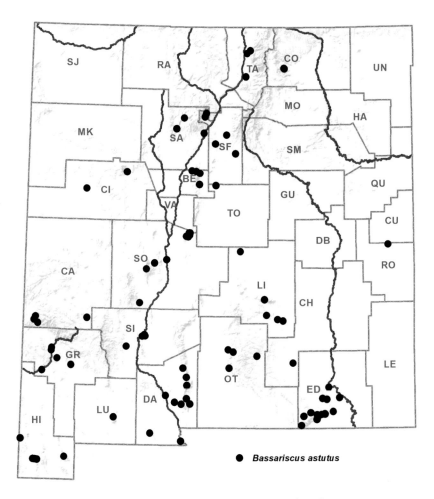

Map 30.1. Distribution of the ringtail (*Bassariscus astutus*) in New Mexico based on museum specimen records.

HABITAT ASSOCIATIONS

Generally associated with areas with vertical relief such as canyons, escarpments, rocky outcrops, or talus slopes, and often absent when such features are not present, ringtails nevertheless occupy a wide range of land-cover types from desert scrublands to pinyon-juniper woodlands and oak forests to temperate rainforests, and are typically viewed as habitat generalists (Leopold 1959; Poglayen-Neuwall and Toweill 1988; Harrison 2012). In Sonora, Mexico, Armenta-Méndez et al. (2020) documented the presence of ringtails in open desert scrub as well as the lush tropical vegetation of both a canyon

bottom and a desert oasis. On the Edwards Plateau in Texas, Taylor (1954) reported that ringtails used every vegetation type available. We have captured ringtails in oak woodlands, riparian areas, and mixed conifer forests in the Chiricahua Mountains of Arizona, and in 11 different plant associations (excluding wash bottoms) in the Guadalupe Mountains (Table 30.1). In Las Cruces in Doña Ana County, we have also personally observed ringtails on the New Mexico State University campus and recovered a carcass within a residential neighborhood that borders creosote bush shrubland (G. W. Roemer, pers. obs.). Ringtails thus appear to be able to utilize a

Photo 30.9. Ringtail among boulders at Aguirre Springs in the Organ Mountains, Doña Ana County, in March 2006. Canyons, rocky out-crops, hills, and ridges, as well as talus slopes all represent typical ringtail habitat. Photograph: © Jeff Kaake.

Photos 30.10a and b (*left to right*). Examples of ringtail habitat in Carlsbad Caverns National Park in Eddy County. Located in the northern part of the Chihuahuan Desert, Carlsbad Caverns National Park is a land of deep, rocky canyons with many cliff ledges, crevices, and caves (a). The dominant vegetation community is desert scrub, as shown in the view of Rattlesnake Canyon (b). Photographs: © Robert C. Lonsinger.

Photos 30.11a and b (*left to right*). Ringtail habitat in Guadalupe Mountains National Park, which straddles the New Mexico–Texas state line. Photographs: © Robert C. Lonsinger.

Table 30.1. Plant associations where 153 individual ringtails (*Bassariscus astutus*) were captured in the Guadalupe Mountains of New Mexico and Texas. Plant associations are based on Muldavin et al. (2003).

Plant Associations	Total Number of Captures
Arroyo Riparian Woodland and Shrubland	7
Chihuahuan Desert Scrub	1
Chihuahuan Semi-desert Grassland and Steppe	2
Desert Shrubland	8
Grassland	17
Madrean Warm Lowland Evergreen Woodland	21
Madrean Warm Montane Forest and Woodland	43
Montane Shrubland	22
North American Warm-desert Xero-Riparian	11
Warm Interior Chaparral	14
Warm Mediterranean and Desert Riparian, Flooded and Swamp Forest	1
Wash	6
Total captures	**153**

wide variety of land-cover types including those disturbed by humans.

Preferences for specific vegetation types nonetheless have been documented in ringtails but may vary geographically. The distribution of telemetric fixes (location estimates) for radio-collared ringtails revealed a non-random pattern of association with vegetation type in southwestern Utah (Trapp 1978). Ringtail locations were distributed equally with respect to expectations based on the proportional availability of riparian and pinyon-juniper woodland. However, ringtails appeared to use brushy meadows less than expected and both blackbrush (*Coleogyne ramosissima*) and human-disturbed habitats more than expected. In northern California, telemetered ringtails tended to use riparian and/or cottonwood forests more than expected, and oak woodlands and river shrub and grassland habitats less than expected (Lacy 1983).

Although researchers initially believed ringtails required access to free-standing water (e.g.,

Photo 30.12. Ringtail at a watering tank in the San Andres Mountains, Doña Ana County. Although ringtails may be drawn to water in some areas, they are adapted to xeric environments. Their kidneys are adapted for water conservation, and ringtails can persist in the absence of free-standing water in some environments. Physiologically, they can meet all of their water requirements through both preformed water and metabolic water derived from their food. Photograph: © San Andres National Wildlife Refuge.

Photo 30.13. Rob Lonsinger processing a captured and anesthetized ringtail in the Guadalupe Ranger District of the Lincoln National Forest, Eddy County in April 2009. Processing of an anesthetized ringtail involved tagging (with either an ear tag or a passive integrated transponder [PIT]) for identification, the collection of an ear snip for genetic analyses, blood collection, and standard physical measures (i.e., mass, standard body measurements). Photograph: © Robert C. Lonsinger and James Doyle.

Cahalane 1954; Taylor 1954), diets composed of high protein prey and/or water-rich vegetation (e.g., berries, cacti) can provide sufficient water resources for ringtails, who, additionally, have kidneys adapted for conserving water (Richards 1976; Chevalier 1984). Ringtails have been photographed at springs and drinkers in New Mexico (Photo 30.12), but the association between ringtails and riparian areas reported in some (though not all) studies may relate more to the availability of denning sites, a potentially essential element

of any habitat for the species (see below). Ringtails typically use rock and tree dens (see below), the latter being more readily available in riparian woodlands compared to many lowland cover types in New Mexico.

LIFE HISTORY

Although common across much of its range, the ringtail is rarely observed in the wild, and information about its life history remains highly incomplete. In the Guadalupe Mountains, ringtails were widespread and presumably abundant, with both relatively high rates of capture (mean=15%, SD ± 10%) and their scats being found frequently on rocky outcroppings (R. C. Lonsinger and G. W. Roemer, unpubl. data). Still, over nearly three years of ecological investigation, we only ever observed one ringtail that we had not first captured, and we captured a total of 153 individual animals (Lonsinger et al. 2015). The elusiveness of ringtails to the casual observer is likely attributable to the species' primarily nocturnal habits, small size and ability to evade detection, scaling rock walls and trees, and slinking through even minimal amounts of vegetation unexposed.

Nightly activities begin within an hour after sunset and conclude before sunrise (Callas 1987; Harrison 2012). Although diurnal activity is rare (Trapp 1978; Callas 1987), we captured a ringtail in southern New Mexico during midday and observed another in a canyon of Guadalupe Mountains National Park in Texas. Daytime use of dens provides relief from temperature extremes, and ringtails may even frequently shift denning substrates during summer, a form of behavioral thermal avoidance (Callas 1987; Tiedt 2011; Harrison 2012). As mentioned, ringtails do not need open water to survive. Nearly 90% of ringtail home ranges in the Sevilleta National Wildlife Refuge lacked free-standing water (Harrison 2012). Furthermore, Chevalier (1991) found that ringtails in low-elevation desert were larger bodied and had reduced minimum resting metabolic rates compared to their montane counterparts, both of which can limit heat gain. Field metabolic rates of desert ringtails were lower than predicted for desert eutherians and may contribute to avoidance of overheating and to water conservation (Chevalier 1991).

Denning

Ringtails do not dig or construct their own dens (Toweill 1976; Trapp 1978), but rather, are opportunistic and exploit any cavity, crevice, or burrow that can accommodate their small bodies. Rangewide they have been commonly found denning in rock crevices, tree cavities, brush piles, artificial nest boxes, burrows of other animals, and human structures (Poglayen-Neuwall and Toweill 1988). Ringtails at the Sevilleta National Wildlife Refuge primarily denned in rocks, secondarily in trees and shrubs, but also selected holes in the ground (Harrison 2012). In western Texas, ringtails in Elephant Mountain Wildlife Management Area used only rock dens (Ackerson and Harveson 2006), while those on the Edwards Plateau favored rock dens, hollow trees, and brush piles (Toweill 1976). Dilapidated rock fences built by early settlers to demarcate pastures have also provided denning opportunities for ringtails in western Texas (Taylor 1954). Ringtails in central Texas commonly favored nest boxes hung for owls over natural dens (Tiedt 2011). In northwestern California, ringtails in riparian areas commonly denned in tree cavities (Callas 1987).

Despite appearing to casually and opportunistically select denning sites, ringtails may use some denning substrates more frequently than expected and, again, may shift use of denning substrates seasonally to secure optimal microclimatic conditions. For example, ringtails used dens located in trees disproportionally more than other substrates in northwestern California (Callas 1987), and nest boxes more often than natural dens in central Texas (Tiedt 2011). Ringtails in New Mexico utilized rock dens more often than other substrates (Harrison 2012), as did those on the Edwards Plateau; use of denning substrates was not equitable between sexes on the Edwards Plateau, with females and males more often selecting rock and tree dens, respectively (Toweill 1976). In several studies, rock dens were generally used less during the coldest months (Callas 1987; Tiedt 2011; Harrison 2012). In contrast, ringtails occupied rock dens more frequently than expected during warmer months, this likely aiding in reducing thermal loads in summer (Callas 1987; Tiedt 2011). Collectively, these findings suggest that ringtails are adaptable and use a variety of denning substrates, but that variation in the selection of substrates likely reflects a propensity to select optimal microclimates.

Unlike species that invest considerable effort and time into excavating or modifying dens, ringtails practice "informal" denning (Taylor 1954), with dens usually having little or no nesting material and being used for only short durations. Although dens may be reused, ringtails rarely occupy the same one for more than one to two consecutive days, and only natal dens tend to be occupied for extended (>10 days) periods (Toweill 1976; Callas 1987; Ackerson and Harveson 2006; Teidt 2011; Harrison 2012). The number of dens reportedly used by an individual ringtail is likely influenced not only by their availability and spatial arrangements of denning sites and food resources, but also the length of monitoring; the record number of dens used by a single individual during a study has been reported as high as 46 (mean = 27.1, SD ± 14.5; Toweill 1976). Where home ranges overlap, dens may be used by multiple ringtails either concurrently or at different times (Toweill 1976; Callas 1987; Tiedt 2011).

Diet and Foraging

Ringtails are omnivorous, apparently consuming animal and plant foods based largely on their availability (Poglayen-Neuwall and Toweill 1988). On Isla San José, in the Gulf of California

Photo 30.14. Ringtail on snow near the Red River fish hatchery in Taos County on 13 December 2020. Ringtails, who remain active in winter, are opportunistic foragers. The animal in the photograph was captured after finding its way to the hatchery's upper outside raceways through the chain-link fence and apparently consuming rainbow trout (*Oncorhynchus mykiss*). The photo was taken after release, but the ringtail would have also traveled through snow to reach the hatchery. Photograph: © Robin Bonner.

in Mexico, at least 36 animal taxa and 12 plant species were represented in 104 scats collected during the spring (Rodríguez-Estrella et al. 2000). Invertebrates occurred in 91.4% of all scats, with mammals (58.7%), plants (56.7%), reptiles (50%), and birds (4.8%) also detected (Rodríguez-Estrella et al. 2000). Most of the animal species identified in scats were small, with the exception of mule deer (*Odocoileus hemonius*) and wild goat (*Capra hircus*), these last two evidently scavenged. Invertebrate prey most often found consisted of insects (mainly Orthoptera and Coleoptera) but also scorpions (Scorpionoidae) (Rodríguez-Estrella et al. 2000). In the Trans-Pecos region in Texas, the diet of ringtails was similarly diverse and included the seeds of many plants, in addition to other vegetation matter; animal food again included small mammals, reptiles, and arthropods (insects, millipedes, and scorpions), with only small seasonal differences in terms of percent volume of these prey taxa (Ackerson and Haverson 2006). On the Sevilleta National Wildlife Refuge in central New Mexico, Harrison (2012) studied the diet of ringtails over a ~2.5-year period and reported that 82.5% of all 314 scats contained

plant remains, 48.1% invertebrates, 40.4% mammals, 26.1% birds, and 2.2% reptiles. At least 49 animal taxa and 9 species of plants were identified, with fruits from juniper (*Juniperus* spp.) being especially prevalent. The observed diet of ringtails in New Mexico, the Gulf of California, and Trans-Pecos Texas thus showed similarities, while contrasting with Taylor (1954), who reported a preponderance of jackrabbits and cottontails among ringtail mammalian prey. Ringtails in New Mexico, the Gulf of California, and Trans-Pecos Texas instead focused on smaller prey such as heteromyid rodents (e.g., kangaroo rats [*Dipodomys* spp.], pocket mice [*Chaetodipus* spp.]), *Peromyscus* spp., and woodrats (*Neotoma* spp.). The number of arthropod taxa in the diet was also very diverse. Ringtails often consumed venomous arachnids, but beetles (Coleoptera) and grasshoppers and their allies (Orthoptera) were the most prevalent arthropods. All the studies mentioned above, recent or not, have revealed that the ringtail is an opportunistic omnivore (see Poglayen-Neuwall and Toweill 1988).

Modes of food acquisition have not been well studied for ringtails, but the prevalence of plant and animal materials in diet studies suggests that searching and hunting both likely play a role. What we do know of ringtail hunting behavior comes from captive settings, where ringtails have been observed rushing and pouncing on prey, rather than stalking prey (Poglayen-Neuwall 1987).

Social Organization

Estimates of ringtail home range size are limited and the methods used vary substantially, making comparisons problematic. On the Sevilleta National Wildlife Refuge, 95% minimum convex polygons (MCPs) for six males averaged 528 ha (1,305 acres), whereas those for seven females averaged only 115 ha (284 acres) or 22% the size of male home ranges (Harrison 2012). Estimates for 95% kernel home ranges were smaller, 462 ha (1,142 acres) vs. 94 ha (232 acres), respectively, but followed the same trend (female ranges were only 20% of those of males). In Zion Canyon in southwestern Utah, home ranges were both much smaller and more similar in size between the sexes, averaging 139 ha (± 53) (343 acres [± 131]) for males (n = 9) and 129 ha (± 62) (319 acres [± 153]) for females (n = 4) (Trapp 1978). In western Texas, 100% MCPs were similar between sexes, but even smaller, at 43 ha (106 acres) for females (n = 4) and 47 ha (116 acres) for a single male (Ackerson and Harveson 2006). The smallest home ranges (n = 4, mean = 8.8 ha [22 acres]; range=5.0–13.8 ha [12–34 acres]) were reported in a riparian forest of northwestern California (Lacy 1983). Sex-biased differences in home range size among studies could be the result of topography, quality of habitat (i.e., availability of food and denning resources), sampling intensity, or method of home range estimation used. They also may simply be a consequence of the ringtail's mating system (see below), whereby in some areas the larger home ranges of males encompass several female home ranges. Lacy (1983) also suggested that a three-dimensional measure of home range size may be more appropriate for ringtails, which have scansorial (climbing) and arboreal

tendencies, and that viewing home ranges in terms of volumes (rather than areas) may reduce the disparity across studies.

Adult ringtails are solitary. Polygynous or polyandrous social structures occur within the Procyonidae with mating systems often characterized as promiscuous (Gompper et al. 1997; Kays and Gittleman 2001; see Chapters 31 and 32). Home ranges of male ringtails often overlapped with those of two or more females (Callas 1987), suggesting a polygynous mating system (Sandell 1989). Intrasexual home range overlap is uncommon (Lacy 1983; Callas 1987), but when it occurs, it is suspected that use of overlapping areas is temporally staggered (Toweill 1976; Ackerson and Harveson 2006).

Maintenance of territories is likely through olfactory cues. Scent marking with urine, scat, and anal gland secretions is common among territorial carnivores. For ringtails, urine may be a less effective olfactory cue than scats due to rapid evaporation of urine compared to the more gradual decay of scats (Barja and List 2006). Latrine usage in ringtails is well documented (Callas 1987; Barja and List 2006; Harrison 2012), with ringtails preferentially and repeatedly defecating on conspicuous and elevated substrates, increasing the visual and olfactory effectiveness of latrine sites. Latrine usage appears common (Barja and List 2006; Harrison 2012), and in addition to communicating spatial domains, latrines may also advertise the reproductive condition of females (Callas 1987).

Mating/Breeding

Ringtails breed in early February to late May when females are receptive for approximately 24 hours. Females have a short gestation of 51–53 days and typically give birth to one to four young from April to July (Poglayen-Neuwall 1987). Males investigate female receptiveness but are not known to demonstrate any courtship behaviors (Poglayen-Neuwall 1987). Newborns are small, helpless, and lightly

Photo 30.15. Ringtail being released following capture and processing in Guadalupe Mountains National Park in February 2009. Following trapping sessions, the authors collected fine-scale land-cover data at each trap location using a point-line intercept method, which was later used to investigate the relationship between land-cover types/plant associations and population genetic structure. Photograph: © Robert C. Lonsinger and James Doyle.

furred with pelage patterns similar to adults but less pronounced (Richardson 1942). Average masses of neonates have been reported between 22.1 and 37.5 g (between 0.8 and 1.3 oz.), or approximately 2–3% the weight of their mothers (Toweill and Toweill 1978; Poglayen-Neuwall 1987). Newborn ringtails consume only milk for about 30 days, their ear canals and eyes open at approximately 22–38 days, and their milk teeth emerge around 26–40 days post-partum (Richardson 1942; Toweill and Toweill 1978; Poglayen-Neuwall 1987). From approximately 30–90 days, the mother delivers prey to the den and young ringtails develop rapidly, increasing their activity by day 60, refining their climbing skills by day 80, and being weaned by day 90 (Toweill and Toweill 1978; Poglayen-Neuwall 1987). Food remains found around occupied maternal dens in western Texas included bobwhite (*Colinus virginianus*), cardinal (*Cardinalis*

spp.), mourning dove (*Zenaida macroura*), fox squirrel (*Sciurus niger*), cottontail (*Sylvilagus* spp.), and white-footed mouse (*Peromyscus leucopus*; Taylor 1954). From approximately 90 to 135 days, young ringtails develop their adult pelage, acquire their permanent teeth, and join their mother on short foraging excursions (Richardson 1942; Toweill and Toweill 1978; Poglayen-Neuwall 1987); they reach ~98% of their adult weight by 160 days of age. Juveniles disperse in early fall to late winter, and males are able to breed in their first year, though this is uncommon for females (Poglayen-Neuwall 1987).

Communication

As mentioned, communication among conspecifics related to territorial maintenance and spacing is accomplished primarily through olfactory cues and fecal scent stations instead of vocalizations (Barja and List 2006). Adult ringtails are nonetheless known to bark, and growl, and these sounds probably are used for defense or to convey hostility or stress. Both the chitter—perhaps the most frequent vocalization of ringtails—and the whistle-grunt appear to be social contact calls (Toweill and Toweill 1978; Willey and Richards 1981). Although males do not appear to court females, females may indicate their receptiveness with nuptial cries (Willey and Richards 1981). Ringtails may also emit contact calls in the form of chirps, which solicit return chirps from nearby conspecifics (Toweill and Toweill 1978). When not in contact with their mother or siblings, newborn ringtails may utter a whimper, which can escalate to a high intensity "shrill squeak" indicating distress (Toweill and Toweill 1978).

Population Genetic Structure

Genetics can provide insights into population health, viability, and structure. Schweizer et al. (2009) identified 15 microsatellite markers from blood and tissue samples in ringtails captured in the Chiricahua (Arizona) and Guadalupe (Texas) mountains. These markers were later employed to assess the population genetic structure of ringtails in the Guadalupe Mountains of southern New Mexico and western Texas (Lonsinger et al. 2015). Based on 75 ringtails from New Mexico, and 78 from Texas, ringtails displayed high levels of genetic diversity across loci (i.e., allelic richness range = 2.98–6.76, number of alleles range = 7–34, observed heterozygosity range = 0.336–0.915; Lonsinger et al. 2015).

Although sample sizes from each mountain range used in the development of the microsatellite markers were small (n = 10 and 11 for the Chiricahua and Guadalupe mountains, respectively), the proportion of the genetic variance in the two subpopulations relative to the total genetic variance was moderately high (F_{ST} = 0.037, SE = 0.005; G. W. Roemer, unpubl. data; Peakall and Smouse 2012), whereas G'_{ST} (Meirmans and Hedrick 2011), a correction of F_{ST} when sample sizes are small, was higher (G'_{ST} = 0.12, SE = 0.065). Given the small body size and presumably limited dispersal of ringtails, we expected the genetic difference between the Chiricahua and Guadalupe mountain populations, separated by a straight-line distance of over 400 km (250 mi), to be higher. However, it is possible that the high mutation rate of microsatellite loci coupled with high dispersal rates, and a prehistoric distribution that may have been contiguous across the southern United States, could have constrained broad-scale genetic structuring (e.g., Kierepka and Latch 2016). Using the larger sample of ringtails captured in the Guadalupe Mountains (n = 153), we detected six unique genetic clusters with pairwise F_{ST} (range = 0.01–0.07) and G'_{ST} (range = 0.04–0.25) similar to that detected between mountain ranges (Lonsinger et al. 2015). These values indicated weak to moderate levels of genetic differentiation over a much smaller geographic region. This level of genetic structuring occurred despite no evidence of barriers to dispersal and no support for either isolation by distance (Wright 1943) or isolation by resistance

(McRae 2006; Lonsinger et al. 2015). Instead, environmental features (e.g., elevation, slope, land-cover characteristics) apparently drove ringtail population genetic structure, a process termed *isolation by environment* (Wang and Bradburd 2014). These patterns may result from natal-habitat biased dispersal (i.e., individuals preferentially dispersing to, and settling in, areas similar to where they were born), or reduced fitness of immigrants that dispersed to habitats different from their natal habitats, or both. Considering the level of genetic differentiation observed across the Guadalupe Mountains, Lonsinger et al. (2015) suggested that managers should consider the importance that habitat heterogeneity may play in maintaining genetic diversity within populations. Further, given the similarity in genetic distance observed between populations inhabiting the Chiricahua and Guadalupe mountains, a comprehensive phylogeographic study of the species should be undertaken to explore

Photo 30.16. A ringtail foraging in a golden eagle (*Aquila chrysaetos*) nest in Utah on 12 January 2019. The nest was used that year by a golden eagle pair, with the female laying the first of two eggs around 25 February and one young later reaching fledging. At least one golden eagle began to use the nest sporadically from 27 December 2018 on, then more regularly starting in the latter part of January 2019. Ringtails and nesting golden eagles likely co-occur in many of the same areas dominated by canyons and cliffs. In New Mexico, ringtails have been documented as prey of golden eagles. Photograph: © US Army Dugway Proving Ground, Dugway, Utah.

rangewide patterns in genetic structure and the potential underlying mechanisms.

Interspecific Interactions

The small size of ringtails makes them susceptible to predation by many larger mammalian and avian predators, in particular great horned owls (*Bubo virginianus*), which are also active at night (Ackerson and Harveson 2006). It is also quite possible for diurnal raptors, such as golden eagles (*Aquila chrysaetos*), to opportunistically prey on ringtails as these eagles are known to kill other nocturnal carnivores such as raccoons, foxes, weasels, badgers, and cats (Watson 1997). Mollhagen et al. (1972) regularly found ringtail remains in occupied golden eagle nests in western New Mexico, as well as in the Capitan and Guadalupe mountains in the southern part of the state. In New Mexico, golden eagles nest most frequently on cliffs (Stahlecker et al. 2010), thus occupying some of the same habitat as ringtails.

Ringtails are sympatric with a suite of mesocarnivores throughout their geographic range. In the United States, ringtails commonly co-occur with foxes (e.g., gray foxes, *Urocyon cinereoargenteus*), skunks (e.g., striped skunks, *Mephitis mephitis*; spotted skunks, *Spilogale* spp.) and raccoons, among others. Although ringtails may compete with sympatric mesocarnivores (Poglayen-Neuwall and Toweill 1988), resource partitioning may facilitate coexistence. For example, despite an overlap in the food items consumed by ringtails and gray foxes based on an analysis of scats, the relative proportion of foods consumed by the two species differed, as did temporal activity, foraging habit, and patterns of habitat use, leading to the conclusion that the two were not strong exploitative competitors (Trapp 1978). Ringtails and other sympatric mesocarnivores generally demonstrate mutual avoidance, but may also be tolerant of one another. Toweill (1976) even detected ringtails sharing dens with white-backed hog-nosed skunks (*Conepatus leuconotus*)

and a western diamond back rattlesnake (*Crotalus atrox*), as well as with armadillos (*Dasypus novemcinctus*) and wasps. Where ringtails are sympatric with the larger *B. sumichrasti*, co-occurrence may be supported by the latter species exploiting more arboreal portions of the tropical forest, whereas the ringtail shows an affinity toward areas with cliffs or rocky outcroppings (Leopold 1959); these spatial, morphological, and/or behavioral differences may act as reproductive isolating mechanisms between the two species, but this has not been investigated.

In captivity, ringtails may live up to 16 years of age (Poglayen-Neuwall 1987), but in the wild, they are susceptible to predation and interspecific killing. Harrison (2012) found that avian predation was the primary cause of death for ringtails in central New Mexico, and that few ringtails lived more than four years (annual survival rate = 0.375). Survival was even lower in western Texas (annual survival rate = 0.191), where predation was primarily by great horned owls (Ackerson and Harveson 2006). Ringtails arch their tail over their back when traversing open areas, and this may be an anti-predator behavior, with the tail serving as a decoy to avian predators (Poglayen-Neuwall and Toweill 1988). In addition to avian predation, intraguild predation and interspecific killing of ringtails has been documented by coyotes (*Canis latrans*), bobcats (*Lynx rufus*), and raccoons (Poglayen-Newall and Toweill 1988; Ackerson and Harveson 2006); one juvenile ringtail was killed by a snake (Taylor 1954), and young ringtails are likely susceptible to an even greater number of predators.

Diseases and Pathogens

The impact of parasites and diseases on ringtail populations is poorly understood but may contribute to regulating populations (Poglayen-Neuwall and Toweill 1988). Ringtails are susceptible to feline and canine distemper virus, as well as rabies (Poglayen-Neuwall 1987; Kuzmin et al.

Photo 30.17. Ringtail visiting a hummingbird feeder at a residence in Los Alamos Canyon on 20 May 2008. Photograph: © Hari Viswanathan.

Photo 30.18. Captured nuisance ringtail in December 2017 at the White Sands Missile Range Main Post, Doña Ana County. Photograph: © WSMR Garrison Environmental Division and ECO Inc., Ecological Consultants.

2012). They serve as hosts to parasitic fleas, mites, ticks, and lice, as well as cestodes and nematodes. One louse species, *Neotrichodectes thoracicus*, and one species of tapeworm, *Taenia pencei*, are host-specific and only known to occur on or in ringtails (Custer 1979; Rausch 2003). The role that ringtails play in disease or parasite transmission is unknown, but ectoparasites of ringtails are believed to result, at least in part, from interactions between ringtails and their mammalian prey (Custer 1979).

STATUS AND MANAGEMENT

The status of ringtails across their range is unclear, but populations are believed to be relatively stable. The International Union for Conservation of Nature (IUCN) classified ringtails as a Species of Least Concern (Reid et al. 2016). In the deserts of the southwestern United States, ringtails are presumably secure and are managed as a furbearer in New Mexico, Arizona, Texas, Colorado, and Utah; Nevada manages ringtails as an

unprotected mammal. Ringtails are regarded as a Species of Special Concern in California (CDFW 2015), a Sensitive Species in Oregon (ODFW 2016), and a Species of Greatest Conservation Need in Oklahoma (ODWC 2016). Ringtails are not classified (i.e., neither as a game or protected species) in Idaho, Wyoming, Kansas, and Arkansas. Louisiana recognizes ringtails as a species of historical occurrence but with no records within the last 30 years (Lester et al. 2005).

Estimates of abundance using robust estimation methods—such as closed-population models or spatially explicit capture-recapture models, as well as measures of fitness (i.e., survival and fecundity) that could be used to model population dynamics—are sorely lacking for ringtails. Density estimates vary substantially across the species' range. The highest estimates of ringtail density, 7 to 20 ringtails/km², were reported in riparian zones in California (Lacy 1983). In arid environments, estimated densities were lower, ranging from 1.5 to 2.9 ringtails/km² in Utah (Trapp 1978)

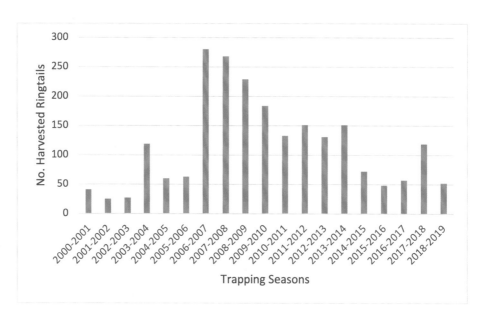

Figure 30.1. Number of ringtails trapped in New Mexico from 2000 to 2019.

and 3.4 to 6.3 ringtails/km² in Texas (Ackerson and Harveson 2006). In New Mexico, the density of ringtails has been estimated at only 0.17 to 0.33 ringtails/km², the lowest density reported rangewide (Harrison 2013). Harrison (2012) suggested that managers in New Mexico should consider the potential impacts of trapping on ringtail populations occurring at low densities.

Ringtails have relatively little economic importance, not being particularly known for damaging crops and attacking poultry or livestock nor being highly prized in the pet or fur trade. Ringtails used to be known for raiding chicken coops in Mexico (Leopold 1959). They have also been occasionally kept as pets, and there is at least one report of ringtails being captured for sale in the eastern US pet trade (Cahalane 1954; ownership of pet ringtails is not legal in New Mexico). Despite their striking appearance, attractive fur and presumed commonality across the southwestern United States and Mexico, ringtails historically have held little economic value as a furbearer and are more often taken incidentally by trappers

targeting more profitable quarry (Cahalane 1954; Leopold 1959). Cahalane (1954:169) reported:

Only in "boom times" on the fur market, is the ringtail trapped intentionally on a large scale. Frequently, its curiosity gets it into traps set for more valuable furbearers. In 1927, good ringtail pelts brought up to three dollars, but ordinarily the price is about fifty cents. The fur, which on a living animal is fluffy and beautiful, mats down and loses its "life" and color and is used as trimming only on the cheapest of cloth coats.

Taylor (1954:57) added,

The ring-tailed cat is less well-known than the more broadly distributed fur animals such as the mink, fox, raccoon, and skunk. An amusing happening based on this was brought to our attention during the last year of World War II (1945). Those were the days of the Office of Price Administration, which, quite logically, had imposed a ceiling price on designated fur animals, including the raccoon, mink, and others, excepting the ring-tailed cat, of whose existence the OPA had apparently never heard. Thus it transpired that while relatively low prices were compulsory for the better-known and normally more expensive furs, the trappers were selling prime ring-tailed cat skins to buyers for $10.00 each, or, indeed, for anything they could get.

Very few ringtails have been taken commercially in New Mexico compared to other furbearers. From 2000 to 2019, an average of 116 ringtails were legally harvested in New Mexico (Figure 30.1). In comparison, an average of 2,413 gray foxes and 1,735 bobcats were legally harvested in New Mexico over the same period, based on trapper harvest reports available from the New Mexico Department of Game and Fish (www.wildlife.state.nm.us). The harvest data for the ringtail also shows a declining trend since the 2006–2007 trapping season, but any inferences drawn from it would be highly speculative because of the multitude of factors that may influence capture rates, such as the number of trappers, trapping effort and procedures, precipitation patterns, and availability of primary ringtail prey. Nevertheless, it might be prudent to further examine any harvest trend by comparing estimates of density in various areas within the state or between protected (e.g., US national parks) and non-protected lands (e.g., US national forests).

Mesocarnivores, among them ringtails, are essential to maintaining ecological processes and can be important drivers of ecosystem function (Roemer et al. 2009). For example, mesocarnivores may serve as reservoirs for pathogens that help regulate wildlife populations, influence the abundance and distribution of prey species, or drive plant communities through direct seed dispersal or interspecific interaction with, or predation upon, seed dispersers (Prugh et al. 2009; Roemer et al. 2009). The ecological role of ringtails has received little study, but their omnivory suggests that they play important roles in seed dispersal, may control small mammal populations, and likely compete with other mesocarnivores for food resources.

Although seldom seen, the ringtail is a widespread member of several diverse ecological communities within New Mexico. Its relative abundance in these ecological communities is not well documented, nor is its ecological role. Robust estimates of abundance are needed to document trends and ensure that this species is not being adversely affected by environmental change. As noted above, ringtails could play a pivotal ecological role that is currently underrecognized but could be revealed through further, more extensive study.

LITERATURE CITED

Ackerson, B. K., and L. A. Harveson. 2006. Characteristics of a ringtail (*Bassariscus astutus*) population in Trans Pecos, Texas. *Texas Journal of Science* 58:169–84.

Ahrens, H. E. 2012. Craniodental characters and the relationships of Procyonidae (Mammalia: Carnivora). *Zoological Journal of the Linnean Society* 164:669–713.

Armenta-Méndez, L., J. P. Gallo-Reynoso, B. T. Wilder, A. A. Gardea, M. M. Ortega-Nieblas, and I. Barba-Acuña. 2020. The role of wild canids in the seed dispersal of *Washingtonia robusta* (Arecaceae) in Sonoran Desert oases. *Revista Mexicana de Biodiversidad* 91:e913129.

Bailey, V. 1931 (=1932). *Mammals of New Mexico*. North American Fauna 53. Washington, DC: US Department of Agriculture, Bureau of Biological Survey.

———. 1936. *The mammals and life zones of Oregon*. North American Fauna 55. Washington, DC: US Department of Agriculture, Bureau of Biological Survey.

Barja, I., and R. List. 2006. Faecal marking behaviour in ringtails (*Bassariscus astutus*) during the non-breeding period: spatial characteristics of latrines and single faeces. *Chemoecology* 16:219–22.

Baskin, J. A. 1982. Tertiary Procyoninae (Mammalia: Carnivora) of North America. *Journal of Vertebrate Paleontology* 2:71–93.

———. 1989. Comments on New World Tertiary Procyonidae (Mammalia; Carnivora). *Journal of Vertebrate Paleontology* 9:110–17.

———. 1998. Procyonidae. In *Evolution of tertiary mammals of North America*, ed. C. M. Janis, K. M. Scott, and L. L. Jacobs, Vol. 1, 144–51. Cambridge: Cambridge University Press.

———. 2004. *Bassariscus* and *Probassariscus* (Mammalia, Carnivora, Procyonidae) from the early Barstovian (middle Miocene). *Journal of Vertebrate Paleontology* 24:709–20.

Bogan, M. A., K. Geluso, S. Haymond, and E. W. Valdez. 2007. *Mammal inventories for eight National Parks in the Southern Colorado Plateau Network*. Fort Collins, CO: Department of the Interior, National Park Service Natural Resource Technical Report NPS/SCPN/NRTR-2007/054.

Cahalane, V. H. 1954. *Mammals of North America*. New York: Macmillan.

California Department of Fish and Wildlife (CDFW). 2015. *California State Wildlife Action Plan*. Sacramento, CA.

Callas, R. 1987, Ringtail (*Bassariscus astutus*) den and habitat use in northwestern California. MS thesis, Humboldt State University, Arcata, California.

Chevalier, C. D. 1984. Water requirements of free-ranging and captive ringtail cats (*Bassariscus astutus*) in the Sonoran Desert. MS thesis, Arizona State University, Tempe.

———. 1991. Aspects of thermoregulation and energetics in the Procyonidae (Mammalia: Carnivora). PhD dissertation, University of California, Irvine.

Custer, J. W. 1979. Ectoparasites of the ringtail, *Bassariscus astutus*, from west Texas. *Journal of Medical Entomology* 15:132–33.

Decker, D. M., and W. C. Wozencraft. 1991. Phylogenetic analysis of recent procyonid genera. *Journal of Mammalogy* 72:42–55.

Dice, L. R., and P. M. Blossom. 1937. Studies of mammalian ecology in southwestern North America with special attention to the colors of desert mammals. *Publication of the Carnegie Institute, Washington* 485:1–129.

Escobar-Flores, J. G., G. Ruiz-Campos, F. Gomis Covos, A. Guevara-Carrizales, and R. Martinez Gallardo. 2012. New records and specimens of three mammals (*Spilogale gracilis*, *Bassariscus astutus*, and *Neotamias obscurus meridionalis*) for Baja California, Mexico. *Western North American Naturalist* 72: 591–94.

Gehrt, S. D. 2003. Raccoons and allies. In *Wild mammals of North America: biology, management, and economics*, 2nd ed., ed. G. A. Feldhamer, B. C. Thompson, and J. A. Chapman, 611–34. Baltimore: Johns Hopkins University Press.

Gompper, M. E., J. L. Gittleman, and R. K. Wayne. 1997. Genetic relatedness, coalitions and social behaviour of white-nosed coatis, *Nasua narica*. *Animal Behaviour* 53:781–97.

Grinnell, J., J. S. Dixon, and J. M. Linsdale. 1937. *Fur-bearing mammals of California*. 2 vols. Berkeley: University of California Press.

Hall, E. R., and K. R. Kelson. 1959. *The mammals of North America*. Vol. 2. New York: Ronald Press.

Harrison, R. L. 2012. Ringtail (*Bassariscus astutus*) ecology and behavior in central New Mexico, USA. *Western North American Naturalist* 72:495–506.

———. 2013. Ringtail (*Bassariscus astutus*) noninvasive survey methods, density, and occupancy in central New Mexico, USA. *Western North American Naturalist* 73:365–72.

Helgen, K. M., R. Kays, L. E. Helgen, M. T. N. Tsuchiya-Jerep, C. M. Pinto, K. Koepfli, E. Eizirik, and J. E. Maldonado. 2009. Taxonomic boundaries and geographic distributions revealed by an integrative systematic overview of the mountain coatis, *Nasuella* (Carnivora: Procyonidae). *Small Carnivore Conservation* 41:65–74.

Helgen, K. M., C. M. Pinto, R. Kays, L. E. Helgen, M. T. N. Tsuchiya, A. Quinn, D. E. Wilson and J. E. Maldonado. 2013. Taxonomic revision of the olingos (*Bassaricyon*), with a description of a new species, the Olinguito. *ZooKeys* 324:1–83. doi:10.3897/zookeys.324.5827.

Hooper, E. T. 1941. *Mammals of the lava fields and adjoining areas in Valencia County, New Mexico*.

Miscellaneous Publications of the Museum of Zoology, University of Michigan, No. 51.

Kays, R. W., and J. L. Gittleman. 2001. The social organization of the kinkajou *Potos flavus* (Procyonidae). *Journal of Zoology* 253:491–504.

Kierepka, E. M., and E. K. Latch. 2016. High gene flow in the American badger overrides habitat preferences and limits broad-scale genetic structure. *Molecular Ecology* 25:6055–76.

Koepfli, K. P., M. E. Gompper, E. Eizirik, C.-C. Ho, L. Linden, J. E. Maldonado, and R. K. Wayne. 2007. Phylogeny of the Procyonidae (Mammalia: Carnivora): molecules, morphology and the Great American Interchange. *Molecular Phylogenetics and Evolution* 43:1076–95.

Kortlucke, S. M. 1984. Variation in *Bassariscus* (Mammalia: Procyonidae). PhD dissertation, University of Kansas, Lawrence.

Kuzmin, I. V, M. Shi, L. A. Orciari, P. A. Yager, A. Velasco-Villa, N. A. Kuzmina, D. G. Streicker, D. L. Bergman, and C. E. Rupprecht. 2012. Molecular inferences suggest multiple host shifts of rabies viruses from bats to mesocarnivores in Arizona during 2001–2009. *PLOS Pathogens* 8:e1002786. https://journals.plos.org/plospathogens/article?id=10.1371/journal.ppat.1002786.

Lacy, M. K. 1983. Home range size, intraspecific spacing, and habitat preference of ringtails (*Bassariscus astutus*) in a riparian forest in California. MS thesis, California State University, Sacramento.

Larrison, E. J. 1967. Guide to Idaho mammals. *Journal of the Idaho Academy of Science* 7:1–166.

Leopold, A. S. 1959. *Wildlife of Mexico: the game birds and mammals*. Berkeley: University of California Press.

Lester, G. D., S. G. Sorensen, P. L. Faulkner, C. S. Reid, and I. E. Maxit. 2005. *Louisiana Comprehensive Wildlife Conservation Strategy*. Baton Rouge: Department of Wildlife and Fisheries.

Liebenthal, S. 2014. Fish and game to get first insight into elusive ring-tailed cat (press release). Idaho Fish and Game. https://idfg.idaho.gov/press/fish-and-game-get-first-insight-elusive-ring-tailed-cat.

Liu, M., S. P. Zack, L. Lucas, D. Allen, and R. E. Fisher. 2016. Hind limb myology of the ringtail (*Bassariscus astutus*) and the myology of hind foot reversal. *Journal of Mammalogy* 97:211–33.

Long, C. A., and H. B. House. 1961. *Bassariscus astutus* in Wyoming. *Journal of Mammalogy* 42:274–75.

Lonsinger, R. C., R. M. Schweizer, J. P. Pollinger, R. K. Wayne, and G. W. Roemer. 2015. Fine-scale genetic structure of the ringtail (*Bassariscus astutus*) in a sky island mountain range. *Journal of Mammalogy* 96:257–68.

Louppe, V., J. Baron, J.-M. Pons, and G. Veron. 2020. New insights on the geographical origins of the Caribbean raccoons. *Journal of Zoological Systematics and Evolutionary Research* 58:1303–22.

McRae, B. H. 2006. Isolation by resistance. *Evolution* 60:1551–61.

Meirmans, P. G., and P. W. Hedrick. 2011. Assessing population structure: F_{ST} and related measures. *Molecular Ecology Resources* 11:5–18.

Mollhagen, T. R., R. W. Wiley, and R. L. Packard. 1972. Prey remains in golden eagle nests: Texas and New Mexico. *Journal of Wildlife Management* 36:784–92.

Muldavin, E., P. Neville, P. Arbetan, Y. Chauvin, A. Browder, and T. Neville. 2003. *A vegetation map of Carlsbad Caverns National Park, New Mexico*. Final report submitted to Carlsbad Caverns National Park, University of New Mexico, Albuquerque.

Oklahoma Department of Wildlife Conservation (ODWC). 2016. *Oklahoma Comprehensive Wildlife Conservation Strategy: a strategic conservation plan for Oklahoma's rare and declining wildlife*. Oklahoma City: Department of Wildlife Conservation.

Oregon Department of Fish and Wildlife (ODFW). 2016. *Oregon conservation strategy*. Salem, Oregon.

Peakall, R., and P. E. Smouse. 2012. GenAlEx 6.5: genetic analysis in Excel: population genetic software for teaching and research—an update. *Bioinformatics* 28: 2537–39.

Poglayen-Neuwall, I. 1987. Management and breeding of the ringtail or cacomistle *Bassariscus astutus* in captivity. *International Zoo Yearbook* 26:276–80.

Poglayen-Neuwall, I., and D. E. Toweill. 1988. *Bassariscus astutus*. *Mammalian Species* 327:1–8.

Prugh, L. R., C. J. Stoner, C. W. Epps, W. T. Bean, W. J. Ripple, A. S. Laliberte, and J. S. Brashares. 2009. The rise of the mesopredator. *BioScience* 59:779–91.

Rausch, R. L. 2003. *Taenia pencei* n. sp. from the ringtail, *Bassariscus astutus* (Carnivora: Procyonidae), in Texas, U.S.A. *Comparative Parasitology* 70:1–10.

Reid, F., J. Schipper, and R. Timm. 2016. Ringtail. *Bassariscus astutus*. The IUCN Red List of Threatened Species 2016:e.T41680A45215881. https://dx.doi.org/10.2305/IUCN.UK.2016-1.RLTS.T41680A45215881.en.

Richards, R. E. 1976. The distribution, water balance and vocalizations of the ringtail, *Bassariscus astutus*. PhD dissertation, University of Northern Colorado, Greeley.

Richardson, W. B. 1942. Ring-tailed cats (*Bassariscus astutus*): their growth and development. *Journal of Mammalogy* 23:17–26.

Rodríguez-Estrella, R., A. R. Moreno, and K. G. Tam. 2000. Spring diet of the endemic ring-tailed cat (*Bassariscus astutus insulicola*) population on an island in the Gulf of California, Mexico. *Journal of Arid Environments* 44: 241–46.

Roemer, G. W., M. E. Gompper, and B. Van Valkengurgh. 2009. The ecological role of the mammalian mesocarnivore. *BioScience* 59:165–73.

Sandell, M. 1989. The mating tactics and spacing patterns of solitary carnivores. In *Carnivore Behavior, Ecology and Evolution*, ed. L. Gittleman, 164–82. Ithaca, NY: Cornell University Press.

Schweizer, R. M., G. W. Roemer, J. P. Pollinger, and R. K. Wayne. 2009. Characterization of 15 tetranucleotide microsatellite markers in the ringtail (*Bassariscus astutus*). *Molecular Ecology Resources* 9:210–12.

Stahlecker, D. W., J.-L. E. Cartron, and D. G. Mikesic. 2010. Golden eagle (*Aquila chrysaetos*). In *Raptors of New Mexico*, ed. J.-L. E. Cartron, 371–91. Albuquerque: University of New Mexico Press.

Taylor, W. P. 1954. Food habits and notes on the life history of the ring-tailed cat in Texas. *Journal of Mammalogy* 35:55–63.

Tiedt, A. R. 2011. Den site selection of ringtails (*Bassariscus astutus*) in west central Texas. MS thesis, Angelo State University, San Angelo, Texas.

Toweill, D. E. 1976. Movements of ringtails in Texas' Edwards Plateau region. MS thesis, Texas A&M, College Station.

Toweill, D. E., and D. B. Toweill. 1978. Growth and development of captive ringtails (*Bassariscus astutus flavus*). *Carnivore* 1:46–53.

Trapp, G. R. 1972. Some anatomical and behavioral adaptations of ringtails, *Bassariscus astutus*. *Journal of Mammalogy* 53:549–57.

———. 1978. Comparative behavioral ecology of the ringtail and gray fox in southwestern Utah. *Carnivore* 1:3–32.

Wang, I. J., and G. S. Bradburd. 2014. Isolation by environment. *Ecology Letters* 23:5649–62.

Watson, J. 1997. *The golden eagle*. London: T & AD Poyser.

Willey, R. B., and R. E. Richards. 1981. Vocalizations of the ringtail (*Bassariscus astutus*). *Southwestern Naturalist* 26:23–30.

Wilson, D. E., and D. M. Reeder, eds. 2005. *Mammal species of the world: a taxonomic and geographic reference*. 3rd ed. Baltimore: Johns Hopkins University Press.

Wozencraft, W. C. 2005. Order Carnivora. In *Mammal species of the world: a taxonomic and geographic reference*, Vol. 1, ed. D. E. Wilson and D. M. Reeder, 532–628. 3rd ed. Baltimore: Johns Hopkins University Press.

Wright, S. 1943. Isolation by distance. *Genetics* 28:114–38.

Wynne, J. J. 2013. Inventory, conservation and management of lava tube caves at El Malpais National Monument, New Mexico. *Park Science* 30:45–55, appendix.

WHITE-NOSED COATI (*NASUA NARICA*)

Jennifer K. Frey and Christine C. Hass

The white-nosed coati (*Nasua narica*; hereafter coati), is arguably one of the most exotic-looking and unusual carnivores found in New Mexico (or the United States for that matter). In the eyes of a novice observer unaware of the species' existence, a coati might seem to be an odd-nosed monkey. The mistaken identification is understandable, because in addition to resembling a primate, this mostly tropical animal is active during the day, often lives in large social groups that communicate using a variety of vocalizations, is adept at climbing trees and cliffs, and is often found in jungly stream bottoms dominated by large broad-leafed trees. But the coati is not a primate. Rather, it is a carnivore in the family Procyonidae, which also includes the more familiar ringtail (*Bassariscus astutus*) and raccoon (*Procyon lotor*).

The white-nosed coati is a medium-sized mammal, with adult males averaging about 5 kg (11 lbs), while females average about 3.8 kg (8.5 lbs; McColgin et al. 2018; Hass 2021). In the American Southwest, coatis are slightly smaller than raccoons in weight (see Chapter 32). However, body shape is quite different in the two species. While raccoons are rather short and squat with a medium-length tail, coatis are lankier in appearance, with a long body, longish legs, and a very long tail. Although no measurements are available for coatis in New Mexico, in the Huachuca Mountains of Arizona, adults average 123 cm (48 in) in total length, with males again larger (126 cm,

50 in) than females (120 cm, 47 in; Hass 1997). A smaller sample of coatis in the Chiricahua Mountains yielded total lengths of 114 cm (45 in) and 103 cm (41 in) for males and females, respectively (Lanning 1976). The long, slender tail makes up almost half of the total length and is usually held upright when the animal is moving about. However, the coati's tail is not prehensile, but rather used for balance when climbing in trees, and as a visual signal to other coatis (Kaufmann 1962; Gompper 1995; Hass 2021). The head of a coati is small and appears pointed due to an elongated

Photo 31.1. Coati photographed on 5 April 2020 by a remote camera in Chihuahuan desert scrub in the Peloncillo Mountains, Hidalgo County, New Mexico. The slender tail, which is typically held upright, is nearly as long as the body. The dark rings on the tail can be faint (as in this coati) or absent. The tail is not prehensile and instead is used for balance or visual communication. Photograph: © Bureau of Land Management, Las Cruces District Office.

(*opposite page*) Photograph: © Sally King.

Photo 31.2 Close-up of the face of a white-nosed coati photographed in October 2020 in the Gila River Bird Area, Grant County. The species is named for the conspicuous patch of white fur encircling the elongated snout. Note also the short, round ears and the white eye spots. Photograph: © Joel Gilb.

and flexible snout. The ears are small. In contrast to body size, the frontal part of the brain is relatively larger in females compared to male coatis and other species of procyonids, which is related to their greater sociality (Arsznov and Sakai 2013). Adult male coatis have dagger-like lower canines that are twice as long as those of adult females (Hass 1997, 2021). The feet have naked soles and a coati's mode of locomotion is plantigrade, meaning that it walks flat-footed on the soles of its feet. The toes of the coati's front feet are tipped with long, almost badger-like claws. The coati's feet resemble those of bears (Photos 31.3 a and b), and so do their tracks, though they are smaller (Photos 31.4 a and b).

The overall body coloration is usually some shade of reddish-brown, though both darker (almost blackish) and lighter (almost blondish) individuals may be found. The neck and chest are usually pale, and this pale coloration may wrap around the upper forearms and up to the shoulders. As with its cousins the raccoon and ringtail,

the tail of the coati is typically banded. However, the bands are not always easy to discern or may even be absent (Allen 1879). The feet and lower legs are usually black. As suggested by the species common name, the muzzle is white. The face has distinctive white spots above and below the eyes, with another spot on each cheek, closer to the ear. The eye spots may serve a communicative function (Hass 2021). The ears are also tipped with white. White lines connect the muzzle to the spots above the eye, but these lines may be incomplete in some individuals. The face is sometimes a darker color, adding additional contrast to the white eye spots and muzzle. The white muzzle of *N. narica* distinguishes it from the brown-nosed coati (*Nasua nasua*; also known as the ring-tailed coati, or South American coati), which among other characteristics has a brown muzzle (Hass 2021).

There has been considerable misunderstanding about the taxonomy of coatis (e.g., Allen 1879; Gompper 1995). Early biologists were confused by the number and kinds of coati species, the result of what is now understood as differences in social behavior and body size between the sexes, in addition to variation in coloration. Today, evidence indicates that there are at least three species of coatis, the white-nosed coati of Central and North America; the brown-nosed coati, which is restricted to the South American continent (Decker 1991; Nigenda-Morales et al. 2019); and one or two species of mountain coatis, found only in the northern Andes of South America (Hass 2021). There is considerable genetic variation across the ranges of both the white-nosed and brown-nosed coatis, which may result in some future new species designations (Nigenda-Morales et al. 2019; Ruiz-Garcia et al. 2020). The first two of the three or four species are sometimes referred to as "lowland coatis" (Hass 2021). The lowland coatis were originally described by Buffon in 1756 as "Le Coati-Mondi" (Allen 1879). Later, Prince Maximillian, a German explorer and naturalist, reported two kinds of coatis in Brazil,

Photos 31.3a and b (*left to right*). Left front foot (a) and left rear foot (b) of an adult male coati, anesthetized. The front feet have longer claws. Photographs: © Christine C. Hass.

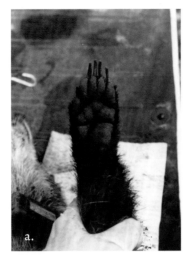 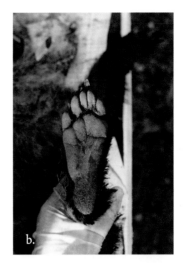

Photos 31.4a and b (*left to right*). Tracks of coati front foot (a) and rear foot (b), Gila River Bird Area, New Mexico, September 2013. Photographs: © Christine C. Hass.

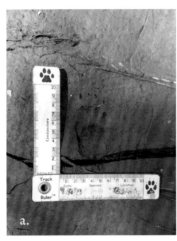 

which were later named *N. socialis* and *N. soliteria* (Allen 1879). This was based on reports by the local people who recognized two distinct forms, "Cuati de Bando," which was small, slender, and social, and "Cuati Mundeo," which was larger and more solitary (Allen 1879). Eventually, it was shown that these two forms were simply different genders of the same species: the smaller, more social form being female, and the larger, more solitary form being male (Allen 1879). Yet today there remains a legacy of taxonomic confusion in that the term "coatimundi" specifically refers to adult male coatis. Throughout much of their range, male coatis are also referred to as "solitarios." The name coati is possibly derived from the Brazilian Tupi Indian words *cua* meaning "belt" and *tim* meaning "nose" (i.e., belted-nose [Collins English Dictionary]; possibly referring to having a flexible snout [Holmgren 1990]). Other colloquial names for the white-nosed coati in the northern part of its range include chulo, chula, chulo bear, cholugo, Mexican monkey, Mexican coon, tejon, and coon-cat (Allen 1879; Bailey 1931; Cahalane 1939; Wallmo and Gallizioli 1954; Gilbert 1973; and various internet hunting blogs).

Four subspecies of the white-nosed coati are traditionally recognized, with animals from New Mexico and elsewhere in the American Southwest and northern Mexico referred to as *N. n. molaris* (Hall 1981; Wilson and Reeder 2005). Coatis from

Arizona, New Mexico, and northwestern Mexico were originally assigned to *N. n. pallida*, on the basis of a paler pelage (Allen 1904), but that subspecies was later synonymized with *N. n. molaris* (Hershkovitz 1951). A recent molecular genetic study found that *N. narica* is composed of five clades (i.e., evolutionary groups) including one representing samples from Arizona, New Mexico, and western Mexico, possibly corresponding to the subspecies *N. n. molaris* (Nigenda-Morales et al. 2019). The same study concluded that *Nasua* likely evolved in South America and dispersed northward to occupy North America.

DISTRIBUTION

The white-nosed coati is a tropical and subtropical species, its distribution extending from northern Columbia northward through Central America and Mexico to the American Southwest in portions of Arizona, New Mexico, and occasionally also Texas (Gompper 1995; Schmidly et al. 2004; Frey et al. 2013; Hass 2021). Prior attempts to synthesize information about the coati's geographic range in the southwestern United States were hampered because museum specimens, which are traditionally used to infer the distributions of mammals, are few in numbers. For instance, Findley et al.'s (1975) classic reference book, *Mammals of New Mexico*, was based almost exclusively on museum specimens. Consequently, their account of the coati in New Mexico was but a single sentence that referenced the sole specimen then known from the state (i.e., from the Animas Mountains, Hidalgo County). That account helped to perpetuate the false notion that coatis had an extremely limited distribution in New Mexico, restricted to the Bootheel region, even though there were reliable records of the species throughout a much wider area of southwestern New Mexico including well into the Gila National Forest of Catron County (Taber 1940).

To this day, there continues to be a paucity of museum specimens of this species, as only nine are known from New Mexico (Map 31.1; Appendix 31.1). To evaluate the distribution of the species, Frey et al. (2013) relied instead on a combination of museum records, photographs, and observations by biologists in addition to Geographic Information System (GIS) computer models. Coatis were thus shown to occur in New Mexico primarily in the highlands of the southwestern quarter of the state, and north into the Mogollon Mountains, including southern Catron, Grant, western Sierra, Hidalgo, and Luna counties (Frey et al. 2013; and see Map 31.1). However, there are occasional records from other areas of southern and central New Mexico including the northern portions of the Gila National Forest (Catron County), the Rio Grande Valley (Doña Ana, Sierra, Socorro, and Sandoval counties), and the Guadalupe Mountains (Eddy County). At least some of these outlying records (e.g., Rio Grande Valley at Hatch) correspond possibly to reproductive troops of females and juveniles (Frey et al. 2013; see under "Life History"). However, the presence of coatis at some locations may only be sporadic (Frey et al. 2013). In addition, solitary adult males disperse to find females or may "wander" (Hass 2021). The northernmost records of coatis in New Mexico and Arizona all appear to be solo males. For instance, the northernmost verified records of coatis in New Mexico were from near the Rio Grande in Sandoval and Socorro counties, and all were single animals observed in March or April of different years (Appendix 31.1), corresponding to the season when males may travel widely in search of mates.

Although there has been no quantitative evaluation of abundance of coatis throughout its US range, anecdotal information suggests the species is most common in the Madrean "sky island" mountains of southeastern Arizona. In New Mexico, coatis also appear to be most abundant in the Madrean Highland region, including the mountains in southern Hidalgo County (e.g., Peloncillo, Animas, and Hatchet mountains) and

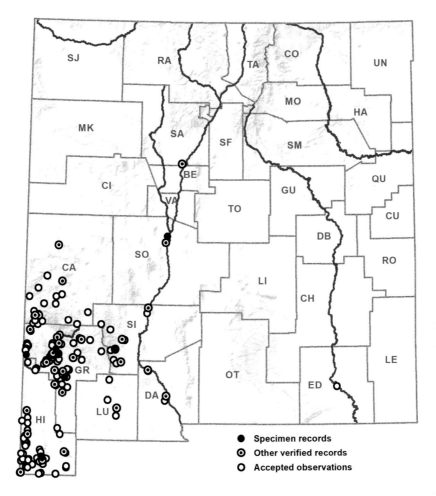

Map 31.1. Distribution of the white-nosed coati (*Nasua narica*) in New Mexico, based on verified records and accepted observations. Adapted and updated from Frey et al. 2013 (specimen and other verified records = Reliability Class A; accepted observations = Reliability Class B + Reliability Class C). See Appendix 31.1 for more details and notes on New Mexico's coati records.

western Grant County (e.g., Big Burro Mountains), as well as near the Gila and San Francisco rivers (western Grant County).

The status and distribution of coatis in the American Southwest have been matters of long-running conjecture (Taylor 1934; Cahalane 1939; Taber 1940; Reeder 1951; Wallmo and Gallizioli 1954; Wetherhill 1957; Kaufmann et al. 1976; Hoffmeister 1986). It is commonly stated that the coati expanded its range into the southwestern United States from Mexico during the early decades of the 20th century (e.g., Davis and

Callahan 1992; Gehlbach 1993; Brown and Davis 1995). As explained herein, there is little evidence to support that belief, and a chronological review of the evidence suggests it stems, at least in part, from scientific zeal to report new distributional records and the overlooking or misinterpreting of some already existing ones. The fervor over the status of the coati in the United States and interpretation of a northward expanding range were products of several confounding factors.

In 1934, Walter P. Taylor published a paper, "Coati added to the list of United States

Photo 31.5. Young coati along the western side of the Big Burro Mountains in Burro Spring Canyon, Grant County, New Mexico, 1999. The species' distribution in New Mexico lies primarily in the southwestern quadrant of the state north into the Mogollon Mountains and east to the Rio Grande. Photograph: © Jason Roback.

mammals," purporting to definitively confirm the presence of the species north of the Mexican border. Four specimens (two taken in 1924, one in 1932, and the last one secured in 1933) served as the basis for Taylor's (1934) paper, in addition to a number of observations of coatis in New Mexico and Arizona, all during the 1920s and early 1930s. Because Taylor (1934) does not reference any specimen records or sightings of the species predating the 1920s, from that report alone the coati appears to represent a recent addition to the fauna of the United States. However, several important points must be made about this report. First, Taylor (1934) overlooked museum specimens collected in the 1890s during early biological surveys at Fort Huachuca, Arizona (Wallmo and Gallizioli 1954; Hoffmeister 1986) and Laredo, Texas (Academy Natural Sciences

of Philadelphia catalog numbers 6096 and 6233; Brown 1898; Bailey 1905), one of which was published by Bailey (1905). He also overlooked a published report of the coati from the southern Animas Valley, Hidalgo County, New Mexico in 1908 (Bailey 1931). Second, Taylor was born in the Midwest and pursued his education as a biologist in California; he only moved to Arizona in 1922, at which time he became professor of economic zoology at the University of Arizona in Tucson (Brown and Babb 2008). Thus, all his coati records postdate the time of his arrival when, presumably, he became aware of the species. Few biologists worked within the range of the coati in the United States before Taylor's arrival, and consequently the paucity of reports about the presence of the species is not surprising. Third, the records of coatis in Taylor's paper (from

Cochise, Santa Cruz, Pima, Greenlee, Maricopa and Yavapai counties in Arizona and Hidalgo County in New Mexico) provide a reasonable approximation of the coati's current, known core reproductive range in the American Southwest (Frey et al. 2013). Lastly, Taylor's paper sparked scientific interest in the coati, which resulted in new published records, often with ad hoc conclusions about a range expansion or introductions (e.g., Cahalane 1939; Taber 1940). The accumulation of historical records demonstrates no consistent progressive northward expansion. In Arizona, occasional sightings and records of coatis in the northern part of the state appear to be solo males. Early papers also interpreted perceived increases in local abundance as additional evidence of a range expansion (e.g., Taber 1940). However, coati populations in the American Southwest fluctuate in abundance from year to year and from place to place, as do the populations of other species (Kaufmann et al. 1976; Hass 2021). Thus, a sudden appearance or apparent increase in abundance does not necessarily equate to a range expansion.

Davis and Callahan (1992) cited both the lack of Pleistocene coati fossils and the absence of reports of the species by early Arizona explorers (see accounts in Davis 2001) as evidence of a recent

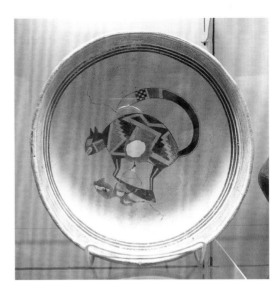

Photo 31.6. Possible coati image on a Mimbres pot. Deming History Museum. Photograph: © Christine C. Hass.

range expansion into the United States. However, there is archaeological evidence of coatis dating back at least 800 years, including images on Mimbres pottery in southwestern New Mexico, and a pot (or mug) possibly depicting a coati that was recovered from Chaco Canyon in the northwestern part of the state (Hass 2021; Photos 31.6–31.8). Petroglyphs possibly depicting coatis also have been found in New Mexico and Arizona

15

11

Photos 31.7a (*left*) and b (*right*). Possible images of coatis on Mimbres bowls. From Fewkes (1923, 1924).

Photo 31.8. Chaco Anasazi vessel fragment (possible coati effigy), Site number Bc 51, 1000–1200 CE, Chaco Culture National Historical Park CHCU 31104. National Park Service Archives. Photograph: © National Park Service.

Photo 31.9. Possible coati petroglyph, Lower Gila Box, Grant County, New Mexico. Photograph: © Christine C. Hass.

(Bernardini 2005; Hass 2021). One of those petroglyphs was discovered at a 14th century Hopi site, Homol'ovi IV, located in northern Arizona along the Little Colorado River near Winslow, Navajo County (Bernardini 2005). Although Homol'ovi is beyond the core reproductive range of the coati in the Southwest as depicted by Frey et al. (2013), solo males occasionally occur in northern New Mexico and Arizona. For example, a coati was reported at Petrified Forest National Park, Navajo County, Arizona, by Bob Housholder, who was the editor of the *Arizona Wildlife Sportsman* magazine and an expert on Arizona's carnivores (Cockrum 1960; Brown 2012). Petrified Forest National Park is located at the same latitude as, but ca. 72 km (45 mi) to the east of, Homol'ovi in the headwaters of the Little Colorado River. One of us (C.C. Hass), specifically looking for glyphs of coatis, uncovered possible petroglyphs and pictographs throughout southwestern New Mexico and southeastern Arizona (Hass 2021; Photo 31.9).

As described by Frey et al. (2013), the current distribution of coatis in the United States is remarkably similar to what it was historically based on records accumulated during early studies (see Taylor 1934 and Taber 1940). Consequently, there is no compelling evidence for a historical range expansion from Mexico into the American Southwest. The overall rarity of the species in the United States, together with fluctuating population numbers (see below), are characteristics of a species at the edge of its distribution (Frey 2006, 2009). That said, anecdotal information suggests more frequent observations of coatis at northern and eastern locations in recent years. However, this could simply be a function of more people accessing coati habitat, the increased ubiquity of digital cameras, and the now-widespread sharing of photographs and information on social media platforms. Whether or not coatis are currently expanding their range due to climate change or habitat alteration is not known and has not been subject to rigorous hypothesis testing.

Frey et al. (2013) studied the influence of climate and habitat variables on the distribution of the coati in New Mexico and Arizona, using occurrence data and GIS-based computer algorithms. The best climatic predictors were: 1) high isothermality (evenness of day-to-night temperature oscillations relative to summer-to-winter [annual] temperature oscillations; indicative of temperature evenness over the course of the year); and 2) low temperature seasonality (i.e., low year-round variation in monthly temperatures). On the basis of that study, coatis in New Mexico and Arizona seem therefore associated with regions of moderate temperatures year-round, which generally correspond to mid-elevations in the southern portions of both states, similar to the conclusions of Kaufmann et al. (1976). Coatis appear to avoid the colder, high elevations in the north, and the hottest deserts in the south. The elevation for 219 of the most reliable occurrence records in New Mexico and Arizona averaged 1,611 m (5,285 ft) and ranged from 704–2,807 m (2,310–9,209 ft; Frey et al. 2013).

Anecdotal evidence suggests that the abundance of coatis varies both geographically and temporally. In Arizona, coatis are thought to be most abundant in the Huachuca, Patagonia, and Tumacacori mountains, south of Tucson, but they are also common in the Santa Rita and Chiricahua mountains (Kaufmann et al. 1976). Based on both the number of sightings and extent of available habitat, coatis in New Mexico may be most abundant in the rugged canyons associated with the San Francisco and Gila rivers, in and below the foothills of the Mogollon Mountains in extreme southwestern Catron and northwestern Grant counties, and in several isolated mountain ranges including the Big Lue (=Brushy) and Big Burro mountains in Grant County, and the Peloncillo and Animas mountains in Hidalgo County. Their broader distribution in the state is probably represented by lower abundance or intermittent occurrence in peripheral areas such as the Rio Grande Valley.

Although the possession of a coati is prohibited by law in New Mexico, in some states (e.g., Texas) it is legal (with proper licenses) to breed

Photos 31.10a and b (*left to right*). Coati (one of a troop of 15–20 individuals) photographed on 31 January 2013, near the Gila River in the Red Rock Wildlife Management Area in Grant County. In New Mexico, areas where coatis can be found most predictably include the rugged canyons associated with the San Francisco and Gila rivers. Photographs: © Mark L. Watson.

Photo 31.11. Coati captured near the Rio Grande in Corrales, Bernalillo County, New Mexico in April 2018. The animal was described as aggressive and appeared wild. It was trapped in response to reports of an animal killing chickens and peacocks. The coati was transported to southern New Mexico where it was released. Photograph: © James N. Stuart.

and sell coatis as exotic pets. Thus, it remains a possibility that some pet coatis could be smuggled into New Mexico. For instance, a juvenile coati, apparently unafraid of humans, was photographed in Rio Rancho, Sandoval County (Roback 2001). This animal was likely an escaped pet as it seems unlikely a reproductive troop of coatis could have gone unnoticed in the Albuquerque-Bernalillo metropolitan area, and the urban setting for the observation is also more suggestive of an escaped pet. However, while it is tempting to attribute seemingly unusual occurrence records to escaped pets (e.g., Taber 1940), this often may be an unfounded assumption that can hamper knowledge about the species. For instance, in April 2018 an adult coati captured in Corrales, Sandoval County, appeared wild and thus could have represented a solo male (Photo 31.11; see also under "Diet and Foraging"). Consequently, great care must be taken in assigning the provenance of unusual distribution records. Coatis originating from the US pet trade are typically

hand-reared (and hence they lack fear of humans) and are often surgically altered to remove claws, canine teeth, and testes (i.e., they are neutered males). In addition, many coatis in the pet trade are brown-nosed, not white-nosed, coatis.

Because coatis are diurnal and vocal, and they may occur in large groups, it may be assumed that their presence in any area would be difficult to miss. However, even during studies that involved radio-collared animals, the investigators rarely saw the animals, likely the result of the surrounding rugged terrain, the dense vegetation, the low density of coatis, and their wide-ranging movements (Valenzuela 1998; McColgin et al. 2018; Hass 2021). Future studies to evaluate the distribution of coatis in New Mexico may benefit from motion-activated cameras that can be deployed to monitor locations over long periods of time. Many records of coati occurrence have been obtained in this manner (e.g., Photo 31.1).

HABITAT ASSOCIATIONS

Throughout their range, coatis are found in thick, wooded, or brushy environments that provide both food in the form of invertebrates and fruits and horizontal and vertical cover for protection from predators (Hass 2021). In the United States and northwestern Mexico, coatis are primarily associated with wooded or forested mountain ranges (Wallmo and Gallizioli 1954; Kaufmann et al. 1976; Hass 1997), especially the Madrean oak-pine woodland biotic community, which occurs elevationally between the lower encinal (= evergreen) oak woodland and the higher montane coniferous forest (pine forest) communities (Brown 1994). In southeastern Arizona, Wallmo and Gallizioli (1954) concluded that the species was most common in the Upper Sonoran (i.e., plains grasslands, and encinal and conifer woodlands) and Transition (i.e., conifer forests dominated by pines) life zones, but that it also occurred in the Lower Sonoran Zone (i.e., desert and desert grasslands). In the Huachuca Mountains

(Cochise County, Arizona), 40% of observations took place in canyon bottoms near water in riparian woodlands dominated by oaks (*Quercus* spp.), sycamore (*Platanus*), walnut (*Juglans*), and often maple (*Acer*); 32% of observations were at or near water sources at higher elevations in canyons dominated by oaks and pines; and the remainder of observations (28%) were in the uplands, either in grasslands, encinal and oak-pine woodlands, or pine forest (Wallmo and Gallizioli 1954). According to Wallmo and Gallizioli (1954), however, the results were skewed toward more observations (72%) in canyons simply because those were used as travel corridors. Wallmo and Gallizioli (1954) also noted that while coatis often visited water sources and commonly used all chaparral, woodland, and forest vegetation types within the Huachuca Mountains, they only infrequently entered the surrounding lower-elevation desert and grassland vegetation (Wallmo and Gallizioli 1954). A subsequent detailed study of habitat use in the Huachuca Mountains (Hass 2021), based this time on tracking of radio-collared animals, found that 85% of locations were in encinal and Madrean coniferous forest, while only 13% were in riparian vegetation. Coatis live in dense vegetation, and they often "freeze" rather than flee when they detect danger (Kaufmann 1962; Hass 2021). Thus, opportunistic coati sightings (as in Wallmo and Gallizioli 1954) may be biased with more detections occurring in areas with greater human access and better sightability.

The availability of water appears to be a key habitat requirement for coatis (Wallmo and Gallizioli 1954; Valenzuela and Ceballos 2000; Valenzuela and Macdonald 2002). In the Chiricahua Mountains in southeastern Arizona, coatis tended to occur near the few permanent springs during dry periods (Lanning 1976). In the Huachuca Mountains, also in southeastern Arizona, the core home range use areas of coatis tended to follow streams with permanent or intermittent water (Hass 2002). Coatis might therefore be selecting core use areas that provide not only drinking water year-round but also a greater abundance of invertebrates and fruits, as tends to be the case along streams (see under "Diet and Foraging").

In the Animas Mountains in southwestern New Mexico, coatis were thought to occur in oak and pinyon-juniper woodland (Cook 1986). Overall, however, and though no detailed field studies of coati habitat have been conducted in New Mexico, Frey et al. (2013) modeled coati habitat suitability using GIS-based ecological niche analyses. In Frey et al.'s (2013) study, the habitat variables most strongly associated with coati occurrence in New Mexico and Arizona were land-cover type and distance to springs, which contributed 28.6% and 26.2% to the best model, respectively. The land-cover types associated with the highest habitat suitability values included Mogollon chaparral, Madrean encinal, and Madrean pinyon-juniper woodland, even though these only account for a small proportion (21.4%) of the total area of predicted habitat in New Mexico and Arizona (suggesting that much of the range of coatis in these states is in inferior habitat). The three land-cover types occur at mid-elevations where they form a natural transition (i.e., from lowest to highest along the elevational gradient: Mogollon chaparral, Madrean encinal, Madrean pinyon-juniper) between the lower-elevation desert and grasslands and the higher-elevation pine forests. These mid-elevation vegetation types are found in areas with mild winters and wet summers, reflecting the model's most important climate variables (i.e., evenness of temperature; see under "Distribution"). They are dominated by a diversity of evergreen shrub and tree species that have the center of their distributions in the Sierra Madre in Mexico, including mountain mahogany (*Cercocarpus* spp.), manzanita (*Arctostaphylos*), diverse oaks (*Quercus*), pinyon pines (*Pinus*), and junipers (*Juniperus*).

Surprisingly, the ecological niche model assigned relatively high habitat suitability values to some of the lower-elevation land-cover types including Chihuahuan mixed salt desert scrub; Apacherian-Chihuahuan mesquite upland scrub; Apacherian-Chihuahuan semi-desert grassland and steppe; and Chihuahuan creosote, mixed desert, and thorn scrub (Frey et al. 2013). These Chihuahuan desert and grassland vegetation types were rated higher than high-elevation land-cover types such as Southern Rocky Mountain ponderosa pine woodland and Colorado Plateau pinyon-juniper woodland. The latter cover types represent northern-affiliated conifer woodland and forest types that exhibited relatively low habitat suitability according to the model. In contrast, their Madrean counterparts (Madrean lower montane pine-oak forest and woodland; Madrean pinyon-juniper woodland) were rated high for habitat suitability (Frey et al. 2013), probably reflecting the influence of more moderate temperatures and

Photo 31.12. Adult male coati on the Gila National Forest at the head of Long Canyon, Catron County on 8 September 2012. It was captured by researchers as part of a gray fox (*Urocyon cinereoargenteus*) rabies study. Atypical for coati habitat in New Mexico, the surrounding vegetation consisted of Petran (Rocky Mountain) mixed coniferous forest dominated by a dense closed canopy of ponderosa pine (*Pinus ponderosa*), Douglas-fir (*Pseudotsuga menziesii*), and aspen (*Populus tremuloides*). Photograph: © Gary W. Roemer.

year-round access to food (Hass 2021). It should be recognized that even land-cover types that were rated relatively low for habitat suitability might still be used by coatis. For instance, on 13 September 2012 a male coati was caught during a study of gray fox (*Urocyon cinereoargenteus*) rabies at about 2,650 m (8,700 ft) elevation in a Southern Rocky Mountain mixed coniferous forest at the head of Long Canyon in the Tularosa Mountains, Catron County (Photo 31.12).

As another way of evaluating the relative importance of various land-cover types to coati populations, Frey et al. (2013) evaluated occurrence records that were suggestive of reproductive troops (e.g., observations of juveniles, pregnant females, or groups of three or more individuals). These records were most frequent in Madrean pinyon-juniper woodland (27.2%), Apacherian-Chihuahuan piedmont semi-desert grassland and steppe (17.5%), and Madrean encinal (13.6%). All the occurrence records likely indicative of reproductive troops within Chihuahuan piedmont semi-desert grassland and steppe were in New Mexico. This finding might reflect the greater extent of this biotic community in New Mexico compared to Arizona (though sightability issues could also contribute). Thus, not all information from Arizona may be directly relevant to New Mexico. The presence of reproductive troops of coatis defines the core distribution of the species in New Mexico and Arizona. Occurrence outside of the core distribution and/or in inferior habitats are most likely to be solo males.

In the ecological niche model, coati occurrence was strongly influenced by distance to springs, but not by distance to streams. The relationship with springs undoubtedly reflects the key importance of drinking water for coatis. The difference between the relative importance of springs and streams likely reflects the fact that most sources of perennial water across the arid Southwest consists of springs rather than rivers. However, this is not to say that other riparian systems are not important to coatis. In fact, several such riparian land-cover types scored high for habitat suitability (North American warm desert lower montane riparian woodland and shrubland, Rocky Mountain lower montane-foothill riparian woodland and shrubland; North American warm desert riparian mesquite bosque; and North American warm desert riparian woodland and shrubland), despite constituting <0.5% of the total predicted habitat. A study of food availability in Arizona found that fruits and invertebrates, which make up most of the coati's diet (see below), were found at higher densities in riparian vegetation (Hass 2021).

LIFE HISTORY

Diet and Foraging

The coati is an omnivore that consumes predominantly arthropods and fruits (Kaufmann 1962; Hass 1997; Valenzuela 1998; Gómez-Ortiz et al.

Photo 31.13. Coati troop on 29 October 2015 in a riparian woodland along the Gila River in Grant County. Fruits and invertebrates, which form the bulk of the coati's diet, are found in greater abundance on the moist soils of riparian areas. Photograph: © Keith Geluso.

Photo 31.14. Coatis along Bear Creek in the Double E Wildlife Management Area in Grant County, New Mexico on 22 September 2015. In the American Southwest, the home ranges of coati troops are usually associated with drainages. Photograph: © Mark L. Watson.

Photo 31.15. Male coati climbing the stalk of an agave in New Mexico's Big Hatchet Mountains in 1998. Photograph: © Jason Roback.

2015). Most of the available data on the diet of the coati comes from tropical regions of Mexico and Panama. In Mexico, analyses of scats (fecal remains) revealed that the most important (>85%) food items were fruits and arthropods (beetles, grasshoppers, and ants), but the observed diet also included smaller amounts of mammals (rodents, cottontails [*Sylvilagus*], raccoons [probably carrion]), and reptiles (Valenzuela 1998; Gómez-Ortiz et al. 2015). Gompper (1996) concluded that the coati's social structure directly influenced patterns of foraging on Barro Colorado Island, Panama, where coatis are mainly frugivorous and insectivorous. The availability of fruit was found to be patchy in distribution, causing intense competition among coatis for access to the fruit. Group living benefited females

as such groups could successfully compete with males for access to fruit, while large males were better able to monopolize fruits when competing with solitary females or small groups (Gompper 1996). These results may not apply to the northern end of the coati's range, where the fruits are smaller and more widely dispersed (Hass 2021).

A detailed study of coati food habits in Arizona's Huachuca Mountains, based on scats and stomach contents, again found a diet of fruits and invertebrates, with only a tiny proportion of lizards and carrion (Hass 1997). Eighty-seven percent of scats contained fruits, and 78% contained invertebrates. The most common fruits consumed year-round were alligator juniper berries (*Juniperus deppeana*) and manzanita berries (*Arctostaphylos pungens*), with seasonally available fruits like buckthorn (*Berberis wilcoxii*), chokecherry (*Prunus virginiana*), prickly pear (*Opuntia phaeacantha*), grape (*Vitis arizonica*), oak (*Quercus* spp.), and madrone (*Arbutus arizonica*) also consumed. Coleopterans were found in 54% of scats; lepidopterans, orthoperans, and gastropods all made up significant parts of the diet. Lizards and carrion were found in fewer than 5% of scats (Hass 1997). In places where alligator juniper does not occur, such as along major streams, coatis consume hackberries (*Celtis pallida*), and may also eat a considerable amount of crickets (Hass 2021). In the Big Burro Mountains, Grant County, New Mexico, five coatis were observed in the canopy of a hackberry tree (*Celtis*) foraging on berries (Geluso 2009). Stomach contents from a coati taken in the Patagonia Mountains, Arizona, contained only what appeared to be manzanita berries (Taylor 1934); while stomach contents of five coatis taken elsewhere in Arizona contained fruits of manzanita, juniper, prickly pear, beetles, and traces of other items (Wallmo and Gallizioli 1954). Feces from two coatis trapped in Arizona were composed of only juniper berries and only manzanita berries (Wallmo and Gallizioli 1954).

In the Chiricahua Mountains, Arizona, coatis

often climbed the stalks of flowering Palmer's agave (*Agave palmeri*) to forage, presumably to access the large amounts of nectar and pollen produced by the plant as well as its associated invertebrates (McColgin et al. 2003). In the Huachuca Mountains, coatis commonly consumed the nectar in Parry's agave (*Agave parryi*) and they leave distinct marks on the yucca stalk, a useful indicator of the presence of coatis in an area (Hass 2021). In the Big Hatchet Mountains in Hidalgo County, New Mexico, an adult male coati was photographed climbing the stalk of an agave and feeding on the nectar from the flowers (Roback 1998–1999; Photo 31.15). Coatis may consume a wide variety of cultivated fruits, poultry, or other anthropogenic foods (Wallmo and Gallizioli 1954; Risser 1963; Kaufmann et al. 1976; Valenzuela 1998). Troops of foraging coatis pose a considerable challenge to fruit growers in some areas of New Mexico, such as the Gila-Cliff valley (S. Davis, pers. comm.).

Although coatis rarely prey on vertebrates (with the exception of sea turtle eggs in the tropics; Valenzuela 1998), Gilbert (1973) observed a male coati attempting to catch a rock squirrel (*Otospermophilus variegatus*), then later observed the same coati with a headless rock squirrel, a sequence of events highly suggestive of predation rather than scavenging. Hass (2021) observed a lactating coati consuming a whiskered screech-owlet (*Megascops trichopsis*). The adult owls were dive-bombing the coati, so it is probable that she killed the fledgling. A coati captured in Corrales (Bernalillo County) in April 2018 had killed a homeowner's poultry (https://www.krqe.com/news/albuquerque-metro/wildlife-officials-coati-mundi-will-be-relocated-to-southern-new-mexico/1115993840; see Photo 31.11). Smith (1977) reported that a captive coati in Arizona killed a woodpecker. Coatis have rather fragile jaws and a weak bite force for their size (Christiansen and Adolfssen 2005); predation on mammals and birds is probably rare and opportunistic.

With the morphology of their forelimbs and forepaws adapted primarily for foraging on the ground, coatis are excellent diggers and shredders (McClearn 1992). Coatis detect invertebrates and carrion using their sensitive nose. They use their forepaws not only to dig holes, shred logs, or overturn rocks but also to hold down food while feeding. When hunting small vertebrates such as lizards, coatis are "cat-like" in that they quickly pounce and hold the prey down with their forepaws and then kill it with a few bites or by biting off its head (Gilbert 1973:157). When eating spiny items such as cactus fruit or tarantulas, coatis roll the item between the leathery pads of their forepaws to remove the bristles; to handle a scorpion, a coati bats it around, apparently avoiding the stinger, until it is dead (Kaufmann 1962; Gilbert 1973).

Although most foraging is conducted on the ground, coatis are also capable of climbing and they often forage up in trees for fruits such as juniper and madrone berries. They have little trouble climbing trees with rough bark. However, smooth-barked trees like sycamore are usually climbed using adjacent trees and vines. Coatis walk on the horizontal surfaces of large branches or sometimes hang suspended by their limbs below the branch (McClean 1992), and they descend trees headfirst, much like a squirrel, by rotating their hind feet (Gompper 1995). The long tail is used for balance. Coatis can stand bipedally for a moment to investigate something above head level or to gain a viewing advantage (McClearn 1992).

In Arizona, coatis are heavier during winter (McColgin et al. 2018), but do not appear to put on large amounts of fat, and they do not hibernate or store food (Gilbert 1973; Hass 2021). Thus, adequate food must be available year-round, and this could serve as a key functional limitation on the coati's potential distribution. Gilbert (1973) found that inclement weather did not seem to prevent coatis from being active as they were observed foraging

Photo 31.16. A troop of coatis' overnight tree nests in a cottonwood along the San Pedro River in Arizona, December 2003. Similar nests have been observed along the Gila River in New Mexico. Photograph: © Christine C. Hass.

in rain and snow, as well as during cold and hot days. The only weather pattern that seemed to reduce daily activity was when windstorms blew through canyons (Gilbert 1973).

Coatis are diurnal (active during the day) and spend most of their day foraging and socializing before retiring to communal cliff dens or tree nests overnight (Kaufmann 1962; Hass 2021). Tree nests are generally small platforms made with branches that coatis bend inward and others they break off before adding to the platform. Some trees may contain many nests to support multiple coatis (Photo 31.16); these may be used repeatedly, and each time more branches are added. Troops typically forage together, though they may break into small and temporary subgroups, and they will usually rest midday. Males

follow the same general activity pattern. Coatis are rarely active at night except during the mating season (Kaufmann 1962), though nocturnal raiding of trash cans has been reported.

Reproduction and Social Behavior

The white-nosed coati is highly social, and its complex social organization influences most facets of its life history. Adult females and their offspring of the past two years live in gregarious groups known as troops (the literature from Central America often refers to them as bands). In contrast, adult males are usually solitary and normally only join troops during the breeding season, though there are exceptions. No studies of coati social behavior have been conducted in New Mexico. Much of the information about coati social structure is derived instead from field studies conducted in tropical forests in Central America (e.g., Kaufman 1962; Gompper and Krinsley 1992; Gompper et al. 1997, 1998; Booth-Binczik 2001; Binczik 2006). However, different behavioral patterns have been described for populations of coatis occurring in tropical dry forests of western Mexico and temperate environments of the southwestern United States (McColgin et al. 2018). Where possible, this chapter presents information drawn from research in Arizona, especially the results of a four-year study involving 73 coatis in the Huachuca Mountains by Hass (1997, 2002, 2021) and a nine-year study of 97 coatis in the Chiricahua Mountains near the New Mexico border by McColgin (2006) and McColgin et al. (2018).

In the Huachuca Mountains, troop size ranged from 5–39 coatis including 2–12 adult females (Hass 2021). In the Chiricahua Mountains troop size ranged from one to six adult females plus their yearlings and juveniles during years of low population density and from three to nine adult females plus their yearlings and juveniles during years of high population density (McColgin et al. 2018). These reported troop sizes are similar to those in the tropics and in western Mexico's tropical dry forest (Kaufmann 1962; Valenzuela 1998). Reports of very large troops of >100 individuals (e.g., see Risser 1963) could be exaggerations or represent instances of several troops congregating at a food source (Gompper 1995), though one study in Guatemala found stable troops of as many as 150 coatis (Booth-Binczik 2001).

Early accounts from Arizona speculated that coatis were nomadic and did not have fixed home ranges (e.g., Wallmo and Gallizioli 1954; Pratt 1962; Kaufmann et al. 1976). However, modern radio-telemetry studies have shown that coatis do indeed restrict their movements to distinct areas or home ranges that are albeit quite extensive. In the Huachuca Mountains, where radio-collared coatis were followed year-round,

the home ranges of males, outside of the reproductive season, averaged 6.1 km² (2.4 mi²) (range 3.5–10.7 km²; 1.4–4.1 mi²) whereas those of troops were twice as large, averaging 13.6 km² (5.3 mi²) (range 7.0–22.4 km²; 3.1–8.7 mi²) (Hass 2002). During the mating season (mid-March to mid-April), both males and females increased their movements, sometimes extending them beyond their normal home range boundaries. In late June, pregnant females left the troop, and established a natal den, usually away from other females, but still within the troop's home range (Hass 2021). During the natal denning season, home ranges of individual females were much smaller, averaging 2.8 km² (1.1 mi²) and centered around the natal den (Hass 2002, 2021). There was considerable overlap of home ranges among both males and troops. Troop home ranges were very stable from year to year, and even as the Huachuca Mountains coati population declined in the late 1990s, troops did not move into adjacent areas that had been vacated.

In the Chiricahua Mountains, the annual home range size for females averaged 4.6 km² (1.8 mi²)

Photo 31.17. Three coatis in Burro Spring Canyon in New Mexico's Big Burro Mountains in 1999. The three individuals belonged to a larger troop consisting of 19 or more coatis. Photograph: © Jason Roback.

(range 3.3–7.7 km² [1.3–3.0 mi²]), though during any given time of the year coatis concentrated their movements in smaller core areas, which were smaller in winter and spring (mean 1.4 and 2.9 km², respectively [0.5 and 1.1 mi²]) compared to summer (mean 4.7 km² [1.8 mi²]; McColgin et al. 2018). Across years, home ranges were stable, though there was some shifting of use areas from season to season within years, and in one instance a troop moved its home range 2 km (ca 1.2 mi) down a canyon. During years of low population density, troops kept separate home ranges, whereas during periods of high population density overlap occurred. Individual dispersion among females within a troop was greater in summer than in winter, likely due to nesting. During some winters, members of different troops commingled, or so it seemed: it is possible that groups interpreted as being different were actually subunits of a larger troop. During years of high density, troops did in fact sometimes break into smaller subgroups for variable lengths of time. On rare occasion adult females were solitary. This occurred after a troop disintegrated during the winter after three adults disappeared and one joined another troop; the remaining two females did not join any troop and remained independent during the following spring (McColgin et al. 2018).

In contrast to the highly social females, males generally lead solitary lives. In Arizona, most males displayed no evidence of gregariousness, though pairs of males sometimes were found in relative proximity (<500 m [<ca. 1,600 ft]) for short periods (McColgin et al. 2018; Hass 2021). In the Chiricahua Mountains, the distance between neighboring adult males averaged 0.8 km (ca 0.5 mi). Despite their independent existence, males maintained home ranges similar in size to those of females. Of 10 adult males that were tracked for over a year, six maintained relatively stable home ranges across seasons, while four shifted the center of their home ranges seasonally by

several kilometers (McColgin et al. 2018). The two maximum reported distances moved by coatis in Arizona were 9 and 18 km (5.6 and 11.2 mi; Hass 2002). These records corresponded to two male coatis that were captured in the Huachuca Mountains and moved after raiding gardens and bird feeders in their home ranges. They were translocated to the opposite side of the mountains but returned to their home ranges within four months (Hass 2002).

In Arizona, the home ranges of coatis (troops and solitary individuals) are huge in comparison with those recorded in the tropics, probably reflecting overall lower productivity and a more seasonal environment in the Southwest (Ratnayeke et al. 1994; Hass 2021). Also, as a consequence of low resource availability, coati densities are low in Arizona, again compared to the tropics, where as many as 70 coatis may occur per km² (Gompper 1995). In Arizona densities range from 0.4–1.2 individuals of all ages per km² in the Huachuca Mountains (Hass 2002) and 0.1–0.5 adults per km² in the Chiricahua Mountains (McColgin et al. 2018). Ratnayeke et al. (1994) documented the home range of a pregnant coati in the Guadalupe Mountains of southwestern Hidalgo County and adjacent areas of Arizona. Otherwise, no home range or density information is available from New Mexico.

In Arizona, coatis mate from mid-March to mid-April (Smith 1977; Hass and Roback 2000). Although male coatis are larger than females, they are almost always subordinate to females (Kaufmann 1962). During most of the year, female coatis are hostile toward adult males. This intolerance disappears during the mating season when females not only become receptive to the presence of males but also initiate mutual grooming sessions and copulation, the latter taking place either on the ground or in a tree (Kaufmann 1962; Smith 1977; Hass and Roback 2000). Estrous is synchronized among the females in a troop, and most copulations occur during a 10–20-day window. Gestation lasts 70–77 days (Gompper 1995), with coatis in Arizona giving birth in late June (Gilbert 1973; Smith 1977; Hass 2021). Throughout the range of coatis, the timing of the birth season is related to the onset of the monsoon (Hass 2021). Pregnant females leave their troop several days to several weeks before giving birth in order to find or build a suitable den, and they remain solitary while they nurse and rear their kits (Kaufmann 1962; Gilbert 1973; Ratnayeke et al. 1994, Hass 2021). In Arizona, crevices in cliffs are common den sites, but caves and old mines are also used. If no rocky areas are available nearby, coatis will build nests in trees (Ratneyeke et al. 1994; Hass 2021). These nests are typically much larger and more substantial than the overnight nests (Kaufmann 1962; Olifiers et al. 2009). Litter size is typically three to six (Smith 1977) and is related to age, as older females give birth to larger litters (Hass 2021). The kits are born with their eyes closed, their fur all black with no markings (Smith 1977). By 19 days of age, their eyes have opened and the tail rings and eye markings are showing. By 45 days the kits have begun to eat solid food (Smith

1977). In late July to early August, the female returns to her troop with her kits; by that time, the kits are about six to eight weeks old and able to travel and commingle with the troop (Gilbert 1973; Kaufmann et al. 1976; Hass 2021).

Coatis engage in a variety of social behaviors (Smith 1980). Hass (2021) analyzed behavior and vocalizations from videotape and audio recordings of captive and wild coatis (mostly white-nosed coatis). She identified three categories of behaviors: affiliative, agonistic, and anti-predator. Affiliative displays included vocalizations, touches, allogrooming, visual displays, and play. Affiliative calls, including contact calls, squeaks, chitters, and chuckles, made up at least 80% of the vocal repertoire. The contact call, issued frequently when coatis were moving, getting ready to move, or anticipating change, is an extraordinarily complex call consisting of multiple parts, and ranges from 40 Hz (near the low end of the human hearing spectrum) into the ultrasonic range at 24,000 Hz. Chitters appear to be a mild distress call and are primarily issued by juveniles and often induce adults to approach (Kaufmann 1962). Chuckles are a riot of sounds produced

Photo 31.18. Adult female coati near her natal den in a sycamore (the entrance is covered with leaves). Huachuca Mountains, Arizona, June 1997. Photograph: © Christine C. Hass.

Photo 31.19. A seven-week-old coati in the Huachuca Mountains in Arizona, July 1997. Photograph: © Christine C. Hass.

when coatis greet each other after an extended absence. Allogrooming is common among troop members and may include adult males during the mating season (Kaufmann 1962). It typically involves grooming with the comb-like front incisors and licking. In addition to maintaining bonds among troop members, it helps reduce external parasites (Gompper 2004; Hass 2021). Coati youngsters play almost constantly, biting, wrestling, and chasing each other up and down trees. Subadults and adults occasionally join in these games (Hass 2021).

Coatis do not communally hunt and tend to be rather defensive of individual food morsels. Juveniles are quite brazen in attempting to obtain food, and adult females will come to their aid during disputes. Loud chittering and squealing by a juvenile bring in the adult females, not just the mother, ready to defend the juvenile or other troop members from any threat (Hass 2021). Visual displays used in agonistic encounters include nose up (a threat) and head down, indicating submission (Kaufmann 1962).

Anti-predator behaviors include jumping ~1.8–3 m (6–10 ft) up into a tree and "freezing" to detect the source of alarm. From this perch, a repetitive, barking alarm call is issued and the coati may wave its tail back and forth. The barking call and waving of the tail both appear to be signals to the predator that it has been seen, but also let nearby coatis know where the source of the threat is and how serious it is (Hass 2021). Coatis have different responses to different types of predators: they run to escape humans, they climb higher to escape terrestrial predators including wild felids, and they drop to the ground and hide under the cover of vegetation to escape birds of prey (Hass 2021).

Interspecific Interactions

Besides humans, known and suspected predators of coatis in the American Southwest include the jaguar (*Panthera onca*), cougar (*Puma concolor*),

bobcat (*Lynx rufus*), ocelot (*Leopardus pardalis*), black bear (*Ursus americanus*), domestic dog (*Canis familiaris*), spotted owl (*Strix occidentalis*), red-tailed hawk (*Buteo jamaicensis*), and golden eagle (*Aquila chrysaetos*). In the tropics, other species (e.g., boa constrictors [*Boa constrictor*] and white-faced capuchins [*Cebus imitator*]) have been shown to prey on coatis (Gilbert 1973; Hass and Valenzuela 2002; Hass 2009, 2021). Across the species range, however, large felids appear to be the coati's main predators (Hass and Valenzuela 2002; Hass 2021). In Arizona's Huachuca Mountains, predation accounted for ≥ 76% of coati mortalities with the main predator being the cougar (Hass and Valenzuela 2002). In studies conducted in both Arizona and central Mexico, coatis ranked as the second most important prey item in the diet of cougars (Hass 2009; Gómez-Ortiz and Monroy-Vilchis 2013). It has been hypothesized that cougar populations today are higher than they were historically due to livestock serving as a prey subsidy (Rominger et al. 2004; but see Chapter 9). An increase in cougar populations could conceivably reduce coati populations, though this idea requires testing. Bobcats are probably too small to successfully prey on adult coatis, but in a study in Arizona, juvenile coatis made up 7% of their diet (Hass 2009). Interestingly, even though coyotes (*Canis latrans*) may be common where coatis occur, there is no evidence that they represent a source of mortality (Hass and Valenzuela 2002). This might be because coatis can climb trees and cliffs to evade attacks by coyotes. Gilbert (1973) observed a troop of coatis chasing a coyote from the carcass of a deer. This report suggests that a troop of coatis might also be able to defend itself from a lone coyote attack.

Coati troops likely experience a reduced risk of predation stemming from a combination of physiological adaptations and group defense behaviors including synchronized birthing; mobbing of potential predators; vigilance and alarm calling; and foraging with juveniles in the middle

Photo 31.20. Adult female with her five kits on 13 August 2021 near Hillsboro in Sierra County. Long-term coati population trends remain unknown in New Mexico and elsewhere in the Southwest. Photograph: © Jake Schoellkopf/New Mexico Department of Transportation.

of the troop. Consequently, it is not surprising that single coatis (solo males and females away from the troop during the natal denning season) experience higher predation rates than do coatis living in groups (Hass and Valenzuela 2002).

Besides predation, other sources of mortality include diseases such as scabies (*Notoedres cati*), canine distemper, and rabies, as well as accidents involving vehicle collisions, electrocutions, and other traumas (e.g., fall from a tree; Hass and Valenzuela 2002). Risser (1963) reported two separate anecdotes of large numbers (30–40 individuals) of coatis found dead in Arizona during the early 1960s: one event occurred beside a spring and the other occurred in an arroyo. Coatis are susceptible to a variety of zoonotic pathogens including the pathogenic bacteria *Escherichia coli* and *Clostridium botulinum*. Coatis are susceptible to the disease canine distemper, which may be facilitated by their social interactions, and was implicated in the population crash documented by Risser (1963). Before the population crash, there were observations of coatis that were blind due to their eyes being sealed shut by a discharge

and infection of the eyes, and one case of distemper was confirmed (Pratt 1962; Risser 1963). Coatis are also susceptible to rabies, and rabid coatis have been observed in Arizona (Krebs et al. 2003; Hass 2021), but it is rare for such a social animal.

STATUS AND MANAGEMENT

The white-nosed coati is designated as a furbearing animal under New Mexico Statute 17-5-2, however, currently there is no open season in the state (NM Administrative Code 19-32-2). Indeed, there was demand for coati pelts in the United States in the 1920s when raccoon coats were popular, not to mention that coati flesh is also said to be tender and tasty (Pratt 1962). The coati is also included in Group IV of the New Mexico Department of Game and Fish "Director's Species Importation List." Its importation into the state is prohibited, and hence it is illegal to obtain and keep a coati even as an exotic pet in New Mexico. This helps prevent the intentional release or accidental escape of pet coatis, which could then introduce diseases or disrupt the genetic integrity of native populations in the state. The coati was listed as Endangered (Group 2 = Threatened) under the New Mexico Wildlife Conservation Act on 24 January 1975 because it was thought to be so rare that disease or harvest could threaten its survival in the state. However, it was delisted on 28 March 1985 based on improved information that indicated it had a larger range in the state than previously known (Jones and Schmitt 1997). The coati was then listed as a Species of Greatest Conservation Need in the 2006 Comprehensive Wildlife Conservation Strategy for New Mexico (NMDGF 2006), but was not included in the most recent version of New Mexico's state Wildlife Action Plan (NMDGF 2016).

In some tropical areas, particularly those with tourist resorts, coatis are easier to observe due to their higher population densities and relative tameness. Easy observability makes tropical coatis attractive for study. In contrast, coatis in the United States exist at very low densities and their habitat is patchy, both across the regional landscape and within mountain ranges. Thus, there are few places in the United States where coatis can be reliably observed, and even in the best conditions where researchers are using radiotelemetry tracking equipment, coatis are still difficult to see (McColgin et al. 2018; Hass 2021). In recent years, however, several factors have improved our knowledge about the distribution and habitat associations of coatis in New Mexico, including the increased presence of humans in poorly studied areas, improved means of wildlife documentation such as remote cameras used in wildlife studies and digital cameras in personal cell phones, and rapid and archivable forms of communication (e.g., email, social media). However, there is still much we do not know about the ecology of coatis in the state, some of which is perhaps different from other areas as New Mexico lies at the very edge of the species' geographic range and ecological tolerances. We also lack information about the coati's population status and trends in New Mexico. Some threats to coati populations can nonetheless be identified, among them hunting, predator control activities, and habitat loss.

Hunting by humans can reduce populations of white-nosed coatis (Glanz 1991). Coatis might be especially vulnerable to hunting because females are the reproductive base of a population, and they occur in groups that can be noisy and conspicuous (Escamilla et al. 2000; Hass and Valenzuela 2002). In tropical regions coatis are regularly taken by poachers and subsistence hunters, and to control damage to poultry or gardens (Escamilla et al. 2000; Wright et al. 2000). Kaufmann (1987) speculated that intensive subsistence hunting for coatis in northern Mexico may have isolated populations in the United States. Currently, no hunting season has been established for the species in New Mexico, but in Arizona, there is a seven-month open season with a bag limit of one coati. Part of the justification for the hunting season in Arizona seems to be the perception that the species has expanded its range and become more common in areas where it was not previously known (but see under "Distribution" for a discussion about the range expansion hypothesis). The belief by many that coatis are also expanding their range and becoming more common in New Mexico, in addition to a demand for additional hunting opportunities, could put pressure on the New Mexico Department of Game and Fish to open a season for the species in New Mexico. However, it seems unlikely that New Mexico has a large enough breeding population of coatis to support sustainable hunting.

There are no data to estimate long-term coati population trends in any area of the United States. However, coati populations are known to undergo year-to-year size fluctuations (Gompper 1995). Anecdotal evidence suggests that coati populations in the United States are highly variable and have occasionally crashed (e.g., Risser 1963; Hass 2021). The extreme variation as to where and when coatis have been observed in Arizona has even led some authors to speculate—wrongly as it turned out (see under "Life History")—that the species is nomadic (Wallmo and Gallizioli 1954). Hass (2021) documented a population crash in the Huachuca Mountains in the early 2000s, attributed to drought and cougar predation, and recovery took more than a dozen years. One other explanation for observed population crashes is due to predator damage control activities. For instance, Pratt (1962) reported a dramatic decline in coatis on Fort Huachuca in Arizona following the 1958–1959 hunting season when hunters were encouraged to kill coatis, which had been blamed for destroying wild turkey (*Melagris gallopavo*) and Montezuma quail (*Cyrtonyx montezumae*) nests the previous year. In the Big Burro Mountains in New Mexico, coatis reportedly disappeared in 1971, about the same time a coyote poisoning program was implemented (Kaufman et al. 1976). Most of the area where coatis occur in New Mexico and Arizona is used for livestock grazing. Historically, there was a concerted effort to eliminate predators from vast areas of rangelands using traps, guns, and various poisons such as strychnine (Robinson 1962; Hawthorne 2004). It is possible that historical poisoning campaigns could have reduced coati populations in some areas. Since the mid-1970s the main poison used for predator control has been the M44 sodium cyanide capsule, which is deployed with a spring-powered ejector specifically designed for coyotes and therefore has a lesser impact on non-target species (see Chapter 3). For the period 2010–2020, Wildlife Services program data reports indicate only a single coati was killed during predator control activities. Thus, the current threat of predator damage management activities to coatis appears to be low. In addition, recently enacted laws limiting trapping on public land may reduce accidental take. It is conceivable that efforts to remove cougars from some desert mountain ranges for bighorn sheep

(*Ovis canadensis*) restoration (Chapter 9) could have a positive impact on coatis as well, though this hypothesis requires testing.

In the United States, coatis tend to occur in remote places with little human activity. However, substantial habitat alterations have occurred throughout this region that could result in reduced coati distribution or abundance. Most areas occupied by coatis in New Mexico and Arizona have long been grazed by livestock. Historical overstocking of rangelands has resulted in desertification of grasslands with concomitant expansion of desert scrub and loss and degradation of riparian zones (Gehlbach 1993). Coatis require a diversity of food items, yet livestock grazing tends to simplify both plant and animal communities upon which coatis depend. Of particular concern is the widespread de-watering of springs, cienegas, streams, and rivers throughout the range of the coati in New Mexico and elsewhere in the United States and northern Mexico (Stevens and Meretsky 2008). The moist soils in riparian zones provide much of a coati's food supply (Hass 2021). Coatis may require access to free water and in many places, those water sources are springs. However, many springs have disappeared due to increasing aridity, groundwater pumping, head-spring alteration, livestock grazing, and changed terrestrial vegetation (Stevens and Meretsky 2008). Artificial waterers, developed for livestock or the conservation of wildlife such as bighorn sheep in desert mountain ranges, may offset some of the losses of springs. However, artificial waters are not necessarily a panacea because they tend to concentrate animals and attract important predators such as cougars (Harris et al. 2020), and potentially draw coatis away from cover. Management efforts to reduce year-round water availability to differentially benefit bighorn sheep as opposed to cougars (Harris et al. 2020) could have the unintended consequence of seasonally eliminating water for coatis. Artificial waterers also do not provide the coati's preferred foods that are associated with natural riparian areas.

The white-nosed coati adds unique diversity to the mammal fauna of New Mexico, and the species garners considerable interest and affection from the public. It is clear that much of southwestern New Mexico is home to this species, though the coati remains rare and uncommonly observed. People fortunate to encounter a coati are sure to remember the event. However, due to their low densities, fragmented habitat, and reliance on vulnerable resources, coatis are always at risk of possible population losses. In order to ensure the long-term persistence of coatis in New Mexico, it is important that more research be conducted to better understand how populations exist on the arid periphery of the species' range. This research should include information on spatio-temporal variation in metapopulation structure, community dynamics around water sources, and habitat use, particularly in Chihuahuan Desert ecosystems. Coatis are thought to be impacted by drought, which makes them even more reliant on water sources (Risser 1963; Hass 2021). Climate change will cause longer and more profound droughts, resulting in additional stresses on water sources and rangeland conditions (see Chapter 5). Research is necessary to project and adapt management to these future changes so that coatis will continue to exist in New Mexico.

LITERATURE CITED

Allen, J. A. 1879. On the coatis (Genus *Nasua*, Storr). *Bulletin of the United States Geological and Geographical Survey of the Territories* 5:153–74.

———. 1904. Mammals from southern Mexico and Central and South America. *Bulletin of the American Museum of Natural History* 20:29–80.

Arsnov, B. M., and S. T. Sakai. 2013. The procyonid social club: comparison of brain volumes in the coatimundi (*Nasua nasua*, *N. narica*), kinkajou (*Potos flavus*), and raccoon (*Procyon lotor*). *Brain, Behavior, and Evolution* 82:129–45.

Bailey, V. 1905. *Biological survey of Texas*. North American Fauna 25. Washington, DC: US Department of Agriculture, Bureau of Biological Survey.

———. 1931 (=1932). *Mammals of New Mexico*. North American Fauna 53. Washington, DC: US Department of Agriculture, Bureau of Biological Survey.

Bernardini, W. 2005. *Hopi oral tradition and the archaeology of identity*. Phoenix: University of Arizona Press.

Binczik, G. A. 2006. Reproductive biology of a tropical procyonid, the white-nosed coati. PhD dissertation, University of Florida, Gainesville.

Booth-Binczik, S. D. 2001. Ecology of social behavior in Tikal National Park, Guatemala. PhD dissertation, University of Florida, Gainesville.

Brown, A. E. 1898. Report of the Board of Directors, Philadelphia, April 28th, 1898. In *The twenty-fifth annual report of the Board of Directors of the Zoological Society of Philadelphia*, 5–15. Philadelphia: Allen, Lane & Scott's Printing House.

Brown, D. E., ed. 1994. *Biotic communities: southwestern United States and northwestern Mexico*. 2nd ed. Salt Lake City: University of Utah Press.

———. 2012. *Bringing back the game: Arizona wildlife management 1912–1962*. Phoenix: Arizona Game and Fish Department.

Brown, D. E., and R. D. Babb. 2008. Walter P. Taylor and Arizona's porcupines. *Arizona Wildlife Views* May–June 2008.

Brown, D. E., and R. Davis. 1995. One hundred years of vicissitude: terrestrial bird and mammal distribution changes in the American Southwest, 1890–1990. In *Biodiversity and management of the Madrean Archipelago: the Sky Islands of southwestern United States and northwestern Mexico*, ed. L. H. DeBano, P. H. Ffolliott, A. Ortega-Rubio, et al., 231–44. Fort Collins, Colorado: US Department of Agriculture, Forest Service, Rocky Mountain

Forest and Range Experiment Station. Gen. Tech. Rep. RM-GTR-264: 669.

Cahalane, V. H. 1939. Mammals of the Chiricahua Mountains, Cochise County, Arizona. *Journal of Mammalogy* 20:418–40.

Christiansen, P., and J. S. Adolfssen. 2005. Bite forces, canine strength, and skull allometry in carnivores (Mammalia, Carnivora). *Journal of Zoology* 266:133–51.

Cockrum, E. L. 1960. *The recent mammals of Arizona: their taxonomy and distribution*. Tucson: University of Arizona Press.

Cook, J. A. 1986. The mammals of the Animas Mountains and adjacent areas, Hidalgo County, New Mexico. *Occasional Papers of the Museum of Southwestern Biology* 4:1–45.

Davis, G. P., Jr. 2001. *Man and wildlife in Arizona: the American exporation period 1824–1865*. Phoenix: Arizona Game and Fish Department.

Davis, R., and J. R. Callahan. 1992. Post-Pleistocene dispersal in the Mexican vole (*Microtus mexicanus*): an example of an apparent trend in the distrbution of southwestern mammals. *Great Basin Naturalist* 52:262–68.

Decker, D. M. 1991. Systematics of the coatis, *Nasua* (Mammalia: Procyonidae). *Proceedings of the Biological Society of Washington* 104:370–86.

Egbert, J. 1980. Observations of wildlife and wildlife habitat in canyon and broadleaf riprarian areas of the Animas Mountains. Unpublished report, Natural History Services, Cliff, NM.

Escamilla, A., M. Sanvincente, M. Sosa, and C. Galindo-Leal. 2000. Habitat mosaic, wildlife availability, and hunting in the tropical forest of Calakmul, Mexico. *Conservation Biology* 14: 1592–601.

Fewkes, J. W. 1923. Designs on prehistoric pottery from the Mimbres Valley, New Mexico. *Smithsonian Miscellaneous Collections* 74:1–47.

———. 1924. Additional designs on prehistoric Mimbres pottery. *Smithsonian Miscellaneous Collections* 76:1–46.

Findley, J. S., A. H. Harris, D. E. Wilson, and C. Jones. 1975. *Mammals of New Mexico*. Albuquerque: University of New Mexico Press.

Frey, J. K. 2006. Inferring species distributions in the absence of occurrence records: an example considering wolverine (*Gulo gulo*) and Canada lynx (*Lynx canadensis*) in New Mexico. *Biological Conservation* 130:16–24.

———. 2009. Distinguishing range expansions from previously undocumented populations using background data from museum records. *Diversity and Distributions* 15:183–87.

Frey, J. K., J. C. Lewis, R. K. Guy, and J. S. Stuart. 2013. Use of anecdotal occurrence data in species distribution models: an example based on the white-nosed coati (*Nasua narica*) in the American Southwest. *Animals* 3:327–48.

Gehlbach, F. R. 1993. *Mountain islands and desert seas: a natural history of the U.S.-Mexican borderlands*. College Station: Texas A&M University Press.

Geluso, K. 2009. Distributional records for seven species of mammals in southern New Mexico. *Occasional Papers, Museum of Texas Tech University* 287:1–7.

Gilbert, B. 1973. *Chulo: a year among the coatimundis*. Tucson: University of Arizona Press.

Glanz, W. E. 1991. Mammalian densities at protected versus hunted sites in Central Panama. In *Neotropical wildlife use and conservation*, ed. J. G. Robinson and K. H. Redford, 163–73. Chicago: University of Chicago Press.

Gómez-Ortiz, Y., and O. Monroy-Vilchis. 2013. Feeding ecology of puma *Puma concolor* in Mexican montane forest with comments about jaguar *Panthera onca*. *Wildlife Biology* 19:179–87.

Gómez-Ortiz, Y., O. Monroy-Vilchis, and G.D. Mendoza-Martinez. 2015. Feeding interactions in an assemblage of terrestrial carnivores in central Mexico. *Zoological Studies* 54, article 16. doi:10.1186/s40555-014-0102-7.

Gompper, M. E. 1995. *Nasua narica*. *Mammalian Species* 487:1–10.

———. 1996. Sociality and asociality in white-nosed coatis (*Nasua narica*): foraging costs and benefits. *Behavioral Ecology* 7:254–63.

———. 2004. Correlations of coati (*Nasua narica*) social structure with parasitism by ticks and chiggers. In *Contribuciones mastozoológicas en homenaje a Bernardo Villa.*, ed. V. Sánchez-Cordero and R.A. Medellín, 527–34. Instituto de Biología e Instituto de Ecológica, UNAM, Mexico.

Gompper, M. E., J. L. Gittleman, and R. K. Wayne. 1997. Genetic relatedness, coalitions and social behaviour of white-nosed coatis, *Nasua narica*. *Animal Behaviour* 53:781–97.

———. 1998. Dispersal, philopatry, and genetic relatedness in a social carnivore: comparing males and females. *Molecular Ecology* 7:157–63.

Gompper, M. E., and J. S. Krinsley. 1992. Variation in social behavior of adult male coatis (*Nasua narica*) in Panama. *Biotropica* 24:216–19.

Hall, E. R. 1981. *Mammals of North America*. Vol. 2. 2nd ed. Caldwell, NJ: Blackburn Press.

Harris, G. M., D. R. Stewart, D. Brown, L. Johnson, J. Sanderson, A. Alvidrez, T. Waddell, and R. Thompson. 2020. Year-round water management for desert bighorn sheep corresponds with visits by predators not bighorn sheep. *PLOS One* 15:e0241131. https://doi.org/10.1371/journal.pone.0241131.

Hass, C. C. 1997. *Ecology of white-nosed coatis in the Huachuca Mountain, Arizona: a preliminary study*. Final report submitted to the Arizona Game and Fish Department, Phoenix.

———. 2002. Home-range dynamics of white-nosed coatis in southeastern Arizona. *Journal of Mammalogy* 83: 934–46.

———. 2009. Competition and coexistence in sympatric bobcats and pumas. *Journal of Zoology*, 278:174–180.

———. 2021. *Nosey beast: natural history of the coatis*. Carson City, NV: Wild Mountain Echoes.

Hass, C. C., and J. F. Roback. 2000. Copulatory behavior of white-nosed coatis. *Southwestern Naturalist* 45:329–31.

Hass, C. C., and D. Valenzuela. 2002. Anti-predator benefits of group living in white-nosed coatis (*Nasua narica*). *Behavioral Ecology and Sociobiology* 51:570–78.

Hawthorne, D. W. 2004. The history of federal and cooperative animal damage control. *Sheep and Goat Research Journal* 19:13–15.

Hershkovitz, P. 1951. Mammals from British Honduras, Mexico, Jamaica and Haiti. *Fieldiana: Zoology* 31:547–69.

Hoffmeister, D. F. 1986. *Mammals of Arizona*. Tucson and Phoenix: University of Arizona Press and Arizona Game and Fish Department.

Holmgren, V. C. 1990. *Raccoons in folklore, history and today's backyards*. Santa Barbara, CA: Capra Press.

Jones, C., and C. G. Schmitt. 1997. Mammal species of

concern in New Mexico. In *Life among the muses: papers in honor of James S. Findley*, ed. T. L. Yates, W. I. Gannon, and D. E. Wilson, 179–205. Albuquerque: University of New Mexico, Museum of Southwestern Biology. *Special Publication of the Museum of Southwestern Biology, University of New Mexico* 3:1–290.

Kaufmann, J. H. 1962. Ecology and social behavior of the coati, *Nasua narica*, on Barro Colorado Island, Panama. *University of California Publications in Zoology* 60:95–222.

———. 1987. Ringtail and coati. In *Wild furbearer management and conservation in North America*, ed. M. Novak, J. A. Baker, M. E. Obbard, and B. Malloch, 501–8. Ontario, Canada: Ministry of Natural Resources.

Kaufmann, J. H., D. V. Lanning, and S. E. Poole. 1976. Current status and distribution of coati in United States. *Journal of Mammalogy* 57: 621–37.

Krebs, J. W., S. M. Williams, J. S. Smith, C. E. Rupprecht, and J. E. Childs. 2003. Rabies among infrequently reported mammalian carnivores in the United States, 1960–2000. *Journal of Wildlife Diseases* 39:253–61.

Lanning, D. V. 1976. Density and movements of the coati in Arizona. *Journal of Mammalogy* 57:609–11.

McClearn, D. 1992. Locomotion, posture, and feeding behavior of kinkajous, coatis, and raccoons. *Journal of Mammalogy* 73:245–61.

McColgin, M. E. 2006. Sociality and genetics of a southeastern Arizona coati (*Nasua narica*) population. PhD dissertation, Purdue University, West Lafayette, Indiana.

McColgin, M. E., E. J. Brown, S. M. Bickford, A. L. Eilers, and J. L. Koprowski. 2003. Use of century plants (*Agave palmeri*) by coatis (*Nasua narica*). *Southwestern Naturalist* 48:722–24.

McColgin, M. E., J. L. Koprowski, and P. M. Waser. 2018. White-nosed coatis in Arizona: tropical carnivores in a temperate environment. *Journal of Mammalogy* 99:64–74.

New Mexico Department of Game and Fish (NMDGF). 2006. *Comprehensive wildlife conservation strategy for New Mexico*. Santa Fe: New Mexico Department of Game and Fish.

———. 2016. *State Wildlife Action Plan for New Mexico*. Santa Fe: New Mexico Department of Game and Fish.

Nigenda-Morales, S.F., M. E. Gompper, D.

Valenzuela-Galvan, A. R. Lay, K. M. Kapheim, C. Hass, S. D. Booth-Binczik, G.A. Binczik, B. T. Hirsch, M. McColgin, J. L. Koprowski, K. McFadden, R. K. Wayne, and K-P. Koepfli. 2019. Phylogeographic and diversification patterns of the white-nosed coati (*Nasua narica*): evidence from south-to-north colonization of North America. *Molecular Phylogenetics and Evolution* 131:149–63.

Olifiers, N., R. C. Bianchi, G. Mourao, and M. E. Gompper. 2009. Construction of arboreal nests by brown-nosed coatis, *Nasua nasua* (Carnivora: Procyonidae) in the Brasilian Pantanal. *Zoologia* 26:571–74.

Pratt, J. J. 1962. Establishment and trends of coati-mundi in the Huachucas. *Modern Game Breeding* 23:10–11.

Ratnayeke, S., A. Bixler, and J. L. Gittleman. 1994. Home range movements of solitary, reproductive female coatis, *Nasua narica*, in south-eastern Arizona. *Journal of Zoology* 233:322–26.

Reeder, W. G. 1951. Breeding record of the coati mundi in the United States. *Journal of Mammalogy* 32:362–63.

Risser, A. C., Jr. 1963. A study of the coati mundi (*Nasua narica*) in southern Arizona. MS thesis, University of Arizona, Tucson.

Roback, J. F. 1998–1999. *Coati survey: Big Hatchet sighting*. Share with Wildlife Newsletter, New Mexico Department of Game and Fish, Santa Fe, Winter 1998–1999.

———. 2001. *Coati surveys: additional records obtained*. Share with Wildlife Newsletter, New Mexico Department of Game and Fish, Santa Fe, Spring–Summer 2001.

Robinson, W. B. 1962. Methods of controlling coyotes, bobcats, and foxes. *Proceedings of the First Vertebrate Pest Conference* 5:32–56. https://digitalcommons.unl.edu/vpcone/5.

Rominger, E. M., H. A. Whitlaw, D. L. Weybright, W. C. Dunn, and W. B. Ballard. 2004. The influence of mountain lion predation on bighorn sheep translocations. *Journal of Wildlife Management* 68:993–99.

Ruiz-Garcia, M., M. F. Jaramillo, C. H. Caceres-Martinez, and J. M. Shostell. 2020. The phylogeographic structure of the mountain coati (*Nasuella olivacea*; Procyonidae, Carnivora), and its phylogenetic relationships with other coati species (*Nasua nasua* and *Nasua narica*) as inferred by mitochondrial DNA. *Mammalian Biology* 100:521–48.

Schmidly, D. J., J. Karges, and R. Dean. 2004. Distribution records and reported sightings of the white-nosed coati (*Nasua narica*) in Texas, with comments on the species' population and conservation status. In *Contributions in natural history: a memorial volume in honor of Clyde Jones*, ed. R. W. Manning, J. R. Goetze, and F. D. Yancey II, 127–147. Special Publications Museum of Texas Tech University, No. 65.

Smith, H. J. 1977. Social behavior of the coati (*Nasua narica*) in captivity. PhD dissertation, University of Arizona, Tucson.

———. 1980. Behavior of the coati (*Nasua narica*) in captivity. *Carnivore* 3:88–136.

Stevens, L. E., and V. J. Meretsky. 2008. *Aridland springs in North America: ecology and conservation*. Tucson: University of Arizona Press and Arizona-Sonora Desert Museum.

Taber, F. W. 1940. Range of the coati in the United States. *Journal of Mammalogy* 21:11–14.

Taylor, W. P. 1934. Coati added to the list of United States mammals. *Journal of Mammalogy* 15:317–18.

Valenzuela, D. 1998. Natural history of the white-nosed coati, *Nasua narica*, in a tropical dry forest of western Mexico. *Revista Mexicana de Mastozoologia* 3:26–44.

Valenzuela, D., and G. Ceballos. 2000. Habitat selection, home range, and activity of the white-nosed coati (*Nasua narica*) in a Mexican tropical dry forest. *Journal of Mammalogy* 81:810–19.

Valenzuela, D., and D. W. Macdonald. 2002. Home-range use by white-nosed coatis (*Nasua narica*): limited water and a test of the resource dispersion hypothesis. *Journal of Zoology* 258:247–56.

Wallmo, O. C., and S. Gallizioli. 1954. Status of the coati in Arizona. *Journal of Mammalogy* 35:48–54.

Wetherhill, M. A. 1957. Occurrence of coati in northern Arizona. *Journal of Mammalogy* 38:123.

Wilson, D. E., and D. M. Reeder, eds. 2005. *Mammal species of the world: a taxonomic and geographic reference*. 3rd ed. Baltimore: Johns Hopkins University Press.

Wright, S. J., H. Zeballos, I. Dominguez, M. M. Gallardo, M. C. Moreno, and R. Ibanez. 2000. Poachers alter mammal abundance, seed dispersal, and seed predation in a neotropical forest. *Conservation Biology* 14:227–39.

RACCOON (*PROCYON LOTOR*)

Jean-Luc E. Cartron

One of New Mexico's three procyonid mammals, the raccoon (*Procyon lotor*; also called northern or common raccoon) is often dubbed the masked bandit, being easily recognized by its black facial mask and notorious for committing nighttime mischief in human-inhabited areas. Little studied yet common in some parts of the state, it is no doubt endearing to many people but also often ranks high on the list of "usual suspects" where human-wildlife conflicts arise. In New Mexico as elsewhere, raccoons can make a nuisance of themselves by raiding crops, chicken coops, campsites, mountain cabins, backyard ponds, bird feeders, and dumpsters. Even the mammalogist working out in the field must occasionally contend with marauding raccoons. At Bandelier National Monument in northern New Mexico, Bogan et al. (2007:29) reported that:

> during [the] summer [of] 2004, we trapped on multiple nights in Cañon de los Frijoles for the meadow jumping mouse (*Zapus hudsonius*). . . . Unfortunately, our efforts to capture this species were hampered by a raccoon that continually molested more than half of our traps each night. In fact, this individual even learned how to remove small mammals from [the] traps.

A medium-sized mammal, the raccoon is distinctive for not only its facial mask but also its stocky build, highly arched back, and ringed tail. The head is broad, with rounded ears and a round face tapering to a pointed muzzle. The facial mask lies across the black eyes and the cheeks, and is fringed by whitish fur. Down the center of the facial mask is a dark brown or black stripe that extends from the forehead to the nose. On the back and along the sides of the body, the raccoon's fur can range in coloration from blackish or silvery gray to pale brown or reddish (see description of subspecies below), while the underside is lighter. Under the long guard hairs is dense, short underfur. Raccoons molt in the spring, and their fur tends to be densest and longest during the winter months.

Although uncommon, albino raccoons have been documented in New Mexico. Among the museum specimens housed at the University of New Mexico's Museum of Southwestern Biology (MSB) is an albino female (MSB 14053) collected on 3 September 1960 16 km (10 mi) northeast of Grants in what is today Cibola County. Rick Winslow (pers. comm.) encountered albino raccoons twice, both in late summer of 1998, one at Narbona Pass (formerly Washington Pass) near Whiskey Creek in the Chuska Mountains, and the other just across the Arizona–New Mexico state line at Red Lake, Arizona, just southwest of Crystal in McKinley County, New Mexico. Both were young-of-the-year animals with normally colored siblings. An adult female albino was admitted to Wildlife Rescue Inc. in Albuquerque on 28 May 2009. On 2 June 2011, the Wildlife Center in Española received a juvenile male albino (Photo

(*opposite page*) Photograph: © Don MacCarter/New Mexico Department of Game and Fish.

Photo 32.1. Adult raccoon photographed in Los Alamos County, New Mexico. Two of the raccoon's most distinctive features are the ringed tail and the black facial mask fringed by whitish fur. Note also the broad head, rounded ears, and dark stripe extending from the forehead down through the face mask to the nose. Raccoons are stocky animals with a highly arched back, short front legs—with dexterous front paws—and longer hind legs. Photograph: © Hari Viswanathan.

32.2) apparently found as an uninjured orphan near Kirtland in San Juan County.

Most North American adult raccoons weigh between 4 kg and 9 kg (~ 9–19 lbs) (Gehrt 2003). Less frequently, adults can reach a weight of up to 18 kg (~40 lbs) or more, the record belonging to a raccoon in Wisconsin weighing 28.3 kg (62.7 lbs) (Scott 1951). Body size varies geographically in North America, with raccoons in the northwestern United States tending to be the largest and those in the southeast tending to be the smallest (Goldman 1950; Ritke and Kennedy 1988; Gehrt 2003). Raccoons also weigh more at the onset of winter than in the spring, and males

are on average larger than females (Gehrt 2003). Few measurements have been taken on raccoons in New Mexico. However, one adult male (MSB 43297) collected at La Joya State Game Refuge on 11 October 1997 weighed 6.75 kg (14.9 lbs). Another male brought into the Wildlife Center in Española (MSB 231084—the raccoon did not survive) weighed approximately 6.4 kg (14.2 lbs). Both of these measurements fall well within the usual range of adult body weights rangewide.

The four subspecies reported by Frey (2004) from New Mexico consist of *P. l. pallidus*, *P. l. hirtus*, *P. l. fuscipes*, and *P. l. mexicanus* (subsumed under *P. l. hernandezii* by Wozencraft in Wilson and Reeder [2005]). As indicated by its name, the subspecies *P. l. pallidus* is characterized overall by its pale coloration, but it also lacks a black

Photo 32.2. Juvenile male albino raccoon brought to the Wildlife Center in Española on 2 June 2011, after being found as an uninjured orphan near Kirtland in San Juan County. Photograph: © Alissa Diane Mundt.

patch behind the ear (Goldman 1950). Raccoons belonging to *P. l. hirtus* tend to be large and dark, with a full, dense pelage often suffused with ochraceous buff; skull measurements also reveal a high, narrow frontal region compared to other subspecies. The subspecies *P. l. fuscipes* is similar in size to *P. l. hirtus*, with a similarly high frontal region; however, its pelage is paler and suffused with gray instead of golden yellow. There appears to be much variation in color in *P. l. hernandezii*, but in New Mexico—where previously it was described as *P. l. mexicanus*—the subspecies is pale. It resembles *pallidus* in size and coloration but with a broader skull particularly in the inter-orbital region (Goldman 1950).

The raccoon has a plantigrade gait, with heels, soles, and digits in contact with the ground while walking. Hind paws and forepaws have hairless soles, five toes or fingers, and non-retractile claws. The forepaws resemble baby human hands as a result of having soles (palms) and long, well-separated fingers, while the elongated hind paws are much like baby human feet. The hind prints are not only longer and wider than the fore prints, they are also deeper, as the raccoon carries more of its body weight posteriorly (Photo 32.3).

Raccoons communicate using a variety of visual, olfactory, tactile, and vocal signals. Sieber (1984) described 13 vocal signals, including whistle, squeal, bark, snort, grunt, and growl. The squeal and a "churr" seem to be uttered exclusively during the nestling period. When facing a conspecific intruder, raccoons may bark, growl and snort while also baring their teeth and laying their ears back.

DISTRIBUTION

Like all other members of the family Procyonidae, the raccoon is native to the Western Hemisphere. Its distribution extends primarily from southern Canada south through the United States, Mexico, and Central America (e.g., Eisenberg 1989; Nowak 1991; De la Rosa and Nocke 2000). Recently,

Photo 32.3. Raccoon tracks in snow at Bandelier National Monument on 10 January 2012. The hind track (left) is large and often shows a long heel pad. The front track (right) is smaller and often compared to that of a baby's hand. As raccoons walk, each hind foot lands next to the opposite front foot. Photograph: © Sally King.

Guzmán-Lenis (2004) reported the raccoon from western Columbia, placing the southern edge of the species' distribution in northernmost South America. Farther south, the raccoon is replaced by its close relative *P. cancrivorus*, the crab-eating raccoon; the two species are sympatric in Costa Rica, Panama, and Columbia (Eisenberg 1989; De la Rosa and Nocke 2000; Guzmán-Lenis 2004). Raccoons were introduced on some islands off the coasts of southeastern Alaska and British Columbia, and raccoon populations on Bahamian islands and on Guadalupe Island are believed to also be the result of introductions (Gehrt 2003). The raccoon is no longer restricted to the Americas, having been introduced into the former Soviet Union for its fur in the 1930s (Redford 1962; Aliev and Sanderson 1966), Germany in the 1930s and 1940s (Röben 1975; Lutz 1984, 1995; Kauhala 1996), and Japan in the 1960s (Ikeda et al. 2004). From Germany, raccoons have steadily expanded their distribution into central and northern Europe, and populations have become established in various areas such as northern France and western Poland (Lutz 1984; Kauhala 1996; Bartoszewicz et al. 2008).

Photo 32.4. Raccoons along the shore of a pond at the Bosque del Apache National Wildlife Refuge in Socorro County. Photograph: © William Horton.

Photo 32.5. Middle Rio Grande Valley in Bernalillo County. Raccoons are common or locally common in the state's river valleys and other areas with proximity to water. Photograph: © Tom Kennedy.

The US and Canada raccoon population has grown manyfold since the 1930s as a result of anthropogenic factors (Sanderson 1987; Zeveloff 2002; Prugh et al. 2009). Raccoons have spread deeper into the Rocky Mountain region, including northwestern Colorado and western Wyoming where before the 1960s the species was apparently scarce or absent (Finley 1995). They have also expanded their distribution northward into parts of Canada where native people previously had no name for the species (Zeveloff 2002). In the Canadian prairies, raccoons have spread northward to now reach the southern limit of the boreal forest biome (Gehrt 2003).

Raccoons are found throughout New Mexico, primarily along the edges of springs, lakes, and perennial streams. Their distribution in the state reaches at least 2,620 m (8,600 ft) in elevation, higher in the Sangre de Cristo Mountains, where according to Berghofer (1967) raccoons occur up to the timberline (~3,660 m [12,000 ft]). The species has been recorded in particular along all five major rivers in the state (Rio Grande, San Juan, Gila, Canadian, and Pecos), and along the Rio Grande in northern and central New Mexico it is by all accounts relatively common to common (Ivey 1957; Hink and Ohmart 1984; Cartron et al. 2008; see below).

Raccoons appear to be comparative newcomers to New Mexico (Chapter 1). The species was identified as part of the late Wisconsin (13,000–11,000 years ago) fauna at Blackwater Draw, near the Texas state line in Roosevelt County (Slaughter 1975), but not anywhere else in the state (Harris 1990). Later, however, it appears among faunal remains at several late prehistoric sites with occupation dates spanning several centuries: two Anasazi sites along the south bank of the San Juan River in San Juan County, LA 126581 (Tommy Site, AD 800–1120; Enright 2008) and LA 16660 (Box B Site, AD 1200–1300; Hogan and Sebastian 1991); one Mogollon site along the upper Gila River (Villareal Pit House 2; AD 1000–1150; Lekson 2002); and possibly also Arroyo Hondo Pueblo south of Santa Fe from between approximately AD 1315 and 1330 (Lang and Harris 1984). During the US-Mexican War, Major Stephen Watts Kearny and a part of his "Army of the West" marched out of Santa Fe on 25 September 1846, following the then-called Rio del Norte (Rio Grande) south to present-day Doña Ana County, before heading west toward California. On 11 October, west of "Jornada Mountain," Captain A. R. Johnston noted raccoon tracks (along with the tracks of other mammals and birds) along the river (Emory 1848). Contrary to Scurlock's (1998: 231) account, however, it appears that Lieutenant James W. Abert (1848) never encountered raccoons during his tour of New Mexico. Abert's narrative does mention raccoons on two occasions, both outside New Mexico. One is a raccoon killed for food on 25 February 1847, four days before reaching Fort Leavenworth, Kansas during the return trip.

Most of the early museum specimen records

Photo 32.6. Raccoon photographed on 16 February 2009 by one of 38–40 camera traps on the San Andreas National Wildlife Refuge in Doña Ana County. The network of camera traps was set up primarily to document distribution, occupancy, and habitat correlates of cougar (*Puma concolor*) and other large carnivores on the refuge. Photograph: © Louis C. Bender.

Photo 32.7. Raccoon at Rattlesnake Springs, a desert oasis with water and tall cottonwoods adjacent to Carlsbad Caverns National Park, Eddy County. Photograph: © Corey Anderson.

from the state are from the upper Rio Grande, from Rinconada downstream to Alcalde in Rio Arriba County: Rinconada on 18 April 1904 (USNM 132483); Velarde on 28 November 1904 (USNM 136308), 30 November 1904 (USNM 136309), 6 January 1905 (USNM 136310), and 11 March 1905 (USNM 136316); and Alcalde on 26 January 1905 (USNM 136317), 5 February 1905 (USNM 136318), 6 February 1905 (USNM 136319), 25 February 1905 (USNM 136320), 1 March 1905 (USNM 136321), 3 March 1905 (USNM 136322), and 12 March 1905 (USNM 136323). Except for the 1904 Rinconada specimen, secured by McClure Surber for the US Biological Survey (see Bailey 1928), all the above specimens are from the same collector, C. C. Beattie. Five more specimens secured by Beattie are from a locality recorded as Senega (sic), likely La Cienaga between Rinconada and Velarde: 29 January 1905 (USNM 136311), 6 February 1905 (USNM 136312), 8 February 1905 (USNM 136313), 28 February 1905 (USNM 136314), and 2 March 1905 (USNM 136315). The 16 raccoon specimens C. C. Beattie collected within a period of less than four months and sent to the Smithsonian Institute were evidently the result of trapping. In a letter published in the October 1905 issue of the magazine *Hunter-Trader-Trapper* (11:59), C. C. Beattie wrote of New Mexico:

> The northern part of the state beats any other place I have ever been, but owing to the strict law just passed, it [trapping] will be [a thing] of the past now. The law prohibits the killing or trapping of any beavers, otters or mink for some time to come. So don't let any other brother trapper come here to trap.

At the time, C. C. Beattie lived in Edith along the New Mexico–Colorado state line. After 12 March 1905 he stopped sending specimens of raccoons or other mammals to the Smithsonian, but his earlier trapping serves to establish the presence of a raccoon population around the turn of the

20th century along the upper Rio Grande in Rio Arriba County.

Other than the 1846 record (Emory 1848), the presence of raccoons farther south along the Rio Grande in early-day New Mexico was never well documented, though from Bailey (1931:348) we know that the species was likely common, as it was found "along practically every permanent stream" at lower elevations. On the basis of extensive surveys, Hink and Ohmart (1984) more recently characterized raccoons as "very common" along a 260 km (160 mi) stretch of the river extending from Española at the southern end of the Rio Grande Gorge—along the upper Rio Grande—south to San Acacia, Socorro County. Along the middle Rio Grande (from Cochiti Dam to Elephant Butte Reservoir), raccoons and their tracks are indeed a familiar sight in many wet areas, including at the Bosque del Apache National Wildlife Refuge (NWR) (Photo 32.4). As elsewhere in the state, however, the likelihood of finding raccoons in the middle Rio Grande valley seems influenced by proximity to concentrated, artificial food resources, and thus the species is particularly common in the Albuquerque area, where the human population increased from around 35,000 to 200,000 people between 1940 and 1960 (Reeve 1961). It was also during that time that bank stabilization efforts led to the formation of a continuous riparian forest along the river's banks, where before open areas would have been predominant (see Cartron et al. 2008). Interestingly, the earliest museum specimens collected along the middle Rio Grande are from no earlier than the 1960s, one from Isleta, Bernalillo County (on 29 June 1963; MSB 22252), the other from the Bernardo refuge in Socorro County (on 3 October 1967; MSB 25609). In his notes on the mammals of Bernalillo County, Ivey (1957) also reported finding raccoon tracks frequently along the Rio Grande.

Although raccoon populations are poorly documented outside the Rio Grande Valley, the species is still known to occur in all parts of the

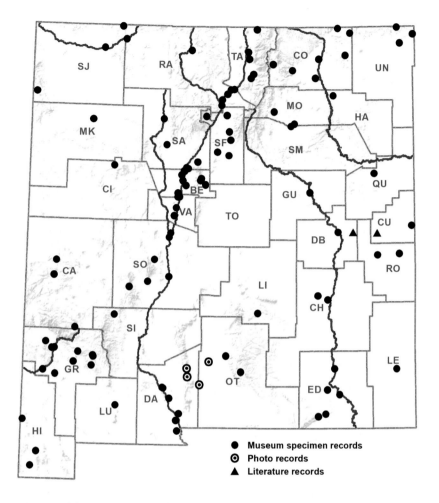

	Museum specimen records
◉	Photo records
▲	Literature records

Map 32.1. Distribution of the raccoon (*Procyon lotor*) in New Mexico, based mostly on museum specimen records. Several atypical records from desert or dry areas of the state are also shown.

state (Map 32.1). Museum specimen records exist from 32 of New Mexico's 33 counties (Appendix 32.1), and harvest of raccoons has been reported from all counties (NMDGF, unpubl. data; Appendix 32.2). Records of occurrence tend to strongly reflect the distribution of lakes and waterways, especially perennial streams and rivers (Findley et al. 1975; Map 32.1). For example, raccoons have been documented in Colfax County both at Maxwell NWR (P. Hoban, pers com) and along the shores of Lake Maloya in Sugarite Canyon State Park (Frey and Schwenke 2012); at Aztec Ruins National Monument (along the Animas River)

and in Frijoles Canyon (along Frijoles Creek) at Bandelier National Monument (Bogan et al. 2007); along Whitewater Creek in Glenwood, Catron County; at Rattlesnake Springs in Eddy County (Photo 32.7); along Animas Creek and on the lower slopes of the Animas Mountains in the Bootheel region of the state (Cook 1986); and along streams in the Chuska Mountains of northwestern New Mexico (and northeastern Arizona) (Halloran and Taber 1965). All these localities are associated with year-round water, as are Bitter Lake NWR in Eddy County, San Simon Cienega in Hidalgo County, and Water Canyon in the

Magdalena Mountains, where raccoon specimens have been secured (MSB 24407, MSB 21474, and USNM 160601, respectively).

Raccoons have also been found in New Mexico where surface water is scarce and/or intermittent. As manager of the then-newly created San Andres NWR, Arthur F. Halloran (1946) did not find any evidence of raccoon occurrence as he recorded his observations of carnivores on the refuge from 14 April 1941 to 26 October 1944. In 2007, however, the refuge's biologist Mara Weisenberger photographed tracks at Upper Ash Spring (Photo 32.8). As part of a study on distribution and habitat correlates of large carnivores conducted by Dr. Louis Bender, New Mexico State University, a raccoon was photographed in 2009 by a remote camera at San Andres Spring, about 10.5 km (7 mi) from Upper Ash Spring. At White Sands National Park, G. W. Roemer (unpubl. data) obtained two raccoon photographs at the same remote camera site, on 21 July and 20 August 2013. No other raccoon photographs exist from six years of extensive remote camera sampling (2009–2020). The site where *Procyon lotor* was documented was located at least 5.4 km (3.4 mi) from any mapped source of surface water (White Sands National Park harbors no permanent source of surface water, though the water table is very shallow in interdunal areas and some animals have been known to dig in order to access it; J. Frey, pers. comm.).

Frey (2003) published five records from the Llano Estacado of southeastern New Mexico, where surface water is also scarce, being generally restricted to playas and springs in addition to irrigation via well water. Of the five records, one is the specimen (ENMUNHM 7629) found dead on 12 October 1979 south of Lovington along New Mexico Highway 483 in Lea County (Map 32.1). Two more are specimens from Roosevelt County, one the skull of a young raccoon near Floyd (ENMUNHM 10537), the other a road-killed adult male discovered near Portales on 6 November

2000 (ENMUNHM 11069). The last two records are from Curry County. One is a road-killed juvenile male discovered near Running Water Draw on 22 October 1998 (ENMUNHM 11115), the other, Frey's observation of a road-killed raccoon north of Melrose (no specimen). Frey (2003) also reported a road-killed raccoon just outside the Llano Estacado, in the drainage of Taiban Creek in Taiban, De Baca County.

Despite the availability of raccoon records from some of the most arid areas of the state, water availability likely represents the primary factor limiting the distribution of raccoons in New Mexico, including in the south and the east. No raccoons were detected during Rollins et al.'s (2009) study on the Armendaris Ranch in Sierra County. The authors of the study used video surveillance technology to evaluate the efficacy of supplemental feeding and water for scaled quails (*Callipepla squamata*) in an arid environment. The lack of raccoon detections by Rollins et al. (2009) stands in contrast to a similar study in western Texas (Henson 2006), where the raccoon was the most frequent visitor at quail feeders, substantially increasing the amount of supplemental feeding lost to non-target species. After Rollins et al.'s (2009) study, however, R. Harrison (pers. comm.) worked at some of the same sites on the Armendaris Ranch and photographed or trapped raccoons at the quail water sources, confirming that New Mexico raccoons need little water to be present in an area.

The upper limit of the raccoon's elevational distribution in Colorado is near 3,000 m (10,000 ft) (Armstrong et al. 2011). Without explicitly giving the elevational range of the raccoon in New Mexico, Bailey (1931: 348) stated that raccoons were locally common in the state "up to the lower border of the Canadian Zone," or only 2,450 m (8,000 ft). Berghofer (1967) described raccoons as reaching the timberline in the Sangre de Cristo Mountains but did not name any specific locality. This is regrettable, in light of the fact that surprisingly few New Mexico records are from localities even approaching 2,450 m (8,000 ft). Among them are the records from Lake Maloya at Sugarite Canyon State Park at an elevation of 2,390 m (7,840 ft) (Frey and Schwenke 2012) and two specimens, one (MSB 156764) collected from Tierra Amarilla in Rio Arriba County on 6 May 2007 at an approximate elevation of 2,295 m (7,525 ft), the other (KUM 42556) from just south of Eagle Nest in Colfax County at an elevation of about 2,510 m (8,240 ft). Specimen and non-specimen records from such localities as Santa Fe and Tesuque in Santa Fe County, Cedar Crest in the Sandia Mountains, Rancho de Taos in Taos County, and Los Alamos Canyon all correspond to lower elevations (i.e., 2,130 m [7,000 ft] or lower).

The highest locality from which a museum specimen record has been secured in New Mexico appears to be Whiskey Creek in San Juan County, at an elevation of approximately 2,740 m (9,000 ft) (Halloran and Taber 1965; USNM 289009). However, there are additional records (but no specimens) from the Sacramento Mountains at the approximate same elevation of 2,740 m (9,000 ft). In particular, R. Winslow found raccoons to be frequent visitors of his yard in Cloudcroft in Otero County from 1992 to 1994. Also notable is J. Frey's observations of raccoon tracks in 2012 along Willow Creek in the Gila National Forest at an elevation of 2,620 m (8,620 ft). Despite spending extensive time in that area since the 1980s, Frey (pers. comm.) had never seen any evidence of local raccoon occurrence. Nor had she received any conclusive reports from cabin owners or campground visitors. In 2012, a massive wildfire swept through some of the Gila National Forest, resulting in loss of vegetation and flooding along streams after the onset of the monsoon season. Although likely always present in the area, raccoons had escaped detection until their tracks were recorded in the layer of silt deposited during flooding along Willow Creek. Since then, remote cameras have detected their presence along the creek on a regular basis, but still Frey has not seen one, a testament to the species' elusiveness.

The subspecies present in extreme northeastern New Mexico (Union and northeastern Colfax counties) is *Procyon lotor hirtus* (Goldman 1950; Frey 2004). This subspecies occurs from the eastern slopes of the Rocky Mountains east to the Great Lakes region south to southern Oklahoma and Arkansas. New Mexico specimens examined by Goldman (1950) and belonging to this subspecies were from Bear Canyon and the mouth of Trinchera Pass, both in the Raton Range. Both specimens (USNM 129104 and USNM 129105) were collected in 1903 by the early naturalist Arthur Holmes Howell (1872–1940).

Procyon lotor pallidus ranges from the Colorado River Delta east and northeast to northeastern Utah, western Colorado, and northwestern New Mexico (Goldman 1950; Frey 2004). In New Mexico, its distribution consists of the Chuska Mountains and the San Juan Basin (Halloran and Taber 1965; Frey 2004). Assigned to this subspecies is the skull of a male collected in 1962 from Whiskey Creek in San Juan County (USNM 289009; see above) (Halloran and Taber 1965). Goldman (1950) gave this subspecies the name of Colorado Desert raccoon.

Procyon lotor fuscipes occurs mainly in Texas and northeastern Mexico, but its distribution also extends eastward to southern Arkansas and parts of Louisiana (Goldman 1950). It is the subspecies found in the Llano Estacado of western Texas and provisionally also New Mexico (Frey 2004).

The distribution of *Procyon lotor hernandezii* includes northwestern Mexico, southeastern Arizona, and New Mexico. In New Mexico, this is the form found in all areas not occupied by the other three subspecies (see Frey 2004).

HABITAT ASSOCIATIONS
Throughout their distribution, raccoons are most abundant in association with permanent water, whether coastal marshes, rivers, or lakes (e.g.,

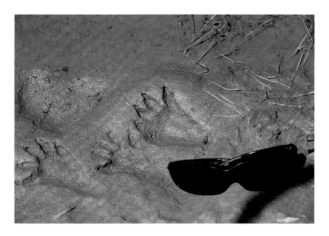

Photo 32.8. Raccoon tracks at Upper Ash Spring on the San Andres National Wildlife Refuge, Doña Ana County, 9 August 2007. Sunglasses are placed next to the tracks for scale. Photograph: © Mara Weisenberger.

Photo 32.9. Two young raccoons sent from White Sands Missile Range to Gila Wildlife Rescue in Silver City in August 2006. The raccoons were rescued from an abandoned building that was being torn down by missile range personnel. Photograph: © Denise Miller.

Photo 32.10. Most observations of raccoons in New Mexico are from near water, but a road-killed juvenile was found in September 2012 on US 70 in the Tularosa Basin near the Doña Ana–Otero county line, in proximity to White Sands National Park and the San Andres and Organ mountains. In addition to the lack of nearby water, the surrounding vegetation (shrub-invaded desert grassland) was also atypical for a raccoon. Photograph: © Jennifer K. Frey.

Photo 32.11. Rare raccoon record from White Sands National Park on 20 August 2013. The site consists of dune field ecotone at least 5.4 km (3.4 mi) from any mapped source of water. Photograph: © Gary W. Roemer.

Photo 32.12a, b, and c. Camera trap project carried out by Matt Farley, Jennifer Miyashiro, and James N. Stuart during the summers of 2007, 2008, 2009, and 2010 at the Bosque del Apache National Wildlife Refuge in Socorro County. a) (*top left*) One of the digital cameras (Moultrie GameSpy I40) used during the project, wired onto a cottonwood tree trunk; b) (*top right*) raccoons photographed at night by one of the project's cameras on 20 October 2008; and c) (*bottom right*) raccoons photographed on 23 July 2010, also during the camera trap project. Non-invasive camera traps have become increasingly popular for carrying out mammal inventories. Photographs: © Matt Farley, Jennifer Miyashiro, and James N. Stuart.

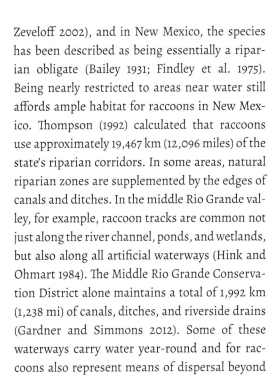

Zeveloff 2002), and in New Mexico, the species has been described as being essentially a riparian obligate (Bailey 1931; Findley et al. 1975). Being nearly restricted to areas near water still affords ample habitat for raccoons in New Mexico. Thompson (1992) calculated that raccoons use approximately 19,467 km (12,096 miles) of the state's riparian corridors. In some areas, natural riparian zones are supplemented by the edges of canals and ditches. In the middle Rio Grande valley, for example, raccoon tracks are common not just along the river channel, ponds, and wetlands, but also along all artificial waterways (Hink and Ohmart 1984). The Middle Rio Grande Conservation District alone maintains a total of 1,992 km (1,238 mi) of canals, ditches, and riverside drains (Gardner and Simmons 2012). Some of these waterways carry water year-round and for raccoons also represent means of dispersal beyond

the levees, into residential neighborhoods and agricultural fields.

Despite being highly adaptable, raccoons are also dependent on the availability of suitable dens for their protection against predators and cold weather, and for giving birth to and sheltering the young. Hollow trees provide suitable sleeping, winter, and litter dens throughout much of the species' distribution, though raccoons also use rock crevices, brush piles, caves, underground burrows, and squirrel nests, and in human-developed areas also drainpipes, attics, barns, crawl spaces under houses, and abandoned homes (e.g., Endres and Smith 1993; Rabinowitz and Pelton 1986; Smith and Endres 2012). Bailey (1931) described New Mexico's raccoons as being mainly associated with cliffs, canyon walls, and boulders along watercourses. In the San Juan Basin of northwestern New Mexico, Harris (1963:

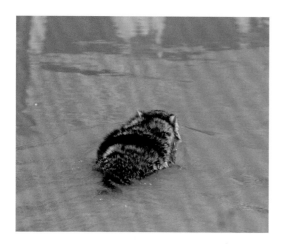

Photo 32.13. Raccoon at the Bosque del Apache National Wildlife Refuge in May 2020. Raccoons are good swimmers and can stay in water for several hours. Photograph: © Sally King.

46) similarly reported raccoons as being most common along streams (e.g., Pine River) "where cliffs and rock debris afford shelter." Other areas of the state are more typical of the raccoon's primary habitat elsewhere (e.g., deciduous woodlands of the eastern United States and wooded bottomlands of the Midwest), with hollow trees likely representing the main source of suitable dens. Hollow tree dens have been documented in woodlots at Maxwell NWR (P. Hoban, pers. comm.) and in riparian woodlands along the Gila River in the Cliff–Gila Valley and the middle Rio Grande (pers. obs.).

There are reports from the Albuquerque area that raccoons use attics and abandoned houses along the Rio Grande. In 2007, University of New Mexico Emeritus Professor and mammalogist Jim Findley wrote a piece for the Sunday, January 7 issue of the *Corrales Comment* newspaper about the multiple generations of young raccoons raised under the roof of his house in Corrales along the west side of the river in Sandoval County:

They never were really any bother. Sometimes a slight coon-odor could be detected in the attic, and that caused the dogs great excitement. Once a baby coon fell down the fireplace chimney and left little sooty coon-prints all over the dining room.

We were quite used to hearing them gallop thunderously across the tin roof in the evening.

Jennifer Frey's observation of a road killed juvenile raccoon on 1 September 2012 on US 70 in the Tularosa Basin is all the more atypical given the species' close association with water and wooded vegetation and/or rocky terrain. The road killed raccoon was found near the Doña Ana–Otero county line in vegetation consisting of shrub-invaded desert grassland, far from water and lacking any conspicuous topographic feature or tree to use as den (Photo 32.10). As it seems unlikely that raccoons would be residents in any such environment, perhaps that individual was dispersing out into the basin from the Organ/San Andres Mountains. Raccoons documented on the Armendaris Ranch by R. Harrison (pers. comm.) occur chiefly in mesquite (*Prosopis*) vegetation adjacent to the Rio Grande bosque, but also occasionally out in black grama (*Bouteloua eriopoda*) grasslands. Although the latter vegetation type occurs well away from the Rio Grande, it is likely that raccoons only forage in those grasslands, returning to the wooded areas along the river. The 2013 White Sands National Park raccoon records are from a site characterized as dune field ecotone (Photo 32.11), again atypical habitat for the species.

LIFE HISTORY

Raccoons are intelligent mammals with excellent long-term memory and strong problem-solving and manipulative skills (they are reported to turn doorknobs), and a highly developed sense of touch. They are excellent climbers and good swimmers, and are mainly nocturnal though

Photo 32.14. Raccoons venturing into a backyard pond in Los Alamos on 20 May 2008. Raccoons are not only known to prey on fish in ponds; there have been reports of raccoons attacking and killing pet turtles (in Albuquerque). Photograph: © Hari Viswanathan.

in New Mexico as elsewhere they can be active during the day (pers. obs.).

Diet and Foraging

Raccoons are omnivorous. Their diet remains poorly documented in New Mexico, but elsewhere typically includes berries, nuts, and grain in addition to aquatic invertebrates, fish, frogs, insects, small rodents, eggs, carrion, pet food, and human garbage (Nowak 1991). This almost certainly captures the essence of the raccoon's diet in New Mexico as well. In his *Mammals of New Mexico*, Bailey (1931:348) mentions crayfish and other crustaceans, freshwater mussels, acorns and "many other nuts," the pods of mesquite (*Prosopis*), and cactus berries as common or favorite foods. Raccoons were reported in Bailey's (1931)

time as raiding watermelon patches and chicken coops in the Carlsbad area in Eddy County and corn fields near Raton (in Colfax County), along the Mimbres River in Deming, and at the mouth of Water Canyon in the Magdalena Mountains. Bailey (1931:349) described corn fields near Water Canyon, where

many of the stalks were pulled down and the green ears eaten, while the stomach of the raccoon caught was found well filled with a pulpy mass of green corn. One was trapped in a field where over an area of several acres more than half of the stalks had been stripped of the ears, which had been partly eaten and left scattered over the ground.

In parts of the raccoon's distribution, the species is described as a major predator of nesting waterfowl and their eggs (Zeveloff 2002), so much so apparently that 108 raccoons were trapped in 1960 in Mora County, New Mexico, all in an effort to protect a local population of nesting geese (Berghofer 1967).

Other, more recent information on the raccoon's diet in New Mexico is only anecdotal. Along the middle Rio Grande, raccoon scats show consumption of Russian olive fruit, garden produce, crawfish, and fish (Cartron et al. 2008). Working in the Gila–Cliff Valley of southwestern New Mexico, S. Stoleson (pers. comm.) observed raccoons foraging along the road for run-over toads and frogs after the onset of the monsoon season. Throughout the top of the Chuska Mountains, R. Winslow (pers. comm.) frequently noted raccoons foraging in perennial ponds, preying in particular on tiger salamanders (*Ambystoma tigrinum*).

Raccoons are opportunistic predators and in some parts of their distribution, they are known to prey on bats hibernating or roosting in caves (Winkler and Adams 1972; Munson and Keith 1984; Geluso and Geluso 2004). Raccoons visit the bat caves, searching for bats that have fallen from roosts or hibernacula, or even seizing them in accessible ceiling and wall areas. Raccoons have preyed on cave-dwelling bats since prehistoric times, and an analysis of raccoon scats revealed estimated predation rates reaching 1,150 bats annually over the last 1,500 years in Wyandotte Cave in Indiana (Munson and Keith 1984). Farther west, raccoon predation on cavern-dwelling bats has been reported from the Edwards Plateau of central Texas, where raccoons were documented visiting bat caves nightly, feeding on fallen bats at the entrances and within the caves (Winkler and Adams 1972). It has also been documented at Carlsbad Caverns National Park in New Mexico (Constantine et al. 1968; Geluso and Geluso 2004). Spanning three decades (from 1973 to 2000),

Geluso and Geluso's (2004) research showed that raccoons visited Bat Cave, King's Palace, Main Corridor, and large natural openings of Carlsbad Cavern. In Bat Cave in particular they fed on fallen bats under the maternity roost of Brazilian free-tailed bats (*Tadarida brasiliensis*).

Reproduction and Social Behavior

Throughout their range, raccoons mate primarily in February and March, and again in May and June in at least some populations (e.g., Gehrt and Fritzell 1996). Temporary pair bonds form while the female is in estrus (Gehrt 2003). Although the raccoon's mating system is traditionally described as polygynous (Macdonald 1984), recent genetic evidence points to promiscuity instead, with most litters sired by multiple males (Nielsen and Nielsen 2007; Hauver et al. 2010). The length of the gestation period averages 63 days (Kaufmann 1982; Sanderson 1987), and the young are typically born from April through June, though births have been reported as early as February in Louisiana (Cagle 1949) and as late as September or even October in Georgia, Texas, and Manitoba (McKeever 1958; Lehman 1968; Cowan 1973; Gehrt and Fritzell 1998a). Litter size is typically three to four but tends to increase with latitude (Gehrt 2003; litter size is given as three to seven in Macdonald [1984]). The young are born with some fur on their body, but their ears and eyes do not open until they are approximately three weeks old, by which time they also have the conspicuous face mask and rings on the tail (Montgomery 1968; Johnson 1970). The young remain in the den until they are approximately six to seven weeks old (Montgomery 1969), but in some cases the adult female may transport them to a new den soon after they are born (see Gehrt 2003). Even after the young have become independent, they may continue to den together with their mother. Very little information is available on reproduction in New Mexico. Data on file at Wildlife Rescue Inc. shows that in central New

Photo 32.15: Tree cavity used as a maternal den at the Maxwell National Wildlife Refuge, Colfax County, in May 2004. Photograph: © Patty Hoban.

Photo 32.16. Raccoon kit, less than three weeks old, found on the ground at the Maxwell National Wildlife Refuge in May 2004. The face mask has not formed yet, and the eyes are still closed. The raccoon might have been dropped by its mother during transport to a new maternal den. Photograph: © Patty Hoban.

Photo 32.17. A six-week-old raccoon from New Mexico. Photograph: © James N. Stuart.

Mexico birth can typically be backdated to mid-March and April; typical litter size is unknown but litters as large as five do occur.

As estimated during studies outside New Mexico, the typical home range of a raccoon seems to vary between 50 and 300 ha (between ~123 and 740 acres), and movements are influenced by the distribution of resources and vary seasonally in some populations (see Gehrt 2003). Greenwood (1982) reported an adult male once traveling over a distance of 14.5 km (9 mi) in one night in North Dakota. Dispersal is male-biased, with most females staying at, or near, natal ranges (see Gehrt 2003).

Much work has been conducted in recent years to better understand the social organization of raccoons. For example, it is now well established that the raccoon social system includes the presence of male, rather than female, social groups (named "coalitions" by some authors). These social groups, which have been described for multiple populations in different parts of the species' range (Gehrt and Fritzell 1998b; Chamberlain and Leopold 2002; Pitt et al. 2008), are particularly interesting in that the combination of male sociality and female asociality is relatively rare among mammals, especially those which are solitary and promiscuous. Adult females typically have overlapping ranges, whereas territoriality in males may vary among populations. In North Dakota, solitary adult males were described as being territorial and maintaining exclusive home ranges (e.g., Fritzell 1978). In southern Texas, however, males formed groups of three or four individuals; their home range overlapped the home ranges of several females but showed little overlap with

Photo 32.18. Juvenile raccoons on a sliding door at a home in Radium Springs, Doña Ana County. Photograph: © Jennifer K. Frey.

Photo 32.19. Two raccoons in a dumpster at Navajo Lake State Park in August 2010. Photograph: © Carol A. Anderson.

neighboring groups of males (Gehrt and Fritzell 1998b). Communal dens involving more than an adult female and her offspring of the year have also been reported. Those dens, often situated underground with several entrances, are used by up to 23 individuals—as reported for a winter den in Minnesota (Mech and Turkowski 1966)—either simultaneously or sequentially. Communal dens may be used by individuals of either or both sexes (see Gehrt 2003). They have not been documented in New Mexico.

Interspecific Interactions

In New Mexico, cougars (*Puma concolor*) are known predators of raccoons both along the Rio Grande and along the eastern foothills of the Black Range (Perry and Upton 2011; Prude and Cain 2021; see Chapter 9). Based on studies elsewhere, New Mexico raccoons might also be included in—but likely represent only a minor component of—the prey base of coyotes (*Canis latrans*), bobcats (*Lynx rufus*), North American red foxes (*Vulpes fulva*), and great horned owls (*Bubo virginianus*). Golden eagles (*Aquila*

chrysaetos) are known to prey on ringtails (*Bassariscus astutus*) in New Mexico (Stahlecker et al. 2010) and their prey base likely also includes raccoons where their territories overlap with water. Where reintroduced wolves (*Canis lupus*) now range in the state (see Chapter 11), they could represent another predator of raccoons (see Finley 1995).

Two types of interspecific competition are generally recognized. In interference competition, one species effectively prevents the other from using a resource. In exploitation competition, individuals of one species use up a resource which is in limited supply, thus reducing the amount available to the other species. In a study in northeastern Illinois, Gehrt and Prange (2007) found little evidence that raccoons avoided areas used by coyotes at a variety of spatial scales. Some raccoons had home ranges that overlapped with coyote core areas, but showed no indication that they avoided those areas. The study thus did not support the existence of interference competition between raccoons and coyotes. Gehrt and Prange (2007) suggested instead that the two

Photo 32.20 (*above*). Raccoons raiding a utility vehicle at Navajo Lake State Park in September 2010. The raccoons were attempting to open a plastic peanut container with a screw-top lid collected earlier during a clean-up detail. Photograph: © Carol A. Anderson.

species may be able to share their habitat by each taking advantage of seasonally or otherwise temporarily available resources. Exploitative competition between raccoons and other mesocarnivores occurs, and in urban centers such as the Albuquerque–Rio Rancho area, it likely involves striped skunks (*Mephitis mephitis*). In Flagstaff, Arizona, Theimer et al. (2015) examined visitation of backyards by striped skunks and raccoons for access to two anthropogenic food sources, bird seeds spilled from bird feeders (control) and cat food left out under the same feeders for feral and domestic cats (*Felis catus*) (treatment). Skunks and raccoons were recorded together in nearly 30 instances. In the presence of just bird seeds, raccoons and skunks ignored one another in a third of all cases; the only other type of interaction between the two species was an aggressive display (for the raccoon, baring of teeth and lunging in the direction of the skunk). When cat food was added, aggressive displays nearly always occurred, and in a few cases, aggressive contact (e.g., nipping and biting) was also recorded (Theimer et al. 2015).

Photos 32.21a and b (*top to bottom*). Adult raccoon on a balcony guardrail (a) in a condo at Angel Fire in Colfax County, New Mexico in July 2010. That same raccoon stands up and reaches for the hummingbird feeder (b). The local elevation is 2,560 m (8,400 ft). Photographs: © Jo Castillo.

Photo 32.22. Raccoon photographed in a chimney in Raton, Colfax County on 23 April 2010. The homeowner called the New Mexico Department of Game and Fish in Raton to have the raccoon removed. With the use of a simple branch, the department's local conservation officer was able to dislodge the raccoon, which then came down the chimney on its own. Photograph: © Clint Henson.

Photo 32.23. Three of four raccoons removed from a chimney in Santa Fe in April 2011. Pest control services may use male raccoon urine to evict adult females with kits from chimneys, attics, and other confined spaces. The repellent is effective in many but not all cases. Photograph: © Alissa Diane Mundt.

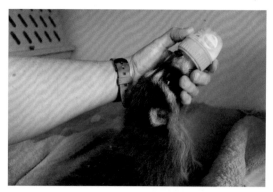

Photo 32.24. Young raccoon being bottle-fed at the Wildlife Center in Española. Photograph: © Alissa Diane Mundt.

STATUS AND MANAGEMENT

No rigorous estimates of raccoon densities—from which to derive population estimates—have been documented from anywhere in New Mexico. From the 1980–1981 through 2019–2020 hunting seasons, some of the largest reported harvest included 233 in Lincoln County in 2007–2008; 153 and 147 in Chaves County in 2017–2018 and 2015–2016, respectively; and 139 in San Juan County in 2019–2020 (Appendix 32.2; and see Map 32.2). In general, however, reported harvest does not lend itself to estimates of densities as it does not just reflect raccoon abundance. To a large extent, it is also influenced by trapping effort, which is simply unknown. Perhaps the only, somewhat reliable—but now dated—density estimate is from La Cueva in Mora County. In a 486 ha (1,200 acre) wetland likely corresponding to high-quality habitat, state trappers caught a total of 108 raccoons in 1960 (Berghofer 1967). In this one instance, trapping was apparently conducted to severely reduce raccoon numbers, if not eradicate the species, in the area (see Berghofer 1967). It might have been intensive enough to justify estimating raccoon density as 22.24 raccoons/km^2 (or 57.6 raccoons/mi^2).

Typical densities reported from rural areas outside New Mexico range from 1 to 27 raccoons

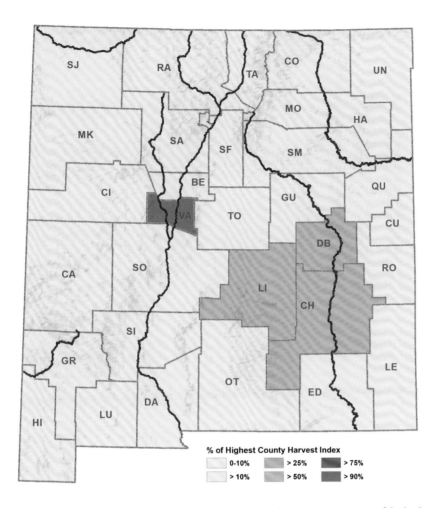

% of Highest County Harvest Index

☐ 0-10%	☐ > 25%	☐ > 75%
☐ > 10%	☐ > 50%	☐ > 90%

Map 32.2. Harvest index by county, 1980–2020. All harvest indices are shown as percentages of the highest harvest index (in Valencia County). See Table 7.1 or 25.1 for details on the calculation of harvest indices (a 40-year record period is used for the raccoon due to the large inter-annual variation in reported harvest at the county level).

per km² (Moore and Kennedy 1985; Kennedy et al. 1986; Gehrt 1988; Seidensticker et al. 1988). Raccoon densities in prairies at the northern end of the species' distribution are lower (i.e., 0.5–1/ km²) (Cowan 1973; Fritzell 1978), whereas those estimated for suburban/urban populations ranged from 55.6/km² in Toronto (Rosatte et al. 1991) to 111/km² in Cincinatti (Schinner and Cauley 1974) or even 125/km² in Washington, DC (Riley et al. 1998). The highest calculated raccoon density (244/ km²) was for a small woodlot in Missouri (Twichell and Dill 1949), with an urban

park in Fort Lauderdale, Florida a close second (238/ km²) (Smith and Engeman 2002). Nowhere in New Mexico is raccoon abundance likely to approach those record high densities (see Figure 32.1). However, simply going by the raccoon density estimate from La Cueva in 1960 suggests that raccoon abundance in some areas of New Mexico may be comparable with—or even exceed—that in other parts of the species' range.

Since the 1940s, raccoon numbers have increased dramatically (15–20 times according to Sanderson [1987]) in North America, while the

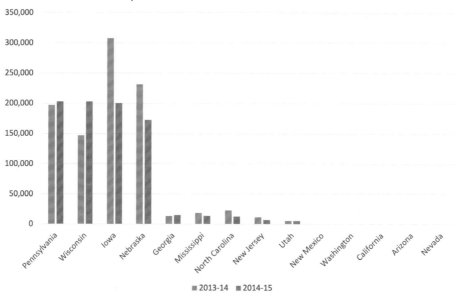

Figure 32.1. Raccoon harvest reported in 14 US states in 2013–2014 and 2014–2015. During those two years, harvest levels in several Midwest states were higher than in New Mexico by nearly three orders of magnitude (totals reported for New Mexico in 2013–2014 and 2014–2015 are 376 and 304, respectively). Although trapping effort was not quantified, such large differences are indicative of the much larger raccoon populations in other parts of the United States compared with New Mexico. Data compiled by the Association of Fish and Wildlife Agencies.

species may have also expanded its distribution by 18% (Laliberte and Ripple 2004). Raccoons were common—along perennial streams and rivers—in New Mexico in Vernon Bailey's time, but according to Berghofer (1967) the species increased notably in numbers in the state after World War II. Even in the more recent past, some of New Mexico's raccoon populations may have grown or continued growing. Hink and Ohmart (1984) cited reports by trappers that the species had increased in numbers during the prior two decades along the Rio Grande. Statewide, however, current or recent population trends are unknown. Total reported harvest from 1980 through 2020 in New Mexico averaged 327 per year (range 156–726) (see Appendix 32.2), fluctuating through time but with no clear trend (Figure 32.2). The highest reported harvest was in

1981–82 (726), followed by 1987–88 (565), 1984–85 (565), and 1999–2000 (504). Annual reported harvest of 200 or fewer raccoons occurred in multiple years, all during the 1990s and 2000s, but here again likely reflecting trapping effort. The number of trapping licenses sold in New Mexico during the 1990s was only about half or less that recorded during the 1980–1981, 1981–1982, and 1982–1983 seasons. After the 1982–1983 season a noticeable drop occurred in the number of trapping licenses sold in the state.

Several explanations have been offered for the raccoon's range expansion and increasing population densities in North America. The main four explanations (Larivière 2004) all center on the impacts of humans and human activities. They can be summarized as 1) greater food availability largely from agricultural development; 2) increased

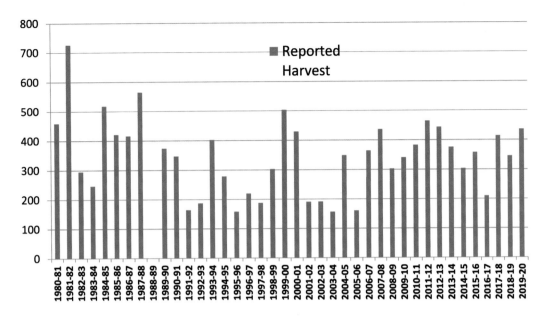

Figure 32.2. Raccoon harvest reported over a 40-year period in New Mexico (no harvest was reported in 1988–1989).

number of suitable winter dens; 3) reduced predation pressure; and 4) raccoon introductions. In support of the reduced predation pressure explanation—Soulé and Terborgh's (1999) "mesopredator release hypothesis"—Finley (1995) argued that the raccoon's range expansion began in earnest in the mid-20th century, or well after the onset of rapid agricultural development (e.g., the 1870s and 1880s in Kansas and Colorado), but closely following the successful eradication of the gray wolf (*Canis lupus*) between 1915 and 1940. However, published studies provide little support for wolves controlling raccoon numbers. Larivière (2004) dismissed the reduced predation pressure/mesopredator release hypothesis because 1) raccoons experience low predation rates even in areas where larger predators exist (but see Prude and Cain 2021) and 2) raccoons now occur in treeless areas also occupied by coyotes. Larivière (2004) instead argued in favor of the increased availability of food as the primary reason explaining the raccoon's range expansion, the explanation that Finley (1995) refuted (and see also Suraci et al. 2016).

In addition to the above four explanations, raccoons were once among the most commercially important mammals. However, the sharp decrease in the price of a raccoon pelt has resulted in reduced commercial harvesting of the species, thus further contributing to what is perceived in some states as raccoon overpopulation (Gehrt 2003). It is possible that in New Mexico at least, most or all the above factors—and others—acted co-jointly to promote higher population densities in the state. Berghofer (1967) argued that reduced trapping was responsible for the state's raccoon population increase since the 1940s, while along the middle and upper Rio Grande Hink and Ohmart (1984) mentioned the release of raccoons by hunters or trappers seemingly to augment the local native population. Although not mentioned by Berghofer (1967) the eradication of the gray wolf in New Mexico, largely achieved by the mid-20th century, might have promoted raccoon population increases, as would some of the most important anthropogenic, landscape-scale alteration of native ecosystems in the state. Today, we know the middle Rio Grande riparian ecosystem as a 280-km (175-mi) (nearly) continuous cottonwood riparian woodland, which in many areas is adjoined by agricultural fields. This continuous woodland did not exist prior to bank stabilization efforts beginning in the 1950s and the construction of Cochiti Dam (completed in 1973). Accounts by early explorers show that the middle Rio Grande valley was a spatially dynamic mosaic of oxbow lakes, wetlands, and isolated, local, and smaller cottonwood groves (Cartron et al. 2008), where tree dens would have been much more limited.

A raccoon's life expectancy in the wild averages approximately two to three years, though one individual is reported to have lived 12 years, 7 months (Haugen 1954). Natural causes of mortality have not been studied in New Mexico, but elsewhere raccoons incur low mortality rates from predators and instead are more likely to die as a result of diseases (Gehr 2003). Overall, most mortality in populations that have been studied are linked to anthropogenic causes, primarily legal fur harvest (but see below) and collisions with road traffic (e.g., Fritzell and Greenwood 1984). In New Mexico, raccoons are designated as "protected furbearers." The open season for trapping and hunting them is from the beginning of September through the middle of May the next year. Signed into law in 2021, New Mexico's Wildlife Conservation and Public Safety Act (also known as Roxy's Law) prohibits trapping but not hunting on public lands; some people still hunt raccoons with dogs. The price of a raccoon pelt has ranged from only about $3 to $20 at the North American Fur Auctions' wild fur sale in recent years, the lesser demand resulting in raccoon fur not even selling in any meaningful quantities in the August 2020 Fur Auction, to the point that it was held with all other unsold goods for future sales (Trapping Today 2020). Today, trapping likely has less of an impact on raccoon mortality in New Mexico, but despite

being protected furbearers, raccoons can still be trapped or removed year-round in areas where they are identified as nuisance wildlife. In such cases, trapping or removal must be conducted by specially permitted wildlife control operators.

Rabies, which is transmitted to humans by the bite of an infected animal, is an important public health concern. Most cases of animal rabies reported from New Mexico occur in skunks, bats, and foxes, but one case of rabies was recorded in a raccoon as recently as 2019 in De Baca County (NMDOH2022). In New Mexico, rabid raccoons have tested positive for the Arizona gray fox rabies strain, not the raccoon strain. The distribution of the raccoon rabies strain—where raccoons serve as a reservoir for the virus—lies primarily in the eastern United States.

Raccoons may look cuddly, but they make difficult and often destructive pets. Possessing a pet raccoon is also essentially prohibited by law in New Mexico. Not only is the species difficult to truly tame, but wild raccoons can also become a problem for homeowners. During the 2014–2015 fiscal year, the New Mexico Department of Game and Fish received 535 nuisance wildlife complaints statewide, with the main species involved consisting of the black bear (*Ursus americanus*), followed by the raccoon (NMDGF 2015). Raccoon-human conflicts are especially likely in urban centers such as the Albuquerque/Rio Rancho area, but some residents remain tolerant of raccoons visiting their garages or attics. In his online blog, for example, Albuquerque's former mayor Jim Baca once humorously described nightly visits by a raccoon fond of cat food. More often, however, the occurrence of raccoons can result in their removal. Raccoons have become a nuisance in Raton in Colfax County, where they establish their dens in the large number of abandoned houses within city limits, while visiting dumpsters or residential yards for food (C. Henson, pers. comm.). Most recently, raccoons increased in numbers to the degree that trapped animals could no longer be relocated and instead had to be euthanized. For a while, raccoons were frequently admitted in New Mexico's wildlife rescue and rehabilitation centers, often as a consequence of raccoon-human conflicts. From May 2004 through June 2011, a total of 78 raccoons were brought into Wildlife Rescue Inc. in Albuquerque. Of those animals, nearly all of them were found in Bernalillo or Sandoval counties—the two counties encompass the Albuquerque–Rio Rancho metropolitan area—and a total of 60 (77%) were immature raccoons with no injuries but too young to survive on their own. Some of them were thought to be orphans; others had been separated from the female parent after she was trapped and relocated or brought into Wildlife Rescue Inc. (WRI, data on file; K. Madden, pers. comm.). New Mexico Department of Game and Fish regulations now prohibit anyone in the state from rehabilitating raccoons (or foxes). As a result, WRI tries to avoid admitting them in the first place and refers all raccoon calls to the city of Albuquerque or to the New Mexico Department of Game and Fish (S. Lindsell, WRI Rehabilitation Coordinator, pers. comm.). When a raccoon is admitted to the clinic, WRI is now required by law to euthanize it. WRI thus avoids admitting raccoons (there have been only six admissions in a six-year period from 2016 to 2021). When it does admit raccoons, they are transferred to the custody of the New Mexico Department of Game or euthanized (S. Lindsell, pers. comm.).

Raccoons are perhaps not emblematic of New Mexico's deserts and mountains. Nor do they rank high on the list of conservation priorities in the state. Like coyotes, they remain an oddity especially among the carnivores of the world, a very adaptable species that overall has benefitted from humans. Despite their secure status, however, New Mexico's raccoons deserve to be studied so we can gain more understanding of their distribution, habitat associations, and natural history in the state.

LITERATURE CITED

Abert, J. W. 1848. *Report of Lieutenant J. W. Abert, of his examination of New Mexico, in the years 1846–47. 30th Congress., 1st session., House Executive Document No. 41.* Washington, DC: Wendell and Van Benthuysen.

Aliev, F. F., and G. C. Sanderson. 1966. Distribution and status of the raccoon in the Soviet Union. *Journal of Wildlife Management* 3:497–502.

Armstrong, D. M., J. M. Fitzgerald, and C. A. Meaney. 2011. *Mammals of Colorado,* 2nd ed. Boulder: University Press of Colorado.

Bailey, F. M. 1928. *Birds of New Mexico.* Santa Fe: New Mexico Department of Game and Fish.

Bailey, V. 1931 (=1932). *Mammals of New Mexico.* North American Fauna 53. Washington, DC: US Department of Agriculture, Bureau of Biological Survey.

Bartoszewicz, M., H. Okarma, A. Zalewski, and J. Szczesna. 2008. Ecology of the raccoon (*Procyon lotor*) from western Poland. *Annales Zoologici Fennici* 45:291–98.

Berghofer, C. B. 1967. Protected furbearers in New Mexico. In *New Mexico Wildlife Management,* 187–194. Santa Fe: New Mexico Department of Game and Fish.

Bogan, M. A., K. Geluso, S. Haymond, and E. W. Valdez. 2007. *Mammal inventories for eight national parks in the Southern Colorado Plateau Network.* Fort Collins, CO: Department of the Interior, National Park Service Natural Resource Technical Report NPS/SCPN/NRTR-2007/054.

Cagle, F. R. 1949. Notes on the raccoon, *Procyon lotor megalodous* Lowery. *Journal of Mammalogy* 30:45–47.

Cartron, J.-L. E., D. C. Lightfoot, J. E. Mygatt, S. L. Brantley, and T. K. Lowrey. 2008. *A field guide to the plants and animals of the middle Rio Grande bosque.* Albuquerque: University of New Mexico Press.

Chamberlain, M. J., and B. D. Leopold. 2002. Spatio-temporal relationships among adult raccoons in central Mississippi. *American Midland Naturalist* 148:297–308.

Constantine, D. G., E. S. Tierkel, M. D. Kleckner, and D. M. Hawkins. 1968. Rabies in New Mexico cavern bats. *Public Health Reports* 83:303–16.

Cook, J. A. 1986. The mammals of the Animas Mountains and adjacent areas, Hidalgo County, New Mexico. *Occasional Papers of the Museum of Southwestern Biology* 4:1–45.

Cowan, W. F. 1973. Ecology and life history of the raccoon (*Procyon lotor hirtus* Nelson and Goldman) in the northern part of its range. PhD dissertation, University of North Dakota, Grand Forks.

De La Rosa, C. L., and C. C. Nocke. 2000. *Guide to the carnivores of Central America*. Austin: University of Texas Press.

Eisenberg, John F. (1989). *Mammal of the Neotropics*. Vol. 1, *The Northern Neotropics*. Chicago: University of Chicago Press.

Emory, W. H. 1848. *Notes on a Military Reconnaissance, from Fort Leavenworth, in Missouri, to San Diego, in California, including Parts of the Arkansas, Del Norte, and Gila Rivers*. Washington, DC: Wendell and Van Benthuysen.

Endres, K. M., and W. P. Smith. 1993. Influence of age, sex, season and availability on den selection by raccoons within the Central Basin of Tennessee. *American Midland Naturalist* 129:116–31.

Enright, E. A. 2008. Faunal analysis of the Tommy Site: subsistence and ritual in the Ancient Chacoan Southwest. MA thesis, Eastern New Mexico University, Portales.

Findley, J. S., A. H. Harris, D. E. Wilson, and C. Jones. 1975. *Mammals of New Mexico*. Albuquerque: University of New Mexico Press.

Finley, R. B., Jr. 1995. *The spread of raccoons (Procyon lotor hirtus) into the Colorado Plateau from the northern Great Plains*. Proceedings of the Denver Museum of Natural History, Series 3, No. 11.

Frey, J. K. 2003. Distributional records and natural history notes for uncommon mammals on the Llano Estacado of eastern New Mexico. *New Mexico Journal of Science* 43:1–24.

———. 2004. Taxonomy and distribution of the mammals of New Mexico: an annotated checklist. *Occasional Papers, Museum of Texas Tech University* 240: 1–32.

Frey, J. K., and Z. J. Schwenke. 2012. Mammals of Sugarite Canyon State Park, Colfax County, New Mexico. *Occasional Papers, Museum of Texas Tech University* 311:1–24.

Fritzell, E. K., and R. J. Greenwood. 1984. Mortality of raccoons in North Dakota. *Prairie Naturalist* 16: 1–4.

Gardner, B. D., and R. T. Simmons, eds. 2012. *Aquanomics: water markets and the environment*. New Brunswick, NJ: Transaction Publishers.

Gehrt, S. D. 1988. Movement patterns and related behavior of the raccoon, *Procyon lotor*, in east-central Kansas. M. S. Thesis, Emporia State University, Emporia, KS.

———. 2003. Raccoon. In *Wild mammals of North America: biology, management, and economics*, 2nd ed., ed. G. A. Feldhamer, B. C. Thompson, and J. A. Chapman, 611–34. Baltimore: Johns Hopkins University Press.

Gehrt, S. D., and E. K. Fritzell. 1996. Second estrus and late litters in raccoons. *Journal of Mammalogy* 77: 388–93.

———. 1998a. Duration of familial bonds and dispersal patterns for raccoons in south Texas. *Journal of Mammalogy* 79:859–72.

———. 1998b. Resource distribution, female home range dispersion and male spatial interactions: group structure in a solitary carnivore. *Animal Behaviour* 55:1211–27.

Gehrt, S. D., and S. Prange. 2007. Interference competition between coyotes and raccoons: a test of the mesopredator release hypothesis. *Behavioural Ecology* 18:204–14.

Geluso, K. N., and K. Geluso. 2004. Mammals of Carlsbad Caverns National Park, New Mexico. *Bulletin of the University of Nebraska State Museum* 17:1–180.

Goldman, E. A. 1950. *Raccoons of North and Middle America*. North American Fauna 60. Washington DC: US Government Printing Office.

Greenwood, R. J. 1982. Nocturnal activity and foraging of prairie raccoons (*Procyon lotor*) in North Dakota. *American Midland Naturalist* 107:238–43.

Guzmán-Lenis, A. R. 2004. Revisión preliminar de la Familia Procyonidae en Columbia. *Acta Biológica Columbiana* 9:69–76.

Halloran, A. F. 1946. The carnivores of the San Andres Mountains, New Mexico. *Journal of Mammalogy* 27:154–61.

Halloran, A. F., and F. E. Taber. 1965. Carnivore notes from the Navajo Indian Reservation. *Southwestern Naturalist* 10:139–40.

Harris, A. H. 1963. Ecological distribution of some vertebrates in the San Juan Basin, New Mexico. *Museum of New Mexico Papers in Anthropology* 8:1–63.

———. 1990. Fossil evidence bearing on Southwestern mammalian biogeography. *Journal of Mammalogy* 71:219–29.

Hauver, S. A., S. D. Gehrt, S. Prange, and J. Dubach. 2010. Behavioral and genetic aspects of the raccoon mating system. *Journal of Mammalogy* 91:749–57.

Henson, K. D. 2006. Species visitation at free-choice quail feeders in west Texas. MS thesis, Texas A&M University, College Station.

Hink, V. C., and R. D. Ohmart. 1984. *Middle Rio Grande biological survey*. Final Report to the US Army Corps of Engineers No. DACW47-81-C-0015. Center for Environmental Studies, Arizona State University, Tempe.

Hogan, P., and L. Sebastian, eds. 1991. *Archaeology of the San Juan Breaks: the Anasazi Occupation*. Albuquerque: Office of Contract Archaeology, University of New Mexico.

Ikeda, T., M. Asano, Y. Matoba, and G. Abe. 2004. Present status of invasive alien raccoon and its impact in Japan. *Global Environmental Research* 8:125–31.

Ivey, R. D. 1957. Ecological notes on the mammals of Bernalillo County, New Mexico. *Journal of Mammalogy* 38:490–502.

Johnson, A. S. 1970. Biology of the raccoon (*Procyon lotor varius* Nelson and Goldman) in Alabama. *Bulletin of the Agricultural Experiment Station, Auburn University* 402: 1–148.

Kaufmann, J. H. 1982. Raccoon and allies. In *Wild mammals of North America: biology, management, and economics*, ed. J. A. Chapman and G. A. Feldhamer, 567–85. Baltimore: Johns Hopkins University Press.

Kauhala, K. 1996. Introduced carnivores in Europe with special reference to central and northern Europe. *Wildlife Biology* 2:197–204.

Kennedy, M. L., G. D. Baumgardner, M. E. Cope, F. R. Tabatabai, and O. S. Fuller. 1986. Raccoon (*Procyon lotor*) density as estimated by the census-assessment line technique. *Journal of Mammalogy* 67:166–68.

Laliberte, A. S., and W. J. Ripple. 2004. Range contractions of North American Carnivores and Ungulates. *BioScience* 54:123–38.

Lang, R., and A. H. Harris. 1984. *The faunal remains from Arroyo Hondo Pueblo, New Mexico: a study in short-term subsistence change*. Arroyo Hondo Archaeological Series 5. Santa Fe, NM: School of American Research Press.

Larivière S. 2004. Range expansion of raccoons in the Canadian prairies: review of hypotheses. *Wildlife Society Bulletin* 32:955–63.

Lehman, L. E. 1968. September birth of raccoons in Indiana. *Journal of Mammalogy* 49:126–27.

Lekson, S. H. 2002. *Salado archaeology of the Upper Gila, New Mexico*. Tucson: University of Arizona Press.

Lutz W. 1984. Die Verbreeitung des Waschbären (*Procyon lotor*, Linné 1758) im mitteleuropäischen Raum. *Zeitschrift für Jagdwissenschaft* 30:218–28.

———. 1995. Occurrence and morphometrics of the raccoon *Procyon lotor* L. in Germany. *Annales Zoologici Fennici* 32:15–20.

Macdonald, D. ed. 1984. *The encyclopedia of Mammals*. New York: Facts on File.

Mech, L. D., and F. J. Turkowski. 1966. Twenty-three raccoons in one winter den. *Journal of Mammalogy* 47:529–30.

Moore, D. W., and M. L. Kennedy. 1985. Weight changes and population structure of raccoons in western Tennessee. *Journal of Wildlife Management* 49:906–9.

Munson, P. J., and J. H. Keith. 1984. Prehistoric raccoon predation on hibernating *Myotis*, Wyandotte Cave, Indiana. *Journal of Mammalogy* 65:152–55.

New Mexico Department of Game and Fish (NMDGF). 2015. *Fiscal year 2015 annual report*. https://www.wildlife.state.nm.us/download/department/annual-report/Annual-Report-Fiscal-Year-2015-New-Mexico-Game-Fish.pdf. Accessed 7 January 2022.

New Mexico Department of Health (NMDOH). [2022]. Animal rabies by county, New Mexico,

2019. https://www.nmhealth.org/data/view/infectious/2222/.

Nielsen, C. L. R., and C. K. Nielsen. 2007. Multiple paternity and relatedness in southern Illinois raccoons (*Procyon lotor*). *Journal of Mammalogy* 88:441–47.

Nowak, R. M. 1991. *Walker's mammals of the world*. Vol. 2. 5th ed. Baltimore: Johns Hopkins University Press.

Perry, T. W., and B. Upton. 2011. *Puma concolor* predation rates and prey selection in riparian and piedmont habitats in New Mexico. Poster presented at the Annual Meeting of the Southeastern Association of Biologists, Huntsville, AL.

Pitt, J. A., S. Larivière, and F. Messier. 2008. Social organization and group formation of raccoons at the edge of their distribution. *Journal of Mammalogy* 89:646–53.

Prude, C. H., and J. W. Cain, III. 2021. Habitat diversity influences puma *Puma concolor* diet in the Chihuahuan Desert. *Wildlife Biology* 2021(4); doi:10.2981/wlb.00875.

Prugh, L. R., C. J. Stoner, C. W. Epps, W. T. Bean, W. J. Ripple, A. S. Laliberte, and J. S. Brashares. 2009. The Rise of the Mesopredator. *BioScience* 59:779 –91. http://www.jstor.org/stable/10.1525/bio.2009.59.9.9.

Rabinowitz, A. R., and R. Pelton, 1986. Day-bed use by raccoons. *Journal of Mammalogy* 67:766–69.

Redford, P. 1962. Raccoon in the U.S.S.R. *Journal of Mammalogy* 43:541–42.

Reeve, F. D., 1961. History of the Albuquerque region. In *Guidebook of the Albuquerque country*, ed. S. A. Northrup, 82–84. New Mexico Geological Society Guidebook, 12th Annual Field Conference, Socorro.

Riley, S. P. D., J. Hadidian, and D. A. Manski. 1998. Population density, survival, and rabies in raccoons in an urban national park. *Canadian Journal of Zoology* 76:1153–64.

Ritke, M. E., and M. L. Kennedy. 1988. Intraspecific morphologic variation in the raccoon (*Procyon lotor*) and its relationship to selected environmental variables. *Southwestern Naturalist* 33:295–314.

Röben, P. 1975. Zur Ausbreitung des Waschbären, *Procyon lotor* (Linne, 1758) und des Marderhundes, *Nyctereutes procyonoides* (Gray, 1834) in der B. R. D. Saugetierkundliche Mitteilungen 23:93–101.

Rollins, D., B. D. Taylor, T. D. Sparks, T. E. Wadell, and G. Richards. 2009. Species visitation at quail feeders and guzzlers in southern New Mexico. In *Gamebird 2006: Quail VI and Perdix XII*, ed. S. B. Cederbaum, B. C. Faircloth T. M. Terhune, J. J. Thompson, and J. P. Carroll, 210–19. 31 May–4 June 2006. Athens, GA: Warnell School of Forestry and Natural Resources.

Rosatte, R. C., M. J. Power, and C. D. MacInnes. 1991. Ecology of urban skunks, raccoons, and foxes in metropolitan Toronto. In *Wildlife conservation in metropolitan environments*, ed. L. W. Adams and D. L. Leedy, 31–38. Columbia, MD : National Institute for Urban Wildlife.

Sanderson, G. C. 1987. Raccoon. *In Wild furbearer management and conservation in North America*, ed. M. Novak, J. A. Baker, M. E. Obbard, and B. Malloch, 486–99. Ontario, Canada: Ministry of Natural Resources.

Schinner, J. R., and D. L. Cauley. 1974. The ecology of urban raccoons in Cincinnati, Ohio. In *Wildlife in an urbanizing environment*, ed. J. H. Noyes and D. R. Progulske, 125–30. Symposium, November 27–29, 1973, University of Massachusetts, Springfield.

Scott, W. E. 1951. Wisconsin's first prairie spotted skunk, and other notes. *Journal of Mammalogy* 32: 363.

Scurlock, D. 1998. *From the rio to the sierra: an environmental history of the middle Rio Grande basin*. General Technical Report RMRS-GTR-5. Fort Collins, CO: US Department of Agriculture, Forest Service, Rocky Mountain Research Station.

Seidensticker, J., A. J. T. Johnsingh, R. Ross, G. Sanders, and M. B. Webb. 1988. Raccoons and rabies in Appalachian mountain hollows. *National Geographic Research* 4:359–70.

Sieber, O. J. 1984. Vocal communication in raccoons (*Procyon lotor*). *Behaviour* 90:80–113.

Slaughter, B. H. 1975. Ecological interpretation of the Brown Sand Wedge local fauna. In *Late Pleistocene environments of the southern high plains*, ed. F. Wendorf and J. J. Hester, 179–92. Rancho de Taos, NM: Fort Burgwin Research Center.

Smith, H. T., and R. M. Engeman. 2002. An

extraordinary raccoon, *Procyon lotor*, density at an urban park. *Canadian Field-Naturalist* 116:636–39.

Smith, W. P., and K. M. Endres. 2012. Raccoon use of den trees and plant associations in western mesophytic forests: tree attributes and availability or landscape heterogeneity? *Natural Resources* 3:75–87.

Soulé, M. E., and J. Terborgh. 1999. *Continental conservation: scientific foundations of regional reserve networks*. Washington DC: Island Press.

Stahlecker, D. W., J.-L. E. Cartron, and D. G. Mikesic. 2010. Golden eagle (*Aquila chrysaetos*). In *Raptors of New Mexico*, ed. J.-L. E. Cartron, 371–91. Albuquerque: University of New Mexico Press.

Suraci, J. P, M. Clinchy, L. M. Dill, D. Roberts, and L. Y. Zanette. 2016. Fear of large carnivores causes a trophic cascade. *Nature Communications* 7, article 10698. https://www.nature.com/articles/ncomms10698.

Theimer, T. C., A. C. Clayton, A. Martinez, D. L. Peterson, and D. L. Bergman. 2015. Visitation rate and behavior or urban mesocarnivores differs in the presence of two common anthropogenic food sources. *Urban Ecosystems* 18:895–906.

Thompson, B. C. 1992. *Ecologically-based management evaluation for sustainable harvest and use of New Mexico furbearer resources*. Las Cruces: New Mexico Cooperative Fish and Wildlife Research Unit, New Mexico State University.

Trapping Today. 2020. *2020–2021 fur prices—trapping today's fur market forecast*. 19 September 2020. https://www.trappingtoday.com/2020-2021-fur-prices-trapping-todays-fur-market-forecast/.

Twichell, A. R., and H. H. Dill. 1949. One hundred raccoons from one hundred and two acres. *Journal of Mammalogy* 30:130–33.

Wilson, D. E., and D. M. Reeder, eds. 2005. *Mammal species of the world: a taxonomic and geographic reference*. 3rd ed. Baltimore: Johns Hopkins University Press.

Winkler, W. G., and D. B. Adams. 1972. Utilization of southwestern bat caves by terrestrial carnivores. *American Midland Naturalist* 87:191–200.

Zeveloff, S. I. 2002. *Raccoons: a natural history*. Washington, DC: Smithsonian Institution Press.

FREE-RANGING AND FERAL DOMESTIC DOG (*CANIS FAMILIARIS*) AND DOMESTIC CAT (*FELIS CATUS*)

David L. Bergman and Scott C. Bender

The most common carnivores in New Mexico are the domestic dog (*Canis familiaris*) and the domestic cat (*Felis catus*). Both species were introduced into New Mexico through human facilitation, though their respective arrivals are separated by approximately 10,000 years or more. The earliest confirmed evidence of domestic dogs reaching the Western Hemisphere dates back between 10,190 and 9,630 years (Perri et al. 2019), whereas domestic cats were introduced much more recently by the first Europeans, as confirmed by recent genetic studies showing New World cats linked to European lineages (Lipinski et al. 2008). In 2016, a survey found that 60.1% of New Mexico households owned a pet, with dogs being the number one species reported (39.4% of households) and cats coming in second (25.2% of households; AVMA 2018). Owned animals are generally kept inside homes or fenced yards, though cats with outdoor access can roam neighborhoods to a variable extent, and dogs, especially in rural settings, are occasionally allowed to roam without restraint. Domestic dogs and cats may also occur as feral animals, those defined as having escaped from domestication and become wild (Merriam-Webster). All feral dogs and cats are free-ranging, whereas owned animals may be either free-ranging or constrained to the owner's residence. Free-ranging dogs and cats go by numerous descriptors, including stray, unowned, community, and feral. The true feral animal is one living in a wild and free state with no direct food

or shelter intentionally supplied by humans (Causey and Cude 1980), and it does not show any evidence of socialization with humans (Daniels and Bekoff 1989a). Daniels and Bekoff (1989c) indicate that feralization occurs through the development of a fear response to humans and does not necessarily involve significant genetic divergence from domestic ancestors. Often the lines are blurred, especially when one considers the provisioning of food as a result of human goodwill. Therefore, the biology and diet of the free-ranging or feral animal may vary from place to place based on different ecological conditions, levels of dependency on humans, and cultural views. Feral dogs and cats represent an important focus of our chapter, though most of what we cover applies more generally to all free-ranging animals, whether owned or not, feral or not. Feral and free-ranging dogs and cats can be found statewide but occur more commonly near or in human communities. Research in New Mexico on these species, after they return to a feral or semi-feral state, has been largely limited to disease-related concerns, ecosystem/environmental damage impacts, and interactions with humans.

HISTORY OF DOMESTICATION

The domestication process can be divided into two distinct phases: 1) animal keeping, the practice of capturing, taming, and keeping animals without any deliberate attempt to regulate their behavior or breeding; and 2) animal breeding, eventually

(*opposite page*) Photograph: © Scott C. Bender.

associated with the conscious, selective regulation and control of the animals' reproduction and behavior (Bökönyi 1969; Clutton-Brock 1992).

Origin and Domestication of Dogs

Gray wolves (*Canis lupus*) are the original progenitor of today's domestic dog, their genetic lineage splitting into three major clades or groups, the North American, Eurasian, and domestic dog lineages (Ramos-Madrigal et al. 2021). A recent, as of yet unpublished genetic analysis has identified what may be one of the closest wild relatives of modern dogs in the extinct Japanese gray wolf (*Canis lupus hodophilax*), with modern Japanese, eastern Asia, and dingo dogs having approximately 5.5% homology (Gojobori et al. 2021). There is still much debate as to when, where, how many times, and from which exact population(s) the ancestor of dogs was domesticated. The earliest proto-wolf/dogs appear to have been especially attracted to permanent or semi-permanent human settlements allowing for multiple persistent dog lineages to arise in Europe, the Middle East, and eastern Asia by the end of the Last Glacial Maximum to early Holocene (Ovodov et al. 2011). The oldest remains considered to be of a domesticated dog, have been dated back to 33,000 years before present (BP) in Russia (Ovodov et al. 2011). Based on Paleolithic remains, the origin of modern dog breeds can be traced to wolf domestication between 20,000–40,000 years ago (Botigué et al. 2017). The divergence was triggered by humans changing from being nomadic hunter-gatherers to adopting a more settled, agriculture-based lifestyle (Vilà et al. 1997). During co-evolution with humans, the modern dog also gained the genetic adaption for the digestion of starches not found in wolf populations (Axelsson et al. 2013). This finding may be useful where domestic dogs and wolves overlap, in genetically "fingerprinting" the species involved with domestic livestock damage or deaths. Europe has played a major role in modern dog evolution and in determining the extent of breed variation seen in the present day (Botigué et al. 2017). Some of the breeds that developed in Europe were introduced by the Spanish conquistadors into the Americas. Earlier, Paleolithic hunter-gatherers migrating from northern Asia over the Bering land bridge had entered North America with their own dogs. The two waves of human arrivals correspond to the first and second major introductions of dogs into the Americas (see below).

Origin and Domestication of Cats

Cats were domesticated on several occasions, all within the Middle East's Fertile Crescent 8,000 to 10,000 years ago, as they began their association with humans as a commensal animal, feeding on the rodent pests that infested the grain stores of the Neolithic farmers. Domesticated cats derive from at least five founders of the African wildcat (*Felis silvestris libyca*) found across the Fertile Crescent region and Northern Africa, as well as their descendants who were transported across the world by human assistance (Driscoll et al. 2007). Despite the five founder events, domestic cats show less genetic diversity than in the domesticated dog, as demonstrated by the lack of synthesis of the amino acid taurine, considered to be essential only in a limited number of feline species, primarily those that evolved in a desert environment (Driscoll et al. 2009; essential amino acids are not produced by the body and instead must be obtained from the diet). Deficiency in the amino acid taurine contributes to degenerative cardiomyopathy and feline central retinal degeneration in domesticated cats that are not fed a strict diet of meat, making them an obligate carnivore like their progenitor, the African wildcat (Hayes and Trautwein 1989). Ancestors of the domestic cat most likely chose to live among humans because of unique ecological opportunities they found for themselves. Natural selection was key for the development of cats through increased tameness and reduced territoriality, with humans only selectively breeding cats in the last few hundred years (Driscoll et al. 2009). It is likely that early humans encouraged cats to remain close by when they realized their ability to dispatch rodent pests (Driscoll et al. 2007).

ARRIVAL OF DOMESTICATED DOGS IN NORTH AMERICA

Domesticated dogs of Siberian lineages arrived in North America with humans after traversing the Bering Strait (Leathlobhair et al. 2018). The earliest confirmed dog remains in North America (from the Koster Site in Illinois) have been radiocarbon-dated to ~10,000 BP (Leathlobhair et al. 2018), approximately 12,500 years after the earliest evidence of humans arriving in North America (Bennett et al. 2021). Pre-Columbus dogs contain an admixture of derived alleles (i.e. variant forms of a gene) from North American wolves (*Canis lupus*) and coyotes (*C. latrans*) (Leathlobhair et al. 2018), evidently the result of crossbreeding when the association with humans remained weak and no formal captivity existed. Through the process of domestication, dogs developed behaviors that allowed them to exist alongside humans, playing a social role that wild canids, such as coyotes and wolves, were not able to fulfill (Morey 2010). Throughout time the size of dogs has varied considerably as verified by Hohokam-era dogs, as well as one individual animal that was even larger, from the Classic period (1150–1450 CE [Common Era date]) (Taylor et al. 2008). Two mummified dogs from previous burials were found in 1916 in White Dog Cave in northeastern Arizona (Fugate 2008). The animals were of light golden color with cloudings of dark brown and dated to 400 BCE (Before Common Era date) during the early Basketmaker II period (1500 BCE to 50 CE). Dogs from the Basketmaker III period (500–750 CE) from the Pueblo I site near Albuquerque had rather heavy bones given the size of their skulls, a shortened and rather

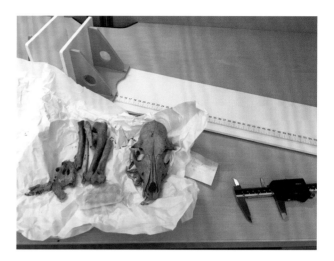

Photo 33.1. Domestic dog (*Canis familiaris*) skull and associated bones from the Spadefoot Toad Site (29SJ 629) in Chaco Canyon, New Mexico, deposited ca. 500–900 CE. Domesticated dogs first arrived in North America with Paleolithic hunter-gatherers from northern Asia after traversing the Bering Strait. Coyote and wolf-derived DNA was present in these pre-Columbus dogs, a legacy of hybridization at a time when the association with humans remained weak. NPS Chaco Museum Collection. Photograph: © Victoria Monagle

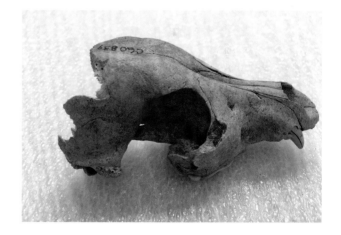

Photo 33.2. Partial domestic dog (*Canis familiaris*) skull from the Spadefoot Toad Site (29SJ 629) in Chaco Canyon, New Mexico, deposited ca. 875–1200 CE. Following the arrival of Europeans, Native American dogs disappeared almost completely and thus left a minimal genetic legacy in modern dog populations. NPS Chaco Museum Collection. Photograph: © Victoria Monagle.

Photo 33.4. Mimbres Classic black-on-white bowl, from the Cienega Ruin, New Mexico, depicting a woman carrying firewood in a burden basket, a man in the background with a log on his head, and their dog. The painted pottery tradition of southwestern New Mexico's Mimbres people dates from about 1000 to 1130 CE. Catalog No. 40.4.276. Photo courtesy: Maxwell Museum of Anthropology, University of New Mexico.

Photo 33.3. Puerco black-on-red dog effigy pitcher, Mogollon, Upper Little Colorado Province dating about 1030 to 1150 CE, from Nutria Canyon in McKinley County, New Mexico. Arizona State Museum catalog no. GP-5973. Photograph: Jannelle Weakly. Photo courtesy: Arizona State Museum, University of Arizona.

Photo 33.5. *The First Thanksgiving 1621*, an oil painting by Jean Louis Gerome Ferris (1863–1930), ca. 1912. Onboard the Mayflower were at least two dogs, an English spaniel and an English mastiff, mentioned in the Pilgrims' journals. Explorers from Spain and later from England, France, and other countries brought their own breeds of dogs from Europe. Those breeds almost completely replaced pre-Columbian dogs, and today the canine gene pool bears almost no trace of the latter. The Europeans also brought with them domestic cats, not found in the pre-Columbian Americas. Photograph in the public domain.

Photo 33.6. Large Navajo pictograph at the Standing Cow Ruin in Arizona's Canyon de Chelly on the Navajo Nation. The polychrome mural (4.5 m x 22.5 m [~15 ft ×75 ft]) depicts mounted Spanish soldiers accompanied by a padre, allied Native Americans (Utes from Colorado), and six dogs during the 1805 Narbona raid into Canyon de Chelly (Dix 1980). Two of the dogs are greyhounds, the other four a different breed with a more compact body. During the Narbona raid, the Spanish fought an all-day battle with a band of Diné (Navajos), killing ninety. The pictograph documents the continued use of war dogs by the Spanish into the 19th century. Spanish war dogs, both mastiffs and greyhounds, were much larger and far more aggressive than the New World's pre-Columbian breeds. Photograph: Robert K. Mark.

narrow muzzle, a moderately steep forehead, and, in adult individuals, jaws that ranged from heavy to very heavy (Lawrence 1966). There is considerable variation in tooth size among dogs of that period, but even in the ones with the smallest teeth, the lower tooth row is crowded and tilted upward posteriorly.

Throughout time in the American Southwest, dog skulls and entire skeletons can be found associated with human burials or buried in a place of honor because dogs were often considered escorts in the afterlife (Fugate 2008; Monagle et al. 2018). Additionally, dogs were buried at the entrances of structures as guardians especially once the location had been desecrated. The dog burial practices decreased over time and ultimately ended after the arrival of Spanish priests. After dog burials ceased, the Pueblo people developed the habit of saying "We bury humans, not animals" (Fugate 2008:4).

Francisco Vázquez de Coronado y Luján was a Spanish conquistador and explorer who led a large expedition from what is now Mexico to present-day Quivira, Kansas, through parts of the southwestern United States between 1540 and 1542, looking for the Seven Cities of Gold in Cibola (Winship 1904). As a member of Coronado's party, the Spanish explorer and navigator Hernando de Alarcón was one of the first Europeans to ascend the mouth of the Colorado River and advance to present day Yuma, Arizona. While in the area, Alarcón's party met a man who stated that he had visited Cibola (Winship 1904), identified as the area of the present-day Acoma Pueblo communities in New Mexico. The same man stated that the Native Americans had a dog similar to the one which accompanied the Spanish explorers. A southwestern breed of dog associated with Native Americans was described by travelers as "fox like," the size of a terrier, heavier of bone, with a shortened rostrum (Allen 1920). Coronado chronicled the Zuni people using varieties of long-haired dogs to make clothes (Winship 1896; Schwartz 1998), as did Captain George Vancouver in 1792 when he encountered tribes of the Pacific Northwest (Hammond-Kaarremaa 2018). Earlier evidence of Native Americans using dog hair was found in a cave on the Navajo Nation in northeastern Arizona. Within the cave were six flat-braided sashes dated between 450 and 600 CE from the Anasazi period (Freer and Jacobs 2008). One individual sash was made with dog hair of white and brown colors (Arizona State Museum catalog number A-21414).

Dogs were chiefly used in the transportation of Native American goods with packs or simple drag-sleds called travois featured prominently in ethnographic and historical accounts of Native American communities (Winship 1904; Welker 2020). Dogs were also used for guarding, as hunting guides, babysitters, bed warmers, and cleanup crew, and for food and fiber resources (Fugate 2008; Taylor et al. 2008; Johnson 2014). The disarticulated remains of dogs found in trash-filled deposits along with the butchered and burned bones of bison (*Bison bison*), pronghorn (*Antilocapra americana*), deer (*Odocoileus* spp.), rabbits (Lagomorpha), and other food debris provides evidence that dogs were also eaten (Bigelow and Speth 2004). Indeed, dogs were consumed on occasion throughout the prehistoric Southwest by Native Americans, and their use as a food is generally seen as a short-term response to stress, a sign that normal or proper foods were not available in sufficient quantity (e.g., Olsen 1990:115; Mick-O'Hara 1994; Bigelow and Speth 2004). At other times, boiled dog meat was often served at social gatherings and feasts, in special ceremonies, and as a food for honored guests, (e.g., Gilmore 1934; Powers and Powers 1990; Snyder 1991; Schwartz 1998). Dogs were a central part of the diet obtained through trade and purchase in the Rocky Mountains and Pacific Northwest by the "Corps of Discovery"—the Lewis and

Clark expedition—between 29 August 1804 and 26 April 1805, as revealed by a quote from the journals of the expedition on 3 January 1805 that "our original aversion to dog meat has long since been overcome" (Clarke 2002:231).

Further evidence of pre-European dogs associated with Native Americans can be found in traditional stories, rituals, and art works produced by tribal members. Oral histories based on traditional ecological knowledge have depicted dogs in positive and negative ways, as both helpers of humankind and the embodiment of evil spirits and witchcraft. The traditional Diné people believe dogs protect homes and families from any external negativity as they can absorb that type of energy without any harm to themselves (Peterson 2017). Peterson (2017) referenced an older Diné man of his acquaintance who stated that the reason dogs were first kept for spiritual protection is that they ate everything, including fecal matter, without any noticeable detrimental effect. However, in very traditional Diné families, dogs are rarely allowed to enter the residence, because the dog is the first cousin to the coyote, who cannot be trusted. Dog impersonators have taken part in Pueblo dances, even as kachinas (Taylor et al. 2008). A Puerco black-on-red vessel depicting dogs identified from Nutria Canyon, near Zuni, New Mexico (ASM catalog number GP-5973) was dated to 1030 to 1150 CE (Ferg 2008; Photo 33.3). Additional art works of dogs are associated with the Mimbres peoples of southwestern New Mexico. The Mimbres were aware of 1,161 unique animals including dogs between 1000 and 1130 CE (Brody 2008). Brody (2008) researched 733 Mimbres painted vessels, of which 26 showed canine paintings, four of them identified as dogs. One of the Mimbres painted vessels consists of a classic black-on-white bowl, from the Cienega Ruin, New Mexico, depicting a woman carrying firewood in a burden basket, a man in the background with a log on his head, and their dog (Maxwell Museum of

Anthropology catalog number 40.4.276; Photo 33.4). A second bowl is a classic black-on-white bowl depicting a lone dog from the Pruitt site, New Mexico (Maxwell Museum of Anthropology catalog number 40.4.133).

After the arrival of Europeans, Native American dogs almost completely disappeared, leaving a minimal genetic legacy in modern dog populations (Leathlobhair et al. 2018). Interestingly, the closest detectable genetic trace of pre-Columbus dogs is found in the canine transmissible venereal tumor, a contagious, relatively unusual cancer clone derived from an individual dog that lived up to 8,000 years ago and still found today (Leathlobhair et al. 2018). The closest living descendants from the original pre-Columbus dogs are believed to be the Xoloitzcuintles and Peruvian hairless dog breeds, which are identified by the Aztecs and Inca, prior to European contact and are described in Columbus's and later European journals. This unique clade has also been confirmed in domestic dog genetic lineage and relationship studies (Parker et al. 2017)

ARRIVAL OF DOMESTICATED CATS IN NORTH AMERICA

Little is known as to when domestic cats reached the Americas. The Spanish are known to have brought cats to the New World (Kiddle 1964), and cats carried on ships were considered in the insurance damage claims made on early maritime shipping. Christopher Columbus and other explorers of his day reportedly carried cats with them on transatlantic voyages (Driscoll et al. 2009). Additionally, passengers onboard the Mayflower and settlers of Jamestown are said to have brought cats with them to control vermin and to bring good luck. Evidence of the introduction of cats by Spaniards can be found in Native American languages with the incorporation of two Spanish words for cats, *miz* and *mozo* (Kiddle 1964). The Diné word for the domestic cat, *Mósí*, is close to the Spanish words, while

Photo 33.7. Mixed-breed sheep dog on the Navajo Nation on 25 August 2017. Mixed-breed dogs have been used since the early 18th century by Diné herders to protect sheep flocks against coyotes (*Canis latrans*) and now also feral dogs. The sheep dogs are fed once a day but otherwise are not handled by their owners. There are no herders accompanying the flocks. The dogs live with the sheep and tend to be wary of humans. Photograph: © David L. Bergman.

Photo 33.8. Free-ranging Navajo Nation dog on 12 February 2017 near Shiprock in San Juan County, New Mexico. The blue eyes and the speckled pattern around the nose are indicative of a mixed breed Australian cattle dog (blue heeler). Australian cattle dogs and rottweilers tended to be the most common recognizable breeds on the Navajo Nation in the late 1990s. Today, however, most Navajo Nation dogs appear to be American pitbull terrier crosses and Chihuahuas. Photograph: © David L. Bergman.

Photo 33.9. Free-ranging cat in Los Alamos County, New Mexico on 9 December 2021. In contrast to dog breeds, domestic cat lineages exhibit little variation in appearance or behavior, differing mostly in the characteristics of their coats (coat color, patterning, texture, and hair length). Photograph: © Hari Wiswanathan.

the domesticated dog in Diné, *Łééchąą'í*, does not have a similar Spanish equivalent. More recently, genetic studies have shown New World domesticated cats closely linked to European lineages (Lipinski et al. 2008).

FREE-RANGING AND FERAL DOGS AND CATS IN PRESENT-TIME NEW MEXICO

With a global population estimated at 800 million and a presence on every continent except Antarctica, dogs now rank second after humans as the world's most widespread large mammal species (Hughes and Macdonald 2013; Lord et al. 2013; Rowan 2020; Wynne 2021). Their success has been linked to a series of adaptations to cohabitation with humans including even a transformation of their facial muscle anatomy (for interspecific facial communication; Kaminski et al. 2019; Wynne 2021). The modeled distribution of dogs has reflected the spatial patterns of human presence on the North American landscape (Pardi and Smith 2015). Indeed, previous investigators have utilized domestic dog fossils as a proxy for humans (Witt et al. 2015). However, not all dogs

live in a state of high intimacy with humans, and in the developing world in particular, where they are often referred to as "street dogs," they are typically not fed intentionally and instead rely on the scavenging of human trash (Butler and du Toit 2002; Coppinger and Coppinger 2002). True feral dogs are relatively rare, representing only an estimated 2.5% of the total free-ranging dog population worldwide (see Beaver 1999). However, feralization can be a lengthy process, and many free-ranging dogs may resemble feral animals behaviorally, depending on needs, locality, and circumstances (Daniels and Bekoff 1989a; Butler et al. 2004; Vanak and Gompper 2009).

In New Mexico as elsewhere, not all dogs are kept as pets inside homes and fenced yards. Free-ranging or feral dogs can occur in many locations, and in a book titled *Wilderness: a New Mexico legacy*, Cory McDonald (1985:14) mentions frequent encounters with them in the Sandia Mountains just east of Albuquerque:

> Twice since then I have seen two dogs chasing deer. . . . The first time was in upper Juan Tabo north of the La Luz trail. Another time was in the upper reaches of Bear Canyon. Many times we have seen tracks of dogs after deer during January and February. I have seen one kill site perhaps a week after the event. . . . The first "sighting" was at night when Phil Tollefsrud and I were sacked out north of the La Luz Trail in Chimney Canyon at about the 8,800-foot elevation level. The big doe came within twenty five meters of our camp and two dogs with collars almost passed through our camp. . . . The second sighting was in mid-morning in Bear Canyon. They looked tired, buck, dogs, and all. Some of the bow hunters said they shot at the dogs, but missed. . . . Some of the feral dogs are feral only during the night. Others, according to some owners living next to the mountain, are sometimes gone for three or four days running.

On 29 December 2012, the Associated Press reported the death of a 12-year-old boy on the Navajo Reservation in Cibola County after he was attacked and mauled by a pack of nine dogs three days earlier. According to the Associated Press article (published in the *Albuquerque Journal*), witnesses indicated that most of the dogs involved in the attack had been stray animals fed by the child's uncle. A prior article also by the Associated Press (in 2011) referred to other humans sustaining severe injuries from bites and a pack of dogs killing 27 sheep all on the Navajo Nation. Although free-ranging and feral dogs are mentioned as a problem around non-tribal communities (e.g., along the US–Mexico border), throughout New Mexico's human history, they have been disproportionally associated with Native American lands including the Navajo Nation, Zuni, Taos, San Juan, Santa Clara, San Ildefonso, Tesuque, Mimbre, Acoma, Cochiti, Santa Domingo, Tiguex Pueblo, Nambre, Arroyo Hondo (Taylor et al. 2008), Lone Kiva, and Pecos Pueblo (DeBoer and Tykot 2007). Along the Arizona–New Mexico border, tribal communities typically have <2,000 residents distributed among multiple small housing areas. Virtually every tribal community can be considered rural in that it provides habitat intermediate between the urban and wild areas. The numerous sandstone canyons are bordered by walls reaching 30 m (~100 ft) or higher. Natural depressions in these walls provide cave-like shelters that have been shown to serve as dens for litters of pups (Daniels and Bekoff 1989a). Feral dogs have resources available within a tribal community and the potential to roam beyond community boundaries and encounter wildlife. Community dumps historically have provided locally abundant food resources that attract feral dogs (Daniels 1988). Additional locations attracting free-ranging or feral dogs include restaurants, grocery stores and food markets, concentrated housing units, higher densities of livestock, flea markets, tourist recreational sites, or other areas where humans aggregate or leave trash.

Dogs have been selectively bred for various behaviors, sensory capabilities, and physical attributes. Humans selected specific characteristics to breed dogs for herding livestock (collies, shepherds, etc.), hunting (pointers, hounds, poodles, etc.), catching rodents (small terriers), guarding (mastiffs, chows), saving fishermen or helping them with their nets (Newfoundland, Labrador retriever), pulling loads (huskies, St. Bernards), guarding carriages and horsemen (dalmatians), and simple companionship. Reflecting the fact that *Canis familiaris* is the world's most phenotypically diverse mammal species (Wayne 1986), dogs are extremely variable in appearance on the Navajo Nation today. Diné people employ dogs for guarding livestock, presumably having learned the technique from early Spanish explorers (Lyman 1844; Dyk 1938; Black 1981; Black and Green 1985). The Spanish brought flocks of sheep into the Western Hemisphere along with dogs of European origin, described as larger than the indigenous varieties (Lyman 1844). The Diné herders may have used mixed-breed dogs with their flocks since the early 1700s, when they became involved in sheep ranching (Black 1981). The number of mixed-breed guard dogs accompanying a sheep flock averages three on the Navajo Nation (Black 1981). In general, the guard dogs are wary of strangers and often the dogs are not handled by their owners to strengthen their bonding to their sheep herd; thus they can best be described as semi-wild (Black 1981), one step removed from being truly feral.

There is no consistent standard for pelage or morphological characteristics within dogs in general, including those free-ranging and/or feral, but the basic morphology is that of the gray wolf, the wild ancestor of dog breeds. Their size in the Southwest can nonetheless vary from that of a Chihuahua to an English mastiff, with the

Photo 33.10. Free-ranging dog on 5 June 2018 in Pacific marten (*Martes caurina*) habitat in the Taos Ski Valley area in New Mexico. Some hikers let their dogs run loose, often unaware of the potential impact their pets can have on wildlife. Dogs can act as predators, prey, competitors, and disease reservoirs or vectors. Photograph: © Brian Jay Long.

majority being between the size of an Australian shepherd and a German shepherd. For New Mexico, sixteen Diné mixed-breed guard dogs were weighed, and they ranged from ~7–27 kg (15–60 lbs) (Black 1981). Daniels (1986) speculated that based on his observations most Navajo dogs are medium-large in size, weighing ~18–23 kg (40–50 lbs). Between 1997 and 2020, the recognizable dog breeds most commonly seen on the Navajo

Nation changed from Australian cattle dogs and rottweilers to American pitbull terrier crosses and Chihuahuas (S. Bender, pers. obs.)

Unlike dogs, which exhibit a relatively wide range of sizes, shapes, and temperaments, domestic cat lineages only show relatively minor distinctions in appearance or behavior, differing mostly in the characteristics of their coats. The uniformity in behavior among cats may be the result of the lack of targeted breeding by humans, or of less plasticity in body shape and size. Overall, cats are small carnivores weighing up to 8 kg (~17.5 lbs), but more commonly 1.5–3.0 kg (3.3–6.6 lbs), when living in the wild, whereas they may be considerably heavier when in a home. The color of cats is extremely variable in domesticated varieties, but feral cats commonly revert to black, tabby, or tortoiseshell with varying extents of white starting from the belly and breast. The tortoiseshell coat coloration is a sex-linked trait associated with the X chromosome, resulting in tortoiseshells being limited to females.

Significantly more research has been conducted on feral or free-ranging dogs than cats. Feral cats have been identified using wildlands and natural areas including around Tucumcari Lake in Quay County (Hoditschek et al. 1985),

Photos 33.11a and b (*left to right*). Free-ranging domestic cats photographed along the Gila River (a) and the Mimbres River (b) in southwestern New Mexico. Five remote cameras deployed from 2014–2016 recorded a total of 76 wild mammal species in the active floodplain and surrounding areas along those two rivers in Grant and Luna counties, respectively (Geluso 2016). Where they occur, free-ranging cats have the potential to interact with all those other mammals, not to mention numerous species of birds, reptiles, and amphibians. Photographs: © Keith Geluso.

Photo 33.12. Golden eagle (*Aquila chrysaetos*) nest with eaglet on 24 May 2010 near the Navajo village of Torreon in Sandoval County, New Mexico. The eaglet is kept covered to keep it calm during the visit of the nest to determine golden eagle reproductive success. The observer is holding a domestic cat flea collar and lower jawbone found in the nest and presumably representing eagle prey remains. Photograph: © Craig Blakemore.

Bitter Lake National Wildlife Refuge (Hafner and Shuster 1996), Carlsbad Caverns National Park (Constantine et al. 1968), Sugarite Canyon State Park, Colfax County (Frey and Schwenke 2007), and along the Mimbres and Gila rivers (Photos 33.11a and b), but they are likely limited, more so than feral dogs in New Mexico, to those areas associated with human habitation or with low mesocarnivore predation pressures. A GIS study conducted on New Mexico's largest city, Albuquerque, determined that the populations of stray cats and dogs were positively correlated with an increasing amount of crime and poverty within the metro areas (Brock 2016).

LIFE HISTORY AND ENVIRONMENTAL IMPACTS

In many locations, free-ranging dogs are the most abundant carnivore and significantly disrupt ecosystems (Feldmann 1974; Bögel et al.1990). A model was developed by the US Geological Survey to map the potential distribution of dogs in the western United States (Hansen and Leu 2008). It assumes that free-ranging synanthropic dogs have the potential to occur in wildlands anywhere within a distance of 0.36 km (0.22 mi) from populated areas or campgrounds (Hanser and Leu 2008). In New Mexico, the model predicted dogs as occurring in and around urban areas, in river valley bottoms, and in rural areas particularly on the Navajo Nation (Map 33.1).

During free-ranging activity, dogs interact with wildlife at multiple levels, including as predators (Lomas and Bender 2007), prey (to larger raptors, coyotes, wolves, and cougars [*Puma concolor*]), competitors (Daniels and Bekoff 1989a), and disease reservoirs or vectors (Eidson et al. 1988; Gould et al. 2008; Gettings et al. 2020). Feral and free-ranging cats impact biodiversity through predation (Hafner and Shuster 1996), fear effects, competition, and disease transmission (Gage et al. 2000), and as prey (including to raptors, bobcats (*Lynx rufus*), and coyotes). Despite this, the impact of dogs and cats on natural environments has not been well documented, especially in New Mexico, with some of what we know based only on anecdotal observations. For example, in many monitored golden eagle (*Aquila chrysaetos*) nests on the Navajo Nation, there has been found an abundance of dog and cat collars, indicating predation by eagles on those species (D. Mikesic, pers. comm.; Photo 33.12).

Diet

Feral and free-ranging dogs are opportunistic feeders or underfed scavengers (Franciscan Fathers 1910; Joseph et al. 1949). The vast majority of populations that have been studied worldwide rely on anthropogenic food sources, including scavenged trash, crops, livestock, poultry, and even human feces, as well as food directly given to them (Vanak and Gompper 2009). Thirteenth-century native dogs had a more homologous diet in the American Southwest. Their diet consisted primarily of maize and occasionally domestic turkey (*Meleagris gallopavo*) due to their association with Native

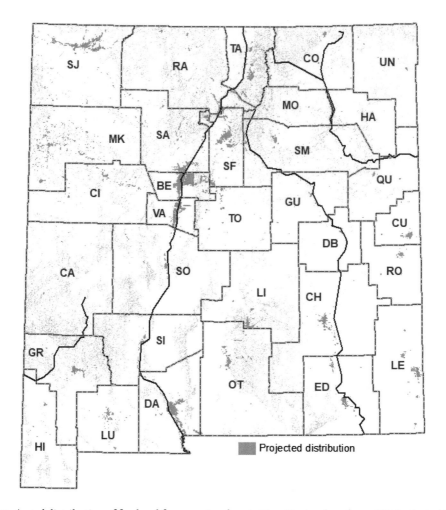

Map 33.1. Projected distribution of feral and free-ranging dogs in New Mexico, based on a US Geological Survey model for the western United States (Hanser and Leu 2008). The model assumes that dogs can range anywhere within a distance of 0.36 km (0.22 mi) from populated areas or campgrounds. Notable in the projected distribution of dogs are urban areas, main river valleys with human development, and scattered rural communities particularly on the Navajo Nation.

American communities (DeBoer and Tykot 2007; Kemp et al. 2017; Monagle et al. 2018), Kemp et al. (2017) even suggesting that dogs that responded well to a maize-rich diet were preferentially bred. By contrast, in more recent times, open and unfenced dumps in New Mexico have provided locally abundant human-derived food resources (Daniels 1988). Populations of feral dogs in northwestern Texas were found to be dependent on cattle carrion found in dead pits near feedlots (Kamler et al. 2003a). On the

Navajo Nation, tourists often report and complain that dogs are observed feeding upon road killed wildlife, horses (*Equus caballus*), and livestock, especially in areas without right-of-way fencing. This feeding behavior has often been observed by both authors. Reported losses due to free-ranging or feral dogs and impacting agriculture include sheep and lambs, watermelons, cantaloupes, milo, sweet and field corn, and wheat (Chapter 3).

Whereas most dogs and other domesticated

Photo 33.13. A free-ranging dog visits a trash can on 7 July 2007 in Aquismón, San Luis Potosí, Mexico. Feral and free-ranging dogs are opportunistic scavengers, though they may also hunt alone or in packs, impacting wildlife and domestic livestock. Most rely, at least partly, on human-generated waste. Photograph: © Scott C. Bender.

Photo 33.14. Domestic cat with orange-crowned warbler (*Leiothlypis celata*). In the United States alone, free-ranging domestic cats kill an estimated 1.3–4.0 billion birds and 6.3–22.3 billion mammals annually (Loss et al. 2013). Throughout the history of their domestication, cats never quite lost their hunting skills. Photograph: © Eastern Ecological Science Center

animals feed on widely available plant foods, felids have a limited ability to digest anything but meat, and in fact, they have lost the ability to taste sweet carbohydrates altogether (Driscoll et al. 2009). Contrary to dogs, cats were mostly left to fend for themselves in the domestication process, thus their hunting and scavenging skills remained sharp. Even today, most domesticated cats are free agents that can easily survive independently of humans. Feeding of feral cats is also a common activity practiced by humans including those that do not own pets (Levy and Crawford 2004). Feral and free-ranging cats can thus have a large impact on wildlife especially birds, small mammals, and lizards. It is estimated that free-ranging domestic cats kill 1.3–4.0 billion birds and 6.3–22.3 billion mammals annually in the United States (Loss et al. 2013), but equivalent research has not been conducted on losses of reptiles and amphibians. Feral cats, as opposed to owned pets, cause most of this mortality (Loss et al. 2013). In southern New Mexico, cats are attracted to human-made water sources (Rollins et al. 2009). The water sources increase the opportunity for cats to prey on wildlife attracted to the same resources. Along with birds, shrews rank among the animals most frequently identified in the scientific literature as prey for cats in New Mexico. Near Tucumcari Lake in Quay County, a cat was documented as having captured a least shrew (*Cryptotis parvus*; state-listed as threatened in New Mexico since 1985) in a heavily grassed area (Hoditschek et al. 1985). Another instance of cat predation on a least shrew was recorded at Bitter Lake National Wildlife Refuge (Hafner and Shuster 1996). At Sugarite Canyon State Park, a cat killed a dusky shrew (*Sorex monticolus*) (Frey and Schwenke 2007). On 22 September 2003, a cat caught a Nelson's sharp-tailed sparrow (*Ammodramus nelsoni*), New Mexico's first recorded observation of that species, near the New Mexico State Fairgrounds in Albuquerque, Bernalillo County (Hubbard and Elliston 2013). Although feral and free-ranging cats are often tolerated, or even promoted, in rural communities, losses have been reported by commercial poultry and game bird rearing operations (USDA, APHIS Wildlife Services, unpubl. data).

Many studies have shown that dogs kill and feed

on wildlife, but they have been mainly focused on prey populations (e.g., Lomas and Bender 2007), reporting rates of mortality caused by different predators. The reliance on human-derived food sources may be a function of both the opportunistic nature of dogs and learned behavior rather than based on ecological constraints. The variance in the diet of dogs may also be a function of location and the availability of resources. However, because of human-derived food resources, dogs can occur at high population densities and thus could potentially outcompete native carnivores, especially when prey is limited (Vanak and Gompper 2009). Population density and environmental carrying capacity are largely influenced by either human intervention such as animal "control" activities, food limitation, and medical reproductive control, or those disease outbreaks limiting population recruitment (Fowler 1981; Morters et al. 2014).

Preying on wildlife by dogs has been documented in particular on mule deer (*Odocoileus hemionus*) fawns in New Mexico (Lomas and Bender 2007). On prairie dog colonies, dogs and cats have been observed digging into burrows to scavenge on prairie dogs that had died of plague (Archibald and Kunitz 1971) or to catch live animals including rabbits (Kartman et al. 1967; Eidson et al. 1988). Feral dogs will cannibalize conspecifics when stressed for resources (Daniels 1987a; D. Bergman, unpubl. data; S. Bender, pers. obs.). Feral or free-ranging dogs have also been documented to kill wild carnivores, such as foxes (Pils and Martin 1974) and coyotes (Kamler et al. 2003b).

Reproduction and Social Behavior

Among all species in the genus *Canis*, dogs are unique for their mating and social systems (Macdonald et al. 2019). The common *Canis* pattern of pair-bonding, social monogamy, extended paternal and alloparental care, and monoestrous seasonal reproduction is conspicuously absent in dogs (Lord et al. 2013). The dog population ratio on the Navajo Nation was 3 to 1 of males over females (Daniels 1986), likely due to human preference for male over females for guarding or costs of preventing reproduction. However, abandoned female dogs outnumber males by two to one with the majority (69%) being puppies in Navajo, New Mexico (Daniels and Bekoff 1989a).

Feral and free-ranging dogs exhibit distinctive social organization characteristics depending on their relationship to human communities. A total of 116 different behaviors have been cataloged among nine different functional categories: investigative, caregiving, care-soliciting, agonistic, sexual, eliminative, ingestive, comfort-seeking, and related to general locomotion (Daniels 1987a). In general, communication among dogs is primarily non-vocal through body language, olfactory means, and pheromones (Daniels 1987b). Dogs can nonetheless utter eight unique types of vocalizations (Yeon 2007). These include the bark for warning, territorial defense, or play solicitation; the growl for threatening; the grunt (as an expression of pleasure); and the howl, whine, snore, groan, and yelp. With the advances in Artificial Intelligence learning, additional vocalizations may be identified (Yeon 2007).

A pack is defined as a group of animals that rest, travel, forage, and hunt together, and pack members usually are related (Mech 1970; Bekoff et al. 1984). Feral and free-ranging dog packs are formed due to dog abandonment (Daniels 1987a), absence or non-enforcement of dog control laws, absence of spay or neuter programs (Daniels and Bekoff 1989b), lack of restraint, escape, natural reproduction, and mixing with other packs. Many advantages accrue for free-ranging and feral dogs to live in packs, including enhanced vigilance resulting in greater protection from potential predators (e.g., humans, larger carnivores) and an increased ability in gaining access to higher-quality food resources (Daniels and Bekoff 1989b). Free-ranging dogs often live in groups, but they do not always form a structured pack (Macdonald and Carr 1995; Kamler et al. 2003a; Majumder et al. 2014). Members of a social group may defend a territory together, but groups can be dynamic in composition, influenced by resource availability, mating interests, and nearness to source populations (Macdonald and Carr 1995; Kamler et al. 2003a; Majumder et al. 2014). Some packs hunt cooperatively (Kamler et al. 2003b; Fleming et al. 2006) while others do not (Macdonald and Carr 1995). Pack sizes vary through time as transients join the pack temporarily, remaining in the pack for days to months before leaving. In small rural human communities, single dogs are observed more frequently than expected by chance alone (as determined by a zero-truncated Poisson distribution); pairs occur less often than expected by chance alone, larger groups more frequently with observed mean group sizes being 1.29 and 1.32 dogs for Navajo, New Mexico, and Window Rock, Arizona, respectively (Daniels and Bekoff 1989a). Community dogs, characterized by their ability to range freely while being passively cared for by one owner or a community (Miklósi 2015), defend territories and are mostly solitary with some loose social grouping. Dogs outside of human communities are more "wolf-like" in behavior, often forming packs and occasionally roaming alone (Boitani et al. 1995). Communal denning occurs with feral dogs. Pack splitting may occur, referring to the temporary emigration from the pack of two or more individuals that travel and feed on their own. The home range of feral dogs on the Navajo Nation ranged from 0.01 km² (2.47 acres) to 5.87 km² (1,450.51 acres) and averaged 0.66 km² (163.09 acres) in two packs and solitary individual animals (Daniels and Bekoff 1989b). Home range size increased tenfold following weaning of puppies within the two packs (Daniels and Bekoff 1989b). Territorial behavior occurs with the incursion of humans or non-pack members and is most pronounced when young pups are around the den site. Feral dog activity peaks in early morning (05:00 to 08:00 hrs) and late in the day (20:00 to 23:00 hrs) with minimal activity performed at midday or during inclement weather. Increases in activity are normally associated with foraging or mating.

Approximately two, and occasionally three, estrus periods occur per year in female dogs, and they are not seasonal, unlike in most other canids (Christie and Bell 1971; Daniels and Bekoff 1989a). Male dogs are able to take advantage of any opportunity to mate year-round (Taha et al. 1981), in contrast to male wild canids who produce sperm only during a single breeding season (Catling et al. 1992). Six types of mating (monogamy, polygyny, promiscuity, polyandry, opportunity, and rape) have been identified in free-ranging and feral dogs (Pal 2011). The dogs' departure from monogamy is an adaptation to resource availability (Macdonald et al. 2019).

Most or all female dogs will at some point in their lives show signs of a pseudo-pregnancy (physiological and behavioral signs associated with pregnancy, but in the absence of any fetuses) (Asa and Valdespino 1998). The demands of gestation and lactation require that a breeding female increase her energy intake from 1.5 to 3 times (Gessaman 1973; National Research Council 1974), which may be difficult to attain under conditions of scarce resources. The gestation period for dogs averages 63 days. The sex ratio at birth is 1:1 (Daniels and Bekoff 1989b). Feral or free-ranging females rear their young apart from the pack for approximately two months after parturition, though regular contact between the females and other pack members can be observed (Daniels and Bekoff 1989a). By four months of age, approximately one-third of pups will survive to maturity (Daniels and Bekoff 1989a). Care for puppies is predominantly provided by the mother (Macdonald and Carr 1995; Lord et al. 2013).

Free-ranging dogs commonly exhibit a promiscuous mating system with no breeding hierarchy (Lord et al. 2013). All adults can have the opportunity to breed, and therefore dog packs can contain multiple lactating females with litters, in addition to other male and female group members (Macdonald and Carr 1995; Pal 2011; Paul et al. 2014). Daniels and Bekoff (1989a) reported a litter of two and a total of ten for two other litters of feral dogs. However, estimates of litter size rely on when the pups are first sighted, which rarely occurs before these are mobile (two to four weeks old), when natal and post-natal mortality might have already contributed to lower estimates. Weaning begins when pups are about five weeks old (Scott and Fuller 1965) and by the age of four months, juveniles are essentially independent (Daniels and Bekoff 1989a). Juveniles then may disperse to another part of the pack's home range, and higher mortality occurs during this period of early independence and exploration. The initiation of pup independence coincides with an increase in pack movement (Daniels and Bekoff 1989b). Feral dogs have a female-skewed sex ratio that is not a bias due to the production of female pups at birth but may be due to the abandonment of females (rather than males) by humans in the vicinity of established packs (Daniels and Bekoff 1989a). Although the probability of survival is low for abandoned animals, the continued addition of abandoned females to an area as a potential source of pack members may maintain feral or free-ranging dog populations (Daniels and Bekoff 1989a).

Female cats are seasonally polyestrous and come into heat several times every year until the onset of pregnancy or a decreasing photoperiod (Gunther and Terkel 2002). Ovulation in cats requires copulation stimulation from barbs that are found on the mature male's penis and develop under the influence of testosterone (Kustritz 2005). A female cat reaches reproductive maturity between 7 to 12 months of age and can be in estrous as many as five times a year (Ogan and Jurek 1997). The gestation period averages 65 days (Nowak 1991) and the average litter is four to six kittens but may range from 2 to 10 (Fitzwater 1994). Kittens are weaned at 35–40 days. Cats can reproduce any month of the year, where food and habitat are sufficient, though they tend to breed more seasonally in more northern latitudes. An adult female can produce up to three litters per year (Fitzwater 1994). In general, the largest males have the greatest mating success overall; alternatively, males that are members of a colony have the greatest mating success within that colony, even if they are small (Yamane et al. 1996).

Felids, in contrast to dogs, tend to be solitary hunters that defend their home ranges fiercely from other cats of the same sex (Driscoll et al. 2009). Feral cats can survive as solitary animals, but sufficient food resources lead to the formation of small or larger social groups referred to as colonies (Crowell-Davis et al. 2004). Colonies of feral cats that are partially dependent upon humans for food can reach over 30 individuals (Macdonald et al. 1987). Behavior within the colony is matrilineal and can be amicable, whereas behavior toward outsiders is often aggressive (Macdonald et al. 1987; Liberg and Sandell 1988; Macdonald et al. 2000). Adult females may be found using communal nests, and will tend, groom, and nurse kittens other than their own (Macdonald et al. 1987). Food resources determine colony size, with large colonies existing where food is abundant and small colonies found where food patches are still clumped but less abundant (Liberg et al. 2000; Crowell-Davis et al. 2004).

Interspecific Interactions

Dogs can be effective competitors, especially with mesocarnivores. They may fill the role of an interactive medium-sized canid within the carnivore community, especially in areas where the native large carnivore community is limited (Vanak and Gompper 2009). Dogs have been reported to suffer from intraguild predation without consumption by larger carnivores. They are killed by

Mexican gray wolves (*Canis lupus baileyi*; D. Bergman, unpubl. data), coyotes (Grinder and Krausman 1998), and cougars (reviewed in Butler et al. 2014). In most cases, predation on dogs occurs in the vicinity of human habitation, or it may happen in wildlands while the dogs are hunting.

Feral dogs have been shown to hybridize with several other *Canis* species, including both coyotes and gray wolves in North America (Gipson et al. 1974; Freeman and Shaw 1979; Bergman et al. 2009). A study based on skull measurements found that coyote-dog hybrids occurred in Oklahoma in Cimarron County, which borders Union County in New Mexico (Freeman and Shaw 1979). Coyote-dog hybrids are also common in other southern states (Gipson et al. 1974). The Calupoh or Mexican wolf dog is a hybrid breed that originated in Mexico and emerged in pre-Hispanic times, the result of the mixture between the Mexican gray wolf and the dog (American Rare Breed Association 2021). Several centuries were needed to achieve the domestication of the Calupoh. Two cases of dogs breeding with Mexican wolves and producing hybrid pups have been documented since the wolves were reintroduced into the Southwest in the late 1990s (Bergman et al. 2009). Despite these two cases of hybridization involving domestic dogs, any significant genetic contamination in New Mexico is at present unlikely (e.g., Wayne and Vilà 2003; see also National Academies of Sciences, Engineering, and Medicine 2019; Chapter 10). Nonetheless, hybridization between wild canine species and domestic dogs has been a concern for conservation and reintroduction programs both in New Mexico—with the Mexican gray wolf—and internationally. With the advent of improved genetic testing, evaluating mitochondrial and microsatellite DNA markers and single-nucleotide polymorphism will provide even better insight on the risk of hybridization (Randi 2008; VonHoldt et al. 2013).

Domestic cats hybridize with wild cats in Eurasia and Africa (Fredriksen 2016). Unlike the domestic dog, however, the domestic cat has not been shown to hybridize in North America, likely because of significant divergence between the African wild cat and North American lynx genetic groups (Slattery and O'Brien 1998). In October 2018, Colorado Parks and Wildlife (CPW) seized an animal believed to be an illegally possessed bobcat. The owner claimed the animal was a bobcat/domestic cat hybrid. Endogenous feline leukemia virus (enFeLV) gene sequences are present in the genome of the domestic cat but not that of North American wild felids, and they can be detected quickly and inexpensively with a polymerase chain reaction assay. Using this assay, CPW was able to confirm that the contested animal lacked enFeLV, and therefore was not a bobcat/domestic cat hybrid (Chiu et al. 2020).

Infectious Diseases, Parasites, and Impacts on Human Public Health and Wild Carnivore Populations

Dogs and cats can be reservoirs of pathogens, largely because most populations of these two species are free-ranging and unvaccinated worldwide, including in New Mexico. Free-ranging and feral dogs and cats have greater interaction with, and endemic and environmental exposures to, sources of infection. Dogs can carry over 40 zoonotic diseases (Bergman et al. 2009). Diseases such as rabies, plague, sarcoptic mange, canine parvovirus, and canine distemper have resulted in severe population declines in several endangered carnivores coexisting with dogs, which can therefore be viewed as pathogen-mediated apparent competitors, capable of facilitating large-scale population declines (Vanak and Gompper 2009). Property, human health and safety, and natural resources were also losses attributed to free-ranging and feral dogs.

The primary public health problem caused by dogs in New Mexico, both those that are truly feral and many free-ranging animals that exhibit some of the same behaviors, are not communicable diseases, but dog bites (Daniels 1986; Burkhart 2018). During the early 1980s, the mean feral dog bite rate on the Navajo Reservation was 605 per 100,000 people, with 87.8% of all cases involving mixed-breed dogs (Daniels 1986). Between 2013 and 2015, the Navajo Area Indian Health Service (IHS) reported 1,474 dog bite cases, 933 being outpatient, 541 being emergency room admissions, while the Navajo Nation Animal Control investigated an additional 1,104 dog bite reports (IHS 2016; S. Bender, unpubl. data).

Besides the physical damage caused by a bite, dogs can transmit pathogens via their saliva because their oral cavities are home to the second richest and most diverse microbiota after their intestinal tracts, harboring potentially over 700 species of bacteria. Depending upon diet and surrounding environment, the oral cavity nurtures numerous microorganisms that include not only bacteria but also fungi, viruses, and protozoa. A noted sequel to dog bites is often bacterial cellulitis, which requires antibiotic or surgical intervention and presents an especially high risk in immune-suppressed and diabetic human patients.

VIRAL DISEASES

Dogs and cats are both susceptible to rabies (*Lyssavirus rabies lyssavirus*). Dogs maintained the canine rabies virus variant in the United States until it was eradicated in the country in 2004 as a result of a mandatory vaccination program started in the 1940s (Pieracci et al. 2019). However, a spillover event occurred in 2006 in New Mexico when a dog in Bernalillo County was determined to be infected with a rabies virus variant associated with Mexican free-tailed bats (*Tadarida brasiliensis*) (Blanton et al. 2007). Cats are also susceptible to rabies with the potential of being exposed to a spillover event such as interactions with bats in caves (Constantine et al. 1968). In New Mexico, there are two primary terrestrial

variants of rabies, the south-central skunk rabies virus variant (RVV) and the Arizona gray fox (*Urocyon cinereoargenteus*) RVV that have resulted in occasional spillover events in both dogs and cats, including an unvaccinated cat in Curry County from south-central skunk RVV in 2018 (Ma et al. 2020). A retrospective review of multiple studies of feral and free-ranging dogs on the Navajo Nation between 2001 and 2017 found that the immunization rate (as determined by antibody titers) was well below the 70% threshold recommended for eradication, and also well below the 59% rate to control or prevent rabies outbreaks (Bender and Bergman 2017). To react swiftly in the case of a re-emergence of rabies on the Navajo Nation, veterinary and wildlife management authorities investigated the potential use of oral rabies vaccination for feral dogs (Bergman et al. 2008; Berentsen et al. 2014, 2016; Bender et al. 2017).

Caused by *Canine morbillivirus*, canine distemper virus (CDV) was first reported in canines in Spain in 1761 (Appel and Gillespie 1972) and likely was introduced into New Mexico with the early Spanish colonists' dogs. The CDV may have played a role in the loss of the North American pre-contact canines, as reflected in the loss of genetics observed in previously discussed studies. The CDV affects a wide variety of mammal species and families, including domestic dogs, coyotes, foxes, wolves, ferrets, skunks (Mephitidae), raccoons (*Procyon lotor*), and large cats within the order Carnivora (Creevy and Evans 2023). The CDV is a respiratory, enteric, and neurotropic virus passed through close contact from domestic and feral dogs, causing epidemics that often result in mass mortalities—it pushes some species to the brink of extinction. The virus contributed to the near-extinction of the black-footed ferret (*Mustela nigripes*; McCarthy et al. 2007; see Chapter 22). Feral dog puppies in New Mexico have been documented to die due to CDV (Daniels and Bekoff 1989a), and this virus continues to be a major cause of juvenile morbidity and mortality in unvaccinated dog populations in New Mexico and the Navajo Nation with occasional spillover to other mesocarnivores like gray and North American red foxes (*Vulpes fulva*) (Perrone et al. 2010; S. Bender, unpubl. data).

Canine parvovirus (*Carnivore protoparvovirus 1*, CPV) is a "new" virus that suddenly emerged in 1978. Antigenically, it is very similar to the long-known feline panleukopenia virus (FPV). Soon after its appearance, CPV was classified as a mutant of FPV and is thought to represent a species jump from cats (Truyen 1994). The CPV is also highly similar to mink enteritis virus and the parvoviruses of raccoons and foxes (Jones et al. 1997). It was first reported in New Mexico in the early 1980s (R. F. Taylor, pers. comm.) as the CPV pandemic swept around the world. Several mutations of CPV continue to circulate in the world's canid species. This disease is highly contagious and is spread from dog to dog by direct contact or indirectly through feces. Following ingestion, the virus replicates in the lymphoid tissue in the throat, then spreads to the bloodstream. From there, the virus attacks rapidly dividing cells, notably those in the lymph nodes, intestinal crypts, and the bone marrow. There is depletion of lymphocytes in lymph nodes and necrosis and destruction of the intestinal crypts (Lobetti, 2003). Vaccines can prevent this infection, but mortality can reach 91% in untreated cases.

Parvovirus has been recognized as a pathogen in several different canid and mesocarnivore species (Holmes et al. 2013). This disease continues to be one of the most important morbidity and mortality factors in unvaccinated populations of juvenile canines in northwestern New Mexico (S. Bender, unpubl. data), and when combined with CDV mortality, may act as a feral dog population self-regulation mechanism by preventing recruitment, because juveniles die from these two diseases when the community carrying capacity is exceeded.

Coronaviruses have been isolated from dogs, cats, horses, cattle, swine, chickens, turkeys, and humans with clinical signs of disease. Most of the diseases caused by coronaviruses are gastrointestinal or respiratory, but encephalomyelitis has also been reported. The alphacoronavirus types are endemic in both dogs and cats within New Mexico, causing intestinal disease. Feline Coronavirus (FCoV; (*Alphacoronavirus alphacoronavirus 1*) corresponds to a common viral infection in cats. It generally causes asymptomatic infections but can cause mild diarrhea. As of yet, poorly understood changes in the virus can give rise to mutants that lead to the development of feline infectious peritonitis from a virus induced antigen-antibody complex causing liver and abdominal pathology. Most cats infected with the FCoV eliminate the virus following infection, but some cats may develop a persistent infection. These cats are generally asymptomatic, can shed large amounts of viruses in feces, and serve as a continual source of infection for other feline species in the environment (Heeney et al. 1990). In recent years, a betacoronavirus called canine respiratory coronavirus has also been reported in some dogs, causing a respiratory disease commonly referred to as "kennel cough" (AVMA 2020). Spillover events of the COVID-19 [SARS-CoV-2], also a betacoronavirus, have also been documented in dogs and cats in close contact with infected humans (Yaglom et al. 2021). There is currently no evidence that cats or dogs play a significant role in human COVID-19 infection; however, reverse zoonosis is possible if infected owners expose their domestic pets to the virus during acute infection (Bosco-Lauth 2020), an event unlikely to occur in feral cats and dogs.

BACTERIAL DISEASES

Plague is a flea-borne zoonotic disease caused by the bacterium *Yersinia pestis* (Davis 2019) and has caused transmissions from animals to humans for decades. Information from human case investigations in New Mexico suggest that peridomestic exposures were due to interactions with dogs and cats (Collins et al. 1967; Mann et al. 1979; Eidson et al. 1988; Gould et al. 2008). Dogs in highly enzootic areas may be repeatedly exposed to plague-infected rodents and fleas, thus maintaining relatively high antibody titers. Exposure arises due to dogs and cats hunting or scavenging in wild areas and becoming infected, especially by rodents. Fleas jump onto the heads or feet of dogs and cats when they investigate holes. Fleas also transfer from carcasses or wildlife recently killed by cats and dogs. In 1965, during a plague epizootic among wild rodents near Fort Wingate, New Mexico, six of eight dogs sampled had positive hemagglutination antibody titers for plague (Archibald and Kunitz 1971). The first documented case of plague in a cat in New Mexico was in 1977 (Eidson et al. 1988). Since 1977, most plague cases associated with cats have been in the states of Arizona, Colorado, New Mexico, and Utah (Gage et al. 2000), with Santa Fe County being the epicenter in New Mexico (Brown et al. 2010). Pneumonic plague was described in 10% of *Y. pestis* infected cats in New Mexico (Eidson et al. 1991). In Bernalillo County, the flea species *Ctenocephalides felis* and *Linognathus setosus* have been identified from dogs (Ford et al. 2004). In Sandoval County, dogs were identified to host *Euhoplopsyllus glacialis*, *Pulex irritans*, and *Spilopsyllus inaequalis* (Ford et al. 2004). Fleas from cats have been identified as *Spilopsyllus inaequalis* in Bernalillo County, *Echidnophaga gallinaceus* in Hidalgo County, and *Euhoplopsyllus glacialis*, *Pulex irritans*, and *Spilopsyllus inaequalis* in Sandoval County (Ford et al. 2004).

Tularaemia is a zoonotic disease caused by the Gram-negative, intracellular bacterium *Francisella tularensis*. It is transmitted from animals to humans through the bites of arthropods (ticks and biting flies), direct contact with infected animal body fluids, ingestion of contaminated food and water, and inhalation. Whereas cats are known to transmit tularemia to humans via bites and other routes (CDC 1982), the role of dogs in facilitating infections is much less understood. Tularemia case investigation records collected through national surveillance during 2006–2016 were reviewed to summarize those involving dogs, characterize the nature of dog-related exposure, and describe associated clinical characteristics (Kwit et al. 2019). Dog-related exposures were classified as direct contact via bite, scratch, or face snuggling/licking (n = 12; 50%); direct contact with dead animals retrieved by domestic dogs (n = 8; 33%); and contact with infected ticks acquired from domestic dogs (n = 4; 17%). Cats are presumably infected through predation on infected lagomorphs and rodents (Magnarelli et al. 2007). Most tularemia isolates from cats belong to the more virulent Type A subspecies (Kugeler et al. 2009).

TICK-BORNE BACTERIAL DISEASES

Dogs and cats are frequently exposed to, and affected by, tick-borne pathogens, with *Borrelia burgdorferi*, the agent for Lyme disease, and *Anaplasma phagocytophilum*, the agent for granulocytic anaplasmosis, being among those most frequently identified in the United States (Gettings et al. 2020). The primary vectors for *B. burgdorferi* and *A. phagocytophilum* are ticks of the genus *Ixodes*. *Anaplasma platys* is also implicated in canine anaplasmosis (Gettings et al. 2020). Tick-borne relapsing fever (*Borrelia turicatae*) is a reportable disease in New Mexico caused by spirochetes in the genus *Borrelia* affecting humans and dogs. The disease was first reported in New Mexico in humans in 1936 with subsequent periodic cases (Dworkin et al. 2002; Esteve-Gasent et al. 2017). Dogs maintain *B. turicatae* as part of the lifecycle in the tick vector *Ornithodoros turicata* (Esteve-Gasent et al. 2017). The newest reported zoonotic tick-borne, dog-associated pathogen is *Rickettsia rickettsia*, which causes

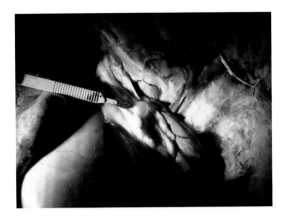

Photo 33.15. Hydatid cyst (at the tip of the scalpel) found during necropsy of a domestic sheep on 22 March 2006 in Chinle on the Navajo Nation in Apache County, Arizona. Hydatid disease or echinococcosis is caused by the tapeworms *Echinococcus granulosus* and *E. multilocularis*. The lifecycle involves both canine species (dogs, wolves [*Canis lupus*], coyotes [*C. latrans*]) as definitive hosts and sheep (wild and domestic) or cervids (deer, elk) as intermediate hosts, with occasional, aberrant transmission to humans, historically documented in New Mexico's northern and western counties. Photograph: © Scott C. Bender.

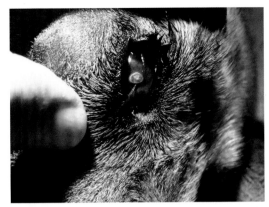

Photo 33.16. Dog with ocular onchocerciasis in Shiprock on the Navajo Nation in San Juan County on 10 May 2013. The white spot present on the eye is a corneal ulceration caused by the vector-borne filarial nematode (roundworm) *Onchocerca lupi*. First identified in the country of Georgia in the Caucasus, *O. lupi* is an emerging zoonotic parasite. It can result in minor to severe eye and skin lesions in both dogs and cats. An increase in human cases has been documented in the southwestern United States, likely via hematophagous (blood-feeding) insects with canine species potentially acting as a reservoir. The adult roundworm visible across the eye of the dog in the photograph is about to be extracted. Photograph: © Scott C. Bender.

Rocky Mountain spotted fever (RMSF), one of the most lethal of all bacterial infectious diseases in the Americas. In the Southwest, the primary vector for RMSF has been the Rocky Mountain wood tick (*Dermacentor andersoni*) in the *Ixodies* tick family. However, in 2004 a new tick vector for RMSF was identified as the brown dog tick (*Rhipicephalus sanguineus*), which had not been previously associated with RMSF transmission in the United States (Demma et al. 2006). As its name indicates, the primary host and fomite for brown dog tick transmission is the dog. The RMSF rarely causes recognizable disease symptoms in dogs, which have not been identified as a reservoir species for RMSF. In response to the outbreaks in east-central Arizona, and to investigate the potential role of dogs as a sentinel species for RMSF outbreaks, a serologic survey of 80 feral and free-ranging dogs was conducted in 2006 in northwestern New Mexico. This survey found that 63% of the dogs studied in San Juan County and 67% in McKinley County carried antibodies to RMSF, indicating a historical (post) exposure to the disease. A second follow up survey in 2012 produced similar serologic results in free-ranging dogs that were owned by humans (S. Bender, unpubl. data).

PARASITIC DISEASES

A 1967 survey of 200 dogs in the Albuquerque area found multiple dogs to be infected with five species of helminths representing cestodes and nematodes (Hathaway 1967). The ascarid (roundworm) nematode helminths found were identified as *Toxocara canis*, *Toxascaris leonina*, *Trichuris vulpis*, and *Ancylostoma caninum*, while helminths belonging to the family Taeniidae were the only cestodes (tapeworms) represented. A total of 13 dogs, or 6.5% of those examined, were found to be infected (Hathaway 1967). The most common cestode identified in cats and dogs by veterinarians in New Mexico is the flea tapeworm *Dipylidium caninum* (R. F. Taylor, pers. comm.).

Hydatid disease or Echinococcosis is caused by two species of very small (2–7 mm [0.08–0.28 in]) tapeworm, *Echinococcus granulosus* and *E. multilocularis*. The lifecycle involves both canine species (dogs, wolves, coyotes) and ruminants, primarily sheep (wild and domestic) and cervids, as sequential hosts (Foreyt 2001; Wen et al. 2019). Occasional severe, aberrant infections in humans are well documented in New Mexico, historically occurring in the northern and western counties. Retrospective surveys of hospital records covering the years 1969–1976 and serologic screening of patients' family members established the diagnosis of 21 human cases acquired locally in New Mexico. Nineteen of these infections occurred on tribal lands, 13 on the Navajo Nation, four in Zuni Pueblo, and two in Santo Domingo (or Kewa) Pueblo. Surveys of the Navajo and Zuni populations revealed a variety of animal husbandry and cultural practices highly conducive to the continued, even possibly increased, transmission of cestodes in the dog-sheep cycle (Schantz 1977). Starting in 1955, the Navajo Area IHS conducted educational outreach and treatment of dogs effectively eliminating this disease on the Navajo Nation by the mid-1980s (IHS 2015). In 2004, however, one of us (S. C. Bender) detected *E. multilocularis* during a domestic sheep necropsy on the Navajo Nation. This appeared to be the result of a reintroduction of the parasite as the number of identified sheep cases increased following the 2004 diagnosis (Photo 33.15), with no sheep cases reported between 1970 and 2004. Further, an age gap was also identified among diagnosed human cases by the IHS, with cases presenting in either those under 10 years of age or those individuals born prior to eradication in the 1970s, approximately 40 years old (IHS 2015). An introduction of this parasite occurred at the Navajo Nation Zoo in Window Rock, Arizona/Tse Bonito, New Mexico, likely as the result of an infected mule deer carcass being donated for feeding to the Mexican gray wolves. All the mule deer in the adjacent enclosure died from acute anaphylaxis over the next three years, due to ruptured hydatid cysts that were found at necropsy. It is suspected that *Echinococcus* was introduced into the mule deer enclosure on equipment and boots that were used by staff during cleaning but not been properly disinfected between enclosures. The zoo immediately instituted procedural and biosecurity changes following the first diagnosed case, including biannual cestode treatment for the wolves, effectively ending this outbreak (S. Bender, unpubl. data).

Chagas disease is an inflammatory, infectious disease caused by the parasite *Trypanosoma cruzi*. Transmitted by triatomine bug vectors, also known as "kissing bugs," this parasite was estimated to infect 6.2 million people worldwide in 2017, mostly in Mexico, Central America and South America, with wildlife and domestic animals acting as reservoirs following infection (WHO 2023). Of the eleven species of triatomine bugs identified in the United States, five are found in New Mexico (CDC 2023). The prevalence of *T. cruzi* in triatomine bugs, feral cats and dogs, and wild animals, was investigated at four different geographical sites along the US-Mexico border, including in El Paso County, Texas, and nearby cities in New Mexico. In total, 66.4% of triatomine bugs, 45.3% of feral and free-ranging dogs, 39.2% of feral and free-ranging cats, and 71.4% of wild animals tested positive for *T. cruzi*. These results underscore the urgent need for implementation of a systematic epidemiological surveillance program for *T. cruzi* infections in insect vectors, feral dogs and cats, and wild animals, and for Chagas disease in the human population in the Southwest (Rodriguez et al. 2021).

Toxoplasma gondii is a parasite of mammals and birds and is considered a zoonotic disease of concern in immunocompromised people and pregnant women. Cats are the only definitive host and thus the only source of infective oocysts, but other mammals and birds can develop tissue cysts (Elmore et al. 2010). Although feline infections are typically asymptomatic, *Toxoplasma gondii* can cross the placenta and cause a severe fetal disease in humans. Marchiondo et al. (1976) found that of the 233 dogs tested on New Mexico tribal reservations (and characterized as feral by the authors of the study), 48% were positive. The estimated seroprevalence for *T. gondii* in cats in the drier climate of Arizona, New Mexico, and Utah is 16.1% (Elmore et al. 2010).

Onchocerca lupi is a filarioid microfilaria species of roundworm related to the canine heartworm *Dinofilaria immitis*. It is of zoonotic concern, infecting dogs and cats, both of which develop ocular and skin lesions ranging from minor to severe. Infected animals do not always display overt clinical signs, rendering the diagnosis of the infection obscure to most veterinarians. Canine onchocercosis is reported as endemic in Mediterranean regions in the Old World (Otranto et al. 2013), whereas its occurrence in the United States, as well as its pathogenesis and clinical management, is still being researched. *Onchacerca lupi* has been diagnosed in both dog and human cases in Bernalillo, McKinley, and San Juan counties in New Mexico (Otranto et al. 2015; S. Bender, pers. obs.). Genetic investigation suggests recent introduction of this nematode into the United States from Europe (Otranto et al. 2015). Unlike cases reported outside the country, cases involving American human patients have not shown subconjunctival (below the lining of the eye) nodules but have manifested more invasive disease (e.g., spinal, orbital, and subdermal nodules). Available evidence suggests that there may be transmission

in the Southwest and that canine species are a reservoir (Cantey et al. 2016).

The mite *Sarcoptes scabiei* var. *canis* causes sarcoptic mange in dogs. This mite can infect a wide range of host species and has been spread across the globe presumably via dogs that accompanied human expansion. While disease caused by *S. scabiei* has been very well studied in humans and domestic animals, there are still numerous gaps in our understanding of this pathogen in free-ranging wildlife. The most common, clinically affected species in North America include red foxes, gray wolves, coyotes, and American black bears (*Ursus americanus*) (Niedringhaus et al. 2019; see Chapter 16). Sarcoptic mange has been diagnosed in New Mexico wildlife (T. Kreeger, pers. comm.) and was the most common skin disease identified in dogs, foxes, and coyotes on the Navajo Nation (S. Bender, unpubl. data).

STATUS AND MANAGEMENT

It is estimated that free-ranging and feral dogs cause more than $620 million in environmental damages and economic impacts annually across the United States, mostly because of predation on livestock (Pimentel et al. 2005). Cats, whether owned or not, are responsible for considerably more damage, estimated at $17 billion per year (Pimentel et al. 2005). Cats have less of a direct economic impact but represent a greater risk to wildlife conservation, primarily due to predation on birds, reptiles, amphibians, and small mammals including bats (Jessup 2004; Pimentel et al. 2005; Oedin et al. 2021). Economic impacts due to feral dogs and cats have not been determined for New Mexico in particular.

Coyotes, cougars, and dogs are the top three predators of cattle in the United States, respectively (USDA, NASS 2011; see Chapter 3). For sheep, coyotes are again the top predator in the United States, followed respectively by dogs and cougars (USDA, NASS 2005). Within New Mexico, dogs were responsible for 1.7% of cattle losses in 2010

(USDA, NASS 2011) and 33.3% of sheep loss (USDA, NASS 2005). Feral and free-ranging dogs caused almost all sheep, calf, and foal damage and deaths seen at Navajo Nation veterinary clinics between 1997 and 2017 (S. Bender, unpubl. data). The federal agency managing feral dogs in New Mexico is the USDA APHIS Wildlife Services. Wildlife Services reported losses due to feral and free-ranging dogs that ranged from just over $1,000 to over $36,000 in any given year (Chapter 3).

Management of feral dogs and cats can be broken into two general categories: removal and reproductive management. In general, both categories are regulated within the state by a variety of political entities. The state of New Mexico has passed the following laws addressing dogs:

- NMSA 1978, § 77-1-12, provides that each municipality and each county shall make provision by ordinance for the seizure and disposition of dogs and cats running at large and not kept or claimed by any person on their premises provided that it does not conflict with state law.

- NMSA 1978, § 77-1-6, provides that state health department shall prescribe regulations for the reporting of animal bites, confinement and disposition of rabies-suspect animals, rabies quarantine, and the disposition of dogs and cats exposed to rabies, in the interest of public health and safety.

- NMSA 1978, § 30-18-1 to 30-18-16; NMSA 1978, § 77-18-2 to 4, constitute New Mexico's anti–animal cruelty provisions. Cruelty to animals occurs when a person mistreats, injures, kills without lawful justification, or torments an animal or abandons or fails to provide necessary sustenance to an animal under that person's custody or control. Extreme cruelty to animals, a fourth-degree felony, consists of a person intentionally or maliciously torturing, mutilating, injuring, or poisoning an animal or maliciously killing an animal.

Across the state, 83 counties and local municipalities have promulgated rules to manage dogs and/or cats. As sovereign entities, many tribes have put into effect their own regulations including the Navajo Nation and the pueblos of Isleta and Laguna. In some cases, the tribes and pueblos receive assistance from the Bureau of Indian Affairs for animal management.

Free-ranging and feral dogs and cats can readily move back and forth across the US-Mexico border and frontier area. Many of these dogs originate in Mexico, and barriers along the border are not a deterrence to their movements (McCallum et al. 2014). Most people in Colonias (rural communities lacking water, sewage, or adequate housing) near the border in El Paso County, Texas, reported that feral and free-ranging dogs were a problem, and nearly 81% stated that these dogs sometimes prevented them from walking outdoors (Poss and Bader 2007).

Feral Dog Management on the Navajo Reservation

The management of feral and free-ranging dogs has been met with mixed success and political angst. The management of dogs on the Navajo Nation provides a good example wherein similarities can be found throughout the tribes in New Mexico. During the 1860s, the United States initiated a scorched-earth campaign to starve the Diné people into submission. In addition to destroying nearly 2 million pounds of crops, US troops slaughtered the domestic animals associated with the Diné people including sheep, goats, horses, and any dogs they could find as they marched through Canyon de Chelly (now Canyon de Chelly National Monument), just a few miles away from today's Chinle, Arizona. The Navajo Reservation's dog problem, while complex, developed from economic and social factors, most of which were and are due to federal policies (Johnson 2014).

Following World War II, the Diné people continued to keep sheep, but others moved to towns for employment or trade and brought their dogs along

Photo 33.17. By some estimates, the Navajo Nation (northeastern Arizona, northwestern New Mexico, and southeastern Utah) is home to 250,000 free-ranging and neglected dogs. In addition to transmitting potential zoonotic diseases, dog bites have become an important public health issue on the Navajo Nation. Between 2013 and 2015, 1,474 dog bite cases were reported. In 2012, a pack of dogs attacked and killed a 12-year-old boy on the Navajo Nation in Cibola County, New Mexico. A 13-year-old girl was also killed by dogs in Fort Defiance, Arizona in 2021. Photograph: © Scott C. Bender.

Photo 33.18. Free-ranging or feral dogs fight in the street in the small Navajo Nation community of Sundance in McKinley County, New Mexico. Photograph: Jim Weber Copyright © 2021 The New Mexican Inc. Reprinted with permission. All rights reserved.

Photo 33.19. Roundup of stray dogs on 15 June 2021 in the Navajo Nation community of Sundance in McKinley County, New Mexico. In a context of staff shortage (as of June 2021, there were only five Navajo Nation Animal Control officers for the more than 100 communities on the 70,000 km² [27,000 mi²] reservation), closure of animal control facilities, and reduction in spaying/neutering services due to the COVID-19 pandemic, the problem of free-ranging dogs was reportedly spreading outside the Navajo Nation (Chacón 2021). Photograph: Jim Weber Copyright © 2021 The New Mexican Inc. Reprinted with permission. All rights reserved.

Photo 33.20. Feral cats tend to live in groups called colonies around available food sources. Feral cat colonies can negatively impact local bird (and other animal) populations and represent a risk of pathogen transmission to domestic pets and wildlife. Photograph: © David L. Bergman.

(Johnson 2014). Most dogs were neither spayed nor neutered, and there were no resources available to hire a veterinarian for the purpose of dog reproduction control until the late 1970s when Navajo Community College (now Diné College in Tsaile, Arizona) started a veterinary assistant certificate program. As the reservation lacked an enforceable leash law, almost all the reservation's dogs roamed and bred freely, and populations skyrocketed (Johnson 2014). The large population of dogs posed a significant threat to human health and safety as dogs could be found in sheep camps, running freely in towns and villages, and raiding garbage cans (Johnson 2014). A representative of the Fort Defiance community and the Window Rock safety committee, Mr. Ed Turner, appeared before Judge Jim Shirley in the Navajo Tribal Court on July 7, 1950, to request that stray and "untagged" dogs be destroyed. The judge was receptive and issued the following ruling: "Since these animals are considered a nuisance among the populated area of [Fort Defiance and Window Rock], therefore, it is the order of the court that any dog straying within these areas, not properly tagged or ownership not properly established, should be disposed in the most humane manner" (Johnson 2014:109).

Today, the Navajo Nation government administers most of the reservation's animal control services under Navajo Nation legal title code (CJY-64-18).

Feral Cat Management in New Mexico

At a local level, the management of free-ranging (typically feral) cats often now relies on trap-neuter-return (TNR) programs, which have replaced the traditional method of impoundment followed by euthanasia when an animal is unadoptable. TNR programs involve the capture of feral cats living in a colony. Those cats are brought to a veterinary clinic to be spayed or neutered, vaccinated for rabies, and ear-tipped (for later recognition), before they are returned to the colony's original territory. TNR programs often pit conservation organizations and scientists against animal-rights activists. From a purely wildlife conservation and disease prevention standpoint, TNR programs are implemented across the United States without widespread public knowledge or consideration of scientific evidence about them, and without first completing the environmental review process often required for actions with potentially harmful consequences (Longcore et al. 2009; Lepczyk et al. 2010). Feral cats encroaching onto private property can intimidate, or even get into fights with pet cats, and steal food (Bradshaw et al. 2012). Colonies attract other cats, including pets (e.g., Swarbrick and Rand 2018). In both situations, and regardless of neuter status, fighting and disease transmission can result. Globally, feral cats can carry diseases and high parasite loads that compromise their health (Crawford et al. 2019). Maintaining or establishing more feral cat colonies increases the likelihood of pathogen

Photos 33.21a and b (*left to right*). Trap-neuter-return (TNR) programs represent a controversial approach to the management of feral cat colonies. Feral cats are captured and brought to veterinary clinics, where instead of being euthanized (if they are not adoptable), they are spayed or neutered, vaccinated for rabies, and ear-tipped (for later recognition), then returned to the colony's territory. TNR programs are popular with animal-rights activists but do not address some of the concerns (wildlife conservation, risk of disease) associated with feral cat colonies. Photographs: © David L. Bergman.

transmission to domestic pets and wildlife (Crawford et al. 2019). TNR cats are therefore potential vectors of these pathogens to other strays, pet cats, wildlife, and humans. Studies from the United States identified reduced populations of native bird species, including the complete absence of ground-foraging species, near sites where unowned cats were fed (Jessup 2004).

Albuquerque's first TNR program was initiated in 1997. Animal Humane New Mexico used TNR on 15,638 cats between 2008 and October 2019 in Albuquerque (Zimmer Feline Foundation 2023). Albuquerque's 1997 TNR program was followed up by the City of Albuquerque's Animal Welfare Department's feral cat program TNR program (Sizemore 2013; Spehar and Wolf 2018). Additional communities that have implemented TNR include Cloudcroft, Deming, Las Cruces, Ruidoso, Santa Fe, and Taos. New Mexico State University's Feral Cat Management Program (Las Cruces) started actively managing the main campus feline population using TNR in 2002 (Corella 2009). A more comprehensive list of resources to assist with dogs and cats is updated annually by Animal Protection of New Mexico (2020).

A community cat program in Albuquerque was initiated in April 2012 and ran for three years until March 2015. It included TNR and a pilot return-to-field program (involving stray cats impounded by animal control or brought to the municipal shelter by individuals) (Spehar and Wolf 2018). Over the course of the program implementation, 11,746 cats were trapped, sterilized, vaccinated, and returned or adopted; consequently, feline euthanasia at the Albuquerque Animal Welfare Department declined by 84.1% and feline intake went down by 37.6%, but the live release rate of cats increased by 47.7% (Spehar and Wolf 2018).

The state of New Mexico took a stand against managing feral cat colonies with TNR programs when it wrote a letter of opposition to the mayor of Albuquerque (NMDGF 2013). In 2019, a federal complaint was filed against the City of Albuquerque and its mayor (Marcy Britton, Plaintiff, v. Mayor Tim Keller, Danny Nevarez, and City of Albuquerque, Defendants), asserting claims for unlawful taking of property under the US and New Mexico constitutions, as well as related claims for trespass and nuisance, all due to the city's catch and release of feral cats and kittens as part of a TNR program. On 16 April 2020, the US District Court for the District of New Mexico (2020) dismissed the case. The court reasoned that colonies of cats did not constitute government occupation of the plaintiff's property. The court went on to state that the TNR policy of the City of Albuquerque ought to be considered an exercise of police powers and that the injuries to the plaintiff and her property were "incidental" to the TNR program, thereby precluding recourse under the US Constitution's protections against the taking of private property.

Damage to wildlife and livestock resources prompted two peer-reviewed studies to document the dentition of feral and free-ranging dogs on the Navajo Nation. The first study found an upper canine (maxillary tip) spread of 36.2 mm (1.43 in) (SE = 1.11 mm [0.04 in], n = 39); a lower canine (mandibular tip) spread of 31.3 mm (1.23 in) (SE = 0.92 [0.04 in], n = 38) (Bergman et al. 2010). The second study measured intercanine width of 197 dogs from 2008 to 2013 after roundups by the Navajo Nation Animal Control Program or in conjunction with wildlife damage management activities (Verzuh et al. 2018). Measurements for 118 male dogs averaged 39.3 ± 5.8 mm ([1.55 ± 0.23 in]; range = 15.9–53.4 mm [0.63–2.1 in])

for upper intercanine width and 34.3 ± 5.0 mm ([1.35 ± 0.2 in] range = 14.7–46.2 mm [0.58–1.82 in]) for lower intercanine width. Measurements for female dogs averaged 36.1 ± 6.0 mm ([1.42 ± 0.24 in; range = 14.1–49.2 mm [0.56–1.94 in], n = 77) for upper intercanine width and 31.8 ± 5.7 mm (1.25 ± 0.22 in; range = 10.8–42.8 mm [0.43–1.69 in], n = 74) for lower canine width (Verzuh et al. 2018). This dataset may be significant for future studies to determine the activity and impacts of feral and free-ranging dog depredation on wildlife and domestic livestock. These forensics data have been utilized in identification of carnivore-caused trauma/deaths and ruling out dogs as the predator in some cases.

The promotion of responsible dog and cat ownership can significantly reduce the numbers of feral and free-ranging dogs and cats, their impact upon the environment, and the incidence of zoonotic diseases. Because feral and free-ranging dog and cat ecology is linked with human activities, control of dog and cat populations can only be effective if it is accompanied by changes in human behaviors. Impact of dogs and cats on natural environments has been minimally documented worldwide and is especially lacking in New Mexico. Feral and free-ranging dogs and cats, their impact and mitigation, truly warrant further investigation and research.

LITERATURE CITED

Allen, G. M. 1920. Dogs of the American aborigines. *Bulletin of Harvard College, Museum of Comparative Zoology* 63:431–517.

American Rare Breed Association. [2021]. *Calupoh Mexican wolfdog breed standard*. https://www.arba.org/PDF%20Files/Group1/calupoh.pdf. Accessed 7 November 2021.

American Veterinary Medical Association (AVMA). 2018. *AVMA pet ownerships and demographics sourcebook*. 2017–2018 ed. Schaumburg, IL: American Veterinary Medical Association.

———. 2020. *Coronaviruses in domestic species*. Updated February 18, 2020. [Schaumburg, IL]: American Veterinary Medical Association. Unnumbered pages.

Animal Protection of New Mexico. 2020. *New Mexico animal resources guide*. Albuquerque: Animal Protection of New Mexico.

Appel, M. J. G., and J. H. Gillespie. 1972. *Canine distemper virus*. Virology Monographs/Die Virusforschung in Einzeldarstellungen No. 11. Vienna: Springer Vienna.

Archibald, W. S., and S. J. Kunitz. 1971. Detection of plague by testing serums of dogs on the Navajo Reservation. *HSMHA Health Reports* 86:377–80.

Asa, C. S., and C. Valdespino. 1998. Canid reproductive biology: an integration of proximate mechanisms and ultimate causes. *American Zoology* 38:251–59. doi:10.1093/icb/38.1.251.

Axelsson, E., A. Ratnakumar, M. L. Arendt, K. Maqbool, M. T. Webster, M. Perloski,O. Liberg, J. M. Arnemo, Å. Hedhammar, and K. Lindblad-Toh. 2013. The genomic signature of dog domestication reveals adaptation to a starch-rich diet. *Nature* 495(7441):360–64.

Beaver, B. V. 1999. *Canine behavior: a guide for veterinarians*. Philadelphia: W. B. Saunders.

Bekoff, M., T. J. Daniels, and J. L. Gittleman. 1984. Life history patterns and the comparative social ecology of carnivores. *Annual Review of Ecology and Systematics* 15:191–32.

Bender, S., and D. L. Bergman. 2017. Evaluating the endemic dog vaccination rate on the Navajo Nation and potential risk to the public (abstract). In *28th Rabies in the Americas Conference*, 77. Calgary, Alberta.

Bender, S., D. Bergman, A. Vos, A. Martin, and R. Chipman. 2017. Field studies evaluating bait acceptance and handling by dogs in Navajo Nation, USA. *Tropical Medicine and Infectious Disease* 2. doi:10.3390/tropicalmed2020017.

Bennett, M., D. Bustos, J. S. Pigati, K. B. Springer, T. M. Urban, V. T. Holliday, S. C. Reynolds, M. Budka, J. S. Honke, A. M. Hudson, B. Fenerty, C. Connelly, P. J. Martinez, V. L. Santucci, and D. Odess. 2021. Evidence of humans in North America during the Last Glacial Maximum. *Science* 373:1528–31.

Berentsen, A. R., S. Bender, P. Bender, D. Bergman, A. T. Gilbert, H. M. Rowland, and K. C. VerCauteren. 2016. Bait flavor preference and immunogenicity of ONRAB baits in domestic dogs on the Navajo Nation, Arizona. *Journal of Veterinary Behavior* 15:20–24.

Berentsen, A. R., S. Bender, P. Bender, D. Bergman, K. Hausig, and K. C. VerCauteren. 2014. Preference among 7 bait flavors delivered to domestic dogs in Arizona: implications for oral rabies vaccination on the Navajo Nation. *Journal of Veterinary Behavior and Clinical Applied Research* 9:169–71.

Bergman, D., S. Bender, K. Wenning, D. Slate, C. Rupprecht, C. Heuser, and T. Deliberto. 2008. Bait acceptability for delivery of oral rabies vaccine to free-ranging dogs on the Navajo and Hopi Nations. *Developmental Biology* 131:145–50.

Bergman, D. L., S. W. Breck, and S. C. Bender. 2009. Dogs gone wild: feral dog damage in the United States. *Proceedings of the Wildlife Damage Management Conference* 13:177–83.

Bergman, D. L., W. Sparklin, C. D. Carrillo, and J. A. Schmidt III. 2010. Depredation investigation: using canine spread to identify the predator species. *Proceedings of the Vertebrate Pest Conference* 24:304–7.

Bigelow, L., and J. D. Speth. 2004. The Henderson site dogs. In *Life on the periphery: economic change in late prehistoric southeastern New Mexico*, ed. J. D. Speth, 221–24. Ann Arbor: Museum of Anthropology, University of Michigan Press.

Black, H. L. 1981. Navajo sheep and goat guarding dogs: a New World solution to the coyote problem. *Rangelands* 3:235–37.

Black, H. L., and J. S. Green. 1985. Navajo use of mixed-breed dogs for management of predators. *Journal of Range Management* 38:11–15.

Blanton, J. D., C. A. Hanlon, and C. E. Rupprecht. 2007. Rabies surveillance in the United States during 2006. *Journal of the American Veterinary Medical Association* 231:540–56.

Bögel, K., K. Frucht, G. Drysdale, and J. Remfry. 1990. *Guidelines for dog population*. World Health Organization, Veterinary Public Health Unit and World Society for the Protection of Animals.

Boitani, L., F. Francisci, P. Ciucci, and G. Andreoli. 1995. Population biology and ecology of feral dogs in central Italy. In *The domestic dog: its evolution, behaviour, and interactions with people*, ed. J. Serpell, 217–44. Cambridge: Cambridge University Press.

Bökönyi, S. 1969. The development and history of domestic animals in Hungary: the Neolithic through the Middle Ages. *American Anthropologist* 73:640–74.

Bosco-Lauth, A. M., A. E. Hartwig, S. M. Porter, P. W. Gordy, M. Nehring, A. D. Byas, S. VandeWoude, S. K. Ragan, R. M. Maison, and R. A. Bowen. 2020. Experimental infection of domestic dogs and cats with SARS-CoV-2: pathogenesis, transmission, and response to reexposure in cats. *PNAS* 117:26382–88.

Botigué, L. R., S. Song, A. Scheu, S. Gopalan, A. L. Pendleton, M. Oetjens, A. M. Taravella, T. Seregély, A. Zeeb-Lanz, R.-M. Arbogast, D. Bobo, K. Daly, M. Unterländer, J. Burger, J. M. Kidd, and K. R. Veeramah. 2017. Ancient European dog genomes reveal continuity since the early Neolithic. *Nature Communications* 8, article 16082. https://www.nature.com/articles/ncomms16082.

Bradshaw, J. W. S., R. A. Casey, and S. L. Brown. 2012. *The behaviour of the domestic cat*. 2nd ed. Wallingford, UK: CABI.

Brock, C. 2016. Urban animals: GIS analysis of stray canines and felines in Albuquerque, New Mexico. MS thesis, University of New Mexico, Albuquerque.

Brody, J. J. 2008. When is a dog in Mimbres art? *Archaeology Southwest* 22:10–11.

Brown H. E., P. Ettestad, P. J. Reynolds, T. L. Brown, E. S. Hatton, J. L. Holmes, G. E. Glass, K. L. Gage, and R. J. Eisen. 2010., Climatic predictors of the intra- and inter-annual distributions of plague cases in New Mexico based on 29 years of animal-based surveillance data. *American Journal of Tropical Medicine and Hygiene* 82:95–102.

Burkhart, G. 2018. Thousands of strays, hundreds of dog bite calls reported in Albuquerque (special assignment). KRQE. Posted 9 November 2018, updated 10 November 2018. https://www.krqe.com/news/investigations/thousands-of-strays-hundreds-of-dog-bite-calls-reported-in-albuquerque/.

Butler, J. R. A., and J. T. du Toit. 2002. Diet of free-ranging domestic dogs (*Canis familiaris*) in rural Zimbabwe: implications for wild scavengers on the periphery of wildlife reserves. *Animal Conservation* 5:29–37.

Butler, J. R. A., J. T. du Toit, and J. Bingham. 2004. Free-ranging domestic dogs (*Canis familiaris*) as predators and prey in rural Zimbabwe: threats of competition and disease to large wild carnivores. *Biological Conservation* 115:369–78.

Butler, J. R. A., J. D. C. Linnell, D. Morrant, V. Athreya, N. Lescureux, and A. McKeown. 2014. Dog eat dog, cat eat dog: social-ecological dimensions of dog predation by wild carnivores. In *Free-ranging dogs and wildlife conservation*, ed. M. E. Gompper, 117–43. Oxford: Oxford University Press.

Cantey, P. T., J. Weeks, M. Edwards, S. Rao, G. A. Ostovar, W. Dehority, M. Alzona, S. Swoboda, B. Christiaens, W. Ballan, J. Hartley, A. Terranella, J. Weatherhead, J. J. Dunn, D. P. Marx, M. J. Hicks, R. A. Rauch, C. Smith, M. K. Dishop, M. H Handler, R. W. Dudley, K. Chundu, D. Hobohm, I. Feiz-Erfan, J. Hakes, R. S. Berry, S. Stepensaski, B. Greenfield, L. Shroeder, H. Bishop, M. de Almeida, B. Mathison, and M. Eberhard. 2016. The emergence of zoonotic *Onchocerca lupi* infection in the United States—a case series. *Clinical Infectious Disease* 62:778–83.

Catling, P. C., L. K. Corbett, and A. E. Newsome. Reproduction in captive and wild dingoes (*Canis familiaris dingo*) in temperate and arid environments of Australia. *Wildlife Research* 19:195–209.

Causey, M. K., and C. A. Cude. 1980. Feral dog and white-tailed deer interactions in Alabama. *Journal of Wildlife Management* 44:481–84.

Centers for Disease Control and Prevention (CDC). 1982. Tularemia associated with domestic cats—Georgia, New Mexico. *Morbidity and Mortality Weekly Report* 31:39–41.

———. [2023]. Triatomine bug FAQs. https://www.cdc.gov/parasites/chagas/gen_info/vectors/index.html. Accessed 14 March 2023.

Chacón, D. 2021. From friend to foe: free-roaming dogs overrun Navajo Nation. *Santa Fe New Mexican*, 19 June 2021.

Chiu, E. S., K. Fox, L. Wolfe, and S. Vandewoude. 2020. A novel test for determination of wild felid-domestic cat hybridization. *Forensic Science International: Genetics* 44:102160. https://doi.org/10.1016/j.fsigen.2019.102160.

Christie, D. W., and E. T. Bell. 1971. Some observations of the seasonal incidence and frequency of oestrus in breeding bitches in Britain. *Journal of Small Animal Practice* 12:159–67.

Clarke, C. G. 2002. *The men of the Lewis and Clark expedition: a biographical roster of the fifty-one members and a composite diary of their activities from all known sources."* Lincoln: University of Nebraska Press.

Clutton-Brock, J. 1992. The process of domestication. *Mammal Review* 22:79–85.

Collins, R. N., A. R. Martin, L. Kartman, R. L. Brutsché, B. W. Hudson, and H. G. Doran. 1967.

Plague epidemic in New Mexico, 1965. Introduction and description of cases. *Public Health Reports* 82:1077–99.

Constantine, D. G., E. S. Tierkel, M. D. Kleckner, and D. M. Hawkins. 1968. Rabies in New Mexico cavern bats. *Public Health Reports* 83:303–16.

Coppinger, R., and L. Coppinger. 2002. *Dogs: a new understanding of canine origin, behavior and evolution*. Chicago: University of Chicago Press.

Corella, M. 2009. On the positive side: NMSU sets example with feral cats. *Las Cruces Sun News* (Las Cruces, NM), 19 April 2009. http://www.hssnm.org/uploads/1/0/5/3/105381009/pawsitive41909.pdf.

Crawford, H. M., M. C. Calver, and P. A. Fleming. 2019. A case of letting the cat out of the bag—why trap-neuter-return is not an ethical solution for stray cat (*Felis catus*) management. *Animals* 9(4), article 171. https://doi.org/10.3390/ani9040171.

Creevy, K. E, and J. B. Evans. [2023]. Canine distemper. *The Merck veterinary manual* (online). https://www.merckvetmanual.com/generalized-conditions/canine-distemper/canine-distemper-overview. Accessed 14 March 2023.

Crowell-Davis, S. L., T. M. Curtis, and R. J. Knowles. 2004. Social organization in the cat: a modern understanding. *Journal of Feline Medicine and Surgery* 6:19–28.

Daniels, T. J. 1986. A study of dog bites on the Navajo Reservation. *Public Health Reports* 101:50–59.

———. 1987a. Conspecific scavenging by a young domestic dog. *Journal of Mammalogy* 68:416–18.

———. 1987b. The social ecology and behavior of free-ranging dogs. PhD dissertation, University of Colorado, Boulder.

———. 1988. Down in the dumps. *Natural History* 97:8–12.

Daniels, T. J., and M. Bekoff. 1989a. Population and social biology of free-ranging dogs, *Canis familiaris. Journal of Mammalogy* 70:754–62.

———. 1989b. Spatial and temporal resource use by feral and abandoned dogs. *Ethology* 81:300–12.

———. 1989c. Feralization: the making of wild domestic animals. *Behavioural Processes* 19:79–94.

Davis, R. G. 2019. Plague in cats and dogs: a public health concern. *Clinicians Brief* (February) 2019:57–60. https://files.brief.vet/2019-01/Plagues%20in%20Cats%20&%20Dogs.pdf.

DeBoer, B., and R. Tykot. 2007. Reconstructing ancestral Puebloan diets in the middle San Juan River, New Mexico, through stable isotope analysis (abstract). In *Proceedings of the 72nd annual meeting of the Society for American Archaeology, Austin, TX.*, 125. https://documents.saa.org/container/docs/default-source/doc-annual-meeting/annualmeeting/abstract/abstract_2007.pdf?sfvrsn=412587ae_4.

Demma, L. J., M. Traeger, D. Blau, R. Gordon, B. Johnson, J. Dickson, R. Ethelbah, S. Piontkowski, C. Levy, W. L. Nicholson, C. Duncan, K. Heath, J. Cheek, D. L. Swerdlow, and J. H. McQuiston. 2006. Serologic evidence for exposure to *Rickettsia rickettsii* in eastern Arizona and recent emergence of Rocky Mountain spotted fever in this region. *Vector-Borne and Zoonotic Diseases* 6:423–29.

Dix, A. S. 1980. Spanish war dogs in Navajo rock art at Canyon de Chelly, Arizona. *Kiva* 45:279–83.

Driscoll, C. A., J. Clutton-Brock, A. C. Kitchener, and S. J. O'Brien. 2009. The taming of the cat. *Scientific American* 300:68–75.

Driscoll, C. A., M. Menotti-Raymond, A. L. Roca, K. Hupe, W. E. Johnson, E. Geffen, E. H. Harley, M. Delibes, D. Pontier, A. C. Kitchener, N. Yamaguchi, S. J. O'Brien, and D. W. Macdonald. 2007. The Near Eastern origin of cat domestication. *Science* 317:519–23.

Dworkin, M. S., P. C. Shoemaker, C. L. Fritz, M. E. Dowell, and D. E. Anderson Jr. 2002. The epidemiology of tick-borne relapsing fever in the United States. *American Journal of Tropical Medicine and Hygiene* 66:753–58.

Dyk, W. 1938. *Son of Old Man Hat: a Navajo autobiography recorded by Walter Dyk*. Lincoln: University of Nebraska Press.

Eidson, M., J. P. Thilsted, and O. J. Rollag. 1991. Clinical, clinicopathologic, and pathologic features of plague in cats: 119 cases (1977–1988). *Journal of the American Veterinary Medicine Association* 199:1191–97.

Eidson, M., L. A. Tierney, O. J. Rollag, T. Becker, T. Brown, and H. F. Hull. 1988. Feline plague in New Mexico: risk factors and transmission to humans. *American Journal of Public Health* 78:1333–35.

Elmore, S. A., J. L. Jones, P. A. Conrad, S. Patton, D. S. Lindsay, and J. P. Dubay. 2010. *Toxoplasma gondii*: epidemiology, feline clinical aspects, and prevention. *Trends in Parasitology* 26:190–96.

Esteve-Gasent, M. D., C. B. Snell, S. A. Adetunji, and J. Piccione. 2017. Serological detection of tickborne relapsing fever in Texan domestic dogs. *PLOS One* 12:e0189786. ttps://doi.org/10.1371/journal.pone.0189786.

Feldmann, B. M. 1974. The problem of urban dogs. *Science* 185:903.

Ferg, A. 2008. A rare breed. *Archeology Southwest* 22:7.

Fitzwater, W. D. 1994. House cats (feral). In *The handbook: prevention and control of wildlife damage*, C45–50. Lincoln: University of Nebraska.

Fleming, P. J. S., L. R. Allen, S. J. Lapidge, A. Robley, G. R. Saunders, and P. C. Thomson. 2006. A strategic approach to mitigating the impacts of wild canids: proposed activities of the Invasive Animals Cooperative Research Centre. *Australian Journal of Experimental Agriculture* 46:753–62.

Ford, P. L., R. A. Fagerlund, D. W. Duszynski, and P. J. Polechla. 2004. *Fleas and lice of mammals in New Mexico*. Fort Collins, CO: US Department of Agriculture, Forest Service, Rocky Mountain Research Station, Gen. Tech. Rep. RMRS-GTR-123.

Foreyt, W. J. 2001. *Veterinary parasitology: reference manual*. 5th ed. Hoboken, NJ: Wiley-Blackwell.

Fowler, C. W. 1981. Density dependence as related to life history strategy. *Ecology* 62:602–10.

Franciscan Fathers. 1910. *An ethnologic dictionary of the Navajo language*. St. Michaels, AZ: Navajo Indian Mission.

Fredriksen A. 2016. Of wildcats and wild cats: Troubling species-based conservation in the Anthropocene. *Environment and Planning D: Society and Space* 34(4):689–705. doi:10.1177/0263775815623539.

Freeman, R. C., and J. H. Shaw. 1979. Hybridization in *Canis* (Canidae) in Oklahoma. *Southwestern Naturalist* 24:485–99.

Freer, R., and M. Jacobs. 2008. Basketmaker dog-hair sashes from Obelisk Cave. *Archeology Southwest* 22:6.

Frey, J. K., and Z. J. Schwenke. 2007. Mammals of Sugarite Canyon State Park, Colfax County, New Mexico. Final report submitted to New Mexico State Parks, Santa Fe, New Mexico, 16 November 2007, 49 pp.

Fugate, D. 2008. Pueblo dogs. *Archeology Southwest* 22:4–5.

Gage, K. L., D. T. Dennis, K. A. Orloski, P. Ettestad, T. L. Brown, P. J. Reynolds, W. J. Pape, C. L. Fritz, L. G. Carter, and J. D. Stein. 2000. Cases of cat-associated human plague in the Western US, 1977–1998. *Clinical Infectious Diseases* 30:893–900.

Geluso, K. 2016. *Mammals of the active floodplains and surrounding areas along the Gila and Mimbres rivers, New Mexico*. Final report, New Mexico Share with Wildlife contract.

Gessaman, J. A. 1973. Methods of estimating the energy cost of free existence. In *Ecological energetics of homeotherms*, ed. O. A. Gessaman, 3–31. Logan: Utah State University Press.

Gettings, J. R., S. C. W. Self, C. S. McMahan, D. A. Brown, S. K. Nordone, and M. J. Yabsley. 2020. Regional and local temporal trends of *Borrelia burgdorferi* and *Anaplasma* spp. seroprevalence in domestic dogs: contiguous United States 2013–2019. *Frontiers in Veterinary Science* 27:1–12.

Gilmore, M. R. 1934. The Arikara method of preparing a dog for a feast. *Papers of the Michigan Academy of Science, Arts and Letters* 19:37–38.

Gipson, P. S., J. A. Sealander, and J. E. Dunn. 1974. The taxonomic status of wild *Canis* in Arkansas. *Systematic Biology* 23:1–11.

Gojobori, J., N. Arakawa, X. Xiayire, Y. Matsumoto, S. Matsumura, H. Hongo, N. Ishiguro, and Y. Terai. 2021. The Japanese wolf is most closely related to modern dogs and its ancestral genome has been widely inherited by dogs throughout East Eurasia (preprint). https://doi.org/10.1101/2021.10.10.463851.

Gould, L. H., J. Pape, P. Ettestad, K. S. Griffith, and P. S. Mead. 2008. Dog-associated risk factors for human plague. *Zoonoses and Public Health* 55:448–54.

Grinder, M., and P. Krausman. 1998. Coyotes in urban areas: conflicts and solutions. In *Cross border waters: fragile treasures for the 21st century*, ed. G. J. Gottfried, 235–43. USDA Forest Service Proceedings RMRS-P-5. Fort Collins, CO. https://www.fs.usda.gov/rm/pubs/rmrs_p005/rmrs_p005_235_243.pdf.

Gunther, I., and J. Terkel. 2002 Regulation of

free-roaming cat (*Felis silvestris catus*) populations: a survey of the literature and its application in Israel. *Animal Welfare* 11:171–88.

Hafner, D. J., and C. J. Shuster. 1996. Historical biogeography of western peripheral isolates of the least shrew, *Cryptotis parva*. *Journal of Mammalogy* 77:536–45.

Hammond-Kaarremaa, L., 2018. Threads, twist and fibre: looking at Coast Salish textiles. In *Proceedings of the Textile Society of America 16th biennial symposium*. Presented at Vancouver, British Columbia, 19–23 September 2018. https://digitalcommons.unl.edu/tsaconf/.

Hanser, S. E., and M. Leu. 2008. Probability of synanthropic feral dog presence in the Western United States. USGS (US Geological Survey, Washington, DC). https://www.sciencebase.gov/catalog/item/542af32fe4b057766eed2885.

Hathaway, R. P. 1967. A survey of intestinal helminths in dogs from Albuquerque, New Mexico. *Journal of Parasitology* 53:1240.

Hayes, K. C., and E. A. Trautwein. 1989. Taurine deficiency syndrome in cats. *Veterinary Clinic North American Small Animal Practice* 19:403–13.

Heeney, J. L., J. F. Evermann, A. J. McKeirnan, L. Marker-Kraus, M. E. Roelke, M. Bush, D. E. Wildt, D. G. Meltzer, L. Colly, and J. Lukas. 1990. Prevalence and implications of feline coronavirus infections of captive and free-ranging cheetahs (*Acinonyx jubatus*). *Journal of Virology* 64:1964–72.

Hoditschek, B., J. F. Cully Jr., T. L. Best, and C. Painter. 1985. Least shrew (*Cryptotis parva*) in New Mexico. *Southwestern Naturalist* 30:600–601.

Holmes, E. C., C. R. Parrish, E. J. Dubovi, V. I. Shearn-Bochsler, R. W. Gerhold, J. D. Brown, K. A. Fox, D. J. Kohler, and A. B. Allison. 2013. Frequent cross-species transmission of parvoviruses among diverse carnivore hosts. *Journal of Virology* 87:2342–47.

Hubbard, J. P., and E. P. Elliston. 2013. New Mexico's first specimen of the Nelson's Sharp-tailed Sparrow, with an assessment of its source and relevance to the species' overall western status. Unpublished report. 12 pp.

Hughes, J., and D. W. Macdonald. 2013. A review of the interactions between free-roaming domestic dogs and wildlife. *Biological Conservation* 157:341–51.

Indian Health Service (IHS). 2015. *The Environmental Health Services Program of the Indian Health Service, U.S. Department of Health and Human Services annual report 2015*. https://www.ihs.gov/sites/dehs/themes/responsive2017/display_objects/documents/annual_reports/DEHS-2015-Annual-Report.pdf.

———. 2016. *The Division of Environmental Health Services annual report 2016*. https://www.ihs.gov/sites/dehs/themes/responsive2017/display_objects/documents/annual_reports/DEHS-2016-Annual-Report.pdf.

Jessup, D. A. 2004. The welfare of feral cats and wildlife. *Journal of the American Medical Association* 225:1377–83.

Johnson, K. A., Jr. 2014. The Chinle dog shoots: federal governance and grass-roots politics in postwar Navajo country. *Pacific Historical Review* 83:92–129.

Jones, T. C., R. D. Hunt, and W. King. 1997. *Veterinary pathology*. 6th ed. Baltimore: Williams & Wilkins.

Joseph, A., R. B. Spicer, and J. Chesky. 1949. *The desert people: a study of the Papago people*. Chicago: University of Chicago Press.

Kaminski, J., B. M. Waller, R. Diogo, A. Hartstone-Rose, and A. M. Burrows. 2019. Evolution of facial muscle anatomy in dogs. *PNAS* 116(29):14677–181. https://doi.org/10.1073/pnas.1820653116.

Kamler, J. F., W. Ballard, and P. S. Gipson. 2003a. Occurrence of feral dogs (*Canis lupus familiaris*) in Northwest Texas: an observation. *Texas Journal of Agriculture and Natural Resources* 16:75–77.

Kamler, J. F., K. Keeler, G. Wiens, C. Richardson, and P. S. Gipson. 2003b. Feral dogs, *Canis familiaris*, kill Coyote, *Canis latrans*. *Canadian Field-Naturalist* 117:123–24.

Kartman, L., A. R. Martin, W. T. Hubbert, R. N. Collins, and M. I. Goldenberg. 1967. Epidemiologic features and results of field studies. *Public Health Reports* 82:1084–94.

Kemp, B. M., K. Judd, C. Monroe, J. W. Eerkens, L. Hilldorfer, C. Cordray, R. Schad, E. Reams, S. G. Ortman, and T. A. Kohler. 2017. Prehistoric mitochondrial DNA of domesticate animals supports a 13th century exodus from the northern US southwest. *PLOS One* 12: e0178882. doi:10.1371/journal.pone.0178882.

Kiddle, L. B. 1964. American Indian reflexes of two Spanish words for cat. *International Journal of American Linguistics* 30:299–305.

Kugeler, K. J., P. S. Mead, A. M. Janusz, J. E. Staples, K. A. Kubota, L. G. Chalcraft, and J. M. Petersen. 2009. Molecular epidemiology of *Francisella tularensis* in the United States. *Clinical Infectious Diseases* 48:863–70.

Kustritz, M. V. R. 2005. Reproductive behavior of small animals. *Theriogenology* 64:734–46.

Kwit, N. A., A. Schwartz, K. J. Kugeler, P. S. Mead, and C. A. Nelson. 2019. Human tularemia associated with exposure to domestic dogs—United States, 2006–2016. *Zoonoses Public Health* 66:417–21.

Lawrence, B. 1966. Early domestic dogs. *Zeitschrift für Saugetierkunde* 32:44–59.

Leathlobhair, M. N., A. R. Perri, E. K. Irving-Pease, K. E. Witt, A. Linderholm, J. Haile, O. Lebrasseur, C. Dimopoulos, M. Eldridge, J. Enk, S. Gopalakrishnan, K. Gori, V. Grimes, E. Guiry, A. J. Hansen, A. Hulme-Beaman, J. Johnson, A. Kitchen, A. K. Kasparov, Y.-M. Kwon, P. A. Nikolskiy, C. Peraza Lope, A. Manin, T. Martin, M. Meyer, K. Noack Myers, M. Omura, J.-M. Rouillard, E. Y. Pavlova, P. Sciulli, M.-H. S. Sinding, A. Strakova, V. V. Ivanova, C. Widga, E. Willerslev, V. V. Pitulko, I. Barnes, M. T. P. Gilbert, K. M. Dobney, R. S. Malhi, E. P. Murchison, G. Larson, and L. A. F. Frantz. 2018. The evolutionary history of dogs in the Americas. *Science* 361:81–85.

Lepczyk, C. A., N. Dauphine, D. M. Bird, S. Conant, R. J. Cooper, D. C. Duffy, P. J. Hatley, P. P. Marra, E. Stone, and S. A. Temple. 2010. What conservation biologists can do to counter trap-neuter-return: response to Longcore et al. *Conservation Biology* 24:627–29.

Levy, J. K., and P. C. Crawford. 2004. Humane strategies for controlling feral cat populations. *Journal of the American Veterinary Medical Association* 225:1354–60.

Liberg, O., and M. Sandell. 1988. Spatial organisation and reproductive tactics in the domestic cat and other felids. In *The domestic cat: the biology of its behaviour*, ed. D. C. Turner and P. Bateson, 67–81. Cambridge: Cambridge University Press.

Liberg, O., M. Sandell, D. Pontier, and E. Natoli. 2000. Density, spatial organization and reproductive tactics in the domestic cat and other felids. In *The domestic cat: the biology of its behaviour*, ed. D. C. Turner and P. Bateson, 119–48. 2nd ed. Cambridge: Cambridge University Press.

Lipinski, M. J., L. Froenicke, K. C. Baysac, N. C. Billings, C. M. Leutenegger, A. M. Levy, M. Longeri, T. Niini, H. Ozpinar, M. R. Slater, N. C. Pedersen, and L. A. Lyons. 2008. The ascent of cat breeds: genetic evaluations of breeds and worldwide random-bred populations. *Genomics* 91:12–21.

Lobetti, R. 2003. Canine parvovirus and distemper. *Proceedings of the 28th World Congress of the World Small Animal Veterinary Association*. https://www.vin.com/apputil/content/defaultadv1.aspx?id=3850237&pid=8768.

Lomas, L. A., and L. C. Bender. 2007. Survival and cause-specific mortality of neonatal mule deer fawns, north-central New Mexico. *Journal of Wildlife Management* 71:884–94.

Longcore, T., C. Rich, and L. M. Sullivan. 2009. Critical assessment of claims regarding management of feral cats by trap-neuter-return. *Conservation Biology* 23:887–94.

Lord, K., M. Feinstein, B. Smith, and R. Coppinger. 2013. Variation in reproductive traits of members of the genus *Canis* with special attention to the domestic dog (*Canis familiaris*). *Behavioral Proceeding* 92:131–42.

Loss, S. R., T. Will, and P. P. Marra. 2013. The impact of free-ranging domestic cats on wildlife of the United States. *Nature Communications* 4, article1396. https://www.nature.com/articles/ncomms2380%C2%A0.

Lyman, J. H. 1844. Shepherd dogs. *American Agriculturist* 3:241–42.

Ma, X., B. P. Monroe, J. M. Cleaton, L. A. Orciari, C. M. Gigante, J. D. Kirby, R. B. Chipman, C. Fehlner-Gardiner, V. Gutiérrez Cedillo, B. W. Petersen, V. Olson, and R. M. Wallace. 2020. Public veterinary medicine: public health: rabies surveillance in the United States during 2018. *Journal of the American Veterinary Medical Association* 256:195–208.

Macdonald, D. W., P. J. Apps, G. M. Carr, and G. Kerby. 1987. Social dynamics, nursing coalitions and infanticide among farm cats, *Felis catus*. *Ethology* 28(Suppl):1–66.

Macdonald, D. W., L. A. D. Campbell, J. F. Kamler, J. Marino, G. Werhahn, and C. Sillero-Zubiri. 2019. Monogamy: cause, consequence, or corollary of success in wild canids? *Frontiers in Ecology and Evolution* 7, article 341. https://www.frontiersin.org/articles/10.3389/fevo.2019.00341/full.

Macdonald, D. W., and G. M. Carr. 1995. Variation in dog society: between resource dispersion and social flux. In *The domestic dog: its evolution, behaviour, and interactions with people*, ed. J. Serpell, 199–216. Cambridge: Cambridge University Press.

Macdonald, D. W., N. Yamaguchi, and G. Kerby. 2000. Group-living in the domestic cat: its sociobiology and epidemiology. In *The domestic cat: the biology of its behaviour*, ed. D. C. Turner and P. Bateson, 95–118. 2nd ed. Cambridge: Cambridge University Press.

Magnarelli, L., S. Levy, and R. Koski. 2007. Detection of antibodies to *Francisella tularensis* in cats. *Research in Veterinary Science* 82:22–26.

Majumder, S. S., A. Bhadra, A. Ghosh, S. Mitra, D. Bhattacharjee, J. Chatterjee, et al. 2014. To be or not to be social: foraging associations of free-ranging dogs in an urban ecosystem. *Acta Ethology* 17:1–8.

Mann, J. M., W. J. Martone, J. M. Boyce, A. F. Kaufmann, A. M. Barnes, and N. S. Weber. 1979. Endemic human plague in New Mexico: risk factors associated with infection. *Journal of Infectious Diseases* 140:397–401.

Marchiondo, A. A., D. W. Duszynski, and G. O. Maupin. 1976. Prevalence of antibodies to *Toxoplasma gondii* in wild and domestic animals of New Mexico and Colorado. *Journal of Wildlife Diseases* 12:226–32.

McCallum, J. W., J. M. Rowcliffe, and I. C. Cuthill. 2014. Conservation on international boundaries: the impact of security barriers on selected terrestrial mammals in four protected areas in Arizona, USA. *PLOS One* 9:e93679. https://doi.org/10.1371/journal.pone.0093679.

McCarthy A. J., M. A. Shaw, and S. J. Goodman. 2007. Pathogen evolution and disease emergence in carnivores. *Proceedings of Biological Sciences* 274: 3165–74.

McDonald, C. 1985. *Wilderness: a New Mexico legacy*. Santa Fe: Sunstone Press.

Mech, L. D. 1970. *The wolf, the ecology and behavior of an endangered species*. New York: Natural History Press.

Mick-O'Hara, L. S. 1994. Nutritional stability and changing faunal resource use in La Plata Valley prehistory. PhD dissertation, University of New Mexico, Albuquerque.

Miklósi, Á. 2015. Intra-specific social organization in dogs and related forms. In *Dog Behaviour, Evolution, and Cognition*, 172–82. 2nd ed. Oxford: Oxford University Press.

Monagle, V., C. Conrad, and E. L. Jones. 2018. What makes a dog? Stable isotope analysis and human-canid relationships at Arroyo Hondo Pueblo. *Open Quaternary* 4:1–13.

Morey, D. F. 2010. *Dogs: domestication and the development of a social bond*. Cambridge: Cambridge University Press.

Morters, M. K., T. J. McKinley, O. Restif, A. J. K. Conlan, S. Cleaveland, K. Hampson, H. R. Whay, I. M. Damriyasa, and J. L. N. Wood. 2014. The demography of free-roaming dog populations and applications to disease and population control. *Journal of Applied Ecology* 51:1096–106.

National Academies of Sciences, Engineering, and Medicine 2019. *Evaluating the taxonomic status of the Mexican gray wolf and the red wolf*. Washington, DC: National Academies Press. https://doi.org/10.17226/25351.

National Research Council. 1974. *Nutrient requirements of dogs*. Washington, DC: National Academy of Science.

New Mexico Department of Game and Fish (NMDGF). 2013. Letter to Albuquerque Mayor Richard J. Berry. 16 September 2013. https://abcbirds.org/wp-content/uploads/2014/01/albequerque_cats.pdf.

Niedringhaus, K. D., J. D. Brown, K. M. Sweeley, and M. J. Yabsley. 2019. A review of sarcoptic mange in North American wildlife. *International Journal of Parasitology and Parasites in Wildlife* 13: 285–97.

Nowak, R. M. 1991. *Walker's mammals of the world*. Vol. 2. 5th ed. Baltimore: Johns Hopkins University Press.

Oedin, M., F. Brescia, A. Millon, B. P. Murphy, P. Palmas, J. C. Z. Woinarski, and E. Vidal. 2021. Cats *Felis catus* as a threat to bats worldwide: a review of the evidence. *Mammal Review* 51:323–37.

Ogan, C. V., and R. M. Jurek. 1997. Biology and ecology of feral, free-roaming, and stray cats. In *Mesocarnivores of northern California: biology, management, and survey techniques*, ed. J. E. Harris and C. V. Ogan, 87–91. Workshop Manual, 12–15 August 1997, Humboldt State University, Arcata, CA. The Wildlife Society, California North Coast Chapter, Arcata, California.

Olsen, J. W. 1990. *Vertebrate faunal remains from Grasshopper Pueblo, Arizona*. Anthropological Paper No. 83. Ann Arbor: Museum of Anthropology, University of Michigan.

Otranto, D., F. Dantas-Torres, E. Brianti, D. Traversa, D. Petrić, C. Genchi, and G. Capelli. 2013. Vector-borne helminths of dogs and humans in Europe. *Parasites and Vectors* 6, article 16. https://parasitesandvectors.biomedcentral.com/articles/10.1186/1756-3305-6-16.

Otranto, D., A. Giannelli, M. S. Latrofa, F. Dantas-Torres, N. S. Trumble, M. Chavkin, G. Kennard, M. L. Eberhard, and D. D. Bowman. 2015. Canine Infections with *Onchocerca lupi* Nematodes, United States, 2011–2014. *Emerging Infectious Disease* 21:868–71.

Ovodov, N. D., S. J. Crockford, Y. V. Kuzmin, T. F. G. Higham, G. W. L. Hodgins, and J. van der Plicht. 2011. A 33,000-year-old incipient dog from the Altai Mountains of Siberia: evidence of the earliest domestication disrupted by the Last Glacial Maximum. *PLOS One* 6:e22821. doi:10.1371/journal.pone.0022821.

Pal, S. K. 2011. Mating system of free-ranging dogs (*Canis familiaris*). *International Journal of Zoology* 2011:314216. https://doi.org/10.1155/2011/314216.

Pardi, M. I., and F. A. Smith. 2015. Biotic responses of canids to the terminal Pleistocene megafauna extinction. *Ecography* 39:141–51.

Parker, H. G., A. Harris, D. L. Dreger, B. W. Davis, and E. A. Ostrander. 2017. The bald and the beautiful: hairlessness in domestic dog breeds. *Philosophical Transactions of the Royal Society B: Biological Sciences* 372:20150488. http://dx.doi.org/10.1098/rstb.2015.0488.

Paul, M., S. S. Majumder, and A. Bhadra. 2014. Grandmotherly care: a case study in Indian free-ranging dogs. *Journal of Ethology* 32:75–82.

Perri, A., C. Widga, D. Lawler, T. Martin, T. Loebel, K. Farnsworth, L. Kohn, and B. Buenger. 2019. New evidence of the earliest domestic dogs in the Americas. *American Antiquity* 84:68–87.

Perrone, D., S. Bender, and S. Niewiesk. 2010. A comparison of the immune responses of dogs exposed to canine distemper virus (CDV)—differences between vaccinated and wild-type virus exposed dogs. *Canadian Journal of Veterinary Research* 74:214–17.

Peterson, H. A. 2017. A Review of *The Navajo and the Animal People: Native American traditional ecological knowledge and ethnozoology*. *IK: Other Ways of Knowing* 3(1):74–76.

Pieracci, E. G., C. M. Pearson, R. M. Wallace, J. D. Blanton, E. R. Whitehouse, X. Ma, K. Stauffer, R. B. Chipman, and V. Olson. 2019. Vital Signs: trends in human rabies deaths and exposures—United States, 1938–2018. *Morbidity and Mortality Weekly Report* 68:524–28. doi:10.15585/mmwr.mm6823e1.

Pils, C. M., and M. A. Martin. 1974. Dog attack on a communal fox den in Wisconsin. *Journal of Wildlife Management* 38:359–360.

Pimentel, D., R. Zuniga, and D. Morrison. 2005. Update on the environmental and economic costs associated with alien-invasive species in the United States. *Ecological Economics* 52:273–88.

Poss, J. E., and J. O. Bader. 2007. Attitudes toward companion animals among Hispanic residents of a Texas border community. *Journal of Applied Animal Welfare Science* 10:243–53.

Powers, W. K., and M. N. Powers. 1990. *Sacred foods of the Lakota*. Kendall Park, NJ: Lakota Books.

Ramos-Madrigal, J., M.-H. S. Sinding, C. Carøe, S. Mak, J. Niemann, J. A. S. Castruita, S. Fedorov, A. Kandyba, M. Germonpré, and H. Bocherens. 2021. Genomes of Pleistocene Siberian wolves uncover multiple extinct wolf lineages. *Current Biology* 31:198–206.

Randi, E. 2008. Detecting hybridization between wild species and their domesticated relatives. *Molecular Ecology* 17:285–93.

Rodriguez, F., B. S. Luna, O. Calderon, C. Manriquez-Roman, K. Amezcua-Winter, J. Cedillo, R. Garcia-Vazquez, I. A. Tejeda, A. Romero, K. Waldrup, D. M. Watts, C. Khatchikian, and R. A. Maldonado. 2021. Surveillance of *Trypanosoma cruzi* infection in Triatomine vectors, feral dogs and cats, and wild animals in and around El Paso

County, Texas, and New Mexico. *PLOS Neglected Tropical Diseases* 15:e0009147. doi: 10.1371/journal. pntd.0009147.

Rollins, D., B. D. Taylor, T. D. Sparks, T. E. Wadell, and G. Richards. 2009. Species visitation at quail feeders and guzzlers in southern New Mexico. *National Quail Symposium Proceedings* 6:210–19.

Rowan, A. 2020. Global dog populations. Well Being International. 30 May 2020. https://wellbeingintl. org/global-dog-populations-2.

Schantz, P. M. 1977. Echinococcosis in American Indians living in Arizona and New Mexico: a review of recent studies. *American Journal of Epidemiology* 106:370–79.

Schwartz, M. 1998. *A history of dogs in the early Americas.* New Haven, CT: Yale University Press.

Scott, J. P., and J. L. Fuller. 1965. *Genetics and the social behavior of the dog.* Chicago: University of Chicago Press.

Sizemore, G. C. 2013. Expert statement of Grant C. Sizemore: trap, neuter and release (TNR) programs harm wildlife, the environment, public health, and the cats they are designed to aid. Invasive Species Programs for the American Bird Conservancy. Unnumbered pages.

Slattery, J. P., and S. J. O'Brien. 1998. Patterns of Y and X chromosome DNA sequence divergence during the Felidae radiation. *Genetics* 148:1245–55.

Snyder, L. M. 1991. Barking mutton: ethnohistoric and ethnographic, archaeological, and nutritional evidence pertaining to the dog as a Native American food resource on the Plains. In *Reamers, Bobwhites, and Blue-Points: Tributes to the Career of Paul W. Parmalee*, ed. J. R. Purdue, W. E. Klippel, and B. W. Styles, 359–78. Illinois State Museum, Scientific Papers No. 23 and University of Tennessee, Department of Anthropology, Report of Investigations No. 52. Springfield: Illinois State Museum.

Spehar, D. D., and P. J. Wolf. 2018. The impact of an integrated program of return-to-field and targeted trap-neuter-return on feline intake and euthanasia at a municipal animal shelter. *Animals* 8, article 55. https://doi.org/10.3390/ ani8040055.

Swarbrick, H., and J. Rand. 2018. Application of a protocol based on trap-neuter-return (TNR) to manage unowned urban cats on an Australian university campus. *Animals* 8, article 77. https:// doi.org/10.3390/ani8050077.

Taha, M. B., D. E. Noakes, and W. E. Allen. 1981. The effect of season of the year on the characteristics and composition of dog semen. *Journal of Small Animal Practice* 22:177–84.

Taylor, T., A. Ferg, and D. Fugate. 2008. Dogs in the Southwest. *Archeology Southwest* 22:1–2.

Truyen, U. 1994. Canine parvovirus: Neuere Erkenntnisse über die Entstehung und Entwicklung eines viralen Pathogens [Canine parvovirus: recent knowledge of the origin and development of a viral pathogen]. *Tierarztl Prax* 22:579–84.

United States Department of Agriculture (USDA), National Agriculture Statistics Service (NASS). 2005. *Sheep and goats death loss.* Washington, DC: USDA, NASS.

———. 2011. *Cattle death loss.* Washington, DC: USDA, NASS.

United States District Court for the District of New Mexico. 2020. Marcy Britton, Plaintiff, vs. Mayor Tim Keller, Danny Nevarez, and City of Albuquerque. Case No. 1:19-cv-01113 KWR/JHR. 17 pp.

Vanak, A. T., and M. E. Gompper. 2009. Dogs *Canis familiaris* as carnivores: their role and function in intraguild competition. *Mammal Review* 39:265–83.

Verzuh, T., D. L. Bergman, S. C. Bender, M. Dwire, and S. W. Breck. 2018. Intercanine width measurements to aid predation investigations: a comparison between sympatric native and non-native carnivores in the Mexican wolf recovery area. *Journal of Mammalogy* 99:1405–10.

Vilà, C., P. Savolainen, J. E. Maldonado, I. R. Amorim, J. E. Rice, R. L. Honeycutt, K. A. Crandall, J. Lundeberg, and R. K. Wayne. 1997. Multiple and ancient origins of the domestic dog. *Science* 276:1687–89.

VonHoldt, B. M., J. P. Pollinger, D. A. Earl, H. G. Parker, E. A. Ostrander, and R. K. Wayne. 2013. Identification of recent hybridization between gray wolves and domesticated dogs by SNP genotyping. *Mammalian Genome* 24:80–88.

Wayne R. K. 1986. Cranial morphology of domestic and wild canids: the influence of development on morphological change. *Evolution* 40:243–61. doi:10.1111/j.1558–5646.1986.tb00467.x.

Wayne, R. K., and C. Vilà. 2003. Molecular genetic

studies of wolves. In *Wolves: behavior, ecology, and conservation*, ed. L. D. Mech and L. Boitani, 218–37. Chicago: Chicago University Press.

Welker, M. H. 2020. Travois transport and field processing: the role of dogs in Intermountain and Plains food transport. *Human Ecology* 49: 721–33.

Wen, H., L. Vuitton, T. Tuxun, J. Li, D. A. Vuitton, W. Zhang, and D. P. McManus. 2019. Echinococcosis: advances in the 21st century. *Clinical Microbiology Reviews* e00075-18. https://doi.org/10.1128/CMR.00075-18.

Winship, G. P. 1896. The Coronado expedition, 1540–1542. In *14th Annual Report of Bureau of American Ethnology*, 339–558. Washington, DC: US Government Printing Office.

———. 1904. *The Coronado expedition 1540–1542*. New York: Barnes.

Witt, K. E., K. Judd, A. Kitchen, C. Grierd, T. A. Kohler, S. G. Ortman, B. M. Kemp, and R. S. Malhiai. 2015. DNA analysis of ancient dogs of the Americas: identifying possible founding haplotypes and reconstructing population histories. *Journal of Human Evolution* 79:105–18.

World Health Organization (WHO). [2023]. Chagas disease (also known as American trypanosomiasis). www.who.int/en/news-room/fact-sheets/detail/chagas-disease-(american-trypanosomiasis). Accessed 14 March 2023.

Wynne, C. D. L. 2021. The indispensable dog. *Frontiers in Psychology* 12, article 656529. https://doi.org/10.3389/fpsyg.2021.656529.

Yaglom, H. D., G. Hecht, A. Goedderz, D. Jasso-Selles, J. L. Ely, I. Ruberto, J. R. Bowers, D. M. Engelthaler, and H. Venkat. 2021. Genomic investigation of a household SARS-CoV-2 disease cluster in Arizona involving a cat, dog, and pet owner. *One Health* 13, article100333. https://doi.org/10.1016/j.onehlt.2021.100333.

Yamane, A., T. Doi, and Y. Ono. 1996. Mating behaviors, courtship rank and mating success of male feral cat (*Felis catus*). *Journal of Ethology* 14:35–44.

Yeon, S. C. 2007. The vocal communication of canines. *Journal of Veterinary Behavior* 2:141–44.

Zimmer Feline Foundation. 2023. Animal Humane New Mexico's trap-neuter-return program for feral cats. https://web.archive.org/web/20210513155508/http://zimmer-foundation.org/pgm3/ccp8.html. Accessed 14 March 2023.

APPENDIX 5.1. MODEL SIMULATIONS OF PROJECTED CLIMATE CHANGE

To assess the potential for species distribution changes resulting from ongoing climate change, temperature and precipitation changes across New Mexico are examined using projections of future climate derived from dynamic models. These models simulate the evolution of global climate changing as the result of increasing atmospheric greenhouse gas concentration imposed on the model. The simulations are part of the Coupled Model Intercomparison Project, Phase 5 (CMIP5; Taylor et al. 2012), which were used as part of the Fifth Assessment Report of the United Nations Intergovernmental Panel on Climate Change (IPCC 2013). The CMIP5 models simulate global climate change using dynamical equations (similar to a weather forecast model) expressed on 3-D global grids filling the volume of the atmosphere, coupled to another three-dimensional model simulating the ocean. The horizontal resolution of the atmospheric grid varies among models, but grid cells are typically about 1 or 2 degrees of latitude and longitude on a side. CMIP5 simulations are forced by one of several standardized future greenhouse gas scenarios that are commonly used in climate research (Meinshausen et al. 2011).

The resolution of global model output is far coarser than the gradients of climate and landscapes generally considered by wildlife biologists, so for many purposes it is useful to downscale the global output to much finer resolution. The results shown in this chapter were derived by statistically downscaling global model output using the Multivariate Adaptive Constructed Analogs procedure (MACA; Abatzoglou 2011), obtained from https://climate.northwestknowledge.net/MACA/. MACA utilizes a statistical algorithm based on historical observations and actual topographic features to introduce realistic (but artificial) high-resolution spatial variability to the coarse-resolution model output.

The spatial scale of the MACA output is 1/24 degree (roughly 5 km), fine enough to resolve the major topographic features of New Mexico: major mountain ranges (e.g., the Sacramento, Sangre de Cristo, Chuska, Black Range, and Manzano ranges) and river valleys (Rio Grande, Pecos, San Juan). But even these landscape features are still represented rather coarsely, and individual topographic features (arroyos, mountain peaks, lesser hills) are at best incompletely resolved. Furthermore, the maps of projected climate change shown in this chapter have been averaged over 20 individual downscaled simulations to reduce uncertainties associated with particular models or simulations. But such averaging also tends to smooth out sharp gradients in climate change projected by any particular model simulation.

APPENDIX 13.1. KIT FOX MUSEUM SPECIMENS FROM NEW MEXICO

Museum institution acronyms: AMNH (American Museum of Natural History); ASNHC (Angelo State Natural History Collections); ENMUNHM (Eastern New Mexico University, Natural History Museum); FHSM (Fort Hayes State University, Sternberg Museum of Natural History); FMNH (The Field Museum of Natural History); HSU (Humboldt State University Vertebrate Museum); KU (University of Kansas, Natural History Museum and Biodiversity Research Center); MSB (University of New Mexico, Museum of Southwestern Biology); MVZ (University of California, Berkeley, Museum of Vertebrate Zoology); NMMNH (New Mexico Museum of Natural History and Science); NMSU (The Vertebrate Museum, New Mexico State University); NMSUVWM (New Mexico State University Vertebrate Wildlife Museum [a.k.a. NMSU Wildlife Museum]); OMNH (University of Oklahoma, Sam Noble Oklahoma Museum of Natural History); TCWC (Biodiversity Research and Teaching Collections [formerly Texas Cooperative Wildlife Collection]); TTU-NSRL (Texas Tech University–Natural Science Research Laboratory); UA (Collection of Mammals, University of Arizona); UMMZ (University of Michigan Museum of Zoology); USNM (United States National Museum of Natural History); UTEP (University of Texas, El Paso, Biodiversity Collections, Mammals Division); UWYMV (University of Wyoming Museum of Vertebrates); WNMU (Western New Mexico University); YPM (Yale University, Peabody Museum of Natural History).

BERNALILLO CO.
MSB:Mamm:101289
Albuquerque, 8 mi SW
27 December 1928
Edgington, E.

MSB:Mamm:421
T11N, R1W, SEC9
10 December 1950
Koster, W.J.

MSB:Mamm:64390
3 mi W Albuquerque
25 July 1951
Ivey, R.D

MSB:Mamm:64391
Albuquerque, near edge of city, E
    mesa by reservoir
6 October 1956
Ivey, R.D

MSB:Mamm:64392
Albuquerque, near edge of city, E
    mesa by reservoir
6 October 1956
Ivey, R.D.

MSB:Mamm:64393
Albuquerque, near edge of city,
    E mesa by reservoir

6 October 1956
Ivey, R.D

MSB:Mamm:16190
2 mi S Gibson, 1 mi E Yale, Albu-
    querque, T9N, R3E
22 October 1962
Young, J.J.

MSB:Mamm:20562
Sandia Base
1 October 1963

Banks, R.C.
MSB:Mamm:24546
Rio Puerco, N of US 66, S of Ber-
    nalillo
8 February 1967
Domenici Taxidermy

CHAVES CO.
MSB:Mamm:54444
10 mi W Hagerman
31 January 1984
Jensen, B.

MSB:Mamm:54442
10 mi W Hagerman
before 19 May 1984
Jensen, B.

MSB:Mamm:54443
10 mi W Hagerman
before 19 May 1984
Jensen, B.

MSB:Mamm:142892
ca. 5.2 mi. W, 0.1 mi. S Roswell; Six
    mile Hill; junction of US 70/380
    (=2nd Street) and Mark Road
5 December 1994
Najier, Denadli-Thompson

MSB:Mamm:232370
0.5 mi S, 5 mi W Bitter Lake NWR
    Headquarters
24 November 1995
Najara, S.

MSB:Mamm:232375
3.2 mi N, 11.3 mi W Hope, DOR on
    US 62 at Jct. with hwy 13.
5 June 1996
Kintigh, K.

MSB:Mamm:232368
8 mi S Roswell, DOR on US 285
19 July 1996
Kintigh, K.

MSB:Mamm:89147
17.4 mi S. (by road) Chaves County
    courthouse, Roswell
24 December 1996
Campbell, M.L., Brunt Jr., J.W.

MSB:Mamm:142648
10 mi E Roswell
14 January 1997
Ballou, J.

MSB:Mamm:142665
12 mi W Roswell, Marley Whitney
    Ranch
25 January 1997
Ballou, J.

MSB:Mamm:142662
12 mi W Roswell, Marley Whitney
    Ranch
27 January 1997
Ballou, J.

MSB:Mamm:142681
12 mi W Roswell, Marley Whitney
    Ranch
27 January 1997
Ballou, J.

MSB:Mamm:232355
12 mi W Roswell, Marley
    Whitney Ranch
27 January 1997
Ballou, J.

MSB:Mamm:142682
18 mi W, 5 mi S Roswell, Hondo
    Canyon Ranch
12 February 1997
Ballou, J.

MSB:Mamm:142658
20 mi. W Roswell, Rio Hondo
    Ranch
15 February 1997
Ballou, J.

MSB:Mamm:142663
20 mi N, 2 mi W Roswell
27 February 1997
Cummings, M.

MSB:Mamm:142664
20 mi N, 2 mi W Roswell
27 February 1997
Cummings, M.

MSB:Mamm:142685
8 mi W Roswell
17 March 1997
Ballou, J.

MSB:Mamm:232369
8 mi W Roswell, 2 Cs Ranch
28 March 1997
Ballou, J.

MSB:Mamm:140117
NM Hwy 285, 18.5 mi S junction w/
    hwy 2 at mile marker 85, south
    of Roswell.
30 March 1997
Campbell, M.L., Brunt Jr., J.W.

MSB:Mamm:231268
Roswell, Sycamore St. in NW part
    of town
10 April 1997
Daw, A.

MSB:Mamm:92716
5 mi E, 7.5 mi S of Mesa on US Rte
    285
5 July 1997
Polechla Jr., P.J.

MSB:Mamm:140707
Junction of Red Bridge road and E
    Pine Lodge road, ca. 4 mi N and
    3 mi E Roswell
22 August 1997
Wells, J.

MSB:Mamm:142680
8 mi E Dexter
27 October 1997
Ballou, J.

MSB:Mamm:232360
5 mi N Roswell, DOR on US 285
2 December 1997
Bell, M.

MSB:Mamm:232366
5 mi NW Roswell; Armstrong
    Ranch
21 January 1998
Cummings, M.

MSB:Mamm:142701
33.5 mi N, 18 mi W Roswell; Gallo
    Draw, 2 mi S Truman Pierce
    Ranch HQ
21 January 1998
Cummings, M.

MSB:Mamm:142683
5 mi NW Roswell; Armstrong
    Ranch
22 January 1998
Cummings, M.

MSB:Mamm:142715
20 mi W Roswell along Rio Hondo
27 January 1998
Ballou, J.

MSB:Mamm:142716
25 mi. E Roswell; Cooper Ranch
28 January 1998
Ballou, J.

MSB:Mamm:142714
25 mi. N Roswell; Hub Corn Ranch
28 January 1998
Cummings, M.

MSB:Mamm:142709
32.5 mi N, 18 mi W Roswell; Gallo
    Draw, 3 mi S Truman Pierce
    Ranch HQ
28 January 1998
Cummings, M.

MSB:Mamm:142704
18 mi. W Roswell; Rio Hondo
30 January 1998
Ballou, J.

MSB:Mamm:142710
25 mi N Roswell; Hub Corn Ranch;
    Round Top Pasture
30 January 1998
Cummings, M.

MSB:Mamm:142708
35.5 mi N, 16 mi W Roswell; 2 mi E
   Truman Pierce Ranch HQ
1 February 1998
Cummings, M.

MSB:Mamm:142702
35.5 mi. N, 16 mi. W Roswell; 2 mi.
   E Truman Pierce Ranch HQ
2 February 1998
Cummings, M.

MSB:Mamm:142707
35.5 mi. N, 16 mi. W Roswell; 2 mi.
   E Truman Pierce Ranch HQ
2 February 1998
Cummings, M.

MSB:Mamm:142717
22.5 mi. N 3.5 mi. W Roswell; 5 mile
   Ranch
6 February 1998
Cummings, M.

MSB:Mamm:142711
22.5 mi N, 3.5 mi W Roswell; 5 mile
   Ranch
12 February 1998
Cummings, M.

MSB:Mamm:232379
22.5 mi N, 3.5 mi W Roswell, 5 mile
   Ranch
23 February 1998
Cummings, M.

MSB:Mamm:232380
17 1/2 mi N Roswell, Eden Valley
   Ranch
23 February 1998
Cummings, M.

MSB:Mamm:92664
2.75 mi N junct. US Rte 70 on US
   Rte 285
23 December 1998
Polechla Jr., P.J.

MSB:Mamm:265446
No specific locality recorded.
27 October 1999
Polechla Jr., P.J.

MSB:Mamm:141304
Bitter Lake National Wildlife
   Refuge, Black-tailed Prairie Dog
   Town, 0.3 mi. NW of middle
   tract of BLNWR
4 April 2002
Butts, K.

MSB:Mamm:284103
US 285, 9.8mi N of Arroyo del
   Macho
26 September 2002
Polechla Jr., P.J.

MSB:Mamm:140706
Junction of U.S 285/70
29 September 2002
Polechla Jr., P.J.

CIBOLA CO.
MSB:Mamm:102111
35 mi W Albuquerque
18 December 1928
Aldous, S.E.

DE BACA CO.
MSB:Mamm:14760
8 mi S Yeso
16 June 1962
Harris, A.H.

ENMUNHM 1468
6 mi S, 2 mi W Yeso
16 April 1968
Collector unknown

ENMUNHM 1469
6 mi S, 2 mi W Yeso
16 April 1968
Collector unknown

ENMUNHM 1470
7 mi N. Yeso
16 April 1968
Collector unknown
MSB:Mamm:232358
10 mi SW Yeso
3 February 1997
Milliron, D.

MSB:Mamm:142705
8 mi W Yeso
10 April 1998
Milliron, D.

MSB:Mamm:142713
1 mi. S Yeso; Aclen Ranch
5 June 1998
Milliron, D.

OMNH 9247
3.2 mi NE Dunlap
No date given
Best, T.L.

DOÑA ANA CO.
USNM 98646—Type Specimen
Baird's Ranch, San Andreas
   Range, ca 50 mi N of El Paso,
   Texas
4 Apr 1899
Barber, C.

USNM 125568
Parker Lake, E,
   Organ Mountains
31 January 1903
Gaut, J.

USNM 125569
Parker Lake, E,
   Organ Mountains
31 January 1903
Gaut, J.

TCWC 4456
1 mi S Parker Lake
19 November 1943
Halloran, A.F.

TCWC 5505
1 mi S Parker Lake
19 November 1943
Halloran, A.F.

OMNH 2
Las Cruces
12 December 1958
Burns, R. D.

UTEP 19
Ca 1.5 mi E Strauss.
14 November 1965
Harris, A.H.

NMSUVWM 967
30 mi N of Las Cruces
1970
Kakaleo, N.

NMSU 4132
11mi N, 1.0 mi E Leasburg
11 November 1971
Carter, J.G.

NMSU 4134
10.0 mi N, 0.2mi E Leasburg
11 November 1971
Carter, J.G.

NMSU 4131
6.0 mi N, 3.0 mi E Leasburg
15 November 1971
Martin, D.F.

NMSU 4133
7.0 mi N, 6.0 mi E Leasburg
16 November 1971
Alberico, M.S.

UTEP 2855
Escarpment on rd to Strauss, from
    Rio Grande Valley
29 November 1971
Mellen, R.

NMSU 14600
3mi N Radium Springs; (exit of I- 25)
7 October 1984
Lawler, R.M.

NMSUVWM 2382
Jornada Exp Range Pasture 12A
24 January1989
Collector unknown

UTEP 8281
Mesilla Valley, T25S, R2E, ca Sec 21
7 February 1990
Redetzke, K.

MSB:Mamm:66086
Jornada Experimental Range
15 November 1990
US Department of Energy

MSB:Mamm:66087
Jornada Experimental Range
15 November 1990
US Department of Energy

MSB:Mamm:66081
Jornada Experimental Range
23 February 1991
US Department of Energy

MSB:Mamm:66083
Jornada Experimental Range
28 February 1991
US Department of Energy

MSB:Mamm:66084
Jornada Experimental Range
28 February 1991
US Department of Energy

MSB:Mamm:66082
Jornada Experimental Range
7 March 1991
US Department of Energy

MSB:Mamm:66088
Jornada Experimental Range
12 March 1991
Gannon, W.L.

MSB:Mamm:66085
Jornada Experimental Range
16 March 1991
US Department of Energy

MSB:Mamm:66079
Jornada Experimental Range
17 March 1991
US Department of Energy

MSB:Mamm:66080
Jornada Experimental
    Range
17 March 1991
US Department of Energy

MSB:Mamm:66078
Jornada Experimental Range
18 March 1991
US Department of Energy

NMMNH 3392
5 km SW of Hatch
2 April 1998
Morgan, G.S.

MSB:Mamm:274056
2 mi SW Hatch
5 November 2006
Hafner, D.J.

MSB:Mamm:274056
2 mi SW Hatch
5 November 2006
Hafner, D.J.

TTU-NSRL 17481
Jornada Experimental Range
No date given
Packard, R.L., Williams, S.L.

TTU-NSRL 17482
Jornada Experimental Range
No date given
Packard, R.L., Williams, S.L.

EDDY CO.
UTEP 338
4 mi. N, 10 mi. W Carlsbad
31 December 1966
Harris, A.H.

UTEP 3151
0.5 mi W Lakewood
11 November 1972
Harris, A.H.

MSB:Mamm:142652
3.4 mi N, 9.4 mi E Artesia
6 March 1996
Derrick, W.

MSB:Mamm:232359
12 mi SW Artesia, 2 1/2 mi NW
    4 Dinkus Ranch Headquarters
21 January 1997
Landrie, B.

APPENDIX 13.1. *continued*

MSB:Mamm:85578
ca. 4.5 mi N and 2 mi W of Artesia
along US Route 285
5 January 1999
Polechla Jr., P.J.

GRANT CO.
USNM 148292
Cliff, Gila River
26 August 1903
Bailey, V.

USNM148293
Cliff, Gila River
9 November 1906
Bailey, V.

USNM 158889
Faywood, 9 mi N
14 September 1908
E. Goldman, E.A.

UAZ 14386
35 mi. S Silver City
26 September 1965
Patton. J.L.

WNMU 1215
2 mi S Silver City, Ridge Rd
11 October 1966
St. Angelo, P.J

NMSU 11777
4 mi W, 3.2 mi S Eagle Pk.
30 September 1979
 McLean, E.

WNMU 6923
9.6 mi N, 21.4 mi E Lordsburg
20 March 2009
Jennings, R.D.

GUADALUPE CO.
MSB:Mamm:214964
Along U.S. Hwy 285, @
 Long Draw; 7 mi SE
 Vaughn
26 January 2007
Campbell, M.L.

HIDALGO CO.
USNM 167995
Animas Valley, Cloverdale Ranch
24 March 1905
Hotchkiss, H.

USNM 167996
Animas Valley, Cloverdale Ranch
Spring 1910
Hotchkiss, H.

FMNH 143709
Lordsburg, SE of
1 May 1971
Dillon, L.S.

FMNH 143710
Lordsburg, SE of
1 May 1971
Dillon, L.S.

FMNH 143711
Lordsburg, SE of
1 May 1971
Dillon, L.S.

FMNH 143712
Lordsburg, SE of
1 May 1971
Dillon, L.S.

YPM MAM 013907
Animas Valley
2 September 1974
Rea, A.M., Johnson, P.C.

MSB:Mamm:54422
Antelope Wells
3 February 1984
Jensen, R

MSB:Mamm:54426
10 mi S Playas Smelter
11 February 1984
Jensen, R

MSB:Mamm:54427
10 mi S Playas Smelter
11 February 1984
Jensen, R

MSB:Mamm:54423
5 mi S Playas Smelter
14 February 1984
Jensen, R

MSB:Mamm:54424
5 mi S Playas Smelter
14 February 1984
Jensen, R

MSB:Mamm:54425
5 mi S Playas Smelter
16 February 1984
Jensen, R

MSB:Mamm:54428
15 mi S Playas Smelter
20 February 1984
Jensen, R

MSB:Mamm:54429
15 mi S Playas Smelter
20 February 1984
Jensen, R

MSB:Mamm:54430
15 mi S Playas Smelter
20 February 1984
Jensen, R

FHSM 23661/MSB:Mamm:213434
10 mi N Antelope Wells
7 January 1985
Jensen, R

FHSM 23662/MSB:Mamm:213435
10 mi N Antelope Wells
7 January 1985
Jensen, R

FHSM 23665/MSB:Mamm:213439
11 mi N Antelope Wells
7 January 1985
Jensen, R

MSB:Mamm:213436
10 mi N Antelope Wells
7 January 1985
Jensen, R

FHSM 23664/MSB:Mamm:213438
11 mi N Antelope Wells
9 January 1985
Jensen, R

FHSM 23663/MSB:Mamm:213437
11 mi N Antelope Wells
10 January 1985
Jensen, R

FHSM 23659/MSB:Mamm:213432
8 mi N Antelope Wells
19 January 1985
Jensen, R

FHSM 23660/MSB:Mamm:213433
9 mi N Antelope Wells
23 January 1985
Jensen, R

FHSM 25098/MSB:Mamm:213440
12 mi S Hachita
16 February 1985
Jensen, R

UTEP 7904
DOR, state hwy 338, ca. 4.6 mi S
    Animas (at jct with state hwy 9)
7 July 1989
Dillon, T.J.

LEA CO.
TTU-NSRL 1678 (possible hybrid)
7.3 mi N Tatum
17 March 1962
Hoddenbach

MSB:Mamm:142666 (possible
    hybrid)
8 mi SW Tatum; Ingles Ranch
5 February 1997
Banman, J.

LINCOLN CO.
USNM 130668
Capitan Mountains, Loveless Lake,
    10 miles NW
25 August 1902
Gaut, J.

USNM 130669
Capitan Mountains, 10 mi NW
    Loveless Lake
26 August 1903
Gaut, J.

MSB:Mamm:22280
No specific locality; T5S, R18E,
    middle SEC 20
8 January 1966
Howell, D.J.

MSB:Mamm:22323
No specific locality; T5S, R18E,
    middle SEC 21
8 January 1966
Boone, D.

MSB:Mamm:22497
No specific locality; T3S, R16E, Sec 20
8 January 1966
Howell, D.J.

MSB:Mamm:22498
No specific locality; T3S, R16E, Sec
    20
8 January 1966
Howell, D.J.

MSB:Mamm:142667
15 mi S Corona, 8 mi W US 54
9 January 1997
Davidson, J.

MSB:Mamm:96188 (possible
    hybrid)
27 mi E Corona; DOR on NM Hwy
    247
30 August 1997
Campbell. M.L., Brunt Jr., J.

MSB:Mamm:142703 (possible
    hybrid)
9 mi N Capitan; Edgar Ranch
17 November 1997
Sisneros, F.

MSB:Mamm:232381
30 mi N, 28 mi W Roswell Gallo Draw
3 February 1998
Cummings, M.

Luna Co.
HSU 1385
Uvas Valley, 10 mi W of Hatch
6 November 1977
Woodroof, W.

MSB:Mamm:142699
Hermanas; W.R. Johnson and Son
    Ranch
6 January 1997
Jones, M.

MCKINLEY CO.
MSB:Mamm:12157
Approximately 3 mi N, 2 mi W
    Estrella; T20N, R6W
17 August 1961
Harris, A.H.

MSB:Mamm:142660
15 mi E of Hospah
14 December 1996
Grant, J.

MSB:Mamm:142684
2.7 mi S, 9.6 mi E of Hospah; Ball
    Ranch
12 December 1997
Grant, J.

OTERO CO.
USNM 119920
Tularosa, 9 mi S
22 November 1902
Gaut, J.

MVZ 74423
White Sands National Monument
8 December 1935
Borell, A.E.

MVZ 74424
White Sands National Monument,
    18 mi SW Alamogordo
8 March 1936
Borell, A.E.

AMNHM-131834
18 miles W of Alamogordo
30 September 1938
Gleaves, L., Hill, J.E.

TCWC 4198
Point of Sands
October 1942
Halloran, A.F.

MSB:Mamm:15115
14 mi N Hueco, Texas; Wallbridge
    Ranch; T26S, R10E
17 April 1962
Collector unknown

FMNH 143713
Alamogordo, W of
1 May 1971
Dillon, L.S.

FMNH 143714
Alamogordo, W of
1 May 1971
Dillon, L.S.

UTEP 4386
White Sands National Monument.
1974
Carraway, L.

UTEP 4387
White Sands National Monument.
1 July 1975
Carraway, L.

UTEP 4385
White Sands National Monument.
1976
Carraway, L.

TTU-NSRL 68785
Fort Bliss, North McGregor Range
30 December 1994
Rodrick, P., Yancey, F.D.

TTU-NSRL 68787
Fort Bliss, North McGregor Range
23 April 1995
Wilson, H., Yancey, F.D.
TTU-NSRL 68783
Fort Bliss, North McGregor Range
3 June 1995
Rodrick, P., Yancey, F.D.

TTU-NSRL 68784
Fort Bliss, North McGregor Range
10 June 1995
Rodrick, P., Yancey, F.D.

TTU-NSRL 68786
Fort Bliss, North McGregor Range
12 June 1995
Rodrick, P., Yancey, F.D.

ROOSEVELT CO.
MSB:Mamm:142649 (possible
    hybrid)
4 mi S, 4 mi E Elida; Lambrith
    Ranch
23 December 1996
Banman, J.

MSB:Mamm:142712 (possible
    hybrid)
8.5 mi S, 1.5 mi E House;
    Hill Ranch
7 March 1998
Banman, J.

SAN JUAN CO.
USNM 289005
S, Fruitland
16 May 1905
Halloran, A.

MSB:Mamm:9295
7 mi S, 1 mi W Bloomfield; T28N,
    R11W
14 March 1960
Sands, J.L.

MSB:Mamm:10842
16 mi S, 1 mi W Farmington; T26N,
    R13W, SEC 17
11 August 1960
Harris, A.H.

MSB:Mamm:10997
13 mi S, 11 mi E Farmington; T27N,
    R11W
7 September 1960
Harris, A.H.

MSB:Mamm:142696
3.3 mi S, 6.8 mi E Aztec (at Jct NM
    44 and US 550)
5 November 1997
Smith, S.

MSB:Mamm:142700
3.3 mi S, 6.8 mi E Aztec (at Jct NM
    44 and US 550)
6 November 1997
Smith, S.

MSB:Mamm:214901/231409 (double
    cataloged)
Navajo Indian reservation; ca.
    5 mi. N. Newcombe along US 491
    (formerly U.S. 666)
9 November 2006
Polechla Jr., P.J.

UTEP 4786
SEC. 24, T20N, R6W
No date given
Harris, A.H.

SAN MIGUEL CO.
UMMZ 41714
Las Vegas
9 March 1905
Atkins, A.

SIERRA CO.
UAZ 12759
40 mi N Truth or Consequences on
    U.S. Hwy 80
2 November 1964
Waddell, E.P.

MSB:Mamm:121838
+/- 4 mi N Engle on main rd.
    to Naval Station, on Pedro
    Armendaris Ranch
11 January 1995
Henry, A.

SOCORRO CO.
USNM 160606
San Augustine Plain, 12 mi NW of
    Monica Spring
16 September 1909
Goldman, E.A.

USNM160607
San Augustine Plain, 12 mi NW of
    Monica Spring
18 September 1909
Goldman, E.A.

USNM160637
San Augustine Plain, 12 mi NW of
    Monica Spring
18 September 1909
Goldman, E.A.

MSB:Mamm:24406
1.5 mi SW jct NM Hwy 60 & Hwy 85
1 January 1967
Mares, M.A.

MSB:Mamm:24744
Ladron Peak
8 July 1967
Domenici Taxidermy

MSB:Mamm:24745
Ladron Peak
8 July 1967
Domenici Taxidermy

MSB:Mamm:24746
Ladron Peak
8 July 1967
Domenici Taxidermy

MSB:Mamm:142697
12.3 mi S, 6.9 mi W Magdalena; 3
    Links Ranch
10 December 1997
Graves, M.

MSB:Mamm:283989
2.2 mi N of Bernardo, Exit no 175
    on I-25
4 March 1999
Polechla Jr., P.J.

TORRANCE CO.
MSB:Mamm:24347
8 mi (by road) W of Vaughn on Hwy
    285
26 December 1966
Mares, M.A.

MSB:Mamm:24401
Moriarty
9 January 1967
Collector unknown

MSB:Mamm:24544
T1N R8E
9 February 1967
Domenici Taxidermy

MSB:Mamm:34963
7 mi S Lucy
9 February 1969
Smith, J.F.

UWYMV 5847
Ca. 4 mi W of Willard
15 June 1977
Collector unknown

UWYMV 5846
Ca. 4 mi W of Willard
15 September 1977
Boyce, M.S.

ENMUNHM 9467
10 mi NW, Encino (Hwy 84)
19 September 1983
Johnson, C.B.

MSB:Mamm:69525
NM Hwy 42, 107 road miles S of
    Willard
7 October 1992
Collector unknown

MSB:Mamm:142670
24 mi N, 4.8 mi E Encino, DOR on
    I-40
16 July 1996
Lang, B.K.

MSB:Mamm:142694
7.5 mi W Vaughn, DOR on US 285
4 November 1996
Teutsch, J.

MSB:Mamm:142669
1.5 mi N, 2.5 mi E Vaughn DOR on
    US 60
8 November 1996
Teutsch, J.

MSB:Mamm:231189
11 mi N Corona, West of US 54
9 November 1996
Davidson, J.

MSB:Mamm:142668
3 mi SE Duran
9 December 1996
Sandoval, L.

MSB:Mamm:142706
25 mi NW Corona
12 March 1998
Davidson, J.

MSB:Mamm:284100
Vaughn, 9.2 mi W of along US 285
1 June 2003
Polechla Jr., P.J.

ASNHC 697
no specific locality
No date given
Carter, J.

VALENCIA CO.
USNM 244509
Los Lunas, 15 mi NE
2 February 1924
Ligon, J.S.

YPM MAM 001751
12 miles E of Los Lunas
20 February 1925
Ligon, J.S.

KU 79090
Belen, 4 mi SE of
11 June 1958
Alcorn, J.R.

MSB:Mamm:24708
1 mi S, 14 mi W Los Lunas on NM 6
27 March 1967
Williams, D.F.

MSB:Mamm:36138
12 mi E Belen
12 September 1975
Elliot, R.

APPENDIX 14.1. SWIFT FOX MUSEUM SPECIMENS FROM NEW MEXICO.

Museum institution acronyms: ENMUNHM (Eastern New Mexico University, Natural History Museum); FHSM (Fort Hays State University, Sternberg Museum of Natural History); KU (University of Kansas, Natural History Museum and Biodiversity Research Center); MSB (University of New Mexico, Museum of South-western Biology); OMNH (University of Oklahoma, Sam Noble Oklahoma Museum of Natural History); TTU (Collection of Recent Mammals, Museum of Texas Tech University); UMMZ (University of Michigan Museum of Zoology); and USNM (United States National Museum of Natural History).

CHAVES CO.
ENMUNHM 1080
2.5 mi N of Caprock
27 December 1967
Collector unknown

ENMUNHM 1127
7 mi W, 23 mi S of Elida
6 December 1967
Collector unknown

COLFAX CO.
MSB 140042
3.2 mi W, 0.5 mi S of Abbott
27 February 2001
Harrison, R. L.

MSB 142692
4.5 mi S of Abbott, NM 139
3 March 1997
Harrison, R. L.

MSB 145872
3.1 mi S, 1.9 mi E of Abbott
20 March 2000
Harrison, R. L.

MSB 145883
0.5 mi E, 4.5 mi S of Abbott
1 August 2001
Harrison, R. L.

MSB 265774
20 February 2001
Carnivore Initiative Project

CURRY CO.
MSB 1238
23 mi NW of Melrose
23 February 1952
Widner, W. J.

MSB 3851
5 mi NE of Clovis
1 August 1957
Harris, A. H.

DE BACA CO.
MSB 50574
Hwy 60 between Ft. Sumner and Vaughn
15 October 1982
Bryne, T.S.

FHSM 1263
8 mi S of Yeso
4 October 1962
Schroeder, M. H.

EDDY CO.
ENMUNHM 8441
T21S R24E SW1/4S8
28 May 1980
Collector unknown

HARDING CO.
MSB 142622
5 mi N, 2.6 mi W of Lake Chicosa St. Park
3 March 2001
Harrison, R. L.

MSB 142693
5 mi E of Roy
15 October 1997
Garcia, J.

MSB 142698
2 mi E of Mosquero
15 January 1998
Garcia, J.

MSB 142894
4.7 mi N, 2 mi W of Lake Chicosa St. Park
16 November 1999
Harrison, R. L.

MSB 145875
12 mi N, 0.5 mi E of Lake Chicosa St. Park
22 June 2000
Harrison, R. L.

MSB 145876
5 mi N, 1 mi W of Lake Chicosa St. Park
27 July 2000
Harrison, R. L.

MSB 145877
7 mi N, 0.6 mi W of Lake Chicosa St. Park
27 July 2000
Harrison, R. L.

MSB 145885
5 mi N of Lake Chicosa St. Park
16 April 2001
Harrison, R. L.

MSB 145886
Spear Hills
21 April 2001
Harrison, R. L.

MSB 232361
1.5 mi N of Roy, NM 120
25 August 1997
Bates, S.

MSB 232372
1.5 mi S of Roy
10 October 1997
Garcia, J.

MSB 232374
4.8 mi N, 1.2 mi W of Roy
1 July 1997
Garcia, J.

MSB 232376
5 mi E of Roy
15 October 1997
Garcia, J.

MSB 267559
4 mi W, 1.3 mi N of Lake Chicosa St. Park
3 November 1999
Harrison, R. L.

MSB 267560
1 mi S, 0.2 mi E of Lake Chicosa St. Park
2 May 2001
Harrison, R. L.

MSB 284081
2 mi E, 0.5 mi S of Lake Chicosa St. Park
30 April 2001
Harrison, R. L.

ENMUNHM 1857
12 mi N, 1 mi W of Roy
Date unknown
Collector unknown

LEA CO.
KU 100643
KU 100644
4 mi S of Lovington
22 January 1965
Clifton, P. L.

KU 100645
KU 100646
4 mi S of Lovington
24 January 1965
Clifton, P. L.

KU 100700
4 mi S of Lovington
17 May 1965
Clifton, P. L.

KU 100721
KU 100722
4 mi S of Lovington
18 May 1965
Clifton, P. L.

MSB 142647
15 mi N, 8.2 mi E of
 Maljamar
17 February 1996
Derrick, W.

MSB 142650
4.8 mi N, 1.6 mi E of
 Maljamar
21 February 1996
Derrick, W.

MSB 142653
16.2 mi N, 5.3 mi E of
 Maljamar
19 February 1996
Derrick, W.

MSB 142654
15.9 mi N, 8.2 mi E of
 Maljamar
22 February 1996
Derrick, W.

MSB 142655
15.9 mi N, 8.2 mi E of
 Maljamar
19 February 1996
Derrick, W.

MSB 142656
15.9 mi N, 3.9 mi E of
 Maljamar
22 February 1996
Derrick, W.

MSB 142657
15.9 mi N, 8.7 mi E of
 Maljamar
17 February 1996
Derrick, W.

MSB 142659
13.3 mi N of Maljamar
29 February 1996
Derick, W.

MSB 232353
12.2 mi N, 3.4 mi E of
 Maljamar
22 February 1996
Derrick, W.

TTU 1678
7.3 mi N of Tatum
17 March 1962
Hoddenbach

TTU 4229
19 mi E, 8 mi S of Mal-
 jamar
21 December 1965
Judd, F.

QUAY CO.
MSB 232433
6 mi W of Jordan, NM 156
21 September 1997
Roberson, W.

ROOSEVELT CO.
MSB 14932
4 mi N, 1 mi E of Portales,
 Oasis St. Park
9 September 1962
Harris, A. H.

MSB 232365
5 mi S, 6 mi E of Elida
15 December 1997
Banman, J.

ENMUNHM 775
8.5 mi S of Melrose
4 November 1966
Collector unknown

ENMUNHM 1083
4 mi S of Elida
17 March 1967
Collector unknown

ENMUNHM 1094
4 mi N of Causey
28 October 1967
Collector unknown

ENMUNHM 4626
6 mi W, 2 mi S of Dora
23 February 1978
Collector unknown

SAN MIGUEL
USNM A 16249
Cabra Springs
March 1879
Pease, W. B.

UMMZ 41714
1895
Atkins, E.

UNION CO.
MSB 142661
19.3 mi N, 8 mi E of Mt.
 Dora (town)
3 March 1997
Jaureguiberry, P.

MSB 142671
19.3 mi N, 8.2 mi E of Mt.
 Dora (town)
3 March 1997
Jaureguiberry, P.

MSB 142672
12.6 mi E of Gladstone,
 US 56
17 October 1996
Schmitt, C. G.

MSB 142673
12 mi S, 3.7 mi W of
 Amistad
19 November 1997
Jones, R.

MSB 142674
12 mi S, 1.9 mi W of
 Amistad
19 November 1997
Jones, R.

MSB 142675
12 mi S, 3.7 mi W of
 Amistad
5 November 1997
Jones, R.

MSB 142676
12 mi S, 2.8 mi W of
 Amistad
19 November 1997
Jones, R.

MSB 232373
8.1 mi E of Gladstone,
 US 56
2 November 1996
Schmitt, C. G.

OMNH 9656
5 mi E of Clayton
24 December 1969
Best, T. L.

OMNH 9657
5.3 mi E of Clayton
27 December 1969
Best, T. L.

ENMUNHM 2464
7.4 mi W, 2.6 mi N of
    Clayton
31 December 1967
Collector unkn

ENMUNHM 2465
4.5 mi W of Clayton
24 March 1968
Collector unknown

APPENDIX 14.2. GOVERNMENT AGENCIES AND OTHER ORGANIZATIONS INVOLVED AS MEMBERS OF THE SWIFT FOX CONSERVATION TEAM (SFCT).

US STATE AGENCIES
Colorado Parks and Wildlife
Kansas Department of Wildlife and Parks
Montana Fish, Wildlife and Parks
Nebraska Game and Parks Commission
New Mexico Department of Game and Fish
North Dakota Game and Fish Department
Oklahoma Department of Wildlife Conservation
South Dakota Department of Game, Fish and Parks
Texas Parks and Wildlife Department
Wyoming Game and Fish Department

US FEDERAL AGENCIES
US Bureau of Land Management
US Fish and Wildlife Service
US National Park Service
USDA-APHIS Wildlife Services
USDA Forest Service
USGS/Biological Resources Division

CANADIAN GOVERNMENT AGENCIES
Alberta Fish and Wildlife Division
Canadian Wildlife Services

TRIBES
Fort Peck Assiniboine and Sioux Tribe
Blackfeet Nation
Oglala Lakota Nation
Lower Brule Sioux Tribe
Kainai Nation

NON-GOVERNMENTAL ORGANIZATIONS
Association of Zoos and Aquariums
World Wildlife Fund
Turner Enterprises
Pueblo (Colorado) Zoo

APPENDIX 15.1. NORTH AMERICAN RED FOX OCCURRENCE RECORDS FROM NEW MEXICO.

Catalog Number/ Reference	County	Locality	Latitude	Longitude
		Museum Specimen Records		
MSB:Mamm:6072	Chaves	1 mi S of Railroad Mt, Crosby Cross B Ranch; T8S, R30E	33.6202018	-103.8837966
USNM271973	Doña Ana	San Andres Refuge, Head of Salt Canyon	36.6045615	-106.5392816
MSB:Mamm:246686	Harding	Harding County (no specific locality)	No precise coordinates	No precise coordinates
MSB:Mamm:142679	McKinley	2 mi S, 2.5 mi E of Gallup (at Jct. I-40 and NM 666)	35.49638	-108.69133
MSB:Mamm:212823	McKinley	1.25 mi W of exit 16 on I-25, 5 mi W of Gallup	35.52539	-108.82441
MSB:Mamm:232356	McKinley	1.5 mi S, 3 mi E of Rehoboth	35.50715	-108.60157
MSB:Mamm:232362	McKinley	1.5 mi S, 3 mi E of Rehoboth, Bruce Williams Ranch	35.50715	-108.60157
MSB:Mamm:232378	McKinley	1.5 mi S, 3.5 mi W [sic] of Rehoboth Bruce Williams Ranch	35.50715	-108.60157
USNM128536	Mora	Pecos Baldy	35.908797	-105.666225
KU 4107	Mora	Truchas Peak; Pecos River	35.9624353	-105.6451795
MSB:Mamm:1919	Rio Arriba	10.5 mi E of Chama; T31N, R5E, SEC 17	36.9227973	-106.4428024
MSB:Mamm:341620	Rio Arriba	3 km WNW of Hopewell Lake, on US Highway 64	36.72213196	-106.2621086
MSB:Mamm:140705	Roosevelt	Portales, Privett Hatchery, 2 mi S of Portales	34.15268	-103.33578
MSB:Mamm:232364	Roosevelt	2.7 mi S, 4.2 mi E of Elida	33.90395	-103.58097
MSB:Mamm:232367	Roosevelt	1 mi W of Portales, Lime Street	34.19363	-103.35238
MSB:Mamm:232371	Roosevelt	3 mi E of Elida, DOR on NM 114	33.93766	-103.60176
ENMNHM 9707	Roosevelt	0.5 mi S, 3 mi E of Portales, T1S, R35E, Sec 32 NE1/4	34.178936	-103.29234
ENMNHM 10533	Roosevelt	New Mexico Highway 267 at Portales city limits	34.185669	-103.387048
ENMNHM 11149	Roosevelt	West-side Portales sanitary Landfill, southern outskirsts of Portales	34.157	-103.326
MSB:Mamm:160217	San Juan	Vicinity of Shiprock	36.78555	-108.68703
MSB:Mamm:267562	San Juan	Bloomfield, 4 mi E of	36.71312	-107.91271
MSB:Mamm:336757	San Juan	Flora Vista, Old Aztec Road	36.78139257	-108.1146304
MSB:Mamm:8683	San Juan	5 mi SE of Blanco Canyon, Canyon Largo; T29N, R9W	36.7271011	-107.7855988

Catalog Number/ Reference	County	Locality	Latitude	Longitude
Museum Specimen Records				
MSB:Mamm:91656	San Juan	1 mi E of Farmington on US 64	36.72854	-108.200073
USNM158896; 158912-158921; 158959	San Juan	Liberty [Waterflow], 12 mi W of Fruitland	36.7522	-108.615
USNM158959	San Juan	Liberty [Waterflow]	36.7607	-108.493
USNM130371-130372	Taos	Twining	36.5947511	-105.4502846
MSB:Mamm:282010	Union	Clayton (edge of town) DOR	36.44866	-103.18104
KU 1628-1636	Unknown	Pecos National Forest	No precise coordinates	No precise coordinates
Photograph Records				
<tabtxt>Remote camera at wildlife road-crossing structure, monitoring by AZGFD and NMDOT	Bernalillo	Tijeras, Tijeras Crosswalk	35.066665	-106.427329
Remote camera at wildlife road-crossing structure, monitoring by AZGFD and NMDOT	Colfax	Raton, Raton Creek CBC	36.897747	-104.425879
NRA Whittington Center Facebook page	Colfax	NRA Whittington Center, SW of Raton	36.7737224	-104.485824
Remote camera at wildlife road-crossing structure, monitoring by AZGFD and NMDOT	Colfax	Raton, unnamed CBC	36.914741	-104.439615
NMDGF Clint Henson (captured silver phase)	Colfax	Raton	36.9033581	-104.4391532
Nancy Cox photo	Curry	North of NMSU ASC Clovis Research HQ. On NM 288 just W of NM 289	34.604354	-103.234604
photo of road kill by Calvin Smith	Lea	1 mi S of Lovington; southern city limits	32.9225663	-103.340016
David Mikesic photo	McKinley	Backyard in Gallup	35.51800556	-108.7280444
Brian Jay Long and Jon Klingel	Mora	Headwaters of Rito de Gascon, S of Gascon Point	35.91439068	-105.4875906
Mark Watson	Rio Arriba	Corkins Lodge Ranch Phase II, 9,643 ft. elev.	36.7431263	-106.4396487

Catalog Number/ Reference	County	Locality	Latitude	Longitude
		Photograph Records		
Mark Watson	Rio Arriba	US 64 ca. 1 mi N of the NM 95-U.S. Hwy 64/84 junction; and ca. 9 miles south of Chama	36.75290	-106.56151
Remote camera, Peale, Long, and Klingel	Rio Arriba	Upper Lagunitas	36.8767704	-106.3203746
Marty Peale	Rio Arriba	Forest Rd 87 near Lagunitas Campground entrance	36.8806624	-106.3151609
Brian J. Long photo	Rio Arriba	Brazos Pass	36.69360757	-106.3640683
NMDGF Remote camera	Rio Arriba	N of Jemez, between Canjilon and El Rito	36.53142	-106.28864
NMDGF Remote camera	Rio Arriba	N of Jemez, between Canjilon and El Rito	36.35546	-106.16588
photos provided to NMDGF (Jim Stuart) by photographer	Rio Arriba	Canjilon, Forest Service office	36.4794611	-106.4378091
Mark Peyton, NPS camera record	Sandoval	Near San Juan Canyon, Jemez Mtns	35.70133	-106.67262
NMDGF Remote camera	Sandoval	Jemez Mtns, between Cuba and San Pedro Parks	36.06173	-106.90649
Mark Watson	San Juan	US 550 ca. 3.5 mi S of the Colorado border	36.955106	-107.887372
Remote camera at wildlife road-crossing structure, monitoring by AZGFD and NMDOT	San Juan	Aztec, UP2 CBC	36.955186	-107.887230
Remote camera, Peale, Long, and Klingel	San Juan	Animas River	36.93145466	-107.8788977
Remote camera, Peale, Long, and Klingel	San Juan	Waterflow, San Juan County	36.7489146	-108.4205058
Albuquerque Journal 31 July 1994	San Juan	Pinon Hills Golf Course, Farmington	36.767909	-108.178216
road kill photo by Jim Stuart on Flicker https:// flickr.com/photos/stu- artwildlife/2803182591 http://flickr.com/photos/ stuartwildlife/2804029248	San Juan	Flora Vista	36.7944478	-108.0803484

Catalog Number/ Reference	County	Locality	Latitude	Longitude
		Photograph Records		
Eric Rominger home	Santa Fe	Santa Fe	35.6869752	-105.937799
Remote camera, Brian J. Long	Taos	Des Montes	36.51857476	-105.6038241
Remote camera, Brian J. Long	Taos	Arroyo Hondo	36.534944	-105.678060
Photo, Brian J. Long	Taos	El Prado (roadkill)	36.4318	-105.5747
Brian J. Long photo	Taos	Ranchos de Taos (roadkill)	36.3586	-105.6095
Brian Long	Taos	Williams Lake	36.556661	-105.429137
Alyssa Radcliff	Taos	Questa	36.7043492	-105.59565
Dan Kuehn	Taos	Questa	36.736667	-105.579167
		Literature Records		
Julyan and Stuever 2005	Bernalillo	Sandia Mountains, Capulin Springs Picnic Area	35.216857	-106.416133
Clothier 1957	Bernalillo	Sandia Mountains, Sandia Crest	35.210102	-106.449593
Frey 2004:7	Cibola, McKinley	"Towns along Interstate Highway 40 west of Albuquerque"	No precise coordinates	No precise coordinates
Frey and Schwenke 2012	Colfax	"Near the [Sugarite Canyon State Park] entrance"	36.936313	-104.378468
Frey 2003	Curry	Clovis, northern outskirts	34.4581634	-103.19664
Frey 2003	Curry	Clovis, southern outskirts	34.3655439	-103.1962967
Frey 2003	Curry	Vicinity Ranchvale	34.4925753	-103.3188406
Halloran 1946	Doña Ana	Ash Canyon	32.6510113	-106.4799728
Halloran 1946	Doña Ana	Lostman Canyon	32.8631377	-106.5175027
Frey 2004:7	Eddy, Chaves, De Baca	"Pecos River Valley from Eddy Co. northward to at least De Baca Co."	No precise coordinates	No precise coordinates
Halloran 1946	Otero	Point of Sands	32.7470311	-106.1958268
Frey 2003	Roosevelt	US Highway 70 and New Mexico Highway 467	34.202778	-103.318212
Frey 2003	Roosevelt	Portales wastewater treatment plant, southern outskirts of Portales	No precise coordinates	No precise coordinates
Frey 2003	Roosevelt	Abandoned airstrip near golf course, western outskirts of Portales	34.173198	-103.377022
Frey 2003	Roosevelt	Southern Roosevelt County	No precise coordinates	No precise coordinates

Catalog Number/ Reference	County	Locality	Latitude	Longitude
Literature Records				
Frey 2003	Roosevelt	3.2 km N of Arch	34.1412196	-103.143596
Halloran 1964	San Juan	Sheep Springs	36.14472	-108.7061
Kamler et al 2005	Union	"Near Clayton"	36.451693	-103.1841039
Credible Observations				
Larry Cordova	Lincoln	Capitan Mountains; near Michallas Canyon	33.6207662	-105.2191544
Navajo Natural Heritage Program's Biotics database	McKinley	6.5 mi E of NFPI, on Asaayi Lake road (N31). E of Navajo NM	35.94806	-108.93
Navajo Natural Heritage Program's Biotics database	McKinley	Hunting Canyon, which is S of I-40 Port of Entry & NE of Manuelito Canyon	35.43222	-108.9181
Navajo Natural Heritage Program's Biotics database	McKinley	US666. Along US 491 at Gamerco, N of Gallup	35.57333	-108.7581
Rhonda Stewart	Otero?	144 on Mescalero Apache Reservation	No precise coordinates	No precise coordinates
Rhonda Stewart	Otero	Pedestrian trail bridge on Hwy 82	32.96107	-105.76914
Rhonda Stewart (she has a photo)	Otero	E of Bluff Springs	32.8295104	-105.739032
Navajo Natural Heritage Program's Biotics database	San Juan	0.5 mi S (E?) of Crystal, along SR134	36.00311	-109.0103
Navajo Natural Heritage Program's Biotics database	San Juan	South Wing of Shiprock, where Red Valley Hwy (N13) passes through dike	36.63816	-108.8257
Navajo Natural Heritage Program's Biotics database	San Juan	Northernmost "headwaters" of Cottonwood Arroyo, ca. 7 mi E of Chaco River	36.54722	-108.4519

AZGFD = Arizona Game and Fish Department; CBC = Concrete box culvert; NMDGF = New Mexico Department of Game and Fish; NMDOT = New Mexico Department of Transportation; NPS = National Park Service.

APPENDIX 21.1. RECORDS OF THE AMERICAN ERMINE (*MUSTELA RICHARDSONII*) IN NEW MEXICO ARRANGED FROM NORTH TO SOUTH IN EACH MOUNTAIN RANGE

Museum acronyms: MVZ (University of California, Berkeley, Museum of Vertebrate Zoology); MSB (University of New Mexico, Museum of Southwestern Biology); NMSU (The Vertebrate Museum, New Mexico State University); NMSUVWC (New Mexico State University Vertebrate Wildlife Museum [a.k.a. NMSU Wildlife Museum]); NMMNH (New Mexico Museum of Natural History and Science); TTU (Collection of Recent Mammals, Texas Tech University Museum); and USNM (United States National Museum of Natural History). Other acronyms: Jennifer Frey's field catalog (FT); North American Datum of 1983 (NAD 83). Mountain range acronyms: SJ (San Juan Mountains); SDC (Sangre de Cristo Mountains); and JZ (Jemez Mountains).

Location	Mountain Range	County	Locality	Date
1	SJ	Rio Arriba	"South of Jawbone Mtn. in GMU 52 (NW of Tres Piedras at UTM 13S 387071E 4066418N (WGS84/NAD83)"	10 Sep 2005
2	SJ	Rio Arriba	24.1mi. W, 3.8 mi. N of Tres Piedras, T29N, R7E, Sec.31, 9800 ft	22 Sep 1973
3	SJ	Rio Arriba	Upper Canjilon Lake	7 Sep 1974
4	SJ	Rio Arriba	5 mi. NE of Canjilon, Canjilon Lakes, (Canjilon Lakes Campground, Lower Lakes)	Unknown
5	SJ	Rio Arriba	"4.25 mi. N, 6.5 mi. E of Canjilon, Canjilon Creek Campground"	25 Jul 1980
5	SJ	Rio Arriba	"4.25 mi. N, 6.5 mi. E of Canjilon, Canjilon Creek Campground"	26 Jul 1980
5	SJ	Rio Arriba	"4.25 mi. N, 6.5 mi. E of Canjilon"	26 Jul 1980
6	SDC	Colfax	"Sugarite Canyon State Park, small tributary to Soda Pocket Creek, 7.9 km N, 3.8 km E of Raton"	12 Jul 2006
7	SDC	Taos	"Southern tributary to Holman Creek, along Forest Road 1950, 1.1 mi (by road) E of junction Comanche Creek Valley, 16.6 km S, 16.8 km E of Amalia"	26 Jul 2006
8	SDC	Colfax	Philmont Scout Ranch, 18 mi NW of Cimarron, 10,000 ft	15 Aug 1968
9	SDC	Colfax	Philmont Scout Ranch, 17.5 mi NW of Cimarron, 10,000 ft	15 Aug 1968
9	SDC	Colfax	"Philmont Scout Ranch, 17.5 mi NW of Cimarron, . . . South Ponil Canyon Elev. 10,000 ft"	19 Aug 1968
10	SDC	Colfax	"Philmont Scout Ranch, 17 mi NW of Cimarron, . . . French Henry Camp. Elev. 9,600 ft"	26 Jun 1968
10	SDC	Colfax	Philmont Scout Ranch, 17 mi NW of Cimarron, 9,600 ft	6 Aug 1968
11	SDC	Colfax	"Philmont Scout Ranch, 15 mi NW of Cimarron, . . . Baldy Town Camp. Elev. 10,000 ft"	19 Aug 1968
12	SDC	Taos	Twining, 10,700 ft ["a small meadow at 10,700 feet altitude a few miles east of the mining camp of Twining in the Taos Mountains, near the headwaters of the Hondo River, at the upper edge of the Canadian Zone" (Bailey 1931:329)]	8 Aug 1904

| Collector | Reference | Geo-referencing | |
		Latitude (NAD 83)	Longitude (NAD 83)
F. Winslow	Observed by F. Winslow, pers. comm.	36.736758	-106.264844
C. Housinger	NMSU 5002	36.703284	-106.243354
James S. Findley	MSB 34315	36.563283	-106.329079
Unknown	OMNH 18415	36.548928	-106.342263
D. J. Warren	D. J. Warren field notes; no specimen	36.541729	-106.317725
D. J. Warren	MSB 53986	36.541729	-106.317725
D. J. Warren	MSB 43503	36.541729	-106.317725
J. K. Frey	FT 503, MSB 140418	36.974734	-104.395183
J. K. Frey	FT 600	36.792687	-105.2636
R. A. Rowlett	USNM 554484	36.640617	-105.198808
R. A. Rowlett	USNM 554483	36.637731	-105.193174
R. A. Rowlett	MSB 29538	36.637731	-105.193174
R. A. Rowlett	MSB 29537	36.635789	-105.184647
R. A. Rowlett	USNM 554482	36.635789	-105.184647
R. A. Rowlett	MSB 29539	36.626379	-105.191256
J. Gaut	USNM 133431	36.612818	-105.428702

Location	Mountain Range	County	Locality	Date
13	SDC	Taos	"Taos Ski Valley"	7 Jul 1963
14	SDC	Colfax	"Philmont Scout Ranch, 7,800 ft, 1 mi. W of Cimarron Reservoir"	11 Oct 1963
15	SDC	Taos	"Rio La Junta, 17 mi S, 6 mi. E of Taos, 9400 ft"	5 Sep 1976
16	SDC	Taos	"Tres Ritos"	Unknown
16	SDC	Taos	"Tres Ritos"	Unknown
17	SDC	Taos	"Rio Pueblo, beaver ponds just below junction Agua Sarca Canyon, 1.1 mi (by road) above La Junta Canyon, 1.6 km S, 1.4 km E of Tres Ritos"	10 Jul 2006
18	SDC	Santa Fe	Near Cowles NM, Parchucla Creek, on forest service trail 106	2 Oct 2004
19	SDC	Santa Fe	"Saddle S of Santa Fe Baldy, 11,000 ft., Santa Fe Range"	
20	SDC	San Miguel	1.5 km N of Elk Mountain, Santa Fe Mtns., 3447 m	21 Sep 2008
21	SDC	Santa Fe	"Tesuque" [Chupadero Canyon]	9 Mar 2004
22	SDC	Santa Fe	"Teseque" (SIC)	11 Feb 1977
23	JZ	Rio Arriba	"San Gregorio Lake"	
24	JZ	Sandoval	9 mi E of Cuba, 9,000 ft	9 Nov 1939
25	JZ	Sandoval	[Rio de las Vacas] 17 km SE of Cuba, Elev. 2600m; T20N R1E S12	14 Jul 1985
26	JZ	Sandoval	"16 mi N of Jemez Spgs"	9 Sep 1979
26	JZ	Sandoval	"2 mi E, 15 mi N of Jemez Springs, T20N, R3E, Sec 7"	6 Sep 1980
26	JZ	Sandoval	2 mi E, 15 mi N of Jemez Springs	6 Sep 1980
26	JZ	Sandoval	"2 mi E, 15 mi N of Jemez Springs (T20N, R3E, Sec 7)"	6 Sep 1980
27	JZ	Sandoval	Seven Springs Hatchery	5 Jul 1961
28	JZ	Sandoval	Jemez Mountains, Santa Fe National forest, W slope of Barley Canyon; T19N R3E SEC6, 2 air miles NW of La Cueva USFS Road 144, 8,600 ft	21 Aug 1992
29	JZ	Sandoval	"Fenton Lake, T19N, R2E, Sec. 10"	Aug 1986
29	JZ	Sandoval	"Fenton Lake, Jemez Mtns." [in catalog: 12.5 mi N of Jemez Springs, Fenton Lake]	4 Aug 1979
30	JZ	Sandoval	"Rio Cebolla, 0.6 mi (by FR 376) SW FR 376 bridge over Rio Cebolla at junction of Lake Fork Canyon, 9.5 km N, 6.5 km W of Jemez Springs, T19N, R2E, W1/2 of NE 1/4 Sec 30"	15 Aug 2006
31	JZ	Sandoval	"7.5 mi N, 6 mi E of Jemez Springs"	22 Jul 1981
31	JZ	Sandoval	"7.5 mi N, 6 mi E of Jemez Springs"	22 Jul 1981
32	JZ	Sandoval	Valles Caldera National Reserve, ca. 3.2 km W. to 4 km. S of Redondo Peak along Redondo Creek	31 Aug 2004
33	JZ	Sandoval	"6.5 mi. N, 5 mi. E Jemez Springs"	24 Jul 1981

| Collector | Reference | Geo-referencing | |
		Latitude (NAD 83)	Longitude (NAD 83)
J. Delso (R. Packard)	TTU 1866	36.594729	-105.45027
B. E. Miller	MSB 18960	36.485832	-105.070844
D. J. Hafner	MSB 36132	36.171705	-105.440178
L. L. Janecek	MSB 57771	36.130568	-105.51584
L. L. Janecek	MSB 57772	36.130568	-105.51584
J. K. Frey	FT 496	36.115537	-105.49873
A. F. (V. L. Mathis)	NMSUVWC 2339	35.834441	-105.667341
	Hall 1951 (ANSP)	35.822747	-105.751196
A. Little	MSB 196418	35.779896	-105.553881
New Mexico Wildlife Center	MSB 143855	35.785732	-105.867401
W. Isaacs	MSB 60780	35.763642	-105.932491
	Findley et al. 1975. TU	36.03756	-106.849657
A. E. Borell	MVZ 89555	35.995331	-106.843222
R. C. Szaro	MSB 67514	35.978186	-106.791979
A. S. Christmas	MSB 41623	35.974946	-106.657043
K. Arganbright	MSB 65695	35.974946	-106.657043
Unknown	MSB 212659	35.974946	-106.657043
W. E. Byers	MSB 65696	35.974946	-106.657043
B. R. Donaldson (J. E. Wood)	NMSU 15386	35.926409	-106.703133
C. W. Painter	MSB 87344	35.907344	-106.666966
K. K. Kleyboecker	NMMNH 1161	35.885167	-106.722779
K. E. Peterson	MSB 41054	35.885167	-106.722779
J. K. Frey	FT 611	35.85209	-106.76908
D. J. Hafner	MSB 46491	35.87711	-106.587193
D. J. Hafner	MSB 46492	35.87711	-106.587193
D. J. Wells	MSB 144524	35.864997	-106.598795
D. J. Hafner	MSB 46493	35.861789	-106.602412

Location	Mountain Range	County	Locality	Date
34	JZ	Sandoval	"Cerros del Abrigo, 11 mi. N, 12 mi. E of Jemez Springs"	15 Aug 1991
35	JZ	Los Alamos	"N-facing rockslide, Pajarto Mtn., 2850 m"	31 Jul 1970
36	JZ	Sandoval	Cerro Grande, off the trail from Highway 4, just inside the VCNP boundary	Unknown
37	JZ	Sandoval	"Between Sierra de los Valles and Frijoles Canyon, 2.4 km SW Cerro Grande 2,710 m"	22 Jul 1970
38	JZ	Sandoval	"Meadow N Hwy 4 & Dome Rd at junction of Baca location #1"	21 Jul 1970
38	JZ	Sandoval	"Between Sierra de los Valles and Frijoles Canyon, 2.5 km S-SW of Cerro Grande 2,690 m"	21 Jul 1970
38	JZ	Sandoval	"Between Sierra de los Valles and Frijoles Canyon, 2.5 km S-SW of Cerro Grande 2,690 m"	21 Jul 1970
39	JZ	Sandoval	"W side ridge between Frijoles and Water Canyons, 3.7 km S-SE of Cero Grande 2,733 m"	5 Nov 1970
40	JZ	Los Alamos	"Near Ponderosa Group Campground"	Summer 1978
40	JZ	Los Alamos	"Near Ponderosa Group Campground"	Winter 1977 1978
41	JZ	Los Alamos	"Near Apache Spring"	Summer 1978
42	JZ	Sandoval	"Jemez Springs 3.25 mi N, 10 mi of E , 35° 48.867' N 106° 31.312 W 2,610 m"	21 Sep 1996
43	JZ	Sandoval	"0.2 mi E of Las Conchas Campground on NM Rt. 4 35 48.418' N, 106 31.287' W 2,160 m"	19 Sep 1998
44	JZ	Unknown	"Jemez Mtns."	Unknown
45		Sandoval	No specific locality recorded	21 Sep 1996
46	SN	Bernalillo	"Sandia Mountains [vicinity road to crest] . . . forest of white and Douglas fir . . . 9,500 feet"	21 Apr 1956
47		Unknown	Unknown	Unknown

| Collector | Reference | Geo-referencing | |
		Latitude (NAD 83)	Longitude (NAD 83)
D.J. Hafner	NMMNH 1763	35.931323	-106.477586
	EIA; Swickard et al. 1971	35.892484	-106.390375
Unknown	B. Parmenter, pers. comm.	35.86814	-106.414123
	EIA; Swickard et al. 1971	35.853624	-106.430543
R. P. Martin	MSB 31223	35.848539	-106.423109
	EIA; Swickard et al. 1971	35.848539	-106.423109
	EIA; Swickard et al. 1971	35.848539	-106.423109
	EIA; Swickard et al. 1971	35.837188	-106.401222
	Guthrie and Large 1980	35.832228	-106.356027
	Guthrie and Large 1980	35.832228	-106.356027
	Guthrie and Large 1980	35.825322	-106.393564
Mammalogy Class 1996	MSB 140947	35.815161	-106.514239
Mammalogy Class 1998	MSB 140861	35.814636	-106.521313
Unknown	UTEP 7163	NA	NA
Unknown	MSB 89652	NA	NA
R. D. Ivey	Ivey 1957	35.211632	-106.428119
R. P. Martin	MSB 33270	NA	NA

APPENDIX 21.2. HABITAT DESCRIPTIONS AT SITES WHERE AMERICAN ERMINES (*MUSTELA RICHARDSONII*) WERE COLLECTED IN NEW MEXICO

Acronyms for mountain ranges: SJ (San Juan Mountains); SDC (Sangre de Cristo Mountains); JZ (Jemez Mountains).

Location	Mountain Range	County	Locality	Habitat Description
1	SJ	Rio Arriba	South of Jawbone Mtn. in GMU 52	"Found in downed, hollowed out aspen logs"; "Spruce-fir with lots of open meadows filled with seeps and glacial moraine . . . Not true riparian, but it might as well be, vole runs everywhere. True riparian w/in 0.5 mi. approximately." "In large log and downed litter, primarily spruce but under an aspen canopy, patch of wooded area was approx. 1/2 acre and surrounded by meadows and similar patches of spruce/fir-aspen. Definitely good herbaceous cover under the aspens and in the wetter portions of the seeps and meadows, open upland grassy areas are overgrazed by cattle and elk. The 'talus' (moraine) areas generally contain some of the seeps, even perennial in some cases."
5	SJ	Rio Arriba	4.25 mi. N, 6.5 mi. E of Canjilon, Canjilon Creek Campground	"60 ft from water near fallen logs."
5	SJ	Rio Arriba	4.25 mi. N, 6.5 mi. E of Canjilon, Canjilon Creek Campground	"5 ft from water's edge in quartz cobble embankment near lush sedge area."
6	SDC	Colfax	Sugarite Canyon State Park, tributary to Soda Pocket Creek	Site was on a bench in Sugarite Canyon with the dominant surrounding vegetation Gambel oak montane scrubland on hills and meadows on flats. Sherman traps were set along a very small (ca. 20 cm) steam in a meadow opening and bordered by tall sedges, hemlock (**Cicuta**), and various fobs; nearby the stream flowed through dense willow edged with locust.
7	SDC	Taos	Southern tributary to Holman Creek, along Road 1950	Site was in a small deep valley that held a small isolated wet meadow of nearly monotypic sedge that was very tall and formed laid-over mats of cover. The Sherman trap was set near a downed dead tree in the meadow. Vertical cover as measured with a Robel pole was 30 in.
12	SDC	Taos	Twining, 10,700 ft	"A small meadow . . . near the headwaters of the Hondo River, at the upper edge of the Canadian Zone." "Set in the runway of [a vole], under the overhanging grass near a little stream."
17	SDC	Taos	Rio Pueblo, below junction of Agua Sarca Canyon	Site was in a broad valley with an extensive complex of beaver workings on the stream, which was in the meadow valley bottom and surrounded by coniferous forest. Sherman traps were set at the upper end of a beaver complex in a beaked sedge wetland on saturated soil.

Reference	Source
R. Winslow, pers. comm.	Observed
Field notes	Field notes
MSB 43503, MSB 53986	Field notes
FT 503/MSB 140418	Field data sheets
FT 600	Field data sheets
USNM 133431	Bailey 1931:329
FT 496	Field data sheets

Location	Mountain Range	County	Locality	Habitat Description
26	JZ	Sandoval	16 mi N of Jemez Spgs	"Montane grasslands in a ponderosa and mixed coniferous forest"; a creek with beaver ponds in the valley bottom; ermine was caught "in a Sherman trap near the beaver ponds."
26	JZ	Sandoval	2 mi E, 15 mi of N Jemez Springs	Site was in a valley along a creek that was bordered by lush sedges and had a beaver workings; "Shermans were set as close to the waters' edge as possible."
26	JZ	Sandoval	2 mi E, 15 mi N of Jemez Springs	Site was in a valley along a creek that was bordered by lush sedges and had a beaver workings; ermine came from "sherman on the hillside"
28	JZ	Sandoval	Jemez Mountains, W of Slope Barley Canyon	un-logged forest
30	JZ	Sandoval	Rio Cebolla, at junction of Lake Fork Canyon	Site was in a broad valley with a stream and extensive beaver complex in the meadow, which was surrounded by coniferous forest. Sherman trap was set 20 cm for edge of a small channel (0.2 m x 1.0 m) bordered by sedge, timothy, other grasses, a diversity of forbs and some cattail. A large gooseberry and 2 large decadent willows were nearby.
32	JZ	Sandoval	Valles Caldera National Reserve, along Redondo Creek	"Along Redondo Creek"
35	JZ	Los Alamos	Pajarto Mtn.	"N-facing rockslide"
37	JZ	Sandoval	2.4 km SW of Cerro Grande	"Meadow"
38	JZ	Sandoval	N Hwy 4 & Dome Rd at junction of Baca location #1	"Meadow"
38	JZ	Sandoval	2.5 km S-SW of Cerro Grande	"Meadow"
39	JZ	Sandoval	3.7 km S-SE of Cerro Grande	"Meadow"
43	JZ	Sandoval	"0.2 mi E of Las Conchas Campground on NM Rt. 4"	"Along south side a stream south of road. A boreal forest type landscape."

Reference	Source
MSB 41623	Field notes
MSB 65695, MSB 212659	Field notes of D. J. Warren
MSB 65696	Field notes
MSB 87344	Field notes
FT 611	Field data sheets
MSB 144524	Specimen tag
Swickard et al. 1971	Swickard et al. 1971
Swickard et al. 1971	Swickard et al. 1971
MSB 31223	Specimen tag
Swickard et al. 1971; 2 specimens, one possibly MSB 31223	Swickard et al. 1971
Swickard et al. 1971	Swickard et al. 1971
MSB 140861	Field notes of L. Roberts

APPENDIX 22.1. TIMELINE OF MAJOR EVENTS IN BLACK-FOOTED FERRET HISTORY, RESEARCH AND MANAGEMENT

(adapted from Hubbard and Schmitt 1984; Clark 1989; Miller et al. 1996; Smithsonian 2011).

1–2 million years BP: Black-footed ferret ancestor crossed the Bering land bridge.

800,000 BP: Earliest fossil record of a black-footed ferret in North America.

2,000–3,000 BP: Estimated age of charred remains of black-footed ferret found in Atlatl Cave, San Juan County, New Mexico.

1851: *Mustela nigripes* described as a species by Audubon and Bachman.

1880s: Poisoning of prairie dogs begins on many large cattle ranches in the West.

1900: Sylvatic plague likely arrives in North America from Asia.

1915: US government begins funding prairie dog poisoning campaigns.

1934: Last verified black-footed ferret specimen collected in New Mexico.

1949: First reported case of human plague in New Mexico.

1960: Prairie dogs are conservatively estimated as occupying 2% of their former range in the United States.

1964: A black-footed ferret population is rediscovered in Mellette County, South Dakota, even as the federal government is about to declare the species extinct.

1966: The Endangered Species Preservation Act becomes law; black-footed ferret listed as Endangered the following year.

1971: Nine black-footed ferrets captured in Mellette County, South Dakota and moved to Patuxent Wildlife Research Center in Laurel, Maryland for captive breeding; four die of vaccine induced canine distemper.

1971: Executive Order no. 11643 is implemented, banning the use of poisons with secondary hazards on public lands.

1973: US Endangered Species Act becomes law, superseding the 1966 Act; Black-footed ferret is listed as Endangered.

1974: Mellette County, South Dakota black-footed ferret population is extirpated.

1975: Ferret listed as State Endangered in New Mexico under the newly enacted Wildlife Conservation Act.

1979: Last captive black-footed ferret from Mellette County, South Dakota dies; the black-footed ferret is declared extinct.

1981: 25 September: A Wyoming dog belonging to John and Lucille Hogg kills a black-footed ferret. 29 October: a live black-footed ferret is spotted near Meeteetse, Wyoming. Conservationist and researchers begin an intensive search and study of wild black-footed ferrets.

1984: The Meeteetse population consists of 129 black-footed ferrets.

1985: Outbreaks of sylvatic plague and canine distemper nearly wipe out the Meeteetse black-footed ferrets. 27 August: The US Fish and Wildlife Service and Wyoming Game and Fish decide to move all wild black-footed ferrets into captivity.

1987: The last known black-footed ferret is captured and removed from the wild at Meeteetse, Wyoming. These 18 captive black-footed ferrets are probably the rarest mammals in the world. A captive breeding program is initiated, and two litters of kits are produced bringing the total population to 25 animals.

1988: Black-footed ferret is removed from New Mexico Endangered list after surveys indicate no populations remain in the state.

1989: Seventy-two black-footed ferret kits are born in captivity.

1991: Shirley Basin, Wyoming becomes the first black-footed ferret reintroduction site.

1992: Two litters of wild-born kits are reported in Shirley Basin, Wyoming.

1994: Sylvatic plague sweeps through Shirley Basin, Wyoming. Black-footed ferrets persist at the site.

1995: Black-footed ferrets are discovered to be highly sensitive to the plague.

1998: The black-footed ferret captive breeding effort produces 442 kits; 339 survive to weaning. The US Fish and Wildlife Service estimates there are more black-footed ferrets in the wild than in captivity.

1999: The Turner Endangered Species Fund completes construction of 12 outdoor preconditioning pens for black-footed ferrets at Vermejo Park Ranch, Colfax County, New Mexico. From 1999 to 2005, 353 black-footed ferrets were preconditioned in the pens and sent for permanent release.

2008: Black-footed ferrets are released onto a black-tailed prairie dog complex at Vermejo Park Ranch.

2009: First wild-born black-footed ferret in New Mexico in perhaps 75 years is documented at Vermejo Park Ranch.

2012: The Turner Endangered Species Fund withdraws from future black-footed ferret releases on the black-tailed prairie dog complex at Vermejo Park Ranch due to drought-related poor survival and degrading habitat conditions, but does initiate releases onto the Gunnison's prairie dog colonies in the mountain meadows surrounding Castle Rock.

2013: A revised federal recovery plan for the black-footed ferret is released.

2015: Plague decimates the Castle Rock prairie dog population, and releases of ferrets at Vermejo Park Ranch are suspended for the foreseeable future.

2018: A new ferret release site on Wagon Mound Ranch in Mora County is established by the US Fish and Wildlife Service, the landowner, and the New Mexico Department of Game and Fish. An initial release of eight ferrets takes place in September on the ranch's black-tailed prairie dog colony.

APPENDIX 25.1. AMERICAN BADGER MUSEUM SPECIMENS FROM NEW MEXICO

Museum institution acronyms: ENMUNHM (Eastern New Mexico University, Natural History Museum); MSB (University of New Mexico, Museum of Southwestern Biology); MVZ (University of California, Berkeley, Museum of Vertebrate Zoology); NMSU (The Vertebrate Museum, New Mexico State University); NMSUVWM (New Mexico State University Vertebrate Wildlife Museum [a.k.a. NMSU Wildlife Museum]); UCM (University of Colorado Museum of Natural History); UTEP (University of Texas, El Paso, Biodiversity Collections, Mammals Division); and WNMU (Western New Mexico University).

BERNALILLO CO.
MSB:Mamm:6578
Albuquerque, Tingley Rd., 0.25 mi
    N of Bridge St.
21 November 1958
Forbes, R. B.

MSB:Mamm:15979
NE mesa, Albuquerque
15 November 1962
Boyle, T. P.

MSB:Mamm:24547
Rio Puerco, N of US66, S of
    Bernalillo
8 February 1967
Domenici Taxidermy

MSB:Mamm:46567
Albuquerque, intersection of Rio
    Bravo and Broadway
23 August 1976
Hafner, D. J.

MSB:Mamm:36885
Albuquerque, intersection of Rio
    Bravo and Broadway
18 May 1977
Scott, N. J., Jr.

MSB:Mamm:46568
~3 mi W of Alameda, on Coors Rd
14 September 1977
Hafner, D. J.

MSB:Mamm:38000
Albuquerque, Bridge St.
27 August 1978
Pennington, D. L.

MSB:Mamm:42177
Albuquerque, 1 mi S. of Intl. Airport
1 March 1980
Hafner, D. J.

MSB:Mamm:50575
Albuquerque, 2 mi N of I-40 on
    Coors Rd.
19 August 1982
Findley, J. S.

MSB:Mamm:65017
Coors Blvd, by Paseo del Norte,
    Albuquerque
3 October 1989
Dunnum, J. L.

MSB:Mamm:140047
Paradise Hills, Albuquerque
2 April 2004
Carnivore Initiative Project

CHAVES CO.
MSB:Mamm:213551
1/3 mi S of Roswell on NM 285
1 April 1996
Campbell, M. L.

CIBOLA CO.
MSB:Mamm:328717
El Malpais National Conservation
    Area, North Pasture
19 January 2018
Cook, J. A.

COLFAX CO.
MSB:Mamm:27524
2 mi W of Raton
3 February 1967
Bogan, M. A. and Studier, E. H.

CURRY CO.
ENMUNHM 1268
1 mi W., 10.5 mi N. of Artesia
2 December 1967
[Collector unknown]

DE BACA CO.
UCM:Mamm:4776
Hung on fence by trapper 50 mi N
    of Roswell
21 March 1948
Noble, A. N., and Burcki, D.

MSB:Mamm:294771
14.4 mi S, 3.2 mi W of
    Fort Sumner
29 August 1967
Best, T. L.

DOÑA ANA CO.
NMSUVWM 184
No locality given
12 December 1961
Wood, J. E.

NMSUVWM 198
No locality given
10 February 1962
Wood, J. E.

NMSUWM 608
No locality given
31 March 1964
Rogers, J.

NMSU 2902
7.6 mi W., 0.4 mi of N Organ
8 April 1967
Thaeler, C.

NMSUVWM 866
No locality given
21 March 1970
Pursley, D.

NMSUVWM 2199
Mesa west of Mesilla Dam
26 September 1970
Griffing, J. P.

NMSU 3866
3mi E. of Las Cruces
20 November 1970
Barkley, R. C.

UTEP:Mamm:3653
7 mi N of US 70 on USDA Rd. to
Jornada del Muerto Exp. Range
No date given
Roueche, W. L.

UTEP:Mamm:6374
Between Hatch and Doña Ana on
I-25
4 July 1980
Bond, M.

NMSU 12064
4 mi N., 2.2 mi E. of White Sands
6 September 1980
Stinnett, S.

UTEP:Mamm:7899
Mesilla Valley Trapper's Dump
1987 (no exact date given)
Redetzky, K. A.

MSB:Mamm:281425
Jornada Experimental Range
18 March 1991
Department of Energy

MSB:Mamm:64932
Jornada Experimental Range, Trap
#157-B
18 March 1991
Davenport, S. R.

MSB:Mamm:281426
Jornada Experimental Range
18 March 1991
Department of Energy

MSB:Mamm:87713
35 km N of Las Cruces, near Radium Springs, DOR on I-25
27 November 1992
Mammalogy Class 1992

EDDY CO.
MVZ:Mamm:80549
Pecos River, 6 mi S, 5 mi E of
Artesia
30 December 1937
Hall, E. R., and Wade, R. E.

MVZ:Mamm:91169
14 mi NW of Carlsbad
4 June 1940
Miller, E. V.

MSB:Mamm:634
Washington Ranch
1 May 1954
Hibler, M. P., and Campbell H.

UTEP:Mamm:2543
Dry Cave, McKittrick Hill
7 June 1970
Brown, S. M.

ENMUNHM 3967
28 mi E., 1 mi N. of Carlsbad
5 September 1975
[Collector unknown]

ENMUNHM 5011
23.5 mi E., 4.5 mi S. of Carlsbad
4 November 1978
Best, T. L.

MSB:Mamm:68564
Carlsbad Caverns NP, Exhibit 13 on
Scenic Loop Rd.
17 August 1990
Geluso, K. N.

GRANT CO.
WNMU:Mamm:1271
Bear Mtn. Rd, ~8 mi NW of Silver City
22 December 1961
Collector unknown

WNMU:Mamm:1217
2 mi S of Silver City
17 October 1962
Smith, W.

WNMU:Mamm:716
1 mi NW of Gila on the Gila River
14 March 1965
Francis, E. E.

WNMU:Mamm:628
13 mi W of Silver City,
US Hwy 180
7 October 1965
Collector unknown

UTEP:Mamm:216
16 mi SW of Silver City, US 181
27 May 1966
Hunt, D. L.

WNMU:Mamm:2906
Ridge Rd, 10 mi SE of Silver City
11 October 1970
Cole, D. C.

NMSU 11778
4 mi W, 3.2 mi of S Eagle Pk.
30 September 1979
McLean, E.

MSB:Mamm:214951
1.1 mi E, 0.6 mi N of Cliff on NM
211 near Radio Tower
1 September 2005
Fugagli, M.

MSB:Mamm:284713
Mangus, U Bar Ranch
March 2014 (no exact date given)
Mammalogy Class 2014

HARDING CO.
ENMUNHM 1802
3 mi N., 6 mi W. of Mills
27 October 1967
[Collector unknown]

APPENDIX 25.1. *continued*

HIDALGO CO.

MSB:Mamm:2570
2 mi W San Luis Pass, T33N, R19W
18 December 1955
Findley, James S.

WNMU:Mamm:2899
15 mi SE of Animas
15-Sep-1973
Baltosser, W. H.

WNMU:Mamm:3366
Alamo Hueco Mtns, Stone Cabin
    Tank; Township 33S, Range 14W,
    Sec. 29
20 January 1979
Collector unknown

LOS ALAMOS CO.

MSB:Mamm:54854
21.7 mi E of NM 4; T19N, R5E
22 August 1984
Thompson, D.

LUNA CO.

MSB:Mamm:140659
Ca. 3.5 mi N, 2.5 mi E of Deming at
    Mile Marker 4 of NM Hwy 26
19 October 1996
Polechla, P. J., Jr.

MSB:Mamm:226046
No locality given
23 February 2010
Polechla, P. J., Jr.

MCKINLEY CO.

UCM:Mamm:21775
Ft. Wingate Depot Activity near
    Buffalo Canal
July 1978 (no exact date given)
Stucky, R. K.

MORA CO.

MSB:Mamm:27518
La Cueva
10 April 1967
Bogan, M. A.

OTERO CO.

UTEP:Mamm:1905
Sleepy Hollow, 2 mi E of Cloud-
    croft, 9,000 ft
4 July 1967
Roueche, W. L.

UTEP:Mamm:7451
5 mi N of Texas state line on US Hwy 54
19 September 1981
Fuller, M. J.

QUAY CO.

MSB:Mamm:7626
Near Tucumcari; T11N, R30E
9 April 1959
Forbes, R. B.

RIO ARRIBA CO.

MVZ:Mamm:81502
T23, 24N R3, 4W
1 January 1938
Maslin, T. P., and Gartin B. H.

MVZ:Mamm:81503
T23, 24N R3, 4W
1 January 1938
Maslin, T. P., and Gartin B. H.

MVZ:Mamm:81504
T23, 24N R3, 4W
1 January 1938
Maslin, T. P., and Gartin B. H.

ROOSEVELT CO.

ENMUNHM 233
Portales
28 July 1966
Roadkill

ENMUNHM 922
3.5 mi S., 1 mi W. of Portales
10 December 1966
[Collector unknown]

ENMUNHM 1504
3 mi W of Floyd
25 May 1968
[Collector unknown]

MSB:Mamm:326949
8.8 mi S of Tolar
29 August 1968
Best, T. L.

MSB:Mamm:268958
7.1 mi S. of Tolar
30 August 1969
Best, T. L.

ENMUNHM 3730
12 mi W. of Portales
5 October 1976
[Collector unknown]

ENMUNHM 8327
5 mi W. of Floyd
17 October 1979
Best, T. L.

ENMUNHM 8455
2.8 mi E. of Portales
24 August 1980
[Collector unknown]

ENMU 9752
5 mi N., 3 mi E. of Portales
6 March 1985
[Collector unknown]

ENMU 9920
30 mi NW of Elida
17 January 1986
Uihlien, W.

ENMUNHM 10800
4 mi S of Floyd on RRAG
2 December 1998
[Collector unknown]

SANDOVAL CO.

ENMUNHM 1798
9.5 mi S., 9.2 mi W. of San Ysidro
    (Armijo Lake)
2 April 1970
[Collector unknown]

MSB:Mamm:89005
Jemez Mountains;
    34D45.644N,106D19.196W
27 September 1998
Klingel, J. T.

MSB:Mamm:92651
2.3 mi W of Counselors, NM Rt 550;
    36 mi W of Cuba
16 May 1999
Dohner, C. Z.

MSB:Mamm:284149
W slope of Sandia Mts. Along
    I-25, ca. 0.1 mi N, 0.4 mi E of
    Bernalillo, Sandia Pueblo Indian
    Reservation
20 August 2005
Carnivore Initiative Project

MSB:Mamm:143769
Santa Ana Indian Reservation,
    along U.S. 550, 3 mi [W] and
    4.5 mi N of Bernalillo
24 September 2005
Carnivore Initiative Project

MSB:Mamm:156766
Along US 550, ca. 8 mi N. of Cuba
27 Aug 2007
Mann, L. J.

MSB:Mamm:214958
Along U.S. 550 at Mile Marker 8
4 July 2009
Elder, K.

SAN JUAN CO.
MSB:Mamm:5921
M A Lucero Ranch, W side of Los
    Pinos River
21 August 1958
Harris, A. H.

MSB:Mamm:7678
San Juan River; T32N, R6W, SE 1/4
    Sec. 22
26 August 1959
Harris, A. H.

MSB:Mamm:10205
Pine River; T31N, R7W, NE 1/4 SE
15 July 1960
Harris, A. H.

MSB:Mamm:123259
~15 mi S of Bloomfield on
    Hwy 44
10 August 1994
Peurach, S. C.

MSB:Mamm:143788
12 mi S of Bloomfield, Navajo Agri-
    cultural Product Industries land
    off US550
8 November 2005
Dohner, C. Z.

MSB:Mamm:195636
Along U.S. Highway 550, near Blan-
    co Trading Post and junction
    with NM Highway 57
21 February 2008
Kaplan Lentz, S.

SANTA FE CO.
MSB:Mamm:22321
15 mi S of Santa Fe, Alamo Creek
6 November 1965
Boone, D.

MSB:Mamm:306211
NM Hwy. 4, about 0.65 mi S of East
    Jemez Rd
7 Aug 2013
Hathcock, C. and Reeves, R. R.

SIERRA CO.
MSB:Mamm:628
2.5 mi S of Truth of Consequences.;
    T14S, R4W, Sec. 10
8 August 1954
Davis, F. W.

MSB:Mamm:60988
2 mi E of I-25 on Hwy 52; 1.5 mi W
    of Elephant Butte
13 November 1988
Snell, H. L.

MSB:Mamm:250041
4.5 km SW of Engle
26 April 2012
Harrison, R. L.

MSB:Mamm:297045
Hunting Camp Road
7 February 2016
Rood, J.

SOCORRO CO.
MVZ:Mamm:55220
Lava Beds, 10 mi SE of Clyde
12 September 1932
Alexander, A. M.

MSB:Mamm:42447
Los Alamos Spring, 4 mi SE of La
    Joya
8 July 1980
Scott, N. J., Jr.

MSB:Mamm:92724
1.25 mi N of Bernardo on I-25;
    0.25 mi N of Mile Marker 176
7 September 2001
Frank, B. D.

MSB:Mamm:89184
6 mi N of Bernardo on I-25, at Mile
    Marker 176
2 October 1996
Friggens, M. T.

MSB:Mamm:85746
I-25 S towards Sevilleta; at Mile
    Marker 183; Sabinal
19 July 1997
Friggens, M. T.

MSB:Mamm:281926
1 mi N of LaJoya/Sevilleta Exit on
    I-25
8 September 1997
Frey, J. K.

MSB:Mamm:85725
I-25, 11 mi N of Bernardo
21 September 1997
Friggens, M. T.

MSB:Mamm:140041
3.4 mi S of Bernardo on I-25
13 August 1998
Polechla, P. J., Jr.

MSB:Mamm:240572
I-25, at Mile Marker 185
13 September 1999
Friggens, M. T.

NMSUVWM 2452
La Joya, E. Hwy 25
21 November 2005
Roemer, G.

MSB:Mamm:269005
I-25, 1.45 mi N. of Exit 169 (La Joya),
    3.8 mi S Bernardo
22 August 2010
Harrison, R. L.

MSB:Mamm:294395
Area surrounding Dusty
December 2015 or January 2016
Rood, J.

MSB:Mamm:297046
Area surrounding Dusty
January 2016 (no exact date
    given)
Rood, J.

MSB:Mamm:327343
Area surrounding Dusty
17 January 2017
Warren, D. J.

MSB:Mamm:327435
Area surrounding Dusty
2017 or early 2018 (no exact date
    given)
Warren, D. J.

MSB:Mamm:327447
Area surrounding Dusty
6 March 2018
Warren, D. J.

MSB:Mamm:327448
Area surrounding Dusty
6 March 2018
Warren, D. J.

MSB:Mamm:327452
Area surrounding Dusty
6 March 2018
Warren, D. J.

MSB:Mamm:327453
Area surrounding Dusty
6 March 2018
Warren, D. J.

MSB:Mamm:327454
Area surrounding Dusty
6 March 2018
Warren, D. J.

MSB:Mamm:327455
Area surrounding Dusty
6 March 2018
Warren, D. J.

TAOS CO.
MSB:Mamm:234704
~20 road mi S of Tres PiedrasUS 85
19 September 2006
Christman, B. L.

TORRANCE CO.
MSB:Mamm:1218
15 mi NE of Corona; T2N, R11E,
    Sec. 11
21 April 1956
Robinson, J. L.

MSB:Mamm:15137
15 mi NE Corona; T2N, R14E
21 April 1956
[Collector unknown]

MSB:Mamm:2031
10 mi E, 2 mi S of Willard
10 November 1956
Harbour, J.

MSB:Mamm:54237
Gran Quivira Nl Monument [sic],
    6,600 ft
8 May 1979
Woodward, B. D.

MSB:Mamm:87771
Roadkill on NM 41; 2 mi S of Macintosh
23 June 1994
Scott, D. T.

UNION CO.
ENMUNHM 711
39 mi E., 3 mi S. of Folsom
6 September 1966
[Collector unknown]

MSB:Mamm:294784
4.4 mi W and 2.2 mi S of Mt. Dora
3 June 1968
Best, T. L.

MSB:Mamm:85753
5 mi E of Clayton on Hwy 56/412
    near Kiowa National Grassland
4 July 1998
Ford, P. L., and Ruiz, E.

VALENCIA CO.
MSB:Mamm:24236
5.2 mi (by NM 6) of W Los Lunas
27 August 1966
Dunham, A. E.

MSB:Mamm:91634
14 mi N of Sevilleta NWR on I-25
25 September 1995
Parmenter, R. R.

NO COUNTY GIVEN
MSB:Mamm:13937
No locality recorded
No date given
Hibben, F. C.

MSB:Mamm:13938
No locality recorded
No date given
Hibben, F. C.

MSB:Mamm:231528
No locality recorded
No date given
Dunn, W. L.

MSB:Mamm:267526
No locality recorded
No date given
Collector unknown

MSB:Mamm:283586
No locality recorded
No date given
Carnivore Initiative Project

UTEP:Mamm:1314
No locality given
1968 (no exact date given)
Rausche, W.

APPENDIX 30.1. RINGTAIL MUSEUM SPECIMENS FROM NEW MEXICO

Museum institution acronyms: AMNH (American Museum of Natural History); ASNHC (Angelo State Natural History Collections); CAVE (Carlsbad Caverns National Park); ENMUNHM (Eastern New Mexico University, Natural History Museum); FHSM (Fort Hays State University, Sternberg Museum of Natural History); KU (University of Kansas, Natural History Museum and Biodiversity Research Center); MSB (University of New Mexico, Museum of Southwestern Biology); MVZ (University of California, Berkeley, Museum of Vertebrate Zoology); NCSM (North Carolina Museum of Natural Sciences); NMMNH (New Mexico Museum of Natural History and Science); NMSU (The Vertebrate Museum, New Mexico State University); NMSUVWM (New Mexico State University Vertebrate Wildlife Museum [a.k.a. NMSU Wildlife Museum]); NYSM (New York State Museum); OMNH (University of Oklahoma, Sam Noble Oklahoma Museum of Natural History); TCWC (Biodiversity Research and Teaching Collections [formerly Texas Cooperative Wildlife Collection]); USNM (United States National Museum of Natural History); WNMU (Western New Mexico University); and YPM (Yale University, Peabody Museum of Natural History).

BERNALILLO CO.
MSB:Mamm:953
Juan Tabo Ranger Cabin; T12N, R4E, Sec. 35
15 August 1953
Clothier, R.

MSB:Mamm:5513
W side of Sandia Mtns.
25 November 1956
Forbes, R.

MSB:Mamm:64401
Sandia Mtns., lower end Doc Long's picnic area
22 December 1956
Ivey, R.

MSB:Mamm:224994
South highway 337 3.55 mi of South Junction NM Highway 337 and I-40
30 April 2008
Dragoo, J.

CATRON CO.
AMNH M-127253
Glenwood
11 April 1937
Mellinger, S.

MSB:Mamm:11201
Black Range, Wall Lake; T11S, R12W, Sec. 10
19 July 1960
Jones, C.; Fleharty, E.

WNMU 2761
9 mi S of Glenwood; T12S, R19W, Sec. 8
8 April 1973
Hayward, B.

MSB:Mamm196611
Catwalk Rd., Glenwood
28 March 2005
Lorentzen, O.

CHAVES CO.
MVZ 94907
Penasco River, 11 mi W of Hope
4 January 1941
Russell, W.; Wade, R.

CIBOLA CO.
MSB:Mamm:54853
11 mi S, 14 mi W of San Rafael; T9N, R12W, SW 1/4 Sec. 32
23 September 1984
Mammalogy Class

MSB:Mamm:281413
Water Canyon, Mt. Taylor
10 August 1987
Yates, T.

COLFAX CO.
AMNH M-131842
8 mi W of Cimarron, Philmont Ranch, mouth Cimarroncito Canyon
24 September 1938
Hill, J.

DOÑA ANA CO.
MVZ 51993
Soledad Canyon, Organ Mtns., 17 mi E of Las Cruces
3 December 1931
Cummings, C.; Benson, S.

MVZ 51994
Soledad Canyon, Organ Mtns., 17 mi E of Las Cruces
3 December 1931
Cummings, C.; Benson, S.

MVZ 51995
Soledad Canyon, Organ Mtns., 17 mi E of Las Cruces
3 December 1931
Cummings, C.; Benson, S.

MVZ 51996
Soledad Canyon, Organ Mtns., 17 mi E Las of Cruces
5 December 1931
Cummings, C.; Benson, S.

MVZ 51997
Soledad Canyon, Organ Mtns., 17
   mi E of Las Cruces
5 December 1931
Cummings, C.; Benson, S.

MVZ 51998
Soledad Canyon, Organ Mtns., 17
   mi E of Las Cruces
7 December 1931
Cummings, C.; Benson, S.

MVZ 51999
Soledad Canyon, Organ Mtns., 17
   mi E of Las Cruces
9 December 1931
Cummings, C.; Benson, S.

MVZ 52000
Soledad Canyon, Organ Mtns., 17
   mi E of Las Cruces
9 December 1931
Cummings, C.; Benson, S.

MVZ 51990
Soledad Canyon, Organ Mtns., 17
   mi E of Las Cruces
23 December 1931
Cummings, C.; Benson, S.

MVZ 51991
Soledad Canyon, Organ Mtns.,
   17 mi E of Las Cruces
28 December 1931
Cummings, C.; Benson, S.

USNM 564259
San Andres NWR, Ash Canyon
23 February 1942
Halloran, A.

TCWC 4192
Ash Canyon, San Andres Mtn.
28 October 1942
Halloran, A.

USNM 274134
San Andres Mtns., Ash Canyon
7 March 1944
Halloran, A.

NMSUVWM 2345
San Andres Wildlife Refuge
26 February 1960
Unknown collector

NMSU 1205
30 mi. NE of Las Cruces, St. Nicho-
   las Canyon
5 December 1960
Donaldson, B.

NCSM 16297
Las Cruces; 10 mi E of University
   Park
19 March 1964
Rippy, C.

UTEP:Mamm:2519
0.9 mi. S of Country Club Rd. on
   NM 273
2 October 1970
Smartt, R.

UTEP:Mamm:7580
2.7 km W of Mt. Riley (T27S,
   R2W, SW 1/4 SW 1/4 Sec. 30) ca.
   4550 ft
26 March 1988
Dillon, T.

UTEP:Mamm:7581
2.7 km W of Mt. Riley (T27S, R2W,
   SW 1/4 SW 1/4 Sec. 30) ca. 4550 ft
26 March 1988
Dillon, T.

UTEP:Mamm:8386
About 2 air mi. N from top of Sug-
   arloaf Peak, Organ Mtns., T22S,
   R4E, Sec. 28, NW 1/4
4 May 1996
Worthington, R.

NMSUVWM 2432
Las Cruces, on Stone Canyon Rd.,
   near junction with Paseo del
   Onate, NAD 27, Z13 E338327,
   N3576534
17 October 2002
Roemer, G.

NMSUVWM 2431
Las Cruces, Mile Marker 7 on HWY 185
8 March 2004
Schemnitz, S.

EDDY CO.*
USNM 244447
Carlsbad Cave, in Walnut Canyon
29 March 1924
Bailey, V.

USNM 244448
Slaughter Canyon, found poisoned
   in cave
20 April 1924
Bailey, V.

USNM 564260
Carlsbad Cave, in deepest and
   farthest room of cave
5 May 1924
Bailey, V.

CAVE 1890
Carlsbad Caverns National Park
   headquarters area
22 January 1951
Gale, B.; Prasil, R.

CAVE 1897
January 1951
Carlsbad Caverns
Prasil, R.

CAVE 1896
Slaughter Canyon Cave
November 1951
Prasil, R; Gale, B.; Young, [F.?]

CAVE 1889
Carlsbad Caverns National Park
   headquarters area
11 August 1952
Gale, B.; Prasil, R.

MSB:Mamm:626
25 mi. SW of Carlsbad; T25S, R24E,
   Sec. 6
16 January 1954
Hibler, M.; Campbell, H.

FHSM 6000
48 mi. NE on State Hwy 137, El Paso
  Gap
25 March 1966
Collector unknown

UTEP:Mamm:2238
Dry Cave, McKittrick Hill, 100 ft
  NE of Balcony Room
Smartt, R.
27 June 1970

UTEP:Mamm:4320
Ca. 4 mi. N of Whites City on HWY
  62-180
31 October 1970
Harris, A.

CAVE 3839
A fenced cave at Carlsbad Canyon
  National Park
24 January 1972
NPS Staff

UTEP:Mamm:3062
T22S, R25E, NE 1/4 SE 1/4 Sec. 29
25 November 1972
Metcalf, A.

MSB:Mamm:68562
Carlsbad Caverns National Park
5 July 1973
Geluso, K.

UTEP:Mamm:6861
Ca. 5.5 mi. NW of Whites City,
  Walnut Canyon Rd., Carlsbad
  Caverns National Park
6 September 2006
Unknown collector

MSB:Mamm:68563
Carlsbad Caverns National Park,
  Walnut Canyon
21 June 1986
Geluso, K.

NYSM zm-10900
9 km E of Queen, Rt. 137
14 September 2014
Steadman, D.

ASNHC 13813
Lincoln National Forest, Guadalupe
  Ranger Dist., 13S 519617, 3544431
Vestal, A.
4 October 2003

ASNHC 13814
Lincoln National Forest,
  Guadalupe Ranger Dist., 13S
  519617, 3544431
Vestal, A.
4 October 2003

ASNHC 13812
Lincoln National Forest, Guadalupe
  Ranger Dist., 13S 522420,
  3557197, 5960 ft elev.
Vestal, A.
8 January 2004

GRANT CO.
USNM 157860
Red Rock
Goldman, E.
2 October 1908
Bassariscus astutus arizonensis

WNMU 3368
2 mi. S of Cliff
23 December 1977
Hayward, B.

WNMU 5194
1 mi W of Silver City, Boston Hill
13 December 1987
Rogers, B.

WNMU 6791
1 mi E of Cliff
10 April 1995
Morris, J.

MSB 156963
Mangas Springs
6 November 2007
Fugagli, M.

HIDALGO CO.
USNM 20360
Big Hatchet Mtn., upper pinyon
  zone
20 May 1892
Mearns; Holzner, F.

MVZ 95829
26 mi S of Animas Mtns.
9 June 1941
Johnson, D.; Reiche, P.
Bassariscus astutus arizonensis

UTEP:Mamm:61
Near Aspen spring, Animas Mtns.
8 April 1966
Harris, A.

MSB:Mamm:46326
Animas Mtns., 2 mi. W of Black Bill
  Spring; T31S, R20W, Sec. 25
27 December 1981
Donald, W.

KU KUM 145287
Rodeo, vicinity of
5 November 1991
Eifler, M.

LINCOLN CO.
MSB:Mamm:5917
Capitan Mtns.
3 July 1958
Findley, J.

MSB:Mamm:160390
San Patricio
21 February 1962
Goodpaster, W.; Goodpaster, L.

OMNH 8080
Sunset
13 January 1969
Geluso, K.

NMSU 5402
1.0 mi. S, 4.0 mi. E of Tinnie
3 November 1975
Boles, P.

MSB:Mamm:76213
Eureka Mine
7 March 1991
Altenbach, J.

LOS ALAMOS CO.
NMSUVWM 970
T19 R6 S#2 1 mi. N of Cumbeas Jr.
    High, elev. 7200 ft
5 October 1971
Hatch, M.

MSB:Mamm:280043
Los Alamos
25 November 2012
The Wildlife Center Inc. staff

LUNA CO.
MSB:Mamm:8619
Florida Mtns., 10 mi. S, 7 mi. E of
    Deming; T25S, R8W
30 January 1960
Gennaro, A.

OTERO CO.
YPM MAM 001965
High Rolls
10 February 1936
Unknown collector

ENMUNHM 649
1.5 mi. E. of La Luz
1 February 1967
Vaughn, T.

UTEP:Mamm:7582
Dog Canyon, Sacramento Mtns.
15 May 1988
Dillon, T.

ENMUNHM 11026
1 mi. E of Mayhill on US 82
10 March 2001
Hope, A.

ROOSEVELT CO.
ENMUNHM 965
6.4 mi. N, 6 mi. W of Portales
21 April 1967
Beard, W.

SANDOVAL CO.
MSB:Mamm:64402
Jemez Canyon, talus slope above
    Jemez Creek
1 March 1949
Ivey, R.

NMMNH 1093
Virgin Creek, Virgin Canyon, Jemez
    Mtns.
11 August 1987
Smartt, R.

MSB:Mamm:141054
Cochiti Canyon near Dixon
    Orchard and Rio Chiquito
8 April 2001
Carnivore Initiative Project

SANTA FE CO.
MSB:Mamm:35094
1/4 mi. N of Canoncito, on Hwy 25,
    9 mi. S and 6 mi. E of
    [Sante Fe]
15 January 1976
Palmer, F.

MSB:Mamm:61903
2086 Calle Navidad, Santa Fe,
    7,000 ft elev.
8 March 1986
Martinez, H.

MSB:Mamm:61904
Bridge crossing at Mile Marker 270
    La Cienega [on I-25]
27 October 1986
Pease, C.

SIERRA CO.
UTEP:Mamm:5314
Elephant Butte Dam
1 September 1978
Messing, H.

WNMU 4907
Truth or Consequences, Rio Grande
1 January 1986
Hunter, K.

NMSUVWM 2343
Hwy 51, E of Elephant Butte Lake
    dam Unknown collector
15 February 2002

MSB:Mamm:263609
Ladder Ranch, 1 km N, 10 km W of
    Caballo
27 October 2012
Mammalogy Class

SOCORRO CO.
MSB:Mamm:82596
12 mi N of San Antonio
25 April 1994
Dickerman, R.

MSB:Mamm:87714
33° 33.102' N, 107° 16.534' W
10 September 1994
Perry, T.

MSB:Mamm:282000
Magdalena Mtns., Langmiur Light-
    ning lab. visit
20 May 1999
Knapp, W.; Ruedas, L.

MSB:Mamm:92615
5.9 mi. W, 2.1 mi. S of Socorro;
    along Hwy 60 on Sedillo Hill; Six
    Mile Canyon
19 April 2000
Mensell, M.

MSB:Mamm:199739
Sevilleta NWR; Goat Draw Well
29 November 2005
US Fish and Wildlife Service

MSB:Mamm:199831
Sevilleta NWR; Goat Draw Water
    Tank
5 January 2006
US Fish and Wildlife Service

MSB:Mamm:224018
Sevilleta NWR; Los Pinos Range
1 September 2009
Harrison, R.

MSB:Mamm:224048
Los Pinos Range, Sevilleta NWR
8 March 2010
Harrison, R.

MSB:Mamm:225060
Sevilleta NWR
22 March 2010
Harrison, R.

MSB:Mamm:225031
Sevilleta NWR
22 March 2010
Harrison, R.

MSB:Mamm:223988
Sevilleta NWR; Los Pinos Range
12 April 2010
Harrison, R.

MSB:Mamm:223990
Sevilleta NWR; Los Pinos Range
17 August 2010
Harrison, R.

TAOS CO.
MSB:Mamm:7643
Red River Canyon, near Taos
7 June 1959
Shockey, M.

MSB:Mamm:85572
Carson NF; Cebolla Mesa trail No.
    102; second switchback from rim
22 November 1998
Polechla, P., Jr.

MSB:Mamm:140657
Confluence of Rio Pueblo de Taos
    and Rio Grande
22 November 1998
Polechla, P., Jr.; Chavez, C.; Frank,
    B.; Knapp, W.

TORRANCE CO.
MSB:Mamm:127029
Edgewood, 35 Lexco Rd.
27 November 2009
Detten, M.

*Two Carlsbad Caverns National
Park specimens (CAVE 2205 and
CAVE 2228) are listed as missing

APPENDIX 31.1. VERIFIED AND PROBABLE WHITE-NOSED COATI (*NASUA NARICA*) OCCURRENCE RECORDS FROM NEW MEXICO

Reliability Class A records are those considered verified. Class B and C records represent accepted observations (see Frey et al. 2013).

Reliability	County	Date	Specific location	Notes	Observer/Collector
A	Catron	Feb–Mar 2001	Glenwood, Blue Front Café's trash bins	1 very large adult photographed	Bucky Allred (Bluefront Café in Glenwood)
A	Catron	7 Dec 2008	Largo Creek, Highway 32	1 coati photographed	Nick Smokovich (NM State Forestry Division, Socorro)
A	Catron	8 Sep 2012	Tularosa Mtns., head of Long Canyon	1 adult male coati captured, photographed, and released at capture location	Dylan Burrus (NMSU)
A	Doña Ana	20 Mar 2006	Hatch, Hwy 185 at Elm St. (released W of Lake Valley)	1 coati captured and photographed	Campanella, James (NMDGF)
A	Doña Ana	Spring 2006 (as well as two earlier sightings in same area)	Hatch	2–3 coatis captured in a tree and photographed	Ray Aaltonen (NMDGF)
A	Doña Ana	28 Nov 2006	3222 Doña Ana Rd., Las Cruces	1 coati on roof of a residential home (photographed)	Ray Aaltonen (NMDGF), Jennifer K. Frey (NMSU)
A	Grant	4 Feb 1993	Lake Roberts	1 coati caught in a foothold trap, photographed, and released	Roy C. Hayes (NMDGF)
A	Grant	8 Jul 1997	Mangas Springs, backyard pond of residence	6 coatis photographed	Ralph Fisher
A	Grant	29 Jul 1999	Across Forest Service Rd. 851 in Burro Mtns.,on the E side of Burro Spring Canyon	19 coatis photographed	Jason Roback
A	Grant	11 Oct 2000	Little Whitewater Canyon near Glenwood	4–8 coatis (including at least 2 young) photographed	Jason Roback
A	Grant	1 Jan 2001	Burro Mtns., Burro Mtn. Homestead RV Park	16 coatis photographed while feeding on seeds under a bird feeder in light snow	Herb Atchinson
A	Grant	1 Feb 2001	Mule Creek residence	4–5 coatis (4 adults and 1 juvenile), video-taped in an open grassy area	Jim Gregory
A	Grant	2 Apr 2001	Iron Knot Ranch, 14.5mi N of Virden [at foot Apache Box]	Specimen-MS-B:Mamm:283633 [skeleton of old adult killed by dog]	Mike Bradfute (Ironknot Ranch), Jason Roback, Paul Polechla (UNM)

Reliability	County	Date	Specific location	Notes	Observer/Collector
A	Grant	21 Jul 2001	Big Burro Mtns., Gila National Forest, 2.4 mi W, 0.1 mi S of White Signal	Specimen-MS-B:Mamm:140072 (juvenile)	Paul Polechla
A	Grant	Between 5–9 Aug 2001	Gila Bird Area, in a stand of hackberry, walnut, and sycamore ca. 0.5 mi N of Bill Evans Rd.	1 coati photographed; vocalizing, likely due to observer's proximity	Sadoti, Giancarlo, Scott Stoleson, Hope Woodward, Rebecca Hunt
A	Grant	27 Mar 2005	Little Burro Mtns., south of SR 90, SE of Round Mtn. south of [SR] 90, SE of Round Mountain. On a dirt road.	1 coati photographed	Patrick C. Morrow
A	Grant	23 Aug 2006	Silver City residence	1 coati photographed in apple tree	http://www.avelinomae-stas.com/2006/08/23/coati-in-a-tree/
A	Grant	18 Jul 2007	Greenwood Canyon, Hwy 180 Mile Marker 90	Specimen-MS-B:Mamm:157858	Mike Fugagli
A	Grant	Mid-Oct 2007	On the trail to Turkey Creek Hotsprings, along the Gila River, Gila National Forest	1 coati photographed	Colin Lee (Bosque del Apache National Wildlife Refuge)
A	Grant	3 Nov 2007	Mangas Springs	Specimen-MS-B:Mamm:156969	Stephen O. MacDonald
A	Grant	24 Feb 2009	Pitchfork Ranch head-quarters, ca. 1 mi NNW of Silver City	1 coati (older male?) eating cat food photographed	A.T. and Cinda Cole (Pitch-fork Ranch)
A	Grant	31 Jan 2013	Red Rock Wildife Management Area	15–20 coatis initially in a netleaf hackberry tree eating berries, photographed	Mark Watson (NMDGF)
A	Grant	11 Feb 2013	Highway 180, Mangas Springs	Specimen-MS-B:Mamm:268390	Stephen O. MacDonald
A	Grant	30 Sep 2013	Gila River Bird Area	Photograph of track	Christine Hass
A	Grant	22 Sep 2015	Double E Wildlife Management Area	3 coatis photographed by remote camera	Mark Watson (NMDGF)
A	Grant	29 Oct 2015	The Nature Conservancy Gila Box	Troop photographed by remote camera	Keith Geluso (University of Nebraska-Kearney)
A	Grant	29 Apr 2016	Double E Wildlife Management Area	1 coati photographed by remote camera	Mark Watson (NMDGF)
A	Grant	4 Jun 2018	Pitchfork Ranch	1 coati photographed	Bob Wilcox (iNaturalist)
A	Grant	1 Oct 2020	Gila River Bird Area, Gila River	1 coati photographed	Joel Gilb (NMSU)
A	Hidalgo	25 Mar 1936	Animas Mtn., Pine Canyon	Specimen-USNM Mammals 261661	W. Echols

Reliability	County	Date	Specific location	Notes	Observer/Collector
A	Hidalgo	15 May 1982	Peloncillo Mtns. Renegade Tank No. 28, 200m SE of Burro Peak	1 coati drinking at small dam photographed by remote camera	Amy Elenowitz (NMDGF)
A	Hidalgo	On or before 1986	Indian Creek Canyon	Specimen-MS-B:Mamm:60680	Marshall C. Conway; reported by Cook (1986)
A	Hidalgo	16 Jul 1998	Upper Thompson Canyon, Big Hatchet Mtns.	1 adult coati climbing agave stalk and feeding on its flowers photographed	Jason Roback
A	Hidalgo	19 May 2009	Peloncillo Mtns., E of Rodeo	16–18 coatis observed, and a group of 2 adults and 3 young photographed	Patrick Mathis and Eric Rominger (NMDGF)
A	Hidalgo	16 Jan 2011	Guadalupe Canyon Wilderness Study Area	2 coatis photographed	Cole Wolf on iNaturalist
A	Hidalgo	5 Apr 2020	Blue Mtn., Peloncillo Mtns.	1 coati photographed by remote camera	Bureau of Land Management
A	Luna	8 Apr 2009	Rockhound State Park, near the campground area	1 coati photographed by visitor	Unknown
A	Sandoval	1 Apr 2018	Corrales, near river in bosque	1 coati trapped, photographed, and relocated to Red Rock Wildlife Area in southern New Mexico; was captured by pest control company after homeowner reported an animal attacking poultry	James Stuart (NMDGF); https://www.krqe.com/news/wildlife-officials-co-atimundi-will-be-relocated-to-southern-new-mexico/; https://magazine.wildlife.state.nm.us/out-of-range/
A	Sierra	2008	13 mi S of Hillsboro, NM	1 coati with injured hind foot photographed	Matilide Holzwarth (NMSU)
A	Sierra	30 Dec 2008	Ladder Ranch	1 coati photographed by a remote camera	Travis Perry (Furman University)
A	Sierra	4 Jul 2018	5 mi W of Hillsboro on 152	Specimen-NMSU	Tricia Rossettie (NMSU)
A	Sierra	13 Aug 2021	NM152 Bridge near Hillsboro	1 adult with 5 kits photographed	Jake Schoellkopf (NMDOT); https://www.newsbreak.com/news/2343395697402/mother-coati-and-her-5-kits-spotted-in-hillsboro; https://www.koat.com/article/mother-coati-and-her-5-kits-spotted-in-hillsboro/37329700#

Reliability	County	Date	Specific location	Notes	Observer/Collector
A	Socorro	8 Mar 2017	I-25 exit 169, just south of the off-ramp for Sevilleta National Wildlife Refuge	Specimen-MS-B:Mamm:300000	Kathy Granillo (Sevilleta National Wildlife Refuge)
A	Socorro	20 Mar 2012	San Acacia	Coati captured and photographed after having become stuck in a fenced yard; released W of Riley and S of Rio Salado	Storm Usrey (NMDGF)
A	Socorro	21 Mar 2018	Upper end of Elephant Butte Reservoir, near Indian Springs	1 coati photographed	Shania Lopez, Dustin Armstrong (Bureau of Reclamation)
B	Doña Ana	11 Feb 2009	Up a power pole at a business on the west mesa of Las Cruces (frontage road going to the airport)	1 coati	Ray Aaltonen (NMDGF)
B	Catron	Numerous times 1990–2000s	Glenwood	Most observations were of single adults (feeding, traveling, escaping up trees, etc.), but two were of family groups (troops) of about 6 and 12 individuals each	Joe Truett (Turner Endangered Species Fund)
B	Catron	Numerous times 1990s–2000s	San Francisco Hot Springs	Most observations were of single adults (feeding, traveling, escaping up trees, etc.), but two were of family groups (troops) of about 6 and 12 individuals each.	Joe Truett (Turner Endangered Species Fund)
B	Catron	Numerous times 1990s–2000s	Pueblo Creek in the Blue Range Wilderness	Most observations were of single adults (feeding, traveling, escaping up trees, etc.), but two were of family groups (troops) of about 6 and 12 individuals each	Joe Truett (Turner Endangered Species Fund)
B	Catron	Numerous times 1990s–2000s	Little Whitewater Creek	Most observations were of single adults (feeding, traveling, escaping up trees, etc.), but two were of family groups (troops) of about 6 and 12 individuals each.	Joe Truett (Turner Endangered Species Fund)
B	Catron	Numerous times 1990s–2000s	Deer Park Creek	Most observations were of single adults (feeding, traveling, escaping up trees, etc.), but two were of family groups (troops) of about 6 and 12 individuals each.	Joe Truett (Turner Endangered Species Fund)

Reliability	County	Date	Specific location	Notes	Observer/Collector
B	Catron	28 Nov 1999	Along the San Francisco River down from the Ranger District Headquarters	2 adults seen from rocky cliffs above river	Scott Posner (Gila National Forest)
B	Catron	Mid-2007	San Francisco Hot Springs on southeast side of river	8-10 coatis	Melinda Benton (Gila National Forest)
B	Eddy	Spring of late 1980s or early 1990s	Roadway close to the Pecos riverbed just east of US Hwy 285, north of Carlsbad. ~1 or 2 mi. N of Living Desert Turnoff, but not as far N as the McNew subdivision or truck bypass	1 male coati dead on road	Mark O. Rosacker
B	Grant	Before 1971	Burro Mtns.	Many troops seen until 1971 coyote poisoning campaign began.	Kaufmann et al. (1976:629), citing Bruce Hayward (WNMU) and David Brown (AZGF)
B	Grant	Mid-1980s	Cliff, NM	1 coati treed by dog in a residential yard	Jerry Monzigo (Gila National Forest)
B	Grant	June or July 1995	San Francisco River	8 or more coatis jumped down from tree and escaped into thick vegetation	Jerry Monzigo (Gila National Forest)
B	Grant	1989–1998	Burro Mtns. and mtns. W of Gila River and N of Red Rock	Signs and sightings increased in frequency between 89–98	Nick Smith (NMDGF)
B	Grant	Spring 1989	Bear Creek, E of Gila	1 coati treed by dogs	Nick Smith (NMDGF)
B	Grant	11 Oct 1993	Gila National Forest, Sheep Corral Canyon Rd.		Bruce Thompson (NMDGF)
B	Grant	3 Jan 1997	Mangas Springs, elm tree along Mangas Creek	Troop of 10 coatis	Ralph Fisher (noted local naturalist)
B	Grant	15 Dec 1999	Burro Mtns. in a piñon-juniper woodland, near rocky outcrop on top of hill ca. 6–10 miles N of Hwy 90.	Troop of 12 coatis	Darrel Weybright (NMDGF)
B	Grant	1997–2000	South of Glenwood where US 180 crosses Big Dry Creek	Multiple sightings of singles and family groups	Nick Smith (NMDGF)
B	Grant	21 Sep 2000	Big Dry Creek at western boundary of Gila Wilderness	1 adult	Jason Roback

Reliability	County	Date	Specific location	Notes	Observer/Collector
B	Grant	10 Jun 2000	Gila National Forest's Gila River Bird and Wildlife Habitat Area, due west of state Highway 180, ~1–3 miles upstream from Gila Middle Box	At least 3 coatis	Scott Stoleson (USFS Northern Research Station)
B	Grant	1994–2000	"In [an] abandoned metal barn at Nichol's Canyon on the east end of the [Gila Lower] Box"	"Usually one at a time"	Bill Merhege (BLM)
B	Grant	9 Mar 2001	"In the Bill Evans Recreation Area"	Troop of 7–9 coatis	Rich Beausoleil (NMDGF)
B	Grant	25 May 2002	Gila River, Apache Box	1 coati seen among large rock outcrops on slope in oak woodland	Bruce Christman
B	Grant	2002	White Signal, NM	1 coati next to house	Donna Stevens (Gila Conservation Coalition and Upper Gila Watershed Alliance)
B	Grant	2003	Gila River near Slate Canyon in Middle Box; described as a "rocky riparian [area] in a steep box canyon"	Troop of 13	Dennis Miller (WNMU)
B	Grant	2003	Gold Gultch in the Burro Mtns; on a roadway	1 coati	Russel Kleinman and Dale A. Zimmerman (WNMU)
B	Grant	1 Jan 2004	Gila River Bird Area, Gila River, Grant, NM	2-3 coatis	Bruce Christman
B	Grant	1 May 2004	Apache Box, S of Mule Creek, NM, Grant	1 coati, possibly male	Bruce Christman
B	Grant	19 Dec 2004	Middle Box of Gila River, below Bill Evans Lake	2 juveniles	Bruce Christman
B	Grant	12 Dec 2004	Black Hawk Canyon in the Big Burro Mtns., 0.6 km E of Saddle Rock Canyon	5 coatis in a hackberry tree	Personal observation in Geluso (2009)
B	Grant	5 Oct 2005	Up Meersham Creek near Lake Roberts	1 coati, probably foraging, moving along the creek (which had only intermittent water)	Kelley Kindscher (Kansas Biological Survey)
B	Grant	2005	Santa Clara, NM in town-captured and released in Birding Area of Gila River	1 coati	Dennis Miller (WNMU)

Reliability	County	Date	Specific location	Notes	Observer/Collector
B	Grant	Ca. 2005	At Mangas Springs, while driving on Hwy 180 (going from Silver City to Cliff) coatis were adjacent to road	6–8 coatis	
B	Grant	6 Oct 2006	Jacks Peak, Big Burro Mtns., just south of the Forest Service Rd. that goes to the radio/communication towers on Jacks Peak.	15–25 coatis, including young of the year (but no adult males, based on size), foraging on slope (nosing and clawing in leaf litter, at logs, etc.) and uttering grunting and other low "communication" noises.	Van Clothier and Donna Stevens (Gila Conservation Coalition and Upper Gila Watershed Alliance)
B	Grant	22 Nov 2006	Gila Lower Box	25 coatis	Jack Barnitz (BLM)
B	Grant	Spring 2007	Gila National Forest on FS Rd. 851, between Oak Grove subdivision and Axel Canyon Rd.	1 coati crossed the road	Donna Stevens (Gila Conservation Coalition and Upper Gila Watershed Alliance)
B	Grant	2 Jun 2007	Glenwood, pasture 100 yards from a riparian area		Martin Frentzel (NMDGF)
B	Grant	1 Mar 2008	Goat Canyon in Big Burro Mtns. of Gila Natnl. Forest.	2 coatis running up rocks to escape observers	Donna Stevens (Gila Conservation Coalition and Upper Gila Watershed Alliance)
B	Grant	Spring or summer 2008	[State Rd. 90] between Lordsburg and Silver City within Burro Mtns. of Gila National Forest	1 adult, dead	Peter Houde (NMSU)
B	Grant	16 Sep 2008	Mangas Springs	1 coati, roadkill	Jerry Monzigo (Gila National Forest)
B	Grant	Sep or Oct 2008	Bill Evans Rd. about 1/2 mile from from Bill Evans Lake in the Burro Mtns.	1 coati	Russel Kleinman and Dale A. Zimmerman (WNMU)
B	Grant	15 Nov 2008	Saddle Rock area, in a really rocky area with thick brush on a steep slope	7 coatis	Tony Ybarra (Gila National Forest)
B	Grant	Fall 2008	Hwy 180 bridge near Cliff	roadkill	Stephen O. MacDonald
B	Grant	2008	Santa Clara, NM in town; captured and released in lower Burro Mtns.	1 coati	Dennis Miller (WNMU)

Reliability	County	Date	Specific location	Notes	Observer/Collector
B	Grant	3 May 2009	Bill Evans Rd, NE of the Gila River Bird Area	1 coati dead on road	David Griffin (Griffin Biological Services)
B	Grant	28 Nov 2008	Tyrone Mine	1 coati	Dennis Miller (WNMU)
B	Grant	No date provided	Gila Bird Area below Bill Evans Lake		Roland Shook (WNMU)
B	Grant	No date provided	Lake Roberts	1 troop	Patrick Mathis (NMDGF)
B	Hidalgo	On or before 1934	Cloverdale country, generally along the boundary line through Hidalgo		John C. Gatlin (US Biological Survey, Albuquerque)
B	Hidalgo	Late Sep or early Oct in 1968 or 1969	Little Hatchet Mtns., north of "bat cave"	Possibly 6 coatis (several adults and several young).	Jolane Culhane (WNMU)
B	Hidalgo	11 Nov 1975	Coronado National Forest, Clanton Canyon	2 adults and at least 3 juveniles	Gregg Schmitt (NMDGF)
B	Hidalgo	Either 1966, 1971, or 1979	Animas Mtns., high up in Indian Canyon	1 coati	Arthur Harris (UTEP)
B	Hidalgo	15 Aug 1980	Guadalupe Canyon, 1 mi S of Hadley Ranch, few hundred meters upstream Mormon Battalion marker	1 coati	William Baltosser (University of Arkansas)
B	Hidalgo	22 Jul 1981	Peloncillo Mtns., 0.5mi. W of Gray Peak, desert shrub–associated area	1 coati	Amy Elenowitz (NMDGF)
B	Hidalgo	12 Oct 1981	Peloncillo Mtns., N side of Skeleton Canyon, in a narrow draw	2 adults and 2 juveniles	Amy Elenowitz (NMDGF)
B	Hidalgo	1979–1985	Antelope Wells, SE corner of Diamond A Ranch (formerly Gray Ranch)	Unspecifiied numbers	Steve Dobrott (Ladder Ranch)
B	Hidalgo	1979–1985	Various sighting in the San Luis Mtns.	Unspecifiied numbers	Steve Dobrott (Ladder Ranch)
B	Hidalgo	Ca. 1985	Peloncillo Mtns., E side, just S of Geronimo Trail	3–4 coatis	Lief Ahlm (NMDGF)
A	Hidalgo	Before 30 June 1986	Animas Mtns., Pine Canyon	1 coati	Personal observation in Cook (1986)
B	Hidalgo	Before 30 June 1986	Animas Mtns., mouth of Pine Canyon	1 coati	Cook (1986), citing Steve Dobrott (Ladder Ranch)
B	Hidalgo	Before 30 June 1986	Animas Mtns., upper Double Adobe Canyon	2 coatis	Cook (1986), citing Steve Dobrott (Ladder Ranch)

Reliability	County	Date	Specific location	Notes	Observer/Collector
B	Hidalgo	Before 30 June 1986	Animas Mtns., 1 km E of Spur Windmill (which is at T32S R18W, NW 1/4 Sec 18)	1 coati	Cook (1986), citing Steve Dobrott (Ladder Ranch)
B	Hidalgo	Before 30 June 1986	Animas Mtns., Walnut Canyon	12 coatis	Cook (1986), citing Steve Dobrott (Ladder Ranch)
B	Hidalgo	1 Aug 1986	Guadalupe Canyon ~1 mi E of the AZ–NM border	2 coatis climbed about a 20 ft cliff to escape the canyon bottom	Mark Hakkila (BLM)
B	Hidalgo	1986	Several sightings within 2 air mi of Cloverdale, NM	Multiple sightings in 1986, including of coatis in trees on two occasions	John Sherman (BLM)
B	Hidalgo	Summer 1989	Peloncillo Mtns., ~ 13 mi southwest of Animas	~15 coatis, females and young, scurried off into a stand of dense oak trees when they saw the observer	Mark Hakkila (BLM)
B	Hidalgo	Late 1980s or early 1990s	Stream area near Gray Ranch	1 coati	Margarita Guzman (BLM)
B	Hidalgo	Since 1990	Observed coatis in Double Adobes Creek		Ben Brown
B	Hidalgo	1 Nov 1990	Peloncillo Mtns., steep canyon on W side of Black Point	Troop of 15, half of them juveniles	Doug Burkett (WSMR)
B	Hidalgo	1990 or later	At the head of Indian Creek	Coati troop	Ben Brown
B	Hidalgo	1990 or later	Upper Deer Creek watershed	Coati troop	Ben Brown
B	Hidalgo	1990 or later	Upper end of Whitewater Creek on the Diamond A Ranch (formerly the Gray Ranch)	Coati troop	Ben Brown
B	Hidalgo	1990 or later	Clanton Draw on the Coronado National Forest	Coati troop	Ben Brown
B	Hidalgo	1990 or later	Creek bed in Guadalupe Canyon	Coati troop	Ben Brown
B	Hidalgo	21 Oct 1997	None given	1 adult and 3 young seen from a helicopter during an aerial bighorn sheep survey	Eric Rominger (NMDGF)
B	Hidalgo	1996 1997	Alamo Hueco Mtns.	1 adult and 3 juveniles	Eric Rominger (NMDGF)

Reliability	County	Date	Specific location	Notes	Observer/Collector
B	Hidalgo	6 Aug 1998	NM portion of Skeleton Canyon	20 coatis	William Baltosser (University of Arkansas)
B	Hidalgo	Ca. 1998	Skeleton Canyon (W side of Peloncillos)	2 coatis	Randy Jennings (WNMU)
B	Hidalgo	Between 1995 and 1998	"Multiple observations on or adjacent to Geronimo Trail in the Peloncillo Mountains . . . best observation was daylight, small troop on rocks in Clanton Draw, adjacent to roadway"	Troop on rocks in Clanton Draw adjacent to roadway	Andy T. Holycross
B	Hidalgo	Late Nov 2001	Nichol's Canyon, BLM Gila Lower Box Wildlife Area, 150 m E of parking under large cottonwood	1 coati emerging from the brush to get a drink of water from the river	Carol Campbell (NMSU)
B	Hidalgo	3 Jan 2002	Brushy Canyon	1 coati	Carroll D. Littlefield (Bioresearch Ranch)
B	Hidalgo	14 Jan 2002	Brushy Canyon, E slope, near S. boundary fence (of Bioresearch Ranch?); later same male (allegedly) at Bioresearch Ranch HQ	1 coati along E slope of Brushy Canyon	Carroll D. Littlefield (Bioresearch Ranch)
B	Hidalgo	24 Mar 2002	Entrance of Katie Draw	Troop (8-10?)	Carroll D. Littlefield (Bioresearch Ranch)
B	Hidalgo	4 Jun 2002	Fox Draw	1 coati	Carroll D. Littlefield (Bioresearch Ranch)
B	Hidalgo	25 Jul 2002	(Coati was heading) into Beehive Canyon	1 coati	Carroll D. Littlefield (Bioresearch Ranch)
B	Hidalgo	8 Aug 2002	Brushy Canyon	Troop (all age groups represented)	Carroll D. Littlefield (Bioresearch Ranch)
B	Hidalgo	24 Oct 2002	Pinyon Draw	Troop(?)	Carroll D. Littlefield (Bioresearch Ranch)
B	Hidalgo	6 Nov 2002	Lower Lee's Canyon (Bear Hollow)	Troop of 12 (including a number of juveniles)	Carroll D. Littlefield (Bioresearch Ranch)
B	Hidalgo	6 Feb 2003	Upper Brushy Canyon	Troop	Carroll D. Littlefield (Bioresearch Ranch)
B	Hidalgo	23 May 2003	NM Hwy 80 N of Rodeo	1 adult dead and decomposing	Mike Hill
B	Hidalgo	25 Sep 2003	Beehive Canyon, ~1 ml NNE of HQ	Troop	Carroll D. Littlefield (Bioresearch Ranch)

Reliability	County	Date	Specific location	Notes	Observer/Collector
B	Hidalgo	1 Oct 2003	Indian Creek, Animas Mtns.	1 large adult crossed the creek bottom on a fallen tree	Doug Burkett (WSMR)
B	Hidalgo	2 Nov 2003	Post Office Canyon, ~800 ft E of HQ(?)	Troop(?)	Carroll D. Littlefield (Bioresearch Ranch)
B	Hidalgo	22 Dec 2003	Upper Brushy Canyon	Troop	Carroll D. Littlefield (Bioresearch Ranch)
B	Hidalgo	27 Sep 2004	Upper Skull Canyon	1 coati	Carroll D. Littlefield (Bioresearch Ranch)
B	Hidalgo	3 Jan 2004	Upper Lee's Canyon	1 adult coati, old (based on gray hairs)	Carroll D. Littlefield (Bioresearch Ranch)
B	Hidalgo	4 Dec 2004	Pond Draw	Coati diggings	Carroll D. Littlefield (Bioresearch Ranch)
B	Hidalgo	2004	Side of road, 2 mi NW of Cloverdale, NM	1 dead adult coati, decomposing	Doug Burkett (WSMR)
B	Hidalgo	28 Mar 2005	Upper Skull Canyon	Many coati diggings on alligator-barked juniper snag	Carroll D. Littlefield (Bioresearch Ranch)
B	Hidalgo	11 Oct 2005	Beehive Canyon	2 adults and 3 juveniles	Carroll D. Littlefield (Bioresearch Ranch)
B	Hidalgo	1 Nov 2005	Brushy Canyon	1 adult with 3 juveniles	Carroll D. Littlefield (Bioresearch Ranch)
B	Hidalgo	23 Jan 2005	Near Owl-Maverick Canyons' Saddle	At least 3 coatis	Carroll D. Littlefield (Bioresearch Ranch)
B	Hidalgo	Between Sep and Nov 2007	Head of Owl Canyon in the Peloncillos	1 coati	Ben Brown
B	Hidalgo	22 Nov 2008	Peloncillo Mtns. S of Geronimo Trail	At least 4 coatis	Larry Kamees (NMDGF)
B	Hidalgo	Oct–Nov 2008	A spring in Weatherby, Peloncillo Mtns.	1 coati, remote camera	Jack Barnitz and Jesse Rice (BLM) remote cameras
B	Hidalgo	Oct–Nov 2008	A spring at Pratt [Benchmark], Peloncillo Mtns.	1 coati, remote camera	Jack Barnitz and Jesse Rice (BLM) remote cameras
B	Hidalgo	2002–2008	Bioresearch Ranch and surrounding areas		Carroll D. Littlefield (Bioresearch Ranch)
B	Hidalgo	13 Feb 2009	South end of Big Burro Mts, Hoodoo Canyon FS rd 841	1 coati	Bruce Christman
B	Hidalgo	24 Feb 2009	Summit of Owl Canyon Ridge	1 coati	Carroll D. Littlefield (Bioresearch Ranch)

Reliability	County	Date	Specific location	Notes	Observer/Collector
B	Hidalgo	19 May 2009	Owl Canyon, southern Peloncillo Mtns.	1 coati	Eric Rominger (NMDGF)
B	Hidalgo	17 Sep 2009	On Owl Canyon Ridge	Several coatis, including 1 small juvenile	Carroll D. Littlefield (Bioresearch Ranch)
B	Hidalgo	No date provided	[Big] Hatchet Mtns.	1 troop	Patrick Mathis (NMDGF)
B	Hidalgo	No date provided	Little Hatchet Mtns.	1 coati	Patrick Mathis (NMDGF)
B	Luna	No date provided	Cooke's Peak	1 coati	Patrick Mathis (NMDGF)
B	Luna	No date provided	Cooke's Peak		Ray Aaltonen (NMDGF)
B	Luna	No date provided	Florida Mtns.	1 coati	Patrick Mathis (NMDGF)
B	Sierra	1 Jan 2009	Ladder Ranch, confluence of Animas Creek and Cave Creek	1 coati (probably male based on size) in tree	Steve Dobrott (Ladder Ranch)
B	Sierra	1979–1985	Various sighting in the Animas Mtns.	Unspecifiied numbers	Steve Dobrott (Ladder Ranch)
B	Sierra	26 Oct 2008	Game Management Unit 21, Pierce Canyon, off of Forest Rd. 888	1 coati	James Campanella (NMDGF)
B	Sierra	17 Dec 1999	Fra Cristobal Mtns., W side of range, in juniper and sage vegetation, near a rocky ridge	Observed by a net-gun crew hunting sheep from a helicopter	Eric Rominger (NMDGF) and Paul Wolf
B	Sierra	15 Sep 2001	White Rock Canyon (Ten Canyons Ranch) in oak savannah bottoms mixed with piñon-juniper	1 coati	Ray Aaltonen (NMDGF)
B	Sierra and/or Grant	No date provided	Black Range		Ray Aaltonen (NMDGF)
C	Catron	Summer 1998	W side of Alma bridge, just past the State Highway 78 turnoff [near Glenwood]	1 coati	John Nielson (Gila National Forest)
C	Catron	1 Oct 1999	3–4 mi N of Willow Creek and 4–5 mi N of Snow Lake	1 coati	Jeff Quick (Gila National Forest)
C	Catron	1 Nov 1999	Mogollon Mtns., near cattle tank in Burnt Cabin Cienega	1 coati	Joe Anderson (Gila National Forest)

Reliability	County	Date	Specific location	Notes	Observer/Collector
C	Catron	18 Jun 2006	West Fork of Gila River, several miles downstream of White Creek Cabin	1 coati, treed by dogs	Kirk Patten and Sean Buczek
C	Catron	8 Aug 2008	Forest Service Rd. 49 near Cross V Ranch, ~7–8 mi NE of Reserve, NM. This area is approximately 3–4 mi from San Francisco River (see notes)	1 coati	Lyndsay Hellekson (Gila National Forest)
C	Catron	11 Oct 2008	W of Eckleberger Canyon, W side of Telephone Canyon	1 coati	Jim Johns
C	Catron	15 Dec 2008	Gila National Forest, Forest Rd. 150 between Beaverhead and Wall Lake (habitat=pinyon-juniper uplands)	1 coati	Ellen Heilecker (NMDGF)
C	Catron	12 Nov 2009	In a canyon in the Eckelberger Canyon area	1 coati	Joseph Aragon (Gila National Forest)
C	Catron/ Grant	No date provided	W fork of Gila River, near Heart Bar Ranch	1 coati, treed by dogs and shot	Taber (1940), citing Jack Pinkerton
C	Grant	Between Dec 1993 and Feb 1994	Lower end of Burro Mtns., Gold Gultch	5–9 coatis	Anthony Romero
C	Grant	Apr 1996 or 1997	Burro Mtns., near Saddle Rock	5 adult coatis	Joan Helen (Gila National Forest)
C	Grant	1 May 1998	Entrance to Red Rock State Wildlife Experimental Area	1 coati chased up a tree by human	Duke Ford
C	Grant	Summer 1998	BLM Apache Box Wilderness Study Area	3 coatis	Brian Lummus (BLM)
C	Grant	8 Sep 1998	Dark Thunder Canyon, Apache N.F.	Troop (not seen directly seen), but diggings and coati-sized diggings observed and grunts heard by J. Roback (able to distinguish coati grunts from other animals' grunts)	Jason Roback
C	Grant	Mar–Apr 1999	Hwy 78, E of Mule Creek	At least 5 coatis	Joanne Kopren

Reliability	County	Date	Specific location	Notes	Observer/Collector
C	Grant	17 Jul 1999	N slope of Lydian Peak facing Ash Creek Canyon, in the Experimental area of Redrock State Wildlife Experimental Area	1 coati	Juanita Graves (Redrock Wildlife Area)
C	Grant	1 Nov 1999	Top of the rocks along Apache Box	30 coatis; observers reportedly took photos, but J. Roback never saw them	Mike Bradfute (Ironknot Ranch)
C	Grant	1 Dec 1999	Burro Mtns., S of Hwy 90, along Forest Service Rd. 837	1 coati chased up tree by dog	Hans and Gabrielle Heynekamp
C	Grant	20–25 May 2000	Hwy 90 [SR 152]	1 coati	Stever Logsdon
C	Grant	May–Jun 2000	Mouth of Clyde Canyon, S of Clark Peak, N of bighorn sheep preserve	1 adult coati	Alton Ford (NMDGF)
C	Grant	Late Aug 2000	Burro Mtns., between Walnut Creek and Copeland Mine	Troop of 15-20 coatis	Hans and Gabrielle Heynekamp
C	Grant	Late Aug 2000	Bitter Creek, between Bar-55 Ranch and the turnoff to the Ironknot Ranch (vic BLM Apache Box Wilderness Study Area)	Ca. 2 adults and 6 young seen several times over a period of a few weeks	Larry Monzigo
C	Grant	First week of Aug 2000	Lincoln Canyon, north of McKnight Canyon, a tributary of the Mimbres River	3 adult coatis	Pete Fust (Black Range Lodge)
C	Grant	Mar or Apr 2001	Steeple Rock Mining District, Mule Creek area	2 coatis	Russel Lummus (BLM)
C	Grant	23 Mar 2001	Pinos Altos residence	1 coati shot by resident	Silver City Daily Press article by Stephen Siegfied
C	Grant	1 Apr 2001	E side of Gila River below Billings Vista in shady sycamore, cottonwood, oak	1 very large coati treed by dogs	Katherine Rowland
C	Grant	1 May 2001	Steeple Rock Mining District, Mule Creek area	2 coatis	Russel Lummus (BLM)
C	Grant	2004	Upper part of Pitchfork Ranch, 35 mi S of Silver City, where cienaga runs above ground	1 coati	A. T. and Cinda Cole (Pitchfork Ranch)
C	Grant	1 Mar 2007	Near Pitchfork Ranch headquarters	2 coatis	A. T. and Cinda Cole (Pitchfork Ranch)

Reliability	County	Date	Specific location	Notes	Observer/Collector
C	Grant	End of Jan 2009	Pitchfork Ranch headquarters front wall, overflow pond close by	1 coati walking toward the overflow pond	A. T. and Cinda Cole (Pitchfork Ranch)
C	Grant	End of Jan 2009	Pitchfork Ranch headquarters	1 coati	A. T. and Cinda Cole (Pitchfork Ranch)
C	Grant	Late Jan 2009	Black Range	1 coati	John Titre (Gila National Forest)
C	Hidalgo/ Grant	Winter 2000	Along Gila River to town site of Telegraph and up the canyons E and W of the river, 4-5 miles; riparian areas only especially in areas with fruiting hackberry	1 troop and several males, via daily sightings	Bill Trussle (Missouri Department of Conservation)
C	Hidalgo/ Grant	21 Nov 2006	Gila River, Nichols Canyon	1 coati in the canyon bottom before running up the rock cliff when it spotted the observer	Jesse Rice (BLM)
C	Hidalgo/ Grant	22 Nov 2006	Gila River, lower Box Canyon, below Nichols Canyon	At least 9 coatis traveling down the river along the rim of the canyon	Jesse Rice (BLM)
C	Hidalgo	Several times since 1955	Peloncillo Mtns., various locations	Either singles or groups with as many as 20 coatis together	Warner Glenn
C	Hidalgo	On or before 1980	Birch Spring, Animas Mtns.	Number not specified	Cook (1996), citing J. Egbert (Egbert 1980)
C	Hidalgo	Fall 2005 or 2006	Guadalupe Canyon	~12 coatis	Jim Scanlon (BLM)
C	Luna	Late spring or summer between 1993 and 1996	Near city limits of Deming, just S of Mimbres River	1 adult	Ray Hewitt (BLM)
C	Sierra	1 Sep 2007	Along Palomas Creek beside the ghost town of Hermosa	5 or more coati tracks	Travis Perry (Furman University)
C	Sierra	Jan–Feb 2000	Upper Southwest Canyon (S of the town of Kingston)	1 coati (and occasional sightings during winters)	John Able (Gila National Forest)

APPENDIX 32.1. RACCOON MUSEUM SPECIMENS FROM NEW MEXICO

Museum institution acronyms: AMNH (American Museum of Natural History); CAVE (Carlsbad Caverns National Park); ENMUNHM (Eastern New Mexico University, Natural History Museum); FMNH (The Field Museum of Natural History); KU (University of Kansas, Natural History Museum and Biodiversity Research Center); MSB (University of New Mexico, Museum of Southwestern Biology); MVZ (University of California, Berkeley, Museum of Vertebrate Zoology); NMMNH (New Mexico Museum of Natural History and Science); NMSUVWM (New Mexico State University Vertebrate Wildlife Museum [a.k.a. NMSU Wildlife Museum]); USNM (United States National Museum of Natural History); WNMU (Western New Mexico University).

BERNALILLO CO.
KU 68055
Tijeras
11 June 1955
Baker, R. H.

MSB 16418
Albuquerque, Rio Grande Zoo
1962 or 1963 (no exact date)
Young, J. J.

MSB 16419
Albuquerque, Rio Grande Zoo
1962 or 1963 (no exact date)
Young, J. J.

MSB 16420
Albuquerque, Rio Grande Zoo
1962 or 1963 (no exact date)
Young, J. J.

MSB 16421
Albuquerque, Rio Grande Zoo
1962 or 1963 (no exact date)
Young, J. J.

MSB 16422
Albuquerque, Rio Grande Zoo
1962 or 1963 (no exact date)
Young, J. J.

MSB 16423
Albuquerque, Rio Grande Zoo
1962 or 1963 (no exact date)
Young, J. J.

MSB 22252
Isleta; T8N, R2E
29 June 1963
Findley, J. S.

MSB 41012
Foothills S of Albuquerque airport
    along Tijeras Arroyo
22 September 1979
Woodward, B. D.

MSB 49876
Cedar Crest
4 May 1982
Ligon, J. D.

MSB 50762
Cedar Crest
17 July 1982
Ligon, J. D.

MSB 50763
Cedar Crest
17 July 1982
Ligon, J. D.

MSB 56870
Cedar Crest
23 July 1982
Collector unknown

MSB 49874
Cedar Crest
17 August 1982
Dickerman, A. W.

MSB 49875
Cedar Crest
20 October 1982
Lewis, S.

MSB 49876
MSB 49877
Cedar Crest
28 November 1982
Ligon, J. D.

MSB 54908
Cedar Crest
25 October 1984
Ligon, J. D.

MSB 55592
Cedar Crest
20 February 1985
Malcolm, C. K.

MSB 55593
Cedar Crest
6 March 1985
Malcolm, C. K.

MSB 64944
Albuquerque, 0.1 mi W Alameda Bridge
25 September 1990
Brunt, M. H.

MSB 89179
Isleta Pueblo, 2.3 mi S, 2.5 mi W of
    Isleta Marsh on Mile Marker 12
16 July 1996
Painter, C. W.

MSB 78999
Rio Grande at Bridge Street, ca.
    0.25 mi N of bridge
14 November 1996
Stuart, J. N.; Miyashiro, J. B. M.;
    Davenport, S. R.

MSB 96189
Cedar Crest
25 June 1997
Collector unknown

MSB 142758
Cedar Crest
19 May 1999
Collector unknown

MSB 142918
Tijeras, 0.6 mi (by road) from intersection of NM 333 and NM 33
16 October 2003
Dragoo, J. W.

MSB 195593
Vicinity of Albuquerque
25 July 2007
Wildlife Rescue

MSB 199713
Albuquerque on Alameda Blvd at
Corvetta Dr.
5 November 2007
Polechla, P. J., Jr.

MSB 231523
0.5 mi S of Juan Tomas
8 October 2009
Degenhardt, W. G.

MSB 199596
Bosque area on Montaño Bridge
just before the river
9 October 2009
Keller, A. E.

MSB 231444
0.5 mi S of Juan Tomas
20 September 2010
Degenhardt, W. G.

MSB 231007
0.5 mi S of Juan Tomas
22 September 2010
Degenhardt, W. G.

MSB 231445
0.5 mi S of Juan Tomas
7 October 2010
Degenhardt, W. G.

MSB 247769
1 km S of Juan Tomas
4 August 2011
Degenhardt, W. G.

MSB 264307
Albuquerque, intersection of
Alameda and Rio Grande
Boulevard
21 October 2012
Schuyler W. Liphardt

CATRON CO.
AMNH 127255
Mogollon Mtns., Sacaton Creek
No date given
Brigham, E. M., Jr.

FMNH 48111
Mogollon Mtns., W Fork of Gila
River
15 June 1938
Osgood, W. H.

MSB 13145
Tularosa Creek just above Apache
Creek
16 May 1960
UNM Mammalogy Class

MSB 87715
NM 32, 5.4 mi S of USFS 23, Gallo
Mtns.
17 August 1994
Kelt, D. A.

CHAVES CO.
MSB 24407
Bitter Lake National Wildlife
Refuge
14 June 1966
Mares, M. A.

MSB 24408
Bitter Lake National Wildlife
Refuge
14 June 1966
Boggs, D. L.

MSB 140702
10.2 mi NW of intersection of
US245 and Macho Draw
10 December 2000
Polechla, P. J., Jr.

CIBOLA CO.
MSB 14053
10 mi NE of Grants
3 September 1960
Wright, J. W.

COLFAX CO.
AMNH 132533
3 mi SE [sic] of Cimarron, Philm-
ont Ranch
No date given
Buchanan, W. J.

USNM 129104
Raton Range, Bear Canyon
1903 (no exact date given)
Howell, A.
Procyon lotor hirtus

USNM 129105
Raton Range, mouth of Trinchera
Pass
12 September 1903
Howell, A.
Procyon lotor hirtus

KU 42556
1 mi S, 2 mi E of Eagle Nest
21 June 1951
Rainey, D.

MSB 27525
25 mi SE of Raton
9 January 1967
Bogan, M. A.; Studier, E. H.

MSB 56919
4 mi N of Maxwell
23 April 1982
Collector unknown

MSB 196678
Vicinity of Springer
24 September 2006
New Mexico Department of Game
    and Fish

CURRY CO.
ENMUNHM 11115
1 mi W of Pleasant Hill, Running
    Water Draw
22 October 1998
Collector unknown

DE BACA CO.
ENMUNHM 573
2 mi S of Ft. Sumner,
February 1964 (no exact date given)
Washborn, G.

DOÑA ANA CO.
NMSUVWM 2340
Ft. Selden Monument, HW 85 and
    Radium Springs Rd.
Vuong, H. B.
2004

NMSUVWM 2451
0.7 mi E of HWY 185, Taylor Rd. on
    Leasburg Bridge
22 January 2006
Berryhill, J.

NMSUVWM 2456
Las Cruces on Snow Rd., 100 yds W
    of junction with Onnies Acres Rd.
4 March 2006
McFarland, J.

EDDY CO.
MVZ 79825
Pecos River, 6 mi S, 5 mi E of
    Artesia
30 December 1937
Wade, R. E., Benson, S. B.

MVZ 79824
Pecos River, 6 mi S, 5 mi E of
    Artesia
31 December 1937
Wade, R. E.; Benson, S. B.

MVZ 80526
Pecos River, 6 mi S, 5 mi E of
    Artesia
3 January 1938
Hall, E. R.; Wade R. E.

MVZ 80527
Pecos River, 6 mi S, 5 mi E of
    Artesia
4 January 1938
Hall, E. R.; Wade R. E.

MVZ 80528
Pecos River, 6 mi S, 5 mi E of
    Artesia
5 January 1938
Hall, E. R., Wade R. E.

MVZ 80529
Pecos River, 6 mi S, 5 mi E of
    Artesia
17 January 1938
Hall, E. R.; Wade R. E.

CAVE-1889
Carlsbad Caverns National Park,
    0.1 m W of Natural Entrance
11 August 1952
Gale, B.; Prasil R.G.
lotor

CAVE-4320
Walnut Canyon, Carlsbad Caverns
    National Park
4 September 1964
Collector unknown

CAVE-7930
Carlsbad Caverns National Park,
    0.17 mi N of the Visitor
    Center
February 1994
Kelly, K.

MSB 156761
Carlsbad
14 October 2007
Wildlife Center

GRANT CO.
USNM 170882
Near Central City
July 1910 (no exact date given)
Hotchkiss, H.
Procyon lotor hernandezii

WNMU 1221
2 mi N of Silver City
21 July 1962
Raught, R.

WNMU 1223
Bear Canyon Dam, ~8 mi N of Lorenzo
16 November 1963
Hayward, B. J.

WNMU 1222
5 mi NNE of Pinos Altos, McMillan
    Forest Camp, Cherry Creek
1 November 1964
Hayward, B. J.

NMSUVWM 968
1 mi N of Mimbres
25 September 1971
Spengler, C.

WNMU 3275
Cliff
7 November 1971
Hunt, D. L., Jr.

WNMU 5785
Gila
April 1990 (no exact date given)
McDonald, S.

MSB 85631
Gila National Forest, Big Burro
    Mtns., Forest Rd. 851, 11.8 mi E
    of junction NM 464
1 April 2001
Polechla, P. J., Jr.

GUADALUPE CO.
USNM 118625
Santa Rosa
7 October 1902
Gaut, J.
Procyon lotor hernandezii

USNM 127592
Santa Rosa
27 May 1903
Weller, A.
Procyon lotor hernandezii

HARDING CO.
ENMUNHM 1800
13 mi N, 6 mi W of Roy,
15 April 1966
Beard, M.

HIDALGO CO.
MSB 14244
30 mi S, 4 mi W of Animas, T32S,
    R20W
27 November 1959
Wright, A. W.

MSB 21474
San Simon Cienega; T26S, R22W
7 February 1965
Culbertson, J. L.

MSB 52950
Animas Valley, ~20 mi S of Animas
9 April 1982
Cook, J. A.

LEA CO.
ENMUNHM 7629
14 mi S of Lovington on NM 483
12 October 1979
Collector unknown

LOS ALAMOS CO.
MSB 195602
Vicinity of Los Alamos
11 July 2007
Alencia, R.

LUNA CO.
USNM 158241
8 mi E of Deming
2 September 1908
Goldman, E.
Procyon lotor hernandezii

MORA CO.
MSB 27527
3 mi E of Watrous
21 March 1967
Bogan, M. A.; Studier, E. H.

MSB 27513
La Cueva
22 March 1967
Bogan, M. A.; Studier, E. H.

MSB 27510
La Cueva
25 March 1967
Bogan, M. A.; Studier, E. H.

MSB 27511
La Cueva
25 March 1967
Bogan, M. A.; Studier, E. H.

MSB 27512
La Cueva
25 March 1967
Bogan, M. A.; Studier, E. H.

MSB 27514
La Cueva
8 April 1967
Bogan, M. A.; Studier, E. H.
Procyon lotor

MSB 27515
La Cueva
5 March 1968
Bogan, M. A.; Studier, E. H.

OTERO CO.
MSB 140701
Alameda Park Zoo, Alamogordo
3 April 2001
Carnivore Initiative Project
Note: wild animal found after
being accidentally electrocuted at
the Alameda zoo.

QUAY CO.
MSB 58982
2 mi E. of Tucumcari, at intersec-
    tion of I-40 and Hwy 339
21 December 1986
Lee, R. M.

RIO ARRIBA CO.
USNM 132483
Rinconada
18 April 1904
Surber, M.
Procyon lotor hernandezii

USNM 136308
Velarde
28 November 1904
Beattie, C.
Procyon lotor hernandezii

USNM 136309
Velarde
30 November 1904
Beattie, C.
Procyon lotor hernandezii

USNM 136310
Velarde
6 January 1905
Beattie, C.
Procyon lotor hernandezii

USNM 136317
Alcalde
26 January 1905
Beattie, C.
Procyon lotor hernandezii

USNM 136311
Senega
29 January 1905
Beattie, C.
Procyon lotor hernandezii

USNM 136318
Alcalde
5 February 1905
Beattie,C.
Procyon lotor hernandezii

USNM 136319
Alcalde
6 February 1905
Beattie, C.
Procyon lotor hernandezii

USNM 136312
Senega
6 February 1905
Beattie, C.
Procyon lotor hernandezii

USNM 136313
Senega
8 February 1905
Beattie, C.
Procyon lotor hernandezii

USNM 136320
Alcalde
25 February 1905
Beattie, C.
Procyon lotor hernandezii

USNM 136314
Senega
28 February 1905
Beattie, C.
Procyon lotor hernandezii

USNM 136321
Alcalde
1 March 1905
Beattie, C.
Procyon lotor hernandezii

USNM 136315
Senega
2 March 1905
Beattie, C.
Procyon lotor hernandezii

USNM 136322
Alcalde
3 March 1905
Beattie, C.
Procyon lotor hernandezii

USNM 136316
Senega
11 March 1905
Beattie, C.
Procyon lotor hernandezii

USNM 136323
Alcalde
12 March 1905
Beattie, C.
Procyon lotor hernandezii

MSB 71725
No specific locality recorded
1 July 1994
Wildlife Center

MSB 156764
Tierra Amarilla
6 May 2007
New Mexico Department of
    Game and Fish

MSB 231084
Española
18 June 2007
Wildlife Center

MSB 232467
Española
22 February 2008
Española Animal
    Control

ROOSEVELT CO.
ENMUNHM 10537
3 mi W, 4 mi S of Floyd
No date given
Collector unknown

ENMUNHM 11069
near Portales, 2.7 mi E of
    South Avenue C on 18th
    Street
6 November 2000
Collector unknown

SANDOVAL CO.
MSB 143727
6 mi S of Bernalillo,
18 March 2002
Dohner, Z.

MSB 231446
12 mi S. of Cuba on 550
21 June 2004
Dohner, Z.

MSB 231138
Corrales bosque
11 June 2008
Cook, J. A.
Procyon lotor

MSB 231473
Placitas
11 March 2009
Wildlife Center

MSB 231518
Placitas
11 March 2009
Wildlife Center

MSB 231056
Corrales
7 August 2009
Cook, J. A.

MSB 231032
Findley residence,
    Corrales
1 September 2010
Collector unknown

MSB 265692
Ojito
January 2013
Long, S.

SAN JUAN CO.
MSB 5390
Los Pinos River, SW 1/4,
    Sec 8, T30N, R7W
30 August 1958
Harris, A. H.

APPENDIX 32.1. *continued*

MSB 7266
Los Pinos River; T32N, R7W, NE 1/4
    Sec 18
Harris, A. H.
27 June 1959

USNM 289009
Whiskey Creek
1962 (no exact date given)
Collector unknown
Procyon lotor pallidus

MSB 142757
3 mi E of Bloomfield on Hwy 64
8 August 1999
Dohner, Z.

MSB 92652
3 mi E of Bloomfield on Hwy 64
8 August 1999
Dohner, Z.

SAN MIGUEL CO.
MSB 32510
2 mi N of 7th Street [no town name
    given]
28 September 1970
Allex, A. H.

MSB 213560
2 mi S of Watrous
12 October 1985
Moore, D. W.

MSB 55829
2 mi S of Watrous
12 October 1985
Aquino, A. L.

SANTA FE CO.
MSB 85776
Santa Fe, Galisteo Street, Bataan
    Building, in tree
17 June 1996
Dunn, W.

MSB 156771
132A 9-mi road, Santa Fe
16 February 2005
Gonzalez, C.

MSB 156763
Galisteo
28 May 2007
Wildlife Center

MSB 156762
Tesuque
4 October 2007
Frosch, R.

MSB 231089
Santa Fe
25 April 2010
Wildlife Center

MSB 231433
Cerrillos
10 May 2010
Wildlife Center

MSB 232468
Cerrillos
10 May 2010
Walding, C., Walding B.

MSB 231085
Nambe
18 July 2010
Wildlife Center

MSB 259044
Santa Fe
15 March 2012
Wildlife Center

MSB 259045
Santa Fe
15 March 2012
Wildlife Center

MSB 259049
Santa Fe
15 March 2012
Wildlife Center

MSB 259050
Santa Fe
15 March 2012
Wildlife Center

MSB 259051
Santa Fe
15 March 2012
Wildlife Center

SIERRA CO.
USNM 213129
4 mi N of Chloride
7 December 1915
Beach, F.
Procyon lotor hernandezii

SOCORRO CO.
USNM 160601
Magdalena Mtns., Water Canyon
30 August 1909
Goldman, E.
Procyon lotor hernandezii

ENMUNHM 574
15 mi W, 25 mi S of Magdalena
January 1964 (exact date not given)
Washborn, G.

MSB 25609
Berrnardo Refuge
3 October 1967
New Mexico Department of Game
    and Fish

MSB 64943
Bosque del Apache National Wild-
    life Refuge, main loop
4 January 1990
Collector unknown

MSB 82505
Near Bernardo overpass, Hwy 60 at
    Rio Grande
27 September 1994
Wilson, W.; Brunt, M. H.

MSB 85729
La Joya State Game Refuge, I-25 at
    Exit 162
11 October 1997
Friggens, M. T.
Procyon lotor

TAOS CO.

NMSUVWM 919
Red River Fish Hatchery
8 March 1975
Harris, J.

MSB 60679
N. bank of San Cristobal
    Creek at San Cristobal
25 October 1979
Schmitt, C. G.

MSB 85577
Rio Castilla on NM Rt 196
17 January 1999
Polechla, P. J., Jr.

MSB 140048
Vicinity of Ranchos de Taos
8 October 2003
Ramsey, K.

MSB 231125
Taos
30 May 2010
Wildlife Center

UNION CO.

ENMUNHM 230
41 mi N, 5 mi W of
    Clayton
19 August 1966
Best, T.

ENMUNHM 229
41 mi N, 5 mi W of Clayton
27 August 1966
Best, T.

ENMUNHM 231
39 mi E 3 mi S of Folsom
7 September 1966
Best, T.

MSB 30892
0.05 mi W of Carrizozo Bridge on
    Rte. 325
1 June 1969
Geluso, K. N.

VALENCIA CO.

NMMNH 3862
Los Chaves
1998
Hafner, D. J.

MSB 142645
2.5 mi S of Belen on I-25
13 January 2000
Polechla, P. J., Jr.

MSB 196618
NM 47 at border between Isleta
    Reservation and Bosque Farms
1 September 2008
Lamb, A.

NO COUNTY GIVEN

USNM 151220
Gila National Forest
1907 (no exact date given)
Hotchkiss, H.
Procyon lotor hernandezii

USNM 158243
Red Rock
29 September 1908
Goldman, E.
Procyon lotor hernandezii

USNM 176640
Gila National Forest, Upper Gila
    River Valley
1911 (no exact date given)
Hotchkiss, H.
Procyon lotor hernandezii

USNM 248503
Lincoln [National] Forest
8 September 1928
Collector unknown
Procyon lotor hernandezii

APPENDIX 32.2A. REPORTED RACCOON HARVEST BY NEW MEXICO COUNTY AND BY HUNTING/TRAPPING SEASON, 1980–2000

Data from New Mexico Department of Game and Fish (unpubl. data).

County	Hunting/Trapping Seasons								
	1980–1981	1981–1982	1982–1983	1983–1984	1984–1985	1985–1986	1986–1987	1987–1988	1988–1989
Bernalillo	3	29	0	0	24	0	1	2	ND
Catron	12	20	6	2	9	2	4	50	ND
Chaves	30	16	49	44	12	66	42	54	ND
Cibola	0	12	0	1	3	0	0	1	ND
Colfax	16	62	38	5	21	19	14	16	ND
Curry	3	12	0	10	1	2	8	10	ND
De Baca	8	22	0	0	45	20	16	24	ND
Doña Ana	18	37	2	50	45	2	17	25	ND
Eddy	46	46	3	19	34	34	34	20	ND
Grant	13	26	2	12	34	45	12	24	ND
Guadalupe	13	4	8	0	5	11	4	3	ND
Harding	11	0	2	0	0	0	2	0	ND
Hidalgo	7	26	1	1	1	1	11	0	ND
Lea	1	0	0	0	0	0	3	1	ND
Lincoln	31	26	27	4	38	29	59	47	ND
Los Alamos	0	0	0	2	0	0	0	0	ND
Luna	4	2	0	2	2	2	0	3	ND
McKinley	3	1	0	0	1	0	0	0	ND
Mora	1	2	12	0	10	23	10	9	ND
Otero	51	72	13	16	13	36	17	52	ND
Quay	35	24	11	14	57	16	15	39	ND
Rio Arriba	34	55	31	10	16	1	15	26	ND
Roosevelt	0	0	0	0	0	0	0	0	ND
Sandoval	29	14	17	12	9	13	5	27	ND
San Juan	34	40	16	18	91	11	30	45	ND
San Miguel	8	10	37	0	13	4	6	6	ND
Santa Fe	2	0	0	0	1	0	4	4	ND
Sierra	0	29	2	0	4	4	8	27	ND
Socorro	7	48	3	7	5	9	44	28	ND
Taos	0	1	12	0	16	0	8	0	ND
Torrance	1	1	1	1	0	1	0	0	ND
Union	14	42	1	15	8	29	25	18	ND
Valencia	14	18	1	1	0	41	2	4	ND
Unspecified	10	29	0	0	0	0	0	0	ND
TOTAL	**459**	**726**	**295**	**246**	**518**	**421**	**416**	**565**	**ND**

ND = no data. No harvest information was collected for the 1988–1989 hunting/trapping year; reported harvest was not recorded/tabulated by county during the 1991–1992 hunting/trapping season.

1989–1990	1990–1991	1991–1992	1992–1993	1993–1994	1994–1995	1995–1996	1996–1997	1997–1998	1998–1999	1999–2000
0	0	ND	0	0	20	0	42	0	0	1
5	4	ND	4	8	9	27	0	0	0	0
90	65	ND	15	120	0	14	23	1	6	16
12	0	ND	0	0	2	0	0	0	0	0
20	4	ND	17	27	0	18	5	10	0	1
0	0	ND	3	2	49	7	0	0	0	0
80	96	ND	15	38	0	0	0	0	0	20
0	20	ND	1	7	14	13	0	10	47	25
14	26	ND	16	6	0	7	10	12	0	40
8	26	ND	1	7	5	2	7	4	1	10
0	0	ND	3	0	0	2	0	0	0	0
0	0	ND	0	0	0	0	0	0	0	0
0	0	ND	3	0	0	2	0	0	0	0
0	0	ND	0	0	5	0	0	0	0	0
36	26	ND	21	70	4	22	8	14	31	12
0	0	ND	2	0	0	0	0	0	0	0
1	1	ND	0	10	0	0	0	0	0	0
0	0	ND	0	0	0	0	0	0	0	0
0	0	ND	0	6	0	0	0	0	5	30
20	1	ND	3	2	74	0	2	5	5	22
0	12	ND	20	30	2	3	10	0	0	0
7	8	ND	18	6	0	1	2	10	0	1
0	0	ND	0	0	1	0	0	0	0	4
11	13	ND	0	0	17	4	0	0	1	0
56	23	ND	14	7	7	19	29	96	59	94
0	0	ND	0	38	0	8	30	0	8	7
0	1	ND	1	0	1	2	0	0	0	0
0	0	ND	0	7	12	4	0	2	10	15
2	10	ND	6	2	32	0	17	5	34	25
0	0	ND	9	0	0	0	0	0	0	0
0	1	ND	0	0	25	0	0	0	0	1
12	6	ND	15	4	0	0	0	0	0	0
0	4	ND	3	0	0	1	35	19	96	54
0	0	ND	0	6	0	2	0	0	0	126
374	**347**	**164**	**187**	**403**	**279**	**158**	**220**	**188**	**303**	**504**

APPENDIX 32.2B. REPORTED RACCOON HARVEST BY NEW MEXICO COUNTY AND BY HUNTING/ TRAPPING SEASON, 2000–2020

Data from New Mexico Department of Game and Fish (unpubl. data).

County	Hunting/Trapping Seasons								
	2000–2001	2001–2002	2002–2003	2003–2004	2004–2005	2005–2006	2006–2007	2007–2008	2008–2009
Bernalillo	3	0	8	9	11	3	6	0	0
Catron	11	8	10	1	0	0	0	0	0
Chaves	14	49	12	54	33	147	11	153	110
Cibola	0	0	0	3	1	2	0	0	1
Colfax	4	4	3	8	7	0	1	0	2
Curry	0	6	0	0	0	0	0	0	1
De Baca	30	24	58	19	32	12	7	10	8
Doña Ana	20	33	67	7	20	5	11	13	19
Eddy	29	78	16	7	36	7	37	11	32
Grant	6	17	1	4	5	12	1	6	2
Guadalupe	0	0	1	0	0	0	0	0	0
Harding	0	0	0	0	0	0	0	3	0
Hidalgo	39	3	0	1	5	4	7	0	0
Lea	0	4	0	10	0	0	0	0	0
Lincoln	62	48	59	50	2	11	37	4	12
Los Alamos	1	0	0	0	0	0	0	0	0
Luna	0	1	7	2	0	0	2	0	0
McKinley	0	0	0	0	0	0	0	0	0
Mora	6	0	0	2	1	0	0	0	1
Otero	3	2	8	15	9	0	6	12	6
Quay	6	13	11	8	19	7	5	3	0
Rio Arriba	18	13	24	17	10	18	11	16	25
Roosevelt	0	0	0	0	0	0	0	0	0
Sandoval	26	18	28	74	17	2	21	56	41
San Juan	0	7	16	9	8	50	0	0	0
San Miguel	7	13	1	6	1	4	0	0	2
Santa Fe	3	0	2	4	4	4	4	7	3
Sierra	3	2	0	11	2	3	0	4	0
Socorro	34	4	7	2	3	2	3	5	5
Taos	14	18	4	1	3	0	1	3	0
Torrance	2	0	1	1	0	1	4	0	5
Union	0	4	1	14	3	11	10	12	4
Valencia	42	23	28	37	72	53	25	97	67
TOTAL	**430**	**184**	**192**	**157**	**349**	**156**	**361**	**434**	**303**

2009–2010	2010–2011	2011–2012	2012–2013	2013–2014	2014–2015	2015–2016	2016–2017	2017–2018	2018–2019	2019–2020
10	3	0	8	9	11	3	6	0	0	10
9	11	8	10	1	0	0	0	0	0	9
72	14	49	12	54	33	147	11	153	110	72
0	0	0	0	3	1	2	0	0	1	0
2	4	4	3	8	7	0	1	0	2	2
0	0	6	0	0	0	0	0	0	1	0
0	30	24	58	19	32	12	7	10	8	0
21	20	33	67	7	20	5	11	13	19	21
41	29	78	16	7	36	7	37	11	32	41
7	6	17	1	4	5	12	1	6	2	7
0	0	0	1	0	0	0	0	0	0	0
0	0	0	0	0	0	0	0	3	0	0
0	39	3	0	1	5	4	7	0	0	0
0	0	4	0	10	0	0	0	0	0	0
19	62	48	59	50	2	11	37	4	12	19
0	1	0	0	0	0	0	0	0	0	0
0	0	1	7	2	0	0	2	0	0	0
0	0	0	0	0	0	0	0	0	0	0
4	6	0	0	2	1	0	0	0	1	4
21	3	2	8	15	9	0	6	12	6	21
0	6	13	11	8	19	7	5	3	0	0
12	18	13	24	17	10	18	11	16	25	12
0	0	0	0	0	0	0	0	0	0	0
139	26	18	28	74	17	2	21	56	41	139
2	0	7	16	9	8	50	0	0	0	2
3	7	13	1	6	1	4	0	0	2	3
0	3	0	2	4	4	4	4	7	3	0
2	3	2	0	11	2	3	0	4	0	2
3	34	4	7	2	3	2	3	5	5	3
0	14	18	4	1	3	0	1	3	0	0
4	2	0	1	1	0	1	4	0	5	4
5	0	4	1	14	3	11	10	12	4	5
60	42	23	28	37	72	53	25	97	67	60
341	**383**	**392**	**373**	**376**	**304**	**358**	**210**	**415**	**346**	**436**

Admixture, genetic: genomic mixing resulting from the interbreeding between species or otherwise genetically distinct groups.

Alloparental care: care provided for the young by individuals other than parents.

Altricial: having young that are born helpless and require significant parental care.

Aposematism: use of conspicuous coloration to warn potential predators that an animal is poisonous, venomous, or otherwise dangerous.

Arboreal: living in trees.

Baculum: bone found in the penis of certain mammals, including the Carnivora.

Biodiversity: variety of life at all levels of biological organization, including number of existing species and associated genetic variation.

Blastocyst: fertilized egg in its second phase of growth.

Body length: distance between the tip of the snout and the posterior end of the body, and corresponding to total length minus tail length.

Bottleneck, population: drastic decline in population size due to environmental factors and resulting in the loss of genetic variation.

Breeding season: see mating season.

Carrying capacity: in a species, maximum population size that can be sustained by a particular location or habitat given the amount of food and other resources that are available.

Clade: all the descendants of one common ancestor.

Competition, exploitative: an indirect form of competition, involving the removal of shared resources by one species to the detriment of others, without direct antagonistic interactions between competitors.

Competition, interference: a direct form of competition wherein the dominant competitor harasses, steals food from, or even kills subordinate competitors as an adaptation to increase its share of limited resources.

Congeneric: belonging to the same genus.

Conspecific: belonging to the same species.

Convergent evolution: independent evolution of a feature with similar form or function by lineages whose most recent common ancestor lacks the feature.

Cross-fostering (wolf recovery): placement of captive-bred pups less than 14 days old into wild dens with similarly aged pups to be raised as wild wolves (see Chapter 11).

Cursorial: adapted for running.

Den: hidden shelter where carnivores and other animals sleep or give birth, raise their young, and/ or hibernate.

Depredation: death or injury of livestock (and poultry) due to predators.

Digitigrade: species that stands and moves with its digits, but not the soles of its feet, in contact with the ground.

Dispersal, natal: movement of an animal from its natal (i.e. birth) area to a new site where breeding may occur.

Dorsal: along the upper side or back of an animal.

Dorsum: upper side or back of an animal.

Ecomorph: suite of morphological, ecological, and behavioral adaptations shared by species that tend to occupy the same niche but are not necessarily closely related (see Chapter 25).

Endemic: restricted to a single geographic area.

Enzootic: disease constantly present in a population but only affecting a small number of individuals.

Epizootic: disease outbreak that affects an unusually large number of individuals.

Estrus: recurring phase of the estrous cycle during which the female is fertile and receptive to mating.

Estrus cycle: recurring hormonal and physiological changes in sexually mature females mammals, preparing them for pregnancy.

Extirpation: local extinction of a species that still exists elsewhere.

Fitness, genetic: capacity of individuals to survive,

reproduce, and pass on their genes to the next generation.

Fomite: inanimate object that can be contaminated with an infectious pathogen and serve in disease transmission.

Fossorial: adapted to digging and life underground (see Chapter 25).

Founder effect: loss of genetic variation that results from a new population being established by a very small number of individuals descended from a once larger population.

Fragmentation, habitat: habitat loss that results in the subdivision of former larger tracks of habitat into smaller, isolated tracks.

Genetic drift: random changes over time in the relative frequency of different genotypes, with more pronounced effects (e.g., loss of an allele) in small populations.

Genetic structure: amount and distribution of genetic variation within and between populations (see Chapter 30).

Gestation: period of time between fertilization and birth, during which the offspring grows and develops inside the uterus.

Guild: group of species all similarly exploiting the same type of environmental resources.

Haplotype: group of genes that tend to be inherited together from a single parent.

Home range: area used regularly by an animal but not necessarily defended against others.

Homology, evolutionary: features inherited from a common ancestor, even if they are do not closely resemble each other and are adapted to different purposes.

Host (infectious disease): an organisms that acts as a refuge in which a pathogenic organism lives.

Hybridization: interbreeding between genetically distinguishable groups (e.g., species).

Hyperphagia: period of intensive eating observed in bears in late summer and fall as they fatten for hibernation.

Inbreeding: breeding of animals that are closely related genetically.

Inbreeding depression: reduced biological fitness within a population as a result of inbreeding.

Induced ovulation: process whereby a sexually mature female ovulates (release of mature eggs from the

ovaries) as a response to an external stimulus, generally copulation.

Inguinal: belonging to the lower abdominal region.

Interspecific: occurring or existing between species.

Intraguild predation (in carnivores): combination of interference competition (in its most extreme form) and predation among carnivores that compete for access to the same prey. One outcome of intraguild predation is that the subordinate competitor is killed but not consumed.

Intraspecific: occurring or existing within a species.

Introgression (or introgressive hybridization): introduction of genetic material from one species into the gene pool of another species through hybridization.

Kleptoparasitism: stealing of resources, usually food, by one animal from another. In carnivores, it may involve the stealing of a kill by a dominant competitor. There is a cost to the subordinate competitor, which expended energy to capture and kill the prey.

Lentic: standing or slow moving water environments such as lakes and ponds.

Lineage, evolutionary: Species or taxa that are connected by a continuous line of descent from ancestor to descendant (see also clade).

Lotic: flowing water environments such as streams and rivers.

Mast: fruits of trees and shrubs.

Mating season: period of the year when the female's estrous cycle is active, signaling animals to mate.

Mesopredator: a predator that occupies a mid-rank trophic level in a food web and generally having a medium body size.

Metapopulation structure: existence of local (sub) populations spatially separated but linked to one another by the dispersal of individuals. Due to asynchronous population dynamics (birth and death rates, immigration, and emigration), meta-populations can persist through time (even though local populations may be small and vulnerable to stochastic extinctions) through recolonization (see Chapter 9).

Monoestrous: species that only have one estrus cycle per year.

Monogamous: characterized by exclusive pair-bonds between a male and a female, with the two mating with each another and raising offspring together (see also monogamy).

Monogamy: Mating system involving one male and one female forming an exclusive social pair-bond, mating with each another, and raising offspring together. Pair-bonds may last for only a breeding attempt or a breeding season, or for life.

Monophyletic group: taxonomic group that includes all the descendants of a common ancestor.

Morphology: the size, shape and structure of an organism.

Mortality, additive: type of mortality resulting in the immediate reduction in total survival within a population.

Mortality, compensatory: type of mortality resulting in no immediate reduction in total survival within a population.

Multiparous: producing more than one young at birth.

Niche, ecological: range of resources (e.g., food sources) and conditions (e.g., climate) allowing a species to maintain a viable population in an area.

Nictating membrane: transparent or translucent inner eyelid that can be drawn over the eye to protect it (e.g., from dirt) without impacting vision.

Paraphyletic: taxonomic group that does not include all the descendants of a common ancestor.

Parturition: process of giving birth to offspring.

Pectoral: belonging to the anterior chest region.

Pelage: fur, hair, or wool of a mammal.

Philopatry: tendency for an animal to remain in, or return to, the area of its birth. Philopatry is usually sex-based in mammals and birds.

Phylogeny: evolutionary relationships among or within groups of organisms.

Phylogeography: in the biological sciences, field of study focusing on the historical processes that are responsible for the past and present geographic distributions of evolutionary lineages.

Piscivorous: feeding on fish.

Plantigrade: species that places the full length of its feet on the ground during each stride.

Polygynous: characterized by males mating with multiple females but females mating with only one male.

Polygyny: mating system in which one male mates with more than one female whereas each female mates with only one male.

Population dynamics: patterns of change in population size and structure over time, based on births, deaths, immigration, and emigration.

Prey base: all the prey species available to a predator in a geographic area.

r- and K-selection theory: theory in population ecology that predicts the evolution of life histories in response to different environments, an unstable, temporary but resource-rich, non-competitive r-environment, and a stable, long-lasting but resource-limited, competitive K-environment. A r-environment selects for traits that maximize population growth rate (r), or early sexual maturation and high reproductive output. A K-environment selects for traits that enhance competitive ability at carrying capacity K (i.e., in crowded populations), or delayed maturity, longer life span, and high investment in each offspring. Whereas r-selected species have a high population growth rate but low individual survivorship, K-selected species are characterized by a low population growth rate but high individual survivorship. Today, the r- and K-selection theory is considered an oversimplification of life history strategies, but some of its elements such as density effects have been retained.

Range, geographic: the spatial area in which a species is found.

Refugium (pl. refugia): during an extended period of continental climatic change (e.g., glaciation), area characterized by relatively unaltered environmental conditions and where widespread species can retreat and persist, and later expand their distribution from.

Reintroduction: release of a species in an area from which it was previously extirpated.

Reservoir (infectious disease): an organisms or environment in which an infectious pathogen naturally lives.

Riparian: relating to vegetation occurring beside, and influenced by, a stream or river.

Rostrum (in mammals): part of the cranium situated in front of the zygomatic arches and encompassing the teeth, palate, and nasal cavity.

Sexual dimorphism: distinct difference in size or appearance between males and females within a species, aside from the sexual organs.

Sink population (metapopulation): population that occurs in low-quality habitat and relies on immigration from other populations for its persistence (see Chapter 9).

Sister taxon: the group representing the closest relative of an evolutionary lineage.

Source population (metapopulation): population that occurs in high-quality habitat and through population growth can provide a net donation of immigrants to other, nearby populations occupying lower-quality habitat (see Chapter 9).

Spillover: cross-species transmission of a pathogen into a host population not previously infected.

Straddle: width of the tracks left by an animal or distance between the outermost edges of the prints (see Chapter 6).

Subspecies: within a species, subpopulations that tend to be distinguishable by morphological characteristics.

Sympatric species: species co-occurring in the same geographic area.

Synanthrope: an undomesticated animal that lives near and benefits from humans.

Take (wildlife management): kill or obtain possession of any wildlife (general definition); harass, harm, pursue, hunt, shoot, wound, kill, trap, capture, or collect wildlife, or attempt to engage in any such conduct (Endangered Species Act definition).

Taxon (pl. taxa): taxonomic unit of any rank, including subspecies, species, genus, or family.

Taxonomy: in the biological sciences, the identification, naming, and classification of living organisms.

Total length: distance between the tip of the snout and the tip of the tail.

Trophic cascade: in an ecosystem, effects resulting from the addition or removal of a top predator and propagating down the food web, with potential inversions in the patterns of abundance among trophic groups.

Trophic level: hierarchical position within a food chain (succession of organisms that eat other organisms and might be consumed by other organisms).

Vector (infectious disease): an organisms that transmits a pathogen from one host to another.

Venter: underside of an animal.

Ventral: along the underside of an animal.

Vibrissae: long, stiff hairs on the face of many mammals and used for the sense of touch.

Whelping season: period of year during which parturition occurs.

Zoonosis: Contagious disease that can be transmitted by animals to humans under natural conditions.

Zygote: fertilized egg.

Echinococcus multilocularis, 995–96

Edgington, Edward, 380, 381

El Malpais National Monument, 163, 351, 899–900

Elk: as prey of the black bear, 494, 495, 496; as prey of the cougar, 227, 229, 230, 231, 232, 233; as prey of the coyote, 280, 282, 283, 286; as prey of the gray wolf, 316–17, 318–19, 320, 321, 325, 329–30; as prey of the grizzly bear, 541; as prey of the wolverine, 594

Endangered Species Act (ESA) and ESA listing, 72, 82–83, 84, 115, 144, 200, 206, 303, 398, 467, 548–51, 598, 600, 612, 626, 722, 834

Evans, G. W. ("Dub"), 537, 539, 559, 568, 569

"Evil Quartet" (main causes of biodiversity loss), 92

Evolutionary Significant Units, in the swift fox, 410

Exploitative competition, 241, 322, 323, 324, 361, 366, 393, 548, 910, 962–63

Extirpation, of the black-footed ferret, 721–25; of the gray wolf, 325–27; of the grizzly bear, 531, 552–71; of the northern river otter, 613–14, 625–26

Extra-pair copulations, in gray foxes, 358; in red foxes, 462

Felidae, and climate niche modeling, 93; evolutionary history, xx, 2; during the Pleistocene, 6–8

Feliformia, during the Pleistocene, 6–8; origin of the, xix-xx

Felinae, 2, 6, 7–8

Feline Coronavirus, 146, 993

Feline distemper. *See* feline panleukopenia virus

Feline immunodeficiency virus (FIV), 146, 173, 255

Feline panleukopenia virus (feline distemper; FPV), 112, 146, 173, 993

Felis catus, xx, 110, 112, 154, 156, 157, 173, 255, 640, 674, 692, 750, 760, 852, 868, 888, 896, 963, 975–77, 981–82, 984–87, 990–97, 1000–1002

Findley, James S., xi, 958

Fire ecology: influence on the distribution of ecosystem types, 21–45. *See also* wildfire threats

Fleas, as vectors of the plague, 111, 365, 394, 423, 657, 720, 726, 812, 993–94

Francisella tularensis, 146, 173, 293, 423, 867, 994

Fur farms, 438, 442, 443, 446, 447, 453, 454–55, 457, 458, 467, 771, 772, 778, 779, 786, 787–88

Fur trapping: history of, xi, xii, 73, 74, 76, 613, 625–26, 658, 952; impacts on carnivores, 173, 174, 290–91, 362–63, 367, 396, 423, 424, 425, 426, 428, 598, 600, 612, 613, 626, 657, 786, 810, 811, 868, 912, 968. *See also* fur farms

Galisteo Basin, 2

Gambel Oak Shrubland (ecosystem type), cross-referenced to Merriam's life zones, 22; description of, 27, 36; distribution, 25; projected change in extent, 100

Game Protection Fund, 70, 72

Game Protective Association (GPA), 71, 72, 558

Giant short-faced bear, 13–14, 484

Gila National Forest, 83, 277, 314, 320, 321, 328, 332, 334, 352, 452, 558, 559, 922, 930, 955

Gila Wildlife Rescue, 156, 956

Gila Wilderness, 52, 100, 328, 329, 507, 563, 748

Glacier bear, 481

Glenn, Warner, 196, 197

Gracile sabertooth, 6

Granulocytic anaplasmosis, 111, 994

Gray fox, xxi, xxxviii, 11–12, 22, 55, 56, 60, 74, 76, 77, 81, 102, 107, 112, 114, 164, 172, 173, 232, 234,

245, 347–67, 377, 383, 409, 427, 437, 438, 444, 449, 452, 453, 459, 464, 465, 466, 494, 760, 786, 850, 869, 910, 913, 930, 968, 992

Gray wolf, xxi, xxii, xxiii, xxxi, 8, 9, 10–11, 16, 52, 55, 56, 57, 62, 69, 72, 74, 81, 83, 85, 91, 102, 114, 195, 219, 231, 236, 238, 245, 246, 269–71, 272, 273, 274, 277, 279, 282, 285, 288, 289, 293, 303–36, 361, 377, 392, 411, 423, 425, 440, 445, 448, 451, 457, 459, 463, 512–14, 547–48, 552, 556, 557, 558, 559, 560, 561, 564, 566, 589, 596, 597, 598, 625, 809, 962, 967, 976, 977, 983, 985, 990, 991, 992, 995, 996, 997

Great American Biotic Interchange (GABI), xx

Great Plains (ecosystem type), cross-referenced to Merriam's life zones, 22; description of, 27, 41–42; distribution, 25; projected change in extent, 100

Grizzly bear. *See* brown bear

Gulo gulo, xx, xxi, xxii, 2, 101, 102, 103, 289, 579–601, 697

Harris, Arthur H., xi, 380

Habitat Management and Access Validation Stamp, 76

Habitat Stamp Program, 76

Harvest levels, 77, 157, 159, 160, 174, 176–78, 251, 362–65, 396, 397, 426, 466, 625, 658, 695, 761, 813, 814, 851, 867, 868, 912, 967

Hemicyon ursinus, 2

Henshaw, Henry W., 52

Hesperocyoninae, 306

Hibernation, in black bears, 496–501; in grizzly bears, 542–44

Hollister, Ned, 192–93

Homestead Act, xxii

Hooded skunk, xxi 55, 82, 359, 823, 824, 825, 833, 841–52, 857, 858, 859, 867, 877

Hunting and trapping regulations, 78, 79, 80–82

New Mexico Wildlife Federation, 72

Newberry, J. S., 585–86

Nimravidae, xx

North American Model of Wildlife Conservation, 73, 292

North American red fox, xxi, 2, 13, 55, 77, 81, 82, 103, 105, 112, 144, 288–89, 323, 347, 348, 349, 353, 359, 360, 361, 362,, 377, 386, 393, 394, 397, 416, 422–23, 437–68, 514, 594, 598, 755, 760, 787, 962, 992, 997, 1029–33

North American spectacled bear, 479

Northern Jaguar Reserve, 190, 199

Northern river otter, xx, xxi, xxii, xxiii, 14, 69, 74, 75, 77, 80, 81, 85, 86–87, 108, 110, 609–33, 769, 770, 782, 785, 786, 795, 952

Nuisance animals. *See* urban carnivores

Number 6 Newhouse trap, 555, 556

Occupancy modeling, to infer black bear habitat, 487, 488

Odocoileus hemionus, as prey of feral and free-ranging domestic dogs, 988; as prey of the black bear, 494, 495, 496; as prey of the bobcat, 167; as prey of the cougar, 220, 227, 230, 231, 232, 233, 234–35, 240; as prey of the coyote, 273, 280, 281, 282, 286; as prey of the gray wolf, 314, 317–18; as prey of the grizzly bear, 541; as prey of the jaguar, 202; as prey of the wolverine, 594. *See also* deer populations

Odocoileus virginianus, xxii, 69, 165, 202, 224, 227, 232, 233, 317, 329, 494, 495

Olson, Christian, 553

Omission errors (distribution maps), XXXV

Onchocerca lupi (onchocercosis or onchocerciasis), 995, 997

Otero, Page, 70

Ovis canadensis, 4, 63, 64, 69, 70, 223, 232, 233, 234–36, 252, 256, 282, 452, 496, 514, 594, 941

Oxyaenidae, xix

Pacific marten, xxi, xxii, 21, 22, 29, 74, 75, 77, 81, 85, 86, 102, 103, 104, 138, 289, 463, 464, 580, 581, 589, 639–60, 671–72, 693, 745–46, 770, 795, 799, 984

Pack formation and organization: in coyotes, 269, 282–83, 285–86, 289; in feral and free-ranging domestic dogs, 982–83, 987, 988–90, 999; in gray wolves, 303, 316, 319–21, 334

Palomas Creek Cave, 4, 13

Panthera atrox, 4, 7

Panthera gombaszoegensis, xx

Panthera onca, xx, xxi, 7, 8, 16, 81, 85, 102, 114, 115, 187–212, 219, 221, 850, 938

Paternity, multiple (or mixed), in black bears, 504; in gray foxes, 358; in grizzly bears, 544; in red foxes, 461, 462

Patton, James L., 380

Patuxent Wildlife Research Center, 724

Pecos Wilderness, 534, 539, 553, 570, 646, 648

Pendejo Cave, 4, 5, 14, 273, 350

Pesticides: impacts on skunks, 868; impacts on the kit fox, 397; impacts on the long-tailed weasel, 761; impacts on the American mink, 786

Philmont Scout Ranch, 85, 518, 519, 688, 800, 860, 864

Pinyon-Juniper Woodlands (ecosystem type), cross-referenced to Merriam's life zones, 22; description of, 27, 34–35; distribution, 25; projected change in extent, 100

Pittman-Robinson Act. *See* Wildlife Restoration Act

Plague, 62, 111, 146, 173, 255, 293, 365, 394, 423, 513, 657, 711, 715, 720–21, 722, 725, 726–27, 730–31, 761, 812, 988, 991, 993–94

Pleistocene Megafauna Extinction. *See* Quaternary Megafauna Extinction

Plesiogulo, 2

Pliotaxidea, 2

Poisoning (early predator control programs), xxii, 52, 75, 132, 192–93, 291, 327, 395, 423–24, 425, 428, 485, 513, 557, 561, 564, 565–66, 568, 711, 721, 722, 731, 762, 835, 941

Polar bear, 14, 112, 117, 460, 531

Ponderosa Pine Forest (ecosystem type), cross-referenced to Merriam's life zones, 22; description of, 27, 31–33; distribution, 25; projected change in extent, 100

Prairie dogs, as important prey of the American badger, 288; as important prey of the kit fox, 382, 386, 387, 388, 395–96, 398; as key prey of the black-footed ferret, 14–15, 710–31

Predator Damage Management (PDM), 51–65, 941

Predatory Animal and Rodent Control (PARC), xxiii, 558–69

Primary constituent elements (PCEs; jaguar habitat), 200

Procyon lotor, xxi, 16, 55, 56, 60, 61, 62, 74, 77, 82, 110, 111, 112, 165, 202, 232, 233, 289, 759, 812, 825, 850, 870, 895, 898, 900, 910, 912, 919, 920, 932, 939, 947–69, 992, 993, 1073–83

Procyonidae, during the Pleistocene, 16; evolutionary history, xx, 2, 825–26, 898; taxonomy, 895, 897–98; typical dentition, 896

Promartes, 2

Pronghorn, 59, 63, 69, 165, 232, 233, 280, 282, 292, 388, 397, 419, 496, 980

Protictis, 1

Pseudaelurus, 2

Puma. *See* cougar

Puma concolor, xv, xx, xxi, xxii, xxxi, xxxii, 8, 16, 21, 52, 54, 55, 56,